International Conference on Materials Science and Technology

(ICMST 2012)

IOP Conference Series: Materials Science and Engineering Volume 73

Kerala, India
10 – 14 June 2012

ISBN: 978-1-5108-0079-3
ISSN: 1757-8981

Printed from e-media with permission by:

Curran Associates, Inc.
57 Morehouse Lane
Red Hook, NY 12571

Printed by Curran Associates, Inc. (2015)

For permission requests, please contact the Institute of Physics
at the address below.

Institute of Physics
Dirac House, Temple Back
Bristol BS1 6BE UK

Phone: 44 1 17 929 7481
Fax: 44 1 17 920 0979

techtracking@iop.org

Additional copies of this publication are available from:

Curran Associates, Inc.
57 Morehouse Lane
Red Hook, NY 12571 USA
Phone: 845-758-0400
Fax: 845-758-2634
Email: curran@proceedings.com
Web: www.proceedings.com

TABLE OF CONTENTS

FROM THE CONVENOR'S DESK

The Department of Physics, St.Thomas College Pala, is highly privileged to organize an International Conference on Materials Science and Technology (ICMST 2012) during 10-14 June 2012, and as Convenor of the conference it is with legitimate pride and immense gratitude to God that I remember the most enthusiastic responses received for this from Scientists all over the world. In a time of tremendous revolutionary changes in Materials Science and Technology, it is quite in keeping with the tradition of a Pioneering Institute that St.Thomas College is, to have risen to the occasion to make this conference a reality. We have no doubt that this proved to be a historic event, a real breakthrough, not only for us the organizers but also for all the participants. A conference of this kind provides a nonpareil, a distinctly outstanding platform for the Scholars, Researchers and the Scientists to discuss and share ideas with delegates from all over the world. This had been most fruitful to the participants in identifying new collaborations and strengthening existing relations.

That experts of diverse disciplines from across the world were sitting under one roof for five days, exchanging views and sharing findings, was a speciality of this conference. The event has evoked excellent responses from all segments of the Materials Science community worldwide. 600 renowned scholars from 28 countries participated in this. We were uniquely honoured to have **Prof. C.N.R. Rao, Chairman, Scientific Advisory Council to the Prime Minister of India, to inaugurate this conference.**

May I take this opportunity to thank all those who have contributed their valuable share, diverse in tone and nature, in the making of this conference.

My whole hearted gratitude is due to the International and National members of the advisory committee for their valuable guidance and involvement. I place on record my heartfelt gratitude to our sponsors. I am sure that this conference has contributed to more innovations and inventions in the field and promoted the advancement of Science and Technology in India.

It is with immense pain and sorrow that I pay **tribute to Prof. Antony Simon** , one of the members of the organizing team, who departed from us after bestowing everything that one could for the success of ICMST 2012.

I wish all the success to all who contributed papers to the conference.

Dr. Ginson P. Joseph

Convenor, ICMST 2012

IOP Conf. Series: Materials Science and Engineering **73** (2015) 011001 doi:10.1088/1757-899X/73/1/011001

ICMST 2012: A few snaps

Prof.C.N.R.Rao, Chairman, Scientific Advisory Council to Prime Minister of India inaugurating the ICMST 2012

Audience of the Inaugural Session of ICMST 2012

His Excellency Mar Joseph Kallarangattu, Patron of our college delivering the keynote address

Prof. Norbert Kroo, Presidential Advisor & Vice-President , Hungarian Academy of Science, giving Plenary talk on *'There is plenty of light at the bottom'*

Dr. Maria Farsari, Institute of Electronic Structure & Laser (I.E.S.L), giving a keynote lecture on *'Beyond the diffraction limit : Diffusion-Assisted Direct fs Laser Writing'*

Dr. John Philip, Head, SMARTS, Metallurgy and Mateials group, IGCAR,Kalppakkam, India giving Plenary talk on *'Nanomaterials-Excitement and challenges'*

A scene from Conference dinner

Delegates at the stalls

IOP Conf. Series: Materials Science and Engineering **73** (2015) 011002 doi:10.1088/1757-899X/73/1/011002

Peer review statement

All papers published in this volume of *IOP Conference Series: Materials Science and Engineering* have been peer reviewed through processes administered by the proceedings Editors. Reviews were conducted by expert referees to the professional and scientific standards expected of a proceedings journal published by IOP Publishing.

Temperature dependent DC characterization of InAlN/(AlN)/GaN HEMT for improved reliability

K Takhar, U P Gomes, K Ranjan, S Rathi and D Biswas

Advanced Technology Development Center, Indian Institute of Technology Kharagpur, Kharagpur -721302, West Bengal, India

Email: takharkuldeep@gmail.com

Abstract. $In_xAl_{1-x}N$/AlN/GaN HEMT device performance is analysed at various temperatures with the help of physics based 2-D simulation using commercially available BLAZE and GIGA modules from SILVACO. Various material parameters viz. band-gap, low field mobility, density of states, velocity saturation, and substrate thermal conductivity are considered as critical parameters for predicting temperature effect in $In_xAl_{1-x}N$/AlN/GaN HEMT. Reduction in drain current and transconductance has been observed due to the decrease of 2-DEG mobility and effective electron velocity with the increase in temperature. Degradation in cut-off frequency follows the transconductance profile as variation in gate-source/gate-drain capacitances observed very small.

1. Introduction

Limitation of mature silicon technology for high power and high temperature operation paves the way for wide-band gap semiconductors like GaN, SiC etc. These semiconductors are extremely attractive for commercial and military applications because of their high power handling capabilities. Low carrier generation by thermal activation and high breakdown field in wide band semiconductor makes them excellent candidate for high temperature applications. Various hetereostructures like AlGaN/GaN, InAlN/GaN, AlN/GaN, etc., are prominent for utilizing the optical and electrical properties of nitride materials. One of the key issues for high power/high temperature applications is the junction and channel temperature which affects the key device parameters like mobility, saturation velocity etc. thus affecting the device characteristics and reliability. Therefore, an in-depth study of device operation at high temperature is required for efficient thermal management and to improve the device reliability. As compared to conventional $Al_xGa_{1-x}N$/GaN HEMT, $In_xAl_{1-x}N$, x~0.19 barrier layer on GaN eliminates the strain induced reliability issues and exhibits a high two dimensional electron gas (2DEG) density of the order of ~$2.5\times10^{13}cm^{-2}$, almost double than normally obtained in $Al_xGa_{1-x}N$ /GaN heterostructure, therefore allows the realization of device with higher power density. The device performance is further improved by inserting thin layer of AlN between the $In_xAl_{1-x}N$ barrier layer and GaN active layer to reduce the alloy disorder scattering thus improving the transport properties in the channel. High sheet carrier concentration (order of 2.8 x $10^{13}cm^{-2}$) is also developed at AlN/GaN interface due to large polarization difference and band gap discontinuity without intentionally doping the barrier layer. Moreover, AlN also possesses highest spontaneous polarization in materials matrix and a Curie temperature well above 1000°C [1]. Further, lattice matched $In_xAl_{1-x}N$ /GaN HEMT is

thermally, chemically and electrically stable at high temperature then $Al_xGa_{1-x}N$ /GaN counterpart, potentially allowing operation at higher temperature and improving robustness.
In the present work, TCAD analysis of $In_xAl_{1-x}N$ /AlN/GaN HEMT in the temperature ranges from 300 K to 800 K is carried out to study the impact of temperature on device characteristics like drain current, transconductance, gate-source/gate-drain capacitance and cut-off frequency.

2. Device structure and Material Model Specification

Figure 1 shows the schematic of the device considered in the analysis. It consist of 13 nm InAlN followed by 1 nm AlN inter-layer separating the active 2 μm thick GaN layer on sapphire substrate. The schottky gate length (Lg) of 0.25 μm with a source-gate distance (Lsg) of 1 μm and gate-drain distance (Lsd) of 1.35 μm are considered.

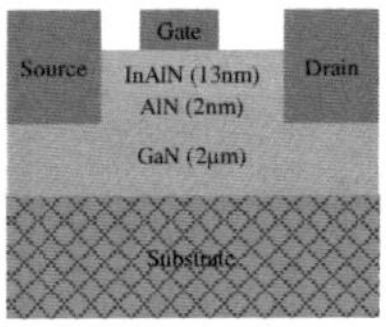

Figure 1. Schematic of InAlN/AlN/GaN HEMT

Great deal of efforts is seen in literature to idealize the value of polarization charges for InAlN/AlN/GaN HEMTs (Table 1). Although, polarization in III-Nitrides is very high, a very small change in this polarization is observed with temperature [2-4]. However, significant changes on various material parameter such as band gap, saturation velocity, thermal conductivity and density of states have been observed with temperature [2-4] and their effect have been incorporated in the device simulation using suitable models. Low thermal conductivity of III-V semiconductors creates the necessity of lattice heating models for high power devices. The GIGA module has been used consistently with BLAZE in order to encounter the effects of lattice temperature on material and transport parameters.

Table 1. Material Parameters

Material	a_0 (A^0)	P_{sp} (C/m^2)	e_{31}-(C_{31}/C_{33})e_{33}	α	β	γ	κ	Ref.
GaN	3.189	-0.029	-0.68	-1.08	745	-0.28	130	[5]
AlN	3.112	-0.081	-0.86	-1.799	1462	-1.64	285	[5]
$In_xAl_{1-x}N$, x=0.19	3.200	-0.071	-	-	-	-	-	-

$$c\frac{\partial T_L}{\partial t} = \nabla(\kappa \nabla T_L + H) \tag{1}$$

Where c is heat capacitance per unit volume, κ is thermal conductivity (Table 1), H is heat generation, and T_L is local lattice temperature.
For realistic TCAD simulation, critical temperature dependent parameters have also been modeled in the present study. The temperature dependence of the energy gap is approximated by Varshini formula which is expressed by the following equation [4].

$$E_g = E_g(0) - \frac{\alpha T^2}{T + \beta} \tag{2}$$

Where E_g (0) is energy band gap at 0 K, α and β are empirical constants. The values used for GaN, AlN and ternary InAlN compounds are obtained through extensive bibliographic review and linear interpolation using Vegard's rule. These values are summarized in table 1 for reference.
The temperature dependence on the electron mobility has been taken into account using the Albrecht mobility model. Along with temperature dependence, this model also takes care of the effects caused

by ionized impurities and compensation ratio dependency of drift mobility. The expression for this model is given as [7]

$$\frac{1}{\mu}=\left\{a\left(\frac{N_I}{10^{17}cm^{-3}}\right)\ln\left(1+\beta^2\right)\left(\frac{T}{300}\right)^{\frac{-3}{2}}+b\left(\frac{T}{300}\right)^{\frac{3}{2}}+c\left(\frac{1}{\exp\left(\frac{\Theta}{T}\right)-1}\right)\right\} \quad (3)$$

Where T is temperature in K; N_I is ionized donor concentration (cm^{-3})

$$\beta^2=3.0\left(\frac{T}{300K}\right)^2\left(\frac{N_I}{10^{17}cm^{-3}}\right)^{\frac{-2}{3}}$$

a, b and c are the fitting parameters whose values are 2.0×10^{-4}, 2.5×10^{-4} and 1.7×10^{-2} respectively. Temperature dependent mobility of minority carriers are modeled using more generic models expressed as:

$$\mu_n=\mu_n(300)\left(\frac{T}{300}\right)^{-1.5} \quad (4)$$

$$\mu_n=\mu_n(300)\left(\frac{T}{300}\right)^{-1.5} \quad (5)$$

3. Results & Discussion

In order to validate the TCAD simulation, the device under study has been calibrated with the experimentally demonstrated device structure both at room and high temperature [8]. Various aspects of the DC characteristics like I_D-V_D, transfer characteristics, threshold voltage have been analyzed in the temperature range of 300 K to 800 K. figure 2(a) plots the I_D-V_D for temperature variation from 300 K to 800 K. The value of maximum drain current decreases as the operating temperature is increased from room temperature to 800 K. At 300K, the device exhibited a saturated drain current (I_D) of 1.6 A/mm, which decreases to 0.83 A/mm as the temperature was raised to 800°C. This significant reduction in maximum drain current can be explained from the degradation in 2-DEG mobility and effective electron velocity with increasing temperature. The inefficient thermal management due to low thermal conductivity of 42 $WK^{-1}m^{-1}$(148 $WK^{-1}m^{-1}$ for Si) of sapphire adds further degradation at high temperatures. Lower value of thermal conductivity results in more lattice heating due to inefficient dissipation of the heat generated in the device. This lattice heating results in additional phonon scattering, further degrading the carrier mobility hence lower drain current.

In order to further explore the thermal heating effects on the device performance, the I_d Vs V_{gs} and transconductance behavior of the device have been plotted in figure 2(b) at various temperatures at 10 V of drain bias. It is observed from the figure that the peak transconductance value also follows downward trend as temperature is scaled up. In contrast to the current and g_m degradation at high temperature, g_m profile becomes flatter (an average g_m ~ 140 mS/mm at 800 K over V_{GS}=-4.4 to 0.3 V) which indicates an improvement in the overall linearity of the device.

A slight positive shift in threshold voltage can also been observed with the increase in temperature, which is depicted clearly in figure 2c. This positive shift with increase in temperature can be explained by the phenomena of trapping effect at the gate-source and gate-drain recess interface. Similar increase in threshold voltage with temperature has also been studied by S. Arulkumaran et al. [9] and can be explained by the following expression

$$V_{th}=V_{bi}-q\left(N_d-N_T\right)\frac{a^2}{2\varepsilon}$$

Where V_{bi} build in potential, ε dielectric constant of barrier layer, N_d is carrier density and N_T is carrier trap density.

Variation in cut-off frequency and total gate capacitance with temperature variation has been plotted in figure 2(d). It can be seen from the figure that there is little variation in capacitance profile while the cut-off frequency shows almost linearly degradation with the increase in temperature. Degradation in g_m and increase in the value of capacitance supports the sharp decay in f_T, but the variation in g_m is almost 3 times greater than that of $c_{gs}+c_{gd}$, so g_m degradation plays a dominant role.

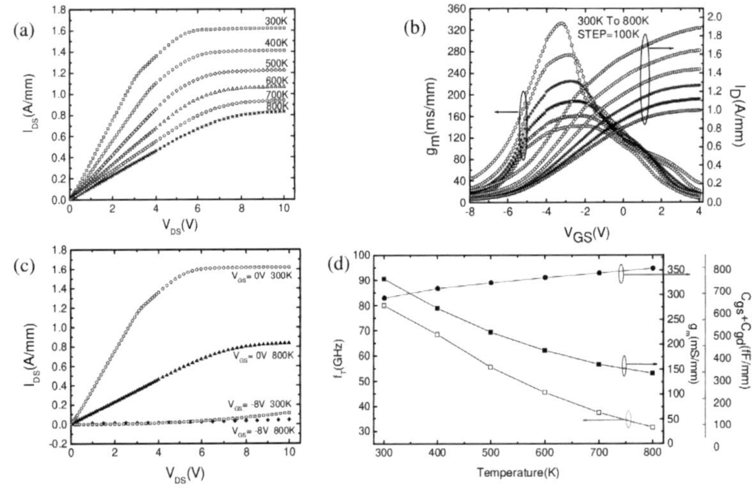

Figure 2. (a-d) Temperature dependent (a) I_D-V_D characteristics at V_{GS}=0 V (b) Transfer characteristics and transconductance variation (c) I_D-V_D characteristics at V_{GS}=0 V and at V_{GS} = -8 V (Pinch off) (d) Variation in cut-off frequency, gate capacitance (C_{gs} + C_{gd}) with temperature from 300 K to 800 K.

4. Conclusion

The degradation in the device performance with the adverse thermal condition is observed through the DC and RF characteristics of the device. This degradation is attributed to the phonon scattering which deteriorates the 2DEG mobility and effective electron velocity. The phenomena of trapping effects may explain the positive V_{th} shift observed with the increase in temperature. The reduction in channel thickness may arrest the degradation observed in the present study by confining the 2DEG channel thus reducing the carrier scattering and dispersion of the carriers. The cut-off frequency follows g_m variation as capacitance maintains a constant profile, which will be studied further using engineered epitaxial design or channel engineering so that degradation in g_m and hence f_T could be avoided. This work is currently under study and presents one of the possible solutions for better device performance even at higher temperatures.

Acknowledgements

The authors acknowledge the Council of Scientific and Industrial Research organization (CSIR) and DIT (ENS project) Govt. of India for providing financial assistance to carry out this work.

References

[1] F Medjdoub, Carlin et al 2008 *Open Electr. Electron. Eng. J.* **2** 1-7
[2] V O Turina, A A Balandin 2006 *Journal of Applied Physics* **100** 054501
[3] S K Islam, H F Huq 2007 *International Journal of Electronics* **94** No.12 1099
[4] Y.Change, K Y Tong, C.Sury 2005 *Semiconductor Science and Technology* **20** 188.
[5] O Ambacher, J Smart et al 1999 *Journal of applied physics* **85** No.6
[6] Silvaco International Jan2012 *Silvaco International*, ATLAS User's Manual.
[7] J D Albrecht, P P Ruden et al 2000 *IEEE Trans. Electron Devices* **47** 2031-2036
[8] F Medjdoub, J F Carlin et al 2006 Int. *Electron Devices Meeting (IEDM)* Tech. Digest **927**.
[9] S Arulkumaran , Z H Liu et al 2007 *Thin Solid Films* **515** 4517-4521

A comparative TCAD assessment of III-V channel materials for future high speed and low power logic applications

U P Gomes, K Takhar, K Ranjan, S Rathi, and D Biswas
ATDC, Indian Institute of Technology Kharagpur, Kharagpur-721302, India

umesh.gomes@gmail.com

Abstract. In this work, by means physics based drift-diffusion simulations, three different narrow band gap semiconductors; InAs, InSb and $In_{0.53}Ga_{0.47}As$, and their associated heterostructures have been studied for future high speed and low power logic applications. It is observed that $In_{0.53}Ga_{0.47}As$ has higher immunity towards short channel effects with low DIBL and sub-threshold slope than InSb and InAs. Also it is observed that for the same device geometry InSb has the highest drive current and lower intrinsic delay but its I_{ON}/I_{OFF} figure of merit is deteriorated due to excess leakage current.

1. Introduction

The narrow bandgap III–V semiconductors (InAs, InSb and $In_{0.53}Ga_{0.47}As$) have attracted considerable attention as alternative channel materials for beyond silicon logic applications due to their very high electron mobility and saturation velocity. The superior electronic properties of III-V materials translate to favorable logic figure of merits for ultra high-speed and low-power logic applications. This paper endeavors to predict the logic performance of these next-generation digital III-V family materials and present a comparative analysis using physics based two-dimensional drift-diffusion simulator. For the first time in order to investigate the potential of these materials for digital applications they are introduced into the channel of their associated heterostructure devices of similar geometry.

2. Device Structure and Model

Reliability of device modelling is based on accurate physical models and material parameters which are not completely available in the simulator database. As a prerequisite for simulation of devices based on III-V compound semiconductors, material parameters have been obtained through bibliographic review. The physical material parameters for the ternary alloys are obtained from experimental data and vegard's interpolation schemes from generally accepted values for the binary compounds. Figure 1 shows the vertical band profile of the devices under study. These structures are designed to evaluate the digital performance of specific channel materials and to bring forward an effective comparison. Therefore for the first time the transistors with different channels are modeled with same device geometry. In order to rationale that gate leakage is mostly due to material properties the heterojunctions have type I configuration which confines the holes in the channel and minimizes leakage current [1]. The structures consist of a 10 nm narrow bandgap channel sandwiched between a 4 nm, high bandgap barrier layer and a 0.16 μm thick buffer layer. The device that employs InAs as the primary channel is inserted between $In_{0.53}Ga_{0.47}As$ sub-channels to realize the lattice-match criteria. Channel charge is supplied through a delta-doped layer at a concentration of 3×10^{12} cm^{-2} located in the barrier just below the gate and a spacer layer of 5 nm separates the channel from the barrier to prevent scattering of electrons. The gate length of the devices is 40 nm and gate-source/drain separation of 50 nm respectively. The drain is biased at 0.5 V in all simulations unless otherwise stated to reduce overall dc power consumption. The simulation temperature is maintained at 300 K. In order to have a fair analysis, the schottky gate work-function is adjusted for a near e-mode threshold voltage of -0.04 V in all the three devices.

The simulation models are tuned by benchmarking the drain and transfer characteristics with experimental data from [2][3][4]. All the simulations have been performed in SILVACO/BLAZE simulator using physical models accounting for the electric-field-dependent and concentration-dependent carrier mobilities, Shockley–Read–Hall recombination/generation, Selberherr's impact ionization and the schottky thermionic emission models. The experimental uncertainties were incorporated using interface charge models and S/D resistances. The effect of compressive biaxial strain on InAs and InSb channels due to lattice mismatched heteroepitaxial growth is also incorporated through quantitative bandgap corrections [5] [6].

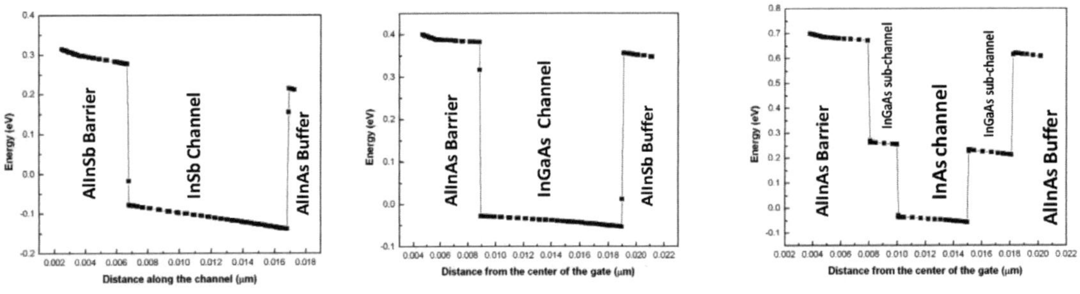

Figure 1. Schrodinger-Poisson solutions for conduction band profile of III-V heterostructures

3. Results And Discussion

The drain characteristics plot shows that the extraordinarily high mobility and saturation velocity of InSb translates to remarkable drive current.

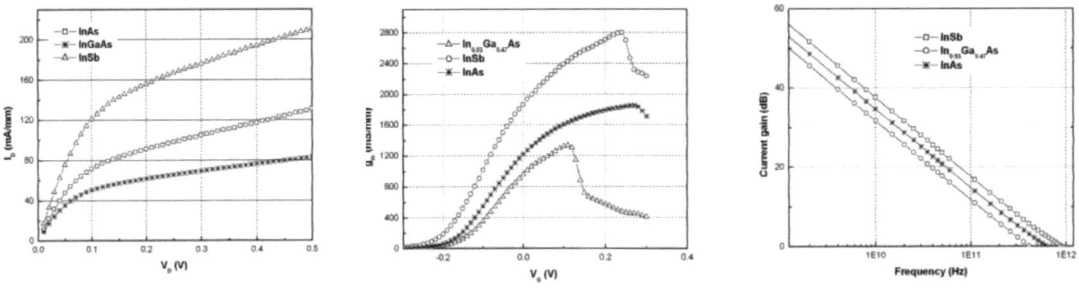

Figure 2. I-V, transconductance and cutoff frequency characteristics

Transconductance measurements show a peak transconductance of 2800 mS/mm for InSb HEMT while InGaAs has maxima at 1336 mS/mm. Although the transconductance of the device is a suitable measure of the device speed, it is more convenient to use the intrinsic delay as the figure of merit. We have investigated the intrinsic delay of the device by measuring the cutoff frequencies through S-parameter simulation. InSb reports the highest cutoff frequency (0.89 THz) and therefore the least delay time of 0.179 ps. This simulation results suggests that THz frequency response may be possible with nanoscale InSb HEMTs. Therefore InSb turns out to be a capable candidate for ultra high-speed logic devices [7].

IOP Conf. Series: Materials Science and Engineering **73** (2015) 012002 doi:10.1088/1757-899X/73/1/012002

Furthermore, a typical curve of drain current versus gate voltage (V_G) in logarithmic scale is plotted in figure 3 to present the various device metrics. The inset in the figure shows the evaluation methodology undertaken in the study proposed by R.Chau et.al and successfully demonstrated in various III-V logic devices. The sub-threshold slopes are relatively similar for the three devices with InSb showing the highest sub-threshold slope confirming poor short channel effects. As shown in Figure 3, InAs and InSb FETs exhibit DIBL=108 mV/V and 204 mV/V respectively, whereas, for InGaAs FETs DIBL=55 mV/V, thereby showing immunity to logic degradation. The increase in DIBL and SS of InSb is due to their high channel dielectric constant and low effective mass that enhances potential barrier lowering. However, InSb exhibits the highest ON-current (324 mA/mm) due to relatively high electron velocity $v_{sat}=5\times10^5$ m/s and mobility of 78000 cm/Vs.

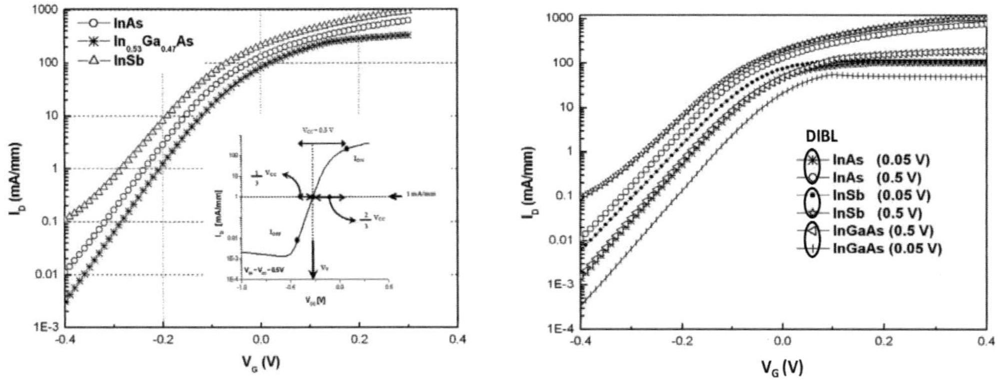

Figure 3. Sub-threshold and DIBL characteristics of the three devices

Although the three devices have shown high ON state current and intrinsic delay, the device performance is dictated by the leakage current behavior. The parameters that determine gate leakage are the schottky barrier height, bandgap, conduction band offset, dielectric constant of the barrier and channel dielectric constant. In InAs HEMT, the InGaAs sub-channel forms an additional barrier for the tunneling electrons and holes in InAs channel reducing the leakage current. The leakage and SCE analysis is carried out by measuring the electric field in the gate and channel regions. A large channel dielectric constant results in the electric field penetrating the channel that causes severe performance degradation [8].The electric field profile along the channel of the three devices is represented in figure 4.InSb shows the highest peak electric field. Because of the strong variation of the electric field along the heterojunction interface impact ionization occurs near the drain end causing additional leakage current and SCEs [9]. Also the increased electric field at the gate edges decreases the tunneling barrier width and increases the gate leakage current [10].The fringing field distribution profile is obtained using a cutline at few angstroms distance below the gate. This needle-shaped electric field is a function of the dielectric constant of the barrier which is highest for AlInSb (16.6).

The chart in figure 5 shows the gate leakage current, the sub-threshold OFF current and the total leakage current. Although the ON current of InSb is the highest among the III-V digital materials, the excess leakage current degrades the I_{ON}/I_{OFF} figure of merit as shown in figure 4. This high leakage current in InSb is attributed due to low confinement in the channel, low bandgap, reduced barrier height and high dielectric constant of the channel and barrier materials causing high field effects as shown in figure 4.The author is aware that the resulting leakage current does not include a complete current leakage assessment, and therefore the OFF state current predictions should be considered as lower limit estimate. The effects of band-to-band tunneling (function of bandgap) are not activated in the simulation which can further magnify the leakage current.

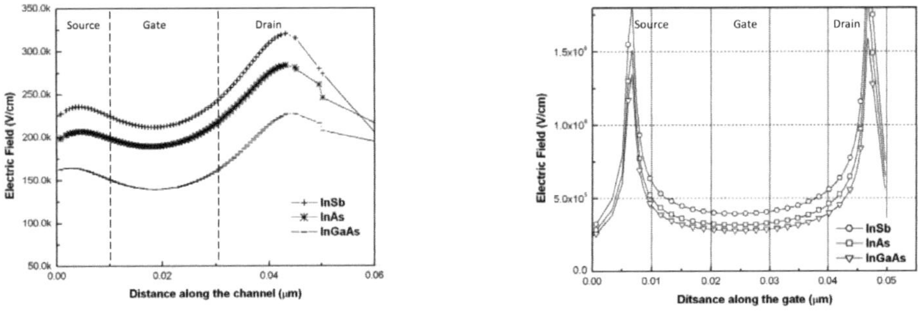

Figure 4. Electric field profiles along the channel and gate regions

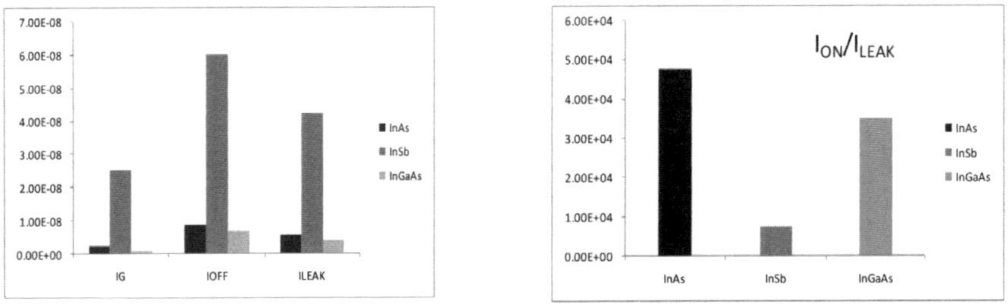

Figure 5 Leakage current analyses

4. Conclusion

As a concluding note, the various figure of merits of the three devices are summarized in table 1.

Table 1. Summary of figure of merits

Channel	SS (mV/dec)	DIBL (mV/V)	I_{ON}/I_{OFF} ($\times10^4$)	f_t (GHz)	T (ps)	V_{CC} (V)
$In_{0.53}Ga_{0.47}As$	84	55	2.0	422	0.377	0.5
InAs	88	108	3.1	640	0.248	0.5
InSb	94	204	0.5	890	0.179	0.5

References

[1] Desplanque L,Vignaud D,Godey S,Cadio E,Plissard S,Wallart X,Liu P and Sellier H 2010 *J. Appl. Phys.* **108** 043704

[2] Kim D H, del Alamo J A, *IEEE IEDM* 2007 629

[3] Ashley T,Barnes A R,Buckle L,Datta S,Dean A B, Emery M T, Fearn M,Hayes D G, Hilton K P Jefferies R,Martin T,Nash K J, Phillips T J,Wilding P J and Chau R, *ICSICT* 2004 **3** 2253

[4] Vasallo B G,Wichmann N,Bollaert S,Roelens Y,Cappy A,Gonzalez T,Pardo D and Mateos J, *IEEE Transactions on Electron Devices* 2007 **54** 2815

[5] Luisier M,Neophytou N,Kharche N and Klimeck G, *IEEE IEDM* 2008 1

[6] Kasturiarachchi T,Brown F,Dai N,Khodaparast G A,Doezema R E,Chung S J and Santos M B, *Applied Physics Letters* 2006 **88** 171901

[7] Serge Oktyabrsky, Peide Ye, "*Fundamentals of III-V Semiconductor MOSFETs*," Springer,2010 and references therein.

[8] Oh Saeroonter, Ph.D dissertation, 2010

[9] Christian H, Buchali F, Prost W,BrockerhoffV W,Fritzsche D,Niskel H,Losch R,Schlapp W, Tegude F J *IEEE Transactions On Electron Devices* 1994 **41** 1685

[10] Sathaiya D M and Karmalkar S *IEEE Transactions On Electron Devices* 2007 **54** 2614

IOP Conf. Series: Materials Science and Engineering **73** (2015) 012003 doi:10.1088/1757-899X/73/1/012003

Fabrication of ZnO and doped ZnO waveguides deposited by Spin Coating

Rosmin Elsa Mohan [1], Neha P R [3], Shalu T [3], Darshana K C [3] and K S Sreelatha[2]
[1] Department of Physics, Amrita Vishwa Vidyapeetham, Kollam, Kerala, India
[2] Department of Physics, Government Polytechnic College, Kottayam, Kerala
[3] Department of Physics, CMS College, Kottayam, Kerala, India

Email: rosminelsa@am.amrita.edu, elzaroz@gmail.com

Abstract. In this paper, the synthesis of ZnO and doped $Zn_{1-x}Ag_xO$ (*where x=0.03*) nanoparticles by co- precipitation is reported. The precursors used were Zinc Nitrate and Potassium hydroxide pellets. For doping, 3% $AgNO_3$ in $ZnNO_3$ was considered as a separate buffer solution. The prepared nanoparticles were subsequently spin coated onto silica glass substrates at a constant chuck rate of 3000 rpm. The substrate acts as the lower cladding of a waveguide structure. The upper cladding is assumed to be air in the present investigation. The nanostructures of the ZnO powders in the doped and undoped cases were studied using X-ray Diffraction patterns. There was a decrease in the grain size with doping which increase the tunability of the powders to be used as photoluminescent devices. The optical characteristics of the sample were also investigated using UV-Visible spectrophotometer at 200-900 nm wavelengths. The photoluminescence peaks also report a dramatic increase in intensity at the same wavelength for the doped case compared to the undoped one.

1. Introduction

The use of nanocrystalline materials as potential candidates in the field of nano device fabrication has gained much attention in the recent past [1-4]. Their unique electrical, optical and piezo electric properties have allowed extensive uses as semiconductor, acoustic wave and accousto-optical devices. Zinc oxide is a direct band gap (3.4 eV) II-VI compound and an n-type semiconductor which has a hexagonal wurtzite crystal structure with a number of alternating planes with tetrahedrally – coordinated Zn^{2+} and O^{2-} stacked alternately . Therefore it has attracted a lot of intense research for its unique properties and versatile applications in transparent electronics, UV light emitters, piezo electric devices, chemical sensors and spintronics [5-8].

The wide band gap of nano zinc oxide has been identified as a promising material for exhibiting the passage of various optical signals .Transition metal doped nanostructures enable intense optical effects such as variation of refractive index, grain size, adjustment of energy level surface states of ZnO leading to varied physical and optical properties [9]. Nano zinc oxide is also non-toxic which offers it an additional edge to being increasingly user friendly.

The method of co- precipitation is an appropriate method for the synthesis of nanopowders especially because of the ease of preparation. It may be treated as a phenomenon whereby the fractional precipitation of a specified ion in a solution results in the precipitation of the target ion and also of other ions existing side by side in the solution. The excess precipitation of the other ions is an added advantage especially when dopants are present, though may be an impediment to the analytical process as well.

ZnO thin films depend critically on several physical and chemical parameters such as deposition technique, drying process, annealing temperature, pH of solution, concentration of solution etc. In the present work, we have fabricated ZnO and doped ZnO ($Zn_{1-x}Ag_xO$) thin films as a prerequisite for a waveguide structure. The nanoparticles were initially co precipitated, coated onto glass substrates by the sol gel process and characterized. The size and the thickness of the films were uniform and controlled.

2. Experimental

For the synthesis of ZnO nanoparticles, AR grade zinc nitrate dehydrate and potassium hydroxide pellets from MERCK laboratory were used without further purification. Polyvinylpyrrolidone (PVP) was used as a capping agent. In the present work, zinc oxide nanoparticles were prepared through chemical co-precipitation method in an aqueous medium at room temperature.

2.1 Co-precipitation of ZnO

The starting materials used for this study are AR grade zinc nitrate dihydrate [$Zn(NO_3)_2$.2H2O] ,the zinc source and Potassium hydroxide pellets. In a typical synthesis of undoped ZnO, 0.1 M $Zn(NO_3)_2$ was dissolved in distilled water. 0.1 M of KOH was also simultaneously prepared by dissolving appropriate amounts of potassium hydroxide. The Zinc nitrate solution was magnetic stirred subsequently heating to a temperature of 50°C. The potassium hydroxide was added drop wise to produce a white, gelatinous zinc oxide complex. Meanwhile 0.05 g of PVP was also added so as to prevent agglomeration of the particles. The mixture was continuously stirred for two hours at room temperature, after addition of the KOH to a pH value of 8. The precipitates were filtered, centrifuged and dried at room temperature.

2.2 Co- precipitation of $Zn_{1-x}Ag_xO$

For the synthesis of ZnO doped with the transition metal silver, a separate buffer solution was made with 5% $AgNO_3$ in $Zn(NO_3)_2$ and dissolved in 50ml distilled water. It was then mixed with KOH following the same procedure as with undoped ZnO. The potassium hydroxide precipitates the silver oxide which appears as a grayish black complex.

2.3 Thin Films

ZnO thin films were prepared and deposited onto glass substrates in a colloidal consistency, the molarity was maintained for the undoped and the doped cases. The glass substrates were precleaned with detergent, and immersed overnight in sulfochromic solution. They were then washed with distilled water and rinsed with acetone and dried. The coating sol was dropped onto the glass substrates, the latter rotated at 3000 rpm for 20 s by using PHRL/TFA/002 Spin coater. After depositing by spin coating, film was dried at room temperature. The procedures from coating to drying were repeated ten times to obtain uniform thin undoped and doped ZnO films. The deposition time and the chuck rotation rate were found to be inversely proportional to the film thickness.

3. Results and discussion

3.1 X-Ray diffraction studies

The nanoparticles obtained were subjected to various analysis studies which gave insight into its various physical, chemical and thermal properties. For identifying the specimen's crystalline phases and measuring its structural properties, homogenous and inhomogeneous the diffraction pattern was used. Consequently, the change in d-spacing, which is the result of the change of lattice constants under a strain [10] was measured. Peak broadening, independent of $\sin\theta$ caused by the finite size of crystallites was observed. Both crystallite sizes and the inhomogeneous strain contribute to the peak width, which may be separately determined by the careful analysis of peak shapes [11,12]. In the case of no inhomogeneous strain, the crystallite size, D, can be calculated from the peak width with the Scherer's formula $D = \frac{K\lambda}{B\cos\theta_B}$, where λ is the X-ray wavelength, B is the full width of height maximum of a diffraction peak, θB is the diffraction angle, and K is the Scherer's constant of the order of unity for usual crystal [12].

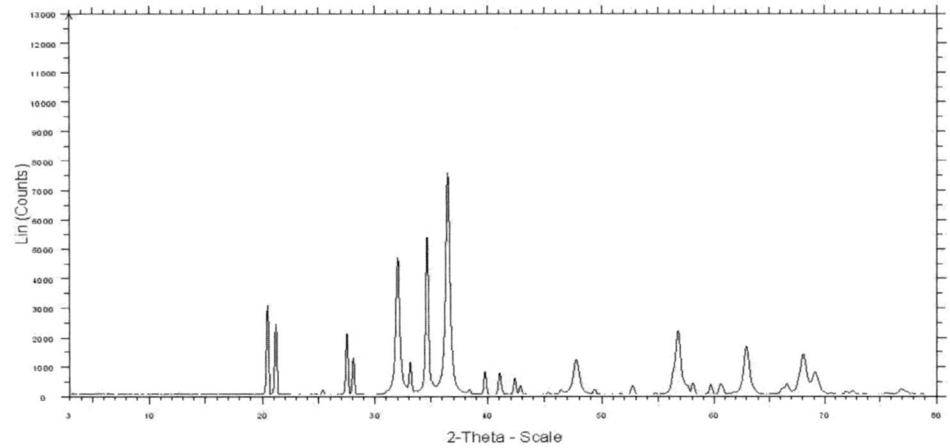

Figure 1.a .XRD pattern of undoped ZnO nanoparticles synthesized peaks observed are at 2θ =36.481, 34.640,31.996, 37.24, 56.800, 63.042 corresponding to the lattice planes (111), (111), (111), (220), (220) respectively (JCPDS card no:77-0191)

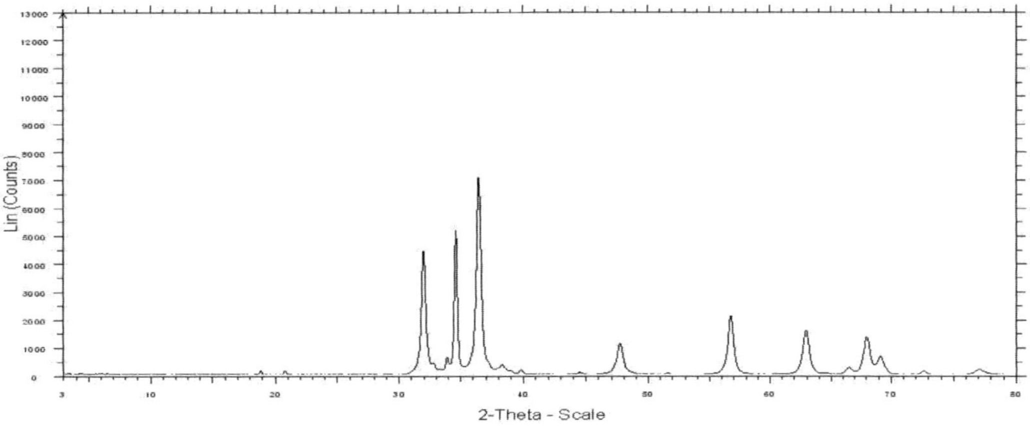

Figure 1.b .XRD pattern of doped ZnO nanoparticles synthesized peaks observed are at 2θ = 36.446, 34.596, 31.957, 37.24, 56.773, 63.008 corresponding to the lattice planes (111), (111), (111), (220), (220) respectively (JCPDS card no:77-0191)

The XRD pattern of undoped ZnO powders show diffraction peaks from various directions of ZnO. the powders being polycrystalline exhibit intense peaks at (111) and (220).

Sample Name	Data observed from XRD		D (nm)	JCPDS data corresponding to observed from XRD pattern		Strain	Dislocation
	2th (degree)	Phase		Cubic phase	Hexgonal phase		
				2th	hkl		
Undoped ZnO	36.481	C	23.495	36.326	111	$8.87x10^{-4}$	$6.561x10^{-4}$
	31.996	C	25.428	36.326	111	$9.890x10^{-4}$	$8.141x10^{-4}$
	56.800	C	20.385	61.199	220	$1.111x10^{-3}$	$1.028x10^{-3}$
	63.042	C	17.921	61.199	220	$9.240x10^{-4}$	$7.107x10^{-4}$

Doped with $AgNO_3$	36.446	C	20.599	36.326	111	$1.682x10^{-3}$	$2.356x10^{-3}$
	31.957	C	21.976	36.326	111	$1.577x10^{-3}$	$2.070x10^{-3}$
	56.773	C	17.567	61.199	220	$1.973x10^{-3}$	$3.240x10^{-3}$
	63.008	C	16.549	61.199	220	$2.094x10^{-3}$	$3.651x10^{-3}$

Table 1. Tabulated XRD values from graph and from the JCPDS card no:77-0191 data. The strain and dislocation values for the corresponding (hkl) planes is also listed in the doped and undoped cases.

The prepared sample as observed, is in the cubic phase of ZnO nanoparticles. The average sizes for the particle was calculated from the Debye Scherer's formula .The average size of the particle D for the doped case was about 18 nm in whereas it is 25 nm in the undoped case. This accounts for the decrease in grain size and thereby increase in the band gap energy which gives more space for excitonic transitions. The crystallite size (D) and the corresponding strain (ε) and dislocation density (ρ) were estimated as tabulated for the doped and undoped variants.

3.2 Optical studies

The optical absorption spectra of ZnO and $Zn_{1-x}Ag_xO$(x=0.05) by UV-viz spectrophotometer in the range 200 to 1000 nm were studied. The figure 2(a) depicts the excitonic absorption peak for both the doped and undoped samples is seen around 230 nm. Doping incorporates a fairly blue shift as can be seen from the curve 2(b) for the doped case. The peak is intense at a particular temperature and evidence suggests that the peaks become broader as the temperature increases [13].

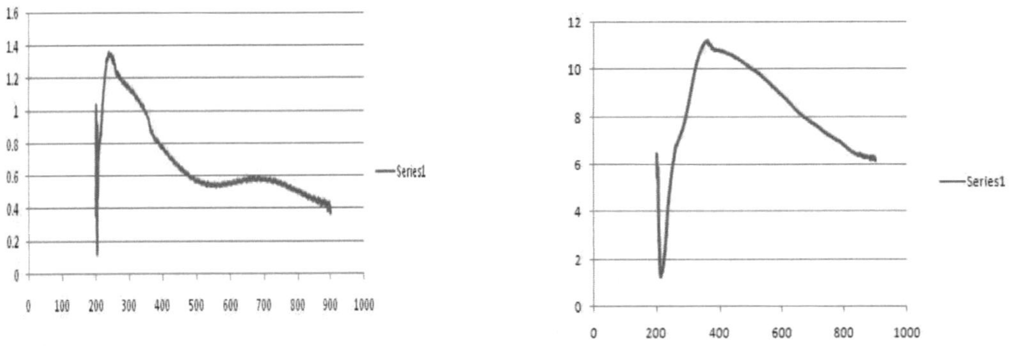

Figure 2. Absorption spectrum measured at various wavelengths under a spectral range of 105 to3300 nm (UV-SHIMADZU 2401 PC) ,and plotted by a recorder between 100 to 1000 nm (a) undoped (b)doped $Zn_{1-x}Ag_xO$

A UV-Visible spectrophotometer was used to measure the amount of light absorbed at each wavelength of the UV and visible regions of the electromagnetic spectrum. This gave information about electronic transitions in the sample. Absorption of radiation by a sample was measured at various wavelengths under a spectral range of 105 to3300 nm(UV-SHIMADZU 2401 PC) ,and plotted by a recorder between 100 to 1000 nm to give the spectrum , a plot of the wavelength of the entire region versus the absorption of light at each wavelength. The samples were dispersed in water to record absorption spectra. The doped ZnO exhibits high transparency in the visible region as shown.

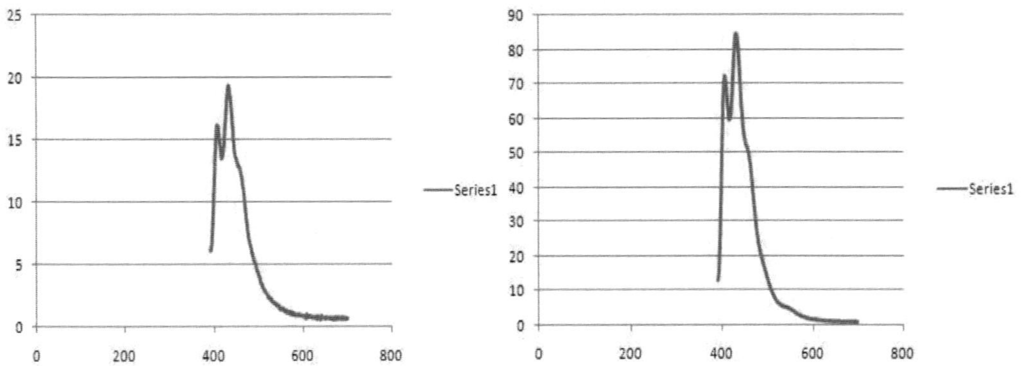

Figure 3.Photoluminescence spectra recorded at the visible range (400-700nm) .Sharp peaks are observed at 420nm and 470 nm in (a) undoped (b) doped

The photoluminescence spectra of undoped and doped ZnO shows two photoluminescent emission spectra at 420 and 470 nm respectively. The doped variant shows a slight blue shift in the emission spectra. The sharp peak corresponds to band to band transition [14]. The less intense peaks if occur correspond to deep level defects, which are uncommon in the current investigation. The intensity peaks in the doped structure shows a dramatic increase at the two wavelength regions. The scope of the doped variant as a good photoluminescent and photovoltaic device thus is evident. The peaks may also be attributed to the excitonic recombination phonon process near the band edge levels in ZnO [15].

4. Conclusion

ZnO is a typical II-VI semiconductor with band gap energy of 3.6 eV at room temperature [12]. Optical properties of ZnO nanoparticles have been studied in the past and evidence for its quantum size effect due to nano dimensions can easily be found in literatures. In the present project, ZnO nanoparticles were prepared by a simple cost effective method. The structural analysis of the fabricated ZnO nanoparticles has been done using XRD techniques. The analysis shows that the prepared samples are in the cubic phase of corresponding nanoparticles .The average size of particles, calculated from the Debye Scherer's formula for the undoped case with respect to most intense peaks (111) and (220) is about 23 nm. The size dramatically decreases with doping to an average of 18 nm increasing the band gap so as to increase the tunability. Also the corresponding strain (ε) and dislocation density (ρ) have been estimated.

It may be also noted that ZnO shows photoluminescent properties in the near ultra violet and visible region making it possible to study the emission properties of nanocrystalline ZnO nanoparticles. The doped variant shows an increase in the photoluminescent emission. The V-I characteristics may be studied which may harness its photo voltaic properties. More optical properties of the synthesized sample may be done which may reveal its optical characteristics such as variation of refractive index with size which can be used for the fabrication of optical waveguides(to be fabricated). By doping ZnO with other metals, it is possible to increase its nonlinear properties and thus it may be used for the production of super luminescence in materials.

References

[1] Golego N, Studenikin S A and Cocivera M 2000 *J.Electrochem. Soc.* 147 1592
[2] Wang X, Song J, Liu J ,Wang Z L 2007 *Science* 316 102
[3]Water W,Yuan S Chu 2002 *Mater.Lett* 55 67
[4]Huang Y,Liu M,Jiang S, Zeng Y, Li C, Liu S, Zhou D 2003 *Microelectron.Eng*. 66 760
[5]Nakada T, Hirabayashi Y, Nokado T, Ohmori D, Mise T 2004 *Sol.Energy* 85 6004
[6]Nomura K, Ohta H,Ueda K, Kamiya T.,Hirano M Hosano H 2003 *Scienc* 300 1269
[7]Konenkamp K Word R C Schlegl C 2004 *Appl.Phys.Lett,* 12 7
[8]Lee S Y, Shim E S, Kang H S ,Pang S S, Kang J S 2005*Thin Solid Films* 437 31
[9]TMckinstry S ,Muralt P 2004 *J.Electroceram.* 12 7
[10] Guozhong C 2004 *Nanostructures & Nanomaterials – Synthesis, Properties & Applications,*

IOP Conf. Series: Materials Science and Engineering **73** (2015) 012003 doi:10.1088/1757-899X/73/1/012003

[11] Ibanga E J et al 2003 *Materials Chemistry and Physics*, 80, 490-495
[12] Kohler M and Fritzsche W, 2004 *Wiley-VCH,* Weinheim
[13] Chauhan R, Kumar A, Chaudhary R P, 2010 *Archives of Applied Science Research* 2(5):378-385
[14] Ghodsi F E ,Ablasan H 2010 *Acta Physica Polonica A*,Vol 118
[15] Studenikin S A ,Cocivera N, Pasher H 2000 *J.Lumin*,91,223

Photoluminescence Property of 1-(2-Hydroxyphenyl)thiourea and its Quenching by Chromium(VI) Ions

A Sunil and S Jagadeeswara Rao

Department of Chemistry, Sri Sathya Sai Institute of Higher Learning, Prasanthi Nilayam, Puttaparthi-515134, Anantapur, Andhra Pradesh, India.

E-mail: asunil@sssihl.edu.in

Abstract. 1-(2-Hydroxyphenyl)thiourea is a fluorogenic compound and it undergoes a very slow autooxidation to form an yellow coloured disulphide, 1-hydroxy-2-[({[N-(2-hydroxyphenyl) carbamimidoyl]disulfanyl} carbonoimidoyl) amino] benzene. In the presence of chromium(VI) ions, the rate of formation of the disulphide increases which results in the quenching of fluorescence. This analytical property has been utilized for chromium(VI) determination. The emission studies were recorded at λ_{Ex}=416nm and λ_{Em}=520nm respectively. The influence of analytical properties, role of surfactants and activators are reported. Fluorescence quenching studies have been experimentally verified using Stern-Volmer plots.

1. Introduction

There is an expeditious requisition for conscientious and reliable analytical methods for the determination of chemical forms of elements that are toxic and pose harm to the environment. Chromium is one such element that most frequently occurs as chromium(VI) or chromium(III) in aqueous solutions. Among these, chromium(VI) is known to be a strong oxidizing agent, posing a high risk to humans and animals due to its carcinogenic and mutagenic properties[1]. That is why the determination of chromium in environmental and biological samples is of great interest.

Chromium has versatile industrial applications and enters the environment as a result of effluent discharge from industries. The major industries that contribute to water pollution by chromium are mining, leather tanning, textile dyeing, electroplating, aluminum conversion coating operations, plants producing industrial inorganic chemicals and pigments, and wood preservatives. Therefore, the level of chromium in expelled industrial wastewater should be monitored[2].

Many sophisticated techniques such as electro-osmosis[3], flame atomic absorption spectrometry[4], electrothermal atomization[5,6], flow injection analysis[7], electro analytics[8] etc. have been utilized for determination in trace levels but they are tedious and require very expensive equipment. Fluorescence based analytical techniques are very attractive because fluorescence measurements are consistently very sensitive (parts per billion/trillion), relatively cheaper and easy to monitor[9]. Catalytic methods have proved to be more economical and impeccable alternative tool to the instrumental analyses for trace determinations[10]. Catalytic methods for chromium sensing are mostly developed based on its effect on the oxidation of organic compounds with appropriate oxidant, most frequently hydrogen peroxide[11-13]. The current method utilizes the auto-oxidation property of

1-(2-Hydroxyphenyl)thiourea (**A**) to form its disulphide (**B**). This special indicator reaction led to the ascendancy of the current proposed method.

2. Experimental

All the chemicals used in the investigation are of analytical grade with high purity (Sigma Aldrich Chemicals Ltd) and HPLC grade solvents (Merck). The buffer solutions for pH studies were prepared as prescribed by Vogel[14]. MICROPRO pH meter was used for all the pH adjustments. Quartz cells of 10mm path length (Hellma make) were used to record the emission and excitation intensities of the chromophore in Perkin Elmer LS-55 fluorescencemeter. Mettler Toledo: AB204-S was utilized for all the weight measurements.

Compound **A** was synthesized as per the standard procedure [15,16]. 1.0mL of potassium phosphate buffer solution (pH=8.0) and appropriate amount (0.3ng/mL-320ng/mL) of chromium(VI) solution were pipetted in to a 10.0mL volumetric flask, followed by 1.0mL of 1% 2,2'-bipyridyl and 1.0mL of 1% sodium dodecyl sulphate (SDS). The temperature was maintained at 25^0C. 3.0 mL of 1-(2-Hydroxyphenyl)thiourea (1mg/mL) solution was then added and millipore water was added up to the mark. The resulting reaction mixture was transferred into 10mm quartz cuvettes of fluorescencemeter. The fluorescence emission measurements were recorded at λ_{Em}=520nm upon excitation at λ_{Ex}=416nm respectively at 20th minute. The blank experiments were repeated by following the same procedure to obtain relative fluorescence intensity I_0 and the value of I_0-I was calculated. The calibration graph was plotted and the method was applied for the determination chromium in waste water samples. Water samples were boiled, treated with concentrated nitric acid and then filtered to remove organic particulate matter before using for the analysis.

3. Results and discussion

Compound **A** undergoes a very slow auto-oxidation to form **B** and based on its auto-oxidation property which was reported[17], determination of cadmium(II), copper(II) and mercury(II) was done[18, 19]. The excitation and emission studies of **A** were studied thoroughly, which reveals that **A** is fluorogenic in nature and the catalytic action of hexavalent chromium leads to the product formation **B** quickly with a visible colour change and a drop in fluorescence intensity. By monitoring both the colour and signal changes the quantification of chromium was done. Various analytical parameters were studied for reporting the ambient reaction conditions. Activators and surfactants were used to enhance the sensitivity to much lower limits of detection. The fluorescence emission studies were done by excitation of compound **A** at λ_{Ex}=416nm, which gave a good emission peak at λ_{Em}=520nm. At this emission wavelength, the intensity of the chromophore starts reducing upon addition of chromium(VI). The experimental observations show that fluorescence quenching of **A** is proportional the concentration of chromium(VI) added.

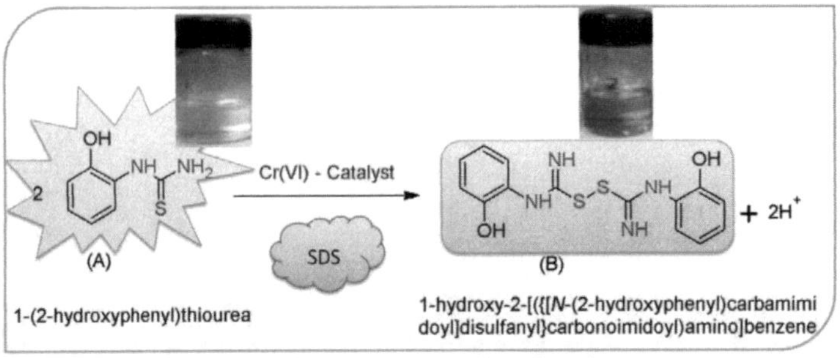

Figure1. Chromium(VI) sensing based on photoluminescence property of 'A'

The investigation of analytical parameters such as pH, reaction time, temperature, concentration of the reagent and chromium(VI) ion were studied. It was evident from the recorded observations that the chromium(VI) catalysed transformation of **A** to **B** is more pronounced at pH~8. Hence, the same pH was maintained throughout the study. The rate of reaction showed no significant improvement upon altering temperature. Therefore, the investigation was carried out at normal laboratory temperature of $25^0\pm5^0$C. Optimum concentration of **A** was determined to be 3mg/mL and the same concentration was maintained throughout the investigation. By varying the concentration of chromium(VI) ion in the range 0.3ng/mL-320ng/mL, fluorescence emission study was done. The fluorescence spectrum [figure2(I)] shows that the intensity quenches linearly with the increase in concentration of chromium(VI) ions, which was supported and proved using the Stern-Volmer plot [figure2(II)]. The Stern-Volmer plot is obeyed in the range 10 ng/mL-250ng/mL of chromium(VI) with Stern-Volmer quenching constant, K_{SV}=0.929 indicating a good linearity in the method.

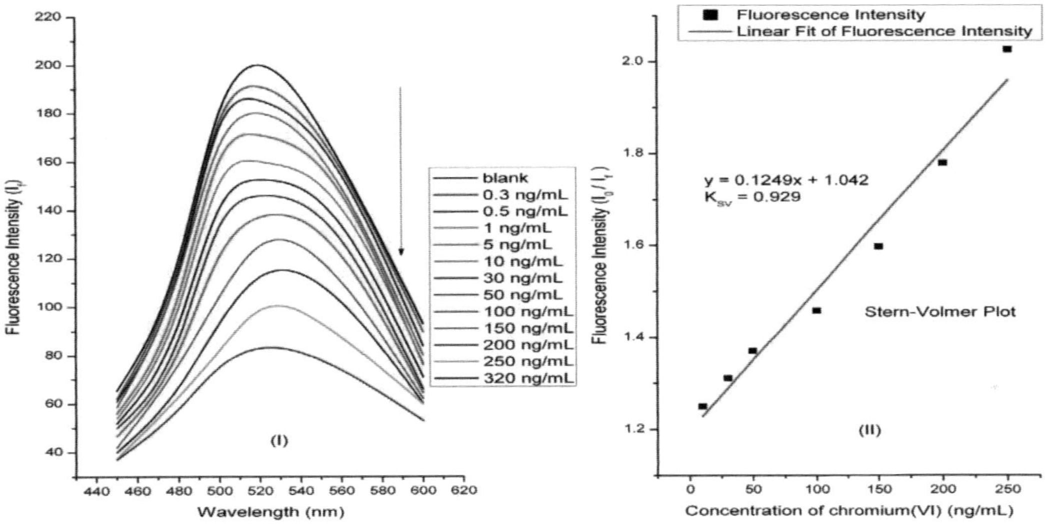

Figure2. (I) Fluorescence spectrum showing the quenching of 1-(2-Hydroxyphenyl)thiourea by chromium(VI) ions. (II) Stern-Volmer plot showing the efficiency of fluorescence quenching

Activators and surfactants have been used to improve the sensitivity of the method,. The activators such as 1,10-phenanthroline, bathophenenthroline, pyridine, 2-aminopyridine, quinoline, 2,2'-bypridyl, nicotinic acid were used and the observations show that 1% 2,2'-bipyridyl enhanced the reaction rate in the formation of product **B**. One can achieve more stable reaction conditions using micellar medium and to observe the effect of surfactants on the reaction, surfactants such as cetyltrimethylammonium bromide(CTAB), Tween-80, Triton-X, sodium dodecyl sulphate(SDS) were used and 1% SDS has showed a promising effect on the reaction rate. In this current investigation, the synergy of 2,2'-bipyridyl and SDS in enhancing the sensitivity is a notable result. Without the activator and surfactant system, the determination was possible only till 30ng/mL but, by using the same, the sensitivity was enhanced up to 0.3ng/mL. In current range of determination many metal ions like manganese(II), cobalt(II), arsenic(V), lead(II), mercury(II), cadmium(II), zinc(II), rhodium(III), iridium(III), platinum(II), silver(I), molybdenum(IV) do not interfere at all. Iron(II) and copper(II) interfere very slightly but, can be eliminated by the addition of fluoride and thiosulphate respectively. The method is fairly sensitive and was applied to determine the chromium(VI) content in waste water samples using the calibration curve. Recovery test was done to validate the method with a satisfactory range of 98%-102%.

4. Conclusions

A simple and sensitive photoluminescence technique has been developed for determination of hexavalent chromium, exhibiting a good number of advantages compared to the various methods reported earlier. The reaction utilized in the investigation is an auto-oxidation reaction and hence, no need for any oxidizing agents unlike other methods. Reagent synthesis is very simple, cost effective and starting products required for synthesis are also inexpensive. The method is bench-stable with no inert conditions required and can be done at normal room temperature. No pre-concentration procedures are involved. The sensitivity levels are greater or equivalent to some of the earlier methods.

Acknowledgements

The authors sincerely acknowledge the inspiration and support rendered by the Founder Chancellor, Bhagawan Sri Sathya Sai Baba and authorities of Sri Sathya Sai Institute of Higher Learning.

References

[1] Barceloux DG 1999 *Clin. Toxicol.* **37**(2) 173-194
[2] Dakiky M, Manassra A, Mereb M 2002 *Adv. Environ. Res.* **6**(4) 533-540
[3] Gan W, Yang L, He Y, Zeng R, Cervera M L and De La Guardia M 2000 *Talanta* **51**(4) 667-675
[4] Anthemidis A N, Zachariadis G A, Kougoulis J S and Stratis J A 2002 *Talanta* **57**(1) 15-22
[5] Ezer M, Elwood S A and Simeonsson J B 2001 *J. Anal. At. Spectrom.* **16**(10) 1126-1130
[6] Nise G and Vesterberg O 1979 *Scand. J. Work Environ. Health.* **5**(4) 404-410
[7] Paleologos E K 1998 *Analyst* **123**(5) 1005-1009
[8] Lin L, Lawrence N S, Thongngamdee S, Wang J and Lin Y 2005 *Talanta* **65**(1) 144-148
[9] Quang D T and Kim J S 2010 *Chem. Rev.* **110**(10) 6280
[10] Yatsimirskii K and Tikhonova L 1987 *Talanta* **34**(1) 69-75
[11] Mohamed A A, Ahmed S A and El-Shahat M F 2001 *J. Trace Microprobe Tech.* **19**(2) 297-311
[12] Stoyanova A M 2004 *Turk. J Biochem.* **29**(2) 204-207
[13] Liu Z and Wang X 1996 *Fenxi Huaxue(Analytical Chemistry)* **24**(2) 164-167
[14] Vogel A I and Jeffery G H. 1989 *Vogel's Textbook of Quantitative Chemical Analysis* (Longman Scientific &Technical) p 268
[15] Beilstein. 1943 *Handbuch Der Organischen Chemie* (Ann Arbor, Michigan) Band XIII 375 (Edwards Brothers Inc., Publishers)
[16] Sunil A and Jagadeeswara Rao S 2012 *Res.J.Chem.Sci* **2**(3) 30-40
[17] Feigl F 1966 *Spot tests in Organic Analysis* (New York: Elsevier)
[18] Sunil A and Jagadeeswara Rao S 2012 *Int. J. Curr. Res.* **4**(2) 210-215
[19] Sunil A and Jagadeeswara Rao S 2012 *Int. J. Chem. Anal. Sci.* **3**(2) 1318- 1321

Influence of Sol Concentration on the Properties of Spin Coated Zirconia Thin Films

A Kavitha [1], N Sankara Subaramanian [2] and S Vadivel [3]
[1, 2] Thinfilm Research Laboratory, Department of Physics, Thiagarajar college of Engineering, Madurai – 625 015, Tamil Nadu State, India.
[3] Department of Physics, Paavai Engineering College, Paavai Nagar, Pachal - 637 018 Namakkal District, Tamilnadu State, India.

[1] kavitha_26sep@yahoo.com , [2] shankersathiya@yahoo.com ,
[3]vadivel.physics@gmail.com

Abstract. Present work involves with the preparation of Zirconia thin films on glass substrate by the cost effective spin coating method at different sol concentration and investigation of its structural, optical and surface properties. XRD profile of Zirconia thin films prepared at different concentration conforms tetragonal phase with preferred orientation along the (111) plane. The average particle size evaluated from the XRD data for solute concentration 0.3 mole is 61 nm. The optical transmittance of the Zirconia thin films have been measured in the visible and NIR region. The band gap energy lies in the range of 3.6 - 3.8 eV for thin films prepared at different concentrations. FTIR analysis shows the peaks 472 and 485cm^{-1} are attributed to Zr-O stretching in tetragonal ZrO_2.The surface morofological features of the Zirconia thin films have been investigated from the 2D and 3D AFM images, which indicates smooth and uniform surface pattern without any pits of the pin holes.

1. Introduction

Ceramic coatings has been applied to high energy applications due to its low thermal conductivity, high band gap, high thermal expansion coefficients and high wear resistance [1-3]. Ceramic material particularly Zirconia thin films are used in thermal barrier coatings, optical coatings , buffer layers oxygen sensors, heat resistant coatings , ion conductors and optical storage elements etc [4-6]. There has been great scientific concentration in Zirconia thin films due to their phase transitions. Pure Zirconia exhibits three polymorphs depending on temperature: monoclinic (m) stable up to1170° C, tetragonal (t) stable up to 1170-2370° C, and above it have cubic (c) phase until it melt at 2706° C [7]. The property of thin films strongly lies on the method of preparation and growth parameters. Zirconia thin films has been prepared by a wide variety of deposition processes Viz., Plasma spraying [8], sol-gel spin coating [9], chemical spraying method [10], electro chemical deposition [11], ion assisted deposition [12], dip coating [13], hydrothermal processing [14], electron beam evaporation method [15] etc., Out of these, sol-gel spin coating method gives us high superiority oxide materials and is well adaptable for thin film preparation. The surface of films prepared by sol-gel spin coating technique is very smooth which is very important to achieve desire properties for many applications. Other advantages of this method include high purity of precursor, high homogeneity of material, low processing temperature, absence of vacuum, low cost and high flexibility. The grain size and thickness generally inclined by the number of coatings and concentration of the sol. The sol concentration affects the thickness via the viscous drag on the substrate by the moving liquid [16, 17]. The present work deals with the influence of sol concentration on the preparation of Zirconia thin films and to investigate its structural, optical and surface properties.

2. Experimental:

Aqueous solutions of ZrO_2 with concentrations of 0.1, 0.2, 0.3 and 0.4 mol dm^{-3} were prepared using Zirconium Oxychloride Octahydrate ($ZrOCl_2.8H_2O$) as starting material. In order to adsorb and remove the acids produced the salt was mixed with 100 ml doubled distilled water. The solution was stirred well in magnetic stirrer for about an hour at a temperature of 85° C. The solution was air cooled in the ambient and then aged in open beakers at room temperature for gelation. After a period of aging, the solution becomes gel. ZrO_2 thin films were developed by dispensing about 2 ml of the as prepared gel on the glass substrate mounted over the turn table. Zirconia thin films have been deposited on glass substrates using the following optimum coating parameters: 1) coating period of the sol - 2-5 day 2) spin rate – 3000 rpm 3) spin time – 15sec 4) baking temperature - 400°C 5) baking duration -7 minutes. The sol-gel coated substrates were baked at a temperature of 400° C for 7 minutes in air after each spin coating. By repeating this step for several times, pin hole free and uniform ZrO_2 films were obtained. The coated films were then annealed at temperature of 600° C for one hour duration each.

X-ray diffraction measurements of the ZrO_2 films were recorded using Philips pananlytical diffractometer. The optical transmittance of the films has been measured in the UV- Visible region, using the Perkin and Elmer spectrophotometer, from which other optical constants were arrived. FTIR spectral data for the ZrO_2 films were obtained using the SHIMADZU FTIR spectrometer in the wave number range of 2000-400 cm^{-1}. The surfaces feature of the ZrO_2 films has been investigated using Atomic Force Microscope (AFM).

3. Result and Discussion

3.1 Micro Structural properties

XRD analyses of spin coated ZrO_2 thin films were prepared at different solute concentrations are shown in Figure 1. The films prepared at different solute concentrations confirms metastable tetragonal phase of Zirconia thin films with preferred orientation along (111) plane. The other orientation observed with relatively lower intensities are (200), (220) and (311). For lower solute concentration the peak intensity is very low, on other hand, when the concentrations are increased, the intensity count and peak sharpness of all planes abruptly increases and this may attributed to increases oriented over growth and presence of large-sized grains. As solute concentration increases the tetragonal phase changes to monoclinic phase and have a sharp peak at 31°. The grain size of ZrO_2 films were calculated from the full width half maximum using the Scherrer equation and tabulated in table 1.

$$D = 0.94\lambda / \beta \cos\theta,$$

where D is the grain size, β is the full width half maximum of the most intense plane, λ is the wavelength of X-ray (CuKα radiation λ=1.5406A°, θ is the Bragg angle. As the solute concentration increases the grains size also increases which may be due to the solute particles becomes larger thereby increasing the chance of more solute to be gathered together forming a grain. The FWHM value decreases linearly as the concentration increases indicating a better crystallinity of the films with higher solute concentration [16,17].The dislocation intensity and micro strain were calculated using the relation $\delta = 1/D\text{^}2$ and $\varepsilon = \beta \cos\theta / 4$. The dislocation density and micro strain tend to decreases as solute concentration increases which may be due to crystalline structure of the Zirconia thin films. At higher solute concentration dislocation density and strain increases due to the change in the preferred orientation of the films [18]. The texture coefficient of the film defined by Barret and Massalcki [19] can be used to describe preferred orientation of the film is given by

$$T.C_{hkl} = (I_{(hkl)} / I_{o(hkl)}) / 1/N \sum_{N} I_{hkl}/I_{ohkl}$$

Where T.C $_{(hkl)}$ is the texture coefficient of the (hkl) plane, $I_{i\,(hkl)}$ is the measured intensity, $I_{o\,(hkl)}$ is the standard intensity of JCPDS power diffraction pattern of the corresponding peak and N is the number of reflection considered for the analysis. From the tabular column the value T.C $_{(hkl)}$ = 1, for 0.3 mole concentration represents films with randomly oriented crystallites, while lower values indicate the, abundance of grains oriented in a given (hkl) direction. The evaluated value of T.C equals to unity implies that randomly oriented crystallites and chemically stable due to its closely packed surface.
Standard deviation σ from the powder diffraction condition was calculated by the following expression and tabulated for different solute concentration.

$$\sigma = (\Sigma I^2_{hkl} - \Sigma (I_{hkl})^2 / N) / N$$

The standard deviation σ is less for 0.3 mole solute concentration compare to another solute concentration which indicates the crystalline structure of thin films.

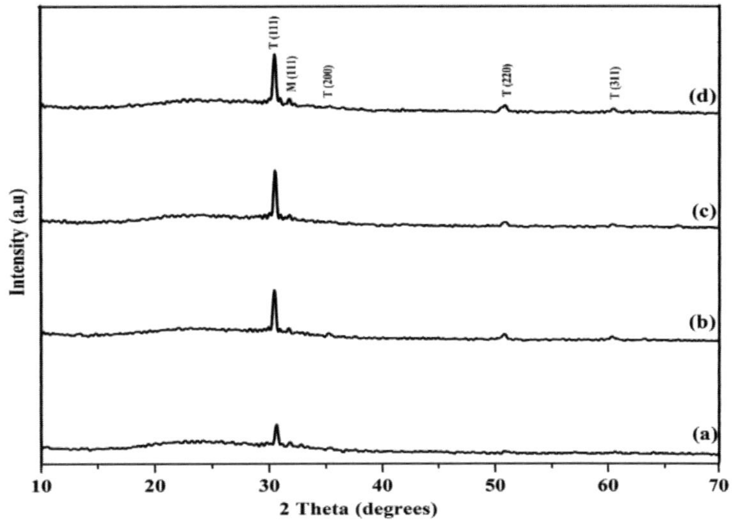

Figure 1 XRD spectra of Zirconium Oxide thin films prepared at different concentration (a) 0.1mole (b) 0.2mole (c) 0.3mole and (d) 0.4mole.

Table 1 Crystallographic parameters of the Zirconium Oxide thin films coated at different concentrations.

Concentration in mole dm^{-3}	Particle size n m	Texture coefficient	Standard deviation σ	Dislocation density δ *10^{15} line/meter^2	Micro Strain ε *10^{-5} line^2/m^{-4}
0.1	51	0.999	0.4358	37	76
0.2	59	0.999	0.4758	28	70
0.3	61	1.0056	0.408	27	69
0.4	57	0.999	0.51187	29	71

The 2D and 3D AFM image of the zirconia thin films spin coated with solute concentration 0.3 mole is shown in the Figure 2 (a, b). The AFM micrograph illustrates smooth uniform surface pattern without any dark pits, pin holes and cracks. The AFM image further shows accumulation of large number of fine grains associated with hump like oriented growth of the crystallites.

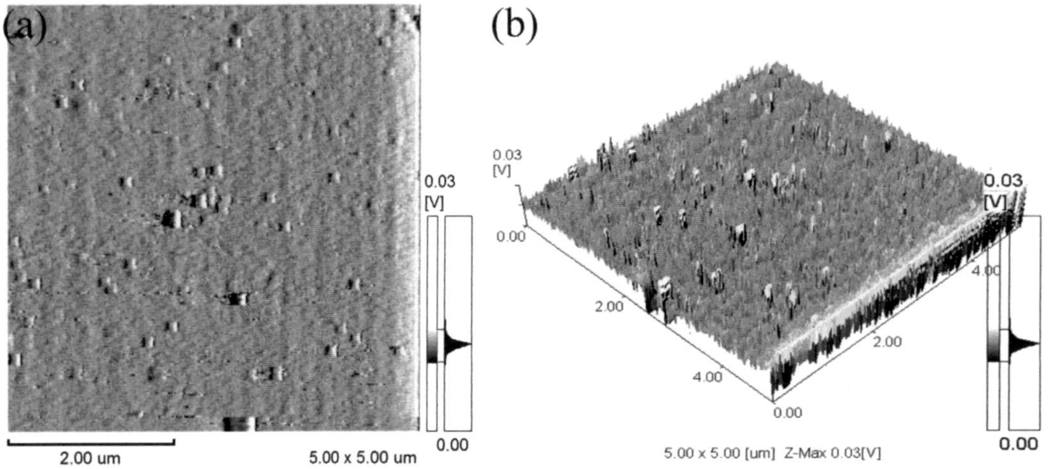

Figure 2 (a) 2D AFM image of the Zirconia thin film spin coated with solute concentration 0.3 moles.
Figure 2 (b) 3D AFM image of the Zirconia thin film spin coated with solute concentration 0.3 moles.

3.2 Optical characterization

Transmission spectra of the ZrO_2 thin films prepared at different solute concentrations are shown in Figure 3. From the figure it is evident that for all the films, the transmittance spectra lie in the UV-Visible region. The transmittance is found to be 65-75 % at the wavelength 400 nm for the Zirconia thin films prepared at different solute concentrations. Further, the figure shows that the transmittance increases with the increase in scattering due to large grain size and crystal aggregate, caused by the increase in film thickness with concentration [16]. The extension coefficient k_f is obtained from the transmission spectra using the relation,

$$Kf = 2.303 log\,(1/T)\lambda \;/\; 4\pi\, t$$

where T is the transmittance, λ is the wavelength of light and t is the thickness of the corresponding film. The absorption coefficient α can be obtained from the relation $\alpha = (4\pi k_f)/\lambda$ and the band gap energy (E_g) proportional to absorption coefficient can be obtained from the following relation $(\alpha h\nu)^2 = k\,(h\nu - E_g)$, k is a constant.

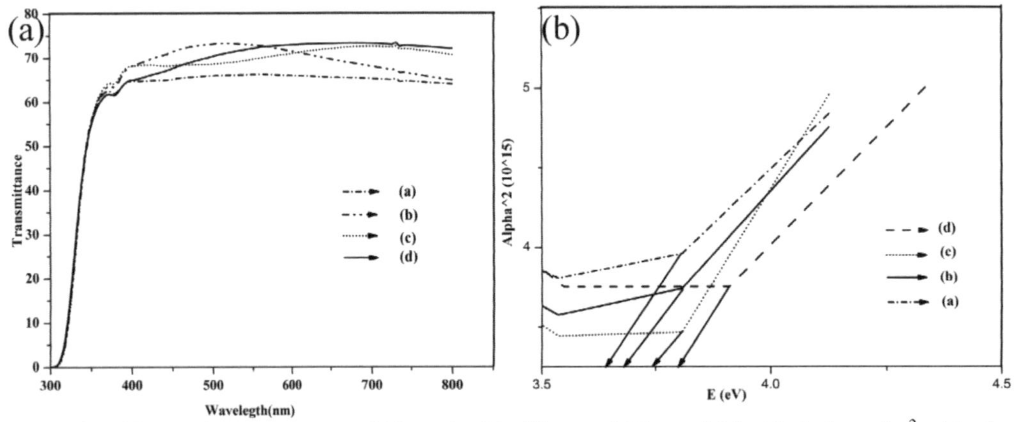

Figure 3(a) Transmittance spectra of Zirconia thin films and Figure 3(b) Variation of α^2 with photon energy prepared at different solute concentration (a) 0.1 (b) 0.2 (c) 0.3 and (d) 0.4 mole.

The variation of square of the absorption coefficient with photon energy of zirconia thin films prepared at different solute concentration is shown in Figure 3(b).The band gap energy value is found to be 3.6 - 3.8 eV. These values are coincides with the values reported by Ramamurty et al [20]. The low value of band gap may due to the non stoichiometric nature of the films. Moreover the band gap energy has been found to increase with the increase in concentration, which may be attributed to the increases in carrier density with increase in solute concentration [7].

3.3 FT-IR spectroscopy

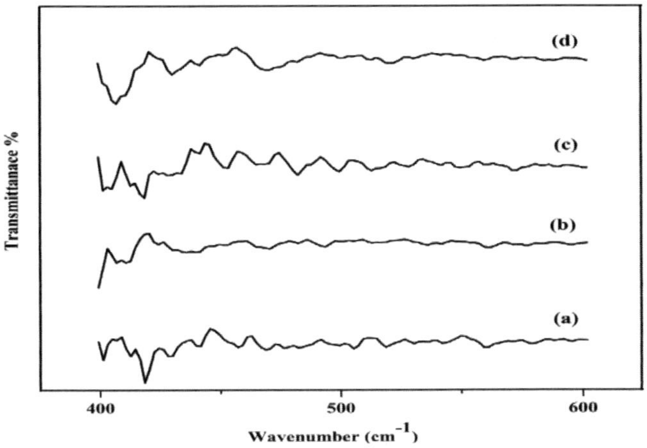

Figure 4 FTIR spectra of zirconia thin films spin coated with (a) 0.1 (b) 0.2 (c) 0.3 and (d) 0.4 moles.

Figure 4 shows the FT-IR spectra of the Zirconia thin films, prepared at different solute concentrations. The FT-IR spectrum, of the thin films prepared at 0.1- 0.3 mole concentration shows absorption band at 472 cm^{-1}. This band was shifted to 485 cm^{-1} on increasing the solute concentration at 0.4 moles. The peaks 472 and 485 cm^{-1} are attributed to Zr-O stretching in tetragonal ZrO_2 .This shift in the absorption band at lower to higher frequency, with increase in solute concentration may be attributed to an increases of the Zr-O bond strength. The band observed around 2500cm^{-1} is characteristic of Zr-O species. The bands at 3500cm^{-1} are characteristic of OH groups [21].

4. Conclusion

Zirconia thin films were developed by cost effective spin coating method with different solute concentrations. The ZrO_2 thin films prepared with different solute concentrations indicates tetragonal phase with preferred orientation along the (111) plane. The tetragonal phase stabilization in the Zirconia thin films arises due to the presence of large-sized grains. The increase in optical transmittance with increase in solute concentration may be attributed to the reduced gain boundary scattering due to the enhanced crystallinity. The evaluated band gap energy lies between 3.6 - 3.8 eV. FTIR analysis shows the peaks 472 and 485cm^{-1} are attributed to Zr-O stretching in tetragonal ZrO_2. The AFM image further shows accumulation of large number of fine grains associated with hump like oriented growth of the crystallites.

References

[1] Yunxia Chen, Weimin Liu, *Thin Solid Films* **516** (2008) 8352–8358
[2] Gurudayal, Amit Kumar Srivastava, Jitendra Kumar, *Materials Letters* **83** (2012) 172–174
[3] Shyankay Jou, Tien-Wei Chi, *Vacuum* **81** (2007) 911–919
[4] Hembram K P S S, Mohan Rao G, *Materials Letters* **61** (2007) 502–505
[5] Kaiqing Luo, Shuxue Zhou, Limin Wu, *Thin Solid Films* **517** (2009) 5974–5980
[6] Sung-Wei Yeha, Hsing-Lu Huangb, Dershin Gana, Pouyan Shen, *J. Cryst. Growth* **289** (2006) 690–702
[7] Sankara Subramanian N, Sornakumar T, Gurumoorthy S, Harihara Sundar M, Vickraman P, *Ionics* (200) **13**, 461-466
[8] PeshevI. Stambolova P, Vassilev S, Stefanov P, Blaskov V, Starbova K, *Material Science and Engineering* **B97** ,(2003) 106-110.
[9] Sue-min Chang, Ruey-an Boong, *Thin Solid Films* **489** (2005) 17-22.
[10] Bhavana Godbole, Nitu Badera, Shrivastav S B and Ganesan V, *Jl. Of Instrum. Soc. of India* Vol. **39** No. **1** March 2009
[11] Brunet M , Mafhoz Kotb H, Bouscayrol L, Scheid E, Andrieux M, Legros C, Schamm-Chardon S, *Thin solid films* **519** (2011) 5638-5644
[12] Ghanashyam Krishna M, Narasimha Rao K and Mohan S, *Thin Solid Films*, **207** (1992) 248-251
[13] Ivanovaa T, Harizanovaa A, Koutzarovab T, Krinsc N, Vertruyenc B, *Materials Science and Engineering B* **165** (2009) 212–216
[14] Bokhimi X, Morales A, Novaro O, Portilla M, Lopez T, Tzompantzi F and Gomez R, *Journal of Solid State Chemistry* **135**, 28-35 (1998)
[15] Hampikian J M and Dohoke N B 1999 *Elevated Temperature Coating*: *Science and Technology* 13-36
[16] Dutta M, Mridha S, Basak D , *Applied Surface Science* **254** (2008) 2743–2747
[17] Joy K , Maneeshya L V, Jijimon K. Thomas, Thomas P V, *Thin Solid Films* **520** (2012) 2683–2688
[18] S. Lalitha, R. Sathyamoorthy, S. Senthilarasu, A. Subbarayan, K. Natarajan, *Solar Energy Materials & Solar Cells* **82** (2004) 187–199
[19] Barrett C S, Massalski T B, 1981 *Structure of Metals*.
[20] Ramamurty S, Sankara Subramanian N , Sarangarajan K and Santhi B, *Bulletin of Electrochemisty* **14** (11) 1998, 446-448
[21] Santos V, Zeni M, Bergmann C P and Hohemberger J M, *Rev.Adv.Mater.Sci.* **17** (2008) 62-70.

IOP Conf. Series: Materials Science and Engineering **73** (2015) 012006 doi:10.1088/1757-899X/73/1/012006

CO_2 Laser irradiation of GeO_2 planar waveguide fabricated by rf-sputtering

S Valligatla[1,2], A Chiasera[2], N Bazzanella[3], L Lunelli[4], A Miotello[3], M Mazzola[2], D Narayana Rao[1] and M Ferrari[2]

[1] School of Physics, University of Hyderabad, Hyderabad 500046, India.
[2] CNR-IFN, CSMFO Lab., via alla Cascata 56/c, Povo, 38123 Trento, Italy.
[3] Department of Physics, IdEA Lab., University of Trento, via Sommarive 14, Povo, 38123 Trento Italy.
[4] FBK, BioSInt Lab., Via Sommarive 18, Povo, 38123 Trento, Italy.

Emial: srihcu08@gmail.com

Abstract: We present a fabrication protocol combining radio frequency sputtering technique and CO_2 laser annealing for GeO_2 based planar optical waveguides. The effects of pulsed CO_2 laser irradiation on the optical and structural properties of pure GeO_2 planar waveguide are evaluated by different techniques as m-line and micro-Raman spectroscopy and AFM measurements. Amorphous GeO_2 planar waveguide was fabricated by Radio Frequency magnetron sputtering system on v-SiO_2 substrate. An increase of the refractive index of approximately 0.04 at 1.5 μm and a decreasing of the attenuation coefficient from 0.9 to 0.5 dB/cm at 1.5 μm have been observed after pulsed CO_2 laser annealing. Raman spectroscopy and AFM results showed that after an adapted pulsed CO_2 laser annealing, the resulting materials showed a crystalline environment in which the phase of the crystalline GeO_2 varies with varying irradiation time.

Introduction

GeO_2-based glasses have demonstrated to be one of the most promising candidates for a high-functionality material platform for integrated optics. Germania glass exhibits a lower Tg (~ 790 K) and a lower phonon energy (~ 900 cm^{-1}) than SiO_2 based glasses. Moreover, GeO_2 glasses are transparent in the near infrared region and their refractive index allows fabrication of confined structures. Finally, Germania allows fabrication of highly photorefractive planar waveguides [1] and drawing fiber with smaller absorption losses in the middle IR range as compared to silica glass [2]. However the tailoring of the synthesis and optimization of a particular phase of GeO_2 crystals is very important as this material exhibits several polymorphs [3]. For these reasons it is of technological and scientific importance to develop a particular fabrication protocol to allow obtaining optical waveguides based on GeO_2 system that exhibits low attenuation coefficients and simultaneously embed GeO_2 nanocrystals with a particular phase.

In order to produce active rare earth nanocrystals in a glass matrix, heat treatment using a furnace has been commonly used, but a potential candidate for reaching future technology nodes is the laser annealing (LA) process. For laser annealing several types of lasers have been used, differing primarily in wavelength (e.g. XeCl 308 nm, frequency doubled Nd-YAG 532 nm, Nd-YAG 1064 nm, CO_2 10.6 lm) [4, 5]. It was reported that CO_2 laser annealing can reduce scattering losses in Corning 7059 glasses [6, 7] and ZnO [8] thin-film waveguides fabricated on thermally oxidized silicon substrates; losses as low as 0.05 dB/cm for Corning 7059 glass waveguide [6] and 0.01 dB/cm for ZnO waveguides [8] have been achieved by this technique.

This paper is devoted to the fabrication by rf-sputtering of amorphous GeO_2 planar waveguide and the investigation of the possibility to derive from this system a glass-ceramic in planar format by sequent CO_2 laser irradiation.

Experimental

Pure GeO_2 planar waveguide was prepared by Radio-frequency Sputtering (RFS) technique. The wave guiding film was deposited on Silica substrate. In order to improve the adhesion of the films, the substrate was cleaned inside the chamber by heating at 120°C for 30' just before the deposition procedure. Sputtering deposition of the film was performed by using 15x3 cm^2 Germania target. The residual pressure, before deposition, was about $1.6x10^{-6}$ mbar. During the deposition process, the substrates were not heated. The sputtering was carried out with an Ar gas at a pressure of $5.4x10^{-3}$ mbar; the applied rf power was 80 W and the reflected power 0 W. The deposition time, necessary to reach the appropriate thickness for one propagating mode at 1.5 μm, was 1h 58 min.

As prepared GeO_2 planar waveguides has been used for the irradiation with pulsed CO_2 laser. We use the S50 TEA CO_2 laser with a mixture of CO_2:N_2:He fixed at 33%:33%:33% percentages and is a pulsed laser with 0.8 J per pulse and have 5 Hz repletion rate, 380 ns pulse width and 10.6 μm wavelength with beam diameter of 1.5 cm. The samples were irradiated in air with 1.3 W medium power and irradiation time varies from 1 to 2 hours. The compositional analysis was performed using energy dispersive spectroscopy (EDS), by using a Noran Instruments mod. Voyager apparatus. Standard AFM technique has been used to measure the surface roughness of the samples before and after laser annealing.

The thickness and the refractive index of the waveguide at 632.8, 1319 and 1542 nm were measured in TE and TM polarizations, by an m-line apparatus (Metricon model 2010) based on prism coupling technique. The losses at 632.8, 1319 and 1542 nm were evaluated by collecting the light intensity scattered out of the waveguide plane for the TE_0 mode using a fiber scanner. More detail about the experimental setup can be found here [9]. Raman scattering measurements were performed at room temperature in a wave number range between 100 and 1000 cm^{-1} by Horiba-Jobin-Yvon, LabRam Aramis Micro Raman set up with exciting radiation at 633nm by using He-Ne laser with power 20 mW. To avoid unwanted laser-induced transformations, neutral filters of different optical densities were used, whenever necessary. The resolution was about 0.35 cm^{-1}/pixel.

Results and Discussions

Table 1 reports the optical parameters for as prepared sample and after CO_2 laser irradiation for 2 h. The waveguide present thickness around 1μm and support one TE and TM mode at 1319 and 1542nm. No change in the thickness was evidenced after the laser irradiation. The refractive indices measured in TE and TM polarizations are equal within the experimental uncertainty, so that film birefringence can be considered negligible. Moreover the laser irradiation do not induce changes between the refractive indices measured in TE and TM polarizations so we can affirm that in these conditions that the CO_2 laser annealing do not induce birefringence in these systems.

Table1: Optical parameters for as prepared sample and after CO_2 laser irradiation for 2h.

Laser wavelength(nm)		Refractive Index		thickness (μm)	Attenuation co-efficient(dB/cm)
		TE	TM		
632	Before LA	1.614±0.001	1.616±0.001	1.1±0.1	1.9±0.2
	After LA	1.652±0.001	1.653±0.001	1.1±0.1	1.1±0.2
1319	Before LA	1.590±0.01	1.590±0.01	1.0±0.1	1.4±0.2
	After LA	1.631±0.01	1.634±0.01	1.0±0.1	0.7±0.2
1542	Before LA	1.585±0.01	1.585±0.01	1.0±0.1	0.9±0.2
	After LA	1.623±0.01	1.624±0.01	1.0±0.1	0.5±0.2

Comparing the refractive indices in GeO_2 waveguide before and after the CO_2 laser irradiation, we observe an increasing of refractive index about 0.04 with the irradiation for all the considered wavelengths. Similar results were obtained in other systems treated with CO_2 laser irradiation [10]. We have observed that laser annealing can lead to waveguides with a lower attenuation coefficient, than the attenuation co-efficient obtained with as prepared sample. In fact, we observe an attenuation coefficient of 0.5 dB/cm at 1542 nm, respectively for the irradiated system while we obtain attenuation coefficient of 0.9 dB/cm for the system before laser annealing. The decrease in the attenuation coefficient on the CO_2 laser irradiated systems, has been attributed by Dutta et al [7] to the reduction of surface irregularities and this effect carry out the reduction of the attenuation coefficient with the CO_2 laser irradiation at all the considered wavelengths as shown in Table 1. EDS analysis was used to monitor the composition of the samples as deposited and after the laser annealing processes. The measurements confirm that the composition of the sample is not affected by the laser treatment and the correct ratio between Germanium and Oxygen is always present.

In figure 1, we have reported the Micro Raman spectra carried out at room temperature for GeO_2 planar waveguide before and after laser irradiation of different time. The Raman spectra observed from the as prepared GeO_2 waveguide before irradiation is typical of GeO_2 amorphous system. In fact, the broad peak at 440 cm^{-1} clearly indicates the amorphous nature of GeO_2 film [11]. After 1 hour CO_2 laser irradiation, the Raman spectra indicate that the presence of rutile like GeO_2 crystalline phase. The peaks observing in the Raman spectra after 1h CO_2 laser irradiation are in good agreement with those reported in the literature [11]. The peak at 302 cm^{-1} corresponds to Ge optical phonons, being related to Ge–Ge bonds [12].

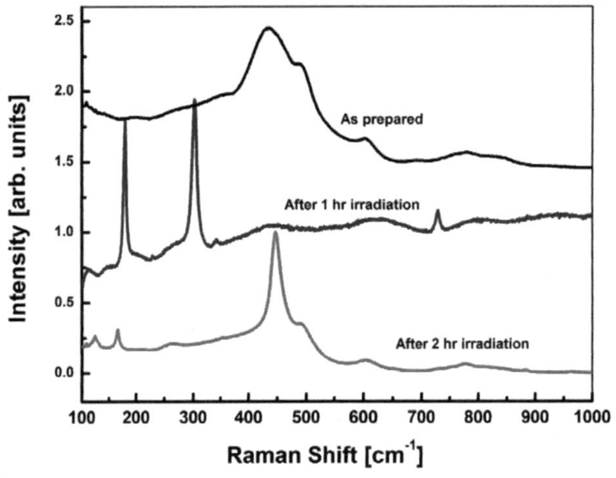

Figure 1: Micro Raman measurements carried out at room temperature for GeO_2 planar waveguide before and after CO_2 laser irradiation for 1 h and 2h.

After 2 hours CO_2 laser irradiation, the Raman spectrum on the GeO_2 planar waveguide shows that the presence of trigonal GeO_2 crystal phase. The peaks at 123, 166 and 263 cm^{-1} correspond, in fact, to the complex translation and rotation of GeO_4 tetrahedra [13]. The peak at 882 cm^{-1} is assigned to Ge-O stretching motion with tetrahedral GeO_4 units [14]. A shift in the band at 447 cm^{-1} is due to symmetric Ge-O-Ge stretching [14]. The peaks observing at 490 and 603 cm^{-1} are D_1 and D_2 defect bands from the Silica substrate. All the bands in the spectrum obtained for the GeO_2 waveguide after 2 hours CO_2 laser irradiation are in good agreement with those of GeO_2 calcinated under T = 1050 °C [15].

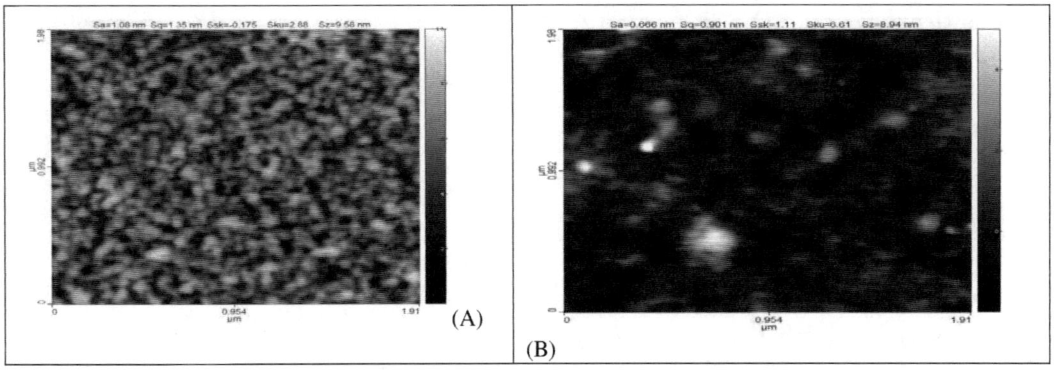

Figure 2. AFM image of a 1.98 x 1.91 μm^2 area on the as prepared sample (A) and after 2h of CO_2 laser irradiation (B).

The AFM image of a 1.98 x 1.91 μm^2 area on the as prepared sample (A) and after 2h of CO_2 laser irradiation (B) are reported in figure 2. AFM analysis put in evidence a roughness of 1.1 nm in the as prepared sample and a roughness of 0.7 nm for the waveguide after 2 h of CO_2 irradiation. This result is in agreement with the reduction of the attenuation coefficient with the laser annealing. Moreover the AFM image of figure 2 (B) show the presence of structures of nanometric scale on the surface of the sample with the dimension ranging from 10 to 50 nm attributed to the presence of GeO_2 nanocrystals. It is interesting to note as although the presence of these scattering points that can increase the attenuation coefficient [16] the protocol develop for the CO_2 irradiation allow to reduce the total value of the attenuation coefficient.

Conclusions

We have developed a fabrication protocol to prepare GeO_2 planar waveguide by Radio Frequency Magnetron sputtering technique. Moreover we note that using an adapted irradiation procedure, the pulsed CO_2 laser irradiation induced optical and structural changes in GeO_2 planar waveguide. The refractive index changes in the GeO_2 planar waveguide after 2 h of CO_2 laser irradiation were 0.04. The technique of laser annealing has been shown to significantly reduce propagation loss on GeO_2 planar waveguide. Low attenuation coefficients of 0.7 and 0.5 dB/cm at 1319 and 1542 nm, respectively, were measured after irradiation. Raman results have shown that after laser treatment GeO_2 planar waveguide shown rutile like and trigonal crystal phase after 1 h and 2 h irradiation respectively. AFM measurements shows that the surface roughness decreased from 1.1 to 0.7 nm after CO_2 laser irradiation and put in evidence the presence of structures of nanometric scale on the surface of the sample with the dimension ranging from 10 to 50 nm attributed to the presence of GeO_2 nanocrystals.

Acknowledgements

This research was performed in the framework of the research project ITPAR Phase – II (2008-2011).

References

1. Sebastiani S, Nunzi Conti G, Pelli S, Righini G C, Chiasera A, Ferrari M, and Tosello C 2005 *Opt. Express* **13** 1696
2. Anan'ev A V, Bogdanov V N, Champagnon B, Ferrari M, Karapetyan G O, Maksimov L V, Smerdin S N and Solovyev V A, 2008 *J. Non-Cryst. Solids* **354** 3049
3. Prakapenk V P, Guoyin Shen, Dubrovinsky L S, Rivers M L and Sutton S R 2004 *J. Phys. Chem. Solids* **65** 1537
4. Zevin M and Reisfeld R 1997 *Opt. Mater.* **8** 37
5. Dutta S, Jackson H E, Boyd J T, Davis R L and Hickernell F S 1982 *JIEEE Quant. Electron.* **18 (4)** 800
6. Dutta S, Jackson H E and Boyd J T 1980 *Appl. Phys. Lett.***37 (6)** 512
7. Dutta S, Jackson H E and Boyd J T 1981 *J. Appl. Phys.* **52 (6)** 3873
8. Dutta S, Jackson H E, Boyd J T, Hickernell F S and Davis R L 1981 *Appl. Phys. Lett.* **39 (3)** 206
9. Ribeiro S J L, Messaddeq Y, Gonçalves R R, Ferrari M, Montagna M, and Aegerter M A 2000 *Appl. Phys. Lett.* **77** 3502
10. Goyes C, Ferrari M, Armellini C, Chiasera A, Jestin Y, Righini G C, Fonthal F, Solarte E 2009 *Opt. Mater.* **31** 1310
11. Trukhin A and Capoen B 2005 *J. Non-Cryst. Solids* **351** 3640
12. Yang Y M, Yang L W and Chu Pal K 2007 *Appl. Phys. Lett.* **90** 081909
13. Mernagh T P and Liu L g 1997 *Phys. Chem. Miner.* **24** 7
14. Atuchin Victor V, Gavrilova Tatiana A, Gromilov Sergey A, Kostrovsky Vitalii G, Pokrovsky Lev D, Troitskaia Irina B, Vemuri R S, Carbajal-Franco G and Ramana C V 2009 *Cryst. Growth Des.* **9** 1829
15. Madon M, Gillet Ph., Julien Ch. and Price G D **1991** *Phys. Chem. Miner.* **18**, 7
16. Berneschi S, Soria S, Righini G C, Alombert-Goget G ,Chiappini A, Chiasera A, Jestin Y, Ferrari M, Guddala S, Moser E, Bhaktha S N B, Boulard B, Duverger Arfuso C and Turrell S, 2010 *Opt. Mater.* **32** 1644

IOP Conf. Series: Materials Science and Engineering **73** (2015) 012007 doi:10.1088/1757-899X/73/1/012007

Size Dependent Magnetic Properties of $Nd_{0.7}Ca_{0.3}MnO_3$ Nanomanganite

K N Anuradha[1], P R Koushalya[2] and S V Bhat[3]

[1]Department of Physics, Dr. Ambedkar Institute of Technology, Bangalore, India
[2]Department of Physics, Don Bosko Institute of Technology, Bangalore, India
[3]Department of Physics, Indian Institute of Science, Bangalore, India

E-mail: anuradhabshankar@gmail.com

Abstract: Nanoscale materials show different properties compared to bulk materials. Due to the size dependent properties the nanoscale materials have potential applications in industry. In this paper the size dependent magnetic properties of $Nd_{0.7}Ca_{0.3}MnO_3$ nanomanganite have been investigated. $Nd_{0.7}Ca_{0.3}MnO_3$ nanoparticles were prepared by low temperature sol-gel method. X-ray diffraction (XRD), Transmission Electron Microscopy (TEM) and EDAX techniques were used to understand the structure, grain size and composition. Nanoparticles prepared were of the sizes 15 nm, 19 nm and 25 nm respectively. SQUID magnetometer was used to study the magnetic behavior of the nanoparticles. Field cooled (FC) and zero field cooled (ZFC) magnetization of all the nanosamples with respect to temperature was studied and compared. We have observed drastic changes in magnetic properties of 15 nm particles compared to the other nanoparticles. The 'charge order peak' was seen to have disappeared in 15 nm particles while it was present in the other nanoparticles. All the nano particles exhibit superparamagnetism whose blocking temperature decreases as a function of decreasing particle size. The possible reasons for the influence of the particle size on the magnetic properties are discussed.
...

1. Introduction

Studies on perovskite rare earth manganites of general formula $A_{1-x}B_xMnO_3$ (where A is a trivalent rare earth ion such as La^{3+}, Pr^{3+} etc. and B is a divalent alkaline earth ion such as Ca^{2+}, Sr^{2+}, Ba^{2+}) have been a very active research area in the last few years in condensed matter physics due to their technological applications[1-3]. These materials show coexisting/ competing electronic and magnetic phases. Charge ordering(CO) is one of the fascinating properties exhibited by manganites. It is an insulating phase that is unstable under various perturbations such as magnetic field, electric field and pressure. We have shown [4-7] that it is also unstable to size reduction to nanoscale. Recent research on manganites focuses on the effect of the particle size reduction on the physical properties of these materials [8-9]. The particles with smaller grain sizes show richer magnetic and electronic properties due to the influence of the disorder on the surface. The published reports on nanomanganites show the interesting effects of the reduction in size in these systems.

Quintela's group [10] has studied the electronic and magneto transport properties of sol-gel derived nanoparticles of $La_{0.67}Ca_{0.33}MnO_3$. They show that the saturation magnetization and magneto calorific effect (MCE) decrease linearly with surface/volume ratio. Zheng *et. al* [11] have shown that $La_{0.67}Ca_{0.33}MnO_3$ nanoparticles with diameter below 60 nm exhibited surface phonon modes which disappeared above that particle size. The softening and hardening of the stretching and bending modes of phonons were observed respectively with the decrease of particle size from 140 to 20 nm. It was also shown that the metal-insulator transition temperature T_{M-I} shifts towards lower temperatures with the decrease in the grain size while the Curie temperature T_c is insensitive to the grain size. Dutta *et.al* [12] have studied the effect of particle size on magnetic and electronic properties on sol-gel derived nanoparticles of $La_{0.875}Sr_{0.125}MnO_3$. It has been shown that no signatures of structural phase transitions are observed with the decrease of particle size at room temperature. There is an increase of T_c and T_{M-I} with the decrease of particle size. Zhangyz et.al., [13] studied the sol-gel prepared nanosized $La_{0.85}Sr_{0.15}MnO_3$ particles. With the grain growth, there is a structural phase transition from cubic perovskite phase to orthorhombic phase. Magnetization increases with the decrease of particle size and a small decrease in the Curie temperature Tc with the grain growth is also reported. Our group has not observed any monotonic increase or decrease in magnetization in $Pr_{0.57}Ca_{0.41}Ba_{0.02}MnO_3$ nano-particles [4-7]. Some other reports have concluded that the magnetization decreases as particle size increases[14]. The published reports on nano manganites show the interesting effects of the reduction in size in these systems though many results are contradictory to each other. To understand the size dependent magnetic properties of manganites more in detail we have investigated the $Nd_{0.7}Ca_{0.3}MnO_3$ nanomanganite.

The bulk $Nd_{0.7}Ca_{0.3}MnO_3$ manganite shows an insulating behavior in a large temperature range (5-240K) due to CO of Mn^{3+} and Mn^{4+} cations (T_{CO}= 240K) but exhibits a ferromagnetic ordering below T_c = 120K . It shows that the ferromagnetic ordering of manganese cations does not induce a transition to a metallic state in contrast to what is observed in many other manganites which exhibit a metallic state due to the double exchange mechanism. Therefore $Nd_{0.7}Ca_{0.3}MnO_3$ is a rare combination of CO and ferromagnetic insulating phase. In this study we present the magnetic properties of $Nd_{0.7}Ca_{0.3}MnO_3$ nanomanganite. We have mainly focused on the charge order region and the ferromagnetic region of the system.

2. Experimental

$Nd_{0.7}Ca_{0.3}MnO_3$ nanoparticles were prepared by sol-gel method. The preparation of a homogeneous polymeric precursor is the most crucial step in this synthesis procedure. The advantage of this method is its versatility and the possibility to obtain high purity materials, control over stoichiometry, homogeneity and the precise size control of the particles. Stoichiometeric proportions of high purity Nd_2O_3, $CaCO_3$, and $MnCO_3$ were dissolved in concentrated nitric acid and an appropriate amount of triple distilled water was added to this mixture. Then an equal amount of ethylene glycol was added with continuous stirring. Initially the mixture was heated to 100^0C and at this stage the water evaporates. Further the mixture was heated to 140^0C. Excess ethylene glycol is then removed by heating to about 180^0 C and at this stage a thick sol was formed. This sol was further heated at 250^0 C in a furnace for about 6 hours to get the polymeric precursor. This precursor was then sintered at 700^0 C, 900^0 C and 1100^0 C for about 6 hours each to obtain particles of different sizes.

3. Characterization

X-ray diffraction experiments were carried out on all the samples using a Philips powder diffractometer (Cu-Kα; λ = 1.54056 A^0) at room temperature to confirm the phase formation and to estimate the crystallite size. Elemental composition analysis was done by Energy dispersive X-ray analysis (EDAX). Further the particle size of 700^0 C sintered samples was also estimated using transmission electron microscopy (TEM). D. C. magnetization measurements were carried out using a commercial SQUID magnetometer at 1000 G on all the three nano samples in the temperature range of 10-300K.

4. Results and Discussions

4.1 Structural details

Figure 1 shows the X-ray diffraction patterns of the three nanosamples. The average grain diameter/size (d) was evaluated indirectly from the XRD data by applying the Scherrer formula d = kλ

/β cosθ (where k = 0.89 is a shape factor, λ is the wavelength of X-Rays, β is the full width at half maximum of the peaks and θ is the angle of diffraction). The average crystallite sizes estimated from XRD for 700° C, 900° C and 1100° C sintered samples are 15 nm, 19 nm and 25 nm respectively Figure 2a shows the X-ray diffraction profiles for the most intense peak of the $Nd_{0.7}Ca_{0.3}MnO_3$ nanosamples. It was observed that for the 15 nm particles the broadening is more compared to 19 nm and 25 nm particles. The structure was studied by the Rietveld powder diffraction profile fitting technique(Fig.2b). All the samples crystallize in the orthorhombic space group P_{bn}

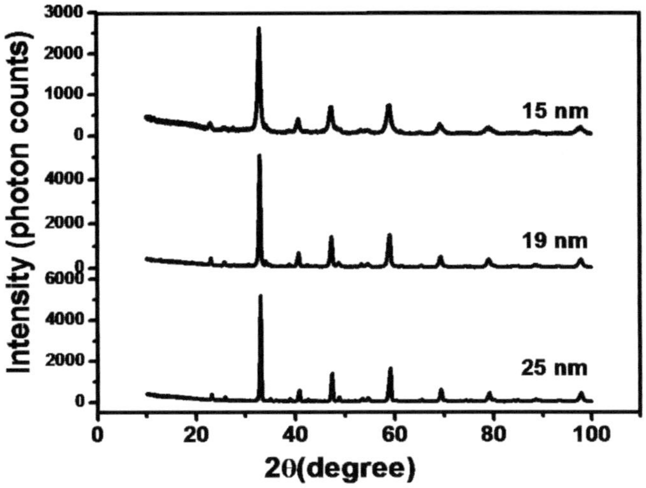

Fig.1 X-ray diffraction pattern of $Nd_{0.7}Ca_{0.3}MnO_3$ showing single phase character of the sol-gel derived samples

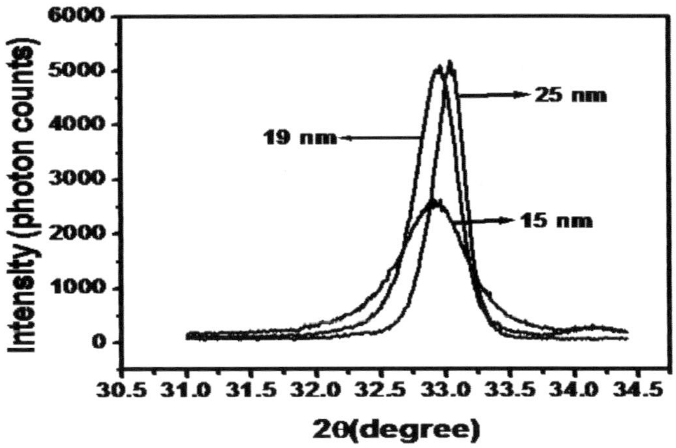

Fig. 2a X-ray diffraction profiles for the (121) reflection of the $Nd_{0.7}Ca_{0.3}MnO_3$ nanosamples

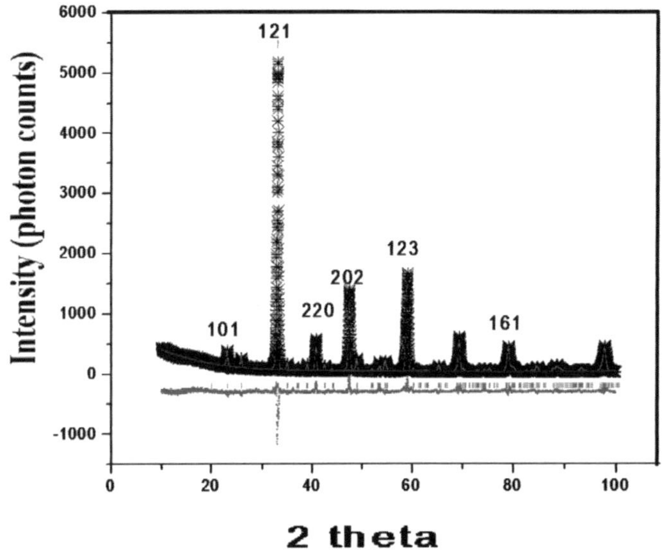

Fig.2b Observed (dots) and Rietveld fitted (continuous lines) powder diffraction pattern of 15 nm particle

The particle size of 700^0 C (NCMO7) sintered sample was also determined directly by employing transmission electron micrscopy. Figire 3(a,b) show the TEM pictures of 700^0 C sintered sample. The average particle size of 700^0 C sintered sample estimated using TEM was 15 nm and is in good agreement with the average grain size estimated by XRD. The TEM observations reveal that the particles are nearly spherical in shape. EDAX measurements have shown the good stoichiometry for all the three nanosamples

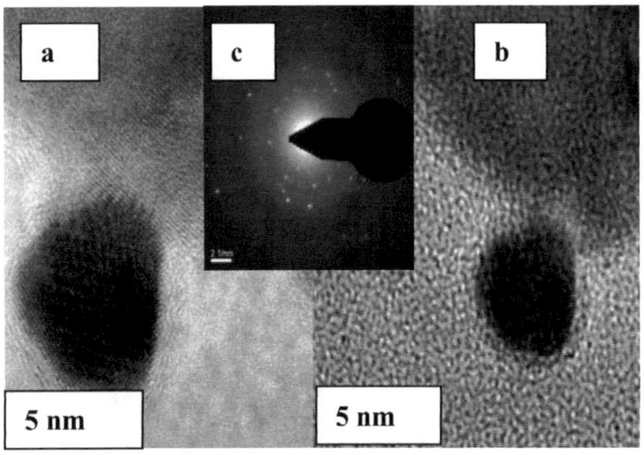

Fig. 3 (a b) High resolution TEM image of 15 nm particles (c) the SAED pattern showing diffraction spots indicating polycrystalline behavior

4.2 Magnetization studies

Figure 4 shows the field cooled (FC) dc magnetization curves for the three nanosamples at 0.1T. The inset shows a clear picture of the melting of charge order in 15 nm particle whereas the CO peak is seen for both 19 nm and 25 nm particles at 235 K. It confirms that size reduction melts the charge ordering in manganites.

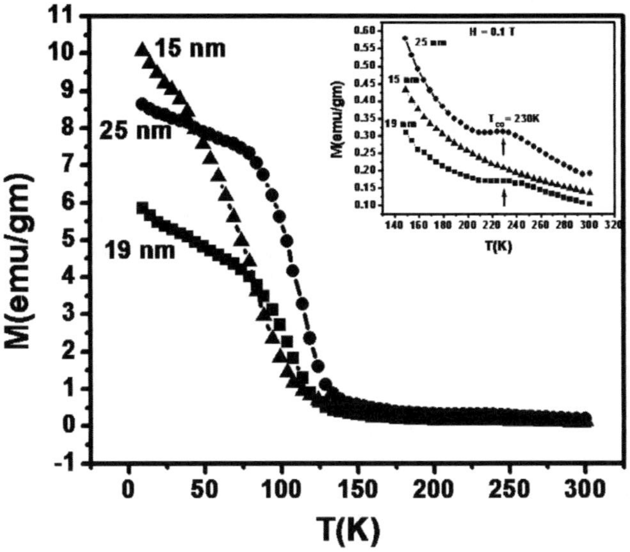

Fig. 4 Comparison of field cooled (FC) dc Magnetization vs. Temperature for all the 3 nanosamples (inset shows the melting of CO in 15 nm particle)

T_C is determined from the inflection point of the field cooled magnetization curve as observed in the dM/dT vs T curve (Fig.5). T_C values increase with an increase in particle size ie. 85 K, 105 K and 110 K for 15 nm, 19 nm and 25 nm particles respectively. Among the three samples maximum magnetization was observed for the lowest particle size ie for 15 nm particles.

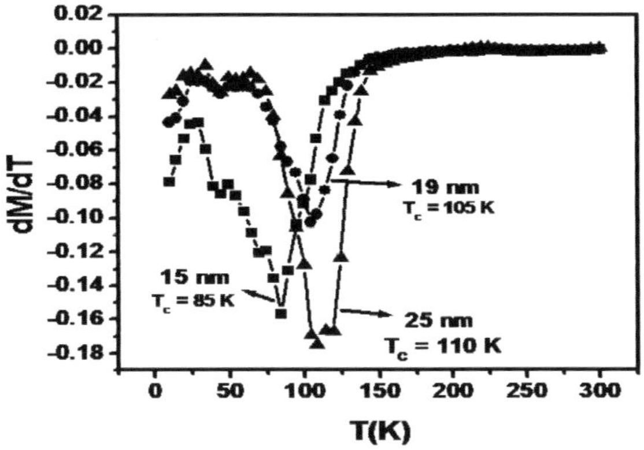

Fig. 5 dM/dT curve showing the T_C of different nano-particles

IOP Conf. Series: Materials Science and Engineering **73** (2015) 012007 doi:10.1088/1757-899X/73/1/012007

Figure 6 (a, b, c) shows the temperature dependence of the field cooled (FC) and zero field cooled (ZFC) magnetization. Zero field cooled(ZFC) magnetization was measured as a function of temperature while warming the sample under a field of 0.1 T.

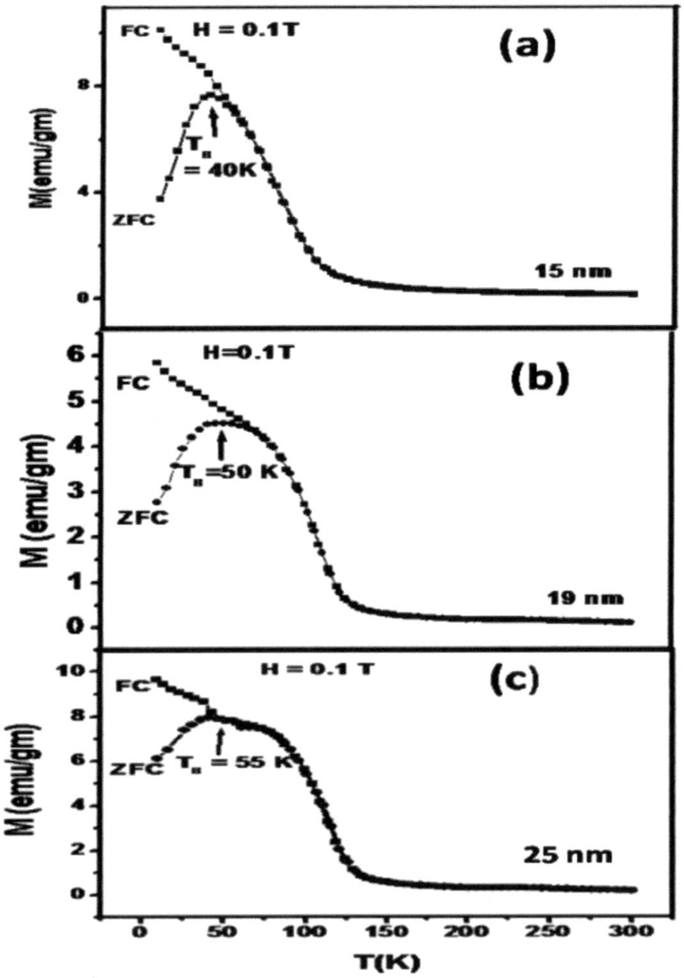

Fig. 6(a,b,c) Temperature dependence of the zero field cooled and field cooled magnetization of the nanosamples showing the decrease in blocking temperature for the 15 nm particles.

In M(T) curves we obtained two different curves for zero-field-cooling (ZFC) and filed-cooling (FC) for super paramagnetic materials. In ZFC curve, the moment increased with the increase in temperature and then decreased, while the moment decreased in FC curve with increase in temperature. All the samples show irreversibility between FC and ZFC curves with ZFC magnetization exhibiting prominent peak at T_B where T_B is called the blocking temperature. The region below T_B ($T< T_B$) is called blocking region, here the net magnetization of the individual particles is zero since the particles are oriented in different directions. Once the temperature is increased in the presence of the field the particles try to align in the field direction and hence magnetization increases with increase in temperature and a peak is observed in the magnetization curve of ZFC ie the blocking temperature[15].

Above T_B the magnetization decreases since the moments no longer align in the field direction. The regionT>T_B<T_C is called superparamagnetic region. Above T_C the moments of individual particles get disordered and the material becomes paramagnetic[16].The variation of T_C and T_B as a function of particle size is as shown in figure 7. The T_C increases with increase in temperature and the blocking temperature decrease with decrease in particle size. Disapperance of charge order in the 15 nm particles is one of the interesting observations of our studies. This is in consistency with our earlier reports.

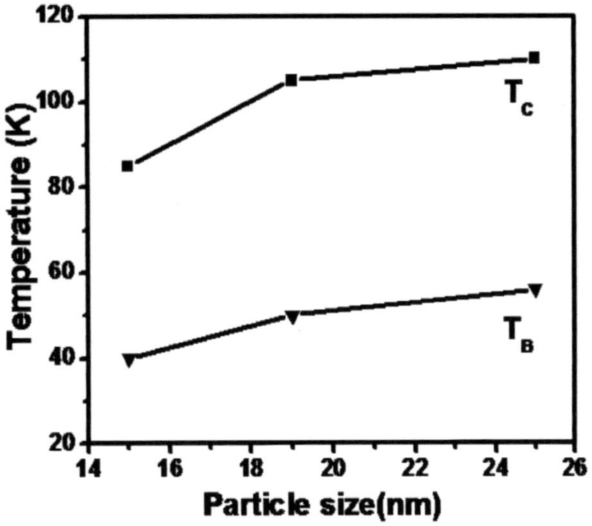

Fig. 7 Variation of Curie temperature(T_C) and blocking temperature(T_B) with respect to particle size.

In nano scale materials a large number of the atoms reside on the surfaces. Such features provide environments that are in general distinct from the bulk. For example, strain, change in co-ordination number and symmetry, reconstructions etc are ubiquitous features in nanostructure materials. It is well known from general considerations that due to the predominance of fluctuation effects in smaller dimensions there is reduced probability for ordering. This explains why the CO transition is absent in the 15 nm particles [9]. We have observed a super paramagnetic behavior in all the nanoparticles since it is more probable for smaller systems.

The blocking temperature is given by,

$$T_B = E_a/ k_B \ln(tf_0) \text{ ------------- } (1)$$
$$T_B = KV/k_B \ln(tf_0) \text{ ----------- } (2)$$

where E_a is the anisotropy barrier and can be obtained by E_a = KV (K = anisotropy energy density constant and V is the volume of particle), t is the experimental measuring time and $k_B\ln(tf_0)$ is treated as constant. [17-20]. From equation 2 it is clear that T_B is directly proportional to volume of the particle For small particles, they have small volume and thus lower energy barrier and lower blocking temperature. Hence T_B decreases with decrease in particle size.

5. Conclusions

We have successfully prepared nanoparticles of NCMO by polymer-assisted sol-gel method. The nanoparticles were characterized by various techniques like XRD, TEM, EDAX and SQUID. Magnetization measurements show interesting temperature and also particle size dependence. Except 15 nm particles all other particles show charge ordering. The disappearance of charge order in nanosystem is quite interesting as it provides a new intrinsic way of melting the CO in addition to the external perturbations. Blocking temperature is seen to decrease with decrease in particle size. We have observed the strong dependence of magnetc properties with respect to particle size.

5. Acknowledgement

Funding from Indian Academy of Sciences, University Grants commission and Visvesvaraya Technological University, Belgaum is gratefully acknowledged.

References

[1] Jonker G H and Van Santen J H 1950 Ferromagnetic compounds of manganese with perovskite structure. *Physica* **16** 337

[2] C N R Rao and B Raveau 1998 Colossal Magnetoresistance Charge Ordering and Related Properties Eds. *World Scientific, Singapore*

[3] C N R Rao et al 2000 Charge ordering in the rare earth manganates: the experimental situation

J. Phys.: Condens. Matter **12** R83

[4] Rao S S et al 2005 Weakening of charge order and antiferromagnetic to ferromagnetic switch over in Pr0.5Ca0.5MnO3 nanowires. *Appl Phys Lett.* **87** 182503

[5] S S Rao,et al 2006 Suppression of Charge order, disappearance of antiferromagnetism and mergence of ferro-magnetism in Nd0.5Ca0.5MnO3 nano-partiucles. Phys. *Rev. B* **74** 144416

[6] Anuradha K N et al 2007 Complete melting of charge order in hydrothermally grown Pr0.57Ca0.41Ba0.02MnO3nanowires. *J. of Nanoscience and Nanotechnology* **7** 1775

[7] Anuradha K N et al 2008 Premonitory charge order fluctuations in Pr0.57Ca0.41Ba0.02MnO3 naoparticlesApplied Magnetic Resonance. **33** 127

[8] *R.D. Sanchez* et al 1996 Giant magnetoresistance in fine particle of $La_{0.67}Ca_{0.33}MnO_3$ synthesized at low temperatures *Appl. Phys.* Lett. **68** *134*

[9] Zhu T et al 2001 *Appl Phys Lett* **78** 3863

[10]. M. A. Lopez-Quintela et al 2003, 14 12 C. Vazquez-Va zquez, et al 1998, *J. Mag. Mag. Mat,* **189**, 321

[11]. Lei Zheng et al 1998 *Phys. Rev B* **58** 8613

[12]. A. Dutta et al 2003 *Phys. Rev. B* **68** 054432

[13]. Ning Zhangyz, et al 1997 *J. Phys.: Condens. Matter*, **9** 4281

[14] N.Zhang et al. 1997 *J. Phys. Cond.Matt.***9** 4281

[15] Dupont F et al 1996 *Phys Rev* B **54** 3007

[16] R un-Wei Li et al. 2000 *J. Phys. Cond.Matt* **13** 141

[17] Sujoy Roy et al. 2004, *J. Aplied Physics* **96** 1202

[18] M.El.Hilo et al. 1992, *J.Mag.Mag.Mate.* **114** 295

[19] D.A.Dimitroo et al 1996 *Phys.Rev.B*, **54** 9237

[20] S.Lee et al 1999 *Phy.Rev.Lett.* **82**, 4508

Spectroscopic studies of xLi_2O-(40- x)Bi_2O_3-20CdO-40B_2O_3 Glasses

R. Vijaya Kumar, A. Edukondalu, P. Muralimohan, M. Srinivas, K. Siva Kumar

Department of Physics, Osmania University, Hyderabad- 500 007, India.

E-mai: kumar_nizamian@yahoo.co.in

Abstract. Glasses with composition xLi_2O-(40- x)Bi_2O_3-20CdO-40B_2O_3 where ($0 \leq x \leq 40$) have been prepared using melt quench technique. Optical absorption spectra of the pure glasses reveled that the cutoff wave length decreases and optical band gap (E_{opt}) and Urbach energy (ΔE) increases with increase in Li_2O content. The E_{opt} values of these glasses are found to be in the range 2.818-3.258 eV where as the values of ΔE are in the range 0.21-0.33 eV. Raman and infrared spectroscopies have been employed to investigate these glasses in order to obtain information about the competitive role of Bi_2O_3 and B_2O_3 in the formation of glass network. IR and Raman spectra show that these glasses are made up of [BO_3], [BO_4], [BiO_3] and [BiO_6] units. The formation of Cd in tetrahedral co-ordination was not observed.

1. Introduction

The interest in heavy metal oxide glasses is due to their long infrared cut-off and optical non-linearity [1]. Bismuth based oxide glasses have attracted the attention of scientific community, due to their important applications in the field of glass ceramics, thermal and mechanical sensors etc. a survey of literature shows that there are many reports available on quaternary bismuth borate glasses [2, 3]. The present work is focused on studying the effect of Bi_2O_3 and Li_2O on cadmium borate quaternary glass system. The presence of the two glass formers Bi_2O_3 and B_2O_3 increases the interest in present study.

2. Expermental

The optical absorption spectra of the present glass samples (thickness ~1 mm) were recorded at room temperature using a double beam Shimadzu spectrometer (model UV-3100) in the wavelength range 400-800 nm. The uncertainty in the observed wave length is about ±1 nm.

The room temperature Raman measurements were performed in the range 100-2000 cm^{-1} on a micro Raman system from Jobin-Yvon Horiba (LABRAM HR-800) spectrometer. The system is equipped with high stability confocal Microscope for Micro Raman 10×, 50×, 100× objective lens to focus the laser beam. Ar^+ laser beam of 488 nm (E = 2.53 eV) was used for excitation. The incident laser power is focused in a dia of ~1–2μm and a notch filter is used to suppress Rayleigh light. In the present system Raman shifts are measured with a precision of ~0.3 cm^{-1} and the spectral resolution is of the order 1 cm^{-1}.

Infrared spectra of the powdered glass samples were recorded at room temperature in the range 400–1500 cm^{-1} using a spectrometer (Perkin-Elmer FT-IS, model 1605). These measurements were made on glass powder dispersed in KBr pellets

IOP Conf. Series: Materials Science and Engineering **73** (2015) 012008 doi:10.1088/1757-899X/73/1/012008

3. Results and Discussion

3.1Optical absorption studies

A distinct cutoff (λ_c) was observed for each glass sample from optical absorption spectra, λ_c decreases with increasing Li_2O content. The optical absorption coefficient, $\alpha(\omega)$ of a material can be evaluated from the optical absorption using the relation [4]

$$\alpha(\omega) = (1/t)\ \ln(I_o/I),$$

where 't' is the thickness of each sample and $\ln(I_o/I)$ corresponds to absorbance. The relation between $\alpha(\omega)$ and photon energy of the incident radiation, $\hbar\omega$ is given by the following relation [5]

$$\alpha(\omega) = A_o\ (\hbar\omega - E_{opt})^p\ /\ \hbar\omega$$

where E_{opt} is the optical band gap energy in eV, A_o is a constant and p is an index which assumes the values 1/2, 3/2, 2 and 3 depending on the nature of the electronic transition responsible for absorption. The values of indirect optical band gap energy E_{opt} can be obtained from expression above by extrapolating absorption coefficient to zero in the $(\alpha h\omega)^{1/2}$ vs $h\omega$ (Tauc plot) as shown in the figure 1. From table 1, It is observed that the values of E_{opt} increase with Li_2O content. This can be understood in the terms of structural changes that are taking place with the introduction of Li_2O into the glass.

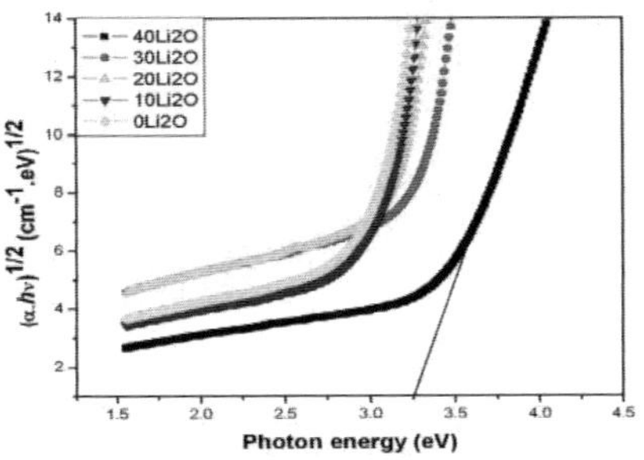

Figure 1. Tauc plots of present glasses.

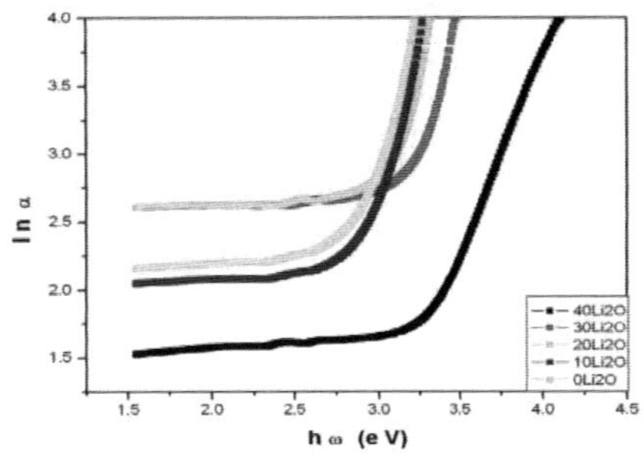

Figure 2. Urbach plots of present glasses.

The relation between $\alpha(\omega)$ and Urbach energy (ΔE) is given by the well known Urbach law given by the relation

$$\alpha(\omega) = \text{const.}\exp(\hbar\omega - \Delta E),$$

ΔE is usually interpreted as the width of the tail of the localized stares in the band gap. The above equation can be rewritten as

$$\ln \alpha(\omega) = (\hbar\omega/\Delta E) - \text{const.}$$

Urbach plots are the plots where the natural logarithm of absorption coefficients, ln α, is plotted against photon energy, $\hbar\omega$. Urbach plot for a typical glass sample is shown in the figure 2. The values of Urbach energy (ΔE) were calculated by determining slopes of the linear regions of the curves and taking their reciprocals. The probable error in (ΔE) is found to be ±0.001 eV. It is found that Urbach energy decreases with increase in the Li_2O content in the present glass system shown in table 1.

Table 1: λ_c, E_{opt}, ΔE values of -(40- x)Bi_2O_3-20CdO-40B_2O_3 where ($0 \leq x \leq 40$) glasses.

Li_2O mol%	x=0	x=10	x=20	x=30	x=40
cutoff wavelength (λ_c)	381	375	370	354	295
optical band gap energy E_{opt}	2.818	2.901	2.999	3.104	3.258
urbach energy (ΔE)	0.21	0.23	0.28	0.30	0.33

3.2 Raman and IR studies

Figure 3 shows the Raman spectra of xLi_2O-$(40-x)Bi_2O_3$-$20CdO$-$40B_2O_3$, $0 \leq x \leq 40$ glasses. In the present glasses the band was observed in the range 140-145 cm^{-1} and its intensity decreases with increasing Li_2O content. Therefore we assume that Bi^{3+} cations are incorporated in $[BiO_3]$ and $[BiO_6]$ groups. This is confirmed by the presence of bands in the present IR spectra (figure 4) around 498cm^{-1} can be assigned to stretching vibrations of Bi-O bonds in strongly distorted vibrations BiO_6 units [3, 6]. In the present study this band shifts from 498cm^{-1} to higher wavenumbers with increasing Li_2O content, which is due to the change of local symmetry in $[BiO_n]$ polyhedra. Similar observations were found in Bi_2O_3-B_2O_3-ZnO-Li_2O, Bi_2O_3-B_2O_3-Li_2O [1, 7] glasses and the band 878cm^{-1} is assigned to symmetric stretching vibrations of Bi-O bonds in BiO_3 units, and stretching vibrations of B-O bonds in BO_4 units from diborate groups. The broad absorption band around 995cm^{-1} in the present IR spectra assigned to stretching vibrations of B-Ø bonds in $BØ_4$ units from tri-,tetra- and pentaborate groups [8, 9]. The broad but strong band occurring in the present Raman spectra centered around 338cm^{-1} can be attributed to the Bi-O bonds strongly distorted in BiO_6 units and the band around 416$cm^{-1}$is assigned to the Bi-O-Bi bending vibrations in BiO_6 units, while the shoulder at 566cm^{-1} can be attributed to Bi-O^- stretching vibrations of $[BiO_6]$ units [1, 7, 8]. In Raman spectra at $Li_2O \geq 20$ glasses, new bands arises around 508cm^{-1}, may be attributed to the Bi-O vibrations in BiO_6 and the band around 725cm^{-1} is attributed to the vibrations of chain type metaborate groups [8].

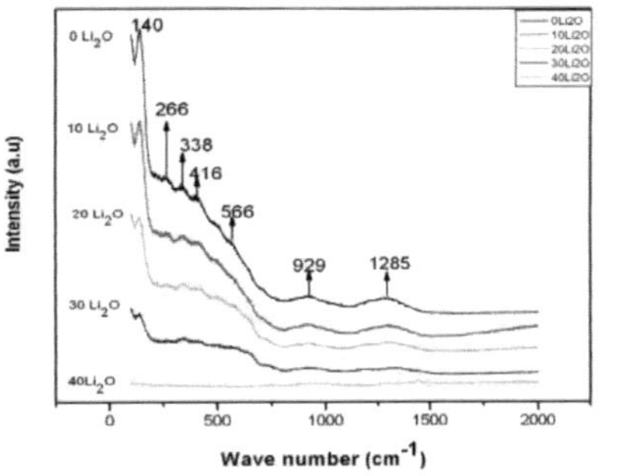

Figure 3. Raman spectra of present glasses

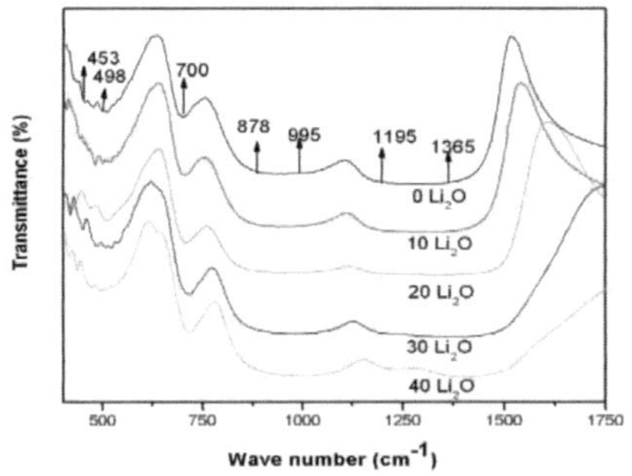

Figure 4. IR spectra of present glasses.

In Raman spectra the strong band observed around 929cm^{-1} can be assigned to vibrations of orthoborate groups [1, 10]. It is observed that the intensity of this band remains constant this is because the B_2O_3 content remains constant in the glass system. Assignment has been suggested for 1274cm^{-1} band in the Raman spectra of pyroborate groups, involving primarily the stretching of the terminal B-O^- bonds [11, 12]. This is confirmed from the fact that in the present glasses, the band observed at 700-712cm^{-1} in infrared spectra is assigned to B-O-B bending vibrations in $[BO_3]$ triangles while the strong bands in the range 1195cm^{-1} is assigned for stretching vibrations of B-O bonds in BO_3 units from meta and ortho-borate groups [8] and 1365cm^{-1} arises from B-O^- bonds stretching in BO_3 units from varied types of borate groups respectively [9, 13, 14].

As the expense of the Bi_2O_3 content i.e at high Li_2O composition of glass $40Li_2O$-$20CdO$-$40B_2O_3$ in IR spectra the band around 508cm^{-1} can be assigned to O-B-O bond bending vibrations of various borate segments the next one centered at 712cm^{-1} is assigned to the O_3B-O-BO_3 bands in bending vibrations [15] and the band 890cm^{-1} which is a characteristic for stretching vibrations of B-O bonds in BO_4 tetrahedra from diborate groups [16, 17] and a band centered at 1025cm^{-1} assigned to B-Ø bonds stretching vibrations in $BØ_4^-$ tetrahedral (Ø: oxygen atom bridging two boron atoms) from tri-, tetra- and penta- borate groups [16, 17]. The bands around 1215cm^{-1} is due to the stretching vibrations of B-O^- bonds in BO_3 units from meta and ortho borate groups and 1397cm^{-1} assigned to asymmetric stretching mode of borate triangles with ($BØ_3$, $BØ_2O^-$) and stretching vibrations of the borate triangles with non-bridging oxygen (NBO) in various borate groups [13, 16]. In Raman spectra the bands at 333cm^{-1}, 419cm^{-1} and 507cm^{-1} are assigned to tetrahedral borate groups mainly pentaborate and tri borate groups [17]. The band situated at 764cm^{-1} has been attributed to the symmetrical breathing vibrations of six-member borate rings with one or two BO_4 tetrahedra [13]. The band at 950cm^{-1} is assigned to vibrations of diborate groups and 1150cm^{-1} is assigned to vibrations of pyro-borate groups [16] and the band 1304cm^{-1} is assigned to B-O$^-$ vibrations from various borate groups the presence in our Raman spectra of the band at 1443cm^{-1} assigned to vibrations of the chain type metaborate units [18].

In all glass samples except in $40B_2O_3$-20CdO-$40Bi_2O_3$ glass sample the weak bands at 438 m^{-1} and 463cm^{-1} in IR spectra were attributed to the Li-O-Li and Li-O-Bi/B bonds [7]. In the present glass system the band at 806 cm^{-1}indicates the absence of boroxal ring formation and the band at 840 cm^{-1} is not observed in which suggest that the formation of tetrahedral co-ordination of Cd (CdO_4) is absent [2, 19]. The low frequency band around 452cm^{-1}(weakly observed) in the IR spectra and Raman peak at 266cm^{-1} observed in all glass samples may be attributed to the Cd-O vibrational modes of metal cation Cd^{2+} [19].

4. Conclusions

In the present glass samples it is observed that cutoff wave length(λ_c) decreases with increasing Li_2O and The values of E_{opt} and ΔE depends on Li_2O concentration and increases with Li2O content. The present glasses are made up of [BiO_3], [BiO_6],[BO_3] and [BO_4] units. In the present glass system there is no formation of tetrahedral co-ordination CdO_4 units and CdO acts as network modifier in present glasses.

Acknowledgements: One of the author R. Vijaya Kumar would like to thank UGC-RGNF for providing financial assistance and thanks to CSR-Indore for providing instrumental facilities.

References

[1]. Bale. S, Rahman. S, 2008 J. Alloys and Compounds 457 545-548.
[2]. Rani. S, Sanghi. S, Anshu, Agarwal. A, Kishore. N, Seth. V.P. 2008 J. Phys and Chem. Solids 69 1855-1860.
[3]. Radu. A, Baia. L, Kiefer. W and Simon. S, 2005 Vibrational Spectroscopy 39 127-130.
[4] Sands. R. H, 1955 Phys. Rev. 99 1222.
[5] Pankove. J, 1971 Optical Processes in Semiconductors. Prentice-Hall, Englewood Cliffs, NJ.
[6]. Culea. E, Pop. L, Pascuta. P and Bosca. M, 2009 J. Molicular Structure, 192 924-926.
[7]. Moustafa. EI. S, Sadeek. Y. B, Shaaban. E. R, 2008 J. Phy and Chem of solids 69 2281-2287.
[8]. Ardelean. I, Simona Cora, Dorina Rasu, 2008 Physica B, 403 3682-3685.
[9]. Pascutta. P, Pop. L, Rada. S, Bosca. M, Culea. E, 2008 J. Mater. Sci: Mater Electron 19 424-428.
[10]. Meera. B. N and Ramakrishna. J 993 J. Non-Cryst. Solids 159 1-21.
[11] Lucacel. R. C, Ardelean. I, 2007 J. Non-Cryst. Solids, 353 2020-2024.
[12]. Griguta. L, Ardelean. I, 2008 J. Optoelectronics and Advanced Materials, 2 256.
[13]. Kamitsos. E. I, Karakassides. M. A, Chryssikos. G. D. 1987 Phys. Chem. Glasses 91 1073.
[14]. Kamitsos. E. I, Patsis. A. P, Karakassides. M. A and Chryssikos. G. D 1990 J. Non-Cryst. Solids, 126 52.
[15]. Fuxi. G, 1991 Optical and Spectroscopic Properties of Glass, Springer-Verlag, Berlin .
[16]. Kamitsos. E. I, Karakassides. M. A, Chryssikos. G. D. 1989 Phys. Chem. Glasses 30 229.
[17]. White K, Hubert T, Reich P and Split C, 1993 J. Raman Spectra, 24 407.
[18]. Konijnendijk. W. L and Stevels. J. M. 1975 J. Non-Cryst. Solids 18 307.
[19]. Alemi. A. A, Sedghi. H, Mirmohseni. A. R and Golsanamlu. V. 2006 Bull. Mater. Sci., 29 55-58.

IOP Conf. Series: Materials Science and Engineering 73 (2015) 012009 doi:10.1088/1757-899X/73/1/012009

Crystalline Indium Sulphide thin film by photo accelerated deposition technique

A.C.Dhanya[1,2], K.C.Preetha[2], K.Deepa[2] ,T.L.Remadevi[1, 2]
[1.] Department of Physics, Pazhassi Raja N.S.S.College, Mattannur, Kannur, Kerala,
[2.] School of Pure and Applied Physics, Department of Physics, Kannur University, 670327, India

acdhanyaac@gmail.com

Abstract. Indium sulfide thin films deserve special attention because of its potential application as buffer layers in CIGS based solar cells. Highly transparent indium sulfide (InS) thin films were prepared using a novel method called photo accelerated chemical deposition (PCD). Ultraviolet source of 150 W was used to irradiate the solution. Compared to all other chemical methods, PCD scores its advantage for its low cost, flexible substrate and capable of large area of deposition. Reports on deposition of high quality InS thin films at room temperature are very rare in literature. The precursor solution was initially heated to 90^0C for ten minutes and then deposition was carried out at room temperature for two hours. The appearance of the film changed from lemon yellow to bright yellow as the deposition time increased. The sample was characterized for its structural and optical properties. XRD profile showed the polycrystalline behavior of the film with mixed phases having crystallite size of 17 nm. The surface morphology of the films exhibited uniformly distributed honey comb like structures. The film appeared to be smooth and the value of extinction coefficient was negligible. Optical measurements showed that the film has more than 80% transmission in the visible region. The direct band gap energy was 2.47eV. This method is highly suitable for the synthesis of crystalline and transparent indium sulfide thin films and can be used for various photo voltaic applications.

1. Introduction

Indium sulphide is a gifted contender for many technological applications in optoelectronic and photovoltaic field [1,2] due to its stability, wider band gap and photo conducting behaviour [3] and a good non toxic substitute for CdS in Copper Indium Gallium Sulphide (CIGS) based solar cells. Teny Therasa John et al deposited indium sulphide thin films using spray pyrolisis technique with different In/S ratios and the characterization showed that in terms of crystalline behaviour, band gap and photo response the samples with In/S ratio of 1.2/8 is very suitable for any photovoltaic application [4]. Amorphous indium sulphide films were synthesized on FTO coated glass substrate and explored the formation of In_2O_3/In_2S_3 multilayer film electrode and its use in photo-electro chemical cell [5]. Photo chemical deposition (PCD) became great successful in depositing semiconductor thin films from aqueous solutions [6-9]. For economic reasons, it should be very interesting if thin films can be deposited using a cheap deposition technique. Nearly stoichiometric CuInS films were deposited in acidic medium for the first time by photo chemical deposition technique [10] and it was observed that, the longer irradiation of $CuSO_4$ solution enhances the Cu content of the films.

2. Experimental details

*2.1. Photo accelerated chemical deposition.*UV light source of 150W was used to irradiate the solution. The energy from the light source has been utilized to increase the kinetic energy and hence the inter diffusion of the adsorbing atoms. Since the substrate was kept vertically very near to the UV source, only a small part of energy gets diffused into the solution. The direct entry of UV rays on to the substrate causes an amplified vibration among the molecules of the glass substrate. These augmented vibrations support the adherence of the incoming atoms, and develops the crystalline behaviour.

The precursor solution was initially heated to 90°C for ten minutes and then deposition was carried out at room temperature for two hours. Films were deposited through double dips, each with one hour duration. All the chemicals used were of analytical grade and the chemical bath contained 0.025M indium chloride, 0.3M thioacetamide and 1drop glacial acetic acid. Glacial acetic acid acts as complex agent which allows slow release of cations. Well cleaned soda lime glass substrates were used to deposit the film. It was kept in detergent solution for many hours and boiled in chromic acid solution for days. A subsection

3. Results and discussion

The film was highly adhered to the substrate and appeared as lemon yellow to bright yellow as the deposition time improves. The thickness of the film using gravimetric technique was 215nm.

*3.1. Structural properties.*The structural aspects of the samples were studied using an X- ray diffractometer (Bruker AXS D8) with Cu- K_α radiation of wavelength $1.5405A^0$ as the source. Figure.1 displays the XRD pattern of the film with mixed phases. The d value of monoclinic structure coincides with the orientations of (130) and (110). (332) and (220) corresponds to cubic structure of indium sulphide in accordance with the standard JCPDS data card [80-0126, 73-0844]. The average grain size from Debye Scherrer's formula was 17nm.

Figure.2 shows the scanning electron microscopic images of the film, a large number of aggregates and discrete nano particles just like a honeycomb structure. The morphology shows that it can be sensitized with dyes for many other applications.

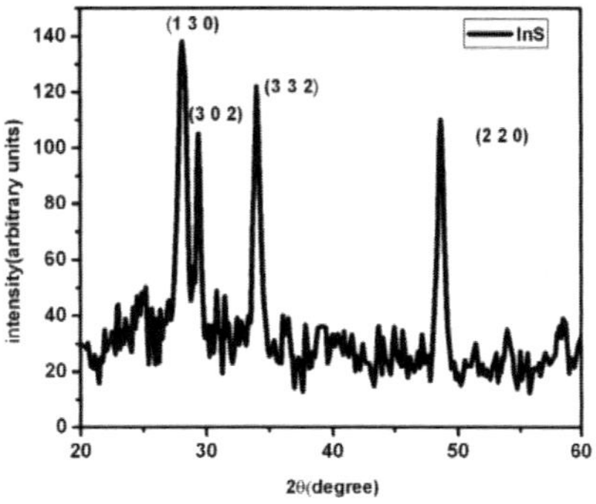

Figure.1.XRD pattern of the Indium sulphide film

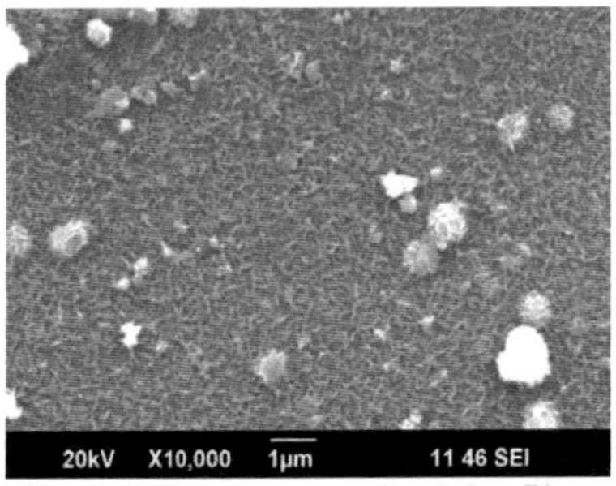

Fig.2.SEM photograph of the film

*3.2. Optical properties.*The optical properties like absorption and transmission spectra were taken using a Varian Carry 5000 UV-Vis-NIR spectrophotometer in the range 200-2000nm. Fig.3 displays absorbance and a transmission spectrum of the indium sulphide films deposited by double dips, each of one hour in freshly prepared chemical baths. The samples exhibit more than 80% optical transmission in the visible region and attain 88% at the wavelength about 300nm, which is noticeably better than previous reports. High transparency in the visible region indicates the absence of adsorbed powdery colloids formed by homogeneous reaction during the growth of the film. A sharp absorption edge is observed in the visible region, signifying good crystallinity and low defect density close to the band edge [11]. Band gap energy calculated from the band edge was 2.47eV

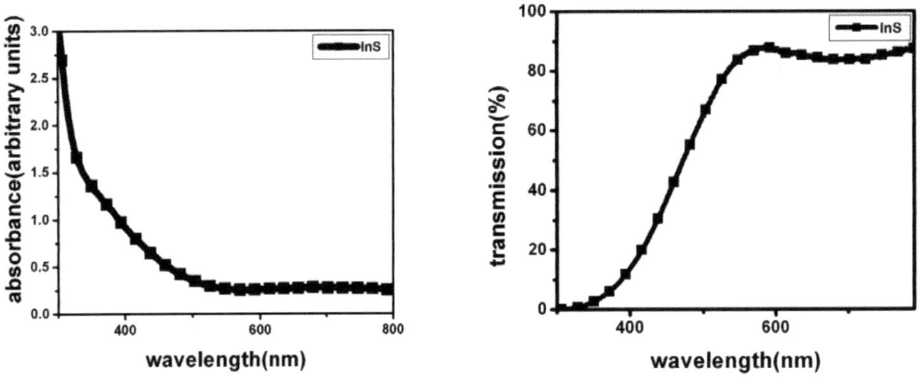

Figure.3.absorbance and tranmittance spectra of the Indium sulphide film

Extinction coefficient (k) measures the rate of diminution of transmitted light through scattering and absorption for a medium and it was calculated from the absorption coefficient by using, $k=\alpha\lambda/4\pi$. The film appeared to be smooth and the value of extinction coefficient was negligible.

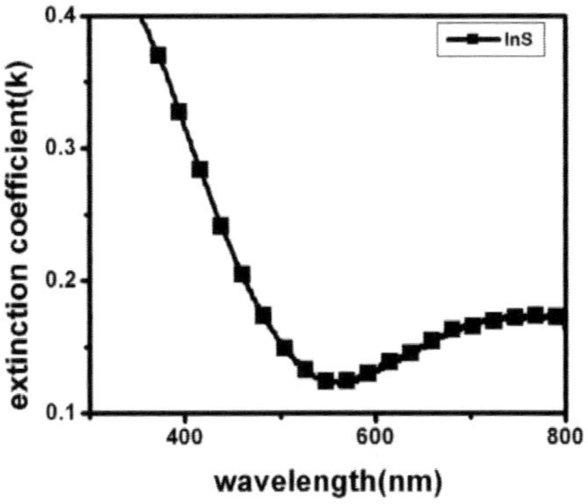

Figure. 4 extinction coefficient of the films as a function of wavelength

4. Conclusion

Crystalline indium sulphide thin films were prepared using photo accelerated deposition technique. Surface morphology of the films appeared to be of honey comb like and is useful in the fabrication of dye sensitized solar cell. Uniform transmission in the visible region signifies its application as window layer in solar cells. Hence this method is highly suitable for the synthesis of crystalline and transparent indium sulfide thin films which can be used for various photo voltaic applications.

References

[1] O Brien P, Octway D J and Walsh J R *Thin Solid Films* **315** 57
[2] Amlouk M 1999 *J. Appl. Phys* **38** 26
[3] Bhira L 2000 *Phys. Status Solidi* **181** 427
[4] Teny Therasa John, S Bini, Y Kashiwaba, T Abe, Y Yasuhiro, C Sudha Kartha and K P Vijayakumar, 2003*Semicond.Sci.Technol.***18** 491
[5] S S Kale, R S Mane, C D Lokhande, K C Nandi and Sung-Hwan Han 2006 *Material Sci. and Eng.*
[6] F Goto, M Ichimura and E Arai *1997Jpn J. Appl. Phys*. 36 L1146
[7] M Ichimura, F Goto, Y Ono and E Arai 1999*J. Cryst. Growth***198/199** 308
[8] M Ichimura, K Takeuchi, A Nakamura and E Arai 2001*Thin Solid Films* **384** 157
[9] R Kumaresan, M Ichimura, N Sato and P Ramasamy 2002*Mater.Sci.Eng B* **96** 37
[10] Jiban Podder, Tetsuya Miyawaki and Masaya Ichimura 2005 *J. Cryst. Growth* **275** e937-e942
[11] J.Cheng, D B Fan, H Wang, B W Liu, Y C Zhang and H Yan 2003 *Semicond. Sci.Technol.***18** 676

Preheated substrate effects on the structural and optical properties of chemically prepared ZnS thin films

K Deepa[1,2], K C Preetha[3], A C Dhanya[1] and T L Remadevi[2]

[1]School of Pure and Applied Physics, Kannur University, Kerala, India-670327
[2]Pazhassi Raja NSS College,Mattannur, Kerala, India-670702
[3]Sree Narayana College,Thottada,Kannur

email:deepa00hari@gmail.com

Abstract. Nanocrystalline zinc sulphide thin films were prepared by an improved chemical bath deposition technique in which glass substrates were thermally treated prior to the deposition process. The microscopic glass slide was heated to 200^0C before mounting into the chemical bath. The films prepared with and without preheating the substrate were characterized using various techniques and their properties compared. Structural analysis of the films was done using powder X-ray diffraction technique. Films deposited without preheating were amorphous whereas preheating the substrate resulted in the deposition of highly crystalline film with wurtzite structure having an intense (100) orientation. Morphological analysis using scanning electron microscope revealed the densification of the films deposited on preheated substrate indicating that the nucleation rate was higher in this case as compared to films deposited without preheating. Gravimetric analysis of the films confirmed that preheating results in the deposition of thicker films. The optical analysis of the samples was done using UV-Vis-NIR spectroscopy. All the films had a very low absorbance in the visible region. A red-shift in absorption edge was observed on preheating the substrate which indicates a decrease in band gap. However the transmittance in the visible region decreased on preheating the substrate. Room temperature photoluminescence spectra exhibited UV and violet emission for both the samples. The luminescence intensity was found to increase on preheating.

1.Introduction

Zinc sulfide (ZnS), which is a II-VI semiconducting compound, has received much attention owing to its potential uses [1-2] in a wide variety of applications. In opto-electronics, it can be used as light-emitting diode in the blue to ultraviolet spectral region due to its wide band gap at room temperature. Furthermore, ZnS films are used as light source for viewing screens and buffer layer for CIGS solar cells. Large area deposition of high quality ZnS films is required for it to be effectively used in solar cells and electroluminescent devices. There are many deposition techniques to fabricate ZnS thin films [3]. Among them chemical bath deposition (CBD), is more popular due to its simplicity, low cost and uniformity for large area deposition of high quality films. It is often difficult to obtain crystalline ZnS thin films of appreciable thickness using CBD technique .In most cases multiple depositions are required to obtain films of appreciable thickness and larger grain size. In this work the substrates were preheated to a temperature of 200^0C before mounting into the chemical bath. The films obtained in this single step deposition process were thicker and highly crystalline. It is confirmed that the nucleation, growth and morphology of thin films is affected by the thermodynamic conditions of the interfaces between the substrates and reaction solution [4].

2. Experimental

Nanocrystalline ZnS thin films were deposited using conventional and very popular chemical bath deposition technique. Aqueous solutions of 0.2M zinc acetate and 0.6M thiourea were used as cationic and anionic precursors. Triethanolamine and trisodium citrate were used as complexing agents. Thoroughly cleaned microscopic glass slides were used as substrates. Two identical substrates were immersed simultaneously into the chemical bath one of which was thermally pretreated to a temperature of 200^0C for 15 minutes. The bath was maintained at a temperature of 80^0C and deposition was carried out for two hours. The slides were removed, thoroughly rinsed and dried using hot air blower. The as prepared samples were characterized using X-ray diffractometer, UV -Vis –NIR spectrophotometer, Scanning electron microscope and spectroflurometer.

3. Results and discussion

The films deposited on both the substrates appeared to be smooth and homogenous. The thickness of these films as determined using gravimetric technique was 43 nm for the one deposited on untreated substrate and 200 nm for the other.

*3.1 Structural analysis.*Figure1 depicts the XRD pattern of the as prepared samples. In general zinc sulphide crystallizes in cubic and hexagonal form. Cubic form is stable at room temperature whereas the hexagonal form is found to be stable at higher temperatures. ZnS film formed on untreated substrate was amorphous owing to very fine grains. The film on preheated substrate showed good crystallinity with an intense peak at 26.806^0 corresponding to (100) orientation of wurtzite structure. Another less intense peak occurred at 53.901^0 corresponding to (1 0 38) orientation. The average grain size using Debye Scherer formula was calculated to be 35nm. The improved crystallinity in the latter case could be attributed to the higher thickness of the deposit.

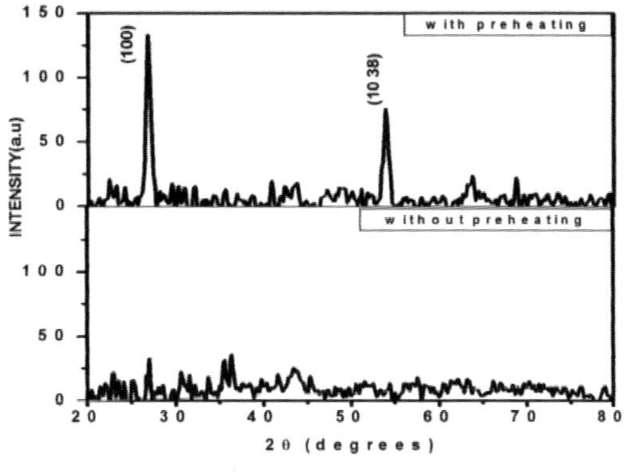

Figure 1.XRD pattern of ZnS samples

*3.2. Morphology.*The SEM images of the samples are shown in figure2.The film on pretreated substrate exhibits more uniform morphology compared to the other film. The grains appear to be tightly packed indicating densification of the film when preheated substrate was used. It is clear that thermal treatment of the substrate has enhanced the nucleation which resulted in a more compact film with decreased porosity.

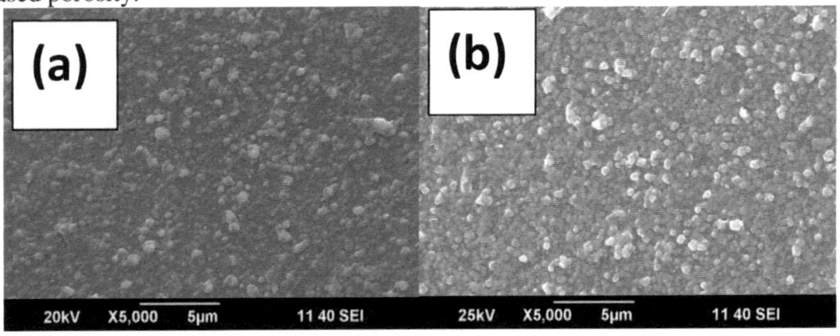

Figure2 SEM micrographs of ZnS thin films (a) without and (b) with preheating

*3.3. Optical analysis.*The transmission and absorption spectra of the samples in the UV-Vis-NIR region are shown in figure 3. The transmittance of the film deposited on preheated substrate in the UV-Vis region was found to be less than that deposited on normal substrate. This can be attributed to the higher thickness of the former sample. The improved thickness of the film has also caused a shift in absorption edge implying a reduction in band gap from 3.65eV to 3.45eV. Figure 4 shows the photoluminescence emission spectra of the samples at an excitation wavelength of 325nm.The samples exhibit emission at 376nm and 400 nm .The photoluminescence spectra of ZnS films depends on preparation conditions, defects present, crystallite size and shape and stoichiometry. The intensity of PL spectra usually decreases with increase in thickness and crystallinity of the film[4]. In this work the PL intensity was found to be more for the film on preheated substrate though it was more crystalline. Probably preheating of the substrate could have resulted in film with improved stoichiometry [5].Poor stoichiometry often results in low intensity luminescence peaks [6].

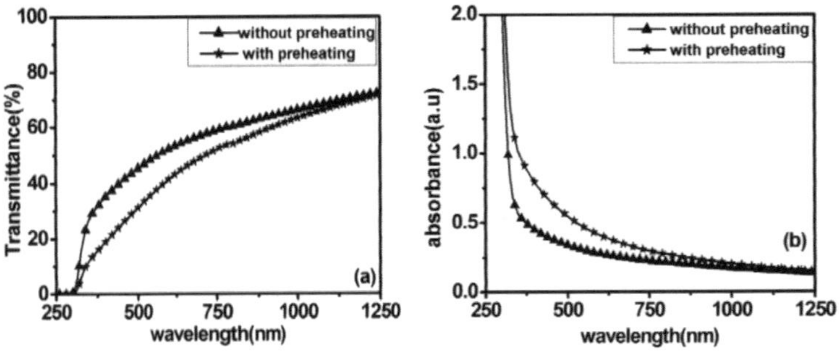

Figure 3 (a) Transmittance and (b) absorbance spectra of ZnS Thin films

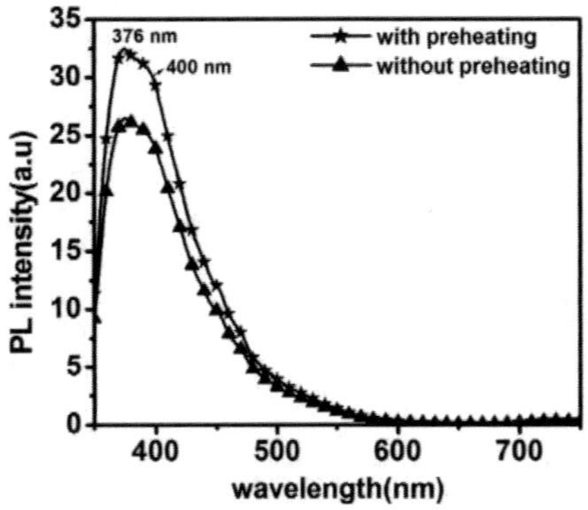

Figure 4.PL emission spectra of ZnS thin films

4. Conclusion

Highly crystalline ZnS thin films were deposited chemically on thermally treated glass substrates. The effect of thermodynamic conditions prevailing at the substrate solution interface is found to affect the properties of the film. The films prepared using this single step facile method exhibited good optical properties and are feasible for photovoltaic and optoelectronic applications. The effect of varying preheating temperature on the properties of ZnS thin films is open to investigation.

References

[1] Mach R and Mueller G O 1991 *Semicond. Sci. Technol.* **6** 305.
[2] Varitimos T E and Tustison R W 1987 *Thin Solid Films* **151** 27
[3] Dedova T, Krunks M, Volobujeva O and Oja I 2005 *Phys. Stat. Sol. (c)* **3** 1161
[4] Prathap P, Revathy N, Venkata Subbaiah Y P and Ramakrishna Reddy K T 2008 *J Phys:Condens.Matter* **20** 035205
[5] Long F,Wei-Min Wang, Zhan-kui Cui, Li-Zhen Fan, Zheng-guang Zou and Tie-kun Jia 2008 *Chem.Phys.Letters* **462** 84-87
[6] Denzler D, Olschewski M and Sattler K1998 *J.Appl.Phy.***84** 2841S

Dielectric properties of nano crystalline $LnTiNbO_6$ (Ln = Ce, Pr, Nd, Sm, Gd, Dy, Er, Yb) ceramics

Fergy John[1], J.K. Thomas[1] and Sam Solomon[1,2*]
[1]Department of Physics, Mar Ivanios College, Thiruvananthapuram 695015
[2]Department of Physics, St. John's College, Anchal, 691306

e-mail: samdmrl@yahoo.com

Abstract. Nano ceramics are potential candidates for various technological applications. Nano sized $LnTiNbO_6$ ceramics are prepared using solution combustion technique. The properties of the nano ceramics are compared with that of micro ceramics. XRD shows that nano crystalline $LnTiNbO_6$ (Ln= Ce, Pr, Nd, Sm) have aeschynite orthorhombic structure and Ln= Gd, Dy, Er, Yb have euxenite orthorhombic structure. The structure is confirmed by FTIR. From the TEM study, it is clear that polycrystalline powder is nano sized. The sintering temperature and sintering time of nano ceramics get lowered as compared to micro ceramics. The SEM image shows the surface morphology of the sintered sample. The variations of dielectric constant (ε_r), conductance (G) and loss factor (tan δ) of the samples are studied in the radio frequency range. The nano ceramics show enhanced dielectric properties than that of micro ceramics.

1. Introduction

Synthesis of nano material is one of the major challenges in material producing technology. Nano materials are the suitable for various technological applications, because they exhibit enhanced properties than bulk materials. Nano crystalline materials have better sinterabilty, high surface to volume ratio, superior phase homogeneity, greater chemical activity, etc. For the communication and satellite broadcasting, microwave dielectric resonators (DR) are essential component. Nano ceramics are potential candidate for developing DR's due to their improved electrical and dielectric properties than micro ceramics. Sebastian et al [1] reported microwave dielectric properties of $LnTiNbO_6$ micro ceramics. Here we reports the enhanced properties of nano sized $LnTiNbO_6$ ceramics.

2. Experimental

$LnTiNbO_6$ (Ln= Ce, Pr, Nd, Sm, Gd, Dy, Er, Yb) is prepared by solution combustion technique [2] using the corresponding metal nitrate (oxidizing agent) and suitable fuel (reducing agent) [3]. Calculations are based on the principles used in propellant chemistry, keeping fuel to oxidant ratio unity in order to produce maximum energy. In this synthesis, high-purity Niobium pentachloride, Titanium isopropoxide and the respective rare earth oxides are used as cation sources and oxidant agents, and urea is used as fuel reagent. Citric acid is used as complexing agent to get precursor complex. Stoichiometric amount of oxidizing agents and reducing agent in a minimum volume of deionized water to obtain transparent aqueous solutions in a glass beaker is heated using a hot plate at 250 °C to forms a viscous gel. The gel thus formed undergoes dehydration on further heating and self-ignites with the evolution of huge quantity of gases. Hence the residual ash that is formed after combustion has a fluffy nature. This ash is further heated up to 600 °C to get the pure phase nano powder. The prepared powder is mixed with one drop of 1 wt% solution of polyvinyl alcohol and uniaxially pressed at room temperature for 1 min at the pressure 190 MPa. The sintering of the pressed pellet is done with the help of furnace.

3. Characterization

The XRD (Model: Philips PW1710 diffractometer) is used to determine the formation of the desired phase, for the powder. The particle size of the prepared powders is also analyzed by the TEM (Model: TM-300, Philips). The Fourier Transform infrared (FTIR) spectra of the samples are recorded in the range 400-4000 cm^{-1} on a Thermo-Nicolet Avatar 370 Fourier Transform Infrared Spectrometer using KBr pellet method. The density of the sintered pellets are measured. The microstructures of the sintered samples are analyzed by SEM (Mode: JEOL JSM-5800) taken on the polished surfaces. The dielectric constant (ε_r), ac conductance (G) and the tangent loss (tan δ) measurements are done at radio frequency range on the polished pellet samples as a function of frequency using LCR meter (HIOKI, model 3532, Japan). Here after we denote $LnTiNbO_6$ (Ln= Ce, Pr, Nd, Sm, Gd, Dy, Er and Yb) as CTN, PTN, NTN, STN, GTN, DTN, ETN and YbTN respectively.

4. Result and discussions

The XRD patterns of nano $LnTiNbO_6$ ceramics are shown in figure 1. The patterns are indexed according to the ICDD files (File nos. 15-0864, 20-1361, 52-1130, 27-0221, 20-1434 and 20-1216). The XRD patterns of Ln= Ce, Pr, Nd, Sm are similar and have aeschynite orthorhombic structure with space group Pnma and that of Ln= Gd, Dy, Er, Yb have euxenite orthorhombic structure with space group Pbcn. The XRD pattern is closely agreement with the result that reported by sebastain et al [1]. The peaks are broad as compared to micro ceramics, which indicates that the particle size gets lowed. This is due to the nano size and micro strain in the crystallites. The particle size calculated using scherrer formula [4] from the main peak of XRD patterns are tabulated in the table 1.

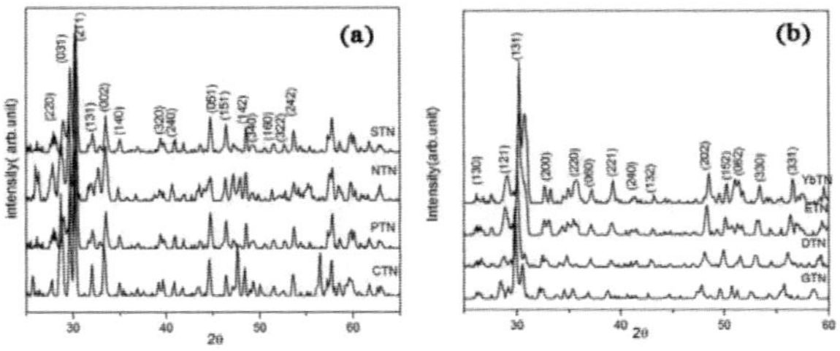

Figure 1.XRD patterns of the nano ceramics $LnTiNbO_6$ (a) Ln=Sm, Nd, Pr and Ce (b) Ln= Gd, Dy, Er and Yb

In FTIR spectra, very strong and broad absorption bands ranging from 400 to 800 cm^{-1} are observed in figure 2. In figure 2(a) the symmetric stretching vibrations of Nb/TiO_6 octahedra give rise to $\upsilon_1 A_{1g}$ mode is observed at 814 cm^{-1}. The $\upsilon_2 E_g$ mode of Ln= Ce, Pr, Nd, Sm is observed at 661 cm^{-1} and Ln= Gd, Dy, Er, Yb at 655 cm^{-1} confirm that figure 2(a) has the aeschynite and figure 2(b) has euxenite structure. In the range of 3000–3500 cm^{-1} a band corresponding to O-H bond [5].This is due to the contact of sample with atmospheric oxygen and water vapour.

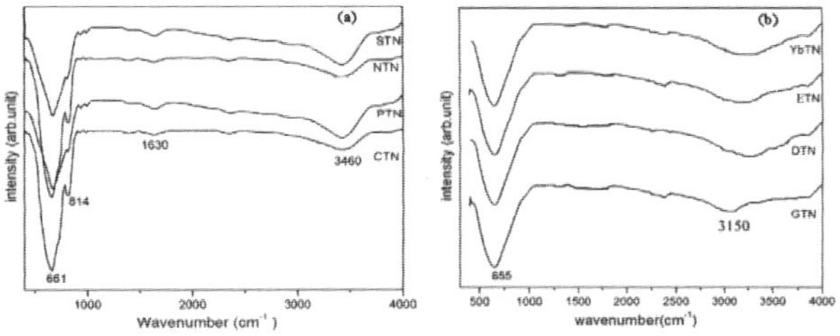

Figure 2. FTIR spectra of the nano ceramics $LnTiNbO_6$ (a) Ln= Ce , Pr, Nd, and Sm (b) Ln= Gd, Dy, Er and Yb.

The particle size is confirmed from TEM image of NTN ceramic is shown in figure 3. The average particle diameters lay in the range of 23 nm for NTN nano powder, which is in close agreement with the crystallite size obtained from XRD. The corresponding selected area electron diffraction (SAED) pattern of the sample NTN showed more "spotty" patterns of an assembly of nano crystallites shown as inset. This indicates the polycrystalline nature of the crystallite.

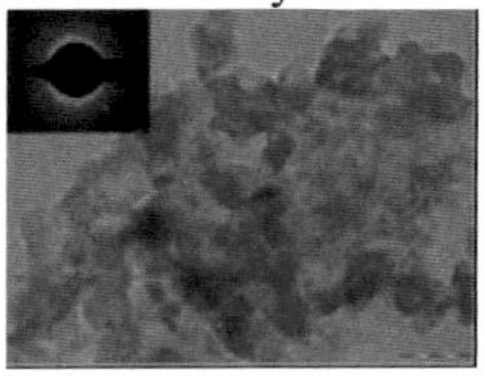

Figure 3. TEM image and SAED pattern of NTN ceramic

Due to fine nature of Nano-sized ceramics, it has high surface area and surface energy which leads to its excellent sintering characteristics. They are sintered below 1300 °C only for 2 hrs which is relatively shorter time and temperature as compared to the conventional micro ceramics. Sebastain et al [1] have reported that micro sized $LnTiNbO_6$ ceramics sintered around 1370 °C for 4 hrs. The sizes of the particles also affect the duration and temperature of sintering. Percentages of densification obtained for all samples are higher than micro ceramics. Table 1 compares the sintering temperature and density attained.

The scanning electron micrographs (SEM) of the sintered samples of STN and GTN are shown in figure 4(a) and 4(b). STN has elongated uniform grains and GTN has equiaxed grains and are appeared to be size less than 4 μm. It is evident from the SEM picture that the samples have achieved high densification with minimum porosity

Figure 4. Surface morphology of (a) STN and (b) GTN

Figure 5 shows variation of ε_r, G and tan δ with frequency range 100 Hz to 5 MHz at room temperature. Dielectric constant decreases initially with the increase in frequency and it reached a constant value at higher frequency range are given in the table 1. This is due to Maxwell-Wagner interfacial polarization [6]. The decrease in dielectric constant is due to the delay in polarization with the application of the electric field because of inertia. Dielectric constant of group I decreases from Ce to Sm depending on its ionic radii and that of group II get low values.

Variation of loss shows that it reached a minimum value and almost constant as frequency increases for all ceramics in accordance with Koops phenomenological model [7]. It means that, material has low loss in radio frequency range because of the relaxation frequency of the sample is out of this frequency range. Therefore, dissipation of electrical energy is low at this range. Loss-factor can be attributed to the presence of impurities like voids in the material and to the electrode effects.

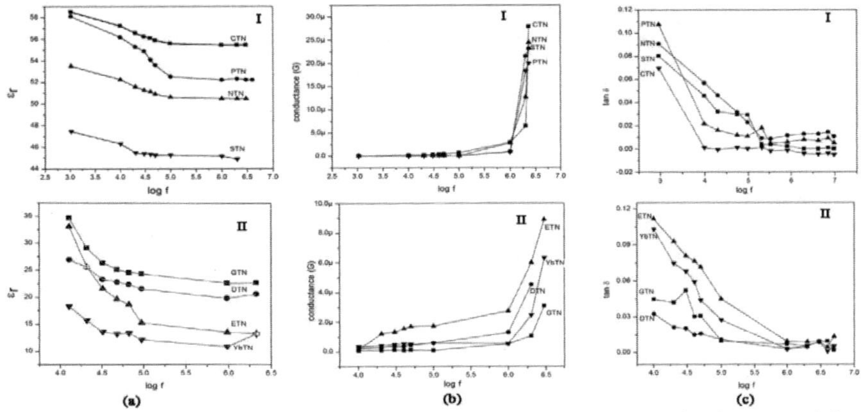

Figure 5. Variation of (a) ε_r (b) G (c) tan δ with radio frequency (I. aeschynite II. euxinite structure)

Table 1. Particle size and dielectric properties of $LnTiNbO_6$ nano ceramics

Ceramic	Particle size (nm)	Sintering temperature (°C)	Density attained (g/cm^3)	ε_r
CTN	25	1200	5.4	55
PTN	27	1230	5.3	53
NTN	23	1260	5.7	51
STN	32	1250	5.6	45
GTN	22	1290	5.7	25
DTN	34	1290	5.9	23
ETN	28	1280	5.8	15
YbTN	29	1280	6.1	13

And at low frequency range the variation of G is very less and almost constant up to 1MHz, and then increases according to the increase in frequency. Thus this material can be useful for capacitive applications in MW communication systems.

5. Conclusion

The Nano $LnTiNbO_6$ ceramics have been prepared using solution combustion technique. The particle sizes calculated from XRD are in the range of 20-35 nm which is agreement with TEM. XRD and FTIR studies shows that ceramics (Ln= Ce, Pr, Nd, Sm) possess aeschynite orthorhombic structure and ceramic (Ln= Gd, Dy, Er, Yb) possess euxenite orthorhombic structure. SEM image clarifies the high densification of nano ceramics. Dielectric constant is in the range 30 to 35.The variation of ac conductivity, dielectric loss gave the evidence that samples can be used as DR application in communication systems. From this study it can be concluded that solution combustion is a potential technique for the preparation of highly sinter-active, phase pure, nano powders of lanthanide oxide ceramics. And also the nano ceramics show enhanced dielectric properties than that of micro ceramics prepared by solid state ceramic route.

References

[1] M.T.Sebastian,Sam Solomon, and Ravindran Ratheesh 2001 Preparation, Characterization and microwave properties of $RETiNbO_6$ (RE=Ce, Pr, Nd, Sm, Eu, Gd, Tb, Dy, Y and Yb) dielectric ceramics *J. Am. Ceram. Soc*, **84** 1487

[2] K.C. Patil, S.T. Aruna, Tanu Mimani 2002 Combustion synthesis: an update *Curr. Opin. Solid. St. M.* 6 507

[3] A.K Tyagi, S.V Chavan & R. D Purohit 2006 Visit to the fascinating world of nano-ceramic powders via solution combustion, *Indian. J. Pure. Ap. Phy.* **44** 113

[4] U. Holzwarth and N. Gibson 2011 The Scherrer equation versus the Debye-Scherrer equation Nature *Nanotechnology* **6** 534

[5] K. Nakamoto 1986 Infrared and Raman Spectra of Inorganic and Coordination Compounds fourth ed. John Wiley & Sons New York 484

[6] J.C. Maxwell 1954 A treatise on Electricity and Magnetism Oxford University Press: Oxford U.K. **2** 328

[7] C.G. Koops 1951 On the Dispersion of Resistivity and Dielectric Constant of Some Semiconductors at Audiofrequencies *Phys. Rev.* **83** 121

IOP Conf. Series: Materials Science and Engineering 73 (2015) 012012 doi:10.1088/1757-899X/73/1/012012

Enhanced electrical and optical properties of iodine doped LDPE films

K Sreelatha and P Predeep[1]
Laboratory for Unconventional Electronics & Photonics, Department of Physics, National Institute Technology, Calicut, Kerala, India 673 601
Condensed Matter Physics Laboratory, Department of Physics, Sree Narayana College, Kollam, Kerala, India

E-mail: ppredeep@gmail.com

Abstract. In this study, an attempt has been made to dope Low Density Polyethylene (LDPE) with iodine and to study the effect of iodine doping on the optical band gap of LDPE films. The DC conductivity measurements and UV/Vis spectral studies are employed to characterize the samples. Iodine treatment induces colour change in the polymer film which supports the interaction between iodine molecules and polyethylene chains. I_2 molecule links the polymer chains electronically and provides conducting path ways by forming DA complexes. The absorption band of complex films has extended to the visible and near infrared region of the spectrum. Further there are discernible shifts found in the energy gap and band edge towards lower energies on doping with iodine. This is essentially due to the formation of strong DA complexes upon iodine doping which improves the conducting behaviour.

1. Introduction

Organic electronics or plastic electronics is a relatively new field of research that involves challenges regarding electronic devices made with carbon-based materials, such as (semi)conducting polymers and other organic materials. Conducting plastics are the best choice to be employed in the fabrication of electronic devices as they possess the electronic properties of a semiconductor and the specific properties of plastics such as light weight, high mechanical strength and flexibility. Another important motivation for the interest in organic (semi)conductors is its easy processability and expected low cost of the end product. The developments in memory materials, optoelectronic devices and applications focused to biomedicine are also emerging as progressing areas. There are many potential applications including, light emitting devices, nonlinear optical devices, photovoltaic devices, plastic field-effect transistors, and electro-magnetic shielding. The ability of the polymer to form thin film from solution allows them to be deposited on many kinds of substrates, providing opportunities [1] for many novel applications.

Low Density Polyethylene (LDPE) with its proven advantages of plastic flexibility and transparency has all the potential to find applications ranging from consumer electronics to optoelectronics, microelectronics and a host of new photonic applications. LDPE has short and long

[1] Corresponding author: P Predeep

IOP Conf. Series: Materials Science and Engineering **73** (2015) 012012 doi:10.1088/1757-899X/73/1/012012

branches of polyethylene [2] made of covalently bonded carbon atoms with hydrogen or methyl (CH_3) groups in a planar zigzag chain. Owing to the branched nature, LDPE has a reduced percentage of crystallinity and this increases the dependence of electrical behaviour on the non-crystalline and interfacial regions. Even though LDPE is chemically unreactive at room temperature it is slowly attacked by strong oxidizing agents such as halogens. Electrical conductivity can be enhanced by doping the polymer with oxidizing or reducing agents. A suitable dopant improves [3, 4] the charge carrier mobility of the polymer. It is known that iodine molecule possess acceptor properties and high electron affinity and it captures electron from the polymer macromolecule, forming donor/acceptor (DA) complexes with LDPE chains which improves the conducting behaviour of the polymer.

In this study, an attempt has been made to dope LDPE with iodine molecule and to study the effect of iodine doping on the optical band gap of LDPE films. UV/Vis spectral studies are employed to characterize the samples. The DC conductivity measurements of the doped films are made by the four probe techniques.

2. Experimental

2.1. Materials
Low Density Polyethylene films of thickness 0.03 mm were purchased from Good Fellow materials; U. K. Iodine (Merck) of analytical grade was purchased from local suppliers and used as received.

2.2. Sample preparation
Thin films of LDPE and iodine crystals are placed in a sealed tube for several days. Iodine is introduced through diffusion of iodine vapour into the PE chains. After this the samples are uncovered and stayed in air for time long enough to remove excess iodine from the surfaces of the samples.

2.3. Characterization
The DC conductivity measurements of the doped films are made by the four probe techniques using the Keithly 6514 Electrometer and Keithly 2000 DMM. The UV-Visible absorption spectra of the samples are recorded at room temperature on Cary 5000 at a scan rate of 600 nm per minute in the wavelength range 300-900 nm. From this data, the optical band gap is determined.

3. Result and discussion

3.1. DC Conductivity
Iodine treatment changes transparent LDPE films to purple and then to black shade with the increase in the doping period. The colour change supports the interaction between iodine molecules and polyethylene chains. Iodine exists in the form of neutral I_2 or as molecular aggregates I_n in the amorphous regions of the polymer [5-8]. I_2 molecule links the polymer chains electronically and reduces the inter chain barrier to charge transfer thereby increasing conduction. The electron transferred to the I_2 molecule moves either as stable iodide ion [5] or is transferred again to another polymer molecule with the iodine acting as a transfer agent [8, 9]. Thus iodine provides conducting pathways through the non-crystalline regions interconnecting the crystallites by forming DA complexes at appropriate sites on the chains. Further electrons captured by iodine pass through iodine states to reach electron vacancy (EV) sites on adjacent PE chains where recombination can occur to maintain conduction [5]. As the time of exposure to iodine is increased the colour, conductivity and iodine content in the film also increases. The conductivity of the doped films reached a maximum value of about 3.5×10^{-5} Scm^{-1}.

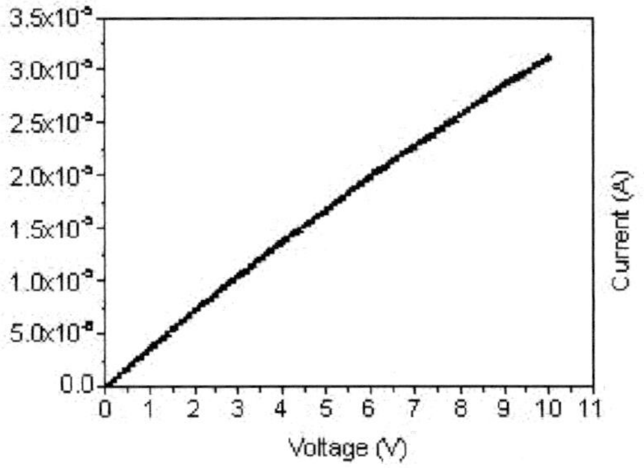

Figure 1. I-V Characteristics of iodine doped LDPE films

The current-voltage characteristic of LDPE iodine films measured at room temperature is shown in figure 1. The conduction observed in these films shows a practically linear plot indicating the Ohmic characteristics of the charge transfer complexes

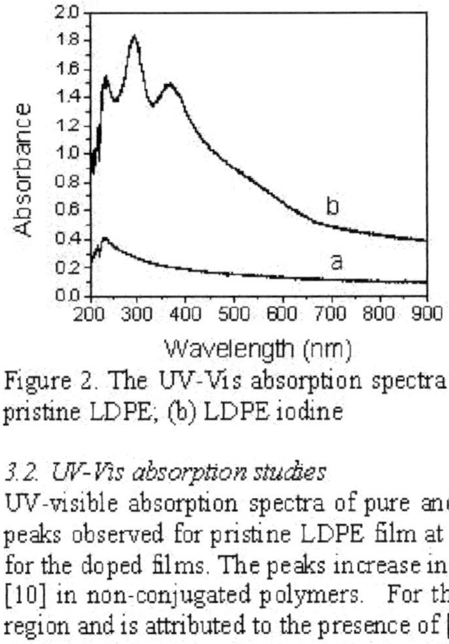

Figure 2. The UV-Vis absorption spectra of (a) pristine LDPE; (b) LDPE iodine

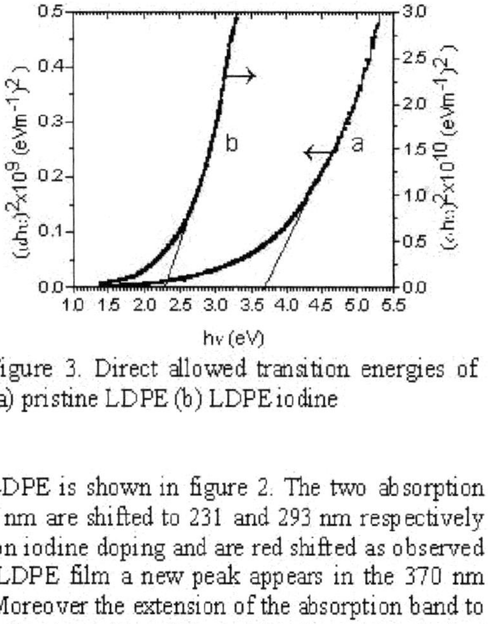

Figure 3. Direct allowed transition energies of (a) pristine LDPE (b) LDPE iodine

3.2. UV-Vis absorption studies
UV-visible absorption spectra of pure and iodinated LDPE is shown in figure 2. The two absorption peaks observed for pristine LDPE film at 215 and 228 nm are shifted to 231 and 293 nm respectively for the doped films. The peaks increase in intensity upon iodine doping and are red shifted as observed [10] in non-conjugated polymers. For the iodinated LDPE film a new peak appears in the 370 nm region and is attributed to the presence of [11] iodine. Moreover the extension of the absorption band to the near infrared region and the overall increase in the intensity of the peaks shows that iodine molecule is well incorporated into polymeric matrix which improves the conducting behaviour.

3.2.1. Determination of optical bandgap. The energy band gap of materials is related to the absorption coefficient by the Tauc's relation

$$\alpha h\vartheta = A(h\vartheta - E_g)^n \qquad (1)$$

Here α is the absorption coefficient, hν the photon energy, A is a constant and the index n is the distribution of the density of states which assumes values 1/2, 3/2, 2 and 3 depending on the nature of the electronic transition responsible for the absorption. n = 1/2 corresponds to the direct allowed transition energy gap.

The direct band gap is determined from plot of $(\alpha h\nu)^2$ as a function of photon energy ($h\nu$) as shown in figure 3. The intercept of these curves on the photon energy axis gives the direct bandgap. For pure LDPE films the direct bandgap lies at 3.65 eV while for doped films, the value decreases to 2.25 eV. The reduction in the band gap due to iodine doping indicates the incorporation of iodine into the polymer chain and, thereby, extending the density of states more into the visible and near IR region of the electromagnetic spectrum as compared to that of the undoped case.

4. Conclusion

Electrical and optical properties of the LDPE-iodine film depend on exposure time and iodine concentration. Energy gap shifts to lower energies on doping with iodine thus increasing the capability of polymers to conduct. In LDPE, iodine exists as neutral iodine in between the polymer chains, acts as electron transfer agents, links the polymer chains electronically and reduces the inter chain barrier to charge transfer thereby increasing conduction.

5. References

[1] Gustafsson G, Cao Y, Treacy G M, Klavetter F and Heeger A J 1992 *Nature* **357** 477
[2] Peacock A J 2000, Handbook of Polyethylene (Marcel Dekker Inc., Basel)
[3] Predeep P, Najitha S, Saxena N S, Sreeja R and Unnithan C H 2005 *Mater. Lett.* **59** 3431
[4] Predeep P, Sreeja R, Najitha S, Remya Jayan S, Maciej Mazurm,and Sharma P D 2006 *Polymer* **47** 617
[5] Lewis T J and D. M. Taylor D M 1972 *J. Phys. D: Appl. Phys* .**5** 1664
[6] Davies D K 1972 *J. Phys. D:Appl. Phys.* **5** 162
[7] Mizutani T, Mitani K and Ieda M 1984 *J. Phys. D: Appl. Phys.* **17** 1477
[8] Swan D W 1967 *J. Appl. Phys.* **38** 5051
[9] Liberman E A and Topaly V P 1968 *Biochim. biophys. Acta* **163** 125
[10] Vippa P, Rajagopalan H, Thakur M 2005 *J. Polym. Sci. B. Polym. Phys.* **43** 3695
[11] Schopov I, Sinigenski V 1992 *Makromol Chem.* **193(8)** 1839

IOP Conf. Series: Materials Science and Engineering **73** (2015) 012013 doi:10.1088/1757-899X/73/1/012013

ELECTRICAL AND PHOTOCONDUCTIVITY STUDIES ON $AgSbSe_2$ THIN FILMS

T Namitha Asokan, K S Urmila and B Pradeep

Solid State Physics Laboratory, Department of physics, Cochin University of Science and Technology, Kochi- 682022, Kerala.

Corresponding author e-mail: **namithaasokan@gmail.com**

Abstract. Silver antimony selenide thin films have been deposited on ultrasonically cleaned glass substrate at a vacuum of 10^{-5} torr using reactive evaporation technique. The preparative parameters like substrate temperature and incident fluxes have been properly controlled in order to get highly reproducible compound films. The polycrystalline nature of the sample is confirmed using XRD. The dependence of the electrical conductivity on the temperature has also been studied. The prepared $AgSbSe_2$ samples show p-type conductivity. The samples show a little photoresponse.

1. INTRODUSTION

The investigations on I-III-VI chalcopyrites have been done extensively because of their strong absorbing properties that make these materials good for photovoltaic applications [1].The compound silver antimony selenide thin film exists in two phases- cubic $AgSbSe_2$ and orthorhombic Ag_5SbSe_4. Garza et al [2] prepared $AgSbSe_2$ thin films by heating sequentially deposited Sb_2S_3, Ag_2Se and Ag thin films in close contact with Se thin film. In this work, using the reactive evaporation technique $AgSbSe_2$ thin films have been deposited successfully by optimizing the deposition parameters such as impingement rates and substrate temperature. The as-deposited samples are structurally and electrically characterized.

2. EXPERIMENTAL DETAILS

Silver antimony selenide thin films were deposited onto ultrasonically cleaned glass substrates using 12'' vacuum coating unit. For the preparation of ternary $AgSbSe_2$, two separate glass crucibles kept in molybdenum baskets were used as sources for the evaporation of 99.999% pure elemental Sb and Se. A molybdenum boat was used for the evaporation of 99.999% pure elemental Ag. The impingement rate of the elements and the deposition temperature used for the preparation of stoichiometric silver antimony selenide thin film was given below [3].

Impingement rate of Silver	$\approx 3x10^{15}$ atoms cm^{-2} s^{-1}
Impingement rate of Antimony	$\approx 2x10^{15}$ atoms cm^{-2} s^{-1}
Impingement rate of Selenium	$\approx 1x10^{14}$ atoms cm^{-2} s^{-1}
Substrate temperature	$=398 \pm 5K$

3. RESULT AND DISCUSION

3.1. *Structural analysis*

In order to see the structural properties, the X-ray diffraction pattern was taken using Rigaku D-Max diffractometer. Figure 1 shows XRD pattern of a typical $AgSbSe_2$ thin film. The indexing of the pattern was done with the standard ASTM card no.12-379 [4].

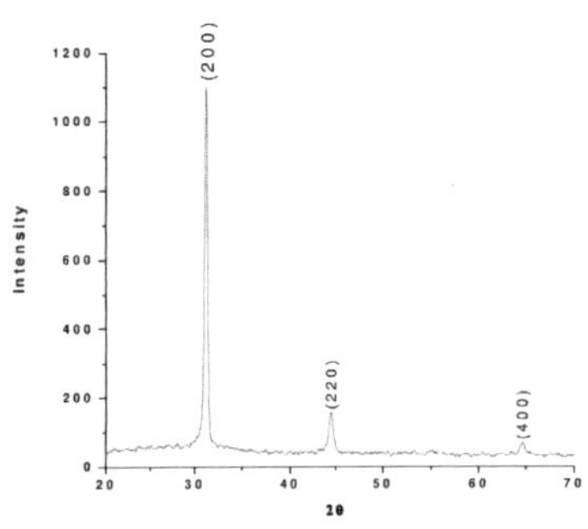

Figure 1. X-ray diffraction of as prepared $AgSbSe_2$ thin film.

The peaks at 2θ=31.15, 44.45, 64.7 and the corresponding planes (200), (220) (400) indicates the films are polycrystalline in nature. The most reported structure is FCC. The average particle size was calculated using the Debye Scherrer's equation [5]:

$$\mathbf{D = 0.9\lambda / \beta \cos \theta}$$

where λ is the wavelength of X-ray used, β the full width at half maximum in radian, and θ the glancing angle in degree. The average particle size calculated using above equation is 22nm. The lattice constant calculated is 5.74Å.

The dislocation density (ρ) is calculated using the relation [6]:

$\boldsymbol{\rho = 1/D^2}$ and it is $2.066x10^{15}$ lines/m^2.

The number of crystallite per unit area (N) calculated as $3.757x10^{15}m^{-2}$ using the relation [6]:

$\mathbf{N = t/D^3}$, where t is the thickness of the film.

3.2. *Electrical studies*

With regards to electrical properties, variations in the conductivity with respect to temperature were studied inside a conductivity cell at a vacuum of 10^{-2} torr. It is shown in figure 2 for three continuous cycles. From the figure it is clear that the film is stabilized after under first heating and cooling. The monotonous increase of current with increasing temperature exhibits the semiconducting nature. All the $AgSbSe_2$ samples showed p-type conductivity. The activation energy calculated is 0.08eV. Patel et al [7] studied growth and electrical characterization of $AgSbSe_2$ thin films. They reported the activation energy as 0.45eV in the high temperature region and 0.091eV in the low temperature region respectively.

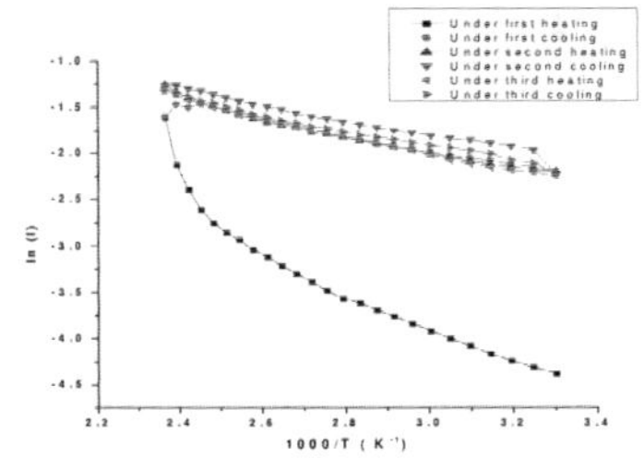

Figure 2. Variation of ln(I) vs inverse of temperature for $AgSbSe_2$ films

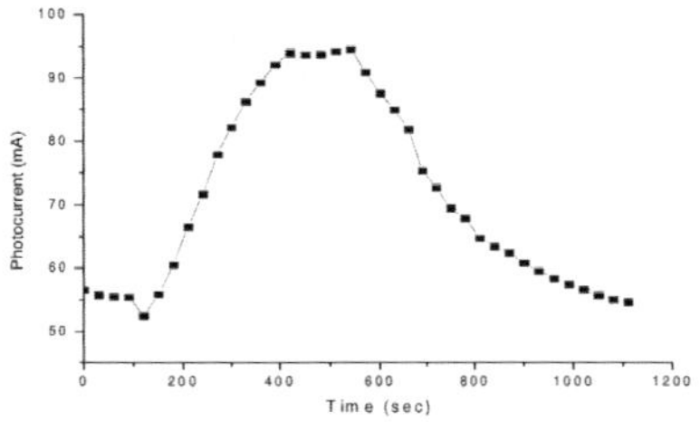

Figure 3. The photoconductivity response of a typical $AgSbSe_2$ with time.

Figure 3 shows the photoconductivity response of a typical $AgSbSe_2$ thin film. The sample is exposed to white light from tungsten halogen lamb of power 100W. At illumination the photocurrent rises gradually with time and attains constant value. It falls down gradually when the illumination is cut off. The $AgSbSe_2$ samples prepared using the reactive evaporation technique showed a little photorespose and can be used as absorber layer in solar cell [8].

4. CONCLUSION

Silver antimony selenide thin films are successfully deposited on glass substrates using reactive evaporation technique. The deposited thin films are structurally characterized by XRD. The average particle sizes, lattice constant, dislocation density and number of crystallite per unit area are calculated. The activation energy calculated is 0.08eV. $AgSbSe_2$ samples show p-type conductivity and all the samples shown a little photoconductivity.

ACKNOWLEDGEMENT

One of the authors (TNA) would like to thank University Grants Commission for the financial assistance in the form of Research Fellowship in Science for Meritorious Students (RFSMS). Second author (KSU) would like to thank Cochin University of Science and Technology for the financial assistance in the form of Research Fellowship.

References

[1] Karaagac H, Kaleli M and Parlak M 2009 J.Phys.D: Appl.Phys. **42** 165413.
[2] Garza J G, Shaji S, Rodriguez A C, Das Royand T K and Krishnan B, 2011 Applied Surface Science **257** 10834
[3] Leon I Maissel and Reinhard Glang 1970 Hand book of thin film technology (McGRAW-HILL) pp 1-26.
[4] Powder Diffraction file 1948 Joint Committee on Powder Diffraction Standards (JCPDS) (Swarthmore, PA), ASTM card no.12-379.
[5] Cullity B D 1956 Elements of x-ray diffraction (Addison-Wesley publishing company Inc, USA).
[6] Dhanam M, Manoj P K and Rajeev R Prabhu 2005 *Journal of Crystal Growth* **280** 425.
[7] Patel A R, Lakshminarayana D and Rao K V 1982 Thin Solid Films **94** 51.
[8] Bindu K, Jose Campos, Nair M T S ,Sanchez A and Nair P K (2005) Semicond. Sci.Techno. **20** 496.

IOP Conf. Series: Materials Science and Engineering **73** (2015) 012014 doi:10.1088/1757-899X/73/1/012014

Raman Spectroscopic Studies on L-histidine, aniline Doped Triglycine Sulphate Single Crystals

A Milton Franklin Benial[1*], V Ramakrishnan[2] and A Parameswari[1]

[1] Department of Physics, NMSSVN College, Madurai-625 019, Tamilnadu, India.
[2] Department of Laser Studies, School of Physics, Madurai Kamaraj University, Madurai 625021, India.

E-mail:miltonfranklin@yahoo.com

Abstract. Single crystals of triglycine sulphate (TGS) doped with L-histidine and aniline were studied by Raman Spectroscopy. The structure and symmetry of molecules, the nature of bonding and the effect of crystalline field on molecular vibrations were studied for pure and doped TGS. The characteristic group frequencies were identified and analysed for H_2SO_4 and glycine. The skeletal motion, lattice vibrational peaks were observed in the low wavenumber region. The site symmetry effect and the correlation field effect were studied from the splitting of vibrational bands. The observed Raman shift towards higher wave number region reveals that the symmetry reduction in doped TGS crystals. The broadening of Raman spectral line showed that a decrease in the hardness value for the doped crystals. Comparative studies of the Raman Spectra of pure TGS and doped TGS were also carried out.

1. Introduction

Triglycine sulphate [TGS] single crystal is one of the most extensively studied material as it possess excellent pyroelectric and feroelectric properties. It exhibits strong absorption in most of the infrared region. This crystal is suitable to make speed broad band infrared pyroelectric detectors and vidicon. The crystal structure of TGS was reported by Hoshina *et al* [1]. The crystal is monoclinic in structure below and above the Curie temperature (49°C) [2]. Pure TGS crystals have some disadvantages over doped TGS crystals such as the ferroelectric domains possess high mobility at room temperature, easy depolarization by electrical and thermal means, microbial contamination with time during the growth and low Curie point. These disadvantages can be overcome by adding suitable impurities to the lattice sites of TGS crystals [3-5].

2. Materials and methods

Pure triglycine sulphate was synthesized [6] by taking glycine and sulphuric acid in the molar ratio 3:1. The required amount of sulphuric acid was diluted with double-distilled water. A calculated amount of glycine was added to the diluted sulphuric acid. The solution was heated until the salt became recrystallised. TGS aqueous solution was equilibrated with the amino acid L-histidine and aromatic compound aniline. FT Raman spectral measurements were carried out using Bruker IFS 66v FT-IR spectrometer attached with FRA 106 Raman module.

3. Results and discussion

In the present study, Raman spectra were recorded for pure TGS, L-histidine doped TGS and aniline doped TGS crystals and shown in Fig.3.1 – Fig.3.3. Raman vibrational modes for pure TGS and doped TGS crystals were identified and assigned.

3.1 Vibrational modes of glycine

In this section, the glycine group frequencies of NH_3^+, COO^- , CH_2, CH and C-C were assigned by using the reported Raman spectral data [7]. In the spectrum of pure TGS (Fig.3.1), CH_2 wagging/twisting is observed as a strong peak at 1302 cm^{-1}, but this mode is observed as a broad and strong peak for L-histidine doped TGS (Fig.3.2), and aniline doped with TGS (Fig.3.3) at 1302 cm^{-1} and 1320cm^{-1} respectively. The observed splitting confirms the symmetry reduction in aniline doped TGS crystals. The Raman spectra show CH_2 rock mode as a strong band at 889 cm^{-1} for all the three samples. The spectrum of pure TGS (Fig.3.1) shows a strong band corresponds to CH_2 bending at 1440 cm^{-1}. The spectrum of L-histidine doped TGS (Fig.3.2) and aniline doped TGS (Fig.3.3) show strong band corresponds to CH_2 bending at 1440 cm^{-1}and 1439 cm^{-1} respectively. The CH_2 scissoring is also observed as weak bands around 1482 cm^{-1} for pure TGS and doped TGS crystals (Fig.3.1-Fig.3.3). The spectrum of pure TGS (Fig.3.1) shows a strong band at 2961 cm^{-1} corresponds to CH_2 symmetric stretching vibrational mode but it is not observed for doped TGS. Another mode is also observed due to the CH_2 asymmetric stretching vibration at 2985 cm^{-1} for pure TGS and L-histidine doped TGS as a strong band, but is observed as a very strong band at 2986 cm^{-1} for aniline doped TGS. The CH_2 symmetric and asymmetric stretching modes are well resolved in pure TGS and L-histidine doped TGS as shown in (Fig.3.1 and Fig.3.2). In aniline doped TGS crystals, the CH_2 symmetric stretching mode is not resolved. In aniline doped TGS crystals, the CH_3 symmetric and asymmetric stretching modes also fall in the same region. The dopant (L-histidine or aniline) causes much broadening of the peaks in the spectra, compared to the pure TGS crystal. The line broadening effect is further confirmed by Aravazhi et.al [8]. This broadening leads to the unresolved CH_3 symmetric stretching and anti symmetric stretching modes in the region 2960-2990 cm^{-1}. The symmetry reduction also causes the broadening of spectral lines.

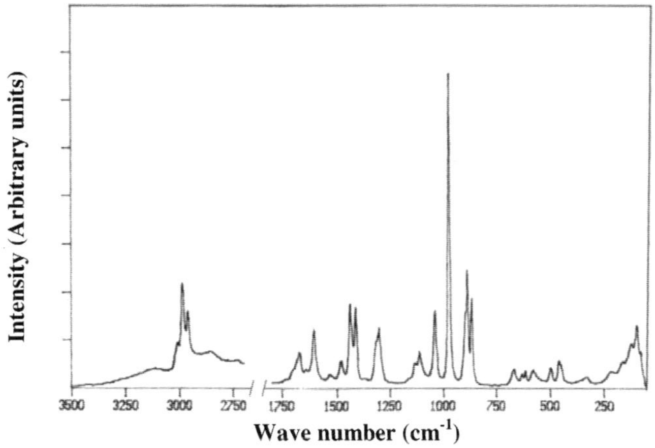

Fig. 3.1. Raman Spectrum for pure TGS

In the present study, the NH_3^+ stretching frequency is appeared as a medium broad band at 3008 cm^{-1} (Fig.3.1) in pure TGS crystal. In L-histidine and aniline doped TGS crystals, (Fig.3.2 and Fig.3.3) show medium band at 3007cm^{-1} and 3018 cm^{-1} respectively. The broadening of spectral line is due to the site symmetry and crystalline field effect in solids. In the pure TGS crystal (Fig.3.1) the asymmetric NH_3^+ deformation is observed at

1674 cm^{-1} as a medium broad band for L-histidine doped TGS and strong band for aniline doped TGS at 1675 cm^{-1} (Fig.3.2 and Fig.3.3). This reveals that the reduction in the crystal hardness as a result of dopant incorporation. The symmetric NH_3^+ deformation mode is appeared as a weak band in doped TGS crystals. In pure and doped TGS crystals, the NH_3^+ torsional mode/COO^- rock mode is appeared around 500 cm^{-1} as a medium band. The previous Raman studies on amino acids revealed that the COO^- asymmetric stretch is observed in the region 1605-1555cm^{-1}, symmetric COO^- stretching is observed in the region 1425-1393 cm^{-1}. In pure and doped TGS crystal (Fig.3.1-3.3), the COO^- symmetric stretching is observed as a strong band at 1414 cm^{-1} and COO^- asymmetric stretch is observed as a strong band at 1607 cm^{-1}. For pure and doped TGS crystals, the COO^- scissoring and COO^- wagging modes are noticed as medium and weak band around 665 cm^{-1} and 579 cm^{-1} respectively.

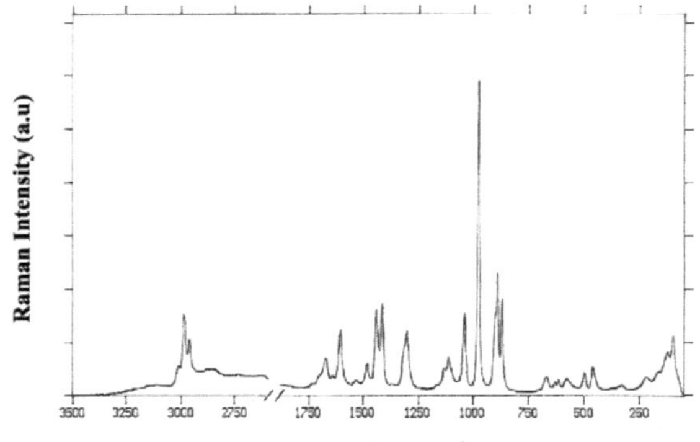

Fig. 3.2. Raman Spectrum for L-histidine doped TGS crystals

3.2 Vibrational modes of H_2SO_4

The spectrum of pure TGS (Fig.3.1) shows the vibrational modes of H_2SO_4 molecule. These modes also appeared in the spectra of L-histidine doped with TGS (Fig.3.2) and aniline doped with TGS (Fig.3.3). The vibrational modes of H_2SO_4 have been assigned in all the three systems. All the three spectra show symmetric stretching mode ($S[OH]_2$), asymmetric stretching mode ($S[OH]_2$) and bending mode (SO_4). In pure TGS and doped TGS crystals, symmetric stretching mode, asymmetric stretching mode are appeared as a strong and very strong peak around 889 cm^{-1} and 978 cm^{-1} respectively.

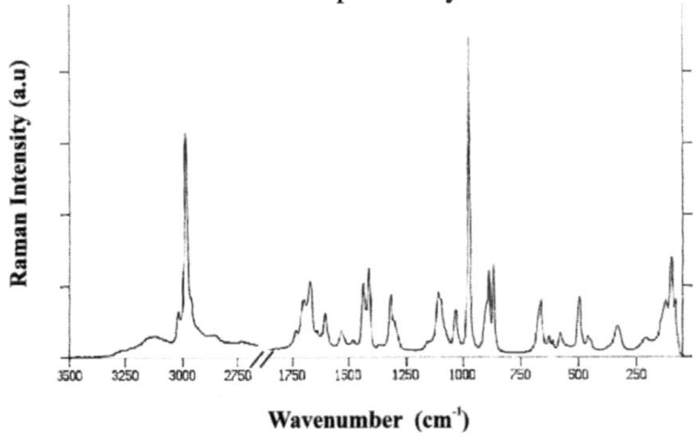

Fig. 3.3. Raman Spectrum for aniline doped TGS crystals

The spectra of pure TGS (Fig.3.1), L-histidine doped TGS (Fig.3.2) and aniline doped TGS (Fig.3.3) show SO_4^- bending modes as weak band around 613 $cm^{-1}$and 629 cm^{-1}. The Raman spectra (Fig.3.1 and 3.2) show out-of-plane SO_4 stretching mode as a medium broad band around 1113 cm^{-1} for pure TGS and L-histidine doped TGS. But it is observed as a strong broad band at 1114 cm^{-1} for aniline doped TGS. In pure TGS and doped TGS crystals, the C-C stretching mode appeared as a strong band around 868 cm^{-1}. The C-N stretching mode is appeared as a strong band for pure TGS and L-histidine doped TGS, but it is observed as a medium band for aniline doped TGS at 1034 cm^{-1}.

The skeletal motions are observed as a weak bands around 331 cm^{-1} and 462 cm^{-1} in all the three systems. The Raman spectra (Fig.3.1-3.3) show the lattice vibrational modes in the region 100-250 cm^{-1}. The combination band is also appeared around 1482 cm^{-1} as a weak band in pure TGS and doped TGS crystals. The combination band arises due to the $S[OH]_2$ asymmetric stretching mode and COO^- rock /NH_3^+ torsion.

4. Conclusion

The Raman vibrational modes for pure TGS and doped TGS have been assigned and analysed. The correlation field effect and site symmetry reduction in solids have been discussed. A comparative study between pure and doped TGS crystals is also carried out. The line broadening was observed for doped TGS crystals compared with pure TGS. The spectral broadening reflected the crystal hardness as a result of dopant (L-histidine and aniline) incorporation. The broadening of the spectral lines in doped TGS crystals is due to the asymmetry of the L-histidine and aniline molecules, which are bigger in size than the glycine molecules and difficult to incorporate into the glycine site. The skeletal motion and lattice vibrational modes are also assigned.

Acknowledgements

The authors A. Milton Franklin Benial and A.Parameswari thank NMSSVN College, Madurai-625 019, India for encouragement and permission to carry out this work. One of the author A.Parameswari thank the University Grant Commission (UGC), New Delhi, India for the financial support (Grant No: **F. No :4-4/2013-14**).

REFERENCES

[1] Guangyong zhu, Xian zhu, QiFan, Xueliang Wan 2011 *Spectrochimica Acta Part A.* **78** 1187-1195

[2] Michalski J, Kucharska E, Sasiadek W, Lorenc J, Hanuza J 2013 *Spectrochimica Acta Part A.* **112** 263-275

[3] Arunmozhi G Jayavel R and Subramanian C 1998 *Mat. Lett.* **33** 251

[4] Krishnakumar V, Rajaboopathi M, Nagalakshmi R 2011 *Advanced Materials Letters.* **2(2)** 163-169

[5] Khanum F and Podder J 2011 *Journal of Crystallization process and technology.* **2** 26-31

[6] Novotny J Prokopova L and Micka Z 2001 *J. Cryst. Growth,* **226** 333

[7] Shanthi N T, Selvaraan P, and Mahadevan C K 2009 *Indian ournal of Science and Technology.* **3** 49-52

[8] Meera K Muralidharan R Santhanaraghavan P Gopalakrishnan R and Ramasamy P 2001 *J. Crystal Growth.* **226** 312

IOP Conf. Series: Materials Science and Engineering **73** (2015) 012015 doi:10.1088/1757-899X/73/1/012015

Synthesis and dielectric studies of poly (vinyl pyrrolidone) / titanium dioxide nanocomposites

Prathibha Vasudevan, Sunil Thomas, Arunkumar K V, Karthika S and Unnikrishnan N V*

School of Pure & Applied Physics, Mahatma Gandhi University, Kottayam- 686 560, Kerala, India.

*Email id: nvu100@yahoo.com

Abstract. In this paper, we present the synthesis of poly vinyl pyrrolidone (PVP) / titanium dioxide nanocomposites via sol- gel technique. The structural and dielectric properties of the samples were also analysed in this work. PVP doped with varying concentrations of TiO_2 are prepared by the sol–gel route. The prepared composites were characterized by X-ray diffraction (XRD), transmission electron microscopy (TEM) and impedance spectroscopy. XRD and TEM confirm the presence of TiO_2 nanoparticles in the composites. The dielectric response and the AC electrical conductivity of the samples are investigated for the frequency range 1 kHz–2MHz at room temperature. The dielectric studies show low values for dielectric constant and loss at high frequencies.

1. Introduction

Organic–inorganic hybrid composites has attracted considerable attention in many applications, such as packaging material, circuit board, interlayer dielectrics, antenna, and passivative protection. Composite materials are typically formed when at least two distinct materials are mixed to form a monolith [1]. Hybrid organic–inorganic materials can be formed by reacting an inorganic alkoxide directly with an organic polymer or an oligomer having the appropriate functional groups, thus providing covalent linkages between the organic phase and the inorganic network [2]. Polymers possess some limitations for much higher performance applications because of their intrinsic organic nature and thus require inorganic materials for improving them. Inorganic materials exhibit excellent thermal stability and functional properties such as dielectric and magnetic properties, but have high brittleness and bad film forming characteristics. Thus, the composite formation of polymers with inorganic material has been suggested to meet the demands of balanced properties for both organic and inorganic materials [3, 4]. The overall properties of a composite material are determined not only by the properties of the parent components but also by the morphologies, volume fractions, and connectivity of the phases as well as their interfacial properties [5, 6].

Titania–polymer nanocomposites represent a new class of potential materials for optoelectronic applications. Titanium dioxide (TiO_2) is a very promising material for a large number of potential applications in many aspects such as solar energy conversion, photocatalysis, sensors, and photochromic devices [7, 8]. TiO_2 has been known to have three crystallographic phases: anatase, brookite and rutile [9]. Among them, anatase has been proved to have excellent photoactivity and optoelectronic properties [10]. In this work, we present the non- hydrolytic sol-gel method of polymer-

TiO_2 nanocomposites preparation without any catalysts and the structural and dielectric characterizations of these polymer nanocomposites.

2. Experimental details

The PVP-TiO_2 nanocomposites were prepared by the non- hydrolytic sol-gel (NHSG) process using tetra isopropyl orthotitanante (TIOT) and poly vinylpyrrolidone with ethanol as a solvent. Measured amounts of TIOT and PVP were stirred magnetically for 20 min with ethanol at room temperature to avoid local inhomogeneities. No catalysts were added. The resulting solution became highly viscous after a few seconds. In this study, the weight ratio of the inorganic precursor in the reaction mixture was taken as 25wt% (Sample A), 50wt% (Sample B) and 75wt% (Sample C) respectively. The TiO_2-PVP composite gels, which are transparent and viscous, are dried in an oven at 60°C for 24 h in air and the final product is yellowish in colour. Finally, the samples were annealed in air at 360°C for 2h to obtain the TiO_2-PVP composite material with TiO_2 nanocrystals.

The FTIR spectra were taken using the Shimadzu FTIR and the X-ray diffraction (XRD) of the sample was recorded using Bruker D8 advance diffractometer. The transmission electron microscopy (TEM) image was obtained using FEI Tecnai TF20 FEGTEM. The AC conductivity and dielectric measurements were done using the Agilent E4980A.

3. Results and Discussion

Fig. 1 exhibits the XRD pattern of the hybrid composite TiO_2 nanocrystals. There exist seven peaks at 25.2°, 37.8°, 48.04°, 55.06°, 62.6°, 68.7°,75.02° respectively, which can be indexed to (101), (004), (200), (211), (204), (116) and (215) planes of anatase phase of TiO_2. The above results confirmed that the optimum calcination temperature for the preparation of pure anatase TiO_2 nanocrystals is 360°C. The reflection peaks were broad, indicating the small size of the crystal in the nanoscale range. Therefore, the average crystallite sizes were measured from the most intense XRD peak (101) using the Debye- Scherrer formula and was estimated to be 4.8 nm. In order to find the effect of strain on the particle size, we use a modified equation with full-widths at half-maximum (FWHM) of the diffraction peaks [11]. The FWHMs (β s) can be expressed as a linear combination of the contributions from the strain (η) and particle size (ε) through the following relation:

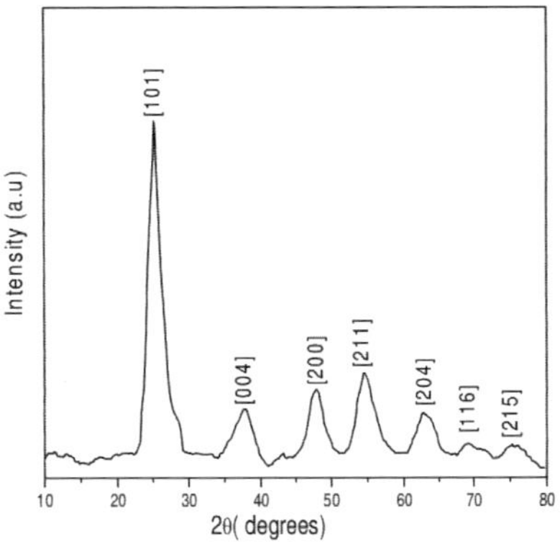

Fig 1: XRD pattern

$$\frac{\beta \cos\theta}{\lambda} = \frac{1}{\varepsilon} + \frac{\eta \sin\theta}{\lambda}$$

Here, ε is the effective particle size and η is the effective strain. Using this equation, the crystal size was evaluated to be 5.58 nm. Comparing the two crystallite size values one can find that the sample does not show the presence of any significant strain.

The TEM image of the hybrid composite containing TiO_2 nanocrystals is shown in Fig 2. It can be clearly observed that a lot of nanoscale particles are dispersed in the spectrum. The image shows several

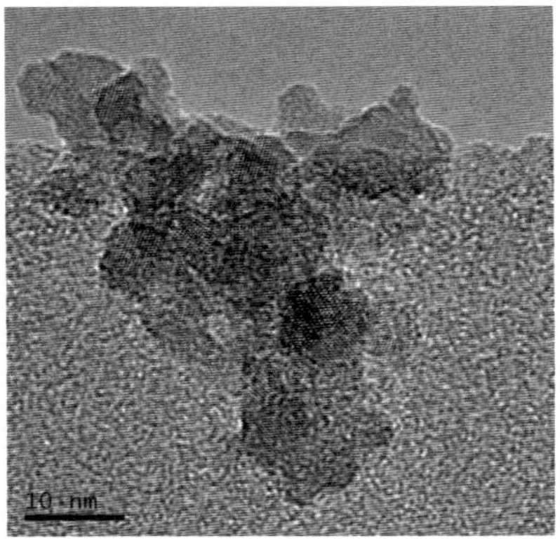

Fig 2: TEM image

nanocrystals oriented randomly, ranging in size from 5 to 7 nm, and not much agglomeration even after heating at 360°C. The nanocrystals mostly had a spherical morphology.

Fig. 3(a) presents the variation of real part of dielectric constant (C) as a function of frequency (ω) at room temperature for different samples. It can be seen from the figure that the value of dielectric constant (ε') is decreased as the frequency increased. It is likely that as the concentration of TiO_2 increases; the distance between the TiO_2 nanoparticles is reduced, thereby increasing the interaction between them [12]. Such an increase in interaction among the nanoparticles, could possibly reduce the diploe- dipole interaction, leading to a reduction in the dielectric constant. The variation of imaginary part of dielectric constant (dielectric loss) ε" with frequency for PVP-TiO_2 nanocomposites at different concentration of TiO_2 nanoparticles is shown in Fig 3(b). It is clear from this graph that dielectric loss

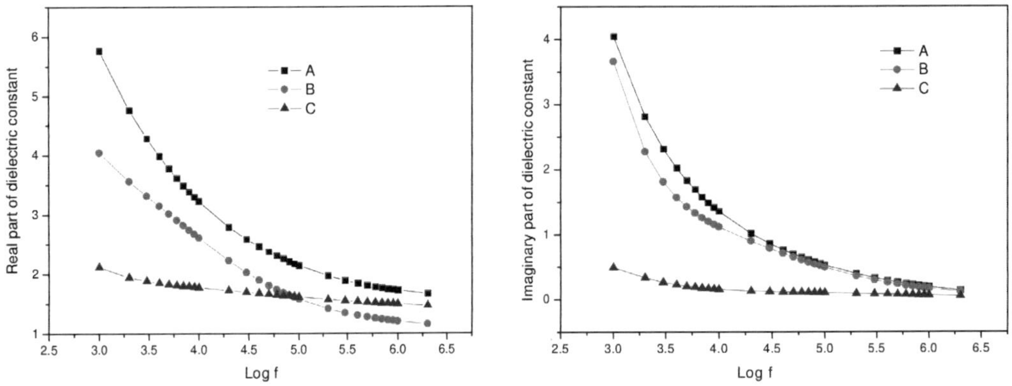

Fig 3(a): Variation of real part of dielectric constant with log f; 3(b) Variation of imaginary part of dielectric constant with log f

decreases with an increase in frequency. The decrease of ε" with frequency can be attributed to the fact that, at low frequencies, the value of ε" is due to the migration of ions in the material. At moderate frequencies ε" is due to the contribution of ions jump, conduction loss of ions migration and ions polarization loss. At high frequencies ions vibrations may be the only source of dielectric loss and so ε" has the minimum value. The large value of ε" at low frequency range is also due to the motion of free charge carrier within the material. As a result of it, power law dispersion in ε" is observed and it does not reveal any peak in the measure frequency range.

Fig. 4 shows the variation of AC conductivity with frequency. The plots of the conductivity as a function of frequency can be divided into a low frequency dispersion region and an almost frequency independent plateau region at high frequency. A plateau region means it is equivalent to the true conductivity value for that particular sample as ions have time to migrate over large distance in polymer

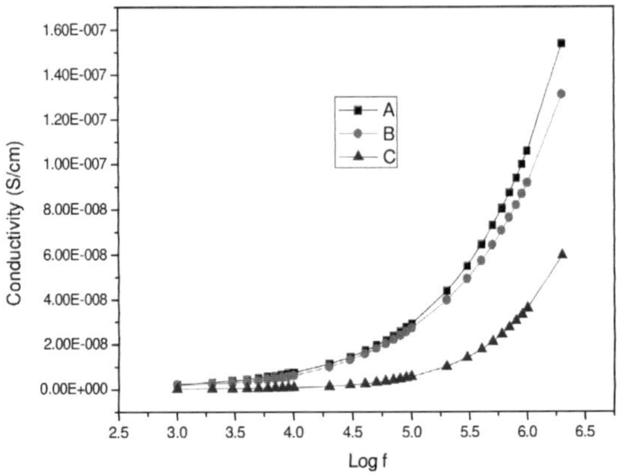

Fig 4: Variation of ac conductivity with log f

chains. The dispersion region is referring to the electrode being polarized as charge accumulates due to the slow periodic reversal of the electric field at low frequency. The low-frequency dispersion may be due to the interfacial impedance or space charge polarization. The conductivity in this low-frequency region is slightly constant owing to the transport of charge occurring at an infinite path. As the frequency decreases, more and more charge is accumulated at the electrode and electrode interface, which will lead to a decrease in the number of mobile ions and eventually a drop in conductivity at low frequency. At the high-frequency region, the conductivity increases with frequency due to the mobility of charge carriers and the hopping of ions from the infinite cluster. As a result, the ion exchange process occurs effectively in the high-frequency region.

4. Conclusion

Poly vinyl pyrrolidone (PVP) / titanium dioxide nanocomposites have been successfully synthesized by means of non hydrolytic sol–gel method. The XRD data confirmed that the TiO_2 nanoparticles in the TiO_2/PVP nanocomposites are crystalline at 360°C with anatase phase having a preferred crystallographic orientation along the (101) direction. The grain size of the TiO_2 nanoparticles estimated from the XRD peak width (and using Scherrer's formula), was about 5.58 nm and that from TEM observations was found to be around 7 nm. The conductivity values show weak frequency dependence at low frequencies but show strong dependence at high frequencies. The conductivity values are found to be in the range 10^{-7}-10^{-8} S/cm. The conductivity decreases steeply with increasing wt% of TiO_2 of the composites. This could be due to an increase in the disorderliness of the composite with increasing amount of TiO_2. Moreover, our as-synthesized nanocomposites contain pure anatase $TiO_{2,}$ without impurities arising from brookite or rutile phases, all of which are desirable characteristics for nanocomposites with potential applications in photocatalysis and other chemical processes. Since there is a stable bonding exists between the organic and inorganic components, these nanocomposites have potential applications in optical and electronic fields.

References

[1] Ahmad Z and Mark J E 2001 *Chem. Mater.* **13** 3320–3330.
[2] Wang B and Wilkes G L 1994 *J Macromol Sci Pure and Appl. Chem.* **A31** 249.
[3] Wang H, Zhong W, Xu P and Du Q 2004 *Macromol. Mater. Eng.* **289** 793–799.
[4] Liu L, Lu Q, Yin J, Zhu Z, Pan D and Wang Z 2002 *Mater. Sci. Eng. C* **22** 61–65.
[5] Shimojima A, Umeda N and Kuroda K 2001 *Chem. Mater.* **13** 3610– 3616.
[6] Wang Z and Pinnavaia T J 1998 *Chem. Mater.* **10** 1820–1826.
[7] Iuchi K, Ohko Y, Tatsuma T and Fujishima A 2004 *Chem. Mater.* **16** 1165- 1167.
[8] Biancardo M, Argazzi R and Bignozzi CA 2005 *Inorg. Chem.* 44 9619- 9621.
[9] Wells A F 1975 *Structural Inorganic Chemistry* Clarendon Press, Oxford.
[10] Yan Q Z, Su X T, Huang Z Y and Gea C C 2006 *J Eur. Ceram. Soc.* **26** 915-921.
[11] Qadri S B, Skelton E F, Hsu D, Dinsmore A D, Yang J , Gray H F and Ratna B R 1999 *Phys. Rev. B Condens. Matter.* **60** 9191-9193.
[12] Jyothy P V, Arun Kumar K V, Karthika S, Rajesh R and Unnikrishnan N V 2010 *J Alloys and Compounds* **493** 223–226.

IOP Conf. Series: Materials Science and Engineering **73** (2015) 012016 doi:10.1088/1757-899X/73/1/012016

Pressure induced phase transition in CeN with NaCl –type structure

R Dubey[1], S Singh[2], M. Sarwan[3]
[1]High Pressure Research Lab,
Department of Physics,
Barkatullah University, Bhopal-462026, India.

E-mail: ritud78@gmail.com

Abstract. We have expressed the Gibbs free energy for CeN compound as a function of pressure and charge transfer through improved interaction potential model (IIPM). The lattice energy in it has been represented by an IIPM which include Coulomb interaction, three body interaction, polarizability effect and overlap repulsive interaction. The phase transition pressure and relative compression reveal that this compound shows a B1-B2 phase transition and this approach are found to be the experimental data in future. The phase transition pressures and volume collapses obtained from this model show a generally good agreement with available results.

1. Introduction

Rare earth nitrides attract much attention due to the intricate electronic and magnetic properties related to the f-electrons. However all rare earth nitrides in rock-salt structure remain almost totally unexplored [1]. These compounds are less studied than chalcogenides and pnictides at the theoretical and experimental level. This is because rare earth nitrides are more difficult to prepare as they are less stable [2] due to the sensitivity to water vapour. The electronic configuration of rare earth nitrides shows that 5d orbital are singly filled and 4f orbitals are progressively filled up with electrons. Among Ce –based compounds CeN exhibit mixed valence behaviour. The 4f electron makes the rare earth and their compounds attractive for applications [3], such as protective materials in reactor, fire resistive materials and spintronics devices.

It seems from above literatures that the present compound is less studied; So that it motivated us to study the structural, stability and phase transitional phenomena at high pressure for CeN using improved interaction potential model (IIPM). The need of inclusion of three body interaction forces was emphasized by many workers for the betterment of results [4, 5].

2. Improved interaction potential model

It is well known that pressure causes an increase in the overlap of adjacent ions in a crystal. Consequently, charge transfer [4] takes place between the overlapping electron shells. These transferred charges interact with all other charges of the lattice via Coulombṡ law and gives rise to many body interactions, of which the most significant part is three body interaction (TBI).The increased effects of TBI, thus obtained, lead to an obvious necessity of their inclusion in the high

[1] R Dubey

pressure study of materials. A phase transition occurs when the solid becomes unstable under a given set of thermodynamic conditions .A phase of the solid is the most stable when its Gibbs free energy G(=U+PV-TS) is minimum. Here U, the internal energy at T= 0K, corresponds to the cohesive energy due to mutual interaction of the ions; S is the vibrational entropy at absolute temperature T, and V is the unit Cell volume at pressure P. At T= 0 K the Gibbs free energies for rock salt (B1) and CsCl (B2) phases can be expressed as [5]

Table 1. Input data and model parameters.

Input Parameter		Model Parameters	
r_i	1.01 A^0	b (10^{-12}erg.)	0.2318
r_j	1.71 A^0	ρ (Å)	0.216
r_0	2.506 A^0	f(r)	-0.0146
B_T	121[a]		

[a] [9].

$$G_{B1}(r') = U_{B1}(r') + PV_{B1}(r') \quad (1)$$

$$G_{B2}(r') = U_{B2}(r') + PV_{B2}(r') \quad (2)$$

$$U_{B1}(r) = \frac{-\alpha_M Z^2 e^2}{r} - \frac{12\alpha_M Z e^2 f(r)}{r} - \frac{e^2(\alpha_1+\alpha_2)}{2r^4} + 6b\beta_{ij}\exp[(r_i+r_j-r)/\rho] + 6b\beta_{ii}\exp[(2r_i-1.414r)/\rho] + 6b\beta_{jj}\exp[(2r_j-1.414r)/\rho] \quad (3)$$

$$U_{B2}(r) = \frac{-\alpha'_M Z^2 e^2}{r'} - \frac{16\,\alpha'_M Z e^2 f(r)}{r'} - \frac{e^2(\alpha_1+\alpha_2)}{2r'^4} + 8b\beta_{ij}\exp[(r_i+r_j-r')/\rho] + 3b\beta_{ii}\exp[(2r_i-1.154r')/\rho] + 3b\beta_{jj}\exp[(2r_j-1.154r')/\rho] \quad (4)$$

With $U_{B1}(r)$ and $U_{B2}(r)$ as the lattice energies for NaCl and CsCl structures, α_M and α'_M as Madelung constants for NaCl and CsCl structure respectively. e as charge of electron and r and r' as inter-ionic separation for NaCl (CsCl). β_{ij} is Pauling coefficients defined as $\beta_{ij}=1+(Z_i/n_i)+(Z_j/n_j)$ with $Z_i(Z_j)$ and $n_i(n_j)$ as the valence and the number of electrons of the i(j)th ion, b is hardness parameter while ρ is range parameter.

These lattice energies consists of long range Coulomb attraction (first term), long range three body or charge transfer interactions (second term), polarizability interaction (third term), and short range overlap repulsive potential (fourth term) extended up to second neighbouring ions is given by Hyfemeister and Flygare (HF) [6]. We are taking the effect of polarization to improve our results in the line of M. Kumar et al. [7] in diatomic molecules of alkali halides. We are taking this effect first time in CeN compound.

3. Computational method and results

The input data and model parameters are given in table 1. We have computed the phase transition pressure and associated volume collapses of CeN with IIPM which includes polarizability along with TBI. The results on phase transition and volume collapse are given in table 2. The lattice energies U

$_{B1}$(r) and U $_{B2}$(r') contain only three model parameters (b, ρ, f(r)) hardness, range and three body parameter, which have been calculated by using input data given in table 1 and have been computed. from the knowledge of the lattice parameter, bulk modulus and equilibrium conditions as described earlier [5] using the self consistent method.

$$[dU(r)/dr]_{r=r0} = 0 \tag{5}$$

and

$$B_T = \frac{e^2}{12r^4}[-4.66Z(Z+12f(r))+(A_1+A_2)-2(B_1+B_2)+27.9612Zrf'(r)] \tag{6}$$

As the stable phase is always associated with minimum of energy, we have followed the technique of minimization of Gibbs free energies of G $_{B1}$(r) (real) and G $_{B2}$ (r') (hypothetical) phases given by equation (1) and equation (2) at different pressures in order to obtain the inter-ionic separations r and r' for B1 and B2 phases associated with minimum energies. ΔG is very important factor in deciding phase transition pressure. We have plotted ΔG against pressure to obtain the phase transition pressure (P_t) at which ΔG approaches zero. Here B1 and B2 are real and hypothetical phases before phase transition and B1 becomes hypothetical phase after phase transition while B2 becomes real. The phase transition pressures so obtained are presented in table 2 and they are compared with the other's value. We have calculated the relative volumes (V (P) / V (0) and plotted them against pressures (P) to obtain the equation of state for B1 and B2 phases. There is a sudden collapse in relative volume indicating the occurrence of first order phase transition.

4. Discussion

The present compound is stable in NaCl structure at normal condition and high pressure it transforms to body centre CsCl structure. We have followed the technique of minimization of Gibbs free energies of NaCl and CsCl phases at ambient conditions corresponding to equilibrium inter-ionic separation r (r') using the IIPM model parameters listed in table 1.We have plotted the values of Gibbs free energy differences ΔG against the pressure (P) as shown in figure 1 for CeN. The phase transition pressure (P_t) is the pressure at which ΔG approaches zero. At P_t CeN undergoes a (B1-B2) transition associated with a sudden collapse in volume showing a first order phase transition. The values of P_t computed by us are in good agreement with the others results [8]. ΔG becomes negative beyond P_t predicting that at high pressure B2 phase is now thermodynamically and mechanically stable as compared to parent B1 phase.

In addition, the pressure volume behaviour of the present compound has also been studied. At the phase transition pressure there is a sudden collapse in volume has been occurred. The computed value of volume collapse for CeN has been given in table 2. The compression curve has been plotted in figure 2 for CeN. The NaCl and CsCl phases have been shown in this figure 2. It is clear that our calculated phase transition pressure is 87.6 GPa with volume collapses – $\Delta V/V_0$ of 11.44 from our IIPM model. Our computed values of P_t and volume collapse are in good agreement with the others results [8] and due to lack of experimental data we could not compare our result.

Table 2. Phase transition pressure (GPa) and volume collapse.

CeN	Phase Transition Pressure	Volume Collapse (%)
Present	87.6	11.44
Expt.	-	-
Other's	88 [7]	5.78

In conclusion, we have applied improved interaction potential model (IIPM) to investigate the structural phase transition of CeN. The results are summarized as follows.

(1) The present compound crystallizes in six-fold coordinated NaCl-type structure (B1) at normal condition and under pressure, they transform to the eight-fold-coordinated CsCl-type structure (B2).
(2) During the NaCl to CsCl phase transition, the volume discontinuity in pressure volume phase diagram identifies the occurrence of first-order phase transition.
(3) The present calculated phase transition pressure and volume collapses are in general in good agreement with the available data.

Finally, we can conclude that an improved interaction potential model (IIPM) has yielded more realistic predictions of the structural phase transition of CeN from NaCl structure to CsCl structure. The experimental result of transition pressure for CeN is not available. So our results may be used as guide to experiments for working range of pressure.

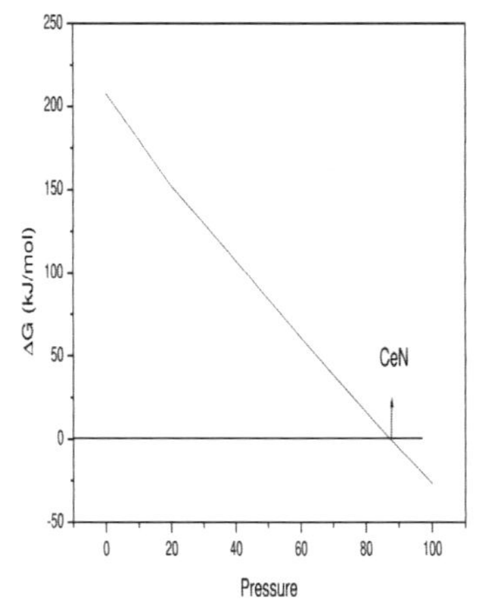

Figure 1. Variation of ΔG (kJ/mol) with Pressure.

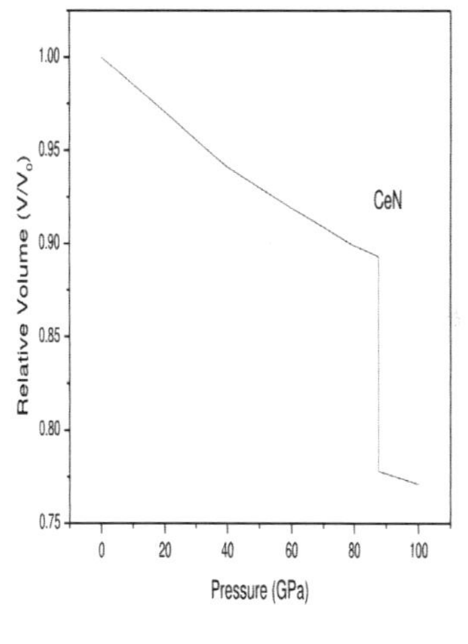

Figure 2. Variation of relative volume with pressure.

5. References

[1] Lee T Y, Gall D., Shin C S, Hellgren N, Petrov I, Greene J E 2003 *J. Appl. Phys.* **94** 921.
[2] Jakobsen J M, Madsen G K H, Jorgensen J E, Staun Olsen J and Gerward L 2002 *Solid state Commun.* **121** 447.
[3] Aerts C M, Strange P, Horne M, Temmerman W M, Szotek Z, Svane A 2004 *Phys. Rev. B.* **69** 045115.
[4] Singh R K 1982 *Phys. Rep.* (Netherlands) **85** 259.
[5] Singh R K, and Singh S 1989 *Phys. Rev. B* **39** 671; 1992 *Phys. Rev. B* **45** 1019.
[6] Hafemeister D W, Flygare W H1965 *J. Chem. Phys.* **43** 795.
[7] Kumar M, Kaur A J and Shankar J 1986 *J. Chem. Phys.* **84** 5735
[8] Rukmangad A, Aynyas M, Sanyal S P 2009 *Indi. J. Pure and Appl. Phys* **47** 114-118.

[9] Yang J, Gao F, Wang H, Gou H, Hao X, Li Z 2010 *Mater. chem. Phys.* **119** 499-504.

Acknowledgments
The authors are thankful to UGC, New Delhi for financial support to this research work. One of us (RD) is thankful to UGC (SAP) for the fellowship.

Effect of applied electric field and grating spacing on performance of ring resonator

Ruchi Singh* R A Yadav and D P Singh
Department of Physics, Banaras Hindu University, Varanasi-221005, India,

*E-mail: *ruchisingh8bhu@gmail.com*

Abstract. In this paper we analyze the effect of grating spacing on relative oscillation intensity and two-beam coupling gain of a ring resonator in presence of an applied electric field. The photorefractive crystal used in the ring resonator is KNSBN. It is found that the relative oscillation intensity increases exponentially with grating spacing in presence as well as in absence of the applied electric field, and approaches to a saturation state after some grating spacing. The increase is however slower under applied field condition, hence, the saturation state is reached at higher value of the grating spacing. The two-beam coupling gain decreases with grating spacing for both conditions on electric field. The decrease is however less abrupt in presence of applied electric field. It is found that the saturation state occurs for the grating spacing over 2 to 4 μm.

1. Introduction

Over the last four decades the photorefractive materials and devices have been a very active field of research because of their enormous applications in telecommunication and sensing. The ring resonator using a photorefractive crystal is an important device because of its rich dynamic behavior and pattern formation characteristics. Several interaction geometries in photorefractive crystal have been studied both theoretically and experimentally [1-4].The ring resonator geometry consists of a photorefractive crystal as an active nonlinear medium with three partially reflecting mirrors M_1, M_2, and M_3 as shown in figure1. Mostly the ring resonators with $BaTiO_3$ crystal have been studied [5-7]. In this work we studied the two-beam coupling performance of a ring resonator with potassium-sodium-strantium-barium niobate (KNSBN) crystal in the resonator cavity. The crystal is pumped by an external light beam. The inhomogeneities and impurities in the crystal produce beam fanning. The scattered light in the direction of signal beam is reflected by the cavity mirrors back into the crystal where amplification occurs by the process of two-wave mixing. The oscillation beam builds up if the two-wave mixing gain is above a threshold value that is when the gain exceeds losses including the absorption in the crystal. High powers can be achieved even with moderate values of two-beam coupling coefficient provided the losses are small. The scattered light away from the signal path will be lost. In the photorefractive ring configuration, the light propagation inside the cavity should be unidirectional as the two-wave mixing is directional determined by the crystal symmetry, alignment and charge transport properties [5]. The two beam coupling gain and the relative oscillation intensity are the two most important parameters that characterize the performances of a ring resonator. In this work we analyze the effect of grating spacing on the oscillation intensity and the two-beam coupling gain both in absence and in presence of an applied electric field. It is seen that an efficient resonator can be realized even with moderate coupling constant if the absorption coefficient is low. These conditions are fulfilled by KNSBN. A He-Ne laser has used as a pump beam.

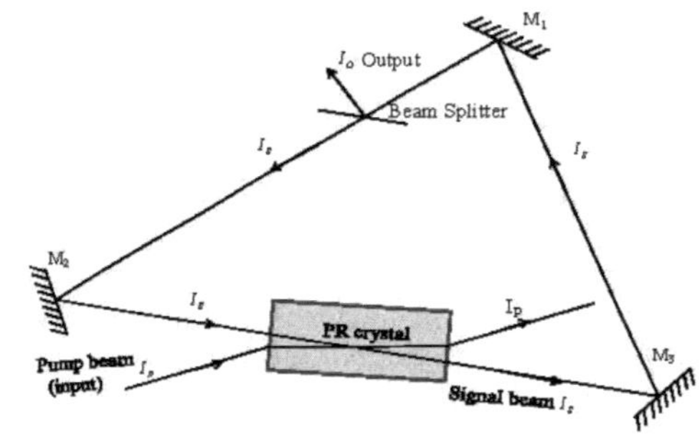

Figure 1. Schematic diagram of ring resonator.

2. Theory

2.1. *Two beam coupling gain*

The two-beam coupling gain is determined by amplification of the signal beam in two-wave mixing. The amplification occurs under conditions that (i) the two-beam coupling overcomes passive losses inside the resonator, and (ii) the phase shift of the oscillating wave in making one round-trip in the resonator is an integral multiple of 2π [5]. That is,

$$GR \geq 1 \tag{1}$$

$$\Delta \psi_c + \int k_g ds = 2N\pi \tag{2}$$

where G denotes parametric two beam coupling gain, R represents the product of reflectivities of mirrors M_1, M_2 and M_3, $\Delta\psi_c$ is the additional phase shift due to two beam coupling, k_g is the grating wave vector, and the integration in the Eq.(2) is over a trip around in the resonator cavity. The parametric two-beam coupling gain can be defined as, and is given by [5],

$$G = \frac{I_{\text{signal}}(\text{with pump beam})}{I_{\text{signal}}(\text{without pump beam})} \equiv \frac{(1+m)\exp(-\alpha l)}{1+m\exp\left(-\dfrac{\gamma_0 l}{1+(\omega\tau)^2}\right)} \tag{3}$$

where m is the ratio of intensities of the two waves at the input, l is the crystal thickness, αl is the absorption strength, $\gamma_0 l$ is the coupling strength, $\omega = \omega_{\text{pump}} - \omega_{\text{signal}}$ is frequency detuning, and τ is the response time (a function of grating spacing) given as [5, Eq.3.6-34]

$$\tau = \tau_{di}\left[\frac{\left(1+\dfrac{E_d}{E_\mu}\right)^2 + \left(\dfrac{E_0}{E_\mu}\right)^2}{\left(1+\dfrac{E_d}{E_q}\right)\left(1+\dfrac{E_d}{E_\mu}\right) + \dfrac{E_0}{E_\mu}\dfrac{E_0}{E_q}}\right] \tag{4}$$

where $\tau_{di} = \varepsilon_r \varepsilon_0 / e\mu n_0$ is the dielectric relaxation time in which ε_r is the relative dielectric constant, ε_0 is the permittivity in vacuum, μ is the mobility and $n_0 = \tau_R s I_0 N_D$ is the average electron density. The other quantities in are: $E_d = k_B T k_g / e$ the diffusion field, $E_\mu = \gamma N_A / \mu k_g$ the drift field, $E_q = eN_A / \varepsilon_r \varepsilon_0 k_g$ is the limiting space charge field, $\tau_R = 1/\gamma_r N_A$ is the recombination life time, s is the photo-ionization cross section, I_0 is the incident light intensity, N_D is the donor density, N_A is the acceptor density, γ_r is the recombination constant, $k_g = 2\pi/\Lambda$ is the grating wave vector with grating spacing Λ and $k_B T$ is the product of Boltzmann constant k_B and temperature T.

2.2. *The relative oscillation intensity*

The ring resonator oscillations start from the noise generated by scattering and quantum fluctuations. In photorefractive crystals the scattering contributions dominate. Initially, there would be a little amount of light scattered in the direction of ring resonator. The two-beam coupling process amplifies the scattered light at frequencies differing from the pump beam by ω ($\leq$ 30 Hz). As the intensity of this light in the resonator cavity builds up the parameter m decreases. The buildup of oscillation intensity leads to gain saturation. The relative oscillation intensity I_r which is same as the input intensity ratio m, given by [7]

$$I_r = \frac{1-R\exp(-\alpha l)}{R\exp(-\alpha l) - \exp\left(-\dfrac{\gamma_0 l}{1+(\omega\tau)^2}\right)} \tag{5}$$

3. Results and discussion

We have used the Eqs. (3) and (5) together with the Eq. (4) for calculating the two beam coupling gain and relative oscillation intensity with pump beam intensity $I_p = 0.5$ W/cm^2. The ring resonator crystal is KNSBN having thickness $l = 5$ mm and absorption coefficient of $\alpha = 10$ cm^{-1}. Figure 2 shows the variation of gain with grating spacing in presence and in absence of the applied electric field. It can be seen that in presence of applied electric field the gain is enhanced, and it decreases less abruptly with grating spacing when compared with no external field condition.

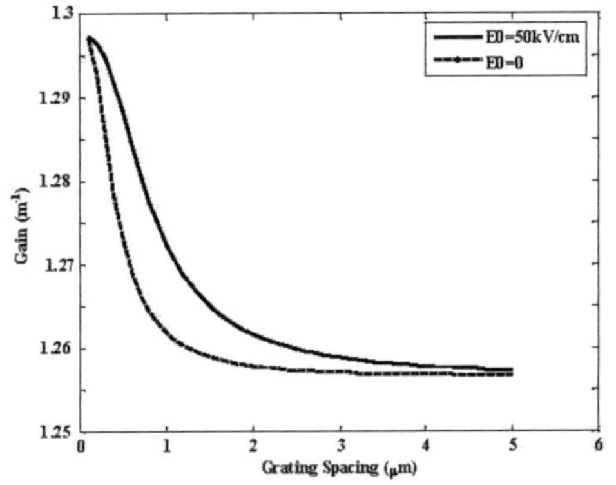

Figure 2. Variation of gain with grating spacing in presence and in absence of the applied electric field.

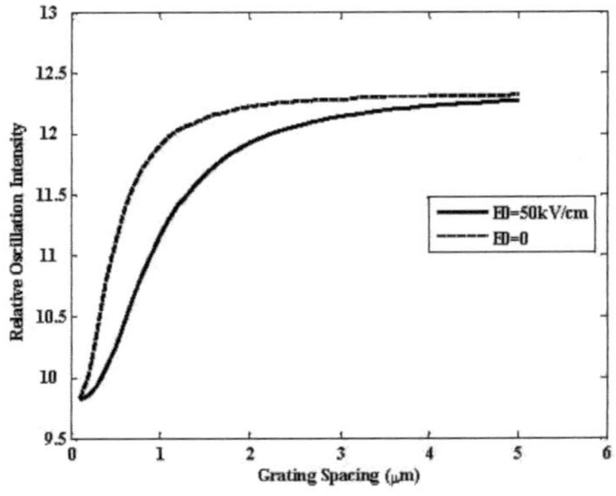

Figure 3. Variation of relative oscillation intensity with grating spacing in presence and in absence of the applied electric field.

Figure 3 shows variation of the relative oscillation intensity with the grating spacing. It is seen that the relative oscillation intensity increases exponentially with grating spacing both in absence as well as in presence of the applied electric field. Saturation occurs for the value of grating spacing of 4 μm in presence of applied field compared to 2 μm under no field condition. It is also clear from the figure that the relative oscillation intensity decreases in the presence of applied electric field for all the values of grating spacing. It means that in the presence of applied electric field the required pump beam intensity for the working of the ring resonator is smaller as compared to that required in the absence of applied electric field.

4. Conclusion

It is concluded that an externally applied electric field improves the performance of the ring resonator for all values of grating spacing. It is also found that the performance of the ring resonator is better for smaller values of grating spacing.

5. References

[1] Feinberg J P, Heiman D, Tanguay A R, Hellwarth R 1980 *J. Appl. Phys.* **51** 1297–1305.

[2] Madsen C K and Zhao J H 1999 *Wiley-Interscience Publications,* New-York.

[3] Maurya M K, Yadav T K, Singh Ruchi, Yadav R A and Singh D P 2010 *Opt. Comm.* **283** 2416-2424.

[4] Solymar L, Webb D J, Grunnet-Jepsen A 1996 *Oxford.*

[5] Yeh P 1993 *New York* John Wiley.

[6] Vahala K J 2003 *Nature* **424**, 839.

[7] Maurya M K, Yadav R A 2010 *Opt. Laser Technol.* **42** 883–93.

IOP Conf. Series: Materials Science and Engineering **73** (2015) 012018 doi:10.1088/1757-899X/73/1/012018

STRUCTURAL AND OPTICAL CHARACTERISATION OF TIN DIOXIDE THIN FILMS BY SOL-GEL DIP COATING TECHNIQUE

S SujathaLekshmy, I John Berlin, L V Maneeshya, Anitha, K Joy*
Thin film Lab, Post Graduate and Research Department of Physics,
Mar Ivanios College, Thiruvananthapuram 695015.
*Corresponding author. e-mail: *jolly2jolly@gmail.com*

Abstract

Tin oxide (SnO_2) thin films were deposited on quartz substrates using sol-gel dip coating technique. X-ray diffraction (XRD) pattern indicated that the film annealed in air at 350°C was amorphous in nature, whereas, the films annealed in oxygen atmosphere at 350°C showed crystalline phase. The films were further annealed in oxygen atmosphere at 450°C and 550°C. All the diffraction peaks can be indexed to the tetragonal phase of SnO_2.The surface morphology (SEM) showed that surface of all films were continuous and without micro cracks. The Energy dispersive X-ray spectroscopy (EDXS) spectra indicated an increase in the concentration of oxygen content with increase in annealing temperature. The energy band gap value for the film annealed in air was 3.88 eV. The optical band gap increased to 4.05 eV when annealed in O_2 atmosphere. The photoluminescence (PL) spectra showed the presence of emission peaks in UV region and visible region of the electromagnetic spectra. Transparent oxide semiconductor SnO_2 film finds potential application as an active channel layer for transparent thin film transistor.

1. Introduction

SnO_2 is a wide direct band gap n-type semiconductor possessing much importance in various fields of science and technology. In the form of thin films, SnO_2 posses interesting applications in opto-electronic devices, gas sensors, flat panel displays[1-2].
There are very few reports on the synthesis and systematic study on the films prepared by sol–gel dip coating technique and annealing in oxygen-rich environment. Sol gel process posses several advantages like lower processing temperature, better homogeneity , controlled stoichiometry and flexibility of forming thin films or nanoparticles. The environment in which the films are prepared plays an important role in controlling defects like oxygen vacancies, which consequently determine the physical properties of the films. In this study, we report the preparation and characterisation of nanocrystalline transparent tin oxide thin films by sol-gel dip coating technique.

2. Experimental details

The SnO_2 solution was prepared and were deposited onto well cleaned quartz substrates by sol-gel dip coating technique[5].The dip coated films were pre-fired at 250°C for 15 minutes in air and O_2 atmosphere respectively. The film pre-fired in air was annealed only at 350°C. The films pre-fired in O_2were annealed at different temperatures 350°C, 450°C and 550°C respectively. All the samples were annealed at the rate 4°/minute, and held at respective temperature for 15 minutes each during the course of annealing. The process of dipping, pre-firing and annealing were repeated seven times.
The phase analysis of the SnO_2 films were characterized by XRD using X-ray diffractometer (Model-PW 1710 PHILIPS). The surface morphology of the annealed films was investigated by using SEM (Model - JSM 5600LV JEOL).Elemental analysis of the films was done by EDXS (JEOL JSM 6390LA).Optical transmittance was studied using a spectrophotometer

(Model – JASCO-V550). PL spectra were recorded by using a Perkin-Elmer Fluorescence spectrometer (Model-LS55) with a 40 W Xenon Lamp as the excitation source.

3.1 Structural studies

3.1.1 X ray diffraction Studies: The XRD patterns of SnO_2 films annealed at 350°C in air and at different temperatures in O_2 atmosphere are shown in the Figure 1.The films annealed in air at 350°C were amorphous in nature (Figure 1a).When the films were annealed in O_2 atmosphere at 350°C, low intensity XRD peaks (110) and (211) were visible, indicating that the crystallisation of the SnO_2 particles have started. At lower annealing temperatures, the SnO_2 particles are in the initial phase of structural formation preventing full crystallization of the films. Since low intensity peaks appeared at 350°C when annealed in O_2 atmosphere, the higher temperature annealing were performed in O_2 atmosphere. When the temperature increased to 450°C and 550°C, the XRD patterns indicated that the intensity of peaks (110) and (211) increased, along with the appearance of (101) phase. All the diffraction peaks can be indexed to the tetragonal phase of SnO_2. Annealing in O_2 atmosphere at higher temperatures causes improvement in crystallinity as well as O_2 chemisorption at the inter grain boundaries [6].The grain size of the films were calculated using Scherrer's equation(Table 1). The increase in crystallite size is mainly due to an increase in the thermal energy for crystallization and growth of grains in the films.

Annealing Temperature(°C)	Crystallite size(nm)	Band gap(eV)
350	8.6	4.05
450	15.7	4.0
550	23.5	3.96

Table 1: Crystallite size and Band gap of films annealed in O_2 atmosphere

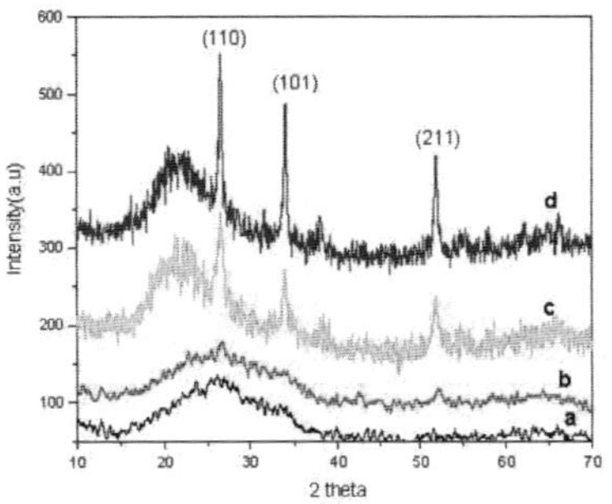

Figure 1: XRD spectra of SnO_2 films annealed in air at a) 50°C and annealed in oxygen at b)350°C c)450°C d)550°C

3.1.2. Scanning electron micrographs: The micrographs of films annealed at 350°C in air and at different temperatures in O_2 atmosphere are shown in the Figure 2.It is evident from the micrographs that all the films display a homogenous crack-free appearance. The micrograph of the film annealed in air at 350°C, shows agglomeration of particles (Figure 2a), whereas, the micrographs of the films annealed in O_2 atmosphere showed the presence of few tiny grains confirming the partial crystalline nature. At 450°C and550°C, some spherical shaped grains have started forming on the surface. It is obvious from the micrographs that by increasing the annealing temperature, grains gradually grown up and the surface of the films became slightly rough in nature. The origin of rough surface can be speculated to be due to the increase of total volume during the transformation from a metallic tin film to a tin oxide film. The roughness of the thin films plays a vital role for developing optical coatings especially in the ultraviolet region for applications such as lithographic uses [8,9].

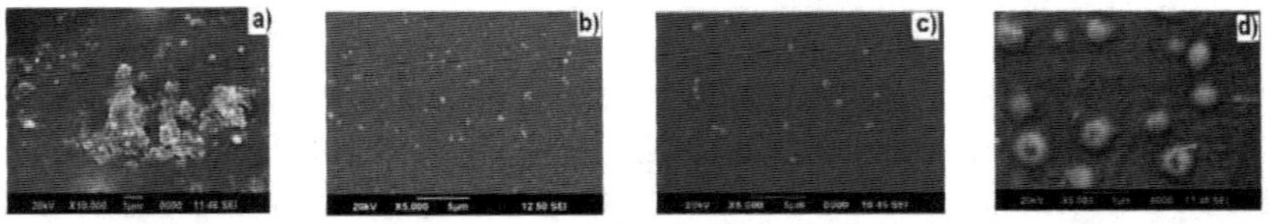

Figure2: SEM micrographs of SnO_2 films annealed in air at a)350 °C and in O_2 at b)350°C c) 450°C d)550°C

IOP Conf. Series: Materials Science and Engineering **73** (2015) 012018 doi:10.1088/1757-899X/73/1/012018

The EDXS(Figure 3) indicates that films were composed of elements Sn and O. It is evident from EDXS, that as the annealing temperature increased, the concentration of O_2 is found to increase for O_2 annealed samples.

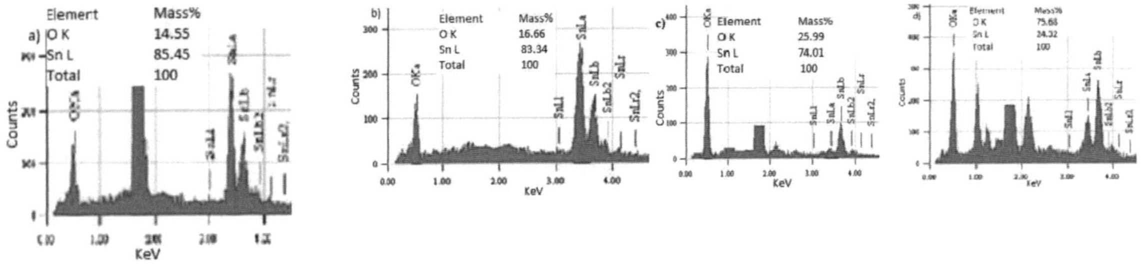

Figure3: EDX Spectra of SnO_2 films annealed in air at a) 350°C and in O_2 at b) 350°C c)450°C d)550°C

3.2 Optical Studies

The transmission spectra of SnO_2 films annealed in air at 350°C and in O_2 atmosphere at different temperatures are shown in Figure 4. It was seen from the spectra, that the transmission of the films were approximately >80%.Thismay be attributed to the formation of the Fermi level in the conduction band[10].The plots of $(\alpha h\upsilon)^2$as a function of energy (inset figure 4)shows that the investigated films have a direct optical band gap(Eg).The calculated Eg [11] is given in Table 1.The estimated Eg of film annealed in air is found to be 3.88eV which is lower than for the films deposited in O_2 environment.The Eg values are sensitive to O_2 content during the annealing process.Hence,O_2 vacancies play an important role in determining the optical properties of thin films.The observed decrease in the energy gap of films, upon increasing the annealing temperature, may be due to the increase in the crystalline size. The decrease in Eg shows that annealing the film causes a strong red shift in the optical spectra, due to the agglomeration of the nanocrystallites into larger crystallites. This change has been attributed to grain size dependent properties of the energy bandgap.

3.2.1. Photoluminescence studies

The room temperature PL spectra of the SnO_2 films annealed in air at 350°C and in O_2 atmosphere at different temperatures are shown in Figure 5. At 350°C in air, PL spectrum (Figure.5a) of the films exhibited a exciton related UV emission peak at 385 nm (3.22eV).The excitonic emissions are sensitive to imperfections and particle size in a crystal. XRD patterns of the films showed amorphous nature(Figure. 1a). Also, it is evident from EDXS profiles(Figure 3a) that the concentration of O_2 is less in films annealed in air. This can trap the electrons from the valence band and can act as luminescent centres. Hence, PL peak may be related to the crystalline defects induced during growth of the film. When the films were annealed in O_2 atmosphere at 350°C, the peak shifted to 382nm(Figure5b). The XRD pattern (Figure1b) of the film indicates partially crystalline nature. This verifies the fact that O_2 atoms flowed from the O_2 atmosphere into the film during the annealing process,leading to decrease of O_2 vacancies in the film.The lower intensity after annealing in O_2 indicates a reduction in the number of defect states.As the annealing temperature increased to 450°C and 550°C,UV peak shifted to 376nm(3.29eV).The weak emission peaks at 456 nm and 480 nm at 450°C lies in the visible region.The visible light emission of SnO_2 is known to be related to defect levels within the band gap of SnO_2 associated with Sn interstitials.At 550°C, the peak at 456 nm shifted to 441nmand peak at 480 nm disappeared. The intensity of the PL peaks decreased with increase in annealing temperatures at 450°C and 550°C.As the temperature increased, the grain size of nanocrystals became larger, in turn,favours a decrease in the concentration of non-radiative defects.This can lead to the decrease in number of the luminescence centers due to reductions on both the ratio of surface area and concentration of O_2 vacancies.Hence, SnO_2 nanoparticles annealed in O_2 atmosphere posses unique property of luminescence characteristics, whose specific emission wavelengths mainly depend on the nature of the semiconductors and the chemical environments, bringing on great application potentialities in the fields of light-emitting devices.

4. Conclusion

This study systematically discussed the effect of annealing atmosphere and temperature on the structural and optical properties of nanocrystalline SnO_2 thin film. The intensity of the XRD peaks increased with annealing temperature leading single structure formation. The surface morphology of the films was found to be strongly dependent on the annealing atmosphere and temperature. The intensity of PL emission peak decreased with increase in annealing temperature. Nanocrystalline SnO_2 films with wide band gap and short wavelength luminescence emission can serve as a better luminescent material for photonic applications.

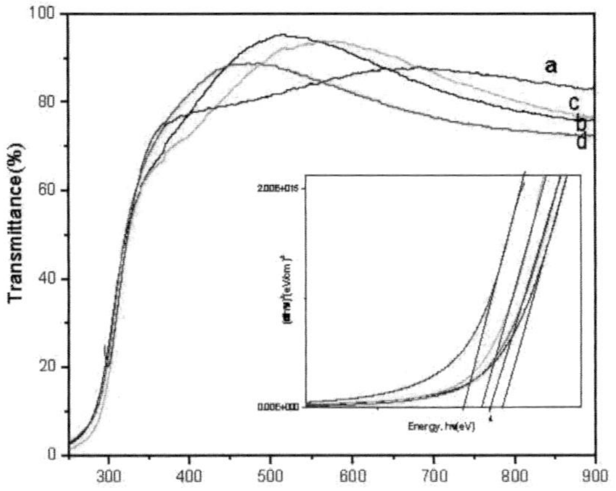

Figure 4: Transmittance spectra of SnO_2 films annealed in air at a)350°C and annealed in Oxygen at b)350°C c)450°C d)550°C

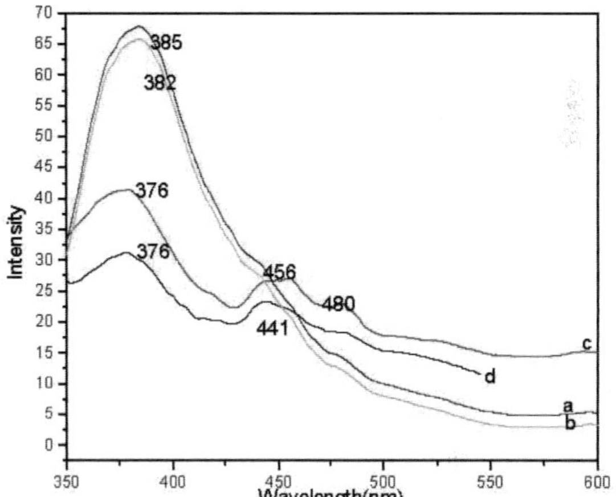

Figure 5: PL spectra of SnO_2 films annealed in air at a)350°C and annealed in Oxygen at b)350°C c)450°C d)550°C

References

[1] Tatsuyama C and Ichimura S 1976 *J. Appl. Phys.* **15** 843–847
[2] Matsui Y, Yamamoto Y and Takeda S 2000 *Materials Research Society SymposiumProceedings* (San Francisco USA) 621 Q 4.9.1
[3] Chatelon J P,TerrierC,BernsteinE,BerjoanR,Rogre J A 1994 *Thin Solid Films* **247** 162-168
[4] Nair M T S, NaikP K, Zingaro R A ,Meyers E A 1993 *J. Appl. Phys.* **74** 1879
[5] Duparre A 1995 Handbook of Optical Properties, **1** CRC Press
[6]Senthilkumar M , Sahoo N K , Thakur S, Tokas R B 2005 *Appl. Surf. Sci.*, 245 114
[7] KilicC,Zunger A 2002 *Phys. Rev. lett.* **88** 095501
[8] Butler, M. A. *J. Appl. Phys.*1977 ***48*** 1914-1920
[9] Wang Y G , Lau S P, Lee H W, Yu S F, Tay B K, Zhang X.H, et al., 2003 *J. Appl.Phys.* **94** 354

Chemically synthesized nano composite (Zinc/Magnesium) Oxide for tunable band gap devices

P P Sharmila [1]& Nisha J Tharayil[2]

1. Dept of Physics, T.K.M.M.College, Nangiarkulangara, Kerala .
2.Dept of Physics.S.N college for Women, Kollam,.Kerala

sharmilavishram@gmail.com

Abstract. Formation of hetro structures in nano structured materials is essential for their potential applications in nano electronics and photonic devices. As a promising candidate for blue and ultraviolet optoelectronic devices, ZnO has attracted much attention due to its wide band gap (3.37eV), large exciton binding energy (60meV), low epitaxial growth temperature and high oxidation resistance. In addition since the ionic radii of $Mg^{+2+}(0.57A^0)$ and $Zn^{2+}(0.60A^0)$ are quite close, they may alloy by replacing each other in the matrix. The doping of Mg in ZnO is done through a simple and novel technique from metal acetates using ammonium carbonate as precipitant. An organic capping agent (EDTA) is used prevent agglomeration and the addition is done under constant stirring. The carbonate precursor obtained is heated on the basis of TGA to obtain the metal nano composite. The effects of different parameters on particle size and morphology of (Zn-Mg)O nano composite is optimized by "one at a time" method. Under optimum conditions, spongy shaped, uniform and homogeneous structured (Zn-Mg)O nano composite powders with particle size few nano meters are obtained. The optical and structural properties of nano composite prepared by solution techniques are investigated by X-ray diffraction, UV-Visible spectroscopy, and PL, FTIR and electron microscopy techniques. The effect of annealing on the optical properties of this nano composite is also studied

1. Introduction

During the last few years, synthesis of nano structured metal oxide has been attracted considerable attention due to its wide range of applications and important properties owing to large surface to volume ratio. ZnO nanoparticles have wide range of application in solar cells, catalysts, gas sensors, luminescent devices etc There are a number of methods to prepare nanoparticles and composites. In recent years, researchers have focussed more on the synthesis of both nanoparticles and nano composite of ZnO and MgO due to their application in advanced technologies. Various synthesis methods have been in use for making nanostructured metal oxides and its composites [1-6]. The (Zn-Mg) O Nano composites were also prepared using different techniques using different metal salts as starting material but most involving costly atmosphere and need high temperature. Here we suggest a

IOP Conf. Series: Materials Science and Engineering **73** (2015) 012019 doi:10.1088/1757-899X/73/1/012019

simple room temperature routine using aqueous solutions of mother chemicals to make porous and spongy Nano structured ZnO,MgO and their composite.

In the present work a wet chemical method is used for preparing nano composite and nano metal oxides. Metal acetates and ammonium carbonates were used as starting materials and EDTA was used to prevent agglomeration to obtain homogeneous structure.

The TGA analysis was performed in the temperature range from 28^{0}C to 800^{0}C at a heating rate of 15^{0}C/minute under nitrogen. The surface morphology of the powdered sample was obtained by scanning electron microscope (SEM) [JEOL/EO JSM-6390] and the chemical composition were found out using the energy dispersive analysis of X-rays(EDAX). XRD study was carried out using XPERT-PRO model powder diffractometer (PAN analytical, Netherlands) employing Cu- K_{α} radiation (λ = 1.54060A^{o}) operating at 40kV,30mA and the optical properties were analyzed using UV-Visible analysis, FTIR and PL analysis.

2. Experimental methods

2.1. Materials used and apparatus

Analytical grade Zinc acetate dehydrate, Magnesium acetate dehydrate, ammonium carbonate EDTA and Sodium hydroxide were obtained from MERCK .All solutions were made in distilled water in different molar concentrations and the whole precipitation technique was done by slow addition under constant stirring using a magnetic stirrer. The pH of the solution was adjusted using sodium hydroxide to aid precipitation. The obtained metal carbonate precursor was washed several times with distilled water, filtered, dried and ground thoroughly to obtain fine powders.

2.2 Results and discussion

Analysis of TGA and derivative of TGA curve show a considerable weight loss in the temperature range 300^{0} c to 450^{0}C ,the maximum weight loss being near 400^{0} C for all the samples. So the decomposition temperature was taken as 500^{0}C to convert the carbonates into oxides.

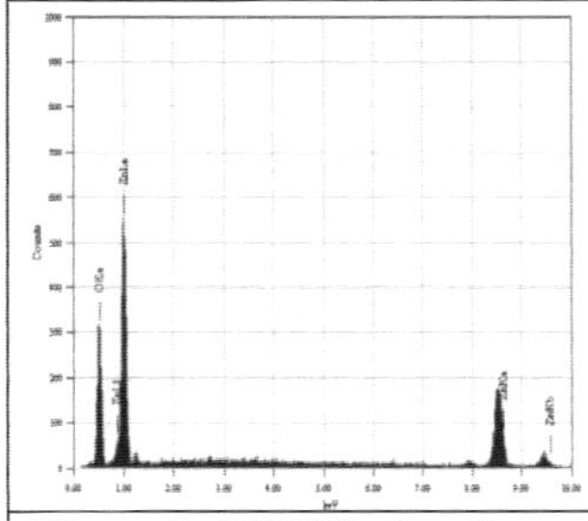

Figure 1a EDAX of ZnO

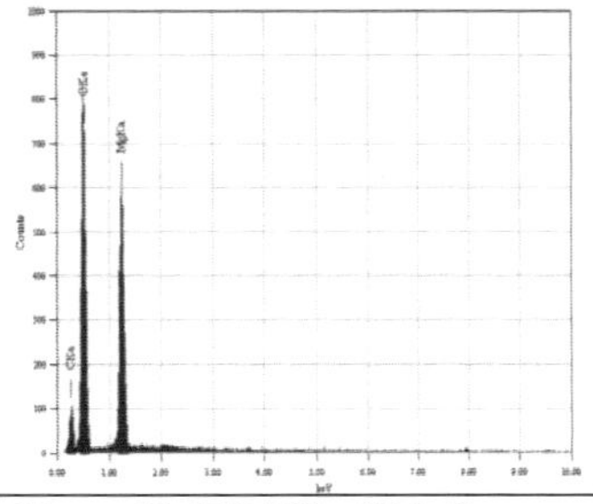

Figure1b EDAX of MgO

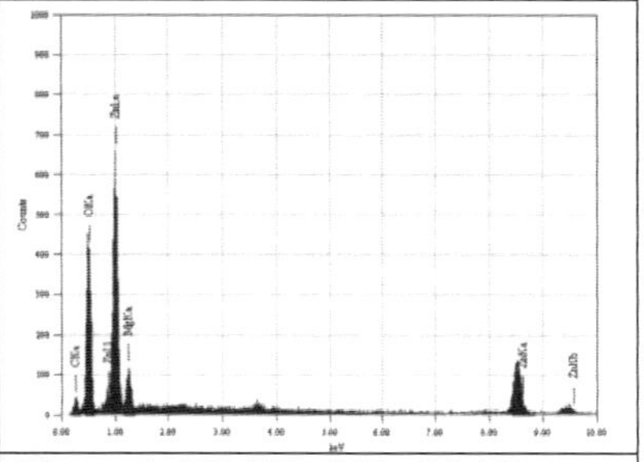

Figure 1c EDAX of (Zn-Mg)O

In the present study samples are also sintered at 700^{0}C and 900 ^{0}C to understand the effect of annealing on the structure and optical properties of these nano metal oxides. Figures 1(a),1(b) and 1(c) depicts EDAX of ZnO,MgO and ZnO/MgO and they show that the samples are free from impurities. Figures 2 (a), 2 (b) and 2 (c) are SEM micrographs and they show the morphology of the samples. The magnesium oxide nanoflakes can be seen in Fig 2(b). The other two shows spherical morphology.

International Conference on Materials Science and Technology (ICMST 2012) IOP Publishing
IOP Conf. Series: Materials Science and Engineering **73** (2015) 012019 doi:10.1088/1757-899X/73/1/012019

Figure 2a SEM of ZnO

Figure 2b SEM of MgO

Figure 2c of (Zn-Mg)O

The XRD spectrum of the samples (given in Fig 3 a,3b,and 3c) reveals that all the oxides are crystalline. The fine particle size is detected in X-ray line broadening. The particle sizes of both samples determined using the Scherrer equation and found that size of particles are of few nm ranging from 8~15nms and the particle size increase with temperature due to agglomeration . The XRD spectra of samples are compared with JCPDS data, and the lattice parameters are calculated. The obtained values are found in good agreement with JCPDS values (ZnO- card no079-0205,MgO-Card no 03-0998). The peaks of the composite itself is a proof for the presence of Mg in ZnO lattice

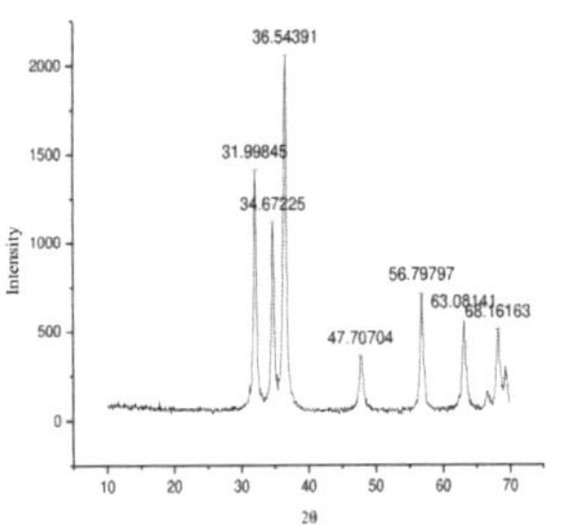

Figure 3a XRD of ZnO

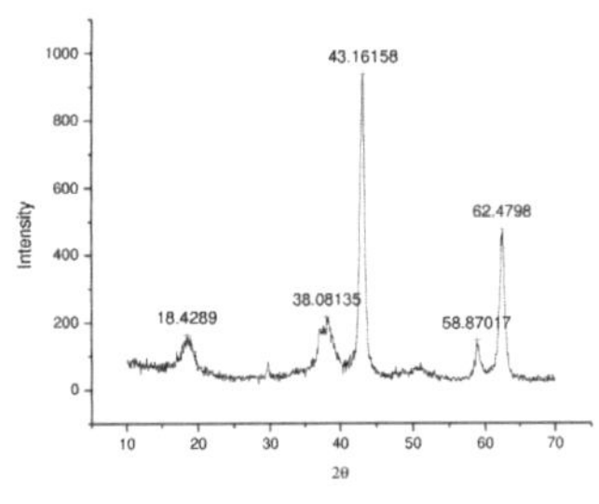

Figure 3b XRD of MgO

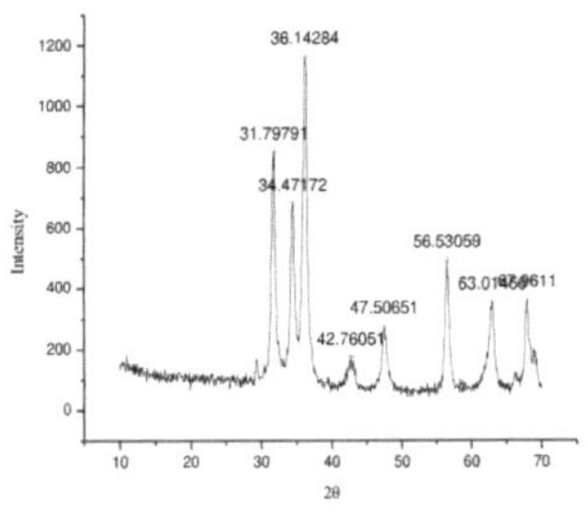

Figure3c XRD of(Zn-Mg)O

UV spectrum of the three samples is shown in figure 4 and the absorption peak of composite shows a small blue shift as compared to that of pure ZnO which clearly shows that the bandgap increases due to the doping of Mg. The peak value of absorbance are used to find the band gap which were obtained as 3.487eV for ZnO ,4.824eV for MgO in the nano forms and 3.624eV for the nano composite.

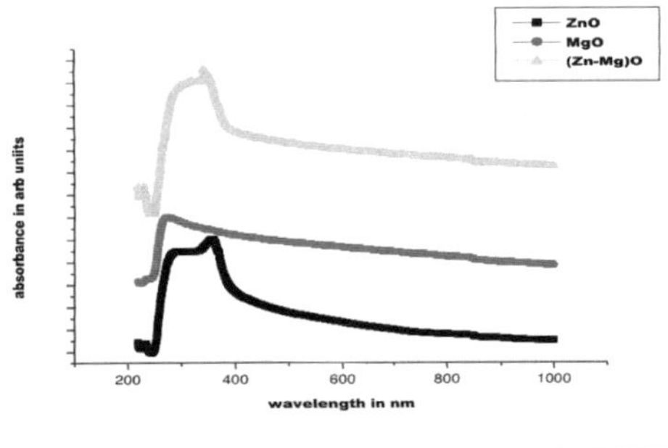

Figure4 absorbance spectrum

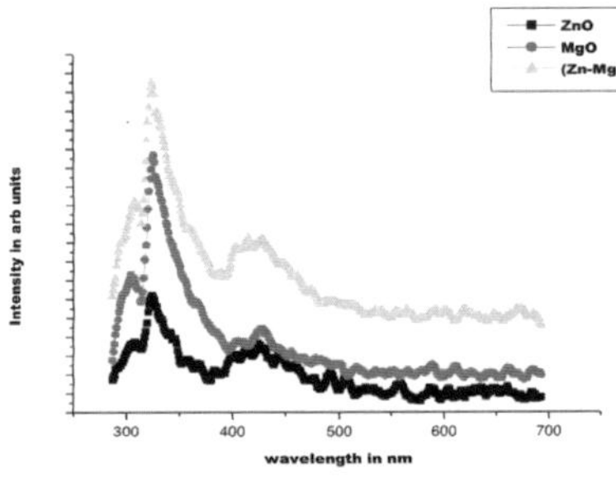

Figure 5 Luminescence spectrum

The bandgap of nano semi conducting metal oxides are found to be size dependent and can be increased slightly by reducing the size even though within limits. But controlling the size of nanoparticles through controlling the parameters in the synthesis method is very difficult. As an alternative the bandgap can also be varied by doping with suitable metal oxides. The luminescence spectrum obtained contains a broad UV emission at 350nm as well as a wide green to red band . The nature and shape of the luminescence spectra also varied with doping. We can also see the changes in the peak intensity caused due to doping. The UV emission is originated from excitonic recombination corresponding to the near band-gap emission of ZnO, while the emission bands in the visible range are due to the recombination of photo-generated holes with singly ionized charge states in intrinsic defects such as oxygen vacancies, Zn interstitials, or impurities [5,8,9]. The FTIR diagrams(Fig 6a,6b and 6c) show stretching and bending bands of OH bands attached to metal ions and to the surfaces. There are two types of OH bonding present in this spectra. First is the OH stretching and bending bonded with Metal ions and secondly, OH stretching and bending attached to the surface of the samples. The OH stretching bonded with Metal ions will be appeared as a sharp peak at 3702cm^{-1} , bending bond at 1490cm^{-1}. For the second O-H stretching appears as a broad band at 3600-3200 cm^{-1} while bending at 1639cm^{-1}. These bands have been characterized by many researchers [7].

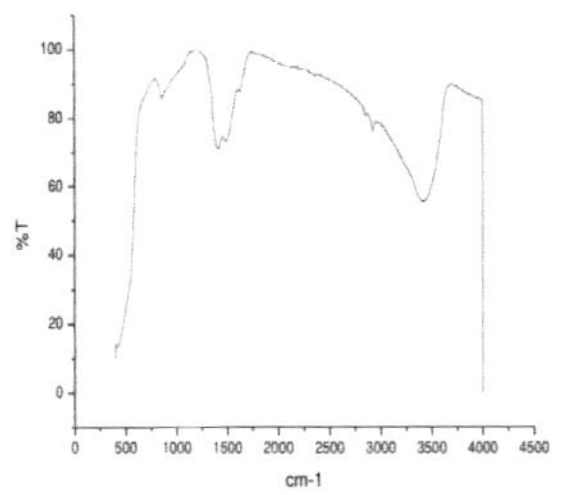	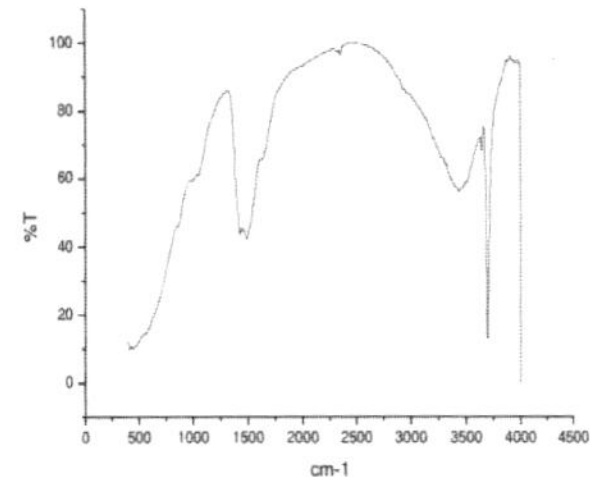	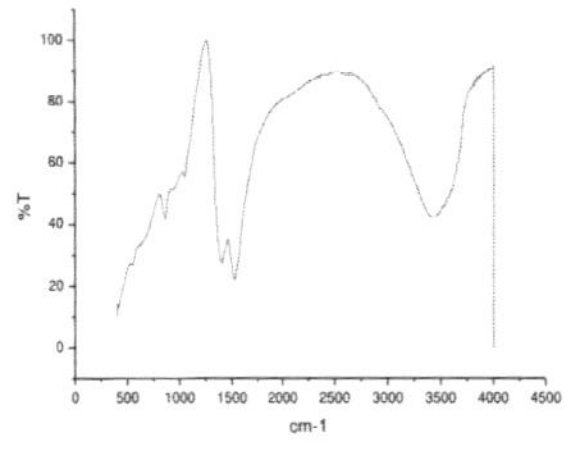
Figure 6a FTIR spectrum of ZnO	Figure 6b FTIR spectrum of MgO	Figure 6c FTIR spectrum of composite

3. Conclusion

The nanocomposite ZnO/MgO and nano ZnO,Nano MgO are prepared using chemical co precipitation technique and there surface morphology and optical preoperties are studied. This nano powders were characterized by EDAX,SEM ,XRD,FTIR,UV-Vis spectroscopy and PL studies. The band gap ,particle size and emissions are studied with doping.FTIR analysis are also done to have a deep insight about the structure. This study confirms the well reported nano size effect and the fact that the band gap can be properly tuned by changing the stoichiometry of (Zn-Mg)O.

Acknowledgement –The first author acknowledges UGC for financial assistance and STIC Cochin, NIST Trivandrum for providing SEM and XRD facilities.

References

[1] Yang Y, Chen H, Zhao B and Bao X 2004 *J.Cryst. Growth.* **2639** 447
[2] Hong Y C and Uhnm H S 2006 *Chem.Phys Lett.* **422** 174

[3] Choi H S and Hwang S T 2000 *J.Mater.Res.* **15(4)** 842-845
[4] Li Y B, Bando Y and Sato T 2002 *Chem Phy Lett.* **359** 141.
[5] Kim T W, Kazawoe T, Yamazaki S, Ohtsu M and Sekiguchi T 2004 *Appl. Phys. Lett.* **84** 3358.
[6] Sunder Manoharan S and Arora Sonia 2009 "Photoluminescent properties of Mg doped ZnO by microwave combustion and microwave polyol method",*Mat sci and eng* **b162**:68-73.
[7] Regragui M,.Addou M, Outzourchi A, Bernede J C, Eldrissi E ,Beneseddik E and Kachouane A 2000 *Thin solid films.* **358** 40
[8] Qiu Z and Wong K S 2004 *Appl. Phys. Lett.* **84** 2739.
[9] Look D C 2001 *Mater. Sci. Eng. B* **80** 383.

IOP Conf. Series: Materials Science and Engineering **73** (2015) 012020 doi:10.1088/1757-899X/73/1/012020

Structural and optical studies of nickel-cobalt-ferric oxides nanocomposite

Dedhila Devadathan, R. Raveendran

Nanoscience Research Laboratory, Department of Physics, Sree Narayana College, Kollam, Kerala

dedhila@yahoo.com

Abstract. In this study single salts of nickel oxide, cobalt oxide, ferric oxide and the nanocomposite of nickel-cobalt-ferric oxide was prepared by the well-known co-precipitation method. The samples were annealed at different temperatures and were characterized using SEM, EDAX, TGA, FTIR, XRD, UV and Photo luminescence (PL). UV analysis showed that the nanocomposite can be suitably tuned to a wide band gap material. The Photo Luminescence analysis showed that the nanocomposite can be used as light emitters in visible region.

1. Introduction

During the last few years, synthesis of nano structured oxide materials had attracted considerable attention [1]. The metal oxides are extremely important technological materials for use in electronic and photonic devices, in chemical industries and in medical fields [2]. An expanding trend for the nanomaterials is the fabrication of composite structures and devices with materials capable of enhancing the properties of the composite material. This can be done either by utilizing the size advantage through templating on the nanomaterials or by enhancing the properties to drive new synergetic properties of the two combined materials. It is also found that the magnetic nanoparticles provide unprecedented levels of new functionality for nano medicine. Such particles commonly consist of magnetic elements such as iron, nickel and cobalt and their chemical compounds [3]. After surface modification these materials provide both biocompatibility and functionality. In the present work a nanocomposite of nickel-cobalt-ferric oxide was prepared and their modifications in various known properties have been studied.

2. Materials and Methods

AR grade chemicals obtained from Merck were used for the preparing single salts of nano nickel oxide, cobalt oxide, ferric oxide and then the composite. Individual nano oxides and the composite were prepared by the co- precipitation method. The samples were annealed at 350^0c, 700^0c and 900^0c for three hours.

The surface morphology of the powdered sample was obtained by scanning electron microscope (SEM) [JEOL/EO JSM-6390] and the chemical compositions were found out using the energy dispersive analysis of X-rays (EDAX). FTIR studies were made using Perkin-Elmer FTIR Spectro Photo Meter in the wavenumber range 400 cm^{-1} and 4000 cm^{-1} by KBr disc method. XRD study was carried out using XPERT-PRO model powder diffractometer (PAN analytical, Netherlands) employing Cu- K_α radiation (λ = 1.54060A^o) operating at 40kV, 30mA. TG analysis was carried out by Perkin-Elmer, diamond TGA/DTA apparatus. UV study was carried out by Schimadzu 160 A

Spectrophotometer and PL study was carried out by Horiba Jobin Yvon Flurolog 3 modular Spectroflurometer.

3. Results and Conclusions

From the SEM analysis it was found that the as prepared sample showed aggregates of uniform, smooth, polished solid like pattern and the sample annealed at 350^0c showed aggregates of scaly layers as shown in Figure 1 (a) & (b). Figure 1 (c) shows the EDAX of as prepared sample which confirms the presence of nickel, cobalt, iron and oxygen in the nanocomposite.

(a) (b) (c)

Figure 1 (a) & (b) *SEM images of nanocomposite in as prepared and annealed at 350^0c* ***(c)*** *EDAX of as prepared sample.*

FTIR spectrum of the nanocomposite annealed at different temperatures showed that the precursor exhibits the basic absorption frequency bands between 4000 and 400 cm^{-1}. The prominent bands observed in the spectrum include: 3418 cm^{-1} due to hydrogen-bonded hydroxyl groups, 1618 cm^{-1} due to bending vibration of water molecules. The weak bands at 538 cm^{-1} and 429 cm^{-1} were attributed to metal oxide vibrations. It was seen that as the annealing temperature increased the bands shifts towards shorter wavelength side.

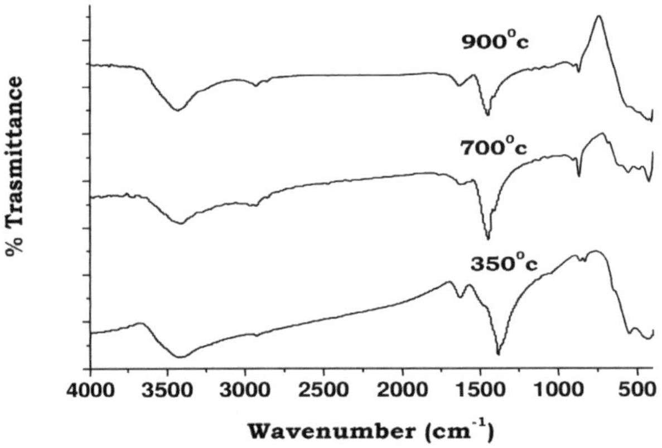

Figure 2 *FTIR spectrum of the nanocomposite*

Figure 3 (a) depicts the typical TGA curve of the nickel oxide [4]. Correspondingly, the TGA curve has two sharp weight losses. The former corresponds to the evaporation of the adsorbed and intercalated water molecules and the latter is associated with the loss of water produced by dehydroxylation of the hydroxide layers. Figure 3(b) depicts the typical TGA curves of the cobalt oxide. The first peak indicates the departure of physically sorbed water molecules; the next peaks correspond to the loss of water of constitution and decomposition. The final weight loss is due to the decomposition of Co_3O_4 into CoO [equation (1)]. The theoretical value of the weight loss in this conversion is 4% which is in close agreement with the result.

$$2Co_3O_4 \rightarrow 6CoO + O_2 \ldots\ldots\ldots\ldots\ldots\ldots\ldots (1)$$

Figure 3(c) shows the TG curve of ferric oxide. From the curve it can be concluded that there is no transformation. Figure 3(d) shows the TG curve for the nanocomposite and it confirms the formation of composite.

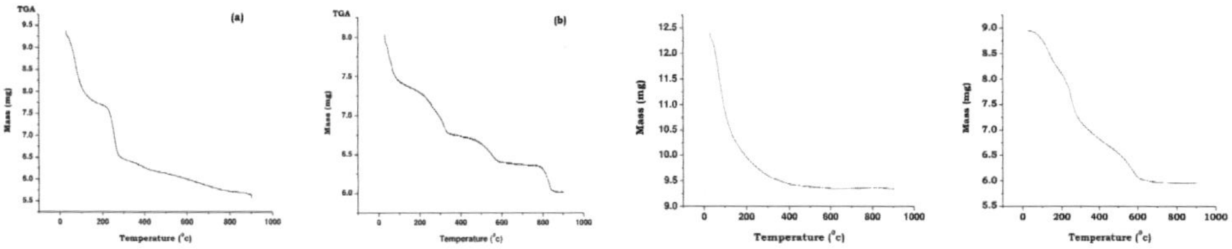

Figure 3 ***(a)*** *Prepared sample of nickel oxide,* ***(b)*** *Prepared sample of cobalt oxide,* ***(c)*** *Prepared sample of ferric oxide and* ***(d)*** *Prepared sample of nickel-cobalt-ferric oxides nanocomposite.*

XRD analyses of all the samples were carried out and except ferric oxide the other two showed crystalline nature. The particle size was calculated using Scherer equation (2)

$$D = k\lambda / (\beta_{hkl})\text{measured} * \cos\theta_{hkl} \ldots\ldots\ldots\ldots (2)$$

The particle size for all the samples are in the range 10-40 nm. The XRD spectrums obtained for nano oxides are in close agreement with the one available in literature, especially the XRD of the prepared sample of nickel oxide is in close agreement with XRD of $\beta Ni(OH)_2$ [5]. The XRD's of the annealed nickel oxide is matching with the JCPDS card no. #895881 and Cobalt oxide with #742120. With respect to these cards the two transition metal oxides are face centred cubic crystals. In the case of composite polycrystalline nature was observed due to the presence of ferric oxide.

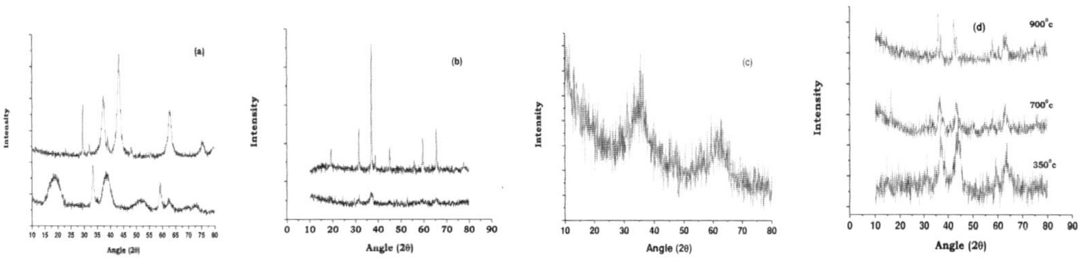

Figure 4 ***(a)*** *Prepared and annealed sample of nickel oxide,* ***(b)*** *Prepared and annealed sample of cobalt oxide,* ***(c)*** *Prepared sample of ferric oxide.* ***(d)*** *Annealed samples of nickel-cobalt-ferric oxides nanocomposite at* 350^0c*,* 700^0c *and* 900^0c *for three hours.*

UV analysis of the nanocomposites annealed at 350^0c and 700^0c for three hours was done. The UV spectrums of both the samples are shown in figure 5 (a) & (b). The absorbance peaks for the samples were at 221.15 nm and 301.91 nm range. The Corresponding band gap curve of the sample annealed at

350^0c is shown in figure 5 (c). The composite is found to have a band gap of 4.2eV. By proper tuning these materials can be converted into a wide band gap material.

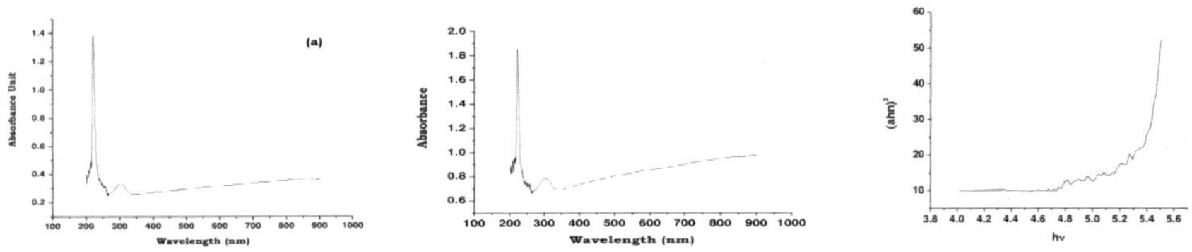

Figure 5 (a) & (b) *UV spectrum of Nickel-cobalt-ferric oxides nanocomposite annealed at 350^0c & 700^0c* ***(c)*** *Tauc plot of nanocomposite annealed at 350^0c*

Photo luminescence study of the nanocomposites was carried out. From the literature it is found that the NiO nanoparticle is a good emitter of 341nm wavelength [6]. The photoluminescence studies are preferred rather than the optical absorption. This is a mechanism where the impurity on absorption of light, gives rise to the bound excited state from which it returns to its ground state abiding in accordance with the colour centre creation mechanism. The room temperature photoluminescence spectra of the nickel-cobalt-ferric oxides nanocomposite annealed at 350^0C and 700^0c are given in figure 6. The PL spectrums of both the samples were almost same with a slight variation in intensity. From PL analysis it could be concluded that this material can be used as light emitters in the visible range.

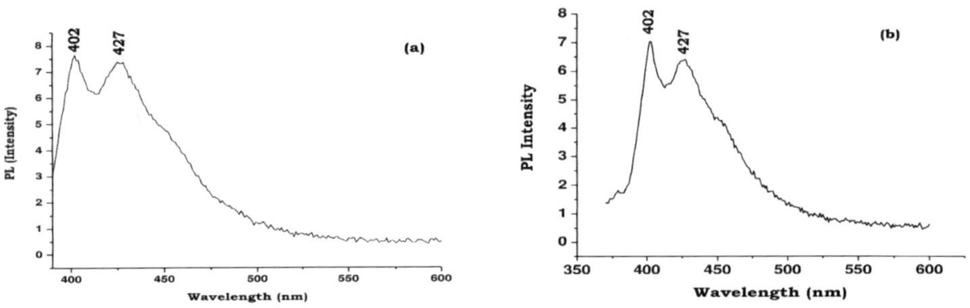

Figure 6 *PL spectrum* ***(a)*** *Nickel-cobalt-ferric oxides nanocomposite annealed at 350^0C.* ***(b)*** *Nickel-cobalt-ferricmagnesium oxides nanocomposite annealed at 700^0c.*

In the present work single salts of nano nickel oxide, cobalt oxide, ferric oxide and their composites were prepared. Using UV analysis the band gap of the nanocomposite was found to be 4.2 eV and if properly tuned the material can be made as a good wide band gap material. The photoluminescence studies showed that the composite can be used as light emitters in the visible region.

References

[1] Lee J, Mahendra S and Alvarez P. 2010 *ACS Nano* **4** 3580–3590.
[2] Karimi M. A. *et al* 2011 *Int. Nano Lett.* **1** 43-51.
[3] Neamtu J and Verga N. 2011 *J. of Nanomat. and Biostru.* **6** 969-978.
[4] Fu G. *et al* 2009, *Int. J. Electrochem. Sci.* **4** 1052 – 1062.
[5] Cheng M and Hwang B. 2007 *Nanoscale. Res. Lett.* **2** 28–33.
[6] Chakrabarty S and Chatterjee K. 2009 *J.Phy.Sci.* **13** 245-250.

IOP Conf. Series: Materials Science and Engineering **73** (2015) 012021 doi:10.1088/1757-899X/73/1/012021

Energy loss for heavy ions in polymers: A possible correlation with the induced changes in optical and electrical properties

Kalpana Sharma[1*], Neetu[1], Vijay Kumar[1], Rishi Pal[1] and Shyam Kumar[1]
[1]Department of Physics, Kurukshetra University, Kurukshetra-136119, India

*Email: kalpanaphysics@gmail.com, +91 9620025345

Abstract. Polymers are being widely used as versatile materials in many scientific and technological applications because of their excellent inherent characteristics along with the possibility to further improve their properties through various treatments like chemical doping, gamma irradiation, ion beam irradiation etc. There are many reports in the literature where the optical, mechanical and electrical properties of the polymers are improved after irradiation to ion beams. Such changes are as a result of the energy transferred by the incident ion leading to the structural changes in the polymers. For the quantitative data interpretation in such experiments it is highly essential to know the exact values of energy deposited by the incident ion in the polymeric material. In the present work, a new approach without any empirical parameterizations for the correct evaluation of energy transfer by the incident ion beam within the polymeric material has been developed. Finally, the observed changes in the optical and electrical properties of PET polymer, as quoted in the literature, have been tried to be explained in terms of the energy transferred by the incident ions.

1. Introduction

In recent days, MeV ion beams are widely being used to tailor the properties of polymers as per requirements for their applications as versatile materials in various fields [1-5]. In this regard, the basic process involved is the linear energy transfer (LET) by the incident ions within the polymeric material leading to the structural changes which are, finally, responsible for the modification in their optical, electrical, mechanical properties etc.[6-11]. To understand these changes in terms of LET, it is highly essential to know the exact values of energy loss by the incident ion while traversing through the polymeric material. Although various formulations [12-21] for energy loss rate calculations are available in the literature but most of these employ empirical/ semi empirical means for effective charge parameterization which is one of the most crucial input parameter. In the present work, we have developed an approach without resorting to any empirical parameterization for correct evaluation of LET for MeV heavy ions. Finally, observed changes in the properties like optical energy gap and conductivity of PET polymer after irradiation to C and Si ions have been tried to be understood in the light of these calculations.

2. Theoretical Calculations

The present calculations are based on the Bohr's equation [22], which is applicable in classical as well as quantum mechanical regime. According to this equation, the energy loss rate (-dE/dx) is given by

$$-\frac{dE}{dX}=\frac{2\pi Z^{*}e^{4}n}{mv^{2}}\left[\sum_{i}\ln\left(\eta_i^2[\chi]^{-2}\right)+\sum_{i}\ln\left(\eta_i^2\left[\frac{\chi}{\eta_i}\right]^{-1}\right)\right] \tag{1}$$

where $\chi = 2Z^{*}v_0/v$ with Z^{*}, v_0 and v as the effective charge, Bohr's velocity and ion velocity respectively and $\eta_i = \left[\frac{2v}{u_i}\right]$ with u_i being the orbital speed of the ith electron of the target atom; n is the number of target atoms per unit volume. The quantities in the square brackets i.e. χ and $\frac{\chi}{\eta_i}$, if less than unity, are to be replaced by unity. In the above equation, the two input parameters involved, i.e. u_i and Z^{*}, are calculated in the following manner.

2.1. Calculation of u_i

The orbital velocity u_i of the ith electron of the target atom is related to I_i (the successive ionization energies) through the relation $I_i = \frac{mu_i^2}{2}$ where m is the mass of the electron. The values of u_i are calculated after modifying the values of I_i, which were initially taken to be x-ray critical absorption energies[23], following the Sternheimer's [24, 25] approach, so as to be consistent with the most accepted values of mean ionization potential I [26]of the target atom.

2.2. Effective Charge Calculations

When a projectile ion at high velocities enters in a material medium, it loses all its electrons. Under this situation the charge on the incident ion is same as that of the nuclear charge. However, during the passage of the ion, its velocity continuously reduces and when it becomes comparable with the velocity of the orbital electrons, it starts picking up of the electrons in a gradual manner. Under this situation, the effective charge Z^{*} is related to the total nuclear charge through the relation

$$Z^{*} = Z - S \tag{2}$$

where S is the combined shielding effect of all inner electrons which is the sum of individual shielding contribution by various electrons. During the passage of an ion, the velocity of the orbital electrons is the resultant of its instantaneous orbital velocity (v_e) and its velocity by virtue of motion of the ion (v). For the electron to escape from the atom, its resultant speed needs to be $\geq (2)^{1/2}$ v_e. The values of v_e can be obtained from the successive ionization energy values quoted in the literature [27, 28]. The shielding effect can be determined using Slater or Clementi and Raimondi rules [27] taking in to account the extent of capture. The total shielding effect S can now be determined from the sum of the individual shielding effects and finally the effective charge Z^{*}can be determined using equation (2). On substituting the values of effective charge Z^{*} and orbital velocities of electron u_i in equation (1), the energy loss rate values of the incident ion can be determined. For the calculations in the composite materials, like polymers, the Bragg's additive rule has been used.

3. Results and discussion

In order to understand the role of energy loss rate by impinging ions resulting in the observed changes in the optical and electrical properties of polymers; the experimental data related to optical band gap and conductivity due to C (90 MeV) and Si(120 MeV) ion irradiation in PET as available in the literature [29, 30, 31] have been considered. For the evaluation of energy loss per unit path length (-dE/dx) for C and Si ions, the presently developed approach as already discussed in the previous section has been followed and the results are presented in figure 1.

Figure 2 presents the values of optical band gap as already quoted in the literature [29] for PET polymer after irradiation to C (90 MeV) and Si (120 MeV) ions as a function of ion fluence. The figure clearly indicates that the reduction in optical band gap as a function of ion fluence is more pronounced in case of Si ions than for C ions. Such a behavior in the reduction of optical band gap as a result ion irradiation is due to induced structural changes within the host polymeric material and is directly correlated to energy loss rate of impinging ions, being higher for Si ions than for C ions throughout the trajectory of these ions. These results are also supported by the increased conductivity of the polymers [30, 31] due to formation of defects and carbon clusters arising as a result of scissoring of polymer chains caused by energy loss after ion irradiation.

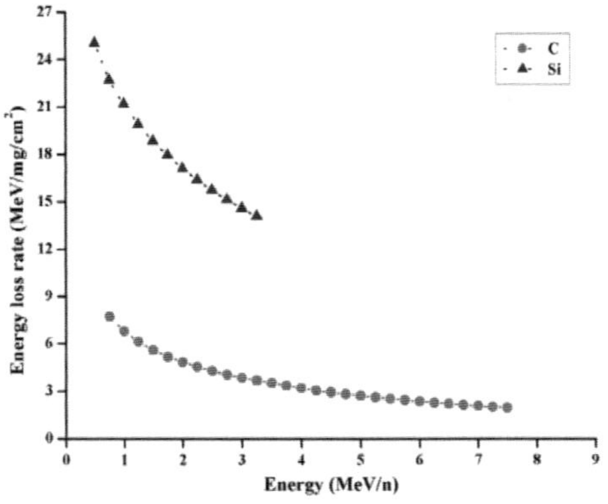

Figure 1.Variation of energy loss rate as a function of energy for C and Si ions in PET polymer.

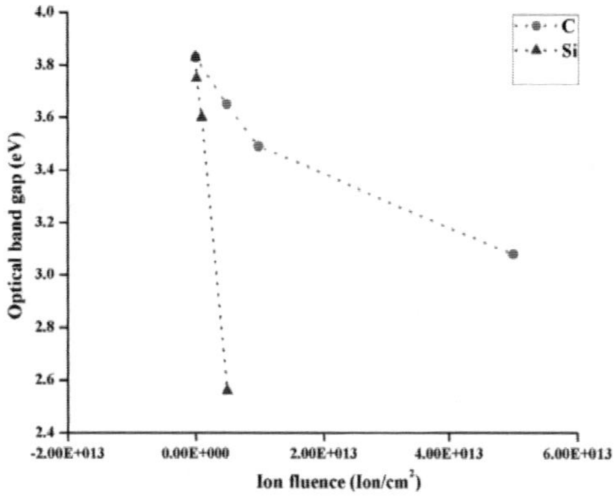

Figure 2. Variation of optical band gap as a function of ion influence for C and Si ions in PET polymer (the experimental data taken from reference 29)

4. Conclusion

A new approach to evaluate energy loss rate without resorting to any major empirical parameterization has been developed. The energy loss along the trajectory of impinging ions seems to be directly correlated with the induced changes in optical and electrical properties of the polymers.

Acknowledgement

Two of the authors (Kalpana Sharma and Neetu) are thankful to the Council of Scientific and Industrial Research(CSIR), New Delhi, India for providing financial assistance by the way of Junior Research fellowship and Senior Research fellowship respectively.

Reference

[1] Rauhala E and Raisanen 1990 *J, Phys. Rev. B* **42** 3877
[2] Bimbot R, Gardes D, Geissel H, Kitahara T, Armbruster P, Fleury A and Hubert F 1980 *Nucl.* Instrum. *Methods Phys. Res.* **174** 231
[3] Raisanen J and Rauhala E 1987 *Phys. Rev. B* **35** 1426
[4] Raisanen J, Rauhala E, Fulop Z, Kiss A Z, Somorji E and Hunyadi I 1994 *Radiat. Measur.* **23** 749
[5] Raisanen J, Rauhala E, Fulop Z, Kiss A Z, Somorji E, Hunyadi I 1992 *Nucl. Tracks Radiat. Measur.* **20** 611
[6] Burroughes J H, Bradley D D C, Brown A R, Marks R N, Mackay K, Friend R H, Burns P L and Holmes A B 1990 *Nature* **347** 539
[7] Fink D 2004 *Fundamentals of Ion-Irradiated Polymers* (Springer-Verlag, Berlin, Heidelberg)
[8] Fink D, Alegonkar P S, Petrov A V, Wilhelm M, Szimkowiak P, Behar M, Sinha D, Fahrner W

R, Hoppe K and Chadderton L T 2005 *Nucl. Instrum. Methods Phys. Res. B* **236** 11
[9] Migahed M D and Zidan H M 2006 *Current App. Phys.* **6** 91
[10] Rizk R A M, Abdul-Kader A M, Ali M and Ali Z I 2008 *J. Phys. D: Appl. Phys.* **41** 205304
[11] Chen T, Yao S, Wang K, Wang H and Zhou S 2009 *Surf. Coat. Technol.* **203** 3718
[12] Lindhard J, Scharff M and Schiott H E 1963 *Mat. Fys. Medd. Dan. Vid. Selsk.* **33** 14
[13] Northcliffe L C and Schilling R F 1970 *Nucl. Data Tables A* **7** 233
[14] Diwan P K, Sharma A and Kumar S 2001 *Nucl. Instrum. Methods Phys. Res. B* **174** 267
[15] Ziegler J F, Biersack J P and Littmark U 1985 *The Stopping and Range of Ions in Solids* Vol. 1 (Pergamon, New York)
[16] Ziegler J F *Computer Code SRIM-2008.05 Available from:* http://www.srim.org/
[17] Paul H and Schinner A 2001 *Nucl. Instrum. Methods Phys. Res. B* **179** 299
[18] Paul H and Schinner A 2002 *Nucl. Instrum. Methods Phys. Res. B* **195** 166
[19] Wambersie A 2005 *J. ICRU* **5** 1
[20] Sigmund P and Schinner A 2002 *Nucl. Instrum. Methods Phys. Res. B* **195** 64
[21] Hubert F, Bimbot R and Gauvin 1990 *At. Data Nucl. Data Tables* **46** 1
[22] Bohr N 1948 *Mat. Fys. Medd. K.Dan.Vidensk. Selsk.* **18(8)** 1
[23] Carlson T A 1975 *Photoelectrons and Auger Spectroscopy* Appendix 1 (Plenum, New York)
[24] Sternheimer R M 1952 Phys. Rev. **88** 851
[25] Sternheimer R M 1956 *Phys. Rev.* **103** 511
[26] Ahlen S P 1980 *Rev. Mod. Phys.* **52** 21
[27] Huheey J E 1978 *Inorganic Chemistry-Principles of Structure and Reactivity* 2nd edition (Harper and Row Publishers, New York)
[28] Weast R C *(Editor in Chief)* 1988 *CRC Handbook of Chemistry and Physics* Ist student edition (CRC Press, Florida)
[29] Srivastava A, Singh V, Aggarwal P, Schneeweiss F, Scherer U W and Friedrich W 2010 *Indian Journal of Pure & Applied Physics* **48** 782
[30] Aggarwal P, Singh V, Singh A, Scherer U W, Singh T, Singla M L and Srivastava A 2012 *Radiation Physics and Chemistry* **81** 284
[31] Singh N, Sharma A and Avasthi D K 2003 *Nucl. Instrum. Methods Phys. Res. B* **206** 1120

IOP Conf. Series: Materials Science and Engineering **73** (2015) 012022 doi:10.1088/1757-899X/73/1/012022

Trace moisture detection in oil filled transformer by ceramic sensor

Debdulal Saha*[1], K Sengupta[2]

[1] Sensor and Actuator Division, CSIR-Central Glass & Ceramic Research Institute,196, Raja S.C. Mullick Road, Jadavpur, Kolkata-700 032, INDIA.
[2]Adjacent Faculty of Engg. & Technology, Jamia Millia Islamia (Central University) Jamia Nagar, New Delhi - 110025, India.

* E-mail: debdulal@cgcri.res.in , debdulal76@yahoo.co.in

Abstract. This paper reports on the suitability of thin film nano porous γ-alumina sensor for sensing parts per million (ppm) moisture present in transformer oil. Transformer oil degrades slowly by weathering, causing dielectric break down voltage of the oil to fall down. For improving this break down voltage, water must be removed from the transformer oil. Flash point of the transformer oil ranges from 150^0C to 200^0C.When the oil is slowly heated up to 75^0C water vapour comes out from oil which is detected by ceramic sensor. The sensor is prepared from organo-metallic precursor by sol-gel process. Gold coated α-alumina substrate was dipped within the alumina hydra-sol and a thin film of γ-alumina formed on the substrate. The sensor capacitance was measured as a function of ppm moisture level. The circuit produces an output voltage which is precisely related to the absolute value of the capacitance of the dielectric material. In order to improve the sensitivity, parallel electrode structure was patterned on the nano porous dielectric. The response is sufficiently linear in extremely low ppm level moisture. A prototype hygrometer was built for detection of trace moisture in transformer oil. Porous alumina can be produced at a relatively low cost and in a variety of structural configurations. Sol- gel processing of alumina allows superior control on pore morphology, phase formation, purity and product microstructure compared to the more traditional techniques like Anodic oxidation of alumina sheets, tape cast by different sizes of alumina powder etc.

1. Introduction

Sol-gel processing can produce uniform distribution of micro-pores (< 2 nm), meso- pores (2-50 nm) and macro-pores (> 50 nm) [1]. For detection of extreme trace moisture range, the pore size should be within (2-50nm) range[2]. The existence of several crystalline alumina phases adds further flexibility in the design of porous alumina relative to other ceramics. Among different grades of aluminas such as γ- Al2O3 is excellent for making porous sensor because of their high surface area [≈ 200-500 m2/gm] [3]and thermal stability up to 10000C [4]. The sensitivity and selectivity to vapours of different volatile components depends on the pore morphology and pore dimension [5]. The pore formation parameter depends on the sintering time, temperature and purity of the sol. These sensors are low cost and suitable for low temperature operation and of integration with electronic circuits [6]. The present paper describes in details on the suitability of thin film porous alumina moisture sensor in ppm level with an accurate capacity measuring system. The preparation of metal oxide thin films on different solid supports has been studied for over 40 years, as a part of the propensity to miniaturize the dimensions of various systems to the meso scale, such films will be beneficial in different industries and fields of research such as sensing tribology and catalysis [6.7]. As more opportunities become available for implementation of ultra thin films, there is a growing need to develop reproducible methods for producing nano-metric dimension for sensitive detection of low relative humidity (RH) level [8-10]. The nano metric dimension can be advantageous due to better control of chemical and physical process across the films [11-14].

2. Experimental

The sol-gel method for preparation of a boehmite sol was similar to that proposed by Yoldas. Hydrolysis was performed at 90^{0}C by introducing aluminium tri-sec.-butoxide (ASB) into excess amount of water (water alkoxide molar ratio = 100/0.8) under vigorous stirring. After 45 minutes of hydrolysis, the colloidal solution was peptized by adding a given amount of 1.6 (N) HNO_3 solution (acid/alkoxide molar ratio = 0.071/1) and then maintained at 90^{0}C for one hour constant stirring. A transparent sol was finally obtained by continuing refluxing the solution at 90^0-95^{0}C for 16 hours. The gold coated alumina support was dip coated with the binder and dispersant mixed boehmite sol by a rotating dipping technique. The support was fixed on the bottom of a rotating shaft and immersed into the solution for 1 minute of slow rotating and then followed by fast spinning to remove the residual sol. After drying in air, the support with xerogel was calcined to 500^{0}C with a heating rate of 2^{0}C/min. and soaking for 3 h to obtain a thin top layer on it. The process was repeated six times with dipping and calcinations in order to get a crack-free supported alumina coating. Response and recovery time were measured by HP 34970A data acquisition system.

3. Results and Discussion

Porous layer absorbed moisture molecules from the water mixed oil and condensed inside the pores shown in figure 1. The molecular dimensions of the vapour molecules must be less then the Kelvin radius. Increase in capacitance due to moisture absorption in dielectric layer. The capacitance of the layer changes due to the difference in Permittivity as a function of vapour uptake and depends on ambient vapour concentration. The sensitivity, linearity and response depends on pore morphology and the thickness of the dielectric and geometry of contracts. Dissipation factor also change when moisture absorb in alumina nano pore. In Figure2, AFM image of surface coating layer, from the image we see that lots of nano pore distribution through out the surface.

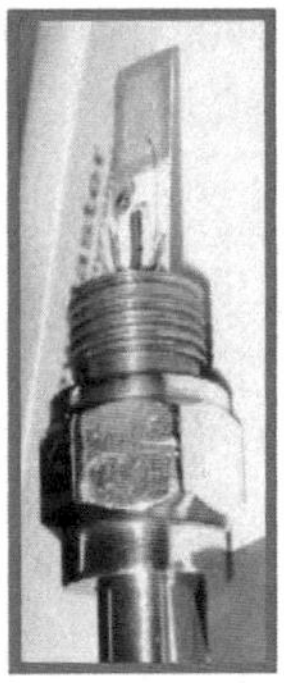

Figure1. Sensor mounted in a holder

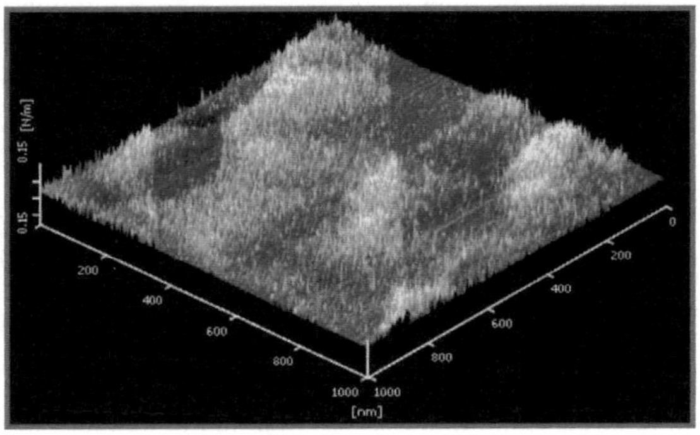

Figure 2. AFM Image of γ-alumina thin film

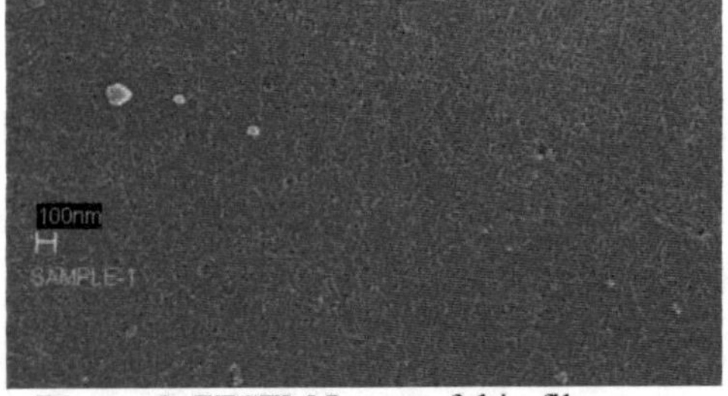

Figure 3. FESEM Image of thin film

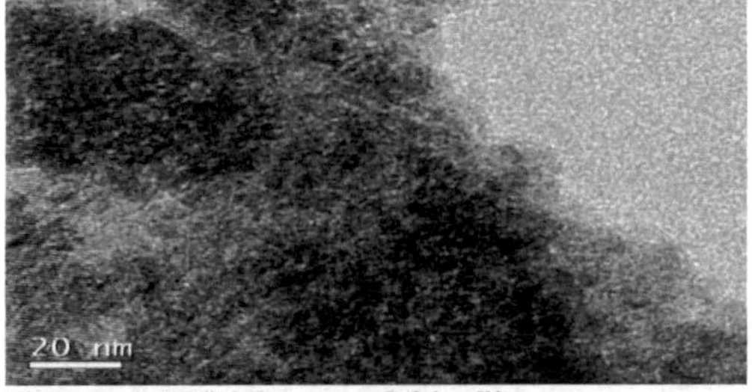

Figure 4. TEM Image of thin film

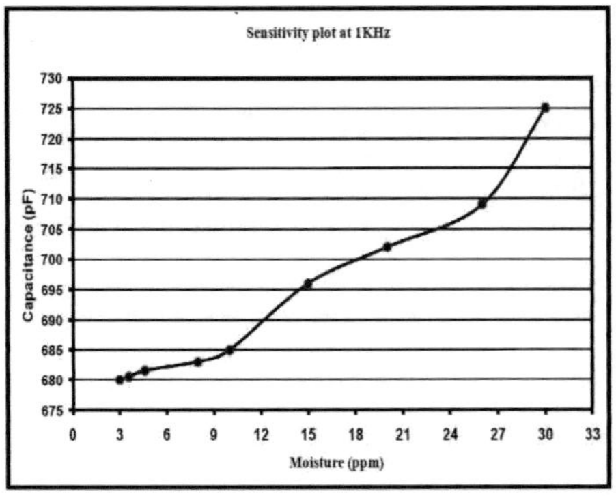

Figure 5. Sensitivity plot in air condition

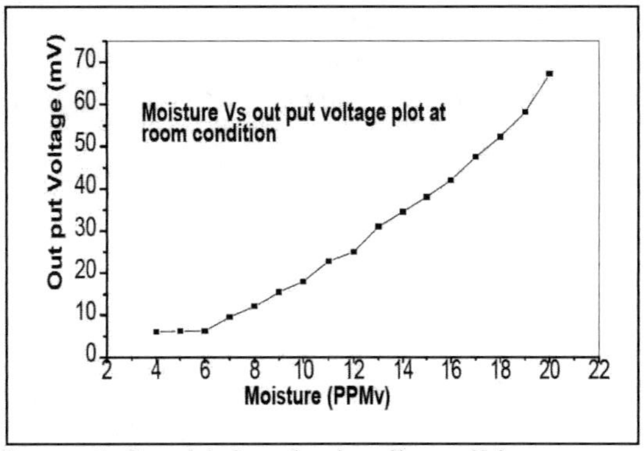

Figure 6. Sensitivity plot in oil condition

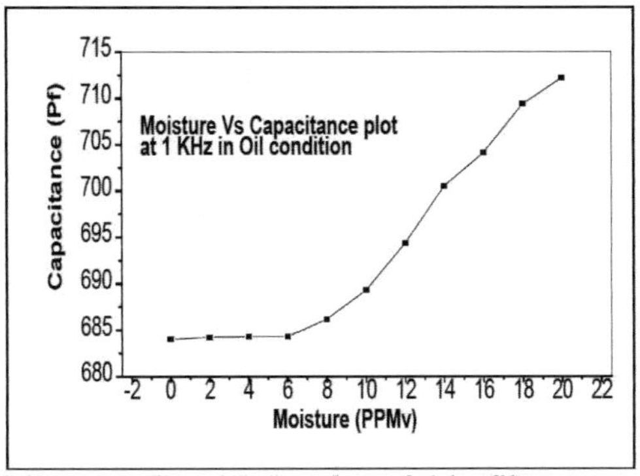

Figure 7. Sensitivity plot of thin film sensor

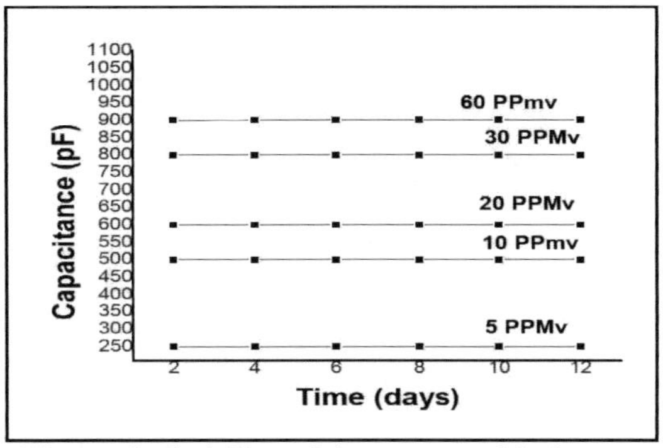

Figure 8. Stability of the sensor with time

Figure 3 and figure 4 are the FESEM and TEM image for thin film gamma alumina. From these figures we see lots of nanopore distributed through out the structure. Figure 5 and figure 7 are the sensitivity plot at 1 KHz without transformer oil condition i.e. sensitivity measurement at moist nitrogen gas medium. Figure 6 is the sensitivity plot of the thin film sensor in transformer oil condition. Form the graph we see that with increasing moisture in ppm out put voltage (mV) increase. Figure 8 is the ageing plot with different ppm moisture level but capacitance values are steady with time.

4. Conclusion

Sol-gel thin film α-Al_2O_3 was developed by dip coating on a gold coated alumina substrate of size (36mmx12mm) as humidity sensor material, their structure was characterized by XRD, FESEM, and AFM. The ordered nanoporous thin film coated structure are highly sensitive to trace moisture level. The sensor characteristics are quasi linier over -64^{0}C to -46^{0}C DP. The reproducibility of the sensors were checked with SHAW dew point meter (UK).

Acknowledgements
The authors are thankful to DST, Govt. of India for financial support and the authors are also thankful to Prof. I Manna, Director, CSIR-Central Glass and Ceramic Research Institute, Kolkata for rendering all sorts of cooperation for conducting the research work.

References

[1] B.E. Yoldas, A transparent porous alumina, Jr. of Mater. Science, **10,** 1856 (1975).
[2] K.Mistry, D.Saha, K Sengupta, " Sol-Gel processed alumina thick film template as sensitive, capacitive trace moisture sensor", Sensors and Actuators B **106** (2005)

[3] Z.M.Rittersma, "Recent achievement in miniaturized humidity Sensors: A review of transduction techniques" Sensors and Actuators A. vol. **96** no 2-3 pp 196-210, 2002.

[4] E. Traversa; "Ceramic sensor for humidity detection: the state of the art and future developments" Sensors and Actuators B **23** (1995) 135.

[5] B. M. Kulwicki; Humidity Sensors. J. Am. Ceramic. Soc. **74**(1991) 697.

[6] J. G. Fagan and Amarakoon; V. R. W. "Reliability and reproducibility of ceramic sensors: part III, humidity sensors", Am. Ceram. Soc. Bull. **72** (1993) 119.

[7] N. J. Arfsten, A. Eberle, J. Otto and A. Reich, J. Sol-Gel Science and Technology **8** (1997) 1099.

[8] G. Gusmano, G. Monterperelli, P. Nunziaute, E. Traversas, A. Montenero, M. Braghini, G. Mattogno, and A. Bearzotti. "Humidity sensitive electrical properties of titania films prepared via sol-gel method". J. Ceram. Soc. Jpn., **101** (1993) 1095.

[9] E.Traversa, A. Montenero and G. Gusmano. "Innovative humidity sensitive electrical properties of sol-gel processed ceramic thin films". Eur. J. Solid State Inorg. Chem, **32** (1995) 719.

[10] Traversa. "Structure and electrical properties of sol-gel processed $CdTiO_3$ powder and films". Appl. Organomettallic Chem. **11**(1997) 137.

[11] J.Chengbin, Z. Xiujian, and Z. Yongheng, "Sol-gel fabrication of compact, crack-free alumina film" Mater. Res. Bull. **42** (2007) 600.

[12] T. Islam, K. Mistry, K. Sengupta and H. Saha, "Measurement of Gas moisture in the ppm range using porous silicon and porous alumina sensor" Sensors and Materials ,Japan Vol 16 No. **7** (2004) 345-356

[13] X.L. Pan, N.Storoh, H. Brunner,G.X. Xiong, S.S. Sheng "Depositation of sol-gel derived membranes on γ-Al_2O_3 hollow fiber by a vacuum-assisted dip-coating process" J. Memb. Sci. **226** (2003) 111-118

[14] X. Ding, J. Zhang, R. Wang and C. Feng "Formation of porous alumina with oriented pores" J. Euro. Ceram. Soc. **22** (2002) 411-414

International Conference on Materials Science and Technology (ICMST 2012) IOP Publishing
IOP Conf. Series: Materials Science and Engineering **73** (2015) 012023 doi:10.1088/1757-899X/73/1/012023

Study of spatial rings in $TPPOH_4$ doped in boric acid glass

Srinivasa Rao Allam[1], Mudasir H Dar[1], N Venkatramaiah[2], R. Venkatesan[2] and Alok Sharan[1,3]
[1] Department of Physics, Pondicherry University, Puducherry-605014, INDIA
[2] Department of Chemistry, Pondicherry University, Puducherry-605014, INDIA

E-mail: aloksharan@gmail.com/alok.phy@pondiuni.edu.in

Abstract. Single beam optical nonlinearity is studied in $TPPOH_4$ doped in boric acid sandwiched films between two microscope glass slides at three different molar concentrations ($4x10^{-4}$M, $1x10^{-4}$M, $5x10^{-5}$M). It shows absorption peak at 700nm with bandwidth of 70nm. We have used diode laser output at 671nm to probe resonant optical nonlinearities. We have observed interesting phenomena of formation of spatial concentric rings centered on the z-axis of the sample. To our knowledge this is first such observation of spatial rings in these systems. We have carried out studies to separate the contribution to the ring formation due to absorptive/refractive optical nonlinearity and the thermal nonlinearity.

1. Introduction

The phenomenon of spatial ring as nonlinear optical phenomenon has been long since a subject of study in different media like gases and solids [1-7]. The rings were observed as the incident intensity increases beyond certain threshold value. Various explanations have been put forward to understand the phenomena in respective media. The spatial concentric rings of same wavelength can occur because of thermal/ absorptive optical nonlinearity or due to both. When medium interacts with the optical field in the non-resonant case, the rings pattern observed is primarily due to thermal in origin. In the case of resonant irradiation we could observe the rings pattern only beyond the threshold intensity which could be attributed to the changes in the intensity dependent nonlinear optical complex refractive index as well as due to thermo-optic coefficient.

S. D. Durbin et al (1981) [1,2] observed rings pattern in the transmitted laser beam, when the irradiation intensity of linearly polarized cw Argon ion laser was above 130 W/cm^2 on the thin film 4-cyno-4'-phenyl-biphenyl (5CB) nematic liquid crystal. They observed that the rings are formed beyond the threshold intensity which coincides with threshold intensity of Fredericksz transition. They further observed that the width of the rings decreases as we move towards the center of the ring pattern. R.Sohrab and N. M. Lawandy [3] theoretically predicted central dark zone formation in pink ruby rod when excited with laser lines from with Argon-ion and dye lasers. Martine Le Berre et al [4] theoretically explained spatial ring formation in cw lasers due to dense absorbing gaseous media. Dense gaseous medium gives conical emission for nearly resonant high intensity laser beams [5-7]. S.J. Mathews et al also observed spatial rings in Phthalocyanines[8].

2. Sample preparation and characterization

[3] To whom any correspondence should be addressed

Organic dye doped in an inorganic host like boric acid glass have been well studied for quite some now. The idea of using organic dyes like porphyrin molecules, (which has large absorption cross-section in the visible band) doped in boric acid glass films helps lower the saturation intensities. This helps us to investigate numerous nonlinear optical phenomena at lower power level. For our study we have chosen $TPPOH_4$ doped in boric acid glass sandwiched between two glass slides at three different molar concentrations ($4x10^{-4}$M, $1x10^{-4}$M, and 5x10-5 M) with corresponding thickness of $176 \pm 5x10^{-6}$ m, $242 \pm 18x10^{-6}$ m, $153 \pm 5x10^{-6}$ m[9] as our nonlinear medium. Samples were prepared by melt quenching method. The absorption spectra of all the sandwiched films were recorded using (HR 4000) ocean optics UV- visible spectrometer with resolution of 0.23 nm in the range 200-1100. The dye free boric acid film was taken as the reference. We observed two absorption peaks at near at 445 nm and 700 nm with bandwidth of 72.5nm and 66nm respectively for $4x10^{-4}$M concentration as seen in figure 1. It shows emission peak at 752nm with Stokes shift of 52nm. We have used diode laser output at 671nm to probe resonant optical nonlinearities

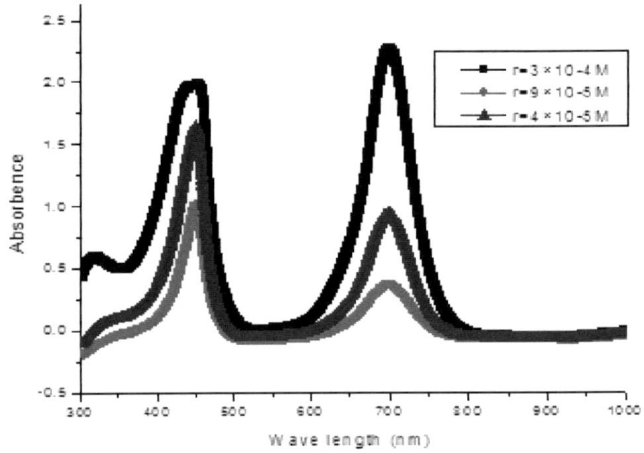

Figure 1: Absorption spectrum of $TPPOH_4$ doped in boric acid (○) $9.7x10^{-5}$M (□) $3.9x10^{-4}$M (Δ) $4.8x10^{-5}$M.

3. Experimental procedure

The experimental setup is in single beam z-scan geometry where the sample is translated along the z-axis for varying intensity as function of position on the axis beyond the focusing lens. It is set in the tight focusing geometry by using lens of focal length 10cm with completely filled aperture with expanded laser beam of 1.5in to achieve the smallest spot-size of 3.8μm and peak intensity of 8.4×10^{8} W/m^2. Laser power of 150mW at 671nm from Diode Laser (Model SDL-671-120T) having a Gaussian TEM_{00} mode is used. Z scan measurements for both open and closed aperture case are carried out for all the samples studied. Z scan plots were obtained after correcting for fluctuations in the laser power by way normalizing (D2) with that of the power monitored through the reference arm of the setup (D1). Reverse saturable absorption is clearly seen in case of the open aperture data wherein the transmission decreases as the intensity increases as it approached the beam waist. The change in transmission is more for higher concentrations. Closed aperture data shows the typical signature of negative optical nonlinearity with peak followed by valley. The asymmetry in the peak height as compared to valley depth is due to the presence of reverse saturable absorption which enhances the valley depth and reduces the peak height.

3.1 Data analysis

Open aperture data collected the z-scan open and closed aperture data of $TPPOH_4$ at three different concentrations are as shown in figure 2. From open aperture data we confirm that our sample has reverse saturable absorption behavior at all three concentrations of interest and this property is increasing with increasing concentration. Closed aperture data tells that sample has negative nonlinear refractive index because of peak followed by valley. The nonlinear refractive index(n_2) of samples because of thermal and optical nonlinearity at three different concentrations were estimated as $5.07\pm0.13\times10^{-5}$ cm^2/W, $1.84\pm0.04\times10^{-6}$ cm^2/W, $9.13\pm0.11\times10^{-6}$ cm^2/W and corresponding third order susceptibilities($\chi^{(3)}$) estimated to be $3.27\pm0.19\times10^{-7}$ cm^2/V^2, $1.18\pm0.05\times10^{-8}$ cm^2/V^2, $5.89\pm0.07\times10^{-7}$ cm^2/V^2. These nonlinear refractive indices, which we measured through above process originated from optical and thermal nonlinear process. The thickness of these sandwiched films is by estimated by finding the first zeroes of the degenerate four wave mixing signal [9].

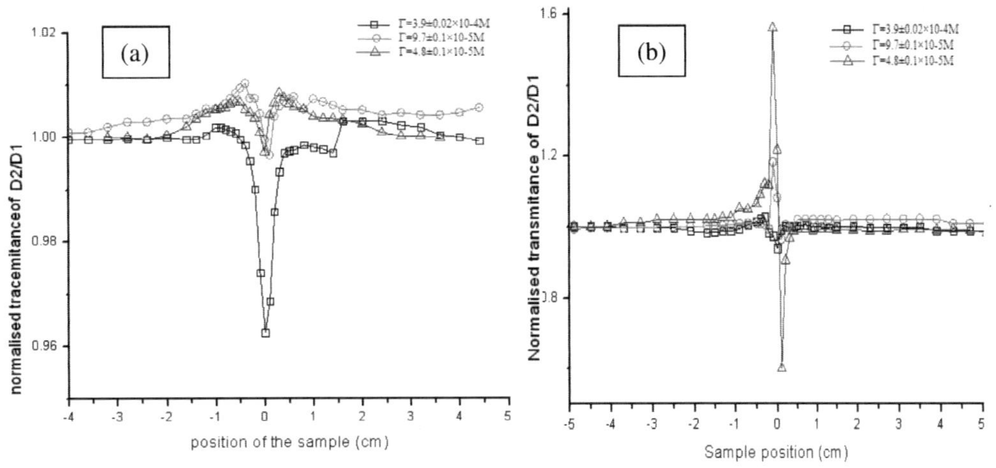

Figure 2. Normalized transmittance of Z Scan data in (a) Open aperture mode and (b) Closed aperture mode of $TPPOH_4$ doped in boric acid glass with different molar concentrations viz (○)$9.7x10^{-5}$M (□)$3.9x10^{-4}$M (Δ)$4.8x10^{-5}$M.

3.2 Analysis of rings

During the Z scan experiment in these samples we observed spatial rings, of alternate dark and bright intensity, which seems to depend upon the position of the sample on the z-axis and hence the intensity. We performed the same z-scan with dye free boric acid films and did not observe any rings in the transmitted beam profile. From this we conclude that $TPPOH_4$ is only responsible for the rings formation. The rings thickness is decreasing from first ring to nth ring as we move radially outwards. Size of the rings & number of rings are only dependents on the beam spot size. Once the rings are formed at a given spot size it seems that changing the intensity levels, while keeping the spot size same does not changes the ring pattern in terms of number and size of rings formed. Possibly we can say that once the rings are formed it is kind of semi permanent in nature. The onset of ring formation at the threshold intensity of the $TPPOH_4$ sample for all three different concentrations samples were measured to be 1.35×10^5 W/m^2, 2.89×10^5 W/m^2 and 2.25×10^5 W/m^2. Further we observe that the rings so formed would take about 10 hours to disappear after the laser irradiation is stopped. Figure 3 shows the rings pattern pictures in the transmitted intensity of $TPPOH_4$ at molar ratio $3.9\pm0.02\times10^{-4}$M.

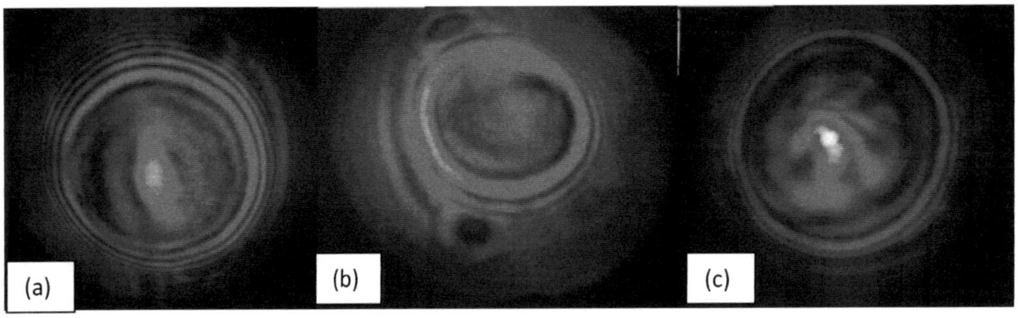

Figure 3. Observed rings in the transmitted intensity of $TPPOH_4$ ($3.9\pm0.02\times10^{-4}$M) doped in boric acid glass at intensity of (a) 1.63×10^{10}W/m^2 (b) 1.21×10^8W/m^2 (c) 1.22×10^6W/m^2

We modulated the input intensity by help of mechanical chopper and observed no rings as the chopping frequency is increased beyond 1600 Hz, 1300Hz and 1400 Hz for the 4×10^{-4}M, 1×10^{-4} M, 5×10^{-5} M samples respectively. As seen in the figure 4 at 60 Hz chopper frequency we could see the thermal decay and at 850 Hz there was no thermal decay in the transmitted pulse. Rings were observed at both these frequencies in the sample with 4×10^{-4}M of $TPPOH_4$ doped in boric acid glass

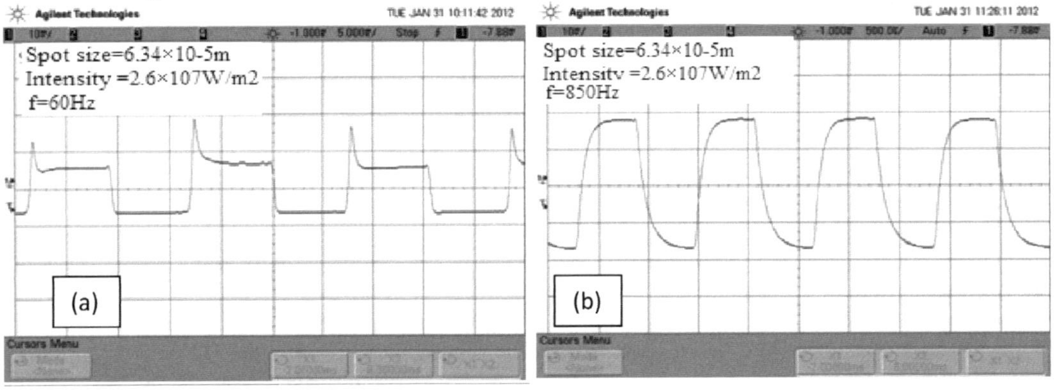

Figure 4. Temporal variation of thermal nonlinearity with modulating frequency at (a) 60Hz (b) 850Hz in the sample with 4×10^{-4}M of $TPPOH_4$ doped in boric acid glass.

We have done the z-scan closed aperture for rings pattern. We maintained the aperture size in such a way that the central spot of rings passes through the aperture. The z-scan has been carried out different chopping frequencies when the rings were formed. Figure 4 is the snapshot for two chopping frequencies, both at low and high frequency. In figure 5 the low frequency curve is due to combined thermal and optical nonlinearity [T_1] and high frequency curve [T_2] gives the estimate of optical nonlinearity. T_1-T_2 transmittance curve gives the effect of thermal nonlinearity. We have measured optical nonlinear refractive indices (n_2) from the graphs as $2.29\pm0.06\times10^{-7}$ cm^2/W, $4.42\pm0.06\times10^{-7}$ cm^2/W, $1.04\pm0.01\times10^{-6}$ cm^2/W and corresponding susceptibilities are $1.82\pm0.52\times10^{-9}$ cm^2/V^2, $3.52\pm0.05\times10^{-9}$ cm^2/V^2, and $8.30\pm0.15 \times 10^{-9}$ cm^2/V^2. Thereby we have succeeded in estimating the contribution of thermal and optical nonlinearity in the formation of the rings. More study are underway to understand the exact mechanism of these phenomena.

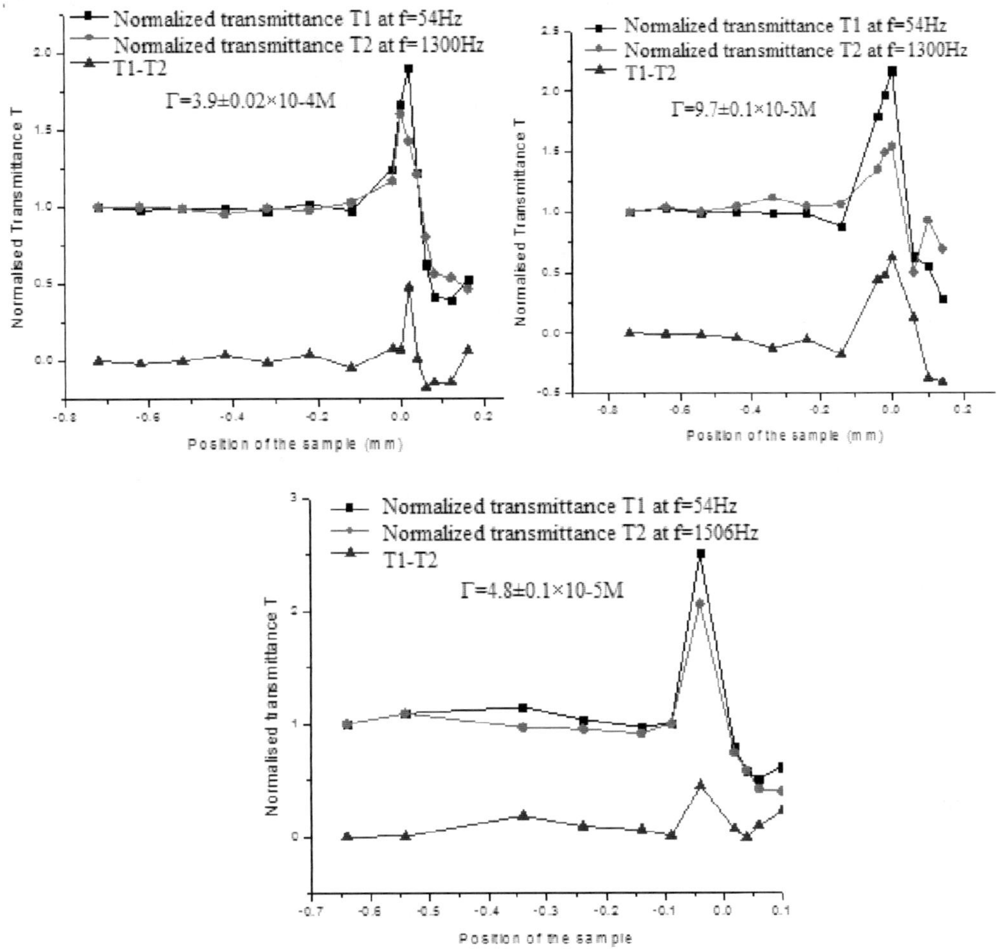

Figure 5. Closed aperture Z-Scan of $TPPOH_4$ at three different concentrations of for low (54Hz) and high(1300 Hz) chopping frequencies.

References

[1] D. Durbin, S. M. Arakelian and Y.K. Shen 1981 *Opt. Lett* **6** 145
[2] D. Durbin, S. M. Arakelian and Y.K. Shen 1981 *Phys Rev Lett* **47** 1411
[3] R. Sohrab Afzal and N. M. Lawandy1989 *Opt. Lett.* **14** 794
[4] Martin Le Berre 1982 *Phys. Rev.* A **25** 1604
[5] W. Chalupczak W. Gawlik and J. Zachorowski 1994 *Phys. Rev.* A **49** R2227
[6] W. Chalupczak W. Gawlik and J. Zachorowski 1993 *Opt. Commun* **99** 49
[7] Y. H. Meyer 1980 *Opt. Commun* **34** 439
[8] S.J Mathews, S.Chaithanya Kumar, L. Giribabu, S.Venugopal Rao 2007 *Opt. Commun.* **280** 206
[9] Mudasir H Dar, M Phil Thesis 2011 Pondicherry University
A Srinivasa Rao M Phil Thesis 2011 Pondicherry University

Low field magneto caloric effect in $(Gd_{1-x}\,Pr_x)_5\,Si_{3.2}\,Sn_{0.8}$ alloys

J Xavier and K V Saban

Smart Materials Analytical Research and Technology (SMART), Department of Physics, St. Berchmans College, Changanassery, Kerala, 686 101, India.

E-mail: kvsaban@gmail.com

Abstract. The structural, thermal and magnetic properties of the arc melted $(Gd_{1-x}\,Pr_x)_5\,Si_{3.2}\,Sn_{0.8}$ alloys with $x = 0$, 0.05, 0.1 and 0.15 were studied. A maximum magnetic entropy change of 1.2034 J/kgK occurs for the $x = 0.1$ alloy at 299 K for a low magnetic field change of 1.58 T. The Arrott plot technique confirms a second order ferromagnetic to paramagnetic phase transition in all these Si rich orthorhombic samples. Tuneable T_c around room temperature, moderate values of ΔS_m in low magnetic fields and absence of magnetic hysteresis make these alloys useful with regard to room temperature magnetic refrigeration.

1. Introduction

The rising concern for the environment has created much interest in using the energy efficient and eco-friendly magnetic refrigeration based on the magneto caloric effect (MCE) as an alternative to the conventional vapour cycle refrigeration and there is an extensive search for magnetic materials with better MCE values following the discovery of giant MCE in $Gd_5Si_2Ge_2$ alloy [1]. The thermal hysteresis and the magnetic hysteresis associated with the first order transition in $Gd_5Si_2Ge_2$ alloy, the high cost of Gd and Ge and the need for a high magnetic field to produce GMCE make the commercialization of this technique difficult. For practical applications, we need magneto caloric materials of little hysteresis, exhibiting high magnetic entropy change around room temperature. Minor alloying additions on the magnetic Gd part and non-magnetic Si/Ge lattice can reduce the hysteresis loss, adjust the Curie temperature (T_c) and enhance the magnetic entropy change in low applied magnetic field. Much research has been conducted on the structural and magnetic properties of the $Gd_5(Si_xSn_{1-x})_4$ alloy system, which exhibits a phase similarity with the $Gd_5Si_xGe_{4-x}$ system [2, 3]. Usually, heavy rare earth metals having high magnetic moment are preferred for MCE. In contrast, this paper presents the preparation and characterization of $(Gd_{1-x}\,Pr_x)_5\,Si_{3.2}\,Sn_{0.8}$ alloys, with $x = 0$, 0.05, 0.1 and 0.15, in which a portion of the heavy Gd is replaced by the lighter Pr.

2. Experimental

Alloy buttons of ~5 g with the nominal composition $(Gd_{1-x}\,Pr_x)_5\,Si_{3.2}\,Sn_{0.8}$, where $x = 0$, 0.05, 0.1 and 0.15, were prepared by repeatedly arc melting appropriate quantities of very high pure constituent elements in a water-cooled copper hearth under Argon atmosphere. The alloys were studied in the as cast condition without any heat treatment. The XRD spectrum was recorded on a Philips XPERT-PRO diffractometer with Cu Kα radiation and calorimetric measurements were performed using Thermal Analysis Differential Scanning Calorimeter (Mettler Toledo DSC 822e). The *M(T)* and *M(H)* studies were carried out using a vibrating sample magnetometer (Lakeshore, Model No: 7407). The isothermal magnetic entropy change ΔS_m *(T, H)* was calculated from the isothermal magnetization measured at discrete temperatures, using the numerical formula:

$$\Delta S_m = \sum_i \frac{1}{T_i+1-T_i} (M_i - M_{i+1})\, \Delta H_i \qquad (1)$$

where, M_i and M_{i+1} are the magnetization values measured in a field H at temperatures T_i and T_{i+1} respectively and ΔH_i is the magnetic field step within each isotherm [4].

3. Results and discussion

3.1. Structural analysis

The powder XRD patterns of the quaternary alloys, $(Gd_{1-x}Pr_x)_5Si_{3.2}Sn_{0.8}$ (x = 0, 0.05, 0.1, 0.15), are shown in figure 1.The diffraction peaks were identified using the JCPDS database and the most intense peaks at ~ 31.9^0, 32.1^0 and 33^0 in all the samples confirmed an orthorhombic Gd_5Si_4-type phase for all the alloys (PDF 87-2319) [5]. However, the peaks at ~ 29.9^0, present in all the samples, are more intense than that is expected for the Gd_5Si_4 orthorhombic system. This peak is indicative of the presence of an orthorhombic GdSi-type minor phase (PDF 80 - 0705) [6]. The crystal structure remained invariant with Pr addition except for the slight shift of the peaks to lower angles, indicating a lattice expansion arising from the differences in the atomic radius of Gd (2.54 Å) and Pr (2.64 Å).

3.2. Thermal studies (Differential scanning calorimetry - DSC)

The heat flow from the samples were measured as a function of temperature from 225 to 375 K, under zero magnetic field and the results are shown in figure 2. Absence of large peaks in the thermal curves arising from the structural transition near the Curie temperature (*T*c) revealed the existence of only a second order magnetic transition (~ 300 K) in the samples.

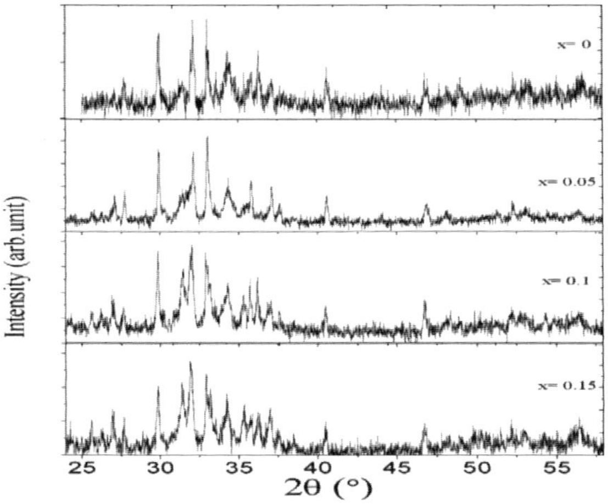

Figure 1. X-ray powder diffraction pattern of the alloys at room temperature.

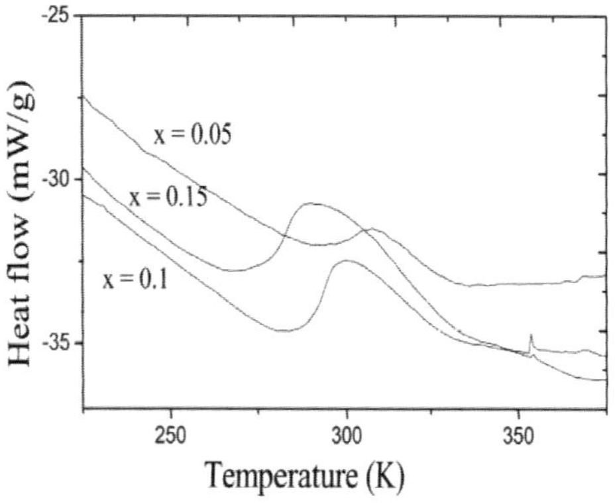

Figure 2. DSC thermograms of the alloys with zero magnetic field applied.

3.3. Magnetic studies

The *M-T* curves for the samples for a temperature range of 400 K to 200 K in a field of 1.6 T, are shown in figure 3. The minimum of the $(\frac{\partial M}{\partial T})$ versus T curve gives the Curie temperature (T_c). The substitution of Pr with Gd has led to a decrease in T_c up to $x = 0.1$ and from there Tc increases with Pr content which is attributed to the change in exchange energy between these magnetic atoms with doping.

In order to calculate the maximum value of the isothermal magnetic entropy change (ΔS_m), the field dependant isothermal magnetization (*M -H*) up to 1.58 T was measured 4 K above and below the ordering temperature (figure 4).

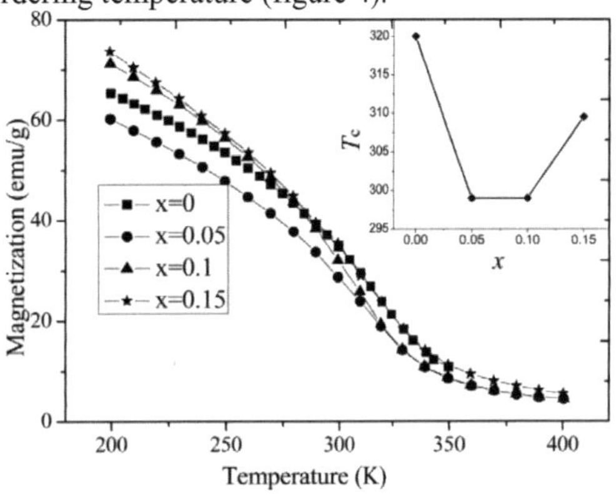

Figure 3. *M-T* curves for the alloys.The inset figure show the variation of *Tc* with *x*.

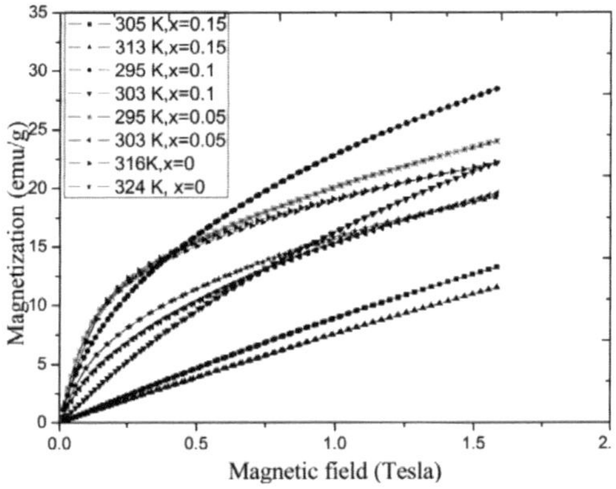

Figure 4. The *M-H* curves for the alloys in a magnetic field of 1.58 T.

The *M-H* isotherms taken around T_c are completely reversible and show a typical ferromagnetic nature below T_c. The magnetization varies nearly linearly with the magnetic field for low fields at $T > T_c$,showing a paramagnetic behaviour. The saturation magnetization was not reached in low magnetic fields due to the magneto crystalline anisotropy of the sample. The ΔS_m values of the $(Gd_{1-x}\,Pr_x)_5\,Si_{3.2}\,Sn_{0.8}$ alloys, at their ordering temperatures are 0.6779, 0.9322, 1.2034 and 0.2138 J/kgK respectively for x = 0,0.05,0.1,0.15 in a low magnetic field of 1.58 T . These values are comparable with that of the $Gd_5\,Si_xSn_{4-x}$ second order transition system (1.69 J/kgK at 301.5 K for $Gd_5\,Si_{2.8}Sn_{1.2}$ and 2.26 J/kgK at 290.5 K for $Gd_5\,Si_{2.6}Sn_{1.4}$ for ΔH = 1.8 T) [3]. It is evident that Pr doping up to $x = 0.1$ would lead to an increase in the maximum magnetization and in the magnitude of the maximum entropy change for these alloys. The lower values of the magneto caloric effect are due to the absence of structural transitions in the samples as inferred from the calorimetric measurements. Although the first order transition (FOT) systems have higher MCE values, the hysteresis and the narrow temperature span of those systems will adversely affect the refrigerant capacity and relative cooling power. Therefore, second order transition (SOT) systems exhibiting MCE over a wide temperature range with little hysteresis are of commercial importance. Figure 5 shows the Arrott plot (M^2- H/M) of the samples which could be used for confirming the order of magnetic phase transition using the Banerjee criterion [7]. When the slope of the M^2 versus H/M curve is negative, the system undergoes a first order magnetic phase transition and a positive slope corresponds to a second order magnetic phase transition. All the curves presented in figure 5 show a positive slope and therefore $(Gd_{1-x}\,Pr_x)_5\,Si_{3.2}\,Sn_{0.8}$ alloys have a second order magnetic phase transition at T_c which is consistent with the result obtained from

M-T, M-H and calorimetric measurements. Thus, the alloys are soft ferromagnetic materials with no magnetic hysteresis, which is highly beneficial for magnetic refrigeration.

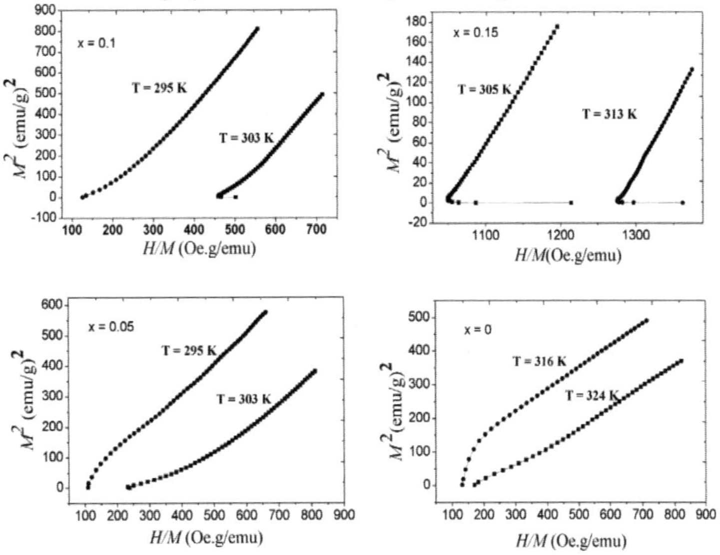

Figure 5. Arrott plots of $(Gd_{1-x}Pr_x)_5Si_{3.2}Sn_{0.8}$ alloys taken around *Tc*.

4. Conclusions

The effect of substituting Pr for Gd, on the crystal structure, magneto caloric effect and order of magnetic phase transition of $Gd_5Si_{3.2}Sn_{0.8}$ alloy has been investigated. A Gd_5Si_4-type orthorhombic structure has been identified in all the samples and the crystal structure remains invariant with the amount of Pr content. All the samples are soft ferromagnetic materials with nearly zero coercivity and show a second order ferromagnetic to paramagnetic phase transition, confirmed by the Arrott plots. The Curie temperatures of the $(Gd_{1-x}Pr_x)_5Si_{3.2}Sn_{0.8}$ alloys are 320 K, 299 K, 299 K, 309.5 K respectively for x = 0,0.05,0.1,0.15 and the corresponding maximum magnetic entropy changes are 0.6779, 0.9322,1.2034 and 0.2138 J/kgK in a magnetic field of 1.58 T. The moderate values of ΔS_m in a low magnetic field near room temperature, absence of magnetic hysteresis and adjustable T_c are the major features of these alloys.

Acknowledgement

The authors thank the University Grants Commission, Govt. of India for the financial support by a major research project under its Xth plan.

References

[1] Pecharsky V K and Gschneidner Jr K A 1997 *Appl. Phys. Lett.* **70** 3299-3301
[2] Wang H B, Altounian Z and Ryan D H 2002 *Phys. Rev. B* **66** 214413 1-6
[3] Zhang E, Chen Y, Zhang T, Tang Y and Tu M 2006 *J. Magn. Magn. Mater.* **305** 410-12
[4] Hu F X, Shen B G, Sun J R, Cheng Z H and Zhang X X 2000 *J. Phys.: Condens. Matter.* **12** L691-704
[5] Pecharsky V K and Gschneidner Jr K A 1997 *J. Alloys Compd.* **260** 98-106
[6] Nagaki D A and Simon A 1990 *Acta Cryst. C* **46** 1197-99
[7] Banerjee S K 1964 *Phys. Lett.* **12** 16-17

IOP Conf. Series: Materials Science and Engineering **73** (2015) 012025 doi:10.1088/1757-899X/73/1/012025

Enhanced response of SnO_2 based thin film sensors towards methane gas due to the collective efforts of catalytic activity and photo-activation phenomenon

D Haridas[1,3] and V Gupta[2]

[1] Department of Physics, Keshav Mahavidyalaya, University of Delhi, Delhi, India.
[2] Department of Physics and Astrophysics, University of Delhi, Delhi, India.

E-mail: divya_h14@rediffmail.com

Abstract. Detection of methane is always a great cause of concern for safety productions in mines and chemical factories. The present study investigates the twin effect of UV illumination and catalytic activity on methane sensing characteristics of SnO_2 sensors. The sensitivity and selectivity of pure SnO_2 thin film sensors are improved by loading it with Pd catalyst clusters (8 nm). Further, optimizing the thickness of Pd clusters leads to an enhanced sensing response of 97% to 99% over a wider temperature range (160°C to 240°C) for 10 nm thick Pd clusters. The room temperature response of SnO_2-Pd (10 nm) sensor increases to 99.7% under UV illumination (which was around 0.6% at room temperature under no illumination) which is attributed to the efficient catalytic dissociation of methane molecules besides the spillover process at room temperature. The present study therefore investigates the effect of UV illumination on methane sensing characteristics of SnO_2 sensors loaded with Pd clusters. Results indicate the possibility of utilizing the sensor structure with novel dispersal of Pd catalyst clusters on SnO_2 film surface for efficient detection of methane at room temperature under the illumination of UV radiations.

1. Introduction

Methane is an explosive gas, used extensively at the domestic and industrial sites. It is the main constituent of natural gas, the fuel, which is supplied at homes and industries including automobiles, is often released from the walls of coal mines and when unmonitored can accumulate, causing dangerous explosions. Methane gas is highly volatile in nature and can cause explosion due to inflammability. It has a lower explosion limit (LEL) of 4.9% and an upper explosion limit (UEL) of 15.4% [1]. Thus, trace level detection of methane gas in the environment is very important to avoid any accidental explosion due to its leakage from automobiles. Continuous efforts are ongoing worldwide towards development of an efficient methane sensor that could work even at low temperatures. Recent concerns over the impacts of methane on global warming (it has a much greater warming effect on the climate than carbon dioxide) have resulted in an increased awareness among the research community towards the development of methane sensors. Therefore, the detection of this potent gas is essential at the environmental, industrial and domestic sites. Continuous efforts are ongoing worldwide towards the development of an efficient methane sensor that could work preferably at lower temperatures with

[3] divya_h14@rediffmail.com

enhanced response. Various analytical techniques are developed since 1990 for the detection of methane including gas chromatography (GC). Gas-chromatography is expensive, bulky, strongly affected by humidity, process limited selectivity and not well suited for on-line monitoring. Alternatively semiconductor gas sensors are becoming popular for the detection of methane gas due to their ability to alter conductance with modest detection limits, small size, low cost and on-line monitoring. Semiconducting Metal oxides have been largely exploited for the detection of such explosive and toxic gases. Among them, tin oxide (SnO_2) is the most explored material as chemical gas sensors, due to its ease in processing and good response to a number of gases at low operating temperature. However, there are problems associated with tin oxide related to its poor selectivity to various gases. In general, suitable catalysts are incorporated with SnO_2 to improve the selectivity and sensing response [2-4]. However a high response is observed only at high temperatures but in real time monitoring it is very difficult to monitor the gas composition in an environment having explosive species using such sensors since high temperature could trigger the explosion. Therefore there is an essential need for the development of gas sensors that can be operated at room temperature with enhanced response and selectivity. One of the additional attractive features associated with the room temperature operated semiconductor gas sensor is that it can lead to a complete integration with silicon substrate, and would not require any heating element. Therefore a new methodology or novel design approach is essentially required in order to fulfill the essential requirements of sensor market. In the present work we target mainly on achieving a high response at low operating temperature to detect low concentration of methane gas (which is below LEL) by exploring the twin effect of catalytic activity and photo-activation process.

2. Experimental

In The SnO_2 thin film of 90 nm thickness was deposited by rf sputtering technique. Palladium catalyst clusters were deposited by rf sputtering using a palladium target in pure argon ambient on the surface of sensing SnO_2 thin film for enhanced response. Pd clusters of varying thickness (2–20 nm) were deposited using a shadow mask having uniformly distributed pores of diameter 600 μm and the SnO_2-Pd sensor structures were prepared [5]. Photostimulation measurements have been carried out at room temperature by enabling a UVlamp (λ = 365 nm) as radiation source (intensity of UV light (2 μWcm^{-2}). Sensor response measurements were carried out in a home-built gas sensing test rig retrofitted with a polished stainless steel bell jar (SS 316L) with a quartz window and calibrated leaks were used for LPG introduction. The sample holder was fitted with a miniaturized heater, controlled through a temperature controller (accuracy ±0.5 °C). The change in resistance was measured using Pt coated spring electrodes fitted to probes of Digital Multimeter DMM (Model No.: Keithley 2002) interfaced to a computer which acted as the data acquisition system.Sensor response (S) at a given temperature T of sensing element (SnO_2) is determined by measuring change in resistance of the sensing element in absence and presence of methane. The sensor response S is defined by eq. (1)

$$S = \left(\frac{R_a - R_g}{R_a} \right)\% \qquad (1)$$

R_a is the sensor resistance in atmospheric air, and R_g the resistance in presence of methane.

3. Results and discussion

The response characteristics of pure SnO_2 thin film and SnO_2–catalyst cluster sensors were studied over a wide temperature range of 60 to 260°C. The resistances of the sensors were measured in both air and in the presence of methane gas at each temperature. The variation of sensing response of both sensor structures were obtained for 200 ppm methane gas and is shown in figure 1 as a function of temperature. The sensing response of both sensor structures prepared in the present study was found to increase with increasing temperature and ultimately reaches the maximum value at a particular temperature (operating temperature). A relatively poor response for sensor having pure SnO_2 film (without any catalyst) with a maximum value (S=57%) was observed at an operating temperature of 200°C. The presence of catalyst on the surface of SnO_2 sensing layer and its nature was identified to be crucial for the enhanced response. A high response (S=97.2%) was observed for SnO_2-Pd clusters (8 nm thin) sensor structure at an operating temperature of 220°C (figure 1). The improved response characteristics of the sensor in the present study may be attributed to the interaction of sensing gas molecules with both the palladium catalyst (dispersed in the form of uniformly distributed clusters) and the uncovered surface of sensing SnO_2 film.

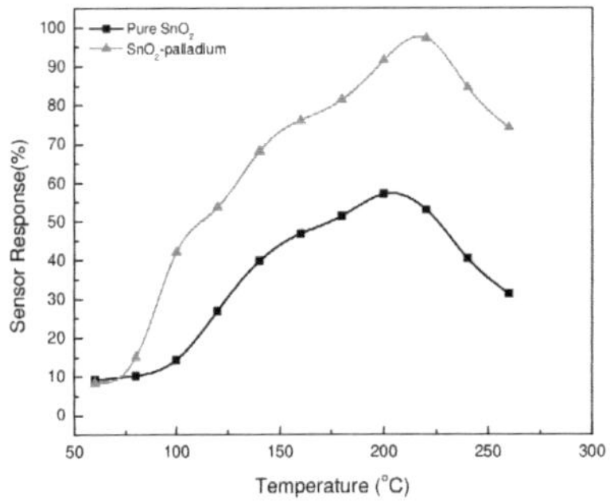

Figure1. Variation of sensor response with operating temperature.

Thickness of the Pd clusters are varied from 2 to 20 nm. The variation of sensor response as a function as a function of the thickness of Pd catalyst cluster loaded on SnO_2 surface at their operating temperature for 200 ppm methane is shown in figure 2.

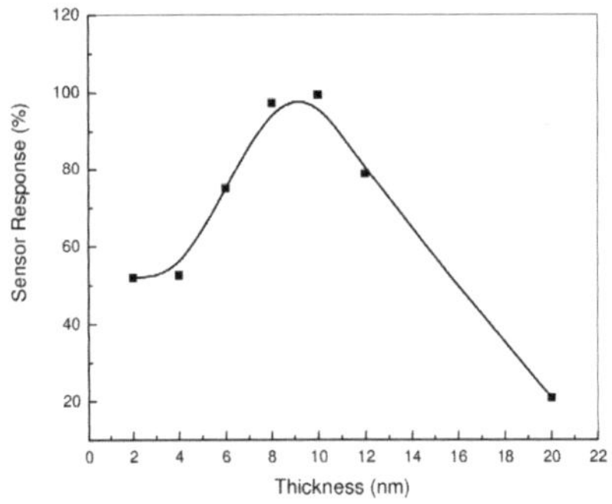

Figure 2. Variation of sensor response vs thickness of Pd clusters at the operating temperature for 200 ppm methane.

It may be noted from figure 2 that the response characteristics depends strongly on the thickness of the Pd clusters loaded on the surface of SnO_2 thin film. The response was found to be relatively higher for the sensor structures having Pd clusters in the thickness range 8–10 nm. The response increases continuously from 52% - 99.2% with increasing thickness of Pd metal clusters from 2 nm to 10 nm. However, response start decreasing with further increase in thickness of the Pd cluster (>10 nm). The maximum response obtained at 10 nm thickness of Pd clusters indicates the importance of the presence of an optimum quantity of catalyst on the surface of the sensing SnO_2 layer.

Since the response of all sensor structures was poor at room temperature, the photostimulation process (under exposure of UV photons) was implemented simultaneously. The sensor structures were exposed initially with UV light and subsequently methane was introduced into the test gas chamber. Figure 3 compares the response of SnO_2-Pd clusters sensor structures (having 8 nm, 10 nm and 12 nm thickness of Pd clusters) obtained with and without UV exposure at room temperature for 200 ppm methane. UV illumination has immensely enhanced the response at room temperature for all prepared sensors. Tremendous increase in the room temperature response 0.19 (=0.6%) to 252.3 (=99.7%) could be seen to occur with dispersal of 10 nm thin Pd clusters over the surface of SnO_2 thin film which is same as that obtained without UV illumination at elevated temperature of 220°C (figure 3). The enhanced response of sensor at room temperature under UV illumination is attributed to the photo-catalytic behavior of SnO_2 towards breakdown of a range of organics

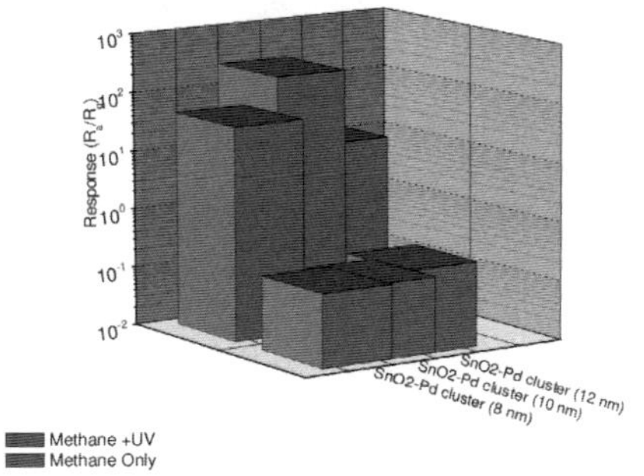

Figure 3. Variation of response for all prepared sensor structures (varying thickness of Pd catalyst) at room temperature.

UV assisted photosensitive activation convert the chemisorbed oxygen available on the surface of SnO_2 film into photoinduced oxygen which has low thermal stability and therefore is amenable to desorption at room temperature. CH_4 gas molecules after interaction with Pd catalyst clusters get dissociated under UV illumination and interact with photoinduced oxygen species available on the uncovered surface of SnO_2 thin film, which thereby briefly explains the gas sensing mechanism.

4. Conclusion

The sensing characteristics of rf-sputtered SnO_2 thin film (90 nm) loaded with palladium catalyst clusters has been studied for methane detection. The SnO_2-Pd cluster structure exhibit enhanced response (97.2%) at a relatively low operating temperature of 220°C. The influence of the thickness of Pd clusters on the response characteristics of SnO_2 thin film for methane is also being studied. The enhanced response (99.2%) at an operating temperature of 160°C was obtained with the presence of 10 nm Pd clusters. Methane detection has been shown to be possible at room temperature under UV illumination using SnO_2-Pd-cluster sensor structure. Room temperature response of the sensor with 10 nm thin Pd clusters is seen to be about 99.7% for 200 ppm methane under UV illumination. UV illumination plays the role of activating reversible adsorption-desorption property in SnO_2-Pd-cluster sensor akin to their operation at elevated temperatures. Obtained results are encouraging for commercial realization of room temperature SnO_2 based thin film sensor for CH_4 gas with enhanced response.

References

[1] Smedt G D, Corte F D, Notele R, Berghmans J 1999 *J. Hazard. Mater.* **70** 105–13
[2] Kim J C, Jun H K, Huh J-S, Lee D D 1997 *Sensors and Actuators B* **45** 271–77
[3] Papadopoulos C A, Vlachos D S, Avaritsiotis J 1996 *Sensors and Actuators B* **32** 61–9
[4] Sun L, Qiu F, Quan B 2000 *Sensors and Actuators B* **66** 289–92
[5] Haridas D, Sreenivas K, Gupta V 2008 *Sensors and Actuators B* **133** 270-75

Study of dc Electrical Conductivity, Magnetic Susceptibility and Thermal Behavior of Pyrrole Aniline Copolymer

H R Wasnik[1]* and D S Kelkar[2]

[1] Yeshwantrao Chavan College of Engineering Nagpur-441 110 , India
[2] Department of Physics, Institute of Science, Nagpur-440 001, India
E-mail: hrwasnik@gmail.com

Abstract: Chemical copolymerization of pyrrole and aniline has been carried out by taking the monomers in equimolar proportion. The structural analysis of the copolymer was done with the help of FTIR, WAXD and SEM techniques, the results confirm the formation of new material involving both pyrrole and aniline units. The dc electrical conductivity of the copolymer has been recorded up to 673k using two probe method. The log σ Vs 1/T plot shows the transition of the sample from metallic to semi conducting behavior and vice versa, at a particular temperature. The magnetic susceptibility was recorded at magnetic field of 0.1KG in the temperature range 313-507 K by using Guoy's method. The Thermal behavior of the copolymer was studied using DTA-TGA revealing variation in the T_g of copolymer when compared to homopolymers.

Key Words: Copolymer, Electrical Conductivity, Susceptibility, Pyrrole.

1. Introduction

The search of improved materials for applications like rechargeable batteries, corrosion protection, and electrochromic display devices, sensors, electromagnetic radiation shielders and many more [1-3] has stimulated the studies of related homopolymers and copolymers. Copolymerization serves as a tool for the synthesis of polyfunctional macromolecules with specific composition, properties, and structure thereby offering a technique to polymerize those monomers that do not polymerize independently [4]. Hence this chemical heterogeneity achieved by simultaneous polymerization of two entirely different monomers in the same reaction mixture is of substantial importance regarding the influence of complexation on the chain formation and propagation. Thus in the present work copolymer of pyrrole and aniline have been copolymerized chemically. Its structural characterization was done by using FTIR, WAXD and SEM. The dc electrical conductivity was recorded at various temperatures using two probe method at a field of 27 V/cm and is reported here. Further the magnetic susceptibility of the copolymer has been recorded by using Guoy's method. The thermal response of the sample was studied using DTA-TGA and the Tg values have been presented herewith.

2. Experimental Aniline (SRL Mumbai) and Pyrrole (SRL Mumbai) were purified by distillation process before use. Also sulphuric acid (Qualigen, Mumbai) was used to generate acidic medium and anhydrous $FeCl_3$ (SRL Mumbai) was used as an initiator for copolymerization. In the present work the copolymerization has been carried out chemically by taking the monomers in equimolar proportion. For this pyrrole and aniline (1 mole each) were mixed together and stirred for about 15 minutes. About 30ml of 0.5M H_2So_4 was added to the reaction mixture and the solution was stirred till it became clear. Finally 0.25M of anhydrous $FeCl_3$ was added to this and the whole reaction mixture was stirred for about one hour, continuously at room temperature i.e. 30°C.The reaction mixture was then allowed to polymerize for a period of 24 hrs. The reaction mixture was then filtered and the precipitate was washed with distilled water and acetone alternately, to remove the impurities. The precipitate was then dried at room temperature and was further studied. The precipitate was then further dried and was further studied.

2.1 Infrared Spectroscopy: This copolymer was then subjected to Fourier Transform Infrared Spectroscopy (FTIR). The transmission spectrum was recorded in the wavenumber region 400-4000cm^{-1}by using a Schimadzu IR spectrometer (FTIR-8101A).

2.2 Wide Angle X-ray Diffraction: The X-Ray diffraction pattern of the prepared copolymer was recorded using PANAnalytical X-Ray B.V. model PW 1710 unit. The X-Ray beam was Cukα radiation from a sealed tube. The specimen sample in the form of powder was fixed on the rotating stage of the diffractometer and was then scanned in the reflection mode over a range of 2θ from 5_o to 40_o.

2.3 Scanning Electron Microscopy: The scanning electron microscopy was utilized for the morphological investigation of the prepared copolymer sample. The scanning was carried out with the help of Scanning Electron Microscope JEOL 6380 A for various magnifications.

2.4 Measurement of dc Electrical Conductivity: The dc electrical conductivity of the sample was measured using two probe method. The conductivity was recorded within the temperature range 313 K to 673 K at a field of 27 V/cm. All the measurements were done with the help of KEITHLY 6514 system ELECTROMETER. The heating rate was maintained constant throughout the measurement i.e. 5_oC per minute.

2.5 Measurement of Magnetic Susceptibility: The magnetic susceptibility of the sample was measured with the help of Guoy's Method. The measurement was done in the temperature range of 313-507K.

2.6 Thermal Analysis: The sample was further investigated for it's thermal properties. The DTA-TGA spectrum was recorded by using THERMOGRAVIMETRIC/DIFFERENTIAL THERMAL ANALYZER (PERKIN ELMER DIAMOND TG/ DTA) in the temperature range of 305 – 873K.

3. Results and Discussion

In order to ensure that during synthesis copolymer is formed and not mixtures of two homopolymers (polyaniline and polypyrrole) solubility test was carried out. Polyaniline and polypyrrole are both completely insoluble in DMF (Dimethyl Formaamide) and DMSO (Dmethyl Sulphoxide); whereas the synthesized material is soluble to a good extent in both these solvants. Also the IR spectrum of the copolymer was compared with that of the physical mixtures of polypyrrole and polyanilne in 1:1 proportion (by weight). The bands obtained in both these cases are quite different from each other. Hence it can be said that copolymer has been formed.

The **FTIR spectroscopy** was used to identify the structure of synthesized copolymer. The IR spectra for copolymer has been shown in figure I. From figure it can be observed the charge delocalization peak appearing at 1114 cm^{-1} in PANI [4-5] is shifted to 1118 cm^{-1} in the copolymer. Also the N-H stretching peak appearing at 1706 cm^{-1} in PANI is shifted to 1743cm^{-1} in copolymer. The C-N stretching band appearing at 1302 and 1385 cm^{-1} in PANI [6-7] and 1250 and 1290 cm^{-1} in PPy [8] have totally disappeared from the spectrum of copolymer. Likewise the benzenoid and quinoid peak appearing at 1493 cm^{-1} and 1599 cm^{-1} for PANI [5, 9] have been vanished from the copolymer spectrum. Similarly the band due to ring stretching vibrations of PPy unit appearing at 1451 cm^{-1} has been disappeared from the copolymer spectrum. However altogether new bands have been observed in IR spectrum of copolymer at 462cm^{-1}, 1020cm^{-1}, 1411cm^{-1}, 1618cm^{-1}, 2804 cm^{-1} and 2924 cm^{-1}. The subsequent shifts as well as appearance of new IR bands in the copolymer can be thought as a result of the hetero association between monomer units. Also due to subsequent attachment of new side groups, the nature of the bond / bond lengths as well as vibrational frequency associated with it are altered. Hence disappearance or shifting of bands can be associated with the interaction between permanent and temporary charges, developed during the co polymerization leading to the polar effects.

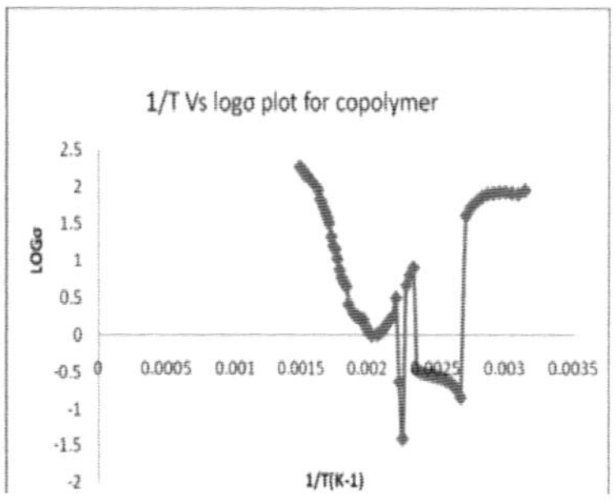

Figure I: FTIR spectra of the copolymer. Figure II: SEM image of the copolymer. Figure III: log σ Vs1/T plot of the copolymer.

The **X-ray diffraction** peaks for PANI, Copolymer and PPy have been tabulated in table I. Further the diffraction peaks of copolymer do not match at all with those of the pure homopolymers indicating that there are structural changes taking place due to copolymerization process. Thus some new material is getting synthesized whose geometrical structure is entirely different from that of the homopolymers. The **Scanning electron micrograph** of copolymer is shown in figure II. When compared with the electron micrographs of parent polymers it has been noticed that the PANI exhibits lamellar structure whereas PPy shows globules [10]. However the copolymer shows agglomerated spheres of approximate 10-20 μm diameter which are dispersed in the matrix. It has been noticed copolymer feed affects the morphology of the copolymer. Also copolymer shows more irregular, disordered spherical agglomerates. [11]

PANI	Copolymer	PPy
9.5	10.2	23.6
14.7	22.5	25.8
20.4	24.2	
25.1	27.4	
	32.8	
	35.7	

Table I: X-ray diffraction peaks of PANI,copolymer and PPy.

The **log σ Vs 1/T plot** for the copolymer is shown in figure III. It is observed that the sample shows a transition from semi conducting to metallic behavior at a certain temperature and vice versa above this transition temperature. The electrical conductivity of copolymer was found to be 1.70105 X 10^{-08} S/cm. As the primary copolymer chain comprises of monomer units of pyrrole and aniline; the charge carriers associated with these units can propagate only upto certain common sequences of either polymer. As soon as the hetero junction is achieved the energy of the charge carriers becomes insufficient for crossing the junction. Thus charge carriers get trapped or their motion is hindered due to the presence of second monomer unit. Hence the conjugation does not remain uniform over the entire polymer chain and overall reduction in the conductivity is observed. Further the steric effects arising due to the attachment of bulky substituents provide torsional twists in the copolymer backbone, reducing the co planarity and average electron delocalization length, and the conductivity [12-13]. Further the **magnetic susceptibility** measurements have been carried out to investigate the nature of charge carriers dominating the conduction process. The susceptibility versus temperature curve shows diamagnetic nature of the sample indicating spinless bipolarons as majority charge carriers.

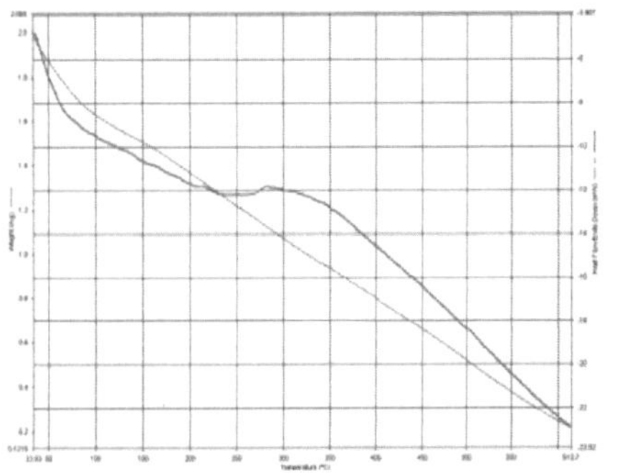

Figure IV: DTA –TGA spectra of the copolymer.

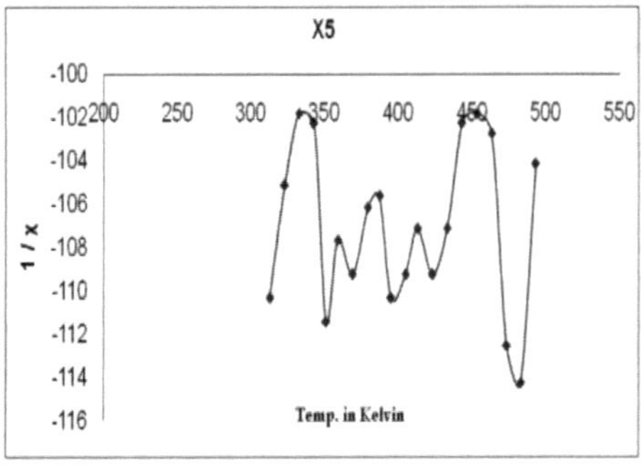

Figure V: Susceptibility Vs temperature curve of copolymer.

The **thermal response** of the copolymer has been studied using DTA-TGA. The sample shows two T_g values at 230°C and 280°C. Both of these T_g values are in between the Tg values for PANI and PPy. From the thermal analysis it can be concluded that the individual monomer units respond differently to the thermal energy. Also at heterojunctions the weakening of two monomers initiates and degradation of the material initiates. As a result first T_g value is obtained. With further supply of heat energy monomeric units became free and the structure collapses hence further dissociation of material takes place and second T_g value is obtained.

4. Conclusion

Copolymer of pyrrole and aniline has been synthesized by using chemical oxidative copolymerization method in equimolar proportion at room temperature. The structural characterization of the prepared copolymer was carried out with the help of FTIR, WAXD and SEM techniques. The results show the formation of new material with modified bands, new IR bands and geometrical structure. The electron micrographs show agglomerated spheres distributed in the matrix. The dc electrical conductivitywas found to be $1.70105X10_{-8}$ S/cm which has been reduced to greater extent when compared with that of the parent polymers. The magnetic susceptibility shows diamagnetic nature of the sample. The thermal property is improved as compared to parent polymers with two T_g values.

5. References

[1] Bonino F, Corradini A, Mastragostino M, Panero S, Scrosati B 1987 *Electrochim Acta*, **32**, 1007.
[2] Otero T F, Cantero I, Azcano B 2000 *Int J Hydrogen Energy*, **25**, 221.
[3] MacDiarmid A G, Yang L S, Huang W S, Humphrey B D 1987 *Synth Met*, **18**, 393.
[4] Langer J 1978 *Solid State Commun*. 26, 839.
[5] Tang J S, Jing X B, Wang B C, Wang F S, 1988 *Synth. Met* **24**, 231.
[6] Yasui T 1985 *Bull. Chem. Soc. Japan*, **10**, 305.
[7] Yang C Y, Smith D, Heeger A J, Cao Y 1994 *Polymer*, **35**, 1142.
[8] Tian B, Zerbi G 1990 *J. Chem. Phys*. **92**, 3886.
[9] Neugebauer C, Wang H C, Fernandez J E 1992 *Macromolecules* **25**, 6179.
[10] Mansouri J, Burford R P 1994 *J Membrane Science*, **87**, 23.
[11] Xu P, Han X J, Wang C, Zhang B, Wang H L 1999 *Synth Met* **104**, 9.
[12] Zhou L and Xue G 1997 Synth *Met*, **87**, 193.
[13] Lemaire M, Buchner W, Garreau R, Hoa H, Guy A, Roncali J, 1990 *J. Electroanal. Chem*., **281**, 293.

IOP Conf. Series: Materials Science and Engineering **73** (2015) 012027 doi:10.1088/1757-899X/73/1/012027

Thickness and oxygen partial pressure dependence on optical band gap of indium oxide by reactive evaporation method

A V Muhammed Ali and Dhananjaya Kekuda
Department of Physics, Manipal University, Manipal, Karnataka,India

E-mail:kekuda@gmail.com

Abstract. Indium oxide film is deposited by reactive evaporation of indium in the presence of oxygen gas onto an unheated glass substrate. It was found that thickness of the film and partial oxygen pressure during the deposition affects the optical properties of the indium oxide thin film. We studied the optical band gap for different thickness and partial pressure keeping a constant annealing temperature. It was found that the band gap varies from 3.5 to 3.8eV, as thickness of the film increased. The band gap energy had also shows the similar trend and it was also studied as a function of annealing temperature. A systematic investigation of the optical band gap as a function of thickness and oxygen partial pressure at different annealing temperature was carried out.

1. Introduction

Indium oxide (In_2O_3) is a wide band gap n-type oxide semiconductor. It has a low resistivity, which is an unusual property for a wide band gap material, compared with conventional semiconductors doped with impurities. The good conductivity of this film is a result of very high doping due to incomplete oxidation. Indium Oxide thin films are characterized by high transparency in the visible region of the spectrum and high reflectance in the infrared region.

Transparent conductive oxide (TCO) material is an interesting semiconductor owing to its unique properties of both high electrical conductivity and high optical transparency [1- 2]. As a matrix for the generation of transparent conducting oxides, indium oxide (In_2O_3) has been the subject of numerous studies for liquid crystal displays [3], solar cell [4], sensors [5], nanowire technology [6], light emitting diodes, and other optoelectronic devices. In the present study, we have investigated the optical properties of InO thin film by reactive evaporation method on a glass substrate.

2. Experimental

The films of Indium Oxide (In_2O_3) were prepared by reactive evaporation technique using HINDHIVAC-12A4D high vacuum coating unit. Resistively heated Molybdenum (Mo) source in the form of boat was used for evaporation of Indium (purity 99.999%). The evaporation was carried out in the presence of oxygen plasma. The vacuum chamber was initially evacuated to base pressure 10^{-3}Pa using diffusion pump. Oxygen was then admitted to the chamber. By adjusting the valve the chamber pressure could be varied between 0.01 and 0.1 Pa during deposition. Optically flat glass slides were used as substrates. The cleaning procedure adopted was as follows: the films were kept in freshly prepared hydrochloric acid overnight and then washed with distilled water. The glass plates were then kept in a detergent solution and cleaned using isopropyl. Finally, the glass plates were cleaned with acetone. The thickness of these films was obtained by quartz crystal monitor. The thickness varied between 500Å -1200Å keeping deposition rate 0.5Å/s.

The films were placed on Muffle furnace and heated for different temperature in open air to investigate annealing effect. The films thus prepared were characterized optically. A UV-VIS double beam spectrophotometer was used to obtain the transmission spectrum of these films.

3. Results and discussion

Figure 1 shows a typical transmission curve of an In_2O_3 thin film with a thickness of 743Å. As can be seen from this figure, the transmission is high in the visible region. However, it depends on the partial pressure of oxygen during deposition [7] and thickness of the film [8].

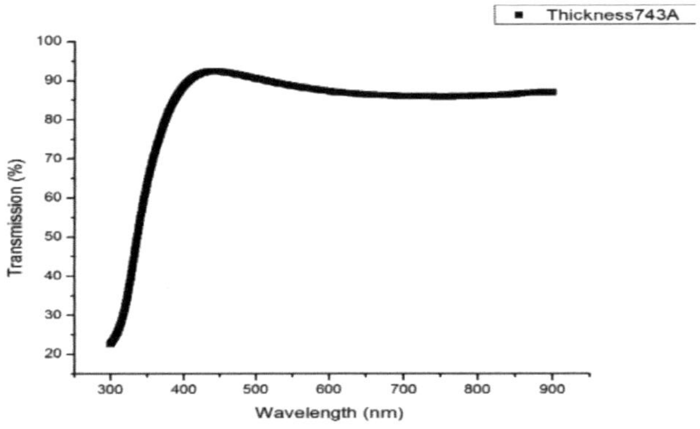

Figure 1. Transmission of In_2O_3 film of thickness 743Å.

From the transmission data, the absorption coefficient as a function of photon energy was calculated and plotted for the direct allowed transition by using the Tauc equation [9], $\alpha(h\upsilon) = A(h\upsilon - E_g)^{1/2}$, Where E_g is the transition energy gap and hυ is the photon energy. Figure 2 shows the $(\alpha h\upsilon)^2$ vs hυ plots for various thicknesses. Extrapolation of linear region of such a curve gives the value of the band gap. The refractive index n of the film was calculated using the formula [10]. Figure 3 shows the variation of refractive index with wavelength

$$n_f = [\frac{n_s(2 - T_{min}) + 2n_s(1 - T_{min})^{1/2}}{T_{min}}]^{1/2}$$

3.1 Influence of thickness on optical properties

The influence of the film thickness on the optical band gap was studied for films deposited 500Å to 1200Å thicknesses, with annealing temperature variation 200^0C - 400^0C. Figure 4 shows the variation of band gap energy as a function of thickness. The calculated band gap energy as a function of thickness, with different annealing temperature are listed in table 1. The results showed that the optical band gap was increased from 3.58-3.74eV by increasing the film thickness and it decreases with increase in annealing temperature. These results are consistent with other published results such as results of M.D Benoy et al [11] who attribute the increase in the band gap as a function of thickness was due to grain size growth and composition changes taking place in the sample.

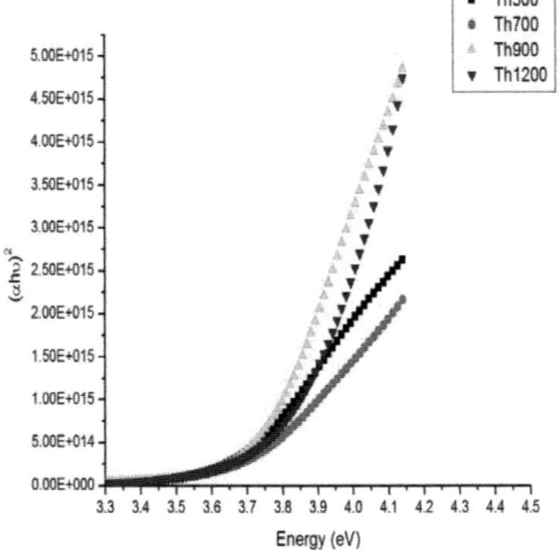

Figure 2. Plot of $(\alpha h\upsilon)^2$ vs hυ for various thicknesses.

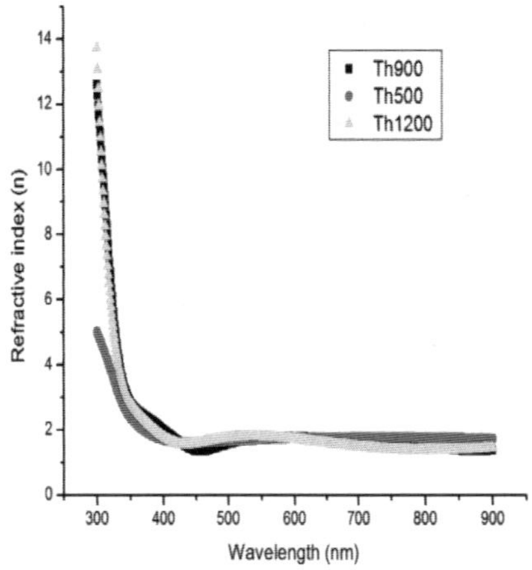

Figure 3. Refractive index vs wavelength for various thickness

3.2 Influence of oxygen partial pressure

Figure 5 shows the plot of $(\alpha h\upsilon)^2$ vs hυ for indium oxide film of thickness 743Å deposited at different oxygen partial pressures and at different annealing temperature. Extrapolation of the linear region of the graph gives the band gap. The values are given in table 2. These values are somewhat higher than the bulk value about 3.5eV [12]. This shift is believed to be due to the Burstein effect, observed in the case of many conducting transparent oxides [13-14]. The band gap is found to increase with oxygen partial pressure. When the oxygen partial pressure is increased, transformation of InO to In_2O_3 is accelerated.

Table 1. Optical band gap dependence on thickness and annealing temperature.

Temp	500Å	700Å	743Å	900Å	1200Å
200^{0}C	3.66	3.68	3.71	3.74	
300^{0}C	3.62	3.63	3.65	3.69	3.71
400^{0}C	3.58	3.60	3.62	3.63	3.65

Table 2. Optical band gap dependence on oxygen partial pressure and annealing temperature.

Temp	3E-5mbar	5E-5mbar	7E-5mbar	9E-5mbar	19E-5mbar
300^{0}C	3.59	3.61	3.63	3.65	3.71
360^{0}C	3.58	3.60	3.62	3.64	3.69
400^{0}C	3.56	3.58	3.60	3.62	3.66

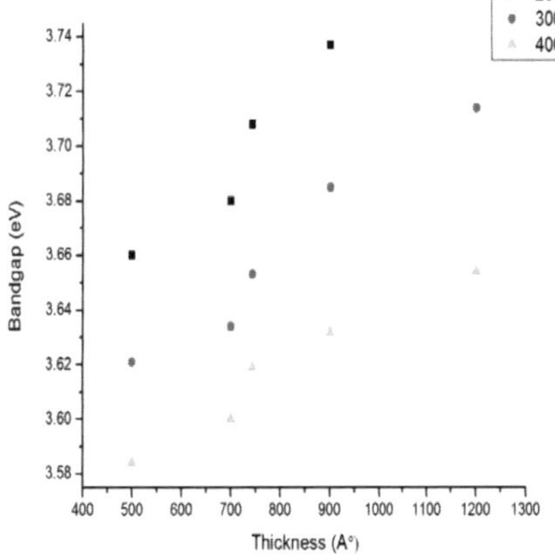

Figure 4. E_g vs thickness for various annealing temperature.

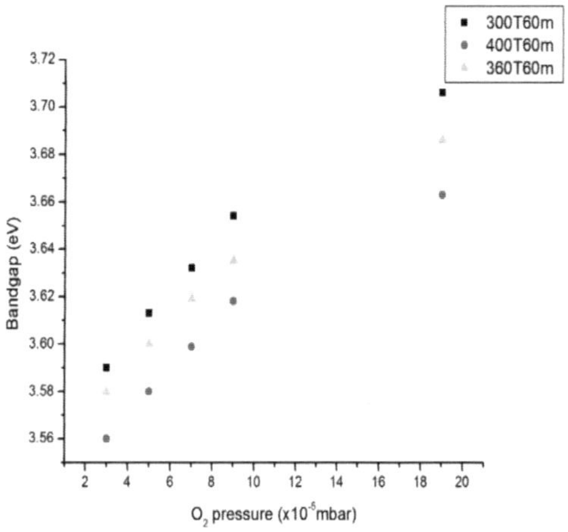

Figure 5. E_g vs oxygen partial pressure for various annealing temperature

4. Conclusion

Indium oxide thin film was deposited by reactive evaporation of indium in oxygen. We found that the thickness, annealing temperature and oxygen partial pressure dependence on the optical band gap of indium oxide film. We also had shown that the transmittance of annealed film was above 95% in visible region.

References

[1] Parthiban S, Elangovan E, Ramamurthi K, Martins R, Fortunato E 2009 *Journal of Applied Physics* **106** 063716
[2] Koida T, Kondo M 2006 *Applied Physics Letters* **89** 082104
[3] Falcony C, Kirtley J R, Dimaria D J, Ma T P, Chen T C 1985 *Journal of Applied Physics* **58** 3556
[4] Granqvist C G 1993 *Applied Physics* A **57** 19
[5] Bender M, Katsarakis N, Gagaoudakis E, Hourdakis E, Douloufakis E, Cimalla V, Kiriakidis G 2001 *Journal of Applied Physics* **90** 5382
[6] Chao Li, Daihua Zhang, Bo Lei, Song Han, Xiaolei Liu, Chongwu Zhou 2003 *Journal of Physical Chemistry* B **107** 12451
[7] Naseem S, Rauf I A, Hussain K and Malik N A 1988 *Thin solid films* **156** 161
[8] Shah M R, Alam M K, Karim M R and Sobhan M A 2005 *Proc. Int. Conf. on Mechanical Enginnering(Dhaka)* ICME05-TH-10
[9] Tauc J 1974 *Amorphous and Liquid Semiconductors* (New York: Plenum)
[10] Mihale Girtan, Rusu G I, Rusu G G , Guluri S 2000 *Applied surface science* **162-163** 492
[11] Benoy M.D, Mohammed E.M, Pradeep B 2009 *Brazilian journal of physics* **39** 629
[12] Jarzebski J.M and Marton J.P 1976 *J. Electrochem. Soc* **123** 333
[13] Burstein E 1954 *Phys. Rev.* **93** 632
[14] Kuku T.A 1986 *Thin Solid Films* **142** 241

IOP Conf. Series: Materials Science and Engineering **73** (2015) 012028 doi:10.1088/1757-899X/73/1/012028

A polymeric route to multielemental boron siliconcarbide ceramic

G Ambadas
Department of Chemistry, Government Arts College, Thiruvananthapuram, Kerala India – 695014, E-mail: ambadasg@yahoo.com

Abstract. A multielemental single source precursor polymer for boron silicon carbide ceramic was prepared by condensation polymerization of boric acid, vinyl silylester and furfuraldehyde. Polymer exhibited excellent thermal stability due the presence of borosiloxane bridges and yield high ceramic residue on pyrolysis. Oxidation behavior of the polymer was studied by heat treating the precursor in air at 1350°C. Morphological analysis of the surface after heat treatment indicates the formation of a uniform glassy layer on the surface. The borosilicate glassy layer on the surface extends superior oxidation resistance to the sample.

Key words: Boron silicon carbide, preceramic polymer, borosilicate ceramic, oxidation resistant material

Introduction

Multielemental ceramics containing boron, silicon, carbon and nitrogen commonly referred as 'Advanced Materials' have unique properties such as extreme hardness, high thermal stability, wear resistance, low friction coefficient and semiconducting properties [1-2]. These ceramics were found to resist crystallization up to 2000°C [3]. Moreover, these materials were found to have interesting electrical, magnetic and optical properties desired for functional applications. The precursor polymer containing boron and silicon are found have excellent oxidation resistant properties [4]. Conventional methods of preparing these materials by high temperature powder mixing result in poor reproducibility as well as inhomogenities in the final composition. Also their thermo mechanical properties start deteriorating at elevated temperatures. In recent years, the preparation of these ceramics from polymeric precursors utilizing both inorganic and organic reagents has gained significant attention [5]. High purity ceramics obtained by this method were found to have homogeneous distribution of elements and are thermodynamically more stable.

Majority of boron containing polymers were prepared from borazine, carborane, decaborane or from other boron hydrides [5]. The high cost of boron reagents is a major impediment in their development. Boric acid is the cheapest source of boron for the synthesis of boron and silicon containing polymers. Using boric acid, Ambadas et. al., has prepared precursor polymers by condensation polymerization with alkyl/aryl substituted silylesters in a high boiling solvent [6]. Although the polymers exhibited good stability at moderate temperatures, their thermo mechanical properties start deteriorates at high temperatures. This is probably due to the presence of residual oxygen in the precursor which at elevated temperature combines with part of the boron and forms volatile of boria (B_2O_3).

The present work deals with the synthesis of a new single source polymeric precursor from boric acid, vinyl silylesters and furfuraldehyde. Furfuraldehyde was incorporated into the polymeric chain so that during pyrolysis carbon can be *in situ* generated at relatively low temperatures which combines with oxygen and form corresponding oxides leaving boron free for ceramic formation.

Experimental

Boron and silicon containing precursor polymer was synthesized by reacting a mixture of boric acid, vinyl triethoxysilane and furfural (in 1:2:1 mole ratio) in presence of 1 ml of concentrated hydrochloric acid and 50 ml diglyme at a temperature of 150°C. The reaction was carried out in an inert atmosphere of nitrogen for 3 hours. Ethanol and water formed as the byproduct were distilled off. Diglyme was distilled off under reduced pressure and the product was dried under vacuum at 80°C for 10 hours. The polymer was obtained as a dark brown solid in excess of 80% yield. The product was found to be sparingly soluble in common organic solvent.

Fourier Transfrom Infra Red (FTIR) spectra were recorded using a Shimadzu FTIR spectrophotometer. Thermo Gravimetric Analysis (TGA) was carried out on a Perkin Elmer STA 6000 thermal analyser at a heating rate of 10°C/min under nitrogen atmosphere. Scanning Electron Micrograph (SEM) were recorded with a Hitachi Model DS 200 machine at variable pressure using an accelerating voltage of 15KV.

Results and Discussion

Organic polymers are one of the primary sources of carbon and furan resins have been used as precursors for new carbon materials [7-8]. Furfural is known to undergo condensation polymerization to form furan resins. Making use of this, a boron and silicon containing resin was synthesized by the condensation polymerization of furfuraldehyde, boric acid and vinyltriethoxysilane in presence of catalytical amount of hydrochloric acid. The product formed is referred as BSCP

The formation of precursor containing B-Si-C units was ascertained by Fourier Transform Infrared spectral studies. The infrared spectrum of BSCP (Figure1) shows characteristic absorption peak due to -Si-O-B- at 771 and 678 cm^{-1} due to stretching and deformation vibrations respectively [9]. The peaks in the region 1120cm^{-1} is attributed to the –Si-O-Si- linkages. The –Si-CH=CH_2 was seen at 1423 cm^{-1}. The absence of peaks at 971 and 881 cm^{-1} due to the in plane and out of plane deformation of furan ring indicate disappearance of the active hydrogen atom and hydroxyl groups of the furan ring and thus confirms the dehydration condensation reaction. The characteristic peak found at 1722cm^{-1} is due to the furan ring.

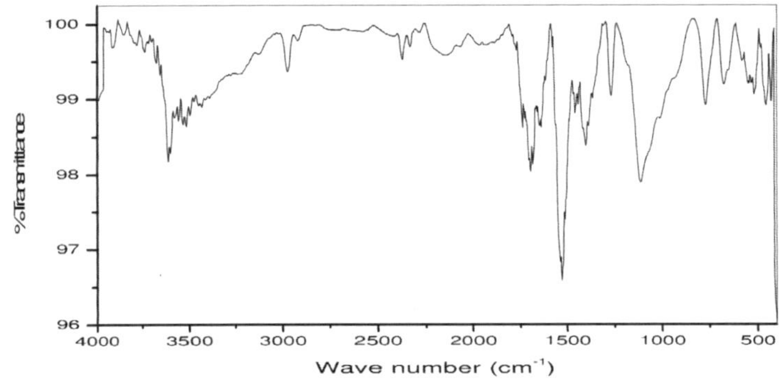

Figure 1. FTIR spectrum of BSCP

Thermo gravimetric analysis of BSCP indicates that polymer is quite stable and gave a ceramic residue of 66% at 900°C (Figure 2).

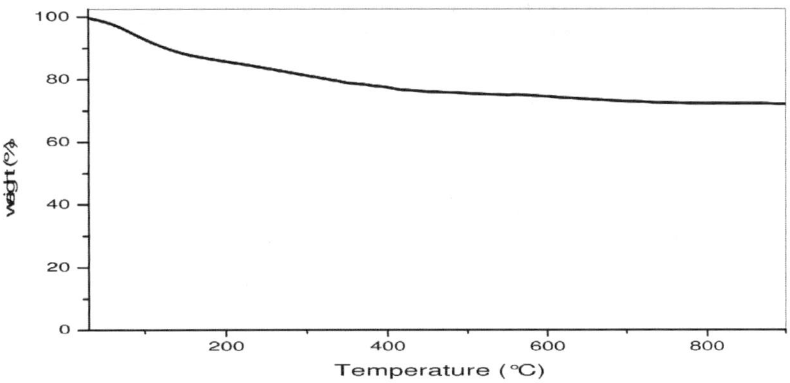

Figure 2.TG curve of BSCP

The high ceramic residue is due to the presence of borosiloxane (-B-Si-C-O-) bridges which prevents the formation of cyclic or cage siloxane formation during pyrolysis. The broad differential thermal analysis peak extending from 180 to 340°C reveals that the curing reaction involving the expulsion of volatiles such as CO, CO_2, CH_4 and H_2O was very slow and the curing time is very long. During this phase the –CH bond of furfual ruptures and the -CH=CH_2 bonds of the ester opens up to form the crosslinking which results in the formation of a three dimensional structure which ultimately results in high ceramic yield.

The oxidation resistance of BSCP was studied by heating the sample to 1350°C in presence of air and the heat treated sample is shown in Figure 3. The surface morphology of the heat treated sample was observed by scanning electron microscopy (SEM)(figure 4). The SEM analysis of the surface shows a uniform glassy layer of borosilicate glass on the surface. The formation of amorphous -BSiOC- glass was confirmed by IR spectroscopy.

Figure 3. BSCP heat treated at 1350°C in air

The glassy layer on the surface was found to protect the inner surface and the high oxidation resistance exhibited by the sample is possibly due to low diffusion rate of oxygen through the –BSiOC- layer. The high oxidation resistance of BSCP ceramic suggests that it is a promising material for many high temperature applications especially for environmental barrier coating and as matrix for composites.

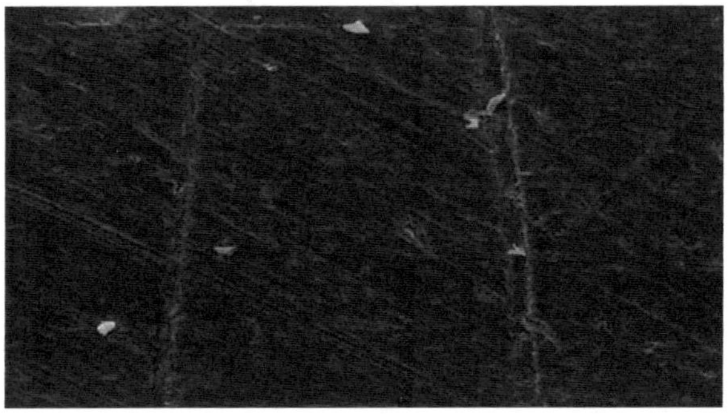

Figure 4. SEM of the surface formed on BSCP on heat treatment at 1350°C in air

Conclusion

A high ceramic yielding precursor polymer was synthesized from boric acid, vinyl triethoxysilane and furfuraldehyde. The precursor exhibited good thermal stability and gave high ceramic residue on pyrolysis due to the presence of highly branched units. The polymer showed superior oxidation resistance when pyrolysed at 1350°C in air which has been attributed to the low infiltration of oxygen through the glassy layer on the surface.

Acknowledgment

The author wish to thank Dr. Hareesh U.S of NIIST, Thiruvananthapuram for carrying out the high temperature heat treatment of the polymeric samples.

References

[1] Bill J and Aldinger F 1995 *Adv. Mater.* **7** 775

[2] Riedel R, Bill J and Kienzle A 1996 *Appl. Organomet. Chem.* **10** 241

[3] Riedel R, Kienzle A, Dressler W, Ruwisch L, Bill J and Aldinger F, 1996 *Nature* **382** A 796

[4] Keller T. M 2002 *Carbon* **40** (30) 225

[5] Packrisamy S 1996 *Prog. Polym. Sci.* **21** 707

[6] Ambadas G, Packrisamy S and Ninan K. N 2002 *J. Mater. Sci. Lett.* **21** 1003

[7] Sevilla M and Fuertes A 2009 *Carbon* **47** 2281

[8] Langang X, Bo Z. H , Xiang X, Lu J and Jian X 2008 *J. Cent. South Univ. Technol.* **15** 753

[9] Kasog A, Misono t and Abe Y 1994 *J. Polym. Sci. Part A Polym. Chem.* **32** 1049

Study of Structural Phase Transitions in $Na_{1-x}Sr_{x/2}NbO_3$

Rajni Bahuguna[14], M.G. Wankhede[2] ,S.K. Mishra[3], A.B. Shinde[3], P.S.R. Krishna[3]

[1]Applied Physics Dept., Thakur College of Engg. and Tech., Mumbai 400101, India
[2]Department of Physics, Yadavrao Tasgaonkar Institute of Engg. & Technology, Karjat, Maharashtra, India
[3]Solid State Physics Division, Bhabha Atomic Research Centre, Trombay, Mumbai 400 085, India

[4] Corresponding author. e-mail: rajni_7000@yahoo.co.in,

Abstract: The solid solution $Na_{1-x}Sr_{x/2}NbO_3$ is prepared by solid state reaction method. Detailed analyses of powder x-ray diffraction data clearly suggest a change of structure from ABO_3 perovskite ($NaNbO_3$ like) to complicated Tungsten Bronze $SrNb_2O_6$ like. The presence of additional reflections clearly suggests that cell multiplicity for $x \geq 0.20$ is different to that of pure $NaNbO_3$. For composition $x \geq 0.20$, the lattice parameters are related with pseudocubic perovskite cell parameters as follows: $A_o = 4a_p$, $B_o = 3b_p$ $C_o = 5c_p$.The lattice parameters and volume increases monotonically with increasing concentration of Sr^{2+} in $NaNbO_3$ matrix.

1. Introduction:

Perovskite oxide family has long been investigated especially in view of the ferroelectric properties that they commonly exhibit. Ferroelectric properties of oxide perovskite have potential applications in piezoelectric transducer and actuators, non-volatile memories etc. These materials also find important application as dielectrics for wireless communication, microelectronics, non-linear optical applications, medical imaging. $NaNbO_3$ is well documented as antiferroelectric at room temperature and have orthorhombic symmetry. But, the presence of substitutional impurities like Li^+, K^+ at the Na^+ site or Mn at the Nb site in $NaNbO_3$, stabilize long range ferroelectric ordering. These structural phase transitions are associated with condensation of zone centre and zone boundary phonon and driven force is strain. The nature of these transitions depends largely on the concentration of impurity. For example, polar phase can be stabilized at room temperature in the $Na_{1-x}K_xNbO_3$ (KNN) and $Na_{1-x}Li_xNbO_3$ (LNN) systems, for x=0.50 and 0.12 respectively. It is also interesting to note that both the systems to exhibit morphotropic phase boundary (MPB) near to these compositions and one of the end members is ferroelectric in nature. In addition to this, solid solutions of alkaline niobate are of great current interest due to their piezoelectric properties and their environmentally friendly nature makes them superior to lead based materials[1,2]. It is believed that substitutional impurities to take off-centre position of cubooctahedron formed by the 12 nearest neighbour oxygen atoms due to mismatch in the radius of ionic radii. In this communication, we present the results of Sr^{2+} substitution at the

Na^{2+} site and find the possibilities of presence of morphotropic phase boundary in this system. It may be noted that similar to $KNbO_3$ and $LiNbO_3$, $SrNb_2O_6$ is a well known ferroelectric.

2. Experimental

Specimens of $Na_{1-x}Sr_{x/2}NbO_3$ (SNN) were prepared by solid state reaction method using powders of $NaCO_3$, $SrCO_3$ and Nb_2O_5 each of minimum assay of 99.0%. These powders were thoroughly mixed in stoichiometric amounts using mortar and pestle using acetone as the mixing medium. Calcination of the mixed powder was carried out at 1150 ^{0}C for 6 hours. The calcined powder was pressed into circular pellets of 10mm diameter using a uniaxial hydraulic press at an optimized load of 10 tons. Sintering of the green pellets was carried out at 1250 ^{0}C - 1300 ^{0}C, depending upon the composition, for 6 hours in air. For powder diffraction experiments, the sintered pellets were crushed to fine powder and subsequently annealed at 500°C to remove strains introduced, if any, during crushing. Rietveld analysis of the powder diffraction data was carried out using structure free model in the program FULLPROF.

3. Results and Discussion:

3.1 Powder diffraction pattern of SNN($0.0 \le x \le 0.5$)

Figure 1 depicts the evolution of the x-ray powder diffraction patterns for SNN in the composition range $0 \le x \le 0.5$ for a limited 2θ range of 20-60 degrees. The powder diffraction patterns in Figure 1 contain the main perovskite and super lattice reflections. These super lattice reflections assume Miller indices represented by odd integers while the main Bragg reflections are represented by even integered indices with respect to the doubled pseudo-cubic unit cell (i.e. $2a_p$, $2b_p$, $2c_p$, α_p, β_p, γ_p, where a_p, b_p, and c_p are the cell parameter of the elementary pseudo-cubic perovskite cell). All-odd integered (i.e., 'ooo' type indices) super lattice reflections correspond to the out of phase (- tilt) tilts of the adjacent octahedra whereas presence of super lattice reflections with two odd and one even integer indices (i.e., 'ooe' type) implies presence of in-phase (+ tilt) tilt of the adjacent octahedral. It is important to mention here that in-phase and anti-phase tilts of octahedral are known to be driven by M (q=0,0, ½,) and R(q= ½, ½, ½) point instabilities of the cubic Brillouin zone respectively. Pure $NaNbO_3$ (x=0) shown in Fig. 1, one observes odd-odd-odd and odd-odd-even type reflections but with additional fractional indices (11 5/2, near 2θ = 37degree) reflection. The fractional indices indicate the quadrupling of one of the elementary perovskite axes.

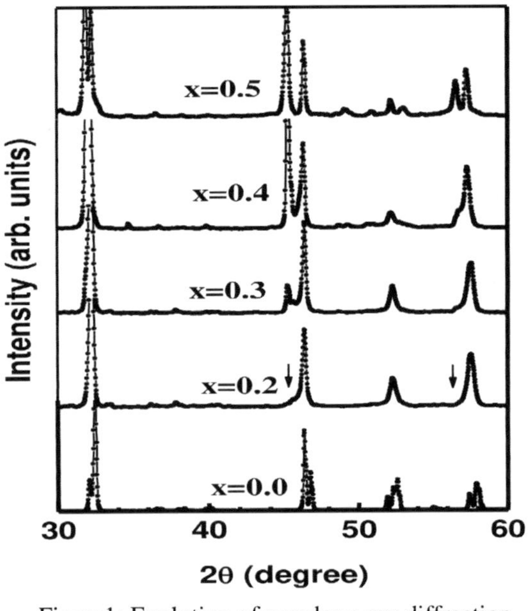

Figure1: Evolution of powder x-ray diffraction patterns of SNN.

3.2 Rietveld analyses:

Cochran and Zia [2] have shown that this quadrupling of the elementary pseudocubic cell in the room temperature phase of $NaNbO_3$ results from a freezing of the Δ point ($q=0,0,{}^1/_4$) soft phonon mode. All superlattice peaks are nicely accounted by orthorhombic structure with Pbcm space group during Rietveld data analysis. A compound tilt system,$[(a^-b^+a^-)_1{}^2(a^-b^-a^-)_2{}^3(a^-b^+a^-)_3{}^4]$, in the space group Pbcm has been found to be consistent with all these features for pure $NaNbO_3$[3,4] although the possibility for slight monoclinic distortion cannot be ruled out. It is evident from the Fig. 1, that there is a drastic change in the profile near 2θ ≈ 46, 52 and 57 degrees in the powder x-ray diffraction pattern as concentration of Sr^2+ increases in $NaNbO_3$ matrix. Appearance of additional peak (marked with arrow) at lower two-theta sides of around 46 and 57 degree Bragg profile and change in Bragg profile at two theta 52 degree are clearly observed in Figure 1. All these features can be explained in terms of a change of crystal structure of $Na_{1-x}Sr_{x/2}$ NbO_3 from ABO_3 perovskite like $NaNbO_3$ to complex Tungsten Bronze type $SrNb_2O_6$ like for x≥0.20 reveals a structural phase transition. The presence of these additional reflections clearly suggest that cell multiplicity for x≥0.20 is different than pure $NaNbO_3$. The lattice parameters are related with pseudocubic perovskite cell parameter as follows: $A_o= 4a_p$, $B_o= 3b_p$ $C_o= 5c_p$. This was further confirmed by structure free Rietveld refinement of SNN using the orthorhombic symmetry with space group P222. In the structure free Rietveld refinements, pseudo-Voigt function was chosen to define the peak profiles. Background was described in terms of a sixth order polynomial.

Figure 2 shows the observed, calculated and difference profiles for $Na_{1-x}Sr_{x/2}$ NbO_3 (x=0.20). As can be seen from this figure, the fit between observed and calculated profiles is quite satisfactory and include the weak super lattice reflections.

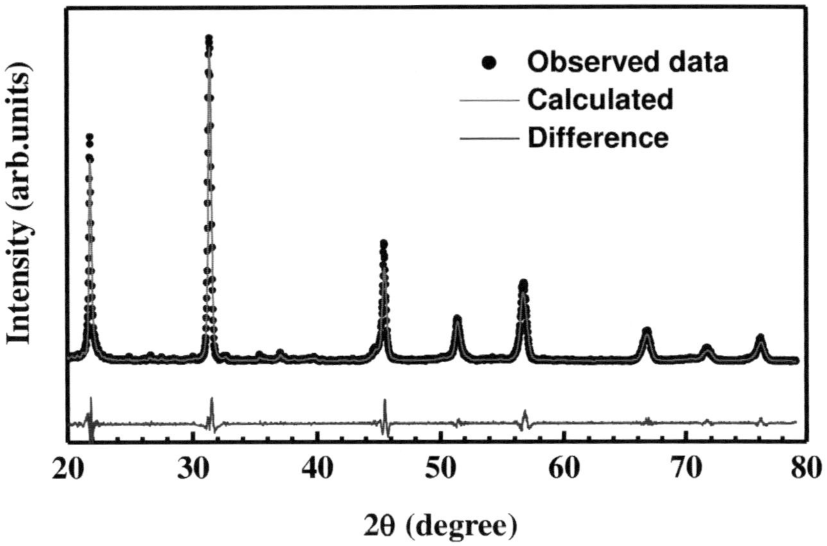

Figure 2: Observed (dots), calculated (continuous line), and difference (bottom line) profiles obtained after the structure free Rietveld refinement of $Na_{1-x}Sr_{x/2}NbO_3$ (x=0.20) using orthorhombic symmetry with space group P222.

3.3 Variation of lattice parameters with composition:

For the sake of easy comparison with corresponding cell parameters of orthorhombic phases, we have plotted the cell parameters of these phases in terms of equivalent elementary perovskite cell parameters. It is evident from Figure 3 that with increasing concentration, the lattice parameters and volume increase monotonically. It is expected that as the Sr^{2+} is substituted at the Na^{2+} site, the lattice expands as ionic radius of Sr^{2+} is larger than ionic radius of Na^{2+}. To explore the possibilities of presence of morphotropic phase boundary in this system, assignments of correct space group and study of the dielectric behaviour is under investigation. Figure 3. Depicts the variation of lattice parameters with the composition.

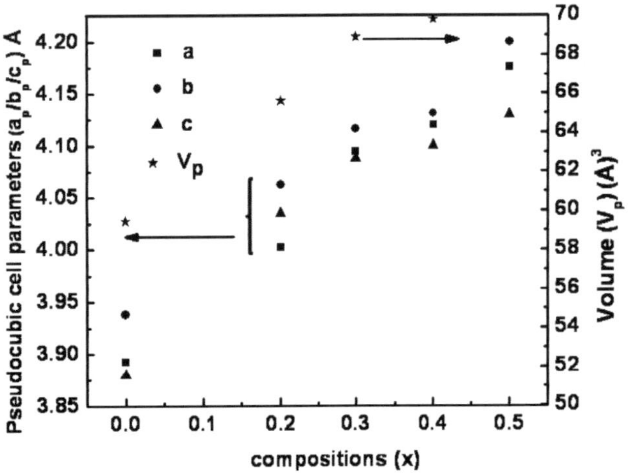

Figure 3: Evolution of lattice parameters of SNN with composition

4. Conclusions:

In conclusion, detailed analysis of powder x-ray diffraction data clearly shows a change of structure from $NaNbO_3$ like to $SrNb_2O_6$ like. The lattice parameters and volume increases monotonically with increasing concentration of Sr^{2+} in $NaNbO_3$ matrix.

5. References:

[1] Yasuyoshi Saito, Hisaaki Takao,Toshihiko Tani, Tatsuhiko Nonoyama, Kazumasa Takatori, Takahiko Homma, Toshiatsu Nagaya & Masaya Nakamura, 2004 Nature 432, 84.

[2] Yu.I. Yuyuk, E. Gagarina, P. Simon, L A Reznitchenko, L Hennet and D. Thiaudiere, Phys. Rev B 69, 144105 2004 Phys. Rev B 69, 144105.

[3] Cochran and Zia W. Cochran and .4. Zia, 1968 Phys. Status Solidi 25, 273.

[4] Darlington C N W, Knight K S 1999 Physica B 266 ,368.

6. Acknowledgements:

One of the author Dr.Rajni Bahuguna gratefully acknowledges Dr. J.V. Yakhmi and Dr. S.L. Chaplot for encouraging and supporting the project in BARC. Help of Mrs Rajul & Chitra is also acknowledged for XRD measurements and Dr Amitabh Das for fruitful discussions.

Effect of sintering temperature on the nature of weak links and flux pinning mechanism in $La_{1.85}Sr_{0.15}CuO_4$ superconductor

Devina Sharma[1,2,3], Ranjan Kumar[1] and V P S Awana[2]
[1]Department of Physics, Panjab University, Chandigarh, India - 160014
[2]Quantum Phenomena & Applications Division, National Physical Laboratory, New Delhi, India - 110012

E-mail: s.sharmadevina@gmail.com

Abstract. *AC* and *DC* susceptibility study is carried out to investigate the granular nature of $La_{1.85}Sr_{0.15}CuO_4$ superconductor. Presence of weak links and flux pinning phenomena are two important implications of granular type II high temperature superconductors. Weak links manifested as Superconductor-Insulator-Superconductor (SIS) / Superconductor-Normal metal-superconductor (SNS) Josephson Junctions are investigated from the temperature and field dependence study of *AC* susceptibility. On the other hand, *DC* susceptibility measurement is used to study the flux pinning mechanism. Mechanism of flux pinning is dependent on the nature and size of the pinning centres as well as on the microstructure wavelength. Thus, the nature of grain boundaries plays an important role in determining the nature of pinning mechanism of flux lines. In the present work, effect of sintering temperature on the nature of weak links and flux pinning mechanism in the bulk polycrystalline sample of $La_{1.85}Sr_{0.15}CuO_4$ superconductor is studied.

1. Introduction

The type II high temperature superconductors (HTS), specially the cuprate superconductors till date are the best potential candidates to be used for generating high magnetic fields, lossless electrical transmission, transportation based on magnetic levitation and so on. But, even after the completion of the twenty five years of their discovery, they have still not been able to put into use extensively. Presence of weak links and poor flux pinning in these materials are among the major grounds of their restricted practical applications.

Cuprate superconductors belong to the class of type II as well as HTS which have a peculiar property of being granular in nature. In granular superconductors, the superconducting grains are separated by non-superconducting grain boundaries which may be either conducting or insulating. Superconducting state in such granular materials is established by tunnelling of cooper pairs between the grains. The grain boundaries act as weak links between the grains, as they tend to decouple them in the presence of ample magnetic field. Therefore, the grain boundaries (weak links) act as Superconductor-Insulator-Superconductor (SIS) / Superconductor-Normal metal-Superconductor (SNS) tunnel or Josephson Junctions [1] assisting the flow of super currents.

[3] To whom any correspondence should be addressed.

In the presence of magnetic field, the flux lines tend to enter inside such superconducting materials in the form of vortices. The motion of vortices causes dissipation of energy which is undesirable for superconductor applications. Therefore pinning of the flux lines or vortices is of utmost importance in polycrystalline bulk type II HTS.

AC susceptibility (ACS) and *DC* susceptibility (DCS) [2] are non-destructive magnetic measurement techniques used for the characterisation and study of superconducting samples. In *AC* (*DC*) susceptibility measurement, sample is magnetised by varying (constant) magnetic field and the induced time dependent (independent) magnetic moment is measured. In the present work nature of weak links/grain coupling and flux pinning mechanism of $La_{1.85}Sr_{0.15}CuO_4$ superconductor is studied using ACS and DCS respectively.

2. Experimental

The superconducting polycrystalline samples of $La_{1.85}Sr_{0.15}CuO_4$ are prepared by sol gel method as described in reference [3]. Three samples are obtained by sintering the precursor powder at 900, 1000 and 1050°C respectively. Phase purity is checked from X-ray diffraction pattern [3] obtained using Rigaku X-ray diffractometer with Cu K_α (λ=1.54Å) at room temperature. Temperature dependent *AC* and *DC* susceptibility measurements are carried out in susceptometer employed in Physical Property Measurement System (PPMS). The amplitude of magnetic field is varied from 0 to 10Oe in the temperature range of 5 to 50K at 333 Hz for ACS study, while DCS measurement is carried out at 5K temperature in the field range of 0 to 30kOe.

3. Results and Discussion

DC susceptibility measurement is carried out to obtain magnetisation (M-H) curves at 5K as described in our earlier report [3]. Flux pinning density curves are obtained from these loops using,

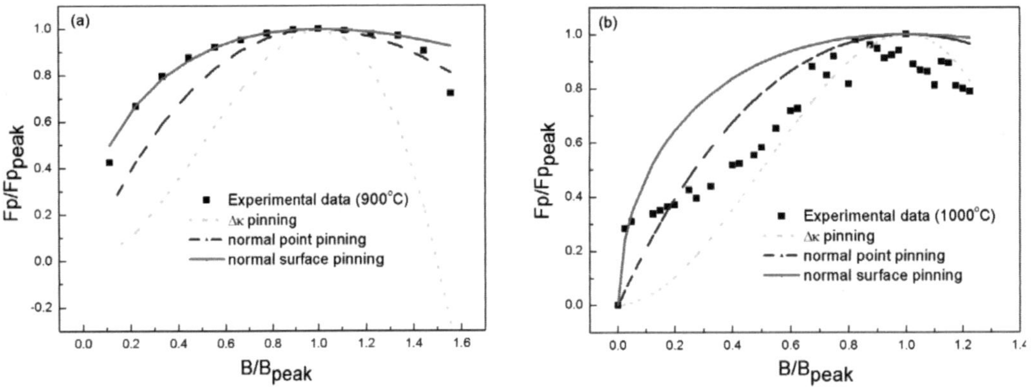

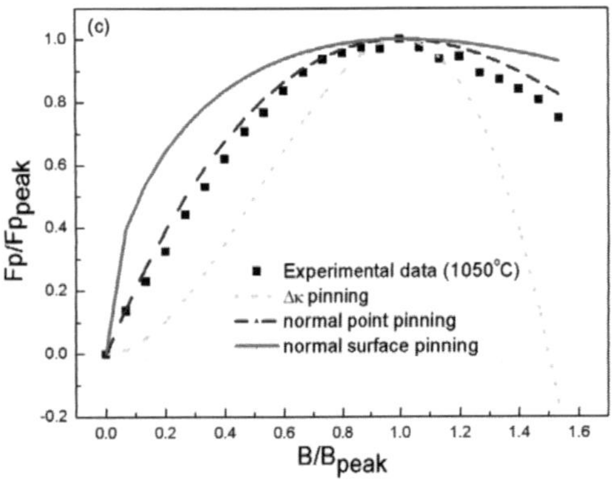

Figure 1. Normalized pinning force F_P/F_{Ppeak} as a function of B/B_{peak} for samples sintered at (a) 900, (b) 1000 and (c) 1050°C fitted to theoretical models given by equation (1), (2) and (3) by green (dot), blue (dash) and red (solid) lines respectively.

J_c(H)=[20×ΔM(H)]/[a(1-a/3b)] [3,4] and $F_P=J_c\times B$ [4]. Figure 1 shows fitting of the flux pinning curves using three theoretical models commonly used for HTS [5]:

$$F_p(B) = 3B^2(1 - 2B/3) \quad \text{; for } \Delta\kappa \text{ pinning} \tag{1}$$
$$F_p(B) = 9/4\,B(1 - B/3)^2 \quad \text{; for normal point pinning} \tag{2}$$
$$F_p(B) = 25/16B^{-1/2}(1 - B/5)^2 \quad \text{; for normal surface pinning} \tag{3}$$

It is clearly visible that sample sintered at 900°C exhibits normal surface flux pinning while that sintered at 1050°C shows normal point pinning. The sample sintered at 1000°C does not follow any single pinning mechanism. It can be concluded that the basic mechanism, i.e. normal pinning mechanism remains same for all the samples but the difference arises due to the change in the size of pinning centres [6] owing to different sintering temperatures.

Figure 2 shows the temperature dependence of imaginary part of AC susceptibility for the three samples measured at different field amplitudes with 333 Hz frequency. It is clearly visible that for sample sintered at 900°C, high inter granular coupling peak has suppressed the intra granular coupling peak. Also the separation of the intra and inter granular dissipation peak decreases with decrease in

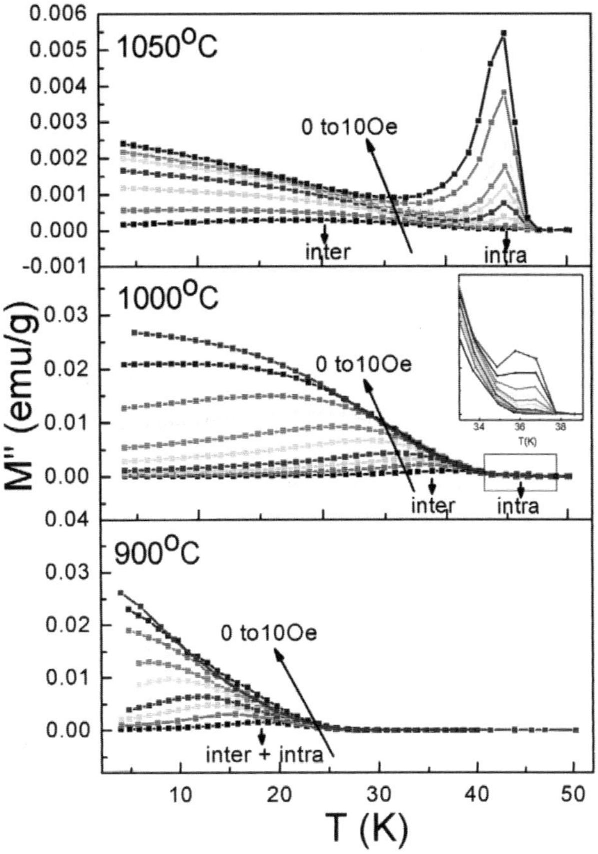

Figure 2. Temperature dependence of imaginary component of AC susceptibility for various samples in the applied field ranging from 0 to 10Oe at 333 Hz. The inset in the middle figure is the magnified view of the intra granular loss peak identified in the figure with rectangular box.

sintering temperature, implying improvement in grain coupling. Improvement in grain coupling with decrease in sintering temperature is also evident from the analysis of the field dependence of inter granular dissipation peak appearing at temperature T_p as shown in Figure 3. According to Herzog *et al* [7], using a critical state model the critical current density (J_c) for thin bar shaped sample at T_p is given by $H_p = J_c d/\pi[1+\ln 2a/d]$, where '*a*' and '*d*' are width and thickness of the sample respectively. Therefore, Figure 3 can be considered as the temperature dependence of J_c which clearly shows linear behaviour as per equation, $J_c = J_c(0)(1-T/T_c)^n$ where, T_c is the superconducting critical temperature. The value of "n" shows that the nature of weak links are of SNS type (n=2) [8] for sample sintered at 900°C and SIS (n=1) [9] for that at 1000°C. Sample sintered at 1050°C is not analysed, since the inter granular loss peak appears below measured temperature range.

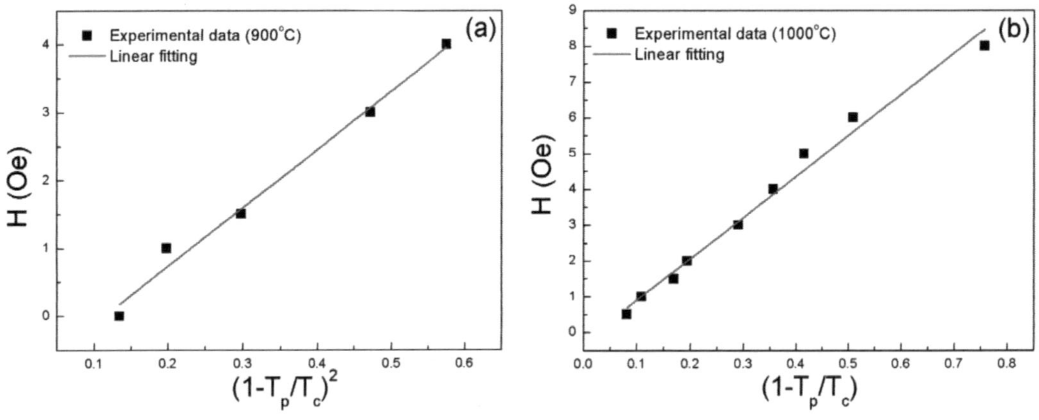

Figure 3. Variation of peak temperature with field as $(1-T_P/T_c)^n$ for samples sintered at (a) 900 and (b) 1000°C.

4. Conclusion

AC and *DC* susceptibility measurements have been carried out to study the effect of sintering temperature on the nature of weak links and flux pinning mechanism in $La_{1.85}Sr_{0.15}CuO_4$ superconductor. It is found that with increase in sintering temperature from 900 to 1050°C, coupling between the grains decreases with their nature changing from SNS to SIS type Josephson Junctions. Also, the flux pinning mechanism changes with sintering temperature due to the change in the size of the pinning centres, though basic mechanism remains same.

References

[1] Josephson B D 1962 *Phys. Lett.* **1** 251
[2] Goldfarb R B, Lelental M and Thomson C A 1991 *Magnetic Susceptibility of Superconductors and Other Spin Systems* ed Hein R A, Francavilla T L and Liebenberg D H (New York: Plenum Press)
[3] Sharma D, Kumar R and Awana V P S 2011 *J. Supercond. Nov. Magn.* **24** 205
[4] Krabbes G, Fuchs G, Canders W, May W and Palka R 2006 *High Temperature Superconductor Bulk Materials* (Weinheim: Wiley-VCH)
[5] Ding Y, Sun Y, Zhuang J C, Cui L J, Shi Z X, Sumption M D, Majoros M, Susner M A, Kovacs C J, Li G Z *et al* 2011 *Supercond. Sci. Technol.* **24** 125012
[6] Huges D 1974 *Philos. Mag.* **30** 293
[7] Herzog T, Radovan H A and Ziemann P 1997 *Phys. Rev.* B **56** 2871
[8] Degennes P G 1964 *Rev. Mod. Phys.* **36** 225
[9] Ambegaokar V and Baratoff A 1963 *Phys. Rev. Lett.* **10** 486

International Conference on Materials Science and Technology (ICMST 2012) IOP Publishing
IOP Conf. Series: Materials Science and Engineering **73** (2015) 012031 doi:10.1088/1757-899X/73/1/012031

Study on the Properties of Fe-substituted $NdMn_2O_5$

K. Saravana Kumar and C. Venkateswaran

Department of Nuclear Physics, University of Madras, Guindy Campus, Chennai 600 025. India

E-mail: cvunom@hotmail.com

Abstract. The oxide, $NdFeMnO_5$, has been synthesized using a two-step synthesis process. First the precursor oxides are high-energy ball milled and the as-milled powders are then sintered to obtain the $NdFeMnO_5$ phase. X-ray diffraction confirms phase formation and the calculated pattern is obtained by the substitution of Mn^{3+} sites by Fe^{3+} sites. Agglomeration of small sized grains is observed from the scanning electron microscopy images. Room temperature magnetization measurement shows a paramagnetic behaviour. A ferrimagnetic ordering is observed to be established below ~110 K from the thermo-magnetization measurements.

1. Introduction

Materials of the RMn_2O_5 (R=Rare earths, Bi) family has Mn at two different crystallographic sites [1]. Mn^{4+} ions occupy the octahedral pyramidal ($4f$) sites where it is coordinated with six oxygen atoms and Mn^{3+} ions at tetragonal pyramidal ($4h$) sites having coordination with five oxygen atoms [1]. They have complex magnetic structures which give rise to magnetodielectric effect around the magnetic transition temperatures [2]. Generally, RMn_2O_5 systems have the orthorhombic structure with Pbam spacegroup and paramagnetic behaviour at room temperature. They have Neel transition temperature around 40 K below which antiferromagnetic ordering exists.

The magnetic ordering in these systems can be investigated by the substitution of Mn^{3+} ions with Fe^{3+} ions [3,4]. The magnetic properties of Fe-substituted RMn_2O_5 in few cases are reported to be ferrimagnetic in nature at low temperatures [5–7]. The Fe substitution has its effect on the structural and magnetic properties due to its high spin only effective magnetic moment. We report the synthesis and properties of Fe-substituted $NdMn_2O_5$. It is considered that Fe^{3+} occupies the Mn^{3+} ($4h$) sites, and the corresponding change in properties observed is compared with the already reported properties of the parent system.

2. Experimental

Powders of neodymium oxide, manganese mono-oxide and iron oxide were taken in stoichiometric proportions and milled for a total of 5 h at 400 RPM with a ball to powder ratio of 28:2. The obtained reddish brown powders were then pressed into dense pellets of 12 mm diameter, sintered at 1073 K for 24 h and furnace cooled. Sintered samples were then ground in a mortar to obtain the black powders of $NdFeMnO_5$. The X-Ray Diffraction (XRD) data was obtained with Cu-Kα radiation at room temperature. High resolution-scanning electron microscopy (HR-SEM) was done in FEI Quanta FEG 200 HR Microscope. Isothermal magnetization and Zero field cooled (ZFC)-Field Cooled (FC) measurements were carried out in the range 20 K-300 K in a

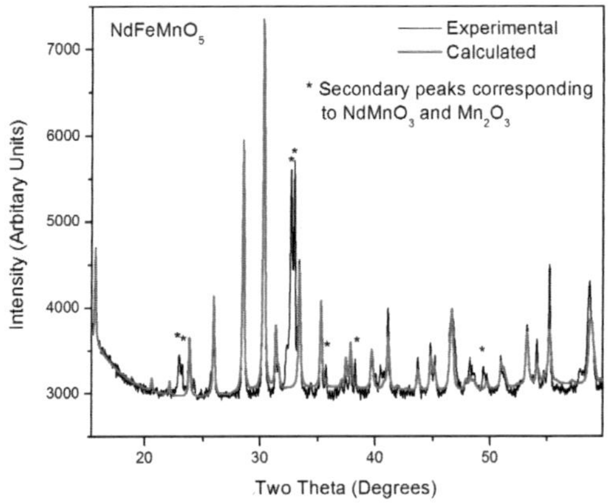

Figure 1. XRD plot with the experimental and generated patterns. * indicates the secondary peaks.

Lakeshore VSM 7410 instrument. A field of 1000 Oe was applied for the ZFC-FC experiment and for iso-thermal magnetization curves a maximum field of 20 KOe was applied.

3. Results and discussion

X-ray diffraction (XRD) pattern is given in fig 1, indicating the formation of stoichiometric $NdFeMnO_5$ which is iso-structural with $NdMn_2O_5$ along with the few secondary phases. The calculated pattern, matching with the experimental pattern, is obtained by using the FULLPROF [8] program by the substitution of Fe^{3+} in Mn^{3+} $(4f)$ sites. From the structural point of view, $NdFeMnO_5$ has the same orthorhombic ($Pbam$) structure as that of its parent $NdMn_2O_5$. Parent system has the Mn^{4+} octahedra network connected through oxygen atoms, extending along $c-$axis [9], and has Mn^{3+} and Nd^{3+} layers between these Mn^{4+} layers. In the case of $NdFeMnO_5$, these Mn^{3+} layers are completely replaced by Fe^{3+} layers connecting Mn^{3+} ions through oxygen atoms. The arrangement of Nd, Mn and Fe atoms in a unit cell of $NdFeMnO_5$ is illustrated in fig 2, along the $c-$axis (a) and $a-$axis (b). HR-SEM image is shown in fig 3, illustrating the random distribution of particles with agglomeration and the EDAX

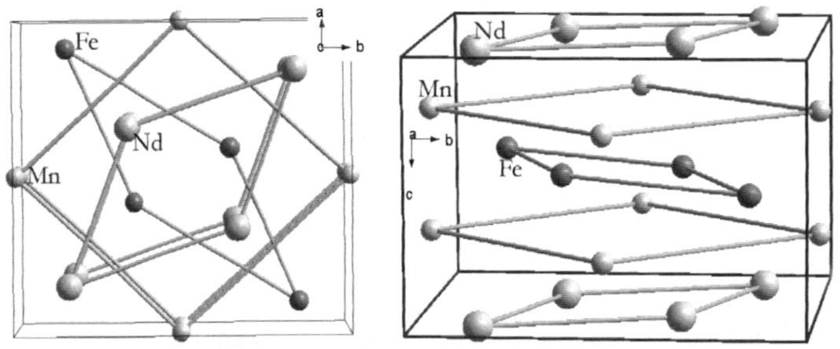

Figure 2. View of Nd, Mn & Fe layers (a) along $c-$axis (left) (b) along $a-$axis(right)

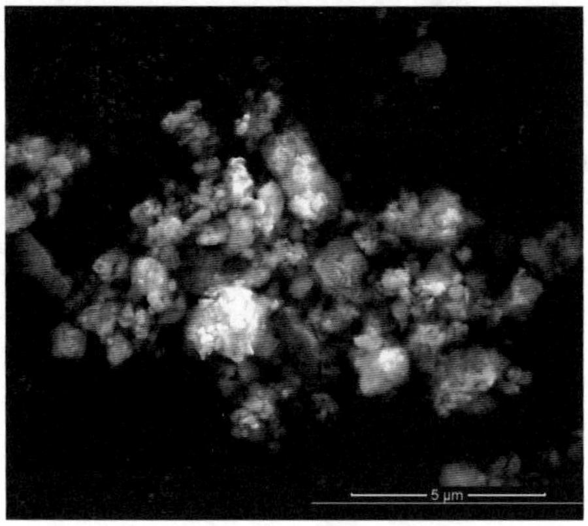

Figure 3. HR-SEM image of $NdFeMnO_5$

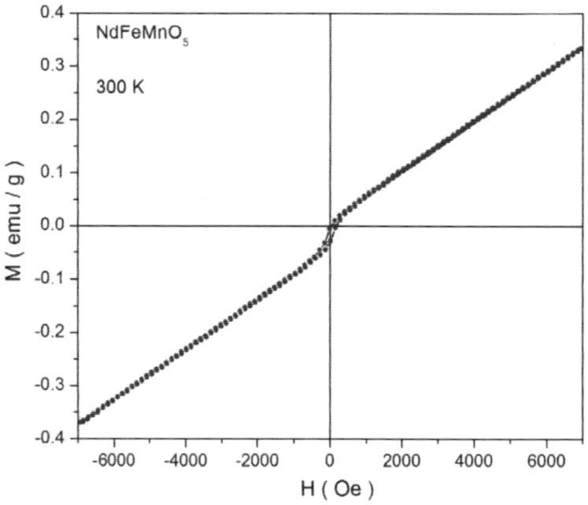

Figure 4. Room temperature magnetization curve

measurements confirm the near stoichiometric presence of all elements in $NdFeMnO_5$.

Room temperature magnetization study, given in fig 4, indicates a paramagnetic behaviour and a weak ferromagnetic property is also observed at initial low fields. The low temperature Zero field cooled-field cooled (ZFC-FC) curves are shown in fig 5, which exhibits a split below ~110 K and an increase in magnetization values till 20 K. This behaviour is indicative of the ferrimagnetic property below ~110 K and is supported by the linear dependence of magnetization on the applied magnetic field obtained at 20 K (inset of fig 5). The parent $NdMn_2O_5$, has a Neel transition (T_N) temperature of ~75K, below which it is reported to have

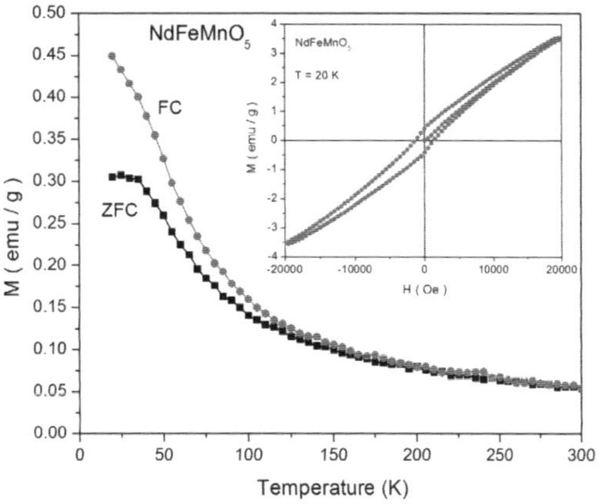

Figure 5. Low temperature ZFC-FC curves showing a transition below ~110 K. [Inset: Magnetization curve at 20 K]

antiferromagnetic ordering [10]. The magnetic transition is due to the ordering of magnetic moments in Nd^{3+}, Mn^{4+} and Mn^{3+} layers. Global magnetic behaviour in RMn_2O_5 is based on the type of magnetic ordering in Mn^{4+}-O-Mn^{4+} and Mn^{4+}-O-Mn^{3+}-O-Mn^{4+} chains determined through superexchange interaction. The spin only effective magnetic moments of Mn^{4+}, Mn^{3+} and Fe^{3+} are 3.87 μ_B, 4.89 μ_B and 5.92 μ_B, respectively [11]. The magnetic ordering arises due to the alignment of magnetic Nd^{3+}, Fe^{3+} and Mn^{4+} moments in the unit cell. Due to Fe^{3+} substitution at Mn^{3+} sites, the magnetic ordering of the chain Mn^{4+}-O-Fe^{3+}-O-Mn^{4+} through the superexchange interaction is affected as the magnetic moment of Fe^{3+} is higher than Mn^{3+}. The interaction in Mn^{4+}-O-Mn^{4+} chains through Nd^{3+} may induce a weak ferromagnetic interaction below the ordering temperature, contributing to the ferrimagnetic ordering observed at low temperatures.

4. Conclusion

The compound, $NdFeMnO_5$, has been synthesized by initial high-energy ball milling of oxides and then sintering of milled powders. Fe^{3+} substitution in the $NdMn_2O_5$ at Mn^{3+} affects the magnetic ordering at low temperatures due to which a ferrimagnetic ordering occurs below $\sim$110 K. This is the result of change in interaction in Mn^{4+}-O-Mn^{4+} chains due to Fe substitution.

5. Acknowledgments

Authors acknowledge SAIF, IIT Madras for low temperature VSM measurements. KSK thanks DRDO, India for support through fellowship.

References

[1] Munoz A, Alonso J A, Casais M T, Martinez-Lope M J, Martinez J L and Fernandez-Diaz M T 2002 *Phys. Rev. B* **65**(14) 144423
[2] Retuerto M, Martnez-Lope M, Muoz A, Ruskov T, Spirov I, Krezhov K, Fernndez-Daz M, Garca-Hernndez M and Alonso J 2010 *Solid State Communications* **150** 1831 – 1836
[3] Retuerto M, Martinez-lope M, Munoz A, Ruskov T, Spirov I, Krezhov K, Fernndez-Daz M, Garca-Hernndez M and Alonso J 2010 *Solid State Communications* **150** 1831–1836
[4] Kumar K S and Venkateswaran C 2011 *Journal of Physics D: Applied Physics* **44** 325001
[5] Muoz A, Alonso J A, Martnez-Lope M J and Martnez J L 2004 *Chemistry of Materials* **16** 4087–4094
[6] Munoz A, Alonso J A, Martinez-Lope M J and Martinez J L 2005 *Phys. Rev. B* **72**(18) 184402
[7] Martnez-Lope M, Retuerto M, Alonso J and Pomjakushin V 2008 *Journal of Solid State Chemistry* **181** 2155 – 2160
[8] Rodriguez-Carvajal J 2001 *Commission on Powder Diffraction (IUCr) Newsletter* **26** 12–19
[9] Alonso J, Casais M, Martnez-Lope M and Rasines I 1997 *Journal of Solid State Chemistry* **129** 105 – 112
[10] Chen Y, Yuan H, Tian G, Zhang G and Feng S 2007 *Journal of Solid State Chemistry* **180** 1340 – 1346
[11] Martnez-Lope M, Retuerto M, Alonso J, Garca-Hernndez M, Krezhov K, Spirov I, Ruskov T and Fernndez-Daz M 2009 *Solid State Communications* **149** 540 – 545

IOP Conf. Series: Materials Science and Engineering **73** (2015) 012032 doi:10.1088/1757-899X/73/1/012032

Mössbauer studies of nanocrystalline $ZnFe_2O_4$ particles prepared by spray pyrolysis method.

Jessyamma Kurian[1], Subin P john [2], Jacob Mathew M[2,*], V R Reddy[3], Abraham K E[2] and V S Prasad[4]

1 Department of Physics, BAM College, Thuruthicad ,Thiruvalla, Kerala, India
2 Department of Physics, Mossbauer Research Group, St. Berchmans College, Changanassery, Kerala, India
3 UGC- DAE Consortium for Scientific Research, Khandwa Road, Indore India
4 Chemical Science Division, National Institute for Interdisciplinary Science and Technology, Thiruvanathapuram, Kerala, India.

* Corresponding author E-mail: jacob.chrisdale@gmail.com Phone +919446385365

Abstract. This paper reports the synthesis and characterization of nanoparticles of Zn Fe_2O_4 by spray pyrolysis method. XRD studies indicates that the particle formed are ultra fine (20 nm) associated with strain. The particle size is confirmed by TEM analysis. The room temperature Mossbauer spectrum shows only a doublet which could be indicative of either paramaganetic or superparamagnetic phase. The value of isomer shift is in the range 0.34±0.01 mm/s while quadruple splitting is in the range 0.48±0.01 mm/s. A series of low temperature studies has been performed and the spectra obtained at 20 K and 5K show the presence of two well resolved sextets, which are attributed to the Fe^{3+} ion in the tetrahedral and octahedral sites of the spinel structure. The application of external magnetic field yields a better fitting resolution of the sub spectra between A and B sites. The temperature dependant and in-field measurements yielded valuable insights about magnetic phase (superparamagnetic), local cation distribution among A sites and B sites, inversion parameter (δ =0.74), canted spin alignment (ψ_A = 29.58, ψ_B= 48.46) and the blocking temperature (T_B= 50 K).

1. Introduction

Magnetic nanoparticle synthesis has received wide research interest due to its unusual magnetic properties compared to other spinel ferrites. Zinc ferrite with nano size shows cation inversion which in turn affects its magnetic characteristics, like super paramagnetic relaxation time, blocking temperature. The inversion parameter δ is used to indicate the degree of inversion in the sample [1-3]. The cation inversion among crystallographically identical sites present in normal spinel ferrites were found to be affected by preparation methods and particle size distribution. The purpose of the present work is to estimate particle sizes and magnetic characteristics of zinc nanoferrites, synthesized by spray pyrolysis of polymer precursor–a simple, rapid and low temperature method and but rarely reported [4-12].

2. Experimental

Synthesis of the zinc ferrite nanoparticles was done by spray pyrolysis of polymer precursor prepared by taking high purity zinc nitrate (Merck), ferric nitrate (Loba) and Polyvinyl Alcohol (PVA) as starting materials. The presence of PVA helps the homogenous distribution of the metal ions in their polymeric network structure and inhibits their segregation/precipitation from the solution. The resulting viscous mixture of PVA metal nitrates is spray dried around 130°C to obtain a brown fluffy powder. The powders were subsequently heat treated at 450°C for 20 minutes for synthesizing desired metal oxides. Five such samples were prepared (Z1, Z2, Z3, Z4, and Z5) with varying molar ratios. The crystalline phase of nanoferrites was analyzed by XRD (Cu K_α radiation λ =1.5406Å) and transmission electron microscope (TEM) operating at 300kV. ^{57}Fe Mossbauer measurements with ^{57}Co radioactive source in constant acceleration mode were done for identifying the magnetic phases in the samples.

3. Results and discussion

3.1 XRD and TEM Studies

The XRD patterns of five molar fractions show well-resolved peaks, which confirm polycrystalline and monophasic nature of the prepared material. The peaks corresponding to planes (220), (311), (222), (400), (422), (511), (440) provide a clear picture for the formation of spinel structure of the ferrite [13]. However, the pyrolytic decomposition process of ferrite powders is a chemically complicated process in which many intermediate products can be formed. The phases that may be present in the sample are ZnO, Fe_2O_3, $Fe_{1-x}O$ and stochiometric $ZnFe_2O_4$ [14]. The XRD pattern of the fifth fraction (Z5) seems to be free from impure phases and hence Z5 has been chosen for further studies. The lattice parameter 'a' of Z5 determined as 0.8447nm matches well with JCPDS (00-022-1012) file for $ZnFe_2O_4$. The grain size of the zinc nanoferrite is determined using Scherrer's equation and is around 20nm. The full width at half maximum (FWHM) value of the peak corresponding to (311) plane is chosen for calculating the particle size. TEM image given in figure 1(a), reveals a narrow particle size distribution and support the average particle size estimated by XRD. Furthermore, the selected area electron diffraction pattern (figure 1(b)) of the same sample shows strong spotty ring patterns, indicating a highly crystalline spinel structure.

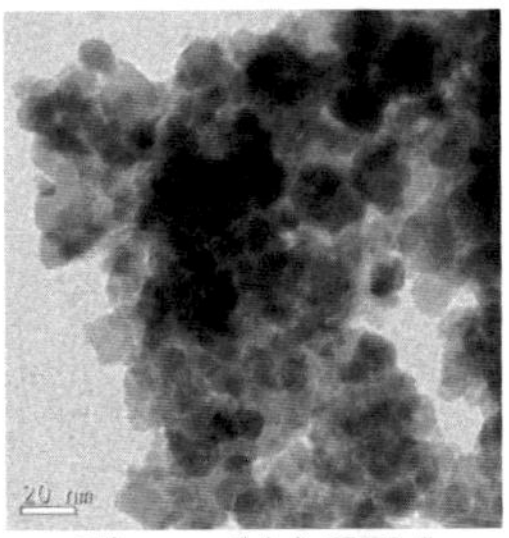

Figure 1(a) TEM picture of Z5

Figure 1(b) SAED pattern of Z5

3.2 Mossbauer study

The temperature depended Mossbauer spectrum (MS) of the 20nm sample has been studied and the nuclear quadruple splitting of ^{57}Fe and there are no magnetic hyperfine interactions. The value of isomer shift is in the range 0.34 ± 0.01mm/s while quadrupole splitting (QS) is in the range 0.48 ± 0.01 mm/s[15]. The QS value indicates a strong electric field gradient around ^{57}Fe nuclei and may be attributed to large surface to volume ratio in nanoparticles, though the bulk material has cubic symmetry [16]. When the temperature is reduced to 90K, a weak sextet pattern appears in addition to the reduction of the intensity of the doublet pattern. The sextet component increases as the spectra obtained are shown in figure 2. The room temperature spectrum shows only a doublet pattern due to temperature decreases to 50K. At 20K the doublet component disappears and only the sextet pattern remains .This typical behaviour of Mossbauer spectra can be attributed to superparamagnetism(SPM)- a unique character of magnetic nanoparticles. MS is a powerful tool to characterize magnetic nanoparticles undergoing SPM relaxation since its experimental measurement time(τ_L) is around relaxation time (τ), determined by the equation

$$\tau=\tau_0\exp(KV/k_BT) \quad (1)$$

where τ_0 (10^{-9} to 10^{-10}s)is the inverse of natural frequency of gyromagnetic precession, K is anisotropic energy constant, V is the volume of the particle, k_B is the Boltzman constant, T is the Kelvin temperature. When the relaxation time τ is smaller than τ_L the fluctuation of the particle magnetization direction is so rapid that the average of internal magnetic hyperfine field is zero at ^{57}Fe nucleus site. Consequently there is only a quadrupole doublet splitting in the spectrum. But, when τ approaches τ_L, a sextet pattern can emerge. However, some ^{57}Fe nuclei still experience a rapid superparamagnetic relaxation and the doublet pattern remains visible. When the temperature is low enough, the relaxation in the nanoparticles are much slower and the spectrum becomes a pure sextet pattern. The characteristic temperature indicating the onset of the blocking process is called blocking temperature (T_B). Its value falls between 50K and 20K & which is not only much higher than that reported for nanoparticles but also the transition occurs over a much wider temperature range. The origin for the large increase in T_B for zinc nanoferrite(ZNF) particles is most likely attributable to redistribution of Fe and Zn cations on A and B sites, a fact which is confirmed by δ –the inversion parameter[17- 20]. The assignment of cation sites and determination of area ratio can be obtained from spectra recorded at 5K with and without field. The data obtained for the cation distribution between the A and B sites are given in Table 1 and those data have been used for extracting the value of δ- an index for cation distribution among A and B sites - using the equation

$$I_{(A)}/I_{(B)}= (f_A/f_B)[\delta/(2- \delta)] \quad (2)$$

where $I_{(A)}$ and $I_{(B)}$ are the sub spectral intensities of A site and B site and the ratio of recoil free fractions of A and B sites ,$f_{(B)}/f_{(A)}$ =1 at 5K. The calculated value of degree of inversion is 0.74.

Based on the δ value, the cation distribution in the spray pyrolytically synthesized zinc nanoparticles are $\{Zn_{0.74}Fe_{0.26}\}$ & $[Zn_{0.26}\,Fe_{1.74}]\,O_4$. The δ value of ZNF supports the redistribution of zinc cation to the octahedral sites. [21-24].

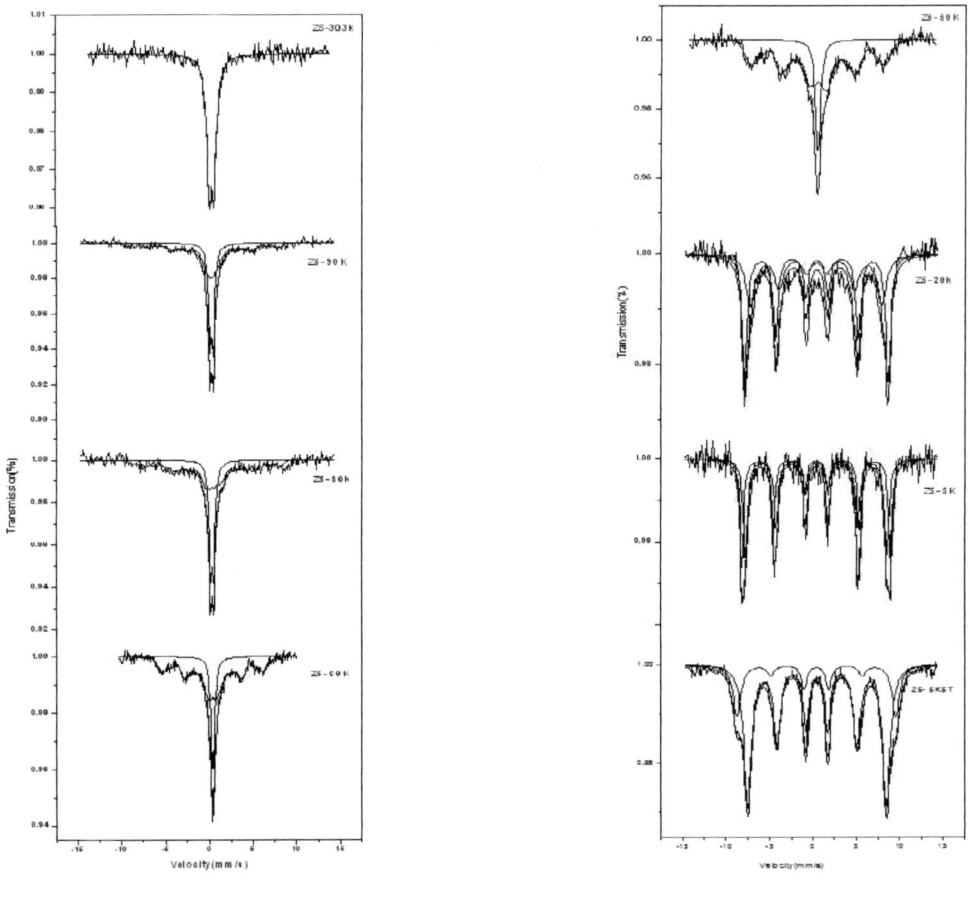

Figure 2. Mossbauer spectra obtained at different temperatures for $ZnFe_2O_4$

Mossbauer spectroscopy is a powerful tool for extracting the orientation of the internal magnetic field. The average canting angle between the external field and the A or B site F^{3+} hyperfine field directions were calculated from the ratio of the intensities of lines 2 and 3 from each sub spectra, I_2/I_3, according to Tuek et al., [25]

$$\psi = \arccos\left[\left(4 - I_2/I_3\right)/\left(4 + I_2/I_3\right)\right]^{1/2} \qquad (3)$$

The value of $<\psi>$ is listed in Table 1. The values are different from the reported values [26] and the variations may be attributed to the dependence of magnetic properties on method of preparation and the particle sizes.

Table 1. Fitted parameters of Mossbauer spectra at 5Kwith and without field H=5T for Z5 sample.

		H=0		H=5T		
Sample	site	Rl.Area %	H_f(T)	H_f(T)	I_2/I_3	$<\psi>$
Z5	A	37	52.8656	56.56	0.55487	29.58
	B	63	50.3024	49.39	1.55602	48.46

4. Conclusion

Spray pyrolysis method has been used to synthesize nanophase zinc ferrite. XRD results indicate that the particles formed are ultrafine (20nm) associated with strain. A temperature dependent Mossbauer spectroscopy has been used for elucidating the magnetic properties of the ZNF particles. MS –a unique technique that is able to distinguish between equivalent crystallographic sites and to reveal the non co-linearity of the spin arrangement present in spinel structure-has been effectively employed and its results gave valuable insight about magnetic phase (super paramagnetic), local cation distribution between A sites and B-sites, inversion parameter (δ= 0.74), canted spin alignment ($\psi_{(A)}$ = 29.58,, $\psi_{(B)}$ =48.46 and the blocking temperature (T_B= 50K).

Acknowledgment
One of the authors, Jacob Mathew thanks UGC-DAE Consortium Indore for providing financial support in the form of collaborative research scheme and VSM, Mossbauer facility. DST, New Delhi is also acknowledged for funding the VSM at UGC-DAE, CSR.

References

[1] Kamiyama T, Haneda K, Sato T, Ikeda S and Asano H January 1992 *Solid State Commun.* **81** p563
[2] Sato T, Haneda K, Seki N and Iijima T 1990 *Appl. Phys.* A **50** p13
[3] Jayadevan B, Tohji K and Nakatsuka K 1994 *J. Appl. Phys.* **76** p 6325
[4] Mahamoud M H, Hamdeh H H, Ho J C, O'shea M J and Walker J C 2000 *J.Magn. Magn. Mater.* **220** pp 139-146
[5] Atif M and Hasanian S K 2006 *Solid state Commun.* **138** pp 416-424
[6] Marykkutty Thomas and George K C 2009 *Indian J. of Pure and applied physics.* **47** pp 81-86
[7] Saez Puche R, Torralvo Fernandez M J, Balanco Gutierrez V, Gomez R, Marquin V, Marquina M L, Perez Mazariego J L and Ridaura R. 2008 *Bol.Soc.Esp.Cream.* **47** pp 3133-137
[8] Pramanik P 1996 *Chemical synthesis of nanosized Oxides. Bull.Mater.sci.* **19** pp 957-961
[9] Ghosh N N 2002 *Mat. Res. Soc. Symp. Proc.***720**
[10] Xinyu Zhao, Baicun Zheng, Hongchen Gu, Chunzhong Li, Shi Chang Zhang, Darrell Ownby P 1999 *J. mat. Res.* **14** pp 3073-3082
[11] Pedro Tartaj, Mar´ del Puerto Morales, Sabino Veintemillas-Verdaguer, Teresita Gonz´alez-Carre ˜no and Carlos J Serna 2003 *J. Phys. D: Appl. Phys.* **36** R182–R197
[12] Drofenik M, Lisjak D and Makovec D 2005 *Materials Science Forum* **494** pp 129-136
[13] Cross W B, Affleck L, Kuznetsov M V, Parkin I P and Pankhurst Q A 1999 *J. Mater. Chem.* **9**

IOP Conf. Series: Materials Science and Engineering **73** (2015) 012032 doi:10.1088/1757-899X/73/1/012032

2545
[14] Yao Li, Jiupeng Zhao, Jiuxing Jiang and Jiecai Han 2003 *Materials Research Bulletin* **38** pp 1393–399
[15] Evans B J, Hafner S S and Weber H P 1971 *J.Chem.Phys.* **55** p 5282
[16] Mandal K, Pan Mandal S, Agudo P and Pal M 2001 *Appl.Surface Sci.* **182** p386
[17] Blanco-Gutierrez V, Torralvo M J, Saez-Puche R and Bonville P 2010 *J.phys* **200** 072013
[18] Krupicka S and Novak P in:E.P.Wohlfarth(ed.) 1982 *Ferromagnetic Materials* **3** p194
[19] Morrish A H 2001, *The Physical principles of Magnetism*, (New York: IEEE press) p 504
[20] Smit J and Wijin H P J 1959 *Ferrites* (New York: Wiley) p 150
[21] Morrish A H and Clark P E 1975 *Phys.Rev*.B **11** p 278.
[22] Sawatzky G A, Van Der Woude F and Morrish A H 1969 *Phys.Rev.***183** p 383
[23] Goya G F and Rechenberg H R 1999 *J.Magn.Magn.Mater* **196-197** 191
[24] Oliver S A, Harris V G, Hamdeh H H and Ho J C 2000 *Appl.Phys.Lett.* **76** p 2761
[25] Tuek J and Zbaril R 2005 *Czech.J.Phys.* **55** p 893
[26] Li F S, Wang L, Wang J B,.Zhou Q G ,Kunkel H P and Williams G 2004 *J.Magn.Magn.Mat.* **268**,pp 332-339

IOP Conf. Series: Materials Science and Engineering **73** (2015) 012033 doi:10.1088/1757-899X/73/1/012033

Growth, optical and thermal studies on N–benzyl-2-methyl-4-nitroaniline

S. Prathap and J. Madhavan
Department of Physics, Loyola College, Chennai -34

E-mail: jmadhavang@yahoo.com

Abstract. In this work we present the growth, optical and thermal studies of the nonlinear organic molecular crystal, N–benzyl-2-methyl-4-nitroaniline (BNA). BNA is also a good THz electromagnetic waves emitter in 0.1 – 15 THz range. Nonlinear optical property of the crystal has been confirmed using the Kurtz powder technique and a study of its second harmonic generation efficiency in comparison with KDP has been made. Knoop hardness test was carried out and its Young's modulus was calculated. Thermal behaviour of the crystal was investigated by TG-DTG analyses.

1. Introduction

Organic crystals exhibit various appropriate properties such as luminescence, photochromism, or high nonlinearoptical (NLO) efficiency due to the high polarizability of π-conjugated molecules [1]. In addition, these organic compounds possess unlimited molecular engineering possibilities, which allow the improvement of these optical properties. Currently, there is substantial activity of research in the development of the new sources in the terahertz region of the electromagnetic spectrum. Coherent tunable terahertz waves have great potential for frequency domain spectroscopy and THz imaging applications. Presently, a number of techniques are employed to access the THz band, based on a large variety of physical mechanisms, and spanning over a size scale of several orders of magnitude from semiconductor-based quantum cascade lasers with a dimension of cm^3 to free-electron lasers and synchrotrons that can be as large as tens of meters.

N-Benzyl 2-methyl-4-nitroaniline (BNA) is an N-derivative of MNA and it easily crystallizes and affords a larger quadratic optical nonlinearity than MNA [2]. Through the optical rectification process, a strong THz electromagnetic wave can be generated. In this paper; first, we describe the growth of a BNA single crystal. Optically transparent single crystal of BNA was grown by slow evaporation technique. Then the crystal was subjected to powder X-ray Diffraction to conform the crystal structure, the TG and DTG were carried out and it shows the single stage weight loss. Nonlinear optical property of the crystal has been confirmed using the Kurtz powder technique and a study of its second harmonic generation efficiency in comparison with KDP has been made. Knoop hardness test was carried out and its Young's modulus was calculated. BNA is also a good THz electromagnetic waves emitter in 0.1 – 15 THz range.

2. Synthesis, solubility and growth

The primary amine (aniline) and substituted amine (2-methyl-4-nitro aniline) were used to prepare

IOP Conf. Series: Materials Science and Engineering **73** (2015) 012033 doi:10.1088/1757-899X/73/1/012033

N-benzyl-2-methyl-4-nitro aniline. The sample was recrystallised from aqueous methanol and the pure N-benzyl-2-methyl-4-nitro aniline is obtained. The solubility of BNA was carried out using aqueous methanol as solvent. The solubility of BNA was measured for various temperatures in the range of 30°C to 50°C and the solution was saturated at 40 °C and the crystals were grown. The as grown crystal is shown in figure 1.

3. Results

3.1 Powder X-ray Diffraction

Powder X-ray diffraction analysis was performed on BNA crystal in the range of 10 – 40° and is shown in Figure 2. All the observed reflections were indexed and the lattice parameters are found to be a=5.624Å, b=7.255Å, c=19.571Å belonging to orthorhombic system with space group of $Pna2_1$, which agrees with the reported values [2].

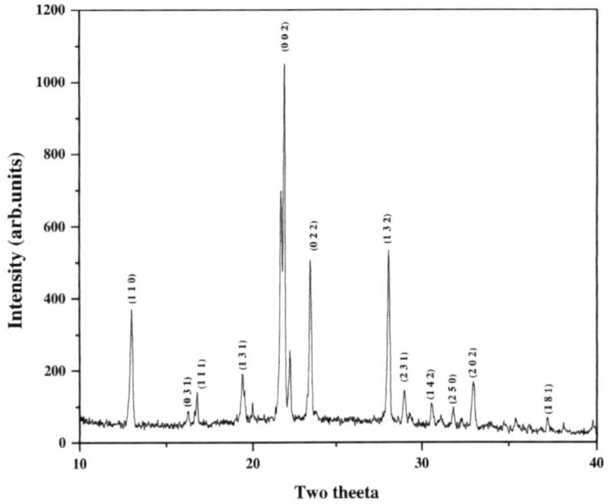

Figure 2. XRD Pattern of BNA

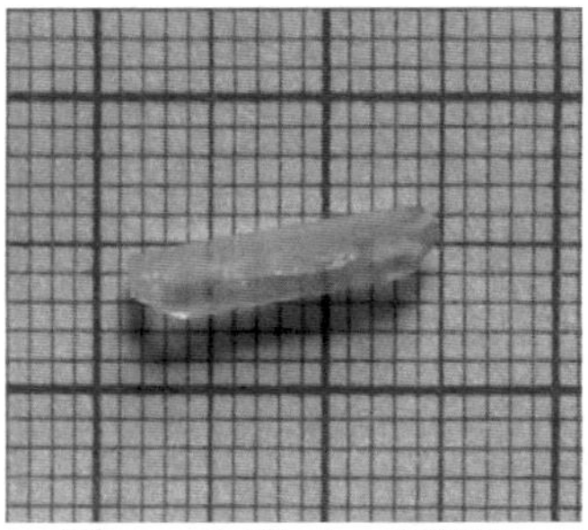

Figure 1. Photo of BNA

3.2 NLO studies

The SHG efficiency of BNA was determined by Kurtz and Perry powder technique[3]. Microcrystalline material of KDP was used as a reference for comparison with BNA for Second Harmonic Generation experiments. For a laser input pulse of 6.2 mJ, the second harmonic signal (532 nm) of 92 mW and 5.2 W, respectively were obtained for KDP and BNA samples. Thus, the SHG efficiency of BNA is fifty six times higher than KDP.

3.3 Thermal studies

Single crystal BNA crystal was subjected to thermo gravimetric analysis (TGA) and differential thermo gravimetric analysis (DTG) simultaneously using STA 409C instrument, in the nitrogen atmosphere at a heating rate of 20 K/min. Figure 3 shows the resulting TGA and DTG traces of the crystal. The material is thermally stable up to 230°C, and the sharp weight loss of the material starts around 230°C. The DTG trace matches with the TGA trace.

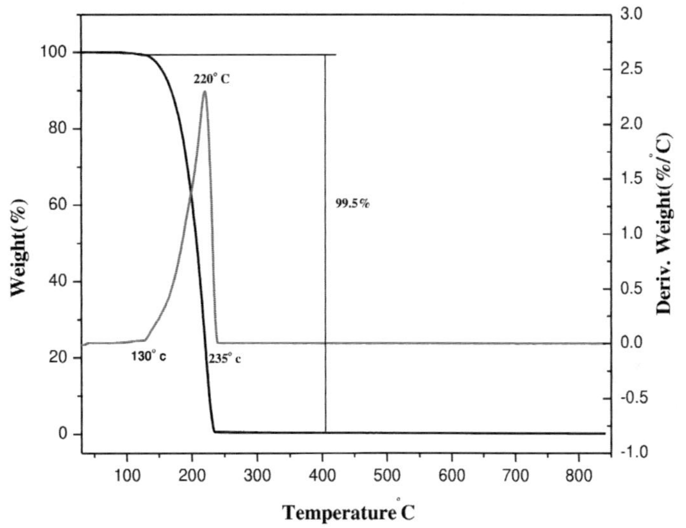

Figure 1. TG/DTG curve of BNA

3.4 Microhardness Studies

Knoop microhardness profiles as a function of applied load is shown in Figure. 4. The decrease of microhardness with increasing load is in agreement with the normal indentation size effect (ISE). Young's modulus (E) of of BNA was calculated using the formula E = 0.45 Hk/(0.1406 – b/a). Where H_k is the Knoop microhardness value at a particular load, b and a are the shorter Knoop indentation diagonal and longer Knoop indentation diagonal, respectively. The Young's Modulus of BNA was found to be 24.9 x 10^8 at a load of 5g.

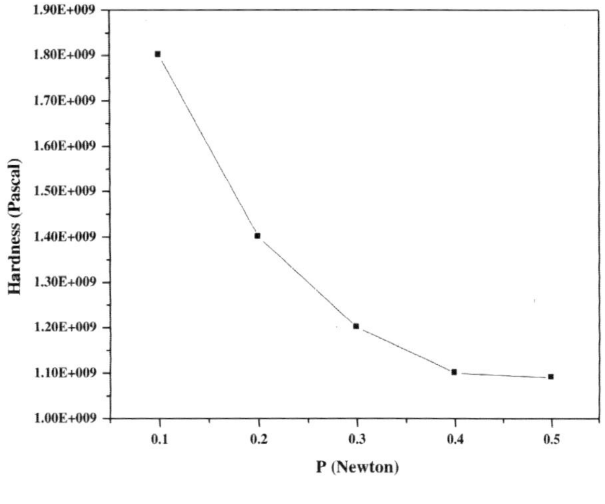

Figure 2. Knoop Hardness Profile

4. Conclusion

Good optical quality single crystal of BNA has been grown successfully by slow evaporation growth technique. Powder XRD studies shows that the crystal belongs to orthorhombic crystal system with space group $Pna2_1$. The second harmonic generation of the grown crystal was measured and compared with KDP. Thermal stability of BNA is also carried out. And Knoop hardness test was carried out and its Young's modulus was calculated.

5. References

[1]. Senthil. S, Pari. S, Sagayaraj. P, Madhavan. J, *Physica B*, 404 (2009) 295.
[2]. Masazumi fujiwara, Kazuhiro Yanagi, Minoru Maruyama, Mitsuru Sugisaki, Kazuyoshi Kuroyanagi, Hironori Takahashi, Shin-ichiro aoshima, Yutaka tsuchiya, Andrew gall and Hideki Hashimoto, *Japanese Journal of Applied Physics,* 45(2006)8676
[3]. S. K. Kurtz, T. T. Perry, *J. Appl. Phys.*,39, 3798 (1968)

Linear and nonlinear optical properties of unidirectionally grown L-Threonine acetate single crystals

G. Prabagaran[1,2] and J. Madhavan[2]
[1] Department of Physics, Govt. Arts College (Men), Nandanam, Chennai 600 035.
[2] Department of Physics, Loyola college, Chennai 600 034. India.

E-mail: jmadhavang@yahoo.com

Abstract. Unidirectional <1 1 0> L-Threonine acetate (LTA) single crystal was grown and reported. The grown crystal was characterized by powder X-ray diffraction (XRD) studies to confirm the crystal structure. The optical band gap (*Eg*) value of the grown crystal is obtained from the tauc's plot of $(\alpha hv)^2$ *vs* *hv* which was determined by extrapolating the linear region of the curve to the *hv* axis where $(\alpha hv)^2 = 0$. Other linear optical parameters such as Extinction coefficient, Reflectance, Refractive index, Complex dielectric constant, and Optical conductivity are calculated for the grown crystal and its variation with incident photon energy is analysed. Second Harmonic Generating (SHG) efficiency of the candidate material was determined by Nd:YAG Q-switched laser Kurtz powder technique.

1. Introduction

Engineering of new nonlinear optical (NLO) materials, both organic and inorganic, structures and devices with enhanced figures of merit over the past two decades has become a major force to drive nonlinear optics from the laboratory to real applications [1]. The optical behavior of materials is an essential parameter to determine its usage in optoelectronic devices [2]. Knowledge of optical constants of a material such as optical band gap and extinction coefficient is vital to scrutinize the materials' potential for optoelectronic applications [3]. The crystal with specific orientation can be grown from solution by Sankaranarayanan–Ramasamy (SR) method [4]. By this method the entire solute can be converted into crystal so that 100% solute–crystal efficiency can be achieved. In this paper we discuss the growth of Unidirectionally grown bulk L-Threonine acetate (LTA) single crystal along with linear and nonlinear optical properties and its variation with incident Photon energy.

2. Experimental Procedure

The experimental setup of SR method consists of growth ampoule made-up of glass with seed mounting pad [4]. The seed crystal from the conventional slow solvent evaporation technique was used for the current study. <110> plane of the seed crystal of LTA was chosen and it was transferred to the saturated solution of LTA. The chemical reaction involved for the grown L-Threonine acetate is given below:

$$C_4H_9NO_3 + CH_3COOH \rightarrow C_4H_{10}NO_3 + CH_3COO^-$$

Growth of highly transparent single crystal of LTA of 10mm diameter and 44 mm length was harvested (Figure 1.) in a period of 27 days and reported for the first time.

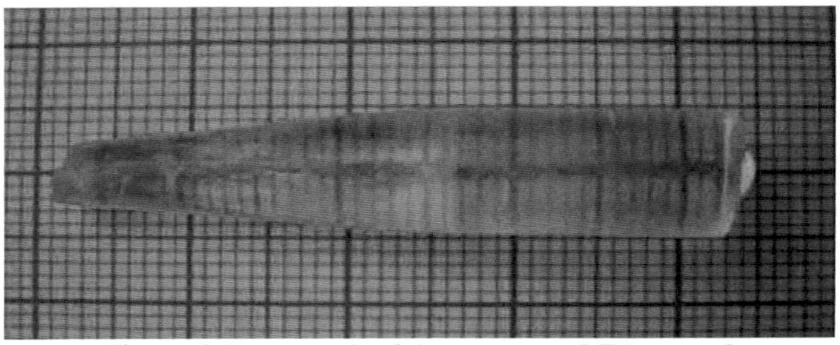

Figure.1 Photograph of the as grown LTA crystal

3. Characterization

3.1 Powder XRD studies

To identify the crystal structure of the grown crystal powder X-ray diffraction studies of LTA crystals were carried out using Siemens D500 X-ray diffractometer with Cu K_α (λ = 1.5418Å) radiation. The samples were scanned for 2θ values from 10° to 70° at a rate of 2°/min. Powder XRD pattern LTA crystals are shown in Figure 2. The diffraction patterns of the LTA crystal are indexed by least square fit method. The lattice parameter values of the LTA crystal have been calculated and were found to be well matched with the reported literature [5]. LTA crystal crystallizes in to the orthorhombic crystal system with the space group of $P2_12_12_1$ and unit cell dimensions were found to be a = 5.542Å, b =8.131Å, c = 13.912Å. The volume of the system is V=626.903Å^3.

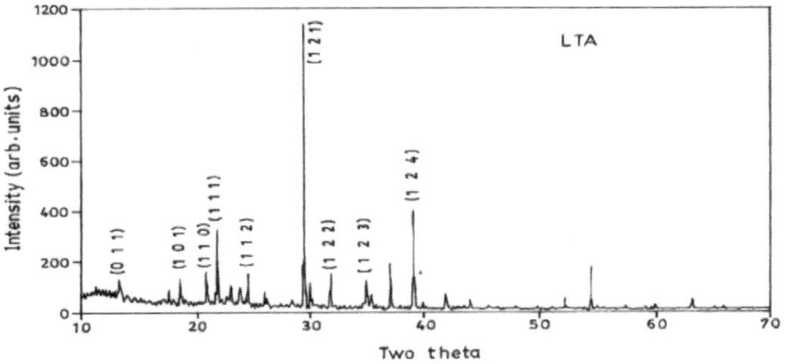

Figure.2 Powder X-ray diffraction pattern of LTA

3. Optical Properties of LTA

The optical absorption spectrum of LTA single crystal is shown in Figure 3. The spectra indicate that LTA crystal has minimum absorption in the entire visible and near infrared region. This data was further used for analyzing optical band gap energy (E_g) using the formula for optical absorption of

$$(\alpha h\nu) = k\,(h\nu - E_g)^{n/2}$$

n is an integer equal to 1 for a direct band gap and 4 for an indirect band gap. The values of the direct optical band gap E_g were obtained from the intercept of $(\alpha h\nu)^2$ versus hν curve plotted in Figure 4. The intercept obtained by the extrapolation of the linear portion of the plot to energy axis gives the band gap energy of the crystal. The band gap is found to be 4.5 eV. The other optical constants were calculated using the theoretical formulae reported elsewhere [6,7]. Figure 5 and Figure 6 shows the plot of extinction coefficient and reflectance against incident photon energy.

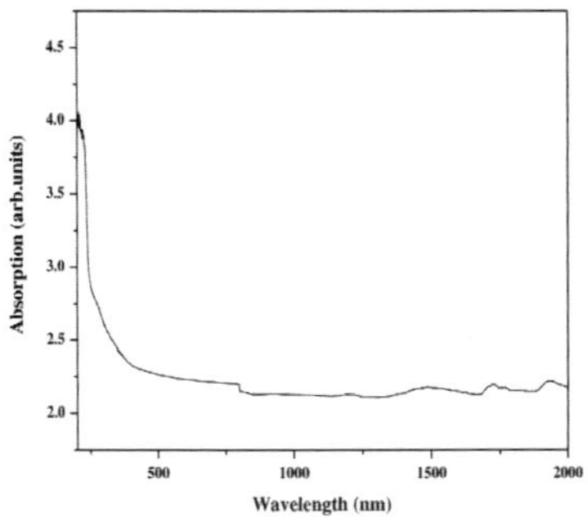

Figure 3. Optical absorption spectrum of LTA

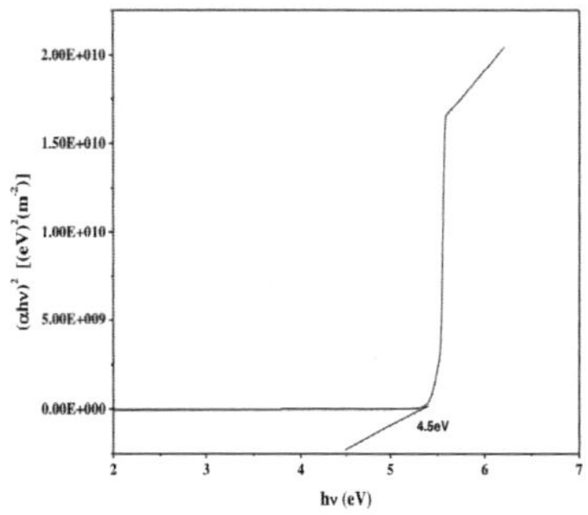

Figure 4. Energy band gap of LTA

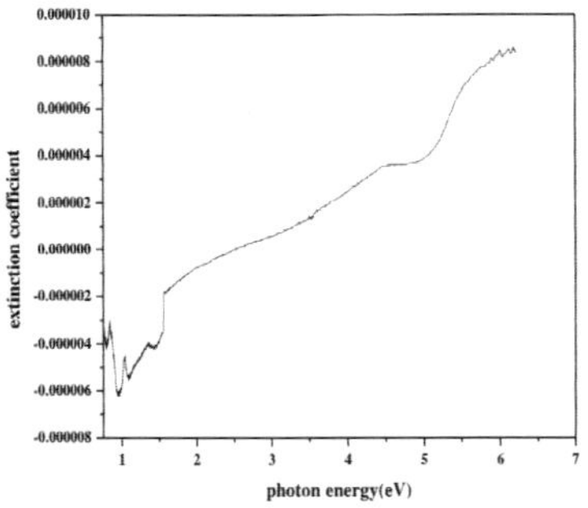

Figure 5. Extinction coefficient Vs Incident Photon Energy of LTA

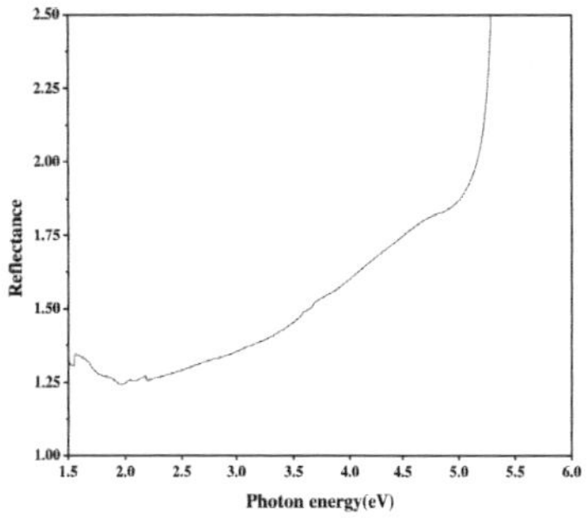

Figure 6. Reflectence Vs Incident photon energy of LTA

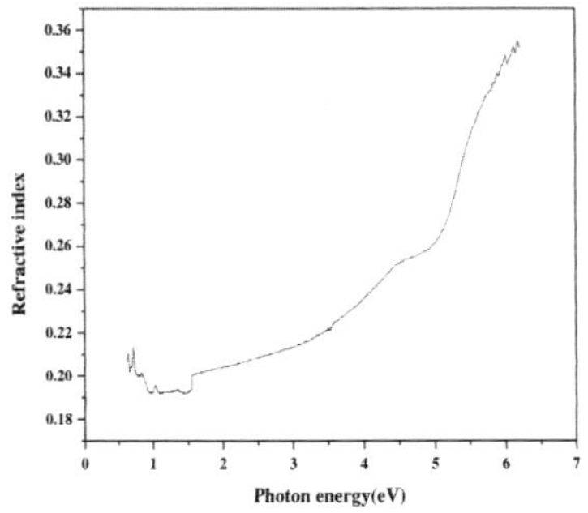

Figure 7. Refractive index Vs Incidient photon energy of LTA

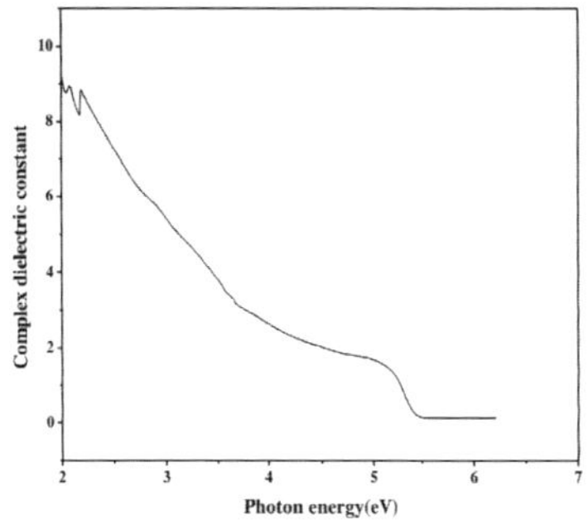

Figure 8. Complex dielectric constant Vs Incident photon energy of LTA

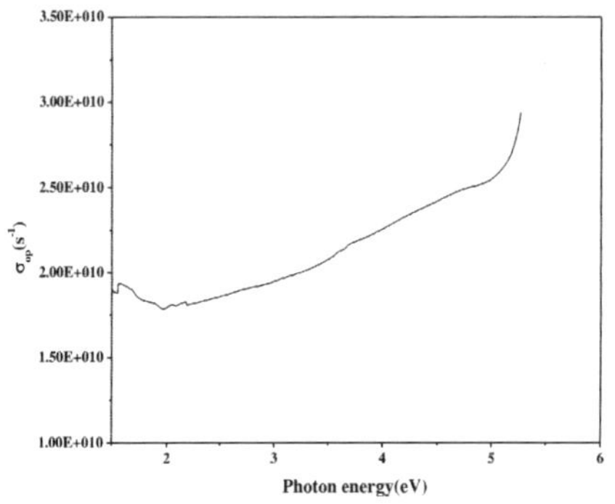

Figure 9. Opticalconductivity Vs Incident photon energy of LTA

It's found that extinction coefficient remains constant for incident photon energy of 1.2 eV to 5.5 eV and reflectance increases linearly with photon energy Figure 7 depicts the variation of refractive index with incident photon energy. Initially the refractive index decreases with increasing photon energy then becomes constant. Variation of complex dielectric constant and optical conductivity of the material with incident photon energy is analysed from Figure 8 and Figure 9. The high optical conductivity (10^9) confirms the high photo response nature of the material.

4. SHG Efficiency Studies

The quantitative work was done on powdered single crystal of LTA and they were graded by the use of standard sieves to desire the range of particle sizes. A relevant comparison was made on KDP with LTA. The reference material was also powdered and used for further studies [8]. The experimental samples reported an input pulse of 6.2 mJ, the second harmonic signal (532 nm) of 89.52 mW for KDP and 268.56 mW for LTA. It is thus elucidated that the SHG efficiency of LTA is 3.0 times higher than that of KDP

5. Conclusion

Good quality single crystal of L-Threonine acetate has been grown successfully by the unidirectional solution growth SR method. It is understood from the present work that the SR technique was found to be suitable method to grow high quality and large- size single crystal of L-Threonine acetate. Optical absorption spectrum was recorded for the given crystal and it is found that it has minimum absorption between 240 -1000 nm. The optical band gap of the material is found as 4.5 eV. Various linear optical constants like extinction coefficient, reflectance, refractive index, complex dielectric constant and optical conductivity are calculated and related with incident photon energy. This optical analysis reveals the NLO behavior of the material.

References

[1]. P.C. Rajesh Kumar, V. Ravindrachary K. Janardhana, H.R. Manjunath, Prakash Karegouda,Vincent Crasta, M.A. Sridhar J. of Molecular Structure 1005 (2011) 1–7.
[2]. P.V. Mehta, N. Tripathi, S.K. Kumar, Chalcogenide Lett. 2 (2005) 39.
[3]. M Dongol, Egypt. J. Solids 25 (2002) 33.
[4]. K. Sankaranarayanan, P. Ramasamy, J. Crystal Growth 280 (2005) 467.
[5]. G. Ramesh Kumar., Gokul Raj., Thenneti Raghavalu, V. Mathivanan., M. Kovendhan., .Bhagavannarayana., Mohan., R. Jayavel., (2007), Materials letters, Vol. 61, No. 27, pp. 4932
[6]. P.A Henikhena, African Phys. Rev., 2(2008) 68.
[7]. J.I.Pankove, Optical Processes in semiconductors, Pentice Hall, New York, (1971).
[8]. S K Kurtz, and T T Perry, J. Appl. Phy., 1968; 39: 3798-3813.

IOP Conf. Series: Materials Science and Engineering **73** (2015) 012035 doi:10.1088/1757-899X/73/1/012035

Molecular structure, First-order hyperpolarizability and HOMO-LUMO studies of L- Histidinium Dinitrate

S. Arulmozhi and J. Madhavan
Department of Physics, Loyola College, Chennai – 34.

E-mail: jmadhavang@yahoo.com

Abstract. The geometric parameters and hyperpolarizabilities of a nonlinear optical material L- Histidinium Dinitrate (LHDN) was investigated by density functional theory and presented. The compound crystallizes in the noncentro symmetric space group $P2_12_12_1$ of orthorhombic system. The first order hyperpolarizability (β) of this molecular system is calculated using B3LYP/6-31G (d,p) method on the finite-field approach. The calculation results also show that the LHDN molecule might have microscopic nonlinear optical behavior with non zero values. The Optical absorption spectrum and second harmonic generation were investigated. The calculated HOMO and LUMO energies showed that charge transfer occurs within the molecule.

1. Introduction
Nonlinear optical (NLO) phenomena have been extensively studied over the last decades; molecules exhibiting large hyperpolarizabilities have a strong NLO potential and could be used, under conditions for optoelectronics and a variety of optical devices [1, 2]. In the present work, LHDN was synthesized by slow evaporation method and the structural properties were calculated by DFT calculation. The experimental spectra are compared with calculated findings. Literature survey also reveals that to the best of our knowledge no DFT calculations have been reported yet. Therefore, the present investigation was undertaken to study the vibrational spectra of this molecule and to identify the various normal modes with greater wavenumber accuracy. In this context, the hyperpolarizability of the title compound was calculated in the present study.

2. Synthesis of LHDN
The starting materials were AR grade L-histidine (99 % Merck) and nitric acid. The solution was prepared by dissolving 1 mol of L-histidine and 2 mols of nitric acid in deionised water. The chemical reaction is as follows:

$$C_6H_9N_3O_2 + 2HNO_3 \rightarrow C_6H_{11}N_3O_2^{2+} . 2NO_3^-$$

The synthesized material was then purified by repeated crystallization process.

3. Results

3.1 Powder XRD Studies
The structural property of the grown LHDN is studied by X-ray powder diffraction technique. Powder X-ray diffraction studies were carried out, using Siemens D500 X-ray diffractometer with Cu K_α (λ = 1.5406Å) radiation. The samples were scanned for 2θ values from 10° to 55° at a rate of 2° /min. The diffraction pattern (Figure 1) was indexed by least square fit method. The grown LHDN crystal belongs to orthorhombic system with the centrosymmetric $P2_12_12_1$ space group. The obtained cell parameters are a = 5.473 (5) Å, b = 8.291 (3) Å, c = 25.612 (4) Å, α = β = γ = 90° and V = 1162.186 (4) $Å^3$. The observed unit cell parameters are in good agreement with the corresponding reported values [3].

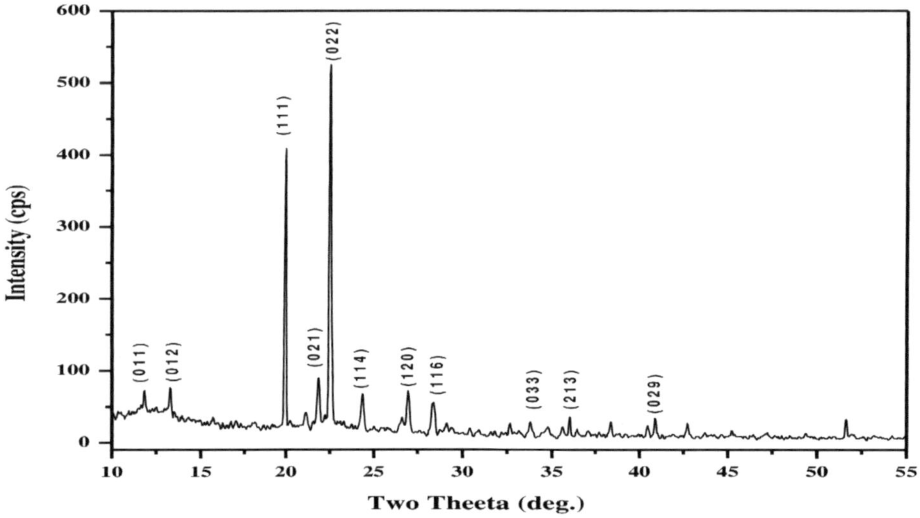

Figure 1. XRD Pattern of LHDN

2.2 Hyperpolarizability

The β values of LHDN compound was calculated by the DFT-B3LYP functional with the 6-31G (d, p) basis set [4]. In this context, the dynamic first hyperpolarizability of the title compound is also calculated in the present study. The first hyperpolarizability (β_o) of this novel molecular system is calculated using B3LYP/6-31G(d) method, based on the finite field approach. In the presence of an applied electric field, the energy of a system is a function of the electric field. First hyperpolarizability is a third rank tensor that can be described by a 3x3x3 matrix. The 27 components of the 3D matrix can be reduced to 10 components due to the Kleinman symmetry [5].

A large value of the first hyperpolarizability (=4.187944 x 10^{-30} esu) is a prerequisite to behave as a good NLO material, and the important parameters influencing β generally are (i) donor–acceptor system, (ii) nature of substituents, (iii) conjugated π system and (iv) the influence of planarity. The results also show that the compounds studied in our research are good NLO materials due to that their β values.

2.3 HOMO – LUMO

The calculations were done with the commonly used exchange-correlation functional B3LYP followed by a comprehensive analysis of the calculated highest-occupied and lowest-unoccupied Kohn-Sham orbital (HOMO and LUMO) energies. The basis set dependence of the DFT results shows that the economical 6-31+G basis set is generally sufficient for calculating the HOMO and LUMO energies (if the calculated LUMO energies are negative) for use in correlating with molecular properties.

IOP Conf. Series: Materials Science and Engineering **73** (2015) 012035 doi:10.1088/1757-899X/73/1/012035

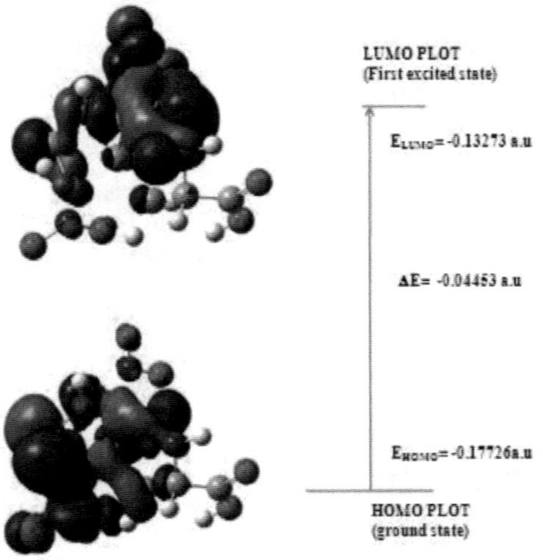

Figure 2. The atomic orbital compositions of the frontier molecular orbital for LHDN

On the basis of fully optimized ground-state structure, DFT/B3LYP/6-31G(d,p) calculations have been used to determine the low-lying excited states of LHDN. The calculated results involving the vertical excitation energies, oscillator strength (*f*) and wavelength are carried out and compared with measured experimental wavelength.

The HOMO-LUMO energy gap of LHDN (Figure 3) was calculated at theB3LYP/6-31(d,p) level reveals that the energy gap reflect the chemical activity of the molecule and found to be -0.04453a.u. LUMO as an electron acceptor represents the ability to obtain an electron, HOMO represents the ability to donate an electron. All of the calculations were performed by using the Gaussian 03.

2.4 SHG Efficiency

The SHG efficiency of the grown crystal was measured by the standard Kurtz and Perry powder technique [6] using a fundamental beam of a Q–switched Nd:YAG laser with a wavelength of 1064 nm. To make relevant comparison with known SHG materials, KDP was also powdered and sieved into the same particle size range. The results obtained by this method shows that the SHG efficiency of LHDN sample is nearly 2.9 times more than that of KDP.

2.5 UV- Vis NIR Spectrum

The UV-Vis-NIR transmission spectrum of the crystal was recorded in a wavelength range of 200 – 1200 nm and is shown in Figure 3. From the spectrum, it is seen that the UV transparency cutoff occurs at 250 nm and there is no remarkable absorption in the entire region of the spectra. This is a favorable character for a NLO material.

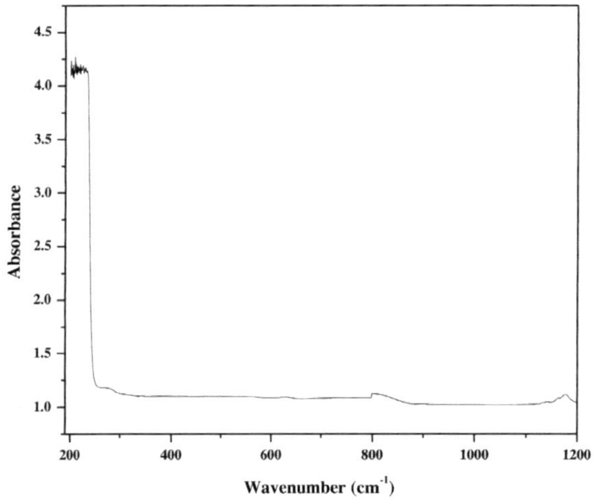

Figure 3. UV Vis NIR Spectrum of LHDN

3. Conclusions

Single crystals of LHDN were synthesized and grown from aqueous solution. The grown crystals were characterized by Powder XRD analysis. The study reveals that the grown crystals belong to orthorhombic system. SHG measurement shows that the crystal has superior properties than KDP crystal. HOMO and LUMO energy gap explains the eventual charge transfer interactions taking place within the molecule. Furthermore, first-order hyperpolarizability of the molecule shows that the title molecule is an attractive object for future studies of nonlinear optical properties.

4. References

[1]. Senthil. S, Pari. S, Sagayaraj. P, Madhavan. J, *Physica B*, 404 (2009) 295.
[2]. S. Senthil, S. Pari, R. John Xavier, J. Madhavan, *Optik,* 123(2011)104.
[3]. S. Aruna, G. Bhagavannarayana, P. Sagayaraj, J. of Crystal Growth 304 (2007) 184.
[4]. Frisch M.J, Trucks G.W, Schlegel H.B, Scuseria G.E, Robb M.A, J.R. Cheeseman J.R, Montgomery Jr J.A, Vreven T, Kudin K.N, Burant J.C, Millam J.M, Iyengar S.S, Tomasi J, Barone V, Mennucci B, Cossi M, Scalmani G,.Rega N, Petersson G.A, Nakatsuji H, Hada M, Ehara M, Toyota K, Fukuda R, Hasegawa J, Ishida M, Nakajima T, Honda Y, Kitao O, Nakai H, Klene M, Li X, Knox J.E, Hratchian H.P, Cross J.B, Adamo C, Jaramillo J, Gomperts R, Stratmann R.E, Yazyev O, Austin A.J, Cammi R, Pomelli C, Ochterski J.W, Ayala P.Y, Morokuma K, Voth G.A, Salvador P, Dannenberg J.J, Zakrzewski V.G, Dapprich S, Daniels A.D, Strain M.C, Farkas O, Malick D.K, Rabuck A.D, Raghavachari K, Foresman J.B, Ortiz J.V, Cui Q, Baboul A.G, Clifford S, Cioslowski J, Stefanov B.B, Liu G, A.Liashenko, Piskorz P, Komaromi I, Martin R.L, Fox D.J, Keith T.,.Al-Laham M.A, Peng C.Y, Nanayakkara A, Challacombe M, Gill P.M.W, Johnson B, Chen W, Wong M.W, Gonzalez C, Pople J.A, Gaussian 03, Revision C.02, Gaussian Inc.,Wallingford, CT, 2004.
[5]. Kleinman. D. A, *Phys. Rev.* 1977, 126, 1962.
[6]. S. K. Kurtz, T. T. Perry, *J. Appl. Phys*.,39, 3798 (1968)

IOP Conf. Series: Materials Science and Engineering **73** (2015) 012036 doi:10.1088/1757-899X/73/1/012036

Synthesis, growth and optical properties of an efficient nonlinear optical single crystal: L-alanine DL-malic acid

R. Kirubagaran and J. Madhavan
Department of Physics Loyola College, Chennai – 34

E-mail: jmadhavang@yahoo.com

Abstract. Single crystals of L-alanine DL-malic acid (LADLMA) have been grown from aqueous solution by slow-cooling technique. Powder X-ray diffraction studies reveal the structure of the crystal to be orthorhombic. The nonlinear optical conversion efficiency test was carried out for the grown crystals using the Kurtz powder technique. The third order nonlinear refractive index and the nonlinear absorption coefficient where evaluated by Z-scan measurements. As the material have a negative refractive index it could be used in the protection of optical sensors such as night vision devices.

1. Introduction

Some organic compounds exhibit large NLO response than widely known inorganic materials. They also offer the flexibility of molecular design and the promise of virtually an unlimited number of crystalline structures. Among organic crystals for non linear optics (NLO) applications, aminoacids display specific features of interest such as (i) molecular chirality, which secures acentric crystallographic structures; (ii) absence of strongly conjugated bonds, leading to wide transparency ranges in the visible and UV spectral regions; (iii) zwitterionic nature of the molecule, which favours crystal hardness [1].

The title compound is the smallest molecule among the amino acids. Although its second harmonic generation (SHG) efficiency is about one-third that of potassium dihydrogen phosphate (KDP), the knowledge of studying the properties is very important since L-alanine can be considered as the fundamental building block of more complex amino acids [2]. Hence, it may be useful to synthesize the amino acid complexes with other carboxylic acids and study their properties. This paper discusses the growth of single crystals of LADLMA by slow cooling technique. Structural, second harmonic efficiency and Z-scan studies were carried out and the results were presented and discussed.

2. Synthesis and Growth

Stoichiometric amount of L-alanine and DL malic were taken with excess of water to synthesize L- alanine DL malic acid [3]. The solubility [4] of LADLMA was measured at 30, 35, 40, 45 and 50 °C. The variation of solubility with temperature is shown in Figure 1. Seed crystals, free from visible defects were used for bulk growth. The rate of cooling was 1 °C/day over a period of 22 days. Single crystal up to size of 11 x 8 x 9 mm^3 was grown and is shown in Figure 2.

3. Results

3.1 Powder XRD

In order to confirm the crystallinity and also to estimate the lattice parameter values, the grown crystals were subjected to powder X-ray diffraction analysis. X-ray powder patterns of the crystal were recorded on a SIEFERT X-ray Diffractometer using CuK α (Kα = 1.5408) radiation. The sample was scanned for a 2θ range of 10–50° and at a scan rate of 2°/ min^{-1}. All the observed reflections (figure 3) were indexed and the unit cell parameters were calculated. The XRD data of the crystal indicates that it crystallizes in orthorhombic system with $P2_12_12_1$ space group with a=5.823 Å, b=6.030 Å and c=12.325 Å which agrees well with the reported values [3].

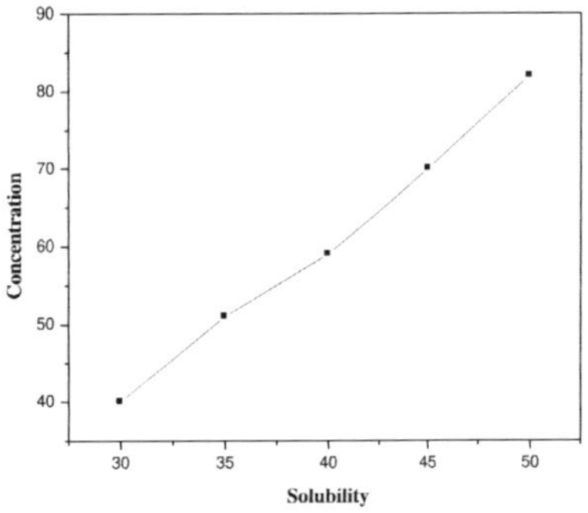

Figure 1. Solubility curve of LADLMA

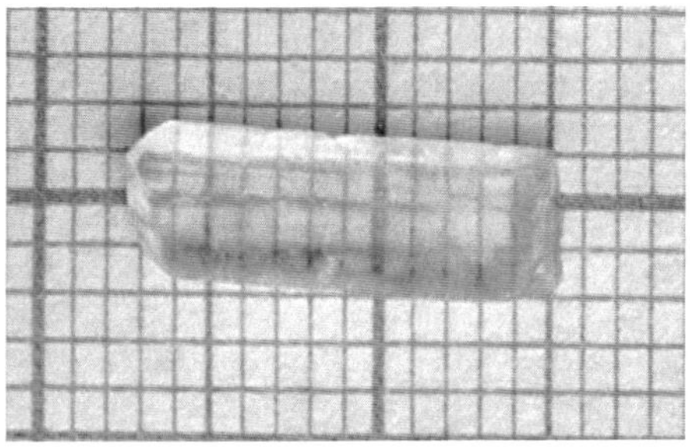

Figure 2. Photograph of LADLMA

3.2 UV-Vis NIR Spectrum

To find the transmission range LADLMA the optical transmission spectrum of LADLMA for the wavelengths between 200-1200nm was recorded using Varian Cary 5E UV-Vis-NIR spectrophotometer. From the graph (figure 4), it is evident that LADLMA crystal has UV cutoff wavelength at 250nm. Moreover the absorption is minimum in the entire visible region and also in the near infrared region. The lower cut-off wavelength of LADLMA crystal is almost comparable to other analogs of L-alanine [2]

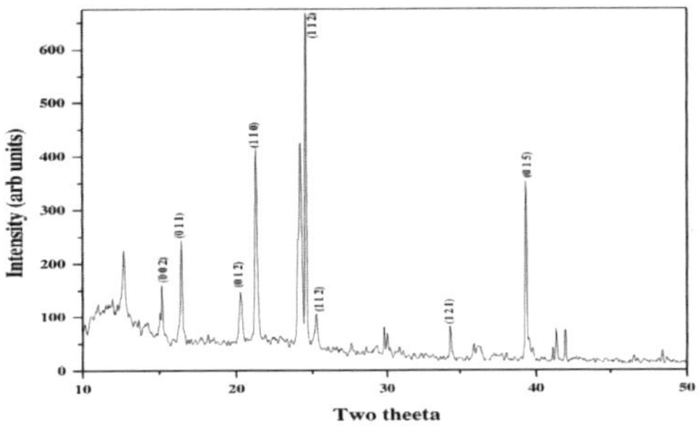

Figure 3. XRD patterns of the powder sample

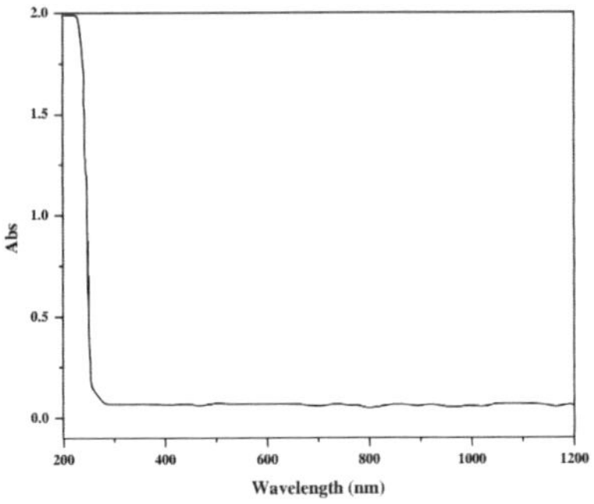

Figure 4. UV-Vis NIR spectrum of LADLMA

3.3 SHG efficiency

Kurtz SHG test [5] was performed on LADLMA crystal to confirm the second harmonic signal generation efficiency. Microcrystalline material of KDP was used as a comparison with LADLMA, for the SHG experiments. For a laser input pulse of 6.2 mJ, the second harmonic signal (532 nm) of 91.66 mV and 90.21 mV were obtained through KDP and LADLMA samples respectively. Thus the SHG efficiency of LADLMA is comparable to that of KDP.

3.4 Z-scan studies

The Z-scan technique is a simple and effective tool for determining the nonlinear properties. The third order nonlinear refractive index and the nonlinear absorption coefficient were evaluated by the Z-scan measurements. The crystals were cut and polished in the z direction and then subjected to Z-scan measurement. Figure 5 and Figure 6 show the open and closed aperture Z-scan for the crystal.

Z-scan parameters are presented in Table 1. As the material have a negative refractive index, it results in defocusing nature of the material, which is an essential property for the application in the protection of optical sensors such as night vision devices [6].

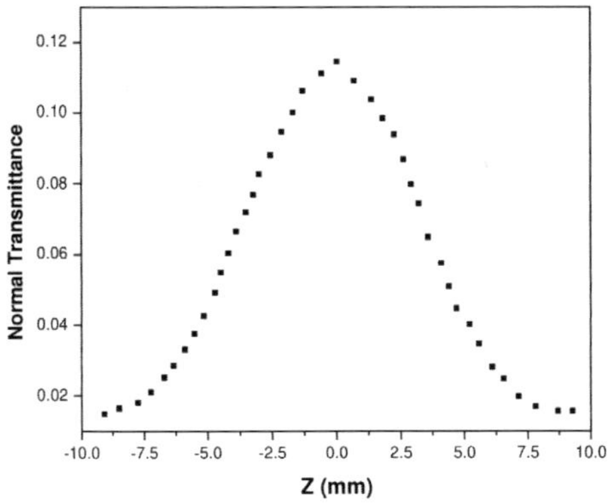

Figure 5. Open aperture Z-scan curve of LADLMA

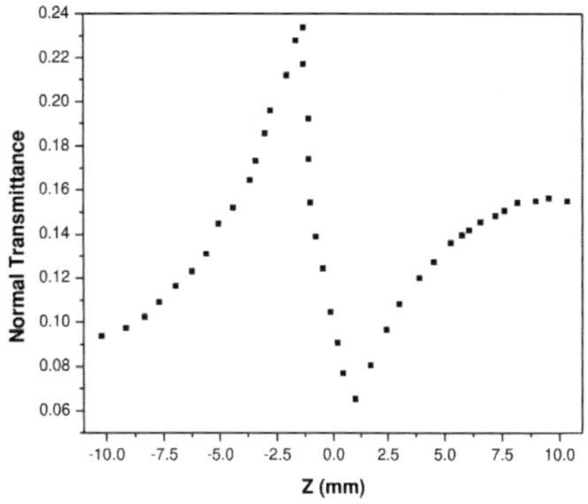

Figure 6. Closed aperture Z-scan curve of LADLMA

Table 1. Z – Scan Data

Laser beam wavelength (λ)	532nm
Lens focal length (F)	10 cm
Spot size diameter in front of the aperture (ω_a)	2 mm
Aperture radius (r_a)	0.087 x 10^{-2}m
Incident intensity at the focus (z=0) (I_0)	2.3915 x 10^5 m W
Effective thickness of the sample (L_{eff})	0.9938 mm
Linear absorption co-efficient (α)	0.0123
Non-linear refractive index (n_2)	-8.1064 x 10^{-12} cm^2/W
Non-linear absorption co-efficient (β)	1.1662 x 10^{-4} cm/W
Real part of the third-order susceptibility (Re$\chi^{(3)}$)	3.6466 x 10^{-10} esu
Imaginary part of the third-order susceptibility (Im$\chi^{(3)}$)	2.222 x 10^{-8} esu
Third-order non-linear optical susceptibility	2.2222 x 10^{-4} esu.

IOP Conf. Series: Materials Science and Engineering **73** (2015) 012036 doi:10.1088/1757-899X/73/1/012036

4. Conclusions

Single crystals of L-alanine DL-malic acid (LADLMA) have been grown from aqueous solution by slow-cooling technique. Powder X-ray diffraction studies reveal the structure of the crystal to be orthorhombic. The nonlinear optical conversion efficiency test was carried out for the grown crystals using the Kurtz powder technique. The third order nonlinear refractive index and the nonlinear absorption coefficient where evaluated by the Z-scan measurements. The results strongly suggest that the material could be used in optical device fabrication.

5. References

[1]. S. Senthil , S.Pari , P.Sagayaraj , J.Madhavan *Physica B* June 2009, pp 1655–1660

[2]. D. Sankar , P. Praveen Kumar , J. Madhavan *Physica B* , February 2010 , pp 1233–1238

[3]. D. Jaikumar, S.Kalainathan, G.Bhagavanarayana *Journal of Crystal Growth* 312 (2009), 120–124

[4]. J. Madhavan, S. Aruna, K. Prabha, J. Packium Julius, Ginson P. Joseph, S. Selvakumar, P. Sagayaraj *Journal of Crystal Growth* , August 2006, pp 409–414

[5]. S. K. Kurtz, T. T. Perry, J. Appl. Phys.,39, 3798 (1968)

[6]. Yun Shan Zhou, En Bo Wang, and Jung Peng, *Polyhedron* 18, 1419 (1999).

Isoconversional and Isokinetic Studies of 2605SA1 Metglass

T Lilly Shanker Rao and T Shanker Rao

Narmada College of Science & Commerce,Zadeshwar,Bharuch,Gujarat-392011(India)

E-mail: ncsclsr@gmail.com

Abstract. Isoconversional methods (KAS,OFW,Li-Tang,and Vyazovkin) as well as isokinetic methods ((IKP) and CR) are used to determine the activation energy for crystallization (E), pre-exponential factor (A) and the conversion function f(α), for Fe-based 2605SA1 metallic glass.The non-isothermal differential scanning calorimetry data obtained at different heating rates ranging from 2 to 10 K/min are utilized for the analysis. The E for the peak I and II obtained by the IKP method are in good agreement with the values obtained from various isoconversional methods. The calculated kinetic triplets from IKP for peak I is: E = 445.6 ± 24 kJ/mol, A = 9.4×10^{22} sec^{-1} and $f(\alpha) = [-\ln(1-\alpha)]^{1/4}$ and for peak II is : E=325.7 ± 8 kJ/mol, A= 7.1×10^{15} sec^{-1} and $f(\alpha) = \alpha^{1/4}$. All the above mentioned methods are summarized and discussed elaborately from the point of view of their applicability and correctness.

1. Introduction

The kinetic analysis of non-isothermal data (DTA, DSC etc.) is based on the rate equation [1]

$$d\alpha/dt = k(T)f(\alpha) = A\exp(-E/RT)\, f(\alpha) \quad (1)$$

where α is the fractional crystallization in the solid reactant, A the Arrhenius pre-exponential factor, T the reactant temperature, f(α) is kinetic function that depends on the reaction models, g(α), E the activation energy for the reaction, and R the gas constant. When the α is constant, for different heating rates β, the methods are called isoconversional [2].These methods are divided into two categories: 1.Linear isoconversional and 2.Non-linear isoconversional methods. The linear isoconversional methods again classified into two groups [3]. One group is called linear integral isoconversional methods and proposed by Kissinger-Akahira-Sunose (KAS) [4,5], Ozawa-Flynn-Wall (OFW) [6,7] and Li and Tang[8].In these methods E values are calculated at a specific α .The other is called non-linear isoconversional method and is described by Vyazovkin [9] which assumes that the reaction model, g(α) is independent of the β.

The isokinetic(model fitting) analysis depends on the reaction model. Among the various methods of this type, the Coats and Redfern (CR) [10] and the invariant kinetic parameters (IKP) [11] are used. Present investigation uses isoconversional and model fitting methods to understand the crystallization kinetics of Fe-based 2605SA1 (wt% Fe=85-95, Si=5-10, B=1-5).

2. Experimental methods

The Fe-based amorphous alloy 2605SA1 was given as a gift from Metglass Inc. Co. USA. The amorphous nature is confirmed by X - ray analysis. The as-quenched samples were heated in DSC (Mettler Toledo) at heating rates 2,4,6,8 and 10 $Kmin^{-1}$ from RT to 840 K in air. The DSC scans were recorded by a thermal analyzer interfaced to a computer.

3. Results and discussion

The DSC thermograms of 2605SA1 amorphous alloy at different β are shown in Figure1. All the DSC traces have double exothermic peaks in the temperature scanned. It is noticed that the T_p shifted to high temperatures with increasing β.This shift in the peak forms the basis for determination of E. Due to increase in the sensitivity, the peak height also increases with increasing β.

3.1. Isoconversional methods.

3.1 (a). Linear Integral isoconversional methods.

(1). Kissinger-Akahira-Sunose (KAS) Method: The method uses the equation

$$\ln(\beta/T^2)=\ln(AR/E\ g(\alpha)) - (E/RT) \quad (2)$$

Figure.2 gives local activation energy $E(\alpha)$, for the peaks I & II .

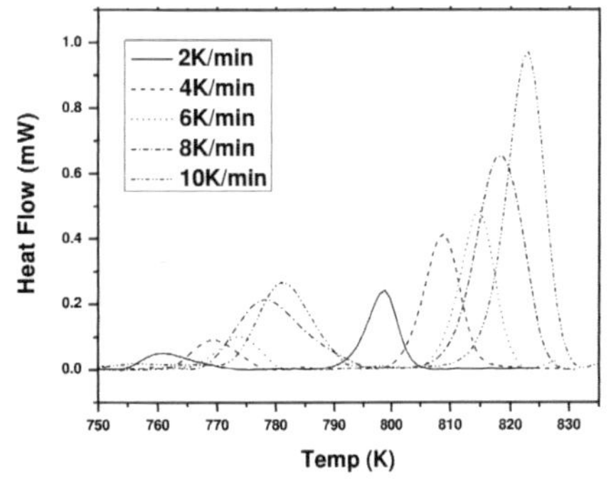

Figure. 1. DSC thermograms at different β

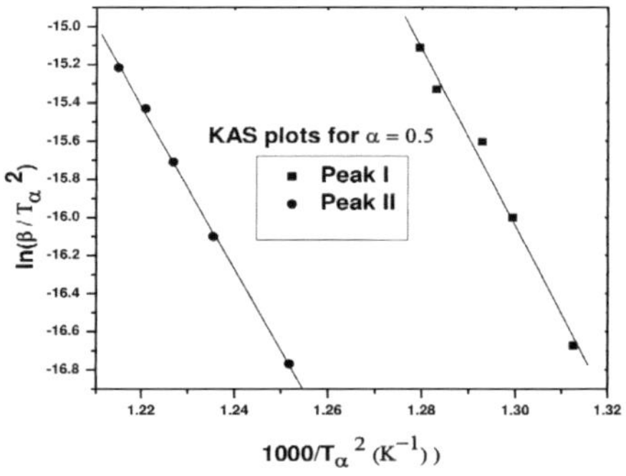

Figure.2 plot of $\ln(\beta/T_\alpha^2)$ vs $1000/T_\alpha$

(2). Ozawa -Flynn-Wall (OFW) method: The expression utilized in this method is

$$\ln\beta=-1.052E(\alpha)/RT_\alpha + \text{constant} \quad (3)$$

Figure.3 gives the slope $-1.0516\ E(\alpha)/R$ from which the $E(\alpha)$ is calculated for both the peaks.

(3). Li andTang method: Figure .4 shows the plot and the equation utilized in this method is

$$\int_0^\alpha \ln(d\alpha/dt)d\alpha = G(\alpha) - E/R \int_0^\alpha (1/T)d\alpha \quad (4)$$

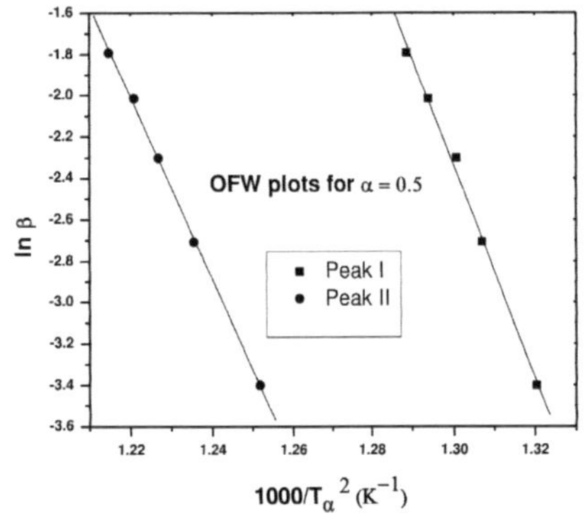

Figure.3 plot of $\ln\beta$ vs $1000/T_\alpha$

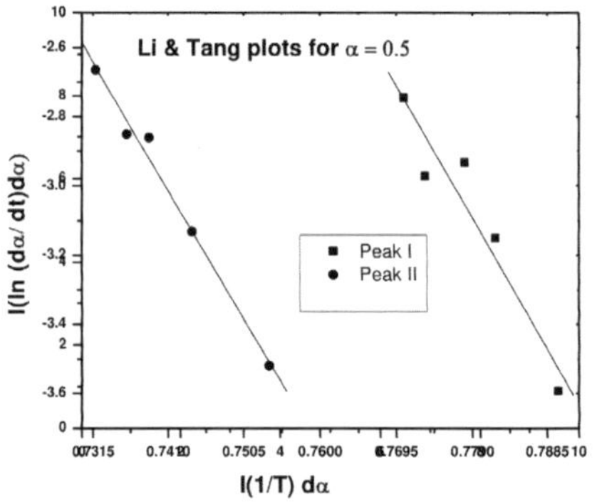

Figure.4 Li and Tang plot

3.1(b) Non-linear integral isoconversional method (Vyazovkin method)

Vyazovkin's method assumes the reaction model, $g(\alpha)$ is independent of β. So, at two different β the ratio of the temperature integral $I(E,T_\alpha)$ to the heating rate β is a constant.For a given α and a set of n experiments performed under different β, the E can be determined at any particular value of α by finding the value of E_α for which the function $\sum_{i\neq j}^{n}\sum^{n}\frac{[I(E_\alpha,T_{\alpha i})\beta_j]}{[I(E_\alpha,T_{\alpha j})\beta_i]}$ is a minimum. Using this procedure E's are calculated and plotted in Figures 5 & 6.

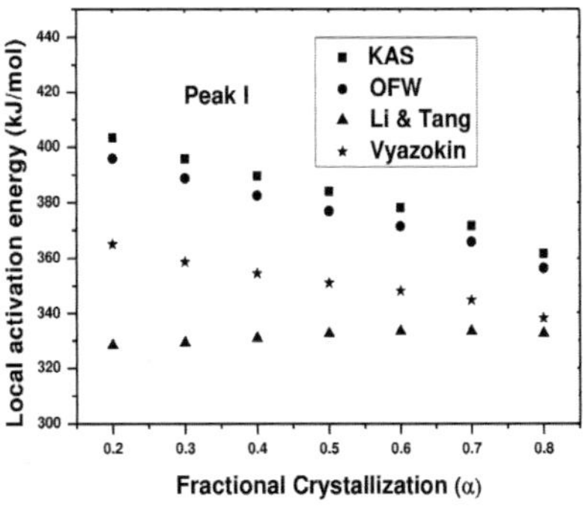

Figure.5 E at different α (peak I)

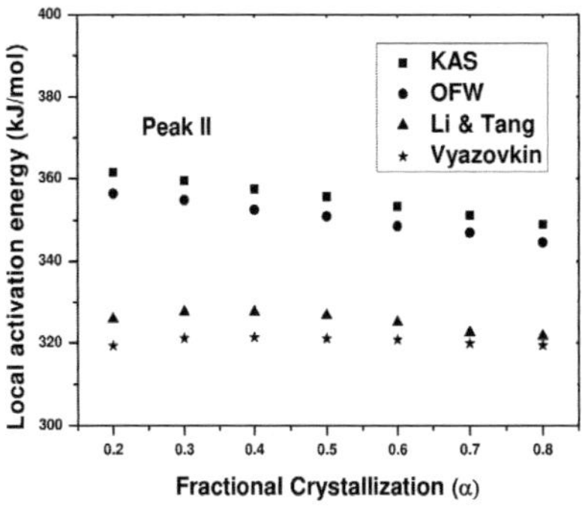

Figure.6 E at different α (peak II)

3.2. Model fitting methods

Model-free isoconversional methods provide accurate estimation of E whereas these methods fail to give correct reaction model. Among the various model fitting methods, CR and IKP methods are used to study the non-isothermal crystallization kinetics.

3. 2 (a). Coats & Redfern (CR) method: This method uses the expression

$$\ln [g(\alpha)/T^2] = \ln [AR/\beta E\ (1\text{-}2RT/E\ )] - E/RT \approx \ln (AR/\beta E) - E/RT \quad (5)$$

CR method uses the above expression and the plot ln $[g(\alpha)/T^2]$ versus 1/T gives E and A for a particular g(α).The method is repeated for all the β's and for different g(α)'s and E's and A's calculated and given in the Table. From the table it is clear that E's and A's are strongly dependent on correlation coefficient |r|.The general rule to determine E is to look for the model corresponding r_{max}. But this corresponds to different E value obtained from other methods. In such cases an r value lower than r_{max} gives the true kinetic model [12].Applying this criterion for peak I, $g(\alpha)= (-\ln(1-\alpha))^{1/4}$ and for peak II, $g(\alpha)= \alpha^{1/4}$ are obtained.

g(α)	β = 2 K/min (Peak I)			β = 2 K/min (peak II)		
	E (kJ/mol)	lnA	|r|	E (kJ/mol)	lnA	|r|
$(-\ln(1-\alpha))^{1/4}$	**321.8±0.7**	**37.8±0.9**	**0.99024**	528.8±0.4	67.0±0.5	0.99896
$(-\ln(1-\alpha))^{1/3}$	434.0±0.9	55.6±1.2	0.99043	710.0±0.5	94.4±0.6	0.99897
$(1-(1-\alpha)^{1/3})$	1175.6±3.1	172.3±4.1	0.98620	1916.4±2.3	275.8±2.8	0.99720
$(1-(1-\alpha)^{1/2})$	1105.8±3.2	161.5±4.2	0.98646	1805.8±2.6	259.5±3.2	0.99591
$\alpha^{1/4}$	190±0.6	76.8±0.7	0.98305	**366.6±0.8**	**42.1±1.0**	**0.98984**

Table 1 Values of E and A obtained for various g(α) by CR method for peak I & II.
Boldface font indicates the most suitable model for the description of the solid state reaction.

3.2 (b). The invariant kinetic parameter (IKP) method

lnA and E are determined from CR method for each β and particular g(α).This process is repeated for different β_i and $g_j(\alpha)$,pairs A_{ij} and E_{ij} are determined. These pairs are given by the compensation effect expression [13]

$$\ln A_i = a_i + b_i E_i \quad (6)$$

where a_i and b_i are called compensation effect parameters. Plotting $\ln A_i$ versus E_i, a_i and b_i, are determined. Again plotting a_i versus b_i, invariant activation parameters E's and A's are determined. These give invariant activation parameters for peak I, E = 445.6 ± 24 kJ/mol, A= 9.4×10^{22} sec^{-1} and for peak II is E=325.7 ±8 kJ/mol and A= 7.1×10^{15} sec^{-1}.

4. Conclusions

Isoconversional methods proposed by KAS and OFW give α dependent E. The values of E calculated from KAS and OFW are very close to each other for both the peaks I and II. This trend is more pronounced for the peak II.The E values remains almost constant with α in the case of peak II and remains at 360 kJ/mol, whereas continuously decreases for the peak I from 460 to 360 kJ/mol. Crystallization Kinetics is a combined process of nucleation and growth. The higher value of E in the initial stage for both the peaks indicates the dominant nature of nucleation and the decreasing trend of E with α reveals the process of saturation of nucleation. That means the nucleation process takes place during the initial stage of transformation and negligible at the end. This is the reason why the isoconversional methods provide E(α) and reveals clearly the crystallization mechanism. The results of Li &Tang for the peak I & II remain at 320 kJ/mol. This may be due to no approximations involved in the derivation and shows clearly the dominant nature of nucleation.

The model fitting isokinetic methods, on the other hand gives single value of E. The CR method is utilized to obtain $g(\alpha)$ as $(-\ln(1-\alpha))^{1/4}$ for peak I and $g(\alpha) = \alpha^{1/}4$ for peak II. The IKP method is also used in conjunction with the CR method to determine E and A values and for peak I, E = 445.6 ± 24 kJ/mol, A= 9.4×10^{22} sec^{-1} and for peak II, E=325.7 ±8 kJ/mol and A= 7.1×10^{15} sec^{-1}.

The evaluated E values from CR and IKP methods are consistent with the values determined from isoconversional KAS and OFW methods for peak I and II. Our analysis concludes that the isoconversional methods gave α dependent E values which reflect the crystallization mechanism. On the other hand isokinetic methods gave single valued E and showed only one mechanism (reaction model) throughout the crystallization process for Fe-based 2605SA1.The present investigation showed that, even though diverse isoconversional and isokinetic methods are used, the results were consistent and complimentary to each other and the same trend is shown in other Co and Fe based metallic glasses [14, 15].

Acknowledgements

T Lilly Shanker Rao acknowledges the financial support rendered by UGC, New Delhi through a minor project no. 47/718/08(WRO) [Project period is over].The authors also acknowledge the experimental facilities (DSC & XRD) utilized from CMCRI, Bhavnagar-364002, Gujarat (India) and UGC-DAE-CSR, Indore (India).

References

[1] Paulik F 1995 Special Trends in Thermal Analysis; (John Wiley & Sons, Ch.10)
[2] Ortega A 2002 Int J. Chem.Kinet **34**193
[3] Sestak J 1979 J. Thermal Analysis **16** 503
[4] Kissinger H E1957 Anal. Chem **29** 1702
[5] Akahira T, Sunose T 1971 Res.Report Chiba.Inst.Technol(Sci. Technol.) **16** 22
[6] Ozawa T1965 Bull. Chemical Society of Japan **38** 1881
[7] Flynn J H, Wall L A 1966 J.Res.Natl.Bur.Standards A Phys.Chem **70**A 487
[8] Li C R, Tang T B 1999 Thermochimca Acta **325** 43
[9] Vyazovkin S, Wight C A1997 J.Phys.Chem,*A* **101**8279
[10] Coats A W , Redfern J P1964 Nature **201** 68
[11] Lesnikovich L , Levchik S V 1985 J. Thermal Anal.**27**83; ibid **30** 667
[12] Budrugeac P et al1996 J.Thermal Analysis **47** 123
[13] Galwey A K 2003 Thermochim Acta **399** 1

[14] Heena Dhurandhar et al 2010 Journal of ASTM International **7** 1
[15] Arun Pratap et al 2007 J.Therm Anal & Calorim **89** 399

Wet chemical synthesis of $Na_{21}(SO_4)_7F_6Cl$:Ce optoelectronics materials

V V Shinde[1] and S J Dhoble[1]
[1]Department of Physics, RTM Nagpur University, Nagpur -440033, India

E-mail: sjdhoble@rediffmail.com

Abstract. A new $Na_{21}(SO_4)_7F_6Cl$ optoelectronics nano phosphor in the Cl–F system doped with Ce^{3+} and synthesized by wet chemical method having particle size up to 100 nm has been reported here and their luminescence properties were investigated by changing the concentration of activators in the host matrix. Polycrystalline $Na_{21}(SO_4)_7F_6Cl:Ce^{3+}$ nanophosphors have been studied for their X-ray diffraction, SEM, TEM, FT-IR and photoluminescence (PL) characteristics. The halo sulphate phosphor $Na_{21}(SO_4)_7F_6Cl:Ce^{3+}$ show an efficient broad band emission at 389 nm in near UV region assigned to $^4f_0{}^5d_1 \rightarrow {}^4f_1$ transition when excited at 355 nm. The investigated halo sulphate phosphors may be suitable for a near UV excited solid state lighting devices.
Keywords: Optoelectronic, Sulphohalite, Schairerite, Photoluminescence.

1. Introduction

Sulphate based inorganic phosphors doped with various rare earth ions form an important family of optoelectronic material for solid state lighting devices. These devices offer many benefits and are tipped to be the next generation lighting devices, in the replacement of conventional incandescent and fluorescent lamps which are Hg pollutants, frangible and high energy consumption devices. In the recent years, research on the sulphate based phosphors has become a hot topic and gained importance. A large number of papers have been published in the related journals about the successful synthesis and luminescence of rare earth doped phosphors with the orientation of exploring novel phosphors suitable solid state lighting applications. Up till now, the group of sulphohalites has not been studied in detail in the field of luminescence. The Cl–F substitution in the system has been investigated by Dhoble et al [1–8]. Klement synthesized halo sulfate based phosphor such as, $Na_{6.45}Ca_{3.55}(SO_4)_6(F_xCl_{1-x})_{1.55}$[9], $Na_6Ca_4(SO_4)_6F_2$[10], $Na_6Pb_4(SO_4)_6Cl_2$[11,12], and $Na_6Cd_4(SO_4)_6Cl_2$[13]. Some investigations are still going on novel halo sulfate based materials for its applications in various purposes. Nanoscale phosphors have a number of potential advantages over traditional micron-sized phosphors providing higher packing densities. In addition, quantum confinement in nanocrystalline materials may result in an enhancement of their luminescence. Therefore, the morphology and size, the stoichiometry and composition, and the surface characteristics must be controlled in order to achieve the desirable objectives of powder phosphors. To our knowledge, no investigations have been reported on studying the photoluminescence properties of rare earth in the hosts of $Na_{21}(SO_4)_7F_6Cl:Ce^{3+}$. Here, we have reported the XRD, SEM, TEM and FT-IR of the undoped and the photoluminescence properties of Ce^{3+} activated alkaline earth $Na_{21}(SO_4)_7F_6Cl$ materials by using the conventional wet chemical method successfully for the possible application in near UV excited solid state lighting devices.

2. Experimental

The pure and Ce^{3+} doped samples nanocrystalline $Na_{21}(SO_4)_7F_6Cl$ phosphors were prepared by a wet chemical method. While preparing $Na_{21}(SO_4)_7F_6Cl$, raw materials Na_2SO_4 , NaCl and NaF of annular grade were taken in a stoichiometric ratio and dissolved separately in double distilled de-ionised water, resulting in a solution of $Na_{21}(SO_4)_7F_6Cl$ stirred for several hours to complete the reaction. The Chemical reaction is as shown below.

$$7\ Na_2SO_4 + NaCl + 6\ NaF = Na_{21}(SO_4)_7F_6Cl$$

To obtain $Na_{21}(SO_4)_7F_6Cl{:}Ce^{3+}$, water soluble Ammonium Hexanitratocerate $(NH_4)_2Ce(NO_3)_6)$ was used which is added drop wise to a pure solution of $Na_{21}(SO_4)_7F_6Cl$ confirming that no undissolved constituents were left behind and all the salts had been completely dissolved in water and thus reacted. $Na_{21}(SO_4)_7F_6Cl{:}Ce^{3+}$ in powder form were obtained by evaporating the solution at 80^0C for 5 hours. The dried samples were then slowly cooled down at room temperature. The resultant polycrystalline mass was crushed to fine particle in agate mortar and then subjected to a post treatment for quenching at 400^0C for 1 hour in a muffle furnace. This powder was then used for further study.The phase purity and structure of the final products of the $Na_{21}(SO_4)_7F_6Cl{:}Ce^{3+}$ optoelectronics nanophosphors was examined by x-ray powder diffraction using Cu Kα radiation on a BRUKER – analytical x-ray diffractometer, at a scanning step size of 0.015, with a continuing step of 5 sec, in the 2 theta range from 0^0 to 80^0. Phosphor morphology was studied by a Geol-6380A scanning electron microscope (SEM). All transmission electron microscopy (TEM) images were obtained on 100 keV and an AMT XR41-B 4-megapixel (2048 x 2048) bottom mount CCD camera. For the TEM measurement, the samples were ground and mixed together with ethylene glycol and dispersed under ultrasonic vibration for 30 minutes. A drop of the dispersed particles was kept over the carbon coated copper grid and evaporated to dryness. Images from the camera are saved as high-resolution TIFF files. Infrared spectra of the pure were recorded on using a Bruker Fourier transform infrared spectrometer. For the measurement of spectroscopic properties, the excitation and emission spectra for all samples were recorded on Shimadzu RF-5301PC spectrophotofluorometer. Emission and excitation spectra were recorded using a spectral slit width of 1.5 nm at room temperature.

3. Results and Discussion

3.1 X-Ray Diffraction

Figure 1 shows the diffraction data pattern for the pure $Na_{21}(SO_4)_7F_6Cl$ phosphor. It shows good agreement with ICDD data file no. #70-1811. The XRD pattern did not indicate presence of the constituents Na_2SO_4 or NaF or NaCl and other likely phases. An excellent match is seen indicating formation of single phase, homogeneous final product $Na_{21}(SO_4)_7F_6Cl$ phosphor.

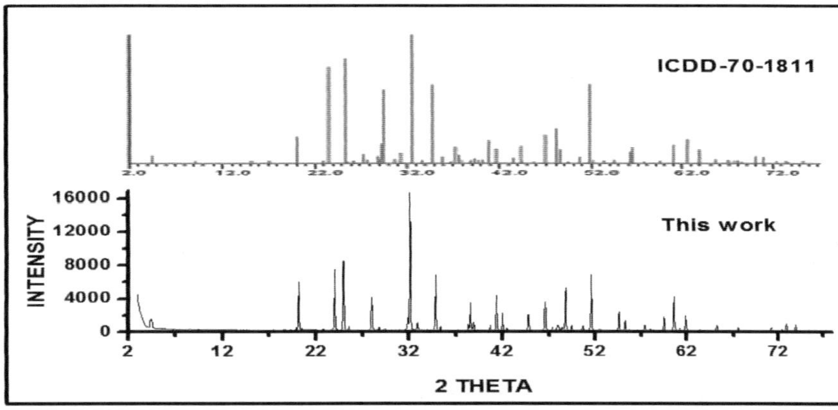

Figure 1. XRD Pattern of synthesized $Na_{21}(SO_4)_7F_6Cl$.

3.2 Crystalline Structure

As per available data the historical name of $Na_{21}(SO_4)_7F_6Cl$ material is Schairerite having hexagonal basis belongs to space group P31m (No. 157) with cell parameter a = 12.19A^o and c = 19.359A^o [14]. A ball (sphere) model of crystal structure of $Na_{21}(SO_4)_7F_6Cl$ (Schairerite) is as shown in figure 2. The crystal structure may be considered as consisting of 7 sheets of Na^+ ions perpendicular to the c axis. These sheets are connected to each other, building up a three dimensional framework. The Na^+ ions in these sheets are arranged in an array of hexagons and triangles. Sulphur atoms are lying in the sheets at the centers of each hexagon. The halogen atoms lie between the sheets midway between the centers of two triangles.

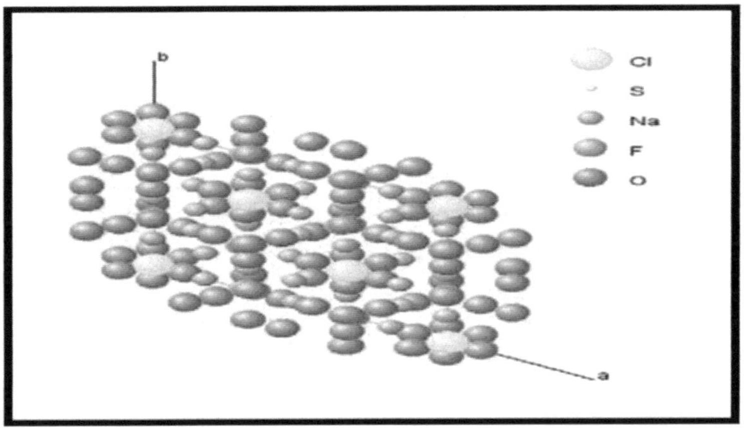

Figure 2. Crystal structure of $Na_{21}(SO_4)_7F_6Cl$.

3.3 *Surface Morphology.*

A typical SEM micrographs of as synthesized $Na_{21}(SO_4)_7F_6Cl$ powders is shown in Figure 3. It is clearly seen that the grains of the phosphor are not uniform they are irregular in shape and size with average grain size is nearly 4-5 μm. The SEM micrograph showed different morphology of the phosphor with agglomerated crystallites. The agglomeration may be due to the prolong stirring of the reaction mixture. However the actual particle size may be even smaller due to the wet chemical synthesis rout, in order to find the actual particle size of the phosphor the phosphor material characterized by TEM by dispersing the powder phosphor in alcohol by sonication. The typical TEM image of the $Na_{21}(SO_4)_7F_6Cl$ phosphor is illustrated in Figure 4. It shows the formation of round shape particles with average size up to 100 nm. Thus TEM study establishes nano crystalline nature of the prepared phosphor material.

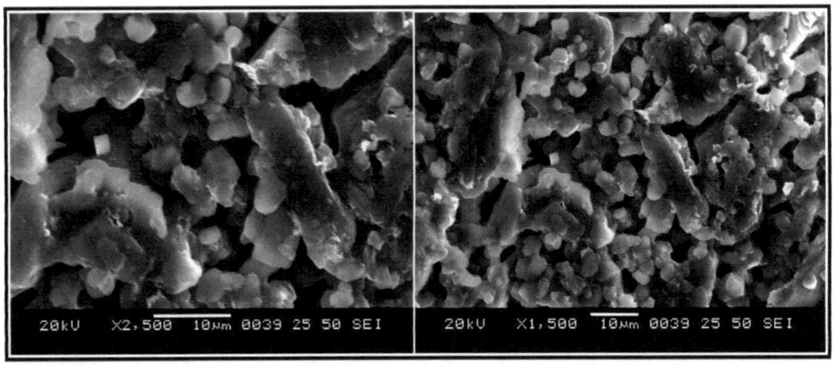

Figure 3. Scanning electron microscope (SEM) image of synthesized $Na_{21}(SO_4)_7F_6Cl$.

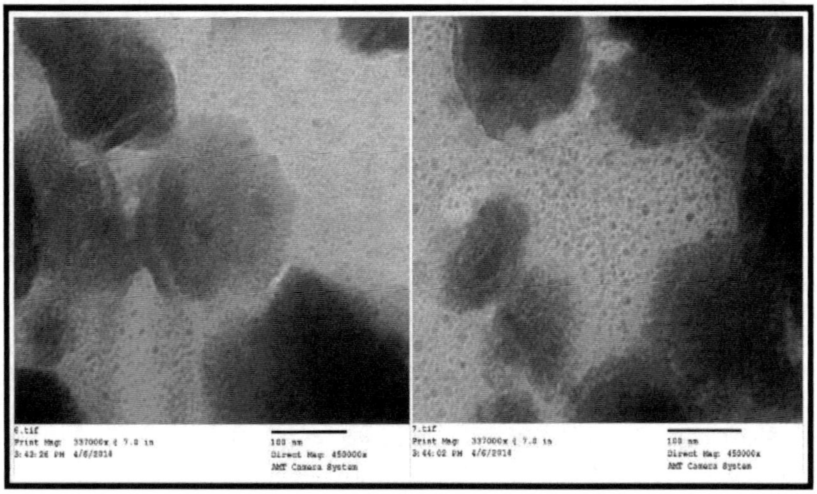

Figure 4. Transmission electron microscope (TEM) image of synthesized $Na_{21}(SO_4)_7F_6Cl$.

3.4. FT-IR Analysis

The infrared spectrum of the synthesized $Na_{21}(SO_4)_7F_6Cl$ is shown in Figure 5. The peak at 3183 cm^{-1} region, the O-H stretching frequency, is significant because it indicates a presence of either OH^- anions or H_2O molecules. The peak appeared at 1313 cm^{-1} is assigned to v_3 asymmetric stretching of $(SO_4)^{2-}$ group. The peak appeared at 1090 cm^{-1} is due to v_1 symmetric stretching of $(SO_4)^{2-}$ group. The peak at 833 cm^{-1} is v_2 out-of-plane bending of $(SO_4)^{2-}$ group. The peaks below 700 could be v_4 in-plane bending of $(SO_4)^{2-}$ group.

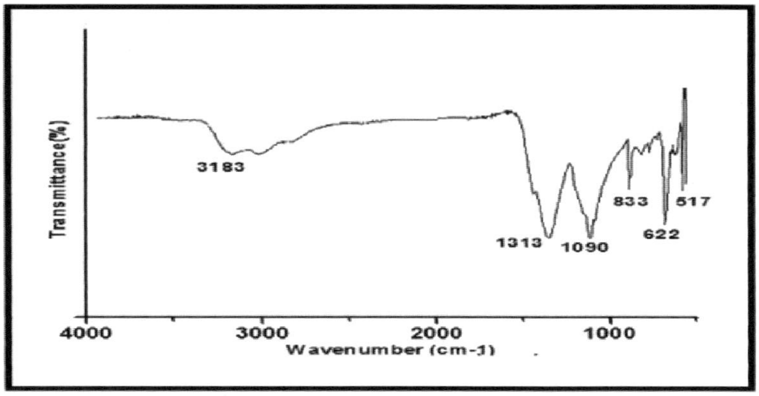

Figure 5. FT-IR Spectrum of the synthesized $Na_{21}(SO_4)_7F_6Cl$ phosphors.

3.5. PL Emission of Ce^{3+} in $Na_{21}(SO_4)_7F_6Cl$

The excitation and emission spectra of Ce^{3+} doped $Na_{21}(SO_4)_7F_6Cl$ optoelectronic phosphor material are shown in Figure 6. The broad excitation peak is observed at 355 nm and broad emission peak in UV range focussing at 389 nm due to the $^4f_0{}^5d_1 \rightarrow {}^4f_1$ transition of Ce^{3+}. The spectroscopy of Ce^{3+} [15] is somewhat different from that of other Ln^{3+} compounds. The promotion of an electron from the ^4f to the ^5d orbitals is due to lowest energy transition of Ce^{3+}. All transitions of Ce^{3+} are spin-allowed because the ground state and the excited states of this ^4f ion are spin doublets. Therefore, the excitation peak appears in the UV spectral region at 355 nm. The emission spectrum centring at 389 nm is due to the $^4f_0{}^5d_1 \rightarrow {}^4f_1$ transition of Ce^{3+} ion from ^5d excited state to ^2F ground state (the lowest level) showing broad emission peak being regarded as the electric dipole allowed one. Generally it is observed that the emission from these f-d state show two bands as the ground state of the Ce^{3+} ion is a doublet ($^2F_{5/2}$ and $^2F_{7/2}$) with a separation of about 2000 cm^{-1}. In the present work, however, such a doublet character of the emission band is not observed. This may be due to lack of symmetry in the host lattice. Thus, the activator Ce^{3+} ion may have experienced a random distribution in local fields of host lattice [16, 17]. The peaks of all concentration of phosphor are centred at 389 nm. With increasing concentration of Ce^{3+} ions, peak intensity of the emission increases as shown in inset of figure 6. Maximum intensity is observed for 2 mole % of Ce^{3+} ions indicating that $Na_{21}(SO_4)_7F_6Cl$:Ce phosphor lattice is more suitable for higher concentrations of Ce^{3+} ions. Beyond this, the emission intensity decreases, this could be attributed to the effect of concentration quenching. This may be due to energy transfer between the Ce^{3+} ions and defects of the $Na_{21}(SO_4)_7F_6Cl$ host and/or due to cross relaxation between the Ce^{3+} ions[18].

Figure 7 shows schematic representation of energy levels, energy transitions and configurational coordinate diagram of Ce^{3+}. The ion Ce^{3+} has $^4f_1$ electronic configuration. The ground state of Ce^{3+} is the doublet $^2f_{5/2}$ and $^2f_{7/2}$. The lower excited states belong to 5d configuration. The Ce^{3+} emission corresponds to transitions from the lowest 5d level to $^2f_{5/2}$ and $^2f_{7/2}$states of the $^4f_1$ configuration. The excited state derived from the 5d state is sensitive to the crystal field and is coupled to the lattice vibrations which results in broader band emission rather than line emission. In configurational coordinate diagram, the two lowest parabolas represent the ground electronic configuration $^4f_1$ by spin–orbit interaction into $^2F_{5/2}$ (the ground state) and $^2F_{7/2}$ (the first excited state). The lowest states of excited electronic configuration $^5d_1$ are represented by two higher parabolas. Electron–lattice interaction in the excited electronic manifold causes the shift of the respective parabolas in the configurational space. The potential energy of the luminescent centre Ce^{3+} is plotted on the vertical axis and the value of single parameter describing the effective displacement of the ion surrounding the activator, Q is plotted on the horizontal axis. Promotion of the inner 4f electron to the outer 5d states perturbs the surrounding ions, the lattice relaxes, and the potential energy curve changes as shown below in figure 7. The good adaptability of the position of the 4f–5d absorptions and the corresponding emissions make Ce^{3+} an important ingredient in solid state light emitting materials. The respective excitation and emission pathways are indicated in figure 7 by arrows [19-20].

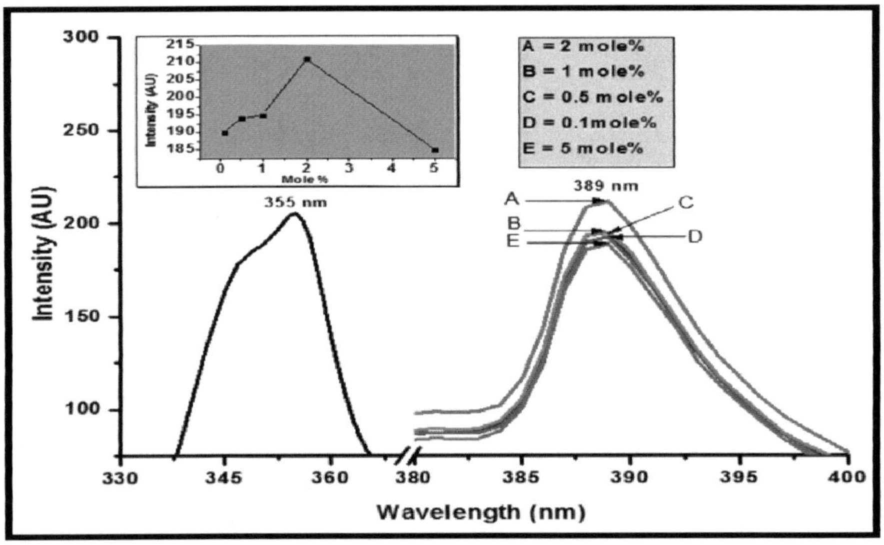

Figure 6. Excitation and emission spectra of $Na_{21}(SO_4)_7F_6Cl$: Ce^{3+} phosphors.

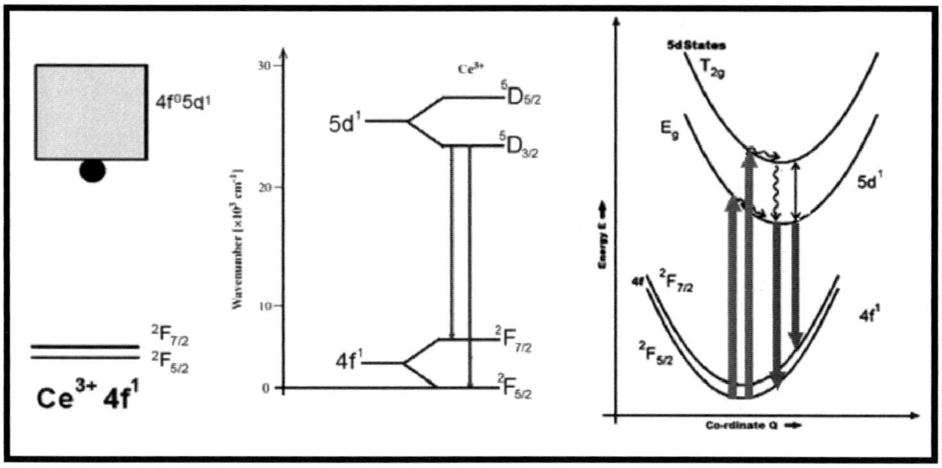

Figure 7. Schematic representation of energy levels, energy transitions and configurational coordinate diagram of Ce^{3+}.

4. Conclusions

A novel nanocrystalline optoelectronic $Na_{21}(SO_4)_7F_6Cl$ nanophosphors in the Cl–F system doped with Ce^{3+} has been prepared by the wet chemical synthesis method and the synthesis were confirmed by XRD. The TEM result shows the formation of round shaped nanomaterial. From the results presented here it can be concluded that Ce^{3+} ions are occupied site with symmetric environment in $Na_{21}(SO_4)_7F_6Cl$ host matrix. Ce^{3+} emission in the host was observed at 389 when excited at 355 nm. Based on the results obtained, we suggest that these optoelectronic luminescent nanophosphors have possible potential applications as a near UV convertible phosphor for light emitting devices and also for energy transfer process in co-activated phosphors to increase the intensity of f-f transitions in near UV region.

References

[1] Gedam S C, Dhoble S J and Moharil S V 2006 *J. Lumin.* **121** 450

[2] Gedam S C, Dhoble S J and Moharil S V 2007 *J. Lumin.* **124** 120

[3] Gedam S C, Dhoble S J and Moharil S V 2007 *Eur. Phys. J. Appl. Phys.* **37** 73

[4] Gedam S C, Dhoble S J and Moharil S V 2007 *J. Lumin.* **126** 121

[5] Gedam S C, Dhoble S J, Omanwar S K and Moharil S V, 2007 *Eur. Phys. J. Appl. Phys.***39** 39

[6] Gedam S C, Dhoble S J, Nagpure I M, Godbole S V and Bhide M K 2007 *J. Phys. D: Appl. Phys.* **40** 6039

[7] Gedam S C, Dhoble S J and Moharil S V 2008 *J. Lumin* **128** 1

[8] Gedam S C, Dhoble S J, Nagpure I M, Godbole S V , Bhide M K and Moharil S V 2008 *J. Mater. Sci.* **43** 3189

[9] Piotrowski A, Kahlenberg V and Fischer R X 2002 *J. Solid-State Chem.* **163** 398

[10] Klement R 1939 *Naturwissenschaften* **27** 568

[11] Schneider W, Jahrb N and Miner F 1967 *Monatschefte* 284

[12] Schneider W, Jahrb N and Miner F 1969 *Monatschefte* 58

[13] Kim H J, Jeong D Y, Zalar B, Blinc R S and Choh H 2000 *Phys. Rev. B* **61** 9307

[14] Pabst A, Sawyer D L and Switzer G, 1963 *J. Mineralogical Society of America* **48** 5

[15] Dorenbos P 2000 *Phys. Rev. B* **62** 15640

[16] Dorenbos P 2000 *J. Lumin.* **87** 970

[17] Van Eijk C W E, Andriessen J, Dorenbos P and Visser R 1994 Nucl. Instr. and Meth. **A348** 546

[18] Dorenbos P 2001 *Phys. Rev. B* **64** 125117

[19] Curie D 1975 *Plenum* 71

[20] Drickamer H G, Frank C W and Slichter C P 1972 *Proc. Nat. Acad. Sci* **69** 933

Thermal conductivity measurement of organic solvents incorporated with silver nanoparticle using photothermal techniques

N Shemeena Basheer [1], B Rajesh Kumar [1], Achamma kurian[1] and Sajan D George[2]

[1]Photonics Lab, Department of Physics, Catholicate College, Pathanamthitta, India, 689645

[2] Centre of Atomic and Molecular Physics, Manipal University, Manipal, Karnataka, India, 576104

E-mail address: shemeenahaffis@rediffmail.com

Abstract.The present paper summarizes our recent work on heat transfer of nanofluid in different organic solvents determined using photothermal techniques. UV- Visible absorption spectrum shows an intense surface plasmon absorption peak at 427 nm. TEM indicates the presence of well dispersed silver nanoparticles having spherical morphology with particle size ranging from 10 to13 nm. The experimental results show that thermal conductivity of organic solvents increases with the incorporation of silver nanoparticles.

1. Introduction.

The Heat transfer fluids such as water, ethylene glycol and mineral oils etc have immense applications in the field of power generation, chemical production, automobiles, computing processes and refrigeration. Nanofluids are colloidal suspensions engineered from conventional base fluids by dispersing solid nanoparticles (less than 100 nm). Metallic nanoparticles, particularly silver nanoparticles have strong absorption in the visible range due to surface plasmon resonance and have got much attention from chemical industry and medicine due to unique properties such as high thermal conductivity, high resistance of oxidation, and antibacterial activity. Nowadays, highly sensitive photothermal techniques such as thermal lens and photoacoustic techniques are utilized to investigate thermal parameters (thermal diffusivity and effusivity) of nanofluids. The thermal diffusivity is the quantity which measures the rate of heat diffusion through the sample and thermal effusivity is a measure of its ability to exchange heat with its surroundings. The thermal diffusivity α and thermal effusivity e can be related to thermal conductivity by $e = k/\sqrt{\alpha}$.

2. Experimental

Inorder to synthesize silver nanoparticles numerous methods have been adopted including chemical reduction method, polyol method and radiolytic process. For the present study citrate reduction method is used for the preparation of silver nanofluid .The silver nanofluid is then diluted to 0.16mM in ethylene glycol and ethanol respectively .The absorption spectrum recorded exhibits an intense absorption peak at 427 nm due to surface plasmon absorption and TEM indicates spherical silver nanoparticles of size ranging from 10-13 nm, which is shown in figure 1 (a and b)

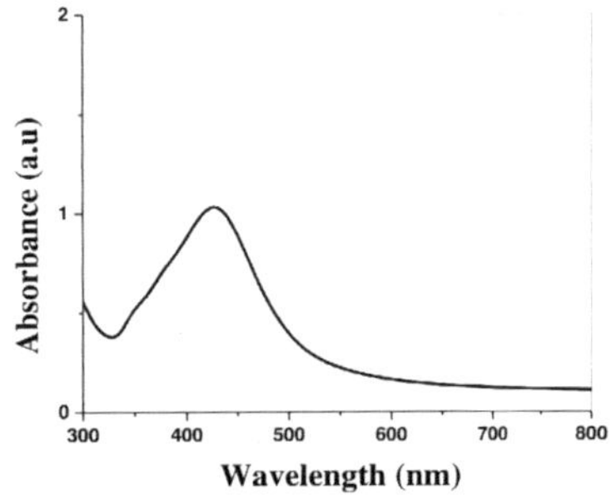

(a)

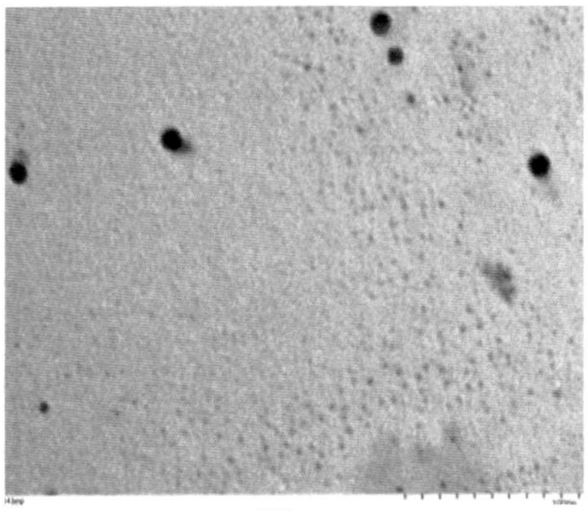

(b)

Figure 1 (a) Absorption spectrum of Ag nanofluid (b) TEM image of silver nanoparticle

3. Result and Discussions

3.1. Thermal lens measurement

Thermal lens technique is highly sensitive and powerful technique enough to measure very small refractive changes due to the temperature variations in the medium. Such refractive changes can be thermally induced by the gaussian laser beam which is the fundamental principle of thermal blooming technique. Due to this modification in refractive index, the medium mimics a lens, called thermal lens (TL).The experimental setup for the thermal diffusivity measurement is shown elsewhere[1]. Gordon et al. [2] reported a detailed theoretical analysis of the formation of thermal lens. The expression for the time dependent probe beam intensity is given by

$$I(t) = I(0)\left[1 - \theta(1 + t_c/2t)^{-1} + \frac{1}{2}\theta^2(1 + t_c/2t)^{-2}\right]^{-1} \tag{1}$$

The parameter θ and t_c can be generated by fitting equation (1). The characteristic time constant t_c is related to the thermal diffusivity D and beam radius ω through the relation

$$t_c = \omega^2 / 4D \tag{2}$$

To eliminate the uncertainty in the determination of beam radius, a reference sample with known thermal diffusivity is used to determine the thermal diffusivity of unknown sample. From this we can calculate the thermal diffusivity D as

$$D = D_{reference}\, {t_{c\,reference}} / {t_c} \,. \tag{3}$$

Here water is used as the reference sample with known thermal diffusivity $14.3\times10^{-8}m^2s^{-1}$. Thermal diffusivity of ethanol /ethylene glycol nano mixtures shows an enhancement with respect to pure solvents [3] and solvent –water mixtures as given in table 1. The thermal lens decay curve of Ethylene glycol/nano mixture is given in figure 2 and the variation of the intensity of the TL signal with time was studied.

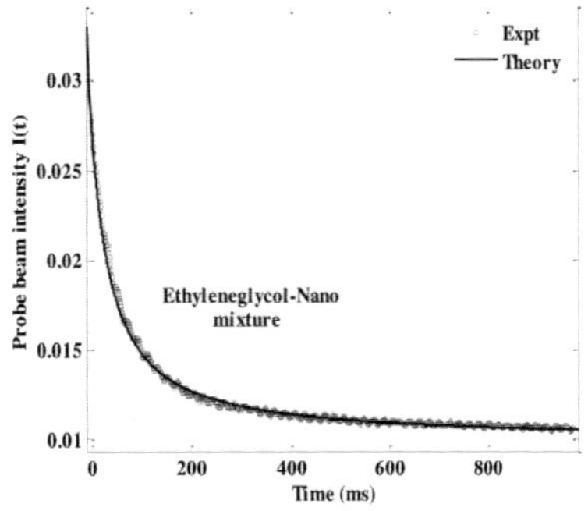

Figure 2. Transient decay curve measured by mode-matched T.L technique.

3.2. Photoacoustic measurement

The microphone based open cell photoacoustic (PA) technique depends upon the detection of pressure fluctuations in a closed cavity created by the periodic thermal waves from the sample. The experimental arrangement and detailed theory for the thermal effusivity measurement is shown elsewhere [4]. The thermal effusivity of organic solvents incorporated with silver nanofluid can be calculated according to Rosencwaig-Gersho theory of PA effect [5]. Sikorska et.al. [6] modified the expression for thermal effusivity e by including the thermal effusivity value of known sample. The expression for thermal effusivity value is rewritten as

$$e_x = \left[\frac{\sqrt{\left\{ \left(e_r \sqrt{d} + 1 \right)^2 + 1 - R^2 \right\}}}{R} - 1 \right] \tag{4}$$

where, $d = \left(e_s l_s \right)^{-2} \dfrac{\alpha_s}{\pi f}$, α represents thermal diffusivity and l denotes the thickness of aluminium foil, subscripts s and x refer to the metal foil and liquid respectively**.** By measuring the ratio of PA amplitudes as a function of modulation frequency f , one can evaluate the thermal effusivity value. Compared to pure solvents [7], solvent-nano mixtures showed an enhancement in thermal effusivity.

The thermal conductivity of organic solvents can be calculated using the relation $e = {k}/{\sqrt{\alpha}}$ by putting the corresponding values of thermal diffusivity and thermal effusivity and is shown in table1. There are several mechanisms proposed for the experimentally observed enhancement in thermal properties of the nanofluids. In the present study when silver nanofluid is added to organic solvents an enhancement in thermal conductivity was observed. Since water has a higher thermal conductivity than organic solvents, any presence of water in organic solvents would lead to an increase in the effective thermal conductivity. The prepared silver nanofluid is water based and therefore the presence of water in solvent-nanofluid mixture could lead one to erroneously conclude that the measured enhancement is due to the presence of nanoparticles. In order to ensure that whether the enhancement is due to the presence of water or nanoparticles, we measured the thermal conductivities of both the solvent-water mixture and solvent-nano mixture. From the result it is clear that thermal conductivity of solvent nano mixture is greater than that of solvent-water mixture. The increase in thermal diffusivity of binary (solvent-water) solution with respect to pure organic solvents may be explained on the basis of cohesive interactions due to hydrogen bonding within the solutions.

Table 1.Thermal parameters of organic solvent-nanofluid mixtures.

Sample	Thermal diffusivity α $(10^{-8} m^2/s)$		Thermal effusivity e $\times 10^4\ Ws^{1/2}K^{-1}m^{-2}$		Thermal conductivity (W/mK)
	Solvent- water mixture	Solvent-nano mixture	Solvent-water mixture	Solvent-nano mixture	Solvent-nano mixture
EG	10.5 ± 0.1	13.2 ± 0.1	0.108 ± 0.001	0.101 ± 0.001	0.3669
Ethanol	10.3 ± 0.2	12.7 ± 0.1	0.072 ± 0.001	0.067 ± 0.001	0.2387

Influence of nanoparticles on the thermal diffusivity of organic solvents can be explained on the basis of nanolayers. Liquid molecules close to a solid surface are known to form layered structures. The layered molecules are in an intermediate physical state or acts as a thermal bridge between a bulk liquid and a solid nanoparticle [8]. This solid-like nanolayer of liquid molecules would be expected to be the reason for thermal conductivity enhancement in organic solvent-nano mixtures.

4. Conclusions

Photothermal techniques have been successfully utilized for the evaluation of thermal conductivity of organic solvent-nanofluid mixtures. Enhancement in thermal conductivity was observed with the incorporation of silver nanoprticles.

References

[1] Joseph S A, Mathew S, Sharma G, Hari M, Achamma Kurian, Radhakrishnan P and Nampoori V P N 2010 *Plasmonics* **5** 63-68
[2] Gordon J P, Leite R C C, Moore R S, Porto S P S and Whinnery J R 1965 *J. Appl. Phys.* **36** 3-8
[3] Bindhu C V, Harilal S S, Nampoori V P N and Vallabhan C P G 1998 *Opt. Eng.* **37** 2791-2794
[4] Rajesh Kumar B, Shemeena Basheer N, Joseph S A, Achamma Kurian and George S D 2010 *Proc.of SPIE Nanophotonic materials V11* **7755** 77550N 1-5
[5] Rosencwaig A and Gersho A 1976 *J.Appl.Phys.* **47** 64-69
[6] Sikroska A, Linde B B J and Wirba W Z 2005 *Chem.Phys.* **320** 31-36
[7] Weast R C 1987 *CRC Handbook of Chemistry and Physics* (CRC press Boca Raton FL)
[8] Yu C J, Richter A G, Datta A, DurbinM K and Dutta P 2000 *Physica B* **283** 27-31

Spectroscopic studies on Rhodamine B intercalated K-10 montmorillonite aqueous dispersions

Lyjo K Joseph* [1], H Suja[2], G Sanjay[3], S Sugunan[4], V P N Nampoori[5] and P Radhakrishnan[5, 6]
[1] Department of Physics, Government College Kottayam 686013- Kerala, India
[2] Department of Chemistry, NSS College Nemmara, Palakkad, Kerala, India
[3] Toc H Institute of Science and Technology, Arakkunnam, Ernakulam 682313
[4] Department of Applied Chemistry, Cochin University of Science and Technology
[5] International School of Photonics, Cochin University of Science and Technology
[6] Centre of Excellence in Lasers and Optoelectronic Sciences, Cochin University of Science and Technology, Kochi 682022 Kerala, India

*Corresponding author email: lyjokjoseph@yahoo.co.in

Abstract: In this paper, the optical absorption and fluorescence studies on rhodamine B (RB) intercalated acid activated K-10 montmorillonite dispersions is presented. The aqueous dispersions were prepared from the dried dye intercalated montmorillonite. The absorption and fluorescence peaks of RB dispersions show a bathochromic shift with respect to the dye concentrations. The samples have a fluorescent emission at 421 nm which is having an intensity variation depending on the concentration of the dyes. The behaviour of samples of same concentration prepared by two different routes is also presented.

1. Introduction:

Clay minerals are attractive hosts for the intercalation of dye molecules. Dye molecules are incorporated in the clay mineral network with strong dye and host matrix interactions. Dye intercalated materials are exploited in modern photochemistry and photophysics due to their transparency in the visible wavelength region and their optical activity [1]

2. Materials and methods:

2.1. Dye adsorption from solution

1g K-10 was treated with 100 ml RB dye solution (concentration 10^{-5}, 10^{-4} and 10^{-3} M respectively) and stirred magnetically for 24 h. It was then centrifuged, Soxhlet extracted with deionised water and dried at 120 °C for 12 h. The samples were designated as KRB–Z where –Z denoted the concentration. To measure the maximum adsorption capacity, the same process was repeated with 10^{-3} M solution for three times. The sample was denoted as KRB–M [2].

2.2. Ultrasonicated dye adsorption

About 100 ml of the dye solution (10^{-3} M) was mixed with 1 g clay and was subjected to ultrasonic treatment. After the required time, the solution was allowed to settle overnight, centrifuged, washed and air dried. It was subjected to Soxhlet extraction with distilled water to remove any weakly adsorbed dye and further dried at 120°C for 12 h. The sample was designated as KRB-S.

2.3. Dispersions of dye intercalated montmorillonite

The aqueous dispersions of the above mentioned samples were prepared by ultra-sonicating 0.1g of each of the samples in 100 ml water for 10 minutes. To check the absorption behaviour of K-10, 0.1g was dispersed in 20ml.

2.4. Optical characterisation

The absorption and the fluorescence spectra of the aqueous dispersions were noted using spectrophotometer (JASCO V-570) and fluorescence spectrophotometer (Varian, Cary Eclipse).

3. Results and Discussions

3.1. Optical Absorption Studies

The main absorption peak of RB at 555 nm in all samples was bathochromically shifted (figure 1). KRB-3 and KRB-S showed the same peak around 580 nm but the intensity of KRB-S was higher. The other main peaks (357 and 257 nm) of RB appear as the RB concentration in the clay mineral increases. KRB-M showed broad absorption spectra with peaks at 537, 560, 572 and 590 nm. The absorption spectra of the KRB-5 and K-10 were similar as the amount of dye loaded was very low in KRB-5. K-10 had a shoulder near 250 nm as in the case of all montmorillonites [3] and the RB had a peak around the same wavelength. The highly concentrated K-10 (0.1g in 20ml water) dispersion had small peaks near 371, 421, 521 and 683 nm (inset of figure 1).

All the dye intercalated samples showed bathochromic shift which may be due to the formation of J- aggregates [4].The more absorbance shown by KRB-S with respect to KRB-3 may be due to the more uniformly distributed dye molecules and the minimisation of cluster formation by ultra sonication. However, they had the same dye concentration and showed the same peak at 580nm. KRB-M showed a broad absorption spectrum due to dye molecular aggregation on the surface of the clay minerals arising from the hydrophobic interactions and the molecular exciton model explains this [4].

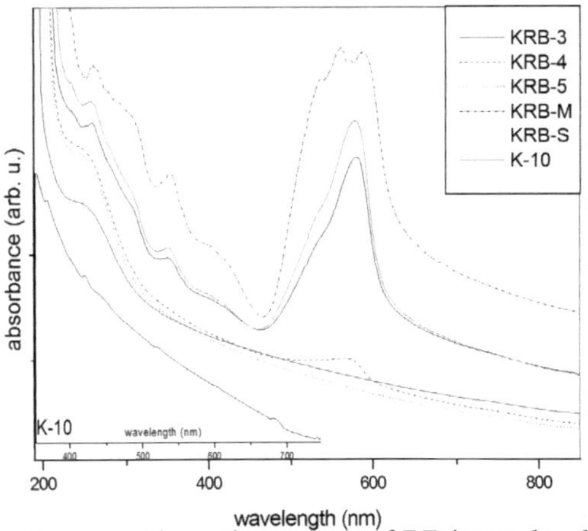

Figure 1. Absorption spectra of RB intercalated K-10 montmorillonite aqueous dispersions (0.1g sample in 100 ml water). Inset: Absorption spectrum of K-10 montmorillonite aqueous dispersion (0.1g K-10 in 20 ml water)

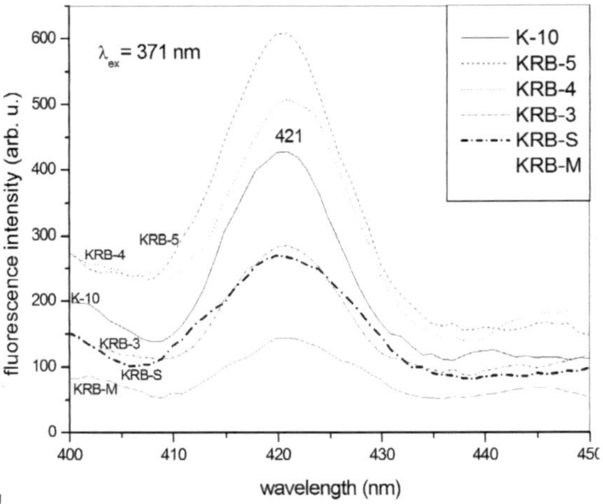

Figure 2. Fluorescence spectra of RB intercalated K-10 montmorillonite aqueous dispersions (0.1g sample in 100 ml water) at 421 nm when excited using 371 nm

Though montmorillonites are weak absorbers in the visible and NIR range [3], the weak peaks of K-10 (figure 1- inset) can be attributed to the silica and alumina layers present in their network. The presence of the dye in the network of the RB intercalated K-10 enhances the absorption at these peaks.

The adsorption of rhodamine onto clays usually leads to the metachromatic effect in the absorption spectra, whereby the main absorption band is shifted to higher energies [1]. In montmorillonite, metachromasy results from π interactions between the adsorbed dye and the oxygen plane of the alumino-silicate [5].

The J-aggregates, which are formed more rarely than the H-aggregates, absorb light of lower energies and exhibit luminescent properties [6]. The bathochromic shift in the absorption spectra of the samples was due to the formation of J-aggregates. But, KRB-M showed the presence of the H-aggregates as evident from the hypsochromic shift and the broadened nature of the main absorption peak of RB.

3.2. Fluorescence studies

The samples showed a fluorescence peak at 421nm when excited with 371nm (figure 2). This may be arising from the silica layer. All the samples were excited at the same conditions and it was observed that the peak intensity increased for KRB-5 than that of K-10. The fluorescence intensity was decreased for all other samples as the dye concentration was increased. Similar results were obtained with 250 nm excitations (figure 3)

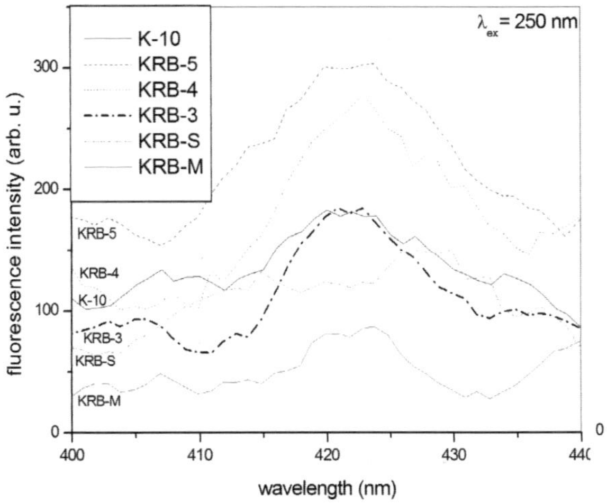

Figure 3. Fluorescence spectra of RB intercalated K-10 montmorillonite aqueous dispersions (0.1g sample in 100 ml water) at 421 nm when excited using 250 nm

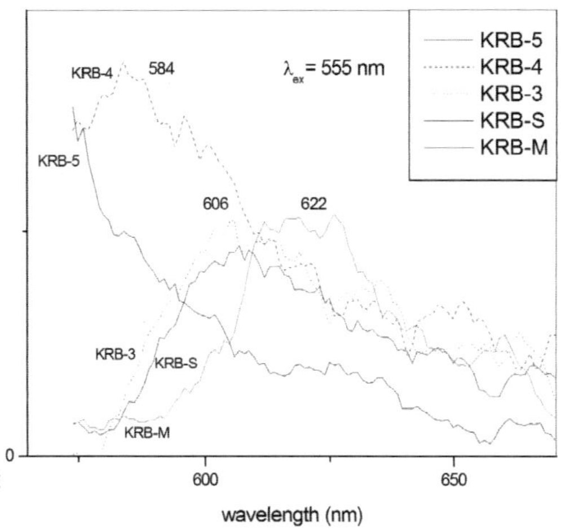

Figure 4. Fluorescence spectra of RB intercalated K-10 montmorillonite aqueous dispersions (0.1g sample in 100 ml water) when excited using 555 nm

Samples excited using 555 nm showed a bathochromic shift as the amount of the intercalated dye increases (figure 4). Since KRB-3, KRB-S and KRB-M were prepared from dye solutions of same molarity, they showed almost same emission intensity, but KRB-M peak was bathochromically shifted.

The fluorescence peak intensity comparison of KRB-3 and KRB-S from figures 2-4 shows that the emission peak intensity for KRB-S is slightly less than that of KRB-3. KRB-S also shows a broadened nature for fluorescence though both KRB-S and KRB-3 have the same absorption peak at 580 nm. But, KRB-S has a higher absorption.

When dye molecules are excited, their excitation energy gets transferred to silica in the matrix so that fluorescence of silica at 421 nm is enhanced as in figure 2. The 421 nm emission of silica showed fluorescence quenching due to the presence of more dye molecules in the clay network which inhibits the emission from the silica layers. As the concentration of the RB is increased the radiative relaxation of dyes become more predominant resulting into fluorescence quenching at 421 nm and enhancement in 580 nm region. The same intensity of KRB-3 and KRB-S also confirms the emission from the silicate layer because they are of the same concentration though they are prepared by different methods. Porous silicon [7], silica spheres [8] and silicon nanoparticles [9] when excited near 370 nm showed an emission around 420 nm. Since, no peak shift is observed for this emission for all the samples we can conclude that the emission arises from the montmorillonite itself. The fluorescence emission of the K-10 colloid also confirms this. However, the presence of a small amount of dye in the network enhances the fluorescence intensity. Moreover, fluorescence quenching can be observed as the amount of dye in the dispersions increases. The fluorescence quenching is due to the energy transfer from the excited state of the monomers to the aggregated species. This transfer can be via either an electron exchange or coulombic interactions. There could also be other effects like light scattering, adsorption by the solid surface [10], deformation to a non-planar conformation upon adsorption etc., on the fluorescence of the dye in the adsorbed state, which may significantly reduce the fluorescence yield [6]. Montmorillonite intercalated with RB cations showed that the bands corresponding to the silicate layer remained unchanged after intercalation [1]. This means that the silicate layers can be treated as rigid bodies during energy minimization.

The fluorescence band of the rhodamine-clay system is shifted to higher wavelength compared with that of the rhodamine in aqueous solution [11]. In addition, the emission band becomes broader as a consequence of the structural inhomogeneity of the rhodamine-clay system, where the surface structure may differ significantly from that in the interlayer space [11]. The bathochromic shift observed in the case of KRB-M (figure 4) is due to the clustering and dense packing of the dye molecules which arises from the repeated adsorption. The fluorescence intensity of the KRB-M was less compared to KRB-3 and KRB-S which was attributed to the usual fluorescence quenching in highly concentrated solutions. The changes in fluorescence band profile with increasing RB^+ concentration reflect the changes in relative proportions of monomers, dimers, and higher aggregates adsorbed on the external surface and intercalated into the interlayer space.

4. Summary

The emission intensity of K-10 montmorillonite can be enhanced by the addition of a small amount of dye in the network. The dye addition increases the absorption peak intensity of K-10. The increase in dye concentration shows bathochromic shift in peaks due to the formation of J-aggregates. The emission intensity decreases as the amount of the dye increases which is due to the fluorescence quenching. KRB-S prepared by ultra-sonication method shows more absorption and less fluorescence than KRB-3 due to the uniform distribution and minimised clustering of dye molecules.

5. References:

[1] Miroslav P, Pavla Č, Helena W, Zdeněk K, Miroslava T, Marta C and Zdeněk W 2003 *J. Mol. Model.* **9** 39. doi: 10.1007/s00894-002-0107-8
[2] Lyjo K J, Suja H, Sanjay G, Sugunan S, Nampoori V P N and Radhakrishnan P 2009 *Appl. Clay. Sci.* **42** 483 ; 2009 *Philos. Mag.* doi: 10.1080/14786430902806652
[3] Laurier L Schramm and Jan C T Kwak 1982 *Clays Clay Miner.* **30** 40.
[4] Juraj B 2006 *Appl. Clay Sci.* **34** 58. doi:10.1016/j.clay.2006.02.011 and the references therein
[5] Dobrogowska C, Hepler L G, Ghosh D K and Yariv S 1991 *J. Thermal Anal.* **37** 1347
[6] Juraj B, Nobuo I and Ryo S 2004 *J. Phys. Chem. B* **108** 4470. DOI: 10.1021/jp037607x
[7] Yue Z, Dongsheng L, Deren Y and Minghua J 2005 *J. Mater. Sci.* **40** 5071.
[8] Adam M Jakob and Thomas A Schmedake 2006 *Chem. Mater.* **18** 3173.
[9] Valenta J, Fucikova A, Pelant I, Kůsová K, Dohnalová K, Aleknavičius A, Cibulka O, Fojtík A and Kada G 2008 *New J. Phys.* **10** 073022. doi: 10.1088/1367-2630/10/7/073022
[10] Chaudhuri R, López Arbeloa F and López Arbeloa I 2000 *Langmuir* **16** 1285.
[11] Pavla Č, Petr M, Miroslav P, Zdeněk K, Helena W and Zdeněk W 2004 *J. Colloid Interface Sci.* **277** 128. doi:10.1016/j.jcis.2004.03.035

Acknowledgments: LKJ is thankful to University Grants Commission- UGC (Govt. of India) for providing experimental facilities through CELOS project.

Study of Electrical Conductivity and Dielectric Behaviour of Molybdenum containing Bismuth Borate Glasses

N. Kishore[1], Sanjay[1], A. Agarwal[1], Sanjay Dahiya[2], Inder Pal[1]

[1]Department of Applied Physics, Guru Jambheshwar University of Science and Technology, Hisar- 125001 (INDIA)

[2]Physics Department, Maharshi Dayanand University, Rohtak-124001

E-mail: nks54@rediffmail.com

Abstract. Bismuth borate glasses containing MoO_3 content have been synthesized by the standard melt-quenching technique. The samples have been characterized by X-ray diffraction and scanning electron microscopy. Frequency and temperature dependent conductivity measurements have been carried out in the range of 20Hz to 1MHz and in the temperature range of 373K to 623K respectively. The frequency dependent conductivity has been studied using both conductivity and modulus formalisms. The dielectric permittivity (ε' and ε''), electrical modulus (M*) and dielectric loss were also studied of the prepared glasses. The values of ε' and ε'' are found to decrease with increase in frequency.

1. Introduction

Heavy metal oxide (such as Bi_2O_3) based glasses have been quite extensively investigated due to their manifold applications in optical and optoelectronic devices such as ultrafast switches, infrared windows and optical isolators [1]. The non-conventional glass forming oxide Bi_2O_3, participates in the glass structure in presence of B_2O_3 with two possible coordinations [BiO_3] pyramidal and [BiO_6 octahedral units [2]. The interest in the present glass system is determined by the presence of network forming oxide, the classical network former B_2O_3 and conditional glass former Bi_2O_3. The introduction of transition metal (TM) oxide like MoO_3 in the glass system may enter either as a network former or as a modifier [3]. The presence of molybdenum ion makes the glasses to be useful for applications in high-density memories, light modulation, large area display devices [4]. Molybdenum ions exist in two valence states viz., Mo^{5+} and Mo^{6+} states. Mo^{6+} ions participate in the glass network with MoO_4 (Td) and MoO_6 (Oh) structural units where as Mo^{5+} ions form $Mo^{5+}O_3^-$ complexes and act as modifiers [5]. The present paper deals with the results of XRD, SEM and dielectric studies such as dielectric constant ε', loss factor tanδ and A.C. conductivity σ_{ac} over a wide range of frequency and temperature apart from the determination of various physical properties of glasses doped with molybdenum ions.

2. Experimental details

Oxide glasses $xMoO_3.(40-x)Bi_2O_3.60B_2O_3$ (x = 0, 2, 4 and 6) have been prepared by the standard melt-quenching technique. The chemicals were thoroughly mixed in appropriate proportions to yield about 15g of the desired compositions. Then the mixture was melted in porcelain crucibles by putting it in an electrical muffle furnace at about 1473K in air for about 30 minutes. The melt was swirled frequently to assure homogeneity. The melt was then quickly poured onto a stainless steel plate and pressed with another stainless steel plate to get the coin shaped samples of 1-2mm thickness. The amorphous nature of the samples is ascertained by XRD patterns as recorded by using Rigaku X-ray diffractometer. The SEM (model JSM-Joel 6510) has been used to study the microstructure of the glasses. The temperature and frequency dependent A.C. conductivity was measured with other dielectric parameters by measuring R_s and X_s values using QuadTech 1920 CE Impedance Analyzer.

3. Results and discussion

3.1. X-ray diffraction and SEM studies

XRD patterns of various glasses are found to be similar and typical XRD pattern for sample with x = 4 is shown in Fig.1. Perusal of this figure shows that the sample exhibits the glassy (amorphous) behaviour with a broad hump. No peak corresponding to MoO_3 is observed which indicates that the added amount of MoO_3 has completely entered the glass matrix. The microstructure of the glasses is identified by scanning electron microscopy (SEM) studies. The Fig.2 shows the typical SEM micrograph of studied glasses for x = 4 which is found to be a glassy (amorphous) nature of the sample.

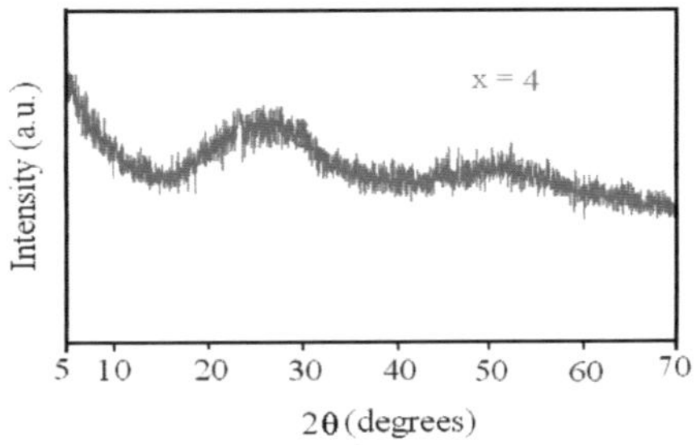

Fig.1 Typical XRD pattern of glass: $4MoO_3.36Bi_2O_3.60B_2O_3$

Fig.2 Typical SEM photograph of glass: $4MoO_3.36Bi_2O_3.60B_2O_3$

3.2. Electrical conductivity

Fig.3 (a) and Fig.3 (b) shows the imaginary and real part of the electrical modulus respectively. The values of the real and imaginary dielectric constants were calculated from the measured capacitance and conductance data. The values of dielectric modulus increase with increase in frequency and decreases with rise in temperature. This suggests that the range in which charge carriers are mobile on long distances i.e. polaron can perform successful hopping from one site to the neighbouring site.The variation of the dielectric constant ε'and dielectric loss, tanδ with frequency of $4MoO_3.36Bi_2O_3.60B_2O_3$ glass sample at different temperature are shown in Fig.4 and 5 respectively. It can be seen that the ε gradually increases with rise in temperature but it decreases with frequency. However, no peaks were observed in this frequency and temperature range for this glass sample. The absence of these peaks is an indication of the non-ferroelectric behaviour [6]. Further, low frequency dielectric dispersion was observed at high temperature. This may be due to the increase in the ionic part (due to Bi^{3+} ions) of conductivity. The value of dielectric constant is higher value at higher temperatures with change in frequency. This may be attributed to the hopping of molybdenum ions through borate glass network [6,7]. Perusal of Fig.5 shows that the dielectric loss decreases with increase in the frequency. The data of all the remaining glasses are being analysed and detail of results will be published elsewhere. Fig.6 shows the frequency dependence AC conductivity of glass (for x = 4 mol%) at different temperature values. This figure shows that the conductivity increases with increase in frequency. Also, it increases with rise in temperature. The AC conductivity in many non-crystalline materials has been considered by many authors and different models were proposed to account for its dependence on temperature and frequency. Owen [8] reported that the electrical properties of glass in a periodic field depend not only on the mobile ions or electrons (which give rise to DC conductivity) but also on other relatively

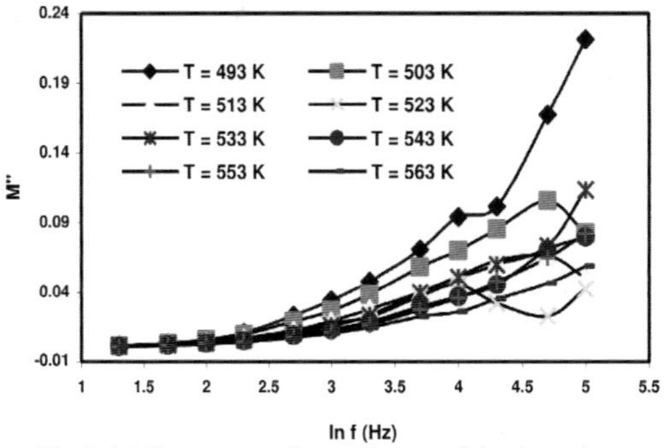

Fig.3 (a) Frequency dependence of the imaginary part of electrical modulus M'' of glass (x = 4): $4MoO_3.36Bi_2O_3.60B_2O_3$.

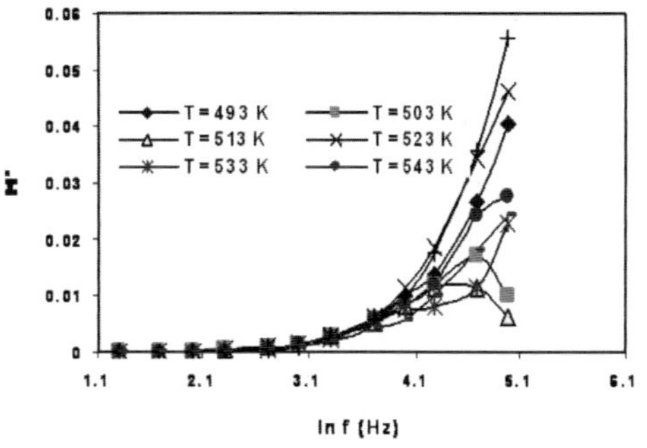

Fig.3 (b) Frequency dependence of the real part of electrical modulus M' of glass (x = 4): $4MoO_3.36Bi_2O_3.60B_2O_3$.

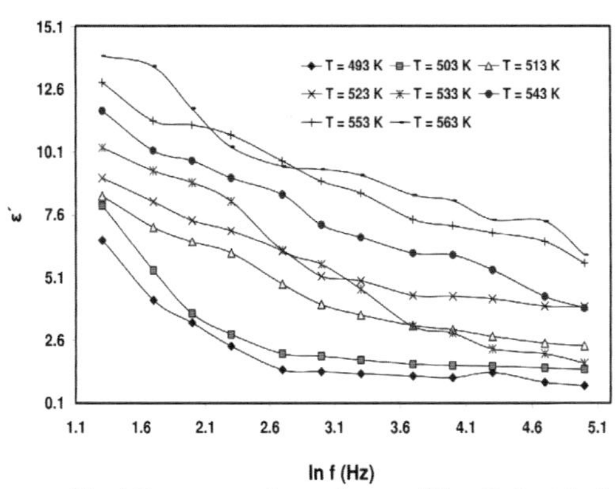

Fig.4 Frequency dependence of the dielectrical constant of $4MoO_3.36Bi_2O_3.60B_2O_3$ glass.

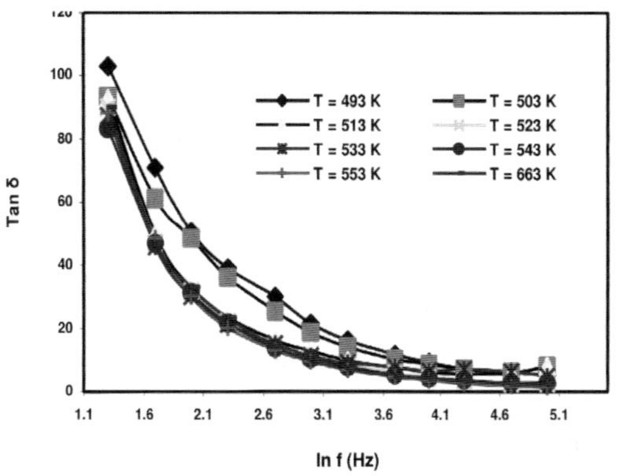

Fig.5 Frequency dependence of the loss tan δ of $4MoO_3.36Bi_2O_3.60B_2O_3$ glass.

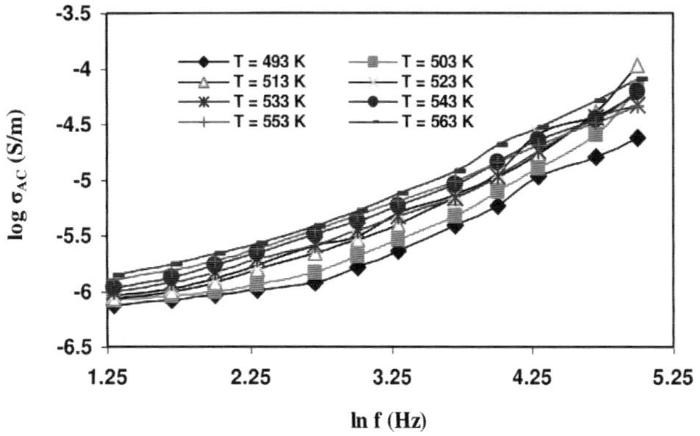

Fig.6 Frequency dependence AC conductivity of $4MoO_3.36Bi_2O_3.60B_2O_3$.glass sample.

immobile ions or on dipole, etc., which form part of the glass network. Hill and Jonscher [9] observed that for the AC conductivity in a wide range of materials, including the ones in which conduction by hopping is expected. The conductivity may be due to the presence of molybdenum ions in Mo^{6+} and Mo^{5+} states, which exhibit in the network former position. There is a possibility for the cross linking between these ions and some BO_3 units (of B_2O_3) to form B-O-Mo bonds in the glass network. Also these cations may replace some Bi ions, where some Mo can act as modifiers [9]. These cross linking bonds weaken the glass network and create pathways suitable for migration of free ions that form space

IOP Conf. Series: Materials Science and Engineering **73** (2015) 012041 doi:10.1088/1757-899X/73/1/012041

charges. Thus, the weaker the network the more is the space charge polarization, and this results in an increase in the dielectric parameters.

4. Conclusions

From XRD patterns, it is confirmed that the given samples are amorphous in nature. The ε' and ε'' decreased with increase in frequency and with decrease in temperature. The dielectric loss for the studied glass is due to electrical conduction with hopping motion of charge carriers between Mo^{5+} and Mo^{6+} states. The change of AC conductivity is attributed to the hopping mechanism of conduction.

Acknowledgements
Authors are highly thankful to DST-FIST, DRDO and UGC New Delhi for financial assistance.

References

[1] Y.Cheng, H.Xiao, W.Guo, Mater. Sci. Eng. A 480 (2008) 56.
Y.Cheng, H.Xiao, W.Guo, Mater. Sci. Eng. A 480 (2008) 56.
[2] P.S. Gahlot, A. Agarwal, V.P. Seth, S. Sanghi, S.K. Gupta, M. Arora, J. Spectrochem. Acta A 61 (2005) 1189.
[3] Jeong-Ryeol Kim, Geun-Kyu Choi, Dong Kyun Yim, Jong-Sung Park, Kug Sun Hong, J. Electroceram. 17 (2006) 65.
[4] Z. Hussain, J.Electron. Mater. 31 (2002) 615.
[5] O. Cozar, D.A. Magdas, I. Ardelean, J. Non-Cryst. Solids 354 (2008) 1032.
[6] S. Furusawa, S. Tange, Y. Ishibashi, K. Miwa, J. Phys. Soc. Jpn. 59 (1990) 2532.
[7] A.K. Jonscher, Dielectric Relaxation in Solids, Chelsea Dielectric Press, London, 1983.
[8] A.E. Owen, Prog. Ceram. Sci. (1963) 77.
[9] R. Hill, A. Jonscher, J. Non-Cryst. Solids 32 (1979) 53.

Development of Rare-Earth Free Mn-Al Permanent Magnet Employing Powder Metallurgy Route

N Singh, R Shyam, N K Upadhyay and A Dhar,

Division of Materials Physics and Engineering
National Physical Laboratory, Council of Scientific & Industrial Research, Dr. K.S. Krishnan Marg, New Delhi-110 012, India

E-mail: adhar@nplindia.org, singhnidhi@nplindia.org

Keywords: MnAl hard magnets, Mechanical alloying, Spark plasma sintering,

Abstract

Most widely used high-performance permanent magnets are currently based on intermetallics of rare-earths in combination with Fe and Co. Rare-earth elements required for these magnets are getting expensive by the day. Consequently, there is a thrust worldwide to develop economical rare-earth free permanent magnets. It is acknowledged that the phase in Mn–Al alloys possesses magnetic properties without the presence of ferromagnetic elements such as Fe, Co, and Ni. In the present study, we report the synthesis of magnetic phase of $Mn_{54}Al_{46}$ alloy synthesized using mechanical alloying followed by solutionizing and annealing to obtain the desired magnetic phase.

It is well known that Al dissolves partially in Mn matrix hence supersaturated solid solution of $Mn_{54}Al_{46}$ alloy powder was obtained by mechanical alloying using a planetary high-energy ball mill. For this purpose elemental Mn and Al powders were ball-milled in Argon atmosphere at 400 rpm using stainless steel bowl with ball to powder ratio of 15:1. These mechanically alloyed $Mn_{54}Al_{46}$ powders were then consolidated using spark plasma sintering at 550°C for 20 min. followed by solution treatment at 1050°C for 5 hrs and then water quenched to retain high temperature phase. Subsequently, the $Mn_{54}Al_{46}$ samples were annealed in the temperature range 450°C-650°C to obtain the magnetic phase. These samples were characterized by XRD and SEM and the magnetic properties were measured using a vibrating sample magnetometer (VSM). It was observed that the magnetization and coercivity of MnAl magnets exhibited strong dependence on annealing temperature and annealing time.

Introduction

MnAl based magnets possess hard magnetic properties without ferromagnetic elements (3d transition metals) and rare-earth metals [1]. The low costs and availabilities of the Mn and Al, good machinability, high specific strength, high modulus of elasticity, as well as the excellent corrosion resistance, make this material particularly attractive. Mn metal is in general antiferromagnetic. By increasing the atomic distance between Mn atoms to 2.96 Å or more, the element becomes ferromagnetic [2]. The hard magnetic properties of MnAl magnets originate from the magnetic τ phase which is a metastable phase that can be obtained either by quenching the high-temperature phase followed by isothermal annealing at temperatures between 400 and 700°C or at a controlled cooling rate from high-temperature non-magnetic

ε–phase (h.c.p.) which transforms into a non-magnetic ε_0-phase (orthorhombic) by an ordering reaction, and then transforms into a ferromagnetic τ-phase (f.c.t.) by a martensitic phase transition, i.e. $\varepsilon-\varepsilon_0-\tau$ Typical properties are saturation magnetization (σ_s) of 88-96 emu/g [1,3] and coercieve force of 1.5 kOe [4]. Prolonged annealing and elevated temperatures result in decomposition of the τ-phase into the equilibrium cubic γ_2- and β-phases.
Mechanical milling have been used to synthesize a number of rare earth permanent magnet alloys, including $Nd_2Fe_{14}B$ [5], $Nd(Fe,Mo)_{12}N_x$ [6] and $SmCo_5$ [7]. In the present study we have done preliminary investigation on the microstructure and magnetic properties of mechanically milled and spark plasma sintered MnAl alloys.

Experimental

Pure Mn and A1 powders were weighed in stochiometric ratio to give compositions of $Mn_{54}Al_{46}$. The powders along with 2% process control agent were then sealed in stainless steel bowl together with 10 mm balls and filled with pure Ar gas. The ball to powder weight ratio was about 15:1. Mechanical alloying of the as-mixed powders was carried out at 400 rpm. Mechanically alloyed Mn & Al powders were sintered employing spark plasma sintering technique followed by two-step heating. The homogenization was carried out at 1050 for 5hrs followed by water quenching to retain the high temperature ε phase. These pellets were annealed in the temperature range 450°C-650°C for the synthesis of metastable magnetic τ phase. As milled, solutionized and annealed specimens were characterized by X-ray diffraction (XRD) measurements. Their magnetic properties were studied using a vibrating sample magnetometer (VSM) at room temperature.

Results and Discussion

Figure 1 shows change in XRD pattern of the MnAl powder with respect to milling time. The XRD pattern of the blended elemental powders shows clear diffraction peaks from α-Mn and Al phases. The intensity of the diffraction peaks from Al phase diminishes with increasing milling time. This indicates diffusion of Al into the α-Mn matrix during mechanical alloying and finally forming α-Mn solid solution.

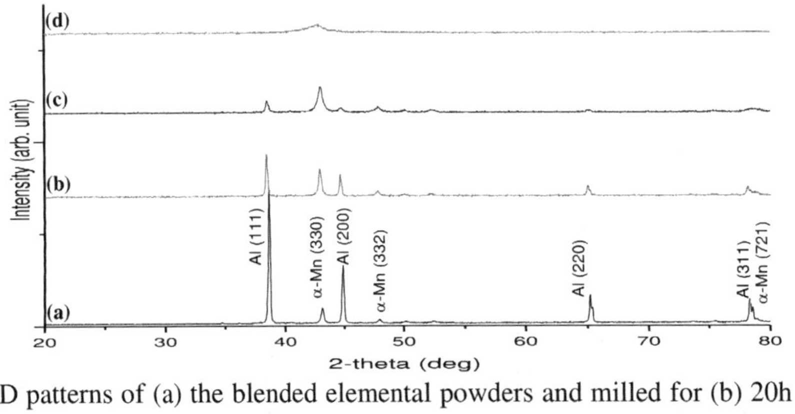

Fig. 1 XRD patterns of (a) the blended elemental powders and milled for (b) 20h (c) 70h, (d) 150h

Figure 2 shows XRD patterns of the solution treated and annealed MnAl pellets exhibiting the high temperature ε phase, metastable τ phase along with stable β and γ_2 phase. It can be clearly seen from fig.2a that solution treatment at 1050°C for 5h and water quenching results in single phase which is high temperature ε phase. As it is well known that the ε phase transform to magnetic metastable τ phase by isothermal annealing at temperatures between 450 and 650°C. To achieve this transformation annealing of water quenched sample was done in this temperature range for varying length of time (10 min. to 40 min.). Present result is from preliminary study and detailed characterization is underway. Annealing treatment given at 550°C for 30 min. results in distinct peaks from magnetic phase τ. Further, magnetic property characterization was done using VSM for all these samples.

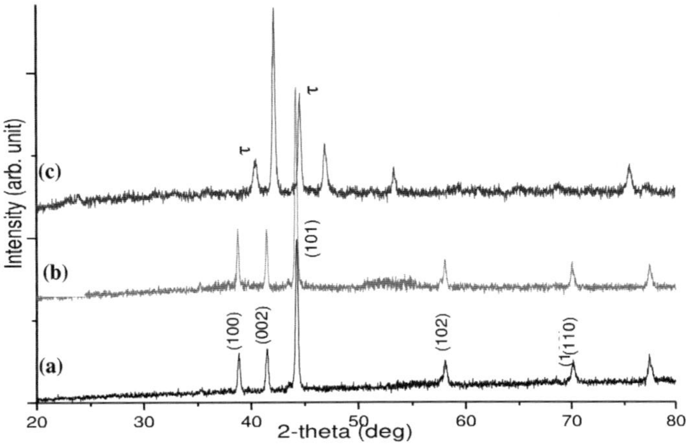

Fig. 2 XRD patterns of mechanically milled and spark plasma sintered MnAl pellet solution treated at 1050°C for 5h (a) Water quenched (WQ) (b) WQ-annealed at 600°C for 10 min. air cooled (c) WQ-annealed at 550°C for 30 min. air cooled

Figure 3 (a - d) shows the hysteresis loop of samples annealed between 550°C and 650°C for 20/30 min. Sample annealed at 550°C for 30 min. exhibits higher coercivity as well as remanence in accordance with the XRD result. The saturation magnetization is strongly dependent on the composition i.e. the amount of τ phase in the present system. Further, transformation of ε phase into τ phase depends on appropriate combination of annealing temperature and time. Optimization of annealing conditions which in turn manipulate the amount of τ phase could lead to further enhancement of magnetic properties.

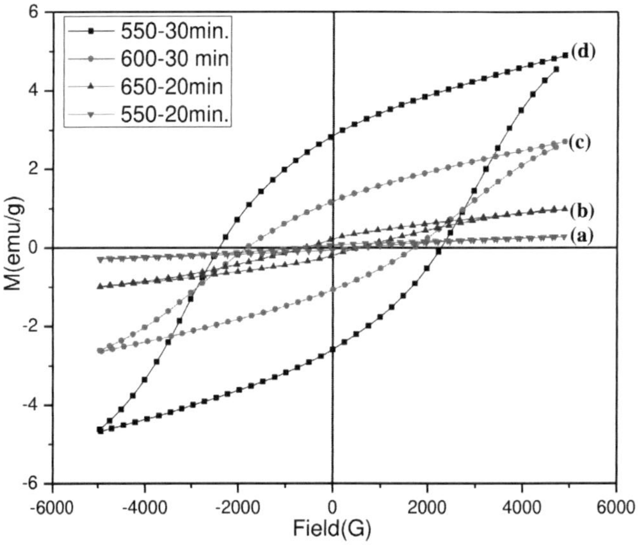

Fig 3 Hysteresis curves of MnAl pellets **(a)** 1050-5h-WQ/550-20min.-AC **(b)** 1050-5h-WQ/650-20min.-AC **(c)** 1050-5h-WQ/600-30min.-AC **(d)** 1050-5h-WQ/550-30min.-AC

Conclusions

Magnetism in rare-earth free Mn-Al system has been demonstrated by the M-H curves, by obtaining the τ phase in $Mn_{54}Al_{46}$ by suitable heat treatments.

- High energy ball milling for extended duration results in complete solid solution of α-Mn.
- High temperature homogenization for 5h followed by water quenching results in ε phase which is required for magnetic phase formation.
- Annealing treatment resulted in transformation of ε phase into magnetic τ phase.

The optimization of annealing process parameters is currently underway for transformation of ε phase of MnAl into magnetic τ phase which would lead to significantly higher magnetic properties.

Acknowledgements

The authors are thankful to Director, National Physical Laboratory, (CSIR), New Delhi, India for providing necessary laboratory facilities and his kind permission to publish this work. Further, authors are also thankful to Dr. R.K. Kotnala for extending his help in magnetic property characterization.

IOP Conf. Series: Materials Science and Engineering **73** (2015) 012042 doi:10.1088/1757-899X/73/1/012042

References

[1] Kono H 1958 On the ferromagnetic phase in manganese-aluminium *J. Phys. Soc. Jpn.* **13** 1444

[2] Bohlmann M A, Koo J C and Wise J H 1981 Mn-Al-C for permanent magnets *J. Appl. Phys.* **52** 2542.

[3] Koch AJJ, Hokkeling P, Sterg MGVD and De Vos KJ 1960 New material for permanent magnets on a base of Mn and Al *J. Appl. Phys.*vol. **31** 75

[4] Sugihara M and Tsuboya I 1962 *J. Appl. Phys.* **33** 1338

[5] Zeng Q, Xiao YF, Liu YB, Dong SZ, Deng YS, Zhang ZY, Wang R, 1999 Magnetic Properties of Nanocrystalline $Mn_{54}Al_{46}$ Powders *J. Mater. Eng. Perform.* **8** 305.

[6] Zeng Q, Xiao YF, Dong SZ, Liu YB, Qiu BQ, Zhang ZY, Wang R, 1999 Influence of milling conditions on magnetic properties of $Nd(Fe,Mo)_{12}N_x$ compounds *J. Magn. Magn. Mater.* **192** 321.

[7] Zeng Q, Zhang Y, Hadjipanayis GC, 2002 *Proceedings of the 17th International Workshop on Rare Earth Magnets and their Applications, Delaware, USA*, 961–966.

IOP Conf. Series: Materials Science and Engineering **73** (2015) 012043 doi:10.1088/1757-899X/73/1/012043

IRRADIATION EFFECTS ON AMMONIUM PENTA BORATE (APB) SINGLE CRYSTALS

K.Prabha[1*], M. Ramesh babu[2], P. Prabhukanthan [3], H.Chen[4], X.L.Chen[5], and P. Sagayaraj[6]

[1]Department of Physics, Mother Teresa Women's University, Kodaikanal-624101, Tamilnadu, India

[2]Department of Physics, Anna Univeristy of Technology Madurai, Dindugal Campus, Dindugal, Tamilnadu, India

[3]Department of Chemistry, Muthurangan Govt. Arts. College, Vellore, Tamilnadu, India

[4]National Laboratory for Condensed Matter Physics, State Key Laboratory, Institute of Physics, Chinese Academy of Sciences, P. O. Box 603, Beijing 100190, China.

[5]National Laboratory for Nanoscale Physics and Devices, State Key Laboratory, Institute of Physics, Chinese Academy of Sciences, P. O. Box 603, Beijing 100190, China.

[6]Department of Physics, Loyola College, Chennai-600034, Tamilnadu, India

E-mail: mtwprabha@gmail.com,

Abstract. Single crystals of ammonium penta borate (APB) were irradiated with 120 MeV Ag^{9+} swift heavy ions (SHI) of fluence 1×10^{13} ions/cm^2. The UV-visible spectrum of irradiated crystal shows slight shift in the absorption edge and it also indicates the non uniformity in its absorbance level. An attempt was made to explain the surface damage caused due to SHI irradiation using AFM and SEM images.

1. Introduction

Borate crystals find commercial application as laser hosts, non-linear optical (NLO) elements, and UV optics [1-2]. In borate crystals, the boron atom usually coordinates with either three or four oxygen atoms forming $[BO3]^{3-}$ or $[BO4]^{5-}$ groups [3]. There are several UV nonlinear optical (NLO) crystals that can produced deep-UV coherent light by harmonic generation. For example, by using fifth harmonic generation (FHG) BBO and KB5 (KB5O8 .H2O) can produce 189 and 195 nm respectively [4]. Ammonium penta borate (APB) is a biaxial crystal of the orthorhombic system and belong to the mm2 (C_{2v}) class [5]. The lattice parameter are a =11.34Å, b = 11.03 Å and c =9.237 Å. The crystallographic elastic and piezoelectric properties of APB crystals have been reported by Cook and Jaffe [6]. We have concentrate swift heavy ion (SHI) effects on APB single crystals.

2. Experimental

Single crystals of ammonium penta borate were harvested using slow – solvent evaporation technique. These crystals were cut in to pieces of dimensions 4 × 4 × 1 mm^3 and polished using alumina paste. 120 MeV Ag^{9+} ions with fluence rate of 1×10^{13} ions/cm^2 from the 16MV Tandem pelletron accelerator at Inter University Accelerator Centre, New Delhi, India. The incident beam current density was about 3 pnA/cm^2. The samples are mounted on a target ladder inside a vacuum chamber pressure 3 × 10−7 Torr.

3. Results and Discussion

3.1 Optical absorption spectrum

The UV-Vis-NIR analysis of non- irradiated and irradiated crystals was carried out between 200 and 2500 nm covering the entire near ultra- violet, visible and near infrared regions, using the VARIAN CARY 5E model spectrometer. From the optical- absorption studies exhibits the band gap of irradiated sample is less when compared to the non-irradiated sample is shown in Fig1. The decrease in band gap of irradiated sample due to the SHI irradiation. It can create some intermediate energy levels due to the structural rearrangements.

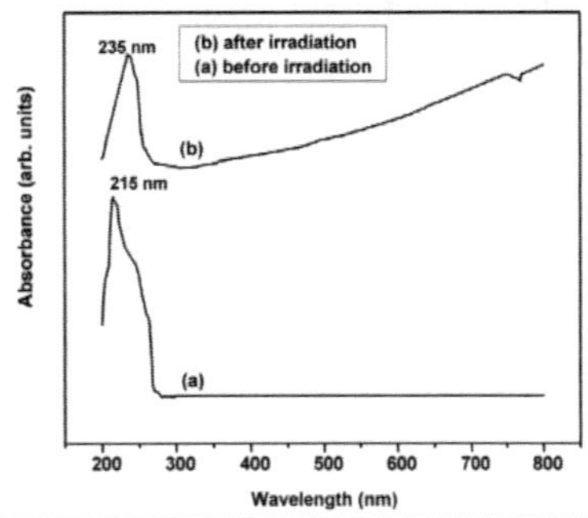

Figure 1. Optical absorption spectra of pure and Ag9+ irradiated APB single crystals

3.2 Atomic Force Microscopy (AFM) studies

The AFM topographical images of as grown and irradiated crystals are shown in Figs. 2 and 3 respectively. The surface roughness of the non-irradiated and irradiated single crystals was also studied. There is no defect are observed by non-irradiated crystal, but the formation of the craters and hillocks were present on the irradiated crystal. The formation of craters and hillocks are due to the SHI irradiation in APB single crystals.

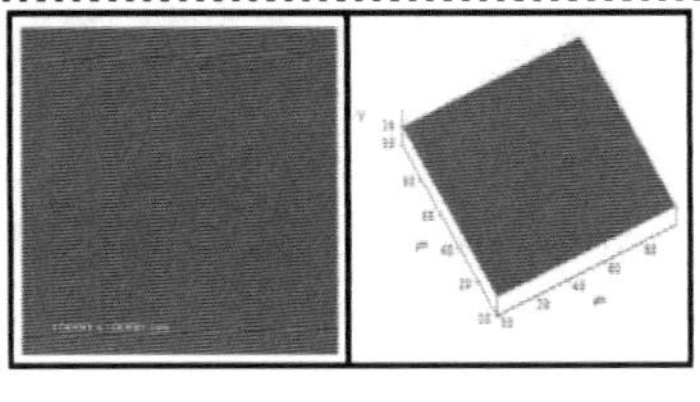

Figure 2. AFM and 3D image of Pure APB single crystal

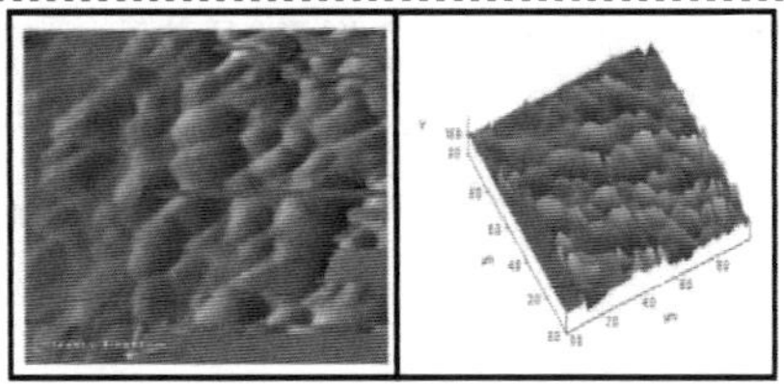

Figure 3. AFM and 3D image of Ag^{9+} ion irradiated APB single crystal

3.3 Scanning Electron Microscopy studeis

Scanning Electron Microscopic (SEM) study on the as grown and irradiated samples were carried out using Hitachi scanning electron microscope model S-4200 operated at 25 kV for different magnifications, after coating the samples with gold layer of 100 Å with Hitachi vacuum evaporator model Oxford 6566. The presence of micro inclusion is identified in the as grown APB single crystal (Fig. 4). When the crystal is subjected to the SHI irradiation it leads to a distorted lattice arrangement and also the generation of shock waves in the medium and this is confirmed by the tiny bubbles and the irregular shapes as observed in Fig. 5.

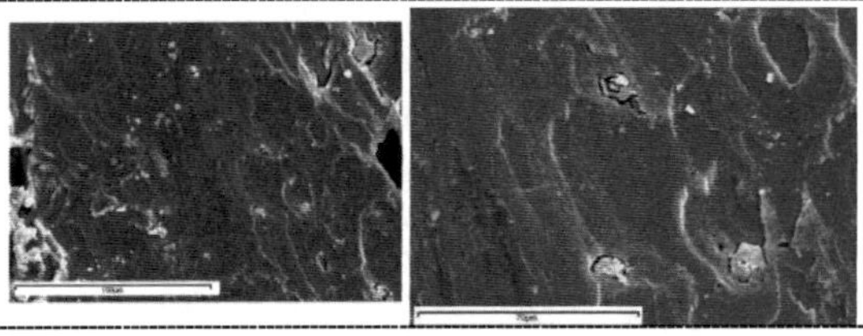

Figure 4. SEM picture of Pure APB single crystal

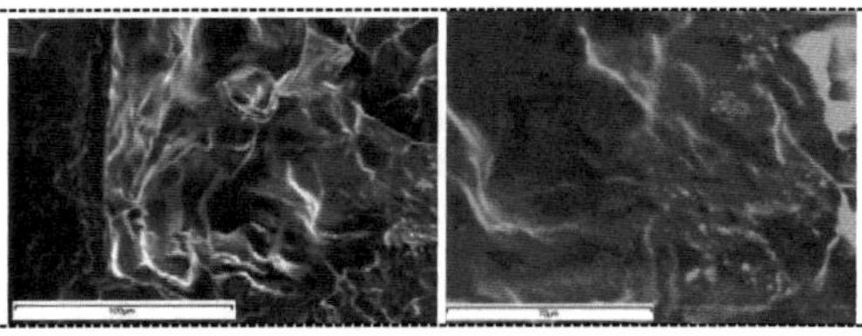

Figure 5. SEM picture of Ag^{9+} irradiated APB single crystal APB single crystal

4. Conclusion

The Ag^{9+} ion induced modification on the optical, structural and dielectric properties of the solution grown APB crystal was studied. The optical absorption study indicates the enhancement in the optical absorbance with the SHI irradiation of the sample and also a decrease in the band gap energy. The decrease in the energy gap is associated with the increased tailing of band edges and production of localized defect states near the Fermi level. The surface damage of the crystal was studied using atomic force microscope (AFM). The AFM image confirms that the SHI irradiated sample has strong presence of raised tails and bulges. The surface roughness of the irradiated crystal is found to be several folds larger than the non-irradiated sample.

5. Acknowledgements

One of the authors (K. Prabha) thanks the CAS-TWAS Fellowship for the financial support to do research in China.

6. References

[1] Keszler D 1996 Current opinions in solid state materials science **1** 204.
[2] Keszler D 1999, Current opinions in solid state materials science **4** 155.
[3] Xue D and Zhang S 1997 Appl. Phys. A **65** 451.
[4] Chen C 2004, Optical Materials **26** 425.
[5] Shahabuddin khan MD, Prasad G, Kumar G.S 1992 , Cry. Res. Technol. **27** K28.
[6] Cook W.R , Jaffer H 1957, Acta crystallogr **10** 705.

Growth and Thermal Studies of Doped and Pure Crystals of L-Argininium Dinitrate

Preema C. Thomas and P Sagayaraj[1]
Department of Physics, CMS College, Kottayam - 1
[1]Department of Physics, Loyola College, Chennai – 600034, India

E-mail: preemacthomas@gmail.com

Abstract Single crystals of pure and Cu^{2+} doped semiorganic L-argininium dinitrate (LADN) crystals were grown by slow solvent evaporation technique, with the vision to improve the physicochemical properties of the sample. Single crystal XRD studies of both pure and doped samples were carried out. The thermal (DTA, TGA and DSC) studies were carried out and the results are compared.

1. Introduction

Within the last few years, a great deal of research work has been carried out in non-linear optical materials. Both the inorganic and organic materials were investigated and it was found that the inorganic crystals were characterized by good stability and high transparency range but their second order coefficients were limited. In contrast organic crystals exhibited higher second order nonlinear coefficients but their transparency domain was reduced [1]. This triggered the search for inorganic-organic hybrid crystals. Complexes of amino acids with inorganic salts proved to be one such material with interesting non linear properties. The salt of amino acid like L-arginine phosphate monohydrate (LAP) is a more recent nonlinear optical material that has been proposed as a replacement for Potassium dihydrogen phosphate (KDP) crystal [2-4]. The structure of LAP was solved by Aoki et al [5]. In view of its encouraging non linear properties scientists tried to grow other salts of L-arginine [6-9].

L-arg.HNO_3 was one among the compounds described by Monaco et al, which precipitated as a fine non-crystalline powder [6]. Interestingly, Petrosyan et al was successful in crystallizing L-arg.2HNO_3 (LADN) from solution containing Arg:HNO_3 in ratios from 2 to 10 [9]. Ramasamy et al have determined the structure of L-argininium dinitrate (space group $P2_1$) and reported that the diprotonated argininium molecule is linked by a strong hydrogen bond to the nitrate anion [10]. Recently, Terzyan et al have also solved the structure of 2(L-arg.HNO_3).H_2O and L-arg.2HNO_3 [11].

Recent experiments reveal that nonlinear performance can be improved by bringing out the compositional modification and metallic substitutions [12-13]. Hence attempts have been made in the present study to investigate the NLO and thermal properties of LADN crystals with Cu^{2+} metallic dopant. The authors have already reported the growth and characterization of L-argininium dinitrate (LADN) crystals [14].

2. Experimental

1.1 Crystal growth

Approximate amount of high-purity L-arginine (Merck 99%) and nitric acid with water were used to prepare the solution of LADN [14]. The following is the reaction

$$(NH_2)\ NHCNH\ (CH_2)_3\ CH\ (NH_2)\ COOH + 4HNO_3 \longrightarrow$$

$$(H_2N)_2^+\ CNH\ (CH_2)_3\ CH(NH_3)^+\ COOH\ (NO_3)_2^- + 2HNO_3$$

The product was purified by repeated re-crystallization typically thrice from double distilled water. The 100 ml saturated solution was kept in a beaker covered with perforated polyethylene lid at room temperature. The crystals were grown by slow solvent evaporation method. The rate of evaporation of the solution was found to be approximately 2 ml per day. In a period of 15 days, seed crystals of LADN were formed due to spontaneous nucleation. Optically clear, defect free crystals with perfect morphological face were chosen for further growth experiment. The pH of the solution was maintained at 2.2. Cu^{2+} Doped LADN crystals were also grown by adopting the same procedure. The concentration of Cu in the doped crystals was maintained as 0.2M. Single crystals of pure and Cu^{2+} doped LADN are grown with dimensions upto 18 x 6 x 2 mm^3 and 52 x13 x 3 mm^3 in a period of 45 days. Photographs of pure and doped LADN single crystals are shown in Figure 1 and 2.

The morphology of LADN crystal has been studied and reported by the author [14] and it is seen that both pure and doped crystals have increased elongation along the crystallographic b- axis.

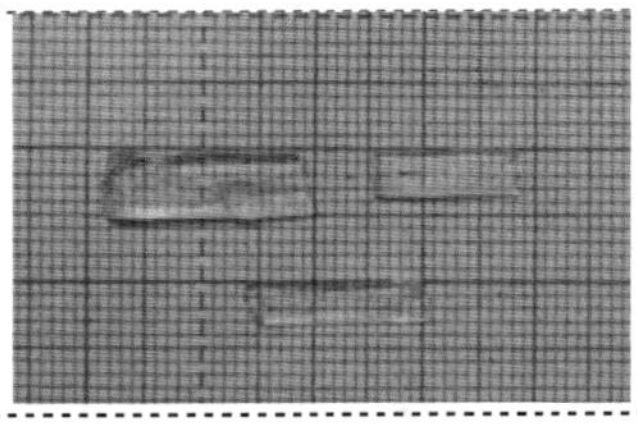

Figure 1. Photograph of as grown single crystal of pure LADN

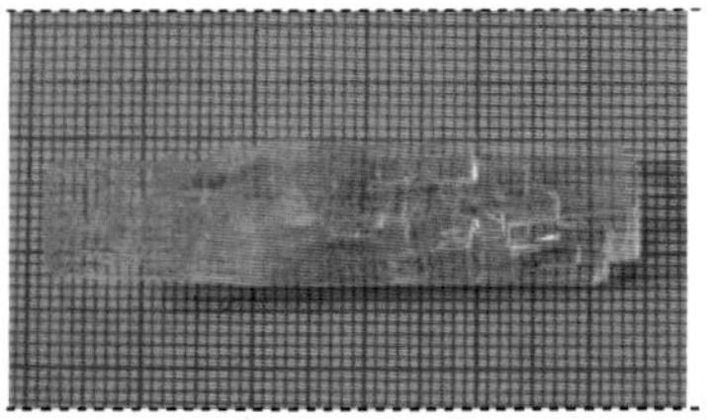

Figure 2. Photograph of as grown single crystal of Cu^{2+} Doped LADN

2.2 Characterization

The X-ray data were collected using an automatic X-ray diffractometer (MESSRS ENRAF NONIUS, The Netherlands) with MoKα (0.717Å) radiation. The structure was solved by the direct method and refined by the full matrix least squares technique using the SHELXL program. The NLO efficiency of pure and doped LADN crystals was evaluated by Kurtz & Perry powder technique using a Q- switched mode locked Nd:YAG laser emitting 1.06nm, 6 ns pulses with spot radius of 1mm [15]. TG / DTA of the crystals were carried out using the instrument NETZSCH STA 409C. DSC analysis of the pure and doped crystals of LADN was carried out using the instrument NETZSCH DSC 204. The TG / DTA of pure and doped LADN was carried out between 28 –1200°C at a heating rate of 10°C /min. The experiment was performed in nitrogen atmosphere. The DSC analysis was done in the range 20 - 120 °C for pure LADN and up to 200°C for doped crystals.

3. Results and Discussion

3.1 X-ray diffraction analysis

The unit cell parameter of pure and Cu^{2+} doped crystals of LADN is presented in Table 1. It is observed from the table that both pure and doped LADN crystals belong to monoclinic system and $P2_1$ space group. The lattice parameters of pure LADN are in good agreement with the reported values [10,11,14].

3.2 NLO Studies

Kurtz SHG test was performed on pure and Cu^{2+} doped crystals of LADN to confirm the second harmonic signal generation efficiency. For a laser input of 6.2 mJ, the second harmonic signal (532 nm) of 91.66, 184.74 and 201.65 mV were obtained through KDP , pure and Cu^{2+} doped LADN samples respectively. This shows that the NLO efficiency of pure LADN crystal have been improved with metallic substitution.

3.3 Thermal Analysis

The TG- DTG traces for pure and Cu^{2+} doped LADN is shown in Figures 3 and 4 respectively. The thermograms of pure and doped LADN crystals nearly appear similar with four stages of decompositions. The first stage of decomposition occurs at nearly 171°C which is increased to 178°C for Cu^{2+} doped LADN. It is followed by two more weight losses between 200 and 350°C. At higher temperatures about 400°C the final stage of decomposition takes place resulting in the total loss of the residue.

Table 1. Unit cell parameters of pure and Cu^{2+} doped LADN single crystals

	Pure LADN	Cu^{2+} LADN
Crystal System	Monoclinic	Monoclinic
a (Å)	7.7532	7.75
b (Å)	7.2779	7.28
c (Å)	11.6754	11.68
α °	90	90
β °	92.7	92.56
γ °	90	90
Volume ($Å^3$)	658.02	659
Z	2	2
Space group	$P2_1$	$P2_1$

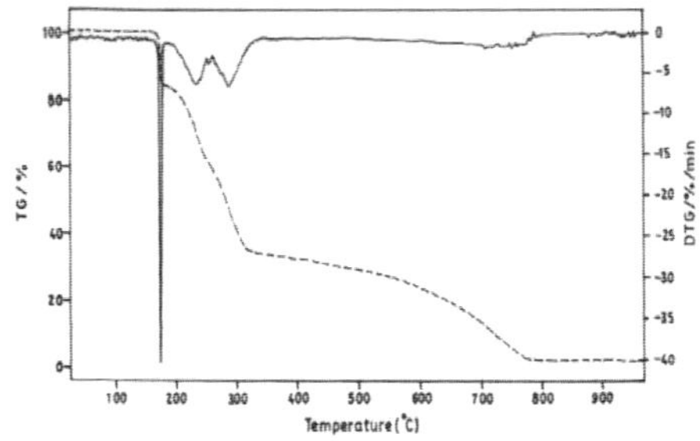

Figure 3. TG – DTG curves of pure LADN

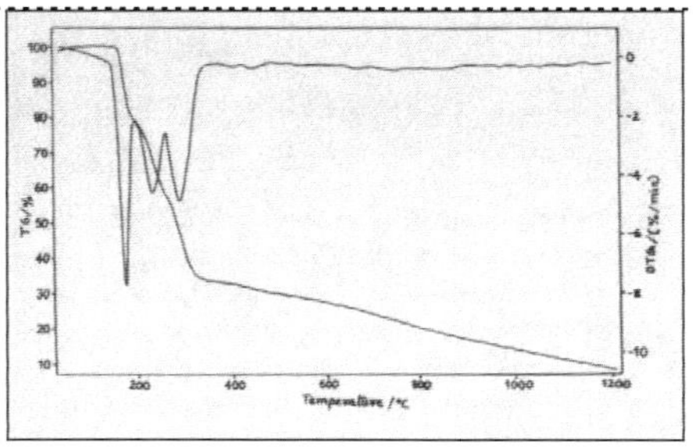

Figure 4. TG – DTG curves of doped LADN

The DTA of pure and doped LADN is shown in Figures 5 and 6 respectively. There is a weak endotherm observed around 145°C in both pure and doped LADN crystals. This corresponds to isomorphic transformation as there is no corresponding weight loss in TGA trace. The melting point of both the pure and doped LADN crystals was found to be nearly equal to 152°C . The melting of the sample is followed by a sharp exotherm starting around 171°C for pure LADN and at 178.2°C for doped sample.

IOP Conf. Series: Materials Science and Engineering **73** (2015) 012044 doi:10.1088/1757-899X/73/1/012044

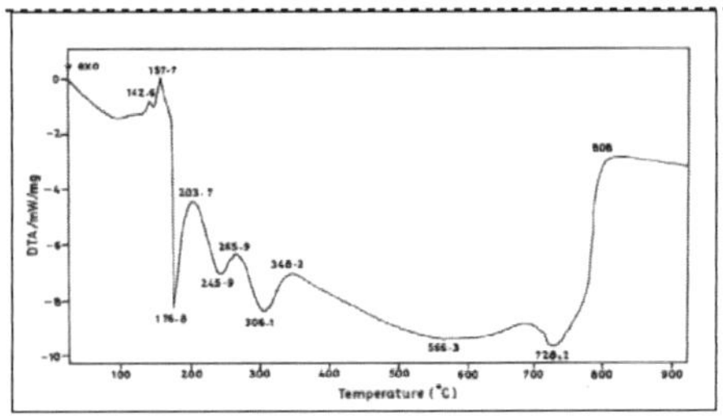

Figure 5. DTA curve of pure LADN

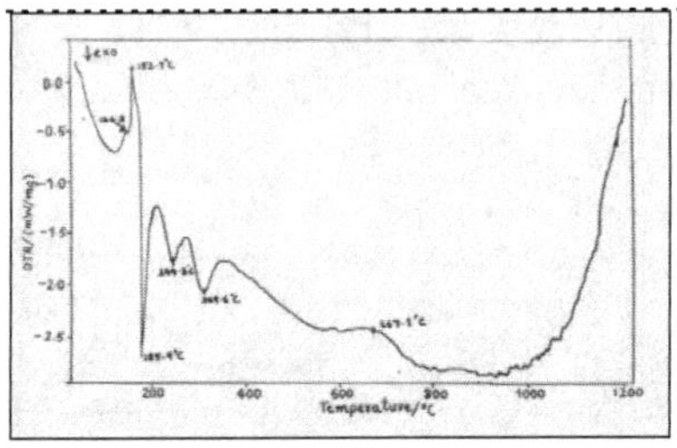

Figure 6. TG – DTA curve of doped LADN

The DSC trace for both pure and Cu^{2+} doped LADN is shown in Figures 7 and 8.The minute endotherm corresponding to isomorphic transformation as mentioned in the DTA trace is seen in the DSC trace. It is seen from the trace that addition of dopants have enhanced the peak value to 142.7°C. This is followed by the endotherm assigned for the melting of the sample. DSC trace shows that decomposition follows immediately after melting. The decomposition temperature for doped LADN sample is enhanced to 185.9°C.

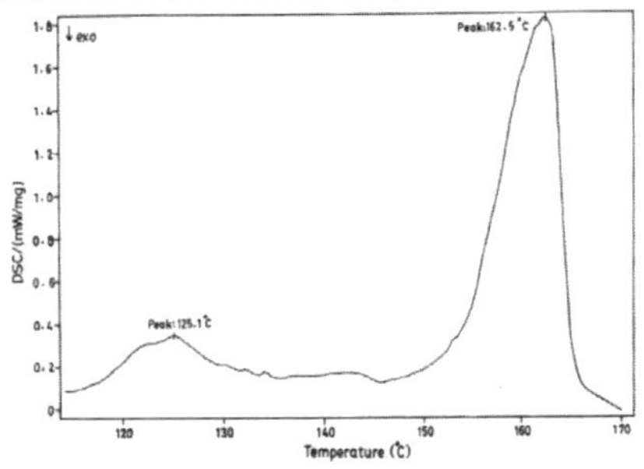

Figure 7. DSC curve of pure LADN

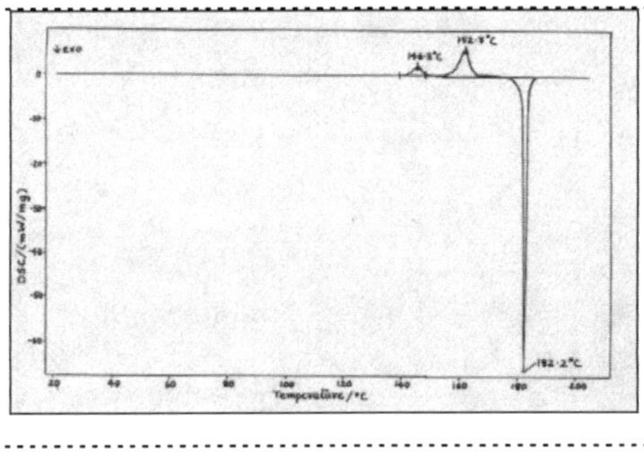

Figure 8. DSC curve of doped LADN

4.Conclusion

Single crystals of pure and Cu^{2+} doped LADN were grown from aqueous solutions by slow evaporation technique at room temperature. The pH of the solution was maintained at 2.2. The grown crystals were characterized by XRD and Thermal studies. The NLO studies show that doping has improved the NLO efficiency of pure LADN crystals. From thermal studies it is observed that the presence of metal dopant (Cu^{2+}) increases the decomposition temperature of LADN.

5.References

[1] C. Owens, K. Bhat, W.S. Wang, A. Tan, M. D. Aggarwal, B. G. Penn, D. O. Frezier, *J. Crystal Growth* 225 (2001) 465.
[2] Xu D., Jiang M. and Tan Z., *Acta Chem. Sinica* 41 (1983) 570.
[3] Sasaki T., Yokotani A., Fujioka K., Yamanaka T. and Nakai S. *Technol. Report of Osaka Univ.* 39 (1989) 25.
[4] Yokotani A., Sasaki T., Yoshida K. and Nakai S. *J. Appl. Phys. Letters*, 55 (1989) 2692.
[5] Aoki K., Kozo Nagano and Yiochi Iitaka, *Acta Crystallogr. B* 27 (1971). 11.
[6] Monaco S.B., Davis L.E., Velsko S.P., Wang F.T. and Eimerl D. *J. Crystal growth,* 85 (1987) 252.
[7] Haussühl S., Chrosch J., Gnanam F., Fiorentini E., Recker K and Wallrafen F. *Cryst. Res. Technol.* 25 (1990) 617.

[8] L. N. Raskovich, B. Yu. Shekunov, *J. Crystal Growth* 112 (1991)183.
[9] Petrosyan A.M., Sukiasyan R.P., Karapetyan H.A., Terzyan S.S. and Feigelson R.S *J.Crystal Growth* 213 (2000) 103.
[10] Ramasamy S., Sridhar B., Rama Krishnan V. and Rajaram R. K. *Acta Cryst. E* 57 (2001) 0872.
[11] Terzyan S. S., Karapetyan H. A., Sukiasyan R. B. and Petrosyan A. M. *Journal of molecular structure* 687 (2004) 111.
[12] Eimerl D., Velsko S., Davis L., Wang F., Loiacono G. and Kennedy G. *IEEE J. Quantum Electronics* 25 (1989)179.
[13] S. A. Rajasekar, K. Tamizharasan, A. Joseph Arul pragasam, J. Packaim Julius, P. Sagayaraj, *J. Crystal growth* 247 (2003) 199.
[14] Preema C. Thomas, Jolly Thomas, Packiam Julius J., Madhavan J., Selvakumar S. and Sagayaraj P. *J. Crystal Growth* 277 (2005) 303.
[15] Kurtz S.K. and Perry T.T. *J. Appl. Phys* 39 (1968) 3798.

IOP Conf. Series: Materials Science and Engineering **73** (2015) 012045 doi:10.1088/1757-899X/73/1/012045

Photo induced effects on the optical properties of $As_{20}Sb_{20}S_{60}$ thin films

Ramakanta Naik[1], E. M. Vinod[2], Sanjit Parida[2], R. Ganesan[2] and K.S.Sangunni[2]

[1]Department of Physics, Utkal University, Bhubaneswar, 751004, India
[2] Department of Physics, Indian Institute of Science, Bangalore, 560012, India

E-mail: ramakanta.naik@gmail.com

Abstract. The thermally evaporated $As_{20}Sb_{20}S_{60}$ amorphous film of 800 nm thickness was subjected to light exposure for photo induced studies. The as-prepared and illuminated thin films were studied by X-ray diffraction, Fourier Transform Infrared Spectroscopy and X-ray Photoelectron Spectroscopy and Raman spectroscopy. The optical band gap was reduced due to photo induced effects along with the increase in disorder. These optical properties changes are due to the change of homopolar bond densities. The core level peak shifting in XPS spectra and Raman shift supports the optical changes happening in the film due to light exposure.

1. Introduction

Chalcogenide glasses are one of the most widely known families of amorphous materials and have been studied for several decades, because of their interesting fundamental properties and wide range of applications [1]. They have been explored as promising candidates for optical memories, gratings, switching devices, acusto-optic devices, holography, and xerography and information storage media [2, 3]. Arsenic sulfide (As_2S_3) is the most studied chalcogenide glass and has applications in infrared optics and optical coatings because of its excellent IR transmission, large glass forming tendency, and resistance to moisture and chemicals [4]. Studies on thin films of Sb_2S_3 are attracting wide attention, for their special applications as a target material for television cameras, microwave devices, switching devices and various opto-electronic devices. Even though As and Sb belong to the same group of the periodic table, As_2S_3 and Sb_2S_3 do not display the same glass forming tendency. However, addition of As_2S_3 to Sb_2S_3 enhances the glass forming ability of the latter and glasses in the mixed As-Sb-S system can be formed [5].

Photo induced changes in amorphous chalcogenides are an object of systematic investigations with a view to better understanding the mechanisms of the phenomena taking place in them as well as their practical applications. It is well established that chalcogenide glasses undergo structural changes under the action of external influences such as laser, neutron, electron and gamma irradiation [6-8]. The knowledge of the optical properties of thin films is very important in many scientific, technological and industrial applications of thin films such as photo-conductivity, solar energy, photography, and numerous other applications. The present communication reports the effect of laser irradiation on optical properties of amorphous thin films of $As_{20}Sb_{20}S_{60}$.

[1] To whom any correspondence should be addressed.

2. Experimental

Bulk glass of $As_{20}Sb_{20}S_{60}$ was prepared by melt quenching technique. Thin films were prepared by thermal evaporation method at a base pressure of 1 $x10^{-5}$ Torr from the prepared bulk glass onto the glass substrates. During the deposition process (at normal incidence), the substrates were suitably rotated in order to obtain the films of uniform thickness. The thicknesses of the films were around 800 nm. The amorphous state of the film was checked by X-ray (Philips, $Cu_{K\alpha}$, λ=1.54 Å) Diffractometer (XRD). To study the photo induced changes, we irradiated the film at room temperature by a diode pumped solid state laser (DPSS) of wavelength 532 nm with a power of 40 mW. The optical absorption spectra of the as-prepared and illuminated films were taken by using the Fourier Transform Infrared (FTIR) spectrometer (Bruker Optics (IFS66v/S) in the visible wavelength range of 400-1200 nm at room temperature. The X-ray Photoelectron Spectroscopy (XPS) core level spectra were obtained with Al K_α X-rays (1486.6 eV) at a base pressure of $\sim 10^{-9}$ Torr in Multilab 2000 Thermo Scientific UK instrument. The C 1s line from adventitious carbon on the surface has been widely used for charge referencing [9] and the BE of the reference C 1s line was set as 284.6 eV. The room temperature Raman spectra were recorded in the 180° backscattering geometry, using a 532 nm excitation from a diode pumped frequency doubled Nd–YAG solid state laser and a custom built Raman spectrometer equipped with SPEX TRIAX 550 monochromator and a liquid nitrogen cooled CCD. Laser power at the sample was ~15 mW and a typical spectrum acquisition took ~ 1-2 min.

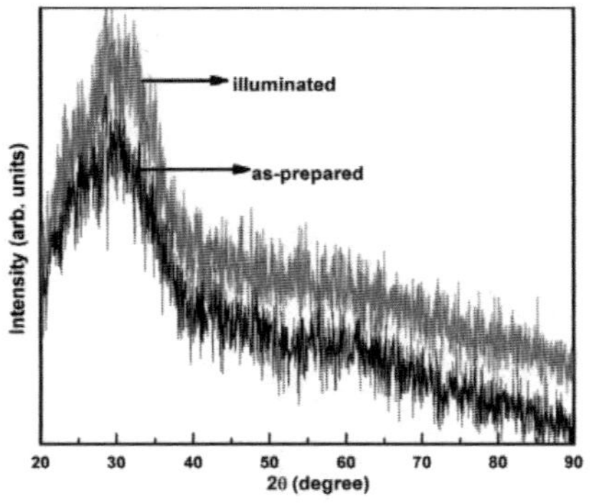

Figure 1. XRD patterns of the $As_{20}Sb_{20}S_{60}$ thin films.

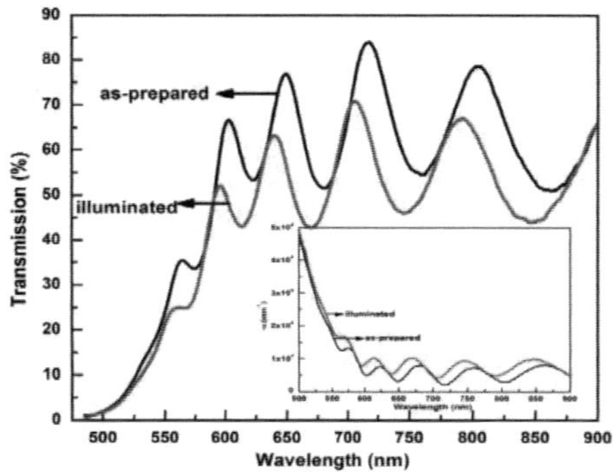

Figure 2. Transmission spectra (inset- α change) of the $As_{20}Sb_{20}S_{60}$thin films.

3. Results and Discussion

Figure 1 shows the X-ray diffraction patterns for the $As_{20}Sb_{20}S_{60}$ thin films. The absence of the diffraction lines in the X-ray patterns indicates that the films have amorphous structures. Transmission spectra corresponding to the amorphous $As_{20}Sb_{20}S_{60}$ thin films are plotted in figure 2, which shows a clear difference between the as-prepared and the illuminated films. The absorption coefficient (α) for the studied films were calculated [10] from the transmittance T and reflectance R using the equation

$$\alpha = \frac{1}{d} ln \frac{(1-R)^2}{T} \quad (1)$$

where d is the thickness of the film and T is the transmission and R is the reflectivity of the film. The absorption coefficient increases with illumination which is shown in fig.2 inset. According to Tauc's relation [11] for the allowed non-direct transition, the photon energy dependence of the absorption coefficient can be described by

$$(\alpha h v)^{1/2} = B^{1/2}(hv - E_g) \quad (2)$$

where B is a parameter that depends on the transition probability and E_g is the optical energy gap. The optical band gaps of the as-prepared and the illuminated films are found to be 1.93 eV and 1.87 eV respectively. So, photo induced optical band gap change was observed due to light illumination on the as-prepared film. It can be suggested that the photo induced changes are caused by chemical bond redistribution [12] according to the equation

$$2(M - S) \Leftrightarrow M - M + S - S \qquad (3)$$

where M stands for As or Sb, S is for sulfer and dash denotes a chemical bond. Films in the virgin state contain a large density of homopolar bonds due to dissociation of the glass in the course of evaporation and quenching of the vapor on the substrate. There is a high degree of disorder in these films; the atoms are partly statistically distributed, and M-S chemical-bond formation is not fully preferred. Thus exposures of $As_{20}Sb_{20}S_{60}$ thin films can shift the equilibrium in both directions depending on the conditions of thin film preparation and/or the conditions of light exposure. In case of illumination, the homopolar bond density increases causing more disorder which can be seen from the $B^{1/2}$ values. The $B^{1/2}$ values for as-prepared and illuminated films are 581 and 551 $cm^{-1/2}eV^{1/2}$ respectively. The $B^{1/2}$ for illuminated film is less than the as-prepared film which indicates the presence of more no of homopolar bonds due to disorder.

A typical XPS spectrum of $As_{20}Sb_{20}S_{60}$ film contains many photoelectrons and Auger peaks of As, S and Sb. But, we have considered only the core peaks such as As 3d, Sb 4d and S 2p for the present study. The XPS spectra of the as-prepared and illuminated film for As 3d core level is shown in figure (3a). From the spectra of As 3d peak, it is found that due to illumination, the peak is shifting towards the lower BE due to the formation of As-As homopolar bonds. The As 3d peak position for as-prepared and illuminated films are at 43.01 and 42.73 eVs respectively. This is in accordance with the reduction in optical band gap of the as-prepared film due to photo induced process. The peak position of Sb 4d core level spectra (figure 3a) for as-prepared and illuminated films are at 33.91 and 33.55 eVs respectively. It is found that the peak is shifting towards the lower BE due to the formation of more Sb-Sb homopolar bonds. The S 2p spectra of the as-prepared and illuminated films are shown in Fig. 3b. The S 2p peak position for as-prepared and illuminated films are at 161.72 and 161.91 eVs respectively. The peak shifts towards higher BE shows the formation of S-S homopolar bonds due to illumination as the electro negativity of S (2.58) is more than that of Sb (2.02).

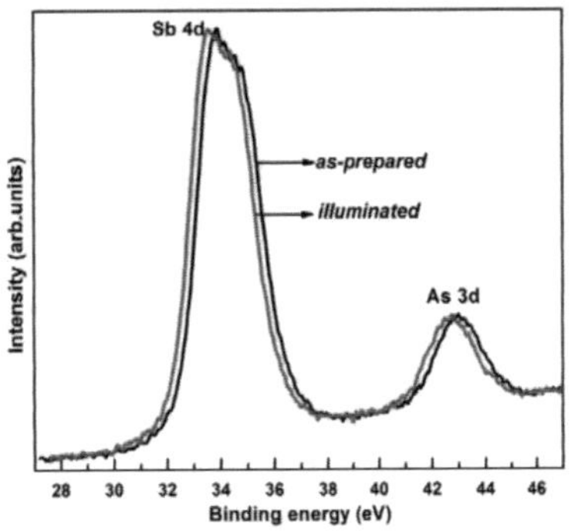

Figure 3a. As 3d and Sb 4d of $As_{20}Sb_{20}S_{60}$ thin film.

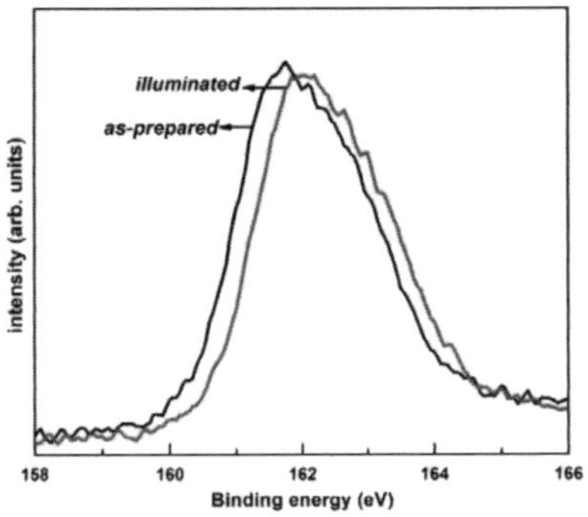

Figure 3b. S2p core level spectra of $As_{20}Sb_{20}S_{60}$ thin film.

Direct evidence of structural changes in $As_{20}Sb_{20}S_{60}$ thin films caused by illumination was obtained from Raman Spectra as shown in figure 4. According to the molecular model [13], each Sb atom in the As-Sb-S ternary glasses is covalently bonded to three S atoms in a pyramidal unit (Sb-S_3), and As atoms are covalently bonded to three S atoms in a pyramidal unit (As-S_3). The basic structural units AsS_3 and SbS_3 are interconnected through bridging S atoms. The coupling between the basic structural groups via S atoms is assumed to be weak, and the vibrational modes are separated into SbS_3 and AsS_3 like vibration in As-S_3, (S_2) As-As (S_2) and the AsS_3 mode is superimposed in that band.

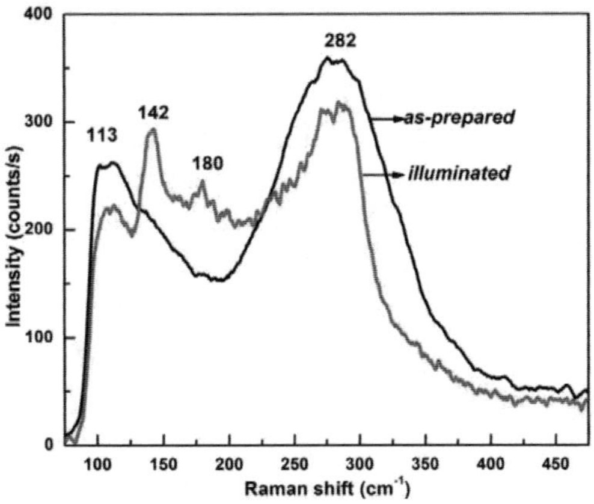

Figure 4. Raman spectra of $As_{20}Sb_{20}S_{60}$ thin film.

The peak at 113 cm^{-1} and 215 cm^{-1} of the as-prepared film corresponds to S-S vibrational modes respectively. The illumination of the as-prepared film causes the increase in homopolar bond density due to which there is shift in the peak position of the various vibrational modes and new peaks at 142 and 180 cm^{-1} are created.

4. Conclusions

In conclusion, thermally-evaporated amorphous thin films undergo structural transformations when exposed to band gap illumination (photo-structural transformations). Photo induced changes of optical transmissivity, optical band gap, $B^{1/2}$ value; XPS core level peaks and Raman spectra change are observed. The optical band gap is decreased due to illumination. Photo induced changes can be assigned to chemical bond redistribution.

Acknowledgements

The authors thank DST for using the National Facility for Optical Spectrometry at Department of Physics and Surface Science facility, IISc for XPS measurement.

References

[1] Shimakawa K, Kolobov A and Elliot S R 1995 *Adv. Phys.* **44** 475
[2] Seddon A. B 1995 *J. Non-Cryst. Solids* **184** 44
[3] Wagner T and Ewen P J S 2000 *J. Non-Cryst. Solids* **266-269** 979
[4] Uhlmann D R and Kreidl N I 1983 *Glass Science and Technology, Vol.1, Glass forming systems*,Academic Press,New York
[5] Cervinka L and Hruby A 1982 *J. Non-Cryst. Solids* **48** 231
[6] Lucas P, King E A, Horner A D and Johnson B R 2006 *J. Non-Cryst.Solids* **352** 2067
[7] Golovchak R, Shpotyuk O, Vlcek M, Ganjoo A, and Jain H 2007 *Philos. Mag.* **87** 4323
[8] Elalaily N A and Mahamed R M 2002 *J. Nucl. Mater.* **303** 44
[9] Barr T L and Seal S 1995 *J. Vac. Sci. Technol. A* **13** 1239
[10] Othman A A, Osman M A, Amer H H and Dahshan A 2004 *Thin Solid Films* **457** 253
[11] Zanatta A R and Chambouleyron I 1996 *Phys. Rev. B* **53** 03833
[12] M.Vlcek, M.Frumar and A.Vidourek 1987 *J. Non-Cryst. Solids* **90** 513
[13] Lucovsky G and Martin R M 1972 *J. Non-Cryst. Solids* **8-10** 185

IOP Conf. Series: Materials Science and Engineering 73 (2015) 012046 doi:10.1088/1757-899X/73/1/012046

Improved Performance of Organic LEDs with Modified Metal-Organic Interface

Ritu Srivastava[1], Omwati Rana[1,2], Razi Ahmad[1,3], C.K.Suman[1], M.Zulfequar[2], M.Husain[2] and M.N.Kamalasanan[1]

[1] CSIR-Network of institute for solar energy (NISE), Physics of Energy Harvesting Division, CSIR-National Physical Laboratory, Dr. K.S.Krishnan Road, New Delhi- 110012, India

[2]Department of Physics, Jamia Millia Islamia University, New Delhi-110025, India

[3]Department of Chemistry, Indian Institute of Technology Delhi, New Delhi-110016 India

Email: - ritu@ mail.nplindia.org

Abstract. Incorporation of thin layers of strong acceptor tetrafluro-tetracyanoquinodi methane (F_4TCNQ) over conducting substrate Indium-Tin-Oxide (ITO) modified its work function. The enhancement depended on the thickness of F_4TCNQ film. These modified ITO substrates were used as hole injection contacts in OLEDs which show enhanced device current, lower operating voltages and enhanced current efficiency.

1. Introduction

In organic based electronic devices like Organic Light Emitting Diodes(OLEDs) and Organic Photovoltaics(OPVs), the basic processes such as charge injection, charge transport, capture and radiative recombination of opposite charged charge carriers are involved[1,2]. In an organic metal interface, if interface dipoles are absent and vacuum level alignment is assumed then the energy barrier for charge injection is given by the difference between the work function of the electrode and the highest occupied molecular orbital(HOMO) or lowest unoccupied molecular orbital(LUMO) of the organic semiconductor (Schottky-Mott Limit). Often the energy barrier off set is modified by interface dipoles which modify the potential barrier [3-5]. The resultant large interface barrier results in injection limited charge transport. Some of our recent studies on electrode/organic interfaces indicate that it is possible to align the Fermi level electrodes with the negative polaron states of thin layers of strong acceptor molecules [6-7]. This process enhances the work function of the substrates and enables to form ohmic contact with many organic molecular films.

To demonstrate the enhancement in work function, we have fabricated OLED devices using indium tin oxide (ITO) modified with different thickness of tetrafluoro-tetracyano-quinodimethane (F_4TCNQ) as the hole injecting anodes, 2,2',7,7'-Tetrakis-(diphenylamino)-9,9'-spirobifluorene (Spiro-TAD) as the hole transport material and a thin film of gold as cathode. The J-V characteristics of the device show pure space charge limited conduction showing the formation of ohmic contact between the modified electrode and Spiro-TAD (HOMO=5.4eV). The OLEDs with modified anodes have shown enhanced efficiency and low turn-on voltages.

2. Experimental Details

The OLEDs were fabricated on ITO coated glass substrates having a sheet resistance of 20 Ω/sq and a thickness of 120 nm. Before fabricating the device, ITO coated glasses were patterned and cleaned by a conventional solution cleaning process using deionized water, acetone, trichloroethylene, and isopropyl alcohol for 20 min each were using an ultrasonic bath. Prior to film deposition, to enhance work function the surface of ITO substrate were treated with oxygen plasma for 5 min. Organic layers were deposited onto glass substrates under high vacuum(4×10^{-6} torr) at a deposition rate of 0.4Å/s. Thickness of the deposited layers were measured in situ by a quartz crystal thickness monitor. Three OLEDs were fabricated with different thickness of F_4TCNQ (x= 0, 5nm and 7nm). The device structure was ITO (12nm)/ F_4TCNQ (x)/ Spiro-TAD (30nm)/ Alq_3(50nm)/ LiF(1nm)/ Al(150nm). The size of each pixel was 5mm×5mm. The emission spectrum has been recorded with a high resolution spectrometer (ocean optics HR-2000 CG UV-NIR). The current density-voltage-luminescence (J-V-L) characteristics have been measured with a luminance meter (LMT-1009) and Keithley 2400 programmable current-voltage digital source meter. All the measurements were carried out at room temperature under ambient conditions.

3. Result and Discussion

The schematic structure of fabricated OLEDs and energy level diagram is shown in Figure 1(a)-(b).

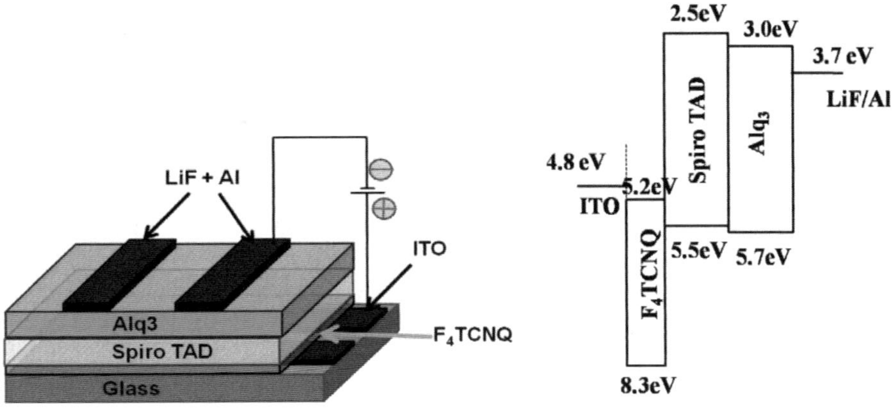

(a) (b)

Figure 1: (a) *Schematic diagram of OLED and (b) energy level diagram of device.*

It has been observed from the energy level diagram that there are large energy barrier (0.7eV) between ITO and Spiro TAD. To reduce the barrier and increase the hole injection into the device it is necessary to modify the ITO/Spiro-TAD interface. We have introduced a thin layer of F_4TCNQ and varied its thickness and observed the effect of the introduction this layer on OLED devices. Three devices (D1(x=0nm), D2(x=5nm) & D3(x=7nm)) have been fabricated and their J-V characteristics have been measured.

Figure 2(a) show the J-V characteristics of the devices with and without anode modification. It is clearly seen that the interface modified devices (D2 & D3) gives higher current density as compared to unmodified device (D1) and as we increase the thickness of F_4TCNQ the current density has also increased and the maximum current density has been observed for device D3.

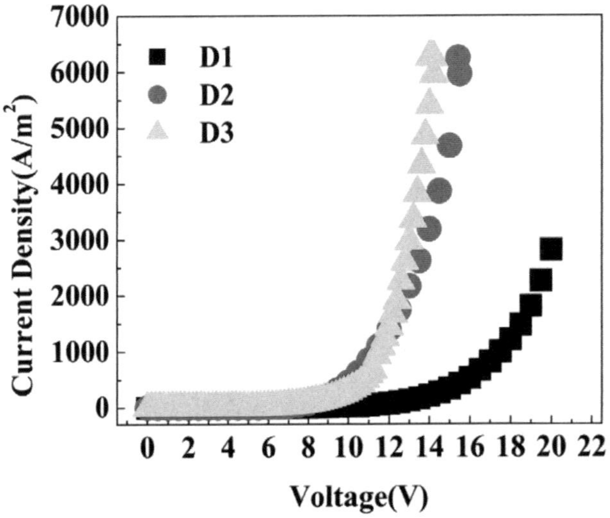

Figure 2(a)*: J-V characteristics of OLEDs with and without anode modification.*

This increase in current density can be attributed to a decreased hole injection barrier at the anode-organic interface and the formation of an ohmic contact for holes at this interface and consequently an enhanced injection of holes from ITO to Spiro-TAD. Since the measured electrical current density depends mainly on the organic electrode injection barrier which depends on the difference in the work function of metal electrode and the HOMO/LUMO of the organic materials, this increase in current density indicates a modification of the ITO's work function . Recently it has been shown by Brown and Li et al [8-10] that in the case organic films grown at low vacuum where the metal surface is passivated by oxide layer formation or hydrocarbon coverage, the metal wave function is not coupled with the organic wave function and interface states are not formed preventing partial charge transfer. Charge transfer take place only through quantum mechanical tunneling by integer charge transfer process. In this process if the metal Fermi level is in between the organic HOMO and LUMO, charge transfer does not occur and the SM limit is followed. However if the metal work function is lower than the negative poloron state of the organic, integer electron transfer occur from the metal in to the organic and positive interface dipoles are formed which pins the metal Fermi level with the negative poloron state. They have also shown that in the case of deposition of F_4TCNQ on to low work function metals, the effective work function is enhanced up to 5.5eV. Similarly if the Fermi level is higher than the positive polaron state of the organic, integer electron transfer occur from the organic into the metal resulting in a negative interface dipole formation and subsequent pinning of the metal work function on the positive poloron state. The increased current density in the present case may be due to this effect

Figure 2(b) shows the Luminescence-Voltage (L-V) characteristics of the devicesD1, D2 and D3.It has been observed that the interface modified devices show higher luminescence and as we increase the thickness of the F_4TCNQ luminescence increases and saturates at about 7nm.

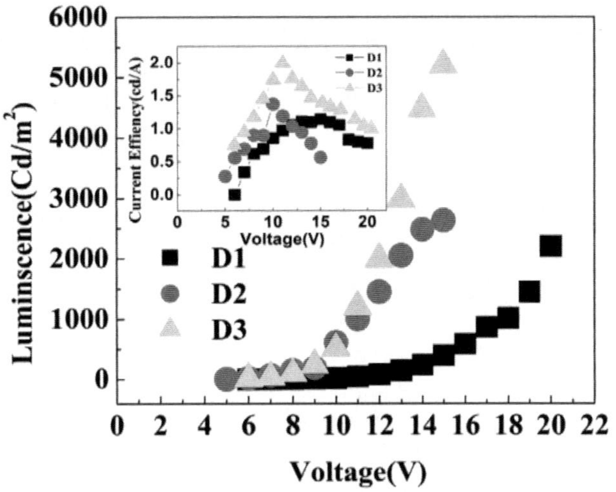

Figure 2(b): *L-V characteristics of OLEDs with and without anode modification. Inset shows the current efficiency-voltage of these OLEDs.*

The highest luminescence (5000 cd/m^2) was achieved by device D3.Also, the operating voltage decreases in case of interface modified devices (D2 and D3).Similar enhancement has been observed for current efficiency(as shown in the inset of Fig2(b).

4. Conclusions

In conclusion, we have explicitly shown that modification of conducting substrates using strong accepter when used as anodes in OLEDs gives higher current density, lower operating voltages and higher luminescence.

Acknowledgements

The authors would like to thank and greatly acknowledge to Director, National Physical Laboratory India to encourage us for novel research, Council of Scientific and Industrial Research (CSIR) for the financial support from the project NWP-55 (TAPSUN). One of the author (OR) acknowledges University Grant Commission (UGC), New Delhi, for providing the Senior Research Fellowship.

References

1. Bruitting W, Berleb S and Muckl A G 2001 *Org. Electronics* **2** pp 1-36.
2. Hoppe H and Sariciftci N S 2004 *J. Mater. Res.***19** 7.
3. Mott N F and Gurney R W 1940 *Electronics Process in Ionics Crystals* (Oxford:Oxford University Press).
4. Blochwitz J, Fritz T, Pfeiffer M, Leo K, Alloway D M, Lee P A, Armstrong N R 2001*Org. Electron* **2** 297.
5. Gao W, Kahn A,2003 *J. Appl. Phys.* **94** 359.
6. Chauhan G, Srivastava R, Kumar A, Rana O, Srivastava P.C.and Kamalasanan M.N. 2012 *Organic Electronics* **13** pp 394–398.
7. Rana O, Srivastava R, Chauhan G, Zulfequar M, Husain M, Srivastava P.C and Kamalasanan M.N. 2012 *Phys. Status Solidi A*, pp 1–7.
8. Braun S, Osikowicz W, Wang Y and Salaneck W R 2007 *Org. Elect.* **8** 14.

9. Braun S and Salaneck W R 2007 *Chem.Phys.Lett.* **438** 259.
10. Li G, Kim C H, Zhou Z, Shinar J, Okumoto K, and Shirota Y 2006 *Phys. Lett.* **882** 253505.

Effect of A-site ionic size variation on TCR and electrical transport properties of $(Nd_{0.7-x}La_x)_{0.7}Sr_{0.3}MnO_3$ with $x = 0, 0.1$ and 0.2

Sudarshan Vadnala[1], Saket Asthana[1,2], Prem Pal[1] and S. Srinath[3]
[1,2]Department of Physics, Indian Institute of Technology Hyderabad, India
[3]School of Physics, University of Hyderabad, India

E-mail: asthanas@iith.ac.in

Abstract. In this work, the structural and transport properties of $(Nd_{0.7-x}La_x)_{0.7}Sr_{0.3}MnO_3$ manganites with $x = 0, 0.1$ and 0.2 prepared by solid state reaction route are studied. These compounds are found to be crystallized in orthorhombic structural form. Experimental results showing a shift in the metal to semiconductor/insulator transition temperature (T_{MI}) towards room temperature (289 K) with the substitution of Nd by La, as the value of x is varied in the sequence $(0,0.1,0.2)$, have been provided. The shift in the T_{MI}, from 239 K (for $x=0$) to near the room temperature 289 K (for $x=0.2$), is attributed to the fact that the average radius $< r_A >$ of site-A increases with the percentage of La. The maximum temperature coefficients of resistance (TCR) of $(Nd_{0.7-x}La_x)_{0.7}Sr_{0.3}MnO_3$ ($x= 0.1$ and 0.2) are found to be higher compared to its parent compound $Nd_{0.7}Sr_{0.3}MnO_3$. The electrical transport mechanisms for $(Nd_{0.7-x}La_x)_{0.7}Sr_{0.3}MnO_3$ ($x= 0$ to 0.2) are explored by using different theoretical models, for temperatures below and above T_{MI}. An appropriate enlightenment for the observed behavior is discussed in detail.

1. Introduction

Pervoskite $AMnO_3$ type manganites have gained wide popularity because of their intriguing physical properties, resulting from the simultaneous presence of spin, lattice and orbital degrees of freedom [1-2]. These manganites have earned momentous attention from many researchers in order to seek their potential for spacious technological applications, e.g. read heads, magnetic information storage, low and high field magnetic sensors, infrared (IR) detectors and numerous other recent spintronic applications [3-10]. The fact that these substituted manganites procure high temperature coefficient of resistance (TCR) in bulk as well as in thin films at room temperature that makes them useful for infrared radiation detectors (i.e. IR detector) for night vision applications [11]. Among all perovskite manganites, $NdMnO_3$ is an antiferromagnetic insulator, characterized by a super exchange coupling between Mn^{3+} sites. This coupling is facilitated by a single e_g electron predominated by strong correlation effects. Partial substitution of Nd^{3+} ions with divalent cations (Sr, Ca and Ba) result in mixed valance states of Mn i.e. Mn^{3+}/Mn^{4+} which is responsible for the ferromagnetic Zener double exchange mechanism [12].

[2] Corresponding author: asthanas@iith.ac.in

The most prevalent experimental way of affecting the physical properties of manganites is by varying the ionic radii of A-sites. This creates structural distortions in manganites due to the size mismatch of A-site, which in turn generates the internal chemical pressure within the lattice. Due to this structural disorder effect, the local oxygen displacement occurs, ensuing into bond angle fluctuations and bond length variations, further leading to carrier localization in pervoskite lattice. This distortion can be controlled by the average size of the A-site cation which in turn modifies the Mn-O-Mn bond angle and Mn-O distances. The Mn-O-Mn bond angle is directly related to the hopping integral between Mn^{3+} and Mn^{4+} degenerate states. Thus, the variations in ionic radii at A-sites lead to competing phases at a particular temperature, hence influencing electrical and magnetic transport properties of the perovskite manganites.

In the present research, the structural and transport properties of Nd-based manganites with composition $(Nd_{0.7-x}La_x)_{0.7}Sr_{0.3}MnO_3$ (where $x = 0, 0.1$ and 0.2), are studied in order to tune the TCR and T_{MI} for IR detector applications.

2. Experimental details

The polycrystalline samples of $(Nd_{0.7-x}La_x)_{0.7}Sr_{0.3}MnO_3$, where x = *0*(NLSMO 0), *0.1*(NLSMO 1) and *0.2*(NLSMO 2) are synthesized by the solid state reaction route, using ingredients Nd_2O_3, La_2O_3, $SrCO_3$ and Mn_2O_3.The mixed powders are calcined, sintered and annealed(oxygen environment) at 1100 °C, 1300 °C and 1000 °C for 24hrs, 5hrs and 5hrs respectively. The structure and phase purity of the samples are analyzed by powder X-ray diffraction (XRD) and Raman spectroscopy. The resistivity measurement, without and with magnetic field of *5* tesla, are carried out using four-probe method in the temperature range from 5 to 300 K on a quantum design Physical Property Measurement System (PPMS).

3. Results & Discussion

3.1. XRD and Raman Spectra

XRD patterns of the polycrystalline manganites $(Nd_{0.7-x}La_x)_{0.7}Sr_{0.3}MnO_3$ ($x = 0, 0.1$and 0.2) exhibiting single–phase orthorhombic unit cell with *Pnma* (No *62*, PCPDF Ref No *861534*) space group are shown in figure1(a). It can be observed from the patterns that peaks are slightly shifted toward lower angle side on substitution of La in place of Nd. This small shift arises due to the mismatch of radius at A-site i.e., larger radius of La-ion (1.36 Å) in comparison to the radius of Nd-ion (1.27 Å) which causes increase in the volume of the lattice. Subsequently, an internal chemical pressure generated within the lattice due to size mismatch at A-site, which results in slight shift in the peaks of XRD pattern.

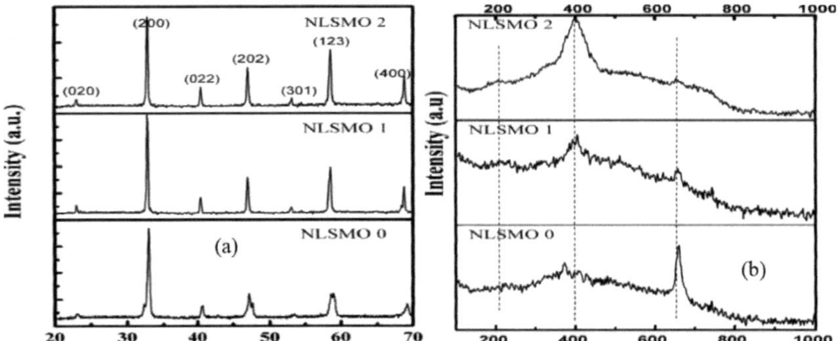

Figure 1. (a) XRD and (b) Unpolarized room-temperature Raman spectra of pervoskite manganites $(Nd_{0.7-x}La_x)_{0.7}Sr_{0.3}MnO_3$ where $x = 0, 0.1$ and 0.2.

The unpolarized room-temperature Raman spectra of $(Nd_{0.7-x}La_x)_{0.7}Sr_{0.3}MnO_3$ ($x = 0, 0.1$and 0.2) are shown in the figure1(b). Several phonon modes have been observed out of 24 allowed Raman active modes in orthorhombic symmetry as per selection rule. The weak modes in the range 205-212 cm^{-1} presumably represent the rotational vibration of the MnO_6 [13]. The presence of two prominent peaks at 656 cm^{-1} and 406 cm^{-1} along with some weaker modes supports the orthorhombic symmetry as observed from XRD results. The modes in the range of 375-406 cm^{-1} occur due to the combined effects from the stretching vibration and Jahn-Teller (JT) distortion of the octahedra [14, 15]. A peak at 656 cm^{-1} is attributed to the bending vibration of octahedral which is also called as breathing mode. Broadening of this peak with La substitution confirms that robustness of JT distortion decreases with increasing $<r_A>$. Based on this argument, it can be attributed that all compounds crystallized with orthorhombic symmetry with slight variation in JT distortion with increasing La-content.

3.2. Electrical transport

The samples average radius of site-A increases from 1.321 Å (for $x = 0$) to 1.339 Å (for $x = 0.2$), T_{MI} shifts from 239 K (for $x = 0$) to 289 K (for $x = 0.2$) is shown in the figure 2a. It is worth to mention as average radius $<r_A>$ increases, the local lattice distortion reduces due to the shifting of Mn-O-Mn bond angle towards symmetrical side (approaches 180°). As a result, hopping amplitude increases which influence the T_{MI} shifts towards higher temperature.

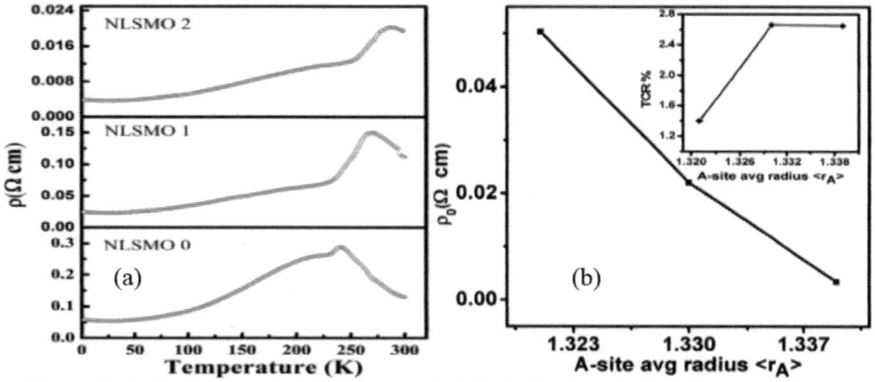

Figure 2. (a)Temperature dependent resistivity behavior of $(Nd_{0.7-x}La_x)_{0.7}Sr_{0.3}MnO_3$ where $x = 0, 0.1$ and 0.2 (b) residual resistivity vs A-site average radius.

The maximum %TCR $(TCR\% = \frac{1}{\rho}\left(\frac{d\rho}{dT}\right)*100)$ vs average radius of A-site $<r_A>$ is given in the inset of figure 2(b). The maximum temperature coefficients of resistance (TCR) of $(Nd_{0.7-x}La_x)_{0.7}Sr_{0.3}MnO_3$ ($x = 0.1$ and 0.2) are found to be higher compared to its parent compound $Nd_{0.7}Sr_{0.3}MnO_3$. It is worth to mention here that %TCR has been optimized to a value of 2.66 (for $x = 0.1$) and 2.65(for $x = 0.2$) which are independent with $<r_A>$. This opens the possibility of further improvement in operating temperature for IR detectors applications without sacrificing an optimum %TCR value.

Below T_{MI}, temperature resistivity behavior is explained by using $\rho = \rho_0 + \rho_2 T^2 + \rho_{4.5} T^{4.5}$ general polynomial equation to fit the experimental data of our samples. In the equation, ρ_0 is the temperature independent residual resistivity which arises due to the grain/domain boundary effects, scattering by impurities, defects and domain walls [16, 17], $\rho_2 T^2$ describes the resistivity due to electron-electron scattering phenomenon [18]. The last terms $\rho_{4.5} T^{4.5}$ ascribes due to the process of electron-magnon scattering process in the ferromagnetic region. The residual resistivity ρ_0 is decreasing with increase of average radius of A-site is shown in figure 2(b).

On the other hand, in semiconducting regime ($T > T_{MI}$), the variation of resistivity with temperature is explained by an Adiabatic small polaron hopping $\rho = \rho_0 \mathrm{T}\exp(\frac{E_P}{k_B T})$. The polaron hopping energy (E_P) decreases with increase of an average A-site ionic radius $<r_A>$ as shown in figure 3(a). A fitted curve on experimental data of NLSMO 0 by polaron hopping model is shown in an inset of the figure 3(a).

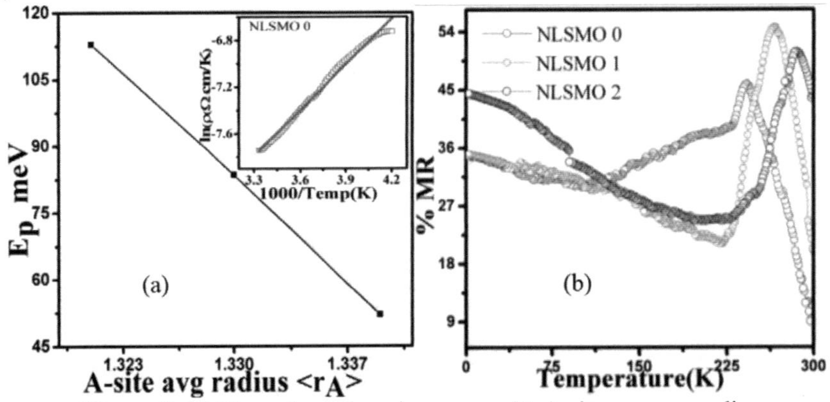

Figure 3. (a) The polaron hopping energy *Vs* A-site average radius
(b) Magnetoresistance *Vs* Temperature with in the field of 5 tesla.

The temperature dependent maximum percentage of magnetoresistance (%MR) at T_{MI} of compounds $(Nd_{0.7-x}La_x)_{0.7}Sr_{0.3}MnO_3$ are *46%* , *52%* and *50%* for $x = 0, 0.1$ and 0.2 with in the field of 5 Tesla. The low temperature MR is effect by the misalignment of adjacent magnetic domains/grains. Large magnetoresistance at T_{MI} is due to the misalignment of adjacent Mn ions and to the double exchange mechanism.

4. Conclusion

The metal-insulator transition (T_{MI}) temperature of $(Nd_{0.7-x}La_x)_{0.7}Sr_{0.3}MnO_3$ for $x = 0.2$ exhibits a shift towards room temperature as the average radius of site-A increases. Conduction mechanism is explained by the polynomial equation and by adiabatic polaron hopping model for low and high temperatures, respectively. Based on the results shown in this paper, it can be concluded that the residual resistivity (ρ_0) and polaron hopping energy decreases with an increase in the average radius of site-A. An interesting observation is that NLSMO 2 provides T_{MI} around the room temperature, and the maximum percentage change in TCR values are independent with average radius $<r_A>$ in $x = 0.1$ and $x = 0.2$.

References

[1] Coey J M D, Viret M and von Molnar S 1999 *Adv. Phys.* **48** 167
[2] Manh D H, Phong P T, Thanh T D, Hong L V and Phuc N X 2010 *J. Alloys Compd.* **491** 8
[3] Seikh Md M, Sudheendra L and Rao C N R 2004 *J. Solid State Chem.* **177** 3633
[4] von Helmolt R, Wocker J, Holzapfel B, Schultz M and Samver K 1993 *Phys. Rev. Lett.* **71** 2331
[5] Jin S, Tiefel T H, McCormack M, Fastnacht R A, Ramesh R, and Chen L H 1994 *Science* **264** 413
[6] Haghiri-Gosnet A M and Renard J P 2003 *J. Phys. D* **36** R127
[7] Heremans J 1993 *J. Phys. D: Appl. Phys.* **26** 1149
[8] Jin S, McCormack M, Tiefel T H and Ramesh R 1994 *J. Appl. Phys.* **76** 6929
[9] Mira J, Rivas J, Hueso L E, Rivadulla F, Lopez Quintela M A, Senaris M A Rodriguez and Ramos C A 2001 *Phys. Rev. B* **65** 024418

[10] Hueso L and Mathur N D 2004 *Nature* **427** 303
[11] Kim J H and Grishin A M 2005 *Appl. Phy. Letters* **87** 033502
[12] Zener C 1951 *Phys. Rev.* **82** 403
[13] Choi K-Y, Lemmens P, Guntherodt G, Pattabiraman M, Rangarajan G, Gnezdilov V P, Balahrishanan G, Mck Paul D and Less M R 2003 *J.Phys.:Condens. Matter* **15** 3333
[14] Kim K H, Gu J Y, Choi H S, Park G W and Noh T W 1996 *Phys. Rev. Lett.* **77** 1877
[15] Li Kebin, Li Xijun, Zhu Kaigui, Zhu Jingsheng and Zang Yuheng 1997 *J. Appl. Phys.* **81** 6943
[16] Snyder G J, Hiskes R, Dicarolis S, Beasley M R and Geballe 1996 *Phys. Rev. B* **53**, 14434
[17] De Teresa J M, Ibarra M R, Blasco J, Garcia J, Marquina C and Algarabel P A 1996 *Phys. Rev. B* **54** 1187
[18] Urushibara A, Moritomo Y, Arima T, Asamitsu A, Kido G and Tokura Y 1995 *Phys. Rev. B* **51**, 14103

IOP Conf. Series: Materials Science and Engineering **73** (2015) 012048 doi:10.1088/1757-899X/73/1/012048

Crystallization kinetics of Fe based amorphous alloy

T Shanker Rao[1] and T Lilly Shanker Rao[2]

[1]Physics Department , [2]Electronics Department
Narmada College of Science & Commerce, Zadeshwar,Bharuch(Gujarat)India 392011

E-mail:ncsctsr@gmail.com

Abstract. Differential Scanning Calorimetry(DSC) experimental data under non-isothermal conditions for Fe based Metglas 2605SA1 (wt% Fe=85-95, Si=5-10, B=1-5) metallic glass ribbons are reported and discussed. The DSC Scans performed at different heating rates showed two step crystallization processes and are interpreted in terms of different models like Kissinger, Ozawa, Boswell, Augis & Bennett and Gao & Wang. From the heating rate dependence of the onset temperature (T_o) and the crystallization peak temperature (T_p), the kinetic triplet, activation energy of crystallization (E), Avrami exponent (n) and the frequency factor (A) are determined. The determined E for peak I is 354.5 ± 2.5 kJ/mol and for the peak II is 348.2 ± 2.2 kJ/mol, respectively. The frequency factor for peak I is $1.1 \times 10^{23}sec^{-1}$ and for peak II is $6.1 \times 10^{20}sec^{-1}$.

1. Introduction

Metallic glasses are amorphous metals. In recent times these materials are known for their high technological applications in various areas. One of the powerful and less cumbersome methods to study the crystallization kinetics in these materials is DSC. Isothermal and non-isothermal methods are utilised in DSC [1-5]. In isothermal method the sample is brought quickly near the glass transition temperature (T_g) and the heat liberated is recorded as a function of temperature, whereas in non-isothermal method the sample is heated with a constant heating rate and the heat evolved is measured as a function of temperature or time. The disadvantage of the isothermal process is to reach the required T_g and record the heat evolved at this point. This disadvantage is removed in non-isothermal method and hence the method is more favoured [6].

The present investigation is aimed at determining the activation energy of crystallization, E, Frequency factor, A and the Avrami exponent, n for the metallic glass 2605SA1 using different methods under non-isothermal conditions.

2. Experimental methods

The Fe-based amorphous alloy 2605SA1 was obtained from Metglass Inc. Co., USA as a gift. The amorphous nature of the sample is confirmed by XRD (figure 1). As-quenched samples were heated in DSC (Mettler Toledo) at five linear heating rates (2, 4, 6, 8 and 10 $Kmin^{-1}$) from RT to 840 K in air. The DSC scans were recorded by a thermal analyzer interfaced to a computer.

IOP Conf. Series: Materials Science and Engineering **73** (2015) 012048 doi:10.1088/1757-899X/73/1/012048

3. Results and discussion

The DSC thermograms of 2605SA1 alloy at different heating rates are shown in Figure 2. T_0 and T_p are given for the peak I & II in the Table 1.It is noticed that the T_p shifted to high temperatures with increasing heating rate. This shift in the peak forms the basis for determination of the activation energy, E. Due to increase in the sensitivity, the peak height also increases with increasing heating rate. The non-isothermal crystallization data of DSC have been analyzed using different models as discussed below.

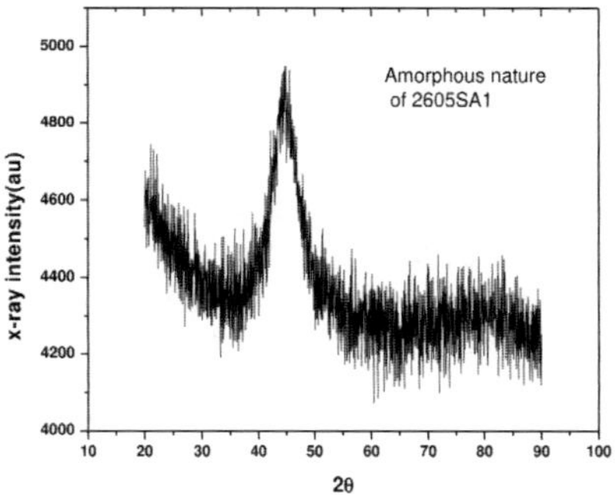

Figure 1. XRD pattern at RT

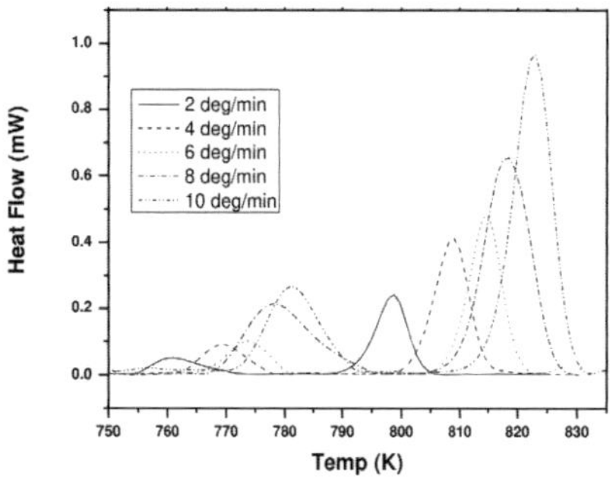

Figure 2. DSC thermograms at different heating rates

Table 1.Onset and peak crystallization temperatures for peak I & II

Heating rate (K/min)	Peak I		Peak II	
	T_0(K)	T_p(K)	T_0(K)	T_p(K)
2	753.14	760.71	786.64	798.88
4	761.16	769.42	797.54	809.04
6	763.60	773.99	803.56	814.71
8	766.56	778.24	804.84	818.43
10	771.16	781.49	807.89	823.02

3.1 Kissinger method. Using Kissinger's equation [7]

$$\ln (\beta/T_p^2) = -E/RT_p + \ln (AR/E) \quad (1)$$

Where β is the heating rate and R is the universal gas constant, the value of E can be obtained from the heating rate dependence of the peak temperature of crystallization. Figure 3 shows $\ln(\beta /T_p^2)$ vs $1000/T_p$ relationship for peak I & II. The value of E for crystallization and the frequency factor (A) are obtained from the slope and intercept of the plot and are given in the Table 2 respectively.

3.2 Ozawa method. According to Ozawa[8], the change of $\ln(1/T_p^2)$ with β is negligibly small compared to the change of ln(β) and therefore, the Kissinger's equation can be written as

$$\ln\beta = -1.0516 \; E/RT_p + \text{const} \quad (2)$$

Plotting lnβ vs $1000/T_p$, E is calculated for peak I & II (Table 2)

3.3 Bosewell method. In order to determine the activation energy Bosewell [9] uses the following equation

$$\ln (\beta/T_p) = - E / RT_p + \text{const} \quad (3)$$

The slope of the graph for peak I & II gives the E and are given in Table 2.

3.4 Augis and Bennett method. The equation given by Augis and Bennett[10] is

$$\ln (\beta/(T_p - T_o)) = - E/RT_p + \ln A \tag{4}$$

Where T_p and T_o are the peak and the onset crystallization temperature respectively. Figure 4 shows the plot $\ln(\beta/(T_p\text{-}T_o))$ vs $1000/T_p$ for peak I & II and gives the E and A values for crystallization and are given in the Table 2.

3.5 Gao and Wang method [11]. This method uses the following expressions to determine the E, A and Avrami exponent n. The determined values are given in the Table 2.

$$\ln\left[\beta\left(\frac{d\alpha}{dT}\right)_p\right] = -\frac{E}{RT_p} + const , \tag{5}$$

where $\beta = dT/dt$ and α is fractional crystallization.

3.6 Matusita and Sakka method. For non-isothermal data, Matusita and Sakka [12] suggested the equation

$$\ln[-\ln(1-\alpha)] = -n \ln \beta - 1.0052\ mE/RT + \text{const} \tag{6}$$

Here *m* is an integer depends on the dimensionality of the crystallization and *n* the Avrami exponent depends on the nucleation process. The plot ln [-ln(1-α)] versus $1000/T_\alpha$ at different heating rates should give a straight line. Figure 5shows such plot at different heating rates for peak I. Figure 6 shows the variation of ln(-ln(1-α)) with lnβ at constant temperatures whose slope gives the value of *n*. The value of *n* shows the scattering trend for both the peaks (Table 2).

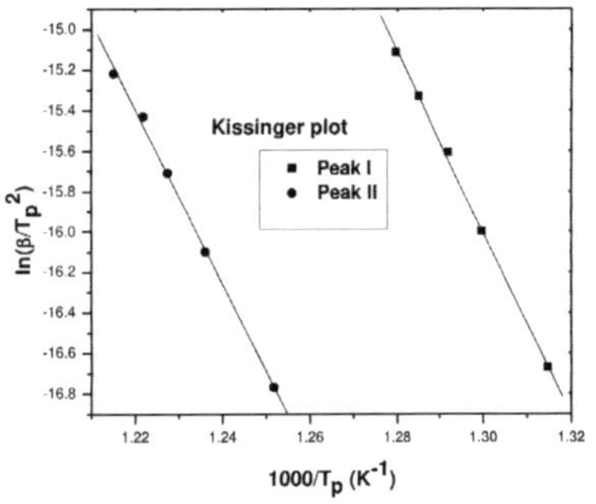

Figure 3 plot of $\ln(\beta /T_p^2)$ vs $1000/T_p$

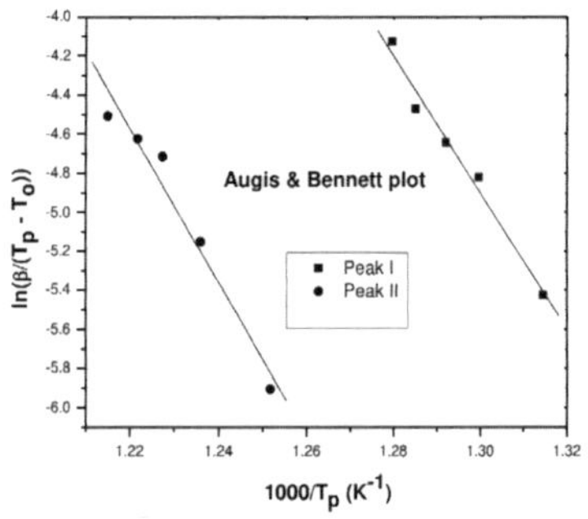

Figure 4. plot of $\ln(\beta/(T_p\text{-}T_0))$ vs $1000/T_p$

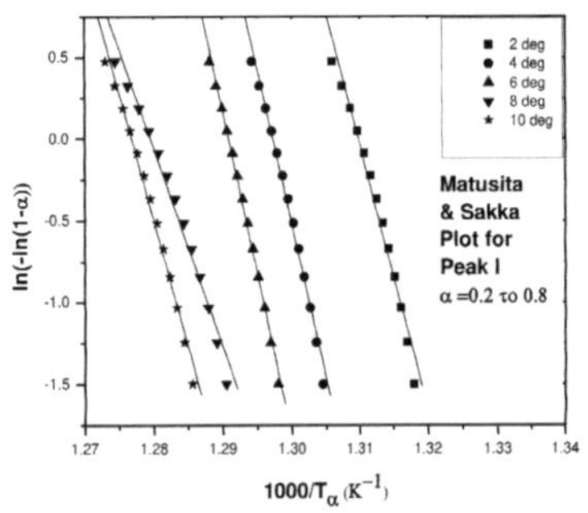

Figure 5. $\ln[-\ln(1-\alpha)]$ vs $1000/T_\alpha$ at different temperatures for peak I

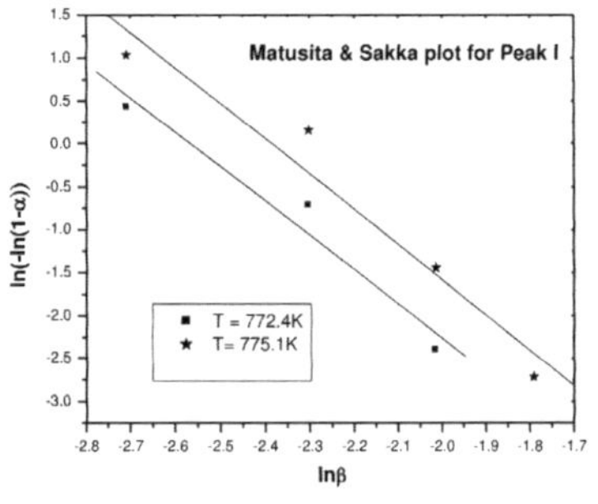

Figure 6. $\ln[-\ln(1-\alpha)]$ vs $\ln\beta$ at different temperatures for peak I

Table 2. Kinetic parameters obtained from different methods

Method	Activation energy(E) (kJ/mol)		Frequency factor(A) Sec^{-1}		Avrami exponent(n)	
	Peak I	Peak II	Peak I	Peak II	Peak I	Peak II
Kissinger	373.3±1.0	358.4±1.2	1.1×10^{23}	$6.1X\ 10^{20}$		
Ozawa	367.0±1.0	353.5±1.2				
Bosewell	379.7±1.0	365.2±1.2				
Augis & Bennett	289.2±2.9	328.2±4.6	3.2×10^{17}	8.6×10^{18}	4.7	6.2
Gao & Wang	363.3±6.4	335.9±2.9	2.3×10^{22}	2.1×10^{19}	3.8	5.8
Matusita & Sakka	350.6±2.0	403.7±1.8			4.0	4.7

4. Conclusions

The crystallization kinetics of Fe based metallic glass, 2605SA1 have been determined under non-isothermal conditions using different methods. The obtained values of E, A and n for the peaks I & II are given in the Table 2. The E and A values by all the methods indicate consistency for both the peaks except in the case of Augis & Bennett. This discrepancy is due to the error in determining the onset temperature of crystallization. It is observed that the frequency factor is very high for both the peaks. The Avrami exponent directly reflects the mode of crystallization. The values obtained for this Fe based 2605SA1 metallic glass for the peaks I and II have a large scattering. The possible reasons may be (i) the sensitivity to alloy composition,(ii) the original state of the sample fully amorphous or partially and (iii) relaxation occurred before the crystallization. Some of these factors may also responsible for the scattering of the Avrami exponent [12] in this sample.

Acknowledgements

T Lilly Shanker Rao acknowledges the financial support rendered by UGC, New Delhi through a minor project no. 47/718/08(WRO) (Project period is over).The authors also acknowledge the experimental help (DSC) taken from Central Salt & Marine Chemicals Research Institute (CMCRI) Gijubhai Badheka Marg, Bhavnagar-364002, Gujarat (INDIA) and UGC-DAE-CSR,Indore for XRD measurements. The authors also wish to thank Dr B M Rawal, Principal, Narmada College for his untiring help and support during the period of the UGC project.

References

[1] A H Moharram, M S Rasheedy, 1998 Phys.Stat.Sol. **A** 16933
[2] J Vazquez,C Wagner, P Villares, R Jimenez-Garay, 1998 J.Non-Cryst.Solids **548** 235
[3] J Vazquez ,C Wagner , P Villares ,R Jimenez-Garay, 1996 *Acta Mater.***44** 4807
[4] Y Q Gao ,W Wang , 1986 J. Non-cryst. Solids **811** 29
[5] R A Ligero ,J Vazquez ,P Villares , R Jimenez-Garay, 1990 Thermochim Acta **162** 427
[6] A H.Moharram , M Abu El-Oyoun, A A Abu-Sehly, 2000 J. Phys. D: Appl. Phys **34** 1.
[7] H E Kissinger, 1957 Annal Chem., **29** 1702
[8] T Ozawa, 1965 Bull. Chem. Soc., **38** 1881
[9] P G Bosewell, 1980 J. Thermal Anal., **18** 353
[10] J A Augis and J E Bennett, 1978 J. Thermal Anal., **13** 283
[11] Matusita and S Sakka , 1979 Phys.Chem. Glasses., **20** 81
[12] X D Wang, Q Wang and J Z Jiang, 2007 J. Alloys & Compounds.,**440** 189

Growth, optical, electrical and photoconductivity studies of a novel nonlinear optical single crystal: Mercury cadmium chloride thiocyanate

S.M. Ravi Kumar[1*], S. Selvakumar[2], P. Sagayaraj[3] and A.Anbarasi[4]
[1]PG & Research Dept.of Physics, Govt. Arts Collge,Thiruvannamalai -606 603, India
[2]Department of Physics, Govt. College, Nandanam, Chennai– 600 035, India
[3]Department of Physics, Loyola College, Chennai – 600 034, India
[4]Department of Physics, Periyar Arts College, Cuddalore- 607 001, India

[*]Email: smravi78@rediffmail.com Mobile: 09884473709

Abstract SCN^- ligand based organometallic non-linear optical mercury cadmium chloride thiocyanate (MCCTC) crystals are grown from water plus methanol mixed solvent by slow evaporation technique. The grown crystals are confirmed by single crystal X-ray diffraction analysis which reveals that the MCCTC belongs to rhombohedral system with R3c space group. MCCTC exhibits a SHG efficiency which is nearly 17 times more than that of KDP. The dielectric constant, dielectric loss measurements of the sample have been carried out for different frequencies (100 Hz to 5 MHz) and, temperatures (308 to 388 K) and the results are discussed. Photoconductivity study confirms that the title compound possesses negative photoconducting nature. The surface morphology of MCCTC was also investigated.
Keywords: Solution growth; Characterization; NLO material; X-ray diffraction

1. Introduction

Nonlinear optics (NLO) has emerged as one of the most attractive fields of current research in view of its vital applications in areas like optical modulation, optical switching, optical logic, frequency shifting and optical data storage for developing technologies in telecommunications and signal processing. Materials with large second order nonlinear optical (SONLO) properties, short transparency cut off wavelengths and stable physicochemical performance are needed to realize many of these applications [1,2]. In this connection, attempts have been made by researchers to develop new organometallic compounds (which contain at least one direct M-C bond between metal and organic ligands) and coordination compounds (in which the metal and the ligands are connected through M-O, M-N, M-S or M-P bonds), these compounds combine the features of both inorganic and organic compounds. Organometallic and coordination compounds offer a variety of molecular structures by changing the metals, ligands, coordination numbers and so on. In metal thiocyanate complexes, the thiocyanate (SCN) plays a crucial role in combining the versatile ambidentate ligand with two donor atoms. Being a ligand with potential S and N donors, thiocyanate (SCN) ion is of interest not only due to the structural chemistry of its multifunction coordination modes, but also because of the formation of complexes with NLO activities [3-7]. Metallic thiocyanates and their derivates are such potentially useful candidates among the organometallic systems because all of them contain –S=C=N– bridges, which connect with metal atoms, forming infinite two dimensional or three dimensional networks. The infinite networks confer a relatively large polarization, which induces relatively large macroscopic nonlinearities in these materials and the same has been proved recently by few research groups [8-10]. Mercury cadmium chloride thiocyanate (MCCTC) is one such potentially useful SONLO crystals with an empirical formula $Hg_3CdCl_2(SCN)_6$. In the earlier works, MCCTC was grown from methanol by slow evaporation; unfortunately, the reported size of the crystal was very small. Hence, the present work addresses the key issue of growing bulk size crystal from a mixed solvent (methanol plus water) by slow evaporation technique. The grown crystal has been confirmed by single crystal XRD. The dielectric, photoconductivity and surface morphology studies are investigated and reported. The nonlinear optical property of the single crystal has been confirmed by SHG test.

2. Experimental procedure

2.1 Synthesis

The starting materials were analytical reagent grade (purity >98.0%) and used as purchased. Appropriate amounts of mercury thiocyanate and cadmium chloride taken in the molar ratio of 3:1 were dissolved in methanol solvent to synthesis MCCTC. The chemical equation is as follows.

$$3(Hg(SCN)_2) + CdCl_2 \rightarrow Hg_3CdCl_2(SCN)_6$$

The synthesized salt was further purified by repeating the crystallization experiment twice or thrice.

2.2 Crystal Growth

The saturated solution of MCCTC was prepared in methanol plus water mixed solvent at room temperature and stirred well to enable homogeniation of the solution. The seed crystals were prepared by the conventional isothermal evaporation method. Good quality seed was chosen and then kept suspended into the supersaturated solution. MCCTC crystal was grown by slow evaporation method at 303 K. Crystal having dimensions up to 10 x 4 x 2 mm^3 (Figure 1)was harvested in a period of 35-40 days.

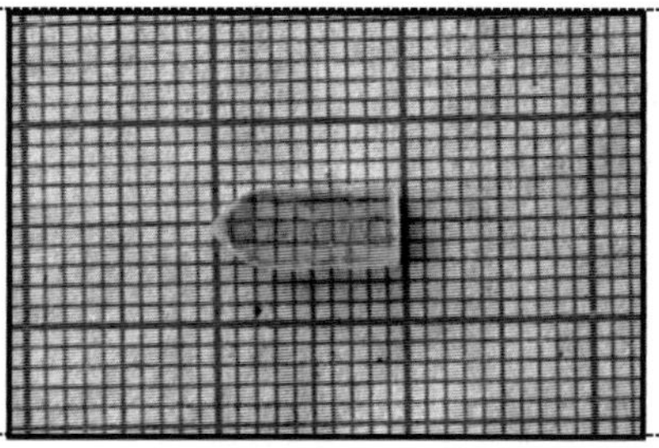

Figure 1. Photograph of as grown crystal of MCCTC.

3. Results and discussion

3.1. X-ray diffraction (XRD) analysis

The lattice parameter values of MCCTC from single crystal XRD are determined as a = 11.184 Å, b = 11.191 Å and c = 59.908 Å. The values of unit cell volume (V) and density (D) are 6493.2 $Å^3$ and 3.47 g/cm^3 respectively. The XRD data is in good agreement with reported value [11], and thus confirm the grown crystal.

3.2 Nonlinear optical (NLO) study

The nonlinear optical property of MCCTC was evaluated by the Kurtz and Perry powder technique [12] and microcrystalline powder of KDP was taken as the reference material. For a laser input pulse of 5.6 mJ, the second harmonic signal of 160 mV and 2.7 V were obtained through KDP and MCCTC respectively and thus confirming its second harmonic efficiency to be 17 times that of KDP.

3.3 Dielectric constant and dielectric loss measurements

Dielectric constant and dielectric loss of the sample have been measured for different frequencies (100 Hz to 5 MHz) at different temperatures (308 to 388 K). Figures 2 and 3 show the variations of dielectric constant and dielectric loss respectively as a function of frequency at different temperatures. It is observed from figure 2 that the dielectric constant (at 308 K) decreases with increase in frequency from 100 Hz to 10 kHz and then attains a constant value of 28.61. The same trend is observed for other temperatures too. It is also observed that the value of dielectric constant increases with temperature. Such variations at higher temperature may be attributed to the blocking of charge carriers at the electrodes [13]. The decrease of dielectric constant at low frequency region may be due to space charge polarization. Figure 3 indicates that as the frequency increases, the dielectric loss decreases exponentially and then attains a lower value of 0.049 at 308 K. The low value of dielectric loss confirms the lesser level of defects present in the grown sample.

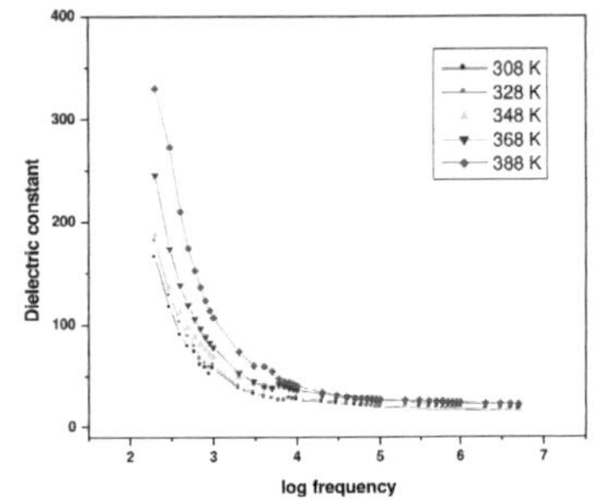

Figure 2. Variation of dielectric constant with frequency for MCCTC at different temperatures

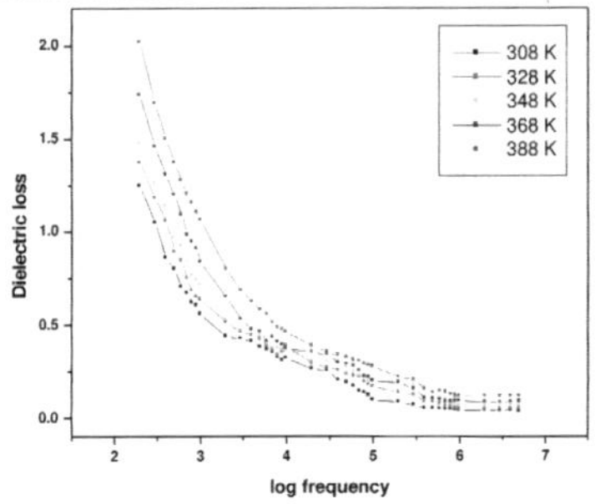

Figure 3. Variation of dielectric loss with frequency for MCCTC at different temperatures

3.4Photoconductivity study

Photoconductivity measurements were made using the Keithley 485 picoammeter. The dark current was recorded by keeping the sample unexposed to any radiation. The experiment was performed at room temperature (303 K). The applied field was varied from 0 to 2500 V/cm. The variations of photo current (I_p) and dark current (I_d) with applied field are shown in Figure 4. It is observed from the plot that the dark current is always higher than the photo current for different applied field. Thus, MCCTC crystal is found to exhibit negative photoconductivity. The negative photoconductivity exhibited by the sample may be due to the reduction in the number of charge carriers in the presence of radiation [14].

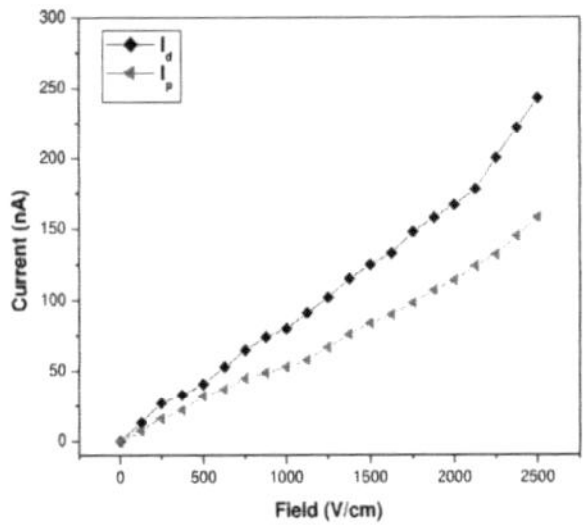

Figure 4. Field dependent photoconductivity of MCCTC

3.5 SEM analysis

The SEM micrographs of MCCTC were taken at room temperature (303 K) with different magnification (x500 and x5000). The micrographs are shown in Figures 5 and 6. The surface appears smooth though it has well defined microcrystals on the surface. The formation of needle shaped microcrystals is clearly evident in figure 6 with higher magnification.

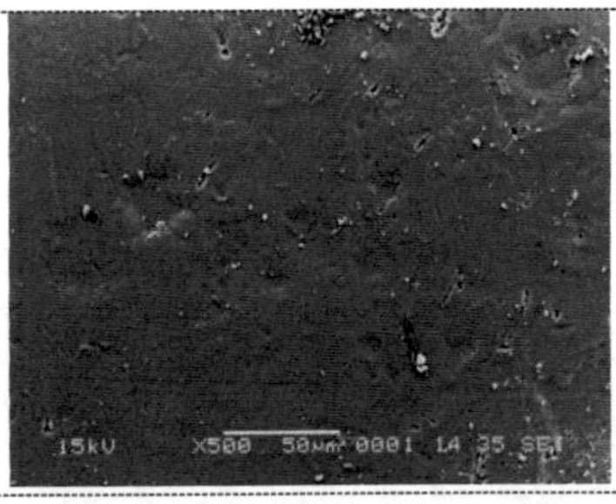

Figure 5. SEM micrograph of MCCTC crystal with x 500

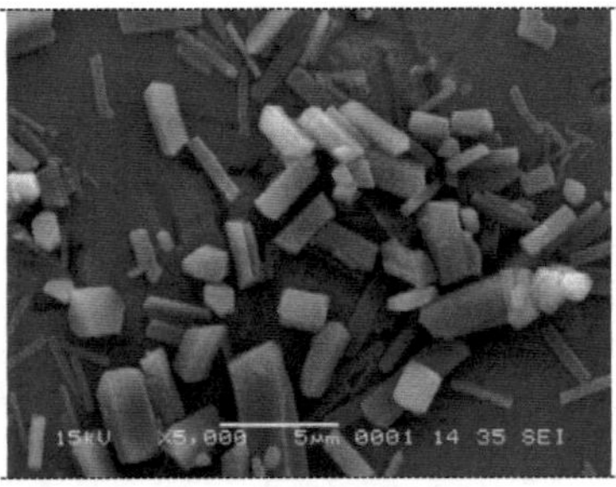

Figure 6. SEM micrograph of MCCTC crystal x 5000

4. Conclusion

Optically clear single crystal of MCCTC was successfully obtained by slow evaporation method. Single crystal XRD study confirms the rhombohedral structure. The dielectric study revealed that the crystal has low dielectric constant and low dielectric loss in the high frequency region. The negative photoconducting nature of MCCTC crystal is confirmed by photoconductivity studies. The surface morphology was analyzed.

5. References

[1] Bergman J G, Mefee J H, and Crane G R 1970 *Mater. Res. Bull.* **5** 913
[2] Sturmer W, and Deserno U 1970 *Phys. Lett. A.,* **32** 539
[3] Xu D, Yu W T, Wang X Q, Yuan D R, Lu M K, Yang P, Guo S Y, Meng F Q, and Jiang M H 1999 *Acta Crst.* **C55** 1203
[4] Wang X Q, Xu D, Yuan D R, Tian Y T, Yu W T, Ying S S, Yang Z H, and Fang Q 1999 *Mater. Res. Bull.* **34** 2003
[5] Duorong Yuan, Zhewu Zhoug, Mingguo Liu, Dong Xu, Qi Fang, Youghong Bing, Suoying Sun and MinhuaJiang 1998 *J. Cryst. Growth* **186** 240
[6] Sun H Q, Yuan D R, Wang X Q, Lu Y Q, Sun Z H, Wei X C, Duan X L, Luan C N, Lu M K and Xu D 2003 *J. Crystal Growth* **256** 183
[7] Wang X.Q., Xu D., Lu M.K., 2001 *Cryst. Res. Technol.* **36** 73
[8] Xinqiang Wang, Doung Xu, Mengkai Lu, Dourong Yuan, Guanghui Zhang, Shouxi Xu, Shiyi Guo, Xuening Jiang, Jiurong Liu, Chunfeng Song, Quan Ren,Ji Huang, Yupeng Tian 2001 *Mater.Res.Bull.* **36** 1287
[9] Rajarajan K, Selvakumar S, Ginson P. Joseph, Samikkannu S, Vetha Potheher I, and Sagayaraj P 2006 *Optical Materials*, **28** 1187
[10] Ginson P. Joseph, Korah I, Raja Rajan K, Thomas P C, Vimalan M, Madhavan J and Sagayaraj P 2007 *Cryst. Res.Technol.*, **42** 295
[11] Alain Mosset, Muriel Bagieu-Beucher, Amelie Lecchi, Rene Masse, Julien Zaccaro 2002 *Solid State Sciences* **4** 827
[12] Kurtz S K and Perry T T 1968 *J. Appl. Phys.* **39** 3798
[13] Arora S K, Vipul Patel, Brijesh Amin and Anjana Kothari 2004 *Bull. Mater. Sci.,* **27** 141
[14] Abraham Rajasekar S, Thamizharasan K, Joe G.M. Jesudurai, Prem Anand D and Sagayaraj P 2004 *Mater. Chem. Phys.,* **84** 157

IOP Conf. Series: Materials Science and Engineering **73** (2015) 012050 doi:10.1088/1757-899X/73/1/012050

Magnetic properties of cobalt ferrite nanoparticles synthesized by sol-gel method

T George[1], A T Sunny[2] and T Varghese[3*]

[1] Holy Kings College of Engineering and Technology, Pampakkuda - 686 667, Kerala, India
[2] School of Chemical Sciences, M G University, Kottayam – 686 560, Kerala, India
[3]Nanoscience Research Centre (NSRC), Department of Physics, Nirmala College Muvattupuzha - 686 661, Kerala, India

[+]E-mail: nanoncm@gmail.com

Abstract. Cobalt ferrite nanoparticles of average size 18 nm are synthesized by sol-gel method and investigated the magnetic properties. The saturation magnetization value calculated from vibration sample magnetometer (VSM) studies for $CoFe_2O_4$ is lower than the reported value for the bulk. The magnetization curves demonstrate a trend towards the superparamagnetic behavior of the as-prepared $CoFe_2O_4$ nanoparticles. The microwave magnetic parameters show a decreasing trend with the increase of frequency.

1. Introduction

Nanosized ferrite magnetic materials are the subject of intense research work because of their numerous applications in various fields such as, magnetic disc drives, spintronics, ferrofluids, catalysts, medical diagnostics, drug delivery systems, pigments in paints and ceramics [1-3]. The high resistivity and low hysteresis loss makes ferrites, suitable to use in microwave applications and radio electronics [4-5]. Various methods have been developed for the synthesis of nanoferrites, like sonochemical reactions, sol-gel, microwave plasma, host template, coprecipitation, microemulsion and chemical vapour deposition [6-8]. We have synthesized cobalt ferrite nanoparticles of average size 18 nm by sol-gel method as described elsewhere [9-10] and investigated their magnetic and microwave properties.

2. Experimental

The precursors are $Co(NO_3)_2\ 6H_2O$ (0.4 M, 99% SD Fine Chem), $Fe(NO_3)_3\ 9H_2O$ (0.8 M, 99 % SD Fine Chem) and aqueous solution of poly acrylic acid (Acros organics), used without further purification, and synthesized by sol-gel method [10]. The phase structure characterization of the synthesized powder was carried out by X-Ray diffraction (XRD) using PANalytical X'Pert PRO diffractometer with CuK_α radiation. Studies on the surface morphology and homogeneity of the powders were carried out using a field emission scanning electron microscope (NOVA NANOSEM 600 FEI).

The magnetic properties of the synthesized nanoparticles were measured using a VSM (Lakeshore) at room temperature and the hysteresis loop was drawn with the magnetic field, cycled between – 10000 G to 10000 G. The experimental set up for the determination of microwave magnetic properties consists of a transmission type S-band rectangular cavity resonator and a network analyzer [11-12]. The rectangular cavity is connected to the Agilent 8714 ET network analyzer (frequency range 300 KHz - 3GHz). To find the complex permeability, the sample powders were tightly filled in thin capillary tubes of low loss material and sealed. The tube was then inserted through the non-radiating slot on the broad wall of the wave-guide cavity, and held with a thin sample holder made of low loss material. Resonant frequencies and the corresponding Q values of the sample loaded cavity and of the air filled cavity were measured in the TE mode. Then the magnetic parameters such as the real and imaginary part of complex permeability (table 1) were evaluated according to the theory explained elsewhere [11-12].

3. Results and discussion

Figure 1(a) shows the XRD pattern of the synthesized cobalt ferrite nanoparticles. All diffraction peaks are indexed to a pure cubic phase (JCPDS Card No 22-1086) of $CoFe_2O_4$ with space group Fd3m. The estimated lattice parameter, a = 8.388 Å well agrees with the JCPDS file. The average crystallite size estimated using the Scherrer equation for $CoFe_2O_4$ is 18 nm. Figure 1(b) shows the FE-SEM images of $CoFe_2O_4$ nanoparticles. It can be seen that the particles are almost spherically shaped and uniformly arranged. The particle size estimated from the FE-SEM analysis agrees well with the XRD results. The extremely fine crystallite size and regular shape of the materials are evident from the FE–SEM studies.

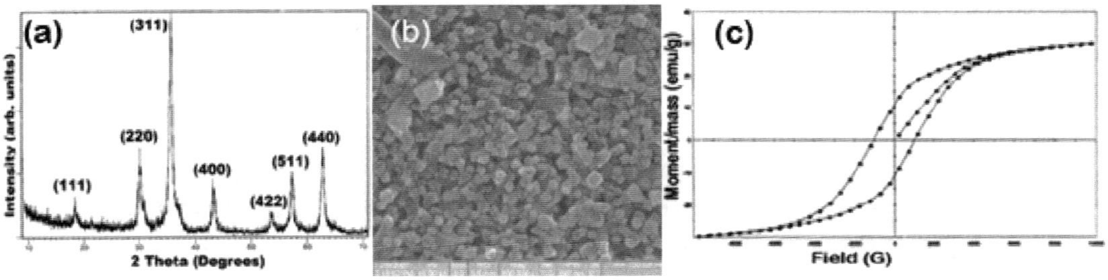

Figure 1. (a) XRD pattern (b) FE-SEM images and (c) hysteresis curves of the $CoFe_2O_4$ nanoparticles.

A hysteresis curve of $CoFe_2O_4$ nanoparticles is shown graphically in figure 1(c). The saturation magnetization (Ms) value calculated for $CoFe_2O_4$ nanoparticles is 30 emu/g, which is lower than the reported value for the bulk samples (80 emu/g) [13], which is a consequence of superparamagnetic nature of the magnetic nanoparticles. The decrease in the density of magnetization of the nanoparticles with respect to the bulk can be attributed to surface defects and morphology. The surface defects are the results of finite-size scaling of nanocrystallites, which in turn leads to a non-collinearity of magnetic moments on their surface. These effects are more intense in ferromagnetic system, where the super-exchange interaction occurs through the oxygen ion, O^{2-} [14]. The coercivity recorded is approximately 1100 G for $CoFe_2O_4$ nanoparticles. The superparamagnetic behavior should be attributed to the extremely fine crystallite size and regular shape of the material [15], which makes it easier for them to be thermally activated to overcome the magnetic anisotropy.

Microwave magnetic parameters of $CoFe_2O_4$ nanoparticles (sample density = $0.46x10^3$ kgm^{-3}) at various resonant frequencies are shown in table 1. The real part of complex permeability μ_r', imaginary part of complex permeability μ_r'' and the corresponding loss tangent of $CoFe_2O_4$ nanoparticles decrease continuously as the frequency increases from 2248 MHz to 2972 MHz. The eddy current loss in a material can be expressed in terms of Tanδ. It is known that ferrites possess feeble eddy currents, which agrees very well with the experimental results. The decrease in loss tangent is due to the fact that, when the frequency of the field increases the dipoles cannot follow the rapid variation of the field. The values of permeability are not really constant for a given material but usually have a strong dependence on frequency [16]. When the frequency increases the magnetic dipoles of the crystallites cannot follow the fast variation of the applied field. The superparamagnetic nature of the $CoFe_2O_4$ nanoparticles might have affected the experimental values of permeability.

Table 1. Variation of magnetic parameters with frequency of $CoFe_2O_4$ nanoparticles.

Frequency (MHz)	Real part of Complex permeability μ_r'	Imaginary part of complex permeability μ_r''	$\tan\delta = \mu_r''/\mu_r'$
2248	8.327	0.4162	0.0500
2440	7.941	0.3861	0.0486
2685	7.612	0.2923	0.0384
2972	7.308	0.2672	0.0365

4. Conclusions

Cobalt ferrite nanoparticles with average particle size 18 nm are synthesized by sol-gel method. The average crystallite size and the particle size are calculated both by XRD and FE-SEM are found to be in good agreement with each other. The saturation magnetization value calculated for $CoFe_2O_4$ is lower than the reported values for the bulk. The magnetization curves demonstrate that $CoFe_2O_4$ nanoparticles at room temperature tend to be superparamagnetic. The microwave magnetic parameters show a decreasing trend with the increase of frequency. The superparamagnetic nature of the $CoFe_2O_4$ nanoparticles might have affected the experimental values of magnetization and permeability.

Acknowledgements

Authors thank University Grants Commission (UGC), New Delhi and Kerala State Council for Science, Technology and Environment (KSCSTE), Thiruvananthapuram for financial support.

References

[1] Emerich D F and Thanos C G 2003 *Expert Opin. Biol. Ther.* **3** 655-663
[2] Ngo A T and Pileni M P 2001 *J. Phys. Chem. B* **105** 53
[3] Didukh P *et al* 2002 *J. Magn. Magn, Mater.* **613** 242-245
[4] Venzke S *et al* 1997 *J. Mater. Res.* **11** 1187
[5] Ahmed M A, Elnimer M K, Tawtic A, and El Hasab 1991 *J. Magn. Magn. Mater.,* **98** 33
[6] Shafi K V P M and Gedanden A 1998 *Chem. Mater.* **10** 3445
[7] Ravikumar G, Vijayakumar K and Venudhar Y C 2012 *Mater. Sci. Appl.,* **3** 87-91
[8] Kohli S, McCurdy P R, Johnson D C, Das J and Fisher E R 2010 *J. Phys. Chem. C* **114** 19557
[9] Chen D H and He X R 2001 *Mater. Res. Bull.,* **36** 1369
[10] George T, Joseph S and Mathew S 2005 *J. Metastable and Nanocryst. Mater.* **23** 141
[11] Raveendranath U and Mathew K T 1998 *Microwave and Opt. Tech. Letters* **18** 241
[12] Harrington R F 1961 *Time Harmonic Electromagnetic fields* (McGraw-Hill: New York)
[13] Roy S and Ghose J 2000 *J. Appl. Phys.* **87(9)** 6226
[14] Caizer C and Stefanescu 2002 *J. Phys. D: Appl. Phys.* **35** 3035
[15] Chen Q and Zang Z J 1998 *Appl. Phys. Lett.* **73** 3156
[16] Wangsness R K 1979 *Electromagnetic Fields* (John Wiley & Sons, New York) 25

International Conference on Materials Science and Technology (ICMST 2012) IOP Publishing
IOP Conf. Series: Materials Science and Engineering **73** (2015) 012051 doi:10.1088/1757-899X/73/1/012051

Enhancement of electrical properties of manganese tungstate nanoparticles by beam irradiation

N Aloysius Sabu[1], K P Priyanka[1], X Sheena[2], E M Mohammed[2] and T Varghese[1*]

[1]Nanoscience Research Centre (NSRC), Department of Physics, Nirmala College, Muvattupuzha - 686 661, Kerala, India
[2]Department of Physics, Maharajas College, Ernakulam - 682 011, Kerala, India

[*]Email: nanoncm@gmail.com

Abstract. $MnWO_4$ nanoparticles were synthesized using simple chemical precipitation method. The dielectric properties of irradiated and non-irradiated samples of monoclinic manganese tungstate nanoparticles have been studied as a function of frequency. It is found that electric properties such as dielectric constant, loss tangent and a.c. conductivity of the irradiated sample were much larger as compared to the non-irradiated sample, which has been attributed to defects and reduction in particle size caused by the beam irradiation. The present investigation observed that the electron beam irradiation is an efficient technique to modify the electrical properties of $MnWO_4$ nanoparticles.

1. Introduction

Nanostructured tungstate materials have aroused much interest because of their luminescence behaviour, structural properties and potential applications. The electrical conductivity of $MnWO_4$ is sensitive to changes in humidity, thereby making it useful as a humidity sensor with potential applications like meteorology, medicine, food production, agriculture, industrial and domestic environment [1-5]. There are a number of processes used to synthesize nanocrystalline $MnWO_4$, such as microwave-assisted synthesis [1], surfactant-assisted complexation-precipitation method [2], melt solution process [6], solvothermal route [5], aqueous salt metathesis reaction [7], sol-gel technique [4,8], ambient template synthesis [9] and solid state metathetic approach [10].

In the present work, we report a systematic study of the dielectric properties and a.c. conductivity of pure and electron beam (EB) irradiated samples of $MnWO_4$ nanoparticles.

2. Experimental details

Manganese chloride ($MnCl_2.4H_2O$) (99.8%, Sigma Aldrich) and sodium tungstate ($Na_2WO_4.2H_2O$) (99.9%, Alfa Aesar) were used for the synthesis of nanocrystalline $MnWO_4$. Manganese tungstate samples were prepared by reacting aqueous solutions of manganese chloride and sodium tungstate (0.1 M each) at room temperature, keeping the p^H=7. The precipitate formed was centrifuged, filtered, washed with distilled water a number of times, and dried in an oven to get fine powders of manganese tungstate. The obtained powder samples were calcined for 3 hours in a furnace at 450^0 C in an aerobic atmosphere. The samples were irradiated at a dose rate of 5 kGy with an electron beam of 8 MeV energy obtained from a variable energy Microtron at Mangalore University, India. S1 is pure sample of $MnWO_4$ and S2 is the irradiated samples with a dose rate of 5 kGy.

The structural characteristics of the pure and irradiated samples have been studied by X-ray powder diffraction (XRD) using Bruker D8 Advance X-ray diffractometer (λ=1.5406 Å). The crystallite size was estimated from the Scherrer's equation [8]. The dielectric properties of pellet samples sintered at 500 ^{0} C are measured at room temperature using a four probe LCR meter (Wayne Kerr H-6500 model) in conjunction with a portable furnace and temperature controller (±1K). The dielectric constant (ε') and a.c. conductivity (σac) were calculated by using equation (1) and (2),

$$\varepsilon' = cd/\varepsilon_0 a, \quad (1)$$

$$\sigma ac = \varepsilon' \varepsilon_0 \omega \tan\delta, \quad (2)$$

where a is face area, c capacitance of the pellet in pF, d thickness of pellet, ε_0 permittivity of vacuum, ω angular frequency and $\tan\delta$ loss tangent.

3. Results and discussion

Figure 1 displays the powder X-ray diffraction (XRD) pattern of pure sample (S1) and EB irradiated sample (S2). Average crystallite size obtained for samples S1 and S2 are 21 nm and 17.4 nm respectively. The principal'd' values taken from the

JCPDS file No. 80-0135 for $MnWO_4$ were in close agreement with the observed 'd' values. The XRD results showed that crystallite size and percentage of crystallinity decreased after irradiation. It confirms amorphization of the sample due to irradiation which is responsible for the change in electrical properties.

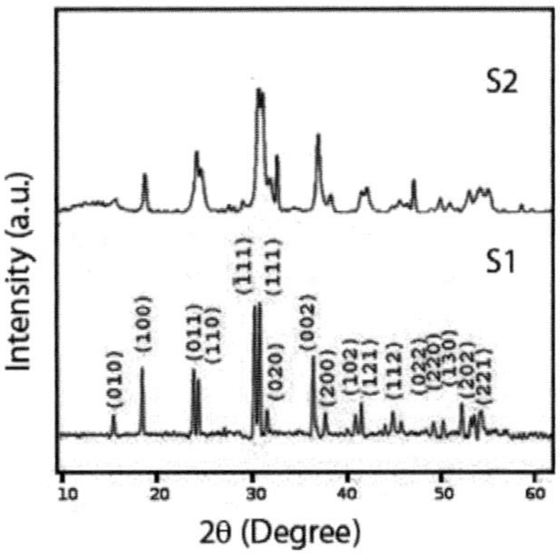

Figure1. XRD spectra of samples S1 and S2.

Figure 2 shows the variation of dielectric constant with frequency at room temperature. It can be seen that the real part of dielectric constant ε' for both the samples S1 and S2 have high values at low frequencies, which decreases rapidly as frequency increases and attains a constant value at higher frequencies. It has been found that the ε' values are shifted upward for the irradiated sample (S2). The value of έ for S1 is 110 at 100Hz while its value for S2 is 215.6. The έ values for S1 and S2 at 10 MHz are 10.7 and 17.2 respectively. The increase in the dielectric constant of the irradiated samples can be attributed to size reduction of the nanoparticles caused by the EB irradiation [11]. When the particle size reduces, the volume percentage of interface boundaries and the amount of defects that cause various kinds of polarizations might increase, which in turn increases dielectric constant.

The variation of tan δ with frequency of samples is shown in figure 3 (a). The loss in $MnWO_4$ can be explained by the electronic hopping model [12]. In the high frequency region, tanδ becomes almost constant because the electron exchange interaction (hopping) between the Mn^{+2} and Mn^{+3} cannot follow the alternatives of the applied electric field beyond a critical frequency. The figure 3 shows that tanδ of irradiated sample S2 has high values for all frequencies as compared to pure sample S1. The value of tanδ for S1 is 9.2 at 100 Hz while its value for S2 is 13.7. The tanδ values for S1 and S2 at 10 MHz are 0.3 and 0.7 respectively. The increase of loss tangent could be due to particle size reduction, defects and imperfections caused by EB irradiation.

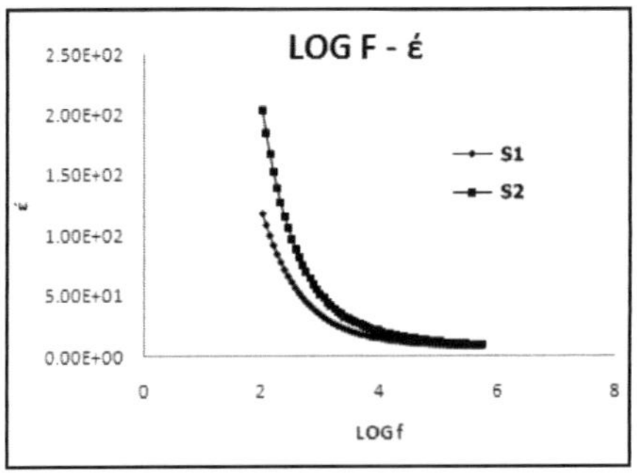

Figure 2. The variation of dielectric constant with frequency.

The variation of a.c. conductivity as a function of frequency is shown in figure 3 (b). At low frequencies, σac has a small value which increases at higher frequencies. The conduction in manganese tungstate is due to the hopping and ionic conduction of Mn^+ ions. When the grain size of the sample is reduced the hopping distance increases which in turn increases the conductivity. The values are shifted upwards for the irradiated sample S2, which might occur due to size reduction caused by the irradiation. The value of σac for S1 is 0.48×10^{-5} S/m at 100 Hz while its value for S2 is 1.57×10^{-5} S/m. The σac values for S1 and S2 at 10 MHz are 3.45×10^{-5} S/m and 6.23×10^{-5} S/m respectively. Thus, electron beam irradiation is found to be an efficient method for the enhancement of the electrical properties of $MnWO_4$ nanoparticles.

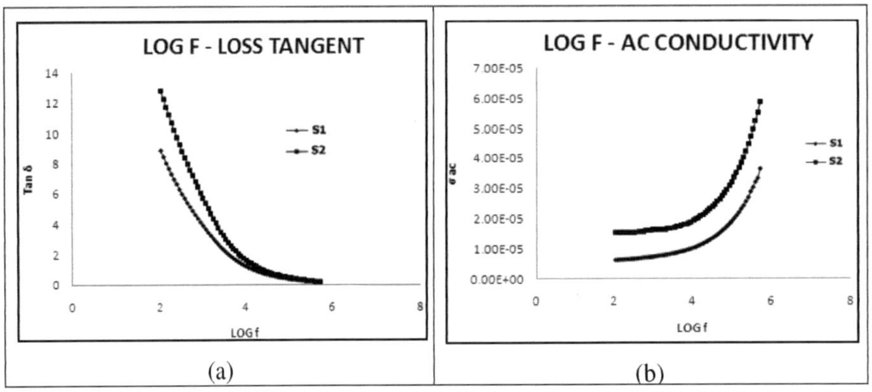

Figure 3. Variation of (a) loss tangent, and (b) a.c. conductivity with frequency.

4. Conclusions

The dielectric properties of pure and EB irradiated samples have been investigated as a function of frequency. At lower frequency regime, ε' and tan δ have higher values while they reach steady lower values at higher frequencies. However, the a.c. conductivity increases with the increase of frequency, which confirms small polaron hopping. It has been found that ε', tan δ and σac values are much larger for the EB irradiated sample. This enhancement of electrical properties might occur due to defects and reduction in particle size caused by the beam irradiation. It is suggested that, the $MnWO_4$ nanoparticles modified with the EB irradiation can be used as a good dielectric material for varied applications.

Acknowledgements

The authors acknowledge their thanks to Nirmala College, Muvattupuzha, Microtron centre, Mangalore University, Mangalore and Maharajas College, Ernakulam for providing the facilities to undertake this study. The financial support from the KSCSTE, Thiruvananthapuram is gratefully acknowledged.

References

[1] Traversa E 1995 *Sens Actuators B* **23** 135-156
[2] Qu W, Wlodarski W and Meyer J U 2000 *Sens. Actuators B* **64** 76-82
[3] Suresh Raj A M E, Mallika C, Sreedharan O M and Nagaraja K S 2002 *Mater. Lett.* **53** 316-320
[4] Jang M, Weakley T J R and Doxsee K M 2001 *Chem. Mater.* **13** 519-525
[5] Chen *et al* 2003 J. Mater. Chem. **13** 1132
[6] Heyer *et al* 2006 *J. Phys: Condens. Matter* **18** L471
[7] Montemayor S M and Fuentes A F 2004 *Ceram. Internat.* **30** 393
[8] Varghese T and Balakrishna K M 2011 *Nanotechnology: An introduction to synthesis, properties and applications* (Atlantic Publishers: New Delhi)
[9] Zhou H, Yiu Y, Aronson M C and Wong S S 2008 *J. Solid State Chem.* **181** 1539
[10] Parhi P, Karthik T N and Manivannan V 2008 *J. Alloys Comp.* **465** 380
[11] Kingon A I, Maris J P and Steiffer S K 2000 *Nature* **406** 1032
[12] Shukla R 1993 *Ind. J. Pure Appl. Phys.* **31** 894-898

Synthesis of PTCDI-C_8 one dimensional nanostructures for photovoltaic applications

T Abhijith[1], M Yoosuf Ameen[1] and V S Reddy[1 *]
[1] Organic and Nano Electronics Laboratory, Department of Physics, National Institute of Technology, Calicut 673601, Kerala, India

[*] Email: sivaji@nitc.ac.in

Abstract. One dimensional organic semiconductor nanostructures have achieved increasing interest in the emerging fields of photonics and electronics in recent years, mainly due to their promising applications in electronic and optoelectronic nanodevices. The solution based self assembly technique is a very convenient method for the fabrication of one dimensional nanostructures with desirable size and shape. Here we report a self assembly of one dimensional N,N′-Dioctyl-3,4,9,10-perylenedicarboximide nanostructures of several micrometer length and 100- 200 nm width. The Ultra Violet Visible near Infrared spectroscopy of nanostructures shows significant improvement in absorption of light in wavelength range 550 to 675 nm. The open aperture Z-scan signature obtained using a Q switched Nd:YAG laser shows strong nonlinear optical properties of these nanostructures.

1. Introduction

In recent decades the one dimensional (1D) semiconductor nanostructures such as nanowires, nanobelts and nanoribbons are of great interest because of their wide applications in optoelectronic and photonic nanodevices. The fabrication of well-defined organic nanowires or nanobelts with controllable size and morphology is not as advanced as for their inorganic counterparts. Molecules with predominant π-π stacking over the lateral side-chain association enables growth of the 1D material [1]. Perylene derivatives such as Perylene tetracarboxylic dianhydride (PTCDA) and Perylene tetracarboxylic diamide (PTCDI) are most widely used n-type organic semiconductor molecules for the fabrication of optoelectronic devices such as transistors, light-emitting diodes and photovoltaic cells [2]. The strong π-π interaction between the planar aromatic skeletons and their non collapsible structure make perylene derivatives ideal candidates for 1D self-assembly. Rapid phase transfer method is one of the widely employed methods for the synthesis of nanoribbons from N,N′-Dioctyl-3,4,9,10-perylenedicarboximide (PTCDI-C_8) molecules [3]. The 1D nanostructures of PTCDI-C_8 show drastic changes in linear and nonlinear optical properties compared to its bulk material due to their strong π-π interactions between perylene backbones along the long axis of nanoribbons [4]. Study of both linear and nonlinear optical properties of 1D nanostructures will provide well rounded, integrated understanding of the optoelectronic properties of the organic materials and their potential applications in devices utilizing nanostructures of materials [2, 5-7].

2. Experimental Details

Ultra-long PTCDI-C_8 nanoribbons were grown using commercially available PTCDI-C_8 powder (Sigma Aldrich, 98%) by solution phase self-assembly. The PTCDI-C_8 nanoribbons were synthesized by using a technique called rapid phase transfer method. In this technique we employed an idea of rapid dispersion of methanol to PTCDI-C_8 chloroform solution to induce self-assembly by π-π interactions of PTCDI-C_8. First, a homogeneous solution of PTCDI-C_8 was prepared by stirring it in chloroform (1.5 mg/ml) for six hours and subsequently equal amount of methanol was added to the solution. After mixing the methanol, complete formation of the nanoribbons was observed within 20 minutes [8].

Field emission scanning electron microscopy (FE-SEM) (HITACHI, SU6600 Variable pressure field emission SEM) and high resolution tunnelling electron microscopy (HRTEM) (JEOL JEM 2100) were performed to determine the shape and size of nanostructures. For FE-SEM studies the film was prepared by dropping the solution of PTCDI-C_8 nanostructures onto a clean cover glass slide, followed by drying at room temperature. The FESEM was performed at accelerating voltages 15-20 kV. One drop of the diluted nanostructure solution was deposited on a carbon coated copper grid for HRTEM measurements. The HRTEM studies were performed at room temperature at accelerating voltages of 80-100 kV. Philips X-Pert PRO MRD X-ray diffractometer with CuKα radiation (λ= 0.15418 nm) was used to investigate the crystallinity of the nanoribbons. To investigate the linear optical properties, the absorption spectra of both bulk and nanostructures of PTCDI-C_8 were recorded using UV–Visible – near IR spectrophotometer (SHIMADZU, UV-1800). Z-Scan technique was employed to study the nonlinear optical properties of films of both bulk and nanostructures. It is a very convenient, fast and well-known experimental method to study the nonlinear optical properties of materials. The experiment was performed using 7 ns pulses from a Q-switched Nd:YAG laser having wavelength 532 nm and a pulse repetition rate of 10 Hz.

3. Results and Discussion

Figure 1 and figure 2 show FE-SEM and HR-TEM images of PTCDI-C_8 nanostructures, respectively. These images reveal well defined ribbon like nanostructures of PTCDI-C_8 molecules. The dimensions of PTCDI-C_8 nanostructures were calculated from the microscopic images and it was understood that the nanoribbons have several micrometer length and 100-200 nm width. The crystallinity of PTCDI-C_8 nanoribbons were studied from X-ray diffractometer using thin film of the material deposited on a glass plate. A strong diffraction peak was observed at 4.2° with a d-spacing of 2.10 nm corresponding to the (001) plane. This result confirms that the nanostructures are crystalline in nature.

Figure 1. FE-SEM image of PTCDI-C_8 nanoribbons and inset: magnified image of nanoribbons.

Figure 2. HR-TEM image of PTCDI-C_8 nanoribbon.

Figure 3 shows the absorption spectra of bulk and nanostructures of PTCDI-C_8 in the wave length range 400 – 750 nm. The bulk material shows main absorption in the wavelength range between 400 – 650 nm with peaks at 456, 490, 525 and 572 nm. On the other hand PTCDI-C_8

nanostructures exhibit slightly red shifted absorption spectra with peaks at 461, 490, 529, and 582 nm. In addition to the red shift, the relative intensity of the peaks at higher wavelengths (530 and 573 nm) became enhanced. Such a strong absorption at higher wavelengths is mainly due to the extended intermolecular π-π stacking along the long axis of the nanostructures which is believed to be more favourable for efficient long-range charge transport due to the effective intermolecular π electron delocalization. This can be utilized for efficient electron transport through the nanostructures to achieve high collection efficiency for photo generated electrons in the photovoltaic devices [9]. The optical nonlinearity of PTCDI-C_8 was measured using standard open-aperture Z-scan method. PTCDI-C_8 shows optical nonlinearities due to a delocalized π electron system and it has potential applications in nonlinear optical devices. Here we observed saturable absorption at the region of tightly focused laser beam as shown in figure 4. Saturable absorption is one of the nonlinear optical properties of material where absorption of light decreases with increasing light intensities. Compared to bulk material the one dimensional PTCDI-C_8 provides large number of electrons to upper energy state at such a rate that there is insufficient time for them to decay back to the ground state before the ground state becomes depleted and the absorption subsequently saturates. This is mainly due to the strong π-π interactions between perylene backbones along the long axis of nanoribbons. In particular, the nonlinear optical properties of PTCDI-C_8 suggest potential applications of the material in mode locking technique inside the laser resonator and in various nanoscale optoelectronic circuitries.

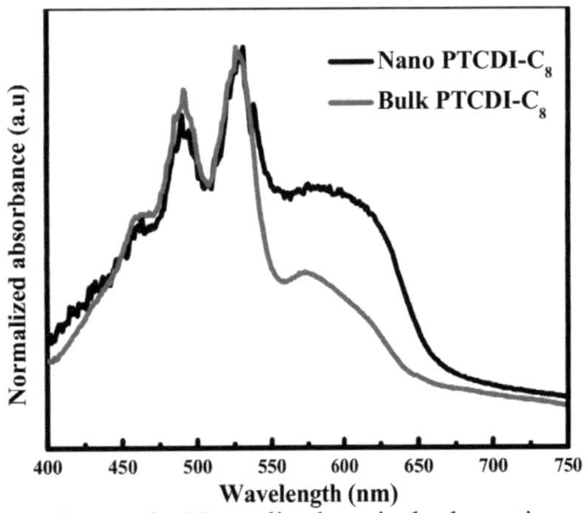

Figure 3. Normalized optical absorption spectra of bulk and nanostructures of PTCDI-C_8.

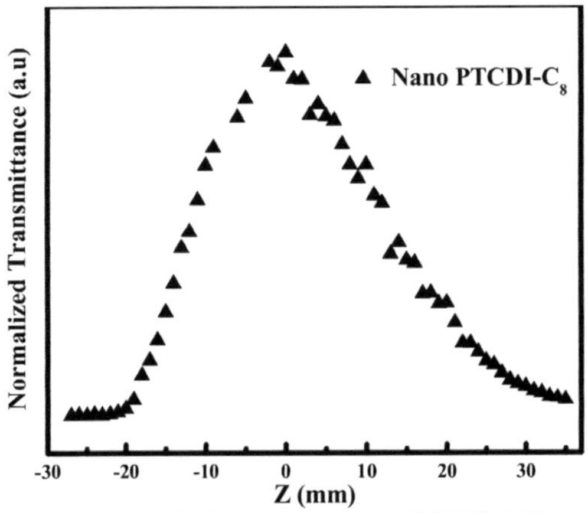

Figure 4. Z-Scan signature of PTCDI-C_8 nanoribbons.

4. Conclusions

In conclusion, we have synthesized ultra long PTCDI-C_8 nanoribbons of several micrometers length and 100-200 nm width. Strong π-π inter chain interactions and its efficient electron transport along the long axis of nanoribbons suggest wide applications of the one dimensional organic semiconductor nanostructures in organic and nanoelectronic devices especially in photovoltaic devices. More interestingly strong nonlinear optical behaviour of these nanostructures provide the material in potential applications as frequency converters or logic/routing elements and as saturable absorbers in laser cavities for Q switching.

References

[1] Balakrishnan K, Datar A, Oitker R, Chen H, Zuo J and Zang L 2005 *J. Am.Chem.Soc.* **127** 10496-10497

[2] Kufazvinei C, Ruether M, Wang J and Blau W 2009 *Org. Electron.* **10** 674-680

[3] Zang L, Che Y and Moore J S 2008 *Acc. Chem. Res.* **41** 1596-1608

[4] Xia Y, Yang P, Sun Y, Wu Y, Mayers B, Gates B, Yin Y, Kim F and Yan H 2003 *Adv. Mater.* **15** 380-389

[5] Supur M, Yamada Y, El-Khouly M E, Honda T and Fukuzumi S 2011 *J. Phys. Chem.C* **115** 15040-15047

[6] Wang J, Fruchtl D and Blau W J 2010 *Opt. Commun.* **283** 464-468

[7] Viviena L, Lancona P, Riehl D, Hache F and Anglaret E 2002 *Carbon* **40** 1789-1797

[8] Che Y , Datar A, Balakrishnan K and Zang L 2007 *J. Am.Chem.Soc.* **129** 7234-7235

[9] Karak S, Ray S K and Dhar A 2010 *Appl. Phys. Lett.* **97** 043306-1-3

International Conference on Materials Science and Technology (ICMST 2012) IOP Publishing
IOP Conf. Series: Materials Science and Engineering **73** (2015) 012053 doi:10.1088/1757-899X/73/1/012053

Doping effect of Ag^{+}, Mn^{2+} ions on Structural and Optical Properties of ZnO nanoparticles

B Sankara Reddy[1], S Venkatramana Reddy[2], P Venkateswara Reddy[3], N Koteeswara Reddy[4] and R P Vijayalakshmi[5]

[1,2,3,5]Department of Physics, Sri Venkateswara University, Tirupati-517 502, Andhra Pradesh, India.
[4]The Center of Excellence in Nanoelectronics (CEN), Indian Institute of Science, Bangalore-12, India.

[2]Email: drsvreddy123@gmail.com

Abstract: Pure ZnO and co-doped (Mn, Ag) ZnO nanoparticles have been successfully prepared by chemical co-precipitation method without using a capping agent. X-ray diffraction (XRD) studies confirms the presence of wurtzite (hexagonal) crystal structure similar to undoped ZnO, suggesting that doped Mn, Ag ions are substituted to the regular Zn sites. The morphology of the samples were studied by scanning electron microscopy (SEM). The chemical composition of pure and co-doped ZnO nanoparticles were characterized by energy dispersive X-ray analysis spectroscopy (EDAX). Optical absorption properties were determined by UV-vis Diffuse Reflectance Spectrophotometer. The incorporation of Ag^{+}, Mn^{2+} in the place of Zn^{2+} provoked to decrease the size of nanocrystals as compared to pure ZnO. Optical absorption measurements indicates blue shift in the absorption band edge upon Ag, Mn ions doped ZnO nanoparticles.

Keywords: Pure ZnO, co-doped ZnO nanoparticles, co-precipitation method, SEM, EDAX, DRS.

1. Introduction:

Zinc oxide (ZnO) is an important electronic and photonic material which has been studied extensively in the past few years due to its technological importance [1]. It has a wide band gap (E_g=3.37eV), large excitation energy of 60 meV, high chemical stability, good piezo-electric properties, nontoxicity and biocompatibility. To synthesize ZnO nanoparticles with transition metals by different routes such as wet chemical methods like sol-gel [2], co-precipitation [3], combustion [4] etc. It has been used in a wide range of applications such as sensors, varistors, piezoelectric, transducers, surface acoustic wave devices, phosphors, transparent conducting oxides, optoelectronic devices, ferromagnetic devices, and heterogeneous photo catalysts [5-9]. ZnO possesses n-type conductivity due to its native defects such as Zinc interstitials and oxygen vacancies, and silver doping is found to be effective for the fabrication of p-type ZnO. On irradiation, the high density free electrons oscillate, which is known as Surface Plasmon Resonance (SPR), and also, recently SPR based Nobel metal semiconductor nanocomposites have become a popular research topic because of their improved optical properties [10]. In this paper, we report that the morphology induced by two different ions such as silver and manganese ions of equal concentrations (x = 0.00 and 0.05) in ZnO nanocrystals was systematically studied. By using chemical co-precipitation method pure ZnO and $Zn_{1-x}Mn_xAg_xO$ nanoparticles were synthesized and their structural, morphology and optical properties have been investigated.

2. Experimental:

All the chemicals used in the experiment are of analytical grade purity and purchased from Merck, Mumbai, India. Pure ZnO and manganese, silver doped ZnO nanoparticles were synthesized by chemical co-precipitation method at room temperature and for silver and manganese doping silver nitrate ($AgNO_3$), and manganese acetate tetra hydrate Mn(CH COO)$_2$·4H$_2$O) have been used. Zn_{1-x} Mn_x Ag_xO (x=0.00, 0.05) nanostructures were prepared at room temperature as the procedure described below. Initially 0.2 M solution was prepared by using Zinc acetate and KOH. For dopants, $AgNO_3$ and manganese acetate have been added drop wise to the above solution in equal concentrations such as 0.00, 0.05 under continuous stirring for 8 hrs. Hence, the precipitate was formed and it was filtered out separately, and repeatedly washed with deionized water to remove unnecessary impurities formed during the preparation process. Ag and Mn doped ZnO nanopowders were obtained after drying at 100 °C for 5 hrs. Then the final products were grinded and annealed at 400 °C in the furnace for 1 hr.

The synthesized nanostructures were characterized by X-ray diffraction technique using BRUKER D8-advance powder X-ray diffractometer (Germany) with source of 2.2 kW Cu anode, Ceramic X-ray tube. Morphology of the samples was analyzed using scanning electron microscopy (SEM) technique. Concentration of dopants in ZnO was estimated by energy dispersive X-ray analysis spectroscopy (EDAX) attached with SEM (Model: CARL-ZEISS EVOMA-15). The optical properties were estimated from the UV-vis diffuse reflectance spectroscopic studies with a model Varian Cary- 4000 spectrophotometer.

3. Results and Discussion:

The prepared Pure and Ag, Mn doped ZnO nanopowders possesses hexagonal wurtzite structure. Fig.1 shows the X-rd pattern of $Zn_{1-x}Mn_xAg_xO$ (x = 0.00, 0.05) powder samples at 400 °C. The diffraction patterns of samples can be indexed to the hexagonal wurtzite structure of ZnO (JCPDS data card No: 36-1451). It should be pointed that in the XRD patterns except diffraction peaks of ZnO, no peak of additional phase was observed. The crystalline sizes of the synthesized powders were determined using Debye-Scherer's equation, $D = \frac{0.89\lambda}{\beta\cos\theta}$, the increasing Mn solubility in ZnO is due to decreasing particle size, as discussed by Straumal et al. [11]. In our study, we observed that for 5 mol% of Ag, Mn doped ZnO nanoparticles of size decreases than pure ZnO. Thus, we may conclude that the lattice constants and particle sizes decreased for the following reason. Silver can also exist in Ag^{2+} Ag^{3+} in addition to Ag^{+}. The ionic radii of Ag^{3+} (0.75 Å), Ag^{2+} (0.94 Å) are nearly equal size of Zn^{2+}, but the ionic radii of Mn^{3+} (0.58 Å) and Mn^{4+} (0.53 Å) are smaller than Zn^{2+}. In the doped ZnO, substitution of Zn^{2+} by Mn^{3+} and Mn^{4+} brings about the lattice constant decreases as noted by Bhatti et al. [12], and for 5 mol% of Ag, Mn doped ZnO nanoparticles sizes are **17 nm**, the corresponding lattice constants are **a = 0.32551 nm, c = 0.52164 nm** respectively, which are lesser than the sizes of pure ZnO nanoparticles (size ~ **23 nm**). The intensity of pure ZnO shows higher diffracted intensity than that of 5 mol% of Mn, Ag doped ZnO.

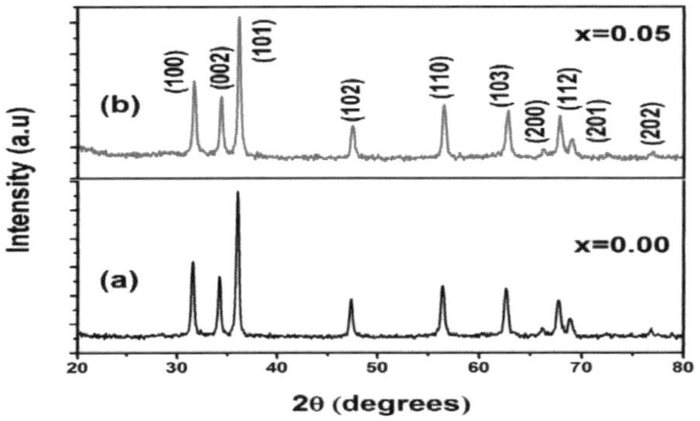

Fig.1. XRD patterns of (a) Pure ZnO (b) 5 mol% of Ag, Mn doped ZnO nanoparticles.

Fig. 2. SEM images of (a) Pure ZnO (b) 5 mol % of Ag, Mn doped ZnO nanoparticles.

Scanning electron microscope (SEM) was used to investigate the morphology of the samples as shown in the Fig. 2. Pure ZnO shows clusters of tiny particles and 5 mol% of Ag, Mn doped ZnO image [Fig. 2(b)] appears like hexagonal structures, these structures agree with the X-rd results. The energy dispersive X-ray spectroscopic analysis of the pure ZnO and co-doped ZnO confirms the presence of Ag and Mn in the prepared oxide as shown in Fig. 3.

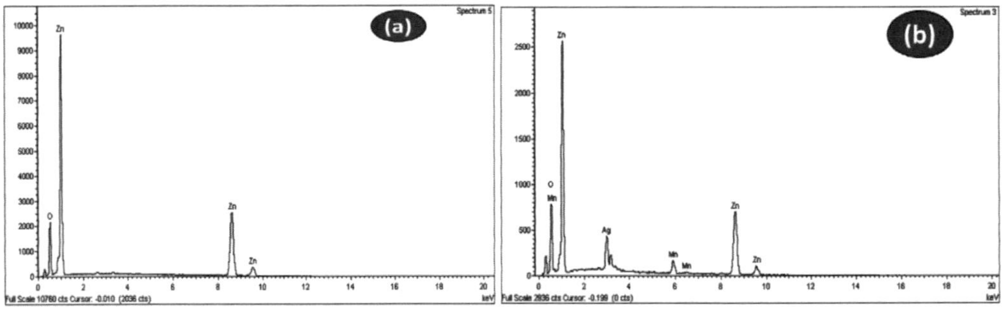

Fig.3. EDAX spectra of (a) pure ZnO, (b) $Zn_{1-x}Mn_xAg_xO$ nanoparticles.

The optical absorption spectra of pure ZnO and $Zn_{1-x}Mn_xAg_xO$ (x = 0.00, 0.05) samples are studied by diffuse reflectance spectrometer (DRS) in the range of 200 to 800 nm. From Fig.4, it can be seen that the excitonic absorption peak for undoped ZnO appears around 363 nm and 5 mol% of Ag, Mn doped ZnO nanoparticles shows that the one excitonic peak at 240 nm and other broad peak is observed, which is centered at 336 nm. It is clearly observed that the absorbance of the co-doped ZnO samples decreased for 5 mol% of Ag, Mn concentration. From Fig. 4, blue shift is observed for 5 mol% of Ag, Mn concentration in ZnO nano-particles, because of the quantum confinement effect [13,14].

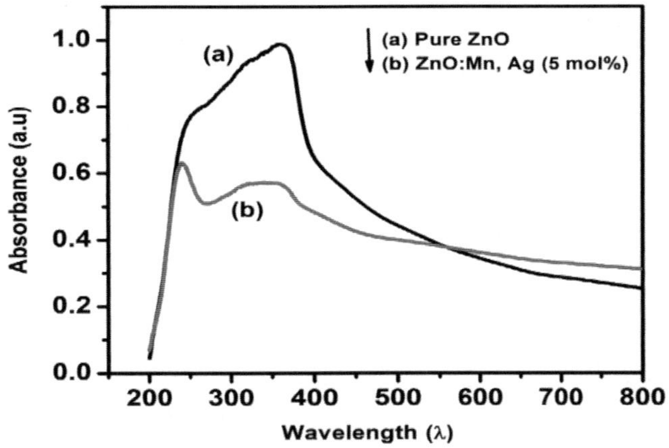

Fig. 4. UV-vis absorption spectra of (a) pure ZnO and (b) 5 mol% of Ag, Mn doped ZnO nanoparticles.

4. Conclusions:

Nanocrystals of pure ZnO and $Zn_{1-x}Mn_xAg_xO$ were successfully synthesized by using a chemical co-precipitation method. The crystalline structure, optical properties were determined by XRD and UV-visible diffuse reflectance spectrophotometer respectively. XRD analysis shows the hexagonal phase and SEM image shows (5 mol% of Ag, Mn doped ZnO) hexagonal structure. The average particle sizes, lattice parameters 'a' and 'c' of nanoparticles decreased for 5 mol% of Ag, Mn doped ZnO nanoparticles as compared to undoped ZnO nanoparticles. Optical absorption measurements indicate the blue shift in the absorption band edge upon 5 mol% concentration of Ag, Mn doped ZnO nanoparticles.

References:

[1] Khorsand Zak A, Majid W H A, Abrishani M E, and Yousefi R 2011 *Solid State Sci.* **13** 251
[2] Abrishami, M E et al 2010 *Phys. Status Solidi* C, **7** 1595
[3] Yadav R S, Pandey A C, and Sanjaya S S 2009 *Chalcogenide Lett.* **6** 233
[4] Riahi-Noori N et al 2008 *J. Ceram. Process. Res.* **9** 246
[5] He J H et al 2007 *Adv. Mater.* **19** 781
[6] Lee C V, Haung Y T, Su W F, Lin C F 2006 *Appl. Phys. Lett.* **89** 231116
[7] Garcia M A et al 2007 *Nano Lett.* **7** 1489
[8] Fonoberov V A and Balandin A A 2006 *J. Nanoelectron Optoe.* **1** 19
[9] Kwon S S et al 2008 *Adv. Mater.* **20** 4557
[10] Lee M K, Kim T G, Kim W, Sung Y M 2008 *J. Phys, Chem.* C. **112** 1009
[11] Straumal B et al 2009 *J. Europ. Ceram. Soc.* **29** 1963
[12] Bhatti K P et al 2005 *Solid State Commun.* **136** 384
[13] Bhargava R N and Gallagher D 1994 *Phys. Rev. Lett.* **72** 416
[14] Moreno D L, Rosa-Curz E D, Cuevas F J, Regalado L E, Salas P, Rodriguez R, Castano V M 2002 *Opt. Mater.* **19** 275

Growth and comparison of physicochemical properties of Lewis base adduct of MMTC and CMTC: Efficient non-linear optical single crystals

I Vetha Potheher[1], K Rajarajan[2], R Jeyasekaran[3] and P Sagayaraj[4]
[1]Department of Physics, Bharathidasan Institute of Technology, Anna University, Tiruchirappalli – 620024, India
[2]Department of Physics, R. V. Govt. Arts College, Chengalpattu – 603001, India
[3]Department of Physics, V. H. N. S. N. College, Virudhunagar – 626001, India
[4]Department of Physics, Loyola College, Chennai – 600034, India

E-mail: potheher11@gmail.com,

Abstract. In the present investigation, single crystals of CMTG and MMTG were conveniently grown by slow evaporation technique. The grown crystals were confirmed by single crystal XRD. The physicochemical properties of the grown crystals were analyzed and compared. The SHG efficiency of CMTG is found to be three times higher than that of urea and for MMTG, it is equal to that of urea. Thermal stability of CMTG and MMTG was found to be 105 °C and 145 °C.

1. Introduction

Organometallic and coordination compounds offer a variety of molecular structure by changing the metals, ligands, coordination numbers and so on. This diversity of molecular structure gives opportunity to tune the electronic properties of the molecules, and hence to exploit the linear and non linear optical properties [1-3]. Chinese scientists have developed a few interesting coordination compounds such as $MHg(SCN)_4 \cdot nH_2O$ (where M = Co, Cu, Zn, Cd, Ni, Fe, Mn) and demonstrated them as the useful SONLO materials [4]. Among these crystals, Cadmium mercury thiocyanate $[CdHg(SCN)_4]$ (CMTC) [5] and manganese mercury thiocyanate $[MnHg(SCN)_4]$ (MMTC) [4] shows high second harmonic intensity and attractive features that are suitable for blue / green laser light generation. Along with these crystals, few Lewis base adducts has also been investigated in recent past due to the interesting variations seen not only in their structures but also in their properties, particularly the relatively high SHG efficiency. Lewis base adducts are ones of the interesting themes of structural chemistry [6]. They have chain structures in which respective side-by-side metal atoms are bridged by thiocyanate ions. Their structural features bestow on them usefulness in nonlinear optics [7], chemical analysis [8], chemical, electrochemical metal sulfide thin film deposition [9]. Recently, some physicochemical properties of CMTG and MMTG have been reported by Duan et al, Xing Qiang Wang et al and Vetha Potheher et al [10,11, 12] respectively. In the present investigation, physicochemical properties of CMTG and MMTG were studied and compared.

2. Experimental

2.1. Synthesis of CMTC

During first stage of experiment, CMTC was synthesized by using $CdCl_2$, $HgCl_2$ and KSCN taken in stoichiometric ratio and dissolved in millipore water. The solution was stirred till the completion of reaction. After the completion of reaction, a white precipitate was formed which was further separated and dried. In order to increase the purity of the synthesized material the product was washed in millipore water three to four times and then dried. The chemical reaction is as follows:

$$CdCl_2 + HgCl_2 + 4KSCN \longrightarrow CdHg(SCN)_4 + 4KCl$$

2.2 Synthesis of MMTC

In first stage, manganese mercury thiocyanate was synthesized by mixing manganese chloride, mercury chloride and potassium thiocyanate in millipore water. The following reaction is used for synthesis:

$$MnCl_2 + HgCl_2 + 4\,KSCN \longrightarrow MnHg(SCN)_4 + 4KCl$$

2.3 Growth of CMTG and MMTG crystals

In the second stage, the synthesized salt of CMTC and MMTC were dissolved in glycol monomethyl ether taken with millipore water in the ratio 3:1. Glycol monomethyl ether (GME) acts as a ligand to react with the CMTC and MMTC salt. The chemical reaction is given as follows:

$$CdHg(SCN)_4 + CH_3OC_2H_5O \longrightarrow CdHg(SCN)_4(CH_3OC_2H_5O)$$

$$MnHg(SCN)_4 + CH_3OC_2H_5O \longrightarrow MnHg(SCN)_4(CH_3OC_2H_5O)$$

To avoid the precipitation, the pH of the solutions was adjusted to be in between 2 and 3 by adding a few drops of HCl to the mixed solutions. The solution was vigorously stirred for about 6 hours and then filtered. The filtered solutions were kept for nucleation. The solutions were allowed to evaporate in room temperature. Single crystals of CMTG and MMTG were grown in a period of 35 – 40 days and 40 – 45 days respectively, with dimensions up to 17 x 18 x 6 mm^3 and 11 x 10 x 6 mm^3 respectively by slow evaporation technique. Figure 1 and 2 shows the photographs of as grown single crystal of CMTG and MMTG. The grown crystals exhibits hygroscopic nature, hence to avoid the loss of optical transparency, the crystals was kept in a dark place.

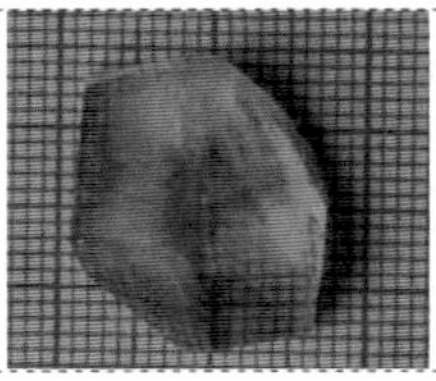

Figure 1. Photograph of as grown single crystal of CMTG

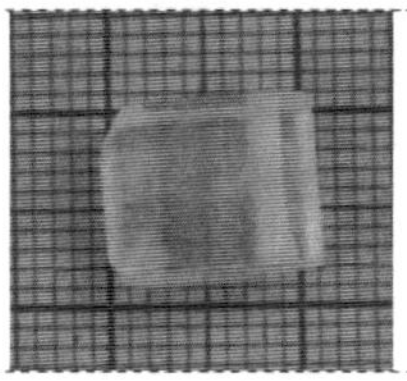

Figure 2. Photograph of as grown single crystal of MMTG

3. Results and Discussion

3.1 Single crystal XRD

Crystal structure of CMTG was confirmed by single crystal X-ray diffraction analysis using ENRAF NONIUS CAD4-F diffractometer. The structure was solved by the direct method and refined by full matrix least square technique using the SHELXL program. From the data, it is observed that CMTG belongs to orthorhombic crystal system with the space group $Pca2_1$. The unit cell parameters are a = 16.6172 Å, b = 7.0143, c = 13.5602 Å and V = 1580.5501 Å^3. There is no change in crystal structure when $CdCl_2$ replaced by $MnCl_2$. But there were changes in the lattice parameter. The unit cell parameter for MMTG are a = 16.2051 Å, b = 7.3012 Å, c = 13.5098 Å and V = 1598.4347 Å^3.

3.2 Optical absorption spectrum

The absorption spectrum of the sample was recorded in the range 200 – 2500 nm using HITACHI U-2800 spectrophotometer. The optical absorption spectrum along with Tauc's plot [13] of CMTG and MMTG is shown in Figure 3 and 4. It is evident from the spectrum that the absorbance of the sample is less than 0.1 units. The UV cut-off wavelength of CMTG and MMTG is about 370 nm and 365 nm respectively, which is comparable to other metal-organic compounds such as CMTC (371 nm) [5], MMTC (373 nm) [14] and MMTD (375 nm) [15]. A plot drawn between the photon energy ($h\upsilon$ = E) and $(\alpha h\nu)^2$ was used to estimate the band gap of the material. The band gap (E_g) is determined by extrapolating the straight line portion of the curve to $(\alpha h\nu)^2 = 0$. From the plot, the

IOP Conf. Series: Materials Science and Engineering **73** (2015) 012054 doi:10.1088/1757-899X/73/1/012054

band gap of CMTG and MMTG is estimated as 3.99 eV and 3.5 eV. MMTG has better optical property compared to CMTG.

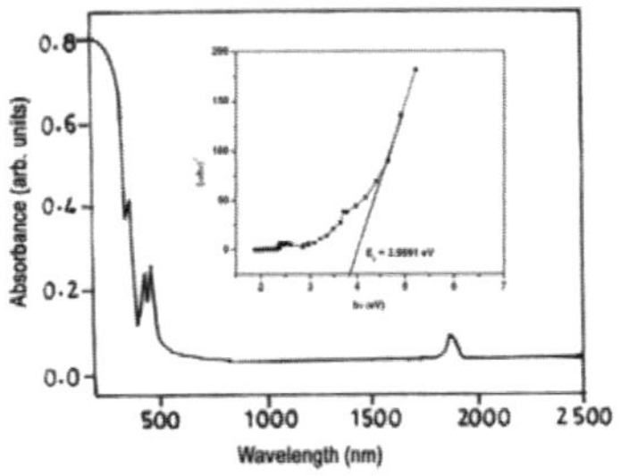

Figure 3. Optical absorption spectrum along with Tauc's plot of CMTG crystal

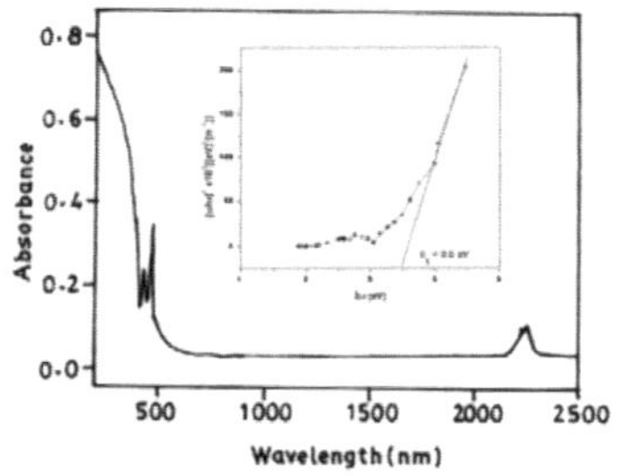

Figure 4. Optical absorption spectrum along with Tauc's plot of MMTG crystal

3.3 NLO Test

The NLO test was performed for CMTG and MMTG using Kurtz and Perry powder technique [16]. The SHG efficiency of the sample was compared with microcrystalline powder of KDP and urea. In the present investigation, the laser pulse of 8 ns with spot radius of 1mm was employed. The input laser beam was passed through the IR reflector and then directed on the microcrystalline powdered sample packed in a capillary tube of diameter 0.154 mm. When a laser beam of 1.35 mJ was passed through the sample, second harmonic signal of 532 nm was generated and the output of 55, 330, 998 and 354 mV were obtained from KDP, urea, CMTG and MMTG respectively. From the experimental data, it was observed that the efficiency of the MMTG is nearly equal to that of urea and the efficiency of CMTG is around 3 times higher than that of urea, which is better than Tetrathiourea mercury (II) tetrathiocyanato zinc (II) (TMTZ) [17], Tetrathiourea mercury (II) tetrathiocyanato manganese (II) (TMTM) [18] and Diaquatetrakis (thiocyanato) cobalt (II) mercury (II) N-methyl-2-pyrolidone (CMTWMP) [19].

3.4 Thermal Analysis

Thermal analysis was carried out by the instrument Perkin Elmer Thermal Analyzer in nitrogen atmosphere at a heating rate of 15 °C/min., from 28 to 1000 °C. Figure 5 and 6 show the TGA-DTA thermogram of CMTG and MMTG respectively. The TGA trace of CMTG and MMTG reveals the different stages of decomposition of the sample. The first stage of TGA shows the weight loss of GME molecules. The melting point of the CMTG and MMTG is 105 °C and 145 °C, which is high when compared with ATZC (58 °C) [20]. Further, the sharp peak observed at 263.14 °C and 358 °C confirms the decomposition of the three dimensional steric structure of CMTC and MMTC. The various DTA peaks observed in the temperature range 400 - 1000 °C correspond to the decomposition of metal sulphides (CdS, HgS), carbon bi sulphides (CS_2) and nitrogen (N_2), dicyanogen ($(CN)_2$).

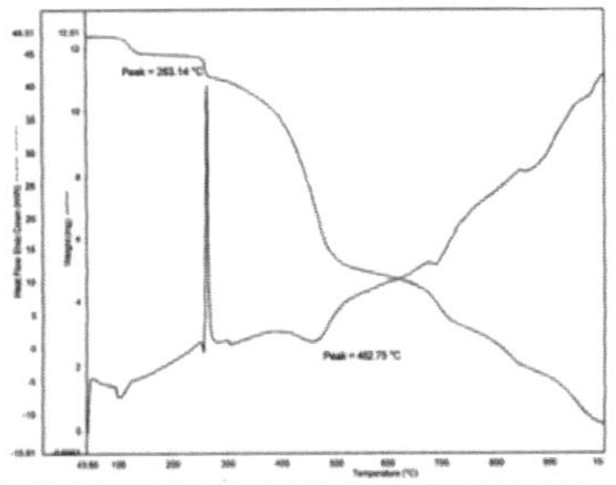

Figure 5. TG – DTA thermogam of CMTG

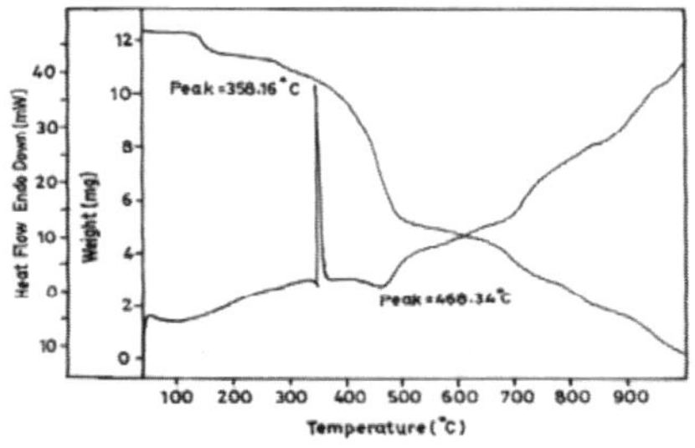

Figure 6. TG – DTA thermogam of MMTG

4. Conclusion

Single crystals of CMTG and MMTG were grown by slow evaporation technique. The crystal systems of the grown crystals were confirmed by single crystal XRD. Both the crystals were belongs to orthorhombic crystal system with non centrosymmetric space group $Pca2_1$. Optical property of the crystals was studied and the band gap energy was found to be 3.99 eV and 3.5 eV. The samples possess good non linear optical property and the SHG efficiency of CMTG is three times higher than that of urea and MMTG is equal to that of urea. The thermal stability of the CMTG and MMTG was found to be 105 °C and 145 °C respectively.

5. References

[1] Duorong Yuan, Zhewu Zhoug, Mingguo Liu, Dong Xu, Qi Fang, Youghong Bing, Suoying Sun and MinhuaJiang 1998 *J. Cryst. Growth* **186** 240
[2] Sun H Q, Yuan D R, Wang X Q, Lu Y Q, Sun Z H, Wei X C, Duan X L, Luan C N, Lu M K and Xu D 2003 *J. Crystal Growth* **256** 183
[3] Xinqiang Wang, Xu D, Lu M K, Yuan D R, Zhang G H, Meng F Q, Guo S Y, Zhou M, Liu J R and Li X R 2001 *Cryst. Res. Technol.* **36** 73
[4] Xinqiang Wang, Dong Xu, Mengkai Lu, Duorong Yuan, Xiufeng Chang, Shigang Li, Ji Huang, Shenglai Wang, Heyi Liu 2002 *J. Crystal Growth* **245** 126
[5] Duorong Yuan, Dong Xu, Mingguo Liu, Fang Qi, Wentao Yu, Wenbo Hou, Yonghong Bing, Suoying Sun and Minhua Jiang 1997 *Appl. Phys. Lett.* **70** 544
[6] Nakamoto K (Translated by Huang D R, Wang R Q) 1986 *Infrared and Raman spectra of inorganic and coordination compounds,* 3rd Edition, Wiley, Beijing, Chemical Industry Press, (in Chinese).
[7] Xinqiang Wang, Wen-Tao Yu, Dong Xu, Meng-Kai Lu, Duo-Rong Yuan, Jiu-Rong, Liu an Guang-Tian Lu 2000 *Acta Cryst.* **C56** 1305
[8] Ptaszynski B, Skiba E and Krystek J 1998 *Thermochemica Acta* **319** 75
[9] Tsukasa Yoshida, Masashi Tochimoto, Derck Schlettwein, Dieter Wohrle, Takashi Sugiura and Hideki Minoura 1999 *Chem. Matter* **11** 2567
[10] Duan X L, Yuan D R, Wang X Q, Cheng X F, Yang Z H, Guo S Y, Sun H Q, Xu D and Lu M K 2002 *Cryst. Res. Technol.* **37** 446
[11] Xinqiang Wang, Dong Xu, Menguai Lu, Dourong Yuan, Ji Huang, Xiufang Cheng, Shenglai Wang, Shiyi Guo, Guanghui Zhang, Mei Pan, Xiulan Duan and Zhaohe Yang 2002 *Phys. Stat. Sol* (a) **191** 106
[12] Vetha Potheher I, Rajarajan K, Vimalan M, Tamilselvan S, Jeyasekaran R and Sagayaraj P 2011 *Physica* B **406** 3210
[13] Tauc J, 1974 *Amorphous and Liquid semiconductors,* J. Tauc Ed. Plenum, New York.
[14] Ginson P Joseph, Philip J, Rajarajan K, Rajasekar S A, Joseph Arul Pragasam A, Thamizharasan K, Ravi Kumar S M, Sagayaraj P 2006 *J. Cryst. Growth* **296** 51
[15] Wang X Q, Xu D, Lu M K, Yuan D R, Xu S X, Guo S Y, Zhang G H, Liu J R, 2001 *J. Cryst. Growth* **224** 284
[16] Kurtz S K and Perry T T 1968 *J. Appl. Phys.* **39** 3798
[17] Rajarajan K, Ginson P Joseph, Ravi Kumar S M, Vetha Potheher I, Joseph Arul Pragasam A, Ambujam K and Sagayaraj P 2007 *Materials and Manufacturing Process* **22** 1
[18] Rajarajan K, Preema C Thomas, Vetha Potheher I, Ginson P Joseph, Ravi Kumar S M, Selvakumar S and Sagayaraj P 2007 *J. Cryst. Growth* **304** 435
[19] Vetha Potheher I, Madhavan J, Rajarajan K, Nagaraja K S and Sagayaraj P 2008 *J. Cryst. Growth* **310** 124
[20] Sun H Q, Yuan D R, Wang X Q, Cheng X F, Gong C R, Zhou M, Xu H Y, Wei X C, Luan C N, Pan D Y, Li Z F and Shi X Z 2005 *Cryst. Res. Technol.* **40** 882

IOP Conf. Series: Materials Science and Engineering **73** (2015) 012055 doi:10.1088/1757-899X/73/1/012055

Thermo-optic nonlinearity of the laser dye LDS 867 under low power CW laser excitation

K A Ann Mary[1,2], E J Sonia Mary[2,3], Viji Vidyadharan[1], Reji Philip[2,4] and N.V Unnikrishnan[1]

[1]School of Pure & Applied Physics, M.G. University, P.D Hills P.O, Kottayam 686560, India

[2]Raman Research Institute, C.V. Raman Avenue, Sadashivanagar, Bangalore 560080, India

3 present address: University of Basel, Petersplatz 1, CH-4003 Basel, Switzerland

[4]email: reji@rri.res.in

Abstract. Thermally induced optical nonlinearity of the laser dye LDS 867 is studied in ethanol solution using the self phase modulation and closed aperture z-scan techniques, employing a continuous wave low power He-Ne laser beam for excitation. The nonlinear optical (NLO) coefficients are obtained by analyzing the z-scan curve on the basis of the thermal lens model. The dye exhibits a negative thermal nonlinearity which can be inferred from the occurrence of a pre-focal peak followed by a post-focal valley in the z-scan. The large nonlinear refractive index (n_2) measured at the excitation wavelength of 633nm reveals that the material is NLO active even at low excitation powers of less than 1 mW. Results indicate that LDS 867 is a promising material for optical power limiting applications.

1. Introduction

Absorptive and refractive nonlinearity of metal nanoparticles [1], organic dyes [2], polymers [3,4] and semiconductors [5] have been extensively studied using the z-scan technique, which is a sensitive tool for measuring the optical nonlinearity of materials. Laser dyes constitute an important group of organic molecules, and because of their intense and broad fluorescence emission spectra, they are becoming increasingly important in spectroscopy, holography, and biomedical applications.

In general, different types of optical nonlinearities can have response times extending from femtoseconds to milliseconds. Thermal nonlinearities are relatively slower, and can be induced by low power laser beams. The absorbed radiation can cause heating of the sample which leads to spatially varying temperature profiles. The temperature dependent induced refractive index leads to the creation of thermal lens as first observed by Gordon *et al* [6]. When a laser beam of intensity higher than a certain threshold value is incident on highly nonlinear materials, a far field diffraction ring pattern may become visible. This is caused by the thermally induced self phase modulation in the medium, and from the number of rings observed; the thermo optic nonlinear refractive index can be calculated [7].

Here we report the thermal nonlinearity measured in an ethanol solution of the laser dye LDS 867 excited by a low power (0.7 mW – 3.5 mW) He-Ne laser beam at 633 nm. The z-scan curve is numerically analyzed using both the Sheik-Bahae formalism and the Thermal Lens model, and the goodness of the resulting fits is compared. The value of nonlinear refractive index obtained numerically from closed aperture z-scan is confirmed from the number of self phase modulation rings observed.

2. Experimental details

Styryl dyes are widely used organic media with various applications in laser dyes [8], nonlinear optics [9], and dye sensitized solar cells [10]. Their derivatives are known for their light emiting properties and are useful in bio-labeling [11], optical recording [12] etc. The laser dye LDS867 belongs to the Styryl family. The UV-Visible absorption spectrum of LDS 867 in ethanol measured using a UV-Vis spectrometer (Lambda 35, Perkin-Elmer) exhibits a peak at 619 nm, with fairly good absorbance at the excitation wavelength of 633 nm (Figure 1).

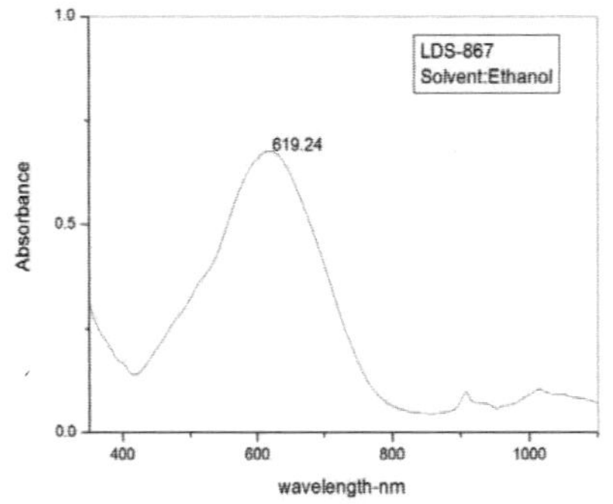

Figure 1: UV-Vis Absorption spectrum of LDS 867 measured in Ethanol.

For closed aperture z-scan measurements (Figure 2), the laser dye solution was prepared in ethanol with 10^{-4} M concentration, and then diluted further to get a linear transmittance of 45% at the probe laser wavelength of 633nm. The sample was taken in a 1 mm cuvette and is gradually translated along the axis of the Gaussian beam (TEM_{00} mode), which was focused to a spot size (ω_0) of 48μm using a 100 mm focal length convex lens. The power transmitted through the far field aperture (2mm diameter) is detected using a photodiode as a function of the sample position z (the focal point is taken as z = 0).

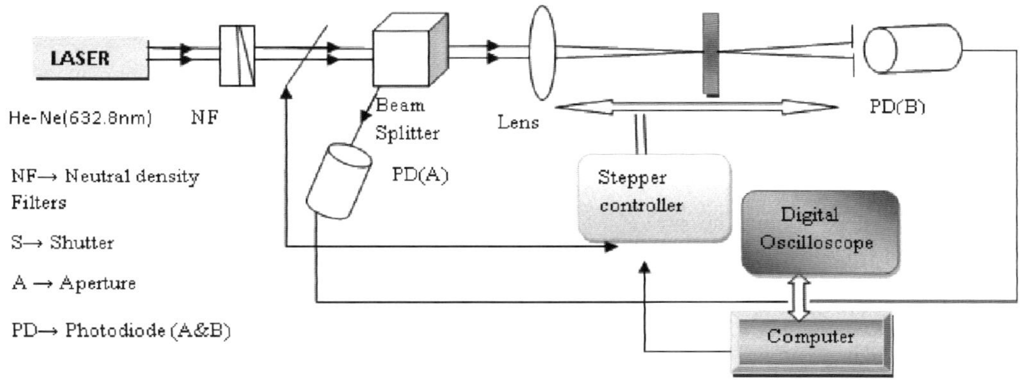

Figure 2: The closed aperture z-scan setup.

For the self phase modulation (SPM) measurements the far field aperture was removed and the beam was observed on a white screen. When the laser beam intensity is higher than a certain threshold value, SPM rings are obtained as a far field diffraction pattern. We found that SPM rings occur when the incident He-Ne laser beam has a power of more than 3.5 mW. The images of the SPM rings were recorded using a CCD camera.

3. Results and discussion

The peak - valley transmittance curve obtained from the z-scan experiment is analyzed using two well known formalisms. According to the Sheik-Bahae Formalism (SBF) [13], the normalized transmittance $T_N^{SBF}(z)$ in the far field for a cubic nonlinearity and small phase shift $\Delta\phi$ is given by the equation,

$$T_N^{SBF}(z) \cong 1 + \Delta\phi \frac{4x}{(1+x^2)(9+x^2)} \tag{1}$$

where x is the dimensionless sample position given by z/z_0, where z_0 is the Rayleigh range. In the Thermal Lens Model (TLM), the light absorbed is converted into heat with a Gaussian radial distribution, and due to the resulting spatial refractive index modification the material behaves as a thermal lens. Steady state is reached at $t \gg t_c$ where t_c is the thermal-diffusion time. In TLM the normalized transmittance $T_N^{TLM}(z)$ is related to nonlinear phase shift θ as [14],

$$T_N^{TLM}(z) \cong \frac{1}{1 + \theta \dfrac{2x}{1+x^2} + \theta^2 \dfrac{1}{1+x^2}} \tag{2}$$

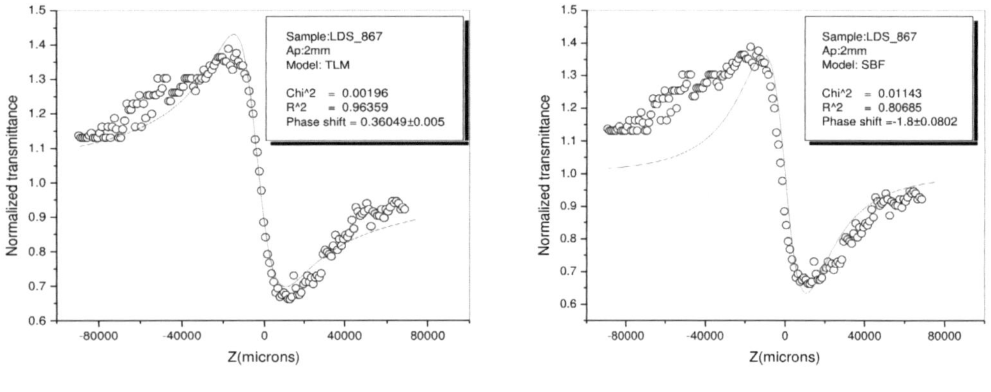

Figure 3: Measured closed aperture z-scan curve, with the data fitted to TLM (left) and SBF (right) respectively.

The experimentally obtained closed aperture z-scan curves are fitted with both models, and we found that TLM gives a better numerical fit to the measured data. For Equation 1 to be valid under SBF the condition $\Delta\phi$ << 1 has to be satisfied, but in the present case the on axis phase shift is 1.8.

Table 1. Fitting parameters for LDS 867

Power 0.75mW	SBF	TLM
Chi-Square Value	0.0114	0.0019
Good Fit		✓
Phase Shift	1.8	0.34

Table 2. Calculated nonlinear parameters

n_2	$1.12\times10^{-6} cm^2/W$
$\frac{\partial n}{\partial T}$	$5.141\times10^{-5} K^{-1}$
$\chi^{(3)}_{therm}$	$4.6\times10^{-5} esu$

The optical nonlinearity arises from the dependence of refractive index on temperature, defined by the thermo-optic coefficient $\frac{\partial n}{\partial T}$. The nonlinear refractive index n_2 is given by [15],

$$n_2^{th} = \frac{dn}{dt}\frac{\alpha\omega^2}{4\kappa} \tag{3}$$

where, α is the absorption coefficient, ω is the beam waist, and κ is the thermal conductivity. The nonlinear refractive index and thermo-optic coefficient are calculated and the thermal contribution to the third order susceptibility is estimated (Table 2). The z-scan curves show a pre-focal peak followed by a post-focal valley indicating a negative nonlinear refractive index, so that $\frac{\partial n}{\partial T}$ also is negative. We performed open aperture z-scans also by removing the far field aperture, but no absorptive nonlinearity could be measured at the low power levels employed.

In a medium exhibiting thermally induced refractive index changes, laser induced self phase modulation can gives rise to far field ring pattern [16]. This effect can be analyzed using the Fraunhofer diffraction integral, and the nonlinear refractive index can be found out from the number of principal SPM rings as [7],

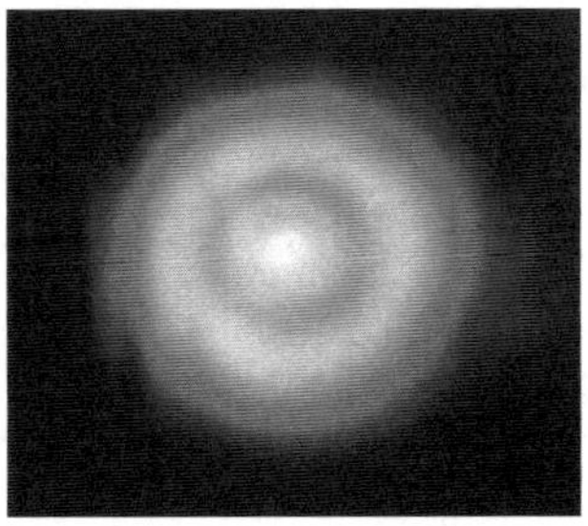

Figure 4: SPM rings observed in the far field.

$$n_2 = \frac{N\lambda}{Id} \tag{4}$$

where N is the number of SPM rings, I is the input intensity, and d is the cell thickness.

The SPM ring pattern in LDS 867-ethanol system shown in figure 4 was obtained above the threshold power of 3.5 mW. We calculated the thermal nonlinear refractive index as 0.94×10^{-5} cm^2/W in LDS 867 from SPM ring measurements. The occurrence of SPM rings and asymmetric nature of closed aperture z-scan curve confirms that the nonlinearity is of thermal origin.

4. Conclusion

Optical nonlinearity due to thermal effects in the laser dye LDS 867 in ethanol under low power cw laser illumination (0.7mW – 3.5mW) is studied using the techniques of closed aperture z-scan and spatial self phase modulation. The closed aperture z-scan curve is analyzed using two models (Sheik-Bahae formalism and Thermal lens model). Due to the high on-axis phase shift exhibited by LDS 867 under these conditions, the Thermal Lens model provides a better fit to the experimental data. The calculated negative nonlinear refractive index and thermo-optic coefficient indicate that LDS 867 is a potential candidate for NLO applications.

5. References

1. He T, Wang C, Pan X and Wang Y 2009 *Phys Lett A* **373** 592

2. Sreekumar G, Louie Frobel P G, Muneera C I, Sathiyamoorthy K, Vijayan C and Mukherjee C 2009 *J. Opt. A: Pure Appl. Opt.* **11** 125204

3. Meyer R K, Benner R E, Vardeny Z V, Liess M, Ozaki M, Yoshino K, Ding Y and Barton T 1997 *Synth. Met.***84** 549–550

4. Jang J I, Mani S, Ketterson J B, Lovera P and Redmond G 2009 *Appl. Phys. Lett.* **95**, 221906

5. Sreejaa R, Jobina Johna, Aneesha P M and Jayaraj M K 2010 *Opt Comm* **283**, 14

6. Gordon J P, Leite R C C, Moore R S, Porto S P S, and Whinnery J R 1965 *J. Appl. Phys.* **36**, 3

7. Binh L N, Dai X and Ja Y H 1989 *Appl.Phys.B* **49** 393-396.

8. Hoffnagle J, Roesch L Ph, Schlumpf N and Weis A 1982 *Opt Comm* **42** 267-268

9. Shettigar S, Umesh G, Poornesh P, Manjunatha K B and Asiri A M 2009 *Dyes Pigm.* **83** 207-210.

10. Dentani T, Ken-ichi Nagasaka, Funabiki K, Ji-Ye Jin, Yoshida T, Minoura H and Matsui M 2008 *Dyes Pigm.* **77** 59-69.

11. Moritz . R, Holzhauser C, Bohlander R P and Hans-Achim Wagenknecht 2012 *Chem. Eur. J* **18** 1299 – 1302

12. Chung-Chun Lee and Hu A H 2003 *Dyes Pigm* **59** 63-69

13. Sheik-Bahae M, Said A A, Tai- Huei Wei, David J Hagan and E W Van Stryland 1990 *Quantum Electron.* **26**, 760.

14. Carter C A and Harris J M 1984 *Appl. Opt.* **23** 3

15. Yu B, Zhu C, Gan F and Huang Y 1997 *Opt. Mater.***7** 103-107.

16. Dabby F W, Gustafson T K, Whinnery J R and Kohanzadeh Y 1970 *Appl. Phys. Lett.* **16** 362-365.

IOP Conf. Series: Materials Science and Engineering **73** (2015) 012056 doi:10.1088/1757-899X/73/1/012056

Study of transport through an electro responsive polymer membrane

D Das[1], A Datta[1] and A Q Contractor[1]
[1]Department of Chemistry, Indian Institute of Technology Bombay Powai Mumbai-400076 India

E-mail: debasreedas@chem.iitb.ac.in, anindya@chem.iitb.ac.in, aqcontractor@iitb.ac.in

Abstract. Conducting polymers have been used widely for development of several electronic, sensing devices because of its electro active nature. In the present work porous polycarbonate (PC) support was coated with a thin gold layer. An electrochemically synthesized polyaniline (PANI) film was deposited on gold coated PC and characterisation was done by field emission gun scanning electron microscopy (FEG-SEM), transmission electron microscopy (TEM) and atomic force microscopy (AFM). For measuring the concentration of potassium ion (K^+) inductively coupled plasma atomic emission spectrometry (ICP-AES) was used. Potassium ion transport across PANI membrane at various potential showed the gradual opening of the coiled PANI. In this work an effort has been given to picture the situation in the membrane electrolyte junction on application of potential.

1. Introduction

Conducting polymers are conjugated polymers, which have an extended p-orbital system, having movement of electrons from one end of the polymer to the other. The most well known are polyaniline (PANI) and polypyrrole (PPY)) whose unique physical and electronic properties [1, 2] have been explored in several technologies. Their unusual high conductivities, have helped in the development of several electronic devices, [3-5] conducting wires, and electrostatic shielding. It is revealed from literature survey that PANI has a distinct position among the conducting polymers, because of its ease of synthesis and several applications.

The different forms of PANI in aqueous solutions are reduced, leucoemeraldine and half-oxidized, emeraldine states. PANI can be reversibly turn on or off in between these states [6]. The transport properties of these membranes are controllable after synthesis, thereby opening up a new route for chemical separation [7]. In our work we have synthesized a polymer membrane and looked in to the transport of potassium chloride across the membrane.

2. Experimental section

Thermal evaporation of gold (200nm) was carried out in HINDHIVAC (model 12A4D) on 1.2 μm diameter pore size polycarbonate (PC) membrane purchased from Scientific Sales Syndicate in vacuum evaporation system. GR grades Aniline, potassium chloride (KCl), hydrochloric acid (HCl), dichloromethane (DCM) were purchased locally. PANI was then deposited on the gold coated membrane electrochemically from a solution containing 0.1 M aniline in 0.5 M H_2SO_4 (Electronic grade). The electrode potential was cycled between -0.2 and +0.8 V vs SCE for 40 cycles. Platinum (Pt) foil was used as counter electrode. The film was then cycled in 0.5M H_2SO_4 to remove the monomer. Potentiostats (model AFCBP1) from Pine Instrument Company and GAMRY Instrument 3000 were used for all electrochemical experiments. The transport measurements were carried out in H type cell shown below in figure 1. The membrane was placed in between the two flanges of the joint. The source compartment contains the counter electrode and the reference electrode. 50ml of 0.1M KCl in 0.1M HCl was filled in source side and the receiver side was filled with 50ml of 0.1M HCl. The potential was changed from 0.1V to 0.4V. The solution in every 10 minutes was withdrawn and K^+ ion transported was determined by inductively coupled plasma atomic emission spectrometer (ICP-AES) (ACROS from M/s. Spectro).

PANI membranes were characterised by field emission gun scanning electron microscope (FEG-SEM) (JEOL JSM 7600F) and AFM Nanoscope IV. A 6-8 nm layer of Palladium was then sputtered on samples before FEG- SEM imaging. For TEM (JEOL JEM 2100F) DCM was added drop by drop on the membrane containing grid to dissolve the PC leaving behind the polymer on the grid.

Figure 1. Cell for transport measurement

3. Results and discussion

To observe the transport of K^+, PANI films of polymerisation charge (28 mC/cm^2) were synthesized by potentiodynamic method on the gold coated PC as shown below in figure 2 & 3.

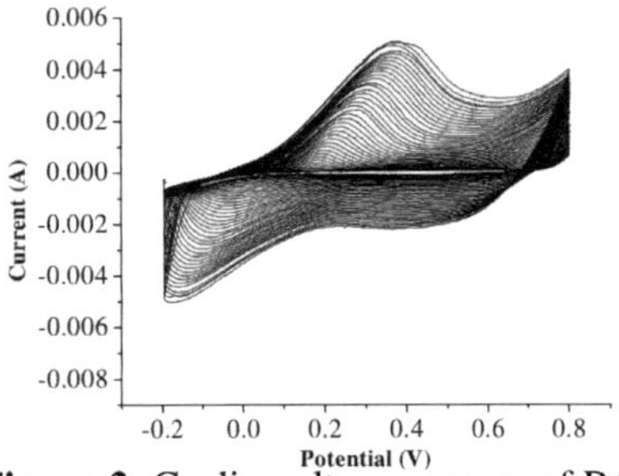

Figure 2. Cyclic voltammogram of PANI from 0.1 M aniline.

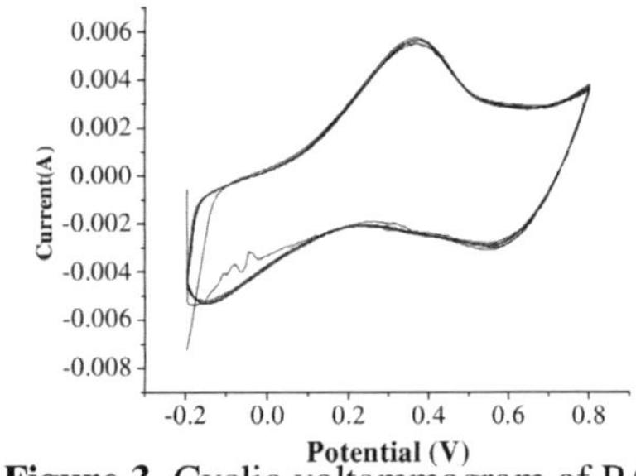

Figure 3. Cyclic voltammogram of PANI in 0.5 M H_2SO_4.

SEM of PC and PANI deposited on gold coated PC are shown in figure 4, 5. The topographic view depicts that due to deposition of the porous PANI film there is a significant decrease in pore mouth. From TEM in figure 6 it is clearly seen there is a PANI film on the PC but it has not completely blocked the pore mouth. Since the polymer has a tendency to grow more on the hydrophobic walls of the PC, so the growth cannot be inhibited resulting in the formation of tubes. The sectional analysis of AFM in figure 7 revealed that there is around 50% coverage of the pore mouths.

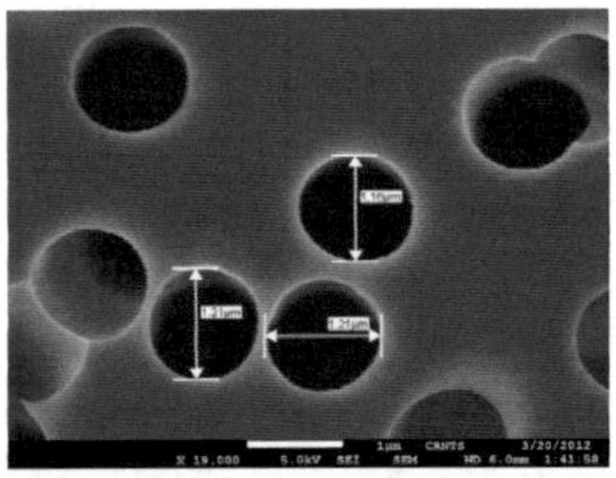

Figure 4. PC.

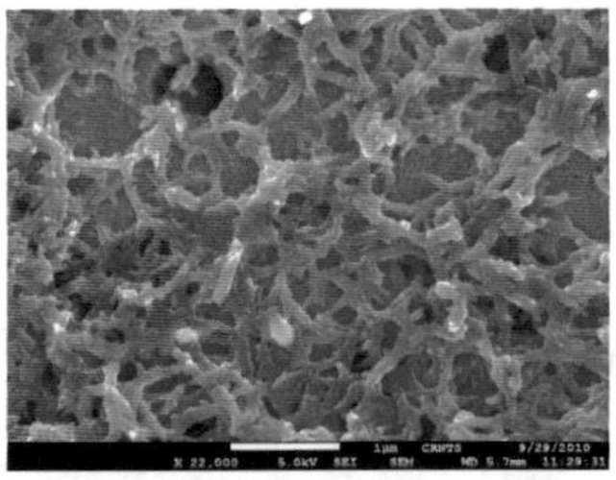

Figure 5. SEM of PANI deposited on PC.

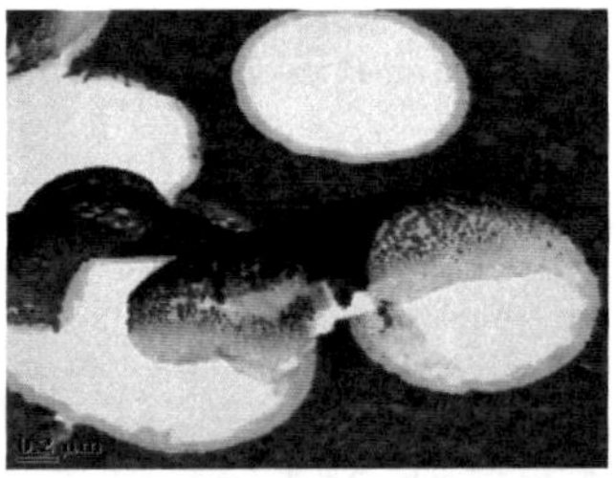

Figure 6. TEM of PANI tubule.

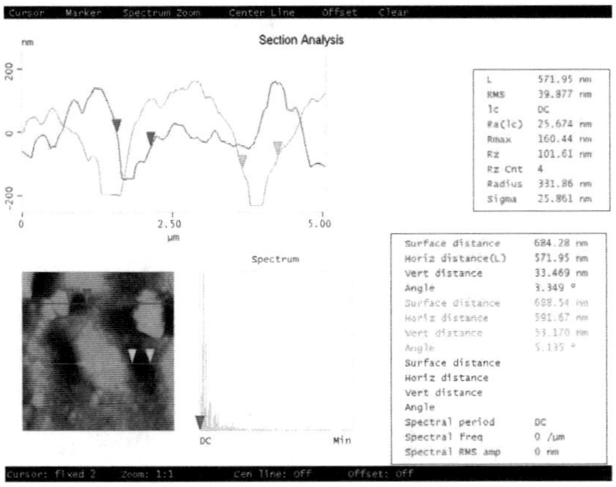

Figure 7. Sectional analysis of PANI (40cycles) on 1.2 μm pores of PC.

Due to the electrochemical processes conducting polymers are often accompanied by transport of ions at the interface. The potential across the double layer plays also a role in electrochemical transport through the conducting polymer.The interfacial properties of a PANI membrane can be switched by electrochemical control of the polymer's oxidation state then the morphology of the polymer undergoes a change on application of electrochemical potential. This change has a huge application in membrane oriented research because the polymer's different redox states can be controlled by potential. The change in conformation is accompanied by the insertion/expulsion of ions in and out of the polymer chain.

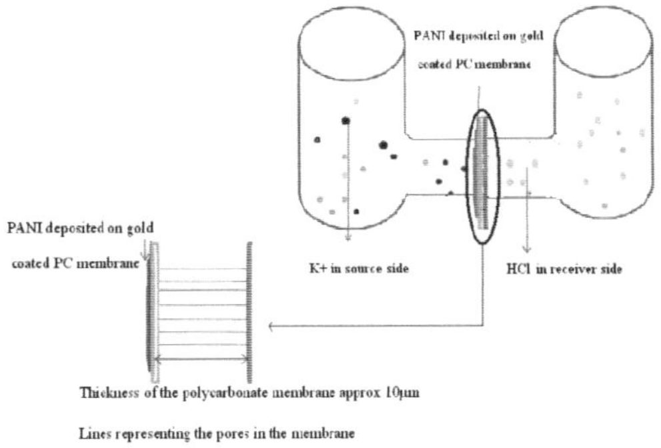

Scheme. Transport across PANI.

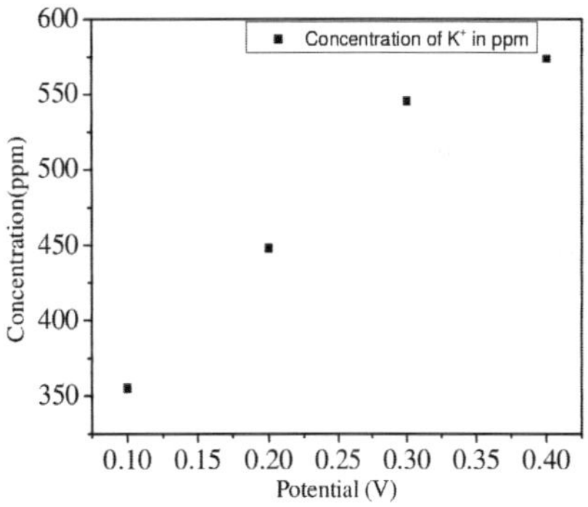

Figure 8. Concentration change of K^+ at different potentials.

As it can be seen from the above scheme the PC has two sides' smooth side and rough side. PANI deposited on the gold coated smooth side faces the source solution and the rough side faces the receiver solution. The auxiliary electrode is placed in the source solution due to which there develops an electric gradient between it and the membrane. The time for which the membrane is held at a particular potential is sufficient for a significant amount of transport to take place. Since the auxiliary electrode is kept in the source solution side the transport is mainly ruled by the ion incorporation and expulsion phenomena [8]. The thickness of the PC is around 10μm. The K^+ ions cross the barrier of the porous PANI membrane, after which it seeps through the walls of the PC to reach the receiver side. The concentration of K^+ was measured as a function of the polymer oxidation state, which was controlled by stepping the potential applied to PANI film from a lower potential to increasingly higher values from 0.1V to 0.4V. The potential was applied for shorter period of time in order to avoid any electrochemical degradation of the polymer. The effective area for transport of K^+ was 1.45 cm^2. In these experiments the driving force is electrochemical stimuli.

Table 1. Flux of K^+ through PANI at different potentials

Potential(V)	Flux of K^+ (10^{-8} mol/cm^{-2}/s)
0.1	0.87
0.2	1.098
0.3	1.336
0.4	1.406

Studies with K^+ were mostly done with atomic absorption spectrometry by the group of Wallace. In our case the concentration of K^+ was measured by (ICP-AES) which improves in précising the quantification. When a potential is applied due to oxidation of the polymer incorporation of anions and release of cations take place. The extent of transport depends on the relative mobility of the cations and anions. Hence for electrochemical transport the movement of the ions play a significant role [9]. Due to this uptake of anions there occurs volume swelling which results in increase in flux through the polymer. The transport is found to depend on the morphology of the polymer. On going from 0.1V to 0.4V the flux is found to be reasonably high near the 0.4V as shown in figure 8 and table 1 because here PANI exists in half oxidised state which as shown in figure 2 corresponds to the highest conductive state of PANI. The increase in flux of K^+ with application of positive potential through PANI membrane with gradual increase of potential shows the opening of the coiled PANI to more expanded and porous structure.

4. Conclusion

SEM, AFM, TEM images show that layer of PANI has been successfully deposited without plugging the pore mouths. The transport of K^+ through porous PANI demonstrated the interfacial responses in the membrane electrolyte interface with the application of electrochemical stimuli. The transport process showed how the polymer undergoes change in conformation from coiled to more expanded and porous structure with the application of potential signifying the electro active nature of PANI.

5. References

[1] Wallace G G Spinks G M and Teasdale P R 1997 *Conductive Electro active Polymers Intelligent Materials Systems* (Lancaster, PA: Technomic Publishing Co., Inc.)
[2] Chandrasekhar P 1999 *Conducting Polymers, Fundamentals and Applications. A Practical Approach* (Boston, MA : Kluwer Academic Publishers)
[3] Hohnholz D and MacDiarmid A G 2001 *Synth. Met.* **121** 1327-1328
[4] Kittlesen G P, White H S, Wrighton M S 1984 *J. Am. Chem. Soc.* **106** 7389-7396
[5] Dabke R B, Singh G D, Dhanabalan A, Lal R and Contractor A Q 1997 *Anal. Chem.* **69** 724-727
[6] Pile D L and Hillier A C 2002 *J. Membr. Sci.* **208** 119-131
[7] Zhao H, Price W E, Wallace G G, 1992 *J. Electroanal. Chem.* **334** 111-120
[8] Zhao H, Price W E, Too C O, Wallace G G and Zhou D 1996 *J. Membr. Sci.* **119** 199-212
[9] Price W E, Too C O, Wallace G G and Zhou D 1999 *Synth. Met.* **102** 1338-1341

Acknowledgements
This work was supported by Department of Chemistry, Nanoelectronics, Physics, Sophisticated Analytical Instrument Facility and IIT Bombay.

Fabrication and characterization of photovoltaic cell with novel configuration ITO/n-$CuIn_3Se_5$/p-CIS/In

R Geethu[1], R Jacob[1], P V Sreenivasan[2], T Shripathi[3], Okram G S[3] and R R Philip[1]*

[1]Thin film research lab, Union Christian College, Aluva, Cochin, Kerala, India.
[2]Department of Chemistry, Union Christian College, Aluva, Cochin, Kerala, India.
[3]UGC-DAE CSR, Khandwa Road,Indore-452 001,M P, India.

E-mail: reenatara@rediffmail.com

Abstract: A novel configuration ITO/n-OVC $CuIn_3Se_5$/p-CIS/In solar cell has been fabricated by multisource vacuum co-evaporation technique on soda lime glass substrates. The pn junction is formed with ordered vacancy compound as the n counter part for the p type $CuInSe_2$. The structural, compositional, hall coefficient, optical and electrical properties of the p and n layers have been studied respectively by X-ray diffraction, Energy Dispersive Analysis of X rays, optical absorbance and conductivity measurements. Current density-Voltage measurements enabled the determination of efficiency of the device.

1. Introduction:

The chalcopyrite semiconductor compound copper indium diselenide ($CuInSe_2$), owing to their good photovoltaic properties, has become a promising candidate for solar-cells applications [1, 2]. The broad range of optical band gaps and carrier mobilities offered by ternary ABX_2 semiconductors, as well as their ability to form various solid solution and to accommodate different dopants, has led to their emergence as technologically significant device materials, including application in photovoltaic solar cells both as single-crystal materials (up to 12% efficient) and as polycrystalline thin films (at least 9.4% efficient), light-emitting diodes, and in various non-linear optical device[3].

Generally Cadmium Sulphide films are used as conducting window layers in the fabrication of heterojuction solar cells with CIS. The present work is based on the preparation of CdS free solarcell with a configuration ITO/n-OVC /p-$CuInSe_2$/In. Here indium tin oxide (ITO) will act as the contact layer at the n region and indium at the active region.

2. Experimental technique:

A modified form of Gunther's three temperature method [4] has been used for the deposition of p-CIS and n-OVC CIS. 99.99 % pure elemental Cu and In were evaporated from two separate electrically heated molybdenum boats and 99.999% Se was evaporated from a glass crucible kept in a molybdenum basket. The deposition was done in the atmosphere of Se at a pressure of 10^{-5} Torr. The evaporation was done in a uniform manner by adjusting the flux through each boat and taking care to adjust the flux conditions and vacuum to the optimised conditions to obtain the films of the required composition. The solar cell fabrication was in three stages. The first stage was the coating of n OVC, on suitably masked ITO coated soda lime glass substrates, in vacuum at substrate temperature of

623K. After ensuring the formation of n-layer by structural, compositional and conductivity type characterization using X-ray diffraction (XRD), Energy dispersive analysis of X-rays (EDAX), Hall coefficient measurement and hot probe method respectively, the films were again suitably masked and p-$CuInSe_2$ was deposited on it at substrate temperature 623K. Indium was deposited at room temperature over p-CIS layer to serve as the electrode contact on the p- side.

X-ray diffraction (XRD) using a Rigaku D MaxC X-ray diffractometer with CuKα1 as source radiation was used to determine the phases and crystallographic structure of the films. Compositional characterization was done by taking EDAX by using a Link 10000, a Cambridge instrument. Hall effect measurements were done placing the film in a magnetic field of 3600 Gauss and by measuring the voltage for different currents by four probe method. The dark conductivity and photoconductivity at room temperature were measured using Keithley 2611A source meter, using silver paste for ohmic contact. Optical absorbance measurements were performed with a Hitachi U-3410 UV-Vis-NIR spectrophotometer. Dark and photo conductivity measurements were done by using a Keithley 2611A source meter and an FHS lamp. The J-V characteristics of the ITO/n-OVC/p-CIS/In solar cell was studied using a Keithley 2611A source meter, using an FHS lamp (82V, 300W) for illumination of the cell from the substrate side.

3. Results And Discussions

The structural characterization has been done by using X-ray diffraction in order to ascertain the formation of n-OVC and p-CIS and the lattice constants has been calculated to check the lattice matching between the two layers. The values obtained showed agreement with the early observed results [5-6]. After ensuring the formation of n and p type layers from hall measurement in conjunction with hot probe technique conductivity has been determined. The n and p type films were found to have conductivity in the order $1.66\Omega^{-1}m^{-1}$ and $3.58\times10^{4}\Omega^{-1}m^{-1}$ respectively. A carrier density of the order of $2\times10^{15}/cm^3$ and $1.9\times10^{19}/cm^3$ has been obtained in the n and p films from Hall effect measurement.

The absorption coefficients of the as prepared n and p type CIS films are determined from the absorbance and the transmittance spectra using equation (1)

$$T=(1-R)^2\exp(-A)=(1-R)^2\exp(-\alpha d) \qquad (1)$$

where A is the absorbance, α the absorption coefficient, d the film thickness (0.1μm), R the reflectance and T the transmittance of the films [7].The absorbance and transmittance of the films for different wavelength are shown in figure 1a and 1b. The band gap has been determined from the equation

$$\alpha h\nu = A_c(h\nu - E_g)^{\gamma} \qquad (2)$$

where α is the absorption coefficient, γ is a constant which is equal to 1/2 and 3/2 for allowed and forbidden direct transitions respectively and equal to 2 and 3 for allowed and forbidden indirect transitions in which the photons are involved, A_c is a constant and E_g is the optical band gap [8].
Copper indium selenide compounds are direct band gap materials [9] so that the extrapolation of straight line portion to x-axis in the graph with hυ along x-axis and $(\alpha h\upsilon)^2$ along y-axis gives the optical band gap value of the material. The optical band gaps of the as prepared films were found to be ~1.3 eV and ~ 0.9 eV respectively for n and p films.

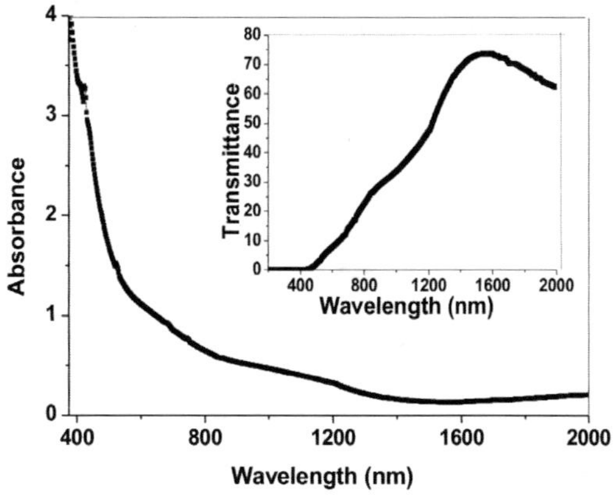

Fig.1. (a) Absorbance spectrum of n OVC, inset shows the transmittance spectrum

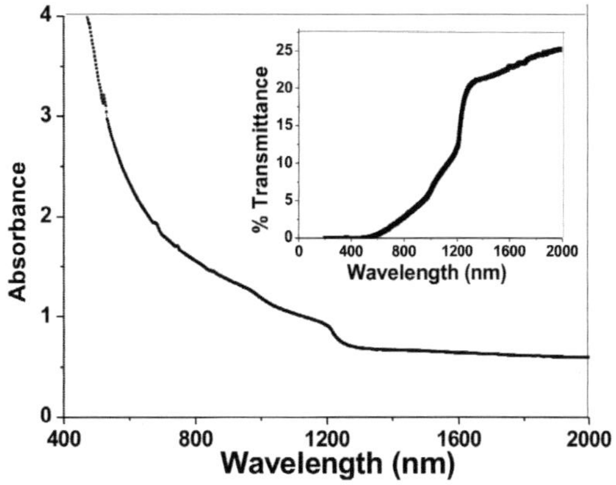

Fig.1. (b) Absorbance spectrum of p-CIS, inset shows the transmittance spectrum

The enhanced band gaps for OVC's indicate that if illuminated from the n-OVC side, photons with lesser energy could pass without absorption or generation of carriers in the OVC to the p-side, thereby resulting in carrier generation in the p-CIS. Carrier separation could then occur at the junction, resulting in external photon generated current. The photosensitive nature of the films has been confirmed by doing photocurrent measurements.

The efficiency of the solar cell has been calculated from the J-V characteristics shown in the figure2. The open circuit voltage and short circuit current ~ 0.05V and ~1.8mA/cm^2 measured in a typical cell of the above configuration at an illumination intensity of 20 W/m^2 is shown in figure 2. The efficiency has been calculated ~0.31%. The cell configuration has the advantage that once optimising conditions are determined, the n- and p-type layers for pn junction formation can be deposited in a single coating simply by adjusting the flux. The possibility for lattice matching and prevention of losses is greater. The cell shows the scope to be developed into a higher efficiency system by optimising the thickness of the junction layers and by reducing the series resistance. Here, the measurement is limited by the low irradiation intensity 12-20W/m^2 that could be given using the lamp used. The results indicate the possibility of obtaining better efficiency of the solar cells under standard measurement conditions.

.

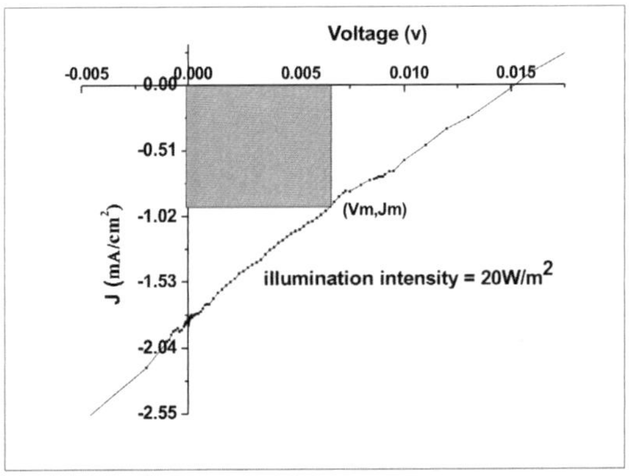

Figure 2. J-V characteristics of a solar cell

4. Conclusion

The structural, compositional and optical analysis confirms the formation of the n OVC and p CIS layers with lattice matching and with an enhanced band gap for n compared to p. The hall and conductivity measurements in dark and under illumination reveal the semiconducting and photosensitive nature of the films. From the J-V characterization the efficiency of the cell is determined.

Acknowledgements

GR and the corresponding author would like to thank Kerala State Council for Science Technology and Environment (KSCSTE) for funding our project through the SRP scheme.

Reference

[1] Guillen C and Herrero J 2002 *Solar energy materials and solar cells* **73** 141.
[2] Rincon C and Bellabarba C1986 *Physical Review B* **33** 7160.
[3] Jaffe J E and Zunger A. 1983 *Physical Review B* **28** 5822.
[4] Gunther K G, *The Use of Thin Films in Physical Investigations*, London Academics, 1966.
[5] Philip R R, Pradeep B 2005 *Thin Solid Films* 472 136.
[6] Philip R R, Pradeep B and Shripathi T 2005 Valence band splitting in CuInSe2 and the related ordered vacancy compounds CuIn3Se5, CuIn5Se8, CuIn7Se12.*Trends in Semiconductor Research*, New York, Nova Science Publishers, p. 33-58.
[7] Philip R R, Pradeep B, 2003 *Semicond. Sci. Technol.***18** 768.
[8] R Jacob, Geethu R, Shripathi T, Okram G S, Ganesan V, Pradeep B, Urmila K S and Philip R R, 2012 *Phys.Status Solidi B* DOI 10.1002/pssb.201147581.
[9] Philip R R, Pradeep B, 2005 *Thin Solid Films* **472** 136.

IOP Conf. Series: Materials Science and Engineering **73** (2015) 012058 doi:10.1088/1757-899X/73/1/012058

Structural and optical characterization of reactive evaporated tin diselenide thin films

K S Urmila, T Namitha Asokan and B Pradeep

Solid State Physics Laboratory, Department of Physics, Cochin University of Science and Technology, Kochi-682022, Kerala, India

E-mail: urmilaks7@gmail.com

Abstract. Tin diselenide thin films with thickness of the order of 300 nm are deposited on glass substrate at a substrate temperature of 523±5 K and pressure of 10^{-5} mbar using reactive evaporation, a variant of Gunther's three temperature method. High purity tin (99.999%) and selenium (99.99%) were used as the elemental starting materials. The deposited film is characterized structurally using X-Ray Diffraction (XRD). The structural parameters such as lattice constant, particle size, dislocation density, number of crystallites per unit area and strain in the film are evaluated. Optical absorption spectrum of the film is analysed using UV-Vis-NIR Spectrophotometer.

1. Introduction

The study of binary IV-VI compound semiconducting thin films has attracted the attention of researchers due to their suitable physical properties, abundance and low environmental impact of their constituent elements. These compounds in bulk and thin film form have been studied for their electro-optical properties. High absorption coefficient makes them good absorbers for thin film photovoltaic devices. $SnSe_2$ is a layered semiconductor of IV-VI family with widespread applications in photovoltaic and infrared optoelectronic devices, radiation detectors, holographic recording systems, electrical switching and polarity dependent memory switching devices [1]. Several methods such as Molecular Beam Epitaxy [2], Selenization of sputtered tin layers [3], Thermal evaporation [4], Elemental layer heating [5] and Chemical Vapour Deposition [6] are used for the preparation of $SnSe_2$ thin films. Compared to other IV-VI compounds $SnSe_2$ thin films have received relatively less attention. Hence in this paper we preset our results concerning the preparation of $SnSe_2$ thin films by reactive evaporation. Results on structural and optical properties of these films are also reported.

2. Experimental details

Thin films of $SnSe_2$ are prepared by reactive evaporation which is a variant of Gunther's three temperature method [7]. For many binary systems it has been found that the deposition of stoichiometric compound layers by simultaneous evaporation of individual components is possible only by selecting a suitable substrate temperature and adequate incident rates for the components [8].

In the present work, a conventional vacuum system operated in the range of 10^{-5} mbar is used for the preparation of $SnSe_2$ thin film. Ultrasonically cleaned optically flat glass slides are used as the substrates and are mounted on a substrate holder with a heating arrangement. The temperature of the substrate is measured using a fine wire chromel-alumel thermocouple. When the required substrate temperature 523±5 K is attained, high purity selenium (99.99%) from a glass crucible kept in a molybdenum basket is evaporated. When sufficient flux of selenium is obtained high purity tin (99.999%) from a molybdenum boat is evaporated in the atmosphere of selenium. Rate of evaporation of tin and selenium is controlled by the amount of current that passes through the sources. When sufficient fluxes of tin and selenium are attained the shutter kept above the sources is removed so that the vapours of individual elements reach the substrate and react to form the compound film on the substrate surface. Unreacted elemental atoms or molecules will re-evaporate from the substrate surface due to its elevated temperature. Source and substrate temperatures are properly controlled throughout the film deposition so as to obtain good quality $SnSe_2$ thin films.

3. Results and Discussion

3.1. Structural Characterization

X-Ray Diffraction (XRD) analysis is carried out using Rigaku D MaxC X-Ray Diffractometer for 2θ varying from 10° to 70° with CuKα (1.5404 Å) as the radiation source. The XRD pattern of the as prepared $SnSe_2$ thin film is shown in figure 1.

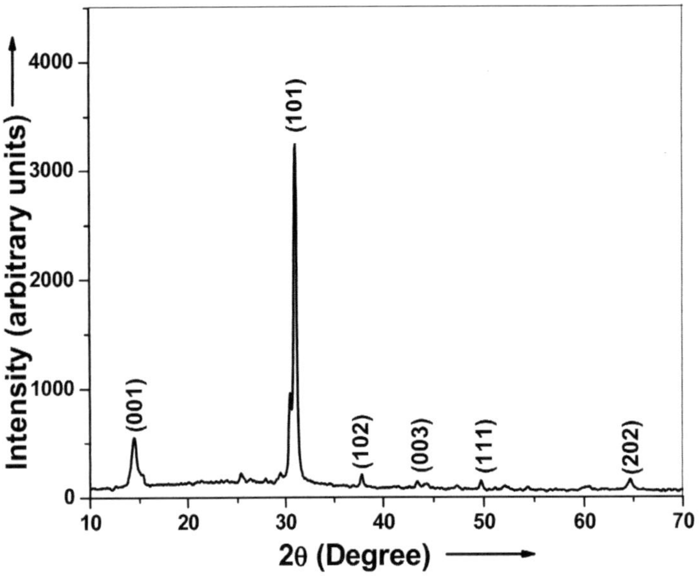

Figure 1. XRD pattern of the as prepared $SnSe_2$ thin film.

The d values of the prepared film obtained from XRD study well matched with that of $SnSe_2$ in standard JCPDS (FileNo.23-0602). Comparison of *d* values of the as prepared film with that of standard *d* values is shown in table 1. The prepared film is polycrystalline in nature with preferential orientation along the (101) plane. The structure of $SnSe_2$ is hexagonal. The calculated values of lattice constants are a = 3.78 Å, b = 3.78 Å, c= 6.13 Å.

Table 1. Comparison of observed *d* values of the as prepared $SnSe_2$ thin film with standard *d* values.

Observed Value		Standard Value (JCPDS File No.23-0602)		(*hkl*)
d (Å)	Relative Intensity	*d* (Å)	Relative Intensity	
6.13	18	6.14	20	(001)
2.89	100	2.91	100	(101)
2.38	7	2.25	45	(102)
2.05	5	2.05	12	(003)
1.83	5	1.82	12	(111)
1.44	5	1.45	12	(202)

Particle size is calculated as 35 nm using Scherrer formula

$$D = \frac{0.9\lambda}{\beta \cos\theta} \qquad (1)$$

where λ is the wavelength of the X-Ray used, β is the full width at half maximum and θ is the Bragg's angle [9]. Dislocation density is estimated as 8.16×10^{10} lines cm^{-2} using the relation [9]

$$\rho = \frac{1}{D^2} \qquad (2)$$

The number of crystallites N per unit area is estimated as 69.97×10^{10} cm^{-2} using the relation [9]

$$N = \frac{t}{D^3} \qquad (3)$$

where t is the thickness of the film.
The strain is calculated as 1.65×10^{-3} using the expression [9]

$$S = \left(\frac{1}{\sin\theta}\right)\left[\frac{\lambda}{D} - \beta\cos\theta\right] \qquad (4)$$

3.2. Optical Characterization
Absorption coefficient α varies with the photon energy $h\nu$ as

$$\alpha h\nu = A\left(h\nu - E_g\right)^n \qquad (5)$$

for direct transitions in which n is a constant that determines the type of electronic transition causing the absorption and can take values 1/2 and 3/2 for direct allowed and direct forbidden transitions respectively and as

$$\alpha h\nu = B\left(h\nu - E_g \pm E_p\right)^n \qquad (6)$$

for indirect transitions in which n can take values 2 and 3 for indirect allowed and indirect forbidden transitions respectively where E_g is the band gap energy, E_p is the phonon energy, ν is the frequency of the incident beam, A and B are constants [10]. Optical absorption of the film is recorded in the wavelength range 2500 nm to cutoff using JASCO V570 UV-Vis-NIR spectrophotometer. Plot of $(\alpha h\nu)^2$ vs $h\nu$ is shown in figure 2. In the absorption data of $SnSe_2$ thin film the best fit is for $n = 1/2$ and it suggests that the transition is a direct allowed one. The straight line portion of the plot is extrapolated to obtain the optical band gap of the film and it is found to be 1.4 eV. Amalraj et al [11] reported an optical bandgap of 1.4 eV for $SnSe_2$ thin films prepared by spray pyrolysis technique at an optimized substrate temperature of 523 K.

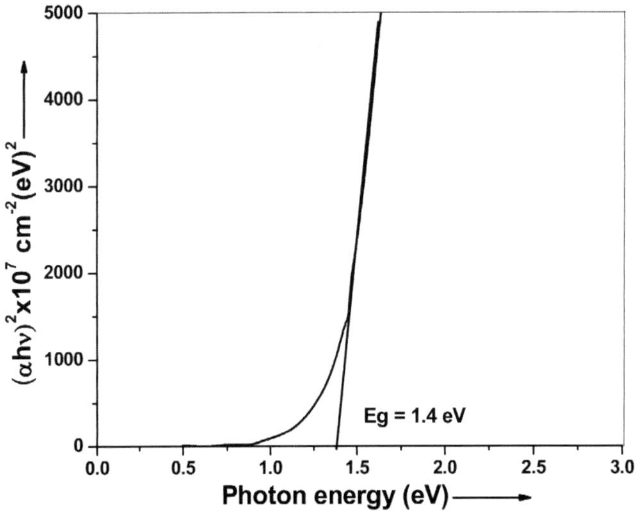

Figure 2. Plot of $(\alpha h\nu)^2$ vs $h\nu$ of a typical $SnSe_2$ thin film.

4. Conclusion

Good quality thin films of tin diselenide are successfully deposited onto glass substrate under optimized deposition conditions using reactive evaporation. XRD study reveals that the films formed are polycrystalline in nature. Structural parameters such as lattice constant, particle size, dislocation density, number of crystallites per unit area and strain in the film are estimated. The film showed a direct allowed transition with a band gap of 1.4 eV.

Acknowledgement

The first author (KSU) would like to thank Cochin University of Science and Technology for financial support in the form of Research Fellowship.

References

[1] Sachdeva R, Sharma M, Devi A, Parihar U, Kumar N, Padha N and Panchal C J 2011 *J.Nano-Electron.Phys.* **3** 507
[2] Schlaf R, Louder D, Lang O, Pettenkofer C, Jaegermann W, Nebesny K W, Lee P A, Parkinson B A and Armstrong N R 1995 *J. Vac. Sci. Technol.* A **13** 1761
[3] Guillen C, Montero J and Herrero J 2011 *Phys. Status Solidi.* A **208** 679
[4] Abdel Hady D, Soliman H, El-Shazly A and Mahmoud M S 1999 *Vacuum* **52** 375
[5] Bindu K and Nair P K 2004 *Semicond. Sci. Technol.* **19** 1348
[6] Boscher N D, Carmalt C J, Palgrave R G and Parkin I P 2008 *Thin Solid Films* **516** 4750
[7] Gunther K G 1996 *The Use of Thin Films in Physical Investigations* vol 1, ed J C Anderson (London: Academic press) p 213
[8] Urmila K S, Namitha Asokan T and Pradeep B 2011 *Proc. Int. Conf. on Light* vol 1391 (New York: AIP) p 770
[9] Dhanam M, Manoj P K and Prabhu R R 2005 *J. Cryst. Growth.* **280** 425
[10] Hankare P P, Khomane A S, Chate P A, Rathod K C and Garadkar K M 2009 *J. Alloys Compd.* **469** 478
[11] Amalraj L, Jayachandran M and Sanjeeviraja C 2004 *Mater. Res. Bull.* **39** 2193

IOP Conf. Series: Materials Science and Engineering **73** (2015) 012059 doi:10.1088/1757-899X/73/1/012059

Shock wave loading of Nickel based superalloy and micro-structural features of the compacts

A D Sharma[1], A K Sharma[2] and N Thakur[1]

[1] Department of Physics, Himachal Pradesh University, Shimla-171005, India
[2] Terminal Ballistics Research Laboratory, Chandigarh-160030, India

E-mail: akash1physics@gmail.com

Abstract. Explosive shock wave loading has been employed to consolidate micro-sized nickel based IN718 superalloy powder. Cylindrical geometry configuring the various critical parameters with optimized detonation pressure has been used to consolidate the powder with desirable means. The thrust on the work is to compact the powder nearer to theoretical density having almost negligible density gradient and without melting the core of the specimen. XRD study indicates that the crystal structure of the post compacts remains the same. Shock wave loading deformed the particles as has been inferred from SEM. The variation in particle size has been measured from Laser Diffraction based Particle Size Analyzer (LDPSA). It is found that this is a rapid fast technique to produce larger and crack free compacts of metal powders without their melting and with less particle size variation.

1. Introduction

It is the concern of every powder metallurgist to compact the powder with efficient and desirable means. For the years shock waves have been employed for the rapid solidification of the hard metal powders which are otherwise difficult to compact using conventional techniques such as hydraulic and iso-static pressure techniques [1]. SWC is a possible alternative for densification and synthesis of micro-sized composites. Experimentalists produce these shock waves through powerful electric discharges; supersonic movement of bodies; gas gun and Laser facility; explosive detonation etc. The main advantage of the explosive shock wave loading to the materials is its rapid time processing and controlled detonation pressure, which directly transmits shock wave to the material [2]. The shock processing of materials utilizes the intense deposition of shock energy preferentially on the particle surface, which gets deformed often melted and re-solidified producing inter-particle bonding without melting the core of the specimen [3]. This paper describes the shock consolidation of IN 718 alloy powder to produce specific dimensional compacts for selected applications. It is well known that IN 718 alloy is a Ni-Cr based superalloy and is widely used in Aerospace industry- cryogenic storage, gas turbines, jet engines, compressor wheels and related applications in critical components such as blades, discs, rocket motors, space crafts, and nuclear reactors etc; due to its very high temperature and corrosion resistant properties.

The work reported here is a part of an extensive program is being undertaken to produce materials of unique industrial applications. The first part of the study comprises the theory and experimental work on shock consolidation of the superalloy powder with maximum compaction and negligible melting using explosive whereas second part provides the information on crystallographic parameters, grain sizes, micro-structure and morphological analysis of post-compacts. Useful information has been obtained on the controlling parameters to produce compacts for desired applications.

2. Experimental

The Various critical parameters such as type of explosive, powder container material and thickness, type of confinement etc, have been optimized keeping in frame of required compacts. Therefore, an explosive mixture (AN/TNT/Al::50/42/8) with moderate detonation velocity (5.8km/s) has been used. The schematic arrangement of the geometry used in the experiment has been illustrated in Figure 1. The cylindrical compaction system shown in figure is a metallic tube made of mild steel. The cylindrical ampoule is placed in a cylindrical plastic container made of PVC having wooden plug at the bottom. The space between the metallic tube and plastic container is filled with the explosive. A wooden plug having an electric detonator and booster covers the plastic container from the top. In order to carry out the experiment, the explosive filled compaction cylinder is placed in an earth pit for easy recovery of the compacted specimen. Figure 1 shows various components of the trial assembly. When the explosive loaded assembly is detonated from the top, an axi-symmetric shock compresses the superalloy powder producing a pressure of about 9.42GPa. Significant in-homogeneity of the structural properties along the ampoule has been observed.

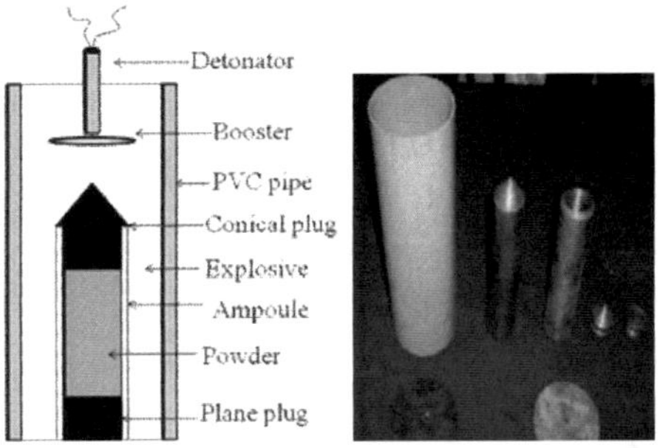

Figure 1. Cross-sectional layout of the configuration used for the consolidation experiment and various trial components.

XRD technique was used for the elementary phase analysis of the pre and post-compacts. The device used was XPERT-PRO X-Ray Diffractometer with a step-angle equal to 0.02° and step-time of 10.33 sec at room temperature. The X-ray beam of Kα lines of Cu with wavelength 1.54Å impinged on the powder sample to get the required XRD pattern. The glancing angle was varied from 30^0 to $80^0(2\theta)$. The particle sizes of the pre and post compacts were determined with Laser Diffraction based Particle Size Analyzer (LDPSA) and correlated to Microscope intercept method. The device used for the particle size distribution was Malvern Mastersizer S-2000. For microstructure analysis of the compacts and to account for the inter-particle melting/bonding SEM technique was adopted and the device used was JEOL JSM6100 system at 15kV.

3. Results and Discussion

Figure 2 corresponds to the X-ray pattern for pre and post shock compacted specimens, respectively. The sharp diffraction peaks indicates pure crystalline structure and after being indexed using ASTM standards, satisfy fcc structure. This means that the crystal phase/structure of the post-compacted specimens remained the same even after the shock-loading. It can be observed from the diffraction pattern that the FWHM values for the post-compacted specimens are increased suggesting a particle size reduction [4].

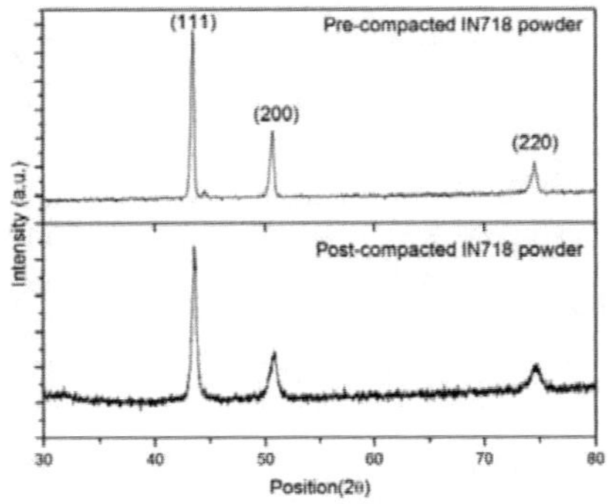

Figure 2. X-ray pattern for pre and post-compacted specimens

The particles size has been validated by using Laser diffraction based particle size analyzer. It has been observed that average particle size for IN718 superalloy powder is close to 105μm. The particle size varies from 30.53μm to 390.80μm. The volume or Mass Moment Mean diameter of the specimen before shock loading was found to be 105μm. The reduction in particle size after shock-loading has been validated by the scanning electron microscopy technique discussed as follows.

Figure 3 represents the SEM images of the pre and post-compacted specimens. The particle are smooth spherical in shape and have the size close to 120μm. The value is in agreement with that of measured by laser diffraction method. After shock-loading, the particle size has been reduced to 70 μm. The particles are welded and interlocked together under the extensive shock energy which preferentially deposits on the particle surfaces resulting into melting and hence responsible for bonding mechanism. The heat generated at the particle boundaries during compaction and consequential microstructure features produced depend on the interparticle friction and local deformation mechanism. The apparent white etched spots in the micrographs at the interparticle regions attributes to a little melting which has been identified by several investigators that produced due to rapid- solidification shrinkage which also responsible for the bonding mechanism[5].

Figure 3. SEM images for pre and post-compacted specimens.

Figure 4 shows surface morphology of pre and post-compacted specimen. It can be seen from the micro-structure that dendritic behavior of the compacted specimens remains the same even after the intensive shock loading.

Density of the compacted specimen was calculated to be close to 97% of the theoretical value.

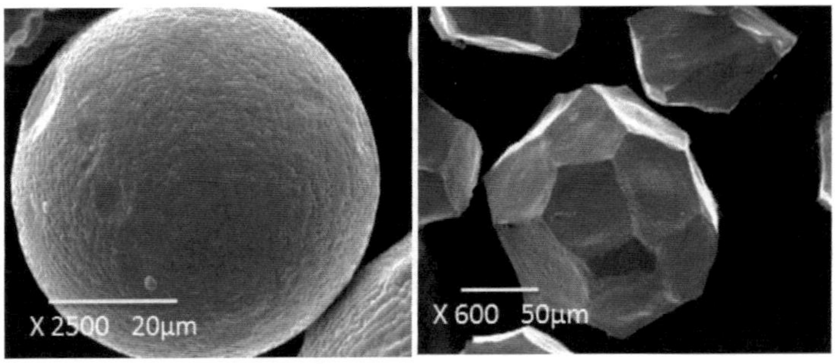

Figure 4. Surface morphology of pre and post-compacted specimens.

Conclusions

Uniform compacts of the superalloy IN718 without the presence of any fracture/micro-voids have been obtained by using explosive having moderate detonation velocity. Cylindrical geometry with optimized critical parameters provided a density 97% of theoretical value. The particle sizes for the compacted specimen were found to be reduced by explosive shock wave loading with a small variation (105-77μm). Shock wave loading deformed the particles micro-structure as has been inferred from SEM but the crystalline structure remained the same (fcc). The technique is now being used to produce denser compacts to achieve tremendous mechanical properties.

Acknowledgement

The authors wish to acknowledge Defence Research and Development Organization (DRDO), India for Grant-in-aid project. Special thanks go to the University Grants Commission (UGC-New Delhi), India, for providing Research Fellowship.

References

[1] Thathani N N, Mutz A H and Vreeland T 1989 *Acta Mettall.* **37** 897-908
[2] Alba-Baena N G, Salas W and Murr L E 2008 *Mater. Charact.* **59** 1152-60
[3] Hokamoto K, Prummer R A, Knitter R and Taira K 2007 *J.Matter. Sci.* **43** 684-88
[4] Cullity B D 1956 *Elements of XRD* (USA- Addison Wesley Publishing Co.)
[5] Wang S L, Mayers M A and Szecket A 1988 *J. Material. Sci.* **23** 1786-1804

Study of post annealing effects on structural and optical properties of sol-gel derived ZnO thin films grown on n-Si substrate

Aniruddh Bahadur Yadav[1], C. Periasamy[2] and S. Jit[1]
[1]Department of Electronics Engineering, Indian Institute of Technology, Banaras Hindu University, Varanasi- 22100, India
[2]Department of Electronics & Communication Engineering NIT Jaipur India

E-mail: abyadav.rs.ece@itbhu.ac.in

Abstract: Zinc oxide (ZnO) thin films were grown on n-type silicon (100) substrates by sol-gel spin coating technique. The prepared thin films were annealed in the presence of Ar at three different temperatures (at 450°C, 550°C and 650°C) to study the impact of annealing temperature on the structural and optical properties of the ZnO thin films. The structural, surface morphology and optical properties of the thin film were investigated using X-ray diffraction (XRD), scanning electron microscopy (SEM) and photoluminescence (PL) measurements respectively. The grown ZnO thin films are polycrystalline in nature with wurtzite hexagonal structure as evident from the XRD and SEM analyses. It further indicates that the crystalline size increases with increasing annealing temperature. The post annealing is also found to influence the optical properties in the terms of band gap energy of the ZnO thin films. The optical energy band gap was found to be decreased from 3.205 to 3.13eV as the annealing temperature is increased from 450°C to 650°C. However, our results concerning the growth of ZnO thin films on Si substrates suggest that there is an intermediate growth temperature allowing for the optimization of the ZnO film growth. The results of the study can be used as a guideline for growing ZnO thin films on n-Si substrates with a homogenous surface morphology, high surface to volume ratio and desired particle size, which are suited for optoelectronic/ gas sensing applications.

1. Introduction

The last decade has witnessed a dramatic development in the field of nanotechnology and nanoscience. The transparent thin films and nanostructures based different materials (e.g., Si, Carbon nanotube, GaN, ZnO) are becoming increasingly attractive for fabrication of optoelectronic and nano-electronic devices based on their thin films and nanostructures [1-3]. Among various nanostructured materials and their thin films, ZnO has drawn maximum attention of the researchers because of its important electrical, optical, magnetic and piezoelectric properties [4]. These properties have made way for many applications ranging from light emitting diodes, UV lasers, gas sensors, transparent field effect transistors, solar cells, piezoelectric nanogenerator and other optical coating applications [5]. Due to the various attractive properties for practical applications of ZnO films, there has been much attention paid on the fabrication of nanocryatalline ZnO thin films in recent years [6]. The ZnO thin film structural and its optical properties strongly depend on growth technique and growth condition.

Many different deposition techniques have already been reported. These include RF sputtering, metal organic chemical vapor deposition (MOCVD), sol–gel, spray pyrolysis, galvanic deposition technique, pulse laser deposition (PLD) and thermal evaporation method [7]. Among several techniques used to obtain high-quality undoped and doped ZnO thin films, the sol–gel processing still offers the possibility of preparing a small as well as large-area coating of ZnO thin films at low cost for technological applications. The main factors affecting the sol–gel film surface morphology and optical properties are dopant concentration, buffer layer on substrate, aging time of sol, the film's thickness and structure, and annealing treatment. Particularly, the annealing treatment is a significant factor to affect the sol–gel process [9]. Therefore, it is still necessary to study the effect of annealing temperature on the structure and optical properties of ZnO thin films. In this work, we deal with structural and optical properties of sol–gel derived nanocrytalline ZnO thin films. As a consequence, the influence of post annealing treatments on the structural surface morphology and optical properties will also be investigated.

2. Experiment

High-purity Zinc acetate dihydrate (Zn $(CH_3COO)_2 \cdot 2H_2O$) was used as starting materials; isopropanol was selected as a solvent, and diethanolamine ($C_4H_{11}NO_2$) was used as stabilizing agent for the fabrication of ZnO thin films by sol-gel technique. The concentration of zinc acetate was chosen to be 0.15mol and the resulting mixture was stirred for 3h at room temperature to yield a clear and homogeneous solution [9] to serve as the coating solution. The coating was usually made 24hr after the solution was prepared. The precursor solution was spin-coated onto n-Si (100) wafer for studying the effect of post annealing treatment on ZnO thin film. Silicon substrates were ultrasonically cleaned with acetone and methanol, then etched for 5 min in 2% HF water solution in order to remove the surface oxide, followed by de-ionized water and dried in vacuum before spin coating. Spin coating was performed at room temperature, with a rate of 1000 rpm for 15s and then of 2000 rpm for another 15s. After depositing by spin coating, the films were dried at 300°C for 10 min on a hotplate to evaporate the solvent and remove organic residuals. This process was repeated for five times to obtained the reasonable thickness of film (~250nm). The films were then inserted into a furnace and annealed in Ar at 450–650°C for 30mintues. The crystal structure and phases of the film were studied by X-ray diffraction (XRD). The surface morphology of ZnO thin films were studied by field emission scanning electron microscope (FSEM). The optical properties of the ZnO thin films were temperature analyzed by the photo luminous spectroscopy (PL). The electrical properties of the ZnO thin film were studied by the four point probe method.

3. Result and discussion

3.1*Crystal structure and surface morphology study*

The XRD patterns of nanocrystalline ZnO thin films deposited on n-Si (100) substrate by spin-coating technique annealed at different temperatures are shown in Figure 1. From Figure 1, (100), (002), (101), (102), (110) and (103) XRD peaks were observed, and it is concluded that all the films were polycrystalline with a hexagonal wurtzite structure and a random orientation, which generally occurs in the growth of ZnO thin films [9]. It is clear that the intensity of (002) peak gradually increases with the increasing annealing temperature from 450°C to 650°C, i.e., the increasing annealing temperature caused the crystallinity of ZnO to enhance. This indicated that the thin films had a high preferential oriented c-axis orientation perpendicular to the substrate surface. The crystallite size of the ZnO films ware calculated by using Scherer's formula [8]

$$D = \frac{.9\lambda}{\beta cos\theta} \tag{1}$$

where, λ is the wavelength of the X-ray, β is the observed angular full-width at half maximum (FWHM) intensity of each diffraction peak located at 2θ position and θ is the Bragg's diffraction angle.

IOP Conf. Series: Materials Science and Engineering **73** (2015) 012060 doi:10.1088/1757-899X/73/1/012060

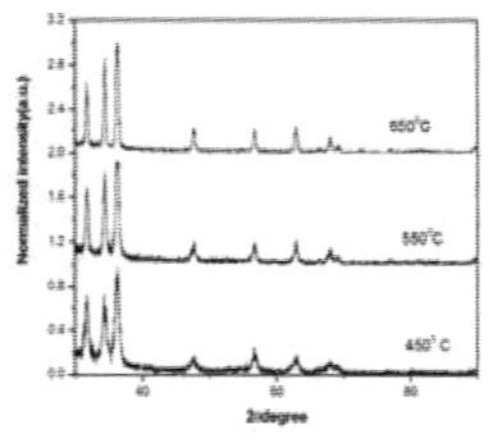

Figure.1: XRD image of ZnO thin films

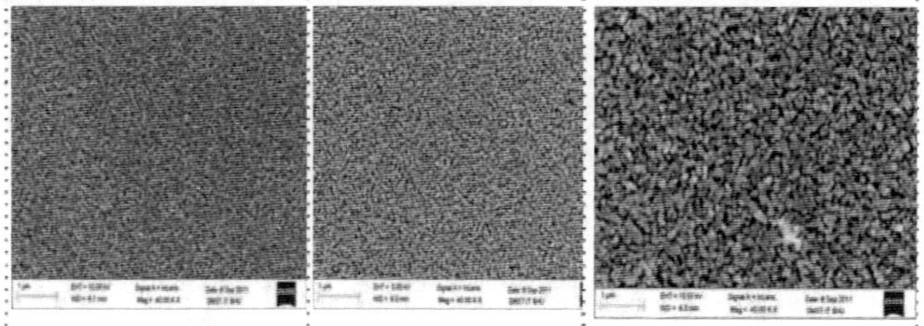

Figure.2: SEM image of ZnO thin films annealed at (a) 450^{0}C, (b) 550^{0}C (c) 650^{0}C temperature

The calculated values of the crystallite sizes ranged between 10 and 20nm. It was observed that crystallite size increased with increasing annealing temperature, which can be understood by considering the merging process induced from post annealing treatment on ZnO thin films. For ZnO nanoparticles, there are many dangling bonds related to the zinc of oxygen defects at the grain boundaries [6]. As a result, these defects are favourable to the merging process to form larger ZnO grains while increasing the annealing temperature

The surface morphology ZnO thin films were studied by field emission scanning electron microscopy (FESEM). Figure 2 shows the FESEM images of ZnO films annealed at various temperatures. The scanning electron microscopy result revealed that uniform ZnO nanoparticles were grown and grain sizes were found to be varied with annealing temperatures in close agreement with the XRD result. The film was granular in nature with good uniformity. Comparing the surface morphologies of ZnO thin films at the annealing temperatures of 450^{0}C, 550°C and 650°C, the grain size of the nanoparticles in the film was found to be increased with the post annealing temperature. In other words, the surface-to-volume ratio was found to be decreased with increased annealing temperature (450 to 650^{0}C). Since, the thin films with a larger surface-to-volume ratio could enhance the surface activity of the film thereby leading to a higher degree of UV absorption in the case of optical detection and gas adsorption in the case of gas sensing, the ZnO films at 450^{0}C has better suitability for both the gas sensing and optoelectronic applications than that of the films annealed at higher temperatures.

3.2. *Optical properties of thin film*

Room-temperature PL spectra of ZnO films annealed at various temperatures are shown in Figure 3. The PL spectra of ZnO thin films exhibit two emissions: a strong UV emission located at ~385nm and a weak broad emission band in the visible range. The UV emissions can be attributed to the near band edge emission and originate from the recombination of free excitons [7]. However, the wider spectrum in the visible band (500 to 650) can be attributed to electron transfer from the conduction band to the defect states in the forbidden band or between different defect levels in the forbidden band. Figure 3 further shows that the UV luminescence is degraded with increased annealing temperature which may be ascribed to excessive oxidized layers formed on the surface and the grain boundary of ZnO thin films after annealing in Ar [5]. Moreover, the UV emission is found to be red-shifted with increased temperature which is likely to be related to Burstein-Moss effect, [8] relaxation of tensile strain [4] and

increase in the grain size. The decrease in optical band gap energy is generally observed in the annealed direct-transition-type semiconductor films. Hong et al (2005) [10] observed an optical band gap shift of ZnO thin films from 3·31–3·26 eV after annealing, and attributed this shift to the increase in the ZnO grain size. Also, the shift of the energy gap was mainly due to both the quantum- size effect and the existence of an amorphous phase in the thin films. In our case, the band gap energy (E_g) value is found to be decreased from 3·205– 3·13eV with increasing annealing temperature from 450–650°C.

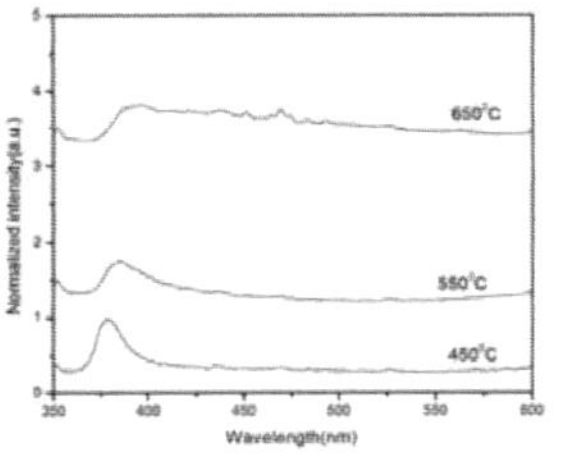

Figure.3 PL spectra of ZnO thin annealed at different temperature.

4. Conclusions

In summary, the effect of post annealing treatment on structural and optical properties of sol–gel derived ZnO films grown on n-Si (100) has been critically examined by XRD, SEM and PL measurements. All the films show polycrystalline hexagonal wurtzite structure ZnO with a high preferential *c*-axis orientation. Morphological and structural studies revealed that the annealing treatment improved crystal quality. The optical band gap energies have been found to be decreased from 3·205eV to 3·13eV as annealing temperature is increased from 450–650°C. The ZnO film at 450°C annealing temperature is observed to be having better suitability for gas sensing and optoelectronic applications than that at other two higher temperatures.

References

[1] Wei L, Xingping P, Xueqin L, Zhiwei H and Front W Y 2007 *Mater. Sci. China* **1(1)** 88-89
[2] Bahadur, Srivastava A K, Rashmi and Chandra S 2008 *IEEE sensors j.* **8** no.6
[3] Hu S Y , Lee Y C, Lee J W , Huang J C, Shen J L, and Water W 2008 *Appl. Surf. Sci.* **254**, 1578-1582
[4] Mandal S, Goswami M L N, Das K, Dhar A, and Ray S K , 2008.*Thin Solid Films* **516**, 8702
[5] Zhao Q X, Klason P , Willander M, Bergman P J , Jiang W L, and Yang J H 2006, *Phys. Scr.* **131**
[6] Teng X, Fan H, Pan S, Ye C and Li 2007 *Mater. Lett.* **61** 201–204
[7] Studenikin S A , Golego N and Cocivera M1998 *J. Appl. Phys.* **84**, 2287
[8] Periasamy C , Prakash R and Chakrabarti P 2010 *J. Mater. Sci. - Mater. Electron.* **21 309**
[9] Ghodsi F E and Absalan H 2010 *Acta Phys. Pol., A* **118**
[10] Kang H S , Kang J S , Kim J W and Lee S Y 2004 *J. Appl. Phys.* **95**, 1246

IOP Conf. Series: Materials Science and Engineering **73** (2015) 012061 doi:10.1088/1757-899X/73/1/012061

Dielectric studies of Fe Doped SmCrO3 Perovskites

Bandi Vittal Prasad[1], B Venugopal Rao[1], K Narsaiah[1], G Narsinga Rao[2], J W Chen[2] and D Suresh Babu[1,*]

[1]**Department of Physics, Osmania University, Hyderabad, India.**

[2]**Department of Physics, National Taiwan University, Taipei, Taiwan, R. O. C.,** [*]

Corresponding author E-mail: s_devarasetty1956@yahoo.co.uk

Abstract. Polycrystalline Fe doped $SmCrO_3$ perovskite ceramics were prepared by sol-gel route and characterized by X-ray diffraction (XRD), Scanning electron microscopy (SEM), dielectric and Impedance measurements. Single-phase perovskite Fe doped $SmCrO_3$ obtained has been confirmed from the XRD pattern with orthorhombic structure. The dielectric properties of perovskite Fe doped $SmCrO_3$ ceramics were studied in the frequency range of 100Hz - 1MHz in the temperature range from 80 K to 300 K. These materials exhibited colossal dielectric constant value of ~ 10^4 at room temperature. The response is similar to that observed for relaxor-ferroelectrics. To get the detailed information, IS data analysis was performed.

1. Introduction

Materials with high dielectric constant and low dielectric loss are of technological importance in the microelectronics industry as they enable device miniaturization. In recent years, materials exhibiting a so called giant dielectric constant have been reported [1-5]. Right from the time this effect was first observed in $CaCu_3Ti_4O12$ (CCTO), attention has been concentrated on materials with perovskite structure (ABO_3), due to the set of specific properties belonged to these materials. Fe doped $SmCrO_3$ (SFCO) materials are semi conductive in nature at room temperature and conductive at high temperatures. In this paper we report details concerning the dielectric response of $SmCr_{1-x}Fe_xO_3$ (x = 0, 0.4, 0.6, 1) (SCFO) as a function of frequency and temperature along with the possible underlying mechanisms.

2. Experimental

Polycrystalline powders of SCFO were prepared by the citric acid route. The powders were annealed at 650 ^oC for 2 hours with intermediate grindings. Circular pellets were pressed at 4000 Kg/cm^2 in uniaxial disc. These pellets were sintered at 1200 ^oC for 4 hours.

Phase purity of sintered pellets was characterized by X-ray diffraction (XRD) using SHIMADZU XRD - 7000 using monochromatic Cu – Kα radiation. Cation ratios were measured using EDAX. Microstructure and the grain size distribution of the sintered pellets were studied using a SEM HITCHI S-3400 N. The surface layer of the sintered pellets was removed by grinding both sides before carrying dielectric measurements. The dielectric constant measurements were carried out in the 100 Hz – 1 MHz frequency range at various temperatures between 80 K to 300 K using HP- 4284A LCR meter and Lakeshore temperature controller with a computer controlled program.

3. Results and Discussions

The XRD patterns obtained at room temperature of the polycrystalline SCFO synthesized by citric acid method are shown in figure 1. It is observed that the XRD spectrum is of a single phase SCFO. All the peaks could be indexed to orthorhombic cell associated with space group Pbnm. The lattice parameters are given in the table 1.

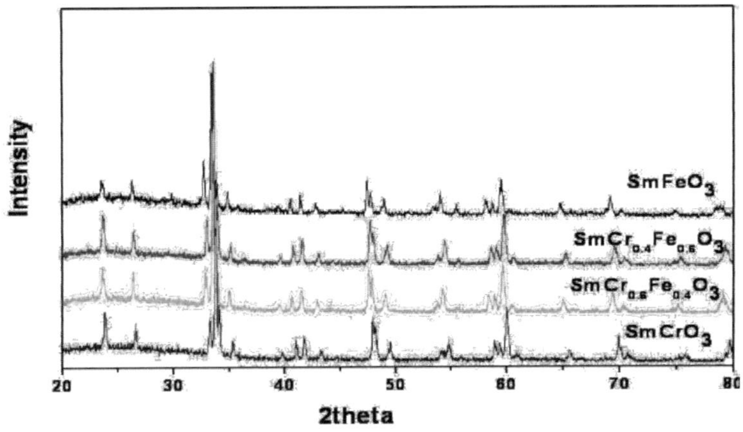

Figure 1. The XRD patterns obtained at room temperature of the polycrystalline $SmCr_{1-x}Fe_xO_3$ (x= 0, 0.4, 0.6, 1) ceramics.

Table 1. Lattice parameters of $SmCr_{1-x}Fe_xO_3$(x= 0, 0.4, 0.6, 1) ceramics.

Sample	a (A^0)	b (A^0)	c (A^0)
$SmCrO_3$	5.373 (3)	5.505(2)	7.655(3)
$SmCr_{0.4}Fe_{0.6}O_3$	5.393(1)	5.544(2)	7.687(2)
$SmCr_{0.6}Fe_{0.4}O_3$	5.404(2)	5.571(3)	7.703(2)
$SmFeO_3$	5.409(2)	5.592 (1)	7.718 (2)

Figure 2. SEM micrograph of the $SmCr_{1-x}Fe_xO_3$ (x= 0, 0.4, 0.6, 1) pellets

The SEM micrograph of the SCFO pellets shown in the figure 2. The microstructure reveals that the pellets contain small grains with considerable porosity and as the Fe content increases the porosity is decreasing and particle size is increasing from 300 nm to 500 nm.The figure 3 and figure 4 presents the temperature dependence of the plots of the complex permittivity (ε') and loss tangent (tanδ = ε''/ε'). It is observed that ε' vs T plots exhibit similar behaviour as that of CCTO, i.e., near room temperature ε'(T) reaches a value as high as 10^4, where the ε'(T) is independent of temperature. With decreasing temperature, ε'(T) displays step like decrease to a lower value of about 7. Below 100K, ε'(T) is almost frequency and temperature independent. The step like increase in ε'(T) consists of two distinct transitions corresponding to two respective peaks in tanδ (T), shifting to high temperature as the measuring frequency increases. This indicates that there exist two sets of thermally activated relaxations in SCFO ceramic sample. It can also be seen that there is an exponential-like increasing background in tan δ curve. It is important to note that the exponential increasing background in loss tangent is associated with the hopping conductivity. This strongly suggests that the observed relaxations might be related to hopping between spatially fluctuating lattice potentials not only produce the conductivity but also give rise to dipolar effects [6].

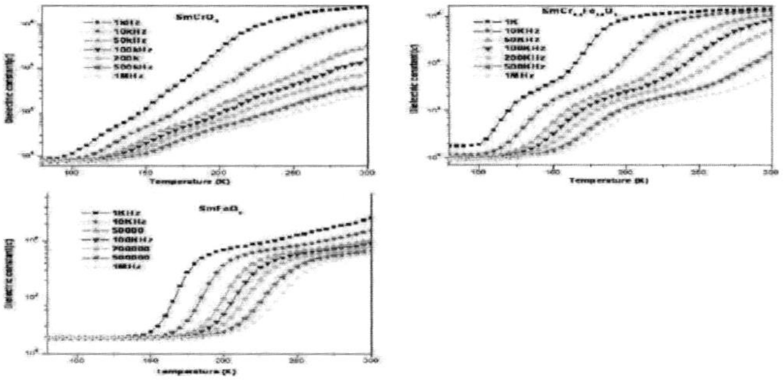

Figure 3. Temperature dependence of the permittivity (ε') of the $SmCr_{1-x}Fe_xO_3$ (x= 0, 0.4, 0.6, 1) ceramics

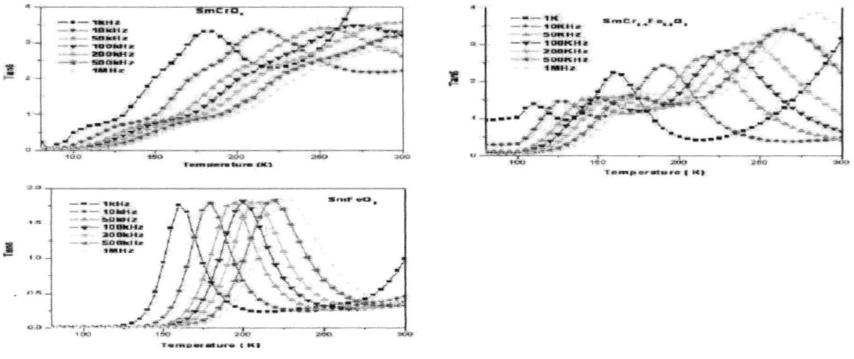

Figure 4. Temperature dependence of the loss tangent (tan δ) of the $SmCr_{1-x}Fe_xO_3$ (x=0,0.4,0.6,1.0) ceramics.

The above discussion suggests that the dielectric behaviour of SCFO ceramics is associated with the carrier polarization. Therefore the dipolar effects of hopping carriers in the grain, IBLC effects related to grain boundary and interfacial effect of electrodes are expected to be responsible for the relaxations observed in figure 4. To clarify this speculation we performed impedance analysis which enables us to distinguish different dielectric relaxations from intragrains, intergrains and electrodes of ceramic material [7-8]. The figure 5 shows an alternative presentation of Z' vs Z''/f at 160 K (5a) and 260 K (5b) as proposed by Abrantes et al [9]. Three well defined regions at 260 K are clearly seen in figure 6. Two regions at 160 K are observed in figure 5. According to Abrantes et al [9] these sequential regions, at 260 K, from low to high frequencies correspond, to the dielectric response from electrodes, grain boundaries and bulk grains respectively, and at 160 K grain boundaries and grain and the electrode effects are not

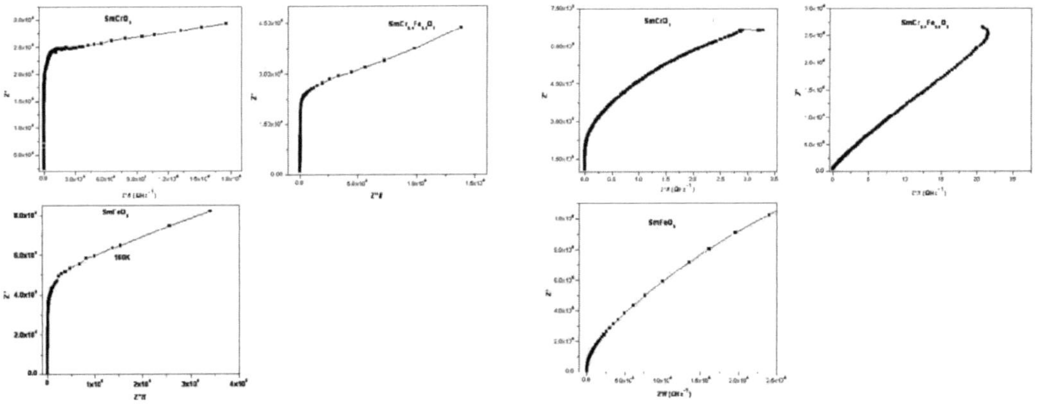

Figure 5. Representation of Z' vs Z''/f at 160 K (5a) and 260 K (5b) for the $SmCr_{1-x}Fe_xO_3$ (x= 0, 0.4, 0.6, 1) ceramics.

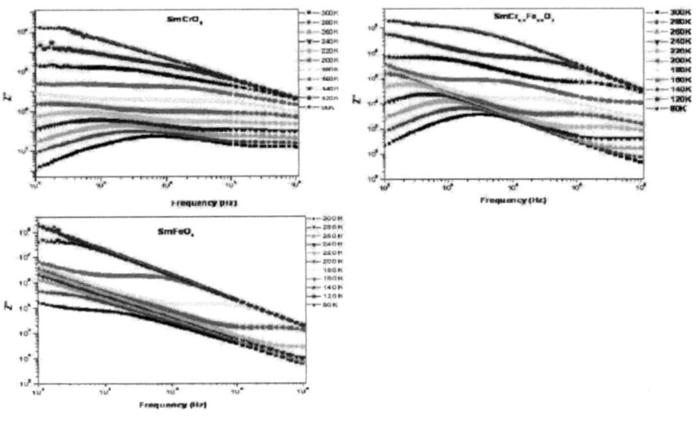

Figure 6. Plot between Z'' vs Frequency at different temperatures for $SmCr_{1-x}Fe_xO_3$ (x= 0, 0.4, 0.6, 1) ceramics.

seen. The figure 6 shows Z'' vs frequency plots for different temperatures. Inspection of these plots over the measured temperature range revealed the presence of a peak which is shifting from low to high frequency with increase in temperature. For temperatures above 200K the peak is moving out of measuring window is corresponds to bulk (grain) and at low frequencies for temperatures above 180K we can observe another peak which is shifting from low to high frequencies is correspond to grain boundary.

4. Conclusion

Ceramic samples of $SmCr_{1-x}Fe_xO_3$ (x= 0, 0.4, 0.6, 1) ceramics characterized using XRD and SEM. With increase in Fe content porosity is decreasing. The effective dielectric constant obtained in frequency range 100Hz to 1MHz exhibited giant dielectric constant value ~10^4 at room temperature. dielectric constant ε' (T) features a step like transition at temperatures ranging from 100 K (< 100 kHz) to 280 K (1MHz). This step like transition was linked with thermally activated dielectric relaxations. With increasing Fe content grain boundary response is decreasing. The observed temperature and frequency dependent high dielectric constant in semiconducting SCFO ceramics is dominated by gain boundary response.

References

[1] M.A.Subramanian, D.Li,N.Duan, B.A.Reisner, and A. W. Sleight, 2000 J.Solid State Chem. 151 323.
[2] T. B. Adams, D. C. Sinclair, and A. R. West, 2002 *Adv. Mater.* (Weinheim, Ger.) 14, 1321.
[3] Bandi Vittal Prasad et al 2011 *Mater. Chem. Phy.* 126, 918.
[4] Bandi Vittal Prasad, et al 2011 *Mater. Res. Bull.* 46, 1670.
[5] Bandi Vittal Prasad, et al 2012 *solid state sci.* 14, 225.
[6] A. K. Jonscher, 1983 *Dielectric Relaxation in Solids*, (Chelsea Dielectrics London).
[7] D. C. Sinclair and A. R. West 1989 *J. Appl. Phys.* 66, 3850 J. Mater. Sci. 29, (1994) 6061.
[8] J. R. Macdonald 1987 *Impedance Spectroscopy* (Wiley New York)
[9] João C. C. Abrantes, João A. Labrincha, and Jorge R. Frade, 2000 *Mater. Res. Bull.* 35 727.

Structural and Optical properties of Brominated Cu thin films

Desapogu Rajesh, C S Sunandana and M Ghanashyam Krishna
School of Physics, University of Hyderabad – 500046, A.P; INDIA

rajesh.desapogu@gmail.com; sunandana@gmail.com; mgksp@uohyd.ernet.in

Abstract. We describe and discuss XRD, FESEM - EDS and optical absorption behavior of nanostructured 50 nm Cu thin films obtained by thermal evaporation and subsequently exposed to bromine vapours. XRD patterns show quasi-amorphous structure due to the nanocrystalline nature of cupper particles. As bromination time increases there is a subtle broadening of the peak that hints at a possible increase in quantities such as particle size. At ambient temperature, bromination time could be used to tune surface morphology, and optical properties of nanostructured Cu thin films. During the short term bromination we observed small Cu Nanoparticles with large density, which exhibit surface plasmon resonance (SPR). Increase of bromination time yields larger Cu Nanoparticles with small density of particles, the particle size and elemental composition assessed by FESEM-EDS. These results offer a promising strategy for situ characterization and monitoring of nanopaticl and for the engineering of Cu nanoparticle with new plasmonic properties even as Cu and brominated Cu films could be promising candidates for plasmonic and nanophotonic applications.

1. Introduction

As micro-electronic devices get smaller and faster, copper interconnects, with their higher conductivity, have replaced other metals such as aluminum to enhance the performance of the devices. Recently, therefore, many researchers have begun to investigate the properties of copper especially in the form of thin films. Copper is considered as a promising material for fabrication of interconnections in integrated circuits based on planar thin films as an alternative to aluminum. Interfacial interactions underpin a range of fundamental phenomena and applications in chemical, biological, and material sciences consequently; significant attention has been directed toward strategies to precisely tune the surface interactions of materials [1]. The reasons are mainly the excellent physical properties of copper and its straightforward and quite cheap deposition using the thermal evaporation process. Film thickness, deposition rate, and method of coating play important role on microstructure as well as macroscopic properties of films [2, 3]. The bromination of Cu thin films opens up the possibility of modifying their optical, structural properties and also strong absorption in the visible to near infrared region thus making them not only feasible for gas sensors, but also as promising organic electro photographic materials, nonlinear optical materials and optical recording media [4–6]. Chemical SPR sensors are based on the measurements of the variations due to adsorption or a chemical reaction of an analyte with the active layer which results in changes of its optical properties. In this paper we reported our work on preparing Cu thin films via thermal evaporation method to study the influence of bromination on the optical, structural properties of Cu thin films. Structure is the most important factor to determine the morphological characteristics of a material. Microstructure properties and grain size were also characterized by using FESEM.

2. Experimental part

The Cu powder purchased from Sigma Aldrich was used as obtained without further purification. Prior to the deposition of Cu thin films, the substrates were thoroughly cleaned by immersing them in soup solution for 15 min and subsequently putting them in the solution of acetone and ethanol sequentially. The pressure during deposition process was upheld at $4.5*10^{-6}$ m bar. The Cu films were sublimed on carefully cleaned glass substrates at room temperature. The thickness of the deposited Cu films and the rate of deposition were maintained at 60nm and 0.01 nm/sec respectively, using DTM quartz-crystal thickness monitor. Thin films of Cu were formed on the cleaned glass substrates by thermal evaporation techniques. The Cu coated films are selected for further bromination treatment and are referred as pristine film. The pristine samples were taken as a reference in comparative experiments. The structural properties of thin films were investigated by X-ray diffraction (XRD) using CuKα (λ = 0.15418 nm) radiation. The optical absorption of the films was measured using JASCO model 7800 spectrophotometer. The Field emission scanning electron microscopic studies were carried out using ZEISS model ULTR 55. Elemental analysis was carried out with an EDAX using energy dispersive X-ray spectrophotometer (EDS).

3. Results and discussion

3.1 XRD

Representative XRD patterns of as-deposited Cu thin films are shown in Fig. 1. It is apparent that thin films in the thickness 50 nm are by and large amorphous-possessing only short range order which is in keeping with the islanded nanostructured quasi-continuous films XRD patterns show quasi-amorphous structure due to the nanocrystalline nature of cupper particles. As bromination time increases there is a subtle broadening of the peak that hints at a possible increase in quantities such as particle size which could be conveniently probed by FESEM to be described later.

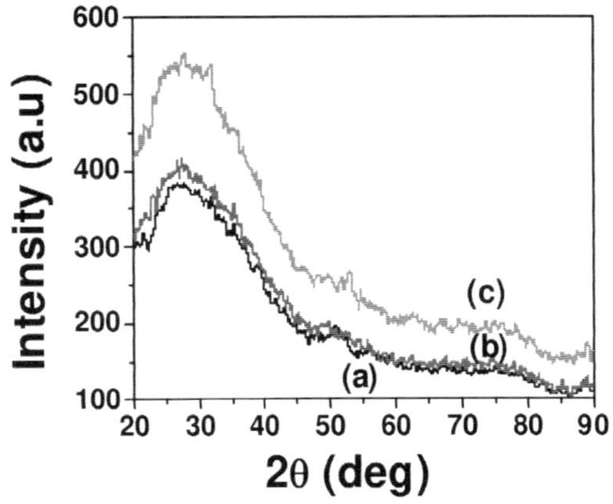

Figure 1. XRD Pattern of (a) Pure Ag thin film (b) 5min bromination (c) 20min bromination

3.2 Optical properties

The optical properties of the copper thin films were measured immediately after the preparation process, and then followed as a function of the exposure time to bromine vapors at room temperature. Figure 2 shows the optical absorption spectra of the films at the different times that elapsed after the reduction process as compared with the spectrum of the film in bromine vapors at room temperature. It was found that the optical density of these samples increase with the increase in bromination time (<5min). The absorption peak centered at about 780 nm is due to SPR absorption of the metallic copper nanoparticles embedded in the glass substrate. After 20, 60min bromination, the optical density spectrum of the reduced film showed stabilized behaviour with no SPR peak similar to the spectrum of the Cu film at 300 ◦C. The optical absorption of the Cu thin films exposed to Br for 20, 60 min did not present any significant absorption feature. We observed that, at ambient temperature, leveraging bromination time it is possible to tune the surface morphology and optical properties of nanostructured Cu thin films.

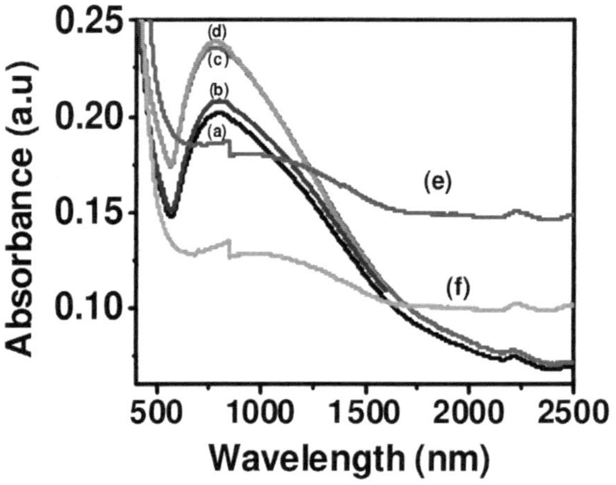

Figure 2. Absorption spectrum of Cu thin films at different bromination times (a) pure Cu film (b) 1min bromine (c) 3min bromination (d) 5min bromination (e) 20min bromination (f) 60min bromination

3.3 FESEM

Surface morphology of the Cu thin films containing copper nanoparticles was studied by FESEM, as presented in fig. 3 for the Cu thin films brominated at room temperature and as compared to the pure Cu films. In fig3.a the growing accumulated Cu Nanoparticles are observed at before bromination, after 5min bromination some quite uniform nano-sized particle features are clearly observed in the fig3.b, the size of the surface particles was in the range of ~10–40 nm (with average size of about 20 nm). Based on the absorption, morphology analysis it was shown that the copper nanoparticles initially accumulated on surface of the films, when we bombard with bromine vapors it could substantially penetrate into the film after 5, 20min bromination we got the different morphologies shown in the fig3.b, fig3.c. Especially in 20min bromination case we got the Cu nano rods, this is we found new observation in thermal evaporation and followed by bromination method. Hence, the particle features observed on the film surfaces could be attributed to the copper nanoparticles, nano rods. Fig3.d shows the elemental analysis of brominated Cu thin films.

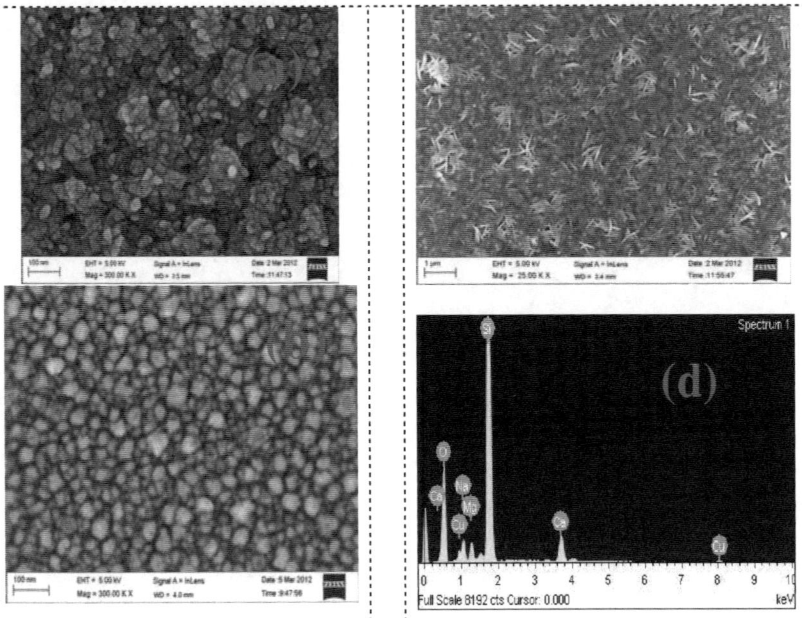

Figure3. FESEM images of (a) pure Cu film (b) 5min bromination (C) 20min bromination (d) EDS spectrum of 5min brominated Cu thin films

4. Conclusions

We describe and discuss XRD, FESEM - EDS and optical absorption behavior of nanostructured 50 nm Cu thin films obtained by thermal evaporation. XRD patterns show quasi-amorphous structure due to the nanocrystalline nature of cupper particles. As bromination time increases there is a subtle broadening of the peak that hints at a possible increase in quantities such as particle size. These results are useful indications for in situ characterization and monitoring of nanopaticle preparation and for the engineering of Cu nanoparticle with new plasmonic properties. These Cu and Cu brominated films could be promising candidates for plasmonic and nanophotonic applications.

Acknowledgement

D Rajesh thanks the University of Hyderabad for providing UGC-BSR fellowship.

References

[1] Chilkoti A and Hubbell J A, 2005 *MRS Bull* **30** 175

[2] Raymond E K, Donald F O, Martin G and David E 1979 *Kirk- Othmer Encyclopedia of chemical technology* third edition **6** 826

[3] Kundu S 1998 *J. Phys. D. Appl. Phys.* **31** L73-L77.

[4] Capone S, Rella R, Siciliano P, Valli L and Troisi L 1998 *Thin Solid Films* **465** 327

[5] Chen Q, Gu D, Shu J, Tang X and Gan 1994 *J. Mater. Sci. Eng.* B **25** 171

[6] Qian S and Wang G, 2001 *Nonlinear Optics – Principles and Developments*, Fudan Univ. Press, Shanghai
[7] Rajesh D and Sunandana C S 2012 *Results in Physics* **2** 22

IOP Conf. Series: Materials Science and Engineering **73** (2015) 012063 doi:10.1088/1757-899X/73/1/012063

Ionic conductivity and electrical relaxation study of polycrystalline $LiNaSO_4$

K N Ganesha and G Govindaraj*
Department of Physics, School of Physical, Chemical and Applied Science, Pondicherry University, Puducherry 605 014, India

*ggraj_7@yahoo.com

Abstract. The temperature dependence of ion conduction and electrical relaxation mechanism of $LiNaSO_4$ has been carried out. DSC curve shows phase transition peak at 520.2 ^{o}C. The temperature dependence relaxation time and dc conductivity follows the Arrhenius relation. We have obtained same activation energy for the hoping and the conduction within the error limit. The activation energy is 1.30±0.04eV in the temperature range between 493 K and 733 K. The dc conductivity varied from 7.30×10^{-10}S/cm at 493K to 3.96×10^{-4}S/cm at 793K. The sample becomes fast ionic conductor at high temperature phase.

1. Introduction

In search for good solid electrolyte materials for applications in high temperature solid state batteries, solid oxide fuel cells and sensors, sulfate based alkali cation systems are vital for achieving a good ionic conductivity during solid-solid phase transition [1]. $LiNaSO_4$ is well known as a high temperature rotator phase and super ionic conductor. The electrical properties were studied by many researchers and the mechanism of conduction has been a subject of discussion [2]. In this paper, we reported the ionic conductivity and electrical relaxation mechanism in β-phase. At high temperature phase, a dramatic increase of conductivity is found. This was due to the combination of "percolation" and "paddle-wheel" mechanism. 'Paddle wheel' model describes the transport of cations is strongly correlated to the rotation of the SO_4 tetrahedra. 'Percolation' model, suggests that there is no direct correlation between the ionic conduction and SO_4, rotations, but that the rotations and vibrations indirectly produce conduction pathways [3].

2. Experimental

$LiNaSO_4$ is synthesized by solid state reaction technique by heating the equimolar amount of $Li_2SO_4.H_2O$ and Na_2SO_4 at 500 ^{0}C. Powder XRD pattern was recorded with 2θ in the range 10° -80° with a 0.01° step size and step scan of 0.5 s at room temperature. Thermal analysis was performed in the temperature range of 40 ^{o}C to 1000 ^{o}C with a heating rate of 10 ^{o}C/min. The mass of the used sample is 17.18 mg. A pellet with diameter 11.40 mm and 1.35 mm thickness was used for impedance measurement. The impedance measurements were carried out using Novacontrol broadband dielectric spectrometer (BDS) in the frequency range 100Hz to 20MHz, with an applied potential of 1 V, between the temperatures ranges 493 K to 593 K under 1 atmospheric pressure.

3. Results and discussion

3.1. Thermal analysis

Thermo-gravimetric curve Figure 1, shows there is no weight loss occurs over the temperature range 30 °C to 1000 °C. The endothermic peak at 793.2K of DSC curve belongs to structural phase transition. The calculated enthalpy change was 89.87J/g (12.26KJ/mol) and the corresponding entropy change was 15.40J/K mol. The value of entropy change is almost equal to Rln6 (=15.40 J/Kmol). Hence it is reasonable to say that the disorder was increased by 6 times at the high temperature phase transition. The second peak at 612 °C belongs to melting of the sample.

3.2 X-ray powder diffraction analysis

The XRD spectrum of $LiNaSO_4$ material sintered at 400 °C as shown in Figure 2. The spectra match well with JCPDS data reference number 01-071-2172. The unit cell volume and lattice parameters are calculated by Rietveld profile fit using FULLPROF software. XRD spectrum is fitted with hexagonal crystal system with P6/mmm space group and the lattice parameter a=7.6249±0.0011 Å, c=9.8545 ±0.0024 Å and the volume v=496.17×10^6 pm^3. The reliable factors χ^2 is 4.17 and Bragg R-factor is 0.15.

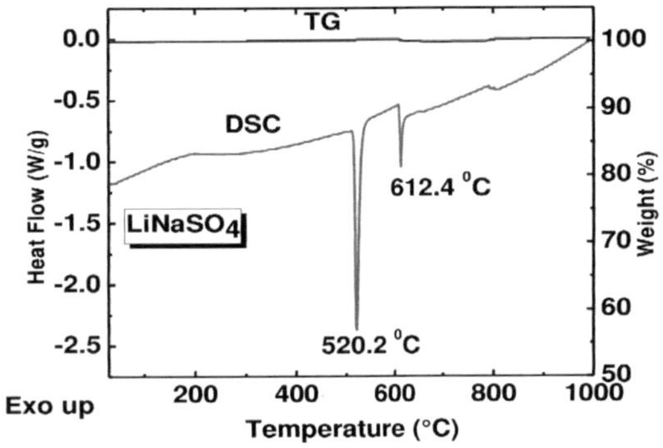

Figure 1. DSC heat flow and TG weight change

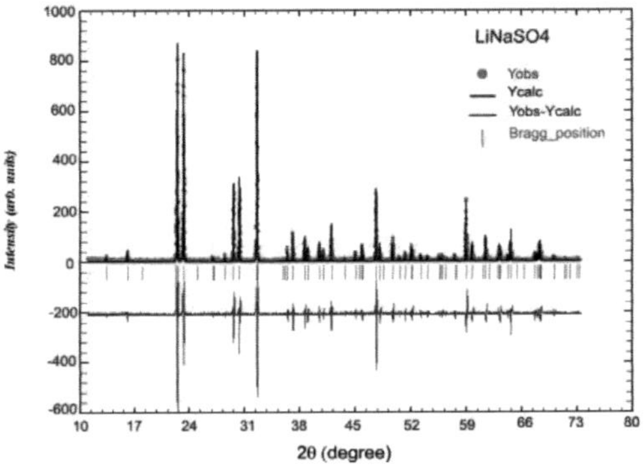

Figure 2. Rietveld fit of XRD spectra

3.3. Impedance spectroscopy study

Figure 3 (a)-(d) shows the complex impedance plane plot of the material from temperature 513 K to 733 K. A depressed semicircle was found for each temperature at higher frequency range and a tail is observed at lower frequency end from 633 K to 713 K. Cole-Cole function form of complex impedance is used to describe the depressed semicircle in complex impedance plane and it is given by:

$$Z^*(\omega)=R_b/(1+(i\omega\tau)^{1-\alpha}) \quad (1)$$

where R_b is the bulk resistance, τ is the mean relaxation time and the parameter α is $0\leq\alpha<1$. The bulk resistance, relaxation time and α were extracted by fitting the above equation in the complex plane using non-linear least square fit software Winfit. At high frequencies ion migration process connected to the sample and at low frequencies the electrode polarization occurs due to space charge accumulation between the electrode and sample. This happens due to the multiple point contacts result to spreading the resistance and the air gap capacitance at electrodes [4]. The bulk resistance is decreases with increasing the temperature. The temperature dependence of relaxation time follows the Arrhenius equation as:

$$\tau=\tau_0 \exp(E_z/k_BT) \quad (2)$$

where τ_0 is the pre-exponential factor, E_z is the activation energy from impedance, k_B is Boltzmann's constant and T is the absolute temperature. The activation energy was calculated from the slope of the straight line plot of $\log(1/\tau)$ versus reciprocal temperature 1000/T as shown in Figure 4. The activation energy obtained for β-$LiNaSO_4$ is 1.30±0.04 eV.

IOP Conf. Series: Materials Science and Engineering **73** (2015) 012063 doi:10.1088/1757-899X/73/1/012063

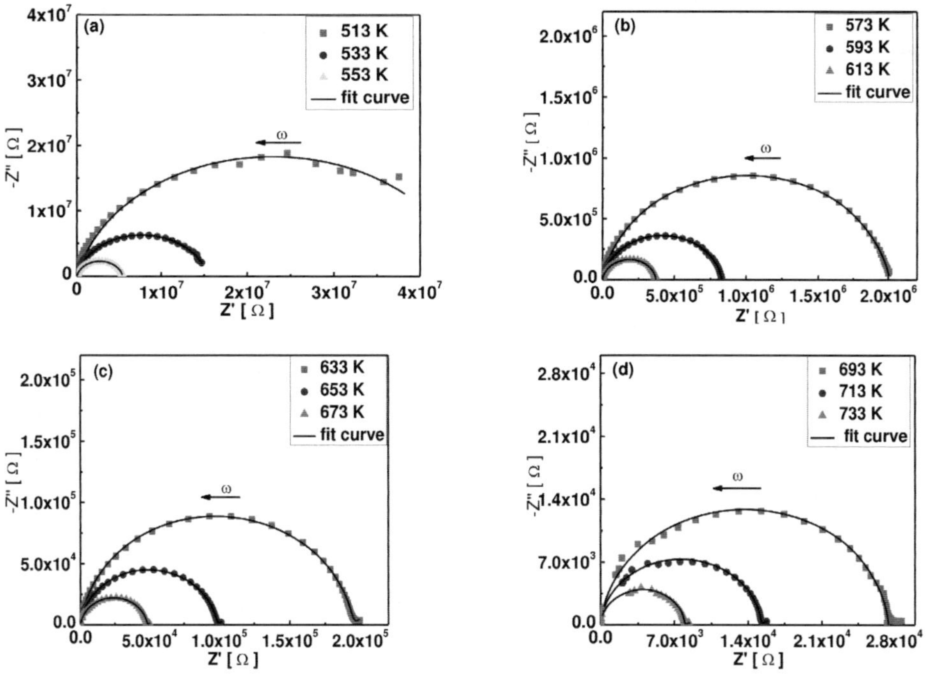

Figure 3 (a)-(d). Cole-Cole fit complex impedance spectra at different temperatures

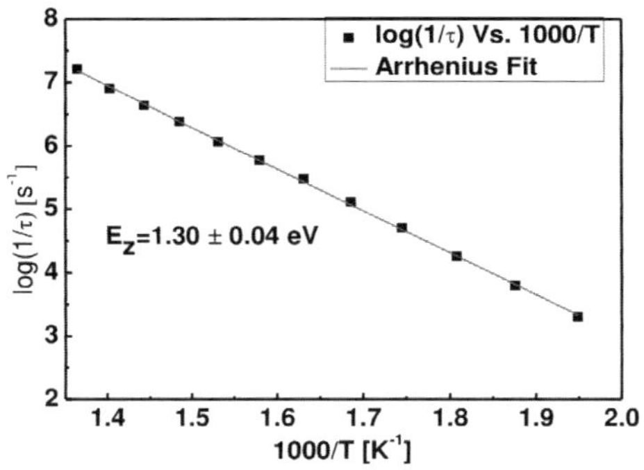

Figure 4. Arrhenius plot of relaxation time

3.4. Conductivity

The conductivity study gives the information of dynamic behaviour of ion conducting materials. Almond and West proposed the following power law to describe the total conductivity:

$$\sigma'(f)=\sigma_{dc}(1+(f/f_H)^n) \quad (3)$$

where σ_{dc} is the dc conductivity, f_H is hoping frequency for conduction and n is the frequency exponent. These parameters are extracted by fitting the conductivity spectra to Equation (3). The frequency dependence of real part of conductivity at temperature range 493 K to 793 K as shown in the Figure 5. From the conductivity spectra we observe, $\sigma'(\omega)$ is independent of frequency at lower frequencies and this gives the dc conductivity. At high frequencies, the conductivity becomes strongly frequency dependent, varying approximately as a fractional power of frequency. The transition point between the frequency independent to frequency dependent region is the signature of onset of conductivity relaxation, which shifts towards high frequency as the temperature increases [4]. The dc conductivity is found to be increases with increasing temperature. This is due to the increase in the thermally activated drift mobility of the ions according to hopping conduction mechanism. The temperature dependence of dc conductivity is well described by Arrhenius law:

$$\sigma_{dc}T=\sigma_0\exp(-E_\sigma/k_BT) \quad (4)$$

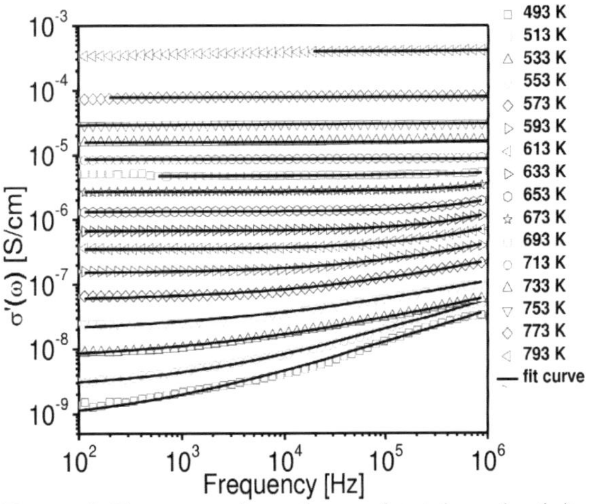

Figure 5. Frequency dependence of total conductivity at different temperature

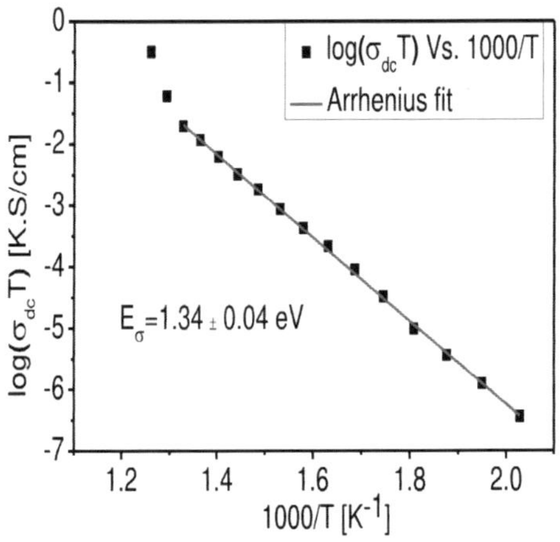

Figure 6. Arrhenius plot of dc conductivity

reflecting the activated nature of ionic hopping processes, where σ_0 is the pre-exponential factor, T is the temperature in K, E_σ is the activation energy and k_B is the Boltzmann's constant. Figure 6 shows the Arrhenius plot of the sample. The activation energy for the conduction process was calculated from the slope of the straight line plot of $\log(\sigma_{dc}T)$ versus reciprocal of temperature. The activation energy obtained for β-$LiNaSO_4$ is 1.32±0.04 eV. At the phase transition region there is a large increase of dc conductivity was found. The dc conductivity obtained at temperature 793 K is 3.96×10^{-4} S/cm.

4. Conclusions

The impedance plots shows non-Debye behaviour and is asymmetric with respect to the peak maximum which indicates the distribution of relaxation time in the conduction process. The activation energy for hoping is obtained from Arrhenius plot and it is found to be 1.30 ± 0.04 eV for the β-$LiNaSO_4$. This is less than the value reported for the $LiNaSO_4$ crystals by Mellander *et al.* (2.15 eV) [5], Kanashiro *et al.* (1.59 eV) [6], Chen *et al.*, (1.63 eV) [2] and comparable to the value obtained Wilmer *et al.*(1.34 eV) [7].

Acknowledgement

Authors would like to acknowledge the UGC-SAP-F.530/15/DRS/2009 for the financial support and Central Instrumentation Facility, Pondicherry University for thermal analysis and BDS facilities.

References

[1] Robertson A D, West A R and Ritchie A G 1997 *Solid State Ionics* **104** 1
[2] Chen R H, Chaw-Ming Tseng, Shern C S and Fukami T 2010 *Solid State Ionics* **181** 877
[3] Karlsson L and McGreevy R L 1995 *Solid State Ionic* **76** 301
[4] Rathan S V and Govindaraj G 2010 *Solid State Sciences* **12** 730
[5] Mellander B E, Graneli B and Roos J 1990 *Solid State Ionics* **40/41** 162
[6] Kanashiro T, Yamanishi T, Kishimoto Y, Ohno T, Michihiro Y and Nobugai K 1994 *J. Phys. Soc.* **63** 3488
[7] Wilmer D, Feldmann H and Lechner R E 2004 *Z. Phys. Chem.* **218** 1439

IOP Conf. Series: Materials Science and Engineering **73** (2015) 012064 doi:10.1088/1757-899X/73/1/012064

Making ceramic- metal composite material by friction stir processing

M Govindaraju[1*], K Balasubramanian[1], Uday Chakkingal[2], K Prasad Rao[2]
[1]Non Ferrous materials Technology Development Centre (NFTDC) Kanchanbagh, Hyderabad – 500 058, India.
[2]Department of Metallurgical and Materials Engineering, IIT Madras, Chennai- 600036, India.

E mail:govindaraju.rajusmg@gmail.com, Ph: 91-9440727051, Fax: 91-40-24342567

Abstract: An innovative method to add ceramic particles in the metal matrix to make ceramic metal matrix composite was experimented and proved with alumina powder as particles and AE42 magnesium alloy as matrix. The alloy was subjected to friction stir processing and alumina particles were added through the processing tool. AE42 magnesium alloy has primary α-Mg phase of 100-150 micron grain size and secondary phase of 10-50 micron size as precipitates. Al_2RE, $Al_{11}RE_3$ and $Al_{17}Mg_{12}$ are main secondary phases in the form of precipitates. Alumina powder was selected with average particles size of 5 micron. For processing parameters of 300-400 rpm tool speed, 15-20 mm/minute traverse speed and a threaded pin geometry; composites with fine distribution of second phase precipitates and alumina particles in the matrix were observed. Mechanical and microstructural characterization revealed uniform properties in longitudinal and transverse directions. Composite material has superior mechanical properties than the magnesium alloy. Distribution of particles was up to the length of tool pin. Tool pin geometry, feed rate and volume percentage of alumina particles, processing speed and tool rpm on the effect of mechanical and micro-structural properties were analyzed in detail.

1. Introduction:

Friction Stir Processing (FSP) is one of the advanced material processing methods based on friction stir welding. It was invented and patented by TWI. – The Welding Institute [1]. This process is used for changing and improving the cast microstructures of metallic materials and many studies were carried out in last 20 years [2]. In FSP, a rotating tool made up of comparatively stronger metal is plunged inside the job and traversed to make processed zone. Many researchers worked on this topic and optimized the process parameters for various materials. Also, literatures are available for FSP of composites [3]. But very few attempts were made to use FSP itself for making composites. These attempts resulted in success for making composites up to a small thickness [3].

[1]M Govindaraju, Sr Engineer, Non Ferrous materials Technology Development Centre (NFTDC) Kanchanbagh, Hyderabad – 500 058, India, govindaraju.rajusmg@gmail.com

For making Metallic Matrix Composites (MMC), various methods including powder sintering, melt addition (squeeze casting, spray deposition, stir casting), diffusion bonding, electroplating are followed [4]. But the limitation of each method results in innovating new methods. All these conventional methods lag the possibility of applying it in-situ or making composites by secondary processes.

For an innovative idea, FSP was experimented to make MMC with alumina as particles, AE42 magnesium alloy as matrix. As this method is expected to make improved surface properties, functionally graded composite and metallic materials, this study was conceptualised and carried out. Due to its light weight, alloys and composites based on magnesium are of special interest for much application including aerospace [5]. But magnesium alloys suffer from brittleness and low strength [6]. Magnesium alloy composites improve them; FSP improves further. So, in the study both were combined. When the composite making by this method is successful, it can be used in-situ composite making, surface modification, and to make functionally graded materials etc. AE42 magnesium alloy was selected for the study as it was specially developed for automotive applications, with better creep resistance.

2. Experimental Details:

Experimental set up for making magnesium-alumina MMC using FSP is shown in figure 1. For powder insertion, a hole was made in the FSP tool and it was inserted by a pressure delivery method. This is shown in figure 2. After powder (alumina particles) filling, the tool is fitted and made to do processing as in conventional FSP. The spring presses and delivers the powder which results in a stir zone (of magnesium) with alumina particles. Alumina particles used in the experiment are of 5-6 micron size. Percentage of alumina in the matrix is controlled by the hole diameter of tool. For comparative purpose, processing of magnesium alloys were carried out with and without alumina (with alumina is composite).

AE42 alloy with 3 mm thickness was used for FSP experiments. Properties and composition of the alloy used are shown in Table1. Experimental parameters are given in Table 2. After the experiments, microstructural analysis by Tescon Scanning Electron Microscope (SEM), hardness testing by Shimadzu Vickers Hardness machine with load of 2 kg, tensile testing by Shimadzu UTM as per ASTM E8 at cross head speed 2 mm per minute were carried out. Samples were prepared for all these analyses as per standards from the processed zone by cutting.

Table 1. Properties of magnesium alloy AE42

Property	Y.S *(MPa)*	UTS *(MPa)*	Elongation %	Hardness *(VHN)*	Composition *(Wt %)*	Form
Value	130	225	6	72	Al :3.9, Ce :1.2, Mn :0.3, La :0.6, Nd :0.4, Pr :0.1, Th :0.2, Mg :Balance	Die cast

Table 2. Experimental parameters

Parameter	Tool speed *(rpm)*	Traverse speed *(mm/minute)*	Tool	Alumina powder feed (Volume)
Value	300-600	15-40	3 mm pin length and diameter	500 mm^3

Figure 1. Experimental set up for FSP

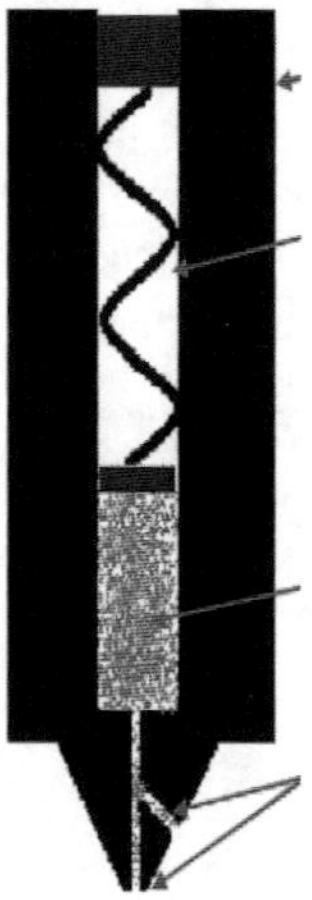

Figure 2. Tool set up for powder delivery

3. Results and Discussions:

3.1. Tensile Testing:

Mechanical properties of magnesium alloy improved by FSP, and further improved for composites. Processed zone with alumina (composite) had 15% higher strength than the processed alloy. Notably elongation of composite was in comparison with processed alloy (without alumina). Table 3 shows the tensile and hardness results of composite achieved for processing parameters of 400 rpm tool speed, 20 mm/minute traverse speed; the strength value was higher for 20 mm/minute traverse speed, which may selected as optimum.

3.2. Hardness:

Hardness of the composite increased to 93 VHN from 72-74 of base metal (magnesium alloy). Increase in hardness may be attributed to refined microstructure and tiny and uniform distribution of second phase particles, and the alumina particles that were highly oriented along with matrix [7]. It is to be noted that the hardness value of processed zone without alumina particles was lower than 93.

Table 3. Tensile Strength and Hardness results

Material condition	YS *(MPa)*	UTS *(MPa)*	Elongation %	Hardness *(VHN)*
Processed zone without alumina (300 rpm, 20 mm/minute)	101	245	11.4	83
Processed zone with alumina (MMC) (300 rpm, 20 mm/minute)	115	281	10.9	93
Processed zone without alumina (400 rpm, 20 mm/minute)	104	255	11	82
Processed zone with alumina (MMC) (400 rpm, 20 mm/minute)	122	291	10.2	90
Processed zone without alumina (600 rpm, 20 mm/minute)	103	251	10.5	80
Processed zone with alumina (MMC) (600 rpm, 20 mm/minute)	120	286	9.9	88

IOP Conf. Series: Materials Science and Engineering **73** (2015) 012064 doi:10.1088/1757-899X/73/1/012064

3.3. Microstructure:

Microstructural results are shown in figure 3 and 4. From the microstructures it is evident that the grain size of the processed zone was refined to 4-5 micron, second phase particles along with alumina particles were distributed evenly to the matrix. Flow of the alumina was as per the tool flow direction. Energy Dispersive X-ray Spectroscopy (EDS) from figure 4 confirmed the presence of alumina particles. Secondary phases and alumina particles were in the range of 1-10 micron in the matrix.

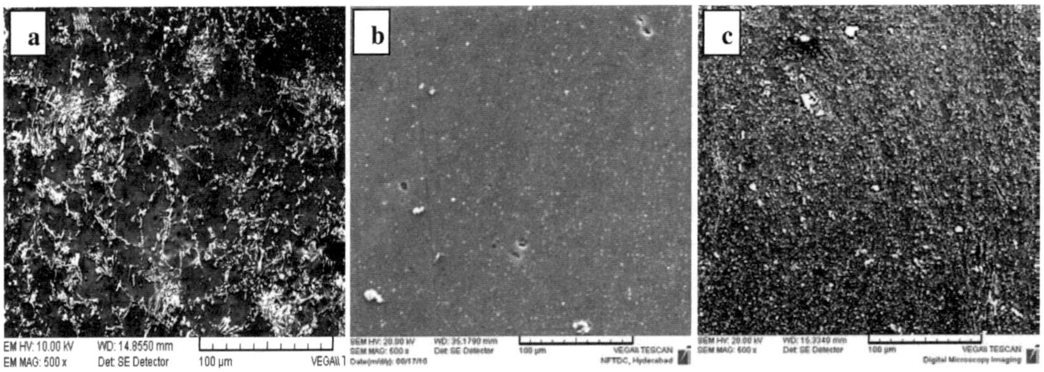

Figure.3. SE Microstructures of (a) PM, (b) processed zone (without alumina) and (a) MMC

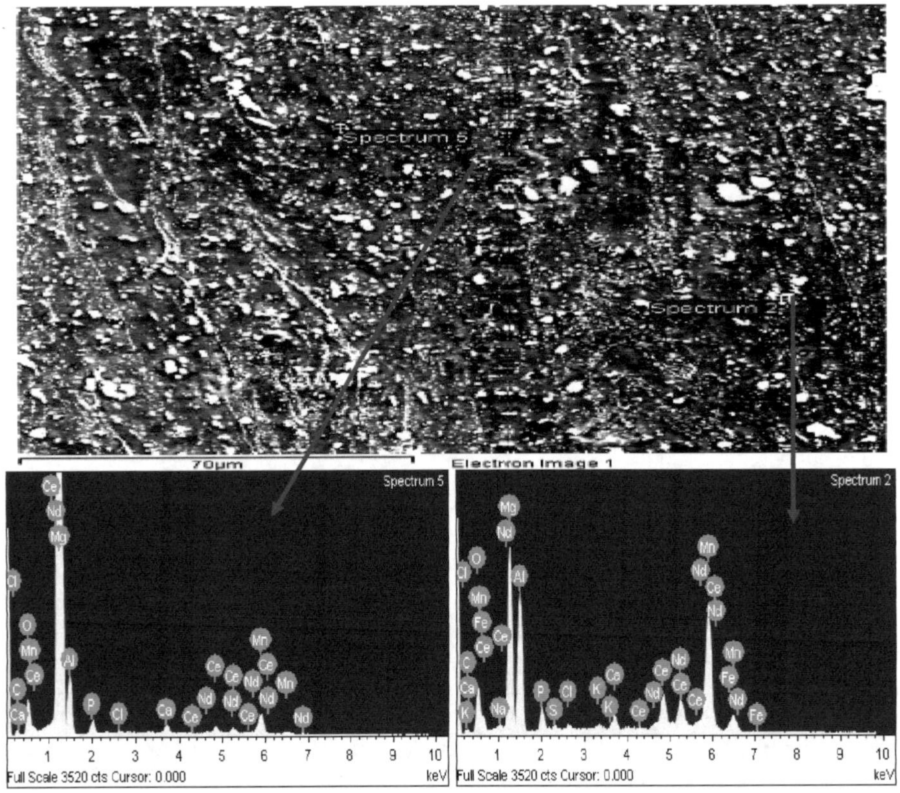

Figure 4. EDS results of MMC, spectrum 2 defines alumina particle

IOP Conf. Series: Materials Science and Engineering **73** (2015) 012064 doi:10.1088/1757-899X/73/1/012064

4. Conclusions:

- A new novel method was experimented and proved to make metal matrix composites by FSP.
- The composite resulted in improving the mechanical properties of magnesium alloy.
- Microstructural results revealed grain refinement, equally distributed alumina particles were confirmed by SEM and EDS.
- This process makes composites up to the pin length of FSP tool, can be used make MMCs from secondary processes.

References:

[1] W.M.Thomas et.al., 1991, US Patent No. 5,460,317
[2] A.H. Feng and Z.Y. Ma. Enhanced mechanical properties of Mg–Al–Zn cast alloy via friction stir processing. *Scripta Materialia.* E 56 (2007), pp.397–400.
[3] Y.Morisada, H.Fujii, T.Nagaoka, M.Fukusumi, WCNTs/AZ31 surface composites fabricated by friction stir processing, *Materials Science and Engineering A* 419 (2006) 344–348
[4] Karl Ulrich Kainer, Metal Matrix Composites. Custom-made Materials for Automotive and Aerospace Engineering, 2006 WILEY-VCH Verlag GmbH & Co. KGaA, Weinheim
[5] Hai Zhi Ye,Xing Yang Liu, Review of recent studies in magnesium matrix composites, *Journal of materials science 39* (2 004) 6153–6171.
[6] A. K i e ł b u s, Structure and mechanical properties of casting MSR-B magnesium alloy; *Journal of achievements in materials and manufacturing engineering* 18, 1-2, 131-134 (2006).
[7] Y.Cai, M. J.Tana, G. J.Shen and H. Q. Su, *Mater Sci.Engng. A* 282 (2000)232.

Effect of Annealing on the Optical Properties of Transition Metal Doped ZnO Thin Films

Jijoy.P.Mathew[1],[*], George Varghese[2],Jacob Mathew[1]

1 Thin Film research Lab, Department of Physics, St.Berchmans College, Changanassery

2 Department of Physics, Calicut University, Thenjippalam, Calicut

* Corresponding Author email ID: mathew.jijoy@gmail.com

Abstract: A group of transition metal (Co,Cu,Ni) doped ZnO thin films were prepared by a low cost dip coating method. To study the effect of annealing temperature on the structural and optical properties of the grown films each film is subjected to three different annealing temperatures. From the calculation of structural parameters from the XRD spectrum, it was found that each film possesses a tensile strain and this tensile strain increases as the annealing temperature increases. This dominant behavior of tensile strain affects the optical properties of the grown film in a greater extent.

Keywords: Band Gap. Stress, Doping, Photoluminescence

Introduction: Thin films of transparent conductive oxides (TCO) have received a considerable attention due to their technological importance in the field of optoelectronic and photovoltaic cells. ZnO is one of the metal oxide semiconductors suitable for these technological applications due to their unique properties such as wide and direct band gap (Eg ~3.2-3.4eV at 300K), large exciton binding energy, excellent thermal and chemical stability etc. the properties of ZnO thin films can be modified by doping it with transition metal ions which motivated intensive research to develop transition metal doped wide band gap semi conductors , dilute magnetic semiconductors etc. The ability to control the fundamental electronic and optical properties of thin films through doping have been the central issue in developing active electronic and optoelectronic devices [1-3]

Transition metal doped ZnO thin films have been prepared by a variety of thin film deposition methods such as pulsed laser deposition[4], sputtering[5] etc. But these deposition technologies need sophisticated and expensive equipments for their operations. In our studies we employed a low cost inexpensive dip coating method for the deposition of doped ZnO thin films. Moreover, most of the research work on transition metal doped ZnO thin films focuses on the studies of magnetic properties. Studies related to optical properties of these films are hardly seen even though they have wide applications in optoelectronic devices . In this paper we report the effect of annealing temperature on the optical properties of transition metal doped ZnO thin film based on single oscillator model

IOP Conf. Series: Materials Science and Engineering **73** (2015) 012065 doi:10.1088/1757-899X/73/1/012065

Experimental: $Zn_{1-x}M_xO$(M=Cu,Co,Ni) thin films were prepared by dip coating method. The coating solution were prepared by mixing appropriate amount of zinc acetate dihydrate [$Zn(CH_3COO)_2.2H_2O$] and appropriate nitrate [$M(NO_3)_2.4H_2O$](M=Cu,Co,Ni) in 2.5wt% of poly vinyl alcohol solution. All the films were prepared for a doping concentration of x=0.06. The obtained solution is stirred for 1 hour to yield a clear solution. The coating is made by dipping precleaned glass substrate to the solution with a withdrawal speed of 10cm/sec. In order to remove the organic residuals and to obtain desired thickness (~700nm) the obtained films were heat treated at 250°C for 10min.The heat treatment process were repeated for 10 times. In order to study the effect of annealing temperature on the structural and optical properties of the films each films were annealed at 350°C,450°C and 550°C for 1 hour in air

The X Ray Diffraction patterns were obtained with XPERT-PRO PANalytical CuKα radiation(λ=1.5406Å).The UV measurements were carried out at room temperature using T90 UV-VIS spectrometer (PG Instruments) in the wavelength region 300nm-700nm with a clean glass substrate in the reference beam

Results and Discussions: Fig 1(a)-1(c) shows the XRD patterns of the doped ZnO thin films with different annealing temperatures. XRD spectrum demonstrates the existence of single phase hexagonal wurzite ZnO structure. There is no evidence for the existence of any secondary phase.

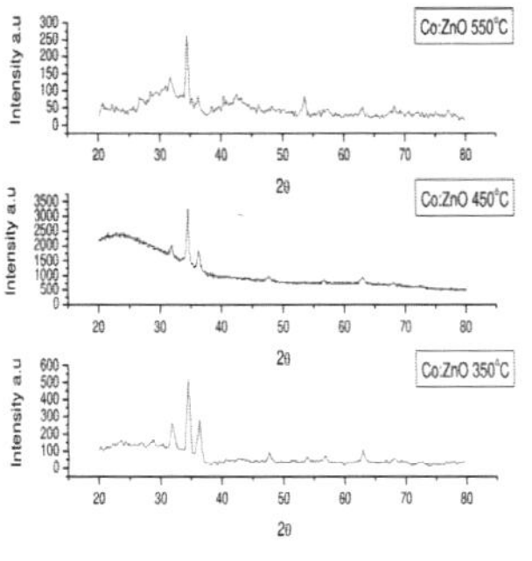

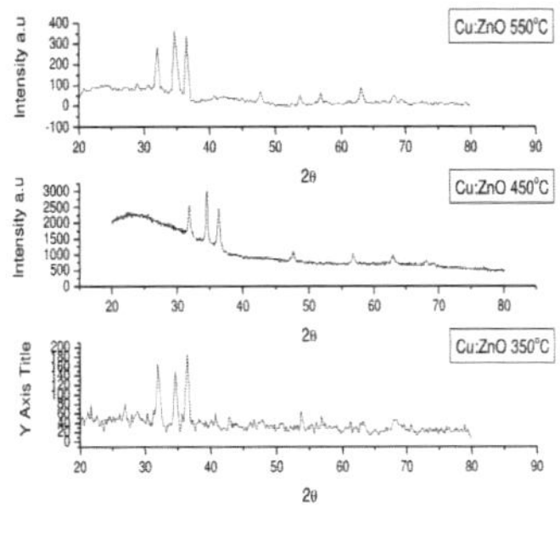

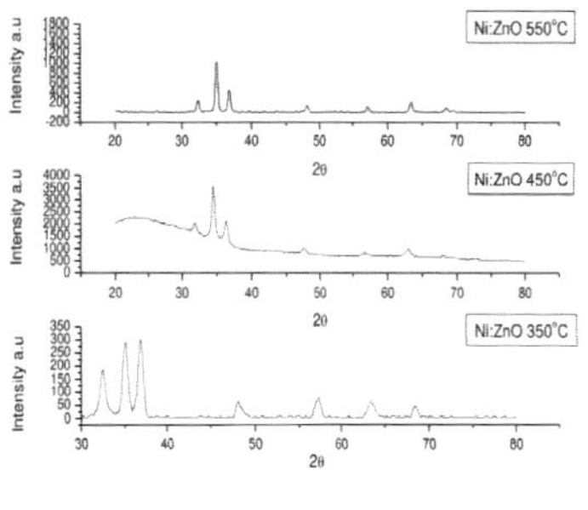

Fig 1(a) Co:ZnO ***Fig 1(b) Cu:ZnO*** ***Fig 1(c) Ni:ZnO***

XRD Spectrum of TM Doped ZnO Thin Films

IOP Conf. Series: Materials Science and Engineering **73** (2015) 012065 doi:10.1088/1757-899X/73/1/012065

Datas obtained from the XRD spectrum are used to determine the c parameter and the grain size of the films. It is seen that the lattice constant 'c' decreases as annealing temperature increases which clearly indicated the relaxation of residual stresses on annealing. The lattice constant 'c' can be further utilized to evaluate the uniform strain εo and biaxial stress in the lattice along the c axis[6-8].From the calculated data it is clear that, there is a dominant behavior of tensile strain in all the films irrespective of the dopant atom. The grain size also increases with increase in annealing temperature. A.Mansingh and V.Gupta[9] suggests that the increase in grain size as annealing temperature increases may also be attributed to the dominant behavior of tensile strain. In their studies on sputtered ZnO thin films they noticed that as annealing temperature increases the residual stress in the film relaxes and tensile strain dominates . This effect consistently accompanied by an enhancement in crystallinity which is manifested by larger grain size

To determine the band gap of the films, we use the equation in which the absorption coefficient α is related to the incident photon energy and band gap as [10]

$$\alpha(h\nu)=A(h\nu-E_g)^{1/2}$$

The direct band gap of the films estimated from the extrapolation of the graph of hv vs.$\alpha h\nu^2$. Fig 2(a)- 2(c) shows the hv vs.$(\alpha h\nu)^2$.plots for the doped ZnO thin films annealed at different annealing temperature.

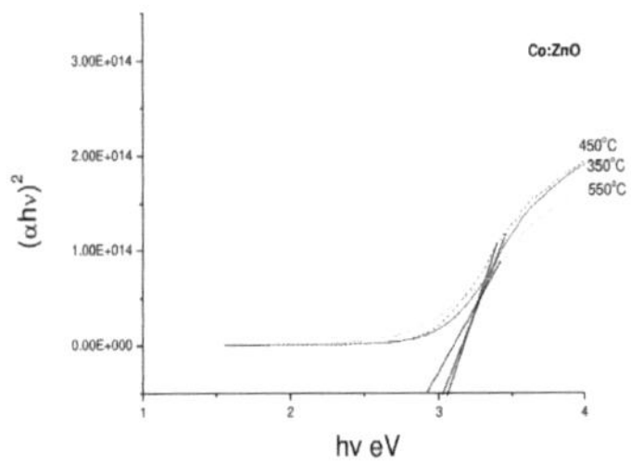

Fig 2(a) Co:ZnO

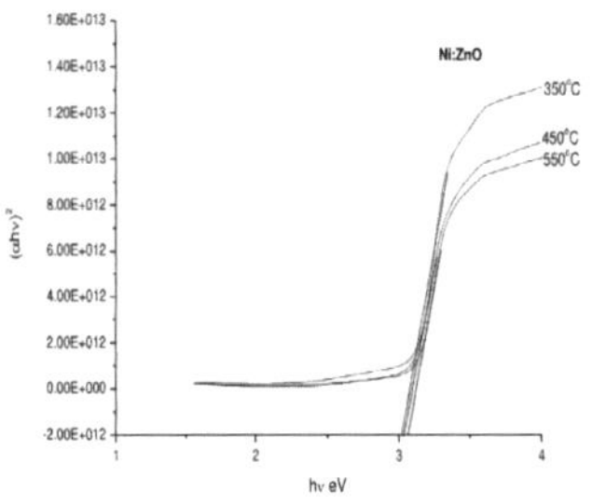

Fig 2(b) Ni:ZnO

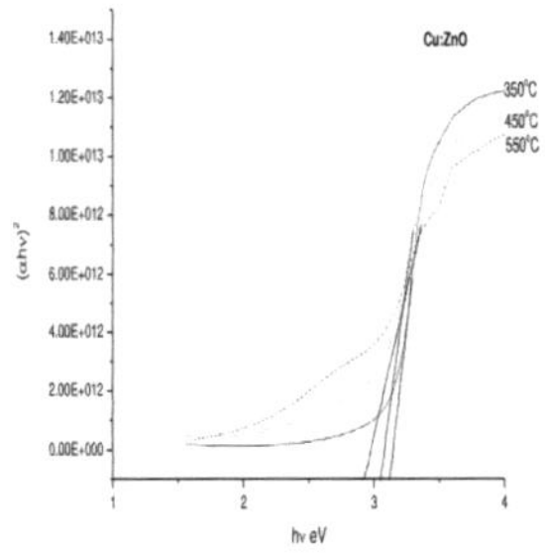

Fig 2(c)Cu:ZnO

$(\alpha h\nu)^2$.plots of TM Doped ZnO Thin Films

The optical band gap is found to be temperature dependent and band gap decreases with increase in annealing temperature. The decrease in band gap can be attributed to the effect of doping and the effect of tensile strain. The narrowing band gap is due to the existence of dopant impurities in the ZnO structures which induce the formation of new recombination structures with lower emission energy [11]. Tang et al.[12],on their studies on Cd doped ZnO thin films, suggests that the tensile strain present in the grown film also cause a decrease in band gap of the alloy thin films

Fig 3(a)-3(c) shows the PL spectra of doped ZnO thin films at different annealing temperatures. All the spectrum shows peaks at UV, blue, blue green region. The peak at UV region is due to radiative recombination process of free excitons. The blue and blue green emission originates from the transition between shallow donor levels of oxygen vacancies to the valance band. Srikantha Reddy et al.,[13] observed a shift in the blue emission peaks to lower wavelength side for Mg doped ZnO nanoparticles, which they states that an identification of band gap modulation. No such shifts were observed in the Cd doped samples, however, we observed a decrease in intensities of PL peaks increases as the annealing temperature increases.

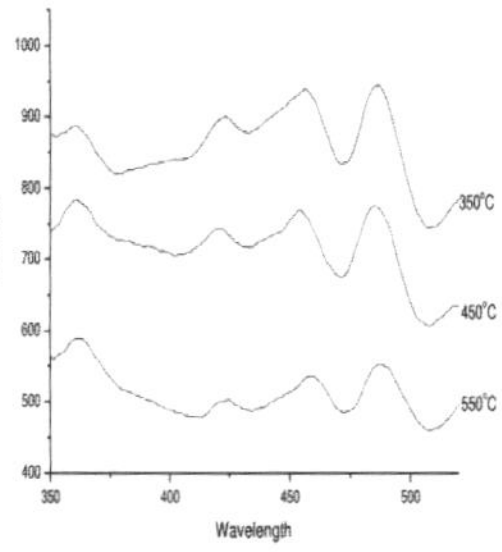

Fig 3(a) Co:ZnO

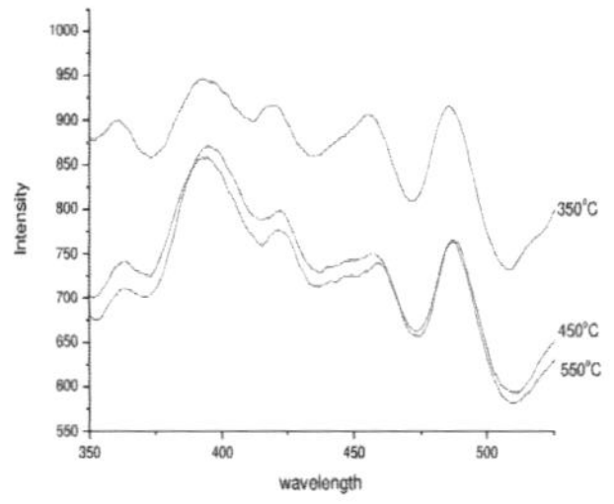

Fig 3(b) Ni:ZnO

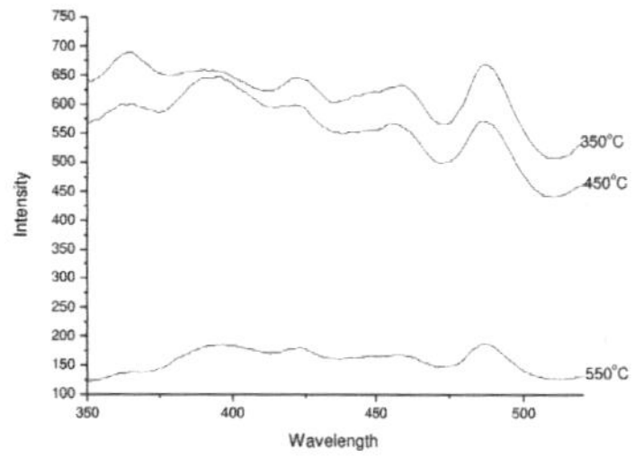

Fig 3(c) Cu:ZnO

PL Spectrum of TM doped ZnO Thin Films

The observed structural and optical constants of the prepared TM doped ZnO thin Films are summarized in Table (1) given below

Sample Name	c(Å)	Strain(ε)	Stress (σ) 10^9Pa	Average Grain Size (nm)	Band Gap (eV)
Co:ZnO (350°C)	*5.192*	*-.00268*	*0.64052*	*34.33*	*3.17*
Co:ZnO (450°C)	*5.189*	*-.00326*	*0.77914*	*36.10*	*3.07*
Co:ZnO (550°C)	*5.180*	*-.00499*	*1.19*	*38.40*	*2.97*
Cu:ZnO (350°C)	*5.189*	*-.00326*	*0.77914*	*27.79*	*3.12*
Cu:ZnO (450°C)	*5.176*	*-.00576*	*1.37664*	*32.36*	*3.04*
Cu:ZnO (550°C)	*5.166*	*-.00768*	*1.835*	*37.02*	*2.94*
Ni:ZnO (350°C)	*5.189*	*-.00326*	*0.77914*	*24.06*	*3.06*
Ni:ZnO (450°C)	*5.180*	*.-.00499*	*1.19*	*30.68*	*3.04*
Ni:ZnO (550°C)	*5.166*	*-.00768*	*1.835*	*35.78*	*3.00*

Table (1) The observed structural and optical constants of TM doped ZnO Thin Films

Conclusion: $Zn_{1-x}M_xO$(M=Cu,Co,Ni) thin films were prepared by a low cost dip coating method. The effect of annealing temperature on the structural and optical properties of the films were using XRD, UV-Vis-NIR and PL spectrum .It is found that in all the doped films there exists a tensile strain and the tensile strain increases with increase in annealing temperature. The dominant behavior of tensile strain also affects the optical properties of the films too. The optical band gap of the films decreases with increase in annealing temperature. The intensities of PL spectra also shows a similar behavior

References

[1] Anderson Jenotti ,Chris G van de Walle 2009 Re.Prog.Phys. 72 126501

[2] Dae-Kue Hwang, Min-Suk Oh, Jae-Hong Lim , Scong Ju Park 2007 J.Phys.D.Appl.Phys. 40 R387

[3] E.V.Monakov, A.Yu Kuzentsov, B.G.Snensson 2009 J.Phys D: App.Phys. 42 153001

IOP Conf. Series: Materials Science and Engineering **73** (2015) 012066 doi:10.1088/1757-899X/73/1/012066

Analysis of ion conduction and relaxation in $Na_2NbCdP_3O_{12}$ glass

N. S. Krishna Kumar, S. Vinoth Rathan, G. Govindaraj*

Department of Physics, School of Physical Chemical and Applied Sciences, Pondicherry University, Puducherry-605014, India

E-mail: ggraj_7@yahoo.com

Abstract. The ac conductivity of ionic conducting $Na_2NbCdP_3O_{12}$ is studied over the frequency range of 100Hz to 1MHz at different temperatures. The conduction mechanism of the material is studied using the concept of mismatch and relaxation (CMR) model and the length scale of the ion transport process is studied from the many particle displacement function $<R(t)^2>$ for various temperature. The non-Debye relaxation mechanism and ionic conduction is analysed through the anomalous relaxation function and it is compared with CMR.

1. Introduction

The dynamics of mobile ion in ionic conducting disordered materials is one of the prime interests of physics and material research. AC impedance spectroscopy is one of the common experimental techniques used to study the dynamics of mobile ion in the applied electric field. It is quite complex to extract the clear picture of dynamics and relaxation of mobile ions. There are several empirical formulas and relaxation models are available to study the conduction mechanism and non-Debye electrical relaxation process [1]. Among them concept of mismatch and relaxation (CMR) [2] and anomalous relaxation function [3, 4] are relatively new models.

Na^+ superionic conductors (NASICON) have a major role in electrochemical research due to their wide applications such as batteries, gas sensors, fuel cells etc [5]. Niobium based phosphate ($Na_4NbP_3O_{12}$) glass have an appreciable ionic conductivity even at room temperature. In the present study the ac electrical data of NASICON type $Na_2NbCdP_3O_{12}$ glass is investigated for ion conduction mechanism using the CMR and anomalous relaxation models, then the physical picture and limitations of these models are studied and compared.

2. Experiment

The ion conducting NASICON type glass $Na_2NbCdP_3O_{12}$ were synthesized by melt quenching method from the analytical grade of Na_2CO_3, Nb_2O_5, CdO and $NH_4H_2PO_4$. The sample was characterized with XRD, FT-IR and DSC and it was reported earlier [6]. The ac impedance measurements were carried out in the frequency range of 100Hz to 1MHz using HIOKI-3532 LCR Hi-TESTER over the temperature range 393K to 513K. The real part of complex conductivity can be obtained from the impedance data using $\sigma'(\omega)=G(\omega)\times d/S$, where d is the thickness and S is the area of cross section of the sample. $G(\omega)$ is the conductance of the sample obtained by using LCR meter.

3. Electrical relaxation and ion transport

Several representations of impedance spectroscopic data are used to study the electric relaxation mechanism worldwide. Many models are formulated based on the ac conductivity behavior, and it is the most used representation for ionic conducting materials. The ac conductivity data of most of the materials shows the similar feature that the real part of conductivity shows a frequency independent character for low frequencies and above certain characteristic frequency the conductivity shows frequency dependence and the characteristic frequency increases with increase of temperature. Correspondingly complex impedance shows a depressed semicircle which is the signature of the non-Debye relaxation [5]. The $Na_2NbCdP_3O_{12}$ glass sample also shows the similar feature.

3.1. Concept of mismatch and relaxation

The CMR model explains the ac conductivity dispersive regime as a consequence of correlated forward backward jumps of mobile ions. Due to the mutual repulsive interaction, the mobile ions will retain some distance each other. A mismatch of arrangement of ions will occur due to the jumping of ion to its neighboring site. This mismatch is reduced either by the rearrangement of neighboring mobile ions or by the backward jump of the ion to its original position. This leads to the correlated forward-backward jumps and consequently the dispersive regime in ac conductivity spectra. The CMR describes the ion dynamics mathematically by two coupled rate equations

$$-\frac{dg(t)}{dt} = A.g^k(t).W(t) \qquad (1)$$

$$-\frac{dW(t)}{dt} = -B.W(t).\frac{dg(t)}{dt}$$

where W(t) is time dependent correlation factor which represent the probability of an ion to be still in the new position occupied directly after the jump and g(t) normalized mismatch function which describes a normalized distance between the actual position of an ion and the position where its neighbors expect it to be. The function g(t) varies with time as g(0)=1 to g(∞)=0. CMR is a model which incorporates the high frequency plateau in the ac conductivity σ(∞). The parameter A is an internal frequency proportional to high frequency limit of the conductivity σ(∞), and B determines the ratio σ(0)/σ(∞)=exp(-B) and K is the shape parameter which is mostly close to 2. The ac conductivity is related with the correlation factor by linear response theory as

$$\sigma'(\omega) = \sigma(\infty).\left(1 + \int_0^\infty \frac{dW(t)}{dt}.Cos(\omega t)dt\right) \qquad (2)$$

The model shows a good microscopic picture of ion dynamics through "single particle route" and "many particle route" [2].

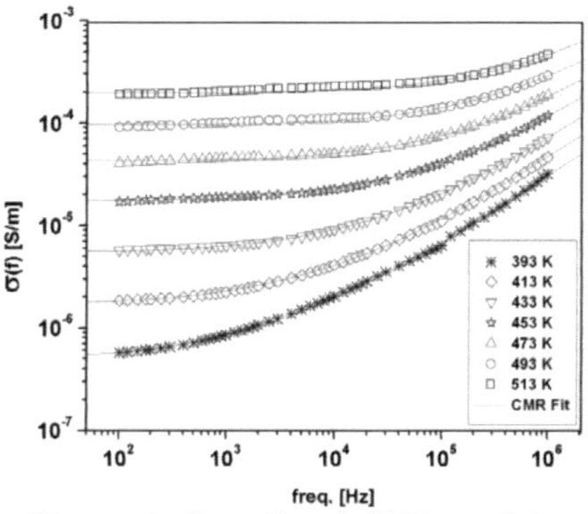

Figure 1. Best fit of CMR model ac conductivities to experimental data.

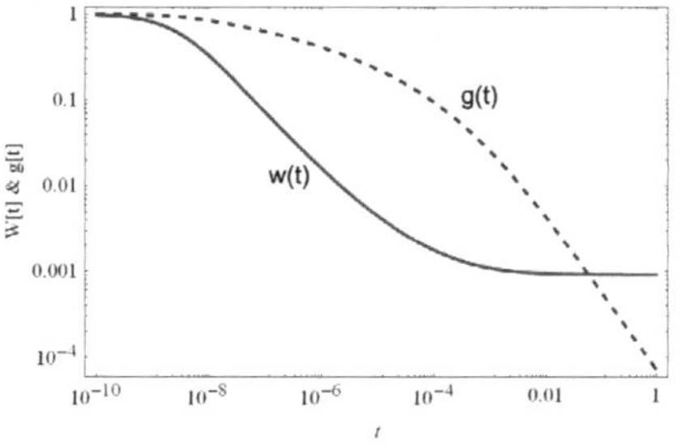

Figure 2. Function w(t) and g(t) with time for K=2.1, B=7 and $A=1.2\times10^8 s^{-1}$.

Experimental result of the ac conductivity of the sample is best fitted for the equation (1) and (2) of CMR model for various temperatures and it is shown in figure 1. The parameter K shows temperature independence with a value K=2.1. σ(∞) and hence parameter A increases with increase of temperature. The high frequency plateau is not observed in glass, since the vibrational contribution dominates to the hopping motion of ions at high frequency. The W(t) and g(t) response with time for the sample at 423K is shown in figure 2. Information on the length scales of ion transport process can be obtained from mean-square displacement. Many particle displacement function $\langle R^2(t)\rangle$ is proportional to the mean-square displacement of the center of charge of the mobile ions by definition. Through linear response theory, the microscopic property of the material $\langle R^2(t)\rangle$ can be obtained from real part of conductivity as [7]

$$\langle R^2(t)\rangle = \frac{12k_BT}{Nq^2\pi}\int_0^t dt' \int_0^\infty \frac{\sigma'(\nu)}{\nu} Sin(2\pi\nu t')d\nu \qquad (3)$$

Where k_B is the Boltzmann constant, N is the carrier density, q is the charge of the carrier and T is absolute temperature. $\langle R^2(t)\rangle$ for the sample $Na_2NbCdP_3O_{12}$ at different temperature is obtained from $\sigma'(\upsilon)$ using equation (3) and it is shown in figure 3. At longer time scale $\langle R^2(t)\rangle$ is linearly dependent on time, this is the characteristics of random walk motion of mobile ions and it corresponds to a frequency independent conductivity. In short time scale $\langle R^2(t)\rangle$ shows sublinear dependency on time which reflects the forward-backward correlations in ion dynamics and it corresponds to the dispersive region of ac conductivity. Crossover length ξ is defined as the average distance that the ions have to cover in order to overcome the forces causing backward correlation. ξ increases with the increase of temperature and its order is a few Angstrom for the sample and it is shown in inset of figure 3.

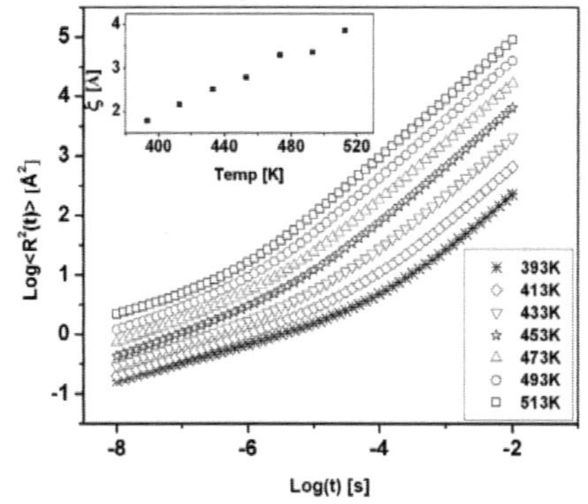

Figure 3. Many particle displacement function with time for different temperature. Inset shows cross over length with temperature.

3.2. Anomalous relaxation function

Anomalous relaxation model has the relaxation function $\phi^*(t)=\exp(-t/\tau^*)$ with $\tau^*=\tau_D/(i\tau_D)^{(1-g)}$, $i=\sqrt{-1}$ and $0<g\leq1$, where τ_D is the Debye relaxation time. The non-Debye response is the consequence of the parameter g. In the impedance representation the model has the simple functional form

$$Z^*=R/(1+\omega(i\tau_D)^g) \qquad (4)$$

where R is the bulk resistance. When g=1 the relaxation mechanism will be Debye type and the imaginary part of the relaxation time vanishes. The model has the functional form $\sigma^*(\omega)=\sigma_0(1+i\omega\tau^*)$ in conductivity representation. According to the model collective motions of hopping ions due to ion-ion interaction are non-random process and it is attributed trapping of charge carriers in disordered host lattice and it is accounted in terms of exponent g. During conduction the ion have to go through several reiterated forward backward ion hopping. The phase factor $i^{(1-g)}$ slows down the τ_D by the factor of $(1/\tau_D)^{(1-g)}$ which incorporate the non-Debye response [3, 4]. The parameters of the model is extracted by fitting complex impedance data with equation (4) and it is shown in figure 4(a)&(b). The

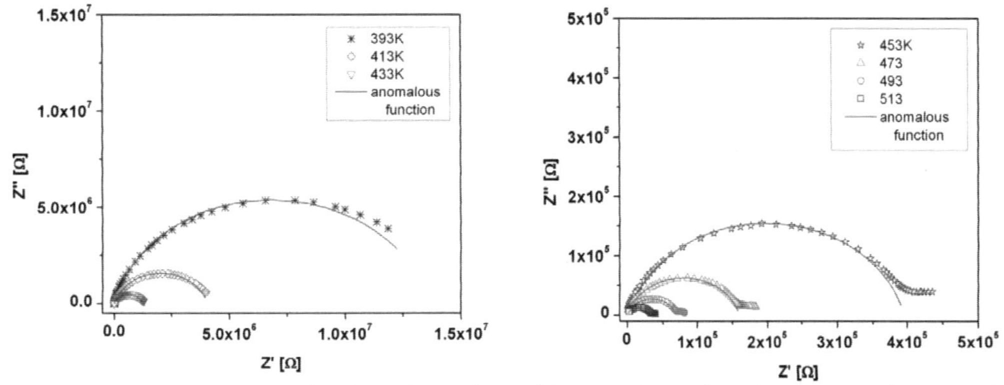

Figure 4 (a) & (b). Complex impedance data with anomalous relaxation function fit curves for different temperatures

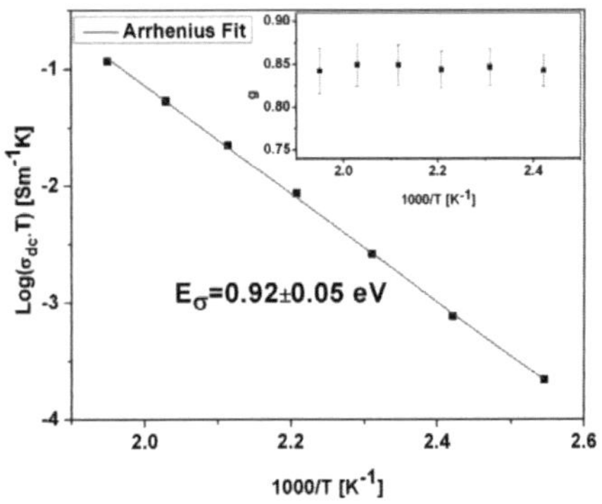

Figure 5. Arrhenius plot of dc conductivity from anomalous function fit of complex impedance. Inset shows the variation of exponent g with temperature

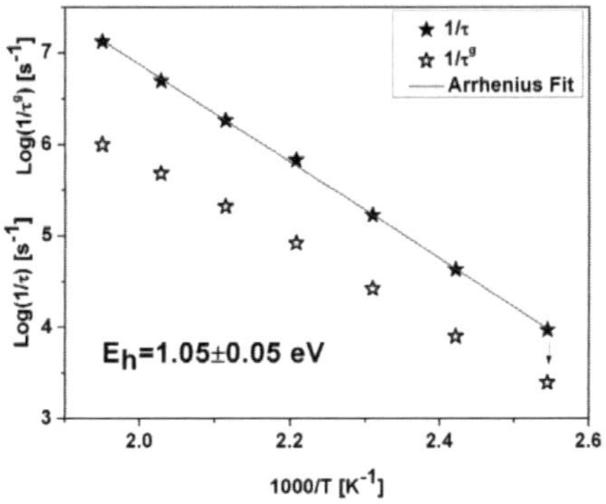

Figure 6. Arrhenius plot of hopping frequency and the temperature dependence of $1/\tau^g$

conductivity and hopping are thermally activated process which shows Arrhenius behavior is shown in figure 5 and figure 6. The activation energy for conduction is 0.92±0.05eV and for hopping 1.05±0.05eV. The exponent g=1 corresponds to the fast relaxation process of mobile charge carriers n_cq and the exponent $0<g<1$ corresponds to slow relaxation process with gn_cq hopping charges and $(1-g)n_cq$ trapped charge carriers. With increase of temperature the number of hopping charge is expected to increase. The sample has g values which are independent of temperature, even though hopping charge may be increasing due to increase of free charge carrier n_c with temperature. For hopping ion have to overcome a barrier energy of E_h, due to the interaction among the charges gn_cq and $(1-g)n_cq$ it alters to $E_h-(gq)b\underline{E}/2$ and $E_h+(gq)b\underline{E}/2$, where $\underline{E}$ is the applied field and b is the hopping distance. The hopping rate of the charge carrier in fast process is given by $1/\tau_p=1/\tau_0 \exp(-E_h/k_BT)$ and for slow process $1/\tau_p{}^g=1/\tau_0{}^g \exp(-gE_h/k_BT)$ where E_h is the activation energy for fast hopping process.

4. Conclusion

The dynamics of Na^+ ion in the $Na_2NbCdP_3O_{12}$ is studied with the CMR and anomalous relaxation function. $\langle R^2(t)\rangle$ and crossover length strengthen the concept of ion dynamics. Both the models are commensurate with the linear response theory and dynamics contains similar features like forward-backward correlated hopping. The microscopic ion transport is mathematically formulated in CMR and shows a good physical picture, but it lacks physical interpretation of some parameter K and it is relatively harder to handle for the experimental data due to mathematical complexities. Anomalous relaxation function gives simple functional form in all representations of complex impedance and it gives better physical picture in terms of hopping and trapped charges. Anomalous relaxation function quantitatively describes non-Debye relaxation and site energy variation through a single exponent g.

Acknowledgement

Financial support from the UGC-SAP F.530/15/DRS/2009 is gratefully acknowledged.

References

[1] Dyre J C and Schroder T B 2000 *Rev. Mod. Phys* **72** 873
[2] Funke K, Banhatti R D, Bruckner S, Cramer C, Krieger C, Mandanici A, Martiny C and Ross I 2002 *Phys. Chem. Chem. Phys* **4** 3155
[3] Govindaraj G and Murugaraj R 2000 *Mat. Sci. and Engg.* B **77** 60
[4] Govindaraj G and Murugaraj R 2005 *Solid State Ionics* **176** 109
[5] Mariappan C R and Govindaraj G 2005 *Solid State Ionics* **176** 1311
[6] Vinod Rathan S and Govindaraj G 2010 *Mat. Chem. Phys.* **120** 255
[7] Roling B, Martiny C and Bruckner S 2001 *Phys. Rev.* B **63** 214203

International Conference on Materials Science and Technology (ICMST 2012) IOP Publishing
IOP Conf. Series: Materials Science and Engineering **73** (2015) 012067 doi:10.1088/1757-899X/73/1/012067

Low velocity impact analysis on glass fiber reinforced composites with varied volume fractions

Rakesh Reghunath[1], Mahadevan Lakshmanan[2], K M Mini

[1]Asst. Professor, Department of Mechanical Engineering, Amal Jyothi College of Engineering Kottayam, Kerala, India

E-mail: rreghunath11@gmail.com

[2] Asst. Professor, Department of Mechanical Engineering, Amrita School of Engineering, Amrita Vishwa Vidyapeetham, Amrita Nagar, Coimbatore, Tamil Nadu, India

E-mail: mlakshmanan90@gmail.com

[3]Professor, Department of CivilEngineering, Amrita School of Engineering, Amrita VishwaVidyapeetham, Amrita Nagar, Coimbatore, Tamil Nadu, India

E-mail: minikmadhavan@gmail.com

Abstract. This paper presents an experimental study to assess the impact response of bidirectional woven type of glass fiber reinforced composite material with varied volume fractions and hence to find out the optimal volume fraction which gives better impact resistance. The specimens are prepared using vacuum bagging process and the volume fraction is estimated by resin burn off method. For getting information regarding surface topography of the impacted specimen, scanning electron microscopy is conducted. The study is done by slightly varying the velocities and it is found that a volume fraction of 43-44% gives a better impact resistance which is also confirmed by the scanning electron microscopy test. From the SEM results it is observed that matrix cracking, fiber breakage, debonding and fiber pull out are the major modes of failure during the impact, which reduces the structural strength and stability of the structure.

1. Introduction

Composites are used as an alternative to metallic material structures where weight is a major consideration. During the last decade fiber reinforced composites have been established as a competitive material for naval, automotive and aerospace industry. Their behavior under impact loading is a major concern, since impacts do occur during manufacture, normal operation, maintenance and so on. The impact behavior of FRP materials is so diverse and complex and the present knowledge is far from complete. The resulting damage due to impact, often in the form of delaminations, matrix cracking and fiber failure may severely reduce the structural strength and stability. Therefore, a current and important design requirement in load bearing composite structures is the ability to tolerate the impact damage. Impact may be defined as a relatively sudden application of an impulsive force, to a limited volume of material or a part of structure.

The impact responses of the composite structures have been studied extensively over the years since the impact damage caused by foreign objects can reduce the load carrying capacity of the composites significantly. Based on different velocity ranges in the impact events, categorization of impact events are reported by Roger L Ellis [1] as low, high, ballistic and hyper velocity impact. Rajesh Mathivan and J Jerald [2] characterized the type and extent of the damage observed in the laminate for a range of thickness subjected to different impact velocity. Studies on fiber reinforced composite plates impacted by standard drop weight with different impact energies are reported by Ramin et al [3], AhemetYapici [5] and Mitvrevski et al [4] investigated the effect of impactor shape on the impact response of thin woven carbon/epoxy laminates. . Low velocity impact tests are further reported by Giovanni Belingrdi & Roberto Vadori [6], Kishore. ZY Zhang and Richardson [16] conducted low velocity impact induced non penetration damage in Pultruded glass fiber reinforced polyester composites material using a instrumented falling weight impact testing machine. A series of low velocity impact test have been carried out on a ($0^{0,}$ 90^{0}) glass fiber reinforced epoxy resin by Yang and Cantwell[9] in order to investigate the influence of varying key parameters on the damage initiation threshold. Energy absorption mechanism in Kevlar multiaxial wrap-knit fabric composite under impact loading is presented by Tae Jin and cheol Kim [11]. Characterization techniques of the impact damage such as cross sectional fractography and scanning acoustic microscopy were employed quantitatively to assess the internal damage of the composite laminate at the subsurface under impact, reported by M S Sohn et al [10]. The repeated impact behavior of self reinforced poly propylene composite has been characterized by tensile impact and instrumented falling weight test by J Aurrekoetxea et al [24].It was reported that near surface damage in glass/epoxy composite beam is sensitive to tensile stress arising from repeated impact load on the undamaged surface by J M Lifshitz and Gandelsman [12]. Indentation and penetration of carbon fiber reinforced laminates under low velocity impact was studied by G Caprino et al[13]. An Overall view of the impact response of woven fabric composite plate is presented by Cesim and Onur Sayman [26]. Evaluation of in plane dimensional effect of the glass fiber/Epoxy composite under low velocity impact was conducted by Zuleyha Aslan et al[14].. Young Shin Lee [15] and his group did numerical analysis using FEM method to predict impact response of laminated composite plates subjected to low velocity impact.. It was also found that the orientation of the damage during impact also have a significant role in determining CAI strength, presented by Mehmet Aktas and Yusuf Arman [17]. G Zhou and O Davies [18] analyzed impact response of thick glass fiber reinforced polyester composites and energy absorbing characteristic have been determined by impact force and absorbed energy histories and by force displacement relations. J N Baucom [20] has conducted an experimental study to obtain a detailed understanding of the effect of reinforcement geometry on the damage progression in woven composites panels under repeated drop weight impact loading conditions. A study of damage mechanism and delamination which are the principle mode of failure and energy absorption in a composite shell and their influence on the impact performance of the helmet was reported by Praveen K Pinnoji and Puneet Mahajan [21]. Low velocity impact damage to composite laminates caused a complicated pattern of matrix cracks, fiber cracks and delamination and was reported by M J Pavier and P Clark [22]. The effect of laminate thickness on the composite strength to repeated low energy impacts has been studied by W A de Morias et.al [23]. His studies proved that below a certain energy level the cross section of the laminate is the most relevant variable that determines the impact resistance. The glass fabric reinforced composite laminates shows the steepest increase in impact resistance with increasing laminate thickness.

Present paper focuses on studying the effect of volume fraction of fibers on the impact strength of GFRP composites under low velocity impact. Composites of different volume fractions are manufactured by vacuum bagging process and percentage of fiber content is determined by resin burn off method. The specimens are subjected to low velocity impact by using the drop/falling weight impact test.

2. Experimental Investigation

Among the different methods available for conducting low velocity impact tests, the drop/falling weight impact test is the widely used one, where the testing can be done under low acceleration. Falling weight impact testing enables to simulate a wide variety of real world impact conditions and collect the detailed performance data.

In the present study, specimens are prepared by using vacuum bagging process [25], which is a clamping method that uses atmospheric pressure to hold the adhesive or resin – coated components of a lamination in place until the adhesive cures. Specimens are manufactured using 310GSM bidirectional E-glass mats. Volume fraction of the prepared specimen is estimated using resin burn off method (ASTM D2584) where specimens of 25 mm x 25 mm size is burned in a muffle furnace to a temperature of 600^0C to burn off the resin, and thus to find the weight of fiber from which the volume fraction of specimen can be calculated . Volume fractions of 41%, 43%, 47%, 48% and 56% are obtained in the present study, thus varying the volume fraction from 40% to 60%. Volume fraction is an important parameter which determines the load carrying capacity of the material. As the volume fraction of composite increases the load carrying capacity by fibers also increases. Rule of mixture [7] is used to plot the effect of varying volume fraction on the load carrying capacity of fibers (figure1) where E_f and E_m represents the Young's Modulus of fiber and matrix respectively.

From Figure1 it is observed that for a given value of E_f/E_m, as the volume fraction of fiber increases the load carrying capacity increases.But practically as the volume fraction of fiber increases beyond a limit,the percentage of matrix phase will go down and the interlaminar shear strength of material will come down and so the load carying capacity should come down. The present paper focuses on this aspect and hence to find the optimum volume fraction.

The tests are conducted on an instrumented falling weight testing machine (figure2) where the impact energy can be controlled by either adjusting the falling height or by adjusting the mass of the falling steel dart. For the present investigation, the maximum height is set as 500 mm and the maximum mass as 2.5 kg which gives maximum impact energy of 15 J approximately. On increasing the mass or height, the potential energy of the dart will keep increasing and on releasing it, the potential energy will be converted to kinetic energy. For the experimental investigation a minimum of three square samples of size 150 mm x 150 mm are tested ASTM D 3029 [8]. Specimens are clamped on a fixture having a central slot of 100 mm x 100 mm which forms the area available for assessing the impact strength and the steel dart is having a hemispherical head of 10 mm radius. Specimen clamping apparatus is specially designed to ensure the constancy of clamping force, through a preloading of hemispherical springs.

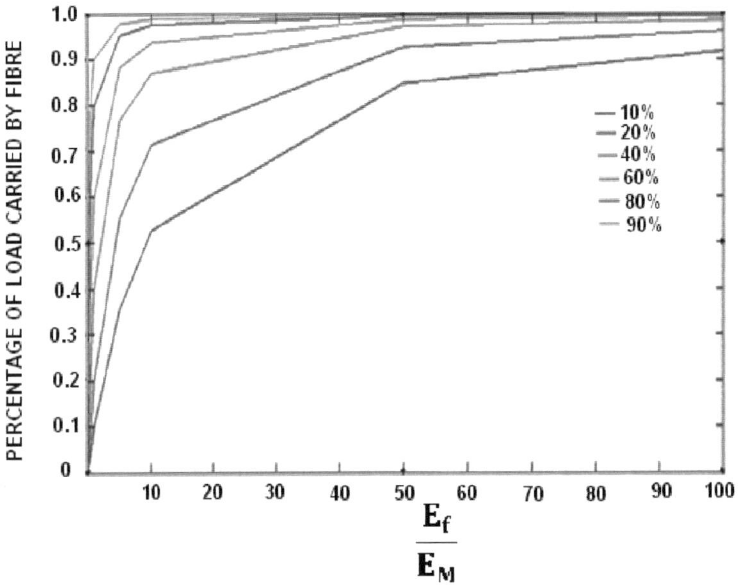

Figure 1. Effect of volume fraction on load carrying capacity of fibers.

Figure 2. Experimental Set Up.

3. Results and Discussion

GFRP composite specimens having a constant thickness of 2mm and volume fractions ranging from 40 % to60% are manufactured using vacuum bagging process. The specimens are clamped on the fixture having a central slot of 100mm × 100mm. The falling height of the impactor is adjusted such that the velocity of impact is attained a value of 2m/s, 2.5m/s and 3m/s. The load taken by the specimens of different volume fractions corresponding to varying velocities are compared and presented in Fig.3.

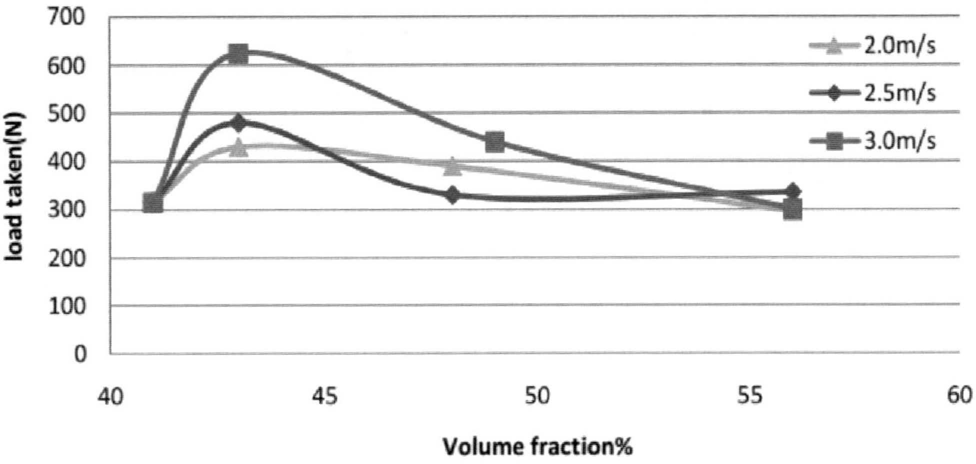

Fig 3 Effect of Volume Fraction on load carrying capacity

From Fig.3 it is observed that as the volume fraction of fiber increases, the load carrying capacity also increases up to a limit and then it starts decreasing. When the volume fraction is 43-44% of fibers it is observed that the load taken by the material is maximum in all the three cases. When fiber content increases beyond a limit the interlaminar shear strength of the fiber will come down so the load taken by fibers will also come down at higher volume fractions. In the present investigation it is observed that a volume fraction 43-44% is the optimum for the load carrying capacity.

Fig. 4 shows effect of volume fraction on the energy absorbed by the specimen. From Fig. it is found that as the volume fraction of the fibers increases the amount of energy absorbed by the specimen increases and reaches a maximum value and then starts decreasing. Same trend of variation is followed for the entire volume fraction. At higher volume fraction more amount of energy is needed for fiber breakage and therefore the amount of energy absorbed by specimen is coming down. At 2m/s maximum amount of energy absorbed by specimen is near to the volume fraction of 43% and as the velocity moves to 2.5m/s and 3 m/s the amount of energy absorbed by specimen is maximum around 49% volume fraction and when impact velocity increases the volume fraction at which max amount of energy is absorbed also increases.

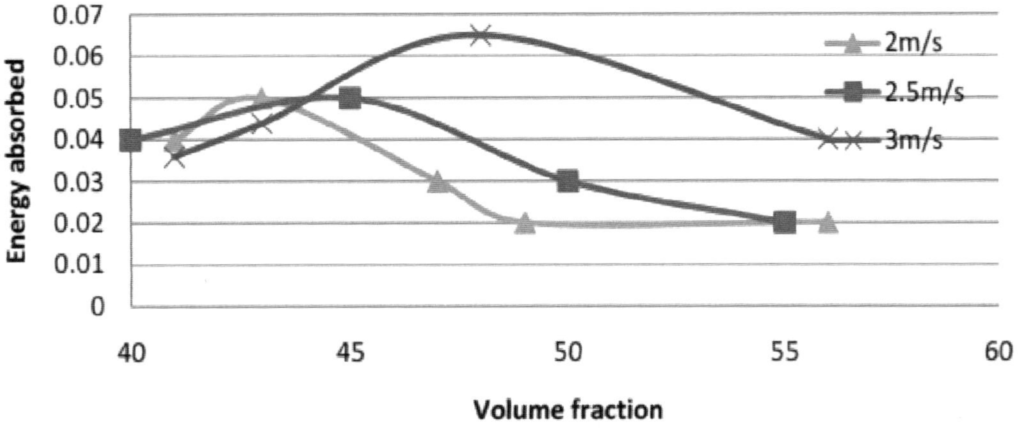

Fig. 4 Effect of Volume Fraction on Energy Absorbed

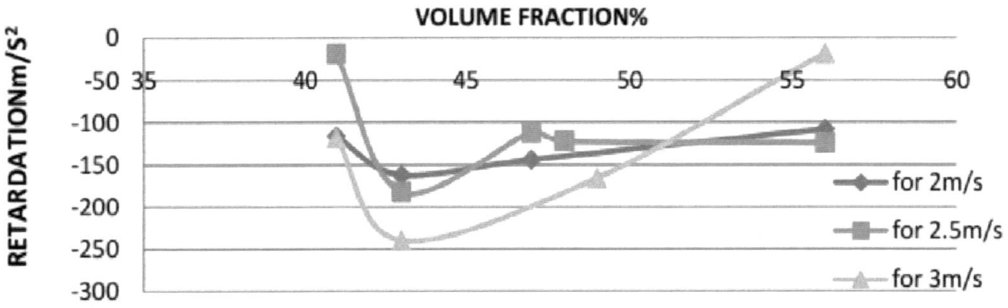

Fig. 5 Effect of Volume Fraction on Retardation

Fig.5 shows the effect of retardation of the indenter on the specimen for varying volume fractions at maximum load, corresponding to varying velocity. From the graph it is observed that retardation offered by the specimen is maximum around 43% which can be considered as the volume fraction of the material which gives better impact resistance. After 43% volume fraction, the retardation offered by specimen is decreasing drastically in all the three cases.

Fig. 6 shows the effect of velocity on the load taken by GFRP composites at different volume fractions. From Fig. it is found that for a volume fraction of 41%, maximum load taken by specimen is 315 N, that is even for the lower velocity of 2m/s velocity, the impactor is completely pierced and the specimen could not take any further load. As the volume fraction is increased to 43% maximum load taken by specimen increases as the velocity increases and it has taken a load of 600N at higher velocity. Load carrying capacity varies slightly linearly in this case. For higher volume fractions of 48% and 56% maximum load taken by material has come down at higher velocity. In this case as the velocity is increasing the load carrying capacity will first increase and then started decreasing.

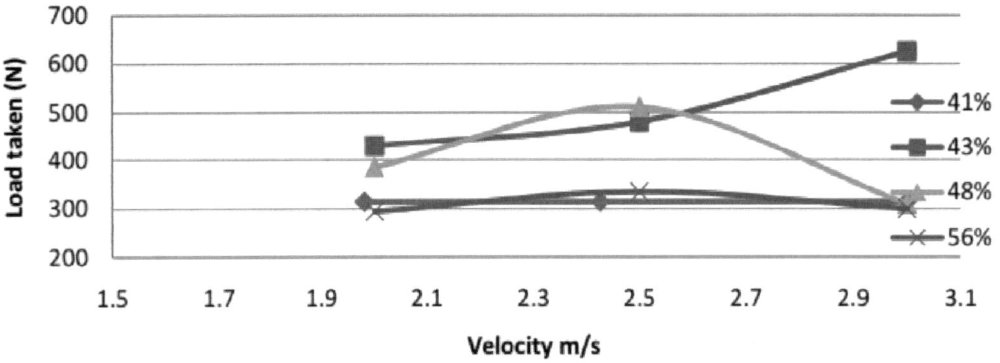

Fig.6 Load Taken Vs Velocity for different Volume Fractions

Conclusion

The present study focused on assessing the impact strength of glass fibre reinforced composite with varying volume fraction and hence to find the optimum volume fraction which gives the maximum impact resistance, for different velocities of impact. Composite specimens of 2 mm thickness are prepared by using vacuum bagging process by varying the fibre content. The volume fraction of the prepared specimens is found using resin burn off method. The specimens are subjected to low velocity impact in a falling weight impact testing machine. By changing the height of the impactor the velocity of impact is varied. Since the investigation is focused on finding the optimum volume fraction corresponding to maximum impact resistance, the velocity of impact is varied slightly from 2 m/s to 4 m/s.From the study it is found that optimal value of fiber volume fraction is around 43% where the composite material is taking maximum load and maximum retardation to the indentation is also found to be around 43%.

Acknowledgment
The authors are very much thankful to Dr. V P Rahupathy and Mr. Rajesh M Mathivan, Professor, PES College of Engineering, Bangalore for their support in conducting the experiments.

References

1) Ballistic impact resistance of graphite epoxy composite with shape memory alloys and Extended chain polythene spectra Hybrid composite (Rigger L Ellis1996, Thesis report submitted at Blacksburg University, Virginia.
2) N Rajesh Mathivan, J Jerald (2010) 'Experimental Investigation of low velocity impact Characteristics of woven glass fiber epoxy matrix composite laminate', 'Journal of materials and design',Vol 31,pp.4553-4560.
3) Ramin ,Mahmood Larry (2010).'Damage behavior of fiber reinforced composites plate subjected to drop weight impacts' Journal of Composite Science and Technology,Vol 66 pp61-68

4) T Mitvrevski et al (2005) 'The effects of impactor shape on the impact response of Composite Laminate 'Journal of Composite structures Vol. 67 pp139-148.

5) AhemetYapici ,Mehmet Metin.'Effect of Low velocity Impact Damage on the Buckling Properties Journal of Scientific Research, 2009 pp 161-166

6) E Milli, B Necib Effect of stacking sequence on the impact induced damage in the cross ply E-glass/ Epoxy Composite Journal of applied Mechanics (2009) Vol 79 pp1019-1031.

7) Giovanni Belingrdi& Roberto Vadori' Low velocity impact test of laminate glass fiber epoxy matrix composite material plate'. International journal of Impact engineering, 27(2002) Pp-213-229

8) Kishore , Ramanthan ' Repeated Drop Weight impact test and post impact ILSS on Glass EpoxyComposite materials 'Bull Mater SciVol 19 PP no 1131-1141

9) Westely J Cantwell et al. Impact response characterization of composite structures –Applied composite materials (2010) pp 463-472

10) P K Mallik- Taylor and Francis group 2007.'Fiber reinforced composites Materials Manufacturing and Design'.

11) Arthur K Kaw –Taylor and Francis Group 2006 'Mechanics of Composite Materials'

12) S R Reid and G Zhou–'Impact behavior of fiberreinforced composite material and structures, CRC Press Washington D C 2000

13) P Gougeon-'A guide to Vacuum Bagging Methodology' CRC Press, Washington D C 7th Edition 2010
14) ASTM D3029- Standards Testing method for impact resistance of flat rigid plastic specimen by means of a Tup or a falling Weight.

IOP Conf. Series: Materials Science and Engineering **73** (2015) 012068 doi:10.1088/1757-899X/73/1/012068

Optical and dielectric studies on tin(iv) tungstate nanoparticles

Beena B and Manoj S

Nanomaterials Research Lab, Department of Chemistry, Devaswom Board College, Sasthamcotta, Kollam, Kerala, India, PIN 690 521
Phone: +919446124229, Fax: +91476-2833323

Email: beenadbc@gmail.com, manojthalavoor@gmail.com

Abstract. Tin(IV) tungstate nanoparticles in the form of disc were synthesized by a novel chemical coprecipitation method. Surface morphology and particle size of the synthesized materials were analysed using Scanning Electron Microscopy (SEM) and Atomic Force Microscopy (AFM). SEM image shows disc like appearance of the nanoparticles. The particle size obtained was found to be ~20nm. Optical absorption for this material arises due to $O_{2p}\rightarrow W_{5d}$ charge transfer in the tungstate structure. The energy band gap determined using optical absorption spectrum shows that it is a direct band gap semiconductor. The extent of disorder determined using Urbach plot was found to be 0.00176 meV. The material showed intrinsic Photoluminescence around 468nm when excited by UV light of 275nm. The variation of dielectric permittivity in the frequency range 316 Hz to 3.16 MHz was studied. The temperature dependence of dielectric permittivity was also studied in the frequency range 10Hz and 32 MHz. The compound exhibit a high dielectric constant at room temperature ($\varepsilon^1>10^3$) for frequencies 10Hz and ($\varepsilon^1>10^5$) for frequencies $3.2x10^7$ Hz. It posses lower dielectric loss ie, ~0.1 at 10 Hz to ~3 at $3.2x10^7$ Hz. The material is very attractive for potential application provides its losses can be minimized.

1. Introduction

Nanodielectrics are the subject of study related to dielectric phenomena of nanoscale materials having morphology of particles, sheets, wires, tubes etc. These materials have enormous scope in power electronics industry, as gate electrodes, capacitors, sensors, electrochemical transducers, fuel injectors for automobile engines etc [1,2,3,4]. The dielectric properties and electrical transport in nanostructured materials differ from micron sized materials due to increase in interfacial atoms or ions and sinking of large amount of defects at or near the grain boundaries. The interfaces of the nanoparticles act as capacitor, thus changing the dielectric permittivity of the materials [5].

Nowadays, there is constant effort for miniaturization of electronic components, which demands the search for new materials with high dielectric constant [6]. The studies on the effect of frequency in the dielectric behavior and ac conductivity give valuable information about the conduction phenomena in nanostructured materials. In this paper, we report the dielectric properties of tin(IV) tungstate nanoparticles synthesized by chemical co-precipitation method. The microstructure and morphological characterizations were carried out using SEM and AFM. The optical absorption spectrum of the material was used to determine the optical energy band gap and the Urbach energy which defines the extent of disorder in the material. The frequency dependence of dielectric properties in the frequency range 316 Hz to 3.16 MHz was studied.

2. Experimental

Tin(IV) tungstate nanoparticles were synthesized and characterized by chemical coprecipitation method as reported earlier [7]. The microstructure of the prepared tin(IV) tungstate was investigated by scanning electron microscopy (JOEL model JSM-6390LV Instrument). The morphology of the nanocrystallites were analysed by AFM taken using NTEGRA (NT MDT) NSG 10 model AFM instrument by contact mode. The absorption spectrum of the tin(IV) tungstate nanoparticles was recorded at room temperature using SHIMADSU UV-2550 UV-Vis spectrophotometer in the wavelength range 200-800 nm. The photoluminescence spectrum of the material was recorded using Fluorolog Spectrophotometer using UV radiation of 275nm excitation wavelength and emission was recorded in the range 300-550nm. The dielectric permittivity measurements were carried out using HIOKI 3532-50 LCR Hi tester in the frequency range 316 Hz to 3.16 MHz.

3. Results and Discussion

3.1 Microstructure Analysis

The microstructural aspects of the nanoparticles of tin(IV) tungstate studied by SEM using an accelerating potential of 25kV is shown in Figure 1. In this case, the particles appear as agglomerates made of tiny crystals of nanodimension. The aggregation of nanoparticles occurs during synthesis and processing and this result in the development of discs. The morphology obtained by AFM shown in Figure 2 reveals the nanosized formation of the prepared material and the particles are more or less spherical in shape. The particle size obtained is ~20 nm.

Fig.1 Secondary Electron Microscopy image (SEM) of tin(IV) tungstate

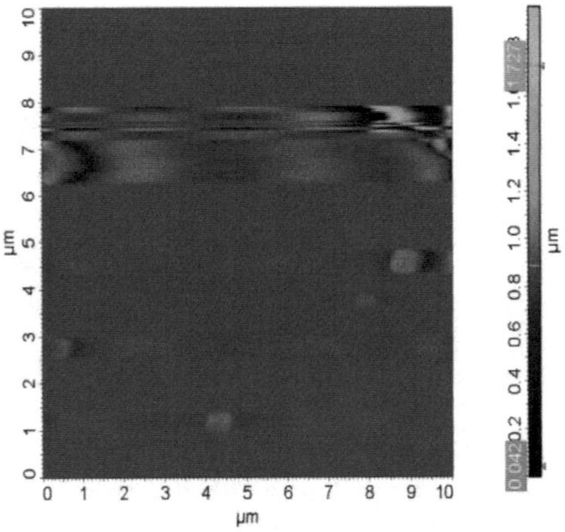

Fig.2 Typical AFM image of tin(IV) tungstate

3.2 Optical Studies

Figure 3 shows the optical absorption spectrum of tin(IV) tungstate nanoparticles in the UV-Vis range. The absorption increases almost vertically when the wavelength is below 240 nm. The characteristic absorption arises due to the $O_{2p} \rightarrow W_{5d}$-$O_{2p}$ charge transfer in the tungstate structure.

According to Tauc relation, the absorption coefficient (α) and the photon energy hν are related as,

$$(\alpha h\nu)^{1/n} = B(h\nu - Eg)$$

where B is a constant, Eg is the band gap, α is the absorption coefficient and hν is the photon energy. The values of n for direct allowed, indirect allowed, direct forbidden and indirect forbidden transitions are respectively ½, 2, 3/2 and 3 [8,9,10]. The direct allowed band gap of this material is obtained by plotting a graph of $(\alpha h\nu)^2$ versus hν as shown in Figure 4. The extrapolation of the straight line portion to the $(\alpha h\nu)^2 = 0$ axis gives the value of optical energy band gap which is found to be ~4.9eV.

The Urbach energy is commonly used to characterize the degree of disorder in amorphous and crystalline systems. The Urbach energy or band tails are developed in the material due to the potential fluctuations of energy levels in the material which extended into the band gap and thus show an exponential behavior. The band tails are characterized by band tail parameter, the Urbach energy (E_u), which is given by the inverse of the slope of ln α versus hν graph (Figure 5) at the lower energy region [11]. The value of E_u for this material is found to be 0.00176 meV. This indicates that the material is disordered due to substitutional defects or oxygen vacancies. This results in the formation of large number of energy levels inside the band gap. This leads to the formation of large band gap in this material [12].

3.3 Photoluminescence Studies

As shown in Figure 6, the material exhibits photoluminescence spectra around 425-525nm up on excitation of the sample with UV radiation of wavelength 275nm. A blue broad emission band located at 468 nm (2.65 eV) is observed, and it is apparent that the broad peak is the result of the convolution of more than one emission lines. These lines arises from the self trapped and localized excitons of the tungstate group [13,14].

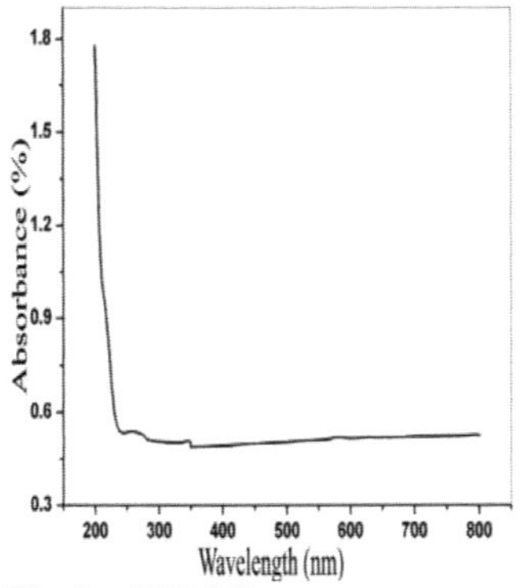

Fig.3 UV-Vis spectrum of tin(IV)tungstate

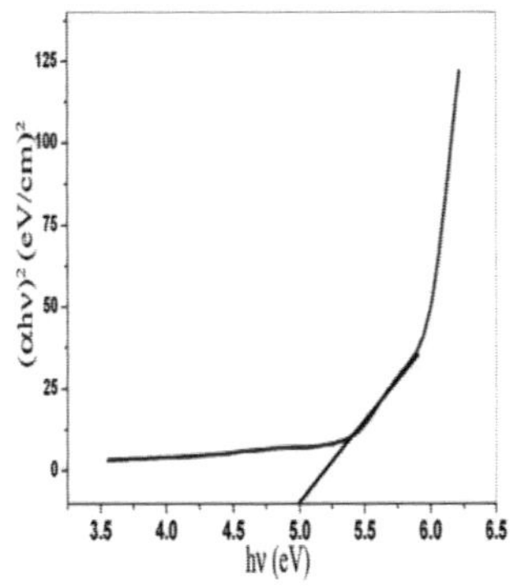

Fig.4 Tauc plot of tin(IV) tungstate

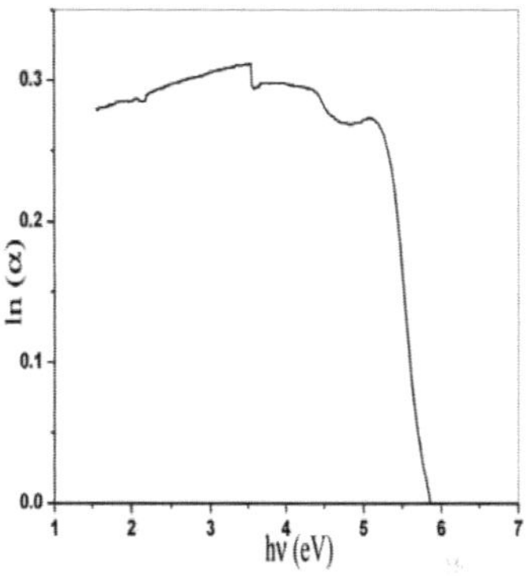

Fig.5.Urbach energy plot of tin(IV) tungstate

3.4 Frequency dependence of dielectric studies

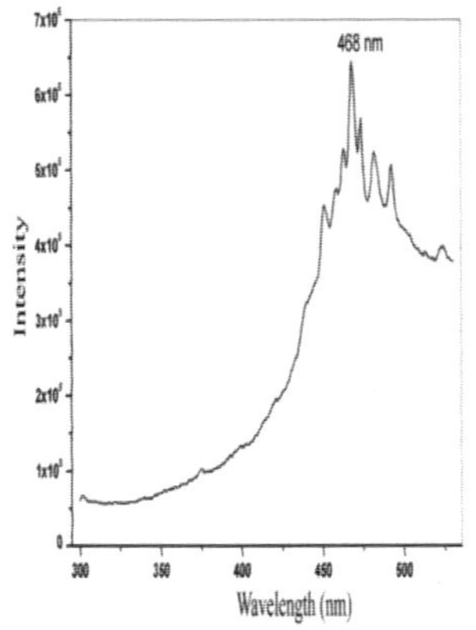

Fig.6 PL of tin(IV) tungstate nanoparticles

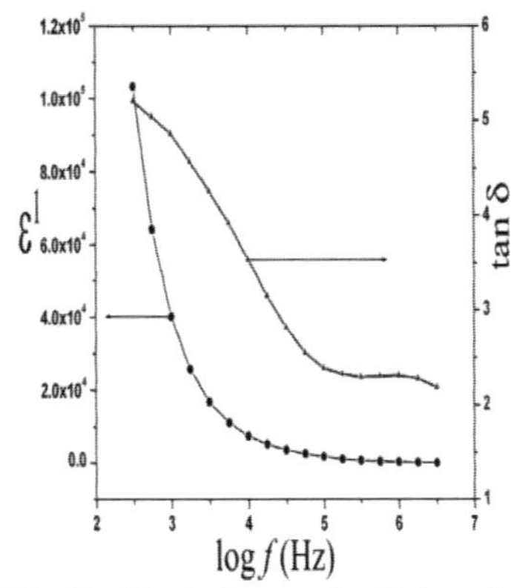

Fig.7 Variation of dielectric constant and loss factor (tan δ) with log frequency

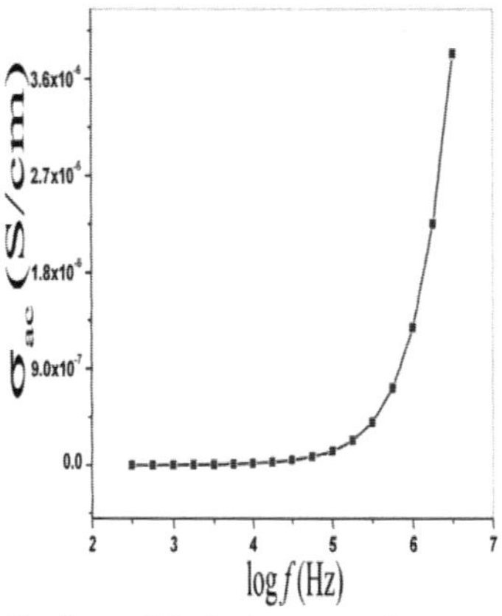

Fig.8 Variation of ac coductivity with log frequency of tin(IV) tungstate

The variation of dielectric constant (ε^1) and the loss factor (tan δ) with frequency measured in the range 316 Hz to 3.16 MHz at room temperature for tin(IV) tungstate nanoparticles as a function of log frequency at 300K is shown in Figure 7. The dielectric constant decreases with increase in frequency and remains almost constant at high frequency region. The dielectric constant and loss factor are found to be $1x10^5$ and 5.2 respectively at 316 MHz. The high value of dielectric constant is because the nanocrystalline grains behave as nanodipoles under the application of voltage. Since the grains are nanometer ordered, the number of grains per unit volume increases and hence the dielectric constant increases [15]. The dielectric loss observed for this material is low compared to the high value of dielectric constant.

Figure 8 shows the variation of ac conductivity as a function of frequency. It is seen from the figure that the ac conductivity increases with increase in frequency. The increase in ac conductivity with increase in frequency is due to accumulation of charge carrier at the boundaries [16]. In most of the disordered solids, ac conductivity is directly proportional to frequency and it results from small polaron hopping [17]. The increase in frequency of the applied field enhances the hopping of charge carriers resulting in increase of conductivity and decrease of resistivity. At higher frequencies, ac resistivity decreases and remains constant because hopping can no longer follow the applied field [16].

4. Conclusion

Morphological studies of the tin(IV) tungstate were carried out using SEM and AFM. Optical study shows absorption characteristics of $O_{2p} \rightarrow W_{5d}-O_{2p}$ charge transfer in the tungstate structure. The UV-Vis spectrum analysis revealed that the material is a wide band gap (Eg = 4.9eV) direct transition semiconductor at room temperature. The exponential band tail in the compound is due to existence of defects or disorders. The photoluminescence spectra exhibited by this material as intense blue luminescence when excited by UV light of 275nm is a promising result for various scintillation applications. The material is found to have high value of dielectric constant with low dielectric loss. The material shows fairly high value of ac conductivity. The dielectric constant of tin(IV) tungstate shows only small variations with change in frequency and temperature in the high frequency region, which is desirable for technological applications.

References

1. Alamdari H.D., Ph.D Thesis, Lavel University, Quebec, Canada, 2000
2. Bazargan A.M, Naghavi M, Mehdi Mazaheri, Sadrnezhaad S.K, Ceram,Int., 35, 1139 (2009)
3. Chang S, Doremus R.H, Ajayan P.M and Siegel R.W, Ceram.Eng.Sci.Proc., 21, 653 (2000)
4. Mukharjee P.K and Chakravorty D, J.Mater,Res., 17, 3127 (2002)
5. Nisha J. Tharayil, Ph.D Thesis, University of Kerala, Thiruvananthapuram, India, 2007
6. Wilk G.D, R.M.Wallace and J.M.Anthony, J.Appl.Phys., 89, 5243 (2001)
7. Manoj Sadanandan and Beena Raveendran., Bull.Chem.React.Eng.Catal., 7(2), 105(2012)
8. Diwakar Chauhan, V.R.Satsangi, Shab Dass, Rohit Srivastav, Bull.Mater.Sci., 29, 709 (2006)
9. Ghosh P.K, Mitra M.K, ChattopadhyayK.K, Nanotechnology, 16, 107 (2005)
10. Mangal R.K, Sing M, Vijay Y.K, Avasthi D.K, Bull.Mater.Sci., 29, 653 (2006)
11. Urbach F, Phy.Rev., 92, 1324 (1953)
12. Huamei Shang, Ying Wang, Brian Milbrath, Mary Bliss, Guozhong Cao, J.Lumin., 121, 527(2006)
13. V.Babin, P.Bohacek, E.Bender, A.Krasnikov, E.Mihokova, M.Nikl, N.Senguttuvan, A.Stolovits, Y.Usuki, and S.Zazubovich, *Radiation Measurements,* **38,** 533(2004)
14. V.V.Laguta, M.Nikl and S.Zazubivich, *Radiation Measurements*, **42,** 515 (2007)
15. Ghosh P.K, Mitra M.K, ChattopadhyayK.K, Nanotechnology, 16 , 107(2005)
16. Tataroglu A and Altindal S, Microelectron.Eng., 85, 1886 (2008)
17. Babar R.A, S.S.Shinde, A.V.Mohalkar, K.Y.Rajpure, J.Alloys Compd., 505, 743(2010)

IOP Conf. Series: Materials Science and Engineering **73** (2015) 012069 doi:10.1088/1757-899X/73/1/012069

Characterization and thermal expansion of $Sr_2Fe_{1-x}In_xMoO_6$ (x=0.0, 0.1 and 0.4 wt. %) double perovskites

Y Markandeya[1]*, V Prashanth Kumar[1], C Vishnuvardhan Reddy[1] and G Bhikshamaiah[1]

[1]Department of Physics, Osmania University, Hyderabad-500007, India.

*E-mail: markphysics@gmail.com

Abstract. Double perovskite oxides of $Sr_2Fe_{1-x}In_xMoO_6$ (x = 0.0, 0.1 and 0.4 wt. %.) were prepared by sol-gel growth followed by annealing under reducing atmosphere conditions of 10% H_2/Ar flow. These samples were characterized by X-ray diffraction, Scanning Electron Microscopy and Electron Spin Resonance studies. High temperature thermal expansion studies were carried out using Netzsch 402PC dilatometer in the temperature range from 40 to 400°C. The coefficient of thermal expansion (α) was calculated at different temperature for all the samples. The value of α was found to increase with increase of temperature while average value of coefficient of thermal expansion found to decrease with indium composition in SFIMO materials.

1. Introduction

The double perovskite oxides of the type $A_2BB\square Q_6$ (where A is divalent cations such as Sr, Ca and Ba; B is tri/divalent cations such as Fe, Cr, Ga, Mg and Mn; B' is penta/hexa valent cations such as Mo, Re and W;) have been extensively studied for the past several years due to their substantial Low Field Magneto Resistance (LFMR) at room temperature that lead to potential applications of these materials as magnetic field sensors, read heads for magnetic hard disk drives and nonvolatile magnetic computer memory[1–6]. Several investigations were carried out in order to understand the magnetic structure and magnetoresistance of Sr_2FeMoO_6 by doping with different elements such as Ga, Ni, Mg and Cr, at Fe site [3– 6]. However, there are no studies done so far on thermal expansion of double perovskites except on stoichiometric Sr_2FeMoO_6 [7]. In view of this, the authors have prepared $Sr_2Fe_{1-x}In_xMoO_6$ (x=0.0, 0.1 and 0.4 wt.%). These materials were characterized by various experiments such as X-ray diffraction (XRD), Scanning Electron Microscopy (SEM), Electron Spin Resonance (ESR) and thermal expansion studies were made.

2. Experimental

Double perovskite samples of $Sr_2Fe_{1-x}In_xMoO_6$ (x = 0.0, 0.1 and 0.4 wt.%) (SFIMO) were synthesized by sol-gel method using stoichiometric amounts of AR grade $Sr(NO_3)_2$, $Fe(NO_3)_39H_2O$, In_2O_3 and H_2MoO_4 were used and powders of SFIMO are obtained. The details of method of preparation of the materials are given in earlier publication [8]. The powders of SFIMO were pressed into rectangular pellets of dimensions 25*mm*×6*mm*×6*mm* using a dies and a hydraulic press by applying a pressure of 5*ton/m*^{2}. These pellets were sintered at 1200°C for 6 hours and then heated at 1000°C in a stream mixture of 10% H_2 / Ar gas for about 3 hours for reducing the Mo^{6+} to Mo^{5+}. These pellets were subjected to XRD studies using Philips PW 1830 generator diffractometer with Cu Kα radiation (40kV×25mA) and graphite monochromator in order to confirm the structure. The surface morphology and micro structural studies of the samples were carried out using the SEM with (Model No. Joel JSM-5600) combined micro analyzer. Room temperature ESR spectra were recorded on a JOEL−1X−ESR X-band spectrometer equipped with a 100 kHz field modulation unit and operating at 9.5 GHz. Thermal expansion measurements were performed using Netzsch 402PC dilatometer in air from 40 to 400°C during heating cycle.

3. Results and discussion

3.1. X-ray diffraction

The XRD patterns of SFIMO samples along with (hkl) values of reflections are shown in Figure. 1. All the XRD profiles of SFIMO were indexed and found to be single phase double perovskite structure [5]. These materials were found to crystallize in tetragonal crystal structure with space group I4/mmm. The lattice parameters a and c of SFIMO compounds were evaluated using 2θ and (hkl) values of XRD profiles shown in Figure 1. These values are given in Table 1. The unit cell volume of SFIMO compounds were also calculated using the relation $V=a^2c$ and included in Table 1. It is found from Table 1 that the values of unit cell parameters of indium doped samples (x=0.1 and 0.4 wt.%) is found to be larger compared to Sr_2FeMoO_6. This may be due to the substitution of the larger In^{3+} (0.79Å) ions into the smaller Fe^{3+} (0.645Å) ionic sites.

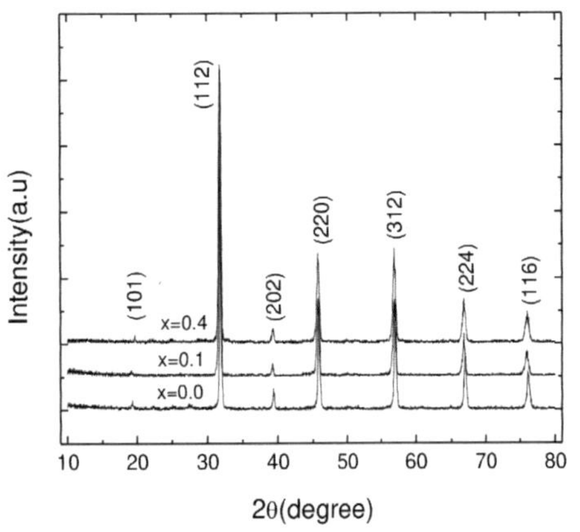

Figure 1. X-ray diffraction pattern of $Sr_2Fe_{1-x}In_xMoO_6$ (x=0.0, 0.1 and 0.4 wt.%) samples.

Table 1.The values of lattice parameters (a and c), unit cell volume (V), g-factor, constants A, B, C values of eq.(2) and average coefficient of thermal expansion ($\bar{\alpha}$) in the temperature range 40-400°C of $Sr_2Fe_{1-x}In_x$ MoO_6 (x=0.0, 0.1 and 0.4 wt.%) double perovskites.

Com(x)	0.0	0.1	0.4
a (Å)	5.569	5.615	5.603
c (Å)	7.898	7.927	7.921
V (Å)^3	244.982	249.987	248.693
g-factor	2.045	2.053	2.047
A	-4.31×10^{-4}	-4.00×10^{-4}	-2.99×10^{-4}
B	8.70×10^{-6}	6.10×10^{-6}	1.61×10^{-6}
C	1.16×10^{-8}	1.68×10^{-8}	2.61×10^{-8}
$\bar{\alpha}$ (°C^{-1})	13.80×10^{-6}	13.50×10^{-6}	13.10×10^{-6}

3.2. Scanning electron microscopy

The SEM photographs of SFIMO are shown in Figure. 2(a-c). The average grain size of each sample was estimated by taking an average of 10 grains. The estimated average grain sizes of SFIMO are varying between 2 – 3.5 μm for all compositions. It is also found that the value of grain size of

Sr_2FeMoO_6 is 2.19 against 1−2 reported by Leilei Zhang et al [7] which is slightly higher. This change may be due to different preparation method and sintering conditions.

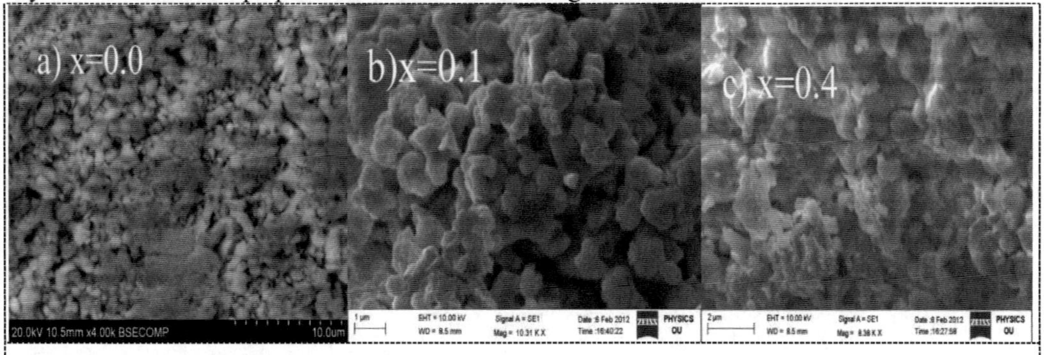

Figure 2(a-c). SEM photo graphs of $Sr_2Fe_{1-x}In_xMoO_6$ (x = 0.0, 0.1 and 0.4 wt.%.) samples

3.3. Electron spin resonance

Figure. 3(a-c) shows the ESR spectrum of SFIMO samples at room temperature. All the figures show a similar spectrum. A peak is resolved at around 310 mT. The g-factor of SFIMO samples were calculated using equation g = hυ / βH, where h = Planck's constant, υ = frequency of magnetic field, β = Bohr magneton and H = Applied magnetic field. The calculated g-factor are given in Table 1. It is found that the g-factor of SFIMO samples vary between 2.045 to 2.053; which is due to the Fe^{3+} ions. The measured g-factor, is clearly identified with a spin-only ground state. This resonance corresponds, in the band picture of Ref. [1], to the localized $3d^5$ Fe cores.

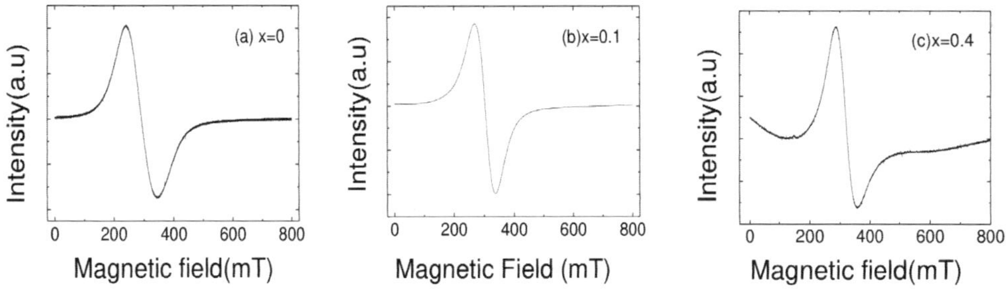

Figure 3(a-c). ESR spectrum of $Sr_2Fe_{1-x}In_xMoO_6$ (x = 0.0, 0.1 and 0.4 wt.%) samples at room temperature.

3.4. Thermal expansion

The thermal expansion characteristics ($\Delta L/L_o$) of SFIMO were obtained using dilatometer in the temperature ranging from 40–400°C in air are shown in Figure. 4. The thermal expansion depends on the electrostatic forces within the lattice, which depend on the concentration of positive and negative charges and their distances within the lattice [9]. The thermal expansion of a lattice with a certain structure and fixed oxygen to metal stoichiometry is characterized by a steady coefficient of thermal expansion (CTE), α, caused by the thermal lattice vibrations. The experimental data of thermal expansion ($\Delta L/L_o$) as a function of temperature (T) of SFIMO in the present study can be fitted to a polynomial of degree 2 as

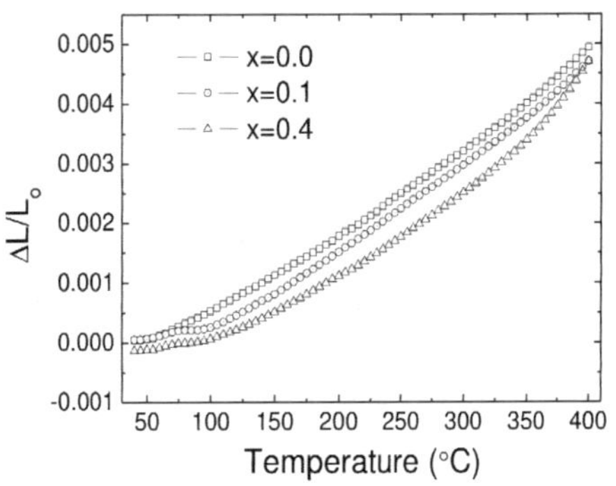

Figure 4. The thermal expansion versus temperature of $Sr_2Fe_{1-x}In_xMoO_6$ (x = 0.0, 0.1 and 0.4 wt.%) samples.

$$\frac{\Delta L}{L_o} = A + BT + CT^2 \qquad (1)$$

where A, B and C are the constants. The value of A, B and C obtained for different compositions of SFIMO were determined by fitting experimental data on $\Delta L/L_o$ with temperature T (°C) using eq.(1) and given in Table 1. The value of α can be determined using the first derivative of eq.(1) with respect to temperature as

$$\alpha = \frac{d}{dT}\left(\frac{dL}{L_o}\right) = B + 2CT \qquad (2)$$

Eq.(2) shows that the value of α at any temperature can be obtained using constants B and C. The average value of CTE ($\overline{\alpha}$) in the temperature range 40−400°C for all SFIMO materials are calculated and given in Table 1. It is observed from Table 1 that the value $\overline{\alpha}$ increases with increase of temperature while decreases with increase of indium content. Falcon et al [10] investigated the oxidation profiles of $Sr_2FeMoO_{6-\delta}$oxide. Their results showed that oxygen was relatively easily incorporated into the perovskite structure. With the increase in the oxygen vacancy concentration, oxygen permeation and lattice expansion become much more pronounced [11]. Therefore, in the present study the value of $\overline{\alpha}$ increase with increase of temperature may be attributed to increase of concentration of oxygen vacancy. The $\overline{\alpha}$ decreases with increase of indium composition may be due to decreases of covalent nature (increase of ionic nature) of SFIMO samples.

4. Conclusions

It has been found that the lattice parameters and unit cell volume increases with increase of indium content in SFIMO samples. The g-factor of SFIMO samples is about 2.04, which confirm the iron is in Fe^{3+} state. The value of $\overline{\alpha}$ increases with temperature and decreases with indium content in SFIMO.

References

[1] Kobayashi K I, Kimura T, Sawada H, Terakura K, Tokura Y, 1998 Nature **395** 677
[2] Ray S, Middey S, Jana S, Banerjee A, Sanyal P, Rawat R, Gregoratti L, Sarma D D, 2011 J of EPL. **94** 47007
[3] Min Feng Lu et al 2007 J. Alloys and Compounds **428** 214
[4] Anurag Gaur et al 2008 J. Alloys and Compounds **460** 581
[5] Falak Sher, Venimadhav A, Blamire M G, Dabrowski B, Kolesnik S, Paul Attfield J 2005 Solid State Sciences **7** 912
[6] Feng X M, Rao G H, Liu G Y, Yang H F, Liu W F, Ouyang Z W, Liang J K 2004 Phy. B **344** 21
[7] Leilei Zhang et al 2010 J. Power Sources **195** 6356
[8] Markandeya Y, Saritha D, Vithal M, Singh A K, Bhikshamaiah G 2011 J. Alloys and Compounds **509** 5195
[9] Prasanth kumar V et al 2008 Materials Chemistry and physics **112** 711
[10] Falcon H et al 2004 Appl. Catal. B-Environ.**53** 37
[11] Onuma S et al 2004 Solid State Ionics **174** 28

Preparation of silver-hydroyapatite/PVA nanocomposites: Giant dielectric material for industrial and clinical applications

Md Jamal Uddin[1*] T R Middya[2] and B K Chaudhuri[1]
[1]Department of Solid State Physics, Indian Association for the Cultivation of Science, Jadavpur, Kolkata-700032, India
[1]Department of Physics, Jadavpur University, Kolkata-700032, India

E-mail: sspjamal@gmail.com

Abstract: Pure hydroxyappatite $Ca_{10}(PO_4)_6(OH)_2$ (or HAP) was prepared from eggshell and potassium dihydrogen phosphate (KH_2PO_4) by a simple self-chemical reaction method. The clean eggshell was heated at 800 °C in air giving the source of CaO. Appropriate amount of CaO was dissolved in KH_2PO_4 solution at 37°C for few days. The PH value decreases with increasing the duration of preparation of HAP. Silver nanoparticles derived from silver nitrate solution using black tea leaf extract had been introduced to hydroxyapatite due to its biocompatibility. The unique size-dependent properties of nanomaterials make them superior and indispensable. In this work, hydroxyapatite-silver nanoparticles/polyvinyl alcohol (PVA) composites with 4 different concentrations of hydroxyapatite (1-4 wt %) were prepared by bio-reduction method. Several techniques like XRD and SEM were used to characterize the prepared samples. Frequency dependent capacitance and conductance of the samples were measured using an impedance analyzer. The results showed a remarkable increase in dielectric permittivity (~5117) with low loss (~0.23) at1000 HZ and room temperature (300K) for 4wt% Hydroxapatie-Silver/PVA nanocomposite. Such nanocomposite might be directly applied in manufacturing clinical devices and also for embedding capacitor applications.

***Keywords*:** Hydroxyapatite, Silver nano-biomaterials, Dielectric properties

1. Introduction

Hydroxyapatite (HAP) with chemical formula Ca $_{10}(PO_4)_6(OH)_2$ is a well known bioactive and biocompatible inorganic material [1]. It is a polarized compound [2] and highly important material for various applications, such as tissue engineering and bone replacement [3], removal of heavy metals form waste water etc. [4]. Several attempts have also been made to prepare HAP from egg shells [5] and the physical properties of HAP are process-dependent. So far different methods have been developed to prepare of HAP, such as sol-gel process [6], hydrothermal syntheses [7] emulsion processing [8], mechano-chemical method [10] and chemical co-precipitation [11]. Most of the methods are little bit complicated and high temperature processing is required. In the present work we have reported a simple method to synthesize high quality uniform size HAP nanoparticles from egg shell by using a chemical reaction of CaO and KH_2PO_4 solution.

It is well known that silver ions exhibit strong inhibitory effects towards a broad spectrum of bacterial strains [12]. Recently (Mostafa et al. [13], studied the effect of two types of polymers polyvinyl alcohols (PVA) and polyvinylpyrrolidone (PVP) on the synthesis of nanohydroxyapatite. The results showed that the polymer affected the crystal morphology and not the phase composition of the product.

In this communication hydroxyapatite-silver nanoparticles/polyvinyl alcohol composites have been prepared where environment friendly technique has been used to synthesize silver nanoparticles derived from silver nitrate solution with the help of black tea leaf extracts which acts as the reducing agent and such Ag derived nanoparticles have been utilized with HAP as the fillers within the polymer matrix. Polyvinyl alcohol (PVA) has been taken as the polymer host matrix which acts as the capping agent due to its advantages having high mechanical strength, water-solubility, good environmental stability and easy processability. The frequency dependent dielectric properties of the composites have been studied over the broad range of frequency (1 kHz-1MHz) at room temperature (RT). This work describes the physical characteristics of silver nanoparticles when it is conjugated with hydroxyapatite polymer matrix composite.

2. Experimental details

Hydroxyapatite having the chemical formula $Ca_{10}(PO_4)_6(OH)_2$ has been prepared in the form of nanoparticles from egg shells collected from the egg shop. The surfaces both inner and the outer were properly cleaned and washed with flowing distilled water. After drying, the shells are first heated around 350°C at the heating rate of 3°C/minute for two hours when smoke comes out. During such heating process, organic substances are removed and the shells become black. The heating temperature of the furnace is increased at the rate of 1^0C/min upto 800^0C, when the dark black shells slowly converted to fine white powder. This heating process is continued for 2 hours and then furnace cooled to room temperature. The white powder is nothing but CaO obtained from the egg shell which is mostly $CaCO_3$ ($CaCO_3$ + heating in air $\rightarrow$ CaO +CO_2).

The highly pure CaO thus obtained is immediately used for the preparation of HAP using KH_2PO_4 solution in deionised water. The reaction concentration was determined considering the fact that Ca/P ratio remains ~1.67. The CaO obtained from the egg shells was slowly added to the solution in a beaker and well mixed for 10 minutes. Subsequently, the beaker was placed in the heating chamber kept at 37°C and keeping the pH of the sample around ~12 by adding NH_4OH solution. The reaction was allowed to complete in different beakers for different period of time viz. 1×24 - 7×24 hrs. The HAP is considered to be produced according to the following reaction:

$$10CaO + 6KH_2PO_4 + 2H_2O \rightarrow Ca_{10}(PO_4)_6(OH)_2 + 6KOH$$

3. Results and discussion

Figure 1(a) shows the variation of pH of the reacting solution with time. The pH decreases with increase of soaking time. It is also noticed that the pH of the reacting solution has a decreasing tendency with time and the grain size decreases with increasing reaction time. It means that acidity of the solution slowly increases with increase of days and the formation of HAP finally stabilizes. Figure 1(b) shows the XRD patterns of CaO obtained from the egg shells and those of the HAP nanoparticles prepared from this CaO powder using various socking time (1 to 7days). No remarkable change is observed in the XRD spectra, with respect to the peak positions. However, notable broadening of the concerned peaks along with reduction in intensity indicates that significant reduction of grain size with time. On comparison of Figure 1(b) and 1(c) we see that the number of peaks in the later one have been reduced because of the addition of polymer which greatly affects the crystal morphology of the composite sample.

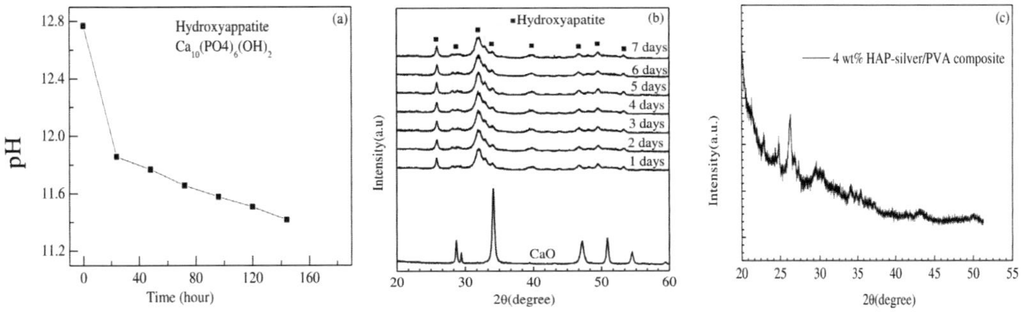

Figure.1(a) Variation of pH of the soaking solution with time (b) XRD patterns of the CaO obtained from the egg shells and those of the HAP nanoparticles prepared from this CaO powder socking in KH_2PO_4 solution for different days and (c)XRD pattern of 4 wt% HAP-silver/PVA nanocoposite.

Figure 2(a & b) show the SEM micrographs of HAP powder and 4 wt% HAP-silver/PVA nanocoposite respectively. Figure 2(a) indicates that the powder exhibits crystalline nanoparticles and are also correspond to the XRD result [see Figure 1 (b)] whereas in Figure 2(b) many white silver nanoparticles of size about100 nm have been observed which are well dispersed within the polymer matrix.

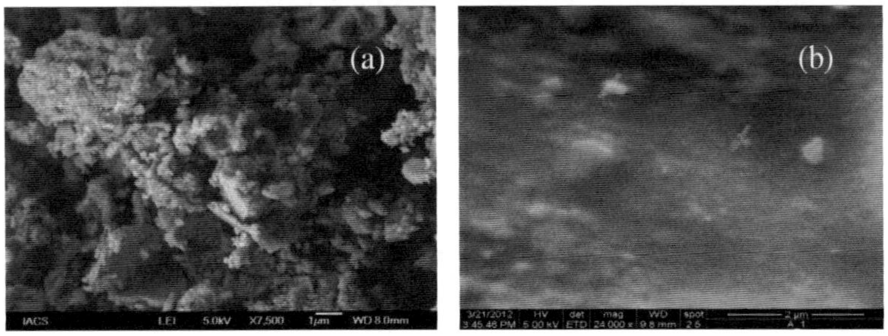

Figure 2. FESEM micrographs of (a) the prepared hydroyapatite powder and (b) 4 wt% HAP-silver/PVA nanocomposite film.

Figure 3(a) represents the frequency dependence of relative permittivity of HAP-silver/PVA nanocomposite with different concentrations of HAP at room temperature. The dielectric permittivity decreases with increasing frequency and we get the maximum value of dielectric permittivity 5117 for 4wt% HAP-silver/PVA nanocomposite at 1 kHz and room temperature (330K). This is because of the facts that at lower frequency dipole moment of hydroxyl ions in HAP follow the variation of field while at higher frequency those ions do not follow the variation of field and dielectric permittivity value decreases. Figure 3(b) shows the frequency dependence of dielectric loss for different composites at room temperature. The dielectric loss increases with increasing the HAP content and we get the value of dielectric loss 0.23 at 1 kHz and RT which is due to the interfacial polarization attributed to the Maxwell Wagner Sailor (MWS) effect.

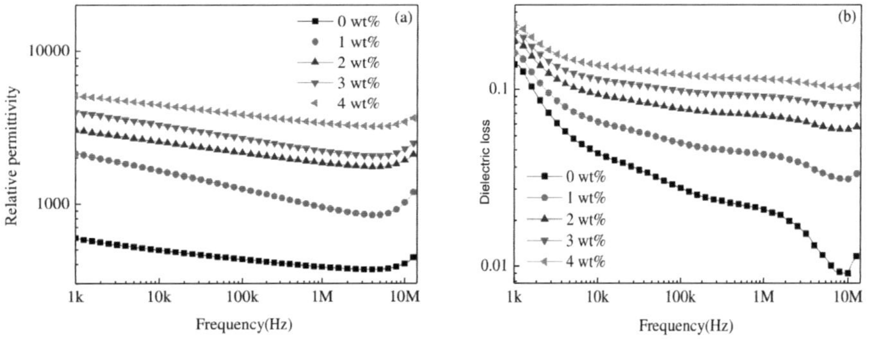

Figure 3. Frequency dependence of (a) relative permittivity and (b) dielectric loss of HAP-silver/PVA nanocomposites with different concentrations of HAP at room temperature.

4. Conclusions

A simplest self-assisted chemical reaction method was used to prepare pure fine particles of HAP from egg shells. HAP-silver/PVA nanocomposites have been prepared using environment friendly technique. Silver nanoparticles about 100 nm diameters were observed well dispersed within the polymer matrix. The addition of polymer to HAP reduced the crystalline nature of the composite. The frequency dependent dielectric studies of the samples have been made exhibited very high relative permittivity value alongwith low dielectric loss at room temperature. Such a nanocomposite scaffold investigated in this study tends to be promising for bone tissue engineering as well in the fabrication of charge storing devices.

Acknowledgements

Authors acknowledges grateful to the Department of Science and technology (DST), Government of India for providing partial financial assistance to carry out this work.

References

[1] Lee S J, and Oh S H 2003 *Mater. Lett.* **57** 4570-.4574
[2] Nakamura S, Kobayashi T and Yamashita K 2002 *J. Biomed. Mater. Res.* **61** 593-599.
[3] Vallet-Regi M and Gonzalez-Calbert J M 2004 *Prog. Solid State Chem.* **32** 1
[4] Layrolle P, Ito A and Tateishi T 1998 *J. Am. Ceram. Soc.* **81** 1421-1428
[5] Balazsi C, Köver Z, Horvath E, Nemeth C, Kasztovszky Z, Kurunczi S and Wéber F 2007 *Mater. Sci. Forum* **105** 537-538

[6] Weng W and Baptista J L1999 *J. Amer. Ceram. Soc.* **82** 27-32
[8] Suchanek W L and Yoshimura M 1998 *J. Amer. Ceram. Soc.* **81** 765-767
[9] Lim G K, Wang J, Ng S C and Gan L M 1999 *J. Mater. Chem.* **9** 1635-1639
[10] Kim W, Zhang Q and Saito F 2000 *J. Mater. Sci.* **35** 5401-5405
[11] Mavis B and Tas A C 2000 *J. Amer. Ceram. Soc.* **83** 989-991
[12] Clement J L and Jarrett P S 1994 *Met. Based Drugs* **1** 467–482
[13] Mostafa A A, Oudadesse H, Mohamed M B, Foad E S , Le Gal Y and Cathelineau G 2009 *Chem. Eng. J.* **153** 187–192

IOP Conf. Series: Materials Science and Engineering **73** (2015) 012071 doi:10.1088/1757-899X/73/1/012071

Fabrication and characterization of ZnO/Al/ZnO multilayers by simultaneous DC and RF magnetron sputtering

K. K. Nagaraja*, A. Santhosh Kumar and H. S. Nagaraja

Materials Research Laboratory, Department of Physics, National Institute of Technology Karnataka, Surathkal-575025, India

Email: nrajkk@gmail.com

Abstract. The present investigation reports the fabrication and characterization of multilayered transparent electrodes by simultaneous DC and RF magnetron sputtering on glass substrates. The multilayer structure consists of three layers (ZnO/Al/ZnO). The influence of Al layer thickness on the electrical and optical properties was investigated. Optimum thickness of Al was determined for high transmittance and good electrical conductivity. High quality films having resistance as low as 25 Ω/sq with optical transmittance upto 65% were obtained at room temperature.

1. Introduction

Transparent conducting oxides (TCOs) are widely used in many technological applications such as solar cells, optoelectronic devices, consumer electronics etc. Tin doped indium oxide (ITO) one of the most used TCOs at present. But, its toxicity and high cost made to look for other alternatives. Zinc oxide (ZnO) is one of the most promising materials for the fabrication of the next generation optoelectronic devices in the UV region and optical display devices. It has many advantages such as less toxicity, less cost and availability in the large scale. It is a direct band gap semiconductor having an energy gap of 3.37 eV at room temperature with 60 meV exciton binding energy [1-3]. It has been subjected to extensive research by the research community due to its attractive properties. The doping effect on its properties is still under investigation. Doped ZnO films have many interesting potential applications such as transparent electrodes, piezoelectric devices, gas sensors, photo diodes and also it can be used as window layer in hetero junction solar cells [4-6]. The main objective of developing new materials for TCO is to achieve less resistivity and good transparency in the visible range of the electromagnetic spectra.

Different methods were employed to develop good TCOs by many researchers. Multilayer stacks of TCOs are one such approach employed to improve the properties of TCOs [7-10]. Kawashima et al. reported improvement of thermal stability in F:SnO_2/ITO stack compared to single layer ITO [11]. Similar stacks of Sn:CdO/$CdIn_2O_4$/Cd_2SnO_4 showed better conductivity than ITO with comparable transparency in the visible region [12]. The deposition of three layer structures with a metallic film embedded between two dielectric layers (D/M/D) exhibits better TCO properties with thickness less than a single layer TCO. These D/M/D structures are widely used in energy saving widows as optical filters [13]. High refractive index zinc oxide makes it a suitable candidate as a dielectric layer. Few early reports on D/M/D include the development of ZnO/Ag/ZnO & ITO/Ag/ITO [14], ITO/Cu/ITO [15], TiO_2/Ag/TiO_2 [16] and ZnO/Cu/ZnO [17]. Ting et al. reported insertion of thin Al interlayer for the improvement of electrical conductivity of ZnO [18].

In the current work, we report the deposition of ZnO/Al/ZnO multilayered structure by simultaneous RF sputtering of ZnO and DC sputtering of Al targets at room temperature. We investigate the effect of Al layer on the properties of multilayers.

2. Experimental

Multilayers of ZnO/Al/ZnO were deposited on pre-cleaned glass substrates. Target materials are compacted zinc oxide (99.99% purity, 2 inch dia., 3 mm thick) and metallic Al (99.999% purity, 2 inch dia., 3 mm thick). The magnetron sputtering system consists of 12 inch vacuum chamber which is equipped with 2 inch magnetron cathodes and RF and DC power supplies. Prior to deposition chamber was pumped down to a pressure less than 5×10^{-6} m bars. High pure argon was used as sputter gas and working pressure was adjusted at 0.001 m bars. Sputtering power was kept 150 Watts for ZnO and 15 Watts for Al. The optical characterizations of alms have been carried out using UV–Vis spectrophotometer (USB2000). Electrical resistance of the deposited multilayers was measured using point probe technique using Keithley 2400 sourcemeter and Keithely 2000 multimeters.

3. Results and discussion

X-ray diffraction patterns of the deposited multilayers are shown in Figure 1. The thickness of the ZnO layer was found around 100 nm and Al layer thickness varies from 5 nm to 30 nm. Thickness of Al layer was calculated by considering the deposition rate of Al. A single peak at $2\theta \approx 34.75°$ confirms the existence of ZnO phase. It is also revealed that the deposited films show preferred orientation along (0 0 2) planes. The intensity of (0 0 2) peak found to decrease with the increase in Al layer thickness which indicates the decrease in the crystallanity. No prominent peaks corresponding to Al were observed. This may be because of the fact that the thickness of the ZnO layer is large compared to that of Al. The observed 2θ positions for multilayer film was bit high compared to that of ZnO film studied here. This is because of residual stressesshows optical transmittance spectra of ZnO/Al/ZnO multilayer with variation in Al deposition time.

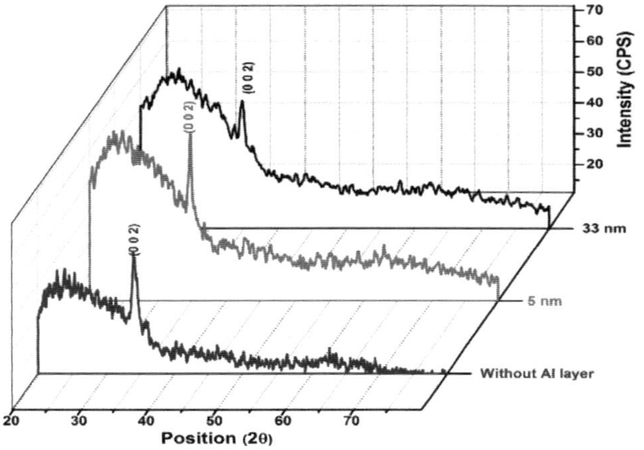

Figure 1. XRD patterns of ZnO and multilayers with different Al interlayer thickness

Optical transmittance spectra of the deposited films are shown in Figure 2. For undoped films a good transmittance over 80% was observed in the visible region of the electromagnetic spectra. The average transmittance of the films in the visible region decreases as the Al layer thickness increases. Absorption edges of the multilayers shift towards higher wavelengths with increase in the thickness of Al interlayer thickness. As the thickness of Al interlayer increases the electrical resistance decreases (Figure 3). Multilayered thin films show low resistance because of Al interlayer. The conduction electrons were supplied from donor sites associated with excess zinc and aluminum ions on substitutional sites of zinc ions in the in the as deposited samples. The lowest resistance obtained for the multilayer thin film with Al interlayer thickness 33 nm.

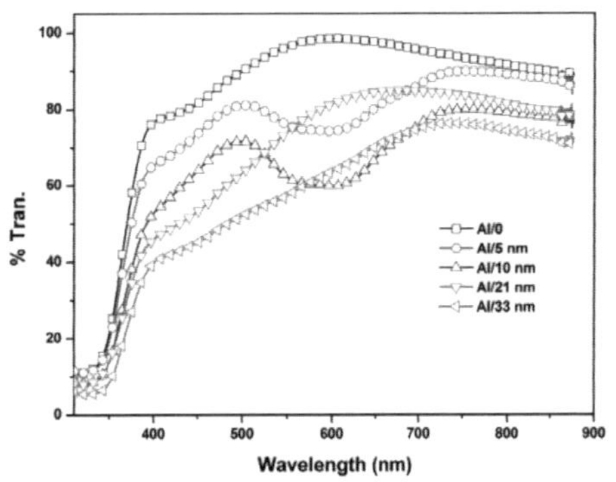

Figure 2. Transmittance spectra of ZnO and multilayer thin films.

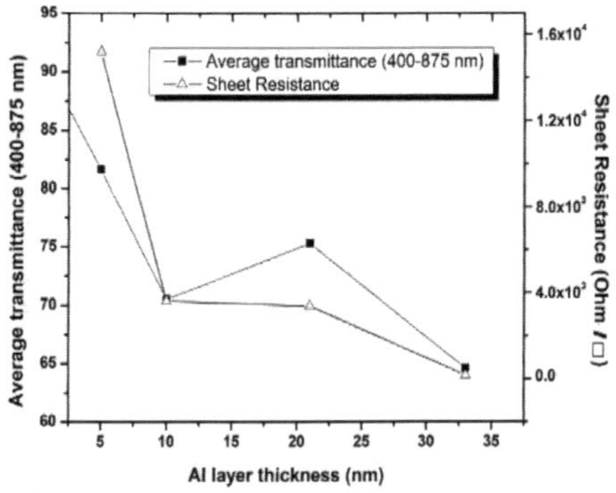

Figure 3. The variation of sheet resistance and average transmittance (400-875nm)

4. Conclusions

We deposited multilayers of ZnO/Al/ZnO were deposited by simultaneous RF and DC sputtering. Obtained multilayers were characterized using optical transmittance spectra and XRD. Sheet resistance of the multilayers found to decrease with the increase in the Al interlayer thickness. XRD patterns of the multilayers show reflections corresponding to Wurtzite ZnO, no peaks corresponding to Al observed due to thicker ZnO layers. Transmittance spectra show a considerable decrease in the average transmittance (in the visible region) of the multilayers.

References

[1] S. J. Pearton, D. P. Norton, K. IP Y. W. Heo and T. Steiner 2004 J. Vac. Sci. Technol. B 22(3), 932-948.
[2] D. M. Bagnall, Y. F. Chen, M. Y. Shen, Z. Zhu, T. Goto and T. Yao 1998 *J. Cryst. Growth* **184,** 605
[3] Y. Chen, D. Bagnall and T. Yao 2002 *Mater. Sci. Engg. B* **75**, 190.
[4] D. C. Look 2001 *Mater. Sci. Eng. B* **80**, 383.
[5] J. M. Lee, K.-K. Kim S.-J. Park and W.-K. Choi 2001 *Appl. Phys. Lett.* **78**, 2842.
[6] K. D. Mitzner, J. Strehagen and D. W. Galipeau 2003 *Sens. Actuators B* **93,** 92.
[7] R. Babar, P. R. Deshamukh, K. Y. Rajpure, R. J. Deokate, D. Haranath, C. H. Bhosale 2008 *J. Phys. D: Appl. Phys.* **41**, 135404.
[8] Z. Ben Ayadi,L. El Mir, K. Djessas and S. Alaya 2007 *Nanotechnology* **18**, 445702.
[9] X. W. Sun, L. D. Wang, H. S. Kwok 2000 *Thin Solid Films* **360,** 75.
[10] H. Kim, J. S. Horwitz, S. B. Qadri, D. B. Chrisey 2002 *Thin Solid Films* 420 –421, 107.
[11] T. Kawashima, T. Ezure, K. Okada, H. Mastui, K. Goto, N. Tanabe 2004 *J. Photochem. Photobiol.-A* **164,** 199.
[12] E. J. J. Martin, M. Yan, M. Lane, J. Ireland, C. R. Kannewurf, R. P. H. Chang 2004 *Thin Solid Films* **461**, 309.
[13] M. Fahland, P. Karlsson, C. Charton 2001 *Thin Solid Films* **392**, 334.
[14] E. Kusano, J. Kawaguchi, K. Enjoiji, 1986 *J. Vac. Sci. Technol. A.* **4(6),** 2907.
[15] M. Bender, W. Seelig, C. Daube, H. Frankenberger, B. Ocker, J. Stollenwerk 1998 *Thin Solid Films* **326**, 67.
[16] G. Leftheriotis, S. Papaefthimiou, P. Yianoulis 2000 *Solid State Ionics*, **136–137**, 655
[17] D. R. Sahu, J.-L. Huang, 2006 Appl. Surf. Sci. **253** 827.
[18] J.-M. Ting, C.-K. Lin 2006 *J. Am. Ceramic Soc.* **89(12),** 3676.

Synthesis, characterization and low temperature electrical conductivity of Polyaniline/$NiFe_2O_4$ nanocomposites

G D Prasanna[1*], V B Prasad[3] and H S Jayanna[2]

[1] PG Department of Physics, JSS College, Ooty Road, Mysore, Karnataka – 570 025

[2] Department of Physics, Kuvempu University, Shankarghatta, Karnataka – 577 451

[3] Department of Physics, Indian Institute of Science, Bangalore, Karnataka – 560 012

E-mail: prasannagd@gmail.com

Abstract. Conducting polymer/ferrite nanocomposites with an organized structure provide a new functional hybrid between organic and inorganic materials. The most popular among the conductive polymers is the polyaniline (PANI) due to its wide application in different fields. In the present work nickel ferrite ($NiFe_2O_4$) nanoparticles were prepared by sol-gel citrate-nitrate method with an average size of 21.6nm. PANI/$NiFe_2O_4$ nanoparticles were synthesized by a simple general and inexpensive in-situ polymerization in the presence of $NiFe_2O_4$ nanoparticles. The effects of $NiFe_2O_4$ nanoparticles on the dc-electrical properties of polyaniline were investigated. The structural components in the nanocomposites were identified from Fourier Transform Infrared (FTIR) spectroscopy. The crystalline phase of nanocomposites was characterized by X-Ray Diffraction (XRD). The Scanning Electron Micrograph (SEM) reveals that there was some interaction between the $NiFe_2O_4$ particles and polyaniline and the nanocomposites are composed of polycrystalline ferrite nanoparticles and PANI. The dc conductivity of polyaniline/$NiFe_2O_4$ nanocomposites have been measured as a function of temperature in the range of 80K to 300K. It is observed that the room temperature conductivity σ_{RT} decreases with increase in the relative content of $NiFe_2O_4$. The experimental data reveals that the resistivity increases for all composites with decrease of temperature exhibiting semiconductor behaviour.

1. Introduction

Among the conducting polymers, polyaniline (PANI) has been extensively studied due to its easy synthesis, low cost, excellent environmental stability, and high electrical conductivity [1]. PANI has attracted much attention because of its potential application in molecular electronics, electromagnetic interference (EMI) shielding, chemical sensors, antistatic coatings, rechargeable batteries, corrosion inhibitors, absorbing materials [2], etc. It is well known that conducting polymers can effectively shield electromagnetic waves generated by an electric source. However, electromagnetic waves from a magnetic source can be effectively shielded only by magnetic materials [3]. Thus, a combination of magnetic constituents and conducting polymeric materials opens new possibilities to provide shielding from various electromagnetic sources.

Many kinds of materials can be used as magnetic part of the composite materials, such as Ni Fe, Co, Fe_2O_3, $ZnFe_2O_4$, $CoFe_2O_4$, $NiFe_2O_4$, Co_3O_4 and $BaFe_{12}O_{19}$ [4,5]. Among magnetic materials, the spinel ferrites exhibit remarkable magnetic properties particularly in radio-frequency region, physical flexibility, high electrical resistivity, mechanical hardness and chemical stability [6]. Their properties are particularly enhanced when the size of the particles reaches the nanometer range [7]. $NiFe_2O_4$ is an important magnetic ferrite which has been found its extensive applications in diverse areas such as permanent magnets, recording media, Ferro fluids and gas sensors.

In this paper, $NiFe_2O_4$ nanoparticles were prepared by citrate-nitrate method, and $NiFe_2O_4$/PANI nanocomposites were then synthesized via an in situ polymerization of aniline monomer in an aqueous. It is pointed out that the $NiFe_2O_4$ and PANI/$NiFe_2O_4$ nanoparticles obtained in our experiment have cubic spinel phase. The electrical, thermal stability and the structural characterization have been investigated through XRD, FTIR and SEM methods. The temperature dependence of dc conductivity on PANI/$NiFe_2O_4$ nanocomposites was investigated in the temperature range 80-300 K.

2. Preparation

Nickel ferrite nanoparticles have been prepared by citrate-nitrate method [5, 8] using nickel nitrate and ferric nitrate in the molar ratio 1:2. A certain amount of citric acid was dissolved in a molar ratio 1:1 of nitrate to citric acid. The combustion reaction of during the process can be described as:

$$Ni(NO_3)_2.6H_2O + 2Fe(NO_3)_3.9H_2O + C_6H_8O_7.9H_2O + H_2O \xrightarrow[pH=7]{Stir\ at\ 120^0C} Sol\ Gel \xrightarrow[250^0C]{Ignition} NiFe_2O_4 + CO_2 + NO + H_2O$$

Polyaniline/$NiFe_2O_4$ nanocomposites were prepared by a simple in-situ polymerization method [5, 8, 9] in the presence of $NiFe_2O_4$ nanoparticles. Different PANI/$NiFe_2O_4$ nanocomposites were synthesized using 0wt%, 20wt% (N1), 40wt% (N2), 60wt% (N3), and 80wt% (N4) of $NiFe_2O_4$ nanoparticles with respect to aniline monomer.

3. Characterization

X-ray diffraction (XRD) pattern of the samples were collected on a Philips X'pert Pro MPD X-ray diffractometer with Cu K_α radiation (λ=1.5418 Å) at a scanning rate of 0.05^0 per step with 15s/step in the range of $10–80^0$. FT-IR spectroscopy was recorded on a PERKIN-ELMER (model-1000) spectrometer in the range of 400–4000 cm^{-1} using KBr pellets. Scanning Electron Micrograph (SEM) was carried out in a Hitachi model S-3200N microscope.

The DC electrical conductivity was carried by four probe method using Keithley source meter (model 220) and a multimeter (Model 2000) in the temperature range 80-300 K by using Lakeshore auto tuning temperature controller (Model-330). The measurements were recorded during cooling cycle.

4. Result and discussion

4.1. Structural and morphological characterization

Diffraction pattern for $NiFe_2O_4$ nanoparticles (figure 1) shows a cubic spinel phase with fcc crystal structure [10] with no extra reflections and are well-indexed to the crystallographic planes of spinel ferrite (220), (311), (222), (400), (422), (511), and (440) with 100% intensity at $2\theta = 31.04^0$ that corresponds to the reflection plane (311). The diffraction result well agrees with the standard XRD data (JCPDS card No. 74-1913). The average particle size was determined by Debye-Scherrer's formula and for pure $NiFe_2O_4$ it is found to be 21.6 nm; and increases with decreasing content of $NiFe_2O_4$ in PANI. It is also evident from the figure that, the pure polyaniline has broad peak in the region of $20-30^0$ with a maximum at around 25^0 indicating its semicrystalline nature and with increasing content of $NiFe_2O_4$ nanoparticles in the composite the relative intensity of the characteristic peaks of PANI/$NiFe_2O_4$ nanocomposites increase regularly, clearly indicating an increase in the $NiFe_2O_4$ content in nanocomposites.

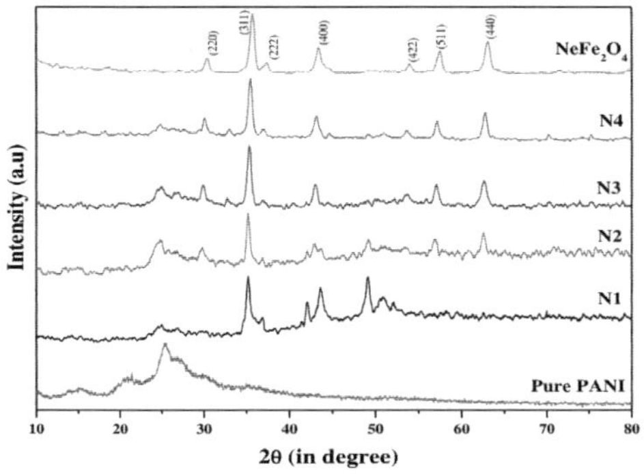

Figure 1. XRD patterns of pure PANI, $NiFe_2O_4$ and PANI/$NiFe_2O_4$ nanocomposites.

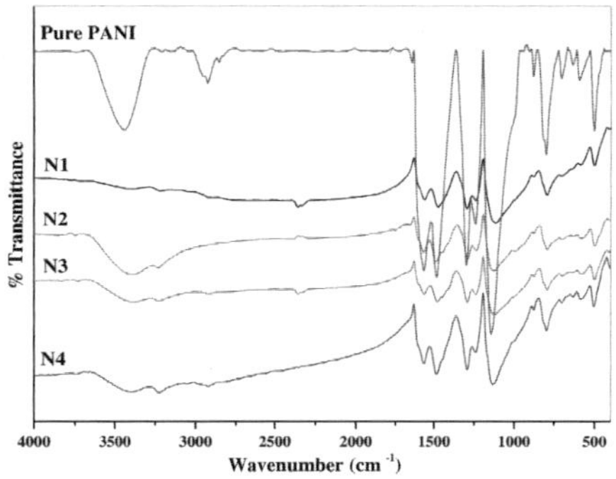

Figure 2. FTIR spectra of PANI/$NiFe_2O_4$ nanocomposites.

Figure 2 shows the FTIR spectra of pure polyaniline and PANI/$NiFe_2O_4$ nanocomposites. The characteristic peaks of PANI occur at 1563, 1483, 1298, 1242, 1143 and 800cm^{-1} [11]. The peaks at 1563 and 1483 cm^{-1} are attributed to the characteristic C=C stretching of the quinoid and benzenoid rings, the peaks at 1298 and 1242 cm^{-1} are assigned to C-N stretching of the benzenoid ring, the broad peak at 1143 cm^{-1} which is described by MacDiarmid *et al.* [12] as the "electronic-like band" is associated with vibration mode of N=Q=N, indicating that HCl doped PANI is formed in our samples, and the peak at 800 cm^{-1} for PANI is attributed to the out of-plane deformation of C-H in the 1,4-disubstituted benzene ring [11]. We can notice that the FTIR spectra for PANI and PANI/$NiFe_2O_4$ nanocomposites are similar and the characteristic peak value of PANI at 1143 and 1483 cm^{-1}. The peaks of the PANI/$NiFe_2O_4$ nanoparticles shift to lower wave numbers. This indicates that the PANI/$NiFe_2O_4$ powders are composed of $NiFe_2O_4$ nanocomposites, which infer there is some interaction between $NiFe_2O_4$ particles and PANI backbone.

Figure 3. SEM images of pure $NiFe_2O_4$, PANI and PANI/$NiFe_2O_4$ nanocomposites (40 wt% of $NiFe_2O_4$).

The SEM images of $NiFe_2O_4$ nanoparticles, PANI and PANI/$NiFe_2O_4$ nanocomposites are shown in the figure 3. It can be observed that the $NiFe_2O_4$ nanoparticles appear uniform sphere-like shape. The SEM micrograph of the PANI indicates the big globular agglomerates with smooth surface. It is noticeable that the SEM micrograph of PANI/$NiFe_2O_4$ (40 wt%) nanocomposites presents different morphology as compared with the bare $NiFe_2O_4$ nanoparticles. The PANI layers are wrapped on the surface of $NiFe_2O_4$ nanoparticles appearing as small aggregated globules [5, 11].

4.2. DC electrical properties

The temperature dependent of resistivity of PANI was measured in the temperature range of 80 to 300K. The resistivity of PANI increases with decreasing temperature exhibiting semiconducting behavior as shown in figure 4. The room temperature conductivity for pure PANI is 5.15 Scm^{-1} and it increases with temperature.

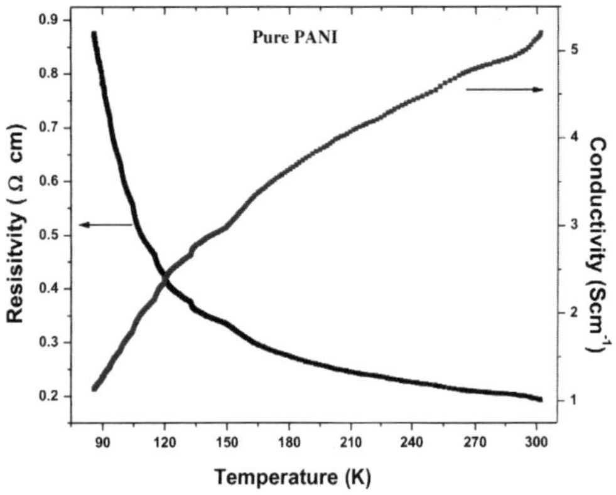

Figure 4. Temperature dependent of resistivity and conductivity of pure Polyaniline.

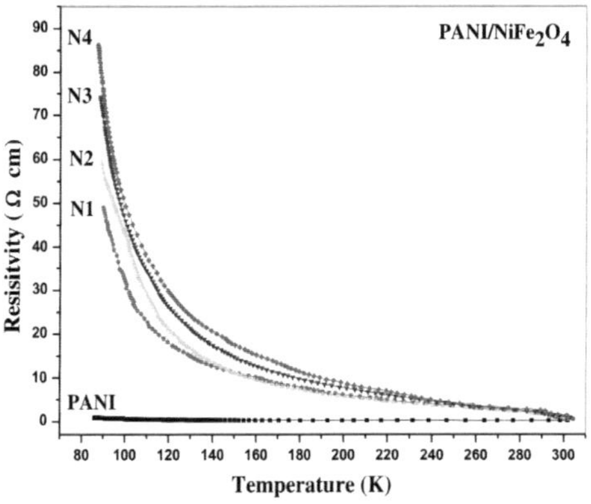

Figure 5. Temperature dependent resistivity of PANI/$NiFe_2O_4$ nanocomposites.

Figure 5 shows, the resistivity of all PANI/$NiFe_2O_4$ nanocomposites which increases with decreasing temperature exhibiting semiconducting behavior [4]. Since the $NiFe_2O_4$ nanoparticles are embedded in the PANI matrix, interactions between the polymer matrix and $NiFe_2O_4$ nanoparticles will increase the charge carrier scattering and thus increase the sample's resistivity resulting in decrease in the conductivity of PANI/$NiFe_2O_4$ composites with increase in $NiFe_2O_4$ content [13, 14]. Other effects like increased charge carrier trapping, either by the nanoparticles themselves or by morphological changes and defects induced by them, could also play a role.

5. Conclusion

In summary, $NiFe_2O_4$ nanoparticles were prepared by sol-gel citrate-nitrate method with an average size of 21.6nm. PANI/$NiFe_2O_4$ nanoparticles were synthesized by a simple general and inexpensive in-situ polymerization in the presence of $NiFe_2O_4$ nanoparticles. The result of XRD, FTIR spectroscopy and SEM shows the formation of the composite and indicate an interaction between PANI and $NiFe_2O_4$ nanoparticles. The effect of $NiFe_2O_4$ nanoparticles on the dc-electrical properties of polyaniline was investigated and the conductivity of the nanocomposite decreases with increase in the doping concentration of $NiFe_2O_4$.

6. References

[1] Gospodinova N and Terlemezyan L 1998 *Prog. Polym. Sci.* **23** 1443
[2] Yuan C L, Hong Y S and Lin C H 2011 *J. Magn. Magn. Mater.* **323** 1851
[3] Yavuz O, Ram M K, Aldissi M, Poddar P and Hariharan D 2005 *J. Mater. Chem.* **15** 810
[4] Prasanna G D, Jayanna H S and Vishnu Prasad 2011 *J. App. Polym. Sci.* **120** 2856
[5] Prasanna G D, Jayanna H S, Lamani A R and Dash S 2011 *Synth. Met.* **161** 2306
[6] Huang X and Chen Z 2005 *Mater. Res. Bull.* **40** 105
[7] Kodama R H 1999 *J. Magn. Magn. Mater.* **200** 359
[8] Hwang C C. Wu T Y, Wan J and Tsai J S 2004 *Mater. Sci. Eng B.* **111** 49
[9] Stejskal J 2002 IUPAC technical report. *Pure Appl. Chem.* **74** 857
[10] Singh K and Usha M E 2007 *Indian J. Chem. Sect A.* **36** 295
[11] Jing Jiang 2008 *J. Macromol. Sci, Part B: Phy.* **47** 242
[12] Ahmad N and MacDiarmid A G 1996 *Synth. Met.* **78** 103
[13] Wang Z H, Javadi H S, Ray A, Macdiarmid A G and Epstein A J 1990 *Phys. Rev.* **42** 5411
[14] Li W and Wan M 1998 *Synth. Met.* **92** 121

Observation of photobleaching and intensity dependent kinetics in $Ge_{22}As_{22}Se_{56}$ thin films under sub-bandgap light illumination

Pritam Khan [1], A R Barik [1], E M Vinod [2], K S Sangunni [2] and K V Adarsh[1]

[1] Department of Physics, Indian Institute of Science Education & Research, Bhopal 462066, India
[2] Department of Physics, Indian Institute of Science, Bangalore 562012, India

Email: adarsh@iiserb.ac.in

Abstract. We experimentally demonstrate photobleaching (PB) in $Ge_{22}As_{22}Se_{56}$ thin films, when illuminated with a diode pumped solid state laser (DPSSL) of wavelength 671 nm, which is far below the optical bandgap of the sample. Interestingly, we found that PB is a slow process and occurs even at moderate pump beam intensity of 0.2 W/cm^2, however the kinetics remain rather different.

1. Introduction

Recently, the demand for chalcogenide glasses (CG) in light wave technology has increased substantially, because of their prominent photoinduced effects. Among the numerous photoinduced effects shown by CG, the most notable effects are photodarkening (PD) in As-based chalcogenides [1] and PB in Ge-based chalcogenide [2]. Besides the fundamental interest of PD/PB in CG, these effects find immense technological applications in high bit rate waveguide writing [3] and also in dense holographic recording [4]. Hence a comprehensive understanding of PD/PB comprising of transient and metastable part is therefore of crucial importance for many of these applications. In contrast, light illumination in Ge-based CG mostly results in PB and is assumed to be of irreversible in nature. However, recent studies on $GeSe_2$ thin films have shown that irreversible PB is accompanied by transient PD. Many models have been proposed to explain the observed PB in Ge-based chalcogenides, but none of them proved to be definite for all compositions. In this paper, we present the interesting results of PB in $Ge_{22}As_{22}Se_{56}$ thin films when illuminated with sub-bandgap light of 671nm. Further, we move on to show that intensity of the illuminating light has a predominant role in determining the kinetics of PB. In our experiments, we observed a 10-fold increase in reaction time when intensity is reduced from 2 to 0.2 W/cm^2. Notably, the kinetic curve of PB follows a stretched exponential function.

2. Experimental

Bulk $Ge_{22}As_{22}Se_{56}$ glass was prepared by melt quenching method from 99.999% pure As, Ge and Se materials. The cast sample was used as the source material for deposition and thin film of average thickness ~1.0 μm was prepared by thermal evaporation in a vacuum of about 1×10^{-6} Torr. PB in these films is studied by a pump probe optical absorption method by using the experimental set up described previously [5]. We have chosen the wavelength of the pump beam as 671nm (from a DPSSL) and kept the intensity of pump beam at 2W/cm^2. The probe beam was a low intensity white light in the wavelength regime of 450-1000 nm. During illumination, the transmission of the sample was recorded using a spectrometer with photodiode array as detector and had the capability to collect the entire optical spectrum in 2ms. In our experiments, the entire optical spectrum was collected in real time of 100 ms/spectrum.

3. Results and Discussions

First we recorded the transmission spectra of as prepared $Ge_{22}As_{22}Se_{56}$ at an intensity of 2W/cm^2. Thereafter we turned on the pump beam and continuously measured the spectrum within an interval of 100ms until the whole effect saturates. After initial PD, PB begins to start nearly at 300 seconds, continuously grow and saturates within an hour. Figure 1 shows the surface plot of transmission as a function of time on selected wavelengths close to the optical bandgap of the sample. To compare the effect of intensity on the kinetics of PB we have done the same experiment at a different intensity of 0.2mW/cm^2 for the same sample. Interestingly, the whole effect is reproducible even at low intensities.

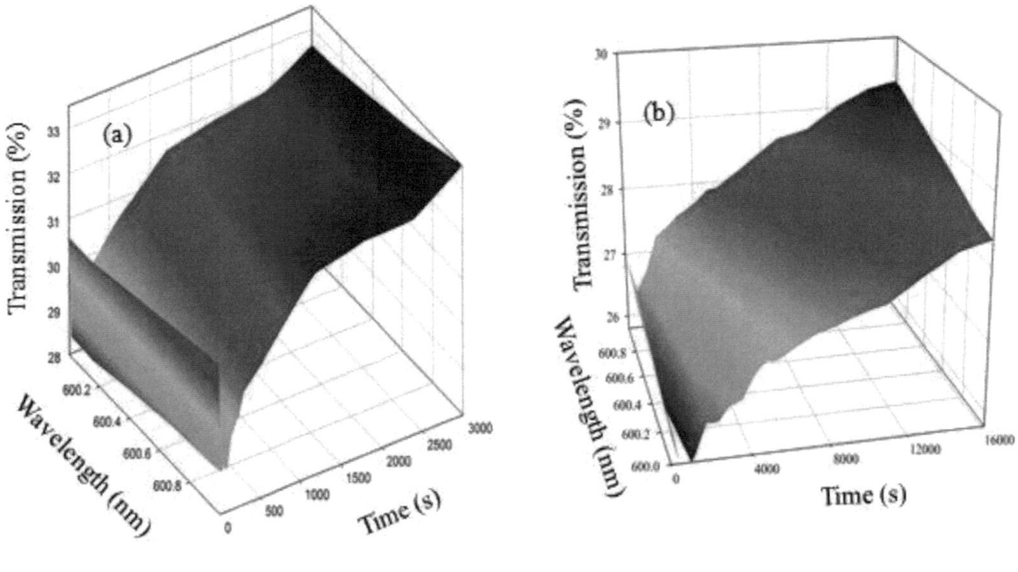

Figure 1. Transmission spectra of $Ge_{22}As_{22}Se_{56}$ thin film at different times when illuminated with a 671 nm DPSSL of intensity (a) 2W/cm^2 and (b) 0.2W/cm^2.

After demonstrating PB in of $Ge_{22}As_{22}Se_{56}$ thin film, it is of great importance to explain the observed effects. At this point, we conjecture that the there exist compositional heterogeneities (As-As, Ge-Ge etc) in the film, which were created from the vapour that contains various atomic fragments. When such a film is illuminated with 671nm, compositional heterogeneities associated with As-Se and Ge-Se responds rather independently and a considerable fraction of metastable homopolar bonds present in the atomic fragments are broken and subsequently converted into energetically favoured heteropolar bonds. The reduction in As homopolar bonds results in PD, however that of Ge atoms leads to photobleaching. Thus we observe an unusual coexistence of PB and PD and the reaction happens in the following way [6]:

$$As_{x-y}Se_z + yAs + h\nu \longleftrightarrow As_xSe_z \quad (1)$$

$$GeSe_a + Se_b + h\nu \longleftrightarrow GeSe_{a+b} \text{ where } a < 2 \leq a+b \quad (2)$$

It is worth mentioning here that PB happens due to internal structural rearrangements due to photo-illumination along with photo-oxidation of the sample forming a new species GeO_2 [7].

Among the reactions described above, rate equation determining PD can be written as

$$\Delta T = C[\exp\{-(\frac{t}{\tau_d})^{\beta_d}\}] + \Delta T_{Sd} \tag{3}$$

and that for PB:

$$\Delta T = \Delta T_{Sb}[1 - \exp\{-(\frac{t}{\tau_b})^{\beta_b}\}] \tag{4}$$

where the subscript 'd' and 'b' corresponds to PD and PB respectively. ΔT_s, τ, β and t are metastable part, effective time constant, dispersion parameter and illumination time respectively. The net rate equation for the whole process is a summation of respective PD and PB as following:

$$\Delta T = C[\exp\{-(\frac{t}{\tau_d})^{\beta_d}\}] + \Delta T_{Sd} + \Delta T_{Sb}[1 - \exp\{-(\frac{t}{\tau_b})^{\beta_b}\} \tag{5}$$

The experimental data at both the intensities fit very well to the stretched exponential function described by equation (5). Further, the fitting parameters calculated from the theoretical fit are listed in table 1. From the table 1, it can be very well understand that PD is a very fast process and the reaction time is of the order of a few 100's of seconds. However on the contrary, PB is a slow process and takes an induction time to start. Interestingly, as we move from higher to lower intensity, magnitude of photoinduced effects remain the same, but the kinetics show a sharp contrast. In our experiments, we observed 10 times decrease in kinetics, when the intensity of the pump beam was decreased from 2 to 0.2 W/cm^2.

Table 1. Fitting parameters obtained from equation 5 corresponding to PD/PB at different intensities for the wavelength 600 nm. The subscript d and b refer to darkening and bleaching respectively.

Intensity (W/cm^2)	τ_d (s)	β_d	ΔT_{Sd}	τ_b (s)	β_b	ΔT_{Sb}
2	12	0.97	0.905	920	0.65	0.185
0.2	260	0.45	0.918	5500	0.72	0.174

4. Conclusions

In conclusion, we observed PB in $Ge_{22}As_{22}Se_{56}$ thin film, when illuminated with a DPSSL of wavelength 671 nm, which is far below the optical bandgap of the sample. Our experimental results clearly demonstrate that intensity of the illuminating beam has a predominant role in determining the kinetics of PD/PB. PB in these samples is explained by considering the compositional heterogeneities associated with the film while evaporation.

5. Acknowledgement

The authors thank Department of Science and Technology (Project no: SR/S2/LOP-003/ 2010) and council of Scientific and Industrial Research, India, (grant No. 03(1250)/12/ EMR-II) for financial support.

References

[1] Ganjoo A and Jain H 2006 *Phys. Rev. B* **74** 024201
[2] Kumar R R *et al* 2013 *Opt. Lett.* **38** 1682
[3] Asatryan K E *et al* 2004 *Appl. Phys. Lett.* **84** 1626
[4] Huges M *et al* 2007 *Appl. Phys. Lett.* **90** 131113
[5] Naik R *et al* 2011 *Opt. Commun.* **284** 5319

[6] Khan P *et al* 2012 *Opt. Express* **20** 12416
[7] Khan P *et al* 2014 Sci. Rep. **4** 4029

Correlation of martensitic transformation temperatures of Ni-Mn-Ga/Al-X alloys to non-bonding electron concentration

M. Ramudu[1], A. Satish Kumar[1], V. Seshubai[1] and T. Rajasekharan[2]

School of physics, University of Hyderabad, Gachibowli, Hyderabad-500 046, India.
Rajiv Gandhi University of Knowledge Technologies, Vindhya C4 building, IIIT-Campus, Hyderabad - 500 032, India.

macrams2@gmail.com

Abstract. The martensitic transformation T_M of the alloys of Ni-Mn-Ga and Ni-Mn-Al show a general trend of increase with electron per atom ratio *(e/a)* calculated from the total number of electrons outside the rare gas shell of the atoms. However prediction of T_M fails among iron substituted Ni-Mn-Ga alloys and those with In doped for Ga, due to the absence of a useful trend. A scheme of computing modified electron concentration is presented considering only the non-bonding electrons per atom $N_{e/a}$ of the compounds, based on Pauling's ideas on the electronic structure of metallic elements. Systematic variation of T_M with $N_{e/a}$ is reproduced for a large number of alloys of Ni-Mn-Ga and the anomaly observed for Fe containing alloys with *e/a* disappears. The non-bonding electron concentration is thus demonstrated to be effective in predicting T_M of shape memory alloys of Ni-Mn-Ga-X system including the isoelectronic compounds of Ni-Mn-Ga-In.

1. Introduction

The study of magnetic shape memory alloys has, of recent, gained momentum due to their potential for applications as sensors and actuators operable at room temperature [1]. Also, their application in magnetic refrigeration is much sought after due to the benefits that could ensue to the environment [2]. Off-stoichiometric compositions derived from ferromagnetic Ni_2MnGa Heusler alloy, are widely studied in this context. The magnitude of magnetic field induced strain in these alloys is governed by key factors like martensitic transformation temperature (T_M), Curie temperature (T_C), saturation magnetization, magnetocrystalline anisotropy and the nature of twin deformation in martensitic phase. Various elements have been substituted to fine tune T_M and the modifications caused to the microstructures and physical properties, is investigated. There are several reports on systematic dependence of T_M on the electron per atom ratio *e/a*, in these alloys. Here, *e/a* is computed as an average of the number of electrons beyond the rare gas shell of the atoms constituting the alloy.

Systematic dependence of T_M on the electron per atom ratio (*e/a*) is reported in Ni-Mn-X (X = Ga, Al, Sn, In and Sb) alloys. However, iron substituted Ni-Mn-Ga alloys have been observed to be anomalous in their behavior; and their deviation from the normal trend has been discussed in several publications [3-5]. Also in the case of isoelctronic substitutions like In- substituted for Ga, where *e/a* remains unaltered, prediction of T_M fails. We discuss in this paper, an alternative scheme of computing effective valence electrons per atom, $N_{e/a}$. We examine the trend of variation in T_M with $N_{e/a}$ for a number of shape memory alloys of Ni-Mn-Ga system and discuss its efficacy in predicting T_M for isoelectronic and iron substituted alloys as well.

2. Present model, an extension of Pauling's approach

In this approach, Pauling's idea of transfer of charge [6, 7] on an atom-pair bond to maintain electroneutrality in metallic alloys is adopted. In this model [7-9], Miedema's parameters Ø and N_{ws} were considered to represent the electronegativity and valence of metallic elements respectively. These parameters were defined by Miedema as work function and the 'electron density at the boundary of the Wigner-Seitz cells' of the elements and were used by him to compute semi-empirically, the heats of formation (ΔH) of binary alloys [10]. This model is shown to accurately predict concomitant and mutually exclusive crystal structure types in metallic binary systems [8]. The versatility of the above approach suggests that Pauling's scheme of filling electrons in orbitals of metallic atoms could be of interest in predicting the magnetic and electrical properties of metallic alloys. Pauling's approach provided a qualitative explanation [11, 12] of many properties of the transition metals such as interatomic distance, hardness, compressibility, and coefficient of thermal expansion, and it accounts satisfactorily for the observed values of the atomic saturation magnetic moments of the ferromagnetic elements iron, cobalt, and nickel and their alloys.

Pauling had proposed [11] that in metals, d-orbitals alone are not especially well suited to use in bond formation, but hybridization of d, s, and p orbitals leads to the best bond orbitals known. There are available a total of nine relatively stable orbitals: five 3d, one 4s, and three 4p for the iron group elements, and corresponding sets for other series. If N_p is the Pauling's valence of an element, N_p electrons would fill up unpaired in the bond orbitals and would be available for the formation of N_p electron-pair bonds with other atoms. Melting and boiling points of elements would therefore be proportional to N_p.

Let O be the total number of orbitals of an atom; Pauling had argued that O = 9, consisting of five 3d, one 4s, and three 4p orbitals for the iron group elements, and corresponding sets for other series [11]. Let E be the total number of electrons beyond the inert gas shell of an atom and N_p be the valence of a metal; Pauling had proposed that N_p orbitals would be filled by an equal number of unpaired electrons which would form resonating covalent bonds with the neighboring atoms. The number of electrons not involved in bonding would be $E- N_p$. Such electrons were referred to as paired non-bonding electrons by Pauling. In the modified Pauling's scheme, we have used Miedema's corrected electron density parameter N_{ws} as the valence of metals. The choice of Miedema's electron density parameter N_{ws} as the valence of metallic elements (instead of N_p) is discussed in detail elsewhere [7]. Then, N_{ws} electrons per atom would occupy N_{ws} orbitals, i. e. one electron per orbital, and form covalent bonds with the neighboring atoms. The residual electrons fill the remaining 9-N_{ws} orbitals in pairs. These electrons are nonbonding electrons and they contribute to magnetism, electrical properties etc. In the case of ferromagnetic elements such as Fe, Co and Ni, a certain number of the nonbonding electrons would remain unpaired, contributing to the magnetic moments of the atoms. The number of non-bonding electrons is calculated as (E-N_{ws}), where E is the number of electrons beyond the rare gas shell of the atom.

In the following, we consider the non-bonding electrons (e_n) to represent the number of electrons per atom relevant in dictating the physical properties of metallic alloys. We compute the effective 'non-bonding electrons per atom' of a compound ($N_{e/a}$) from a knowledge of en contributed by each atom. The values of N_{ws} for the elements Ni, Mn, Ga, Co, Fe, Al, and In are 5.36, 4.17, 2.25, 5.36, 5.55, 2.69, and 1.6, respectively [10]. As a typical example demonstrating how $N_{e/a}$ for an alloy is calculated, we consider the calculation for $Ni_{50}Mn_{26}Ga_{24}$. The values of (E, N_{ws} and e_n) for Ni, Mn and Ga are (10, 5.36 and 4.74), (7, 4.17 and 2.83) and (3, 2.25 and 0.75), respectively. $N_{e/a}$ for the compound can be computed as {50 x (10-5.36) + 26 x (7-4.17) + 24 x (3-2.25)}/100 = 3.24.

3. Systematics in the dependence of T_M on electron concentration

The martensitic transformation temperatures T_M of 171 off-stoichiometric alloys in the Ni-Mn-Ga system, 84 quasi-ternary alloys with Co substitution, 76 alloys with Fe substitution, and 7 alloys with indium substitution in Ni-Mn-Ga system, were collected from compilations [1, 2] and individual reports in the literature [3, 4, 13, 14, 15-18, 19-32]. The e/a values of the compounds were computed taking the E values to represent the valences of Ni, Mn, Ga, Co, Fe and In which are 10, 7, 3, 9, 8, 3 and 3 respectively; the individual electron contributions of all atoms are added up in the formula unit and divided by the total number of atoms [33].

The variations in the martensitic transformation temperatures of off-stoichiometric alloys and Co- doped alloys of Ni_2MnGa with *e/a* and $N_{e/a}$ are given in figure 1. Figure 1 shows that the correlations of T_M with *e/a* and $N_{e/a}$ are equally good for the above alloys. Figure 1 (a) and (b) for off-stoichiometric Ni_2MnGa, shows three regions. 141 alloys out of a total of 171 alloys with $e/a \leq 7.72$ show a linear variation of T_M with *e/a*. The compounds with $e/a > 7.72$ are the Ni-rich ones having $T_M > T_C$, with potential for high temperature shape memory applications. Most of these compounds, which have non-modulated tetragonal structure, show a faster increase in T_M with *e/a*. This has been attributed to a decrease in their unit cell volumes due to smaller Ni atoms replacing the larger Ga atoms in the crystal lattice. A combination of increase in *e/a* and a decrease in cell volumes causes a faster increase in effective electron concentration and hence in the T_M of those alloys [22]. At Ni contents in excess of 57 at.%, γ-phase precipitates out and the concentration of the alloy remains a constant, leading to a constant T_M for *e/a* values greater than 8.0 (see figure). The *e/a* values were computed from the nominal composition of the alloys. Figures 1 (c) and (d), show a linear trend of T_M with *e/a* as well as $N_{e/a}$ for Co-doped Ni-Mn-Ga alloys.

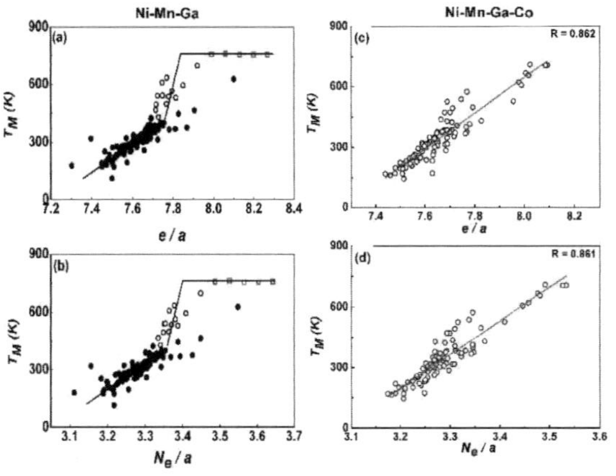

Figure 1. (a) and (b) variation of T_M in off-stoichiometric Ni-Mn-Ga alloys with *e/a* and N_e/a. (c) and (d) T_M variation with *e/a* and N_e/a in off-stoichiometric Ni-Mn-Ga-Co alloys.

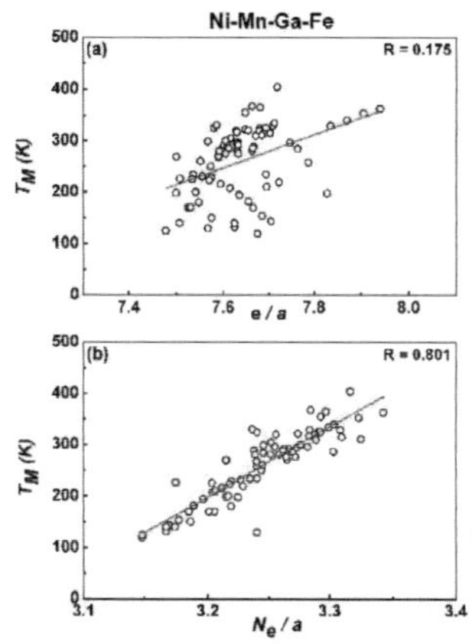

Figure 2. (a) and (b) variation of T_M in off-stoichiometric Ni-Mn-Ga-Fe alloys with *e/a* and N_e/a.

On the contrary, we note from figure 2(a), that the variation of T_M with *e/a* for Fe-doped compounds of Ni-Mn-Ga distributed at random. The anomalous behavior of iron-containing alloys has been discussed by several authors [3-5]. It has been attributed to factors other than *e/a* also contributing to the stability of those phases, like the site occupied in ordered structure [29], volume effects associated with atomic sizes [5, 31], magnetic contributions [3] etc. We now examine the variation of T_M with the non-bonding electrons per atom, $N_{e/a}$ in figure 2 (b) for the iron-substituted alloys of Ni_2MnGa. The regression factor (R) shows a remarkable improvement from R = 0.175 for T_M with *e/a* to R = 0.80 for the plot of T_M with $N_{e/a}$. Hence the correlation of T_M with N_e/a can be used to predict T_M of new compositions in the Ni-Mn-Ga-Fe system. The anomalous behavior seen in Fe-containing compounds when using the earlier electron counting scheme (using *e/a*) is thus resolved.

We consider below the case of isoelectronic substitutions like $Ni_2MnGa_{1-x}In_x$ alloys. In figure 3 (a) we observe that the T_M versus *e/a* plot for $Ni_2MnGa_{1-x}In_x$ alloys does not exhibit any useful trend since the valences assumed conventionally are the same for the isoelectronic elements like Ga and In, viz. 3. In figure 3 (b) we show the dependence of T_M on the non-bonding electrons per atom ($N_{e/a}$) which

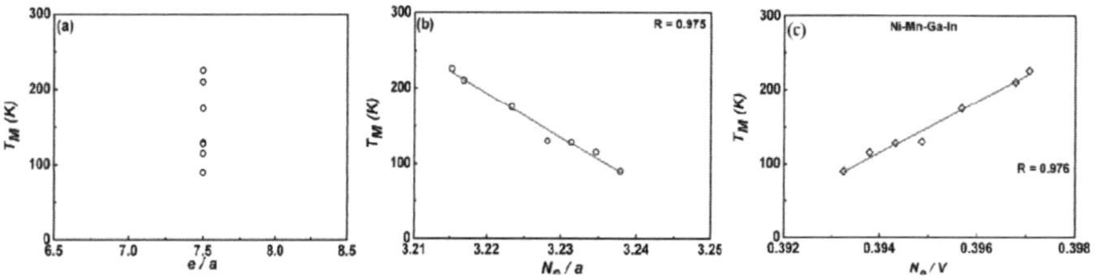

Figure 3. Variation of T_M in $Ni_2MnGa_{1-x}In_x$ alloys (a) with *e/a* showing no trend, (b) linear trend with N_e/a with decreasing trend and (c) shows role of volume effects on the variation of T_M with N_e/V.

shows a systematic variation, but a decreasing trend rather than an increasing trend of T_M with $N_{e/a}$ as observed in other compounds. We examine below the possible role of volume effects discussed in literature [34], associated with large difference in the atomic sizes of Ga and In. We computed N_e/V that defines the electron concentration per unit volume. Here, V is calculated as the average (atomic) volume per atom. Figure 3 (c) shows that the trend of variation in T_M with $N_{e/a}$ reverses with N_e/V for $Ni_2MnGa_{1-x}In_x$ alloys confirming the role of volume effects. The various results discussed above suggest that computing non-bonding electron concentration is worth exploring to understand the mechanisms determining the martensitic transformation temperatures in shape memory alloys.

4. Conclusions

Significance of computing the non-bonding electron concentration ($N_{e/a}$) to predict the martensitic transformation temperatures of T_M of shape memory alloys of Ni-Mn-Ga system is demonstrated. A plot of T_M with $N_{e/a}$ shows a systematic variation for alloys derived from Ni-Mn-Ga and resolves the anomaly observed for Fe containing alloys with *e/a*.

Acknowledgements: MR acknowledges UGC-RFSMS for financial support in the form of fellowship. ASK and VS thank CSIR, India, for research funding (03(1065) 106/EMR-II).

References

[1] Vasil'ev A N, Buchel'nikov V D, Takagi T, Khovailo V V and Estrin E I 2003 *Physics – Uspekhi* **46** 559.
[2] Planes A, Manosa L and Acet M 2009 *J. Phys.: Condens. Matter.* **21** 233201.
[3] Liu Z H, Zhang M, Wang W Q, Wang W H, Chen J L, Wu G H, Meng F B, Liu H Y, Liu B D, Qu J P and Li Y X 2002 *J. Appl. Phys.* **92** 5006.
[4] Kikuchi D, Kanomata T, Yamaguchi Y, Nishihara H, Koyama K and Watanabe K 2004 *J. Alloys Compd.* **383** 184.
[5] Chen F, Wang H B, Zheng Y F, Cai W and Zhao L C 2005 *J. Mater. Sci.* **40** 219.
[6] Pauling L 1950 *Proc. Natl Acad. Sci. USA* **36** 533.
[7] Rajasekharan T and Seshubai V 2011 *J. Mater. Res.* **26** 1539.
[8] Kameswari V L, Seshubai V and Rajasekharan T 2010 *J. Alloys Compd.* **508** 55.
[9] Rajasekharan T and Seshubai V 2010 *Intermetallics* **18** 666.
[10] Miedema A R, De Chatel P F and De Boer F R 1980 *Physica* **100B** 1.
[11] Pauling L 1938 *Phys. Rev.* **54** 899.
[12] Pauling L, *The Nature of the Chemical Bond*, 1960 Chapter 11, *The metallic Bond, Cornell University Press* 393.
[13] Lanska N, Söderberg O, Sozinov A, Ge Y, Ullakko K and Lindroos V K 2004 *J. Appl. Phys.* **95**

8074.
[14] Richard M, Feuchtwanger J, Schlagel D, Lograsso T, Allen S M and O'Handley R C 2006 *Scripta Mater.* **54** 1797.
[15] Glavatskyy I, Glavatska N, Söderberg O, Hannula S –P and Hoffmann J –U 2006 *Scripta Mater.* **54** 1891.
[16] Chen X Q, Yang F J, Lu X and Qin Z X 2007 *Phys. Stat. Sol. (b)* **244** 1047.
[17] Sánchez-Alarcos V, Pérez-Landazábal J I, Recarte V, Gómez-Polo C and Rodríguez-Velamazán J A 2008 *Acta Mater.* **56** 5370.
[18] Chen X Q, Lu X, Wang D Y and Qin Z X 2008 *Smart Mater. Struct.* **17** 065030.
[19] Chernenko V A, Cesari E, Kokorin V V and Vitenko I N 1995 *Scripta Metall. Mater.* **33** 1239.
[20] Jiang C, Muhammad Y, Deng L, Wu W and Xu H 2004 *Acta Mater.* **52** 2779.
[21] Khovaylo V V, Buchelnikov V D, Kainuma R, Koledov V V, Ohtsuka M, Shavrov V G, Takagi T, Taskaev S V and Vasiliev A N 2005 *Phys. Rev. B* **72** 224408.
[22] Ma Y, Jiang C, Li Y, Xu H, Wang C and Liu X 2007 *Acta Mater.* **55** 1533.
[23] Zhou X, Kunkel H, Williams G, Zhang S and Desheng X 2006 *J. Magn. Magn. Mater.* **305** 372.
[24] Richard M L, Feuchtwanger J, Allen S M, O'Handley R C, Lazpita P, Barandiaran J M, Gutierrez J, Ouladdiaf B, Mondelli C, Lograsso T and Schlagel D 2007 *Phil. Mag.* **87** 3437.
[25] Pons J, Chernenko V A, Santamarta R and Cesari E 2000 *Acta Mater.* **48** 3027.
[26] Ma Y, Yang S, Liu Y and Liu X 2009 *Acta Mater.* **57** 3232.
[27] Khovailo V V, Abe T, Koledov V V, Matsumoto M, Nakamura H, Note R, Ohtsuka M, Shavrov V G and Takagi T 2003 *Mater. Trans.* **44** 2509.
[28] Satish Kumar A, Ramudu M and Seshubai V 2011 *J. Alloys Compd.* **509** 8215.
[29] Soto-Parra D E, Alvarado –Hernandez F, Ayala O, Ochoa-Gamboa R A, Flores-Zúñiga H and Rios-Jara D 2008 *J. Alloys Compd.* **464** 288.
[30] Kikuchi D, Kanomata T, Yamaguchi Y and Nishihara H 2006 *J. Alloys Compd.* **426** 223.
[31] Soto D, Hernandez F A, Zuniga H F, Moya X, Manosa L, Planes A, Aksoy S, Acet M and Krenke T 2008 *Phys. Rev. B* **77** 184103.
[32] Luo H, Meng F, Feng Z, Li Y, Zhu W, Wu G, Zhu X, Jiang C and Xu H 2010 *J. Appl. Phys.* **107** 013905.
[33] Chernenko V A 1999 *Scripta Mater.* **40** 523.
[34] Khan M, Dubenko I, Stadler S and Ali N 2004 *J. Phys.: Condens. Matter.* **16** 5259.

IOP Conf. Series: Materials Science and Engineering **73** (2015) 012075 doi:10.1088/1757-899X/73/1/012075

Enhanced Stability of Carbon Doped Boron Nanotubes: An ab-initio Approach

Sandeep Kumar Jain* and Pankaj Srivastava

Nanomaterials Research Group, Computational Nanoscience & Technology Lab
ABV-Indian Institute of Information Technology & Management, Morena Link Road, Gwalior-474015 (MP) India

*Corresponding Author: Email Id: sandeepjain@iiitm.ac.in

Abstract. We have performed ab-initio calculations for investigating structural stability of carbon doped boron nanotubes. The considered structures are Armchair (3,3) ,Zigzag (5,0), and Chiral (4,2) nanotubes consists of 12,20,and 56 atoms respectively. The ground state energies of relaxed nanotubes are calculated, thereby cohesive energy and formation energy. The cohesive energy is increased to a significant extent after carbon doping. Thus, carbon doped boron nanotubes are energetically more stable than pristine nanotubes. The enhancement of energetic stability is also supported by formation energy. It is found that the formation energy is higher for the higher stable structure in all the three nanotubes. It is also observed that carbon doped zigzag nanotube is most favourable nanotube among all the three considered nanotubes because of high formation energy.

1. INTRODUCTION

The nanoscale materials possess intriguing physical and chemical properties thereby scientist and engineers have been fascinated. Nanotube, one of the most important nanostructures is extensively studied and grouped into organic and inorganic nanotubes. Organic and inorganic nanotubes have been studied theoretically as well as experimentally. The organic nanotube i.e. Carbon nanotube (CNT) was investigated by Iijima [1] and subsequently it was investigated that the electronic properties depend on diameter and chirality [2]. Similarly, the nanotubes of inorganic materials include Boron Nitride [3-4], Silicon [5], Gold [6], etc. The boron nanotube was first proposed by Boustani [7] and subsequently various structural conformations of BNTs were proposed. The various conformations include Triangular [8], α-BNT [9, 10] and Hexagonal [11, 12] in contrast to only one conformation of CNT i.e. Hexagonal. It has been predicted that all the BNTs are found to be metallic and independent of diameter and chirality. The experimental synthesis of BNTs by Ciuparu et al [13] has confirmed the existence of boron nanotubes. Victor et al. [14] also synthesized the BNT and studied the field emission properties. The study of transport properties [15] revealed that BNTs are more conducting than that of CNTs. However, the atomic structure of BNTs has not been revealed yet. Moreover, it is predicted that α-BNTs are found to be highly stable while hexagonal to be least. In view of comparison with hexagonal CNTs, it becomes extremely essential to study CNT-like-BNT. Further, hexagonal BNTs of small diameter were found to be stable. However, the stability of hexagonal BNTs can also be enhanced by doping with foreign atoms i.e. carbon. It is our natural choice to prefer substitutional doping rather than other available methods of doping. We have used carbon atoms as impurity because of its size comparable to boron.

2. COMPUTATIONAL METHODOLOGY

The pseudopotential plane wave calculations for determining structural stability of ultrathin BNTs doped with carbon have been performed in the framework of density functional theory (DFT). All the calculations have been carried out using CASTEP simulation tool [16]. The exchange and correlation effects are described by generalized gradient approximation (GGA) [17] proposed by Perdew-Burke-Ernzehrof (PBE). The Vanderbilt [18] ultrasoft pseudopotential in the reciprocal space with plane wave cutoff energy 320eV has been used for geometry optimization. The integration is performed in the first brillouin zone [19] by using k-points generated by 1x1x8 grid parameters. The geometry is optimized until force on each atom becomes < 0.03 eV/ Å. The intertubular distance is set to be 10 Å in order to avoid periodic image interaction. The bond length B-B is chosen to be 1.67 Å. In addition, non-spin-polarized calculations are performed. The considered conformations of BNTs are of the three types viz. Armchair (3,3), Zigzag (5,0), and Chiral (4,2) consisting of 12, 20,and 56 atoms respectively. The concentration of carbon impurity is restricted to two atoms. The diameters of these tubes are 4.60Å, 4.78Å, and 4.87 Å for (5,0), (3,3), and (4,2) BNTs respectively.

3. RESULTS AND DISCUSSION

The effect of chemical doping of carbon atoms in BNTs have been investigated in the framework of density functional theory (DFT). We have put forth our efforts to explore the structural stability of carbon doped BNTs. In this study, three model of BNTs have been considered namely Armchair (3,3), Zigzag (5,0), and Chiral (4,2). The content of carbon impurity is limited two atoms in each model of BNTs. In our previous study [12], it has been investigated that pristine chiral (4,2) BNT is highly stable followed by zigzag (5,0) and armchair (3,3).

In this study, the structural stability is predicted on the basis of cohesive energy and formation energy. The cohesive energy E_{coh} of the SWBNT was obtained from the total energy (E_{tot}), the energy of a single free boron atom (E_1), and the energy of the carbon atom (E_2) using the following relation:

$$E_{coh} = \frac{(E_{tot} - xE_1 - yE_2)}{x + y}$$

Where x and y are the numbers of boron and carbon atoms in the supercell respectively.

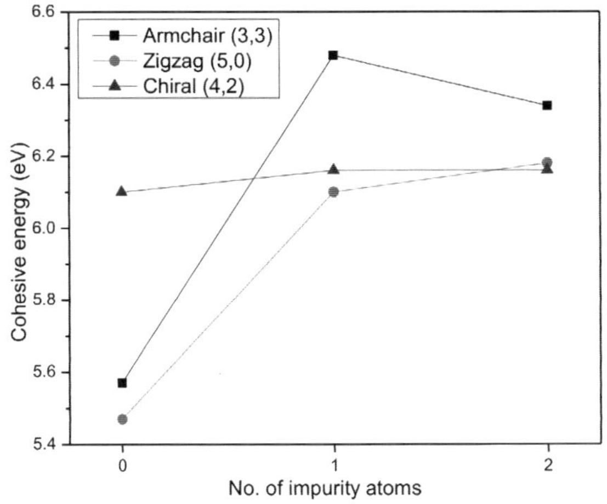

Figure (1): Variation of cohesive energy with impurity concentration

When single atom of carbon substitutes boron atom in armchair (3,3) BNT, the cohesive energy is found to be 6.48 eV while 6.34 eV correspond to two atom doped BNT. Thus, there is a very slight variation in cohesive energy. Consequently, carbon doped BNTs are energetically more stable as compared to pristine one. Similarly, substitution of carbon atoms in zigzag and chiral BNTs causes' slight variation in cohesive energies correspond to single and double atom doped nanotubes. The variation of cohesive energies with impurity atoms is shown in figure (1). Thus, it is revealed from figure (1) that all the three models are equally stable.

Moreover, the formation energy of carbon doped boron nanotube is calculated by using ground state energy of relaxed structures. We determine the formation energy:

$$E_{formation} = \frac{E_{doped\,BNT} - E_{pristine} - xE_{carbon}}{n}$$

Where $E_{doped\,BNT}$ is the total energy of our supercell with doped carbon atoms, $E_{pristine\,BNT}$ is the total energy of pristine BNT, E_{carbon} is the total energy of an isolated carbon atom, x is the total number of carbon atoms, and n is the total number of atoms in the supercell.

Table 1: Variation of formation energy with impurity concentration

Nanotubes	Number of impurity atoms	Formation energy (eV)
Armchair (3,3)	1	10.71
	2	9.20
Zigzag (5,0)	1	12.35
	2	14.14
Chiral (4,2)	1	2.90
	2	3.14

The calculated formation energy for these BNTs is shown in table (1). It is noteworthy that single atom doped armchair BNT possess high formation energy correspond to high cohesive energy as compared to two atom doped BNTs. The high formation energy reflects more favourable structure. Thus, for armchair (3,3) BNT, the formation energy is in good agreement with cohesive energy. Similarly for zigzag (5,0) chiral (4,2) BNTs, the cohesive energy as well as formation energy increase with the concentration of carbon impurity. However, there is very slight variation in energies. Moreover, it is shown in table (1) that the formation energy for zigzag (5,0) BNT is found to be highest which indicates that zigzag (5,0) BNT is highly favourable while chiral (4,2) is least stable among considered conformations .

4. CONCLUSION

Ab initio calculations have been performed for exploring structural stability of carbon doped hexagonal ultrathin boron nanotubes with conformations namely, armchair (3,3), zigzag (5,0) and chiral (4,2) BNTs. It is found that carbon impurity gives rise to higher stability to boron nanotubes. The formation energy is found to be highest for zigzag (5,0) followed by armchair (3,3) and chiral (4,2). The structural stability is found to be highest for armchair (3,3) followed by zigzag (5,0) and chiral (4,2) BNTs.

ACKNOWLEDGEMENTS

The authors are thankful to Computational Nanoscience & Technology Lab (CNTL), ABV-Indian Institute of Information Technology & Management, Gwalior (MP), for providing computational facility.

References

[1] Iijima S (1991) Helical microtubules of graphitic carbon *Nature* **354** 56-8
[2] Saito R and Dresselhaus G Physical Properties of Carbon Nanotubes, (Imperial College Press, London 1998)
[3] Blasé X, Robio A, Louie S G,Cohen M L (1994) Stability and band gap constancy of boron nitride nanotubes *Europhys. Lett.* **28** 335
[4] Rubio A,Corkill J,Cohen M L (1994) Theory of graphitic boron nitride nanotubes *Phys. Rev. B* **49** 5081-5084
[5] Yang X and Ni J (2005) *Phy. Rev.B* **72** 195426; Crescenzi M D et al (2005) *Appl. Phys. Lett.* **86** 231901
[6] Sengar RT et al (2004) *Phys. Rev.Lett.* **93** 196807; Oshima Y et al (2003) ibid **91** 205503
[7] Boustani I and Quandt A (1997) Nanotubes of bare boron clusters, Ab initio and density functional study *Europhys. Lett.* **39** 527
[8] Kunstmann J and Quandt A (2006) Broad boron sheet and boron nanotube: An ab initio study of Structural, electronic, and mechanical properties, *Phys. Rev. B* **74** 035413
[9] Singh A K,Sadrzadeh A and Yakobson B I (2008) Probing properites of boron α-tubes by ab initio calculations *Nano Lett.***8** 1314-1317
[10] Yang X, Ding Y,and Ni J (2008) Ab initio prediction of stable boron sheets and boron nanotubes: Structures, stability, and electronic properties *Phys. Rev. B* **77** 041402(R)-1
[11] Zhang D, Zhu R and Liu C (2006) Density functional theory study on the geometrical and electronic structures of a new thinnest boron nanotube *J. Mater. Chem.* **16** 2429-2433
[12] Jain S K and Srivastava P (2011) Electronic and optical properties of ultrathin boron nanotubes-

an ab initio study *Comp. Mater. Sci.* **50** 3038-3042

[13] Ciuparu D, Klie R F,Zhu Y M, Pfefferle L (2004) Synthesis of pure boron single wall nanotubes J. *Phys. Chem. B* **108** 3967-3969

[14] Bezugly V, Kunstmann J, Grundkotter-Stock B, Frauenheim T, Niehaus T and Cuniberti G (2011) Highly conductive boron nanotubes: transport structural stabilities, *ACS Nano* **5** 4997-5005

[15] Lau K C, Pandey R, Pati R, and Karna S P (2006) Theoretical study of electron transport in boron nanotubes *Appl. Phys. Lett.* **88** 212111

[16] Clark S J, Segall MD, Pickard C J, Hasnip P J, Probert M J, Rafson K, Payne M C (2005) First–principles methods using CASTEP, *Zeitschrift feur Kristallographie* **220 (5-6)** 567-570

[17] Perdew J P, Burke K, Ernzerhof M (1996) , Generalized gradient approximation made simple *Phys. Rev. Lett.* **77** 3865-8

[18] Vanderbilt D (1990) Soft self-consistent pseudopotentials in a generalized eigen value formalism *Phys. Rev. B* **41** 7892-7895

[19] Monkhorst H J andPack J D (1992) Special points for brillouin-zone integrations. *Phys. Rev. B* **13** 5188-92

Synthesis of nanosized platinum based catalyst using sol-gel process

S V Ingale[1,44], P B Wagh[1], D Bandyopadhyay[2], I K Singh[1], R Tewari[3] and S C Gupta[1]

[1] Applied Physics Division, Bhabha Atomic Research Centre, Mumbai, India, 400085
[2] Heavy Water Division, Bhabha Atomic Research Centre, Mumbai, India, 400085
[3] Material Science Division, Bhabha Atomic Research Centre, Mumbai, India, 400085

Abstract. The nano-sized platinum based catalysts using high surface area silica support have been prepared by sol-gel method. Tetramethoxysilane (TMOS) diluted in methanol was hydrolyzed to form a porous silica gel. Platinum (2%) was loaded at sol state using platinum chloride solution. After gelation, the solvent from the gel pores was extracted at ambient temperature which resulted in porous silica matrix incorporated with nanosized platinum. X-ray diffraction studies indicated the presence of elemental platinum in the silica-platinum composites. Transmission electron microscopy of the platinum –silica composites revealed that nanosized platinum particles of about 5-10 nm are homogeneously dispersed in silica matrix. Chemisorptions studies showed high dispersion (more than 50%) of platinum on silica support with specific surface area of 400 m^2/g which puts them as promising candidates as catalyst in heterogeneous reactions.
Keywords: Sol-gel technique, Mesoporous silica, Platinum catalyst, Chemisorption

1. Introduction

The platinum (Pt) based catalysts are highly applicable as catalyst in heterogeneous (solid-gas) reactions. The hydrogen isotope exchange reaction between hydrogen gas and water using Pt catalyst are exceedingly favourable to remove tritium from water [1] in nuclear industries. For such applications, the major requirements of catalyst are large surface area and high dispersion of active sites. The high surface area can be achieved by reducing the particle size of active catalyst to nanometer scale. The high dispersion requirement can be met by choosing proper support matrix for dispersion of catalyst material. The hydrophobic nature of the support matrix is also a critical requirement [2]. The sol gel process is a promising method for preparation of nano-sized particles that can be homogenously dispersed on gel matrix, mostly of silica. The sol-gel process allows the preparation of metal nanoparticles in oxide matrices with narrow particle size distributions and adjustable metal loadings [3]. In the present works, we developed a method to obtain the nano sized platinum dispersed on porous silica matrix prepared by sol-gel process. The micrographs obtained from transmission electron microscopy showed formation of 5-10 nm platinum particles where as the chemisorptions studies revealed high dispersion of platinum in silica matrix.

2. Experimental

A platinum-silica nano composite was prepared by sol-gel process [4]. To prepare silica gel, tetramethoxysilane (TMOS) diluted in methanol was hydrolysed. The molar ratio of acetone/TMOS

[4] Corresponding author : Email: svingale@barc.gov.in, Tel.: 912225591808

was kept as 12. The platinum solution ($PtCl_4$) was added as Pt source to the silicate solution. The molar ratio of $PtCl_4$ to TMOS was chosen so as to achieve 2% platinum loading in silica-platinum composite. Methyltrimethoxysilane (MTMS) was added as hydrophobic reagent to this solution in molar proportion of TMOS/MTMS as 4. The gelation of silicate solution was accelerated using ammonium hydroxide (0.5 M NH_4OH). After gelation, the solvent from the gel pores was extracted at ambient temperature and pressure which resulted in porous silica matrix incorporated with nanosized platinum.

The crystalline data for silica/platinum composite was obtained on a Philips X-ray diffractometer using a PW 1710 goniometer (CuKα, 30 kV, 20 mA). For transmission electron microscopy (TEM), the powdered sample diluted in methanol was dispersed on a carbon coated copper grid of 200 mesh size. Electron micrographs were obtained using JEOL 2000FX transmission electron microscope operating at 160 kV. Hydrophobicity of the silica support system was evaluated by using Karl-Fischer technique [5]. Chemisorptions studies on the samples were carried out using TPDRO 1100, Thermo Scientific, Italy. Samples were pretreated by heating in a stream of Argon (20 ml min^{-1}) for 2 h at 300°C and then cooled to room temperature. Helium with 5% H_2 was passed over the sample. The catalysts were then exposed to pulses of H_2 gas. From the amount of chemisorbed H_2 the percentage of the accessible Pt atoms has been calculated.

3. Results and discussion

The x-ray diffraction spectrum of platinum-silica nanocomposite with 2 % platinum loading is shown in figure 1. The broad peak around 22^0 is assigned to amorphous silica. The formation of nano particles of platinum is indicated by the characteristic peaks at 2θ values of 40, 46 and 67^0 in the diffractogram obtained by using Cu-Kα (1.54 Å) x-rays. The peaks could be assigned to 111, 200 and 220 plane, respectively [3]. The average metal particle size calculated from the width of the [111] peak using Scherer equation is 3 nm.

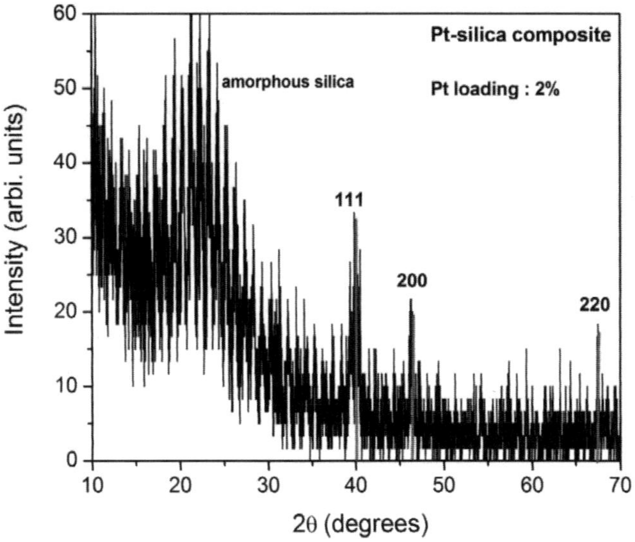

Figure 1 X- ray diffractogram for Pt-silica nanocomposite.

Transmission Electron Microscopy (TEM) image (figure 2 a) shows that nanosized platinum particles of about 5-10 nm are homogeneously dispersed in silica gel matrix. The selected area electron diffraction (figure 2 b) obtained for the platinum-silica composite confirmed the presence of nanocrystalline platinum. The inter planer d-spacing were deduced from the concentric rings and were attributed to platinum.

IOP Conf. Series: Materials Science and Engineering **73** (2015) 012076 doi:10.1088/1757-899X/73/1/012076

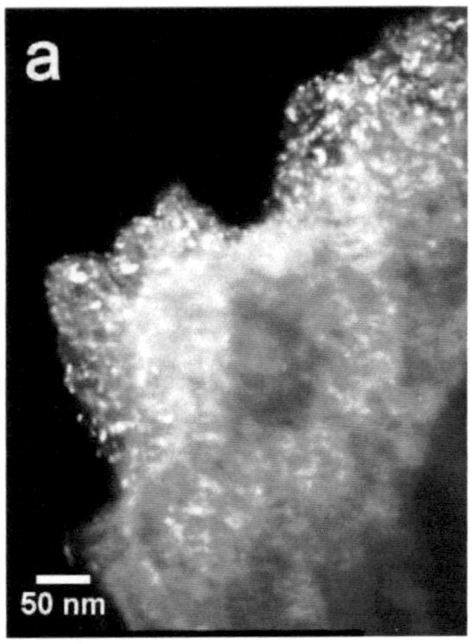

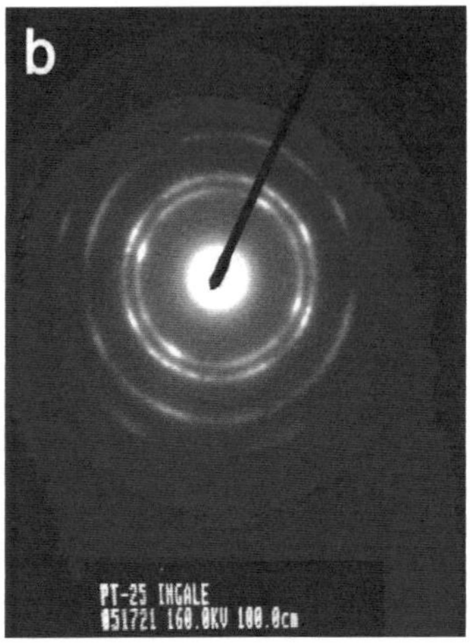

Figure 2 a) TEM image and **b)** SAED for Pt-SiO_2 nano-composite.

The Karl-Fischer titration method showed only 0.08% water adsorption for the Pt-silica nanocomposite sample that was placed on water for 3 days before measurement. It indicated super-hydrophobic nature of the silica support system [5].

The catalytic properties of the nanocomposite are tested through exposure to hydrogen gas. Pulse studies performed on the catalyst to evaluate the Pt dispersion on the support. The typical pulse profile of adsorbed H_2 on the nanocomposite is shown in figure 3.

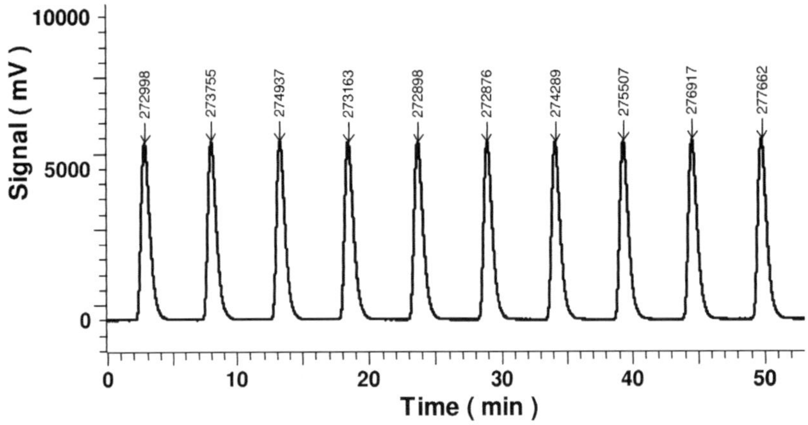

Figure 3 Typical pulse profile of adsorbed H_2 on the Pt-silica nano-composite.

The specific surface area of the Pt-silica nanocomposite was found to be 400 m^2/g where as the surface area of metal was found to be 140 m^2/g. Chemisorptions studies conducted using H_2 showed 50.6 % of accessible platinum atoms on the silica matrix that indicate high dispersion of platinum on silica support.

4. Conclusions

The nano sized platinum particles have been prepared by sol gel method using silica support. The high dispersion (50.6 %) of Pt nanoparticles has been achieved on hydrophobic silica gel matrix with surface area of 140 m^2/g for the metal. The well dispersed Pt nanoparticles on silica support system could be potential catalyst in heterogeneous reactions.

Acknowledgements
The authors thank to R. P. Patel, P. C. Das, Sonu Gavit and Sandip Virnak, APD, BARC and P. R. Golatkar, HWD, BARC, for their assistance in experimental work.

References

[1] Izawa H, Isomura S and Nakane R 1979 *Journal of Nuclear Science* **16** 741
[2] Choi H J, Lee H, Kim K R and Ahn D H 2002 *J. Ind. Eng. Chem.* **8** 586
[3] Lembacher C and Schubert U 1998 *New J. Chem.* **22** 721
[4] Ingale S V, Wagh P B, Tripathi A K, Kamble V S, Rataneshkumar, Gupta S C 2011, *Journal of Porous Materials* **18** 567
[5] Wagh P B, Ingale S V and Gupta S C 2010 *J. Sol-Gel Sci. Technol.* **55** 73

Polymer assisted preparation and characterization of ZnO and Sn doped ZnO nanostructures

A. Santhosh Kumar*, K. K. Nagaraja and H. S. Nagaraja

Materials Research Laboratory, Department of Physics, National Institute of Technology Karnataka, Surathkal-575025, India.

Email: saanthosh.phy@gmail.com

Abstract. Zinc oxide (ZnO) and tin doped ZnO are wide band gap semiconducting materials with excellent optoelectronic properties. In the present study ZnO and Sn: ZnO films are prepared using polymer assisted sol gel process. The thermal behaviour of the dried gel sample studied using DTA and TG analysis. TG-DTA result shown that most of the organic of PVA and CH_3COO group of zinc acetate and other volatiles are removed below 500 °C. The effect of Sn on the crystallinity, microstructral properties of the deposited films was investigated. XRD patterns of undoped and Sn doped ZnO films indicate enhanced intensities for the peak corresponding to (002) plane, resulting preferential orientation along the c-axis. The SEM images confirm that the grown films are composed of nanorods.

1. Introduction

Zinc oxide is one of the most promising materials for the fabrication of the next generation optoelectronic devices in the UV region and optical display devices. Zinc oxide (ZnO) is a wide band gap semiconductor having an energy gap of 3.37 eV at room temperature with 60 meV free exciton binding energy [1, 2]. Recently, one-dimensional (1D) ZnO nanostructures such as nanowires, nanorods, nanobelts, nanoneedles and nanosheets are promising building blocks for optical applications in the fields of blue-light emitting diodes, short wave length laser diodes with low thresholds in the UV light-emitting diodes, transparent and conducting electrodes in solar cells and gas sensors. In order to develop material for special applications, doped ZnO films have been fabricated and investigated by many research groups. The main reason for choosing tin as the dopant in ZnO is one of the most efficient combination to improve the opto-electrical and gas sensing properties [3]. ZnO and Sn doped ZnO (Sn: ZnO) nanostructure films have been deposited by several methods such as thermal evaporation [3], pulsed laser deposition [4], successive ionic layer adsorption and reaction (SILAR) [5], chemical deposition [6], spray pyrolysis [7] and sol gel [8]. Recently, an alternative method namely polymer assisted deposition has been developed, and was implemented successfully to deposit metal oxide thin film [9]. In this process, polymer controls the viscosity and binds metal ions, resulting in a homogeneous distribution of metal precursors in the solution. In the present work we used water soluble polyvinyl alcohol (PVA) as complex polymer and served as a carrier, associated with Zn^{2+} ligands on polymer side chain. Moreover, it is used as a spacer and promoter for ZnO nano rod growth [10]. In this work, undoped and Sn doped ZnO thin films were prepared by PVA assisted sol gel spin coating onto glass substrates. The prepared films were characterized and analyzed by different characterization techniques.

2. Experimental

Polyvinyl alcohol (PVA) with molecular weight of 1,40,000 g/mol [Merck], Zinc acetate dehydrate [$Zn(CH_3COO)_2.2H_2O$, Merck] and Tin tetrachloride pentahydrate [$SnCl_4.5H_2O$, Aldrich] were used as the precursors. The polymer gel for deposition was prepared by dissolving 1Wt% of PVA in 50ml distilled water and stirred at 80 °C for 6 h. 1M zinc acetate was added to the resulting solution, while stirring at room temperature for 6 h. A clear homogeneous gel was obtained. Above prepared gel solution pH=6.4 adjusted with citric acid. Sn doped ZnO gel was obtained by introducing 1% of tin tetrachloride into the above prepared solution. The gel solution was aged for 24hr in order to make the PVA chains fully complex with the metal ions. Subsequently, a pure and Sn doped ZnO polymer gel was two times spin coated onto the cleaned glass substrates. The intermediate drying between successive coatings was done at 180 °C for 30 min. The prepared films were initially heat treated at 250 °C for 60 min, to promote pyrolysis of the polymer resin, resulting in an amorphous film, and later crystallized by annealing at 450 °C for 1 h. The prepared dried precursor and films were characterized and analyzed by different characterization techniques. The decomposition of PVA was confirmed by TG-DTA results. The crystallographic studies were performed using JEOL X-ray Diffractometer (JDX-8030) using Cu-Kα wavelength (λ=1.54059 Å) and scanning range from 10° to 70°. The surface morphology of the films was observed with JEOL scanning electron microscopy (JSM-6380 LV).

3. Results and discussion

3.1. Thermal analysis

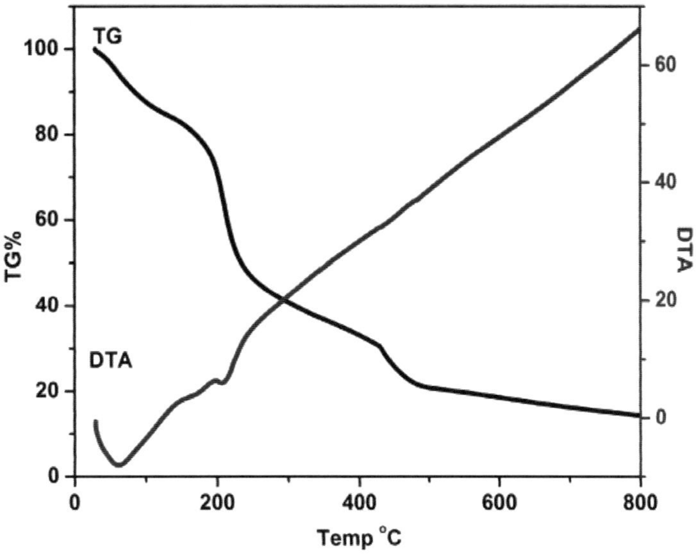

Figure 1. TG-DTA curves recorded on PVA zinc acetate dried gel

The thermodecompostion behavior of the dried PVA -zinc acetate dried gel was studied by TGA and DTA analysis. Fig. 1 shows TG-DTA curves measured on PVA-zinc acetate dried gel, which shows that most of the organic belonged to PVA and CH_3COO group of zinc acetate and other volatiles (H_2O, CO_x, etc.) were removed at temperature less than 500 °C.

3.2. X – Ray Diffraction (XRD) studies

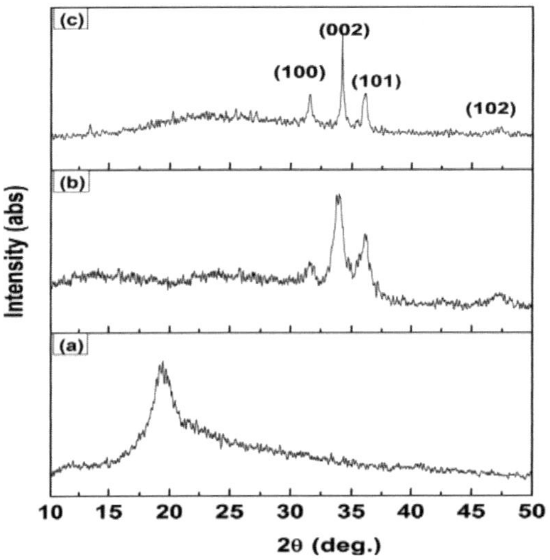

Figure 2. XRD patterns of PVA assisted undoped and Sn doped ZnO films (a) PVA zinc acetate film dried at room temperature (b) PVA zinc acetate film annealed at 450°C / 1hr (c) Sn doped PVA zinc acetate film annealed at 450°C /1hr.

In Fig. 2(a), a broad peak around 2θ=20° corresponds to the (101) plane of semi-crystalline PVA [11, 12] in PVA-zinc acetate film dried at room temperature. This shows that the crystallinity of PVA was largely influenced by the presence of zinc acetate in the PVA-zinc acetate film dried at room temperature, indicating an interaction between PVA and zinc acetate molecules. Notably, after the PVA-zinc acetate thin films were annealed at 450 °C for 1hr (Figure 2 (b)), crystalline peak of PVA disappeared, and three main reflection peaks appeared at 2θ=31.9° (100), 34.5° (002), 36.3° (101) corresponding to pure ZnO (JCPDF: 079-0206; a=b= 3.2499 Å, c= 5.2066 Å) crystalline phase [12]. The XRD patterns of all the ZnO and Sn: ZnO samples indicate enhanced intensities for the peak corresponding to (002) plane, resulting preferential orientation along the c-axis. The strain $\varepsilon = (c_{film} - c_{bulk}) / c_{bulk}$ in c-axis was calculated by XRD data, where c_{film} and c_{bulk} are the lattice constants (c) of ZnO film and bulk respectively. The lattice mismatch between ZnO, Sn: ZnO films and substrates can result in varying degrees of stress during the growth process of the films. To derive stress in the film, following formula is used, which is valid for a hexagonal lattice [13]

$$\sigma_{film} = \frac{2C_{13}^2 - C_{33}(C_{11}+C_{12})}{C_{13}} \varepsilon \qquad (1)$$

Where Cij are bulk ZnO elastic stiffness constants C_{11} = 209.7 GPa, C_{12} = 121.1 GPa, C_{13} = 105.1 GPa, C_{33} = 210.9 GPa. The estimated values of stress σ_{film} in undoped and Sn doped ZnO thin films, have been listed in Table 1.

Table 1. Lattice parameters (a & c), average crystallite size (D), Stress (σ_{film}) evaluated from XRD

	Lattice parameters (A°)	Avg Crystallite size(D) nm	Stress (σ_{film}) GPa
ZnO	a= 3.255; c=5.207	38	-1.129
1%Sn: ZnO	a=3.235; c=5.204	66	-0.860

3.3. Scanning electron microscopy (SEM)

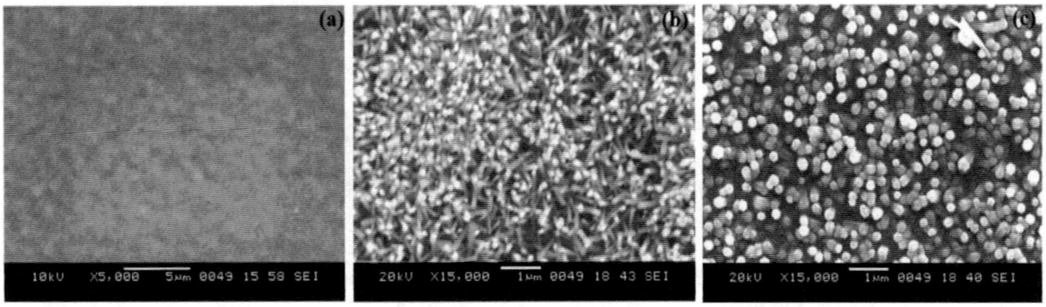

Figure 3. Surface morphology of PVA assisted ZnO and Sn doped ZnO films. (a) PVA zinc acetate film dried at room temperature (b) PVA zinc acetate film annealed at 450°C / 1hr (c) Sn doped PVA zinc acetate film annealed at 450°C /1hr.

The micro structural analysis of the deposited ZnO and Sn: ZnO films were carried out using scanning electron microscope. The samples prepared in presence of 1 wt% PVA at room temperature, seen to be brown and smooth (Fig. 3(a)). When the deposited films are calcined at 450 °C , the surface morphology of the undoped ZnO and Sn doped ZnO film as shown in Fig. 3 (b and c), displays vertically oriented hexagonal rods.

4. Conclusion

The ZnO and Sn doped ZnO thin films have been deposited by the PVA assisted sol-gel process. Thermal behavior of the dried gel shown that most of the organic of PVA and CH_3COO group of zinc acetate and other volatiles were removed at about 450 °C. The undoped and Sn doped ZnO thin films are composed of nanorods, with (002) preferred orientation. The crystallinity of the films improved with Sn doping.

References

[1] Özgür Ü, Alivov Ya I, Liu C, Teke A, Reshchikov M A, Doğan S, Avrutin V, Cho S J, Morkoc H 2005 *J Appl Phys* **98** 041301-103pp
[2] Anderson J, Chris G Van de Walle 2009 *Rep Prog Phys* **72** 126501-29pp.
[3] Shalaka C. N, Mulla I. S. 2009 *Mater Sci Eng C* **29** 1317-1320.
[4] Holmelund E, Schou J, Tougaard S, Larsen NB 2002 *Appl Surf Sci* **197-198** 467
[5] Shishiyanu S T, Shishiyanu T S, Lupan O I 2005 *Sensors Actuators B*, **107** 379
[6] Vaezi M R, Sadrnezhaad S K 2007 *Mater Sci Eng B* **141** 23
[7] Caglar Y, Ilican S, Caglar M, Yakuphanoglu F 2007 *Spectrochim Acta* A **67** 1113
[8] Yung K C, Liem H, Choy H S 2009 *J Phys D: Appl Phys* **42** 185002
[9] Jia Q X, Mccleskey T M, Burrell A K, Lin Y, Collis G E, Wang H, Li A D Q and Foltyn S R 2004 *Nature Mater* **3** 529-532.
[10] Sang W, Fang Y, Fan J, He Y, Min J, Qian Y 2007 *J Cryst Growth* **299** 272-276.
[11] Nishio Y and Manley RS 1988 *Macromolecules* **21** 1270-1277
[12] Santhosh K A, Nagaraja K K, Nagaraja H S 2013 *J Mater Sci: Mater Electron* **24** 3812–3822
[13] Cullity B D, Stock R S 2001 *Elements of X-ray Diffraction*, (3rd ed.), (New Jersey: Prentice Hall)

Observation of confinement effects through liner and nonlinear absorption spectroscopy in cuprous oxide

H Sekhar*, Y Rakesh Kumar and D Narayana Rao
University of Hyderabad, Hyderabad – 500 046, India

*Email address: halubaisekhar@gmail.com

Abstract. Cuprous oxide nano clusters, micro cubes and micro particles were successfully synthesized by reducing copper (II) salt with ascorbic acid in the presence of sodium hydroxide via a co-precipitation method. The X-ray diffraction studies revealed the formation of pure single phase cubic. Raman spectrum shows the inevitable presence of CuO on the surface of the Cu_2O powders which may have an impact on the stability of the phase. Transmission electron microscopy (TEM) data revealed that the morphology evolves from nanoclusters to micro cubes and micro particles by increasing the concentration of NaOH. Linear optical measurements show that the absorption peak maximum shifts towards red with changing morphology from nano clusters to micro cubes and micro particles. The nonlinear optical properties were studied using open aperture Z-scan technique with 532 nm, 6 ns laser pulses. Samples exhibited saturable as well as reverse saturable absorption. The results show that the transition from SA to RSA is ascribed to excited-state absorption (ESA) induced by two-photon absorption (TPA) process. Due to confinement effects (enhanced band gap) we observed enhanced nonlinear absorption coefficient (β_{eff}) in the case of nano-clusters compared to their micro-cubes and micro-particles.

1. Introduction

The past few years have witnessed an increasing trend toward micro- and nano structures of Cu_2O with well-controllable size and shapes. Controlling is being pursued with great interest in current material synthesis and device fabrication. Two crystalline forms of copper oxides are cupric oxide (CuO) and cuprous oxide (Cu_2O). Cuprous oxide having simple and highly symmetric cuprite structure (space group O_h) with six atoms in the unit cell. Cu_2O is low cost, with low toxicity and the natural abundance of its source materials and important inorganic p-type semiconductors with a direct band gap of 2.17 eV. It is believed to be a promising application in multiple technical fields including, solar cells, electrocatalytic, electrode material for lithium ion batteries, dilute magnetic semiconductor (DMS) and bio and gas sensors. Tuning the band gap and influencing the physical, chemical and electronic properties of the semiconductors are possible by the use of changing particle size from nano to even submicron scale or doping with transition metal ions [1-3] leading to properties that are different from those of their bulky counterparts, makes them candidates for various important applications in the field of material research. Cu_2O with different morphologies have been prepared by many methods, such as electrochemical, sonochemical, microwave irradiation, solution phase, sol–gel, etc. Chemical coprecipitation is simple and easy method to prepare Cu_2O with different morphologies. Some of the studies on Cu_2O show that Cu_2O phase was stabilized by a controlled oxidation and the formation of a thin protecting film of CuO [4].

Intense research is currently being pursued in the field of new nonlinear optical material preparation, with high optical nonlinearities [5] and fast response time is gaining interest both from the research as well as industrial point of view. To find the nonlinear optical parameters such as nonlinear absorption (NLA), we utilized the standard Z-scan technique, which is a simple single beam nonlinear transmission experiment [6].

The present work is focused on the cost-effective synthesis and characterization of its linear and nonlinear optical absorption of Cu_2O with different morphologies at nano second excitation.

2. Experimental Section

2.1 Synthesis

Cu_2O with different morphologies were prepared by simple chemical co-precipitation method. By reducing copper (II) salt with ascorbic acid in the presence of sodium hydroxide. In the typical synthesis, $CuSO_4$. $5H_2O$ (2.5g) was dissolved in milli Q water (20ml). Then NaOH (0.8g, 0.5 $molL^{-1}$) was dissolved in 40 ml milli Q water and added to the 20 mL aqueous solution of $CuSO_4$ by constant stirring. A clear blue $Cu(OH)_2$ precipitate appeared immediately. Then 0.88 g ascorbic acid dissolved in 50 mL milli Q water was added drop wise to the above solution with vigorous stirring at room temperature. After 60 minutes of stirring, the precipitate particles were isolated from the solution by centrifugation at 2000 rpm for 15 min. The product was washed by milli Q water and absolute ethanol through multiple cycles of centrifugation and redispersion. The final product was dried at 80 °C for 7 hours. Cu_2O powders with different morphologies prepared by simply changing the concentration NaOH with similar procedure (0.75, and 1.5 M). The sample codes of the Cu_2O powders obtained by varying the concentration of NaOH as 0.5, 0.75, and 1.5 M are referred as Cu_2O-I, Cu_2O-II, and Cu_2O-III, throughout this script.

2.2 Instruments

X-ray powder diffraction (XRD) patterns of the as prepared powders were recorded by using Inel - C120 X-ray diffractometer, which is equipped with a curved position sensitive detector and data was collected using Co-Kα radiation of wavelength 0.17889 nm. Raman spectra of as-synthesized solid powders were recorded using Horiba Jobin Yvon LabRam HR high resolution micro-Raman spectrograph with 514 nm of Argon ion laser as the excitation source. UV–Visible spectroscopy analysis was carried out on a JASCO UV-Visible absorption spectrophotometer. The morphology and the particle sizes of as prepared powder samples were measured with TECNAI G^2 FEI F12 model TEM. For TEM measurement, the samples were deposited on carbon coated copper grids. Second harmonic from the Nd:YAG laser with 6 ns (Spectra-Physics INDI-40) pulse duration and 10 Hz repetition rate was used as the excitation sources for nonlinear absorption (NLA) measurements.

3. Results and Discussion

Fig. 1 shows XRD pattern of as synthesized Cu_2O powders, all of the peaks match well with Bragg reflections of the standard cubic cuprite structure and is compatible with the literature standard value (standard JCPDF file no. 05-0667). One can observe that increasing the concentration of NaOH from 0.5 to 1.5 M, Bragg reflections (peaks) narrowing indicates particle size increase. The average crystallite size of Cu_2O-I (0.5M), was calculated using the (111) diffraction peak as reference according to the line-broadening method using the Scherrer's formula. The average crystallite size was found to be around 20 nm.

TEM samples were prepared by placing a drop of an iso-propanol dispersion of Cu_2O powders on carbon coated copper grid was used for particle morphology and as shown inset of Fig. 1 Cu_2O-I particle size varies from 15-20 nm and agglomerated with 100 to 150nm size. Cu_2O-II powder are cubic in nature and it size varies from 100-150nm. Cu_2O-III powders are in micron size. OH^- ions affect the stability of Cu_2O with different morphologies [7].

The Raman spectrum of as synthesized powders is shown in Fig. 2(a). The peaks at 218, 620 cm^{-1} are assigned to Cu_2O phase, whereas the peak at 280 cm^{-1} belongs to CuO phase [8]. Linear optical absorption spectra of as prepared Cu_2O powders of different morphologies were recorded by

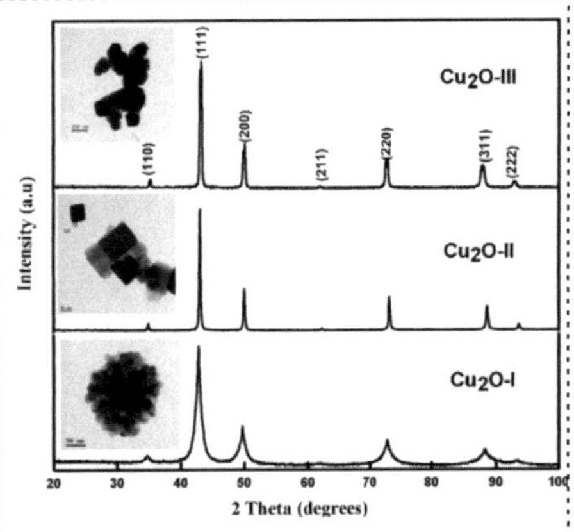

Figure 1. XRD and TEM picture of Cu_2O powders.

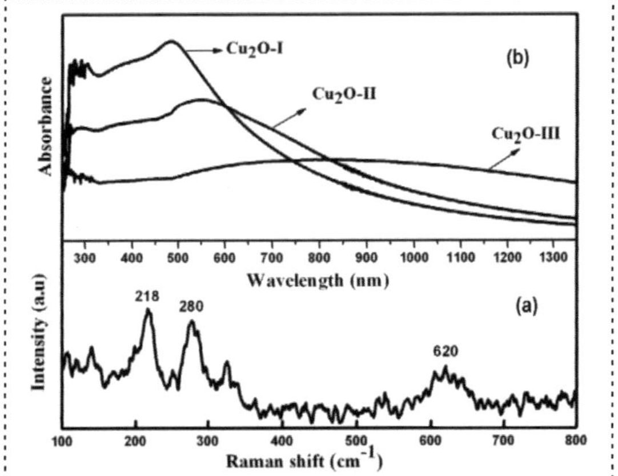

Figure 2(a). Raman spectrum of Cu_2O powders **2(b)** Uv-Vis absorption spectrum of Cu_2O powders.

dispersing them in iso-propanol by ultrasonication for 15 minutes and as shown in Fig. 2(b). From UV-Vis absorption spectra, the optical band gaps of Cu_2O-I, Cu_2O-II, and Cu_2O-III are calculated as 2.6, 2.3 and 2.1 eV respectively. Cu_2O-I, Cu_2O-II optical band gap shifts towards higher energy compared to the bulk Cu_2O band gap at 2.1 eV. This can be ascribed to the quantum confinement effect.

Fig. 3 shows open aperture Z-scan traces of Cu_2O powder ($3X10^{-4}$ M) in iso-propanol. All samples show SA to RSA behaviour. Since 532nm excitation falls in the absorption region, all samples show SA behaviour first (at lower intensities) and then due to strong excited state absorption (ESA), the RSA behaviour dominates at higher intensities or nearer to the focal point. During ESA, the nonlinear process dominates taking the first excited electron in the lower conduction band region to the higher excited conduction band states through a second photon. Such a process is referred to as free-carrier or excited-state absorption.

Switchover from SA to RSA on increase of the input intensity has been observed in various materials under different pulse excitation. In order to model [9] this type of behavior of RSA within SA, the observed experimental data is fitted with the following equation that combines the saturation behavior with saturation intensity of I_S and the effective excited state absorption (β_{eff}) coefficients that included both the excited state absorption and two photon absorption (TPA)

$$\alpha(\mathrm{I}) = \frac{\alpha_0}{1+\frac{I}{I_s}} + \beta_{eff} I \qquad (1)$$

Here, the first term describes the saturable absorption and the second term describes reverse saturable absorption including possible direct two photon absorption (β). α_0 is the linear absorption coefficient at 532nm and is derived from the absorption spectrum. I and I_S are laser intensity at each position in the Z-scan and saturation intensity, respectively. In this case, the propagation of the laser beam through the medium can be given by

$$dI/dz = -\alpha(I)I \qquad (2)$$

where α(I) is the total nonlinear absorption coefficient and z is the propagation distance in the medium. In Fig. 3 solid lines show theoretical fit to the experimental data using above equations, where β and I_S have been used as fitting parameters to match the valley and peak, respectively, in the experimental data. We can find that the theoretical simulations are in good agreement with the experimental results. One can observe that β_{eff} values are higher for nano-clusters and micro-cubes compared to their micro-particles.

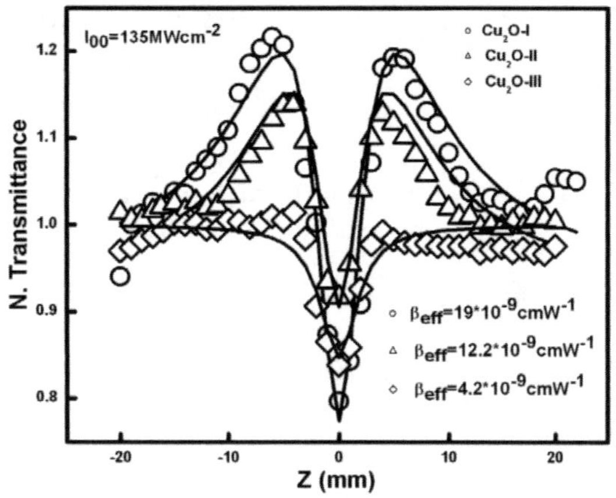

Figure 3. Open aperture Z- scan curves of (a) Cu_2O-I (b) Cu_2O-II (c) Cu_2O-III (open circles, open triangles and open squares are experimental curves and solid line theoretically generated curves.)

4. Conclusions

We observed a strong dependence of the band gap with the change in the particle size inferred from the XRD line shapes, and TEM data. We also observed a strong variation in the effective nonlinear absorption coefficient with change in the particle size. Effective nonlinear absorption coeiffient (β_{eff}) values in the case nanoclusters are higher compare to the micro-particles.

References

[1] Zhang L and Wang H 2011 *ACSNano* **5** 3257
[2] Sekhar H and Narayana Rao D 2012 *J. Mater. Sci.* **47** 1964.
[3] Sekhar H and Narayana Rao D 2012 *J. Alloys and Compd.* **517** 103
[4] Borgohain K, Murase N and Mahamuni S 2002 *J. Appl. Phys.* **92** 1292
[5] Sekhar H, Prem kiran P and Narayana Rao D 2011 *Mater. Chem. Phys.* **130** 113
[6] Sheik-Bahae M, Said A A, Wei T, Hagan D J and VanStryland E W 1990 *IEEE J. Quan. Electron.* **26** 760
[7] Zhang X, Song J, Jiao J and Mei X 2010 *Solid State Sci.* **12** 1215
[8] Wang S, Huang Q, Wen X, Li X Y and Yang S 2002 *Phys. Chem. Chem. Phys.* **4** 3425
[9] Gao Y, Zhang X, Li Y, Liu H, Wang Y, Chang Q, Jiao W and Song Y 2005 *Opt. Commun.* **251** 429

Preparation, structural and linear optical properties of Zn doped CdS nanopowders

H Sekhar*, Y Rakesh Kumar and D Narayana Rao
University of Hyderabad, Hyderabad – 500 046, India

*Email address: halubaisekhar@gmail.com

Abstract. A series of Zn doped cadmium sulfide (CdS:Zn) nanopowders were prepared by a simple co-precipitation method at room temperature by mixing the stoichiometric amount of reactants in a Milli Q water solvent. The composition of nanopowders was accurately adjusted by controlling the molar ratio of Cd, Zn acetate in the mixed reactants. Spectroscopic studies on as prepared nanopowders were investigated by using XRD, Raman, UV-Vis absorption, FE-SEM-EDAX and photoluminescence. Extremely broad reflections of XRD peaks of as prepared powders establish the nanometer scale dimensions and cubic structure. Doping with Zn in CdS does not lead to any structural phase transformation but introduces a decrease in the lattice constants. Two characteristics of LO phonon peaks were observed in pure and Zn doped CdS samples. Raman peaks of Zn doped CdS nanopowders shifts slightly towards higher energy side compared to the pure CdS nanopowders. Exciton-phonon confinement factor (S) varies in between 0.3-0.4. At lower wavelength excitation we observed a broad emission peak maximum centered at 404 nm is attributed to localized band edge emission.

1. Introduction

Colloidal semiconductor nanoparticles or quantum dots (QDs) have generated enormous attention in the past decade because their optical and electronic properties, which are dramatically different from those of the corresponding, bulk crystals, as well as their wide applications in the fields of optoelectronic and biological technology. Chalcogenide semiconductors such as cadmium sulfide is a II-VI compound semiconductor with a wide direct band gap of 2.42 eV (bulk CdS) and a small exciton Bohr radius of 2.5 nm. When the nanoparticle radius is comparable to the Bohr exciton radius in the corresponding bulk material, it leads to splitting of continuum of electronic energy levels into discrete states with the effective band gap energy blue shifted from that of the bulk. II-VI semiconductor nanostructures have been investigated widely and demonstrated for potential applications in solar cells, light emitting diodes, IR photodetectors, electrically driven lasers, optical limiters biological fluorescent labels and upconversion luminescent materials.

Most of the efforts are focused on equal valence charge doping. Recently, transition metal ions doped II-VI nanoparticles have attracted keen interest because transition metal ions incorporated into II-VI nanoparticles can improve significantly the optical properties of nanoparticles. The optical properties of doped nanomaterials differ from the corresponding host nanomaterials [1]. Thrust of recent research has been the synthesis of quantum-dot sized doped CdS nanoparticle powders. Recently, attempts were made on the use of $Cd_{1-x}Zn_xS$ nanoparticles for production of hydrogen from water under visible light irradiation with greater efficiency than using pure ZnS and CdS nanoparticles [2-4]. $Cd_{1-x}Zn_xS$ film was shown to have good photovoltaic efficiency [5].

The present work is focused on the cost-effective synthesis and its physical properties of the cubic structure Zn-doped CdS nanoparticles at room temperature. The synthesis method presented is an easy, inexpensive, and single-step method.

2. Experimental section

2.1 Preparation technique

Pure and doped CdS nanoparticles were prepared by the colloidal chemical precipitation method at room temperature [1]. Three different molar ratios of cadmium acetate to zinc acetate, 10:0, 9:1, and 8:2 were prepared. About 15 ml of polyvinyl pyrrolidone (PVP MW-55,000) solution in methanol (0.5M) was added to 30 ml of the aqueous solution of Cd - Zn mixture solution with vigorous stirring. At room temperature 0.1 M Na_2S aqueous solution was added slowly drop wise to the Cd - Zn mixed solution under vigorous stirring. After 1hour and 30 minutes stirring precipitate particles were isolated from the solution by centrifugation at 6000 rpm for 15 min. The product was washed by milli Q water and absolute ethanol through multiple cycles of centrifugation and redispersion. The yellow precipitate obtained was air dried.

2.2 Characterization

Crystal structure of the powder sample was determined using powder X-ray diffractometer (Inel C120) which is equipped with a curved position sensitive detector and data was collected using Co-Kα radiation of wavelength 0.17889 nm. Raman spectra, was obtained using Horiba Jobin Yvon LabRam HR high resolution micro-Raman spectrograph with 514 nm of Argon ion laser as the excitation source. FE-SEM-EDAX analysis was carried out using Carl Zeiss Ultra 55 model. UV-Visible spectral analysis was carried out on a JASCO UV-Visible absorption spectrophotometer with a resolution of 1 nm in the range of 200 to 1200 nm with a scanning speed of 200 nm/min. Steady-state photoluminescence (PL) characteristics of as prepared powders dispersed in chloroform placed in a 1 cm path non-fluorescent quartz cuvette, were measured by Jobin Yvon Inc., Fluorolog 3 spectrofluorometer with a spectral resolution of 1 nm.

3. Results and Discussions

3.1 XRD

X-ray diffraction (XRD) peaks of pure and doped CdS powders reflection were broadened and as shown in Fig. 1 and its lattice parameters were calculated. The broadening can be attributed to a very small grain size of the particles. The reflection peaks of as synthesized powders can be indexed to be cubic phase, with JCPDS data card no 800019 at Co-Kα radiation of wavelength 0.17889 nm. Since we did not observe any characteristic peaks of the impurity phases in the XRD spectrum, we have concluded that the dopant is distributed homogenously without clustering or segregation with increase

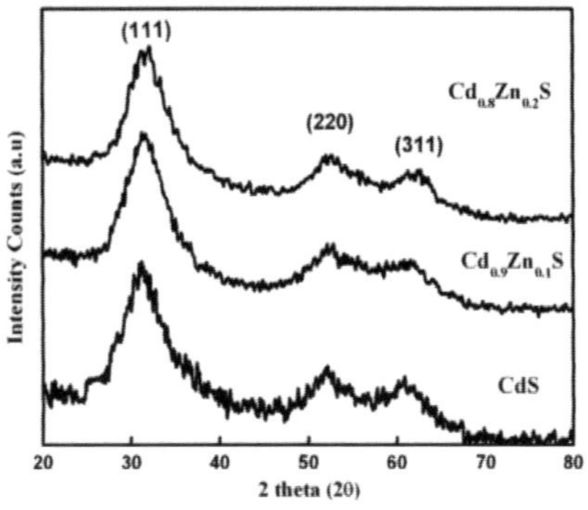

Figure 1. XRD pattern of pure and Zn-doped CdS powders.

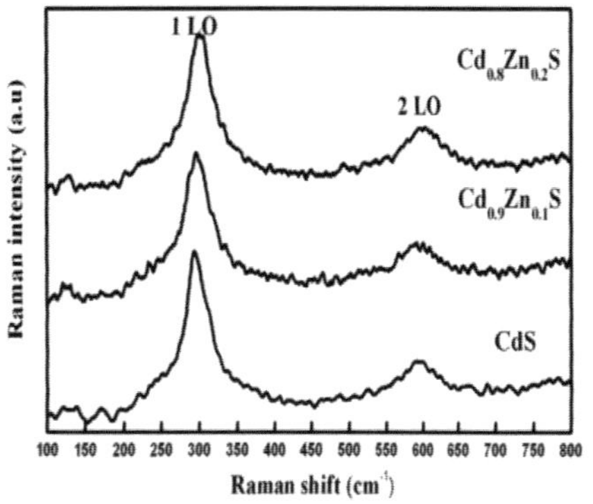

Figure 2. Raman spectrum of pure and Zn-doped CdS powders.

of the doping percentage, the XRD peak position shifts slightly towards higher angle relative to that of the pure CdS powders. Similar patterns were observed in Mn, Zn co-doped CdS nanowires [6]. Lattice parameters of pure and doped (Zn 10%, 20%) CdS were 5.7941, 5.7836 and 5.7001 Å. The ionic radius of Zn^{2+} (0.74 Å) ion are smaller than Cd^{2+} (0.97 Å) ion. So doping with Zn in CdS does not lead to any structural phase transformation but introduces a decrease in the lattice constants.

3.2 Raman Studies

A Raman spectrum of as synthesized powders is shown in Fig. 2. Bulk CdS has two characteristic LO phonon peaks, 1-LO (300 cm^{-1}) and 2-LO (600 cm^{-1}) vibrations [7]. These two characteristic LO phonon peaks were observed in pure as well as in Zn doped CdS samples, however at lower energies (pure CdS 293, 593cm^{-1}) compare to their bulk CdS, which could be due to quantum confinement. With increase of Zn content, 1-LO and 2-LO modes shift towards higher energy side compared to their pure CdS nanopowders which is consistent with the Cd→Zn substitution in the nanopowders. The intensity ratio of the 2-LO to 1-LO modes (I_2^{LO}/I_1^{LO}) is the parameter used to specify exciton-phonon coupling strength (S) in semiconductors. Calculated exciton-phonon coupling strength (S) for pure and doped (Zn 10%, 20%) CdS powders were 0.3914, 0.3414 and 0.3005. Our observations are in agreement with an earlier reported decrease in ratio from 0.88 to 0.42 from bulk to nanospheres at 514 nm excitation [7]. The phenomenon was attributed to lowering of the exciton- phonon coupling due to quantum confinement.

3.3 FE-SEM-EDAX Studies

Field Emission Scanning Electron Microscope (FE-SEM) images and Energy-Dispersive X-ray Analysis (EDAX) were obtained from as synthesized powders mounted on double sided carbon tape. Fig. 3 shows a FE-SEM image of as synthesized pure CdS powders. The grains are found to have aggregated to form clusters. Fig. 4 shows EDAX analysis of as synthesized powders. From the clear peaks of Cadmium (Cd), Zinc (Zn) and Sulphur (S), it can be noted that our sample did not show any trace elements other than cadmium, zinc and sulphur. The compositional ratio of Cd:Zn in the nanocrystals is approximately equal to that reactant mixture over a large composition range.

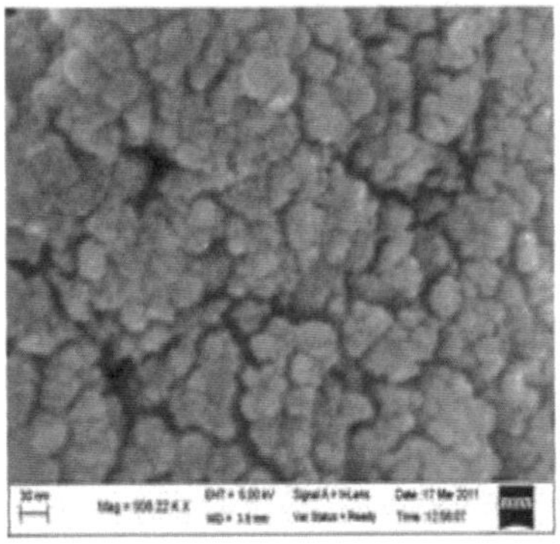

Figure 3. FE-SEM image of pure CdS

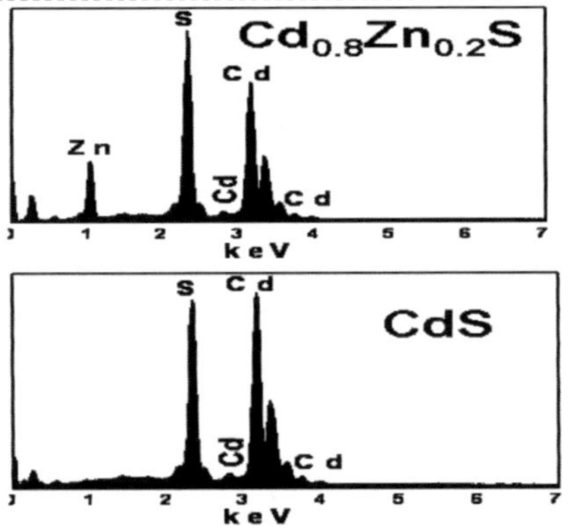

Figure 4. EDAX analysis of a CdS and $Cd_{0.8}Zn_{0.2}S$ powders

3.4 UV-Vis absorption spectroscopy

UV-Vis Absorption spectrum of pure and doped CdS powders dispersed in chloroform at same concentration is as shown in Fig. 5. The band gaps for pure and Zn doped CdS powders are blue-shifted relative to bulk CdS, which has a band gap at ~ 2.42 eV (512 nm), suggesting the formation of nanometer-sized CdS particles. The broad absorption peak around 462 nm is attributed as due to the surface states of the aggregated particles [7] and the narrow peak around 280 nm is attributed as due to the exciton peak of CdS. Absorption peak at 280 nm shows broadening in the case of Zn doped CdS compared to their pure CdS, which could mean that the exciton peaks energies in the doped system are distributed over wide energies.

3.5 PL studies

Emission spectra were recorded by exciting as prepared solid nanopowder dispersed in chloroform (4×10^{-4}M) through 15 minute ultrasonication and as shown in Fig. 6. At lower wavelength excitation we observed a broad emission centered at 404 nm from all samples. It was reported that this kind of band-edge luminescence is caused by the recombination of shallow region trapped electron-hole pairs and/or excitons [7]. The absence of emission from trap (defect) states, which normally appears around 635 to 675 nm range, suggests the stoichiometric nature of CdS, without a surface excess of Cd^{2+} or S^{2-} vacancies. Another interesting feature is that at low doping concentrations of ~ 10 % of Zn, the emission intensity increases but at higher dopings of 20%, the emission intensity slightly decreases as shown Fig. 6. The increase in the emission intensity is attributed as due to ZnS. The attenuation of the fluorescence is attributed to possible self quenching.

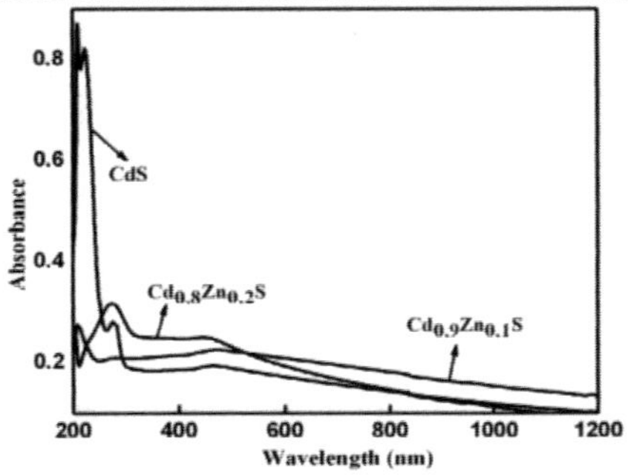

Figure 5. UV-Vis absorption spectrum of pure and Zn-doped CdS powders.

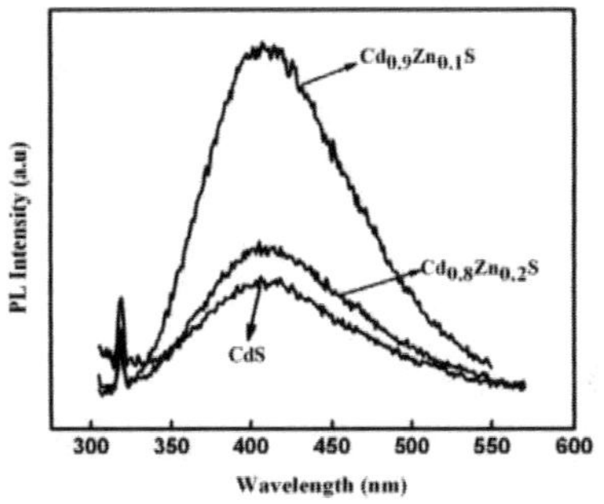

Figure 6. Photoluminescence spectrum of pure and Zn-doped CdS powders.

References

[1] Sekhar H and Narayana Rao D 2012 *J. Alloys and Compd.* **517** 103
[2] Wang X, Liu G, Chen Z, Li F, Qing (Max) Lu G and Cheng H 2009 *Electrochem. Commun.* **11** 1174
[3] Xing C, Zhang Y, Yan W and Guo L 2006 Int. *J. Hydrogen Energy* **31** 2018
[4] Zhang K, Jing D, Xing C and Guo L 2007 Int. *J. Hydrogen Energy* **32** 4685
[5] Chandra T and Bhushan S 2004 *J. Mater. Sci.* **39** 6303.
[6] Kim DS, Cho YJ, Park J, Yoon J, Jo Y and Jung M 2007 *J. Phys. Chem. C* **111** 10861
[7] Zeiri L, Patla I, Acharya S, Golan Y and Efrima S 2007 *J. Phys. Chem. C* **111** 11843

Effect of F ions on physical and optical properties of fluorine substituted zinc arsenic tellurite glasses

Shaik Kareem Ahmmad [1], Edu kondaul [2] Syed Rahman[2]

[1] Physics department, Muffakham jah college of Engineering and Technology, Hyderabad,India

[2] Department of Physics Osmaia universities, Hyderabad, India

Email: kareem.physics@gmail.com

Abstract. The effect of substitution of fluoride ions for oxide ions on the physical and optical properties of glass system (20-x) ZnO-xZnF$_2$-40As$_2$O$_3$-40TeO$_2$ where x = 0, 4, 8,12,16,20 mole % were investigated. The samples prepared by melt quenching method under controlled condition. The amorphous nature of these glasses was checked by X-ray diffraction technique. The density was measured according to Archimedes principle. The room temperature absorption spectra of all glass samples were determined using UV-Vis-NIR spectrometer. The thermal behaviour, glass transition temperature and stability of glass samples were studied by a differential scanning calorimetric (DSC). The density reduction of present glasses with ZnF$_2$ concentrations may be due to the low density of ZnF$_2$ compared with that of ZnO. Breaking the oxide network, the cross linking degree of the glass former could be reduced which results in decrease of both T_g and T_x. In the present glass system when F ions replaced by oxygen ions UV-Vis absorption cut-off wavelength decreases. This resulted form the conversion of structural unit in the glass from TeO$_4$ to Te(O,F)$_4$ and then to Te(O, F)$_3$.

Key words: zinc arsenic tellurite glass , XRD, density, DSC, and UV-Vis-NIR.

1. Introduction

In recent years the study on tellurium oxide based glasses has been the subject of high interest due to their potential applications as IR domes, optical fibers, modulators, memories and laser windows. These glasses are well known for their high density, high refractive index, high dielectric constants, a wide band infrared transmittance and large third order non-linear optical susceptibility and numerous devices including optical amplifiers, planar waveguides and nano wires have been fabricated using tellurite glasses [1-6] . A variety of studies on mechanical, electrical and optical properties of tellurium based glasses with different modifiers are present in literature [7-10]. Since tellurium oxide is a conditional glass former, the addition of arsenic oxide (As$_2$O$_3$) to tellurium oxide glasses improves the glass forming ability and the optical transparency in the blue region [11] and this addition affect the infrared transmission of TeO$_2$ glasses to a less extent, since the frequencies of some of the fundamentals modes of vibrations of As$_2$O$_3$ structural groups lay in the region of vibrations of TeO$_4$ structural groups [12].

The TeO$_2$-ZnO system shows good and stable glass forming stability with a broad region. However this glass formation strongly depends on the cooling rate and the size of the melt, especially in the TeO$_2$ –rich region [13]. Several studies [14-18] on infrared and optical absorption were reported on zinc tellurite glasses. It has been reported that absorption properties in TeO$_2$-ZnO glasses were observed to be very much dependent on the ZnO content. The ZnO-TeO$_2$ glass was used as a basis for multi-component optical glass system. Since fluorine has a higher electronegativity than oxygen, the addition of the modifiers like ZnF_2 to ZnO-TeO_2 glass matrix or the substitution of fluorine ions for oxide ions is expected to increase the glass forming range and glass stability, lower the viscosity and improves the transparency to a substantial extent and makes the glass more moisture resistant [19, 20] Morever, it would be expected that that the glass transition temperature and liquidus temperature decreases with substitution of fluoride ions for oxide ions, since two fluoride ions are substituted for an oxide ion, which causes the breakdown of the glass network structure. Several studies [19, 21-23] on introduced fluorine ions into oxide ions were reported with zinc tellurite glasses. To the best our knowledge there are no reports on fluorine substituted quaternary zinc tellurite glasses. Recently, Kitamura *et al* [24] reported the effects of substitution of fluoride ions for oxide ions on the thermal and optical properties of ZnO-Bi_2O_3-P_2O_5 glasses. In the present study, we prepared xZnF_2-(20-x)ZnO-40As_2O_3-40TeO_2 glasses and evaluated the effect of fluorine substitution on physical and optical properties.

2. Experimental

2.1 Glass preparation

In the present study, the glass samples of composition xZnF_2-(20-x)ZnO-40As_2O_3-40TeO_2 (x= 0, 4, 8, 12, 16 & 20 mole%) were prepared by melt quench technique. High purity (99.99%) zinc fluoride (ZnF_2), zinc oxide (ZnO), tellurium oxide (TeO_2) (all Sigma Aldrich) and arsenic oxide (As_2O_3) (May and Baker) were used as starting materials. A batch of 15 g of the above high purity chemicals in powder form was weighted, well mixed and melted in a porcelain crucible in the temperature range 650-700°C depending on the glass composition in an electrical furnace for about 30 minutes. For samples taken from different regions of the bulk specimen, the absence any Bragg peaks in the x-ray diffraction pattern confirmed that the prepared glasses are amorphous. The room temperature density (ρ) of the glass samples was determined by the Archimedes principle. The thermal behavior of the glass samples was investigated using a differential scanning calorimeter Mettler-Toledo: TGA/DSC1/1600HT. The glass transition temperature (T_g) and the crystallization beginning temperature (T_x) were determined based on the DSC curves using the onset method. The optical absorption spectra for all glass samples were recorded on a jasco model V-570 UV-Vis-NIR spectrometer in wavelength region of 200 to 1000 nm.

Results and discussion

X-ray diffraction patterns of all the present glass samples are shown in Fig.1. The above figure shows no continuous or discrete sharp peaks, which reflects the characteristics of amorphous glass structure.

The room temperature density (ρ) of the glass was determined to an accuracy of ±0.001 by standard principle of Archimedes' using xylene as the buoyant liquid. The density was calculated by the formula

$$\rho = a \times 0.86 / (a-b)$$

Fig. 2 presents the room temperature density of present glasses measured by Archimedes method as a function of ZnF_2 content . From the above figure, it is clear that the density decreases non-linearly. The density of the present glasses varies from 5.374 to 5.516 g/cc.

The density reduction of oxy-fluoride zinc arsenic tellurite glasses with ZnF_2 concentrations may be due to the low density of ZnF_2 compared with that of ZnO. An increase in the ZnF_2 concentrations in the present glass system increases the F/O ratio, and the density reduction is due to the replacement of the Zn=O double bonds by two Zn-F single bonds, the structure of glasses becomes loose. At high ZnF_2 concentrations some ZnF_2 molecules transform into ZnO molecules during the melting process in an atmosphere of air. Similar results were absorbed in TeO_2-ZnO-ZnF_2-Na_2O-La_2O_3 glasses [21]. The molar volume of the glass samples were calculated

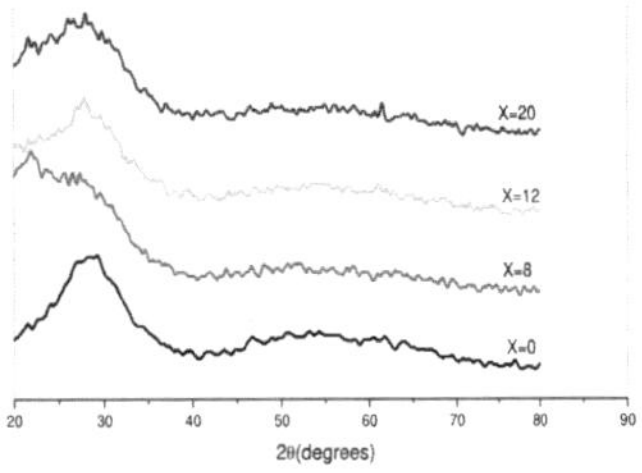

Figure 1

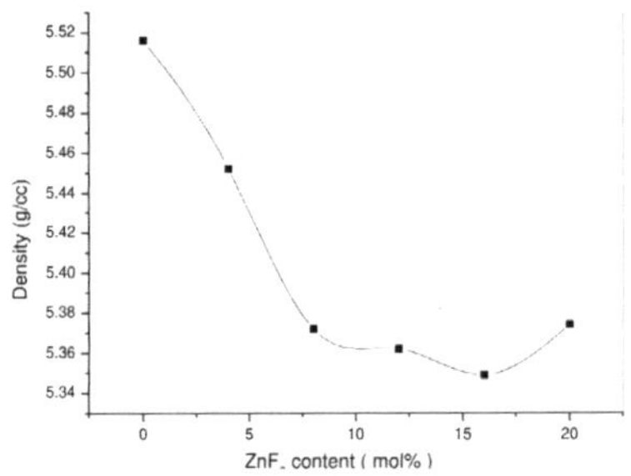

Figure 2

The DSC thermograms of all the glass samples are shown in fig. 3 and values decreasing form 338.6 to 297 ^{0}C The glass transition temperatures (T_g) and onset of crystallization temperate (T_x) were determined based on the DSC curves using the onset method. The uncertainty in T_g and T_x is $\pm 0.1^{\circ}$C. Fig. 4 plots the variation of T_g, T_x and S as a function of ZnF_2 content. The reduction in glass transition temperature and onset of crystallization temperate may be due to the addition of fluorine in higher proportion and consequently, the fluorine atoms are introduced in oxide network can be considered not to act as bridge between two Tellurium atoms because of single valence of fluorine. Breaking the oxide network, the cross linking degree of the glass former could be reduced which results in decrease of both T_g and T_x . Similar results were obtained in TeO-ZnO-ZnF_2-Na_2O-La_2O_3 and TeO_2-ZnO-ZnF_2 glasses [19, 21].

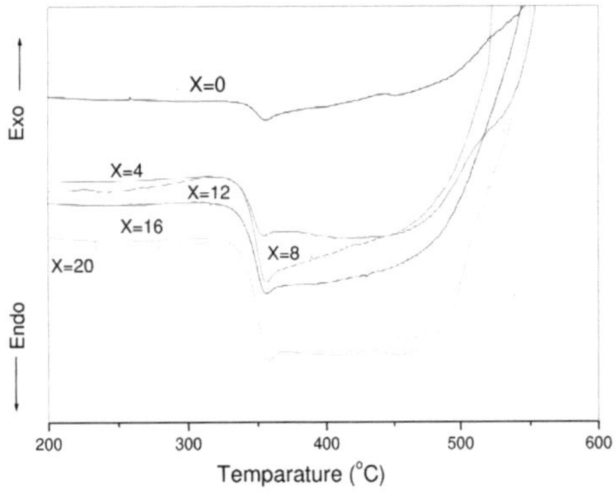

Figure 3

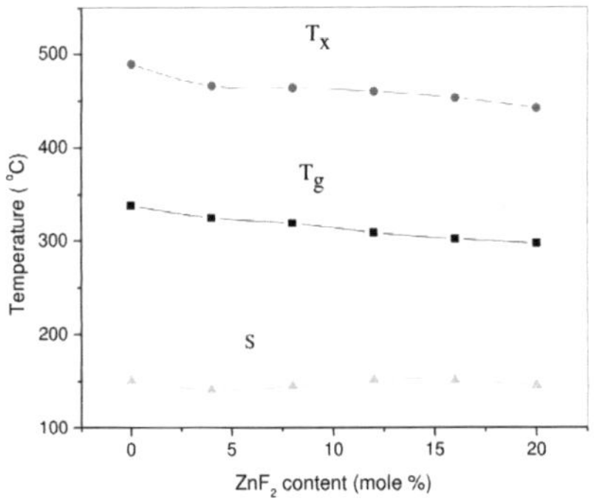

Figure 4

The study of optical absorption is a useful method for the investigation of optical transitions, electronic band structure in crystalline and non-crystalline materials.. The optical absorption spectra of of present glass system is shown in Fig 3. The variation of cutoff wavelength as a function of ZnF_2 content is illustrated in Fig 4. Cutoff of values decreasing from 465nm to 443nm .The UV absorption edge of all glasses fabricated in the present study is blue shifted with increasing ZnF_2 content from approximately 465 nm to 443.5 nm. This trend agrees with other results [19, 29-31] on oxyhalide tellurite and silicate glasses but differs from that observed for some lead germanium and lead silicate glass samples [32, 33], where a typical red shift of the absorption edge with the addition of glass

3 Conclusions.

Glasses based on present system were fabricated and characterized. The addition of ZnF_2 in the tellurite glass changed significantly glass properties such as density, T_g, T_x as well as the glass optical properties as Uv-Vis absorption etc. The density, glass transition temperature T_g decreased with increasing ZnF_2 content. It is noted that the glass thermal stability (S) was significantly increased, particularly when the ZnF_2 concentration was X=12%, which indicates could be a promising candidate for optical fiber fabrication against divitrifiaction. Due to optical loss would allow them to good candidate materials for this application

4. Acnowledgment

One of trhe authors (Shaik Kareem Ahmmad) whishes to thank Dr.M.G.Krishna, UOH for providing UV-Vis-NIR spectctrometer, Priciple Scientifc Officer,CIL for providing DSC and EDAX and Priciple MJCET for suggestions and hleps with this work.

References

[1] A. Mori 2008 *J.Ceram.Soc.Japan.* **116** 1040-1051
[2] A. Mori, Y.Ohishi, S.Sudo, 1997 *Electron.Lett.* **33** 863-864
[3] H. Masuda, A. Mori, K Shikano, M. Shimzu, 2006 *J.Lightwave Technol.* **24** 504-515
[4] G. Qin, H. sotobayashi, M. Tsuchiya, A. Mori, T. Suzuki, Y. Ohishi 2008 *J.Lightwave Technol* **26** 492-498
[5] P.Nanid, G. Jose, C. Jayakrishnan, S. Debbarma, K. Chalapathi, K. Alti, A.K. Dharmadhikari, J.A. Dharmadhikari, D. Mathur, 2006 *Opt. Express.* **14** 12145-12150.
[6] L. Tong, L. Hu, J. Zhang, J. Qiu, Q. Yang, J. Lou, Y. Shen, J. He, Z. Ye, 2006 *Opt. Express* **14** 82-87
[7] M. Pal, Y. Tsujigami, A. Yoshikado, H. Sakata, 2000 *Phys. Stat. Sol.* A **182** 727
[8] G. El-Damrawi, 2000 *Phys. Stat. Sol.* A **177** 385
[9] G. El-Damrawi, S. Abd-El-Maksoud, 2000 *Phys.Chem. Glasses.* **41** 64
G. El-Damrawi, S. Abd-El-Maksoud, 2000 *Phys.Chem. Glasses.* **2** 56
[10] R. A. Montani, S.E Giusia, 2001 *Phys. Chem. Glasses.* **42** 12
[11] B.V Raghavaiah, N. Veeraiah, 2004 *J. Phys. Chem. Solids.* **65** 1153
[12] A. Bishay, C. Magarabi, 1969 *Phys.Chem. Glasses.* **10** 1
[13] V. Kozhukharov, S. Neov, I. Gerasimova, P. Mikula, 1986 *J. Mater. Sci.* **21** 1707.
[14] A. Narazki, K. Tanaka, K. Hirao, T. Hashimoto, H. Nasu, K. Kamiya, 2001*J. Am. Ceram. Soc.* **841** 214.
[15] S. Rosmawati, H.A.A. Sidek, A.T. Zainal and H. Mohd Zobir, 2007 *J. Appl. Sci.* **20** 3051.
[16] A. Nukui, T. Taniguchi, M. Miyata, 2001 *J. Non-Cryst. Solids* **260** 293.
[17] J.C. Boyer, F.Vetrone, J.A. Capobianco, A. Speghini, M. Bettineli, 2003 *J. Appl. Phy.* **93** (12) 9460.
[18] H.A.A. Sidek, S. Rosmawati, Z.A. Talib, M.K. Halimah, W.M. Daud, Am. 2009 *J. Appl. Sci.* **6** (8) 1489.
[19] V. Nazabal, S. Todoroki, A. Nukui, T. Matsumoto, S. Suehara, T. Hondo, T. Araki, S. Inoue, C. Rivero, T. Cardinal, 2003 *J. Non-Cryst. Solids* **325** 85
[20] M.M. El-Desoky, M.M. Abo-Nafs.M , 2004 *J. Mater. Sci. Mater. Electron.* **15** 425
[21] G. Wang, S. Dai, J. Zhang, S. Xu, L. Hu, Z. Jiang, 2005 *J. Luminuscence* **113** 27
[22] G. Wang, S. Dai, J. Zhang, S. Xu, L. Hu, Z. Jiang, 2005 *J. Non-Cryst. Solids* **351** 2147.
[23] M.D. O'Donnell, C.A. Miller, D. Furniss, V.K. Tikhomirov, A.B Seddon, 2003 *J. Non-Cryst. Solids* **331** 48
[24] N. Kitamura, K. Fukumi, J. Nakamura, T. Hidaka, T. Ikeda, H. Hashima, J. Nishii, 2011 *J. Non-Cryst. Solids* **357** 1188
[25] S. Inaba, S. Fujino, 2010 *J. Am. Ceram. Soc.* **93** 217
[26] S. Toyoda, S. Fujino, K. Morinaga, *2003 J. Non-Cryst. Solids* **355** 821
[27] R.D. Shannon, C.T. Prewitt, 1969 *Acta Crystallogr.* B **25** 925
[28] R.D. Shannon, C.T. Prewitt, 1970 *Acta Crystallogr.* B **26** 1046
[29] G. Liao, Q. Chen, J.Xing, H.Gebavi, D. Milanese, M.Fokine, M. Ferraris, 2009 *J. Non-Cryst. Solids* **355** 447
[30] M. D. O'Donnell, K. Richardson, R. Stolen, 2007 *J. Am. Ceram. Soc.* **90** 1448
[31] K.Saito, A. J. Ikushima, 2002 *J. Appl. Phys.* **91** (8) 4886
[32] H. Jia, G. Chen, Z. Hou, 2003 *J. Non-Cryst. Solids 319* 322
[33] D. Lezal, J. Pedikova, J. Hoaka, 1996 *J. Non-Cryst. Solids* **196** 178.

IOP Conf. Series: Materials Science and Engineering **73** (2015) 012081 doi:10.1088/1757-899X/73/1/012081

Theoretical study of methanol as inhibitor and cyclopentane as stabilizer of dodecahedron methane hydrate cage

Snehanshu Pal[1], T K Kundu[2]
[1]Department of Metallurgical and Materials Engineering, National Institute of Technology Rourkela, Rourkela –769008, India, [2]Department of Metallurgical and Materials Engineering, Indian Institute of Technology Kharagpur, Kharagpur – 721302, India

E-mail: snehanshu.pal@gmail.com

Abstract. Density Functional Theory (DFT) based simulations have been performed to explain the role of methanol as an inhibitor and the role of cyclopentane as a promoter for methane hydrate. Interaction energy, Mullikan charges and electrostatic potential parameters for combined system of methanol and dodecahedron methane hydrate as well as cyclopentane and dodecahedron (5^{12}) methane hydrate cage are calculated using B3LYP functional (with and without dispersion function) and 6-31G(d) basis set. Methane hydrate formation inhibition by methanol and methane hydrate stabilization by cyclopentane is critically analyzed based on electrostatic potential and Mullikan charge. It is observed that hydrogen bond between water molecules of clathrate 5^{12} cages become stronger in presence of cyclopentane and weaker in presence of methanol. It is also found that methanol breaks some hydrogen bonds of water molecules.

1. Introduction

Gas hydrates are non-stoichiometric compound of clathrate group having complex cage like crystal structure consist of hydrogen bonded water molecules and stable at high pressure and low temperature. Accommodations of hydrocarbon and other guest molecules in cavities stabilize the gas hydrate structure. Gas hydrate is considered as the largest alternative source of energy because large amount of natural gas is trapped in it. Since gas hydrate formation in oil pipe line has hazardous impact, inhibition of gas hydrate formation is most important in petroleum industry. Gas hydrate stabilizer is also very significant in case of safe transportation of gas hydrates. Gas hydrate evolves as an important research focus globally. There are five types of unit clathrate cages such as 5^{12}, $5^{12}6^{2}$, $5^{12}6^{4}$, $4^{3}5^{6}6^{3}$ and $5^{12}6^{8}$ [1]. Stability analysis of various clathrate unit cages and absorption possibilities of methane, carbon dioxide and hydrogen in those unit cages are theoretically studied by Román-Pérez et. al. [2]. Van der Waals and Platteeuw's model based gas hydrate phase diagrams calculation is performed by Lundgaard et. al [3]. Chemical affinity concept dependent simulation of methane hydrate decomposition has been carried out by Varaminian et.al. [4]. Molecular dynamics study of clathrate-methane-water system for various proportions of methanol / water / $5^{12}6^{2}$ hydrate cage has been performed by Gualdrón et. al.[5]. Theoretical analysis of different antifreeze proteins as inhibitors of gas hydrate formation have been reported by several researchers [6, 7]. Electronic structure based

[1] To whom any correspondence should be addressed.

detail theoretical analysis of gas hydrate inhibitor and stabilizer effect is yet to be performed. In this work we study the electronic structure alteration of pentagonal dodecahedron (5^{12}) cages in presence of cyclo-pentane and methanol separately using Density Functional Theory (DFT) with dispersion function.

2. Simulation Details

One molecule of methane encapsulated pentagonal dodecahedron (5^{12}) clathrate cage structure in presence of methanol or cyclo-pentane is optimized by B3LYP [8] functional, 6-31G(d) basis set and DFT- D [9] functions using Gaussian 09 [10]. Interaction energies (ΔE) are calculated using following equations,

$$\Delta E = E_{com} - (E_{cage} + E_{cyclopentane}) \quad (1)$$

$$\Delta E = E_{com} - (E_{cage} + E_{methanol}) \quad (2)$$

Where, ΔE_{com} is energy for combined configuration of cluster and methanol or cyclopentane. E_{cage}, $E_{methanol}$ and $E_{cyclopentane}$ are energies of empty 5^{12} cage, methanol, cyclopentane respectively;

3. Result and Discussion

Optimized structures of $1CH_4@5^{12}$, one combined configuration having methanol placed and another combined configuration having cyclopentane placed near one face of $1CH_4@5^{12}$ clathrate using B3LYP/6-31G (d) and DFT-D are shown in figure 1, figure 2 and figure 3 respectively. It is observed

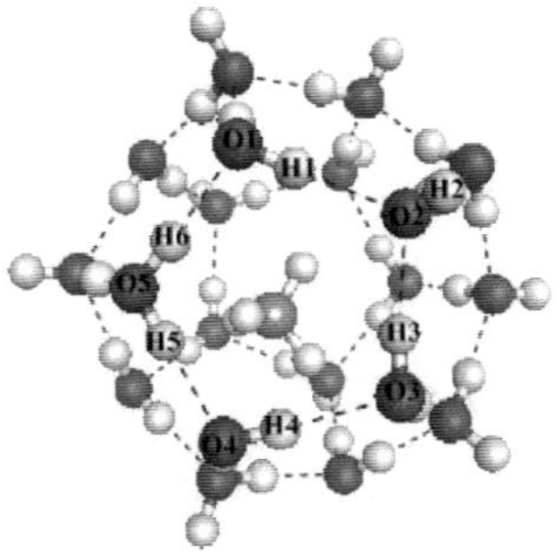

Figure 1. Optimized structure of $1CH_4@5^{12}$ cage structures B3LYP DFT-D / 6-31G(d)[legend: red=oxygen, white=hydrogen, black = carbon, dotted line: hydrogen bond].

in figure 2 that hydrogen bonded network of water molecules of $1CH_4@5^{12}$ cage is broken in presence of methanol. On the contrary, it is found from Fig.1 and Fig.3 that $1CH_4@5^{12}$ cage retains its original structure without any rupture of hydrogen bonded network. In presence of cyclopentane, $1CH_4@5^{12}$

Table 1. Interaction energy for optimized combined systems

	ΔE *(kcal/mole)*		ΔE *(kJ/mole)*	
	B3LYP/6-31G(d)	B3LYP DFT-D / 6-31G(d)	B3LYP/6-31G(d)	B3LYP DFT-D / 6-31G(d)
$1CH_4$ @ 5^{12} +Cyclo-pentane	-13.93	-10.26	-58.27	-42.95
$1CH_4$ @ 5^{12}+Methanol	-16.95	-7.74	-70.93	-32.40

clathrate is more stable because interaction energy of the combine system of cyclopentane and unruffled $1CH_4@5^{12}$ cage is negative as per table 1 for both B3LYP/6-31G(d) and B3LYP / 6-31G(d) with DFT-D based calculations. Presence of methanol causes reverse effect because combined system of methanol and broken $1CH_4@5^{12}$ clathrate has negative interaction energy according to table 1.

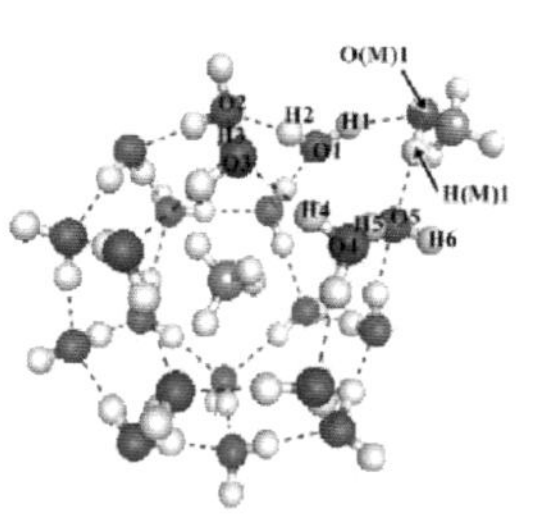

Figure 2. Optimized structure of $1CH_4$ @ 5^{12} + Methanol B3LYP DFT-D / 6-31G(d) [legend: red=oxygen, white=hydrogen, black = carbon, dotted line: hydrogen bond].

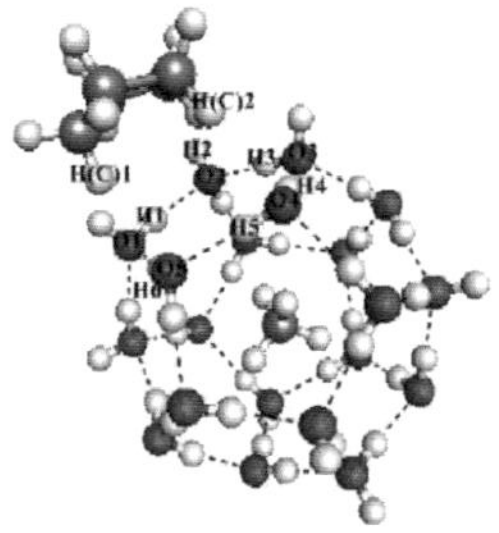

Figure 3. Optimized structure of $1CH_4$ @ 5^{12} + cyclopentane B3LYP DFT-D / 6-31G(d) [legend: red=oxygen, white=hydrogen, black = carbon, dotted line: hydrogen bond].

Calculated values of Mullikan charge, electrostatic potential are summarized in table 2 and table 3 respectively. It is observed that Mullikan charge difference as well as electrostatic potential difference of hydrogen bond forming oxygen and hydrogen atoms for $1CH_4@5^{12}$ cage (e.g H1 and O2) increases in presence of cyclopentane (i.e. absolute charge difference and electrostatic potential difference between H1 and O2 from 1.19 a.u. to 1.27 a.u. and 21.23 a.u. to 21.32 a.u.) and decreases in presence of methanol (i.e. absolute charge difference and electrostatic potential difference between H1 and O2 from

Table 2. Mullikan charges for optimized structure of $1CH_4$ @ 5^{12} and $1CH_4$ @ 5^{12}+ Cyclopentane

Atom	Mullikan charges					
	$1CH_4@5^{12}$		$1CH_4@5^{12}$ + Cyclopentane		$1CH_4@5^{12}$ + Methanol	
	B3LYP /6-31G(d)	B3LYP DFT-D / 6-31G(d)	B3LYP /6-31G(d)	B3LYP DFT-D / 6-31G(d)	B3LYP /6-31G(d)	B3LYP DFT-D / 6-31G(d)
O1	-0.808	-0.808	-0.860	-0.861	-0.800	-0.800
O2	-0.812	-0.813	-0.834	-0.834	-0.786	-0.786
O3	-0.862	-0.858	-0.873	-0.873	-0.790	-0.790
O4	-0.809	-0.810	-0.814	-0.814	-0.819	-0.819
O5	-0.868	-0.862	-0.861	-0.861	-0.808	-0.808
O(M)1					-0.651	-0.651
H1	0.385	0.406	0.434	0.434	0.401	0.407
H2	0.434	0.439	0.451	0.451	0.403	0.403
H3	0.426	0.423	0.412	0.412	0.398	0.398
H4	0.427	0.429	0.442	0.442	0.410	0.410
H(C)1			0.184	0.184		
H(C)2			0.130	0.130		
H(M)1					0.437	0.437

1.19 a.u. to 1.18 a.u. and 21.23 a.u. to 21.22 a.u.). It is also found that absolute Mullikan charge difference between hydrogen atom of methanol (i.e. H(M)1) and oxygen atom of one water (O5) is 1.24 a.u. It is inferred bases on above mentioned facts that cyclopentane helps to strengthen the hydrogen bonded water molecules cluster of $1CH_4@5^{12}$ cage and methanol itself form a stronger hydrogen bond with nearby water molecules of $1CH_4@5^{12}$ cage and causes distortion and disintegration of $1CH_4@5^{12}$ clathrate cage.

Table 3. Electrostatic Potentials for optimized structure of $1CH_4$ @ 5^{12} and $1CH_4$ @ 5^{12} + Cyclo-pentane

Atom	Electrostatic Potentials					
	$1CH_4@5^{12}$		$1CH_4@5^{12}$ + Cyclo-pentane		$1CH_4@5^{12}$ + Methanol	
	B3LYP /6-31G(d)	B3LYP DFT-D / 6-31G(d)	B3LYP /6-31G(d)	B3LYP DFT-D / 6-31G(d)	B3LYP /6-31G(d)	B3LYP DFT-D / 6-31G(d)
O1	-22.29	-22.30	-22.35	-22.35	-22.27	-22.27
O2	-22.26	-22.25	-22.27	-22.27	-22.23	-22.23
O3	-22.32	-22.32	-22.26	-22.26	-22.20	-22.20
O4	-22.26	-22.26	-22.29	-22.29	-22.26	-22.26
O5	-22.35	-22.36	-22.35	-22.35	-22.24	-22.24
O(M)1					-22.34	-22.34
H1	-1.03	-0.99	-0.95	-0.95	-1.01	-1.01
H2	-1.04	-1.00	-0.96	-0.96	-1.02	-1.02
H3	-1.03	-1.01	-0.95	-0.95	-1.00	-1.00
H4	-1.03	-1.02	-0.96	-0.96	-1.05	-1.05
H(C)1			-1.16	-1.16		
H(C)2			-1.13	-1.13		
H(M)1					-1.02	-1.02

4. Conclusion

It is concluded from the rigorous DFT study that cyclopentane stabilizes $1CH_4@5^{12}$ clathrate cage structures and methanol disintegrates the $1CH_4@5^{12}$ cage structure. This electronic structure based detail stability analysis of one methane hydrate cage in presence of methanol and cyclopentane separately will a significant step towards electronic structure property correlation study of inhibition and stabilization process of gas hydrates.

Acknowledgement

We acknowledge Ministry of Earth Science, Government of India for financial support, (Project No. MoES/16/48/09-RDEAS (MRDM5)).

References

[1] Sloan E D 2003 *Nature.* **426** 353
[2] Román-Pérez G, Moaied M, Soler J M and Yndurain F 2010 *Phys. Rev. Lett.* **105** 145901
[3] Lundgaard L and Mollerup J 1992 *Fluid Phase Equilibria.* **76** 141
[4] Varaminian F and Nia Z A 2010 *Iran. J. Chem. Chem. Eng.* **29** 125
[5] Gualdrón D A G and Balbuena P B 2007 *J. Phys. Chem. C.* **111** 15554
[6] Cruz-Torres A, Romero-Martínez A and Galano A 2008 *ChemPhysChem.* **9** 1630
[7] Zeng H, Wilson D L, Walker V K and Ripmeester J A 2006 *J. Am. Chem. Soc.* **128** 2844
[8] Lee C, Yang W and Parr R G 1988 *Phys. Rev. B.* **37** 785
[9] Goerigk L and Grimme S 2011 Phys. Chem. Chem. Phys. **13** 6670
[10] Frisch M J et. al. 2010 Gaussian 09 Revision B.01 Gaussian Inc. Wallingford CT

Effect of Gd substitution on structure and magnetic properties of $BiFeO_3$

Pittala Suresh and S. Srinath
School of Physics, University of Hyderabad, Gachibowli, Hyderabad, 500 046, India.
E-mail: srinath@uohyd.ac.in

Abstract. Multiferroic ceramic samples of $Gd_xBi_{1-x}FeO_3$ (x = 0 - 0.35) have been prepared by solid state reaction method. The results of XRD show a transition from rhombohedral to orthorhombic structure with increase in composition. Ferromagnetic order is observed with an enhanced magnetization (M) and coercivity (H_C) for all the doped samples. Maximum magnetization is observed at the field of 5T for x = 0.3. The temperature dependant magnetization shows the anomalous magnetic behaviour in this system. A minimum in the M-T curve and double hysteresis loop behaviour were observed for the samples with x=0.1 and 0.2

1. Introduction

Multiferroic materials have attracted the focus of many researchers recently because of the co-existence of the spin order as well as the charge order. The coupling between charge and spin ordering drives these materials for device application such as multiple state memory elements. Bismuth ferrite (BFO) is one of the most widely studied multiferroic material with T_N = 370°C and T_C = 810°C. The magnetic order of BFO is G-type ordering with a cycloidal spiral with a period of 620A° [1]. Unfortunately, the aforementioned magneto electric coupling is very small in BFO and its spin cycloid structure constraints the magnetic properties. To enhance the magnetic properties different strategies are adopted in terms of partial chemical substitution, reducing the particle size etc. Partial substitution of rare earth element at Bi site, $Bi_{1-x}R_xFeO_3$ (R= La, Gd, Ho, Sm etc.), is one such ways to improve the magnetic properties as it is expected to modify the spin structure due to the difference in ionic radius. Also, it has an additional advantage that it stabilizes pure phase and reduces the leakage current. Moreover, the introduction of rare earth element suppresses the spin cycloid and improves the magnetic properties [2]. To this end, as Gd^{3+} has a magnetic moment of 7.9 μ_B, effect of doping on the magnetic properties of BFO has been investigated.

2. Experimental details

$Bi_{1-x}Gd_xFeO_3$ (BGFO) (x = 0 - 0.3) powders were prepared using solid state reaction method. Stoichiometric amounts of Bi_2O_3, Fe_2O_3 and Gd_2O_3 oxides were milled for 5 hours in ball mill and pre-calcinated at 600°C. These powders are further milled and calcinated at 820°C for 30 minutes. The powders were then pressed into pellets and finally sintered at 820 °C for 2 h. Phase analysis was carried out by X-ray diffraction (XRD) using Cu Kα radiation. Field and temperature dependent magnetic properties were measured using Quantum design Physical Property Measurement System (PPMS).

3. Results and discussion

3.1. X-ray Analysis:

The *X*-ray diffraction spectra for BGFO samples are shown in Figure.1. For $x = 0$ the structure is rhombohedral with R3c space group and a shift in the peak positions is observed with increase in x. The doublets seen in pure BFO disappear with doping, signifying the change in the structure from Rhombohedral R3c to Orthorhombic Pnma due to the difference in the ionic radii of Gd^{3+} and Bi^{3+} as reported in literature [3, 4].

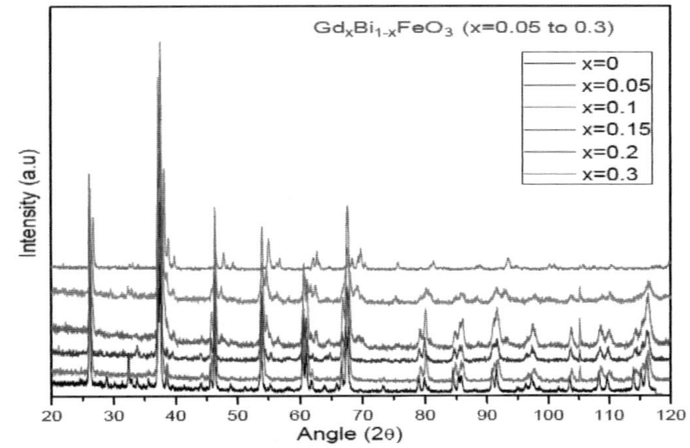

Figure.1. XRD pattern of Gd doped BFO.

There are small traces of non-magnetic $Bi_{25}FeO_{40}$ [5] and $Bi_2Fe_4O_9$, antiferromagnetic (AFM) below 264 K, secondary phases were observed. However, the influence of $Bi_2Fe_4O_9$ phase is not significant in the magnetic properties of these samples.

3.2. Magnetic properties:

The temperature variation of zero-field-cooled (ZFC) and field-cooled (FC) magnetization at a field of 500Oe is depicted in Figure 2. Both ZFC and FC magnetization (for x = 0, 0.1, 0.2, 0.3) increase gradually with decrease in temperature from 300 K. For $T \leq 25$ K there is a sudden increase in magnetization which is due to the ordering of Gd^{3+} magnetic moments [4]. A bifurcation in the ZFC and FC magnetization below 200K indicates the presence of magnetic frustration. There is a linear increase in the moments with x at low temperatures.

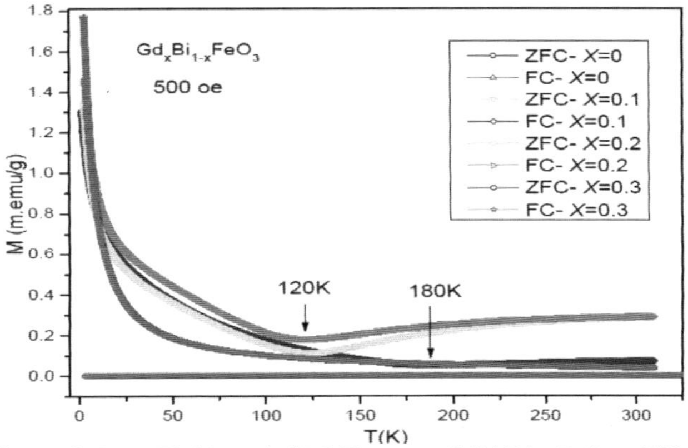

Figure.2. Zero Field cooled M-T curve of $Gd_xBi_{1-x}FeO_3$ at 500 Oe

Only for samples x = 0.1 and x =0.2, the magnetization shows a minimum in both FC and ZFC at 180 K and 120 K respectively [Figure.2]. A similar kind of result was observed in single crystal BFO with a minimum at 100K, which indicates a change in spin ordering [6]. The minimum observed in *x*=0.1 and *x*=0.2 samples may be due to spin re-orientation, typically seen in orthoferrites [7]. This behavior is more prominent in the M-H loops at 300K.

Figure 3 presents the magnetic hysteresis loops recorded at 300 K for BGFO. In pure BFO (*x* = 0), almost no net magnetization is observed due to the spiral spin structure but has a small residual magnetic moment results due to a canted spin arrangement. Gd doping at Bi site in BFO results in the appearance of ferromagnetism at room temperature. Pure BFO shows a weak magnetism with M_r = 0.0005 emu/g and H_C = 8.3Oe.The values of magnetization and coercivity get enhanced remarkably with the doping. This magnetization increase is due to the breaking of spin cycloid structure of BFO with the introduction of Gd.

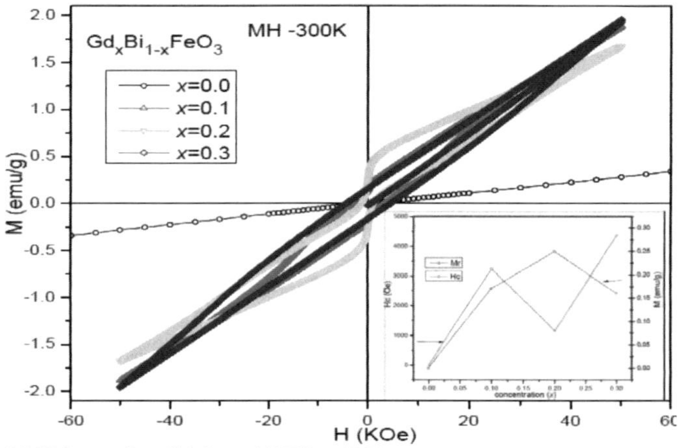

Figure.3. M-H loops for Gd doped BFO at RT. Inset shows H_c and M_r Values with *x*.

The antiferromagnetic structure of pure BFO thus changes to ferromagnetic state at room temperature with doping. Gadolinium doping not only results in structural modification, affecting the Fe-O-Fe antiferromagnetic (AFM) exchange interaction and therefore modifying the magnetic structure but also brings in additional magnetic interaction involving Gd. The samples with *x*=0.1 and 0.2 have a magnetization (at H=5T) of 1.85 and 1.69 emu/g and H_C is 3.2 KOe and 1.1 KOe at 300 K respectively. H_C and magnetization is maximum for *x* = 0.3 with a value of ~ 4.2 KOe and 1.95 emu/g respectively at 300 K. M-H loops for *x*=0.1 and 0.2 show an interesting double hysteresis kind of loop. This effect is more pronounced for *x*=0.2 thereby decreasing the H_C to 1.1 KOe. This may be attributed to the interactions between Fe^{3+} and Gd^{3+}. A similar kind of result was observed in previous studies, which is typical to orthoferrite structure [8] and also in Ho doped BFO [9].

M-H loops measured at 5 K are shown in figure 4. Although they are still unsaturated at 5T, exhibit a huge increase in magnetization reaching a value of 19.8 emu/g and 36.9 emu/g for *x* = 0.1 and *x* = 0.3 respectively compared to the pure BFO of 0.22 emu/g at 5T. The coercivity is maximum for *x* = 0.2 which has a value of ~ 0.95 KOe with the magnetization of 21.76 emu/g. The value of M_r increases linearly with *x* and reaches a maximum value of 0.25 emu/g and 0.97 emu/g for *x*=0.2 both at RT and 5K respectively and then starts decreasing for further doping of *x*=0.3.

The contribution to the magnetism occurred in this system is basically due to the Fe-O-Fe AFM interaction, contribution from Gd^{3+} ions and canted Fe sub lattice [4] As the doping level changes, the contributions from Gd and Fe sub lattice vary and hence results in the nonlinear variation in the M_r with x both at RT and 5K . However, a detailed magnetic structure analysis is required to understand this phenomenon more clearly.

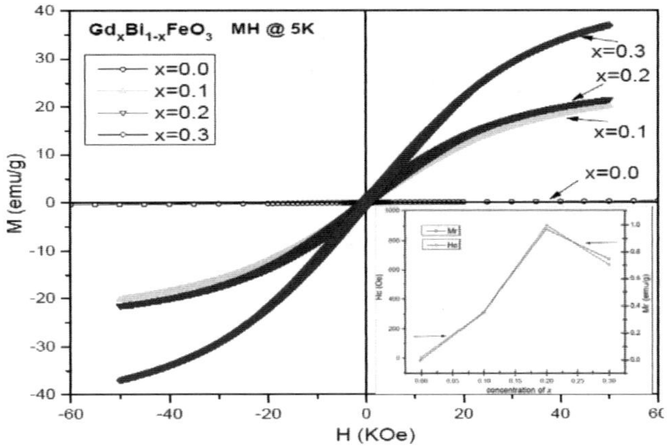

Figure.4. M-H loops for Gd doped BFO at 5K. Inset shows H_c and M_r values with x

4. Conclusions

In summary, we have prepared Gd doped BFO samples using solid state reaction method. Gd doping causes a structural change and accordingly enhance the magnetic properties. The magnetization reaches a maximum values for $x = 0.3$. High coercivity observed for x=0.15. The M-T curves for x = 0.1 and $x = 0.2$ have a minimum in the magnetic moment and a double hysteresis like M-H curve with a low coercive field was observed for the same samples, which is a characteristic of orthoferrites.

5. Acknowledgments

Authors acknowledge the financial support from CAS-RFSMS and Centre for Nanotechnology, University of Hyderabad for PPMS measurements.

6. References

[1]. Kanai T, Ohkoshi S I, Nakajima A, Watanabe T and Hashimoto K 2001 *Adv. Mater.* **13**, 487.
[2]. Julian A *et al* 2010 *Journal of Physics: Conference Series* **200** 012134
[3]. Khomchenko V A *et al* 2008 *Appl.Phys. Lett* **93** 262905
[4]. Weiwei Hu *et al* 2011 *J. Phys. Chem. C* **115** 8869–75
[5]. Chen Y and Zhao J 2009 *J. Alloys Compd.*, **487** 599
[6]. Singh M K *et al* 2008 *Phys. Rev. B* **77**, 144403
[7]. Koshizuka N and Ushioda S 1980 *Phys. Rev. B* **22**, 5394
[8]. Lazenka V V *et al* 2012 *J. Phys. D: Appl. Phys.* **45** 125002
[9]. Jeon N *et al* 2011 *Appl.Phys. Lett* **98** 072901

IOP Conf. Series: Materials Science and Engineering **73** (2015) 012083 doi:10.1088/1757-899X/73/1/012083

A Gold Sensors Array for Imaging The Real Tissue Phantom in Electrical Impedance Tomography

Tushar Kanti Bera[1] and J. Nagaraju
Department of Instrumentation and Applied Physics, *Indian Institute of Science*, Bangalore – 560012, INDIA

E-mail: tkbera77@gmail.com

Abstract. Surface electrodes in Electrical Impedance Tomography (EIT) phantoms usually reduce the SNR of the boundary potential data due to their design and development errors. A novel gold sensors array with high geometric precision is developed for EIT phantoms to improve the resistivity image quality. Gold thin films are deposited on a flexible FR4 sheet using electro-deposition process to make a sixteen electrode array with electrodes of identical geometry. A real tissue gold electrode phantom is developed with chicken tissue paste and the fat cylinders as the inhomogeneity. Boundary data are collected using a USB based high speed data acquisition system in a LabVIEW platform for different inhomogeneity positions. Resistivity images are reconstructed using EIDORS and compared with identical stainless steel electrode systems. Image contrast parameters are calculated from the resistivity matrix and the reconstructed images are evaluated for both the phantoms. Image contrast and image resolution of resistivity images are improved with gold electrode array.

1. Introduction

The design and development errors in the Electrical Impedance Tomography (EIT) [1-2] sensor array attached to the boundary of the EIT phantoms [3-5] usually produces boundary data errors which reduces the signal to noise ratio (SNR) of the boundary potentials. Low SNR of the boundary potential data leads to the errors in reconstruction process providing images with poor quality. Gold electrodes are found suitable for improved image reconstruction of NaCl phantoms in EIT [6]. A gold electrode array with high geometric precision is developed for studying the resistivity imaging of real tissue phantoms in electrical impedance tomography. The layout and required mask of the electrode array containing sixteen identical square electrodes is designed in CADSTAR. The flexible gold electrode array is developed by depositing a gold thin film (2 μm) on a flexible FR4 [7] sheet (acting as a flexible as the substrate) using electro-deposition process [8]. The gold electrode array is put inside a shallow polypropylene tank and the real tissue phantom is developed with chicken muscle tissue paste and fat tissue. An EIT instrumentation [9] developed with a constant current injector, signal conditioners and a USB based high speed data acquisition system is used for constant current injection and boundary data collection. The constant current injector consists of a sinusoidal signal generator and a voltage controlled current source. Signal conditioner block is developed with two filter circuits: a 50 Hz notch filter circuit and a narrow band pass filter circuit with a center frequency of 50 kHz. 1 mA 50 kHz sinusoidal constant current is injected to the phantom boundary with opposite current

[1] To whom any correspondence should be addressed.

injection protocol and the boundary data are collected using the high speed data acquisition system in a LabVIEW platform for different inhomogeneity positions. Resistivity images are reconstructed and compared with identical stainless steel electrode systems. Image contrast parameters are calculated and the images are evaluated for both the phantoms.

2. Materials and Methods

2.1. Sensors development

The flexible gold electrode array is developed by depositing a gold thin film (2 μm) on a flexible FR4 sheet using electro-deposition process [8]. The layout and required mask of the electrode array containing sixteen identical square electrodes (width - 10 mm, height - 10 mm) is designed (Fig. 1a) in CADSTAR software. Copper cladded flexible FR4 sheet (thickness = 200 μm, copper thickness = 35 μm) is chosen as the electrode array substrate (Fig. 1a) and the electrodes patterns are made using UV light lithography. Using photolithography required electrode patterns of the copper layaers are made on the FR4 sheet. All the electrode patterns are electroplated by a Ni layer of 2 μm thickness (Fig. 1b) and the Ni patterns are electroplated with a 2 μm thick gold layer (Fig. 1b). Ni layer is introduced between Au layer and Cu layer for improving the adhesion. All the gold electrodes (Fig. 1c) are made up of a 39 μm thick composite coating (35 μm Cu + 2 μm Ni + 2 μm Au) on the 200 μm flexible FR4 sheet (Fig. 1c). All the gold electrodes in the gold electrode array are connected with the connecting wires by soldering process (Fig. 1c). A stainless steel electrode array [5] is also developed with sixteen square electrodes identical to the electrodes in the gold array. Sixteen electrodes are cut from a stainless steel sheet (type-304, thickness = 50 μm) and the sixteen identical electrodes are made with equal electrode area (10 mm × 10 mm) and the SS electrodes are connected with crocodile clips [5].

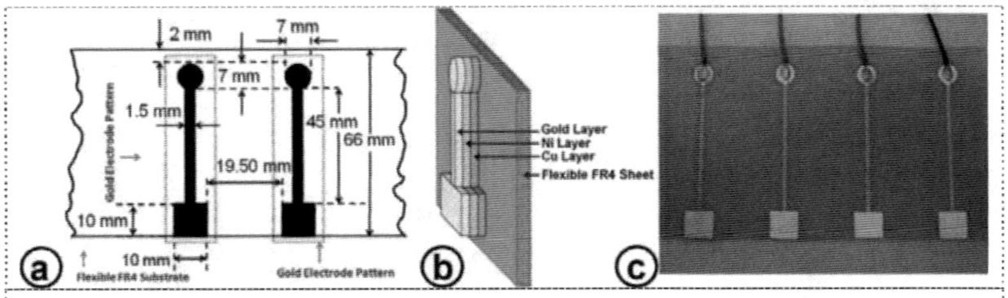

Figure 1. (a) electrode design, (b) electrode array schematic, (c) gold electrodes in the electrode array.

2.2. Phantom development

The performance of the gold electrode array is studied with a real tissue phantom developed with chicken tissues and the results are compared with a similar real tissue phantom containing identical stainless steel electrodes. Two identical phantoms (Fig. 2) are developed with 16 square (10 mm × 10 mm) thin film gold electrodes and stainless steel electrodes respectively and the resistivity imaging is studied with chicken muscle and fat tissues. Chicken muscle tissue paste [5] is poured into the phantom tanks (Fig. 2a-2d) a cylindrical fat tissue (inhomogeneity) of 35 mm diameter is placed in the muscle tissue and the boundary data are collected for different inhomogeneity positions. The height of the bathing medium in the phantom is kept as 10 mm so that an effective electrode area becomes 10 mm ×10 mm. A CME of 25 mm diameter is placed at the phantom center for both the phantoms (Fig. 2). The SS electrode phantom is developed with a stainless steel electrode array (Fig. 2b and 2d) with sixteen square electrodes identical to the electrodes in the gold array. Sixteen identical electrodes with same contact area (10mm × 10 mm) are cut from a stainless steel sheet (type-304, 50 μm thick).

IOP Conf. Series: Materials Science and Engineering **73** (2015) 012083 doi:10.1088/1757-899X/73/1/012083

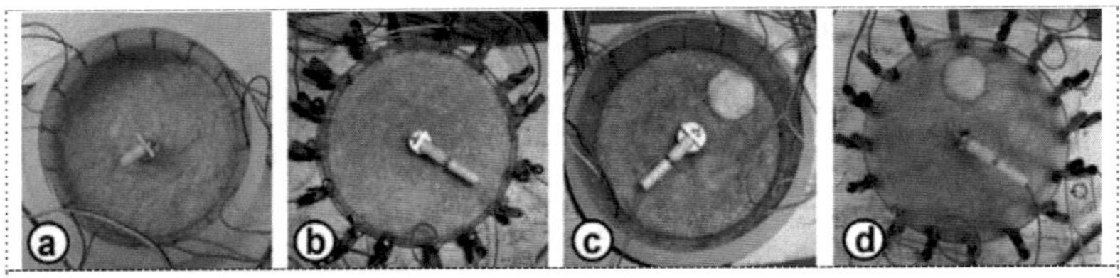

Figure 2. Chicken phantoms (a) gold array and without object, b) with SS array and without object, c) gold array and with fat tissue near elec.- 3, d) SS array and with fat tissue near elec.- 5.

2.3. Data collection and image reconstruction

An EIT instrumentation is developed with a constant current injector [10], signal conditioners, an electrode switching module and data acquisition system for boundary data collection. A USB based high speed data acquisition system is developed with NI USB-6251 DAQ interfaced with LabVIEW software. 1 mA, 50 kHz sinusoidal current signal is injected to the phantom boundary and the surface potentials are collected for all the projections using opposite current injection protocol [11]. Boundary potentials are measured for both the chicken phantoms with different inhomogeneity positions and resistivity images are reconstructed in Electrical Impedance and Diffuse Optical Reconstruction Software (EIDORS) [12]. CNR [13], PCR [13], COC [13] and DRP [13] are studied and the reconstructed images are evaluated for both the phantoms.

3. Results and Discussion

Results show that the contrast and the resolution of the resistivity images of the chicken tissue phantoms are improved with gold sensor array. CNR, PCR and COC of the resistivity images of gold electrode chicken tissue phantom are found more compared to the SS electrode phantom (Table 1). It is observed that the resistivity images obtained from chicken tissue phantom with gold electrode array are less noisy (Fig. 3) compared to the images reconstructed from the identical SS electrode system.

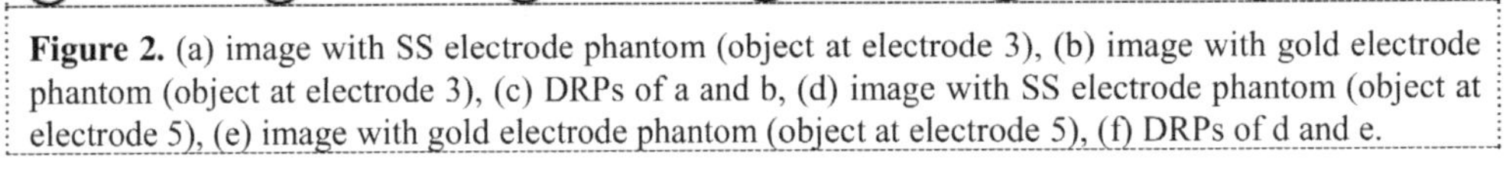

Figure 2. (a) image with SS electrode phantom (object at electrode 3), (b) image with gold electrode phantom (object at electrode 3), (c) DRPs of a and b, (d) image with SS electrode phantom (object at electrode 5), (e) image with gold electrode phantom (object at electrode 5), (f) DRPs of d and e.

Table 1. CNR, PCR and COC of the images of the chicken phantom.

Parameters	Inhomogeneity at Electrode 3		Inhomogeneity at Electrode 5	
	SS Electrode	*Gold Electrode*	*SS Electrode*	*Gold Electrode*
CNR	2.31	2.78	2.28	3.55
PCR	40.41	71.22	51.20	78.36
COC	2.95	3.38	2.96	5.27

It is observed that, for SS electrode chicken tissue phantom with fat tissue at electrode number-3, the image blurring encountered in the reconstruction (Fig. 3a) is found more compared to the blurring encountered in the images with gold electrode system (Fig. 3b). DRPs of the images (Fig. 3c) show that the reconstructed resistivity profiles are more acceptable compared for gold electrode system compared to the profiles obtained for SS electrode array. This is because of the high SNR of the gold electrode system. Similarly it is also observed that, for SS electrode chicken tissue phantom with fat tissue at electrode number-5, resistivity image quality is poor (Fig. 3d) compared to the gold electrode system (Fig. 3e). DRPs of the images (Fig. 3f) with gold electrode chicken tissue phantom show that the reconstructed resistivity profiles are more acceptable compared to SS electrode system.

4. Conclusions

Gold sensors array is developed to improve the resistivity images of real tissue phantoms in EIT. Resistivity images of gold electrode real tissue phantoms are reconstructed in EIDORS and the results are compared with the identical SS electrode systems. Results show that the electrodes performance is improved for the gold electrodes with precise geometries obtained with electrodeposition process. The image quality is improved with gold electrode array compared to the SS electrode systems. Results show that the image contrast and image resolution of resistivity images are improved with gold electrode array compared to the SS electrode system. Results also show that, in gold electrode system, the CNR, PCR, COC and DRP of the reconstructed images are improved. Hence the quality of the boundary data and the images can be improved using a gold electrode array. The entire resistivity imaging studies on real tissue phantoms showed that the gold electrode system improves the resistivity image and hence the gold electrode array can be suitably used for improved EIT image reconstruction.

5. References

[1] Webster J G 1990 *Electrical impedance tomography* (Adam Hilger Series of Biomedical Engineering, Adam Hilger, New York, USA).

[2]. Holder D S 2005 *Electrical impedance tomography: methods, history and applications (Series in Medical Physics and Biomedical Engineering)* (Institute of Physics Publishing Ltd.).

[3]. Holder D S Hanquan Y and Rao A 1996 Some practical biological phantoms for calibrating multifrequency electrical impedance tomography, *Physiol. Meas.* **17** A167–A177.

[4]. Bera T K and Nagaraju J 2011 Resistivity Imaging of A Reconfigurable Phantom With Circular Inhomogeneities in 2D-Electrical Impedance Tomography, *Measurement* **44(3)** 518-526.

[5]. Bera T K and Nagaraju J A 2011 Chicken Tissue Phantom for Studying An Electrical Impedance Tomography (EIT) System Suitable for Clinical Imaging, *Sens Imaging*, **12(3-4)** 95

[6]. Bera T K and Nagaraju J 2011 Gold Electrode Sensors for Electrical Impedance Tomography (EIT) Studies, *Proc. Int. Conf. on IEEE Sensors Application Symposium* (USA) p 24-28.

[7]. Ataman C, Urey H 2009 Compact Fourier transform spectrometers using FR4 platform, *Sensors and Actuators A* **151** 9–16

[8]. Paunovic M and Schlesinger M 2006 *Fundamentals of Electrochemical Deposition*, The Electrochemical Society Series, (2nd Ed., John Wiley & Sons, Inc.)

[9]. Bera T K and Nagaraju J 2009 A Study of Practical Biological Phantoms with Simple Instrumentation for Electrical Impedance Tomography (EIT), *Proc. Int. Conf. on IEEE Instrumentation and Measurement Technology* (Singapore) p 511-16.

[10]. Bera T K and Nagaraju J 2010 A Multifrequency Constant Current Source for Medical Electrical Impedance Tomography, *Proc. Int. Conf. on IEEE Systems in Medicine and Biology* (India) p 278-283

[11]. Bera T K and Nagaraju J 2012 Studying the Resistivity Imaging of Chicken Tissue Phantoms with Different Current Patterns in Electrical Impedance Tomography (EIT), *Measurement* **45** 663

[12]. Vauhkonen M, Lionheart W R B, Heikkinen L M, Vauhkonen P J and Kaipio J P 2001 A MATLAB package for the EIDORS project to reconstruct two dimensional EIT images, *Physiol. Meas.* **22** 107-111

[13]. Bera T K and Nagaraju J 2011 Studying the Elemental Resistivity Profile of Electrical Impedance Tomography (EIT) Images to Assess the Reconstructed Image Quality, *Communications in Computer and Information Science* **157 (CCIS-157)** 621-630

IOP Conf. Series: Materials Science and Engineering **73** (2015) 012084 doi:10.1088/1757-899X/73/1/012084

DC electrical conductivity of nanocrystalline Mn_3O_4 synthesized through a novel sol-gel route

Vipin C Bose[1], K Maniammal[1], G Madhu[1,2], C L Veenas [1], A S Aiswarya Raj [1] and V Biju[1]

[1]Department of Physics, University of Kerala, Kariavattom Campus, Thiruvananthapuram, Kerala, India, Pin – 695581
[2]Department of Physics, University College, Thiruvananthapuram, Kerala, India
Email: vipincbose@yahoo.co.in

Abstract. Manganese oxides have generated considerable interest in recent years due to their electronic and magnetic properties with potential industrial applications. Among manganese oxides, Mn_3O_4 is one of the stable oxides with normal spinel structure and it has variety of important applications. In the present work nanocrystalline Hausmannite, Mn_3O_4 with phase purity and average crystallite size of 8 nm was synthesized through a novel sol-gel route at a temperature of 0-5°C. The samples were annealed at different temperatures to understand the thermal stability and the structural characterization was done using techniques such as X-ray powder diffraction (XRD), Transmission Electron Microscopy (TEM) and Fourier Transform Infrared Spectroscopy (FTIR). The UV visible absorption spectra of the samples show that there are two types of transitions corresponding to Mn^{2+} ions and Mn^{3+} ions. The dc electrical conductivity of the samples increases in comparison with that of bulk.

1. Introduction

Manganese oxides have generated considerable interest in recent years due to their electronic and magnetic properties with potential industrial applications[1]. Manganese (Mn) is a transition metal having multiple valencies +2, +3, +4, +5, +6, +7 etc and hence it forms a number of oxide systems such as MnO, MnO_2, Mn_2O_3, Mn_2O_5, MnO_3, Mn_2O_7 and many nonstoichiometric phases beside Mn_3O_4. Among these Mn_3O_4 is one of the stable oxides with normal spinel structure [2]. Herein, we report the synthesis of nanocrystalline Mn_3O_4 at a temperature of 0-5°C through a sol-gel route and studies on the characterization techniques such as X-ray powder diffraction, Transmission Electron Microscopy(TEM), Surface Area Electron Diffraction(SAED), UV visible absorption spectroscopy, Fourier Transform Infrared Spectroscopy (FTIR) and temperature dependent dc electrical conductivity.

2. Experimental Methods

Nanocrystalline Mn_3O_4 sample was synthesized by a novel sol-gel route, using manganese acetate and lithium hydroxide as the starting materials and ethanol as the reacting medium. The reaction was carried out in the temperature range 0 to 5°C. The reaction yielded highly dispersed nanocrystalline samples in one step[3]. The sample obtained was washed several times with ethanol and water and finally dried at 40°C in a hot air oven (Sample Code Mn40). In order to investigate the stability of the sample against temperature the as prepared sample was annealed at different temperatures, viz.100°C (Sample code Mn 100), 300°C (Sample code Mn 300) and 400°C (Sample code Mn 400) for two hours

IOP Conf. Series: Materials Science and Engineering **73** (2015) 012084 doi:10.1088/1757-899X/73/1/012084

each. The X-ray diffraction (XRD) patterns of the samples were recorded using a Philips XPERT-PRO power diffractometer employing Cu K_α (1.540560Å) radiation in the 2θ range 0° to 80° with a step size of 0.05°. The TEM and SAED patterns were recorded using PHILIPS CM 200 transmission electron microscope, to measure the particle size, surface morphology and crystalline nature. The room temperature UV Visible absorption spectra of the samples were recorded using a SHIMADZU UV-2550 double beam UV visible spectrophotometer in the range 200-800nm. Infrared transmission spectrum of the sample was recorded using Perkin Elmer Fourier Transform Infrared (FTIR) spectrometer for the frequency range 200-4000 cm^{-1}. The dc electrical conductivity of the samples in the temperature range 313-423K was measured with an interval of 10K using KEITHLEY 2400 Source meter have an accuracy of ±.01.

3. Results and Discussion

3.1. XRD Analysis

Figure 1 shows the XRD patterns of all the samples. The broadened nature of the XRD peaks indicates the nanocrystalline nature of the samples. The interplanar spacing (d_{hkl}) and relative intensity (I/I_0) values are in agreement with JCPDS-ICDD Pattern number 80-0382 for tetragonal haumannite Mn_3O_4. The full width at half maxima of the peaks was estimated by a nonlinear curve fitting routine using a Pseudo-Voigt function. The average crystallite size of the samples was estimated using Scherrer equation. The crystalline size of the sample Mn 40 is 8 nm. The crystallite size of the samples Mn 100, Mn 200, Mn 300 and Mn 400 are shown in table 1. Thus annealing of the as prepared Mn_3O_4 sample in air in the temperature range 100–400°C does not lead to any phase change and there is considerable grain growth on annealing. But there is no appreciable change in size of the sample as the temperature increases from 100 to 200°C.

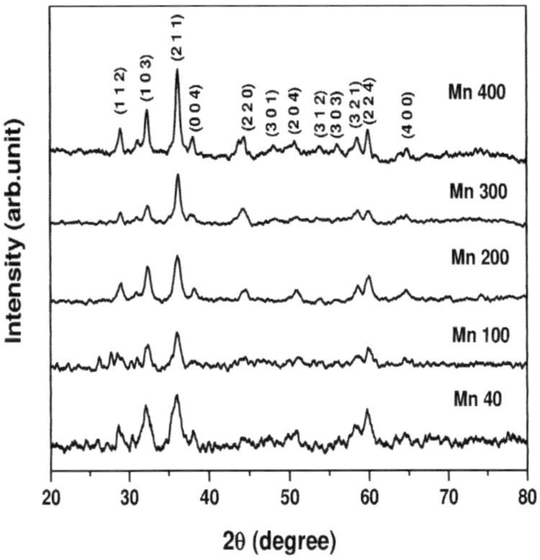

Figure 1. XRD Patterns of all the Samples

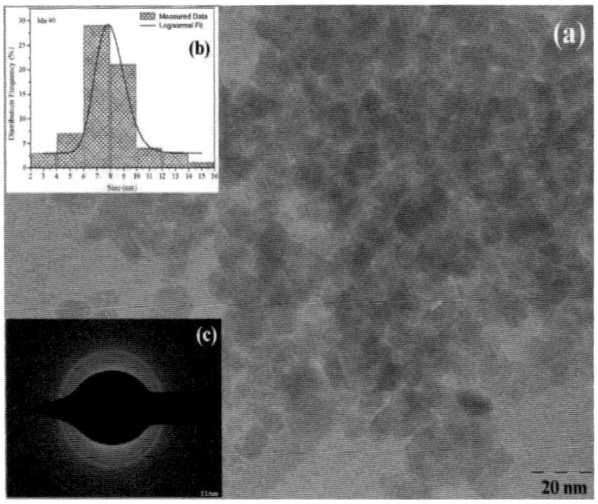

Figure 2. (a) TEM (b) Lognormal fit and (c) SAED pattern of the sample Mn 40

3.2. TEM Analysis

The TEM image of the sample Mn 40 is shown in figure 2. The particle size distribution was obtained by fitting a lognormal profile on the data obtained from TEM and is shown in the inset of figure 2. The average particle size obtained from the XRD pattern for all samples is in agreement with the particle size distribution obtained from TEM, which is shown in table1. The inset of figure 2 also shows SAED pattern of the sample Mn 40. SAED pattern shows multiple intense rings which reveal the crystalline nature of the nanocrystals.

Table 1. Comparison of size from XRD and TEM and Band gap energy.

Sample Code	Size from XRD (nm)	Standard deviation	Size from TEM (nm)	Standard Deviation	Band Gap Energy	
					Eg1 (eV)	Eg2 (eV)
Mn 40	8.66	1.229	8.07	1.110	3.07	2.72
Mn 100	11.26	2.286	8.30	2.078	3.05	2.62
Mn 200	10.67	0.805	8.43	1.498	2.92	2.65
Mn 300	13.08	1.436	11.09	1.591	2.42	1.95
Mn 400	13.66	1.66	15.09	2.607	2.06	1.77

3.3 Analysis of UV Visible Absorption Spectrum

Figure 3 shows the UV Visible absorption spectra of the samples recorded in the range 200-800nm. The spectra show several absorption peaks which was obtained by curve fitting. The absorption peaks were observed at 220, 260, 325 and 500 nm and the edges were located at 325 and 500 nm. The bands at lower wavelength region are attributed to the allowed $O^{2-}\rightarrow Mn^{2+}$ and $O^{2-}\rightarrow Mn^{3+}$ charge transfer transitions and that at higher wavelength region may be related to d-d crystal field transitions [4,5]. The optical band gap energy (Eg) of the sample was calculated using the equation

$$(\alpha h\upsilon)^n = A(h\upsilon - Eg)$$

where α is absorption coefficient, and A is a constant. The values of band gap energy of Mn_3O_4 with an average crystallite size was calculated by assuming the direct transition ie, n=2 because the optical absorption in Mn_3O_4 is direct allowed. The exact values of the band gap energies were obtained by extrapolating the straight line portions of the $(\alpha h\upsilon)^2$ versus hυ graph which is shown in table 2. The band gap energies for sample Mn 40 are 3.07 and 2.72 eV[6].

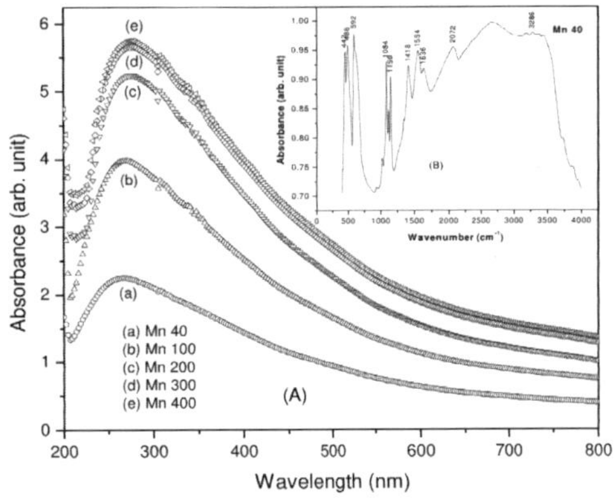

Figure3. (A) UV Visible Absorption spectra (B)FTIR Spectrum of the Sample Mn 40

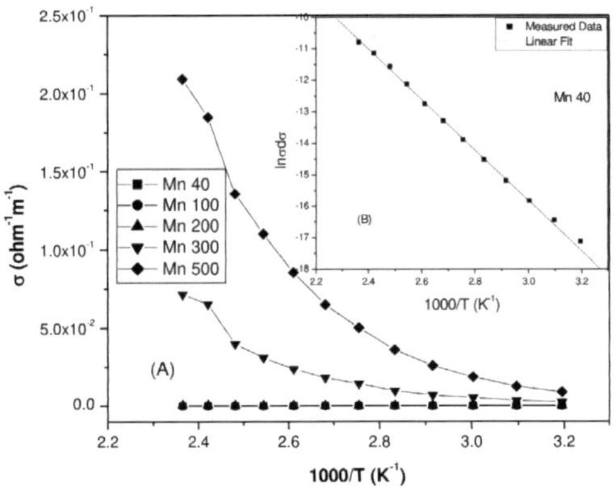

Figure 4. (A) Variation of conductivity with inverse temperature, (B) Arrhenius plot

3.4. Analysis of FTIR Spectrum

The FTIR spectrum of the sample Mn 40 is shown in the inset of figure 3. Several vibrational modes were identified by curve fitting the data in the region 400-800 cm^{-1}. The vibration frequencies of Mn-O modes are 684, 637, 592, 518, 481 and 441 cm^{-1}[6, 7]. The peaks in the range 1084-1554 cm^{-1} correspond to the vibration of O-H bonds connected with Mn atoms[8]. The other peaks in the 1636-3286 cm^{-1} are due to the presence of moisture in the Mn_3O_4 sample[8].

3.5. Analysis of conductivity

The dc electrical conductivity of all the samples in the temperature range 313K to 423K was measured using the source meter. The variation of conductivity with inverse temperature is shown in figure 4. The conductivity of the sample Mn 40 increases from a value (3.68±.03)x10^{-8}ohm^{-1}m^{-1} at 313K to (2.04±0.001) x10^{-5} ohm^{-1}m^{-1} at 423K. The conductivity of all other samples increases with increase in temperature which are listed in the table 2. The electrical transport in Mn_3O_4 is due to the thermally activated polaronic hopping of electron from Mn^{3+} ions in the tetrahedral site to Mn^{2+} ions in the octahedral sites. This is due to interaction of autolocalized holes with lattice vibrations[9]. The inset of figure 4 shows the variation of lnσdσ with inverse temperature (1/T). It is in good agreement with Arrhenius law. The activation energy of the sample Mn 40 is 0.69eV which is smaller than the reported value of 1.35eV of bulk Mn_3O_4 phase[10]. It is found that the activation energy of the samples decreases with increase in crystallite size which is listed in table 2.

Table 2. Variation of conductivity with crystallite size and temperature.

Sample Code	Conductivity (σ) (ohm^{-1}m^{-1})		Activation Energy
	313 K	423 K	(eV)
Mn 40	(3.683 ± 0.037) x 10^{-8}	(2.044 ± 0.001) x 10^{-5}	0.69
Mn 100	(2.036 ± 0.025) x 10^{-8}	(1.093 ± 0.001) x 10^{-5}	0.69
Mn 200	(6.810 ± 0.006) x 10^{-6}	(6.061 ± 0.005) x 10^{-4}	0.45
Mn 300	(2.412 ± 0.002) x 10^{-3}	(7.082 ± 0.008) x 10^{-2}	0.34
Mn 400	(6.426 ± 0.005) x 10^{-3}	(13.772 ± 0.11) x 10^{-2}	0.32

4. Conclusion

Nanocrystalline Mn_3O_4 with an average crystallite size of 8-15nm was synthesized through a sol gel route. The phase purity of the sample was confirmed from the X-ray diffraction analysis. The as prepared sample exhibited grain growth on annealing while no phase transition was observed till 400°C. The optical band gap estimated from UV- Visible absorption exhibits quantum size effect. The Mn-O vibrational modes of the sample were identified through the FTIR spectroscopy. The activation energy of the samples was found to decrease with increase in crystallite size, which shows that the electrical conductivity of the nanocrystals increseases in comparison to that of bulk.

Acknowledgement

The authors Vipin C Bose and K Maniammal acknowledge University of Kerala for financial assistance in the form of Junior Research Fellowship. One of the authors G Madhu acknowledges Univesity Grants Commission (UGC), Government of India for Financial Assistance in the Form of FIP teacher fellowship. The authors C L Veenas and A S Aiswarya raj acknowledge University Grants Commission, Government of India for Financial Assistance in the Form Junior Research Fellowship.

References

[1] Dubal D P, Dhawale D S, Salunkhe R R, Fulari V J and Lokhande C D 2010 *J. Alloys and Componds* **497** 166
[2] Yang L X, Liang Y, Chen H, Meng Y F and Jiang W 2009 *Mater. Res. Bull.* **44** 1753
[3] Chang H J, Lu Cheng Zhu, Wang Yongsheng , Son Chang-Sik, Kim Seong-II, Kim Young-Hwan and Choi In-Hoon 2004 *J. of the Korean Phys. Soc.* **45** 959
[4] Boyero Macstre J, Fernandez Lopez E, Gallardo-Amores J M, Ruano Casero R,et al. 2001 *Int. J. Inorg. Mater.* **3** 889
[5] Vazquez-Olmos A, Redon R, Rodriguez-Gattorno G, Mata-Zamora M E, Morales-Leal F, Fernandez-Osorio A L, et al. 2005 *J. Colloid Interface Sci.* **291** 175
[6] Dubal D P, Dhawale D S, and Pawar S M 2010 *Appl. Surf. Sci.* **256** 4411
[7] Shengtao Xing, Zicheng Zhou, Zichuan Ma and Yinsu Wu 2011 *Mater. Lett.* **65** 517
[8] Ananth M V, Pethkar S and Dakshinamurthi K 1998 *J. Power Sources* **75** 278
[9] Yamashitha J J and Kurosava T 1959 *J.Phys. Chem. Solids* **5** 43
[10] Metselaar R, Van Tol R E J and Piercy P 1981 *J. of Solid State Chem.* **38** 335

Effect of Cu substitution on magnetic behaviour and electrical properties of bilayered manganite $La_{1.2}Sr_{1.5}Ca_{0.3}Mn_2O_7$

A Saxena*, A Krishnamurthy and B K Srivastava

Department of Physics, University of Rajasthan, Jaipur – 302 004 (India)

Email: arpitasaxenaphy@gmail.com

Abstract. DC magnetization and electrical resistance measurements have been made on 0.5 atomic% Cu substituted bi-layered system $La_{1.2}Sr_{1.5}Ca_{0.3}Mn_{1.99}Cu_{0.01}O_7$ for studying the effect of low level substitution of Cu at Mn site. Rietveld analysis of X-Ray diffraction pattern of the sample confirms the single phase $Sr_3Ti_2O_7$ type structure. On cooling below room temperature the system shows magnetic orderings – ferromagnetic (FM) ordering in ab plane below ~270K and 3D FM ordering below ~45K. Also at an intermediate temperature of ~210K there is a signature of another transition which could be attributed to onset of an antiferromagnetic state. Substitution of Cu weakens ferromagnetism and the system shows 96% magnetoresistance (MR) at the field of 8 Tesla. The system remains electrical insulator down to 8K. Resistance versus temperature curves has been fitted following Variable range hopping and semiconducting models. Estimated values of density of states at Fermi energy and of Debye temperature are in reasonably good agreement with those reported for other perovskites. Magnetoresistance versus field curves show hysteresis and discontinuities.

1. Introduction

Bi-layered manganites $La_{2-2x}Sr_{1+2x}Mn_2O_7$ have attracted considerable attention due to their novel properties such as the colossal magnetoresistance effect, tunneling magnetoresistance and fascinating electronic and magnetic properties. A reported study on the effect of substitution of calcium for strontium in $La_{1.2}Sr_{1.8}Mn_2O_7$ showed that the system with 0.3 Ca exhibits maximum magnetoresistance (MR) of 71% at a field of 1.5 T [1]. Different workers have examined the effect of substitution of Cr, Fe and Co, for Mn in $La_{1.2}Sr_{1.8}Mn_2O_7$, on magnetic behaviour and electrical transport [2-5]. We have undertaken a study of the effect of the substitution of Cu for Mn in $La_{1.2}Sr_{1.5}Ca_{0.3}Mn_2O_7$ which is reported to exhibit highest MR. In this paper we present our results on 0.5 atomic % Cu substituted system $La_{1.2}Sr_{1.5}Ca_{0.3}Mn_{1.99}Cu_{0.01}O_7$.

2. Experimental details

Sample of $La_{1.2}Sr_{1.5}Ca_{0.3}Mn_{1.99}Cu_{0.01}O_7$ has been prepared following standard ceramic technique [2, 3]. Structural characterization was done through powder X-ray diffraction measurements made at room temperature on PANalytical make X'Pert PRO MPD diffractometer with CuK_α radiation. Magnetization measurements were made in both zero field cooled (ZFC) and field cooled (FC) conditions, as a function of temperature (20K – 300K) and magnetic field (upto 1 Tesla) using a vibrating sample magnetometer (Lakeshore make Model 7403). Closed helium cycle refrigerator cryostat was used for attaining temperatures down to 20K. Electrical resistance was measured using four probe method. Measurements have been made with a Cryogen Free High Field measurement system (Cryogenics make Model SMS 120C) in the temperature range 8K – 300K and in magnetic fields ± 8 Tesla.

3. Results and discussion

Figure 1 shows X-ray diffraction (XRD) pattern of the prepared sample. The pattern has been subjected to Rietveld analysis using High score plus software. It gets fitted in single phase tetragonal structure (space group I4/mmm) with lattice constants a = 3.8660(1)Å and c = 19.9721(8)Å.

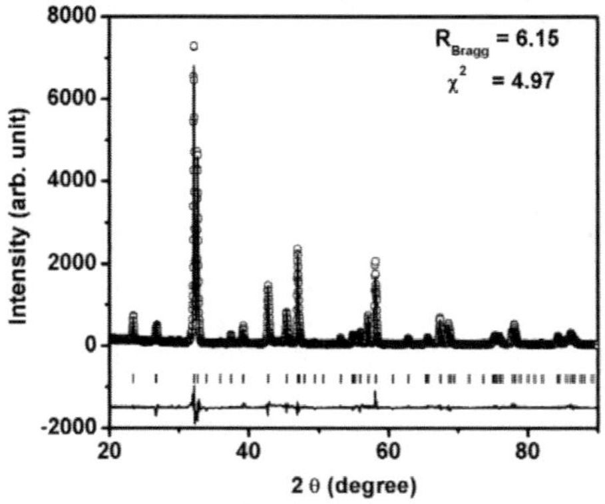

Figure 1. Powder XRD pattern of the sample. Dots and the continuous line show observed points and calculated (Rietveld) patterns respectively. Vertical markers denote the angles of calculated Bragg reflections. The lowest line represents the difference between calculated and observed intensities.

Figure 2 shows magnetization – temperature (M – T) curves recorded in ZFC and FC modes under external magnetic fields of 50 Oe and 500 Oe. Three maxima show up. The prominent one below 50K is attributed to transition to 3D ferromagnetic (FM) ordering [1, 3]. Curie temperature (T_c = ~44K) has been taken as one at which the rate dM/dT (in the M – T plot in ZFC mode) is maximum. In addition to this main transition, two other transitions show up at higher temperatures – T_1 at ~267K (to onset of FM order in *ab* plane) and T_2 at ~208K (to onset of an antiferromagnetic (AFM) state arising due to superexchange interactions). Application of magnetic field causes the prominent low temperature peak, associated with transition to 3D FM state, to shift slightly to lower temperatures. The maxima around T_1 and T_2 are quite broad and it is difficult to visualize any change in their positioning with magnetic field. Wide departure between ZFC and FC curves also exhibits. The departure is maximum at the lowest field of 50 Oe and decreases with increasing field. This departure is a manifestation of large energy barriers indicative of magnetic disorder and high magnetic anisotropy.

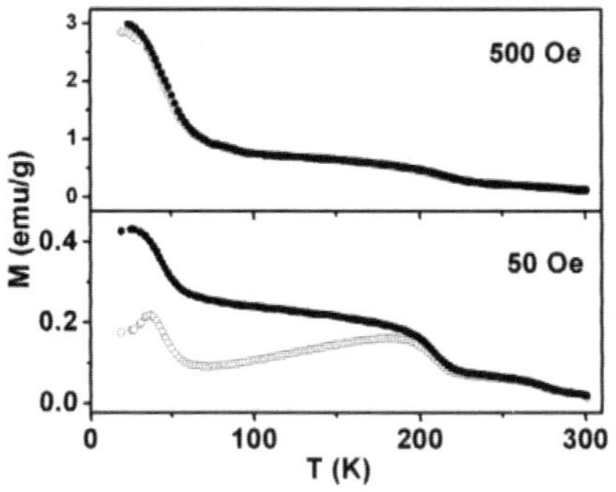

Figure 2. M – T curves recorded in ZFC (open circles) and FC (solid circles) modes under 50 Oe and 500 Oe external fields.

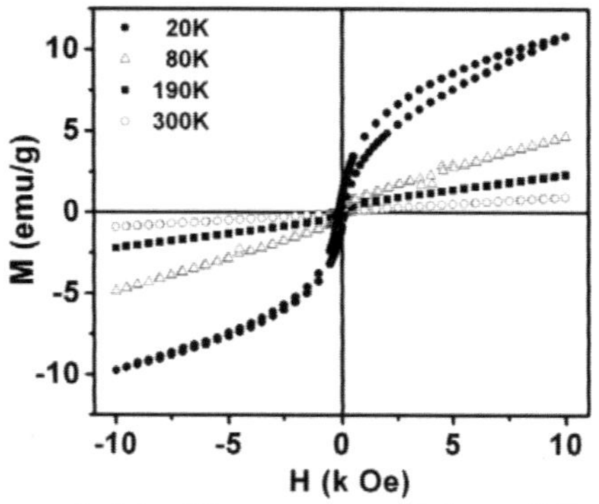

Figure 3. M – H curves recorded at different temperatures.

Our earlier studies gave T_c of ~75K, ~84K and ~70K for the systems with only Mn and with equal amounts of Cr and Fe (as of Cu in the sample under study) respectively [3, 5]. This shows that low level substitution of Cu weakens ferromagnetism more than that of Fe does; contrary to this substitution of Cr enhances ferromagnetism.

Figure 3 shows some typical M – H curves recorded at different sample temperatures. Values of coercivity (H_c) and remanance (M_R) show a general trend of decrease with increasing temperature. The fact that even at 300K, H_c does not go to zero and has a finite value shows that the system does not attain paramagnetic state up to 300K.

Figure 4 shows resistance – temperature (R – T) curves measured in zero field and in external fields of 4 T and 8 T. The electrical state remains insulating down to 8K. Resistance decreases with magnetic field resulting in a large magnetoresistance ($MR = (R_0 - R_H)/R_0)$). This occurs due to increase in hopping conductivity (due to double exchange) resulting from increased ferromagnetic alignment of Mn spins and also increased intergranular tunneling [6, 7]. Steep increase in resistivity at low temperature, could be explained on the basis of a model of parallel conduction channels of better and poor conductivity [3, 7]. Inset to Fig. 4 shows a high negative MR of ~96% at 8K with the field of 8T.

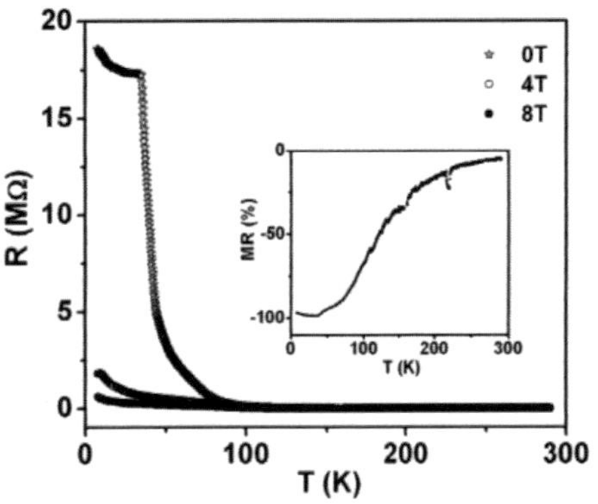

Figure 4. R – T curves recorded at different fields. MR – T plot is shown in the inset.

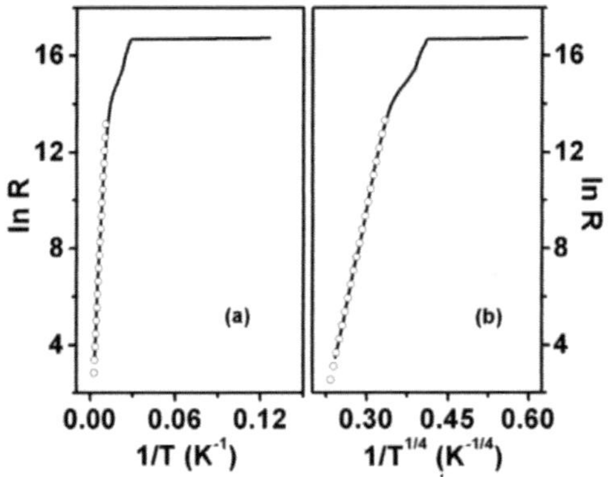

Figure 5. (a) Plot of ln (R) vs T^{-1}. Open circles show the best fit to the equation $\rho = \rho_0 \exp(E_a/k_BT)$ (SC model). (b)Plot of ln (R) vs $T^{-1/4}$. Open circles show the best fit to the equation $\rho = \rho_\infty \exp(T_0/T)^{1/4}$ (VRH model).

Figure 5 shows fitting of R – T curve in (a) semi conducting (SC) model characterized by the equation $\rho = \rho_0 \exp(E_a/k_BT)$ [3, 7-9] and (b) variable range hopping (VRH) model characterized by the equation $\rho = \rho_\infty \exp(T_0/T)^{1/4}$ [3, 7-9]. Estimated values of activation energy (E_a) (in SC model) and T_0, a characteristic temperature (in VRH model) are 111.15 m eV and 135.75 x 10^6 K respectively. The value of the density of states at Fermi energy $N(E_F)$ obtained from T_0, using the equation $T_0 = 16\,\alpha^3/ k_B N(E_F)$, is 1.50 x$10^{19}$ eV^{-1} cm^{-3}. Debye temperature θ_D, estimated as taking the lower limit of temperature region in which VRH Model holds as $\theta_D/2$, is 160K. These values of E_a, T_0, $N(E_F)$ and θ_D are in good agreement with those reported for other bi-layered perovskites [1,3,7-9].

Figure 6 shows MR – H curves at different temperatures. The curves in general show symmetricity in both the field directions and also hysteresis. Measurements at 8K, 30K and 150K exhibit large discontinuities. The features of hysteresis and discontinuities would owe to magnetization reversal by nucleation and domain wall propagation and to discontinuities in anisotropy. Such features have been reported in nano-wires [10]. It is being noted that the temperatures of 150K and 30K lie in the region of AFM transition and 3D FM transition respectively.

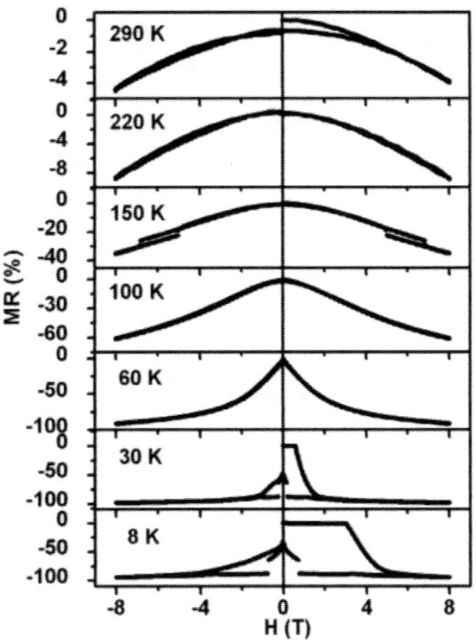

Figure 6. MR-H curves recorded at different temperatures

4. Conclusion

Ferromagnetic (FM) and antiferromagnetic (AFM) states simultaneously manifest in the studied bi-layered system. Cooling from 300K downwards, the system $La_{1.2}Sr_{1.5}Ca_{0.3}Mn_{1.99}Cu_{0.01}O_7$ undergoes a transition to FM order in *ab* plane (~267K), to an AFM state (~208K) and to 3D FM order (~44K). Wide departure seen between M – T curves recorded in ZFC and FC modes are suggestive of disordered magnetic state and high magnetic anisotropy. Substitution of Cu causes weakening of ferromagnetism. Sample does not attain paramagnetic state up to room temperature.

A low amount of Cu (0.5 atomic% for Mn) causes the metallic nature of the sample to vanish. The system does not attain metallic state down to 8K. Estimated values of the density of states at Fermi energy and Debye temperature are in reasonably good agreement with those reported in other perovskites. Sample shows a large MR of ~96% under field of 8T. MR – H curves show interesting features of hysteriesis and discontinuities.

5. Acknowledgement

Authors thank Prof. N. Lakshmi, Department of Physics, M L S University, Udaipur for permission to record magnetization data on their magnetometer.This work has been done under the sponsorship of a University Grants Commission (UGC) sanctioned research project. One of us (AS) thanks UGC, New Delhi, for the award of research fellowship.

6. References

[1] Reddy Y S et al 2007 *Phys. Stat. Sol.* (b) **244** 3719.
[2] Zhang J, Yan Q, Wang F, Yuan P and Zhang P 2000 *J. Phys.: Condens. Matter.* **12** 1981.
[3] Saxena A et al 2011 *Solid State Sciences* **13** 1191.
[4] Gundakaram R et al 1999 *J. Phys.: Condens. Matter.* **11** 5187.
[5] Saxena A, Krishnamurthy A and Srivastava B K 2010 *AIP Conf. Proc.* **1313** 247.
[6] Raychaudhuri P et al 1999 *Phys. Rev.* B **59** 13919.
[7] Saxena A, Krishnamurthy A and Srivastava B K 2011 *AIP Conf. Proc.* **1349** 1209.
[8] Sharma M P et al 2008 *J. Phys.: Condens. Matter.* **20** 425220.
[9] Raychaudhuri P et al 1999 *Physica* B **259-261** 812
[10] Vaz C A F et al 2003 *J. Appl. Phys.* **93** 8104

Role of complexing agents on chemicalbath deposited PbS thin film characterization

K.C. Preetha [*1], K. Deepa [2], A.C. Dhanya[2] and T.L.Remadevi[2,3]

[1]Sree Narayana College,Thottada,Kannur
[2]School of Pure and Applied Physics, Kannur University, Kerala, India-670327
[3]Pazhassi Raja NSS College,Mattannur, Kerala, India-670702
email:k.cpreetha@yahoo.com.com

Abstract In the present work, lead sulphide thin films have been grown on soda lime glass substrates by chemical bath deposition technique using three different complexing agents, namely triethanolamine, diethanolamine and hexamine. Aqueous solutions of lead acetate complexed with the above mentioned ligands were used as cationic precursor and thioacetamide as anionic precursor. The structural, morphological, optical and electrical properties of the as-prepared samples were investigated and compared by means of X-ray diffraction (XRD), Scanning electron microscopy (SEM), optical and electrical studies. The films were well crystallized and highly adherent to the substrate with face centered cubic structure. The average crystallite size was found to be in the range 15-20nm. The SEM micrographs showed variations in morphology. Optical studies revealed that the direct band gap energy was in the range of 1.65-2.27eV. The transmittance and reflectance in IR region was respectively in the range of 14%-47% and 25-45%. Optical constants of all films were determined by absorbance and reflection measurements. The room temperature conductivity of the PbS thin films were in the range $9.67x10^{-9}$ $(\Omega\text{-Cm})^{-1}$-$3.29X10^{-6}(\Omega\text{-Cm})^{-1}$.The optical band gap energy has inverse relation with grain size and electrical conductivity is closely related to structural parameters like grain size, crystallinity and microstrain. PbS thin film using hexamine as complexing agent was proved to be the best in structural, optical and electrical properties. In this attempt we establish that the use of different complexing agents have immense effect on structural, morphological , optoelectronic and transport properties of PbS thin films.

1.Introduction

The semiconductor chalcogenides of IV-VI group have recently attained considerable attention due to their simplicity in preparation and wide applications [1-3]. PbS with a narrow direct band gap energy of 0.41eV can be used as a very good IR detector [4-6]. CBD is a very apposite method for the preparation of poly crystalline PbS thin films [7]. This process uses a controlled chemical reaction to achieve thin film deposition by precipitation. It is necessary to eliminate spontaneous precipitation in order to form a thin film by a controlled ion-by-ion reaction. This can be achieved by using a fairly stable complex of the metal ions which provides a controlled number of the free ions. A complexation reaction with a metal ion involves the replacement of one or more of the coordinated solvent molecules by other nucleophilic groups. In the metal complexes, the metal ions are bonded to one or more neutral molecules or anions so as to define the integral structural unit. The molecules or ions bonded to the central metal ion are called ligands.

In CBD, an aqueous solution of a metal complex when mixed with a solution of chalcogen bearing compound, precipitation of chalcogen occurs under certain conditions. When the precipitation is controlled, the compound gets deposited on clean substrates or other nucleating centers present in the solution. By choosing an appropriate complexing agent, the concentration of the metal ions is controlled by the concentration of the complexing agent. The kinetics of growth of a thin film in this process is determined by the ion-by-ion deposition of the chalcogenide on nucleating sites on the immersed surfaces. Initially, the film growth rate is negligible because an incubation period is required for the formation of critical nuclei from a homogeneous system onto a clean surface. Once nucleation occurs, the rate rises rapidly until the rate of deposition equals the rate of dissolution. Consequently the film attains a terminal thickness. The metal (M^{2+}) ion concentration decreases with increasing concentration of the complexing ions. Consequently, the rate of reaction and hence precipitation is reduced leading to a larger terminal thickness of the film.

In this research work, our main goal is to demarcate the role of complexing agents in the CBD grown PbS thin films. As complexing agents are one of the crucial chemical additives in preparation of thin films, we used a range of organic complexing agents with carboxylate and ammonia groups. The complexing agents under investigation are Hexamine (HA), Triethanolamine (TEA) and Diethanolamine (DEA). TEA $N(CH_2CH_2OH)_3$, is an organic compound that is both a tertiary amine and a triol. A triol is a molecule with three alcohol groups. Like other amines, triethanolamine is a strong base. DEA, is an organic compound with the formula HN $(CH_2CH_2OH)_2$. This colourless liquid is polyfunctional, being a secondary amine and a diol. Like other organic amines, diethanolamine acts as a weak base. Reflecting the hydrophilic character of the alcohol groups, DEA is soluble in water, and is even hygroscopic. HA is a heterocyclic organic compound with the formula $(CH_2)_6N_4$. This white crystalline compound is highly soluble in water and polar organic solvents. Hexamine forms additive or complex compounds with metal salt. These additives are chemically neutral for those areas in which these grades are used.

The role of three complexing agents on structural, optical, surface morphological and electrical properties of as-deposited PbS thin films is highlighted in this work. We have carried out X-ray diffraction (XRD), Optical spectroscopy, Scanning electron microscopy (SEM) and two probe conductivity measurements for thin film characterization

2. Experimental methods

PbS thin films were prepared on ultrasonically cleaned soda lime glass substrate by CBD technique. The reaction bath comprised of an aqueous solution containing 0.3M Lead acetate solution as metal source and equimolar solution of thioacetamide as sulphur source. To avoid precipitation of the bath, HA, TEA and DEA were used as complexing agents for metal source. All the reagents used were of analytical grade. Metal complex and sulphide ions migrate to the substrate surface, where they react to form PbS thin film. Highly adherent PbS thin films were taken out from the reaction bath after three hours. The films prepared by using different complexing agents, HA, TEA and DEA are denoted by PbSHA, PbSTEA and PbSDEA respectively.

The structural studies were carried out using X- ray diffractometer (Bruker AXS-8 advance) having CuKα radiation of wavelength 1.5406 Å as the source. The optical absorption was recorded using Hitachi-U-3410 UV-Vis-NIR spectrophotometer. Scanning Electron Microscopy (SEM) of the samples was recorded using a Hitachi (JEOL Model JSM6490) operating at an accelerating voltage of 25 kV. Conductivity measurements were done with the help of two probe using Keithley Source Measurement Unit (Model SMU Keithley 2400).

3. Results and discussions

3.1. Structural Studies. According to XRD pattern in Figure 1, all the PbS thin films were polycrystalline and have face centered cubic structure with (111) as preferred orientation plane. The average size of the crystallites obtained by Debye scherer formula is in the range of 15-20 nm. Strong and narrow diffraction peaks with maximum intensitiy and crystallite size are obtained for PbSHA. DEA complexed thin film has very low crystallinity. The variation in crystallinty is due to the fact that the resulting concentrations of Pb^{2+} and S^{2-} in solution will be different for different complexing agents, thus leading to different deposition rates. When the effective concentrations of precursor ions are high and that of PbS particles is low, ions-by-ions deposition is dominant in CBD process; and it leads to uniform films as in PbSDEA. When HA is used, the clusters-by clusters deposition is dominant, which leads to the formation of thicker films. Micro strain of the samples is calculated using equation

$$\epsilon = (a0 - a)/a0 \qquad (1)$$

Where 'a_0'=5.936nm.Thin films prepared from HA exhibits minimum strain (0.012) and DEA based film has maximum strain (0.018) , which leads to reduction in crystallinity. The above results confirm that the deposition using HA proved more favourable.

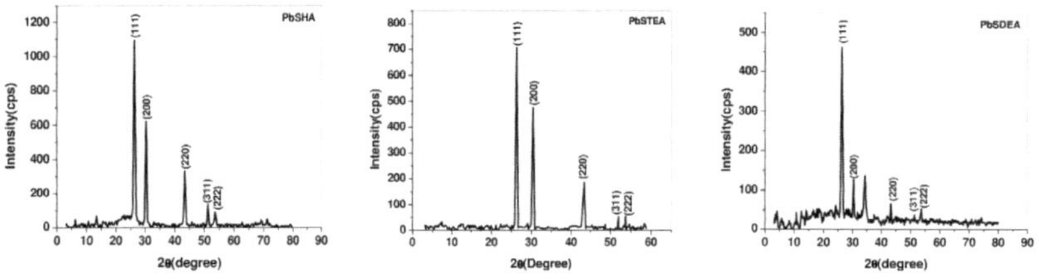

Figure 1. X-ray diffraction pattern of as- prepared samples-PbSHA, PbSTEA ,PbSDEA

3.2. Morphological studies

Figure 2 shows the SEM micrograph of the PbS thin film samples. The micrograph reveals that the film surface is found to be densely packed with bigger grains for PbSHA and PbSTEA, indicating fast growth rate and size of the nano grains. Whereas the surface becomes less dense for PbSDEA. Maximum grain size is noticed in PbSHA and minimum for PbSDEA. Thus SEM results are in accordance with the XRD data.

Figure 2. SEM Image of (a) PbSHA (b) PbSTEA (c) PbSDEA

3.3. Optical Properties

The optical absorbance, band gap, transmittance/ reflectance of the films were shown in Figure 3. According to Figure 3(a), the optical absorption edge shifts to lower wavelength with improvement in crystallinity ,which leads to a blue shift in band gap energy for PbSHA. The direct band gap energy calculated using Tauc's relation is shown in Figure 3(b) and is in the range of 1.65-2.27 eV. The decrease in band gap of HA complexed PbS films may be attributed to the improvement in the crystalline quality of the films along with reduction in porosity, strain relief and increase of grain size.

As it can be seen in Figure 3(c), the transparencies of the PbS films in the visible region strongly depend on the complexing agent used in the CBD. HA complexed PbS film displays low transmission in the visible region; on the other hand, when the complexing agent is DEA, the transmission in the visible region is very high. PbS is normally used as the solar control coating material that is required to have low transmission in the visible region so that not much visible light can penetrate solar cells and appreciable reflectance in IR region as in Figure 3 (d). PbSHA is a very acceptable material for this purpose. Morphologies of three PbS thin films are quite different, which affects the transmission of visible light. Evidently, the rough surface morphology will reduce the transmission in visible region. Moreover the variation in film thickness due to difference in deposition rates also contributes to different transmission and reflection.

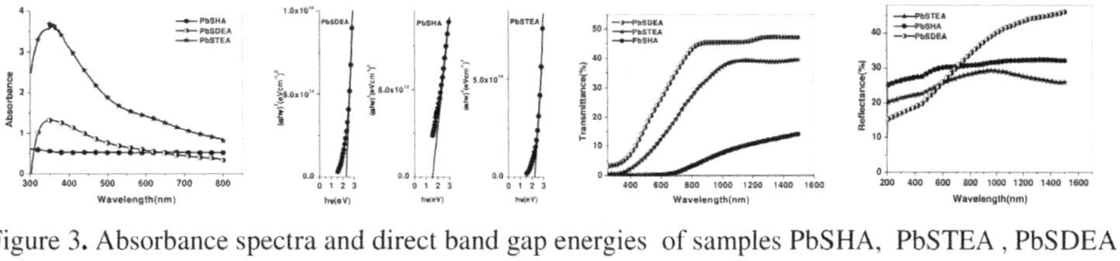

Figure 3. Absorbance spectra and direct band gap energies of samples PbSHA, PbSTEA , PbSDEA

3.4. Electrical characterization

Figure 4 shows the typical I-V characteristic of the PbS thin films prepared with different cationic complexes. The measured conductivity of the PbS thin films are found to be in the range $9.67X10^{-9}$ $(\Omega\text{-cm})^{-1}$- $3.29X10^{-6}$ $(\Omega\text{-cm})^{-1}$. The electrical conductivity is directly connected to crystallinty and grain size and it is inversely related to band gap and micro strain. Thus PbSHA shows maximum electrical conductivity. Lead sulphide thin film with hexamine as complexing agent has least micro strain, minimum optical band gap and maximum electrical conductivity.

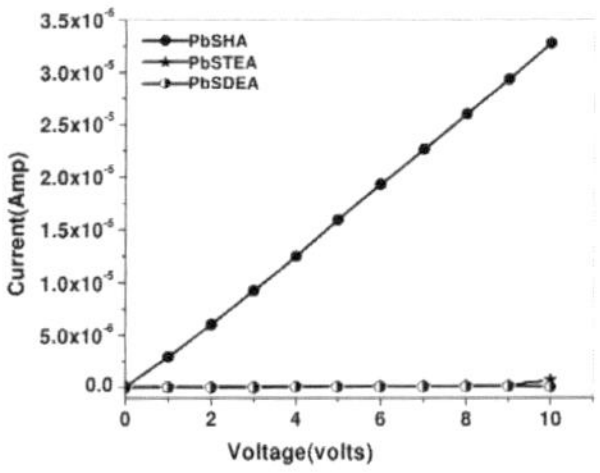

Figure 4. Typical I-V characteristic plot

Conclusion

In this study, we have presented chemical bath deposition method to synthesize PbS thin films using three different metal complexes. Structural analysis specifies that the best crystallinity and high adhesion of PbS on glass substrate is obtained by employing hexamine as complexing agent. The film gown in this condition was well crystallized, with minimum strain and optical band gap and maximum electrical conductivity. As an outcome of this work, the HA is proven to be very feasible for the preparation of PbS films which finds application as solar control coating

Acknowledgments

The financial assistance from the University Grants Commission of India is gratefully acknowledged. The authors also wish to acknowledge SAIF-STIC, CUSAT, Kochi for technical support

References

1. Zogg H, Fach A, Maissen C , Masek, J , Blunier S 1995 *Optical Eng.* **34** 1964.
2. Hicks L .D, Harman T C , Sun X , Dresselhaus M S 1996 *Phys. Rev.* B **53** R10493.
3. Katzir A , Rosman R , Shani Y , Bachem K H , Böttner H , Preier H M 1989 *Handbook of solid state lasers*. Peter K. Cheo (ed.), Marcel Dekker, New York,Basel.
4. Bauer G , Kriechbaum M , Shi Z , Tacke M 1995 *J. Nonlin. Opt. Phys.Mater.* **4** 283.
5. Tacke M Helm M (ed.), 2000 *Optoelectronic properties of semiconductors and superlattices.* Gordon & Breach Science Publishers SA, Amsterdam.
6. Alex P Gaiduk, Peter I Gaiduk, Arne Nyland sted Larsen 2008 *Thin Solid Films.* **516** 3791.
7. N Choudhury and B K Sarma, 2008 *Indian Journal of Pure and Applied Physics.* **46** 261.

IOP Conf. Series: Materials Science and Engineering **73** (2015) 012087 doi:10.1088/1757-899X/73/1/012087

Structural, thermal and dielectric properties of cadmium malate single crystal grown in limited diffusion media

A Lincy, V Mahalakshmi, J Thomas and K V Saban
SMART, Department of Physics, St. Berchmans College, Changanassery 686101, Kerala, India.

E mail: smartlabindia@gmail.com

Abstract. Single crystals of cadmium malate trihydrate have been grown by the controlled diffusion of ionic species in hydrosilica gel. Single crystal X-ray diffraction studies show that the crystal belongs to the monoclinic system with space group $P2_1/n$. The functional groups elucidated from the FTIR spectrum are in conformity with the information derived from the X-ray diffraction studies. The thermal behaviour of the material has been investigated using TG-DTA techniques. The optical band gap of the material has been estimated using diffuse reflectance spectroscopy. The dielectric constant, dielectric loss and a.c. conductivity of the crystal are studied over wide temperature and frequency ranges.

1. Introduction

Extensive interest on metal coordinated complexes of carboxylic acids is generated due to the unique properties exhibited by them. The crystals formed by the coordination of various metals with certain dicarboxylate ligands like oxalates, malonates and maleates are found to possess immense potentials in the realms of magnetism and non linear optics. Due to the outstanding properties exhibited by the above class of materials, recently we have grown certain metal malates of fairly good size using limited diffusion media [1]. This paper is a report on the structural, thermal and dielectric properties of cadmium malate crystals grown in hydrosilica gel.

2. Materials and methods

The silica gel was prepared in single tubes, out of sodium metasilicate solution acidified with malic acid [2]. Experiments were conducted corresponding to different densities and pH values of the gel matrix. Concentrations of the inner malic acid and supernatant cadmium chlorides also were varied in different trials. The single crystal X-ray diffraction studies were carried out using the CAD4 diffractometer with graphite monochromated Mo-Kα radiation. The FTIR spectrum was recorded using a Thermo Nicolet Avtar 370 model FTIR spectrophotometer in the range 400 cm^{-1}-4000 cm^{-1}. Thermal behavior of the material was analyzed using the Perkin Elmer made Pyris Diamond TG-DTA analyzer by heating the sample in nitrogen atmosphere. Optical behaviour of the material in the 200 nm - 2100 nm range was evaluated using the Cary 5 E UV-VIS-NIR spectrophotometer. Dielectric measurements were conducted on silver electroded pellets of diameter 1.3 cm and thickness 1.90 mm employing the Hioki 3532-50 Hi tester.

3. Results and discussion

Well faceted, transparent crystals of the title compound began to grow at the gel matrix-solution interface in about two weeks. Better crystals were grown in a gel medium of density of 1.05 gm/cc at pH 6 by the controlled diffusion of 1M malic acid and 0.25 M cadmium chloride.

X-ray diffraction studies on single crystals revealed that the compound is cadmium malate trihydrate and it crystallizes in the monoclinic system with space group $P2_1/n$. The unit cell parameters are a = 10.578(2) Å, b = 7.289(2) Å, c= 10.799(2) Å, α = 90 °, β = 95.570(10) °, γ = 90 °. These data match well with an earlier report on tiny crystallites synthesized by precipitation method [3]. The ORTEP of the molecule with thermal ellipsoids at 50% probability is shown in figure 1.

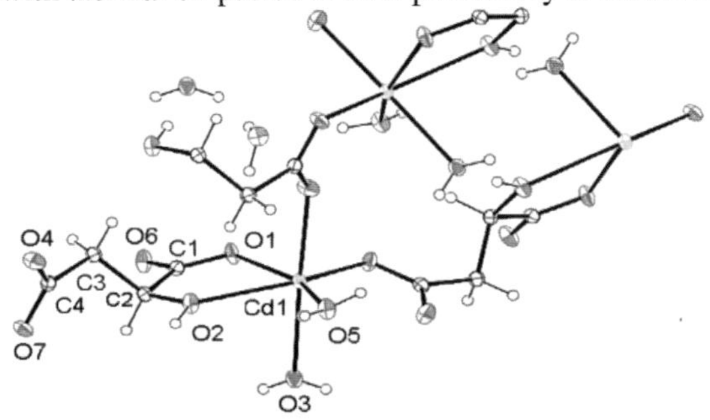

Figure 1. ORTEP diagram of cadmium malate.

In the structure, each cadmium ion is coordinated to six oxygen atoms, four of which coming from three different malate ligands and two from coordinated water molecules. One molecule of water is seen trapped in the lattice too. Each malate ring is linked to three cadmium atoms. The Cd-O distances range from 2.217 Å to 2.495 Å.

The recorded FTIR spectrum (figure 2) contains the signatures of all the functional groups in the title compound. The broad intense peak at 3445 cm^{-1} is due to -OH stretching mode. The asymmetric and symmetric vibrations of the carboxylate groups appear at 1570 $cm^{-1}$and 1343 cm^{-1} respectively. The weak band at 1212 cm^{-1} is due to the asymmetric stretch of the C-C bond while the band at 940 cm^{-1} is due to its symmetric stretch. The rocking and wagging modes of water are represented by the two bands at 841 cm^{-1} and at 553 cm^{-1}.

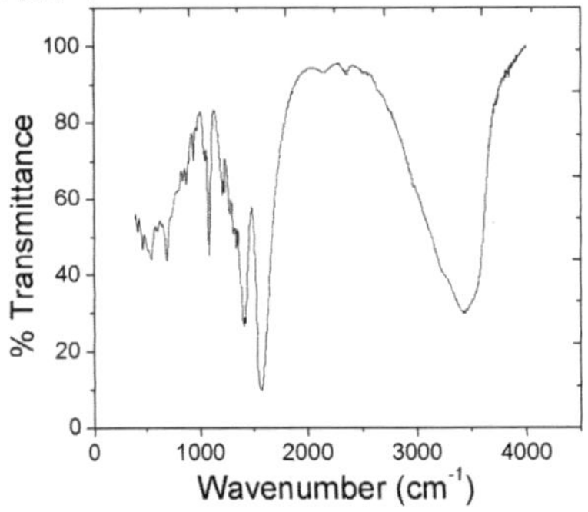

Figure 2. FT-IR spectrum of cadmium malate trihydrate.

The TG-DTA trace (figure 3) of the material reveals a four stage decomposition pattern. Table 1 shows the various steps in the thermal decomposition scheme. All the peaks in the DTA curve give additional support to the TG results.

Table 1. The decomposition process of $CdC_4H_4O_5 .3H_2O$

Stage	Decomposition Temperature (°C)	Product after decomposition	Molecules Evolved	Mass loss Observed (%)	Mass loss Calculated (%)
I	30 - 200	$CdC_4H_4O_5$	$3H_2O$	18±1	18
II	200- 380	CdC_2O_4+ CO+CH_4	CO+CH_4	10±1	10
III	380-400	$CdCO_3$+CO	CO	9±1	9
IV	400-800	CdO+ CO_2	CO_2	18±1	18

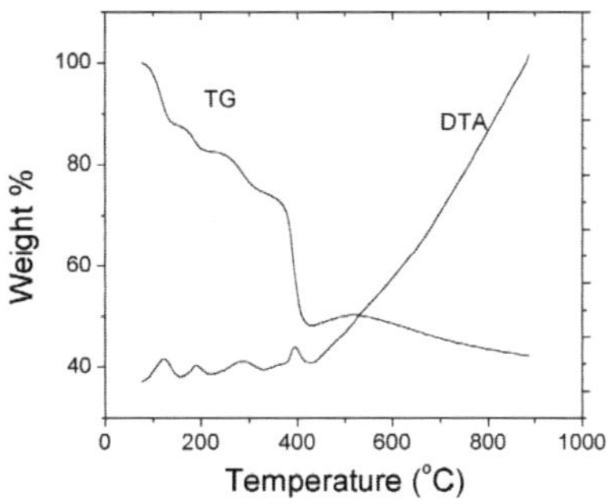

Figure 3. Thermograms of cadmium malate trihydrate.

The diffuse reflectance from the sample in the UV-visible-NIR region was recorded (figure 4). From the Kubelka-Munk function, $F(R) = \frac{(1-R)^2}{2R} = \frac{k}{s}$, the optical band gap of the material is determined by plotting $((k/s)hv)^2$ versus hv (inset of Fig. 3) where R, k and s are the reflectance, absorption and scattering coefficients respectively [4]. The linear fit of the curve on the energy axis yielded the band gap as 1.114 eV. The smaller band gap value indicates the relatively dense crystal packing.

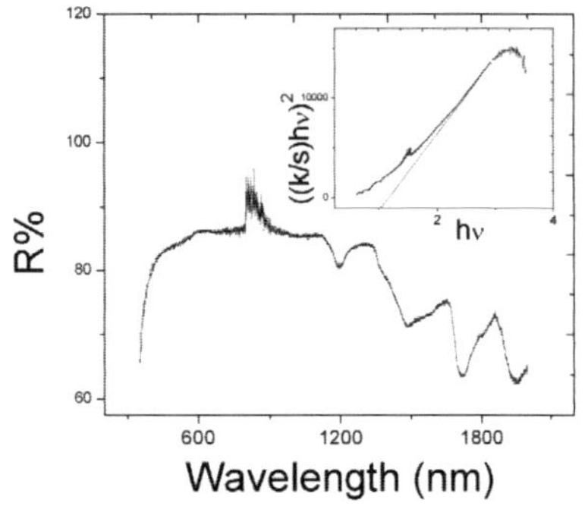

Figure 4. DRS spectrum of cadmium malate trihydrate.

The temperature dependence of dielectric constant at various frequencies are plotted in figure 5. The enhanced values of ε_r at high temperature region is attributed to the space charge polarization arising from the combined effects of ionic mobility and imperfections in the material. The dielectric loss also is found to exhibit strong frequency dependence (figure 6). The low value of dielectric loss at high frequencies indicates the enhanced optical quality of the crystals. The inset of figure 6 gives a magnified view of the loss tangent variation in the mid frequency region. The peaks observed in this region indicate the relaxing dipoles present in the sample. Furthermore, the loss peak shows a trend of shifting towards higher frequencies with temperature, indicating a thermally activated relaxation [5].

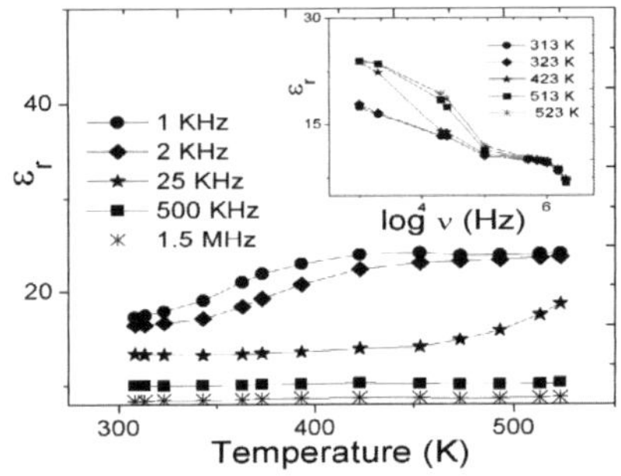

Figure 5. Thermal and frequency response of dielectric constant.

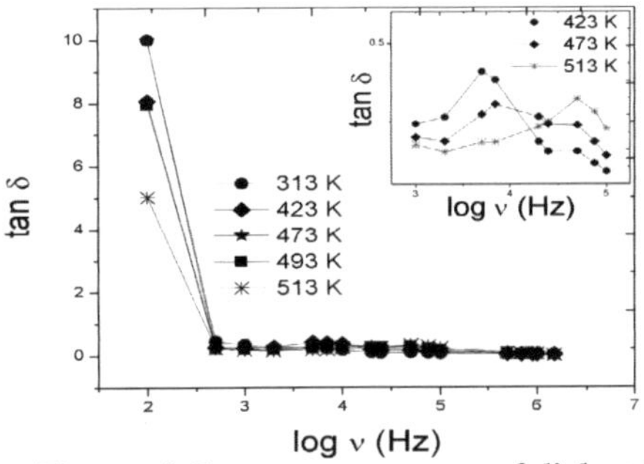

Figure 6. Frequency response of dielectric loss

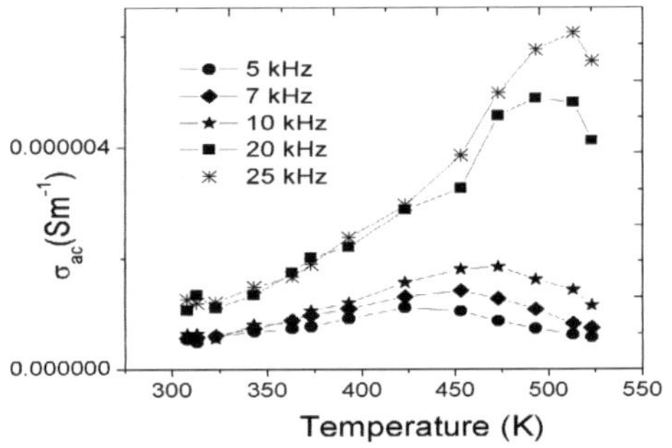

Figure 7. Temperature dependence of a.c. conductivity.

The a.c. conductivity (σ_{ac}) was calculated from dielectric data using the empirical relation, $\sigma_{ac} = \varepsilon_0\epsilon' \omega tan\delta$ where ε_0 is the permittivity of free space, ε'relative permittivity, ω is the angular frequency and $tan\delta$ is the dielectric loss. The conductivity curves (figure 7) show sharp changes in slope around 450 K, suggestive of the dehydration proposed in the TG-DTA analysis. The rupture of hydrogen bonds present in the structure, due to loss of water molecules, is responsible for the diminishing conductivity.

4. Conclusion

Well faceted crystals of cadmium malate trihydrate were grown in inert hydrosilica gel. Crystals of fairly good size and quality were obtained in a gel of density 1.05 gm/cc at pH 6. X-ray diffraction studies revealed that the compound crystallizes in the monoclinic system with space group symmetry $P2_1/n$. Various functional groups in the material were identified from the FTIR spectrum. The thermal decomposition scheme of the material explored using TG-DTA studies are correlated with the dielectric measurements. The optical band gap energy of the material is 1.114 eV.

References

[1] Jini T, Saban K V, Varghese G, Naveen S and Sridhar M A 2007 *J. Alloys Compd.* **433** 211
[2] Henisch H K 1970 *Crystal Growth in Gels* (Pittsburg: Pennsylvania University Press)
[3] Fleck M, Tillmanns E and Bohaty L 2001 *Z. Kristallogr.* **216** 633
[4] Kubelka P and Munk F 1931 *Zh. Tekh. Fiz.* **12** 593
[5] Wang C C and Zhang L W 2007 *New J. Phys.* **9** 210

AC conductivity studies of Fe doped TiO_2 nanotubes

P.P. Vijayan[1] M. Thomas[2] and K.C. George[1*]
1. Department of Physics, S.B.College, Changanacherry, Kerala 686101,India

2. Department of Physics, B.C.M College, Kottayam, Kerala 686001,India

E-mail: drkcgeorge@gmail.com

Abstract. Fe-doped TiO_2 nanotubes are prepared by the combination of sol-gel process with hydrothermal treatment. The morphology and crystalline structure of TiO_2 nanotubes are characterized by transmission electron microscopy (TEM), X-ray diffraction respectively (XRD). Fe doping induces a structural transformation from anatase to rutile. The temperature dependence of the ac electrical conductivity is investigated in the temperature range 303-413 K. Positive temperature coefficient of resistance is observed in the Fe doped TiO_2 nanotubes. PL spectrum shows the presence of oxygen vacancies and self trapped excitons in Fe doped TiO_2 nanotubes and undoped samples.

1. Introduction

TiO_2 is a wideband gap (3.32ev) [1] n type semiconductor with a variety of promising applications in optical and electrical devices. One dimensional nanostructure, such as nanotubes, nanowires and nanorods has attracted a great deal of attention as functional units for mediating the transport of electrons or optical excitation. Therefore, synthesizing novel one dimensional materials and probing their intrinsic properties are critical to assess their possible role in future nano device technology. In nanotubular geometrics, the electronic properties of TiO_2 are more important, because they determine how efficiently electrons can be transferred along the one dimensional path. These features make one dimensional structures promising for applications in photo catalytic and solar energy conversion [2, 3].

Influence of various dopants (Cr, Co, Fe) on the properties of titanium dioxide has been studied by various researchers [4-6]. Bally et al reported electrical conduction in Fe doped TiO_2 thin films [6]. To the best of our knowledge the ac conductivity studies of Fe doped TiO_2 nanotubes have not been exploited yet. Electrical properties of the TiO_2 semiconductors are influenced by oxygen vacancies and extrinsic impurities. The main objective of the present work is to investigate the ac electrical properties of the Fe doped TiO_2 nanotubes for different Fe concentrations.

2. Experimental

Synthesize of Fe doped TiO_2 nanotubes were performed by sol gel method followed by hydrothermal method [7]. Ethanol solution of tetra butyl titanate was added drop wise to the mixture of 20 mL ethanol and 0.045 mol/L ferric nitrate. The mixture was hydrolyzed at room temperature for 2 hours under vigorous stirring, a transparent sol obtained. Gel formed after aging the sol for 12 hours at room temperature. The resulting gel was dried at 80 °C in an oven for 12 hours. After calcination of gel at 500 °C in a muffle for 3 hours Fe doped TiO_2 powder was obtained. Fe doped TiO_2 powder (1.2 g) mixed with 58 ml 10 mol/L NaOH aqueous solution in Teflon lined bottle and maintained at 150 °C for 12 hours. Obtained materials were washed with 0.1 mol/l HNO_3 solution and distilled water successively and dried at 80 °C, Fe doped TiO_2 nanotubes were obtained. Calculated Fe concentrations were 1, 7 and 16 at. % and named as Fe a, Fe b and Fe c respectively.

3. TEM analysis

The surface morphology of the TiO_2 nanostructures formed is shown in Fig.1, which reveals the tubular structure of synthesized Fe doped TiO_2 nanotubes.

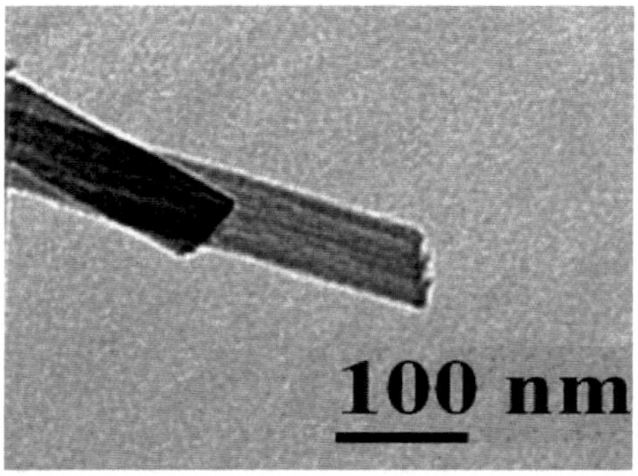

Fig.1. TEM images of Fe doped TiO_2 nanotubes

4. Structural analysis

EDX data reveals the existence of Fe in the Fe doped TiO_2 nanotubes. The calculated weight percentage of Fe in the Fe doped TiO_2 samples are 1at.% (Fe a), 7 at.%(Fe b) and 16 at.% (Fe c).

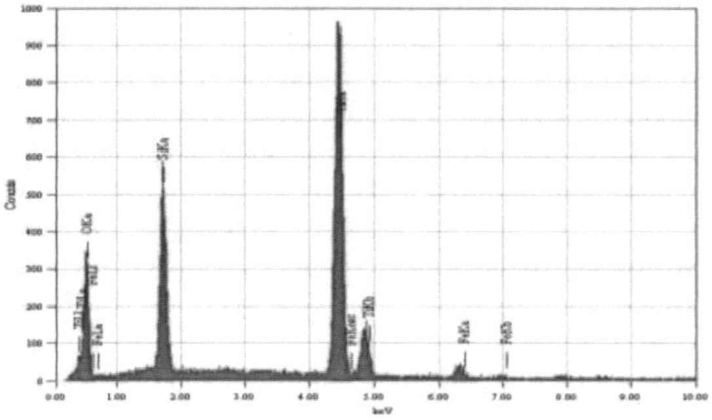

Fig.2. EDX spectrum of 16 at. % Fe doped TiO_2 nanotubes

5. X-ray Diffraction Analysis

The XRD patterns of the undoped and the Fe doped TiO_2 nanotubes are as shown in Fig.3.The undoped TiO_2 samples show only an amorphous structure. Crystallization can be seen for the Fe doped samples. 1at.% Fe doped samples have the peaks of anatase TiO_2. But as the Fe concentration increases the rutile peaks can be seen. From the XRD pattern it is clear that Fe doping induces anatase to rutile transformation. As the concentration of Fe increases from 1at.% to 16 at.% a slight shift of the XRD peaks towards smaller diffraction angles can be observed. The radii of the Fe $^{2+}$ and Fe $^{3+}$ are 0.75 A^0 0 .64 A^0 respectively. The possibility for the increase in the d spacing is the presence of Fe^{2+} formed by the reduction of Fe^{3+} [8]. The mean crystallite size is calculated using Scherrer's formula to be 39-63 nm.

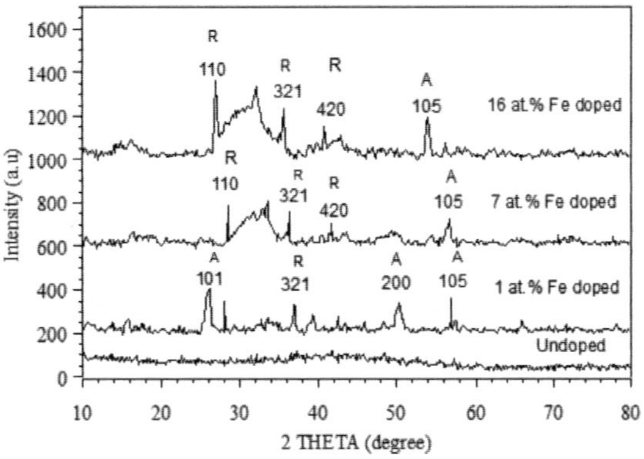

Fig.3. XRD patterns of as synthesized TiO_2 nanotubes & TiO_2 nanotubes doped with different concentration of Fe

6. Photoluminescence Spectrum

The PL spectrum gives information on the properties of the energy levels lying within the band gap. The room temperature PL spectra of the as prepared Fe doped TiO_2 nanotubes under excitation with the 300 nm are displayed in Fig.4.

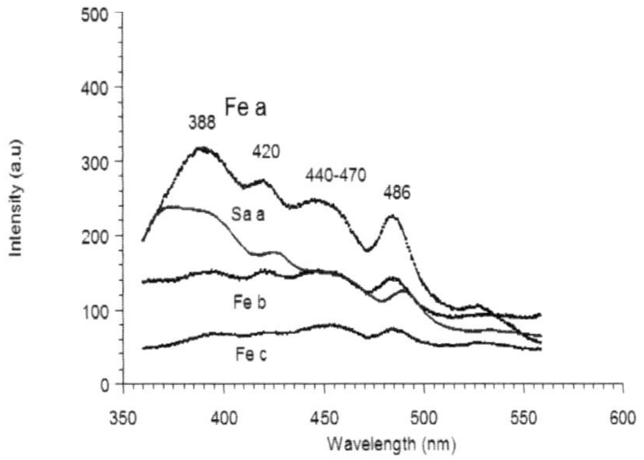

Fig.4. PL spectra of as synthesized TiO_2 nanotubes and Fe doped TiO_2 nanotubes with different Fe concentrations

The spectrum mainly consists of four emission bands at 388, 420, 440-460 and 486 nm. The peak at 388nm corresponds to the direct allowed transition X_2-X_1 [9]. The peaks at 420, 445, 486 and 528 nm in the Co doped TiO_2 nanotubes represent the defect level emissions. Peaks at 420 and 440-470 nm corresponds the emission due to the recombination of self trapped excitons [10]. Broad blue emission ranging from 440-470 nm is due to the self trapped excitons localized on TiO_6 octahedra. The emission peak due to self trapped excitons for the samples Fe b and Fe c are very weak possibly resulting from the formation of rutile for the high iron concentration. The blue green emission at 486nm is be assigned to oxygen vacancies [10]. The more the oxygen vacancies the stronger the blue green emission intensity is. The sample of 16 at.% Fe concentration have low intensity peak for oxygen vacancies. Fewer oxygen vacancies in rutile than in the anatase structure were reported [6]. 1at.% Fe doped samples have higher oxygen vacancies while those of 16 at.% Fe doped samples have lower oxygen vacancies due to the formation of rutile at higher Fe concentration.

7. AC conductivity studies

To study the effect of iron doping on the ac conductivity studies of TiO_2 nanotube log σ_{ac} with inverse temperature are plotted for as synthesized TiO_2 nanotubes and for Fe doped samples at 100Hz. We have reported the variation of ac conductivity of TiO_2 nanotubes with temperature and a negative temperature coefficient of resistance were observed [12]. The ac conductivity of the 7 and 16 at.% Fe doped samples are low as compared to as synthesized undoped TiO_2 nanotubes. In Fe doped samples the ac conductivity increases up to a particular temperature. A decrease in ac conductivity is observed for a further rise in temperature (positive temperature coefficient of resistance) in all compositions of Fe doped TiO_2 nanotubes. The range of temperature for transition to the positive temperature coefficient of resistance is different for different Fe concentrations. For the highest Fe concentration transition is observed at 323K and it extends up to 363K. For the undoped TiO_2 nanotubes this type of conductivity is not observed. So the decrease in ac conductivity with temperature can be related to the Fe concentration and the concentration of oxygen vacancies.

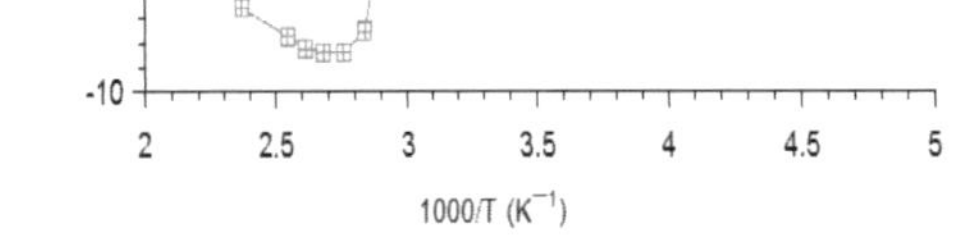

Fig 7 Temperature dependence of ac conductivity of TiO_2 nanotubes and Fe doped samples at 100Hz

In TiO_2, oxygen vacancies are known to be the most common defects. The positive temperature coefficient of resistance observed in Fe doped TiO_2 nanotube can be attributed to the formation of oxygen vacancies and the effect of Fe doping.1 at.% Fe doped titania nanotube (Fe a) has more oxygen vacancy than the undoped samples as is confirmed by the Pl. Each oxygen vacancy gives two electrons in the titania conduction band, leading to an improvement in the ac electrical conductivity

[12]. But in Fe doped TiO_2 nanotubes Fe impurities act as acceptors, which decreases the electron concentration produced by oxygen vacancies [6]. For samples having higher iron concentration (Fe b& Fe c) the rutile formation lowers the oxygen vacancies, which increases the concentration of holes. The increase in concentration of holes leads to the decrease in ac conductivity after a particular temperature in Fe doped TiO_2 nanotubes.

Conclusions

Fe doped TiO_2 nanotubes with different Fe concentrations are synthesized by a sol gel method followed by a hydrothermal method. An ac conductivity data analysis is carried out in the temperature range 303-423K for the TiO_2 nanotubes and Fe doped TiO_2 nanotubes. Positive temperature coefficient of resistance is observed for Fe doped TiO_2 nanotubes. The range of temperature for transition to the positive temperature coefficient of resistance varies with Fe concentrations.

Acknowledgment

Authors are grateful to Central instrumentation facility (CIF), St. Berchmans College and IIT Kharagpur.

References

[1] H Tang, F Levy and H Berger 1995 Urbach tail of anatase TiO_2 Phys. Rev. B 52 7771
[2] Y Yu and D Xu Single –crystalline TiO_2 nanorods: 2007 Highly active and easily recycled photocatalysts. Applied catalysis B: Environmental 73 166
[3] D Kuang, J Brillet, P Chen, M Takata, S Uchida, H Miura, K Sumioka, S M Zakeeruddin and M Gratzel 2008 Application of highly ordered TiO_2 nanotube arrays in flexible dye-sensitized solar cells *ACS Nano* 2 1113
[4] R.C dasilva, E Alves and M M Cruz 2002 Conductivity behaviour of Cr implanted TiO_2 Nucl.Instrum.Methods.Phys.Res.b 191 158
[5] Y Matsumoto, M Murukami, T Shono, T Hasegawa, T Fukumara, M Kawasaki, P Ahmet, T Chikyov, S Koshihara and H.Koinuma 2001 Room temperature ferromagnetism in transparent transition metal doped TiO_2 Science 291 854
[6] A R Bally, E N Korobeinikova, P E Schimid, F Levy and J Bussy 1998 Structural and electrical properties of Fe doped TiO_2 thin films J.Phys.D 31 1149
[7] L Deng, S Wang, D Liu, B Zhu, W Huang, S Wu and S Zhang 2009 Synthesis, characterization of Fe doped TiO_2 nanotubes with high photocatalytic activity Catal.Lett 129 513
[8] J Arana, O Gonzalez Diaz, J M Dona Rodriguez, J A Herrera Melian, C I Garriga Cabo, J Perez Pena, M Carmen Hidalgo and Jose A Navio-Santos 2003 Role of Fe^{3+}/Fe^{2+} as TiO_2 dopant ions in photocatalytic degradation of carboxylic acids Journal of Molecular Catalysis A: Chemical 197 157
[9] N Daude, C Gout and C Jounanin 1977 Electronic band structure of titanium dioxide Physical Review B 15 3229
[10] Y Lei and L D Zhang 2001 Fabrication and characterization of highly ordered TiO_2 nanowire array J. Mater. Res 16 1138
[11] W Siripala and M Tomkieviez 1982 Interactions between photoinduced and dark charge transfer across n-TiO_2-aqueous electrolyte interface J. Electrochem.Soc 129 1240
[12] P P Vijayan, M Thomas and K C George 2012 Effect of calcinations on ac conductivity studies of TiO_2 nanotubes J.Appl.Phys 112 104308

IOP Conf. Series: Materials Science and Engineering **73** (2015) 012089 doi:10.1088/1757-899X/73/1/012089

Graphical modelling of carbon nanotube field effect transistor

R Sahoo[1], R R Mishra

Physics group, Birla Institute of Technology and Science, Pilani, Rajasthan-333031, India.

E-mail: sahoorasmita@yahoo.com

Abstract. Carbon nanotube Field Effect Transistors (CNTFET) are found to be one of the most promising successors to conventional Si-MOSFET. This paper presents a novel modelling for planar CNTFET based on curve fitting method. The results obtained from the model are compared with the simulated results obtained by using the nanohub simulator. Finally the accuracy of the model is discussed by calculating the normalized root mean square difference between the nanohub simulation results and those obtained from the proposed model.

1. Introduction

Among many different potential applications of carbon nanotubes (CNT), we have focused on its use in CNTFET, which is found to be one of the most promising alternatives for current field effect transistors. The first CNTFET was reported in 1998 [1] and after that different types of structures are proposed for CNTFET [2-9], out of which we have chosen the planar CNTFET. Further different models are also developed to study the characteristics of different CNTFETs [10-12]. In this paper we have presented a model for planar CNTFET which reproduces almost same results as obtained from the simulation which is done by using the nanohub simulator [13-15]. The main aim is to report a simple model which produces the results very similar to that obtained through very rigorous methods. The paper is organized as follows. Characterization of planar CNTFET is done based on the simulation results which are discussed in section 2. After that the developed model is presented in section 3. Section 4 is devoted for results and discussions and finally a conclusion is given in section 5.

2. Characterization of CNTFET

The planar CNTFET structure is almost the same as the conventional Si-MOSFET, the only difference being the use of a CNT in the channel region, inside which the carriers move from source to drain. Different CNTFETs having planar structure were simulated to understand how the characteristic curves depend on different tube parameters, like the length of the tube and its chirality. This simulation was done by using the nanohub simulators [13-15]. First of all a CNTFET was simulated in which a (10,0) CNT is used as the channel. For this simulation we have kept all other parameters fixed and changed the channel length as well as the top gate length proportionally. For the second phase of simulation, the diameter of the CNT was changed by changing the chirality, and the channel length was kept constant. The basic characteristic curves for a planar CNTFET based on the above simulation results are shown in figure 1 (a) and 1 (b). Further a (10,0) CNT is considered as the channel and the channel length is taken to be 10 nm. A (10,0) CNT was chosen because the band gap for this is comparable with that of silicon. The range of drain voltage for the output characteristic curves, as shown in figure 1 (a), is 0V - 0.4V and the different curves in the plot are for gate voltages from

0.225V to 0.4V in steps of 0.025V. The drain current is very small for the first plot which is for a gate voltage 0.225V, as this gate voltage is just above the threshold voltage. Moreover, the current increases with applied gate voltage. Figure 1 (b) shows the transfer characteristic curves for drain voltages 0.2V, 0.3V and 0.4V respectively. As the saturation is achieved well before these drain voltages, we have same drain current for all the three drain voltages for any applied gate voltage and hence all the three curves coincide.

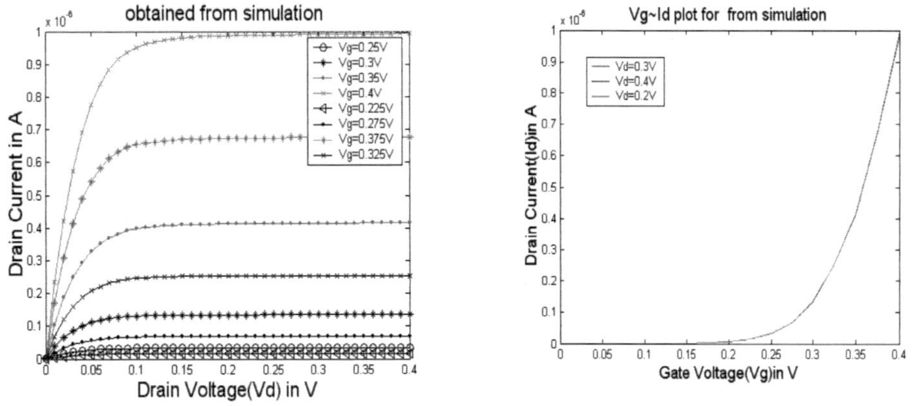

(a) (b)

Figure 1: (a) V_d~I_d plot and (b)V_g~I_d plot for planar CNTFET.

3. Modeling of CNTFET

For the modeling we have first obtained the simulation results from the nanohub simulator as discussed in the previous section. The simulation was repeated for different planar CNTFETs with different physical dimensions (length and diameter) to study the dependence of drain current on those parameters. In the next step we have developed certain empirical relations between the drain current (I_d), drain voltage (V_d) and the applied gate voltage (V_g) , that we have obtained, based on the previous simulation results, using the best-fit method. In this we have used a best-fit schemes based on exponential fit, where the functional relation is chosen to be an exponential function. The best fit is in the sense of minimal root-mean-squared (RMS) difference between the function chosen and the simulation data. This has been done for planar CNTFETs, over the entire range of V_d & V_g values chosen. In order to check the closeness of the fits, we have calculated the root-mean-squared difference (RMSD) and normalized-root-mean- squared differences (NRMSD).This fit also includes the length modulation of the CNT forming the channel and the chirality of the CNT is chosen as (10,0). The choice of a particular chirality fixes the diameter of the CNT used in the channel. The exponential function chosen is:

$$I_d = f(V_g, L)\times\left(1 - e^{d(V_g)\times V_d}\right) \quad (1)$$

Here d is a function of gate voltage only but f is a function of both gate voltage (V_g) and the length of CNT (L) which are given as:

$$f(V_g, L) = a_1 V_g^{a} \quad (2)$$

$$d = d_1 V_g^2 - d_2 V_g - d_3 \quad (3)$$

Further the coefficients a_1 and a in f are considered as functions of length and modeled as:

$$a = c_1 \exp(c_2 L) + c_3 \exp(c_4 L) \quad (4)$$

$$a_1 = c_5 \exp(c_6 L) + c_7 \exp(c_8 L) \tag{5}$$

where c_1, c_2, ..., c_8 are different fitting parameters, which, in the present calculation, have been taken to be constants. However, these coefficients can be modeled further to include dependency on other parameters like diameter, doping, dielectric constants etc.

4. Results and Discussions

The output characteristic curves, obtained with the above fit, for different planar CNTFETs with different lengths of CNT are shown in figure 2 and the results are compared with the simulation results obtained from the nanohub simulator. Figure 2(a)-(d) are for different lengths of the CNT used as the channel. Further for each length of the CNT we have plotted the characteristic curves for different applied gate biases. In the figures the dotted lines are for the results obtained from the nanohub simulator and the solid lines represent the result obtained from our model equations. As can be seen from the graphs, we are getting very similar result as the previous simulation results. We have also calculated the NRMSE for each curve to have a quantitative measure of the error, which are listed in table 1.

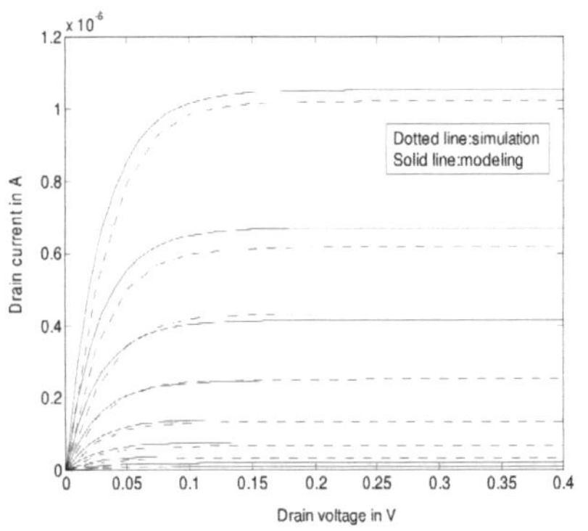

(a)

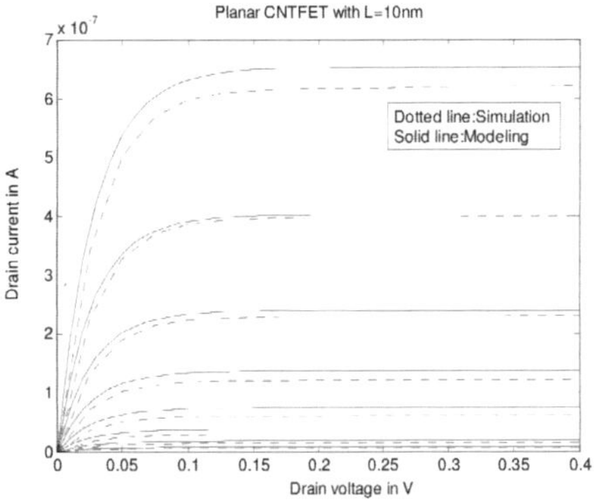

(b)

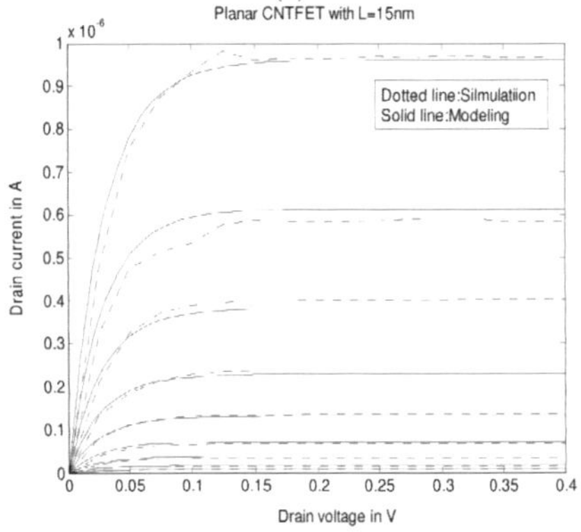

(c)

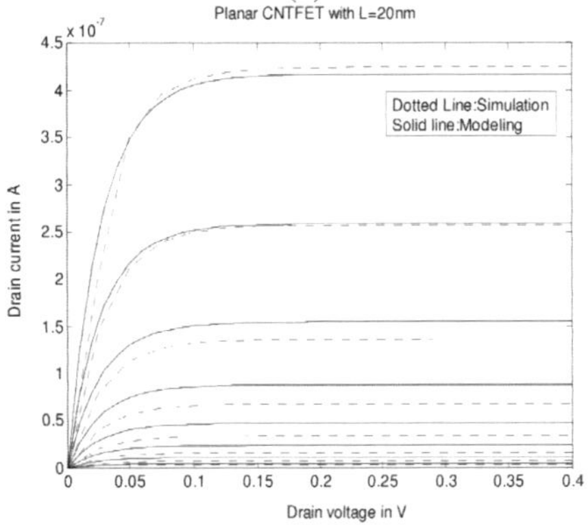

(d)

Figure 2: Comparison between the results obtained from the nanohub simulator and our model: (a) L = 5nm, (b) L = 10nm, (c) L = 15nm and (d) L = 20nm.

Table 1. Calculated NRMSE for different applied gate voltage and different length of the CNT used in the channel of the device.

L_{CNT} in nm	Vg in volt(V)	NRMSE	L_{CNT} in nm	Vg in volt(V)	NRMSE
5nm	0.275	0.1444	15nm	0.275	0.0603
	0.3	0.0600		0.3	0.0284
	0.325	0.0167		0.325	0.0342
	0.35	0.0384		0.35	0.0479
	0.375	0.0806		0.375	0.0398
	0.4	0.0399		0.4	0.0233
10nm	0.275	0.2157	20nm	0.275	0.2793
	0.3	0.1219		0.3	0.1276
	0.325	0.0433		0.325	0.0108
	0.35	0.0133		0.35	0.0292
	0.375	0.0533			
	0.4	0.1594			

Using our model equations as given in equation (1), the transfer characteristic curves are also plotted for two different lengths of the CNT used in the channel region. The plots are shown in figure 3, in which the dotted lines are for the results obtained using our model and the solid lines show the results obtained from the nanohub simulator. The drain voltage considered for both the cases is equals to 0.3V. From the figures it is clearly seen that the fits are quite good.

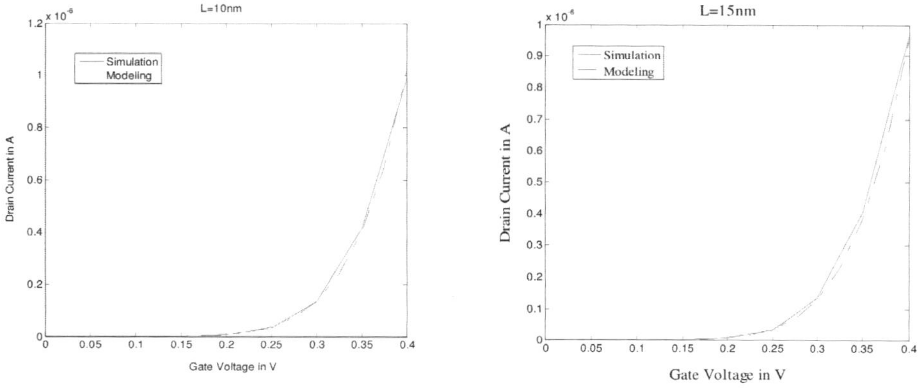

(a) (b)

Figure 3: Transfer characteristic curves for planar CNTFETs with CNTs of length (a) 10 nm and (b) 15 nm.

5. Conclusion

This paper presents some model equations which reproduces output as well as transfer characteristics, for planar CNTFET, very close to some earlier results. Both qualitative as well as quantitative comparisons are done which show the accuracy of the results obtained from the proposed model equations.

References

[1] Tans S J, Alwin R M, Verschueren and Dekker C 1998 *Nature* **393** 49-52.
[2] Avouris P, Appenzeller J, Martel R and Wind A J 2003 Proceedings of the IEEE **91** 1772-84.
[3] Wind S J, Appenzeller J and Avouris Ph 2003 *Phys. Rev. Lett.* **91-5** 058301-1-4.

[4] Guo J, Datta S and Lundstrom M 2002 *IEDM* 711-14.
[5] Mizutani T, Nosho Y and Ohno Y 2008 International Symposium on Advanced Nanodevices and Nanotechnology *J. of Physics: Conference Series* **109** 012002.
[6] Svensson J, Sourab A A, Tarakanov Y, Su Lee D, Park S J, Baek S J, Park Y W and Campbell E E B 2009 *Nanotechnology* **20** 175204.
[7] Pourfath M, Kosina H and Selberherr S 2007 *J. Comput Electron* **6** 243–46.
[8] Park J Y 2007 *Nanotechnology* **18** 095202.
[9] Kordrostami1 Z, Hassaninia I and Sheikhi M H 2008 Proceedings of the 3rd IEEE Int. Conf. on Nano/Micro Engineered and Molecular Systems Sanya China 529-31.
[10] Raychowdhury A, Mukhopadhyay S and Roy K 2004 *IEEE Trans on Comput Aided Design of Integ Circuits And Syst* **23-10** 1411-20.
[11] Pourfath M, Kosina H and Selberherr S 2006 *J. of Physics: Conference Series* **38** 29–32.
[12] Jimenez D, Cartoixa X, Miranda E, Sune J, Chaves F A and Roche A 2006 *J. Comput Electron* **5** 361–64.
[13] Neophytou N, Guo J and Lundstrom M 2006 *IEEE Trans on Nanotechnology* **5** 385-92.
[14] Neophytou N, Ahmed S, Polizzi E, Klimeck G and Lundstrom M 2006, "cntFET," 10254/nanohub-r1091.5.
[15] Neophytou N, Ahmed S and Klimeck G 2007 *J. Comput Electron* **6** 317-20.

Two photon absorption in TeO_2-ZnO glass at different laser irradiances

Rose Leena Thomas[1], Misha Hari[1] ,V P N Nampoori[1] ,P Radhakrishnan[1,2], Sheenu Thomas[2]

[1.] International School of Photonics, Cochin University of Science and Technology, COCHIN,

[2.] Centre of Excellence in Lasers and Optoelectronic Sciences, Cochin University of Science and Technology, COCHIN

KERALA, INDIA-682022.

Ph : 09497023250

E-mail: roseleenat@gmail.com

Abstract: TeO_2-ZnO glasses were prepared and their optical properties were determined. In the present work we measured the optical non-linearity of the glass using a single beam z-scan technique given by Sheike Bahe et al. The open aperture z-scan curve of TeO_2-ZnO proves that nonlinear optical properties are highly fluence dependent.

1.Introduction

Tellurite glasses are very much used in the optoelectronic research due to its attractive properties such as high refractive index, low phonon energy, low transition temperature, excellent infrared transmission [1-6] and very high optical nonlinearities [7].

2.Experimental

TeO_2-ZnO glasses were prepared by melt quench method with different ZnO concentration. Samples were taken in a platinum crucible , melted at a temperature of 900 °C for two hrs , quenched to a brass mould in air and annealed at a temperature 300 °C, using an electric furnace manufactured by High heat furnace and refractories.XRD was taken to verify the amorphous nature of the sample. Glasses were cut and polised for optical measurements such as optical absorbance using spectrophotometer, refractive index using Brewster angle measurement etc.

In this research, we have employed the single-beam z-scan technique with nanosecond laser pulses to measure the nonlinear optical absorption properties of ZnTe glasses. A Q-switched Nd:YAG laser

(Spectra Physics LAB-1760, 532 nm, 7 ns, 10 Hz) is used as the light source. The sample is allowed to move in the direction of the light incidence near the focal spot of the lens with a focal length of 20 cm The transmitted beam energy, reference beam energy, and their ratio are measured simultaneously by an energy ratio meter (Rj7620, Laser Probe Corp.). At the focus the power output of the laser beam was measured in GW/cm^2. The effect of fluctuations of laser power is eliminated by dividing the transmitted power by the power obtained at the reference detector.The experiment is repeated for different powers. The data are analyzed by using the procedure described by Bahae et al[8]. and the two photon absorption coefficients are obtained by fitting the experimental z-scan plot with the theoretical plots.

3.Result and Discussion

The amorphous nature of the prepared glasses were verified using Xray Diffraction technique.The optical band gap is determined from the absorption spectra of the glass samples.

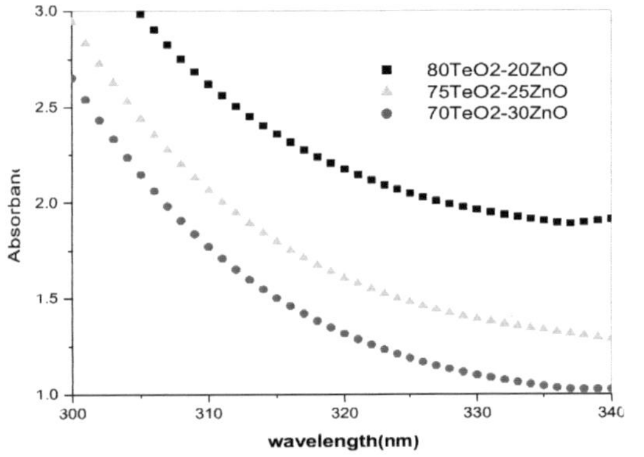

Figure 1: Optical absorption spectra of Tellurite glass at three different ZnO concentrations.

Addition of modifier ions like ZnO increases the density of tellurite glass samples,by creation of non bridging oxygen ions .Zinc ions also occupies interstitials of the glass network and therefore density increases due to the formation of dangling bonds[1].
Band gap reduction and refractive index increase of of telurite glass samples is given in fig.2

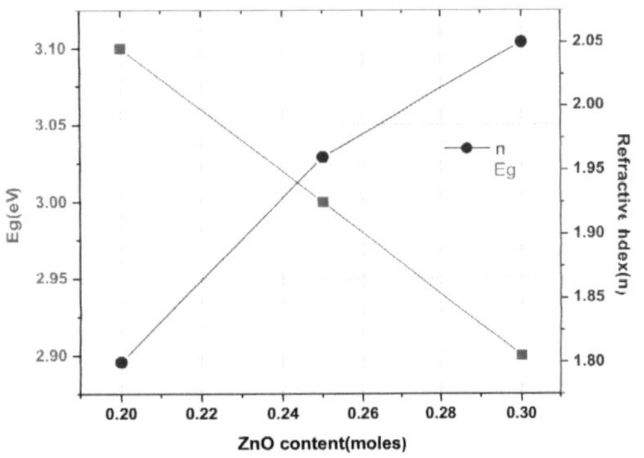

Figure 2: Variation of Eg and refractive index with ZnO concentration

Two photon absorption coefficient is obtained by fitting the experimental z-scan plot to the expression [8],

$$T(Z,S=1)=\sum_{m=0}^{\infty}\sum_{m=0}^{\infty}\frac{[-q_o(z)]^m}{[m+1]^{3/2}}\left|q_o(z)\right|<1 \quad (1)$$

where: $q_0(z)=\dfrac{[I_0\beta L_{eff}]}{1+(Z^2/Z_0^2)}$ and $Z_0 = k\omega_0^2/2$, is the diffraction length of the beam,k = $2\pi/\lambda$ α is the wave factor , ω_o = the beam waist radius at the focal point,
L_{eff} = (1-exp(-αL))/ α is the effective thickness of the sample

On analyzing nonlinear transmittance of the tellurite glass sample, it is clear that the material exhibits a large nonlinear absorption at the focus.Such materials are named reverse saturable absorbers and are well known for their optical limiting property[9].

Two Photon Absorption at different powers is considered to study various nonlinear phenomena taking place in the glass.At lower laser powers, nonlinear absorption increases as photocarriers are absorbed from the ground state,via Two Photon Absorption.At higher fluences,third order susceptibility becomes a nonlinear function of intensity given by the equation,

$$\chi^{(3)} = \chi_0^{(3)} + \chi_1^{(3)} I + \chi_2^{(3)} I^2 \quad (2)$$

At a particular power,ground state absorption is reduced,still there is an increase in the nonlinear absorption indicating the occurrence of higher order nonlinear processes such as free-carrier absorption[10], from the excited state.

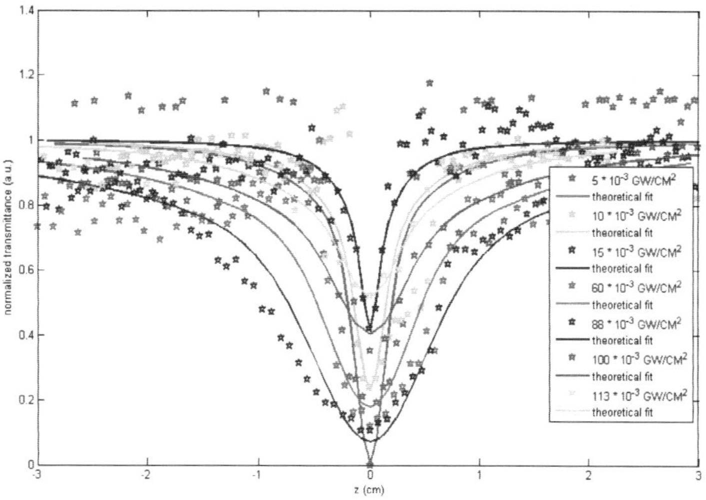

Figure 3: Variation of Two Photon Absorption with laser power

4.Conclusions

Nonlinear behaviour of Tellurite glass with ZnO is studied as a function of laser intensity.It is found that the depth of Reverse Saturable Absorption curves vary with laser irradiance.The result give an evidence to the existence of another nonlinear process called free carrier absorption in the glass sample.

5.Aknowledgements

We acknowledge UGC(RFSMS) ,KSCSTE and DST for providing financial support through research grant.

6.References

[1] Raouf A. H. El-Mallawany 2002 *Tellurite glasses Hand book-Physical properties and data* CRC press LLC .

[2] A. Zakery S.R. Elliott 2007 *Optical Nonlinearities in Chalcogenide Glasses and their Applications* Springer.

[3] A.A. Sidek, S. Rosmawati, Z.A. Talib, M.K. Halimah and W.M. Daud, 2009 *Am. J. Applied Sci.* **6 (8)** 1489-1494.

[4.] D.L. Sidebottom, M.A. Hruschka, B.G. Potter, and R.K. Brow 1997 *J. Non-Cryst. Sol.* **222** 282.

[5] B Erariah 2010 *Bull. Mater Sci* **33** 391–394.

[6] Sun K 1988 *In: Preparation and characterization of rare earth glasses*Brown University.

[7] Rose Leena Thomas, Vasuja, Misha Hari, V. P.N. Nampoori, P. Radhakrishnan, and Sheenu Thomas 2011 *JOAM* **13 5** 523-527.

[8] Mansoor Sheike Bahe,AliA.Said,Tai-Huei Wei,David J Hagan,E.W.Van Stryland, 1990 *IEE j. quant. electron* **26** 760-769.

[9] Rose Leena Thomas, Vasuja, Misha Hari, B. Nithyaja, S. Mathew, I.Rejeena, Sheenu Thomas, V. P. N. Nampoori and P. Radhakrishnan 2011 *JNOPM* **20 3** 351-356.

[10] Litty Irimpan, V. P. N. Nampoori, and P. Radhakrishnan. 2008 *J. Appl. Phys.* **103** 033105 .

Vickers microhardness studies on solution-grown single crystals of potassium boro-succinate

M Lakshmipriya, D Rajan Babu, R Ezhil Vizhi*
Crystal Growth and Crystallography Division , School of Advanced Sciences , VIT University , Vellore, TamilNadu, India

* revizhi@gmail.com, rezhilvizhi@vit.ac.in

ABSTRACT

The semiorganic crystals of potassium boro-succinate (KBS) were grown by slow evaporation method. KBS crystallizes in monoclinic system which was confirmed by powder XRD analysis. Vickers microhardness study has been carried out over a load range of 25–100 g. The Vickers hardness numbers (H_v) of the material increases as the load increases so the material is suitable for device fabrication. The Meyer index '*n*' is estimated to be greater than 1.6, the crystal system belongs to the soft material category. The elastic stiffness coefficient, c_{11}, has also been calculated using Wooster's empirical relation from the hardness data. The fracture toughness values 'K_c', determined from measurements of crack lengths, were estimated to be 0.15166 $MN/m^{3/2}$. The brittleness indices 'B_i' were estimated as 276 $m^{-1/2}$.

Key words:hardness, mayer index, fracture toughness

1. INTRODUCTION

Potassium boro-succinate crystals were grown by slow evaporation method. The crystal structure is determined by powdered XRD, and it is found that the crystal belongs to monoclinic system. FTIR spectrum was recorded to confirm the presence of functional groups and it is found that the presence of potassium and borate ions in the crystal lattice of succinic acid. UV-Vis-NIR was carried out and the spectrum shows the maximum absorption at UV region. The melting point of the material is found to be 195.1°C by thermal analysis. In dielectric studies the dielectric constant decreases with increase in frequency, low value of dielectric loss reveals the high purity of crystal [1].

Mechanical properties, hardness testing provides useful information on the strength and deformation characteristics of the material [2]. The chemical forces in a crystal resist the motion of dislocations as it involves the displacement of atoms. This resistance is the intrinsic hardness of a crystal. As hardness properties are basically related to the crystal structure of the material and hardness studies are carried out to understand the plasticity of the crystal [3]. So far the mechanical properties of KBS were not studied. So, in this investigation, we have measured the hardness and the related physical constant of the solution grown Potassium boro-succinate using Vickers microhardness tester.

2. EXPERIMENT

Potassium boro-succinate (KBS) was grown by slow evaporation method. Potassium hydroxide, boric acid, succinic acid, was dissolved in millipore water in equimolar ratio and the solution was stirred well by using a magnetic stirrer for 4 hours to get saturated solution. The pH of the solution is found to be 3.The solution is filtered using Whatman filter paper .It is poured in a petridish, covered with a perforated cover and kept in dust free atmosphere. Small seeds were observed after 5 days and transparent crystals are harvested after 45 days. The grown crystal is powdered well and powder XRD was taken and the analysis of the peak using POWDERX software, it is confirmed that the crystal belongs to monoclinic system which agrees with the reported values [1].

The microhardness of KBS crystals was determined using Vickers tester. For the static indentation test, loads varying from 25 to 100 g were applied on the grown crystal using Vickers diamond pyramid indenter connected to an incident ray research microscope. For each load *P*, an average of three impressions were recorded and the average of diagonal lengths (*d*) of the indentation mark after unloading was measured using a calibrated micrometer attached to the eyepiece of the microscope.

The Vickers hardness numbers (*H*v) were calculated using the formula

$$ _{v} = 1.8544\, X\, {}^{P}\!/_{d^2} \ \text{kg/mm}^2 \qquad (1)$$

P – applied load in kg
d – diagonal length of the indentation in mm

3. RESULTS AND DISCUSSION

The hardness value calculated by using equation (1) and plotted against the applied load and the graph is shown in figure 1. The hardness test could not be carried out above 100 g because crack initiation and materials chipping become significant beyond this load. The hardness of the material is found to be increases with increase in the applied indentation load, due to the softer superficial layers on the specimen surface. The observed increase in hardness with increase in load is usually termed as reverse indentation size effect. Because of this KBS can be used for device fabrication.

The Meyer's law, provides an expression regarding load and size of indentation [4]:

$$P = k_1 d^n \qquad (2)$$

Where k1 is the material constant and n is the Meyer index

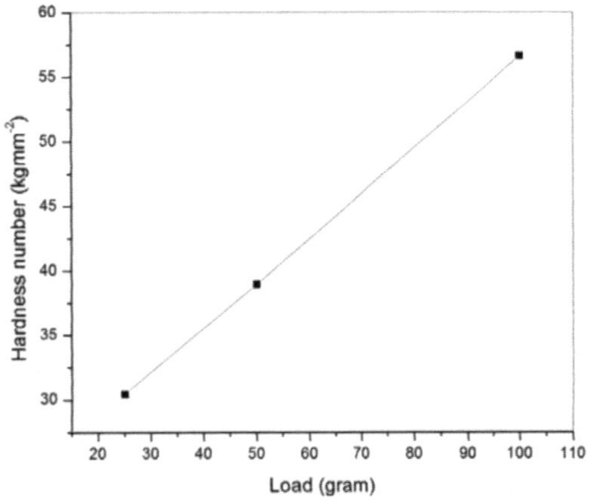

Figure.1 Load vs. Hardness

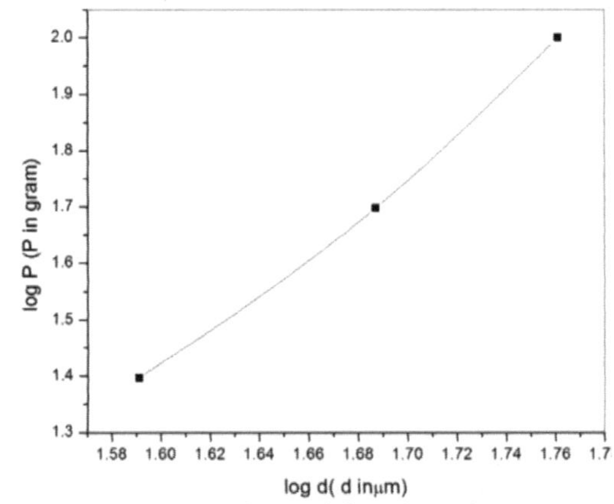

Figure.2 log d vs. log p

The plot of log *P* against log d results in straight-line as shown in figure 2 and from the graph the value of n is found to be 3 which belong to soft material [5]. Material constant was calculated from the plot figure.3 and k_1 was found out to be 4.545 kg/mm.

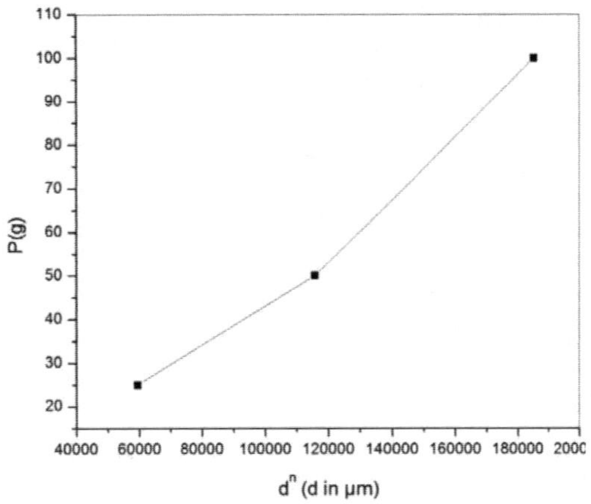

Figure.3 d^n vs P

3.1 ELASTIC STIFFNESS CONSTANT (C_{11})

The elastic stiffness constant (C_{11}) was computed by Wooster's empirical relation [6].
$C_{11}=Hv^{7/4}$ (4)

Table.1 Elastic stiffness constant of KBS

S.No	Load (g)	C_{11} X 10^{14} (Pa)
1	25	3.94
2	50	6.07
3	100	7.03

The elastic stiffness increase with increase in load (Table.1), which authenticates the tightness of bonding between neighbour atoms.

3.2 FRACTURE MECHANICS

The resistance to fracture indicates the toughness of a material and the fracture toughness *Kc* determines how much fracture stress is applied under uniform loading and is given by a relation [7]

$K_c = P/\beta_0 c^{3/2}$ for $c \geq d/2$. (5)

where β_0 is a constant that depends upon the indentation geometry. For Vickers indenter β_0 is equal to 7. For the KBS crystal the value of c/a was 2.034 and the calculated K_c was 0.15166 $MNm^{-3/2}$.

3.3 BRITTLENESS INDEX

Brittleness is an important property that affects the mechanical behavior of a material and gives an idea about the fracture induced in a material without any appreciable deformation. The value of brittleness index B_i is computed using relation
$B_i = H_v / K_c$. (6)

The calculated value of B_i was 276 $m^{-1/2}$.

4. CONCLUSION

The KBS crystals were grown by slow evaporation method. The Vickers microhardness, *H*v was carried out different load. It was observed that the hardness increases with increasing load, termed as reverse ISE. As Vickers hardness number is calculated as 3, so the material belongs to soft material category. The fracture toughness of the material is found to be 0.15166 $MNm^{-3/2}$. The B_i value is computed as 276 $m^{1/2}$. The value of C_{11} gives an idea of tightness of bonding between neighboring ions. The hardness measurements may be useful in indicating the order of magnitude to be expected for the elastic constant in a new material.

ACKNOWLEDGEMENTS

The authors are thankful to DST (SERB), Govt. of India for the research project (SR/FTP/PS-122/2012) and VIT University for their constant support. One of the authors (M.L) is grateful to VIT University for Research Associateship.

REFERENCES

1) Chithambaram.V , Jerome Das S, Arivudai Nambi R , Krishnan S 2011 Synthesis Growth and characterization of novel semiorganic nonlinear optical potassium boro-succinate (KBS) single crystals *Opt Laser Technol* **43** 1229–1232
2) Deepthy and Bhat H L 2001 Growth and characterization of ferroelectric glycine phosphite single crystals *J. Cryst. Growth* **226** 287–293.
3) Ezhil Vizhi R , Rajan Babu D, Sathiyanarayanan K 2010 Study of Microhardness and Its Related Physical Constants of Ferroelectric Glycine Phosphite (GPI) Single Crystals *Ferroelectr. Lett* **37** 23–29.
4) Susmita Karan and Sen Gupta S P 2005 Vickers microhardness Studies on solution-grown single crystals of magnesium sulphate hepta-hydrate. *Mater. Sci. Eng., A* **398** 198–203.
5) Vineeta Gupta, Bamzai K K, Kotru, Wanklyn B M 2005 Mechanical characteristics of flux grown calcium titanate and nickel titanate crystals, *Mater. Chem. Phys* **89** 64–71
6) Wooster W A 1953 Physical properties and atomic arrangements in crystals *Rep.Progr. Phys.* **16**, 62–82.
7) Bamzai K K, Kotru P N, Wanklyn B M 2000 Fracture mechanics, crack propagation and microhardness studies on flux grown ErAlO3 single crystals. *J. Mater. Sci. Technol.* **16**, 405–410.

IOP Conf. Series: Materials Science and Engineering **73** (2015) 012092 doi:10.1088/1757-899X/73/1/012092

Fabrication and characterization of p-type SiNW/n-type ZnO heterostructure for optoelectronics application

Purnima Hazra[1], P Chakrabarti and S Jit
Centre for Research in Microelectronics, Department of Electronics Engineering, Institute of Technology, Banaras Hindu University, Varanasi-221005, India

[1]E-mail: phazra.rs.ece@itbhu.ac.in

Abstract: Semiconductor hybrid structure, known as core-shell heterostructures was fabricated and optical properties were analyzed to make it applicable in future optoelectronic and photonic devices. Large-area, high density, vertically oriented silicon nanowire arrays, synthesized by means of metal-assisted chemical etching of p-type silicon (100) substrate was used as the core and zinc oxide (ZnO) layer, deposited on the SiNW arrays by atomic layer deposition (ALD) was used as shell. The XRD peaks of the heterostructure confirmed the subsequent growth of ZnO film on the template of SiNW arrays having similar crystalline quality. The photoluminescence (PL) spectra showed a very sharp peak at 378 nm, corresponding to the band gap of ZnO material and another broad emission band almost throughout the entire visible range with a peak around 550 nm. The structure also showed a very good antireflection property. The results present that the SiNW/ZnO heterostructure can have potential application in future nanoscale electronic and photonic devices.

1. Introduction

In recent years, high-aspect ratio silicon nanowire based heterostructures attracting interest due to their potential application in broad areas, ranging from nanoelectronic , photonic to biomedical sensing applications [1,2] and their compatibility with present day Si based optoelectronic nano-devices and Si-integrated circuits. In present day optical industries, ZnO is considered as a promising material for applications including blue and ultraviolet optical devices, and gas sensing [3]. ZnO, a II-VI compound semiconductor having wurtzite crystal structure with wide and direct bandgap of ~3.37eV at room temperature and large exciton binding energy (60 meV) can prepare nanostructures with diversified morphology [3]. Um *et al.* for the first time reported radial heterojunction nanowires (NWs) using ZnO (shell)/Si (core) coaxial structure for a novel photodiode application [4]. They showed that the approach can promote surface enhanced electron-hole separation in ZnO nanostructure and the charge carrier separation in radial heterojunction which leads to the improvement in photosensitivity with spectral response enhancement. Although ZnO nanostructures as well as SiNW structures have been introduced in various previous studies to obtain a high conductivity gain with ZnO NWs [3] or strong light absorption with SiNWs [5], the investigation of properties of combined or heterojunction nature of these two is a relatively new field and needs mare attention. In this paper SiNW/ZnO heterostructures were studied for their optoelectronic applications.

The fabrication of SiNW/ZnO heterostructure has to be done in two steps. In first stage SiNWs should be fabricated and the ZnO deposition on SiNW arrays should be done in second stage. There are a number of techniques to produce Si-nanostructures. Currently SiNWs are produced by different methods such as laser ablation, thermal evaporation , chemical vapor deposition , chemical etching , and solution growth [6].These techniques have their own limitations, as complex 3D shapes are difficult to produce and it is very hard to maintain high aspect ratio structures with varying degrees of complexity as feature sizes shrink into the nanoscale. Nevertheless, controllable and repeatable large-area SiNW array fabrication is still a problem. WE synthesized SiNWs on silicon wafer through a simple and conventional nanoelectrochemical etching technique without the use of a template near room temperature, as demonstrated by Kuiquing Peng *et. al.* for the first time [5]. In this process, the simultaneous growth of metal dendrites accompanied with the etching of silicon wafer is a very important factor in the formation of nanostructures. Si nanostructures with different morphologies and properties can be produced by varying several parameters, such as, type of deposited metal (Pt, Ag, Au etc.), variation of etching time duration and temperature, concentration of etchants, wafer orientation and morphology of the metal film [7]. There is a lot of technique available also for ZnO deposition, such as pulsed laser deposition (PLD), atomic layer deposition (ALD), molecular beam epitaxy (MBE) and sol-gel method [3]. Among all the above described methods, ALD is the cost-effective and most promising method, which guarantees conformal deposition of ZnO film with extremely uniform and reproducible thickness, low stress and low defect density. Nanocrystal films are well suited for ALD, since their open structure allows diffusion of gaseous precursors, and thus allows an easy layer-by-layer filling of interstices. In this study, fabrication of SiNW/ZnO heterostructure was done by depositing of ZnO film on metal-assisted chemically etched SiNWs by atomic layer deposition method.

2. Experimental Details

2.1. Synthesis of SiNW

P-type, boron doped silicon (100) wafer was cut into small rectangular strips and properly cleaned. The growth of SiNW was achieved at atmospheric pressure and low temperature (50ºC) using ultrasonication method. Analytical grade (99.99% pure) 4.6 mol/l hydrofluoric acid (HF) and 0.02 mol/l silver nitrate ($AgNO_3$) were mixed in teflon coated vessel and Si stripes were dipped in the solution for 1 hour for the etching purpose. The as grown SiNW substrates with clapped silver dendrite film were dipped in the 30 wt. % HNO3 aqueous solution for removing the clapped silver. Then the samples were rinsed with de-ionized water, dried at oven, and subjected for further deposition.

2.2. ZnO deposition

After SiNW synthesis, ultrathin ZnO film was deposited onto as prepared SiNW by atomic layer deposition (ALD) technique (equipment model TFS 200 from BENEQ). In this process two precursors, diethylzinc [Zn $(C_2H_5)_2$, DEZN] and DI water were used as the source of zinc and oxygen respectively to produce ZnO with pulse duration of 100 ms each and the purging time of 1 sec. Nitrogen was used as both carrier gas and purging gas with flow rate 200 sccm. The substrate and precursor temperature was 150°C and 20°C respectively. The thickness of ZnO film was about 100 nm for 500 cycles of deposition. The microstructures of the films were analyzed by scanning electron microscope (SEM), atomic force microscope (AFM) on contact mode and X-ray diffraction spectra (Bruker D8 Advance XRD). Optical properties of the nanostructure templates were showing by photoluminescence spectra with a laser source with wavelength of 325 nm and a PerkinElmer Lambda 950 spectrophotometer.

3. Results and discussion

The surface morphology of the wafer-scale SiNW arrays before and after deposition of ZnO was studied by FESEM and AFM imaging. Figure [1(a)] shows that the as-prepared blank SiNWs formed bunches with average diameter and length of single nanowire varied around 200-500 nm and 6-8 μm respectively. The formation of nanowire is uniform and aligned vertically in a large area, also covering

IOP Conf. Series: Materials Science and Engineering **73** (2015) 012092 doi:10.1088/1757-899X/73/1/012092

the whole sample. The figure [1(b)] shows the 2-D AFM image of SiNWs coated with ZnO by ALD. The figure clearly depicts that the morphology of SiNWs remain intact after the deposition of ZnO.

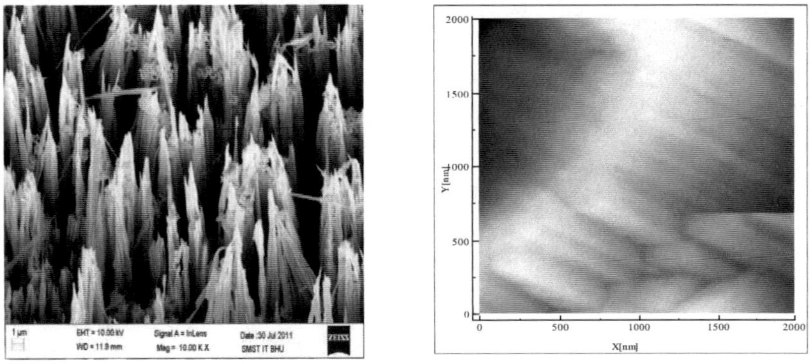

Figure 1: (a) SEM image of SiNW arrays, (b) 2D AFM image of ZnO film grown on SiNW arrays

XRD spectra in Figure [2] showed that the SiNWs had very good crystalline quality and predominantly having (002)-orientation with full-width at half-maximum of only 0.15°. Though, there were some other weak peaks, originating from other crystallographic planes. The spectrum also showed the polycrystalline nature of ZnO layer as grown on SINW template. Although the (002) peak at 2θ=34.45° with FWHM value of 0.7° was predominant in nature, indicating that most of the c-axis of ZnO grains were arranged perpendicular to the substrate surface. Indeed, the results ensured the subsequent grown of ZnO films were having similar crystalline quality as of the SiNW templates.

Figure 2: *θ–2θ* X-ray diffraction pattern of SiNW arrays (blue dashed line) and ZnO film deposited on SiNW arrays (red line)

After the detailed analyses of the microstructures, the point of discussion of our work is optical properties of the Si/ZnO core–shell nanowire arrays. In the PL spectrum shown in figure [3] shows two peaks. First of all there is a sharp peak at 378nm, corresponding to the band gap of ZnO (3.28 eV), presumably due to the direct electron-hole recombination across the intrinsic gap of ZnO [1]. There is also a broad spectrum, from 450 nm to 650 nm, centered at around 550 nm and covering almost the entire visible region, consistent with the previous reports [1, 2,]. SiNW/ZnO heterojunction exhibits a type II core–shell band alignment based on Anderson's model as shown in energy band diagram in Figure [4]. The conduction band minimum of ZnO is 0.34 eV higher than that of silicon. The electrons injected from ZnO into the neighboring one, to recombine with the holes in the valence band of silicon, responsible for the emission on 2.46 eV (~500 nm). Though this Type-I peak was very weak as the absorption depth of the excitation beam depends on the excitation source as well as the absorption coefficient of ZnO and the penetration depth of laser [8]. A comparatively strong peak at 550nm (~2.25 eV) was induced due to the Type-II defect present in ZnO because of the deep level states in ZnO induced by various impurities and those defects which existed close to the valence band maximum of silicon. Here the holes in p-Si can tunnel into ZnO and may be trapped at deep level states in ZnO, resulting in the green emission [8]. Hence, devices based on a n-ZnO/p-Si heterostructure can emit green and red light simultaneously. The weak PL peak ranging from 450 nm to 650 nm originated from the trap state emission in ZnO might be detected due to the low quality of the interface between Si and ZnO [2]. The very good antireflection property of the device (in the order of 0.2% to 2.6% for the wavelength ranging from 300 to 900 nm) may also contribute some significant impacts to the marked enhancement of the observed visible PL emissions.

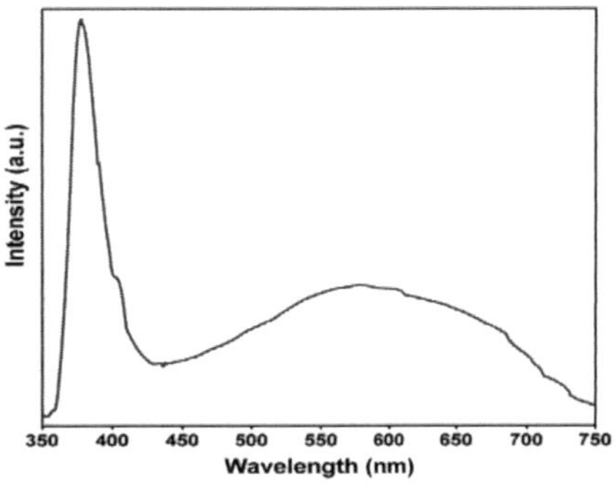

Figure 3: Photoluminescence spectra of SiNW/ZnO heterostructure at room temperature

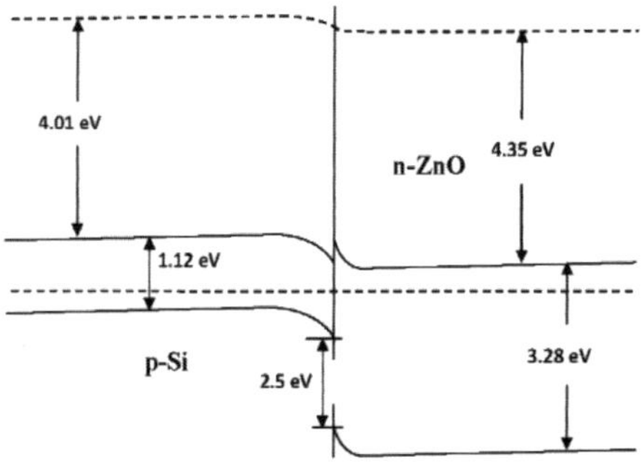

Figure 4: Anderson's model of band diagram of SiNW/ZnO heterostructure

4. Conclusion

A simple method for the preparation of high aspect ratio quasi one dimensional SiNW-ZnO core-shell nanowire arrays is reported. Here conformal deposition of ZnO was done on metal-assisted chemically etched highly-ordered vertical SiNW by ALD technique. The detailed study of structural and optical properties of core-shell SiNW/ZnO heterostructure is represented through FESEM, AFM, PL and reflectometry analyses. The Si/ ZnO core-shell nanowire arrays can perform simultaneously in dual band of the spectrum namely ultraviolet and visible region respectively. The high aspect ratio, anti-reflective characteristics inherent to nanowire structure can be exploited for fabricating future nanoelectronic and optoelectronic devices.

Acknowledgement

The authors are deeply thankful to Laser Materials Processing Division, DAE-RRCAT, Indore for availing facilities to complete a part of the above project. They are also thankful to UGC-DAE CSR Centre, Indore for availing some of the characterization facilities.

Reference

[1] Chang Y M, Jian S R, Lee H Y, Lin C M and Juang J Y 2010 *Nanotech.* **21** 385705 (7pp)
[2] Sun L, He H, Liu C, Lu Y and Ye Z *CrystEnggComm* **13** 2439-2444.
[3] Mende L S, Drisscol J 2007 *Materials Today* **10** 40-48
[4] Um H D, Park K T, Jung J Y, Jee S W, Moiz S A, Ahn C H, Cho H K, Lee E, Kim D W, Park Y C, Yang J M, Lim S K and Lee J H 2010 *IEDM IEEE* 222-225
[5] Peng K Q, Yan Y, Gao S, and Zhu J 2003 *Adv. Funct. Mater.* **13** 127-132
[6] Peng K Q and Lee S T 2011 *Adv. Mater.* **23** 198-215
[7] Peng K Q, Fang H, Hu J, Wu Y, Zhu J, Yan Y and Lee S T 2006 *Chem. Eur. J.* **12** 7942-7947
[8] Tay Y Y, Tan T T, Liang M H, Boey F and Li S 2010 *Phys. Chem. Chem. Phys.* **12** 6008-6013

Structural and electrical properties of neodymium substituted cobalt ferrite nanoparticles

S Xavier, S Thankachan, B P Jacob and E M Mohammed
Research Department of Physics, Maharaja's College, Ernakulam, India

E-mail: emmohammed_2005@yahoo.com, sheena.xavier1@gmail.com

Abstract. A series of polycrystalline spinel ferrites with composition $CoFe_{2-x}Nd_xO_4$(x=0.0, 0.05, 0.1, 0.15, 0.2, 0.25) have been synthesized by sol gel method. The structural characterizations of the prepared samples were done using XRD and TEM. The crystallite size shows an increase with the increase in the concentration of neodymium. The activation energy has been calculated from the temperature dependent DC conductivity measurements. The dielectric properties were studied and analyzed as a function of frequency. All the samples exhibit normal dielectric behaviour which is attributed to Maxwell– Wagner interfacial polarization.

1. Introduction

Polycrystalline ferrites are very good dielectric materials and has many technological applications ranging from microwave to radio frequencies [1]. Cobalt ferrite is an important member of ferrite family and is characterized by moderate saturation magnetization, high coercivity, mechanical hardness and chemical stability [2, 3]. The high electrical resistivity and good magnetic properties make this ferrite an excellent material for electronic and telecommunication applications.

The structural and electrical properties of cobalt ferrite are very sensitive to the method of preparation, the amount and the type of substitution [4]. The sol-gel technique is an excellent method to synthesize nanoparticles with maximum purity and narrow size distribution. It has been reported that partial substitution of Fe^{3+} ions by rare earth in ferrite can modify its magnetic and electrical properties [5]. In the present study, structural and electrical properties of neodymium doped cobalt ferrite prepared by sol gel method have been presented.

2. Experimental methods

AR grade cobalt nitrate, neodymium nitrate and ferric nitrate were dissolved in ethylene glycol in the proper molar ratio using a magnetic stirrer. The solution was then heated at 60°C until a wet gel of the metal nitrates was obtained. The gel was dried at 120°C and finally self ignites to form a highly voluminous and fluffy powder. The obtained powder was ground well and sintered for 4 hours in a muffle furnace at 400°C. For electrical measurements, cylindrical disc-shaped pellets of the samples were made using a hydraulic press by applying a uniform pressure of 5 tons. The pellets were sintered at 500°C for 10 hours in a muffle furnace.

The neodymium doped cobalt ferrite samples were characterized by using X-ray powder diffractometer (Bruker AXS D8 Advance). The particle size was determined by subjecting the sample to Transmission Electron Microscopy. The DC conductivity studies were done using Keithley 6221

DC current source and 2182A Nano volt meter. Dielectric measurements were carried out using precision impedance analyzer Wayne Kerr 6500B in the frequency range 20Hz to 30 MHz.

3. Results and discussion

3.1 X-ray diffraction analysis

The XRD pattern of $CoFe_{2-x}Nd_xO_4$ nanoparticles sintered at 400°C is depicted in figure 1. Comparing the XRD pattern with the standard data (ICDD file No. 22-1086), the formation of spinel phase in all the samples was confirmed.

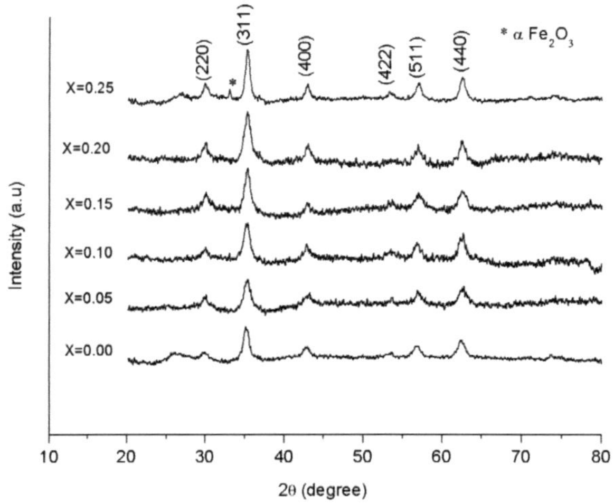

Figure 1. XRD pattern of $CoFe_{2-x}Nd_xO_4$.

The crystallite size of the samples has been estimated from the broadening of XRD peaks using the Scherrer equation [4]. The crystallite sizes were increased by the gradual increasing in neodymium content and it is in agreement with the reported results [5]. The lattice parameter is calculated for prominent peak (311) using Bragg's equation and the actual (X-ray) density were calculated. The bulk density was calculated from mass and bulk volume of the sample. Calculated values of lattice parameter, X-ray density, bulk density and porosity of all the samples are listed in table 1. The porosity of the samples was calculated from the X-ray density and bulk density.

Table 1. Lattice parameter, crystallite size, density and porosity of $CoFe_{2-x}Nd_xO_4$ samples

Nd content x	Lattice parameter(Å)	Crystallite Size (nm)	X-ray density (gcm^{-3})	Bulk density (gcm^{-3})	Porosity (%)
0.00	8.399	11.26	5.260	2.662	48.96
0.05	8.422	13.00	5.315	2.716	48.27
0.10	8.424	13.38	5.408	2.690	50.23
0.15	8.425	14.04	5.504	2.581	53.12
0.20	8.428	14.69	5.598	2.461	56.02
0.25	8.432	17.64	5.689	2.665	53.14

3.2 TEM analysis

The TEM images of the samples $CoFe_{2-x}Nd_xO_4$ (with x =0, 0.1) are shown in figures 2(a) and (b). In these figures, most of the particles appear with almost spherical shape and some are agglomerated. The particle size estimated from the different images of these samples lies in the range of 15 to 25 nm.

IOP Conf. Series: Materials Science and Engineering **73** (2015) 012093 doi:10.1088/1757-899X/73/1/012093

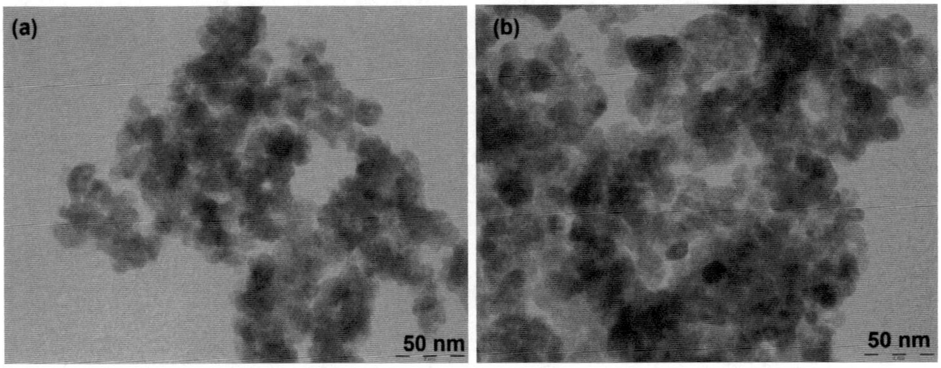

Figure 2. TEM image of (a) $CoFe_2O_4$ and (b) $CoFe_{1.9}Nd_{0.1}O_4$.

3.3 Electrical characterization

The temperature dependence of DC conductivity in cobalt ferrite is studied for different concentrations of neodymium and is shown in figure 3. The variation of conductivity with temperature shows the typical semiconducting nature of the ferrites. With the neodymium doping the DC conductivity of the samples varies from 1.698×10^{-4} to 5.287×10^{-5} S/m. According to the electron hopping model [2], the conductivity is mainly due to the hopping of electrons between Fe^{2+} and Fe^{3+} ions present at the octahedral B sites. The activation energy for conduction calculated from the Arrhenius plot is found to be very high (0.759eV), indicating the high resistive nature of the samples.

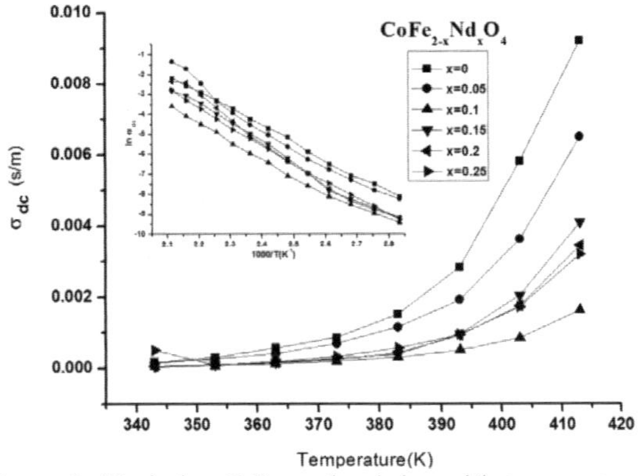

Figure 3. Variation DC conductivity with temperature
(ln σ_{dc} vs 1000/T given as inset).

The variation of dielectric permittivity (ε') with frequency is depicted in figure 4. It can be seen that dielectric permittivity exhibit an inverse dependence on frequency as reported for various ferrites. It decreases with frequency and remains a constant at higher frequencies. The decrease in permittivity with frequency can be explained on the basis of Koop's theory [6] which considers the dielectric structure as an inhomogeneous medium of two layers of the Maxwell – Wagner type [7]. The space charge polarization occurring at the interfaces contribute to the enhanced ε' values at lower frequencies. At higher frequencies, electron exchange between Fe^{2+} and Fe^{3+} ions will not be able to follow the applied electric field thus resulting in a decrease in polarization. When neodymium is doped in the system, it occupies the octahedral site in the ferrite system, decreasing the Fe^{3+} ion number responsible for polarization. Hence the value of dielectric permittivity decreases with lower concentrations of neodymium content. However an increase in ε' is noticed for higher neodymium content.

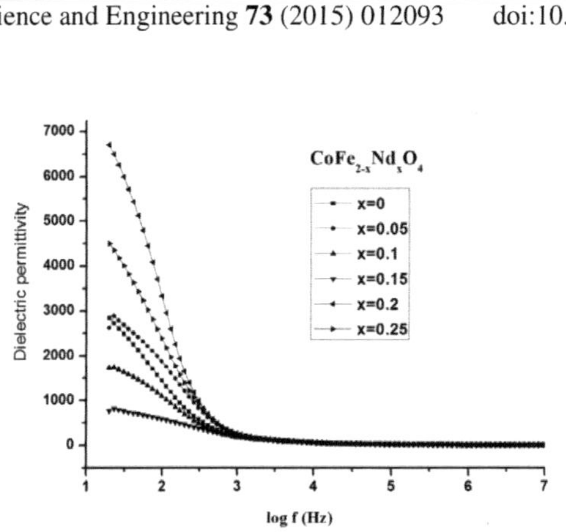

Figure 4.Variation of dielectric permittivity with frequency.

The variation of loss tangent with frequency for various neodymium concentrations is shown in figure 5. All the doped samples in the series exhibit relaxations at specific frequencies. Dielectric relaxation occurs when the hopping frequency of the charge carriers is approximately equal to that of external applied field [8].

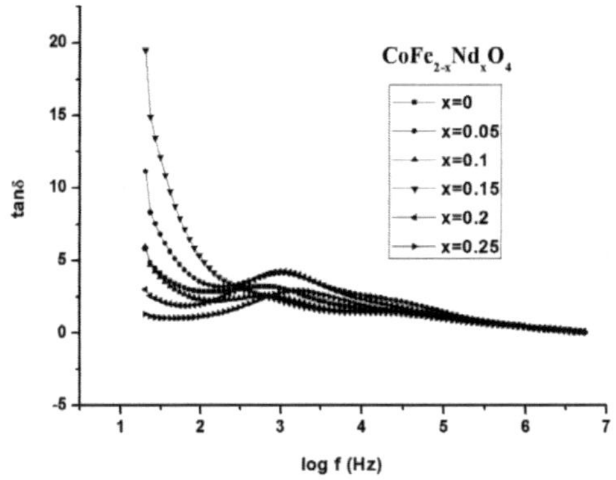

Figure 5.Variation of dielectric loss ($\tan \delta$) with frequency.

4. Conclusions

Neodymium substituted cobalt ferrite nanoparticles were prepared successfully using the sol gel technique. The increase in crystallite size with the increase in the concentration of neodymium can be attributed to the large radii of the Nd^{3+}ions. The temperature variation of DC conductivity indicated the semiconducting behavior of the ferrites. The variation in the dielectric properties as a function of frequency and composition has been explained on the basis of Maxwell-Wagner theory of interfacial polarization. The high resistivity and low dielectric behavior make this ferrite useful in high frequency applications.

References

[1] Kharabe R G, Devan R S, Kanamadi C M and Chougule B K 2006 *Smart Mater. Struct.* **15** N36
[2] Veena G E, Al-Omari I A, Malini K A, Joy P A, Sakthi K D, Yasuhiko Y and Anantharaman M R 2008 *J.Magn.Magn.Mater.***321** 1092
[3] Cedeno – mattei Y and Perales-Perez O 2009 *Microelectronics Journal* **40** 673
[4] Gul I H, Abbasi A Z, Amin F, Anis-ur-Rehman M and Maqsood A 2007 *J.Magn. Magn. Mater.* **311** 494
[5] Binu P J, Smitha T, Sheena X and Mohammed E M 2011 *Physica Scripta* **84** 045702
[6] Koops C G 1951 *Phys.Rev.* **83** 121
[7] Wagner K W 1913 *Ann.Phys.* **40** 817
[8] Shigeki I and Toshio T 1971 Japan. J. Appl. Phys. **10** 260

Electrical and magnetic properties of nano-sized magnesium ferrite

Smitha T, Sheena X, Binu P J and E M Mohammed

Research Department of Physics, Maharaja's college, Ernakulam, India

emmohammed_2005@yahoo.com, smithathankachan11@gmail.com

Abstract. Nano-sized magnesium ferrite was synthesized using sol-gel techniques. Structural characterization was done using X-ray diffractometer and Fourier Transform Infrared Spectrometer. Vibration Sample Magnetometer was used to record the magnetic measurements. XRD analysis reveals the prepared sample is single phasic without any impurity. Particle size calculation shows the average crystallite size of the sample is 19nm. FTIR analysis confirmed spinel structure of the prepared samples. Magnetic measurement study shows that the sample is ferromagnetic with high degree of isotropy. Hysterisis loop was traced at temperatures 100K and 300K. DC electrical resistivity measurements show semiconducting nature of the sample.

1. Introduction

Spinel magnesium ferrite is having a wide range of applications as catalyst, humidity sensors and magnetic memory devices. Nowadays nanostructured spinel ferrites has invoked more attention due to its novel material properties when compared with its bulk counter parts [1-2]. The interesting physicochemical properties of ferro-spinels arise from their ability to distribute the cations among the available tetrahedral (A) and octahedral [B] sites[3-4]. Magnesium ferrite is a pertinent magnetic material showing inverse spinel structure for wide applications owing to its high resistivity, high Curie temperature and environmental stability. The degree of inversion of spinel structure depends upon the heat treatment due to high diffusibility of Mg^{2+} ions[3]. At present, several chemical methods including the solid state reaction, co-precipitation, sol–gel, citrate method have been developed to synthesize small particle sized ferrites [5-7]. Due to good stoichiometric control and production of ultrafine particles in nanorange at relatively low temperature, sol-gel technique is an attractive method of preparation [8]. In addition to this, sol–gel-derived nanoparticles generally possess good chemical homogeneity, high purity and lower calcination temperature [9-10].

2. Experimental methods

2.1 Synthesis

Sol-gel combustion method was used for preparation of magnesium ferrite nanoparticles. AR grade magnesium nitrate and ferric nitrate were used as chemical precursors in ethylene glycol as solvent. Nitrates in the required stoichiometric ratio were dissolved in minimum amount of ethylene glycol at room temperature and the sol was heated at 60^0C to obtain a wet gel. This gel was dried at 120^0C, which self ignites to form a fluffy product. This is then ground to form fine powder and was fired at

Published under licence by IOP Publishing Ltd

400^0C for 3 hours. For doing the d.c. conductivity studies the samples were pelletized in cylindrical disc-shape using a hydraulic press by applying a uniform pressure of 5 tons and were sintered at 500^0C in the furnace for 24 hours.

2.2 Characterization
Structural characterization was done using XRD analysis. X-ray powder diffractometer (Rigaku make RINT 2000) with Cu-Kα radiation (λ=1.54059A^0) at 40kV and 30mA performed a scanning from 20^0 to 80^0 at a step size of 0.02^0 per second for each sample. The crystal structure, lattice parameter, crystallite size, X-ray density, bulk density and porosity were determined from the XRD pattern. The Fourier Transform Infrared (FTIR) absorption spectra of the samples were recorded using FTIR spectrometer (Thermo Nicolet, Avatar 370) in the wave number range 4000 to 400 cm^{-1} with Potassium bromide (KBr) as solvent. Magnetic characterization of the sample was carried out by vibration sample magnetometer at temperatures 40K and 303K with maximum applied field of 20kOe using Lakeshore VSM 7410. Keitheley 6221 current source and 2182A voltmeter was used to determine the d.c. conductivity of the samples at temperature from 300K to 400K.

3.Results and discussion

3.1 Structural characterization
3.1.1 X-ray diffraction analysis The XRD pattern of the prepared samples S1 is shown in figure 1. The XRD pattern confirms the formation of single phase fcc spinel structure with no extra lines corresponding to any other crystallographic phase (compared with JCPD card no: 88-1935). The lattice constant 'a' calculated using Bragg's equation for prominent (311) peak is 8.40A^0 which is comparable with reported value of 8.37A^0 [11]. The average crystallite size calculated from the XRD peak broadening of the peaks using Scherrer formula [14] corrected for micro strain estimated from the Hall- Williamson plot (figure 2)[12] is 19.8nm. Calculated value of x-ray density is 4.47 gm cm^{-3}.

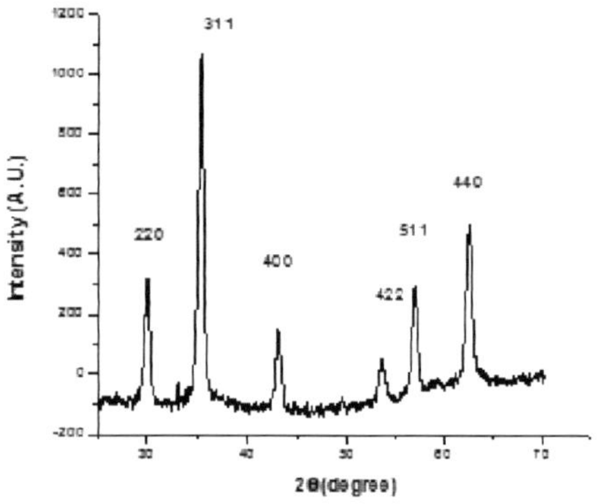

Figure1. XRD pattern of $MgFe_2O_4$

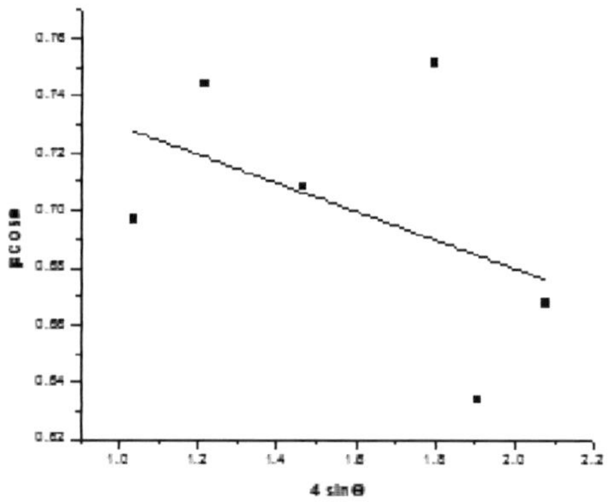

Figure2.Hall- Williamson plot of $MgFe_2O_4$ from XRD

3.1.2 FTIR analysis IR spectra of $MgFe_2O_4$ ferrite system is shown in figure 3. FTIR spectrum analysis helps us to confirm the formation of spinel structure in ferrite ferrites. Two major absorption bands are found in the range 600-550 cm^{-1} and 450-385 cm^{-1} which confirm the spinel structure formation of the prepared samples [13]. The bands corresponding to 3400cm^{-1} and 1400 cm^{-1} represent stretching and bending vibrations of H-O-H which indicates the presence of free or absorbed water in the samples. This is confirmed in the d.c. conductivity studies also.

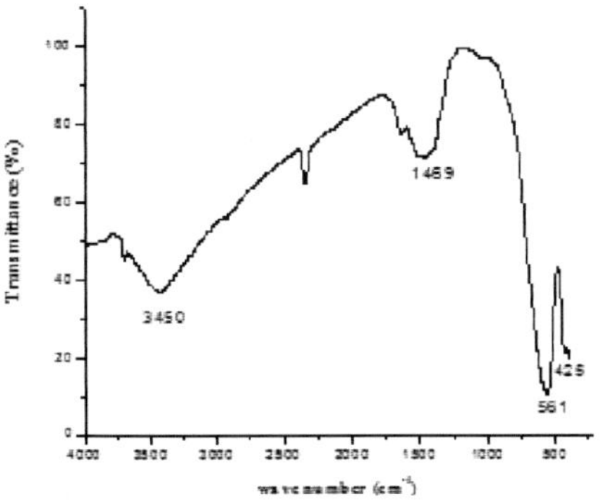

Figure3. FTIR spectrum of $MgFe_2O_4$

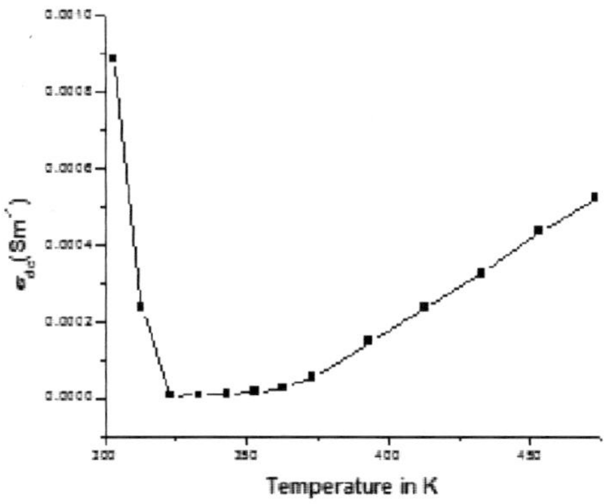

Figure 4. Variation dc conductivity with temperature

3.2 Electrical characterization

3.2.1 DC conductivity studies The temperature dependence of DC conductivity in magnesium ferrite is studied in the temperature range 300K to 473K and is shown in figure 4. The variation of conductivity with temperature shows semiconducting nature of the ferrite. But initially till the temperature reaches323K from room temperature there is a sudden decrease in σ_{dc} value which is due the absorbed water in the sample. The dc conductivity of the sample varies from 5×10^{-7} to 5×10^{-4} S/m in the temperature range 323K 473K.

The conductivity in ferrites can be explained based on the popular electron hopping model [18]. According to this model, the conductivity is mainly due to the hopping of electrons between Fe^{2+} and Fe^{3+} ions present at the octahedral B sites.

3.2.2 Magnetic properties Magnetic characterizations of the samples were carried out by vibration sample magnetometer at room temperature with maximum applied field of 20kOe. Typical magnetic hysteresis loop of sample S1 at temperatures 303K and 40K is shown in figure 5. The saturation magnetization (M_s), coercivity (H_c) and remanance (M_r) of the samples are shown in table 1.

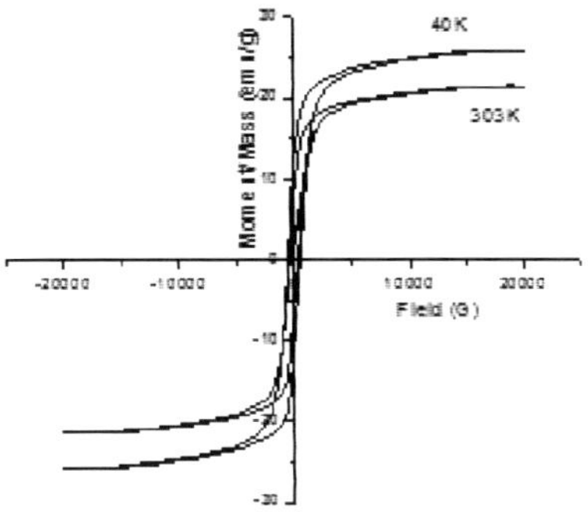

Figure 5: Hysterisis loop of $MgFe_2O_4$ traced at 40K and 303K

Sample at Temperature	Saturation magnetization (M_s)in emu/gm	Coercivity (H_c)in Oe	Remanance (M_r)in emu/gm	M_r/ M_s
303K	21.4	349.9	8.2	0.38
40K	25.8	461	12.7	0.49

Table 1. Saturation magnetization (M_s), coercivity (H_c) and remanance (M_r) of $MgFe_2O_4$ at temperatures 40K and 303K

The remenant ratio $R=M_r/M_s$ shows the ease with which the direction of magnetization reorients to the nearest easy axis direction after the field is removed. The low value of R indicates the isotropic nature of the material [17]. According to Stoner Wohlfarth model [19], a theoretical value of Mr/Ms is 0.5 for non interacting uniaxial single domain particles with the easy axis being randomly oriented. Mr/Ms value observed is found to be 0.49 for sample at low temperature which shows its isotropic nature.

4.Conclusions

Magnesium ferrite particles were prepared successfully using the sol gel method. The XRD pattern confirms the formation of single phase fcc spinel structure with no extra lines corresponding to any other crystallographic phase. FTIR spectrum confirms the spinel formation and presence of water in the sample. The temperature variation of dc conductivity indicated the semiconducting behavior of the ferrite. Decrease in conductivity till temperature 323K is due to water absorbed in the sample. The magnetic study reveals that the sample is ferromagnetic at room temperature and low temperature (40K). At low temperature the sample is highly isotropic in nature.

Acknowledgements
The authors wish to acknowledge SAIF, IIT Madras for providing VSM measurement facility.

References

[1] Santi M, Montana S,Amporn W, 2009 *Nanoscale Res. Lett.* **4**:221-228.
[2] S.Wang, Y.Zhou,.Guan, B.Ding, 2008 *Nanoscale Res. Lett* . **3**, 289
[3] G. Blasse, 1965 *Philips Res. Rept* **20** pp 528.
[4] J.B. Goodenough, 1963 *Mag. Chem. Bond*, John Wiley, New York, London **120**.
[5] A. Verma, T.C. Goel, R.G. Mendiratta, R.G. Gupta, 1999 J.*Magn. Magn. Mater.* **192** pp 271–276.
[6] P.P.Hankare, P.D.Kamble,M.R.Kadam,K.S.Rane,P.N.Vasambekar,2007*Mater.Lett.* **61**2769–71.
[7] Cui-Ping L,Ming-Wei L,Zhong C,Juan-Ru H,Yi-Ling T,Tong Lin,Wen-Bo Mi,2007*J.Mater.Sci.***42** 6133–38.
[8] Manish Srivastava, S. Chaubey, Animesh K Ojha, 2009, *Materi.Chem.Phys.* **118:1**, 174-180
[9] Abdel-Mohsen FF, Emira HS 2005, *Pigment Resin Technol* **34** pp 312
[10] Cui H, Zayat M, Levy D 2005 *J Sol–Gel Sci Technol* **35**:175
[11] Viswanathan B, Murthy VRK, Ferrite material science & technology, 1990Narosa Publishing, p6.
[12] Klug H P & Alexander L e 1974 X-ray diffraction procedures for polycrystalline & amorphous materials *(Wiley: New York)*
[13] Waldron R D, Infrared Spectroscopy of ferrites, *Phy rev* 1955,**99**,1727-35.
[14] Binu PJ, Smitha Thankachan, Sheena X , E M Mohammed, 2011 *Phys. Scr.* **84** 045702 (6pp)
[15] Franco, Jr. and M. S. Silva, 2011 *Journal of Applied Physics* **109**, 07B505
[16] S.S.Khot,N.S.Shinde, B P.Ladgaonkar, B.B.Kale and S C.Watawe, *Advances in Applied Science Research*, **2 (4)**,460-471
[17] Shirsath S E, Toksha B G and Jadhav K M 2009 *Mater. Chem. Phys.* **117,** 163-8
[18] Veena G E, Al-Omari I A, Malini K A, Joy P A, Sakthi Kumar D, Yasuhiko Y, Anantharaman M R 2008 *J.Magn.Magn.Mater.***321** 1092-9
[19] E. C. Stoner , E. P. Wohlfarth, Philos. 1948 *Trans. R. Soc.* A **240,** 599-642

IOP Conf. Series: Materials Science and Engineering **73** (2015) 012095 doi:10.1088/1757-899X/73/1/012095

Upconversion luminescence in Er^{3+}/Yb^{3+} codoped PbO-Bi_2O_3-Al_2O_3-B_2O_3 glasses

B Appa Rao*, Y Raja Rao, K Krishnamurthy Goud and M Srinivas
Department of Physics, Osmania University, Hyderabad-500 007, India

apparao.bojja@gmail.com Mobile: +91 9391060524

Abstract. Er^{3+}/Yb^{3+}co-doped glasses with the composition PbO-Bi_2O_3-Al_2O_3-B_2O_3 were prepared by melt quenching method. Optical absorption (OA) and photoluminescence (PL) and FT-IR spectra of all the glasses were recorded at room temperature. Green and red up-conversion emissions centered at 525, 545 and 654 nm corresponding to $^2H_{11/2} \rightarrow {}^4I_{15/2}$, $^4S_{3/2} \rightarrow {}^4I_{15/2}$ and $^4F_{9/2} \rightarrow {}^4I_{15/2}$ transitions of Er^{3+} ions have been obtained under the excitation of 980 nm laser at room temperature. From emission spectra it was observed that the intensity of red emission increases with increasing the concentration of Er^{3+} ions upto 0.8 mol% and decreases with the further increase in the concentration of Er^{3+} ions. The spectra have been analyzed in the light of structural modifications in the vicinity of the dopant rare earth ions, due to relative variations in the concentration of Al_2O_3 and B_2O_3 in the glass network.

1. Introduction

In recent years, the upconversion of infrared light to visible light by rare earth ion doped glasses has been investigated extensively because of their potential applications in optoelectronics, such as three dimensional volumetric display, optical data storage, medical diagnostics, sensors, undersea optical communication. Among various rare earth ions Er^{3+} is identified as one of the most efficient rare earth ions for obtaining NIR laser emission, frequency upconversion in the glass hosts [1].

Among various glass systems, heavy metal oxide based glass systems (like lead bismuth oxide based systems) find potential applications in non-linear optical devices, because of their high refractive index and low phonon energy [2, 3]. In view of such qualities these glass systems offer suitable environment to rare earth ions for laser emission with high quantum efficiency. In the present study, we have investigated luminescence spectra and up-conversion phenomenon in Er^{3+}/Yb^{3+} co-doped PbO–$Bi_2O_3Al_2O_3$–B_2O_3 glass system. Al_2O_3 is added because it participates in the glass network with AlO_4 and AlO_6 structural units and alternate with BO_4/BO_3 units; strengthen the glass network and increase the mechanical strength of the glasses.

For the present study, glasses of composition $[100-(x+y)][0.5PbO-0.25B_2O_3-0.20Bi_2O_3-0.05Al_2O_3]-xEr_2O_3-yYb_2O_3$ with $y= 0$ for $x=0$; $y=0.2$ for $x=0.2$ to 1.0 (step 0.2) are chosen and the glasses are labelled as A0, A2, A4, A6, A8 and A10 respectively. The details of methods of preparation of glasses, techniques adopted for recording OA, PL and IR spectra were reported in our earlier paper [4].

2. Results and discussion

2.1 Optical absorption

The optical absorption spectra of the co-doped glasses (Fig. 1) exhibited a prominent overlapping band at 980 nm due to $^4I_{15/2} \rightarrow {}^4I_{11/2}$ and $^2F_{7/2} \rightarrow {}^2F_{5/2}$ transitions of Er^{3+} and Yb^{3+} ions respectively. The spectra also exhibited other absorption bands in the visible and the NIR regions due to at $^4I_{15/2} \rightarrow {}^4F_{7/2}$, $^2H_{11/2}$, $^4S_{3/2}$, $^4F_{9/2}$, $^4I_{9/2}$ and $^4I_{13/2}$ transitions of Er^{3+} ions at 525, 554, 660, 808 and 1540 nm respectively. From Fig. 1 it can be seen that the absorption edge shifts towards longer wavelength side with doping of the rare earth ions. From Fig. 1 it can be seen that, for almost all the glasses, there is an exponential rise of spectra towards the edge. The spectra further indicated that the cut-off wavelength shifts towards longer wavelength with increasing the concentration of Er^{3+} ions up to 0.8 mol%. From the absorption edges we have evaluated optical band gap (E_o) by drawing the Tauc plots between $(\alpha\hbar\omega)^{1/2}$ vs $\hbar\omega$; the value of E_o is found to decrease with increasing the concentration of Er_2O_3 up to 0.8 mol% and found to increase for further increase in the concentration of Er_2O_3.

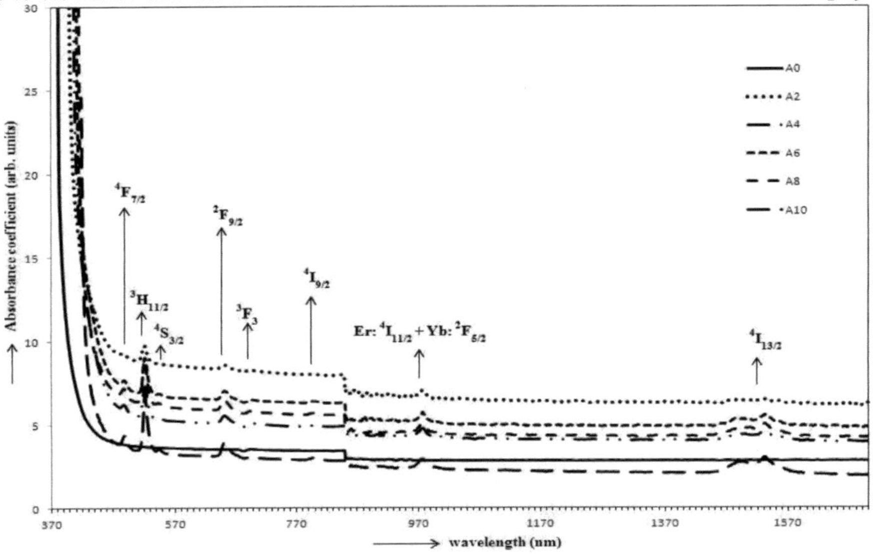

Fig. 1 Optical absorption spectra for $[100-(x+y)][0.5PbO-0.25B_2O_3-0.20Bi_2O_3-0.05Al_2O_3]-xEr_2O_3-yYb_2O_3$ glass system

2.2 FTIR spectra

To have some structural assessment of these glasses we have also recorded the IR spectra of these glasses. The IR spectra of the titled glasses recorded in 1600-400 cm^{-1} region exhibited four characteristic groups of bands (I-IV). The first group of bands in the region (400-650cm^{-1}) has been identified PbO_4, as well as borate rings deformation vibrations. This region may also consist of bands due to AlO_6 structural units. The second group of bands (II) observed in the region 650-850cm^{-1} is attributed to B–O–B bending and AlO_4 structural units [5]. The third group of bands (III) is associated with B-O stretching vibration of tetrahedral BO_4 group (850-1050cm^{-1}) [6-8]. The fourth group of bands (IV) centered at about 1300 cm^{-1} is due to the antisymmetric B-O stretching vibration of the trigonal BO_3 group. Out of these bands the AlO_4 participates in the network forming and alternate with BO_4 structural units, whereas, AlO_6 structural units act as modifiers and induce non-bridging oxygens. With increase in the concentration of Er_2O_3 (up to 0.8 mol%) all the asymmetrical bands are observed to grow at the expense of symmetrical bands. Thus the analysis of IR spectra reveal that the degree of disorder in the glass network increases with increase in the concentration of Er_2O_3 up to 0.8 mol%.

Table 1 Frequencies and their assignments for FT-IR spectra

Wavenumber (cm^{-1})	IR assignments
~490	B-O-B and Pb-O-B bending vibration as well as borate rings deformation. This region may also consist of bands due to AlO_6 structural units
~706, ~730	Due to the B–O–B bending and AlO_4 structural units
~935, ~970, ~990	B-O stretching vibration of tetrahedral BO_4 group
~1270, ~1280	Antisymmetric B-O stretching vibration of the trigonal BO_3 group.

2.3 Upconversion spectra

The room temperature up-conversion spectra of Er^{3+} / Yb^{3+} co-doped [100-(x+y)][0.5PbO-0.25B_2O_3-0.20Bi_2O_3-0.05Al_2O_3]-xEr_2O_3-yYb_2O_3 glasses under 980 nm excitation are shown in Fig. 2. The spectra exhibited three up- conversion bands centered at 525, 544 and 654 nm corresponding to $^2H_{11/2} \rightarrow {}^4I_{15/2}$, $^4S_{3/2} \rightarrow {}^4I_{15/2}$ (green) and $^4F_{9/2} \rightarrow {}^4I_{15/2}$ (red) transitions of Er^{3+} ions, respectively [9,10]. The close examination of the spectra indicated that the intensity of red emission was much stronger than that of the green emission.

Fig. 3 shows the up-conversion mechanisms of green and red emission. Up-conversion excitation mechanisms has been proposed to be involved in the population of the relevant excited state emitting levels of the Er^{3+} / Yb^{3+} co-doped glass as shown in Fig. 3. Firstly Er^{3+} ions excited from the ground state $^4I_{15/2}$ to the excited state $^4I_{11/2}$ through one of the following three processes under 980nm excitation : (i) Ground-State Absorption (GSA) $^4I_{15/2}$ (Er^{3+}) + a photon → $^4I_{11/2}$ (Er^{3+}), (ii) Phonon-Assisted Energy Transfer (PAET) from the Yb^{3+} $^2F_{5/2}$ state: $^2F_{5/2}$ (Yb^{3+}) + $^4I_{15/2}$ (Er^{3+}) → $^2F_{7/2}$ (Yb^{3+}) + $^4I_{11/2}$ (Er^{3+}), and (iii) Energy Transfer (ET) from the $^4I_{11/2}$ state of adjacent Er^{3+}. Among the above three processes, PAET from Yb^{3+} is the dominant one, due to the larger absorption cross section of Yb^{3+} at around 980 nm. Next the ions at the populated $^4I_{11/2}$ state are promoted to the $^4F_{7/2}$ state by the one of the following processes.

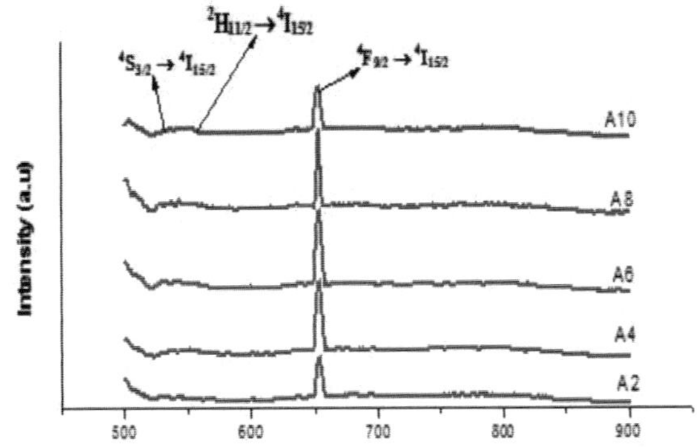

Fig. 2 Upconversion fluorescence spectra of PbO-B_2O_3- Bi_2O_3-Al_2O_3: Er_2O_3/Yb_2O_3 glass system recorded at room temperature

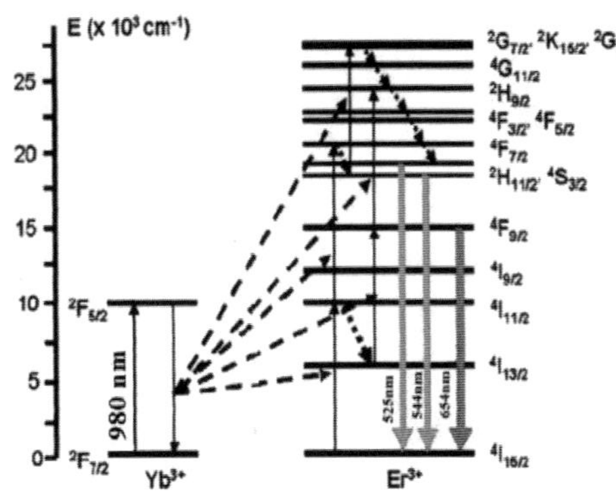

Fig. 3 Energy level scheme of a glass system co-doped with Er^{3+}/Yb^3 ions

(i) Excited state absorption (ESA): $^4I_{11/2}$ (Er^{3+}) + a photon → $^4F_{7/2}$ (Er^{3+}), (ii) Phonon-Assisted Energy Transfer (PAET) from the Yb^{3+}: $^2F_{5/2}$ (Yb^{3+}) + $^4I_{11/2}$ (Er^{3+}) → $^2F_{7/2}$ (Yb^{3+}) + $^4F_{7/2}$ (Er^{3+}), and (iii) Energy Transfer (ET) from the $^4I_{11/2}$ state of adjacent Er^{3+}: $^4I_{11/2}$ + $^4I_{11/2}$ → $^4F_{7/2}$ + $^4I_{15/2}$. Er^{3+} ions at the populated $^4F_{7/2}$ state can relax non-radiative relaxation (NR) very fast to the intermediate state of $^2H_{11/2}$ due to a multiphonon relaxation process (MPR), the $^2H_{11/2}$ → $^4I_{15/2}$ transition shows the 525 nm green emission. Er^{3+} ion at the $^4F_{7/2}$ state can also decay rapidly down to the $^4S_{3/2}$ state, and finally, the $^4S_{3/2}\rightarrow{}^4I_{15/2}$ transition reveals green upconversion at 545 nm. For the red-upconversion emission at 654 nm, Er^{3+} ion is first excited from the ground state $^4I_{15/2}$ to the excited $^4I_{11/2}$ through the above mentioned three processes under 980 nm excitation.

Upconversion luminescence intensities of red emission as a function Er_2O_3 concentration at fixed Yb_2O_3 (0.2 mol%) is shown in Fig. 4. From Fig. 4 it is clearly observed that the glass containing 0.8 mol% of Er^{3+} ions is the best upconversion emission intensity of present glass system. When Er^{3+} concentration is less than 0.8 mol%, the intensity of red emission is decreased. This stronger red emission is due to the Energy Transfer between the excited Er^{3+} ions, when the distance between the two excited Er^{3+} ions is short at higher concentration. When Er^{3+} concentration is beyond 0.8 mol%, the intensity decreases. This may be due to increase in back energy transfer from Er^{3+} to Yb^{3+}, i.e., $^4I_{11/2}(Er^{3+}) + {}^2F_{7/2}(Yb^{3+}) \rightarrow {}^4F_{15/2}(Er^{3+}) + {}^2F_{5/2}(Yb^{3+})$.

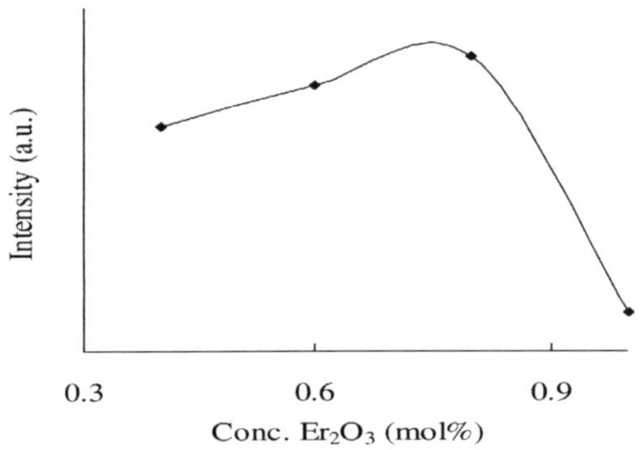

Fig. 4 Upconversion luminescence intensities (I_R) *vs* Er_2O_3 concentration with fixed Yb_2O_3 (0.2 mol%)

3. Conclusions

The studies on up-conversion phenomenon in Er^{3+}/Yb^{3+} co-doped lead bismuth alumina borate glasses have indicated that the up-conversion is more prominent in the glasses containg 0.8 mol% of Er_2O_3. When Er^{3+} concentration is beyond 0.8 mol%, the intensity decreases. This may be due to increase in back energy transfer from Er^{3+} to Yb^{3+}.

Acknowledgements
One of the authors Y. Raja Rao gratefully acknowledge the financial support extended by UGC, Govt. of India for carrying out this work under Faculty Development Program. The authors would like to thank the Director, Central Instrumentation Facility, Pondicherry University for PL measurements.

References

[1] Jiang S B, Luo T, Hwang B S, Smekatala F, Seneschal K, Lucas J and Peyghambarian N 2000 *J Non-cryst Solids* **202** 364-368
[2] Yang G F, Shi D M, Zhang Q Y and Jiang Z H 2007 *J. Fluoresc* **18** 131-137
[3] MacFarlane D, R Newman P J, Plathe R and Booth D 1999 *J. Proc SPIE* **3849** 94-102
[4] Chandra Shekhar Reddy M, Appa Rao B, Brik M G, Prabhakar Reddy A, Raghava Rao P, Jayasankar C K and Veeraiah N*Appl. Physics B* DOI: 10.1007/s00340-012-4983-z
[5] Reddy M S Nagaraju G. Nagarjuna G and Veeraiah N 2007 *J. Alloys Compd.* **438** 41-51.
[6] Shankar Ram and Kanik Ram J 1988 *Mater. Sci. Eng.* **23** 4541
[7] Alaily N A and Mohamed R M 2003 *Mater. Sci. Eng.* **B 98** 193
[8] Batal F H and Ashour A H 2002 *Mater. Chem. Phys.* **77** 677
[9] Davis E A and Mott N F 1970 *Phil. Mag.* **22** 903
[10] Ross S D 1972 *Spectrochem Acta* **28A** 1555

International Conference on Materials Science and Technology (ICMST 2012) IOP Publishing
IOP Conf. Series: Materials Science and Engineering **73** (2015) 012096 doi:10.1088/1757-899X/73/1/012096

Structural characterization of graphene layers in various Indian coals by X-Ray Diffraction technique

B. Manoj *, A.G. Kunjomana

Dept.of[a]Physics,Christ University, Bangalore-29, Karnataka, 5600029
manoj.b@christuniversity.in

Abstract. The results of the structural investigation of three Indian coals showed that, the structural parameters like *fa* & *Lc* increased where as interlayer spacing d_{002} decreased with increase in carbon content, aromaticity and coal rank. These structural parameters change just opposite with increase in volatile matter content. Considering the 'turbostratic' structure for coals, the minimum separation between aromatic lamellae was found to vary between 3.34 to 3.61 A^{o} for these coals. As the aromaticity increased, the interlayer spacing decreased an indication of more graphitization of the sample. Volatile matter and carbon content had a strong influence on the aromaticity, interlayer spacing and stacking height on the sample. The average number of carbon atoms per aromatic lamellae and number of layers in the lamellae was found to be 16-21 and 7-8 for all the samples.

KEYWORDS: Coal, Graphene layer, Aromaticity, Turbostratic structure.

1. Introduction

Allotropes of carbon have been a subject of great interest for researchers several years. Of these best known are graphite, diamond and amorphous carbon. Amorphous carbon is the name used for coal, soot and other impure forms of the element. Amorphous carbon has a wide range of properties that are primarily controlled by the different bond hybridizations possible in such materials. Graphite consists purely of sp^2 hybridized bonds, whereas diamond like carbon is a metastable form of amorphous carbon containing significant fraction of sp^3 bonds. This gives different properties like high strength, flexibility etc. Due to these properties they are used in thin film technology and in nanoscale electronic devices. The existence of crystallites in coal structure is confirmed by the appearance of the peaks corresponding to the (002), (100) and (110) reflections of graphite in coal. It is suggested that carbon in coal has an intermediate structure called turbostratic between graphite and amorphous structure. Coal also contains significant amount of highly disordered material, amorphous carbon, which is responsible for the back ground intensity of the diffractions [1-9].

The aim of the present study is to determine the stacking structure of selected coal samples with carbon content of 72 – 77.4%. The parameters include the interlayer spacing and crystalline size along the c-axis (L_c) and in the layer sheet direction (L_a), the average number of carbon atom (n) per aromatic lamellae and packing.

2. Experimental

The samples were de-mineralized with strong base like NaOH and KOH to avoid the effect of mineral matter on the quantitative analysis before X-ray diffraction studies [1-4]. From the finely powdered coal about 10g of the sample was dispersed in 50ml of NaOH and KOH separately and the mixture was stirred for 1hr at 27°C. Finally the treated coal was washed and dried in the air at 70°C for 3 hrs and allowed to cool slowly in a dessicator.

2.1. X-ray diffraction Analysis

The XRD data collection was performed by a Bruker AXS D8 Advance X-ray powder Diffractometer. The broad hump in this region was fitted with two Gaussian peaks around 20° and 26°, namely γ- band and Π-band (d_{002}) respectively. Theoretically, the areas under the γ and Π-peaks are believed to be equal to the number of aromatic carbon atoms (C_{ar}) and aliphatic carbon atoms (C_{al}) respectively [1-6].

3. Result and Discussion

The X-ray diffraction profile of the demineralized coal sample is shown in Figure 1. The samples exhibited high background intensity indicating that the coals contained a proportion of highly disordered materials in the form of amorphous carbon [1-13].

The diffraction profiles show the presence of a clear asymmetric (002) band around ~25.5°, which suggests the existence of another band (γ) on its left. The (γ) band around 20° was reported by many authors [1-9, 11-13]. It was attributed to the presence of saturated structures such aliphatic side chains, attached to the edge of the coal crystallites. The (002) band indicates the spacing of aromatic ring layer, while (γ) band reflects the packing distance of saturated structures. In addition, the coals also contained some graphite-like structures (crystalline carbon) indicated by the presence of a clear (002) band at ~ 25.5° and (10) weak band at ~ 42.3°. These observations suggest that, the crystallites in all the coal samples have intermediate structures between graphite and amorphous state called turbostratic structure or random layer lattice structure..

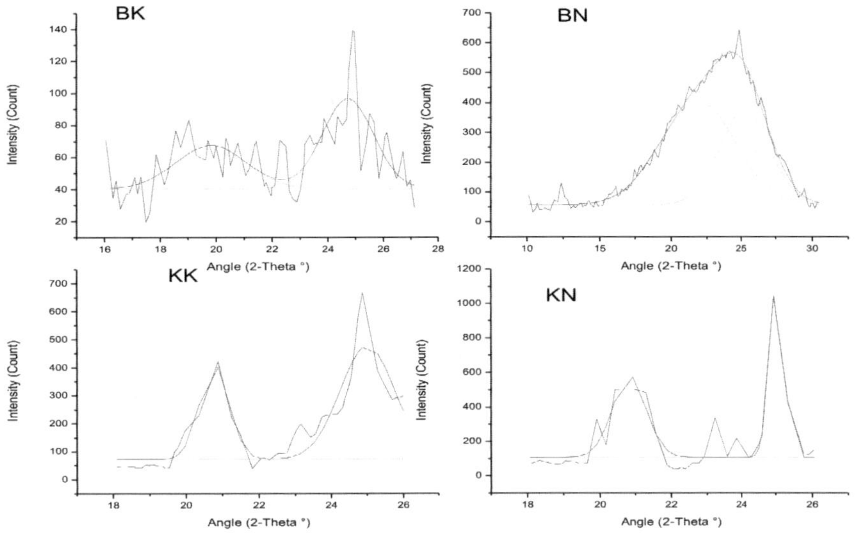

Fig.1. XRD pattern of demineralized coal sample with NaOH and KOH (BN and KN)

3.2. Determination of number of carbon atom (n) and packing aromatic lamellae

The average number of carbon atoms (n) per aromatic lamellae was calculated from the equation given elsewhere [3-4, 11-14]. It was found to be varying from 16 to 21 and also increasing with increase of coal rank. This result is in good agreement with the study carried out by Binoy et al. [2] in Assam coals. There was a high degree of correlation exist between these two parameters (R^2=0.99).

Tuinstra and Koenig [10] noted that the intensity ratio of the D and G modes, I_D/I_G, varies inversely with the in-plane correlation length or grain size of the graphite. This equation was based on the Raman spectrum graphite which is highly ordered carbon. In the present study, the intensity ratio I_{20}/I_{26} from X-ray diffraction, is plotted against *La* values and is shown in Fig.2. It is found that, the sample is agreeing with Tuistra-Koenig relationship for nano crystalline Carbon. The I_{20}/I_{26} ratio is proportional to the number of rings at the edge of the grain. As the *La* increases, the intensity ratio decreases quickly.

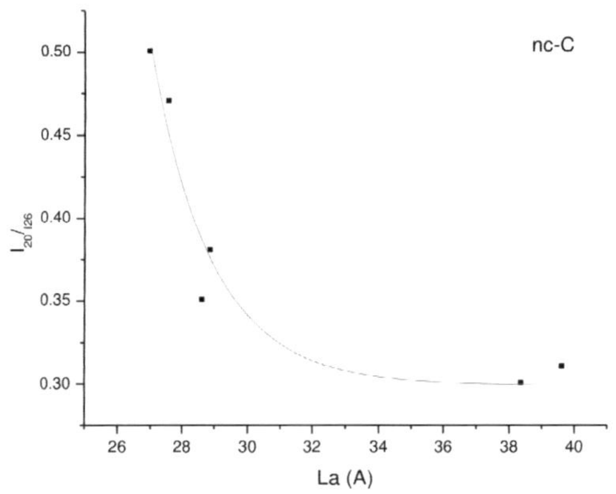

Fig.2. Relation between *Lc* and intensity ratio I_{20}/I_{26}

3.3. Packing of aromatic –lamellae

The average interlayer distance of the lamellae were computed from the position of '002' reflection, where as the average dimension of the packets of the lamellae in the direction perpendicular to their planes were estimated from its half width [12-13]. The number of layer as determined by this method using Lc was found to be 7 to 8 for coal samples. The aromaticity *(fa)* and rank (I_{26}/I_{20}) ratios for the samples ranged from 0.61 to 0.71 and 2.007 to 3.275, respectively. The average lateral sizes (*La*) and average stacking heights (*Lc*) of the layer structures in the coals samples, measured using the Scherrer equation ranged from 27.61 A° to 38.43A° and 23.73 A° to 20.16 A° respectively. Interlayer spacing (d_{002}) of the crystallite structure ranges from 3.60 A° to 3.34A^0. The percentage of ordered carbon and the degree of parallel stacking of the lamellae increased with rank of the sample. The volatile matter (V.M) changed from 36.2wt % to 24.9 wt% where as oxygen content varied from 21.37-16.64 wt% for the studied samples

Conclusions

The results of the structural investigation of three Indian coals showed that the structural parameters like *fa* & *Lc* increased with increase in carbon content where as interlayer spacing d_{002} decreased with increase in carbon content, aromaticity and coal rank. These structural parameters change just opposite with increase in volatile matter content. Considering the 'turbostratic' structure for coals, the minimum separation between aromatic lamellae was found to vary between 3.34 to 3.61 A° for these coals. As the aromaticity increased, the interlayer spacing decreased an indication of more graphitization of the sample. Volatile matter and carbon content had a strong influence on the aromaticity, interlayer spacing and stacking height on the sample. The average number of carbon atoms per aromatic lamellae and number of layers in the lamellae was found to be 16-21 and 7-8 for all the samples. The carbon in coal is having the structure of nanocrystalline carbon.

References

[1] Takagi H et al 2007, Fuel. **83** 2427.

[2] Binoy K.S et al 2009, J.Chem.Sci. **12** 103.

[3] Klug PH and Alexander, LE 1974, X-ray diffraction procedures (John Wiley & Sons, New York)

[4] Manoj B and Kunjomana 2012, Trends in applied science research . **7(6)** 434.

[5] Genetti D et al 1999, Energy Fuels, **13** 60.

[6] Lu L et al 2001, Carbon, **39** 1821.

[9] Xiajiang L et al 2006, Fuel, **85** 1700.

[10] Tuinstra F and J.L. Koenig 1970, J.Chem.Phys.**53** 1126.

[11] .Manoj B and Kunjomana A G 2012, Int J. of Min, Met. and Mats. **19(4)** 279.

[12] .ManojB and Kunjomana A G 2012. International journal of electrochemical sciences **7** 3127.

[13] ManojB and Kunjomana A G 2012 International journal of electrochemical sciences **7** 3215.

[14] Manoj B and AE Jose 2011, Raman Spectrum of graphite layers in Indian coal, AIP Publishing **1391(1)** 140.

Effect of Ga doping on structural, magnetic and transport properties of ordered Ba_2FeMoO_6 double perovskite

Y Markandeya[1], Y Suresh Reddy[2], A Manjula Devi[2], K Suresh[3] and G Bhikshamaiah[2]*,

[1]Department of Physics, Nizam College, Osmania University, Hyderabad-500001, India.

[2]Department of Physics, Osmania University, Hyderabad-500007, India.

[3]International Advanced Research Centre for Powder Metallurgy and New materials (ARCI), Hyderabad-500005, India.

*E-mail: gbhyd08@gmail.com

Abstract. Different compositions of double perovskite oxide powders of $Ba_2Fe_{1-x}Ga_xMoO_6$ (x=0.0, 0.1, 0.25 and 0.3) were prepared by sol-gel method. These materials were subjected to X-ray diffraction studies to confirm the crystal structure. Magnetization studies were performed using VSM and saturation magnetization found to decreases by addition of Ga. Electrical resistivity data were collected while varying magnetic field from 0 to 80 kOe keeping temperature constant at 150 K by four-probe method. Resistivity studies were also performed by varying temperature from 5 to 300 K keeping magnetic field constant at 0 and 80 kOe. Magneto resistance (MR%) was calculated using resistivity data. The MR (%) was found to increase with increase of Ga concentration in BFGMO compounds.

Keywords: Sol-gel, Crystal structure, Electrical resistivity, Magnetoresistance
PACS: 81.20.Fw, 61.05.cp, 81.40.Rs, 75.47.De

1. Introduction

A number of compounds are known with double perovskite structure. Among them, compounds of formula A_2FeMO_6 (A= Sr, Ba or Ca and M=Mo, Re or W) were attracted the attention of researchers since the low-field magnetoresistance (LFMR) at room temperature was observed in the polycrystalline samples of Sr_2FeMoO_6 by Kobayashi et al [1] in 1998. The magnetic structure of Ba_2FeMoO_6 is attributed to the ordered arrangement of Fe and Mo ions arranged alternatively on the B sites in the cubic perovskite ABO_3 [2]. A large antiferromagnetic interaction between the S=5/2 spins of Fe^{3+} and the S=1/2 spins of Mo^{5+} induces an ideal saturation magnetization (M_s) of $4\mu B$ per formula unit (f.u.). However, the observed M_s of Ba_2FeMoO_6 is less than $4\mu B$/f.u. The reduction of M_s is ascribed to the disorder between Fe and Mo sites[3]. Min Feng Lü et al [4] investigated structural, magnetic and magnetoresistance studies on Sr_2FeMoO_6 and found that enhancement of MR(%) by doping of Ga at Fe site. However, there are no studies on Ga doping at Fe site in Ba_2FeMoO_6 in literature. In view of this, the authors have taken up the study of structural, magnetic and magnetotransport properties of $Ba_2Fe_{1-x}Ga_xMoO_6$ (x=0.0, 0.1, 0.25 and 0.3) double perovskites with an objective to see the effect of Ga addition particular on MR(%) in Ba_2FeMoO_6.

2. Experimental

Double perovskite powder samples of $Ba_2Fe_{1-x}Ga_xMoO_6$ (x = 0.0, 0.1, 0.25 and 0.3) (BFGMO) were synthesized by sol-gel method using stoichiometric amounts of AR grade $Ba(NO_3)_2$, $Fe(NO_3)_3 9H_2O$, $Ga(NO_3)_3$ and H_2MoO_4 and pellets were made using powders. The details of method of preparation of the materials are given in our earlier publication [5]. These pellets were sintered at 1200 °C for 6 hours and then heated at 1000 °C in a stream mixture of 10% H_2 / Ar gas for about 3 hours for reducing the Mo^{6+} to Mo^{5+}. These pellets were subjected to X-ray diffraction studies using Philips PW 1830 generator diffractometer with Cu Kα radiation (40kV×25mA) and graphite monochromator in order to characterize crystal structure. The magnetization measurements of BFGMO samples were performed as a function of magnetic field from −20 to 20 kOe at room temperature using Digital Measurement System Model 880 USA make of Vibrational Sample Magnetometer. Electrical resistivity measurements were carried out at constant magnetic field of 0 and 80 kOe in the temperature range of 5 to 300 K employing the standard four-probe method using home made resistivity insert along with OXFORD superconducting magnet system. Isothermal magnetoresistance measurements were also carried out at 150 K while varying magnetic field from 0 to 80 kOe. For all the infield measurements magnetic field direction were kept parallel to current direction. The relevant MR (%) was evaluated using the resistivity data as a function of temperature as well as magnetic field.

3. Results and discussion

3.1. Crystal structure

Figure. 1 shows X-ray diffraction patterns for BFGMO samples at room temperature. Analysis of X-ray diffraction data indicates that these samples are in single phase and crystalizes in cubic double perovskite structure with space group $Fm\bar{3}m$ [6]. The lattice parameter 'a' of BFGMO samples have been evaluated using 2θ and (hkl) values of X-ray diffraction profiles shown in Figure. 1. The unit cell volume ($V=a^3$) of BFGMO has been calculated using lattice parameter. The values of lattice parameter 'a' and unit cell volume V of BFGMO are given in Table 1. It is found from Table 1 that the value of 'a' and V decreases with increase of Ga composition in BFGMO samples. This may be due to the substitution of the smaller Ga^{3+} (0.62Å) ions into the larger Fe^{3+} (0.645Å) ionic sites.

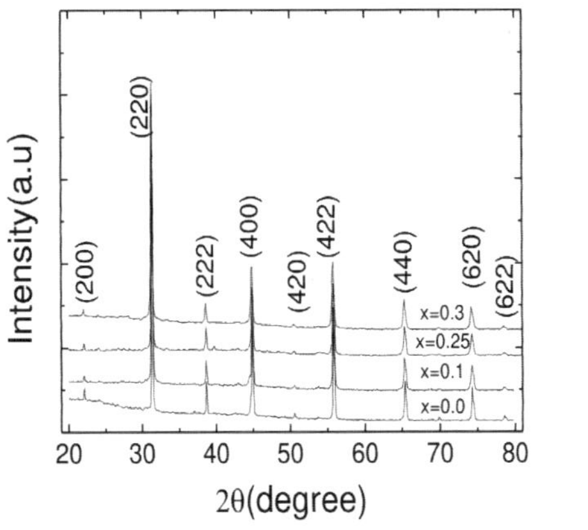

Figure 1. X-ray diffraction pattern of $Ba_2Fe_{1-x}Ga_xMoO_6$ (x = 0.0, 0.1, 0.25 and 0.3) samples.

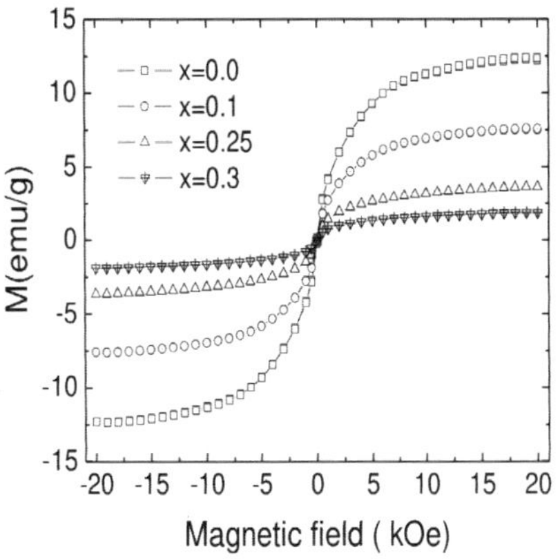

Figure 2. Magnetic field dependence magnetization cure of $Ba_2Fe_{1-x}Ga_xMoO_6$ (x = 0.0, 0.1, 0.25 and 0.3) samples.

IOP Conf. Series: Materials Science and Engineering **73** (2015) 012097 doi:10.1088/1757-899X/73/1/012097

3.2. Magnetic properties

Magnetic field dependence of magnetization of BFGMO samples taken at 300 K is shown in Figure. 2. The value of saturation magnetization (M_s) was obtained from Figure. 2 and are given in Table 1 for all the compositions. It is noticed from Table 1 that the M_s decreases with increase of Ga content in BFGMO samples. The value of M_s was found to reduce from 12.57 e.m.u /g for x = 0 to 1.96 e.m.u /g for x = 0.3. The decrease of M_s by the addition of Ga can be related to the model as developed by Sleight and Weicher [7] and detailed by García-Landa et al [8] where the ferromagnetic coupling between distant B sites in the ordered $A_2BB'O_6$ double perovskites is promoted by the antiparallel alignment of the itinerant electrons of the B' site cations. The cations disorder for the Ga-doped compounds is annihiling this mechanisim due to the presence of significant amounts of Fe and Ga cations on the B' site. i.e. the reduction in M_s with Ga content in BFGMO samples may be due to decrease in degree of Fe/Mo ordering or increase of anti-site defects in double perovskite structure.

3.3. Resistivity and Magnetoresistance studies

The variation of electrical resistivity (ρ) of BFGMO samples as a function of temperature in the range 5 to 300 K keeping magnetic field constant at 0 kOe is shown in Figure.3. It can be seen from Figure. 3 that there is no systematic variation of resistivity with Ga content in BFGMO samples. This may be attributed to disorder in the grain boundaries which varies randomly among the sample [9]. However, the values of resistivity of Ga doped samples ($Ba_2Fe_{1-x}Ga_xMoO_6$, x=0.1, 0.25 and 0.3) have larger compared to resistivity of Ba_2FeMoO_6. This may be expected that Ga^{3+} acts as a barrier to electron transport, reducing the number of pathways for electron percolation [10]. It is found from Figure. 3 that all the samples exhibit semiconductor behavior at 5 K and transform into metallic character as temperature is increased indicating a semiconductor–metallic transition. MR (%) is defined as $MR(\%) = [\{\rho(H,T)-\rho(0,T)\} / \rho(0,T)] \times 100$, where H denotes the applied field and ρ (0, T) and ρ (H, T) are the resistivity at zero field and H fields respectively at temperature T. The temperature dependence magnetoresistance (MR %) at a field of 80 kOe for all BFGMO samples are shown in Figure. 4. It can be seen from Figure. 4 that the variation of magnitude of MR (%) at 80 kOe are found to increase with decrease of temperature suggesting that the grain boundary effect dominates the MR as temperature decreases below Curie temperature of the samples [10].

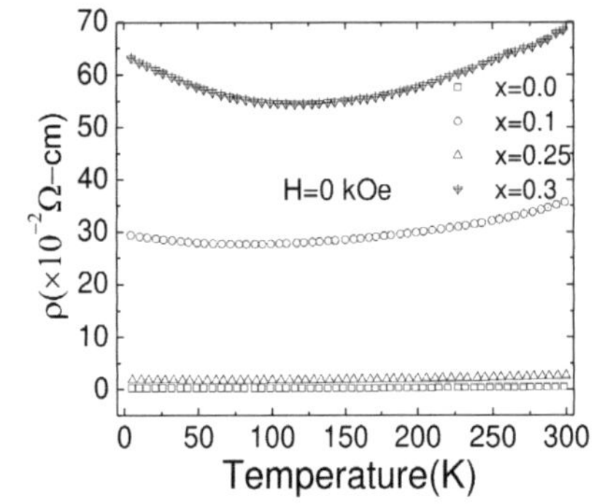

Figure 3. Resistivity versus temperature at magnetic field 0 kOe of $Ba_2Fe_{1-x}Ga_xMoO_6$ (x = 0.0, 0.1, 0.25 and 0.3) samples.

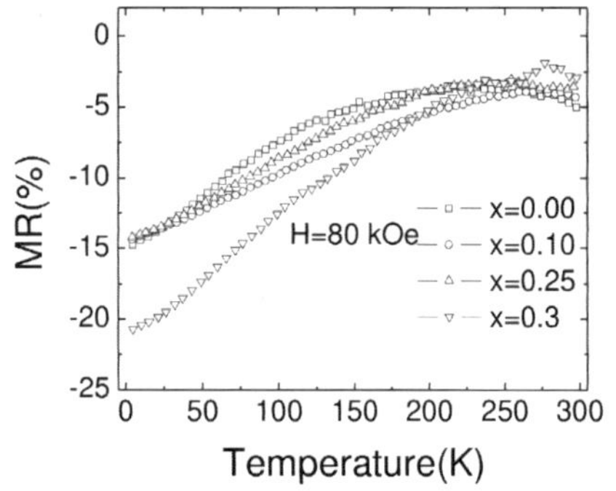

Figure 4. MR (%) versus temperature at magnetic field 80 kOe of $Ba_2Fe_{1-x}Ga_xMoO_6$ (x = 0.0, 0.1, 0.25 and 0.3) samples.

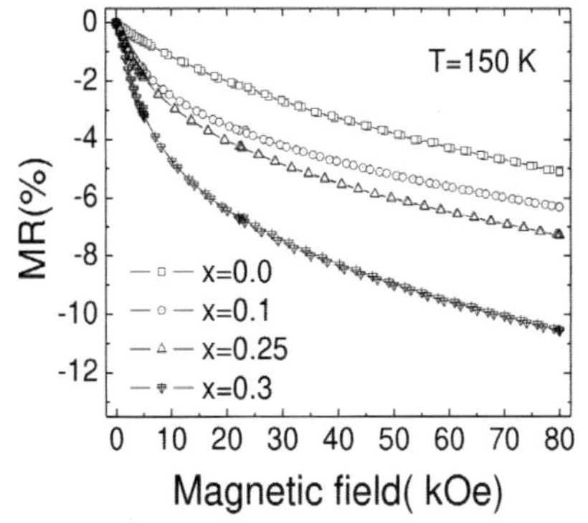

Figure 5. MR(%) versus magnetic field at temperature 150 K of $Ba_2Fe_{1-x}Ga_xMoO_6$ (x = 0.0, 0.1, 0.25 and 0.3) samples.

Table 1. The values of a, V and M_s of BFGMO samples.

Com(x)	0.0	0.1	0.25	0.3
a (Å)	8.073	8.056	8.037	8.011
V (Å)^3	526.14	522.83	519.18	514.12
M_s(emu/g)	12.55	7.72	3.78	1.96

The magnetic field dependence of MR (%) for BFGMO samples in the magnetic field range 0 to 80 kOe at constant temperature 150 K are shown in Figure. 5. It can be seen from Figure. 5 that the MR (%) increases with increase of magnetic field at temperature 150 K. The increase in MR (%) is due to the lowering of spin scattering at grain boundaries and the quenching of spin scattering of the charge carriers by localized spins upon the application of magnetic field [10]. It can be seen from Figure. 5 that the MR (%) increases by addition of Ga content in BFGMO samples at 150 K. The appreciable increase in MR (%) for Ga-doping system is thought to weaken the double exchange barrier after Ga doping in Ba_2FeMoO_6 system [4]. In such a case, Ga ions may act as a barrier for electron transport along the chain in the ferromagnetic segregation and weaken the ferromagnetic exchange.

4. Conclusions

It has been found that the value of lattice parameter, unit volume and saturation magnetization decreases while MR (%) increases by addition of Ga content in BFGMO samples.

Acknowledgments

The authors thank Dr. Rajeev Rawat of UGC-DAE, Consortium for Scientific Research, Indore for providing magnetoresistance facilities. One of the authors (GB) wish to thanks DST for providing financial assistance to carry out this work through a project OU–DST–Scheme No.A-60.

References

[1] Kobayashi K I, Kimura T, Sawada H, Terakura K and Tokura Y 1998 Nature **395** 677
[2] Patterson F K, Moeller C W and Ward R 1963 Inorg. Chem. **2** 196
[3] García-Hernández et al 2001 Phys. Rev. Lett. **86** 2443
[4] Min Feng Lu et al 2007 J. Alloys and Compounds **428** 214
[5] Markandeya Y, Saritha D, Vithal M, Singh A K and Bhikshamaiah G 2011 J. Alloys and Compounds **509** 5195
[6] Sung Baek Kim et al 2003 J. Magn. Mater. **245** 580
[7] Sleight A W and Weicher J F 1972 J. Phys. Chem. Solids **33** 679
[8] García-Landa B, Ritter C, Ibarra M R, Blasco J, Algarabel P A, Mahendiran R and García J 1999 Solid State Commun. **110** 435.
[9] Hemery E K, Williams G V M and Trodal H J 2007 Physica B **394** 74
[10] Falak Sher, Venimadhav A, Blamire M G, Dabrowski B, Kolesnik S and Paul Attfield J 2005 Solid State Sciences **7** 91

Synthesis, characterization and Transport properties of $Co_xZn_{1-x}Fe_2O_4$ Nanoparticles

Jaison Joseph[1], R B Tangsali[2], S M Gurav[3]

[1]Department of Physics/Chemistry[3], Govt. College, Khandola, Goa, India.403107

[2]Department of Physics, Goa University, Taleigao Plateau, Goa, India.403206

Email:jaisonjosephp@gmail.com

Abstract. Spinel ferrites are commercially important materials due to their excellent electrical and magnetic properties. Mixed metal spinel ferrites, essentially with low hysteresis loss and small particle sizes, have found new applications in areas like biosensors, drug delivery systems, and hyperthermia. The electrical conductivity, which gives credible information about conduction mechanism, is one of the important properties of ferrites. The order of magnitude of conductivity greatly influences the dielectric and magnetic behavior of the ferrites and depends on the preparative method. The present work deals with investigations into electrical, dielectric and thermo power parameters of $\mathbf{Co_xZn_{1-x}Fe_2O_4}$ ferrite nano particles synthesized by co-precipitation method. The effect of variation of the cobalt concentration in the sample on transport properties of nano ferrites has been analyzed.

1. Introduction

Nano spinel ferrites are becoming important candidates as far as newer high end technological applications are concerned in fields such as microwave devices, high speed digital storage systems, ferro-fluids, catalysis, and magnetic refrigeration [1-3]. Mixed spinel ferrites, essentially with low hysteresis loss and small particle sizes, have found new applications in areas like drug delivery systems and hyperthermia [4]. Apart from magnetic applications of Co ferrites, they are used in thermistors [5] where resistivity plays an important role. The electrical conductivity, which gives credible information about conduction mechanism, is one of the important properties of ferrites. The order of magnitude of conductivity greatly influences the dielectric and magnetic behavior of the ferrites and depends on the type of preparation method [1,5]. This has provided the motivation to synthesize, characterize and carry out electrical, dielectric as well as thermo power measurements on $\mathbf{Co_xZn_{1-x}Fe_2O_4}$ ferrite nano particles. The nanoparticles under investigation were synthesized by co-precipitation method, which were then analyzed to study various transport properties.

2. Experimental

Nanoparticle $\mathbf{Co_xZn_{1-x}Fe_2O_4}$ ferrite material with (x=0.2, 0.4, 0.6) were prepared in our laboratory by co-precipitating aqueous solutions of $ZnSO_4$, $CoCl_2$ and $FeCl_3$ mixtures in alkaline medium. $ZnSO_4{:}7H_2O$, $CoCl_2{:}6H_2O$ and $FeCl_3{:}6H_2O$ were taken in their respective stoichiometry dissolved in distilled water and heated up to 65^0C.The respective metal ions were precipitated by adding a suitable precipitating reagent in boiling solution state within 10 seconds under constant stirring. After maintaining the solution along with the precipitate at 80^0C for one hour, the precipitate was filtered and dried at 100^0C. The dried powder was pelletized and sintered at 1000^0C for 15 hours, resulted in formation of $\mathbf{Co_xZn_{1-x}Fe_2O_4}$ nanoparticles

The three powdered samples were characterized using high intensity rotating anode Rigaku X-ray diffractometer (XRD). FT-IR absorption spectra of the samples were recorded on a Shimadzu 8900 FT-IR spectrometer. The resistivity measurements were carried out by two probe method using Keithley's electrometer (model 6514), by maintaining a constant supply voltage across the sample. The frequency dependent dielectric measurements were carried out on Wane Kerr 6440B LCR-Q meter at room temperature. Thermo electric power (TEP) measurements were carried out on an indigenously built, lab based set up with variations in ΔT ranging from 0^0C to 150^0C.

3. Results and Discussion

The X-ray diffraction peaks shown in figure 1, indicative of single phase spinel crystal structure were identified and indexed in all three samples. The lattice parameter is found to vary between 8.32A° and 8.34 A° with increasing Co content in the sample. This variation can be ascribed to the difference in ionic sizes of the component ions. The Co2+ ions have larger ionic radius (0.82 A°) than Zn 2+ (0.74 A°), and Fe3+ (0.67 A°) ions. The average particle size calculated from XRD line width of (311) peak using Debye Scherrer's formula, $t = 0.9\lambda/\beta\cos\theta$, where t = particle size, λ = wavelength of X-ray, β=full width at half maxima(FWHM) and θ is Bragg angle works out to 44nm,which indicate formation of nano particle material.

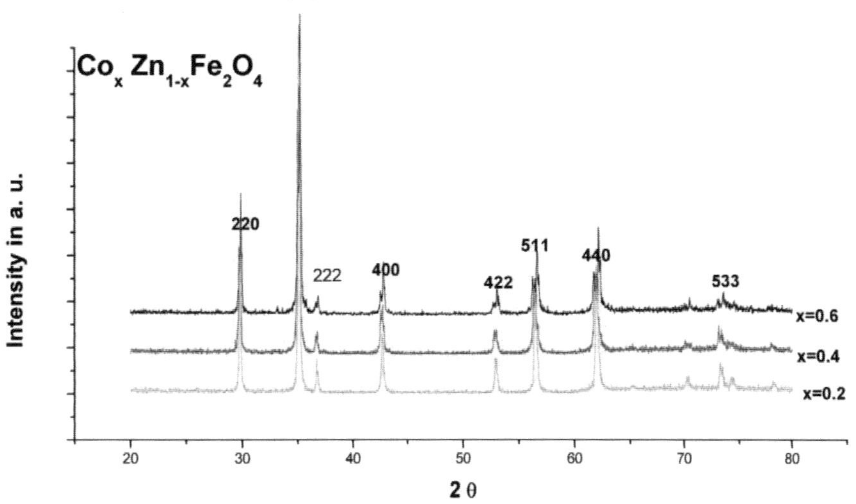

Figure 1. XRD pattern of $Co_xZn_{1-x}Fe_2O_4$

FT-IR absorption spectrum shown in figure 2,with two broad absorption bands (610-550cm^{-1}) which corresponds to intrinsic stretching vibrations of the metal at the tetrahedral site and (450-375 cm^{-1}) due to octahedral metal stretching, confirmed spinel phase formation [6]. The absorptions noticed at higher wave numbers are due to the presence of water molecules and traces of other contaminating molecules present in air.

The temperature-dependent dc resistivity measured from room temperature till 550^0C is as shown in Figure 3. It can be seen that for all the samples variation of resistivity with increasing temperature follow the same trend. The change in slope indicates change in conduction mechanism. The prominent change in slope occurs at 426 K for sample x=0.6, at 393K for sample x=0.4 and at 473K for sample x=0.2. Mild slope change indicative of shift in conduction behavior is noticed at 743K, 683K and 643K for sample x=0.6. The trend is repeated for samples x=0.4 and x=0.2 for temperatures 773K,743K,683K and 803K,763K,663K respectively.

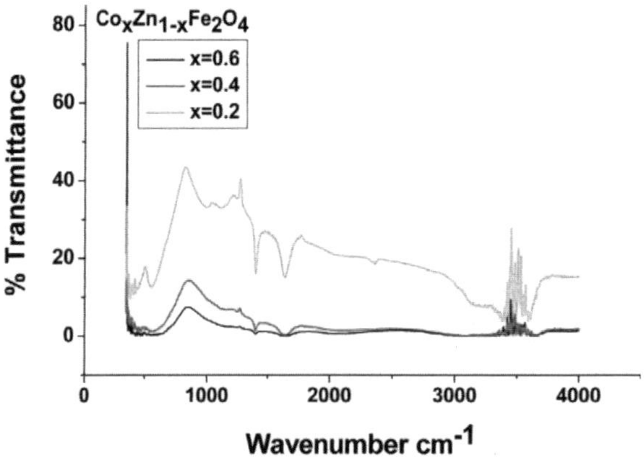

Figure 2. FT- IR spectrum of $Co_xZn_{1-x}Fe_2O_4$

The first low-temperature region (up to about 423 K) is attributing to the conduction due to impurities, voids, defects, etc. The second temperature region (up to Curie temperature) is attributed to ferrimagnetic region. Third temperature region (above Curie temperature) is attributed to paramagnetic region [7–9]. With respect to increase in Co concentration a systematic decrease in temperature at which the change in conduction mechanism occurs is also noticed at higher temperatures.

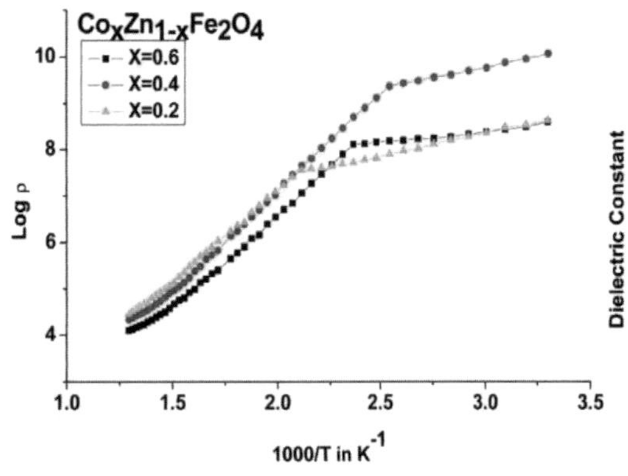

Figure 3. DC resistivity of $Co_xZn_{1-x}Fe_2O_4$

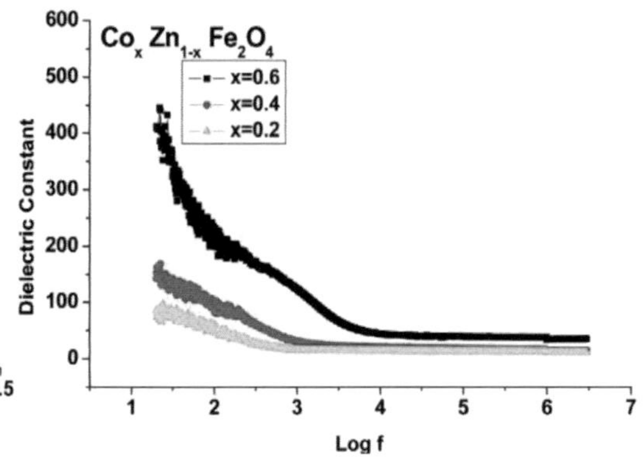

Figure 4. Dielectric constant of $Co_xZn_{1-x}Fe_2O_4$

Figure 4 shows variation of dielectric constant (Є) with frequency for the three samples. Initially Є decreases nearly exponential with increasing frequency and reaches constant value for all the three compositions. According to the model based on dispersion due to Maxwell–Wagner [10,11] type interfacial polarization, which is in agreement with Koop's phenomenological theory [12], a heterogeneous dielectric material is considered containing well conducting grains, separated by highly resistive thin grain boundaries. In such situation, the applied voltage on the sample drops mainly across the grain boundaries resulting in creation of a space charge polarization. According to Koop the effect of grain boundaries is prominent at lower frequencies and the grain boundary thickness is inversely proportional to the dielectric constant value. The observed decrease of Є with increase in frequency can be attributed to the fact that the electron exchange between Fe^{+2} and Fe^{+3} ions cannot follow the change of the external applied filed beyond a certain frequency [13,14].It may be seen from Figure 4 that, an enhancement of dielectric constant occurs with the corresponding increase of Co ions concentration in the matrix.

Thermo electric power (TEP) measurements carried out on the samples are as shown in Figure 5.

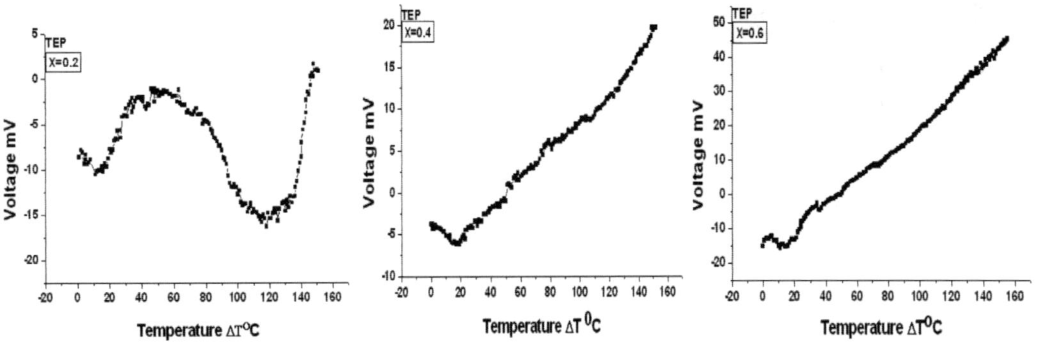

Figure 5. Thermo Electric Power measurements of $Co_xZn_{1-x}Fe_2O_4$

TEP analysis confirm semiconducting nature of the material, with the presence of n-type charge carriers for x=0.4 and 0.6 at lower values of ΔT ^0C, which shifts to p-type at larger values of ΔT ^0C. However for sample with x=0.2, n-type charge concentration is predominant with large amplitude fluctuations of the charge concentration as ΔT ^0C approaches 150^0C.

4. Conclusion

The $Co_xZn_{1-x}Fe_2O_4$ nanoparticles with (x=0.2, 0.4, 0.6) were synthesized by Co-precipitation method and characterized using (XRD) & FT-IR spectroscopy. The decrease in resistivity with increase in temperature indicates semi conductor-like behaviour. The changes observed in dc resistivity at various measuring temperatures provide adequate testimony to the idea of the changing conduction behaviour or mechanism in the samples with rising ambient temperature. An increase in dielectric constant with increasing Co concentration in the sample suggests that the addition of cobalt is responsible for introduction of more dipoles in the sample.TEP analysis confirm semiconducting type behaviour of the samples, with generation of n-type charge carriers in the sample with x=0.2 and dual type, n-type at low temperature difference ΔT ^0C and p-type subsequently for higher values of ΔT ^0C in samples with x=0.4 and x=0.6.

Acknowledgment

This work is supported by University Grants Commission, Government of India, under research project UGC/47-820/04 (WRO)

References

[1] Urcia Romero S and Perales Perez O 2011 *Journal of Applied Physics* **109 07B** 512
[2] Yan S, Yin J, and Zhou E 2008 *J.Alloys Compd.* **450** 417–20
[3] Kurlyandskaya G V, Cunanan J and Bhagat S M 2007 *J.Phys.Chem.Solids* **68** 1527–32
[4] Rudolf Hergt 1998 IEEE *Transactions on Magnetics* **vol.34** 3745-54
[5] Shaikha P A and Kambalea R C 2010 *Journal of Alloys and Compounds* **492** 590–96
[6] Evans B J and Hafner S S 1968 *J. Phys. Chem. Solids* **29** 1573
[7] Liu Y L, Liu Z M,Yang Y and Yang H F 2005 *Sens Actuators B* **107** 600
[8] Solyman S 2006 *Ceram. Int.* **32** 755
[9] Patil M G, Mahajan V C, Ghatage A K and Patil S A 1996 *Ind. J. Pure Appl Phys* **34,** 1665
[10] Devan R S,Kolekar Y D and Chougule B K 2006 *J. Phys. Condens. Matter* **18** 9809–21
[11] Wagner KW 1973 *Am. Phys.* **40** 817
[12] Koop CG 1951 *Phys. Rev.* **85** 121
[13] Sheikh A D and Mathe V L 2008 *J. Mater. Sci.* **43** 2018–25
[14] Jadhav S S and Patange S M 2010 *J. Biomedical and Bioengineering* **Vol.1** Issue **1** 21-29

IOP Conf. Series: Materials Science and Engineering **73** (2015) 012099 doi:10.1088/1757-899X/73/1/012099

La_2O_3 Nano powders by mixture of fuels approach through chemical combustion for dielectric studies

G Nithesh Sharma[1*], K Venkateswara Rao[1], V Sesha Sai Kumar[1], Ch Shilpa Chakra[1], V Rajendar[1], P Ranjith reddy[2]

[1]Center for Nano Science and Technology, IST, JNTUH, India-500085.

[2]Department of Physics, College of Engineering JNTUH, India-500085.

Presenting and Corresponding author: Mobile: +91 9290558256

Email: nithesh.sharma530@gmail.com

Abstract: La_2O_3 which is considered as a promising substitute for present SiO_2 Gate oxide material in CMOS devices has been synthesized by mixture of fuel method . The method involves the mixing of the Lanthanum Nitrate with, mixture 1:1 of Urea and Glycine. The fuel to oxidizer ratio (ψ) has been maintained as 1.25. The powders were studied by several physical characterization techniques such as X-ray diffraction (XRD), Thermo gravimetric analysis (TGA) and Differential Thermal analysis (DTA) and dielectric properties have been studied using LCR meter. Surface morphology has been studied by SEM. Particle size has been analysed using Particle analyzer.

1. Introduction: MOSFETs have become ubiquitous with their wide range of applications spanning over almost all fields starting from automobiles to communications. Miniaturization has played key role in this. Further scaling of transistors requires reducing Gate oxide layer thickness making SiO_2 Gate dielectric layer becoming so thin (1.4nm)[1]. This arises new issues such as Oxide breakdown, leakage current and light atom penetration through the film. Introducing physically thicker high-k dielectric materials while maintaining optimum value of capacitance needed for controlling the current flow in the channel, can resolve the afore mentioned problems[2].La based dielectrics are considered as a promising substitute for present SiO_2 owing to their high dielectric constant(k~30) and better crystallization temperatures than other rare earth oxides[3-5]. La, has been prepared so far by several physical and chemical approaches such as sol-gel method [6], reverse micelle synthesis techniques [7], sputtering, SHS, Thermal decomposition [8] etc. In the present work, La, has been synthesized by mixture of fuel method through chemical combustion.

2. Experimental Procedure:

Lanthanum Oxide nano powders have been synthesized using Chemical Combustion by employing mixture of fuels approach. The initial precursor materials taken were Lanthanum nitrate (LaN_3O_9) by Merck Ltd., Urea (CH_4N_2O) by Finar and Glycine ($C_2H_5NO_2$) by Merck . All reagents used were of Analytical grade and mixed in double distilled water. Here LaN_3O_9 acts as Oxidizer and Urea, Glycine as Fuels. The Fuel to Oxidizer ratio has been maintained as 1 (Ψ = 1.25). The fuels Urea and Glycine were taken in equimolar ratio. Stoichiometric amounts of reagents were taken as proposed by Jain et.al.[9]. Ψ value has been chosen such that the there exists a trade-off between combustion temperature and particle size since ψ value has shown to have effect on the microstructure [10]. The balanced equation obtained for the above combustion is as follows.

$$6LaN_3O_9 + 7.5CH_4N_2O + 7.5C_2H_5NO_2 \rightarrow 3La_2O_3 + 33.75H_2O + 22.5CO_2 + 40.5N_2$$

The calculated amounts of reagents were weighed precisely using electronic weighing balance, shimadzu, Japan. Lanthanum nitrate precursor was taken in a Borosilicate flask and double distilled water was added to it. Calculated amounts of Urea and Glycine were added to the beaker containing solution of LaN_3O_9 and stirred for 15 min using magnetic stirrer till clear solution is observed.

The clear solution comprising LaN_3O_9 , CH_4N_2O and $C_2H_5N_2O$ was kept on hot plate . Initially the solution started boiling , followed by frothing , fumes and instead of flames smoldering occurred indicating that combustion has been completed leaving behind white colored deposit. The sample was Calcinated to 600°c for 1 hour in a muffle furnace for removing any carbonaceous impurity present in the as prepared sample. The samples Before calcination and After calcination were characterized using Particle Analyzer for average particle size The crystal structure and crystallite size of the synthesized powder was determined by X-ray diffractometer (XRD), the X-ray diffraction analysis revealed that the synthesized La_2O_3 Nano powder has the perovskite structure having average crystallite size 30nm. Morphological studies were carried out by scanning electron microscopy (SEM), Energy Dispersive X-ray analysis (EDAX) was investigated by Scanning electron microscopy (SEM) coupled with an energy dispersive X-ray spectroscopy(EDAX) . The phase purity of the Nano powder has been confirmed using differential thermal analysis (DTA), thermo gravimetric analysis (TGA) and the particle size was estimated from Nanoparticle size analyser.

3. RESULTS AND DISCUSSION

3.1 X-Ray diffraction (XRD):

X-Ray powder diffractograms (XRD) were recorded by means of a Bruker's MODEL D8 Advance with Ni filtered Cu Kα radiation(λ=1.54056A°). The device operating controls were maintained at 40KV, 30 mA and scan rate of 1° per minute. The powder sample was packed into the depression in a sample holder mounted in a horizontal position. The intensity(I/I°) versus d- spacing graphs representing diffraction pattern have shown resemblance with reported values.

.

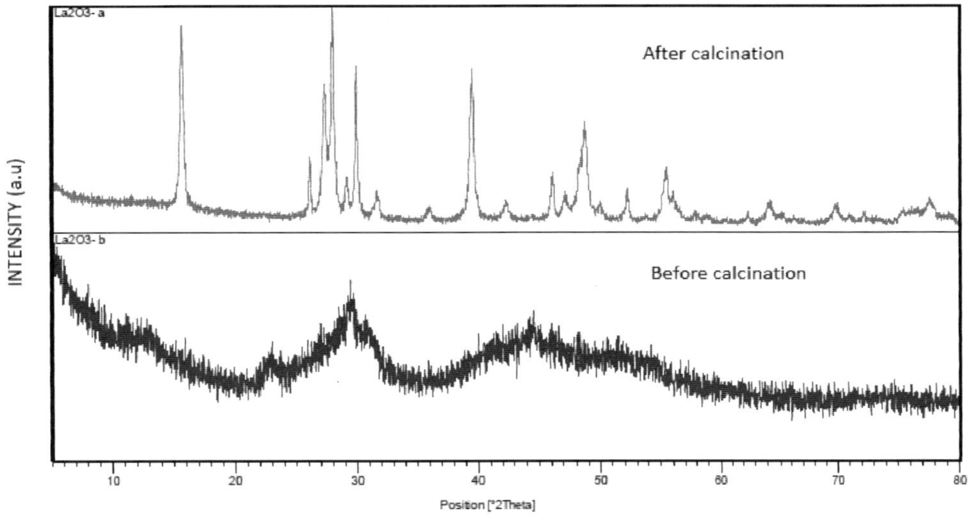

Figure 1. XRD diffraction patterns of La_2O_3 powders Obtained Before and After calcination.

The crystallite size was calculated using the scherrer's formula [11],

$$d = \frac{0.9\lambda}{\beta COS\theta} \qquad (1)$$

Where d is the crystallize size , λ is the wavelength of incident X-ray, β is full width at half maximum of diffracted peak and θ is diffracted angle.

S.No.	Intensity	d-spacing [Å]	2Theta [°]	h k l
1	34	3.41000	26.111	1 0 0
2	31	3.06300	29.131	0 0 2
3	100	2.98000	29.961	1 0 1
4	58	2.27800	39.528	1 0 2
5	63	1.96800	46.085	1 1 0
6	52	1.75300	52.133	1 0 3
7	4	1.70500	53.717	2 0 0
8	24	1.65600	55.441	1 1 2
9	17	1.64200	55.955	2 0 1
10	3	1.53200	60.372	0 0 4
11	5	1.49000	62.260	2 0 2
12	2	1.39800	66.871	1 0 4
13	7	1.30900	72.096	2 0 3
14	2	1.28900	73.396	2 1 0
15	12	1.26100	75.304	2 1 1
16	6	1.20900	79.157	1 1 4

Table 1: peak parameters of the obtained diffractogram matched with JCPDS data card number 05-0602

3.2 Scanning Electron Microscopy (SEM) :

(a)

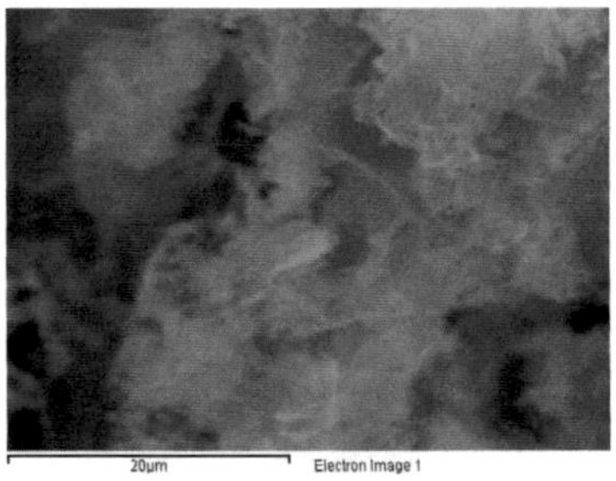

(b)

Figure 2. SEM images of La_2O_3 nanopowder a) before calcinations 2 b) after calcinations

The surface morphology of La_2O_3 nanopowder was investigated by SEM and images are shown in Figure 2(a) and Figure 2(b) with different magnifications. The SEM shows that, the clusters of nanoparticles like porous structures. It has been observed that powder of the combustion reaction has a morphology forming a porous and fuzzy network of Nano crystalline La_2O_3 , which may be due to the rapid release of gaseous byproducts during the combustion reaction. the La_2O_3 agglomerated particles which consists of tiny nanoparticles having average size 40nm, which is in agreement with XRD results.

IOP Conf. Series: Materials Science and Engineering **73** (2015) 012099 doi:10.1088/1757-899X/73/1/012099

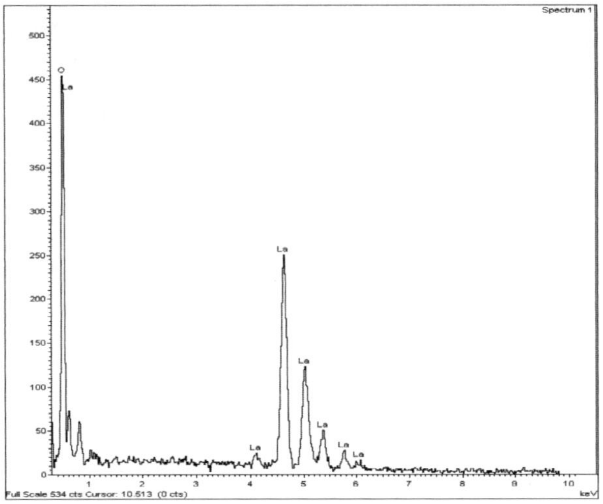

Figure 3. EDAX spectrum of La_2O_3 Nano powder

Element	Weight%	Atomic%
O	36.05	83.03
La	63.95	16.97
Total	100%	100%

Energy Dispersive X-ray analysis (EDAX) was investigated by Scanning electron microscopy (SEM) coupled with an energy dispersive X-ray spectroscopy (EDAX) unit. Figure 3. Shows the chemical compositional analysis of La_2O_3 nanopowder is nearly in stoichimetriy .

3.3 Particle Analyser studies

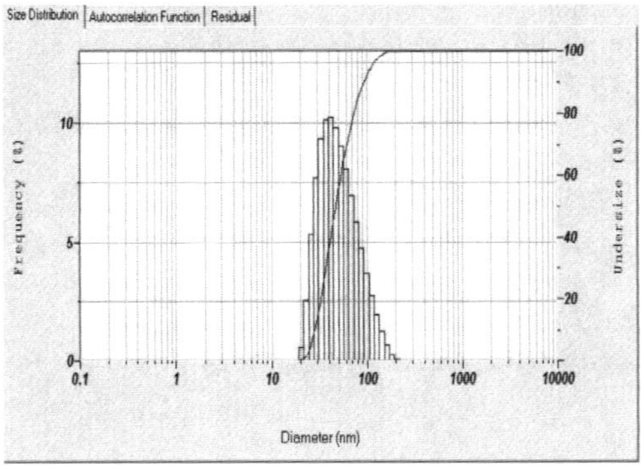

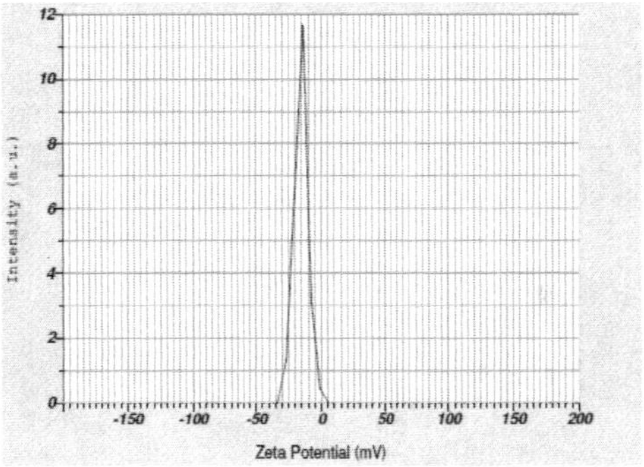

Figure 4(a) **Figure 4(b)**

Figure 4. (a)Particle size analysis of La_2O_3 solid solution (b) Zeta potential of La_2O_3 solid solution

The particle size of nano powders is done by particle size analyzer (Nano particle size analyzer SZ100). Figure 4 shows the distribution of particle size of La_2O_3 nano powders. The size is estimated as 37.5nm from the graph. and Zeta potential as (-15.7) mV.

3.4 LCR Meter Analysis

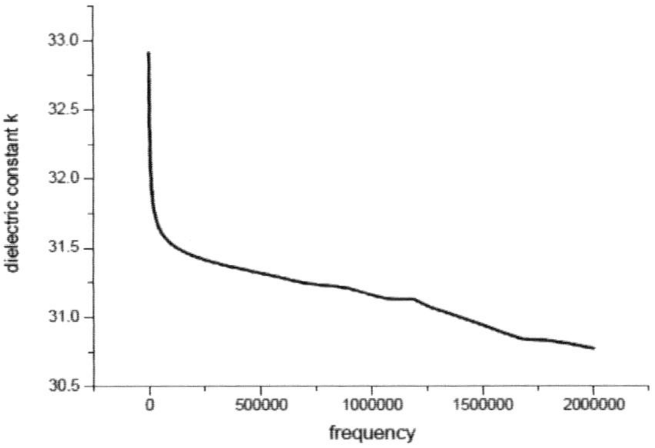

Figure 5. frequency Vs K curve for La2O3 nanopowder

LCR analysis shows that the dielectric constant of La_2O_3 nano powder sharply decreases at low frequencies and stabilizes at higher frequencies. The average per gram Dielectric constant is measured to be K= 31.

4. Conclusions

La_2O_3 nanopowder have been successfully synthesized via mixture of fuel approach in solution chemical combustion. It is observed that from the XRD analysis, the obtained crystallite size varies from 30 nm to 50 nm . Structural properties were examined by SEM,. The average particle size is deduced as 47.2nm from Particle size analyser and Zeta Potential as -15mV indicating that the obtainded sample is chemically less reactive and stable .The LCR meter results show high-k dielectric nature of material such that it can be employed as gate dielectric material.

References:

[1 International Technology Roadmap for Semiconductors, **2011** update.

[2] Robertson J , Eur. Phys. J. Appl. Phys. **28**, 265–291 (2004), DOI: 10.1051/epjap:2004206

[3] Triyoso, D H , Li, R.I. Hegde, Z. Yu, K. Moore, J. Grant, B.E. White Jr., P.J. Tobin, J. Vac. Sci. Tech. B **23** (2005) 2480

[4] A. Chin, Y.H. Wu, S.B. Chen, C.C. Liao, W.J. Chen, in: 2000 Symposium on VLSI Technology, IEEE, Piscataway, 2000, p. 16.

[5] S. Guha, E. Cartier, M.A. Gribelyuk, N.A. Bojarczuk, M.C. Copel, Appl. Phys. Lett. **77** (2000) 2710.

[6] Xingang Wang, Maolin Wang, Hua Song, Bingjun Ding, A simple sol–gel technique for preparing lanthanum oxide nanopowders, Materials Letters, **Volume 60**, Issues 17–18, August 2006, Pages 2261-2265, ISSN 0167-577X, 10.1016/j.matlet.2005.12.142.

[7] Ying et al. SYNTHESIS OF NANOMETER SIZED PARTICLES BY REVERSE MICELLE MEDIATED TECHNIQUES, US PATENT NO.: US 6,413,489B1

[8] Gamal.AM. Hssain, HAMDI M ISMAIL, POWDER powder technology **84**(1995)185-190`.

[9] S.R.JAIN, K.C. ADIGA AND V.R. PAI VERNICAR, Combustion and flame **40**: 71-79(1981)

[10] K.Venkateswara Rao, C.S.Sunandana,Journal of Nano Science and Technology **vol-8,** 4247-4253, 2008

[11] A. WESLEY and B. D. CULLITY, Elements of X-ray Diffraction, (1978).

IOP Conf. Series: Materials Science and Engineering **73** (2015) 012100 doi:10.1088/1757-899X/73/1/012100

Effect of Γ-X band mixing on the donor binding energy in a Quantum Wire

R Vijaya Shanthi, K Jayakumar and P Nithiananthi[1]

Department of Physics, Gandhigram Rural University, Gandhigram – 624 302, Tamilnadu, India.

E-mail:- nithiadhi@yahoo.com

Abstract. To invoke the technological applications of heterostructure semiconductors like Quantum Well (QW), Quantum Well Wire (QWW) and Quantum Dot (QD), it is important to understand the property of impurity energy which is responsible for the peculiar electronic & optical behavior of the Low Dimensional Semiconductor Systems (LDSS). Application of hydrostatic pressure P>35kbar drastically alters the band offsets leading to the crossover of Γ band of the well & X band of the barrier resulting in an indirect transition of the carrier and this effect has been studied experimentally and theoretically in a QW structure. In this paper, we have investigated the effect of Γ-X band mixing due to the application of hydrostatic pressure in a GaAs/$Al_xGa_{1-x}As$ QWW system. The results are presented and discussed for various widths of the wire.

1. Introduction

Confinement of hydrogenic impurity states in nanostructured systems and changes in the energy bands due to external perturbations lead to innovations in the field of band engineering and electronics. So far, a lot of theoretical and experimental investigations on hydrogenic impurity states in Low Dimensional Semiconducting Systems (LDSS) have been carried out worldwide by many researchers [1-3].The influence of position of donor impurity in quantum wire have been investigated theoretically for ground state and some excited states by A. Latge [4] et. al., and a comparative study between donor binding energy in cylindrical and rectangular quantum wire was made by Bryant [5]et. al., Brown and Spector [6] discuss confinement of carrier in Quantum wire with infinite and finite potential barriers. Navaneethakrishnan et. al., have investigated the effect of dielectric screening on the donor states [7] and on the diamagnetic susceptibility [8] of the donor. Tuning the band offsets is possible by applying external perturbations like electric [9] and magnetic field [10], laser field [11], pressure, etc.; Application of pressure on LDSS not only changes the dimensions of the system, but also alters the band gap through the mixing of Γ and X conduction bands in the well and in the barrier regions [12]. Experimental realization on Γ –X mixing in GaAs/$Al_xGa_{1-x}As$ coupled quantum wells under pressure was achieved by J.H.Burnett [13]. Recently the same effect has been investigated theoretically in single quantum well including the effect of non parabolicity of the conduction band by Nithiananthi [14] et.al.,

[1] To whom any correspondence should be addressed.

In the Present work we investigate theoretically the effect of Γ-X on the ground state binding energy of a donor under hydrostatic pressure which is confined in a GaAs/$Al_xGa_{1-x}As$ Quantum wire by using variational method in the effective mass approximation, in view of exploring the technological applications in optics and electronics due to band tailoring.

2. Theory

The pressure dependent Hamiltonian of the donor electron in a GaAs-$Ga_{0.7}Al_{0.3}As$ quantum wire in atomic units is given by

$$\boldsymbol{H} = \frac{-\nabla^2}{2m^*_{w,b}(P)} - \frac{1}{\varepsilon_{w,b}(P)\boldsymbol{r}} + V_B(r,P) \tag{1}$$

where $m^*_{w,b}$ is pressure dependent effective mass of electron in well and barrier is given by[14]

$$m^*_W(P,T) = 1 \Big/ \left[1 + 7.51\left\{2/\Gamma_W(P,T) + \left[\Gamma_W(P,T) + 0.341\right]^{-1}\right\}\right] \tag{2}$$

$\Gamma_w(P,T)$ is the pressure dependent energy gap of GaAs at the Γ point and is given by[14]

$$\Gamma_w(P,T) = 1.519 + \alpha^{\Gamma}_{w} P - 5.405 \times 10^{-4}\, T^2 / (T + 204) \qquad (T = 4\text{ K}) \tag{3}$$

where α^{Γ}_{w} is the pressure coefficient of GaAs at the Γ point

The barrier $Al_xGa_{1-x}As$ effective mass[15] $m^*_b(P,T) = m^*_w(P,T) + 0.083\,x$, x being Al composition.
The pressure dependent dielectric constant for GaAs and $Al_xGa_{1-x}As$ are given by

$$\varepsilon_w(P) = \varepsilon_w(0)\exp(\delta P) \tag{4}$$

where $\varepsilon_w(0) = \varepsilon_0(T_0)\exp[\gamma_0(T-T_0)]$ and

$$\varepsilon_b(P) = \varepsilon_w(P) - 3.12\,x \tag{5}$$

Pressure dependent potential energy of $Al_xGa_{1-x}As$ barrier is

$$V_B(r,P) = \begin{cases} 0 & |x|,|y| \le L(P)/2 \\ V_0(r,P) & |x|,|y| > L(P)/2 \end{cases} \tag{6}$$

$$V_0(r,P) = \begin{cases} \Gamma_b(P,T) - \Gamma_W(P,T) & P \le P_1 \\ X_b(P,T) - \Gamma_W(P,T) + S_{\Gamma X}(P) & P_1 < P < P_2 \end{cases} \tag{7}$$

The pressure dependent Γ - X band mixing strength coefficient

$$S_{\Gamma X}(P) = S_0 x\,(P-P_1)/P \tag{8}$$

where S_0 (=250 meV) is the adjustable parameter. P_1 and P_2 are the critical crossover pressures between X_b - band and Γ_b - band and X_b - band and Γ_w - band respectively.

The variation of X_b band with pressure is $X_b(P) = X_b(0) + \alpha^X_b P$

α^X_b being the pressure coefficient for the barrier

Choosing the trial wave function of the donor impurity in its ground state as

$$\Psi_{1s} = N_{1s} \begin{cases} \cos\alpha x \cos\alpha y\; e^{-\alpha_{1s}\boldsymbol{r}} & |x|,|y| \le L(P)/2 \\ Be^{-\beta(x+y)} e^{-\alpha_{1s}\boldsymbol{r}} & |x|,|y| > L(P)/2 \end{cases} \tag{9}$$

$$\alpha = \sqrt{\frac{2m_W E(P)}{2}} \quad \text{and} \quad \beta = \sqrt{\frac{2m_b(P)[V_0(P) - E(P)]}{2}}$$

The variation of the width of the well as a function of the pressure[14] is given by

$$L(P) = L(0)[1-(S_{11}+2S_{12})P] \tag{10}$$

where S_{11} (=1.16 x 10^{-3} $Kbar^{-1}$) and S_{12} (= -3.7 x 10^{-4} $Kbar^{-1}$) are the elastic constants of GaAs and the pressure dependent subband energy is obtained by solving the transcendental equation

$$\tan\alpha\, L(P)/2 = \left[\frac{m_w^*(P,T)}{m_b^*(P,T)}\left(\frac{V_0(P)}{E(P)} - 1\right)\right]^{1/2} \tag{11}$$

The normalization constant B is obtained by applying the boundary conditions at $x = y = \pm L(P)/2$ on the wavefunction.

The pressure dependent binding energy is given by

$$E_B(P,T) = E(P,T) - \langle H(P,T)\rangle_{min} \tag{12}$$

3. Result and Discussion

The variation of the binding energy of the ground state donor with pressure in a quantum wire has been calculated and it is given in figure 1.

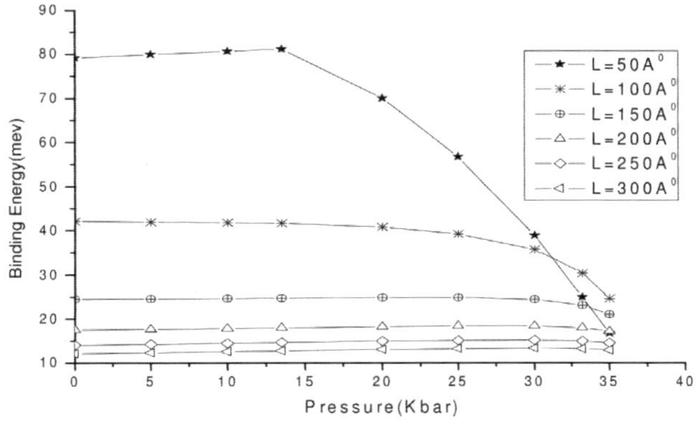

Figure 1:- Variation of binding energy as a function of pressure for various well widths.

From the figure, it is clear that up to $P \leq 13.5$ Kbar, binding energy of the donor increases with pressure due to the increase of effective mass and decrease of static dielectric constant. These factors decrease the kinetic energy and increase the potential energy of the donor respectively. And also, the potential barrier height remains constant in this region. Hence, the donor binding energy increases with pressure in this region for all sizes of the wire. After crossing the first critical pressure (P_1=13.5kbar), for smaller wire size, (L(P)=50Å), the donor impurity is found to be less bound with its host. This is due to the crossover of X band of barrier below the Γ band of barrier, which reduces the potential barrier height. When the size of the wire is increased, the rate of decrease of binding energy is reduced, which manifests the fact that the effect of Γ –X band mixing in the barrier is significant for smaller wires, even after pressure P_1. There is further decrease in the binding energy of the donor after second critical pressure (P_2=33.2kbar) due to the Γ –X band crossover in the well region also. Thus the system enters into an indirect band gap regime.

Figure 2 gives the behavior of binding energy with wire size for different pressures (P=0, P_1, P_2). The figure shows that the confinement of the donor impurity in Quantum wire is more for wires of smaller size for all pressure values. The characteristic behavior of a low dimensional wire is reflected in this figure with turnover in the value of binding energy, and shift in the peak value towards the lower width region.

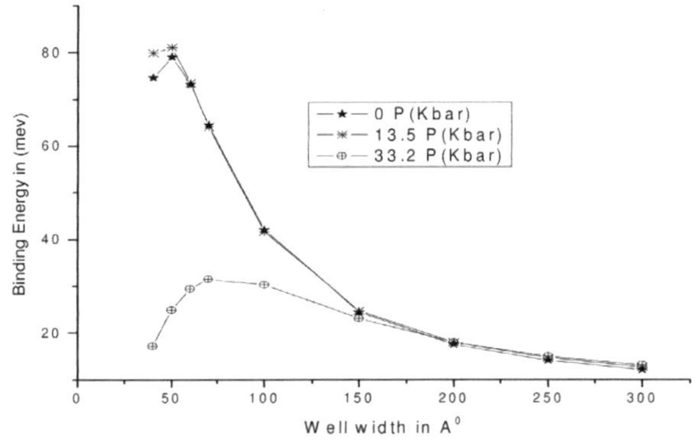

Figure 2:- Variation of binding energy as a function of well widths for various pressures.

To conclude, the effect of Γ –X band crossover is very significant in the lower width region even after P_1 which is seen from the drastic reduction in the binding energy for L(P)=50Å. The effect is very less in the higher well width region, (L (P) >200Å) and the localization of the donor is uniform which reflects the bulk behaviour. For wire sizes 50Å<L (P) <200Å, there is reduction in the donor binding only after P_2, which is similar to the behaviour observed in Quantum wells [14].

References

[1] Nithiananthi P and Jayakumar K 2009 *Phys. Stat. Sol.(b)* **246** 1238
[2] Pokatilov E P Fonoberov V A Balaban S N and Fomin V M 2000 *J.Phys.:Condens. Matter* ***12*** 9037
[3] Merwyn Jasper D Reuben A and Jayakumar K 2006 *Phys. Stat. Sol.(b)* **243** 4020
[4] Latge A deDios-Leyva M and Luiz Oliveira E 1994 *Phy.Rev.* ***B49*** 10450
[5] Bryant G W 1984 *Phys. Rev.* ***B29*** 6632
[6] Brown J W Spector H N and 1986 *J. Appl. Phys,* **59(4)** 1179
[7] Sukumar B and Navaneethakrishnan K 1990 *Solid State Commun.***74** 295
[8] Latha M Rajashabala S and Navaneethakrishnan K 2006 *Phys. Stat. sol.*(b) **243** 1219
[9] Morales AL Montes A Lopez SY and Duque CA 2002 *J.Phys.: Condens. Matter* **14** 987
[10] Jayakumar K and Nithiananthi P 2011 *J. Nano and Elect. Phy* **3** 383
[11] Merwyn Jasper D Reuben A, Nithiananthi P and Jeyakumar 2009 Superlatt. and *Microstruct.* **46** 710
[12] Elabsy A M 1994 *J.Phy.: Condens. Matter* **6** 10025
[13] Burnett J H Cheong H M Paul W,.Koteles E S and Elman B 1993 *Phys. Rev.* B **47** 1991
[14] Nithiananthi P and Jayakumar K 2005 *Int.J.Mod.Phys.*B **19** 3861
[15] Adachi S 1985 *J.Appl.Phys* **58** R1

IOP Conf. Series: Materials Science and Engineering **73** (2015) 012101 doi:10.1088/1757-899X/73/1/012101

Preparation and characterization of ceria magnesium aluminate nano powder for thermal barrier application

N Sankara Subaramainian[1], R Ramalingam[2,3] and M Vijayaraghavan[2,4]

[1] Thin film Research Laboratory, Department of physics, Thiagarajar College of Engineering, Madurai - 625 015, Tamilnadu State, India.
[2] Department of Mechanical Engineering, Thiagarajar College of Engineering, Madurai - 625 015, Tamilnadu State, India.

E-mail: [1] shankersathiya@yahoo.com , [3] ramalingamapk@gmail.com and [4] vijayaraghavantce@gmail.com

Abstract. The present work deals with the preparation of $CeMgAl_{11}O_{19}$ nano powder by sol gel citric acid route and to investigate its structural, surface and thermo-mechanical properties, to access its suitability for thermal barrier application. The XRD profile of the $CeMgAl_{11}O_{19}$ powder heat treated at different temperatures confirms polycrystalline structure. From the XRD data, mean grain size, lattice strain energy, dislocation density and other structural parameters have been investigated and reported. Thermal properties of the as synthesized $CeMgAl_{11}O_{19}$ powder have been studied from the TG-DTA and DSC measurement and the results are presented. Thermal conductivity of the $CeMgAl_{11}O_{19}$ specimen heat treated at different temperatures has been evaluated using the physical property measurement system and the results have been discussed. Hardness of the pelletized $CeMgAl_{11}O_{19}$ specimen heat treated at different temperatures has been measured in the Vickers scale and the results have been discussed. The influence of annealing temperature on the tensile strength and wear resistance of the $CeMgAl_{11}O_{19}$ specimen have been investigated and reported. The 2D and 3D AFM image of the $CeMgAl_{11}O_{19}$ specimen heat treated at 750^0 C shows the uniform surface pattern without any dark pits and pinholes. The surface roughness and grain size evaluated from the AFM data have been analyzed and presented.

1. Introduction

The concept of thermal barrier coatings (TBCs) is well-known among gas-turbine designers. The experimental results of thermal barrier coating indicate that the higher combustion temperature generated by adiabatic engine generally reduces the heat rejection rate [1]. The fundamental idea of a TBC is that an insulating ceramic top coat is used for thermal protection of components which are exposed to hot gas (*e.g.*, combustor linings, turbine guide vanes and turbine blades). Between the substrate and the top coat there is a bond coat, which is a metallic alloy [2]. In these days partially stabilized zirconia (PSZ), used as thermally insulating material on the piston crown face and reported a 19% reduction in heat loss through the piston [3]. Normally, the TBC system is required to have significantly increased phase stability, lower lattice and lower radiation thermal conductivity and improved sintering and thermal stress resistance under the engine high-heat flux and severe thermal cycling conditions. The development of low conductivity thermal barrier coatings requires test techniques that can accurately and effectively evaluate coating's thermal conductivity at high surface temperatures (typically at 1300 to 1400 °C). In particular hot section components of gas turbine engines must function at peak efficiencies at 3000 ^{0} F to near stoichiometric temperature and at high pressure with minimal cooling and without degradation of component life [4]. Since this low thermal conductivity requires at high temperature environment, the evaluation of materials has to take initial and after certain thermal exposure. $CeMgAl_{11}O_{19}$ has been selected as TBCs material because of their ability to with stand at higher operating temperature, high temperature stability, high hardness, wear resistance and tensile strength at higher temperature, low cooling requirements and improved performance achievements.

2. Experimental procedure

The Sol-gel citric acid method was used for the powder synthesis of $CeMgAl_{11}O_{19}$. In Sol-gel citric acid method a stoichiometric amount of citric acid, about 1.2moles per mole of metals in the nitrate solution, was separately dissolved 500 ml of water. Appropriate amounts of cerium nitrate ($Ce(NO_3)_3.6H_2O$), Aluminum nitrate ($Al(NO_3)_3.9H_2O$) and magnesium nitrate ($Mg(NO_3)_3.6H_2O$) was dissolved in 1.2 molar aqueous citric acid solution. The entire mixture was well stirred in a magnetic stirrer at 90 °C. A viscous substance (sol) was formed in a short while from the start of the stirring action. Further stirring lead to the formation of a jelly layer and then a foam. This foam on further stirring resulted in the formation of an amorphous porous material, which on heating resulted in the formation of $CeMgAl_{11}O_{19}$ crystallites, which was washed well and then ball milled for two hours, to get fine grained $CeMgAl_{11}O_{19}$. The prepared $CeMgAl_{11}O_{19}$ powder material was annealed in the atmospheric condition between 400 °C and 900 °C for ten hours and then compacted into pellets of desired geometry. Thermal conductivity studies have been carried out using Techno four meter. The micro hardness testing of the specimen has been conducted in Vickers hardness scale, using Zwick micro-hardness tester.

3. Result and discussion

3.1. Structural analysis

The X- Ray Diffraction pattern of unannealed and annealed $CeMgAl_{11}O_{19}$ powder specimen at 300 °C and 750 °C is shown in the figure 1, The XRD profile of the entire $CeMgAl_{11}O_{19}$ powder specimen confirms the formation of the polycrystalline structure, constituted by nanocrystallites. The $CeMgAl_{11}O_{19}$ powder exhibits the amorphous state at the annealed temperature of 300 °C. The $CeMgAl_{11}O_{19}$ powder heat treated at 450 °C shows hexagonal structure with the preferred orientation along the (2 1 10) plane. The calculated lattice constants and all the orientations in the XRD profile are well in agreement with the standard JCPDS PDF data file number 04-010-0269 for $LaMgAl_{11}O_{19}$ Powder specimen. On the other hand the $CeMgAl_{11}O_{19}$ powder heat treated at 750 °C shows Cubic structure with the preferred orientation along the (1 1 1) plane. The other orientation observed corresponds to (220) and (311) plane.

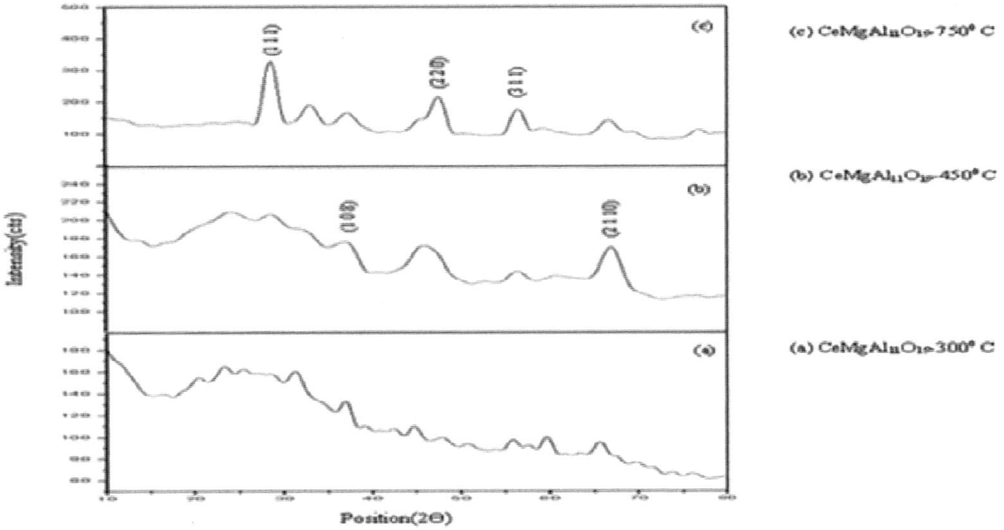

Figure 1. The XRD profile of $CeMgAl_{11}O_{19}$ powder

3.2. *Electrical Characterization*

Variation of resistance (R) and resistivity (ρ) of the pelletized $CeMgAl_{11}O_{19}$ specimen annealed at different temperatures is shown in the figure 2, The plot clearly illustrate that the R and ρ value increases, with the increase in annealing temperature and this may be attributed to the increased lattice scattering effect with the increase in the annealing temperature. The resistance and resistivity is found to be maximum with the value 6 MΩ and 9×10^{-6} Ω-cm respectively, for the $CeMgAl_{11}O_{19}$ specimen annealed at the temperature 750 °C.

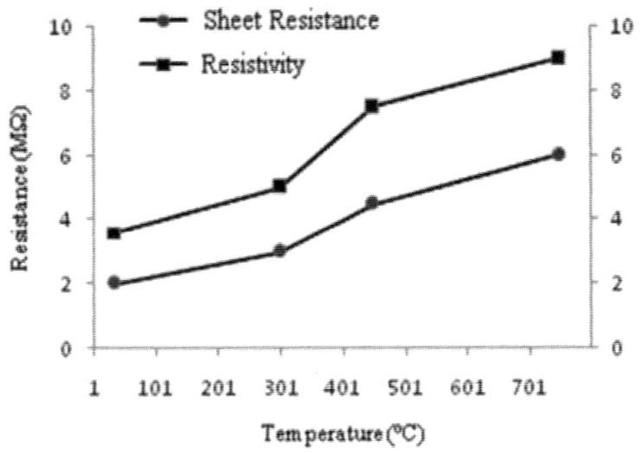

Figure 2. Variation of resistance and resistivity with annealing temperature

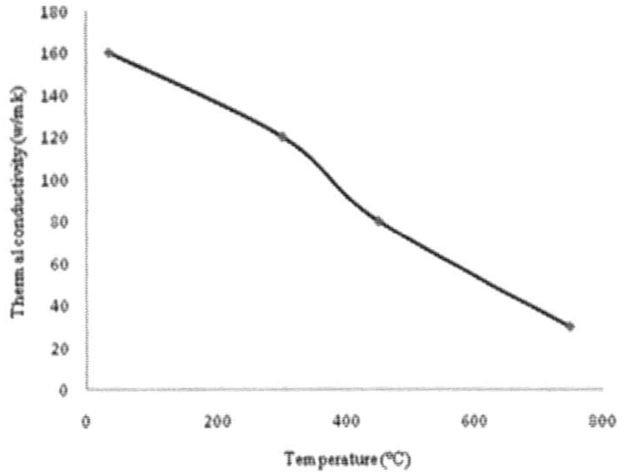

Figure 3. Variation of thermal conductivity with annealing temperature

3.3. *Thermal conductivity studies*

The thermal conductivity of pelletized $CeMgAl_{11}O_{19}$ specimen has been evaluated using the Wiedemann – Franz law stated below.

$$L = \frac{k}{\sigma t} \tag{1}$$

Where **L** is the Lorentz number (2.44×10^{-8} $\Omega W/K^2$), **k** is the thermal conductivity in Watt /mK, **σ** is the electrical conductivity in mho / m and **T** is the absolute temperature in K. Using the above relation, thermal conductivity of the unannealed sample have been estimated as 160 Watt / mK whereas for the pelletized $CeMgAl_{11}O_{19}$specimen annealed at different temperatures a steady fall in thermal conductivity have been observed. The variation of thermal conductivity as a function of annealing temperature is shown in figure 3, The plot shows a non – linear decrease in thermal conductivity with the increase in annealing temperature and this may be attributed to the micro structural variation and gradual raise in thermal diffusivity with the increase in annealing temperatures. The thermal conductivity is found to be maximum with the value 160 W/mK for the unannealed $CeMgAl_{11}O_{19}$ specimen and a minimum thermal conductivity of 18 W/mK is evaluated for the $CeMgAl_{11}O_{19}$ specimen annealed at 750 °C.

3.4. Micro hardness Studies

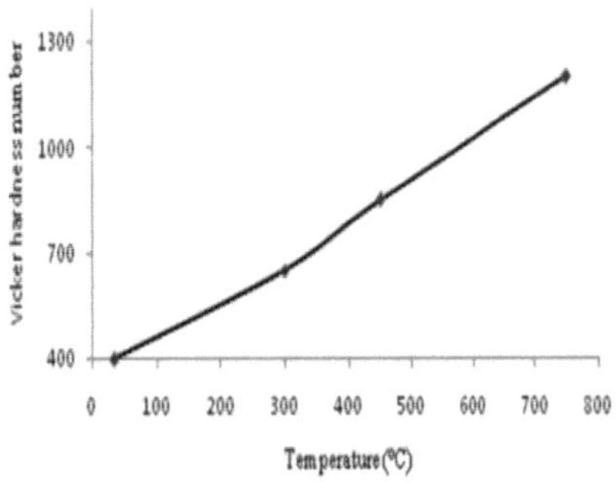

Figure 4. Variation of micro hardness with annealing temperature

The variation of VHN with annealing temperature of the $CeMgAl_{11}O_{19}$specimen is shown in the figure 4, The plot shows a non – linear increase in VHN with annealing temperature, indicating an improvement in the hardness of the films. This may be attributed to increase in the densification of compacted particulates at higher elevated temperatures, which ultimately increases the mechanical and thermal shock resistance of the $CeMgAl_{11}O_{19}$ specimen annealed at higher temperatures.. The hardness is found to be maximum with the value 1140 VHN and minimum with the value 400 VHN, for the $CeMgAl_{11}O_{19}$ specimen annealed at the temperatures 750 °C and unannealed specimen respectively.

3.5. Surface Characterization

The 2D and 3D AFM image of the pelletized $CeMgAl_{11}O_{19}$ specimen annealed at 450 °C and 750 °C are shown in the figure 5, and figure 6, respectively. The $CeMgAl_{11}O_{19}$ specimen annealed at 750 °C confirms the uniform distribution of fine grains throughout the surface without any dark pits and pin holes. The $CeMgAl_{11}O_{19}$ specimen annealed at 450 °C indicate more surface roughness and the presence of unevenly distributed coarse grains and small valley.

Figure 5(a). 3D AFM image of $CeMgAl_{11}O_{19}$ specimen annealed at 450 °C

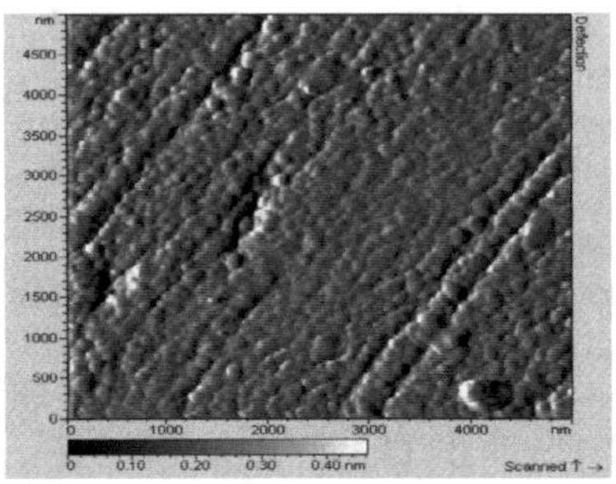

Figure 5(b). 2D AFM image of $CeMgAl_{11}O_{19}$ specimen annealed at 450 °C

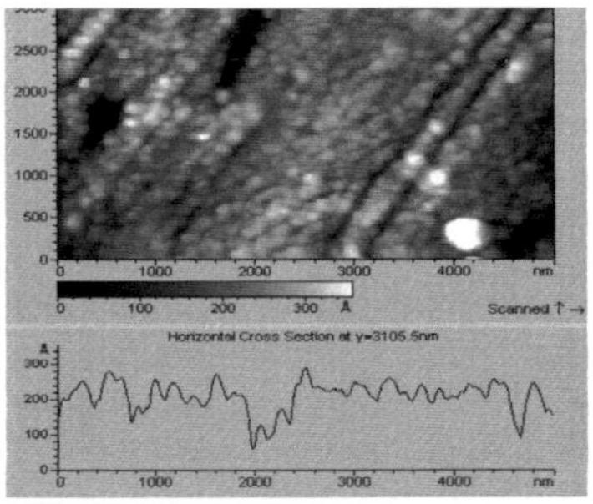

Figure 5(c). 2D AFM Roughness chart of $CeMgAl_{11}O_{19}$ specimen annealed at 450 °C

IOP Conf. Series: Materials Science and Engineering **73** (2015) 012101 doi:10.1088/1757-899X/73/1/012101

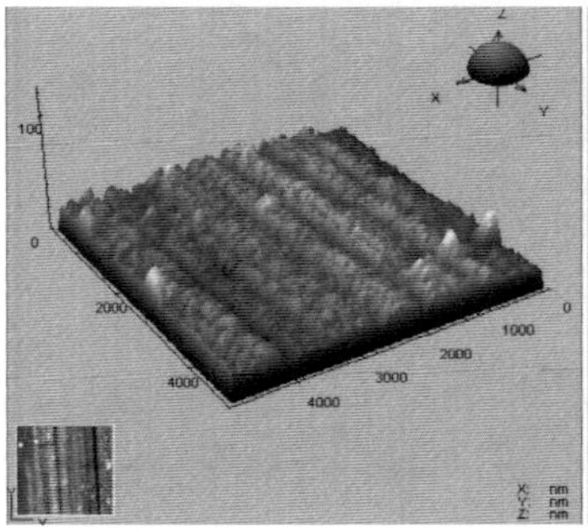

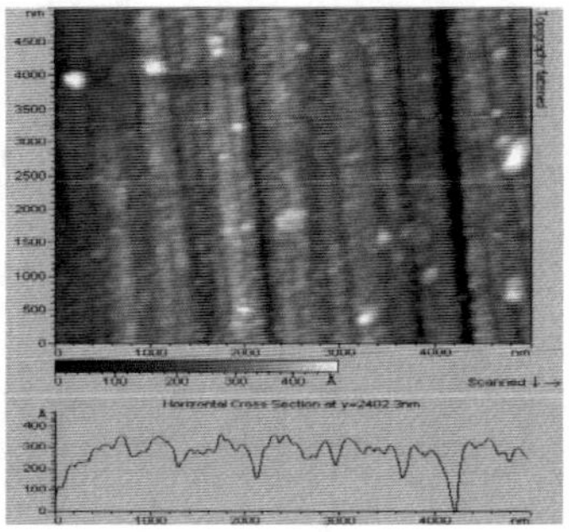

Figure 6(a). 3D AFM image of $CeMgAl_{11}O_{19}$ specimen annealed at 750 °C

Figure 6(b). 2D AFM image of $CeMgAl_{11}O_{19}$specimen annealed at 750 °C

Figure 6(c). 2D AFM roughness chart of $CeMgAl_{11}O_{19}$ specimen annealed at 750 °C

4. Conclusion

Ceria Magnesium aluminate ($CeMgAl_{11}O_{19}$) nano powder was prepared by sol-gel method. The $CeMgAl_{11}O_{19}$ powder heat treated at 450°C shows hexagonal structure with the preferred orientation along the (2 1 10) plane. On the other hand the $CeMgAl_{11}O_{19}$ powder heat treated at 750 °C shows cubic structure with the preferred orientation along the (1 1 1) plane. A sudden transition from the hexagonal to the cubic face is observed for the $CeMgAl_{11}O_{19}$ powder when annealed beyond the temperature 450 °C. The mean grain size evaluated from the XRD data lies between 16.5 nm and 44 nm. The resistance and resistivity is found to be maximum with the value 6 MΩ and 9 x 10^{-6} Ω-cm respectively, for the $CeMgAl_{11}O_{19}$ specimen annealed at the temperature 750 °C. Thermal conductivity of the $CeMgAl_{11}O_{19}$ specimen decreases non – linearly with the increase in annealing temperature and this may be attributed to the micro structural variation and gradual raise in thermal diffusivity with the increase in annealing temperatures. The thermal conductivity is found to be maximum with the value 160W/mK for the unannealed $CeMgAl_{11}O_{19}$specimen and a minimum thermal conductivity of 18W/mK is evaluated for the $CeMgAl_{11}O_{19}$ specimen annealed at 750 °C. The hardness is found to be maximum with the value 1140 VHN and minimum with the value 400 VHN, for the $CeMgAl_{11}O_{19}$ specimen annealed at the temperatures 750 °C and unannealed specimen respectively. The AFM micrographs of $CeMgAl_{11}O_{19}$ specimen clearly illustrate an increase in surface smoothness, grain boundaries, close packing of crystallites and grain size with the increase in the annealing temperature. The $CeMgAl_{11}O_{19}$ specimen annealed at 750 °C confirms the uniform distribution of fine grains without any dark pits and pin holes.

5. References

[1] Narottam P, Ohio D and Maryam E May 2007 Thermal properties of oxides with magnetoplumbite structure for advanced thermal barrier coatings. *NASA/TM* 214850.

[2] Gadow R and Lischka M 2002 Lanthanum hexaaluminate—novel tbc for gas turbine applications *Materials and Process Development. Surface Coatings Tech.* 151 to152. (2002) 392–399.

[3] Dongming Z and Miller A 2004 Development of advanced low conductivity thermal barrier coatings. *NASA/TM* (2004) 212961.

[4] Dongming Z and Miller A M 2002 Thermal conductivity and sintering behavior of advanced thermal barrier coatings. *NASA/TM* (2002)-211481.

Fabrication and characterization of Er^{+3} doped SiO_2/SnO_2 glass-ceramic thin films for planar waveguide applications

S Guddala[1], A Chiappini[2], C Armellini[2,3], S Turell[4], G C Righini[5], M Ferrari[2,6] and D Narayana Rao [1,6]
[1] School of Physics, University of Hyderabad, Hyderabad 500 046, India
[2] CNR-IFN, CSMFO Lab., via alla Cascata 56/c, Povo, 38123 Trento, Italy
[3] Fondazione Bruno Kessler FBK, via Sommarive 18 Povo, 38123 Trento, Italy
[4] LASIR (CNRS, UMR 8516), Université des Sciences et Technologies de Lille (USTL), 59655 Villeneuved'Ascq, France
[5] CNR-IFAC, MDF Laboratory, 50019 Sesto Fiorentino, Italy

Abstract: Glass-ceramics are a kind of two-phase materials constituted by nanocrystals embedded in a glass matrix and the respective volume fractions of crystalline and amorphous phase determine the properties of the glass-ceramics. Among these properties transparency is crucial in particular when confined structures, such as, dielectric optical waveguides, are considered. Moreover, the segregation of dopant rare-earth ions, like erbium, in low phonon energy crystalline medium makes these structures more promising in the development of waveguide amplifiers. Here we are proposing a new class of low phonon energy tin oxide semiconductor medium doped silicate based planar waveguides. Er^{3+} doped (100-*x*) SiO_2-$xSnO_2$ (x= 10, 20, 25 and 30mol%), glass-ceramic planar waveguide thin films were fabricated by a simple sol-gel processing and dip coating technique. XRD and HRTEM studies indicates the glass-ceramic phase of the film and the dispersion of ~4nm diameter of tin oxide nanocrystals in the amorphous phase of silica. The spectroscopic assessment indicates the distribution of the dopant erbium ions in the crystalline medium of tin oxide. The observed low losses, 0.5±0.2 dB/cm, at 1.54 μm communication wavelength makes them a quite promising material for the development of high gain integrated optical amplifiers.

1. Introduction

Glass-ceramics (GC) are nanocomposite materials and have been an intensive topic of research with their specific characteristics of capital importance in photonics. These systems, depending on their glass host and crystalline phase compositions, show improved mechanical, thermal, electrical and optical properties. The applications of GCs have spread over several areas like cook-top panels, electronics, medicine and dentistry, tough GCs and optical materials. The silica based glass ceramic (GC) systems has attracted great attention with the idea of Tick et al., [1] on the possibility of using such GCs as low-loss optical waveguides. The interesting optical property of these systems is the crystal like spectroscopic properties for the segregated optically active dopant ions in the low phonon energy crystalline phase. Especially, reduction in the inhomogeneous line width and therefore to an increase in the emission and absorption cross-sections of the active ions [2-3]. This behavior would thus make these materials attractive host for the rare-earth ions, which play an important role at the

[6] To whom correspondence should be addressed. E-mail: mferrari@science.unitn.it; dnrsp@uohyd.ernet.in

telecommunication wavelength of 1.54μm. In this regard erbium doped fiber amplifiers (EDFA) and erbium doped waveguide amplifiers (EDWA) are well-known for the past few decades, with the erbium ions' intra *4f* emission at 1.54μm. From the past few decades several GC phase systems with low phonon energies of crystalline phase were developed like silica-titania, silica-oxyfluoride, silica-hafnia etc., by our group,[4-8] with the objective of meeting the low optical losses and higher band width and gain. Here we would like to propose a GC phase of erbium doped silica-tin oxide matrix with the low optical losses for 1.54 μm wavelength propagation towards applications in the EDWAs.

2. Fabrication of planar waveguide structures

Once the ethanolic colloidal suspension prepared by using precursor $SnCl_2 \cdot 2H_2O$ becomes transparent, erbium ions were introduced through $Er(NO_3)_3.5H_2O$. The molar concentration of Er^{3+} in the total composition was maintained as 0.3 or 1 mol%. The silica solution obtained by mixing tetraethylorthosilicate (TEOS), ethanol (EtOH), deionized water and hydrochloric acid (HCl) as a catalyst was prehydrolyzed for 1h at 65 °C. A ratio of 1:0.01:2, was used for the TEOS: HCl: H_2O mixture. The quantity of ethanol during the preparation of the two solutions was chosen such that we maintain a concentration of (SnO_2+SiO_2) in ethanol as 0.448 mol/L. The two solutions were mixed and left at room temperature under stirring for 16 h. The resulting sol was filtered and then deposited on pure vitreous SiO_2 (v-SiO_2) by dip-coating, with a dipping rate of 40 mm/min. Each layer was annealed at 900 °C for 3 min prior to the application of the next coat. After every five the resulted film was heat treated again at 900 °C for 5min. Finally the films with Si/Sn molar ratios 90/10, 80/20 and 75/25 were heat treated at 900 °C for 30h, 20h and 5min respectively for attaining the densified and crack-free uniform thin films.

3. Results and Discussion

X-ray diffraction (XRD) spectra showed in figure 1a was measured with grazing angle of incidence of the beam at 1°. The spectra is composed of the diffraction peaks corresponding to single crystals of SnO_2 in tetragonal (cassiterite) phase [9] and a broad band centered at 21.5° corresponding to amorphous silica phase. The peak corresponding to (211) plane of the XRD spectrum (figure 1a) was used to determine the nanocrystals size by using the well-known Debye-Sherrer formula [9]. The size of nanocrystals was found to be ~4nm for the 25% SnO_2 content waveguide. Figure 1b shows the high resolution transmission electron microscope (HRTEM) image for x= 25 mol% SnO_2 content waveguide. Figure 1b also shows the narrow size distribution and lattice fringes of SnO_2 ~4 nm nanocrystals. The selected area electron diffraction (SAED) pattern shown in the inset of figure 1b indicates the polycrystalline nature of SnO_2.

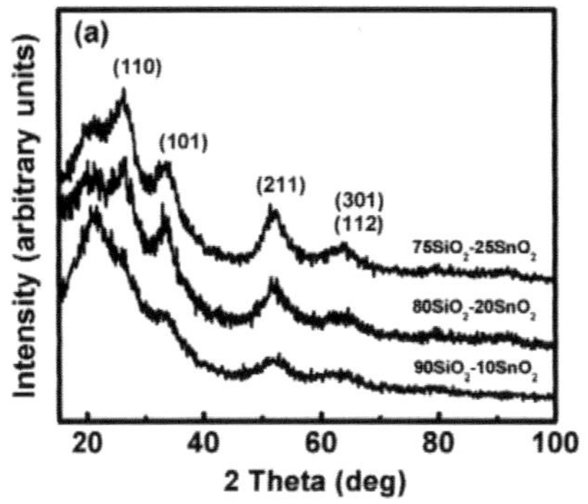

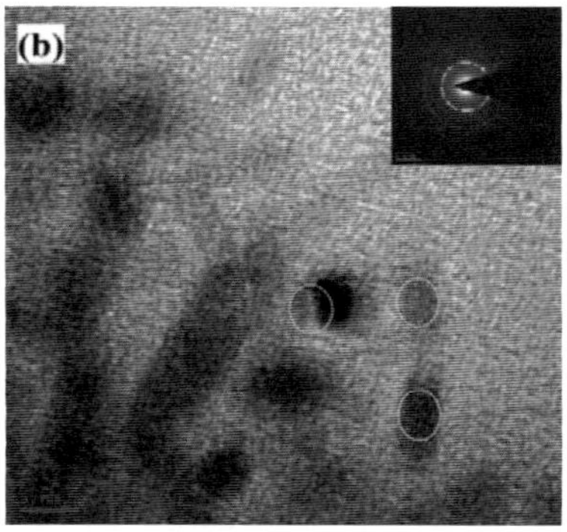

Figure 1.(a) GIXRD spectra measured with grazing angle of incidence at 1°. (b) HRTEM image and corresponding SAED pattern (in the inset) shows the tin oxide nanocrystals distribution in a thin film.

The energy band gap of embedded semiconductor SnO_2 nanocrystals could be obtained by plotting the well-known Tauc relation [10]. SnO_2 is known for its direct band gap and dipole forbidden transition in the UV region [10-11]. The plot of $(\alpha h\nu)^{2/3}$ vs $h\nu$ is shown in the figure 2a. The estimated energy band gap for all the three compositions, which is around 3.75eV, indicates the quantum confinement effect [12] and unique size distribution in all three compositions with a variation in the density of the particles.

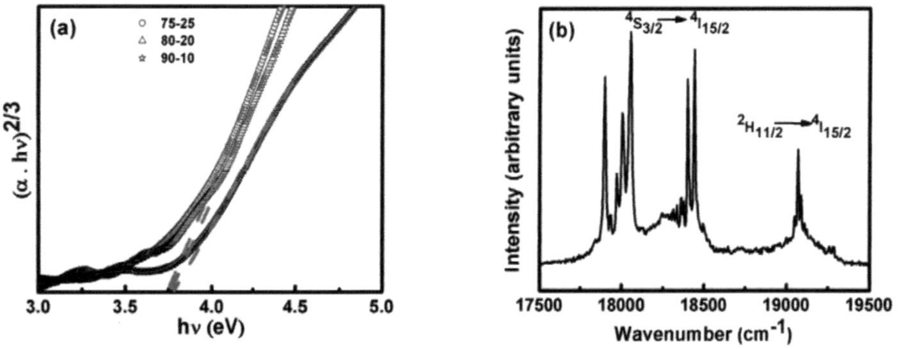

Figure 2: a) Plot of $(\alpha h\nu)^{2/3}$ vs $h\nu$ of Tauc plot for three compositions of SiO_2:SnO_2. The extrapolation of dashed line to $(\alpha h\nu)^{2/3} = 0$ axis gives the value of the band gap of the sample. (b) Photoluminescence spectra in the visible range of the 0.3mol% Er^{3+}-activated $75SiO_2$-$25SnO_2$ planar waveguide. Excitation was performed at 488nm, in the TE_0 mode configuration.

Figure 2b shows the photoluminescence (PL) in the visible range of erbium ions on excitation with the 488nm of Argon ion laser. The presence of narrow structures and the important reduction in the usual inhomogeneous line width of the erbium ions emission in the luminescence spectra (figure 2b) indicates a crystalline local environment for the Er^{3+} ion [12]. The crystal-field strength experienced by the optically active ions directly reflects the maximum energy splitting of the stark components of the electronic level. In fact, the narrowing is more evident in the visible PL spectra for the $^4S_{3/2}\rightarrow{}^4I_{15/2}$, and $^2H_{11/2}\rightarrow{}^4I_{15/2}$ transitions. The later transition is hypersensitive and therefore particularly suitable to detect the change in the environment immediately around the Er^{3+} ion [3]. So the presence of nanocrystals is a factor responsible for the reduction in the inhomogeneous line width, as observed in the present case figure 2b.

Table 1: Optical parameters of the 1mol% Er^{3+} activated $75SiO_2$-$25SnO_2$ planar waveguides. The thickness of the film found to be 1.15(±0.05) microns.

Wavelength (nm)	632.8	1319	1542
Refractive index (TE) (±0.001)	1.566	1.551	1.547
No. of modes	2	1	1
Attenuation coefficient (±0.3dB/cm)	0.7	0.6	0.5

The thickness and refractive index of the thin films were measured by m-line technique based on prism coupling method [8]. An increase in the refractive index value was observed as a function of increase in SnO_2 content. The maximum percentage of SnO_2 in silica matrix to support the propagation modes was found to be x=25%. The thickness and refractive index values for x=25% of SnO_2 are reported in table 1 for three different wavelengths of propagation. The attenuation coefficient (losses for the guided mode) was determined by collecting the scattered light along the propagating TE_0 mode. The observed refractive index values for TE and TM modes are in good agreement with the Lorentz-Lorentz equation with n= 1.457 and n= 1.94 for silica and tin oxide respectively. From the TE and TM measurement of refractive index values it is evident that birefringence is absent in the system [13]. The maximum confinement of the TE_0 mode at 1542nm was found to be about 82%. The attenuation coefficient measurements were performed by scanning the intensity of the signal along the propagating mode with the help of a fiber probe coupled to a detector. Observed low optical losses of 0.5±0.2dB/cm at 1.54μm wavelength propagation makes this 75/25 silica-tin oxide nano structure waveguides as the most promising system for integrated optical amplifiers.

4. Conclusions

In summary, we have fabricated erbium doped SiO_2-SnO_2 glass-ceramic waveguides with a maximum molar composition of SnO_2 at 25 mol%. From the structural and optical assessment the size of SnO_2 nanocrystals was found to be ~4 nm in diameter and energy band gap of 3.75eV. The spectroscopic assessment in the visible and NIR range indicates the local SnO_2 crystalline environment for the erbium ions. From the optical and spectroscopic results, RE-activated SnO_2 glass-ceramic waveguides with low attenuation coefficient of 0.5±0.2dB/cm at 1.54μm appears to be a viable system for integrated optics.

Acknowledgements

This research work is carried out in the frame work of 'India-Trento Program for Advanced Research' (ITPAR) Phase-II research project. The author S. Guddala would like to acknowledge the financial support of UGC- Basic Science Research (BSR) fellowship.

References:

[1] Tick P A 1998 *Opt. Lett.* **23** 1904
[2] Mortier M 2002 *Phil. Mag. B* **82** 745
[3] Gonçalves M C, Santos L F and R.M. Almeida 2002 *Comptes Rendus Chimie* **5** 845
[4] Chiasera A, Tosello C, Moser E, Montagna M, Belli R, Gonçalves R R, Righini G C, Pelli S, Chiappini A, Zampedri L and Ferrari M 2003 *J. Non-Cryst. Solids* **322** 289
[5] Tikhomirov V K, Furniss D, Seddon A B, Reaney I M, Beggiora M, Ferrari M, Montagna M and Rolli R 2002 *Appl. Phys. Lett.* **81** 1937
[6] Jestin Y, Armellini C, Chiasera A, Chiappini A, Ferrari M, Moser E, Retoux R and Righini G C 2007 *Appl. Phys. Lett.* **91** 071909
[7] Guddala S, Alombert-Goget G, Armellini C, Chiappini A, Chiasera A, Ferrari M, Mazzola M, Berneschi S, Righini G C, Moser E, Boulard B, Arfuso C D, Bhaktha S N, Turrell S, Narayana Rao D, and Speranza G, *2010 Proc. 12th Int. Conf. on Transparent Optical Networks (Munich)*, 1–4.
[8] Peled A, Chiasera A, Nathan M, Ferrari M and Ruschin S 2008 *Appl. Phys. Lett.* **92** 221104
[9] Ningthoujam R S, Sudarsan V, Godbole S V, Kienle L, Kulshreshtha S K and Tyagi A K 2007 *Appl. Phys. Lett.* **90** 173113
[10] Lorenzi R, Lauria A, Mochenova N, Chiodini N, and Paleari A 2011 *J. Non-Cryst. Solids* **357** 1888
[11] Jacquemin J L and Bordure G 1975 *J. Phys. and Chem. Solids* **36** 1081
[12] Brovelli S, Chiodini N, Meinardi F, Monguzzi A, Lauria A, Lorenzi R, Vodopivec B, Mozzati M C and Paleari A 2009 *Phys. Rev.* B **79** 153108
[13] Von Rottkay K and Rubin M 1996 *MRS spring meeting Proc. symp. J. Thin films for photovoltaic and Related Device* **426** 449

IOP Conf. Series: Materials Science and Engineering **73** (2015) 012103 doi:10.1088/1757-899X/73/1/012103

Eu and Dy activated $(Sr_{0.90}\,Ca_{0.10})_3(PO_4)_2$ phosphors for near-UV LED

K N Shinde[1] and S J Dhoble[2]
[1] N.S. Science & Arts College, Bhadrawati, Chandrapur 442902, India
[2] Department of Physics, RTM Nagpur University, Nagpur 440033, India

[1]E-mail: kartik_shinde@rediffmail.com

Abstract. The new efficient trivalent Eu and Dy activated $(Sr_{0.90}\,Ca_{0.10})_3(PO_4)_2$ phosphors were prepared by the conventional solid state method. X-ray powder diffraction analysis confirmed the formation of $(Sr_{0.90}\,Ca_{0.10})_3(PO_4)_2$ host lattice. The Eu^{3+} PL emission spectrum was observed in $(Sr_{0.90}\,Ca_{0.10})_3(PO_4)_2$ phosphors at 590 (orange) and 614 nm (red) region, the spectrum due to $^5D_0$-$^7F_1$ and $^5D_0$-$^7F_2$ transitions at near UV excitation, respectively. $(Sr_{0.90}\,Ca_{0.10})_3(PO_4)_2$: Dy^{3+} excited by 350 nm near-ultraviolet (NUV) light, the phosphors show an efficient blue and yellow band emissions, which originates from the $^4F_{9/2}$-$^6H_{15/2}$ and $^4F_{9/2}$-$^6H_{13/2}$ transitions of Dy^{3+} ion, respectively. The excitation spectra of the phosphors are broadband extending from 340 to 400 nm, which are characteristics of NUV excited LED. The investigated prepared phosphors are suitable for a NUV excited LED.

1. Introduction:

White-light emitting diodes are one of the most promising eco-friendly light sources, with low energy consumption [1]. As environmental subject has become serious, the need for mercury-free fluorescent lamps for general lighting has become an important subject for light source manufacturers. As it is possible to improve the luminous efficiency of white light LEDs excited by near-UV LEDs in the near-UV to blue part of the spectrum [2] around 380–420 nm, researchers worldwide have been investigating other suitable chemical compounds for solid-state lighting. Recently, Dhoble and co-workers reported some phosphate based $Eu^{3+}$and Dy^{3+} activated $M_5(PO_4)_3F$ (M= Ba, Sr, Ca) [3], $K_3Al_2(PO_4)_3$ [4], $Na_2X(PO_4)F$ (X = Mg, Ca, Sr) [5] and $NaCaPO_4$ [6] phosphors for lamp industry.

With respect to the phosphors presently used in white LED systems, most do not meet the optimum requirements. Accordingly, novel luminescent materials with improved properties are greatly in demand. In this article we report on the synthesis and characterization of Eu and Dy activated $(Sr_{0.90}\,Ca_{0.10})_3(PO_4)_2$ phosphors were prepared by the conventional solid state method. The X-ray diffraction (XRD) and photo luminescent (PL) properties of these phosphors were investigated at room temperature.

2. Experimental:

Trivalent Eu and Dy activated $(Sr_{0.90}\,Ca_{0.10})_3(PO_4)_2$ phosphate based phosphors were synthesized by the solid state method. The starting materials $SrCO_3$, $CaCO_3$ and $NH_4H_2PO_4$ of analytical grade (pure) were used. These materials were weighed in the proper molar ratio and then Dy_2O_3 was introduced as a dopant followed by mixing and grinding homogeneously in an agate mortar. The mixturewas heated at 500^0C in a silica crucible for 2 h. The vapors extraneous to the desired product that was evolved during the process were allowed to be released. After grinding, the mixture was heated at 800 ^0C for 24 h, thereby obtaining the white phosphor powder. Similar procedure was followed for Eu activated $(Sr_{0.90}\,Ca_{0.10})_3(PO_4)_2$.

IOP Conf. Series: Materials Science and Engineering **73** (2015) 012103 doi:10.1088/1757-899X/73/1/012103

The prepared host lattice was characterized for their phase purity and crystallinity by X-ray powder diffraction (XRD) using PAN-analytical diffractometer (Cu-Ká radiation) at a scanning step of 0.01^0, continue time 20s, in the 2θ range from 10^0 to 80^0; the average crystallite size was calculated from the broadening of the X-ray line (311) using Scherer's equation. The photoluminescence measurement of excitation and emission were recorded on the Shimadzu RF5301PC spectrofluorophotometer. The same amount of sample (2g) was used for each measurement. Emission and excitation spectra were recorded using a spectral slit width of 1.5 nm.

3. Results and discussion:

3.1 XRD

Figure 1 shows the XRD pattern and XRD did not show the presence of the starting constituents and other likely phases which is an indirect evidence for the formation of the desired compound. The XRD pattern of $(Sr_{0.90}\,Ca_{0.10})_3(PO_4)_2$ and it agrees well with JCPDS no. 00-14-0205. This shows final product was formed in crystalline, homogeneous form and solid state reaction of the mixtures took place well.

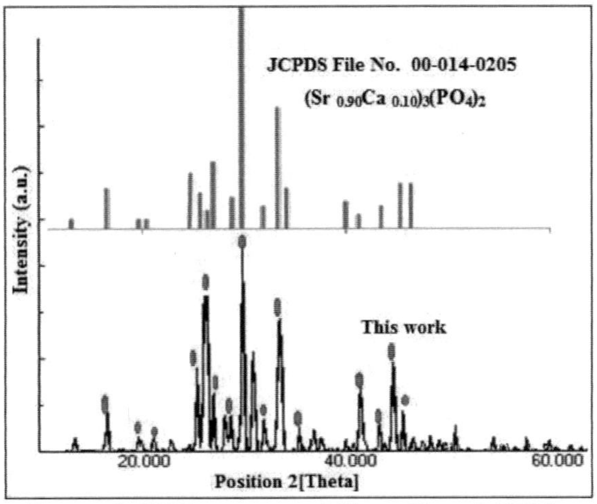

Figure 1: XRD pattern of $(Sr_{0.90}Ca_{0.10})_3(PO_4)_2$

3.2 The excitation and emission spectra of $(Sr_{0.90}\,Ca_{0.10})_3(PO_4)_2{:}Eu^{3+}$

Figure 2(a) shows the PL excitation spectrum of the $(Sr_{0.90}\,Ca_{0.10})_3(PO_4)_2{:}Eu^{3+}$ phosphor. There are two excitation bands are observed at 342 and 394 nm, which are all caused by the f–f transitions. Both are observed with a maximum intensity at 394 nm (stronger excitation band). Under the excitation of 394 nm, the phosphor of $(Sr_{0.90}\,Ca_{0.10})_3(PO_4)_2{:}Eu^{3+}$ has two sharp orange/red-emission bands at 590 and 614 nm shown in figure 2(b).

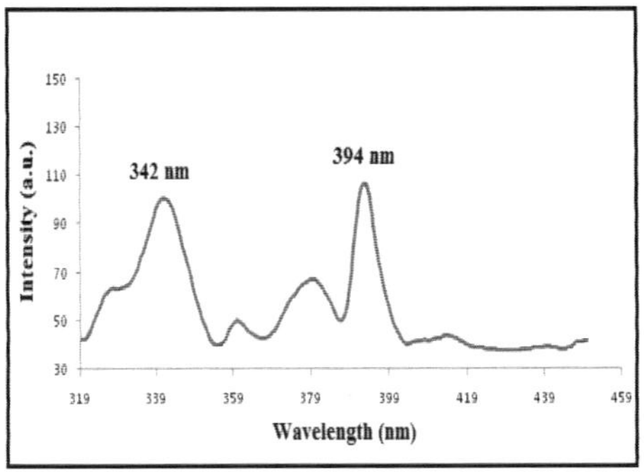

Figure 2(a): Excitation specrtrum $(Sr_{0.90}Ca_{0.10})_3(PO_4)_2:Eu^{3+}$

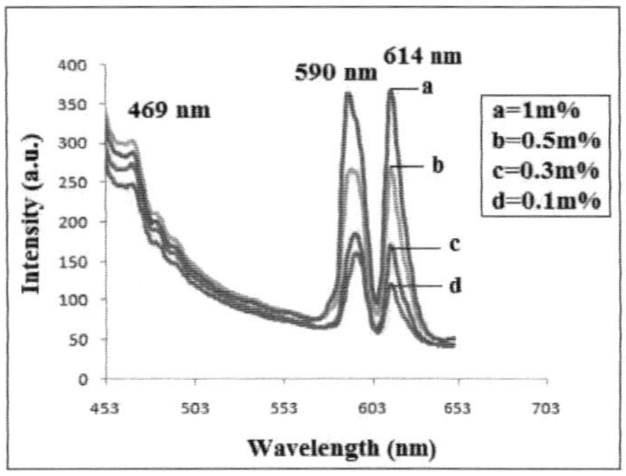

Figure 2(b): Emission spectra of $(Sr_{0.90}\,Ca_{0.10})_3(PO_4)_2{:}Eu^{3+}$

IOP Conf. Series: Materials Science and Engineering **73** (2015) 012103 doi:10.1088/1757-899X/73/1/012103

Among these two emission bands the 614 nm is corresponding to the electric-dipole transition $^5D_0 \rightarrow {}^7F_2$ of Eu^{3+} whereas other is corresponding to the magnetic dipole transition $^5D_0 \rightarrow {}^7F_1$ of Eu^{3+} ion which is the less prominent, these two peaks are obtained due to the splinting of Eu^{3+} ion emission. The luminescent properties of Eu^{3+} ion in the crystalline $(Sr_{0.90}\,Ca_{0.10})_3(PO_4)_2$ phosphor are in good agreement with those obtained through other processes indicating that Eu^{3+} ions have been effectively doped into the host lattice of $(Sr_{0.90}\,Ca_{0.10})_3(PO_4)_2$. The PL intensity increased with concentration from 0.1–1 mol% and it decreased at more than 1 mol% probably due to concentration quenching effect. In present case, the low contributions of the orange $^5D_0 \rightarrow {}^7F_1$ emissions and the high intensity of the red $^5D_0 \rightarrow {}^7F_2$ emission results in high colour purities that is adequate for near UVLED applications.

3.3 The excitation and emission spectra of $(Sr_{0.90}\,Ca_{0.10})_3(PO_4)_2{:}Dy^{3+}$

The intense excitation peak is observed at 349 nm and other weak peaks are also observed at 325 nm to 387 nm shown in figure 3(a) and only emission intensity vary with respect to all three excitation wavelengths. In this context, excitation curve 350 nm peak is near UV excitation which is more applicable (i.e. 385 to 395 nm) for UVLED phosphor. Accordingly, out of all three peaks we select the 350 nm peak for excitation in the experimental work. The emission spectrum show in figure3(b) and all the samples have the two emission bands: one is centered at 485 nm (blue) and another is at 575 nm (yellow).They are assigned to the Dy^{3+} electronic transitions of $^4F_{9/2} \rightarrow {}^6H_{15/2}$ and $^6H_{13/2}$ energy levels, respectively. A series of *$(Sr_{0.90}\,Ca_{0.10})_3(PO_4)_2{:}Dy^{3+}$*phosphor with various Dy^{3+} concentrations (x=0.1mol % – 1 mol %) were prepared and the effect of doped Dy^{3+}concentration on the emission intensity of was investigated.

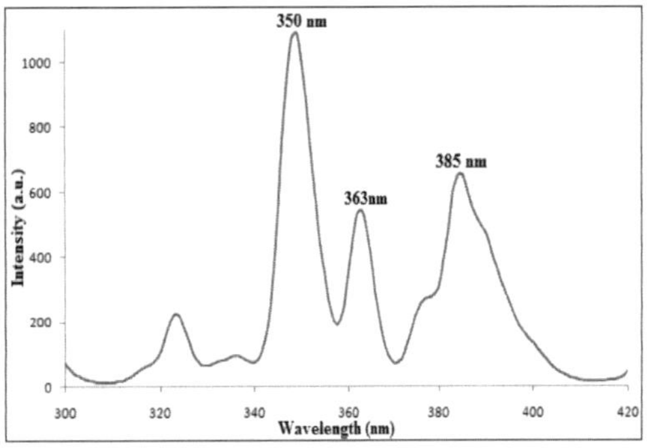

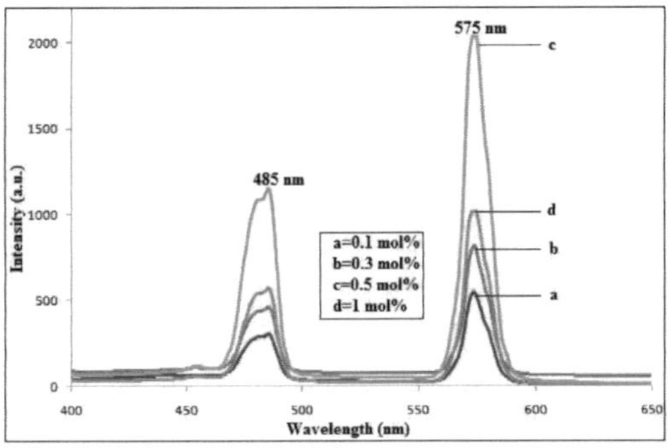

Figure 3(a): Excitation spectrum of $(Sr_{0.90}\,Ca_{0.10})_3(PO_4)_2{:}Dy^{3+}$

Figure 3(b): Emission spectra of $(Sr_{0.90}\,Ca_{0.10})_3(PO_4)_2{:}Dy^{3+}$

Thus the combination of colors gives BYR (blue–yellow–red) emissions which can produce white light, by near UV excitation. $(Sr_{0.90}\,Ca_{0.10})_3(PO_4)_2$ activated with Dy^{3+} and Eu^{3+} ions phosphors exhibit a strong absorption in the range of 340–400 nm and chromatic properties indicated that present phosphor is a promising candidate for producing near ultraviolet white light-emitting diodes (near-UV LED).

4. Conclusions:

In summary, the Eu^{3+} and Dy^{3+} ions activated $(Sr_{0.90}\,Ca_{0.10})_3(PO_4)_2$ phosphors have been synthesized by the solid state method. The formation of these compounds was confirmed by the XRD technique. The photoluminescence characterization of prepared phosphors show role of rare ions in the host lattice. The PL spectroscopic characterizations of the prepared phosphors are done using excitation and emission spectra. Under near UV excitation (350 nm) $(Sr_{0.90}\,Ca_{0.10})_3(PO_4)_2{:}Eu^{3+}$ the emission spectrum shows peaks at 590 nm (orange) while other are at 614 nm (red). When $(Sr_{0.90}Ca_{0.10})_3(PO_4)_2$:Dy^{3+} phosphor excited at 350 nm the emission spectrum shows intense bands at 485 nm (blue), and 575 nm (yellow). Both Eu^{3+} and Dy^{3+} activated $(Sr_{0.90}Ca_{0.10})_3(PO_4)_2$ phosphors show the excitation in the range of 340-390 nm of near UVLED excitation and emission observed in the red and blue/yellow region of the spectrum. This results show the $(Sr_{0.90}Ca_{0.10})_3(PO_4)_2$: Eu^{3+} and $(Sr_{0.90}Ca_{0.10})_3(PO_4)_2$:Dy^{3+} phosphors could be applicable for near UVLED phosphors.

References:

[1] Mikami M , Watanabe H , Uhedra K , Shimooka S , Shimomura Y , Kurushima T, 2009 *IOP Conf Ser: Mater Sci Eng* **1** 012002.
[2] Yamada M , Narukawa Y , Mukai T , 2002 *Jpn. J. Appl. Phys.,* **41** L246.
[3] Nagpure I M , Shinde K N , Dhoble S J , Kumar Animesh, 2009 *J Alloys Compd.,***481** 632.
[4] Shinde K N , Dhoble S J , 2011 *J. Fluoresc* **21** 2053.
[5] Dhoble S J , Pawade V B ,Shinde K N , 2010 *Eur. Phys. J. Appl. Phys.* **52** 11104.
[6] Shinde K N, Dhoble S J , Kumar Animesh, 2011 *J. Rare Earths,* **29(6)** 527.

Oriented grain growth in ZnO thin films by Iodine doping

Deepu Thomas[1,2], Sunil C Vattappalam[2] , Sunny Mathew[2] and Simon Augustine[2*]

[1]*Research and Development Centre, Bharathiar University, Coimbatore, India-641046.*
[2]*Department of Physics, St.Thomas College, Pala, Kottayam, India-686574.*
E-mail: simonstcp@gmail.com

Abstract. ZnO thin films were prepared by Successive Ionic Layer Adsorption Reaction (SILAR) method. Oriented grain growth in Iodine doped ZnO thin films were studied. The oriented grain growth in samples was studied by comparing the peak intensities from X-ray diffraction data and surface morphology by scanning electron microscopy. It is found that oriented grain growth significantly enhanced by Iodine doping. When the oriented grain growth increases, crystallinity of the thin film improves, resistance and band gap decrease. ZnO thin films having good crystallinity with preferential (002) orientation is a prerequisite for the fabrication of devices like UV diode lasers, acoustic-optic devices etc. A possible mechanism for the oriented grain growth is also investigated. It is inferred that creation of point defects is responsible for the enhanced oriented grain growth in ZnO thin films when doped with iodine.

Key words: ZnO thin film, SILAR method, oriented grain growth, annealing ZnO thin films, Iodine doping.

1. Introduction

For nearly half a century, the synthesis and characterization of ZnO thin film is an active area of research. ZnO with wurtzite structure [1] is an n-type semiconductor with direct band gap of 3.37eV [2] and high electronic mobility. As a wide and direct band gap semiconductor, nanostructred ZnO thin films have attracted more attention in optoelectronic devices [2]. Different methods have been applied to obtain ZnO thin films. These include magnetron sputtering [3], chemical bath deposition [4], sol-gel [5], spray pyrolysis [6] etc. Chemical deposition techniques are relatively low cost processes and can be easily scaled up for industrial application. Among the thin films deposition methods, double dip technique from aqueous solutions is the simplest and the most economical one. There are reports that oriented grain growth (textured grain growth along c-axis) in ZnO thin films is enhanced by doping. Controllable n-type doping is easily achieved by substituting Zn with group-III elements like Al [7, 8] or substituting oxygen with group-VII elements such as iodine . There are reports on enhancement of oriented grain growth in ZnO by Al doping. But a study on the effect of Iodine doping on oriented grain growth in ZnO by iodine doping and its possible mechanism is not yet reported.

2. Experimental

ZnO thin films were prepared by Successive Ionic Layer Adsorption Reaction (SILAR) method. In which, the ZnO thin film was coated on the glass substrate (26 × 76 mm) by alternately dipping the substrate in sodium zincate bath at room temperature and then in hot water maintained at 90- 95°C .Iodine doping was done by adding 6 atom.% iodine in Sodium Zincate bath. The pure and doped samples were annealed at 450° C for half an hour in air. The structural analyses of the thin films were done by X-ray diffraction (XRD) and surface morphology by scanning electron microscopy (SEM).

*To whom any correspondence should be addressed. simonstcp@gmail.com

Chemical elemental stoichiometry was examined from energy dispersive X-ray analysis (EDAX) linked with SEM unit. The resistance of ZnO thin film at room temperature is measured by Keithley 2100 Digital Multimeter. The optical absorbance was measured in the wavelength range of 190-1100 nm by means of a UV-VIS spectrophotometer. Oriented grain growth of the samples was measured by using an orientation index (O.I.), which is the ratio of the intensities of (002) reflection ($2\theta = 34^{\circ}$) to (101) reflection ($2\theta = 36^{\circ}$) in the XRD.
In the course of the present investigations, oriented grain growth in ZnO thin films doped with iodine is studied. A possible mechanism for the enhanced oriented grain growth is also investigated

3. Results and Discussion

3.1 Structural and Morphological Studies

XRD patterns of ZnO thin films of pure and doped samples are depicted in fig1.(a-b).

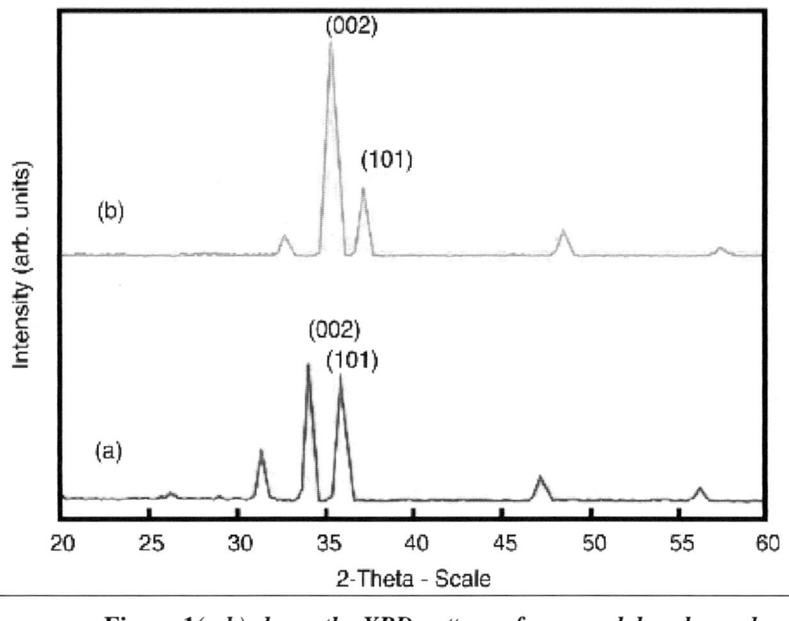

Figure 1*(a-b) shows the XRD pattern of pure and doped sample*

It is seen from figures that the (002) peak appears with maximum intensity at $2\theta = 34^{\circ}$. The other peaks at 31°& 36° can be associated with (100), (101) reflections of ZnO, as expected for wurtzite hexagonal ZnO structure (JCPDS no.36-1-1451). In particular, it is clearly observed that the intensity of the (002) peak enhanced by doping with iodine. The (002) diffraction peak exhibited a higher intensity, showing a fact that the oriented grain growth is along (002) direction. The figure-2 shows the SEM micrograph of the iodine doped sample. The flowered structure of ZnO is evident from the SEM micrograph.

IOP Conf. Series: Materials Science and Engineering **73** (2015) 012104 doi:10.1088/1757-899X/73/1/012104

Figure 2 *shows 2 shows the SEM micrograph of the iodine doped sample*

The average grain size of the ZnO crystals in the films was calculated by using Scherrer's formula: $D = 0.9\lambda / \beta \cos\theta$, where D is the grain size, λ- the wavelength of X-ray used, β- the FWHM(Full width half maximum) and θ- Bragg diffraction angle of the XRD peak. From the studies, the grain size of the iodine-doped sample was found to be 35.05nm (Table-1). The dislocation density $\delta = 1/D^{2}$, where D is the grain size. Dislocation density gives the degree of crystallinity. The larger values of D indicate better crystallization of the films. As the values of D increase, the orientation index is found to increase. Increase in orientation index indicates the increase in the crystallinity of ZnO thin films. ZnO thin film having good crystallinity with preferential (002) orientation is a prerequisite for the fabrication of devices like UV diode lasers [2].

3.2 Resistance Measurements

The grain size and resistance of ZnO thin films measured at room temperature is given in table-1. It shows that as the grain size increases, the resistances of the film decreases. One of the reasons for this is due to the decrease in grain boundary when the grain sizes increase. Another is due to the contribution of iodine ions on the substitutional site of oxygen ions.

Table-1 gives the orientation index (O.I), grain size, resistance, and band gap of samples

Sample No.	Sample details	O.I.	Grain size (D) in manometer	Resistance in KΩ at room temp.	Band gap in eV
1	Pure	1.21	18	9.01	3.25
2	Iodine doped	7.1	35.05	3.51	3.12

3.3 Optical Properties of ZnO Films

Fig.3 shows the optical absorbance spectrum of the pure ZnO thin film using UV-visible region from 200-800nm.The corresponding optical band gap of the thin film is estimated by extrapolation of the

IOP Conf. Series: Materials Science and Engineering **73** (2015) 012104 doi:10.1088/1757-899X/73/1/012104

linear relationship between $(\alpha h\upsilon)^2$ and $h\upsilon$ according to the equation: $h\upsilon = A\ (h\upsilon\text{-}Eg)^{1/2}$ where α is the absorption coefficient, $h\upsilon$ is the photon energy Eg is the optical band gap and **A** is a constant.

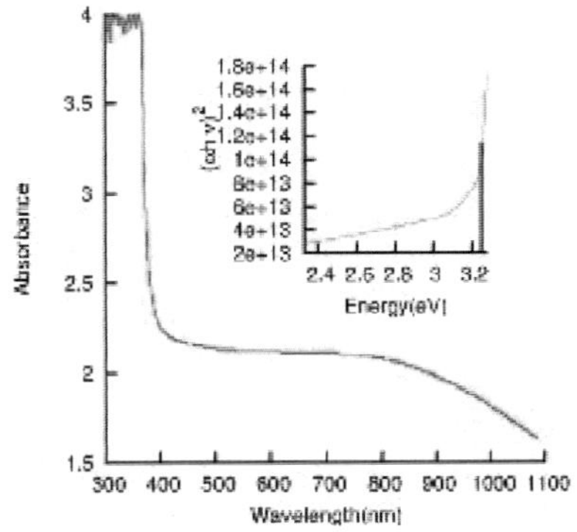

Figure 3 shows absorbance for the sample annealed in air. The inset indicates optical band gap energy of the sample.

The presence of a single slope in the plot suggests that the film has direct and allowed transition. The band gap of the pure sample is found to be 3.25 eV which is slightly smaller than that of bulk ZnO (3.37 eV). It is also reported that the band gap difference between the thin films is due to the grain boundaries and imperfections of the polycrystalline thin films [9]. The corresponding band band gap of iodine doped sample is reported in the table-1. The table-1 shows that band gap decreases with increase in grain size of the film.

3.4 Mechanism

From EDAX analysis it is observed that the atom. % of oxygen in pure and doped samples 27.34% and 30.97 whereas Zn is 72.66 and 69.03 respectively.

Native or intrinsic defects are imperfections in the crystal lattice that involve only the constituent elements. They include vacancies (missing atoms at regular lattice positions), interstitials (extra atoms occupying interstices in the lattice) and anti sites (a Zn atom occupying an O lattice site or vice versa) [10]. Iodine has one more electron than oxygen and when inserted on the oxygen site, it acts a shallow donor in ZnO. Therefore, the atom. % of Zn is low compared to pure sample. From this it is clear that the creation of point defect is the mechanism behind the oriented grain growth.

4. Conclusion

Oriented grain growth in ZnO thin film is enhanced by iodine doping. When the oriented grain growth increases, crystallinity of the thin film improves, resistance and band gap decrease. ZnO thin film having good crystallinity with preferential (002) orientation is a prerequisite for the fabrication of devices like UV diode lasers, acousto-optic devices etc. A possible mechanism for the oriented grain growth was also investigated. It is inferred that creation of point defects is responsible for the enhanced oriented grain growth in ZnO thin film by iodine doping.

Acknowledgement

The authors are thankful for financial support from University Grants Commission, Government of India through a research grant (MRP(S) - 986/ 10 - 11/ KLMG027/ UGC - SWRO).

IOP Conf. Series: Materials Science and Engineering **73** (2015) 012104 doi:10.1088/1757-899X/73/1/012104

Reference

[1] Xiangdong G, Xiaomin L and Weidong Y 2005 Preparation and characterization of highly oriented ZnO film by ultrasonic assisted silar method *J. of Wuhan University of Technology Materials Science Edition* 20 **23– 6.**

[2] Shelke V, Sonawane B, Bhole M and Patil D 2012 Electrical and optical properties of transparent conducting tin doped ZnO thin films *J. of Materials Science: Materials in Electronics* 23 **451–6.**

[3} Yen T, Strome D, Kim SJ, Cartwright A and Anderson W A 2008 Annealing studies on zinc oxide thin films deposited by magnetron sputtering. *J. of Electronic Materials* 37 **76499.**

[4] Luo W H, Tsai T K, Yang J C, Hsieh W M, Hsu C H and Fang J S 2009 Enhancement in conductivity and transmittance of zinc oxide prepared by chemical bath deposition. *J. of Electronic Materials* 38 **2264–9.**

[5] Kokubun Y, Kimura H, Nakagomi S 2003 Preparation of ZnO thin films on sapphire substrates by sol-gel method. Japanese Journal of Applied Physics 42 (Part 2, No. 8A) **L904–6.**

[6] Lee Y, Kim H and Roh Y 2001 Deposition of ZnO thin films by the ultrasonic spray pyrolysis technique *Japanese J. of Applied Physics* 40 (Part 1, No. 4A):**2423–8.**

[7] Pathan H and Lokhande C 2004 Deposition of metal chalcogenide thin films by successive ionic layer adsorption and reaction (silar) method *Bulletin of Materials Science* 27 **85–111.**

[8] Chandramohan R, Dhanasekaran V, Ezhilvizhian S, Vijayan T, Thirumalai J and Peter A 2012 Spectral properties of aluminium doped zinc oxide thin films prepared by silar method *J of Materials Science: Materials in Electronics* 23 **390–7.** .

[9] Saleem M, Fangl L, Wakeel A, Rashad M, Kong CY. Simple preparation and characterization of nano-crystalline zinc oxide thin films by sol-gel method on glass substrate. World Journal of Condensed Matter Physics 2012; 2(6):10–5.

[10] Jannotti A and G C. Fundamentals of zinc oxide as a semiconductor 2009 *Reports on Progress in Physics* 72 (126501) **126501a–126506a30.**

Studies on effect of oxygen flow rate in textured grain growth of ZnO thin films

Deepu Thomas[1,2,*], Sunil C Vattappalam[2] , Sunny Mathew[2] and Simon Augustine[2]

[1]Research and Development Centre, Bharathiar University, Coimbatore, India-641046.
[2]Department of Physics, St.Thomas College, Pala, Kottayam, India-686574.
E-mail: deepuskariankal@gmail.com

Abstract. ZnO thin films were deposited on glass substrate by Successive Ionic Layer Adsorption Reaction (SILAR) method. Effect of oxygen flow rate in textured grain growth, resistance and band gap of the thin films have been done. Textured grain growth of the samples were measured by comparing the peak intensities from XRD. Textured grain growth was found to be maximum when the oxygen flow rate is 2.5 litre/minute. It is found that as the oxygen flow rate increases above this limit, textured grain growth decreases and resistance the samples increases. The optical band gap of ZnO film was found to be increased with the increase of oxygen flow rate.

Key words: ZnO thin film, SILAR method, textured grain growth, Oxygen flow rate.

1. Introduction

ZnO thin films possess very interesting properties in the electrical and optical application fields. They have low electric resistance and high transparency in the visible wavelength range [1- 4]. The material is composed of cheap and abundant elements, and is readily produced for large-scale coatings. These films have attracted interest in many applications such as solar cells, gas and optical sensors, ultrasonic oscillators, transducers, optical waveguides and photoprotective coatings. Various deposition techniques can be used to prepare ZnO thin films, such as reactive and ion-assisted evaporation, laser ablation, spray pyrolysis [6], sol–gel processing [5], chemical vapour deposition [4] and sputtering[3]. In this work, ZnO films were deposited by chemical deposition techniques which is relatively low cost process and can be easily scaled up for industrial applications [7] . Even though, there are reports on studies of enhanced textured grain growth in ZnO under different experimental conditions , a study on the variation of textured grain growth in ZnO annealed in different oxygen flow rates were not yet studied widely. In course of the present investigations, the variations in the textured grain growth in ZnO thin films annealed in different oxygen atmosphere is studied. ZnO thin films having crystallinity with preferential orientation is a prerequisite for the fabrication of devices like UV diode lasers, acoustic-optic devices etc. [1-4].

2. Experimental

ZnO thin films were prepared by Successive Ionic Layer Adsorption Reaction (SILAR) method. In which, the ZnO thin film was coated on the glass substrate (26 × 76 mm) by alternately dipping the substrate in sodium zincate bath at room temperature and then in hot water maintained at 90- 95°C. The samples were annealed at 450° C for half an hour in different oxygen flow rates.The structural

*To whom any correspondence should be addressed. deepuskariankal@gmail.com

analysis of the thin films were done by X-ray diffraction . X-ray diffraction was performed on Bruker AXS-8 using CuK_α radiation. The resistance of ZnO thin film at room temperature was measured by Keithley 2100 6 1/2 Digital Multimeter. The optical absorbance was measured in the wavelength range of 190-1100 nm by means of a UV-VIS spectrophotometer. Textured grain growth of the samples was measured by using an orientation index (O. I.), which is the ratio of the intensities of (002) reflection ($2\theta = 34^\circ$) to (101) reflection ($2\theta = 36^\circ$) in the XRD.

i.e., O.I. = I_{002}/I_{101}

The average grain size of the ZnO crystals in the films was calculated by using Scherrer's formula: $D = 0.9\lambda/\beta \cos\theta$, where D is the grain size, λ- the wavelength of X-ray used, β- the FWHM(full width half maximum) and θ- Bragg angle [8].

3. Results and Discussion

3.3 Structural and Morphological Studies

The figure 1 (a) and (b) show the typical XRD patterns of ZnO thin films annealed at oxygen flow rates of 2.5 and 4.5 litre/minute.

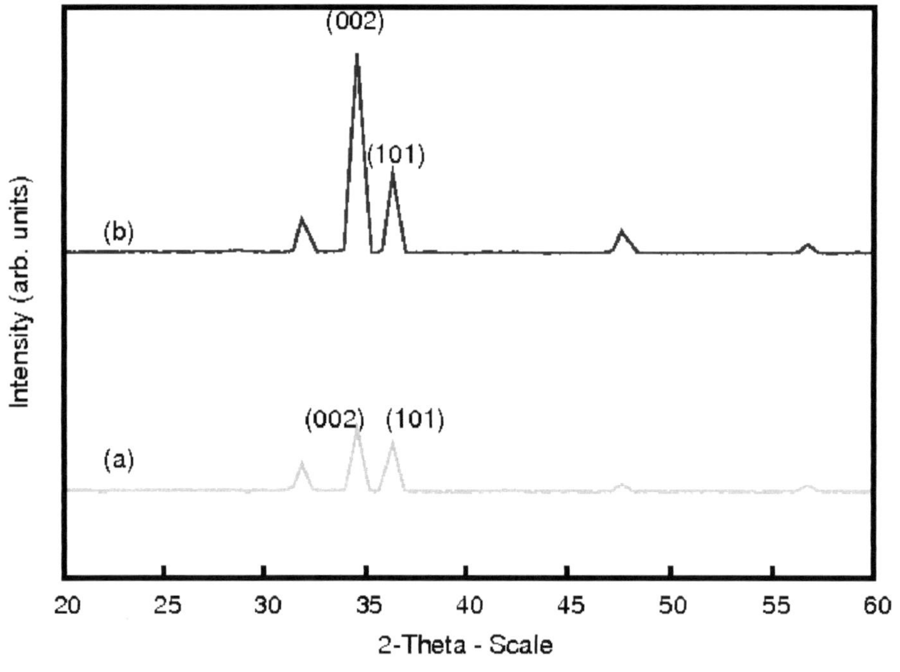

Figure 1*(a-b) shows the XRD pattern of sample annealed in oxygen atmosphere with flow rate at 4.5 litre/min and with flow rate2.5 litre /min. respectively.*

From fig.1 it can be seen that, the (002) peak has maximum intensity, which indicates that the film is c-axis-oriented perpendicular to the surface.The sample under the flow rate 2.5 liter/min has maximum orientation index. When the oxygen flow rate exceeds this limit, the textured grain growth was found to be decreased. Table 1 gives the flow rate,orientation index,grain size, resistance and band gap of different samples.

Table 1 gives the flow rate,orientation index, grain size, resistance and band gap of different samples.

Sample No.	Oxygen flow rate	Orientation Index (O.I.)	Grain size (D) in nm	Resistance in MΩ at room temp.	Band Gap in eV
1	Oxygen flow rate 4.5 litre/minute	1.31	26.5	6	3.26
2	Oxygen flow rate 3.5 litre/minute	1.72	27.2	5.5	3.23
3	Oxygen flow rate 2.5 litre/minute	2.23	28	4.8	3.21

3.2 Resistance Measurements

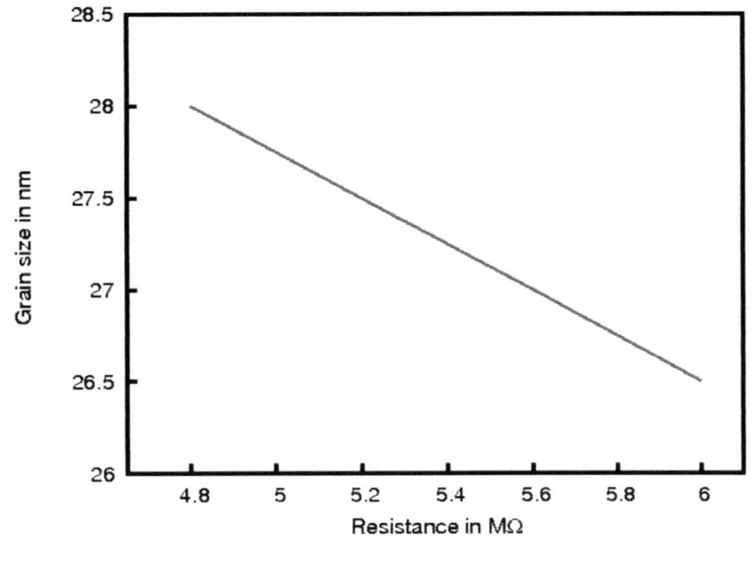

Figure 2 shows variation of resistance with grain size of different samples.

3.3 Optical Properties of ZnO Films

The figure 3 shows the optical absorbance spectrum of ZnO thin film (annealed in oxygen atmosphere with flow rate 4.5 litre/min) using UV-visible region. Figure 3 (inset) depicts the plot of $(\alpha h\nu)^2$ *vs* energy, $h\nu$.

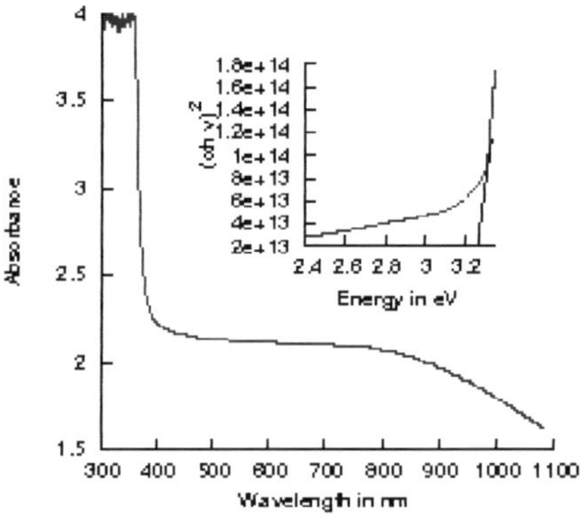

Figure 3 shows absorbance for the sample annealed in oxygen atmosphere with flow rate 4.5 litre/min. The inset indicates optical band gap energy of the sample.

The corresponding optical band gap of the sample is estimated by extrapolation of the linear relationship between $(\alpha hv)^2$ and hv, according to the equation $hv = A\ (hv - E_g)^{1/2}$, where α is the absorption coefficient, hv is the photon energy, E_g is the optical band gap and A is a constant. The values of direct band gap (E_g) are calculated from the intercept of $(\alpha hv)^2$ *vs* hv curve. The presence of a single slope in the plot suggests that the film has direct and allowed transition. The figure 4 shows the optical absorbance spectrum of ZnO thin film (annealed in oxygen atmosphere with a flow rate of 2.5 litre/min). Figure 4 (inset) depicts the plot of $(\alpha hv)^2$ *vs* energy, hv. Table-1 shows that band gap decreases with increase in grain size of the film [9].

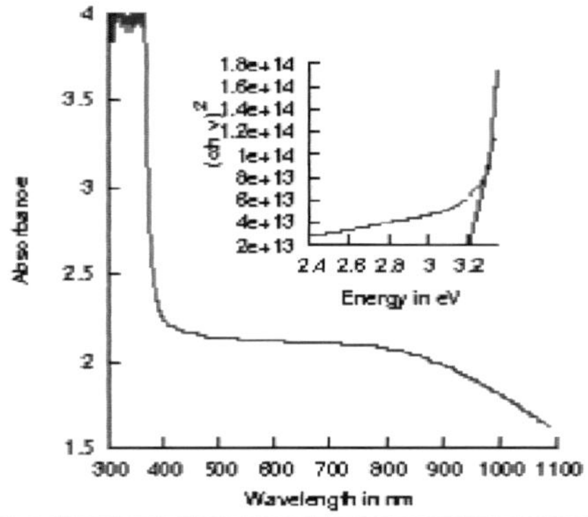

Figure 4 shows absorbance for the sample annealed in oxygen atmosphere with flow rate at 2.5 litre/min. The inset indicates optical band gap energy of the sample.

4. Conclusion

From the studies following conclusions can be drawn.Textured grain growth in ZnO thin film was found to be maximum when the oxygen flow rate is 2.5 litre/minute. It is found that as the oxygen flow rate increases above this limit, textured grain growth decreases and resistance the samples increases.The optical band gap of ZnO film was found to be increased with the increase of oxygen flow rate.

Acknowledgement

The authors are thankful for financial support from University Grants Commission, Government of India through a research grant (MRP(S) - 986/ 10 - 11/ KLMG027/ UGC - SWRO).

Reference

[1] Xiangdong G, Xiaomin L and Weidong Y 2005 Preparation and characterization of highly oriented ZnO film by ultrasonic assisted silar method *J. of Wuhan University of Technology Materials Science Edition* 20 **23– 6.**

[2] Shelke V, Sonawane B, Bhole M and Patil D 2012 Electrical and optical properties of transparent conducting tin doped ZnO thin films *J. of Materials Science: Materials in Electronics* 23 **451–6.**

[3} Yen T, Strome D, Kim SJ, Cartwright A and Anderson W A 2008 Annealing studies on zinc oxide thin films deposited by magnetron sputtering. *J. of Electronic Materials* 37 **76499.**

[4] Luo W H, Tsai T K, Yang J C, Hsieh W M, Hsu C H and Fang J S 2009 Enhancement in conductivity and transmittance of zinc oxide prepared by chemical bath deposition. *J. of Electronic Materials* 38 **2264–9.**

[5] Kokubun Y, Kimura H, Nakagomi S 2003 Preparation of ZnO thin films on sapphire substrates by sol-gel method. Japanese Journal of Applied Physics 42 (Part 2, No. 8A) **L904–6.**

[6] Lee Y, Kim H and Roh Y 2001 Deposition of ZnO thin films by the ultrasonic spray pyrolysis technique *Japanese J. of Applied Physics* 40 (Part 1, No. 4A):**2423–8.**

[7] Pathan H and Lokhande C 2004 Deposition of metal chalcogenide thin films by successive ionic layer adsorption and reaction (silar) method *Bulletin of Materials Science* 27 **85–111.**

[8] Chandramohan R, Dhanasekaran V, Ezhilvizhian S, Vijayan T, Thirumalai J and Peter A 2012 Spectral properties of aluminium doped zinc oxide thin films prepared by silar method *J of Materials Science: Materials in Electronics* 23 **390–7.** .

[9] Saleem M, Fang1 L, Wakeel A, Rashad M, Kong CY. Simple preparation and characterization of nano-crystalline zinc oxide thin films by sol-gel method on glass substrate. World Journal of Condensed Matter Physics 2012; 2(6):10–5.

Band gap and conductivity variations of ZnO thin films by doping with Aluminium.

Sunil C Vattappalam[1], Deepu Thomas[1, 2], Raju Mathew T, Simon Augustine[1] and Sunny Mathew[1,*]

[1]Department of Physics, St.Thomas College, Pala, Kottayam, India-686574.
[2]Research and Development Centre, Bharathiar University, Coimbatore, India-641046.
*E-mail: sunmatn@gmail.com

Abstract. Zinc Oxide thin films were prepared by Successive Ionic layer adsorption and reaction technique(SILAR). Aluminium was doped for different doping concentrations from 3 at.% to 12 at.% in steps of 3 at.%. Conductivity of the samples were taken at different temperatures. UV Spectrograph of the samples were taken and the band gap of each sample was found from the data. It was observed that as the doping concentration of Aluminium increases, the band gap of the samples decreases and concequently conductivity of the samples increases.

Key words: ZnO thin film, SILAR method, Band gap.

1. Introduction

Zinc Oxide is a group II-VI compound semiconductor with a wide band gap of 3.37 eV [1] at room temperature. Thin films of ZnO are widely used in solar cells, surface acoustic wave devices(SAW), heat mirror coatings and recently as Solid state gas sensors. It is reported that structral, elctrical and optical properties are changed by doping with various metal ions like In,Ga,Al [2] etc. Zinc Oxide thin films can be prepared by various methods such as spray pyrolisis, magnetron sputtering, plasma enhanced chemical vapour deposition, pulsed laser deposition, sol-gel process, vacuum evaporation etc. In the present investigation, ZnO thin films were prepared by SILAR [3] method which is simple and involves low cost equipments.

Though studies on dc and ac conductivity of ZnO thin films have received considerable attention, there have been only a few studies on the variation of conductivity with doping concentrations. In the present investigation, variation of band gap and conductivity of the samples with different doping concentrations is studied. The doping concentrations are from 3at.% of Aluminium[4] to 12 at.% varied in steps of 12 at.% and the conductivity of the samples were taken from room temperature to 210°C.

2. Experimental

ZnO thin films were prepared by successive ionic layer adsorption and reaction (SILAR) method. Doping was done by adding adequate amount of Aluminium Sulphate into sodium zincate bath before dipping. The films thus prepared were annealed at 400° C for 1 hour. The conductivity of the samples were taken by two probe method by placing the sample inside an oven and consecutively heating to 210° C from room temperature. Conductivity at different temperatures are noticed and a conductivity curve was obtained. UV spectrograph of the same same samples were taken and band gap of each sample was found from the data.

*To whom any correspondence should be addressed. sunmatn@gmail.com

IOP Conf. Series: Materials Science and Engineering **73** (2015) 012106 doi:10.1088/1757-899X/73/1/012106

3. Results and Discussion

The figures 1 to 4 shows the variation in band gap of the samples at different doping concentrations starting from 3at.% to 12 at.%.

.

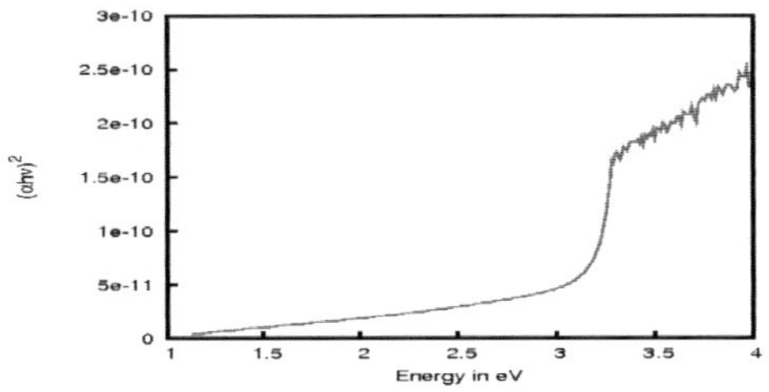

fig 1

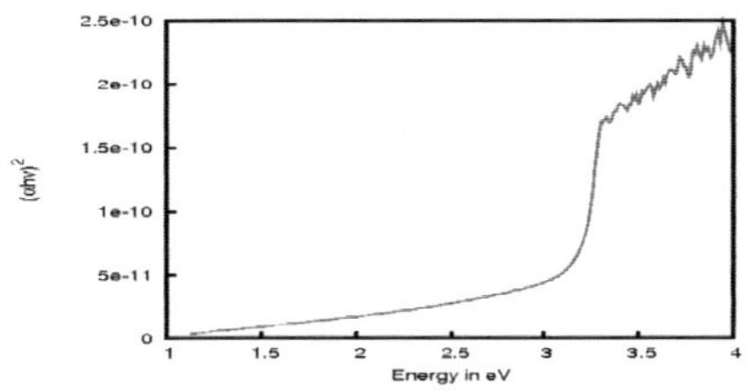

fig 2

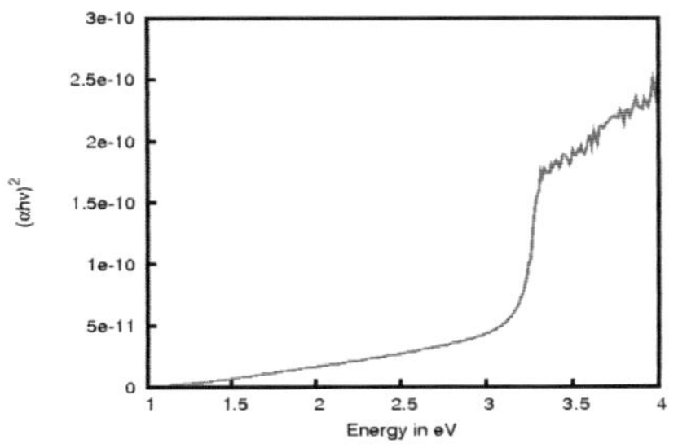

fig 3

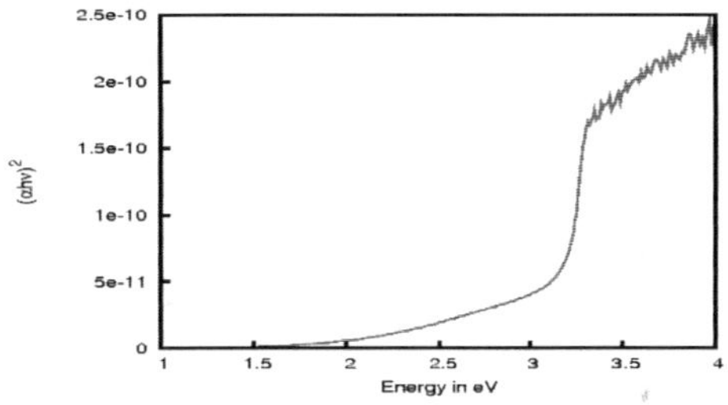

fig 4

The figures 1 to 4 shows the variation in band gap of the samples at different doping concentrations. Fig (1) 3% Al, fig (2) Al 6 %, fig (3) Al 9% and fig (4) Al 12.%

Table 1 gives the Band Gap Values of the samples

Name of the sample	Doping percentage	Band gap observed
Sample 1	3 % Aluminium	3.11 eV
Sample 2	6% Aluminium	3.09 eV
Sample 3	9% Aluminium	3.084 eV
Sample 4	12% Aluminium	3.04 eV

IOP Conf. Series: Materials Science and Engineering **73** (2015) 012106 doi:10.1088/1757-899X/73/1/012106

The following figures from fig 5 to fig 8, gives the conductivity of the samples in the order of band gap and according to doping concentrations.

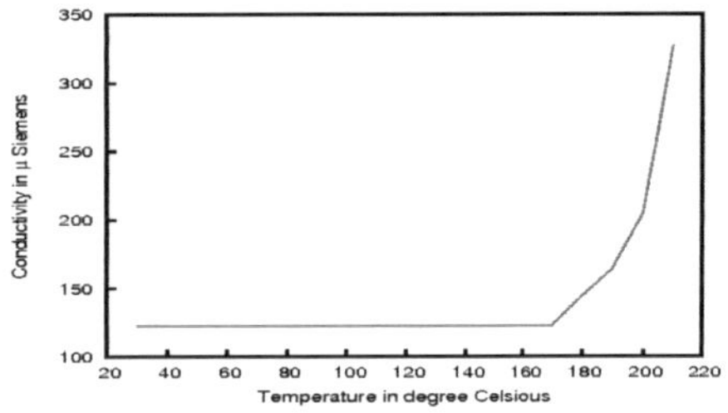

fig 5

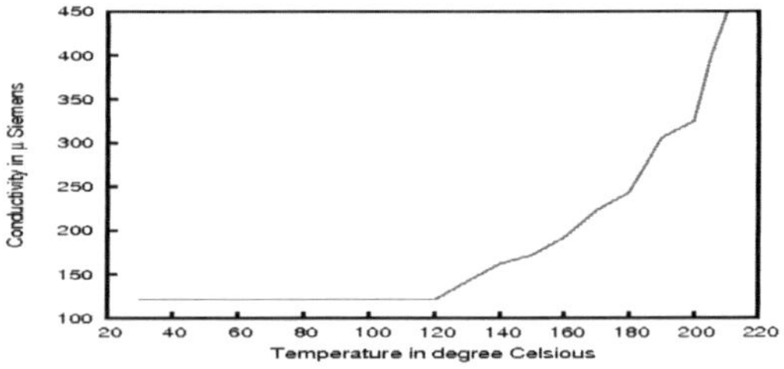

fig 6

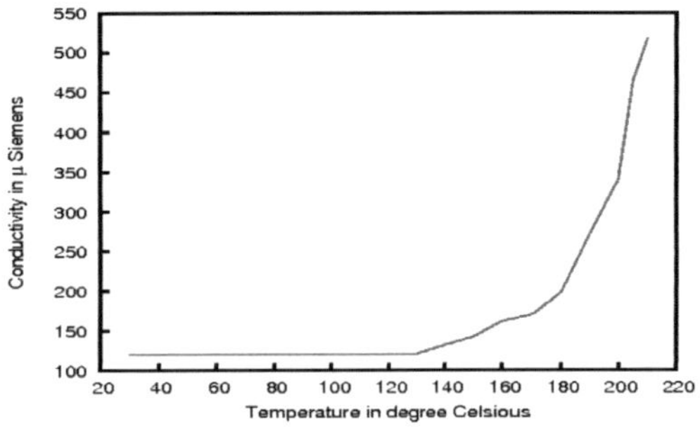

fig 7

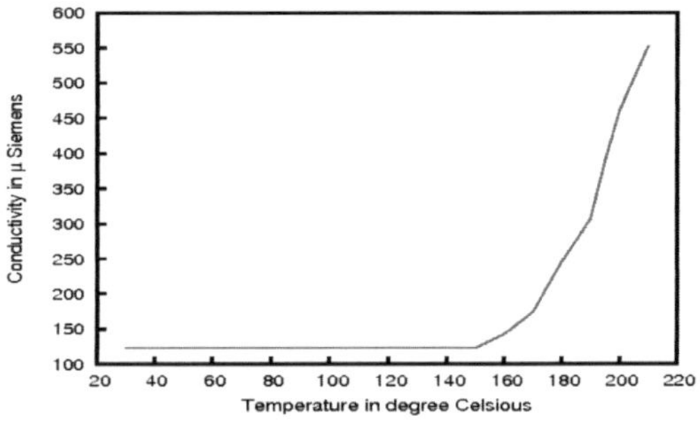

fig 8

The fig 5 to fig 8, gives the conductivity of the samples in the order of band gap and according to doping concentrations. Fig (5) 3% Al, fig (6) Al 6 %, fig (7) Al 9% and fig (8) Al 12.%

Table 2 gives the conductivity of the samples at 210 °C.

Name of the sample	Doping percentage	Maximum conductivity observed at 210 °C
Sample 1	3 % Aluminium	553micro Siemens
Sample 2	6% Aluminium	518micro Siemens
Sample 3	9% Aluminium	448 micro Siemens
Sample 4	12% Aluminium	327.5micro Siemens

IOP Conf. Series: Materials Science and Engineering **73** (2015) 012106 doi:10.1088/1757-899X/73/1/012106

4. Conclusion

Band gap in ZnO thin films prepared by silar method can be decreased by increasing the doping concentrations of the transition metal. Also a decrease in band gap cause an increase in conductivity. So in ZnO thin films prepared by silar method, the doping concentration is directly proportional to conductivity.

5. Acknowledgment

Raju Mathew T is greatful to University Grants Commission for providing financial assistance for this work.

Reference

[1] R Hong, H Qi, J Huang, H He, Z Fana, J Shao, Thin Solid Films 473 (2005) **58 – 62.**

[2] Zhibing Zhan,Jiye Zhang, Qinghong Zheng, Danmei Pan, Jin Huang, Feng Huang and Zhang Lin crystal growth and design communication Vol. 11(2011) **21–25.**

[3] T A Vijayan, R Chandramohan, Santhiyagu Valanarasu, Jagannathan Thirumalai Science and technology of advanced materials.9(2008) 035007 (5pp)

[4] R Chandra Mohan, V Dhanasekharan, S Ezhilvizhiyan, T A Vijayan matter Sci: Mater Electron(2012)23:**390-397**

[5] S Mondal, K P Kanta and P Mitra J.of physical sciences, Vol.12,(2008),**221-229**.

IOP Conf. Series: Materials Science and Engineering **73** (2015) 012107 doi:10.1088/1757-899X/73/1/012107

Band gap and conductivity variations of ZnO nano structured thin films annealed under Vacuum.

Sunil C Vattappalam[1], Deepu Thomas[1,2], Raju Mathew T, Simon Augustine[1] and Sunny Mathew[1,*]

[1]*Department of Physics, St.Thomas College, Pala, Kottayam, India-686574.*
[2]*Research and Development Centre, Bharathiar University, Coimbatore, India-641046.*
*E-mail: sunmatn@gmail.com

Abstract. Zinc Oxide thin films were prepared by Successive Ionic layer adsorption and reaction technique(SILAR). The samples were annealed under vacuum and conductivity of the samples were taken at different temperatures. UV Spectrograph of the samples were taken and the band gap of each sample was found from the data. All the results were compared with that of the sample annealed under air. It was observed that the band gap decreases and concequently conductivity of the samples increases when the samples are annealed under vacuum.

Key words: ZnO thin film, SILAR method, Band gap.

1. Introduction

Due to wide band gap(3.37 eV) [1-4] and large exciton binding energy(~60 meV), the importance of ZnO thin films in opto-electronic devices have attracted more attention. ZnO is a promising candidate material for advanced devices applications due to the above mensioned charecteristics and unique combination of its physical properties-optical , electric, magnetic, piezoelectric and ferroelectric. Zinc Oxide is a group II-VI compound semiconductor and a large number of methods are used for the fabrication of ZnO thin film such as, sputtering, molecular beam epitaxy, metal-organic chemical vapour deposition, pulsed laser deposition etc[5-8]. Semiconducting metal oxides have been known for decades to be good gas sensing materials. Ethanol sensors based on SnO2 thick films have been commercialized for years . Thin films of ZnO are widely used in solar cells, surface acoustic wave devices(SAW), integrated optics, liquid crystal display, light emiting diode, laser diode, heat mirror coatings and recently as Solid state gas sensors. It is reported that structral, elctrical and optical properties are changed by doping with various metal ions like In,Ga,Al [2, 9, 10] etc
Though studies on dc and ac conductivity of ZnO thin films have received considerable attention, there have been only a few studies on the variation of conductivity with annealing under vacuum. In the present investigation, variation of band gap and conductivity of the samples with different annealing atmospheres is studied. The samples are annealed under vacuum and are compared with that annealed in air.

2. Experimental

ZnO thin films were prepared by successive ionic layer adsorption and reaction (SILAR) method by successively dipping the glass substrate in Sodium zincate solution and nearly boiling water. The films were deposited on a cleaned microscopic glass slide of known thickness and weight. The films thus prepared were annealed under vacuum at 400° C. for 30 minutes. The conductivity of the samples

*To whom any correspondence should be addressed. sunmatn@gmail.com

IOP Conf. Series: Materials Science and Engineering **73** (2015) 012107 doi:10.1088/1757-899X/73/1/012107

were taken by two probe method by placing the sample inside a PID controlled oven and consecutively heating to 210° C from room temperature. Conductivity of the samples, annealed in air and in vacuum, at different temperatures are noticed and a conductivity curve was obtained. UV spectrograph of the same same samples were taken and band gap of each sample was found from the data. The UV and conductivity data of the samples annealed in air and that annealed under vacuum are compared.

3. Results and Discussion

The following figures from fig 5 to fig 8, gives the conductivity of the samples according to annealing in air and under vacuum.

.

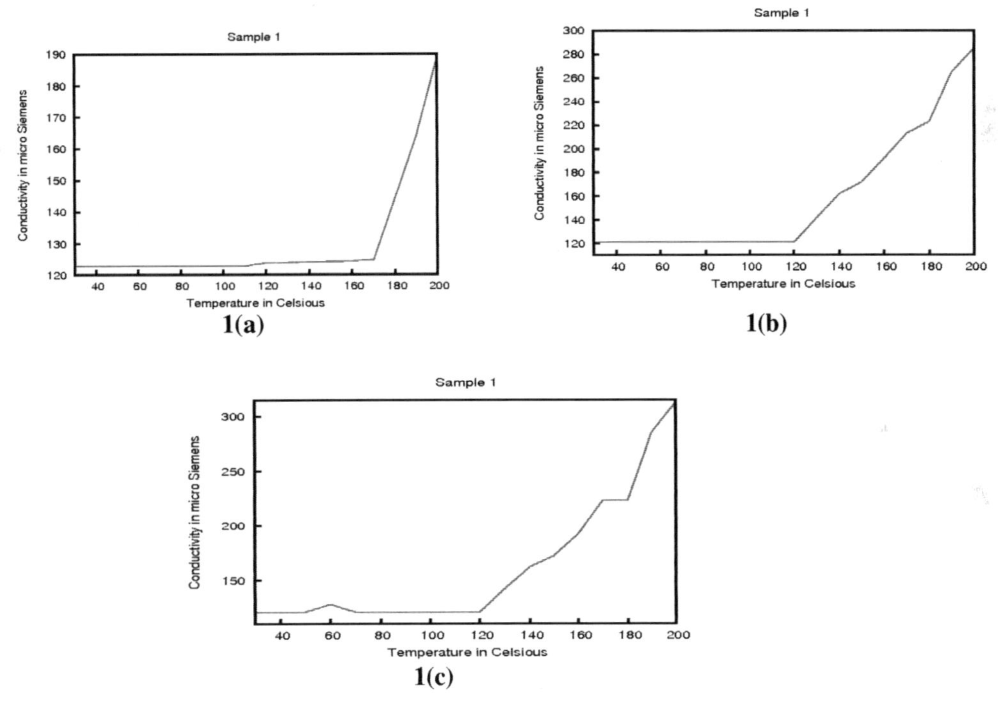

Figure 1(a-c) ***shows the (a) Sample annealed in air ,(b) annealed under vacuum at 400° C. for 30 minutes, (c) annealed under vacuum at 400° C. for 1 Hour.***

IOP Conf. Series: Materials Science and Engineering **73** (2015) 012107 doi:10.1088/1757-899X/73/1/012107

Table 1 shows the condctivity of the different samples

Name of the sample	Annealing	Maximum conductivy observed at 200 °C
Sample 1	Air	188.5 micro Siemens
Sample 2	30 min. in vacuum	285 micro Siemens
Sample 3	1 hr in vacuum	315 micro Siemens

The following figures shows the UV, and band gap graph of samples annealed in air and in vacuum respectivrly.

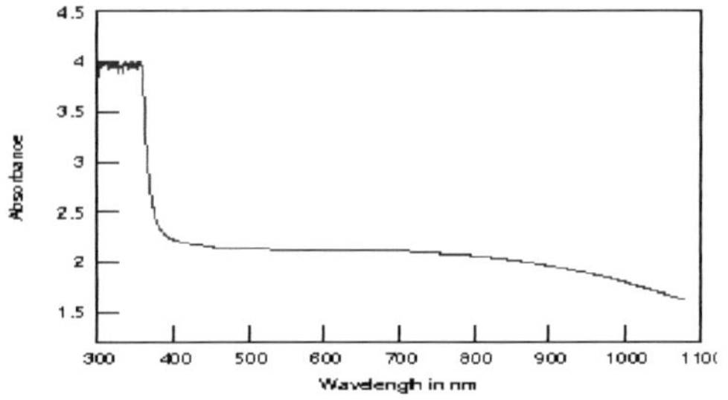

fig 2

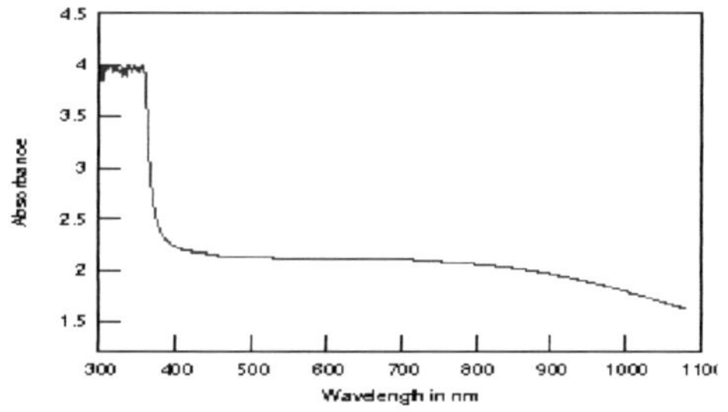

fig 3

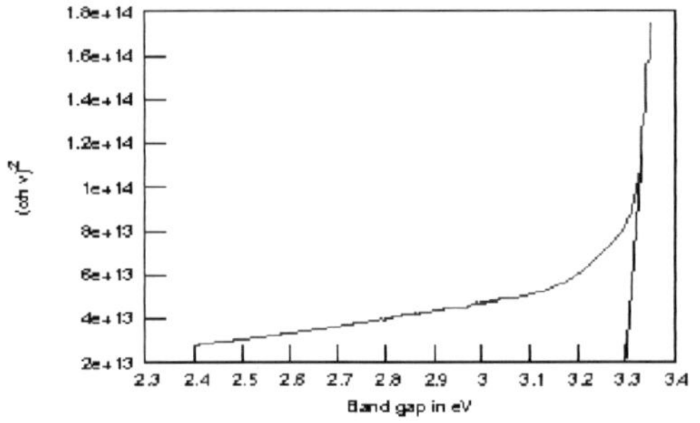

fig 4

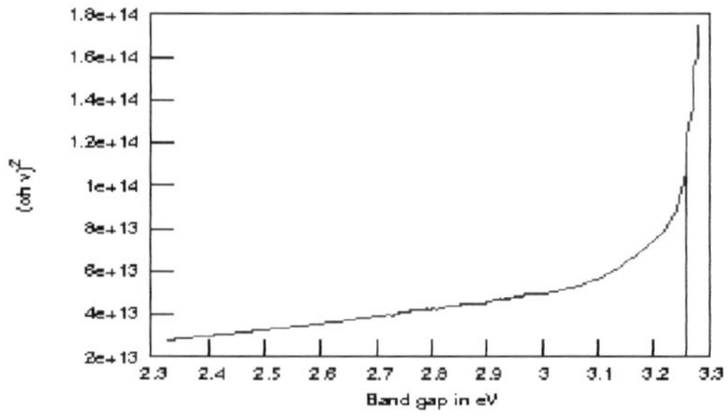

fig 5

Fig 2 and fig 3 gives the UV of the samples annealed in air and in vacuum respectively. Fig 4 and fig 5 gives the band gap of the samples annealed in air and under vacuum respectively.

Table 2 gives the Band Gap Values of the samples according to annealing.

Name of the sample	Annealing atmosphere	Band gap
Sample1	Annealed in air	3.30eV
Sample 2	Annealed under vaccum	3.26eV

4. Conclusion

Band gap in ZnO thin films prepared by silar method can be decreased by annealing the sample in vacuum instead of air. Also a decrease in band gap cause an increase in conductivity. So in ZnO thin films prepared by silar method, the conductivity of the sample incerase as the sample is annealed under vacuum.

5. Acknowledgement

Raju Mathew T is greatful to University Grants Commission for providing financial assistance for this work.

Reference

[1] Ruijin Hong, Hongji Qi, Jianbing Huang, Hongbo He, Zhengxiu Fana, Jianda Shao Thin Solid Films 473 (2005) **58 – 62.**

[2] O Lupan, T Pauporte, L Chow, B Viana, F Pelle, Applied Surface Science jourbnal, 256(2010) **1895-1907**

[3] S.J, Chen J Crystal growth 254(2003) 86

[4] S T Tann, B J Chen X W Sun, W J Fan, H S Kwok, S J Chua, Applied Physics 98(2005) 013505

[5] C Bundesmann, N Ashkenov, R Sauer, B Mensching, C LIu, applied Physics Letter 72(1998) 2589

[6 D C Look, D C Reynold, J W Hemskey, Applied Physics Letter 75(1995) 811

[7] Zhibing Zhan,Jiye Zhang, Qinghong Zheng, Danmei Pan, Jin Huang, Feng Huang and Zhang Lin crystal growth and design communication Vol. 11(2011) **21–25.**

[8] T A Vijayan, R Chandramohan, Santhiyagu Valanarasu, Jagannathan Thirumalai Science and technology of advanced materials.9(2008) 035007 (5pp)

[9] R Chandra Mohan, V Dhanasekharan, S Ezhilvizhiyan, T A Vijayan matter Sci: Mater Electron(2012)23:**390-397**

[10] S Mondal, K P Kanta and P Mitra J.of physical sciences, Vol.12,(2008),**221-229**.

IOP Conf. Series: Materials Science and Engineering 73 (2015) 012108 doi:10.1088/1757-899X/73/1/012108

Microhardness studies of sulfamic acid single crystal

A. Santhosh Kumar[1], Cyriac Joseph [1], Reshmi Paulose[1], Rajesh R [2],Georgekutty Joseph[3] and Godfrey Louis[4]

[1]School of Pure & Applied Physics, Mahatma Gandhi University, Kottayam, Kerala, India

[2] Department of Physics, N S S College, Nenmara,Palakkad, Kerala, India

[3]Department of Physics, S H College, Thevara, Ernakulam, Kerala, India

[4]Department of Physics, Cochin University of Science and Technology, Cochin - 682024, Kerala, India

Email: skumarspap@yahoo.com

Abstract. Vicker's microhardness study of (100), (010) and (001) faces of a non-linear optical crystal sulfamic acid have been reported. Single crystals of sulfamic acid have been grown by slow evaporation method. The load dependence of the Vickers microhardness of sulfamic acid crystal were investigated and analyzed from the stand point of various theoretical models. Crystal samples in a, b and c-axes exhibit reverse indentation effect which is best described by Meyer's law, Hays-Kendall's approach and proportional specimen resistance (PSR) models. The negative values of load dependent quantities in Hays-Kendall's approach and PSR model suggest that the origin of indentation size effect is associated with the process of relaxation of indentation stresses.

1. Introduction

The wide range of industrial applications of the salts of sulfamic acid (NH_3SO^3) necessitates a thorough study of their various physical properties in the crystalline state. The crystal structure of sulfamic acid was reported to be orthorhombic with lattice parameters a=8.066, b=8.115 and c=9.255A^o and have space group Pbca [1]. It has been suggested that it exist as a zwitterion, $NH_3^+SO_3^-$, in the solid state [1, 2] and have nonlinear optical properties [3]. The elastic properties and thermal expansion of sulfamic acid and its sulfamates was reported by Haussuhl and Haussuhl [4]. Sulfamic acid undergoes a single stage irreversible endothermic transition at 448 K, where decomposition starts and the material was fully decomposed at 477 K [3]. In an earlier study we reported the dc electrical conductivity anomalies near 340 K and around 410 K in this crystal along a, b and c-axes and these anomalies are attributed as due to a phase transition and the onset of the thermal decomposition respectively [5]. Differential scanning calorimeter and differential thermal analysis also supported the possible phase transition near 340K [5]. Ramesh et al [6] reported the dielectric constants of sulfamic acid crystal in the frequency range 10 - 103 Hz along with the Micro-Raman spectrum. The dielectric behavior was also studied by Lenin et al. [7].

One of the important properties of any device material is its mechanical strength, represented by its hardness [8]. The fastest and simplest type of mechanical testing is hardness measurement. Various other mechanical properties of the materials such as toughness, brittleness, yield strength, etc. have specific correlation with hardness [9-11]. Among the different testing methods, the Vickers hardness test method is more commonly used. The microhardness have been measured on the a, b and c-axes of sulfamic acid crystals at room temperature in the load region (5 to 100 g) in steps of 5 gm. where hardness is found to be a function of indenter load. The Meyer index otherwise known as the work-hardening exponent is computed using least squares fit analysis for different load regions and the results are discussed.

2. Experimental

Single crystals of sulfamic acid have been grown by slow evaporation method from an aqueous solution at 308 K with threefold recrystallization from distilled water. Transparent crystals of size 33 × 30 × 20 mm with good optical quality are obtained after a period of 60 -70 days. The grown crystals were examined by X-ray diffraction analysis. The powder diffraction pattern of the sample crystal is agreed well with that reported in the literature [JCPDS file No. 74-0342]. The natural faces of the crystal are identified by measuring the interfacial angle of the crystal and comparing it with calculated angles. Bulk samples have been cut into required thickness using a slow speed diamond wheel saw (MTI Corp. USA). The indentations were performed along a, b and c-axes at a very slow rate and for all indentations; care was taken to see that the rate was nearly the same. The indentation mark was square in shape. The diagonals of indentation mark were measured using micrometer eyepiece. The indentations using Vickers' pyramidal diamond indenter were made at different loads ranging from 1 to 100 gm. Indentations were carried out in the low load region at different sites such that the distance between any two indentations was greater than three times the diagonal length, in order to avoid any mutual influence of the indentations. The measurements were made at room temperature and the time of indentation was kept at 10 sec throughout the work. Subsequent impressions were made after a time lapse of 10 min. to allow for any elastic recovery. The two diagonals of the indentation left in the surface of the crystal after removal of the load were measured using a micrometer eyepiece and their average length was calculated. The Vickers hardness is defined as the ratio of applied load to the pyramidal contact area of indentation and it is calculated as

$$H_V = 1.8544\ (p/d^2)\ \mathrm{kg/mm^2}$$

Where H_v = Vickers' microhardness in Kg/mm^2, P = Applied load in gm. d = Mean diagonal length of the indentation mark in μm. The indentation mark is geometrically similar whatever be its size. This would imply the hardness to be independent of load. However, this is not the case and except for loads exceeding about 200 gm. in general, the measured hardness value has been found to depend on load in almost all cases and hence the hardness values measured in the low load region, are known as microhardness values. Though, the limit load is not sharply defined, practically the hardness may achieve a constant value for loads in the range 20–50 gm and beyond, depending on the material [12].

3. Results and discussions

3.1. Load variation of hardness

The variation of microhardness with applied indentation test load is shown in figure 1.The plots indicate clearly that the hardness varies with load in a complex manner. Starting from smallest load used, the hardness increases up to a load of about 50 gm. Beyond 50 gm., it reaches saturation. In

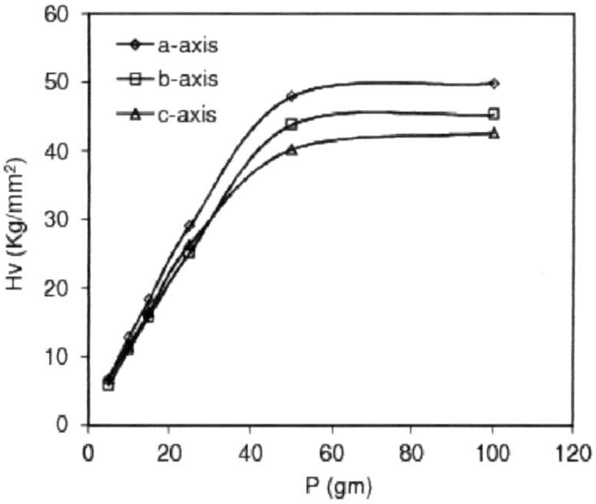

Figure 1. Microhardness vs Load plot for different axes

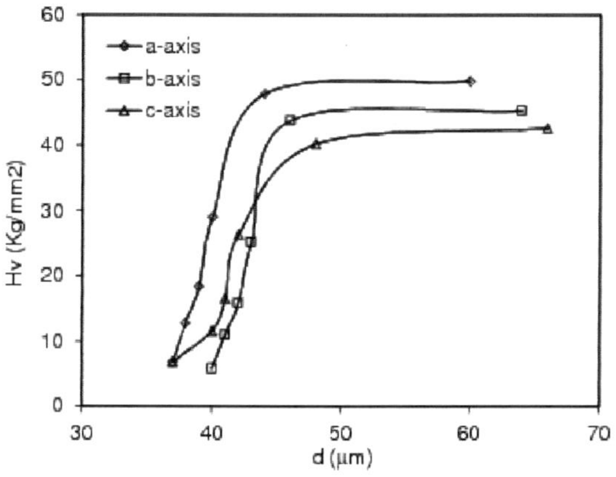

Figure 2. Reverse ISE for a, b and c-axes of sulfamic acid crystal

general the hardness varies considerably in the low load region as the work hardening capacity and elastic recovery of a particular material are dependent on the load, type of surface receiving the load, and the depth to which the surface is penetrated by the indenter. For example, the low load hardness behavior in the case of silicon single crystal has been explained on the basis of elastic recovery and piling up of material around the indentation mark [13]. Both the magnitude of work hardening and the depth to which it occurs depend on the properties of the material and are the greatest for soft metallic materials which can be appreciably work hardened. Since the penetration depth at high loads is usually greater than that of the work hardened surface layer, the hardness value at high loads will be representative of the undeformed bulk of the material and hence independent of load [12].

The apparent microhardness of solids depends on the applied indentation test load. This phenomenon, known as the indentation size effect (ISE), usually involves a decrease in the microhardness with increasing applied test load, i.e. with increasing indentation size [14- 18]. In contrast to the above normal ISE, a reverse type of indentation size effect (called reverse ISE), where the apparent microhardness increases with increasing applied test load, is also known [19-21].

It is observed that the sulfamic acid crystal exhibited reverse indentation size effect (ISE), in which an increase in the apparent microhardness with increasing indentation size. The variation of microhardness with increasing indentation size is shown in figure 2. Although the ISE has been explained by dislocation behaviour during indentation [14, 22], the reverse ISE has been explained in terms of the existence of a distorted zone near the crystal medium interface, effects of vibration, and indenter bluntness at low loads [19, 21].

3.2. log P vs log d plots

In order to describe the ISE behaviour of materials, several models for the relationship between applied indentation test load and indentation diagonal length have been reported in the literature. The most common explanation of the ISE found in the literature is directly related to the intrinsic structural factors of the test material [23, 16-18]. According to this theory, in order to analyze the ISE in the hardness testing, one needs to fit the experimental data according to Meyer's law, which correlates the applied load P and the resulting indentation size d with each other:

$$P = Ad^n$$

where the exponent n, called as the Meyer number (or index), is a measure of the ISE and A is a constant. The values of n and A of above equation may be obtained from the plots of log P against log d. Figure 3 shows the log P vs log d plots. The ISE is usually related to the deviation of the n value from 2, for n equal to 2 is the absence of an ISE or the hardness is independent of the applied test load [24]. When n > 2, there is the reverse ISE behaviour. The graph is more linear for loads greater than

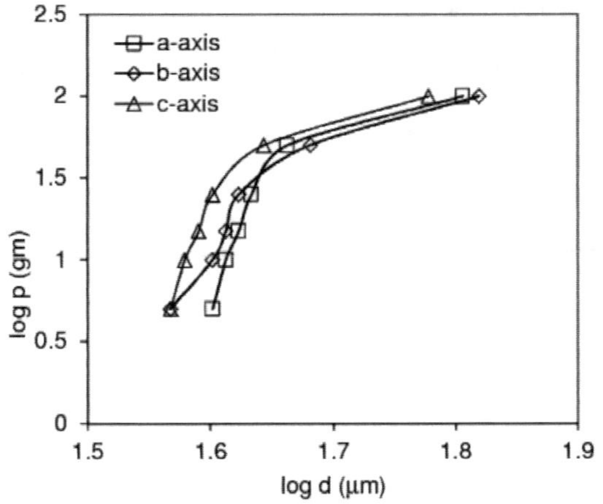

Figure 3. log P vs log d plots for a, b and c-axes

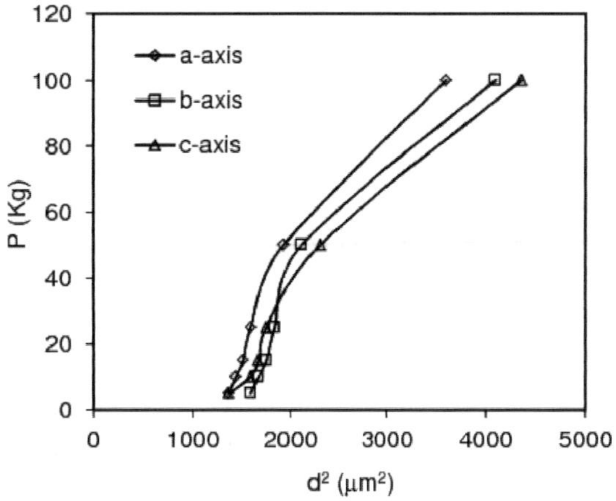

Figure 4. Plots of P against d^2 for a, b and c-axes

50 gm. So we consider the best fit plots for the data at P≥ 50 gm. The values obtained for the Mayer's index are 2.1212, 2.099, and 2.176 for a, b and c axis respectively, for loads greater than 50 gm. The experimental data show that value of n is always greater than 2. This reveals that the crystal samples exhibit reverse indentation size effect. Onitsch [25] and Hanneman [26] from careful observations on various materials pointed out that n lies between 1 and 1.6 for hard materials and it is more than 1.6 for soft materials. The 'n' values observed in the present studies are above 1.6 suggesting that Sulfamic acid crystal is a soft material.

3.3. Hays and Kendall's approach

According to this approach [27] there is a threshold load W necessary to initiate permanent deformation, and the load dependence of hardness may be expressed by

$$P = W + A_1 d^2$$

where W is the minimum load to initiate plastic (permanent) deformation, A_1 is a load-independent constant. The values of W and A_1 may be calculated by plotting the experimental P(d) data as P against d^2 plots. Figure 4 shows the plots of P against d^2.

Table 1. Values of A_1 and W for a, b and c-axes

Axes	Load range	A_1(g/μm²)	W(g)
a-axis	Low P	0.08079	-104.63
	High P	0.02801	-4.23
b-axis	Low P	0.08983	-138.73
	High P	0.02525	-3.424
c-axis	Low P	0.05066	-64.35
	High P	0.03183	-38.65

It is observed that the plots of P against d^2 for a, b and c axes are very similar and consists of two linear parts for d < dc and d > dc, respectively. The calculated values of W and A_1 for different samples are listed in Table 1. It may be noted from Table 1 that the value of W is negative for all loads. The negative values of W express the reverse ISE. This suggests that the crystal samples exhibit reverse indentation size effect. In the range of d < dc there are high negative values of W for a- and b-axes. Very high negative values of W suggest that a sudden increase in the value of Hv for d < dc does not occur by this mechanism [28].

3.4. Proportional specimen resistance model

Several workers [16-17, 29-30] have proposed that the normal ISE behaviour may be described by the relation

$$P = ad + bd^2$$

where the parameter a characterizes the load dependence of hardness and b is a load-independent constant. The term ad has been attributed to the specimen surface energy [29-30], the deformed surface layer, the indenter edges acting as plastic hinges, and the proportional specimen resistance [16-17]. The constants a and b may be obtained from the plots of P/d against d for the samples. Figure 5 shows the plots of P/d against d for a, b and c-axes

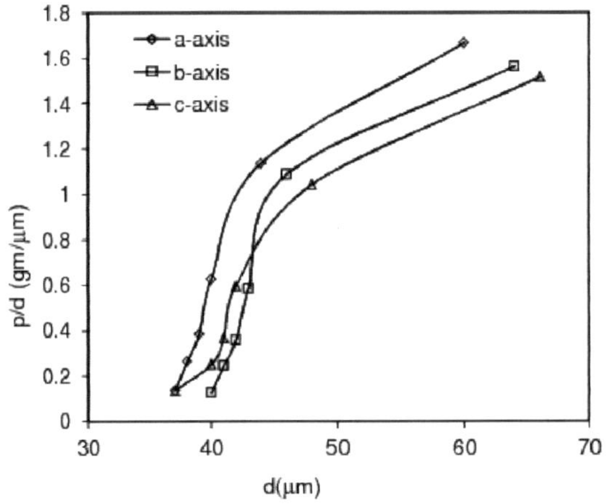

Figure 5. Plots of *P/d* against *d* for a, b and c-axes

It should be noted that $a > 0$ results in the appearance of normal ISE in which H_v decreases with an increase in indentation diagonal d. However, when $a < 0$ a reverse ISE, in which H_v increases with an increase in indentation diagonal d, is produced. According to Li and Bradt [16], the value of *a* is positive when elastic surface stresses are compressive. This leads to a reduction in the value of d upon unloading. This implies that the value of *a* is negative when surface stresses are tensile, which lead to a relaxation of these surface stresses introduced by indentation. There are several cases of the occurrence of normal and reverse ISE for different materials [14]. It was suggested that the value of *a* can be negative both due to the development of indentation cracks as well as motion of dislocations in the form of dislocation rosettes.

It was observed that, the [P/d] vs. (d) plots for a, b and c-axes are very similar and consists of two practically linear parts for $d < d_c$ and $d > d_c$, respectively. Values of *a* and *b* for different samples are calculated and is listed in Table 2. It should be noted that, as expected for reverse ISE, the values of *a* are negative. In the case of reverse ISE similar observations were made earlier for other materials [14]. According to Li and Bradt [16] when elastic surface stresses are compressive the sign of *a* is positive. The negative values of *a* suggest that the surface stresses are tensile, which implies that reverse ISE is associate with the relaxation of these surface stresses introduced by indentation. However, in the present case of sulfamic acid crystals relaxation of the surface stresses occurs as a result of development of indentation cracks.

Table 2. Values of a and b for different samples

Axes	Load range	a (g/μm)	b(g/μm^2)
a-axis	Low P	-5.24	0.14539
	High P	-0.1656	0.02959
b-axis	Low P	-6.44	0.16404
	High P	-0.1284	0.02642
c-axis	Low P	-3.08	0.08691
	High P	-0.2193	0.0628

4. Conclusion

The growth and microhardness study of sulfamic acid crystal are carried out. The grown crystal was almost defect-free and was identified by X-ray diffraction technique. Vickers microhardness has been studied along a, b and c-axes and the results obtained were on predicted lines. The values of microhardness show an increase with increasing the test load. This phenomenon gives the idea of reverse ISE. The values of Meyer's index obtained for a, b and c- axes of the crystal are also in agreement with the above observation. Analysis of the experimental data on hardness as a function of indentation size from the stand point of Hays- Kendall's approach and the proportional specimen resistance model revels that these models satisfactorily explain the reverse ISE on the different axes of sulfamic acid crystal.

References

[1] Kanda FA and King AJ 1951 *J Am Chem Soc* **73** 2315

[2] Sass RL 1960 *Acta Cryst* **13** 320

[3] Valluvan R, Selvaraju K and Kumararaman S 2006 *Mat Chem Phys* **97** 81

[4] Haussuhl E and Haussuhl S 1995 *Zeitschrift fur kristallographic* **210** 269

[5] Kumar AS, Varughese G, Iype L, Rajesh R, Joseph G and Louis G 2010 *Cryst Res Technol* **45** 879

[6] Babu RR, Sethuraman K, Vijayan N, Gopalakrishnan R and Ramasamy P 2007 *Mat Letts* **61** 3480

[7] Lenin M, Balamurugan N and Ramasamy P 2007 *Cryst Res Technol* **42** 39

[8] Subhadra KG, Rao KK and Sirdeshmukh D B 2000 *Bull Mater Sci* **23** 147

[9] Bergner F, Bergmann U, Schaper M, Hammer R and Jurisch M 2001 *Material Prüfung* **25** 117

[10] Lal B, Bamzai KK and Kotru P N 2003 *Materials Chem & Phys* **78** 202

[11] Wooster W A 1953 *Rep Prog Phys* **16** 62

[12] Desai C F, Maunik J, Soni P H and Pandya GR 2009 *J Mater Sci* **44** 3504

[13] Walls MG, Chaudhri MM and Tang TB 1992 *J Phys D Appl Phys* **25** 500

[14] Sangwal K and Surowska B 2003 *Mat Res Innovat* 7 91

[15] Bull SJ, Page TF and Yoffe E H 1989 *Phil Mag Lett* **59** 281

[16] Li H and Bradt RC 1993 *J Mater Sci* 28 917

[17] Li H, Han YH and Bradt RC 1994 *J Mater Sci* **29** 5641

[18] Jain A, Razdan AK, Kotru P N and Wanklyn B M 1994 *J Mater Sci* **29** 3847

[19] Berzina I G, Berman I B and Savintsev P A 1965 *Kristallografiya* **9** 483

[20] Westbrook J H 1967 *Environment-Sensitive Mechanical Behaviour* In: Westwood ARC, Stoloff NS (eds) (New York: Gordon-Breach) p 247

[21] Hanneman R E and Westbrook J W 1968 *Phil Mag* **18** 73

[22] Gane N and Cox J M 1970 *Phil Mag* **22** 881

[23] Mott B W 1956 *Microindentation Hardness Testing* (London: Butter worths)

[24] Pal T and Kar T 2003 *Mater Sci Eng* A **354** 331

[25] Onitsch E M 1947 *Mikroskopie* **2** 131

[26] Hanneman M 1941 *Metallurgia Manchu* **23** 135

[27] Hays C and Kendall E G 1973 *Metallography* **6** 275

[28] Sangwal K and Klos A 2005 *Cryst Res Technol* **40** 429

[29] Frohlich F, Grau P and Grellmann W 1977 *Phys Stat Sol (a)* **42** 79

[30] Michels B D and Frischat G H 1982 *J Mater Sci* **17** 329

Temperature dependent dielectric constant/loss investigations in annealed polyetherimide

R Singh[1,3], B K Kaushik[2] and J K Quamara[2]
[1]Physics Department, SKIET, Kurukshetra-136118, India
[2]Physics Department, NIT, Kurukshetra -136119, India.

[1]E-mail: rskalas@gmail.com

Abstract. Temperature dependent dielectric constant/loss behavior has been investigated for annealed Polyetherimide. For investigating the effect of annealing, the samples were kept at desired temperature (100^0C) for a predetermined time (varying from 2-10 hrs). The samples were then cooled to room temperature and stored in a desicator. The dielectric constant/loss were measured in the temperature region varying from 30^0-250^0C using a precision LCZ meter, Keithley 3321 at different frequencies 120 Hz, 1 kHz, 10 kHz and 100 kHz. A significant decrease in dielectric constant ε' in annealed samples below 120^0C, is due to the suppression of dipolar relaxation process lying in this temperature region due to annealing. The enhancement in ε' above 120^0C, is due to the dominance of α-relaxation process. The dielectric behaviour of Polyetherimide in the temperature region below 120^0C is mainly governed by the dipolar relaxation processes owing either to ether linkages or to >C=O groups. Annealing has resulted in a loss of dipolar relaxation processes which is further confirmed from the fact that ε' varies scarcely with temperature.

1. Introduction

Aromatic polyimides are versatile engineering polymers that have found applications in a number of advanced fields of science and technology owing to their outstanding thermal stability, mechanical and electrical properties [1]. Polyetherimide (PEI), one of the important members of polyimide family is a polymeric material introduced in early 1980s at the General Electric Corporation Research Centre under the commercial name ULTEM [2-7]. The material is an amorphous thermoplastic consisting of ether units and aromatic imide rings. Ether units provide the chain flexibility, processability and good melt flow characteristics while aromatic imide rings provide stiffness, thermal resistance and mechanical properties to the polymer [5, 6]. Polyetherimide has found applications in the fields of microelectronics and aerospace industry besides many other industrial applications owing to its heat and radiation resistance [8, 9]. In the present paper we are reporting the temperature dependent dielectric behaviour of annealed samples.

2. Experimental details

Polyetherimide used in the present investigations was procured from Good Fellow (U.K.) in thin film form of nominal thickness 85 μm. The samples were metallized on both the surfaces by vacuum evaporation of aluminium to make electrical contacts. The sample holder designed for this purpose was suspended in a specially prepared temperature controlled furnace, shielded against stray pickups. The dielectric constant and dielectric loss measurements were made in the temperature region varying from 30^0-250^0C using a precision LCZ meter (model Keithley-3321) at different frequencies. For investigating the effect of annealing, the samples were kept at desired temperature (100^0C) for a predetermined time, then cooled to room temperature and stored in a desicator.

[3]Randhir Singh, Physics Department, SKIET, Kurukshetra-136118, India, E-mail: rskalas@gmail.com

3. Results and discussion

Figures 1 and 2 represent the temperature versus dielectric constant ε' and dielectric loss factor ε'' for pristine and annealed PEI samples at a frequency of 100 kHz. A significant decrease in ε' in case of annealed samples below 120^{0}C has been observed, however, above this temperature the effect appears in an opposite manner. This indicates that the different mechanisms governing ε' in different temperature regions are affected differently by annealing. In the temperature region above 100^{0}C, the occurrence of multiple relaxation processes in PEI, has been reported earlier [4-5, 7, 10]. These relaxations are (i) β- relaxation process due to localized rotational fluctuations of the dipole vectors of >C=O groups (ii) α- relaxation process due to interfacial polarization or space charge formation. The dielectric behaviour of PEI in the temperature region below 120^{0}C is mainly governed by the dipolar relaxation processes associated either with ether linkages or with carbonyl (>C=O) groups.

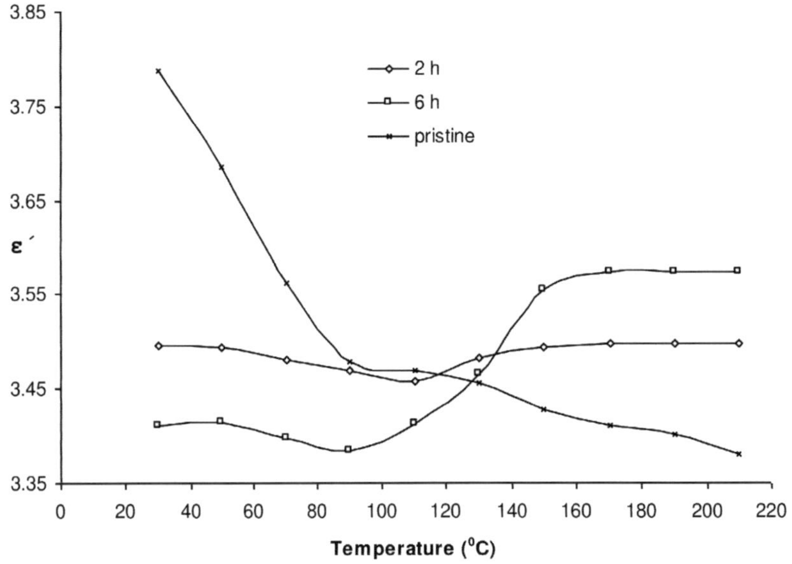

Figure 1. Temperature versus dielectric constant for pristine and annealed PEI samples at 100 kHz frequency.

It appears that annealing has resulted in a loss of these dipolar relaxation processes, which is further confirmed from the fact that ε' varies scarcely with temperature. Cheng et al. [11] have shown that a decrease of the cell edges occurs due to annealing leading to an increase in density in case of poly (p-phenylene sulphide) samples. It appears that in PEI also, an enhancement in the density may result from the annealing which would suppress the dipolar relaxation processes resulting in a decrease in ε'. The increase in annealing time further reduces ε'. The high temperature region above 120^{0}C, is usually dominated by space-charge relaxation processes. Since, in case of annealed samples, there is a significant loss in the β-relaxation process, we are left with α- relaxation process alone. An enhancement in the α-relaxation process due to annealing is evident from the increased magnitude of ε'. The enhancement in the α-relaxation can be associated to the increased density of the annealed samples. An increase in density due to annealing is also supported by the fact that the dielectric loss factor values are much more in annealed samples as compared to pristine one [figures 1 and 2].

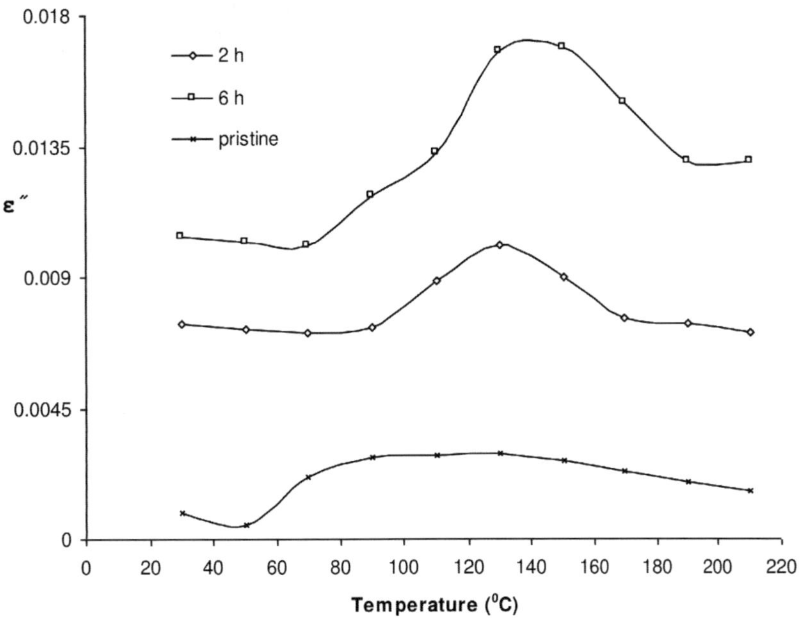

Figure 2. Temperature vs. dielectric loss factor (ε'') for pristine and annealed PEI samples at 100 kHz frequency.

3. Conclusions

1. The dielectric behaviour of pristine and annealed PEI is mainly governed by β′-, β- and α- relaxation processes associated with dipolar nature of ether linkages, dipolar nature of >C=O groups and interfacial polarization or space charge formation respectively.
2. Annealing may result in increased density of PEI, which can suppress the dipolar relaxation processes, thereby resulting in a decrease in ε' in the lower temperature region lying between 30°C – 90°C.
3. The increase in annealing time further reduces ε'.

References

[1] Chen B K, Su C T, Tseng M C and Tsay S Y 2006 *Polym. Bull.* **57** 671
[2] Brydson J A 1995 *Plastics Materials* 6th Edn (Butterworth, Heinenmann) p 508
[3] Suh K S, Nam J H and Lim K J 1996 *J. Appl.. Phy.* **80** 6333
[4] Diaz-Calleja R, Friederichs S, Jaimes C, Sanchis M J, Belana J, Canadas J C, Diego J A and

Mudarra M 1998 *Polym. Intl.* **46** 20

[5] Belana J, Canadas J C, Diego J A, Mudarra M, Diaz-Calleja R, Friederichs S, Jaimes C and Sanchis M J 1998 *Polym. Intl.* **46** 11

[6] Sanchis M J, Diaz-Calleja R, Jaimes C, Belana J, Canadas J C, Diego J A, Mudarra M and Sellares J 2004 *Polym. Intl.* **53** 1368

[7] Singh R, Kaushik B K and Quamara J K 2008 *Indian J. Pure & Appl. Phys*, **46** 127

[8] Serfaty W I 1984 *Polyimides: Synthesis, characterization and applications* Vol 1 ed K L Mittal (Plenum Press, New York) p 149

[9] Long Jr E R and Long T S A 1987 *Nasa Technical Paper* 2663

[10] Matsuoka S 1992 *Relaxation Phenomena in Polymers* (Hanser Publishers, New York) p 212

[11] Cheng Stephen Z D, Wu Z Q and Wunderlich B 1987 *Macromolecules* **20** 2802

X-ray diffraction, spectroscopic and mechanical studies on potential organic NLO materials of metaNitroaniline and N-3-Nitrophenyl Acetamide single crystals

S. Senthil[1] and J. Madhavan[2]
[1] Department of Physics, Govt. Arts College (Men), Nandanam, Chennai 600 035.
[2]Department of Physics, Loyola college, Chennai 600 034. India.

E-mail: jmadhavang@yahoo.com

Abstract. In the present paper, attempts were made to grow good quality metaNitroaniline (mNA) and N-3-Nitrophenyl (3-NAA) single crystals. The lattice parameter values from the Powder X-ray diffraction pattern confirms that mNA belongs to orthorhombic crystal system with the unit cell parameter values of a = 6.501 Å, b = 19.330 Å and c = 5.082 Å with space group $Pbc2_1$. Similarly the powder XRD data indicates that 3-NAA crystal retained its monoclinic structure with lattice parameter values a = 9.762 Å, b =13.287 Å, c =13.226 Å, and β = 102.99°. Investigation has been carried out to assign the vibrational frequencies of the grown crystals by Fourier Transform infrared spectroscopy technique. The SHG efficiency of mNA and 3NAA was determined by Kurtz and Perry powder technique. The Optical absorption study confirms the suitability of the crystals for device applications. The mechanical properties of the grown crystals have been studied using Vickers microhardness tester.

1. Introduction

The emergence of new materials with superior quality is often responsible for major advances in new technologies. In that respect, materials with a nonlinear optical response are expected to play a major role in enabling optoelectronic and photonic technologies [1]. Materials with large nonlinear optical susceptibilities are of current interest in the area of harmonic generation and optical modulation. In recent years, some polar organic crystals, which form a non-centrosymmetric structure exhibit second-order nonlinear optical properties that far surpassed those of the conventional materials [2]. Organic molecules with significant nonlinear optical activity generally consist of a π – electron conjugated structure. The conjugation π -electron moiety provides a pathway for the entire length of conjugation under the perturbation of an external electric field. Fictionalization of both ends of the π bond systems with appropriate electron donor and acceptor group can increase the asymmetric electronic distribution in either or both the ground and excited states, thus leading to an increased optical nonlinearity [3].

2. Synthesis Solubility and Growth of mNA and 3NAA Single Crystal

The purified mNA material was used for solubility studies and to grow bulk single crystal. Since mNA is less soluble in water, acetone was chosen as the solvent. The selected solvent and an excess amount of mNA were added to a 250 ml glass container. The solution was gently stirred for an optimum period to attain saturation. The solution was saturated at 40 °C and the crystals were grown. The habits of the grown crystals were pyramidal in shape. The 3NAA compound was synthesized by the direct combination of N-(3-nitrophenyl) acetamide and acetic anhydride in methanol at room temperature. According to the solubility curve, the saturated solution of 3NAA was prepared by dissolving the source material in the mixed solvent of acetone and methanol (1:1). On reaching saturation, the beaker containing the growth solution was optimally closed for controlled evaporation. Transparent single crystals were obtained from the growth solution after 10 days.

3. Characterization

3.1 Powder X-ray diffraction studies

Finely crushed powder of mNA and 3NAA crystals were subjected to powder X-ray diffraction analysis using CuK_α (λ = 1.5418 Å) radiation. The sample was scanned at a scan rate of 2 degree/minute. The recorded X-ray diffraction pattern of mNA and 3NAA are shown in figure 1 and 2. It is confirmed that mNA belongs to orthorhombic crystal system, the unit cell parameters were found to be a = 6.501 Å, b = 19.330 Å, and c = 5.082 Å with point group mm2, space group Pbc21 and Z = 4. Using the simulated hkl values and the experimental d values, the lattice parameters were estimated and this indicates that 3NAA crystal retained its monoclinic structure with lattice parameters values a = 9.762 Å**,** b =13.287 Å, c =13.226 Å, and β = 102.99°. The results were in good agreement with the reported values [4, 5]

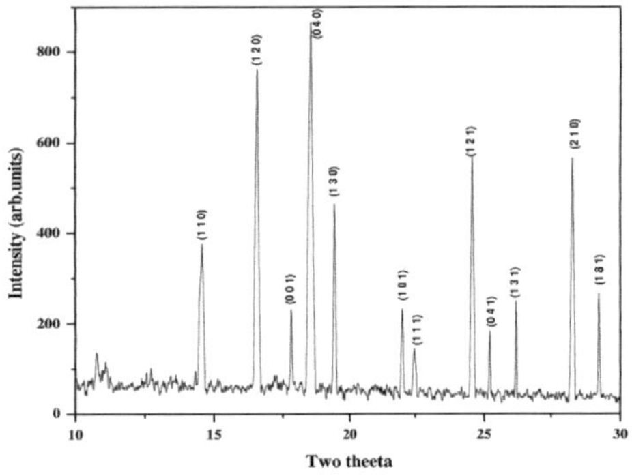

Figure 1. Powder X-ray diffraction pattern of mNA

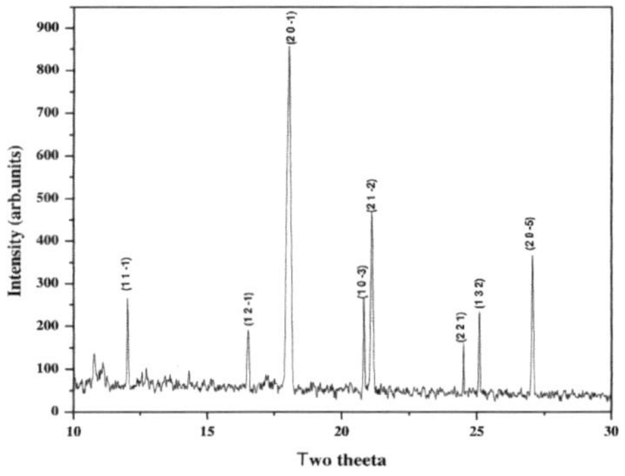

Figure 2. Powder X-ray diffraction pattern of 3NAA

3.2 FT-IR spectral analysis

The mNA sample was prepared by mixing powdered mNA with potassium bromide and a pellet was made. The FT-IR spectrum obtained for mNA sample is shown in Figure 3. Symmetric and asymmetric stretching modes of free NH_2 group are observed at 3328 and 3430cm^{-1}. The peaks at 3201, 3094 and 2923cm^{-1} are due to very weakly bonded N–H stretching modes. The peaks at 2359 cm^{-1} may be assigned to overtone of bands at 1264 and 1087cm^{-1}. The NH_2 band is observed at 1623cm^{-1}. The peaks that appear at 1520 and 1348cm^{-1} are due to the vibrations of NO_2 stretching modes. As either the N–H stretching modes or the NO_2 stretching modes are not much broad, their interaction with the neighboring molecules was presumed to be weak bonds. The vibrations between 909 and 650cm^{-1} are usually associated with the presence of benzene rings in the mNA molecule. The peaks between 750 and 810 cm^{-1} show the meta position of the substituted molecules in the benzene ring of mNA.

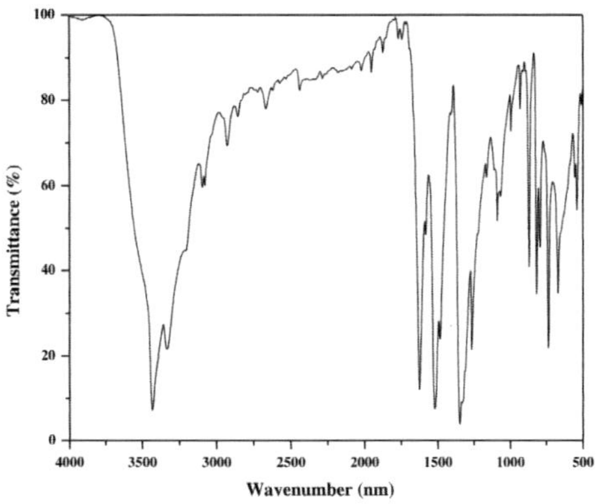

Figure 3. FT – IR Spectrum of mNA

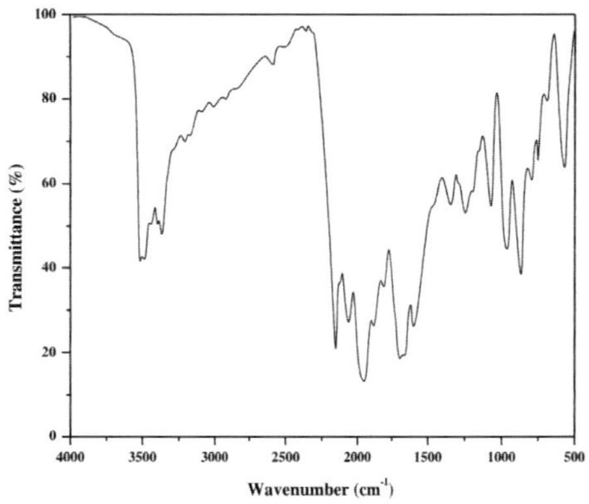

Figure 4. FT – IR Spectrum of 3NAA

The FT-IR spectrum was recorded on the grown 3-NAA crystal at room temperature in the range of 500 – 4000 cm^{-1} and is shown in the figure 4. The spectrum shows the symmetric and asymmetric stretching modes of free NH_2 group, which are observed at 3303 and 3263 cm^{-1}. The absorptions at 1673 cm^{-1} are due to ketone stretching vibration. The aromatic skeletal ring is observed at 1600 cm^{-1}. A well-resolved peak at 604 cm^{-1} represents the C-N-O stretching vibration. Peak at 1260 cm^{-1} is due to N-H bending and C-N stretching vibrations. The vibration observed between 886 and 670 cm^{-1} are usually associated with the presence of benzene rings in the 3NAA molecule. The peak between 741 and 805 cm^{-1} shows the meta position of the substituted molecule (NO_2) in the benzene ring of 3NAA. The peaks at 1531 and 1350cm^{-1} are due to the vibration of NO_2 stretching modes. As either the N-H stretching modes or NO_2 stretching modes are not much broad, their interaction with the neighboring molecule is presumed to be weak bonds.

3.3 Optical assessment

Optical absorption spectra were recorded and the recorded spectra are shown in Figure 5 and Figure 6. From the spectra, it is seen that the mNA crystal has a lower cut-off wavelength of 240 nm. The UV–Vis–NIR spectrum gives limited information about the structure of the molecule because the absorption of UV and visible light involves the promotion of the electron in s and p orbitals from the ground state to higher energy states. It can be seen that 3NAA is transparent in the range 270 – 1200 nm and is the required important property in nonlinear optical applications. The wide transparency in the visible region is one of the additional key requirements for having efficient NLO character [6].

3.4 Second harmonic generation efficiency measurement

The SHG efficiency was estimated using Kurtz and Perry technique [7]. SHG was confirmed by the emission of green light. Using the potassium dihydrogen phosphate (KDP) crystalline powder as reference material, the output of SHG signal was compared and found that the SHG conversion efficiency of mNA is 28 times and 3-NAA is 4 times that of KDP.

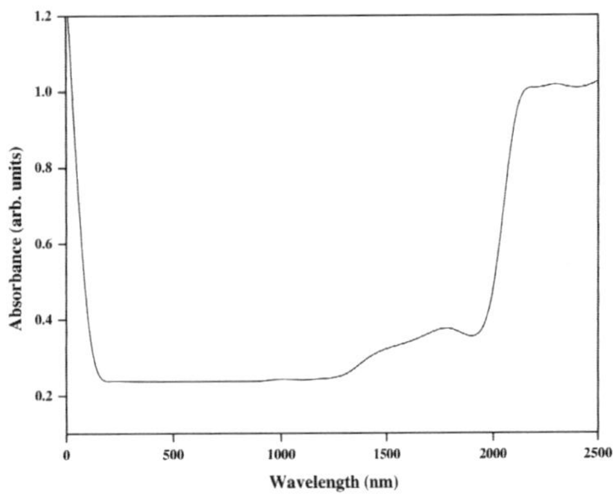

Figure 5. Optical absorption spectrum for mNA

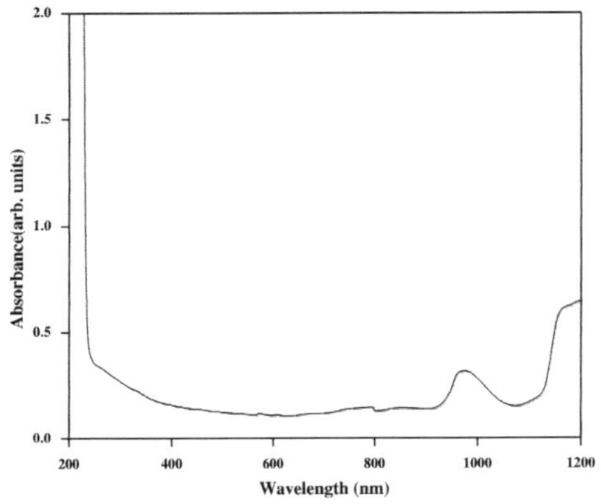

Figure 6. Optical absorption spectrum for 3NAA

3.5 Vickers microhardness studies

Microhardness studies were carried out on mNA and 3-NAA single crystals using Vickers hardness tester fitted with pyramidal indentor. The diagonal length of the indentation in mm for various applied loads in kg is measured for a constant indentation time of 15 s. The Vickers hardness number is calculated using the relation $Hv = 1.8544\ p/\ d^2$ kg/mm^2. According to Onitsch for $1.0 \leq n \leq 1.6$, the material is regarded as hard material and $n > 1.6$ for soft materials [8]. For mNA and 3-NAA, n is found to be greater than 1.6, thus confirming that both the crystals are soft materials.

4. Conclusion

The title compounds of mNA and 3-NAA were successfully synthesized and the single crystals have been grown by solution growth technique. Its cell parameters have been determined by the powder XRD analysis. The presence of functional groups has been identified from the FTIR analysis. Optical absorption study reveals the transparency of the grown crystals. Second harmonic efficiency of the grown crystals has been confirmed by Kurtz and Perry powder method. The microhardness measurements prove that both belongs to the soft category of materials.

5. References

[1]. Sankar. D, Praveen Kumar. P, Madhavan J, *Physica B*, 405(2010) 1233.

[2]. T. Sujatha , A. Cyrac Peter , M. Vimalan , J. Merline Shyla , J. Madhavan, *Physica B,* 405(2010)3365

[3]. J. Madhavan, S. Aruna , A. Anuradha, D. Premanand, I. Vetha potheher ,K. Thamizharasan , P. Sagayaraj, *Optical Materials,* 29(2007A) 1211

[4]. Senthil. S, Pari. S, Sagayaraj. P, Madhavan. J, *Physica B*, 404 (2009) 295.

[5]. S. Senthil, S. Pari, R. John Xavier, J. Madhavan, *Optik,* 123(2011)104

[6]. P. Praveen kumar, V. Manivannan, P. Sagayaraj, J. Madhavan, *Bull. Mater. Sci.*,32(2009), 431

[7]. Kurtz. S.K and Perry. T. T, (1968), *J. Appl. Phys.* 39, 3798–3813.

[8]. J. Madhavan, S. Aruna, K. Prabha, J. Packium Julius, Ginson P. Joseph, S. Selvakumar, P. Sagayaraj, *Journal of Crystal Growth* , 293(2006) 409

Blue-green luminescence in Hg free excited $Sr_2P_2O_7:Tb^{3+}$ pyrophosphate phosphor for NUV excited LEDs

R L Kohale[1], O P Chimankar[1] and S J Dhoble[1]*
[1]Department of Physics, R.T.M. Nagpur University, Nagpur-440033, India

*E-mail: sjdhoble@rediffmail.com

Abstract. Tb^{3+} activated $Sr_2P_2O_7$ phosphor was prepared by modified solid state diffusion technique at 700°C. The XRD pattern of $Sr_2P_2O_7$ is in well argument with the standard ICDD File (24-1011) available. Surface morphology of the present phosphor has been studied by scanning electron microscope (SEM). In photoluminescence investigation, the excitation spectra of the phosphor extends from 200 to 400 nm gives optimum absorption at 352 nm, which is mercury free excitation and characteristics of NUV excited LED. Under the excitation of 352 nm, $Sr_2P_2O_7$ doped with trivalent terbium ions produces weak blue emission between 400–500 nm owing to the $^5D_3$ to the ^7F_J transitions of Tb^{3+} and strong green emission in 500–650 nm region of the visible spectrum due to the 4f–4f transitions from the $^5D_4$ to ^7F_J (J=6, 5, 4, 3) states of Tb^{3+}. The entire study reveals that the present phosphor have promising applications in the lamp industry especially for solid state lighting (mercury free excited lamp phosphor) and NUV LEDs.
Keywords: Photoluminescence; Phosphor; Pyrophosphate; XRD.

1. Introduction

In the recent era of solid-state lighting the demand for mercury free excited phosphor-converted LEDs has significantly increased. Literature study authenticates that phosphate base phosphors are renowned for their thermal and chemical stability [1]. Since, they are extensively used as hosts doped with different rare earth ions to obtain phosphors for verity of lighting applications. Consequently the research on phosphors suitable for fabricating NUV excited light-emitting diodes has attracted more attention. Compared to the conventional incandescent lamps, LEDs have numerous advantages such as less energy consumption, optimum luminosity, and extended lifetime [2]. They have been used mainly for LCD backlighting, traffic lights, and information boards. In sight of green emission at around 545 nm trivalent terbium ions based phosphors play a vital role in the field of displays and enlightenment. Particular phosphors such as $GdMgB_5O_{10}$, Y_2SiO_5 and $LaPO_4$ based on energy transfer from Ce^{3+} to Tb^{3+} found diversified applications for their green emission in lamps [3]. Herein, the main objective of this work is to develop an efficient pyrophosphate phosphor that absorbs energy in the near-UV range and emit in the visible range and to examine the photoluminescence and structural properties of the newly synthesized $Sr_2P_2O_7$ pyrophosphate phosphor. The prepared phosphor has been characterized by XRD and SEM. The photoluminescence investigations are performed to explore the luminescence characteristics of trivalent terbium doped strontium pyrophosphate. The observed blue and intense green emission from $Sr_2P_2O_7:Tb^{3+}$ with 352 nm (mercury free) excitation suggests a possibility of applying the newly synthesized phosphors for phosphor- converted NUV-LEDs and solid state lighting.

IOP Conf. Series: Materials Science and Engineering **73** (2015) 012111 doi:10.1088/1757-899X/73/1/012111

2. Experimental

The powder samples of $Sr_2P_2O_7:Tb^{3+}$ pyrophosphor were prepared by high temperature modified solid state diffusion. The starting materials used in the preparation were high purity strontium carbonate $SrCO_3$ (Merck, 99.995%), ammonium dihydrogen phosphate $NH_4H_2PO_4$. (Merck 99.99% pure) and the molar ratio of rare earth, terbium oxide Tb_2O_3 (Aldrich, 99.999%) was varied from 0.1 to 1 mol% in $Sr_2P_2O_7$ phosphor with respect to Sr ions. After being ground thoroughly in stoichiometric ratios by using an agate mortar, to ensure the best homogeneity and reactivity, powder was transferred to silica crucible, and corresponding mixtures were initially heated in a muffle furnace at 350 °C for 1 hr in order to decompose $NH_4H_2PO_4$. Afterwards, the mixture was reground for 30 min before heating at higher temperatures and then heated in furnace at 700 °C for 12 hrs .The temperature was cooled to 500°C and air-quenched to room temperature. All samples were found out to be purely white and various techniques were used to characterize the synthesized phosphor. The prepared phosphor was characterized by XRD, SEM and PL.
The basic chemical reaction for $Sr_2P_2O_7$ pyrophosphor can be described as follows:
$Sr(CO_3) + 2(NH_4H_2PO_4) \rightarrow Sr_2P_2O_7 + 2NH_3 + 3H_2O + 3CO_2$.

3. Results and discussion

3.1. XRD of $Sr_2P_2O_7:Tb^{3+}$ pyrophosphate phosphor. The phase purities of the sample was checked by powder X-ray diffraction using PAN-analytical diffractometer (Cu-Kα radiation) at a scanning step of 0.010, continue time 20s, in the 2θ range from 10^0 to 60^0 . Figure 1 illustrates the XRD pattern for $Sr_2P_2O_7:Tb^{3+}$ prepared at 700 ^{0}C.

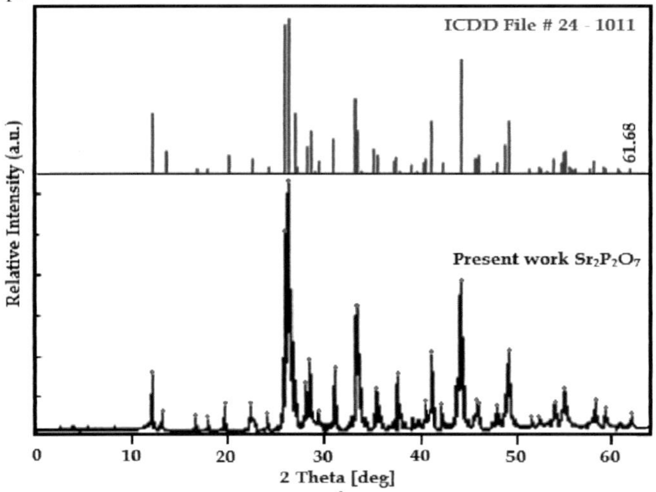

Figure 1. XRD of $Sr_2P_2O_7:Tb^{3+}$ pyrophosphate phosphor.

The above sample was heated in air and the obtained XRD pattern is similar to that reported for the compound strontium pyrophosphate $Sr_2P_2O_7$ (ICDD file No.24-1011), indicating the formation of the crystalline phases.XRD pattern (Figure 1) reveals that the trivalent terbium ions (dopant) have no noticeable influence on the crystalline structure of the host $Sr_2P_2O_7$ [4]. Considering the effect of ionic sizes of the cations, it is predictable that Tb^{3+} can effectively occupy Sr^{2+} sites, since the ionic radii of Tb^{3+} (1.04 Å) and Sr^{2+} (1.12 Å) are in the close vicinity with one another. [5].This shows solid state reaction of the mixtures took place well and final product acquires crystalline phases.
3.2 SEM of $Sr_2P_2O_7:Tb^{3+}$ pyrophosphate phosphor. The particle morphology of the present pyrophosphate was studied by (SEM, JEOL 6380A). An average crystallite size is in sub-micrometer range seen in SEM images. The microstructure of the phosphor prepared at 700°C reveals an average size of about 5 to 10 μm. Particles possess foamy like morphology formed from extremely

agglomerated crystallites and sample $Sr_2P_2O_7$:Tb^{3+} shows formation of porous structure. The sample revealed sintered structures in which individual particles could barely be recognized. However the particles were attached with each other all the way through necking and possibly they may be crushed to obtain separate particles. Most particles showed smoothed rounded surfaces and sizes of a few micrometers. All these characteristics were beneficial when considering their application as phosphor powders for fabrication of NUV- LEDs (coating purpose) [6]. The typical morphological images are represented in Figure 2.

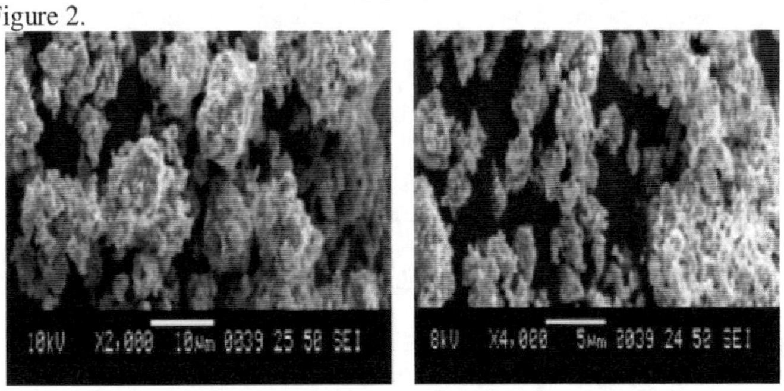

Figure 2. SEM photographs of $Sr_2P_2O_7$:Tb^{3+} pyrophosphate.

3.3 Photoluminescence (PL) spectra of $Sr_2P_2O_7$:Tb^{3+} pyrophosphate phosphor. The photoluminescence (PL) investigations were performed on a Shimadzu RF5301PC spectrofluorophotometer using a spectral slit width of 1.5 nm. The equal amount of powder sample was used (i.e. 2 g, each) for the measurement of emission and excitation spectra. Figure 3(a), demonstrates the PL excitation and emission spectra of $Sr_2P_2O_7$:Tb^{3+} pyrophosphate phosphor. The excitation peak observed at 352 nm is highly intense amongst other excitation peaks which are observed at 262 nm to 280 nm shown in Figure 3(a). The energy level transitions of Tb^{3+} ions are represented in Figure 3(b).

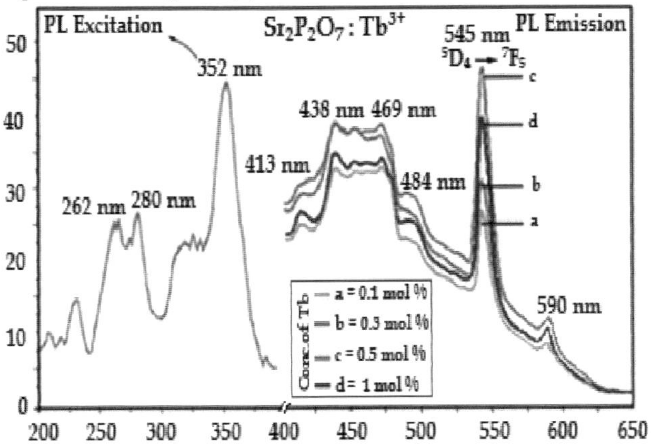

Figure 3(a). PL excitation and emission spectra for $Sr_2P_2O_7$ doped with Tb^{3+} activators.

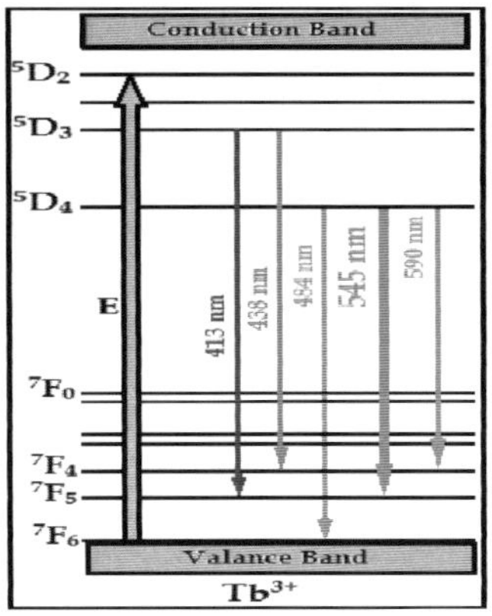

Figure 3(b). Energy level transitions of Tb^{3+} ions.

Only the emission intensity is variable with respect to all excitation wavelengths; however there is no effect on the emission wavelengths. In this framework, we select 352 nm wavelengths for our study since it is more suitable for Hg free excited lamp phosphors and NUV excited LEDs. In the PL emission spectrum $Sr_2P_2O_7:Tb^{3+}$, the broad emission lines are observed at 413,438, 469, 484, 545 and 590 nm. The weak e mission peaks at 413, 438 nm are due to the transition from excited level $^5D_3$ to$^7F_5$ and $^7F_4$ levels of Tb^{3+} ions, respectively. The emission peaks observed at 469 nm is due to $^5D_3 \rightarrow {}^7F_2$ and peak at 484 nm is ascribed to the $^5D_4 \rightarrow {}^7F_6$ transitions of Tb^{3+} [7]. The small peak observed at 590 nm arise due to the $^5D_4 \rightarrow {}^7F_4$ transitions .Sample showed intense green emission of light at 545 nm wavelength corresponding to the strongest $^5D_4 \rightarrow {}^7F_5$ transition of Tb^{3+} ions stimulated probably from both electric-dipole and magnetic-dipole transitions [8].

In general at lower doping concentrations of Tb^{3+} ions ,it shows more intense emission in blue region ,as a result the emission around 380 - 385 nm due to $^5D_3 \rightarrow {}^7F_6$ transition can also observe in the Near UV region, however with increasing concentrations of Tb^{3+} ions, the green emission becomes most intense because of cross relaxation. A series of $Sr_2P_2O_7:Tb^{3+}$ pyrophosphate phosphors have been synthesized with the dopant concentration ranging from 0.1 to 1 mol % and an optimum doped molar concentration of Tb^{3+}ions is 0.5 mol % attained in our study.

4. Conclusions

In conclusion, the $Sr_2P_2O_7:Tb^{3+}$ pyrophosphate phosphor prepared by the conventional high temperature modified solid state diffusion. The formation of crystalline phases was confirmed by the X-ray diffraction technique (XRD). In the SEM study, an average crystallite size is found out to be 5 to 10 μm range with foam like morphology. The photoluminescence (PL) spectra exhibit intense green light emission at 545 nm wavelength when excited by NUV lights at 352 nm (mercury free excitation).In view of the extended excitation band and admirable luminescent properties, Tb^{3+} activated $Sr_2P_2O_7$ pyrophosphate phosphor can be utilized as blue-green phosphors in tricolor lamps for solid state lighting and useful for NUV excited LEDs because of its uncomplicated synthesis, lower manufacturing cost, unhygroscopic nature and mercury free excitation .

Acknowledgement

One of the authors SJD is grateful to UGC, New Delhi for providing financial assistance to carry out this work under research project.

References

[1] Nagpure I M , Shinde K N , Kumar V, Ntwaeaborwa O M , Dhoble S J , Swart H C 2010 *J. Alloys Compd.* **492** 384
[2] Jang H S , Jeon D Y 2007 *Appl. Phys. Lett.* **90** 041906
[3] Kamiya S , Mizumo H , Shionoya S , Yen W M 1998 *Phosphor Handbook, CRC Press, Boca Raton* 393
[4] Lin C C , Xiao Z R , Guo G Y , Chan T S , Liu R S 2010 *J. Am. Chem. Soc.* **132** 3020
[5] Guo C F , Luan L , Ding X , Huang D X 2008 *Appl. Phys. A.* **91** 327
[6] Kohale R L , Dhoble S J 2011 *AIP Conf.Proc.* **1391** 203
[7] Dieke G H 1968 Interscience Publishers, 253
[8] Lei B , Li B , Zhang H , Zhang L , Cong Y , Li W 2007 *J. Electrochem. Soc.* **154** H623

IOP Conf. Series: Materials Science and Engineering **73** (2015) 012112 doi:10.1088/1757-899X/73/1/012112

Polyaniline/clay Nanocomposites: Preparation, Characterization and Electrochemical Properties

B Vijayakumar[1,*], K O Anjana[2] and G Ranga Rao[1]

[1] Department of Chemistry, Indian Institute of Technology Madras, Chennai-600036, India
[2] Department of Chemistry, National Institute of Technology Calicut, Kozhikode-673601, India

E-mail: badathala@rediffmail.com

Abstract. Preparation of polyaniline (PAni)/clay nanocomposites by reacting aniline with ammonium persulphate in presence of clay at 0-5 °C for 12 h. The composites were characterized by X-ray diffraction, infrared spectroscopy, UV-vis diffuse reflectance spectroscopy and thermogravimetric analysis for their physicochemical properties. Morphology was studied by scanning electron microscopy. The electrochemical studies of PAni/clay nanocomposites were carried out using cyclic voltammetry in 0.5 M H_2SO_4 aqueous electrolyte. The PAni/clay nanocomposites showed specific capacitances of 415–455 Fg^{-1} at a scan rate of 10 mVs^{-1}.

1. Introduction

Polymer/clay nanocomposites have attracted great interest, because they exhibit remarkable improvement in materials properties such as mechanical, thermal, electrical and optical properties when compared with virgin polymer or conventional micro- and macro-composites. The physical mixture of a polymer and clay may not form a nanocomposite. This situation is analogous to polymer blends, and in most cases separation into discrete phases takes place. In immiscible systems, which typically correspond to the more conventionally filled polymers, the poor physical interaction between the organic and the inorganic components leads to poor mechanical and thermal properties. In contrast, strong interactions between the polymer and clay in polymer/clay nanocomposites lead to the organic and inorganic phases being dispersed at the nanometer level. As a result, nanocomposites exhibit unique properties not shared by their micro counterparts or conventionally filled polymers [1–4]. The natural clays are hydrophilic in nature and hence incompatibility between clay and hydrophobic polymer matrix to obtain a polymer/clay nanocomposite. However, in this state, clays are miscible with hydrophilic polymers, such as poly (ethylene oxide) (PEO) [5], or poly (vinyl alcohol) (PVA) [6]. To render clays miscible with other polymer matrices, one must convert the normally hydrophilic silicate surface to an organophilic one, making the intercalation of many engineering polymers possible. Generally, this can be done by ion-exchange reactions with cationic surfactants including primary, secondary, tertiary, and quaternary alkylammonium or alkylphosphonium cations. Alkylammonium or alkylphosphonium cations in the organosilicates lower the surface energy of the inorganic host and improve the wetting characteristics of the polymer matrix, and result in a larger interlayer spacing. Additionally, the alkylammonium or alkylphosphonium cations can provide functional groups that can react with the polymer matrix, or in some cases initiate the polymerization of monomers to improve the strength of the interface between the inorganic and the polymer matrix [7, 8]. The polymer/clay nanocomposite is obtained when a polymer intercalates in the interlamellar region of clays separates and gets dispersed in polymer matrix [9].

* Corresponding author. Current address: Department of Chemistry, Vel Tech High Tech Dr. Rangarajan Dr. Sakunthala Engineering College, Avadi, Chennai-600 062, India. E-mail: badathala@rediffmail.com, badathala@yahoo.com

Polyaniline is known for more than a century and is potentially one of the most useful conducting polymers because of its facile synthesis, environmental stability, and simple acid/base doping/dedoping chemistry [10]. The present research work deals with the modification of clays to hydrophobic and latter use them in the insitu polymerization to obtain polyaniline/clay nanocomposites. The polyaniline/clay nanocomposites are characterized for their physico-chemical and electrochemical properties.

2. Experimental

2.1 Materials and methods

Smectite rich clay sample from Bhuj area, Gujarat has been used in this study. The composition of the smectite rich clay sample was found to be 53.44% SiO_2, 16.12% Al_2O_3, 13.65% Fe_2O_3, 2.84% MgO, 2.31% Na_2O, 0.27% K_2O, 1.28% CaO, 1.24% TiO_2, 0.02% Cr_2O_3, 0.11% MnO, 0.01% NiO, 0.04% P_2O_5, 0.16% S, and 8.03% ignition loss. The cation exchange capacity (CEC) of the clay was found to be 0.98 meq per g [11]. The other reagents were purchased from Sd-fine, Loba, AVRA, and SRL, India and used as received. X-ray powder diffraction (XRD) data were collected on a Bruker AXS D8 Discover diffractometer with Cu Kα radiation (λ = 1.5418 Å). The IR spectra were recorded in the range of 450-4000 cm^{-1} on a Perkin-Elmer FTIR spectrometer using KBr pellet. UV-Vis diffuse reflectance (DR) spectra were recorded on JASCO V-530 UV-Vis spectrophotometer. The thermogravimetric analysis (TGA) was conducted on a Perkin-Elmer TGA-7 analyzer in air with a heating rate of 20 °C min^{-1} from room temperature to 800 °C. Field emission scanning electron microscopic/energy dispersive spectroscopic (FE-SEM/EDS) studies were done in Jeol microscope model JSM-6700F.

For electrochemical characterization, electrodes were fabricated on a high purity nickel as a current collector. The nickel was polished with successive grades of emery paper, sonicated and washed thoroughly with detergent and distilled water. 80 wt % of polyaniline/clay nanocomposite, 15 wt % of acetylene black, and 5 wt % of polyvinylidene fluoride (PVDF) were ground in a mortar. Few drops of *n*-methyl pyrrolidinone (NMP) were added to make syrup. This was then coated on to the pretreated nickel foil and dried at 60 °C under reduced pressure for 12 h. The electrochemical studies were performed on a CHI 7081C electrochemical workstation using a three electrode-configuration-cell consisting of polyaniline/clay nanocomposite as the working electrode, platinum foil (1 × 2 cm^2) as a counter electrode and SCE as reference electrode, all dipped in 0.5 M H_2SO_4 aqueous electrolyte.

3. Results and discussion

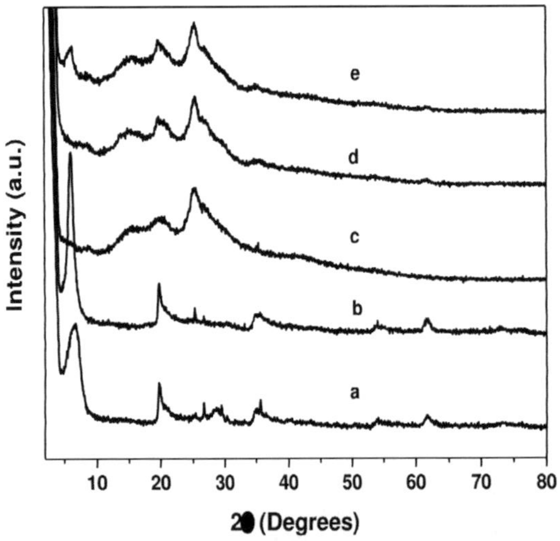

Fig. 1 PXRD patterns of (a) IB, (b) OMIB, (c) PAni, (d) PAni/IB and (e) PAni/OMIB.

3.1 *XRD*

Fig. 1 shows powder XRD patterns of Indian bentonite (IB), organically modified Indian bentonite (OMIB), PAni, PAni/IB, and PAni/OMIB. The peak at 6.75° in Fig. 1a corresponds to the periodicity in the direction of (0 0 1) of the IB sample. The XRD pattern of OMIB showed peak at 18.64 Å which is due to the grafting of CTAB to silanol groups within the interlayer where the silica framework is in contact with the clay layers [12]. The peak is shifted due to the interaction between PAni and the modified clay in PAni/OMIB (Fig.1e). The d-spacing in the direction of (0 0 1) of the IB sample is 13.07 Å, OMIB is 18.64 Å and that of PAni/OMIB is 14.60 Å respectively. The PAni/IB (Fig. 1d) showed very small peaks corresponding to clay due to its lower weight percent in the sample. These results demonstrate that the conducting PAni/clay nanocomposites are of the order of nanoscale size.

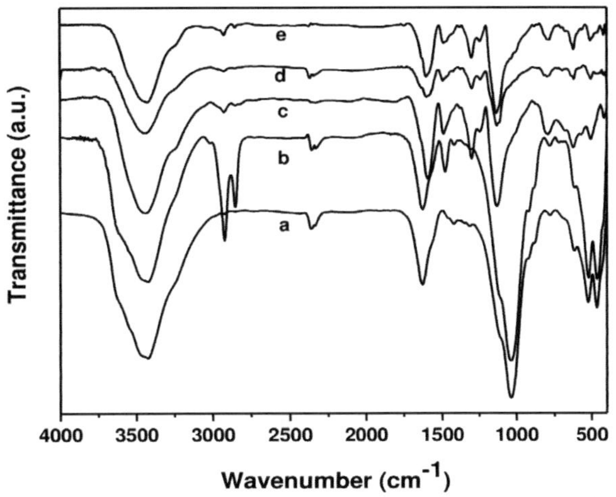

Fig. 2 FTIR spectra of (a) IB, (b) OMIB, (c) PAni, (d) PAni/IB and (e) PAni/OMIB.

3.2 *FTIR*

Fig. 2 shows FTIR spectra of IB, OMIB, PAni, PAni/IB, and PAni/OMIB. The infrared spectrum of PAni (Fig. 2c) shows broad band centred at 3451 cm^{-1} corresponds to N-H stretching with hydrogen bonded amino groups and free O-H stretching vibration and is attributed to the N-H stretching vibrations of the leucoemeraldine component. The characteristic absorption band observed for PAni/IB and PAni/OMIB (Fig. 2d & e) at 3451 cm^{-1} and 1597 cm^{-1} are assigned to the N-H stretching vibration mode, and NH_2 deformation in aniline unit respectively. The absorption bands at 2915 cm^{-1} and 2829 cm^{-1} are assigned to the aromatic sp^2 hybridized C-H stretching vibration mode and aliphatic hydrocarbon C-H stretching due to -CH_2- group in OMIB, PAni, PAni/IB and PAni/OMIB. The absorption bands observed at 1597 & 1479 cm^{-1} in PAni are assigned to the non-symmetric vibration mode of C=C in quinoid and benzenoid ring system in PAni. The C-N stretching vibration mode in aromatic amine nitrogen (quinoid system) in doped PAni is found at 1297 cm^{-1}, corresponding to the oxidation or protonation state. The IR peak at 1236 cm^{-1} is attributed to C-N stretching vibration mode in benzenoid ring system of PAni due to the conducting protonated form. In plane vibration of C-H bending mode in N=Q=N, Q-N+H-B or B-N+H-B (where Q = quinoid and B = benzenoid) is observed at 1113 cm^{-1}. The presence of this absorption band is expected due to the polymerization of PAni, i.e., polar structure of the conducting protonated form. The absorption band at 799 cm^{-1} is attributed to the aromatic ring and out of plane C-H deformation vibrations for 1, 4-disubstituted aromatic ring system [13].

The characteristic bands corresponding to clay are 1038 (Si-O), 911 (Al-OH) and 525cm^{-1} (Si-O-Al). The Si–O–Si stretching vibration of the clay matrix (1038 cm^{-1}) present in the montmorillonite [14] is merged with the peak at 1113 cm^{-1} in the composites and this peak appeared as a broad one. The FTIR peak at 1038 cm^{-1} for the clay corresponds to the internal SiO_4 tetrahedra, especially the Si-O-Si chain structure. The decrease in broadening of FTIR bands in the range 3062 - 3451 cm^{-1} may be due to covalent and hydrogen bonding between $-NH_2$ and –OH group of PAni and clay respectively.

3.3 *UV-Vis DR study*

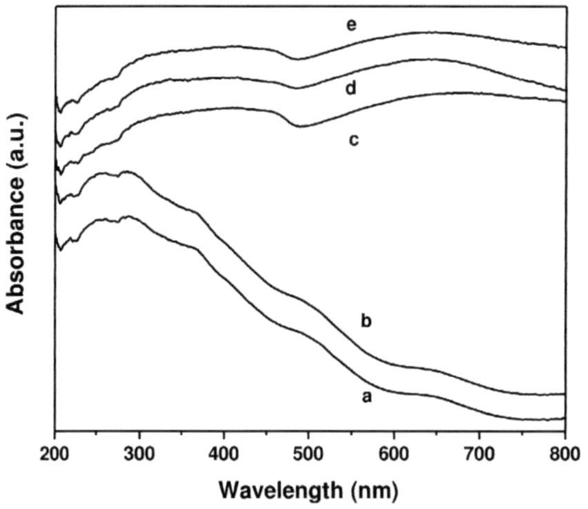

Fig. 3 UV-Vis spectra of (a) IB, (b) OMIB, (c) PAni, (d) PAni/IB and (e) PAni/OMIB.

Fig. 3 shows the UV-vis absorption spectra of IB, OMIB, PAni, PAni/IB, and PAni/OMIB. Two absorption bands are observed in the wavelength region from 315 to 350 nm and a small band at 578 to 712 nm for the PAni (Fig. 3c). PAni always exhibits a π-π^* transition, usually closer to 315 nm [54]. Partially oxidized PAni and its oligomers display an additional absorption at around 712 nm associated with the quinoid (oxidized) units [15]. These peaks are characteristic of the PAni emeraldine base [14, 16] and indicate that nanostructured PAni composites are stabilized in the emeraldine base redox state. The peak at 315 nm is attributed to $\pi - \pi^*$ transition of benzenoid rings and the peak at 712 nm is attributed to the charge transfer excitation of the quinoid structure. In the spectra of IB and OMIB, peaks are observed in the regions at 248-262 nm related to Fe^{3+} ions in octahedral coordination (Fig. 3a & b). PAni/IB and PAni/OMIB nanocomposites show clear similarity in their UV-vis spectra particularly with the presence of the absorption maxima at 315 and 610 nm which is associated with the stabilization of the composite in the emeraldine form. Comparison of the PAni and PAni/clay nanocomposites spectra shows that clay stabilizes the polyanilines in its emeraldine form.

3.4 *TGA*

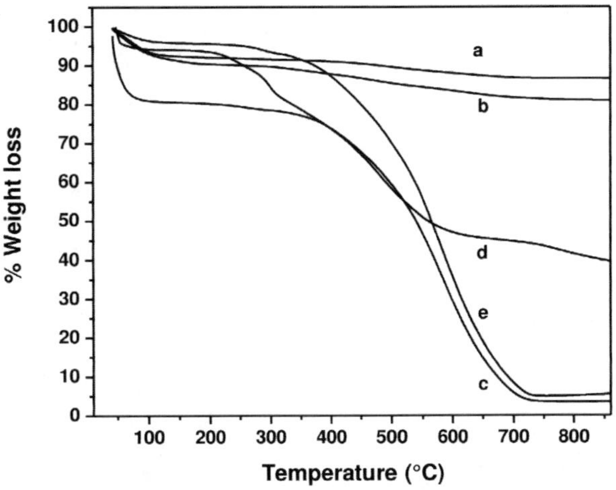

Fig. 4 TG plots of (a) IB, (b) OMIB, (c) PAni, (d) PAni/IB and (e) PAni/OMIB.

Fig. 4 shows TG plots of IB, OMIB, PAni, PAni/IB and PAni/OMIB. The thermogram of PAni indicates three major stages of weight loss (Fig. 4c). In the first stage, 3-4% weight loss at temperature up to 100 °C is associated with the loss of water molecules from the polymer matrix [17]. The weight loss at second stage that commences after 100 °C until 400 °C (about 19%) is due to the removal of the acid dopant bound to the polyaniline chain and low molecular weight oligomers. A slow and gradual weight loss profile observed starting at 400 °C onwards, represents degradation of the skeletal polyaniline chain structure after the dopant has been removed [18]. Above 700 °C, the results obtained are associated with the residues. The thermogram of IB shows an initial sharp decrease in weight ~ 5-6% within the temperature range of 80–150 °C is attributed to the loss of adsorbed water and interlayer water. A steady weight loss of about 6-7% in the temperature range of 150–800 °C which is attributed to the loss of dehydroxylation caused by breaking of structural hydroxyl groups of the montmorillonite (Fig. 4a). The thermogram of OMIB shows an initial sharp decrease in weight ~ 7% within the temperature range of 80–150 °C is attributed to the loss of adsorbed water and interlayer water. A steady weight loss of about 9% in the temperature range of 150–800 °C which is attributed to the loss of dehydroxylation caused by breaking of structural hydroxyl groups of the montmorillonite and decomposition of the organic moiety present in the interlayers of the clay (Fig. 4b). PAni/IB and PAni/OMIB exhibit similar pattern, with a small variation in degradation temperature. In the first stage, 3-4% weight loss at temperature up to 120 °C is associated with the loss of water molecules [19]. In the second stage corresponding to temperature zone 120-350 °C, the weight loss (about 12-19%) is due to the evolution of thermally labile compounds and the breaking of aliphatic structures with low dissociation bonds in the clay matrix. In the third stage that commences after 350 °C until 700 °C, maximum weight loss occurs (33-38%) due to release of soot particles and the loss of dehydroxylation caused by breaking of structural hydroxyl groups of the clay [20,21]. Above, 700°C gradual decreases were observed due to thermal degradation of mineral matter of clay.

3.5 *SEM*

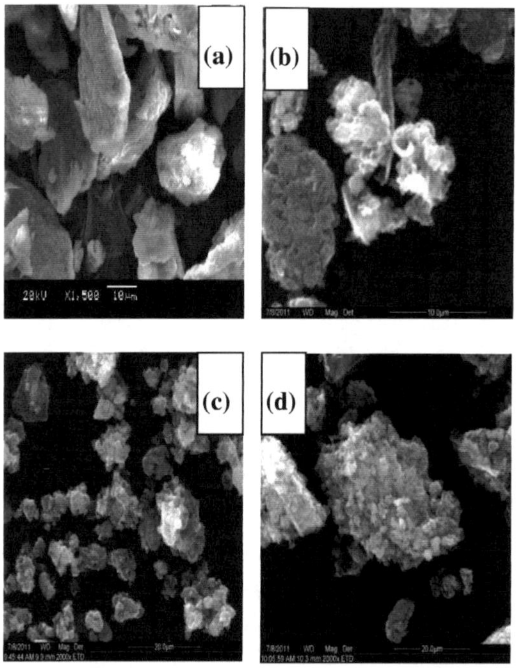

Fig. 5 SEM images of (a) IB, (b) PAni, (c) PAni/IB and (d) PAni/OMIB.

Fig. 5 shows SEM images of IB, PAni, PAni/IB and PANI/OMIB. SEM image of IB (Fig. 5a) shows morphology of clay particles having different irregular shapes, where as PAni/IB (Fig. 5c) has flake like morphology and PAni and PAni/OMIB have particles of both smaller and bigger sizes with irregular shapes (Fig. 5b&d).

3.6 *Electrochemical characterization*

Cyclic voltammetry. Fig. 6-8 display the cyclic voltammograms (CV) of PAni, PAni/IB and PAni/OMIB measured in the potential range of –0.2 to 0.6 V employing the scan rates of 10, 20, 50 and 100 mVs^{-1}. The corresponding specific capacitances (on y-axis), obtained from the CV curves are also shown. Generally, the charge storage occurs in an electrical double-layer capacitor by electrostatic reversible adsorption of ions of the electrolyte. This type of capacitance behavior gives CV curves close to ideal rectangular shape. But these samples showed CV curves distinctly different from rectangular shape. This pseudocapacitance behavior can be attributed to the charge storage mechanism essentially by redox reactions. There is an inverse relation observed between the scan rate and the total amount of charge stored. In other words, there are kinetic limitations associated with the diffusion of ions through the electrode matrix which limits the full storage of charge at higher scan rates. For example, at 10 mVs^{-1} scan rate, PAni electrode reaches maximum specific capacitance of 457 F g^{-1} while at 20 mVs^{-1} it reaches 388 F g^{-1}, whereas at 10 mVs^{-1} scan rate, PAni/IB electrode reaches maximum specific capacitance of 455 F g^{-1} while at 20 mVs^{-1} it reaches 430 F g^{-1} and at 10 mVs^{-1} scan rate, PAni/OMIB electrode reaches maximum specific capacitance of 415 F g^{-1} while at 20 mVs^{-1} it reaches 323 F g^{-1}. The reduced supercapacitance of PAni/OMIB may be attributed to the presence of non-conducting cetyltrimethylammonium moiety. From the results, it is clear that PAni/clay nanocomposites show better supercapacitance properties over PAni alone even at higher scan rates.

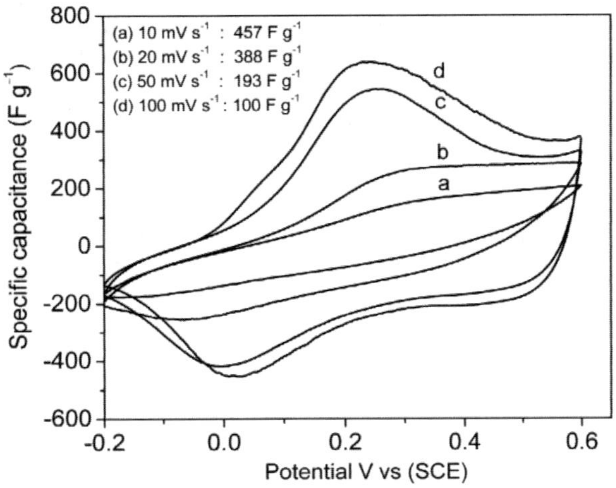

Fig. 6 Cyclic voltammograms of PAni at different scan rates in 0.5 M H_2SO_4 electrolyte.

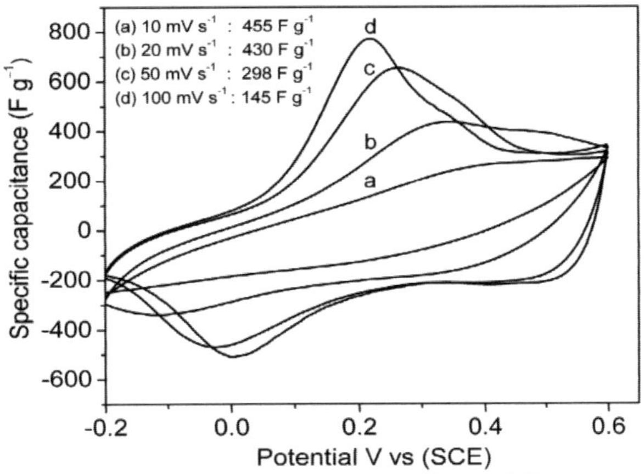

Fig. 7 Cyclic voltammograms of PAni/IB at different scan rates in 0.5 M H_2SO_4 electrolyte.

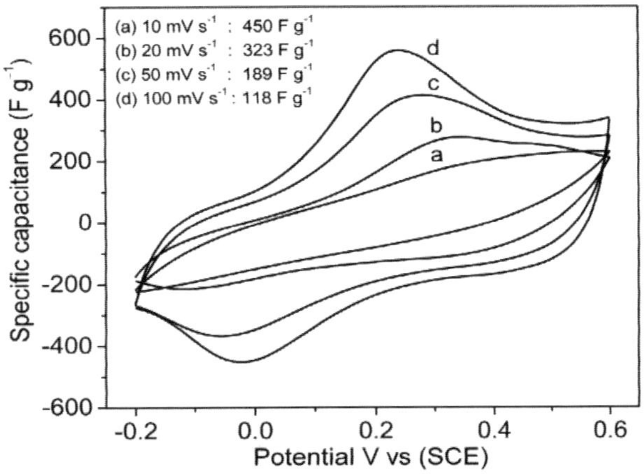

Fig. 8 Cyclic voltammograms of PAni/OMIB at different scan rates in 0.5 M H_2SO_4 electrolyte.

4. Conclusions

Modified clays have been prepared with cetyltrimethyl ammonium bromide. Polyaniline/modified clay and polyaniline/clay nanocomposites have been prepared and characterized for their physico-chemical properties. The polymer/clay nanocomposites are found to be thermally stable compared to pure polyaniline. The electrochemical studies revealed that polyaniline upon modification with clays can be used as supercapacitors for better storage compared to pure polyaniline.

Acknowledgements

This work is funded by the Department of Science and Technology, New Delhi, India, under Fast Track Scheme for Young Scientists (Project No.SR/FTP/CS-11/2006). BVK is grateful to Prof. G. Ranga Rao for his support and encouragement. K. O. Anjana thanks the Indian Academy of Sciences, Bangalore for awarding a summer research fellowship.

References

[1] Giannelis E P, Krishnamoorti R, Manias E 1999 Adv. Polym. Sci. **138** 107
[2] LeBaron P C, Wang Z, Pinnavaia T J 1999 Appl. Clay Sci. **15** 11
[3] Vaia R A, Price G, Ruth P N, Nguyen H T, Lichtenhan J 1999 Appl. Clay Sci. **15** 67
[4] Biswas M, Ray S S 2001 Adv. Polym. Sci. **155** 167
[5] Aranda P, Ruiz-Hitzky E 1992 Chem. Mater. **4** 1395
[6] Greenland D J 1963 J. Colloid Sci. **18** 647
[7] Blumstein A 1965 J. Polym. Sci. A **3** 2665
[8] Krishnamoorti R, Vaia R A, Giannelis E P 1996 Chem. Mater. **8** 1728
[9] Ray S S, Okamoto M 2003 Prog. Polym. Sci. **28** 1539
[10] Li D, Huang J, Kaner R B 2009 Acc. Chem. Res. **42** 135
[11] Vijayakumar B, Nagendrappa G, Jai Prakash B S 2009 Catal. Lett. **128** 183
[12] Krishna B S, Murty D S R, Jai Prakash B S 2000 J. Colloid Interface Sci. **229** 230
[13] Rees C A, Provis J L, Lukeya G C, van Deventer J S J 2007 Langmuir **23** 8170
[14] Cho M S, Park S Y, Hwang J Y, Choi H J 2004 Mater. Sci. Engg.: C **24** 15
[15] Libert J, Cornil J, dos Santos D A, Bredas J L 1997 Phy. Rev. B, **56**(14) 8638
[16] Glasel H -J, Hartmann E, Hormes J 1999 J. Mater. Sci. **34** 1
[17] Chandrakanthi R L N, Careem M A 2002 Thin Solid Films **417** 51
[18] La Puente G D, Marban G, Fuente E 1998 J. Anal. Appl. Pyrolysis **44** 205
[19] Arenillas A, Rubiera F, Pis J J, Cuesta M J, Iglesias M J, Jimenez A, Suarez-Ruiz I 2003 J. Anal. Appl. Pyrolysis **68-69** 371
[20] Malhotra B D, Chaubey A, Singh S P 2006 Anal. Chim. Acta, **578** 59
[21] Kaushik A, Khan R, Gupta V, Malhotra B D, Singh S P 2009 J. Nanosci. & Nanotechnol. **9** 1792

Corrigendum: Polyaniline/clay Nanocomposites: Preparation, Characterization and Electrochemical Properties

B Vijayakumar[1,*], **K O Anjana**[2] **and G Ranga Rao**[1]

[1] Department of Chemistry, Indian Institute of Technology Madras, Chennai-600036, India
[2] Department of Chemistry, National Institute of Technology Calicut, Kozhikode-673601, India

E-mail: badathala@rediffmail.com

CORRIGENDUM TO: 2015 IOP Conf. Ser.: Mater. Sci. Eng. **73** 012112

In Experimental, after 2.1 Materials and methods the following text should be inserted:

2.2 *Preparation of polyaniline/clay nanocomposites*

5 g (13.71 mM) of cetyltrimethylammonium bromide (CTAB) was dissolved in 200 ml of distilled water and 10 g of Indian bentonite (IB) was added. The contents were stirred using a mechanical stirrer for 1 h, the mixture was centrifuged and the centrifugate was discarded. The solid was washed several times with distilled water to remove superficially held adsorbate. The resulting modified clay (OMIB) was dried at 100 °C and ground to a fine powder.

The oxidant solution (0.11 M ammonium persulphate in deionised water) was added dropwise to a mixture of 0.5 g of modified clay, 0.1 M aniline and 0.1 M dopant (*p*-toluene sulphonic acid) in 75 mL deionised water. The reaction was carried out at a temperature of 0-5 °C with ice around the reaction vessel for 12 h under stirring. The controlled low temperature is chosen to prevent high increase in temperature due to the exothermic reactions. The polymer obtained was in the conducting state (emeraldine salt; green colour of doped PAni). The reaction product was separated from the solution by filtration and washed with deionised water several times, followed by washing with ethanol to remove residual monomers. The powder (PAni/OMIB) was then dried at 40 °C under vacuum. Following the same procedure, polyaniline/clay nanocomposite (PAni/IB) and polyaniline (PAni) were prepared by adding parent clay and without clay respectively for comparison.

Structural and morphological properties of Ag ion irradiated SnO_2 thin films

K M Abhirami[1], P Matheswaran[1], B Gokul[1], R Sathyamoorthy[1], K Asokan[2]

[1]PG and Research Department of Physics, Kongunadu Arts and Science College, Coimbatore, Tamil Nadu- 641 029, India.
[2]Materials Science Division, Inter University Accelerator Centre, Aruna Asaf Ali Marg, New Delhi- 110 067, India.
E-mail: abhirami.km@gmail.com

Abstract. SnO_2 thin films of thickness 300 nm were prepared by reactive thermal evaporation and subjected to sintering at 600 °C for 2 hr. The annealed films were irradiated using silver (Ag) ions with energy of 120 MeV at different fluences (1×10^{11}, 5×10^{11}, 1×10^{12}, 5×10^{12} and 1×10^{13} ions/cm^2). The effect of swift heavy ion (SHI) irradiation on structural, morphological and optical properties were studied using X-ray diffractometer (XRD), Scanning electron microscopy (SEM) and UV-visible spectrophotometer. XRD studies showed formation of tin oxide with tetragonal structure. Morphology analysis revealed uniform deposition of the material with increase in grain size after irradiation upto 1×10^{12} ions/cm^2 and beyond that size tends to decrease. In addition, agglomeration of nanocrystalline grains was observed after Ag ion bombardment with varying fluence. The transmittance decreases from 90% to 80% as the fluence increases.

1. Introduction

Ion irradiation is a useful tool for impurity doping and defect production in materials, which helps to alter their structural, electrical, optical, and magnetic properties with a high spatial selectivity [1, 2].The dynamic development of interest in wide band-gap semiconductors thin film and its widespread utilization in the fields of technology have led to the increased interest in the various problems of the properties of these semiconductors thin film in the recent years. Metal oxides are the best suited for wide band-gap semiconductors. Oxides are the basis of smart and functional materials. The synthesis and device fabrication using functional oxides have attracted a lot of attention due to the functionality of the physical properties of these oxides. Functional oxides have two structural characteristics: cations with mixed valence states, and anions with deficiencies (vacancies) [3]. SnO_2 is an n-type semiconductor with a wide band gap of approximately 3.7 eV. SnO_2 film shows the best thermal and chemical stabilities. It has extensive applications such as transparent conducting oxide in solar cells and other optoelectronic devices and remains active material for gas sensor.

In this paper, in an attempt to synthesize SnO_2 thin films with different morphology, we have made use of Ag-ion irradiation at room temperature and have studied changes in structural, optical, electrical, and morphological properties for a wide range of fluence.

2. Experimental details

SnO_2 thin films were prepared by means of reactive thermal evaporation of thickness 300 nm on glass substrates at 300 °C. The vacuum of the chamber during deposition was maintained at about 5x 10^{-6} mbar. The as grown films were sintered at 600 °C for 2 h. The 120 MeV Ag^{9+} ion beam irradiations at fluences of $1x10^{11}$, $5x10^{11}$, $1x10^{12}$, $5x10^{12}$ and $1x10^{13}$ ions/cm^2 were first carried out for the structural phase transformation and bandgap modification. SHI irradiations were carried out with 120 MeV Ag ions using 15 UD tandem Pelletron accelerator at IUAC, New Delhi. The projected range of 120 MeV ions in the films as calculated using TRIM95 code is about 9 μm, which is greater than the total thickness of the film (0.2 μm). Thus, the bombarding ions pass through the entire film and come to rest in the substrate.

Irradiated and pristine films have been characterized using different techniques. Structural modifications were analyzed by X-ray diffraction (XRD) Philip PW1700 diffractometer (using Cu Kα). Surface structures were characterized by field emission-scanning electron microscopy (FE-SEM) [MIRA, TESCAN] at IUAC, New Delhi. Electron beam of energy 20 keV was used to capture the SEM images in the present experiment. UV-visible spectroscopy is performed to study the modification of band gap due to SHI irradiation. Optical absorption spectra were recorded with the conventional two-beam method using the U-3300 UV-visible spectrophotometer of Hitachi.

3. Result and discussion

Structure of the pristine and irradiated films was investigated by XRD technique. Figure 1 shows the XRD patterns of pristine and irradiated films. The diffraction patterns corresponding to irradiated films and pristine films match well with the JCPDS data [4], whereas the diffraction angle and intensity of the (101), (200) peak varies with ion fluence. No additional peaks related to other crystalline phases or metallic tin were observed. The grain size was calculated using Debye–Scherrer's formula and the estimated grain size for the as-grown film is about 17.3 nm. The full width of half maximum (FWHM) of this peak found to dwindle upto 1x 10^{12} ions/cm^2, implying that irradiation at low fluence increases the grain size. Thus, the crystallinity and quality of the film were getting improved with ion irradiation at low fluence. While at higher ion fluence beyond 1x10^{12} ions/cm^2, the FWHM of the peak seems to be augmented, implying that the size of coherent domains decreased. The irradiation results to an increase in the grain size and a maximum grain size of 27.3 nm was estimated for film irradiated at a fluence of 1×10^{12} ions/cm^2. At higher ion fluence of 1×10^{13} ions/cm^2 the grain size decreased to a value of about 11.7 nm. The increase in grain size might be due to the annealing effect of SHI below the threshold S_e value and the growth of small crystalline inclusions present in the film, leading to local crystallization. Similar grain growth was observed in SHI irradiated thin films [5, 6]. However, the observed decrease in the apparent grain size at higher fluences could be due to an increase of the micro-strains generated by irradiation-induced lattice defects.

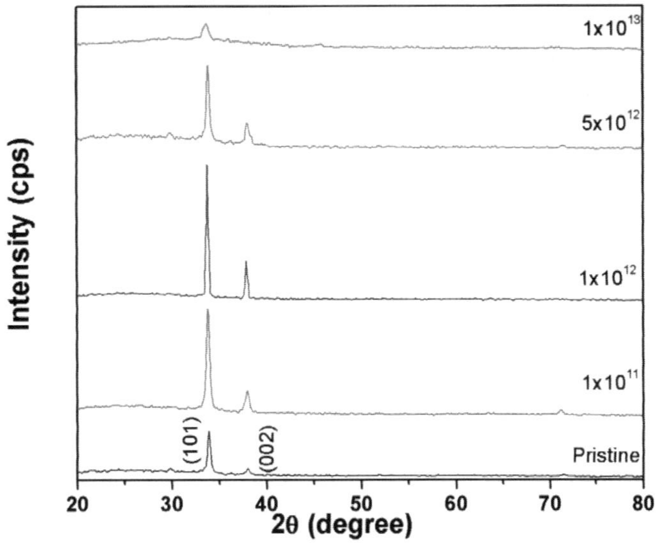

Figure 1. XRD pattern of pristine and irradiated SnO_2 thin films

Surface morphology of the pristine and irradiated films has been characterized by FE-SEM. The SEM image of the pristine sample (figure. 2), contains tetragonal shaped structures grown perpendicular to the surface. As the fluence increases the grains melt due to irradiation induced annealing and form bigger grains. At higher fluence of about $1x10^{12}$ ions/cm^2, the tetragonal like structure gradually disappears and grains were observed to be agglomerated under disorder induced by swift heavy ions.

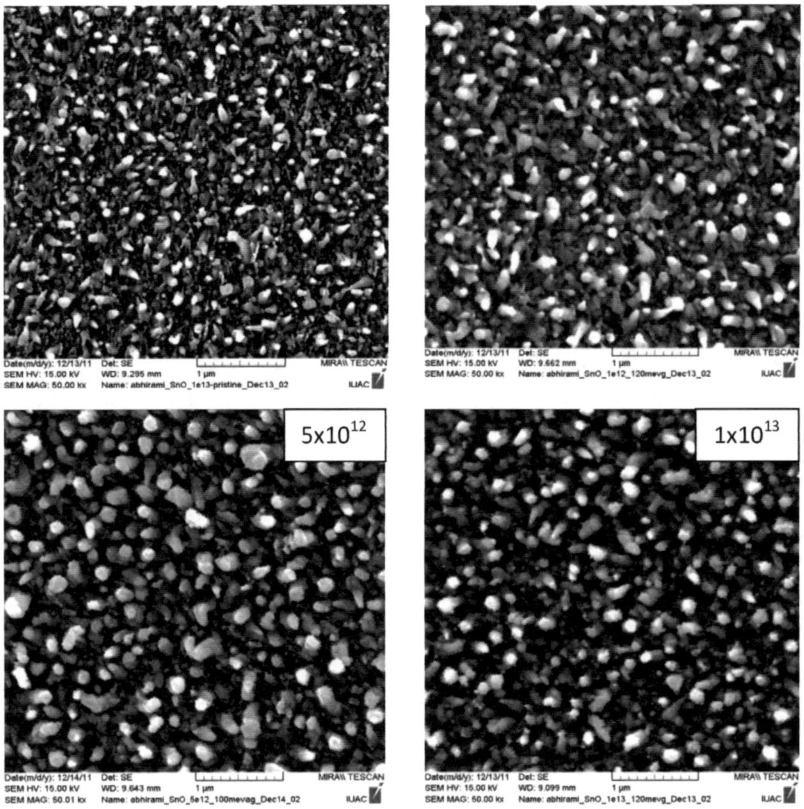

Figure 2. SEM image of pristine and irradiated SnO_2 thin films

UV–Visible spectroscopy provides useful information about the optical band gap of the semiconductors. Transmittance spectrum is shown in figure 3. The optical absorption study of the irradiated and pristine films was carried out and the band gap of the films has been calculated using Tauc's plot by plotting $(\alpha h\nu)^{1/2}$ vs. hν and extrapolating the linear portion of the absorption edge to find the intercept with energy axis. The structural transformation (upto $1x10^{12}$) generally leads to the formation of larger grains, which results in the reduction of band gap of the semiconductors. We also notice an increase in optical bandgap (beyond $1x10^{12}$) in the irradiated film, which may be due to the irradiation provoked defect creation like anti-site oxygen and oxygen vacancies in the film. The calculated values of *Eg* are 3.55, 3.40, 3.62 and 3.84 eV for films irradiated at fluences of 0, 1×10^{11}, 1×10^{12} and 1×10^{13} ions/cm^2, respectively. This shows a consistent decrease of Eg with an increase in ion fluence up to 1×10^{12} ions/cm^2, and at higher fluence (1×10^{13} ions/cm^2) a definite increase in the band gap was observed. The shift in Eg upon irradiation can be attributed to several reasons. The variation in the optical band gap can be candidly related to the change in the grain size because; we have observed a consistent increase of grain size up to 1×10^{12} ions/cm^2 and a sudden decrease at 1×10^{13} ions/cm^2. The drop off in the energy gap with increasing ions fluences is due to creation of midway energy levels [7] and quantum confinement effect [8, 9].

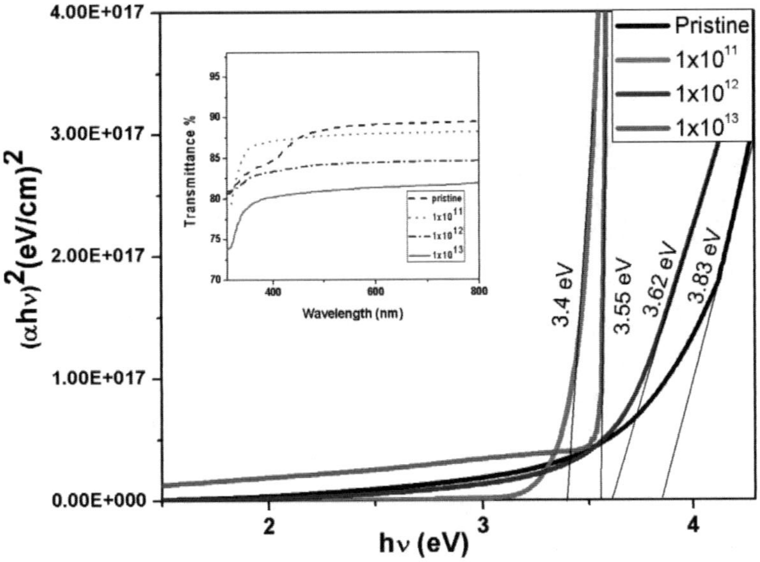

Figure 3. Transmittance graph of pristine and irradiated SnO_2 thin films

4. Conclusion

The experimental data show that Ag ion irradiation (at lower fluences) induces formation of crystalline phase, whereas at higher fluence the crystallinity declines. In addition, we find that irradiation at higher fluences gives rise to inhomogeneous strain due to fluctuations in size and shape of individual grains. The grain size increases upto $1\text{x}10^{12}$ and decreases beyond that forming agglomerated structures at high fluence. The value of bandgap at $1\text{x}10^{12}$ is 3.62 eV which is very near to the bulk bandgap value. The change in optical band gap follows the change in grain size and hence the observed effect could be results of irradiation persuade modifications in the grain size.

5. Reference

[1] Cogan S F, Andersson E J, Plante T D, and Rauh R D 1985 Appl. Optics **24** 2282
[2] Isidorsson J and Granqvist C G 1996 Sol. Energ. Mat. Sol. C **44** 375
[3] Wang Z L and Kang Z C 1998 Functional and Smart Materials, Plenum, New York
[4] JCPDS No. 721147
[5] Sreekumar R et al. 2008 J. Appl. Phys. **103** 023709
[6] Agarwal D C et al. 2006 Nucl. Instrum. Meth. B **244** 136
[7] Chandramohan S et al. 2007 Nucl. Instrum. Meth. B **254** 236
[8] Chaudhary Y S et al 2004 Nucl. Instrum. Meth. B **225** 291
[9] Mohanta D, Mishra N C and Choudhury A 2004 Mater. Lett. **58** 3694

Acknowledgement

The author (KMA) sincerely thanks IUAC, New Delhi for the project and the help during experiment and characterization.

IOP Conf. Series: Materials Science and Engineering **73** (2015) 012114 doi:10.1088/1757-899X/73/1/012114

Ferromagnetic behaviour at room temperature of Co doped CdS nanoparticles by chemical precipitation method: Synthesis, optical and electrical properties

T Lavanya[1], K Satheesh[2], N V Jaya[1*]

[1] Department of Physics, Anna University, Chennai 600 025, India
[2] Research and Development, Dhanalakshmi College of Engineering, Tambaram, Chennai - 601 301, India

E-mail: victor@annauniv.edu

Abstract. Cobalt doped CdS nanoparticles were prepared by chemical precipitation method. The prepared powder were characterized by various analytical techniques such as Powder XRD, SEM, EDAX, Magnetic and spectroscopic analysis of nanoparticles as a function of cobalt concentration have been systematically studied. The XRD study confirms the high crystalline nature of prepared compounds and the presence of hexagonal phase of the CdS nanoparticles. The particle size of CdS nanoparticles are calculated by Scherer's formula and it is found that the particles are in the range of 8 to 10 nm.

1. Introduction and experimental

Semiconductor nanoparticles, as one of the most technologically important class of materials, CdS nanoparticles has the great potential applications as nanoelectronics and photocatalytic materials are have a relatively large nonlinear optical response and photo catalytic activity. Since their properties and structures are closely related to size, shape and dimension play important role to determine their physical properties. Doping is important for semiconductors, which plays a critical role in tunning their optical and electrical properties for the potential applications in wavelength-controlled lasers, biotechnology , and solar cells.

IOP Conf. Series: Materials Science and Engineering **73** (2015) 012114 doi:10.1088/1757-899X/73/1/012114

In the present work, we have prepared the Co-doped CdS nanoparticles by co-precipitation method in air atmosphere.

Co-doped CdS nanoparticles were prepared by chemical precipitation method using $CdSO_4$ CH_4N_2S $Co(CH_3COOH)_2$as a precursor. For the preparation of CdS nanoparticles, initially 0.1M of $CdSO_4$ and CH_4N_2S was dissolved with water under vigorous strriring above the room temperature.Then 1 mole% of $Co(CH_3COOH)_2$ was added to it. After that NH_3 solution was used as a reducing agent for adjusting the pH value to ~10. The yellowish brown color precipitates were obtained in the bottom of the solution after 4-5 hrs strriring and filtered. Finally the as prepared material was sintered at 200°C in the high temperature conventional furnace. Similar procedure was followed for other two concentrations of cobalt (3 and 5 mole %).

2. Results and discussion

The powder X-ray diffraction analysis was performed to study the crystalline structure and avg. particle size of the cobalt-doped cds samples calcined at 300°C in air atmosphere shown in Figure 1A. It is evident that the XRD patterns show an excellent match with hexagonal phase peaks of the system (PCPDF #75-1546). The maximum diffraction peak

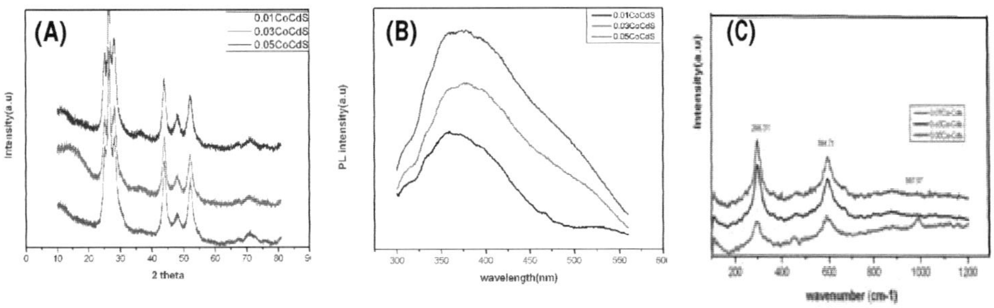

Figure 1 (A) XRD (B) PL, (C) Raman spectra of Co doped CdS nanoparticles with different concentration.

observed for (002) plane along with (100),(001),(110),(102) and (112) planes for their particular angles.The Lattice parameters were calculated as a=b= 3.701 Å and c=6.813Å and confirmed its exhibit Hexagonal structure. According to the Scherer's equation the avg. particle size was calculated to be as ~10 nm for the synthesized Co doped CdS nanoparticles. The PL spectra were employed to study the emission characteristics of CdS nanoparticles. The PL Spectrum was observed in the range from 300 nm to 600 nm excitation wavelengths of the CdS nanoparticles with Co concentrations (0.01 mole%, 0.03 mole %, and 0.05 mole%) is shown in Figure 1B. The emission peaks at 382 and 375 nm are usually observed from the excitonic emission luminescence and blue emission band was associated with the emission due to electronic transition from the conduction band to an acceptor level due to Co concentration. Figure 1C shows Raman spectra of the Co doped CdS nanoparticles. The peaks are obtained at 301, 603 and 909 cm^{-1} corresponding to 1 LO, 2 LO and 3 LO optical phonons respectively. The first order LO Raman line is not only broadened, but also it shows an asymmetric broadening towards the low frequency side with the increase in Co-dopant level. It is now well known that confinement of phonons, optical as well as acoustic, influences the phonon spectra conspicuously when the grain size falls to a few nanometers.

The surface morphology of the CdS nanoparticles were analyzed by using Field Emission-Scanning Electron Microscopy (FE-SEM) measurement From the Figure 2A.

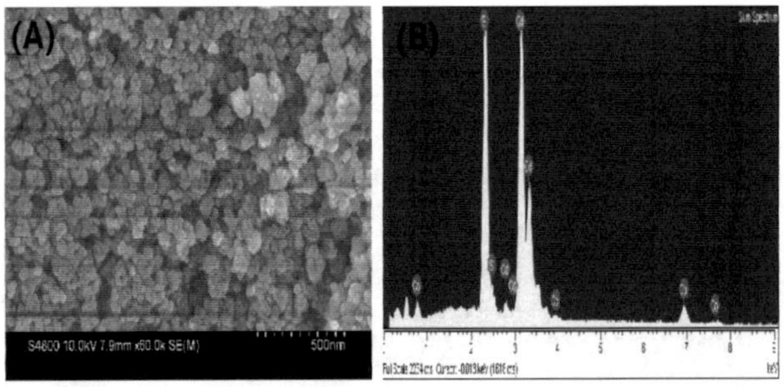

IOP Conf. Series: Materials Science and Engineering **73** (2015) 012114 doi:10.1088/1757-899X/73/1/012114

Figure 2 (A) FE-SEM and (B) EDAX image of Co doped CdS nanoparticles.

It is observed that the morphology of the particles is spherical. The SEM micrographs shows the particles are homogenously distributed without any pores in the lower nanometer range.The morphology having fine nanostructure is detected by Field Emission- scanning electron microscopy. An EDX result (Figure 2B) gives the strong K diffraction peaks at 3.2,3.4 and 2.3 keV, corresponding to the elemental S and Cd and the diffraction peaks at 0.8 and 6.9 keV corresponding to the cobalt element Present in the materials.

The magnetic properties of the Co doped CdS nano particles (Co concentrations 0.01 mole% (Figure 3A), 0.03 mole % (Figure 3B), and 0.05 mole% (Figure 3C)) at roomtemperature were evaluated by vibrating sample magnetometer (VSM). It is clear that the magnetization (M–H curve) of doped samples is greatly sensitive to Co dopant concentration. 0.01 mol% Co-doped CdS and the decrease of magnetization at larger magnetic field strength part can be ascribed to the paraiamagnetic contribution as shown in the Figure 3A. These paramagnetic ions are distributed homogeneously in the samples and the transferred hyperfine interactions between cobalt and Cd nuclei. The wurtzite hexagonal 0.05 mo

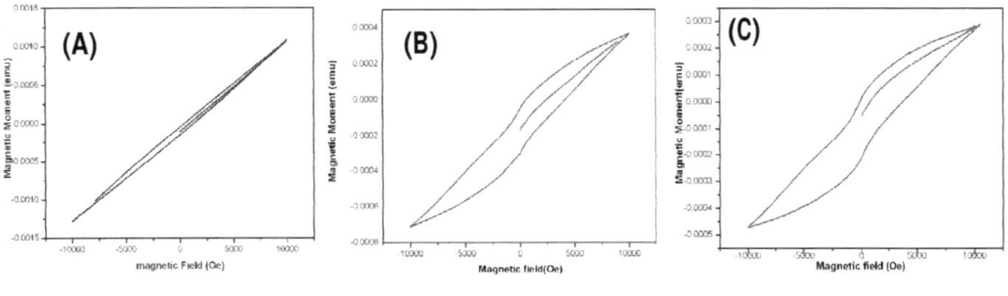

Figure 3(A), (B) and (C) M-H curve of 0.01, 0.03 and 0.05 mol% Co doped CdS nanoparticles.

Cobalt doped CdS sample Figure 3B shows significant ferromagnetic ordering which informs that more number of Cobalt ions are incorporated in to CdS lattices and more CoS clusters are also attached to the surface of CdS: Co. This room temperature ferromagnetism arises from the creation of a spin-split impurity band at the Fermi level, below the conduction band due to the hybridization between the charge carriers of cobalt and cadmium. Figure 3C shows 0.03 mol% Co-doped CdS exhibits a weak ferromagnetic order even at room temperature.Ferromagnetic order is either suppressed or disappeared because it may be due to theparticle size decreases with the increasing Co concentration.

3. Conclusion

In the present work, the Co CdS nanoparticles were synthesized by chemical precipitation method. The XRD results confirm the CdS particles are in the nanometer range and exhibit in the hexagonal crystalline structure. The Raman signals from CdS nanoparticles have longitudinal optical (LO) phonons peaks.. The short emission peak in the PL spectrum which is due to the smaller grain size of the particles. The uniform particle size were confirmed in the SEM analysis. EDX result confirmed the cobalt element Present in the materials. room temperature ferromagnetism. Its magnetic properties are changed from para to ferromagnetic with increasing Co concentration.

4. References

[1] Cahn R W1990 Nanostructured materials *Nature* **348** 389.
[2] Qingqing W, Gang X and Gaorong H 2005, Solvothermal synthesis and characterization of uniform CdS nanowires in high yield *Solid State Chem.* **178** 2680.
[3] Archana, Maurya and Pratima Chauhan 2011 Structural and optical characterization of CdS/TiO2 nanocomposite *Materials Characterization* **62** 382.
[4] MathieunH, Richard T, Allegre J, Lefebvre P and Arnaud G 1995 Quantum confinement effects of CdS nanocrystals in a sodium borosilicate glass prepared by the sol- gel process *J. Appl. Phys.***77** 287.
[5] Chawla P, Lochab S P and Singh N 2010 Photoluminescence, thermoluminescence and Raman studies of CdS nanocrystalline phosphor *J. Alloys & Comps* **492** 662.

[6] Coey J M D , Venkatesan M and Fitzgerald CB 2005 Donor impurity band exchange in dilute ferromagnetic oxides *Nature* **4** 173.
[7] Xu W, Liao Y and Akins DL 2002 Formation of CdS nanoparticles within modified MCM-41 and SBA-15 *J Phys Chem B* **106** 11127.

Microwave assisted synthesis of copper oxide and its application in electrochemical sensing

S Felix[1], R Bala Praveen Chakkravarthy[2] and A Nirmala Grace[1*]
[1]Centre for Nanotechnology Research, VIT University, Vellore – 632 014, Tamil Nadu, India.
[2] Research & Advanced Engineering Division (Materials), Renault Nissan Technology & Business Center India (P) Ltd., Chennai - 603 002, Tamil Nadu, India.
[*]Corresponding author's e-mail:balapraveen2000@yahoo.com, anirmalagladys@gmail.com

Abstract. Copper oxide nanopowders were prepared using copper acetate as the precursor and polyethylene glycol (PEG) as stabilizer in ethanol medium. The mixture containing copper acetate, sodium hydroxide and PEG was irradiated with microwave and nanometric copper oxide particles were formed within 8 min. The prepared nanoparticles were characterized using x-ray diffraction, UV-vis spectroscopy and scanning electron microscopy. The average particle size was found to be ~ 4 nm. This was used to modify glassy carbon electrode with PVDF & DMF as binder and used for sensing of carbohydrates (glucose and sucrose) and H_2O_2. The copper oxide nanoparticles showed excellent sensitivity in the range of 0.1 mM to 1 mM when choronoamperometry was carried out at 0.6 V Vs. Ag/AgCl. The observed sensitivity is much higher when compared with conventional micron sized copper oxide particles.

1. Introduction

Metal oxides are versatile materials with electrical conductivities ranging from insulating to metallic, optical transmission from opaque to transparent, magnetic susceptibility from para to ferromagnetic, etc., [1]. In recent years research focus has been increasing in producing nanostructured metal oxides for a variety of applications like transparent conductors, super paramagnetic materials, battery materials, catalysts etc., Compared with the bulk metal oxide materials, nanoscale metal oxides have many advantages due to their extremely reduced size and large surface to volume ratio provided the crystallinity is maintained [2]. Diabetes is a worldwide public health problem. An important clinical test for diagnosing diabetes is accurate measurement of blood glucose level. Many types of glucose sensors are known, viz., chemiluminence, fluorescence and optic sensors [3, 4]. Sucrose is the important carbohydrate in most of the drinks and foods. Fast, simple and efficient electrochemical method used to determine sucrose in food industry [5]. Hydrogen peroxide (H_2O_2) can induce functional and morphological disturbances and because of its extreme toxicity in cells can cause cancer and its detection is important to clinical, industrial, and environmental analyses. Although technologies such as titrimetry, fluorescence, and absorption spectra have been used to detect H_2O_2 and new sensors based on electrochemical technology have proven simple to use and highly sensitive for real-time detection [6]. Electrochemical sensors have many advantages compared to the optical methods like relatively low cost, high selectivity and sensitivity, and ease of integration with electronic systems.

CuO is a p-type semiconductor and its narrow band gap 1.24 eV makes it suitable for applications in gas sensors, magnetic storage media, lithium batteries and solar cells due to its photoconductive and photochemical properties. The properties of CuO are closely related to its crystal size, orientation and morphology [7]. Microwave irradiation as a heating method has found a number of applications in chemistry. The microwave solvothermal synthesis is popular as it is faster, simpler and efficient in energy for metal oxide nanopowder synthesis. Good phase purity is also reported when microwaves are used [8, 9]

In this work CuO nanoparticles were synthesized by microwave irradiation method in ethanol medium. The prepared CuO nanoparticles were characterized by using XRD, UV-vis spectroscopy and SEM. The electrocatalytic activity of CuO nanoparties with modified electrodes towards the glucose, sucrose and hydrogen peroxide sensing in 0.1 M NaOH was investigated using cyclic voltammetry and chronoamperometry. It was found that sucrose can also be sensed using these modified electrodes with comparable sensitivity to that of glucose, thus opening avenues to use it as a chemical oxygen demand sensor.

2. Experimental procedure

2.1. Chemicals

Copper acetate, sodium hydroxide, ethanol and polyethylene glycol (PEG) were purchased from S. D. fine-chemicals. D-(+)-glucose, Sucrose, H_2O_2, N, N-Dimethyl formamide and polyvinyllidene fluoride were purchased from Sigma Aldrich.

2.2. Synthesis of CuO nano particles by Microwave irradiation method

1.59 g of $Cu(CH_3COO)_2$ was dissolved in 50 mL ethanol and mixed with 0.016 g of NaOH added to 50 mL ethanol. 5 g PEG was then dissolved in the solution mix. The mixture was placed in a domestic microwave oven (Sanyo, EM-G3686WY) and irradiation was carried out in ambient air for a total of 8 min at 80 % power. The solution mixture boils in about 10 s of microwave irradiation. Therefore, it was cooled in an ice bath after every 10 s and heating was continued till. Therefore, it was cooled in an ice bath after every 10 s and heating was continued till. The dark-green precipitate was formed at the end of the reaction. After cooling to room temperature the precipitate was centrifuged, washed with de-ionized water repeatedly and finally with absolute ethanol, dried in a hot air oven at 100 ° C for 3 h. The final product (dark brown in color) was collected for characterization [9].

2.3. Characterization of CuO nanoparticles

The phase purity and crystallinity of the prepared nanoparticles was determined by XRD (D8 Advanced, Bruker diffractometer, Germany). The optical absorbance of the powder characterized in the wavelength between 200 to 800 nm using UV-Visible spectrophotometer (Shimadzu, UV-3600, china). The surface morphology and size of CuO nanoparticles were investigated using Scanning Electron Microscope (JEOL SM 840, Spain) and Electrochemical studies were carried out using an electrochemical analyzer (CHI 660C, CH Instruments, Switzerland).

2.4. Electrode modification and electrochemical measurements

The glassy carbon working electrode had a nominal diameter of 3 mm; it was polished with 0.05 micron alumina powder mixed with deionized water before testing and washed thoroughly with acetone and de-ionized water. 3 mg of CuO was dispersed in 150 μL (PVDF&DMF) solution in the ratio of 1:50 and then sonicated for 10 min. 10 μL of above dispersion was coated on the surface of glassy carbon electrode and dried at room temperature for 5 h. This gave a stable electrode with copper oxide immobilized on it. Electrochemical measurements were performed using a conventional three electrode system with CuO modified glassy carbon as the working electrode, platinum wire as counter electrode and Ag/AgCl as reference electrode. During chronoamperometry measurements, the voltage was maintained at 0.6 V and the current was observed after successive addition of (1000 μL of 5.4 $mg\,cm^{-3}$) glucose to the electrolyte solution with constant stirring. The same condition was followed for sucrose at a constant potential of 0.6 V in a successive addition of sucrose to the electrolyte under stirring condition. In case of hydrogen peroxide, the concentration was varied from 0.5 mM to 10 mM with a constant stirring at 0.6 V.

3. Results and Discussion

3.1. XRD-analysis

XRD analyses were used to study the phases and crystalline nature of CuO nanoparticles as shown in figure 1. All diffraction peaks located at values between 30° and 65° (35.51, 38.71, 48.81 and 58.41) can be attributed to the Miller indexes of the (-1 1 1), (2 0 0), (-2 0 2) and (2 0 2) planes of monoclinic CuO, according to (JCPDS 89-5895). The size of the nanoparticles was calculated by Debye-Scherrer formula.

$$D = \frac{0.9\lambda}{\beta \cos\theta} \tag{1}$$

Where λ is the wavelength of x-rays (Cu kα = 1.54178 Å), β is the full width half maximum value in terms of radians, θ refers to diffraction angle, the calculated particle size was 4 nm.

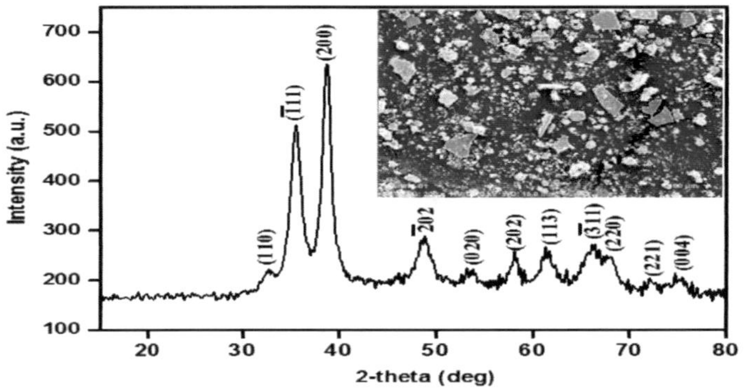

Figure 1. XRD patterns and SEM images (Inset) of synthesized copper oxide nanoparticles.

3.2. Optical properties

The UV-vis spectrum was used to determine the optical absorption properties of CuO nanoparticles. Figure 2 shows the UV-Vis absorption spectrum of CuO nanoparticles. A broad absorption peak at 297 nm confirmed that the band gap is due to the intrinsic transition of the CuO.

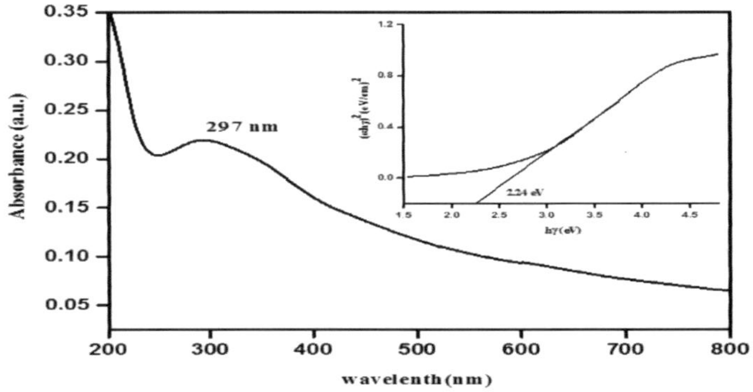

Figure 2. UV–vis absorption spectrum of monoclinic CuO. *Inset:* shows the plots of $(\alpha\ E_{photon})^2$ vs. E_{photon}.

The energy gap of these nano materials were estimated with Tauchi's equation [12] for direct transition in semiconductors given by

$$(\alpha h\gamma) = c(h\nu - Eg) \tag{2}$$

Where hγ, α, c, Eg are respectively incident photon energy, absorption co-efficient, constant and the energy gap. The plot drawn using $(\alpha h\gamma)^2$ on the y-axis against photon energy (E=hγ) on the x-axis show the value of the band gap given by the x-intercept as shown in figure 2 (inset), is 2.24 eV. The blue-shift behavior of the peak position in comparison with that of the bulk CuO (1.85 eV) [13] could be attributed to the enhancement of the quantum confinement effect resulting from the decrease in the size of the nanoparticles [12-14].

3.4. Electrochemical characterization

The sensing behavior of the modified electrode was observed at an applied potential of 0.6 V with successive addition of carbohydrates (glucose, sucrose) ranging from 50 μM to 1000 μM in 0.1 M NaOH solution using chronoamperometry. The first addition of glucose was made after 150 s with 50 μL added to the electrolyte. Then, 50 μL was added at 210 s. After that 100 μL was added once in every 60 s. A near linear behavior was observed. The sensitivity was calculated as 17.2 μA/mM and it gives a linear amperometric response in the glucose concentration from 100 μM to 1000 μM of glucose.

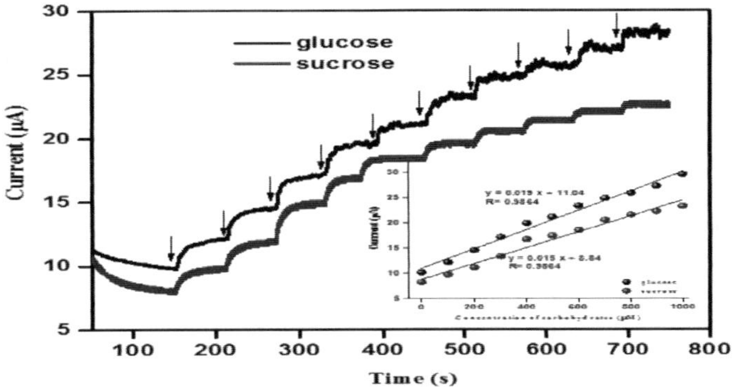

Figure 3. Choronoamperometry response of glucose and sucrose with glassy carbon modified with CuO in 0.1 M NaOH solution at a regular interval of 60 s. *Inset (right)* the corresponding calibration plot of net current vs. concentration.

The linear regression equation for glucose detection was calculated from net current, which is defined as the the total current minus the blank current i.e. current at ~145 s, $I_{glucose}$ =0.019x + 11.04C with R = 0.9864. Figure 3 displays the chronoamperometric behavior of sucrose with 50 μL sucrose added after 150 s. A fast, stable and stepwise increase in current was observed at 0.6 V. The sensitivity was calculated in the case of sucrose as 10.482 μA/mM. The linear regression equation is $I_{sucrose}$ =0.015x + 8.84C with R=0.9864. These values are comparable to those in the literature for copper oxide nanostructure [15, 16] and its sensitivity was calculated to be 6.53 μA/mM. Figure 4 (Inset) showed a poor linear range denoting a slow electron transfer speed and adsorption of the intermediate product produced during the glucose oxidation process. The same condition is followed for H_2O_2. But in the case of H_2O_2 concentration varying from 0.5 mM to 9 mM, a step current change occurred with its linear regression equation I(μA)=0.503x + 2.423, R = 0.9986. In general, insufficient stability of enzyme electrode is the result of the loss and deactivation of enzyme. The fabrication of carbohydrate sensors has the advantages like long term stability, low detection limit, reproducibility, simplicity and inexpensive [15-16] Figure 4 showed that the step response is distorted at higher concentrations. At higher concentrations, the detection is not diffusion limited and more side reactions are possible. Therefore, it may be concluded that this electrode detects better when the concentration of the peroxide is less than 10 mM.

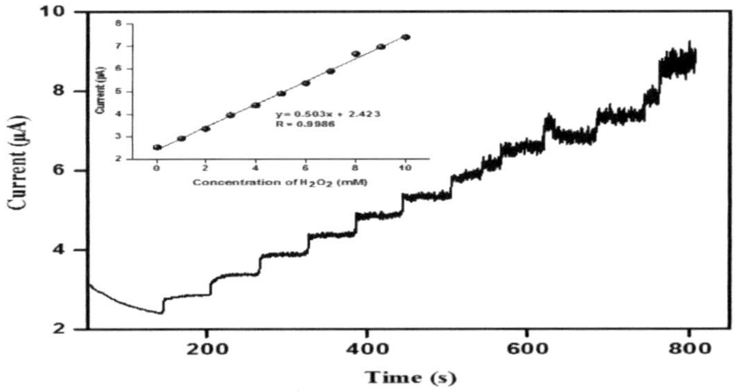

Figure 4. Amperometric response for the detection of H_2O_2. *Inset* shows the corresponding calibration plot.

4. Conclusions

CuO nanoparticles were successfully synthesized by microwave irradiation with phase purity and high crystallinity as seen by XRD; its optical properties and morphological studies were analyzed by UV-Vis spectroscopy and SEM analysis respectively. Using these nanoparticles, a sensitive amperometric electrochemical sensor for detecting glucose, sucrose and hydrogen peroxide is demonstrated.

References

[1] Kang X, Mai Z, Zou X, Cai P and Mo J 2007 *Anal. Biochem* **363** 143
[2] Prathab M U A, Kaur B and Srivastava R 2012 *J. Colloid Interface Sci* **370** 144
[3] Ding Y, Wang Y, Su L, Bellagamba M, Zhang H and Lei Y 2010 *Biosens. Bioelectron.* **26** 542
[4] Lu L M, Zhang X B, Shen G L and Yu R Q 2012 *Anal. Chim. Acta* **715** 99
[5] Surareungchai W, Worasing S, Sritongkum P and Tanticharoen M *Anal. Chim. Acta* **380** 7
[6] YiHu M, Chen H C, Tsai R Y and Lai C S 2011 *Talanta* **85** 631
[7] Chang L, Shao M, Chen D and Zhang Y 2010 *Mater. Res. Bull* **45** 235
[8] Chen G, Zhou H, Wei M, Zhang D and Qiu G 2012 *Solid State Sci* **13** 2137
[9] Wang H, Xu J Z, Jiezhu J and Chen H Y 2002 *J. Cryst. Growth* **244** 88
[10] Li Y, Liang J, Tao Z and Chen J 2008 *Mater. Res. Bull* **43** 2380
[11] Wang F, Tao W, Zhao M, Xu M, Yang S, Sun Z, Wang L and Song X 2011 *J. Alloy. Compd* **509** 9798
[12] Zhang X, Zhang D, Ni X and Zheng H 2008 *Solid-State Electron* **52** 245
[13] Vaseem M, Umar A, Hann Y B, Kim D H, Lee K S, Jang J S and Lee J S 2008 *Catal. Commun* **10** 11
[14] Liu Q, Liang Y, Liu H, Hong J and Xu Z 2006 *Mater. Chem. Phys* **98** 519
[15] Wang X, Hu C, Liu H, Du G and He X 2010 *Sens. Actuators, B Chem* **144** 220
[16] Khatib K M E, Hameed R M A 2011 *Biosens. Bioelectron* **26** 3542

IOP Conf. Series: Materials Science and Engineering **73** (2015) 012116 doi:10.1088/1757-899X/73/1/012116

Low-threshold Optical limiting and Nonlinear refraction in Nanocomposite films of Light Green dye-polymer system

S Sreeja*, P G Louie Frobel, S Mayadevi, S R Suresh, C I Muneera

Department of Physics, University of Kerala, Kariavattom, Thiruvananthapuram, Kerala-695581

Email: sreejasreedhar83@gmail.com

Abstract. Poly(vinyl alcohol) (PVA) /Light Green (LG) dye nanocomposites were fabricated and their structure, microstructure, linear and nonlinear optical properties were investigated. The samples were characterized as nanocomposites with dye molecules encapsulated between the larger molecules (molecular chains) of the polymer host PVA. The nonlinear refraction behaviour was investigated employing the Z-scan technique, using a continuous wave (cw) He-Ne laser operating at 632.8 nm, as the excitation source. The samples displayed negative nonlinear refraction (self-defocusing) under the experimental conditions. Low-threshold optical limiting behaviour exhibited by the PVA based nanocomposites was demonstrated using an aperture limited geometry. The low limiting thresholds (~4 mW and ~2.4 mW) observed for the samples indicate that these low cost and durable nanocomposite films are potential media for optical limiting application under cw laser light excitation.

1. Introduction

The growth of photonic technology during the past decade has intensified research activities for developing new materials that display unusual and interesting NLO properties. This has lead to the design and synthesis of new molecular architectures and hybrid materials with nanoscale dimensions, including nanocomposites with various structures and multiple functions. Polymer matrix nanocomposites (PMNC) have opened up enormous scope for utilizing NLO active materials in various diverse applications ranging from optical limiters and optical wave guides to polymer based nanodevices and OLEDs and, also in many recent significant applications in medical and biological sciences, without causing much health and environmental hazards. Polymer matrices reinforced with micro and nanoparticles have been proven to possess properties superior to those of the starting materials and exhibit good thermal stability, high optical damage threshold, micro-hardness, etc. Organic dye polymer nanocomposites have been generating significant research interest by way of their high degree of compatibility, easy mode of preparation, flexibility in molecular design strategy, versatility and diverse functionality. The vinyl hydroxy polymer PVA has been proven to be a high performance matrix for polymer composites and it is largely in use in medical, biomedical and industrial fields because of its many interesting physical properties such as high transparency, good gas-barrier property against ambient gases, high dielectric strength, bio-degradability, biocompatibility etc. In the present work, we report the results of the structure, morphology, linear and nonlinear optical properties of a PVA based nanocomposite system comprising an organic dye Light Green (LG).

2. Experimental

The PVA based nanocomposite films for the present work were prepared by the solution-cast method, incorporating very small wt% of the organic dye Light Green (LG) (*Loba Chemie, Mumbai*) in the host polymer PVA (molecular weight = 125,000 g/mol s.d. Fine chemicals, Mumbai, India), as has been adopted by many researchers [1]. However, the method was improved to allow film formation in a dark, dust free environment, for NLO study. Good quality transparent films with uniform surface finish and, thickness in the range 30-35 μm were obtained within about five to six days.

The structure and crystallinity of the composite films were examined by X-ray diffraction (XRD) technique. A Bruker AXIS-D8 Advance diffractometer with Ni-filtered CuK_α radiation of wavelength 1.54 Å was used to record the X-ray diffractograms of the samples. The composite films were characterized for details of their microstructure and surface homogeneity employing atomic force microscopy (AFM) (Digital Instruments Nanoscope E, with Si_3N_4 100 μm cantilever, 0.58 N/m force constant in the contact mode). The linear absorption spectra of the nanocomposite films were recorded using a UV-visible spectrophotometer (SHIMADZU UV-2450). The Z-scan technique [2] was employed to investigate the nonlinear refraction behaviour exhibited by LG-PVA composite films. For this, a continuous wave (cw), He-Ne laser operating at 632.8 nm was used as the excitation source. Optical limiting based on nonlinear refraction is demonstrated with an aperture limited geometry [3].

3. Results and Discussion

The XRD patterns of the LG-PVA nanocomposite films along with that of an unfilled PVA film and LG dye molecules in the powder form (inset) are shown in figure 1. The observed XRD pattern of the PVA film characterizes a semicrystalline polymer possessing a clear crystalline peak at $2\theta=19.6^0$, corresponding to 101 spacing. This semicrystalline nature is due to the inter-molecular hydrogen bonding between the PVA chains [4]. The crystalline phase of this polymer may be treated as an imperfect crystalline lattice, in which free volumes/interstitials are filled with amorphous phase [4]. Even though the dye LG is crystalline in nature, it loses its crystallinity when incorporated into the polymer PVA. The dye molecules when incorporated into this matrix may occupy the interstitial space between the polymer chains and lose the crystallinity of their powder form. For the PVA based nanocomposites, the intensity of the peak at $2\theta=19.6^0$, characteristic of the host matrix PVA increases, suggesting an enhancement in the degree of crystallinity of PVA. The increase in crystallinity of the composite films with respect to that of clean (unfilled) PVA may be due to the interactions of LG dye molecules probably with the hydroxyl groups of PVA.

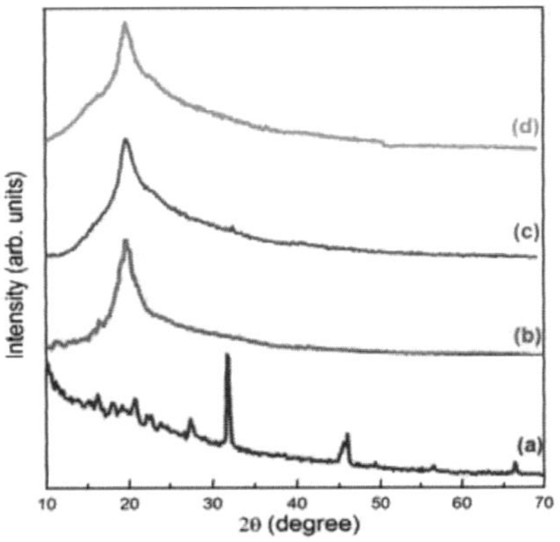

Figure 1. XRD pattern of (a)LG dye in powder form (b) PVA film (c & d) LG-PVA nanocomposite films.

Figure 2. AFM images of 1x1μm^2 regions of LG-PVA nanocomposite film (dye concentration: 5.1×10^{-4} M)

The results of AFM analysis are also in agreement with the above observations. Figure 2 shows the 2D and 3D AFM images of LG-PVA composite films with dye concentration 2.02×10^{-4} M. The

IOP Conf. Series: Materials Science and Engineering **73** (2015) 012116 doi:10.1088/1757-899X/73/1/012116

average roughness of the surface was measured to be ~1.3 nm for the LG-PVA composite film. The surface roughness was found to be much less than those reported for many nanostructured materials and nanocomposite films. This suggests that by dispersing the dye molecules/particles in the PVA matrix, the dye molecules get inside the voids/free volume present in the polymer PVA, as a result of which the free volume of the matrix decreases. Encapsulation of dye molecules in these microscopic/nanoscopic free volumes/interstitials of molecular dimension available between the polymer chains reduces the porosity, and hence the surface roughness also decreases.

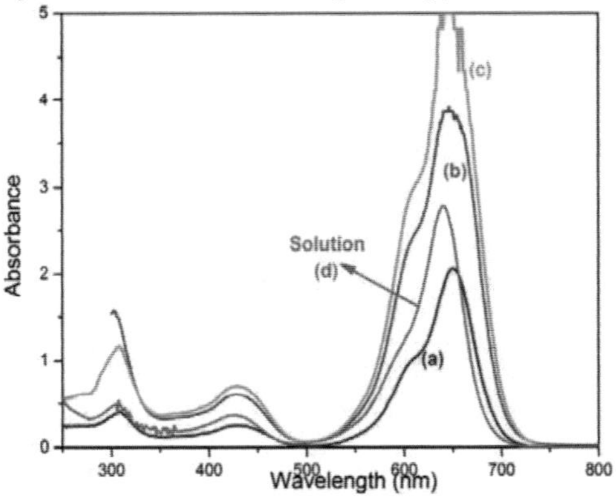

Figure 3. Linear absorption spectra of (a-c) LG-PVA composite films for different concentrations and (d) aqueous solution of LG dye

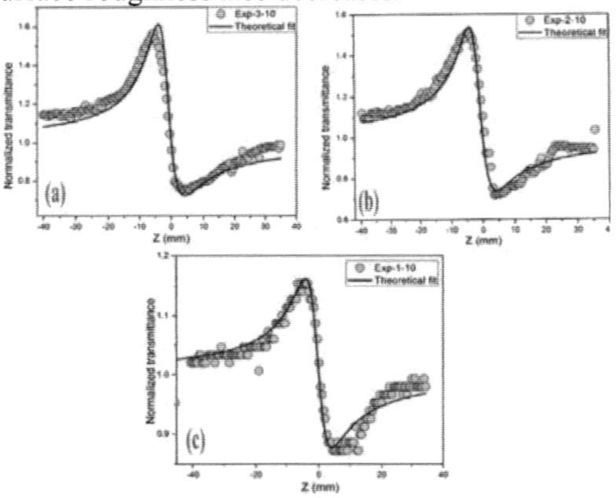

Figure 4. CA Z-scan profiles for different dye concentrations of LG-PVA nanocomposite films (a) 1.0×10^{-3}M (b) 3×10^{-3}M (c) 5×10^{-3}M

The linear absorption spectra of LG-PVA nanocomposite films for different concentrations of the dye content are illustrated in figure 3. The spectra indicate well structured and wide absorption band centered at 649 nm, red shifted by ~11 nm compared to that of aqueous solution of LG dye, as has been observed for other solvatochromic dyes.

Closed aperture (CA) Z-scan experiments were performed to study the nonlinear refraction behaviour of the PVA based nanocomposite films. The normalized closed aperture Z-scans (Fig. 4) exhibit a pre-focal peak followed by a post-focal valley, indicating self-defocusing effect (negative nonlinear refraction). Nonlinear refraction in a medium can be due to different physical mechanisms (Kerr effect, thermal self-focusing, defocusing, photo refraction, molecular reorientation etc). The peak-valley separation of the CA Z-scan curves for the composite films is found to be ~2.3 – 2.4 z_0, in contrast to 1.7 z_0 that is observed for Kerr nonlinearities. According to the literature on nonlinear refraction, the positive deviation of the peak-valley separation from the Kerr value of 1.7 z_0 indicates the presence of thermal contributions [5]. Localized absorption of tightly focused laser beam propagating through an absorbing dye medium produces spatial distribution of temperature and consequently a spatial variation of refractive index, that act as a thermal lens resulting in the phase distortion of the propagating beam. This is a predominant mechanism of nonlinearity for excitation under cw laser light where the experimental data were analysed using Thermal lens model [5], good fitting was obtained for the theoretically obtained curves (shown as solid lines of fig. 4) with experimental ones. The on-axis phase shift θ can be obtained from the best theoretical fit. Assuming θ<<1, the on-axis normalized transmittance can be expressed as:

$$T_N^{TLM}(z) = \frac{1}{1 + \theta \frac{2x}{1 + x^2}} \qquad (1)$$

where, $x = z/z_0$, and θ, the nonlinear phase shift. From the value of θ, the nonlinear refractive index of the sample can be calculated using the expressions,

$$\theta = kn_2 L_{eff} I_0 \tag{2}$$

where, $L_{eff} = (1-\exp(-\alpha L))/\alpha$, I_0 is the on-axis irradiance at the focus, k, the wave vector ($k=2\pi/\lambda$), L, the sample thickness and α, the linear absorption coefficient at the wavelength used for excitation. The estimated values of n_2 for LG-PVA composite films are found to be 0.5 - 1.36 $\times 10^{-6}$ cm^2/W.

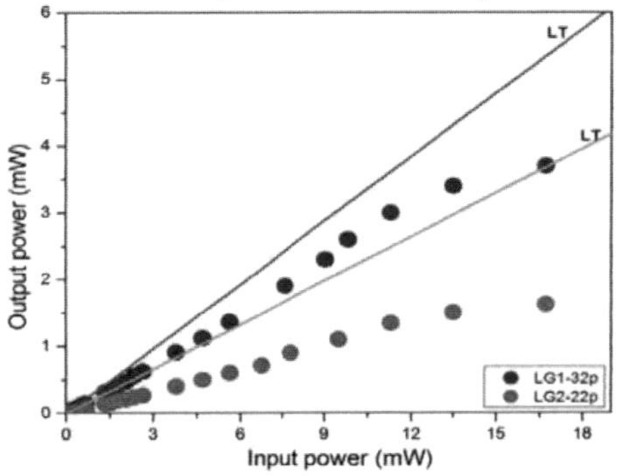

Figure 4. Optical limiting action of LG-PVA nanocomposite films at 632.8 nm cw laser light.

The defocusing effect under cw laser irradiation, associated with nonlinearity of thermo-optic origin, can be used for the design of an optical limiting device [3].Optical limiting based on nonlinear refraction is studied keeping an aperture (2 mm) in front of the detector, which is placed at a distance of 60 cm from the lens of focal length 5 cm. Figure 5 shows the optical limiting performance of LG-PVA samples for two different dye concentrations. The samples were placed at the valley position of the Z-scan trace and the output power from the sample was monitored varying the input power in steps. At low input powers, the output varies linearly with input power, obeying Beers law. With further increase in the input power, the transmitted power get saturated and the limiting thresholds were found to be ~4 mW and ~2.4 mW for LG-PVA composite films (Dye concentrations: 1.1×10-3M and 3.1×10-3M) under cw laser excitation at 632.8 nm. This indicates that the nanocomposite film samples are good candidates for optical limiting of 632.8 nm cw laser light.

4. Conclusion

Good quality nanocomposite films comprising Light Green dye and PVA were prepared by a simple processing technique and their structure, morphology, linear and nonlinear optical properties were investigated. Employing the techniques of XRD and AFM, the samples were characterized as nanocomposites with dye molecules encapsulated between the larger molecules (molecular chains) of the polymer host PVA. The low-threshold nonlinear refraction behaviour exhibited by the PVA based nanocomposites on excitation with 632.8 nm, cw He-Ne laser light were investigated employing the Z-scan technique. Since the nanocomposite films exhibited self-defocusing under cw laser light excitation at 632.8 nm, nonlinear refraction based optical limiting of cw laser light was demonstrated with an aperture limited geometry. The low threshold optical nonlinearities exhibited by these LG-PVA composite films signify the scope for utilizing these nanocomposites as a potential material for photonic device applications.

References

[1] Kumar G S, Frobel P G L, Muneera C I, Sathiyamoorthy K, Vijayan C and Mukerjee C 2009 *J. Opt. A: Pure Appl. Opt.* **11** 125204.
[2] Mansoor S B, Said A A, Wei T-H, Hagan D J, Stryland E W V 1990 *IEEE J.Quantum Electron.* **26** 760
[3] Sathiyamoorthy K, Vijayan C 2008 *J. Phys. Chem. C* **112** 14336
[4] Bhajantri R, Ravindrachary V, Vincent C, Suresh N P, Poojary B 2006 *Polymer* **47** 3591
[5] Cuppo F L S, Neto A M F, Gmez S L, Pal y-Muhoray P 2002 *J. Opt. Soc. Am. B* **19**(6) 1342

Renormalization of Molecular Energy Levels in Single-Molecule Nanojunctions: An Ab-initio Approach

Sweta Parashar[a], Pankaj Srivastava and Manisha Pattanaik
Nanomaterials Research Group, Computational Nanoscience &Technology Lab
ABV-Indian Institute of Information Technology & Management
Morena Link Road, Gwalior (MP)-474015 India

[a]E-mail: sweta@iiitm.ac.in

Abstract. Using first-principle calculations based on Density Functional Theory (DFT) in conjunction with the Non-Equilibrium Greens Functions (NEGF) technique, we have studied the effect of electrostatic environment in molecular energy levels of single-molecule transistors. This approach is applied to three polycyclic aromatic hydrocarbons (PAHs) anthracene, tetracene and pentacene, for calculating charging energies of molecular systems weakly coupled to an electrostatic environment. The result shows renormalization of molecular energy levels in an electrostatic environment. Further, it is observed that on increasing the number of aromatic rings, the addition energy value decreases which increases the conductivity of the system. Subsequently charge stability diagram for PAHs has been obtained.

1. Introduction

The dimensions of conventional silicon-based electronic devices will soon be so small that quantum effect, such as electron tunneling and energy quantization will begin to influence and eventually limits the device functionality. On the other hand, molecular Single-electron transistor (SET) constructing from the bottom up approach using organic molecules as active component have been considered as potential building block for future nanoelectronic systems [1-3]. Aviran and Ratner in 1974, first proposed theoretically the concept of electron transport in single molecule devices [4] and attracted significant interest due to its recent progress in experimental characterization [5-8]. Theoretically, electron transport can be generally distinguished within two limiting regimes, namely, coherent transport (CT) for strong coupling between the molecule and electrodes and coulomb blockade (CB) for weak coupling. However, in case of molecular SET, transport is incoherent [9].

First principle calculations usually based on Non-Equilibrium Green's Function (NEGF) methods in combination with Density Functional Theory (DFT) [10, 11] have been successfully applied in CT regime. However, in the CB regime, ab-initio approach renewed interest recently [12, 13]. Transport in CB regime is characterized by the so called charge stability diagram indicating frontiers between low and high conductivity domains in the source/drain bias gate voltage co-ordinates [14, 15]. Key quantities in this charge stability diagram is addition energy defined here as the difference between molecular ionization potential (IP) and electron affinity (EA) of single molecule.

Despite the great progress achieved, outstanding questions about fundamental issues such as energy levels, influence of electrodes are still not fully explored. Therefore, a more realistic and quantitative theoretical description is needed. Also, Martin-Lasanta et al recently found that on increasing the number of rings in polycyclic aromatic hydrocarbons (PAHs) will give higher conductivity as well as mechanical and current stability [16]. Motivated by this, we use the model to calculate charging energy of three PAHs namely anthracene, tetracene and pentacene of acene series in the SET environment. The purpose of the present study is to examine the influence of electrostatic environment on the positions of molecular energy levels of PAHs. Further, the effect of number of aromatic rings on addition energy has been demonstrated. Thus, we obtain charge stability diagram in the SET environment by calculating the charging energies as the function of external gate potential.

2. Computational method

All the calculations are carried out by using DFT based NEGF formalism which is implemented in Atomistix ToolKit [17]. For the calculations we use non-polarized DFT in the local density approximation (LDA) and expand the wave functions in double-zeta polarized (DZP) basis set. We setup benzene, anthracene, tetracene and pentacene molecule in the SET geometry as shown in the Figure 1. It consists of a metallic back-gate, and above the gate there is 3.8 Å of dielectric material with dielectric constant 10 ϵ_0. The molecule is positioned 1.2 Å above the dielectric. To the left and the right of the molecule, there are metallic source-drain electrodes respectively. Moreover, the distance between the molecule and the electrode is kept 2.8 Å.

Figure 1. A pentacene molecule lying flat on the top of dielectric substrate between source and drain electrodes with metallic back-gate resembling SET environment at zero gate and source-drain bias

3. Results and Discussion

Initially, DFT calculations have been performed in gas phase and in SET environment for benzene molecule. The gas phase values denoted as "isolated" in the Table 1 are obtained from the total energies of the charged isolated molecule. While the "SET" values are obtained by calculating the total energies of the charged molecule in the electrostatic surrounding at zero gate and drain bias. The calculated charging energies for benzene are in excellent agreement with available results [9].

Table 1. The calculated and experimental values of IP and EA for anthracene, tetracene and pentcene

Anthracene	E_I^{+1} (eV)	E_I (eV)	E_A (eV)	E_A^{-1} (eV)
Experiment[a]	-	7.439	0.530	-
isolated	11.56	6.86	0.05	-4.48
SET	7.18	5.68	2.02	0.62
Tetracene	E_I^{+1}	E_I	E_A	E_A^{-1}
isolated	10.46	6.29	0.69	-3.37
SET	6.39	5.20	2.56	1.44
Pentacene	E_I^{+1}	E_I	E_A	E_A^{-1}
isolated	9.68	5.89	1.14	-2.58
SET	5.95	4.91	2.90	1.91

[a]NIST server (http://cccbdb.nist.gov/)

Further, we apply the same approach to anthracene, tetracene and pentacene. Table 1 show the calculated values of charging energies for anthracene in isolated case that are in good agreement with experimental results [18]. These energies are also evaluated for tetracene and pentacene. While in SET environment, the charging energies of all the three molecules are strongly reduced as compared to its gas phase values. The reduction in the charging energy is caused by screening of the charged molecule by surrounding dielectric and metal electrodes [19].

We have compared our findings with the GW calculations [20]. They found gas phase values E_{LUMO} - E_{HOMO} = 6.91 eV and absorbate gap values E_{LUMO} - E_{HOMO} = 4.29 eV for anthracene. Similarly, these values are E_{LUMO} - E_{HOMO} = 6.09 eV and E_{LUMO} - E_{HOMO} = 3.62 eV for tetracene and E_{LUMO} - E_{HOMO} = 5.24 eV and E_{LUMO} - E_{HOMO} = 2.88 eV for pentacene respectively. These results are comparable with our values E_I - E_A = 6.81 eV, E_I - E_A = 3.66 eV for anthracene, E_A = 5.60 eV, E_I - E_A = 2.64 eV for tetracene and E_I - E_A = 4.75 eV, E_I - E_A = 2.01 eV for pentacene in gas phase and in SET environment respectively. Thus, we conclude that the addition energy in an electrostatic environment is heavily reduced compared to their corresponding gas phase values. It can also be noted that in SET environment, on increasing the number of aromatic rings, the addition energy value reduces which indirectly affect the conductivity of the system. This is in correspondence with HOMO-LUMO gap. Since the barrier of electron transfer is approximately proportional to the HOMO-LUMO gap [21], therefore in SET environment, the current increased with increasing the number of aromatic rings.

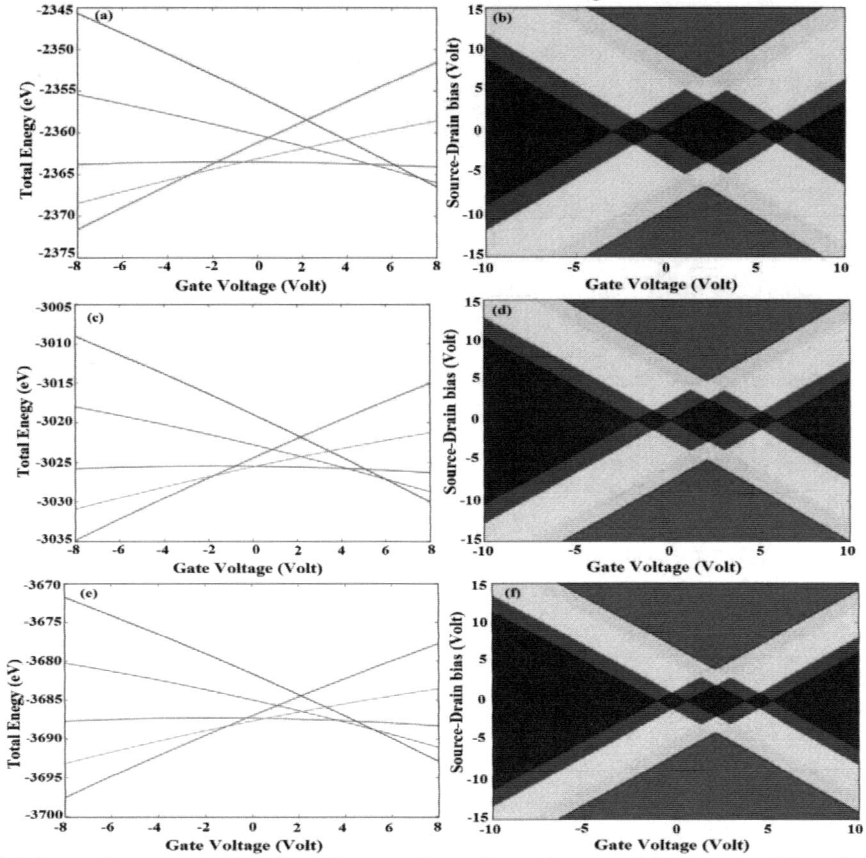

Figure 2. The total energy as function of gate voltage for anthracene (a), tetracene (c), and pentacene (e) in the SET environment. Different colors for different charge states: blue (-2), green (-1), red (0), turquoise (1), violet (2). The charge stability diagram for antracene (b), tetracene (d), and pentacene (f) in SET environment. The color shows the number of charge states dark blue (-2), blue (-1), red (0), green (1), orange (2) within the bias window for a given gate voltage and source-drain bias

We next calculate the total energy of different charge states in SET environment as a function of gate voltage as shown in Figure 2. (a), (c), (e). The total energy includes reservoir energy qW, where q represents charge of organic molecule and W (=5.28eV) is the work function of gold electrode. At zero gate voltage, the neutral molecule has the lowest energy and hence it is stable charge state. Negative charge states lowers energy by positive gate voltage while positive charge states lowers energy at negative bias. This is in agreement with HOMO and LUMO levels follow $-eV_G$, thus, at positive voltage the LUMO level gets below the electrode Fermi level therefore attracts an electron and the molecule gets negatively charged. At negative bias, the HOMO level gets above the electrode Fermi level hence electron escapes organic molecule and the molecule become positively charged.

Further, we calculate the charge stability diagram for anthracene, tetracene and pentacene molecules by evaluating total energies as shown in Figure 2. (b), (d), (f). The different color shows the number of molecular energy levels inside the bias window for given values of gate and source-drain bias. As we know that the height of central diamond related to the addition energy [20], reduces on increasing the number of aromatic rings which indirectly increase the conductivity of the system. This effect is also observed in recently reported results [16]. It can also be noted that, charge stability diagram depends only on the difference in energies between charge states and hence in charge stability diagram the non-linear dependence of the total energy on gate voltage is not observed.

4. Conclusions

We performed an ab-initio calculation to obtain the charge stability diagram of anthracene, tetracene and pentacene in the coulomb blockade regime by calculating the charging energies as a function of external gate potential. We found that the electrostatic environment reduces the charging energies of the molecule. Further, it is also observe that number of rings in PAHs decrease the addition energy and hence increase the conductance of the system.

Acknowledgements

Authors are thankful to the computational nanoscience & Technology Lab, ABV-Indian Institute of Information Technology & Management, Gwalior (India) for providing computational facilities.

References

[1] Cuniberti G, Fagas G and Richter K 2005 Eds. *Springer* p-519
[2] Joachim C, Gimzewski J K and Aviram A 2000 *Nature* **408** 541
[3] Moth-Poulsen K and Bjornholm T 2009 *Nat. Nanotechnol.* **4** 551
[4] Aviram A and Ratner M A 1974 *Chem. Phys. Lett.* **29** 277
[5] Kubatkin S, Danilov A *et al* 2003 *Nature* **425** 698
[6] Danilov A, Kubatkin S *et al* 2008 *Nano Lett.* **8** 1
[7] Osorio E A, O'Neill K *et al* 2007 *Adv. Mater.* **19** 281
[8] Thijssen J M and Van der Zant H S J 2008 *Phys. Status Solidi B* **245** 1455
[9] Stokbro K 2010 *J. Phys. Chem. C* **114** 20461
[10] Brandbyge M, Mozos J L, Ordejon P, Taylor J and Stokbro K 2002 *Phys. Rev. B* **65** 165401
[11] Xue Y, Datta S and Ratner M A 2002 *Chem. Phys.* **281** 151
[12] Stadler R, Geskin V and Cornil J 2008 *Phys. Rev. B* **79** 113408
[13] Chowdhury R, Adhikari S and Rees P 2012 *Physica B* **407** 855
[14] Kouwenhoven L P, Austing D G and Tarucha S 2001 *Rep. Prog. Phys.* **64** 701
[15] Nitzan A and Ratner A 2003 *Science* **300** 1384
[16] Martin-Lasanta A, Miguel D *et al* 2012 *Chem. Phys. Chem.* **13** 860
[17] AtomistixToolKit version 2011.2.3, QuantumWise A/S (http://quantumwise.com)
[18] http://cccbdb.nist.gov/
[19] Kaasbjerg K and Flensberg K 2008 *Nano Lett.* **8** 3809
[20] Neaton J B, Hybertsen M S and Louie S G 2006 *Phys. Rev. Lett.* **97** 216405
[21] Seminario J M, Zacarias A G and Tour J M 2000 *J. Chem. Soc.* **122** 3015

DFT study of CdS-PVA film

Vaneeta Bala[a1], **S K Tripathi**[a] **and Ranjan Kumar**[a]

[a]Department of Physics, Panjab University, Chandigarh, India

E-mail:vaneetabala@yahoo.com, surya@pu.ac.in, ranjan@pu.ac.in

Abstract. Density functional theory has been applied to study cadmium sulphide-polyvinyl alcohol nanocomposite film. Structural models of two isotactic-polyvinyl alcohol (I-PVA) chains around one cadmium sulphide nanoparticle is considered in which each chain consists three monomer units of [–($CH_2CH(OH)$)–]. All of the hydroxyl groups in I-PVA chains are directed to cadmium sulphide nanoparticle. Electronic and structural properties are investigated using ab-intio density functional code, SIESTA. Structural optimizations are done using local density approximations (LDA). The exchange correlation functional of LDA is parameterized by the Ceperley-Alder (CA) approach. The core electrons are represented by improved Troulier-Martins pseudopotentials. Densities of states clearly show the semiconducting nature of cadmium sulphide polyvinyl alcohol nanocomposite.

1. Introduction

In recent years, polymer nanocomposites of II-VI semiconductor compounds have attracted much research interest due to their size dependent structural, electronic, optical and catalytic properties. As the size of nanoparticle approaches to excitonic radius, quantum confinement effect arise which results in size dependent unique properties. Among II-VI inorganic semiconductor compounds, cadmium sulphide (CdS) is regarded as an attractive candidate for optoelectronic devices and solar cells [1]. CdS has a wide direct band gap ~ 2.4 eV with excitonic Bohr radius ~ 3 nm which make it possible to tailor its electronic and optical properties with size. The incorporation of CdS nanoparticles in polyvinyl alcohol matrix has shown significantly enhanced photoluminescence with long term stability [2,3]. Polyvinyl alcohol (PVA) is a water soluble polymer which provides a good control over growth and morphology of embedded nanoparticles. PVA acts as a capping material which binds to nanoparticle through -OH groups of polyvinyl alcohol chains [4]. Various in situ synthetic methods have been used to synthesize CdS nanoparticle/polymer composites [5-7]. Extensive experimental work has been done on CdS Nanocomposites but theoretical studies of these or related systems are very few.

In this paper, we propose three structural models of CdS nanoparticle incorporated within two isotactic-polyvinyl alcohol (I-PVA) chains in different ways utilizing the interaction between Cd and OH group of PVA. Each I-PVA chain contains three monomer units of [-($CH_2CH(OH)$)-]. We have performed density functional based calculations to get optimized structures. We also focus our attention on the electronic properties of these proposed structural models.

[1] To whom any correspondence should be addressed.

IOP Conf. Series: Materials Science and Engineering **73** (2015) 012118 doi:10.1088/1757-899X/73/1/012118

2. Computational Details

We employ first principles calculations on structural models of CdS nanoparticle embedded in polyvinyl alcohol chains using SIESTA [8] code which is based on density functional theory. It implements the standard Kohn-Sham self-consistent method and valence electrons are described by linear combination of numerical atomic-orbital basis set and the atomic core by norm-conserving pseudopotentials. The pseudopotentials generated using the Troullier and Martins scheme [9] are applied to describe the interaction of valence electrons with atomic core and their nonlocal components are expressed in the full separable form of Kleinman and Bylander [10,11]. The Ceperley-Alder (CA) form local density approximation (LDA) corrections are employed for the exchange-correlation potential [12]. We have used a split valence double- ξ polarized (DZP)-basis set throughout the work. To calculate the self-consistent Hamiltonian matrix elements, the numerical integrals are performed and projected on a real space grid with an equivalent cut-off of 250 Ry. Structural models are relaxed using conjugate gradient algorithm until maximum atomic force is less than 0.04 eV/Å.

3. Results and Discussion

We discuss three structural models of CdS nanoparticle within I-PVA chains. Each PVA chain consists of three repeating units of [–($CH_2CH(OH)$)–] with periodicity of 2.52 Å. C-C and C-O bond distances in zigzag PVA chain are 1.53Å and 1.43 Å respectively with tetrahedral bond angles. Cd-S bond length is observed to be 2.53Å. The -OH groups in PVA chains are directed towards cadmium sulphide and act as coordination sites for nanoparticles.

3.1 Structural Models

In model A, CdS nanoparticle is suspended within two I-PVA chains (figure 1(a)). Cadmium is surrounded by four sulphur atoms and each sulphur atom is saturated with hydrogen. Cadmium is in the plane of two oxygen atoms of opposite chains. The methyl groups of two chains are separated by 6.61 Å. The incorporation of nanoparticle decreases the intermolecular hydrogen bonding between two polymer chains.

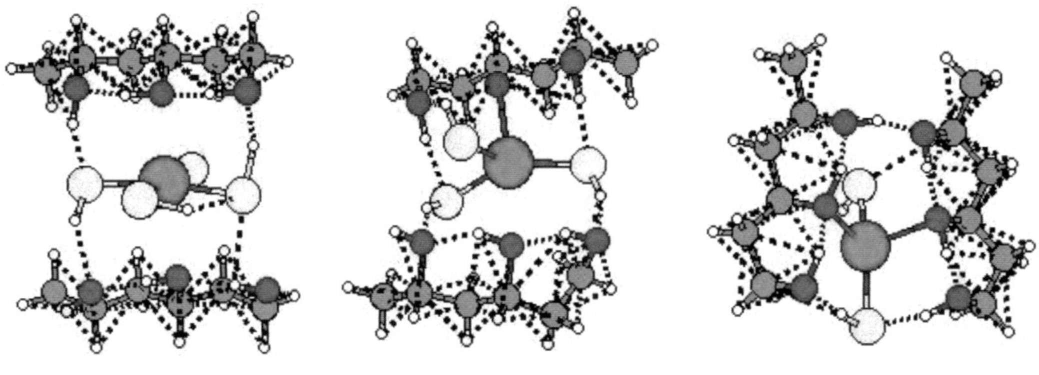

(a) **(b)** **(c)**

Figure 1. (a) Model A, (b) Model B and (c) Model C. In these figures red, yellow, grey and green circles represent O, S, C and Cd atoms respectively. Hydrogen bonding is shown by dotted lines.

In model B, Cd is bonded with three sulphur atoms and to one of the PVA chains through interaction of Cd ion with central OH group of the polymer chain (figure 1(b). Cd-O bond length is observed to be 2.32 Å. Central -OH group of second PVA chain is at a distance of 2.72 Å from Cd. In model C, Cd is coordinated to two hydrogenated sulphur atoms and two polyvinyl alcohol chains through central -OH groups of I-PVA chains (figure 1(c)). O-Cd-O bond angle is observed to be 90°. Intermolecular hydrogen bonding between I-PVA chains increases as the distance between polymer chains decreases in this model.

3.2 Electronic Structure

We focus our attention on the band gap for these structural models which is very important factor for optoelectronic materials. We have plotted total densities of states (DOS) for above mentioned structural models as shown in figure 2. From DOS, band gap is calculated as 3.90 eV, 3.78 eV and 4.02 eV for models A, B and C respectively. These plots clearly show the semiconducting nature of the composite.

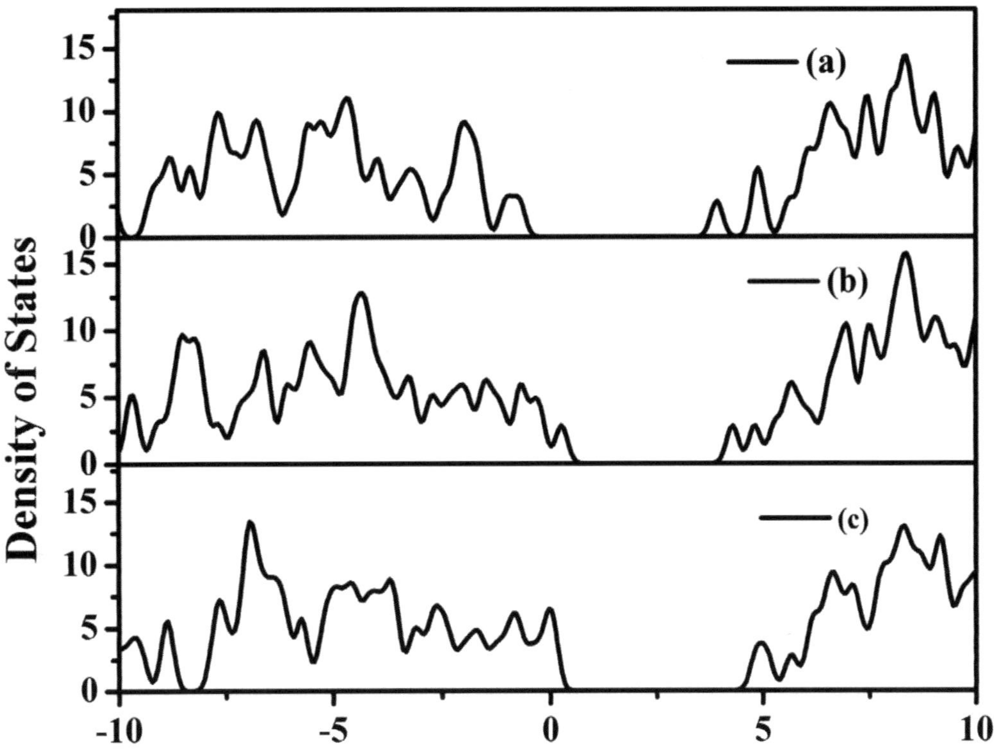

Figure 2. Density of states: (a) Model A, (b) Model B and (c) Model C.

4. Conclusion

We have proposed structural models for CdS nanoparticle embedded in PVA chains. All the structures are optimized using density functional theory. We have considered different situations in which nanoparticle can bind to polymer chains through interaction between Cd and -OH groups of PVA. DOS have been plotted for these structural models which reveal semiconducting nature of CdS-PVA composites. In the present study, we have calculated band gap from DOS which is comparable to experimental results [13,14]. Wide band gap values ensure the use of CdS-PVA composites as a promising material for optoelectronic applications.

Acknowledgements
We are highly thankful to the Department of Physics, Panjab University for providing facility High Performance Computing Centre (HPCC). One of the authors, Vaneeta Bala thanks Department of Science and Technology, Government of India for INSPIRE fellowship.

References

[1] Han Z, Zhang J, Yang X and Cao W 2011 *Sol. Energ. Mat. Sol. Cells* **95** 483
[2] Jung D, Kim J, Nahm C, Nam S, Kim J I and Park B 2012 *Mater. Res. Bull.* **47** 453
[3] Pattabi M, Saraswathi Amma B and Manzoor K 2007 *Mater. Res. Bull.* **42** 828
[4] Wang H, Fang P, Chen Z and Wang S 2007 *Appl. Surf. Sci.* **253** 8495
[5] Mbhele Z H, Salemane M G, van Sittert C G C E, Nedeljković J M, Djokovic V and Luyt A S 2003 *Chem. Mater.* **15** 5019
[6] Mondal S P, Mullick H, Lavanya T, Dhar A, Ray S K and Lahiri S K 2007 *J. Appl. Phys.* **102** 06433
[7] Saikia D, Saikia P K, Gogoi P K, Das M R, Sengupta P and Shelke M V 2011 *Mater. Chem. Phys.* **131** 223
[8] Ordejon P, Artacho E and Soler J M 1996 *Phys. Rev.* B **53** R10441
[9] Troullier N and Martins J L 1991 *Phys. Rev.* B **43** 1993
[10] Kleinman L and Bylander D M 1982 *Phys. Rev. Lett.* **48** 1425
[11] Bylander D M and Kleinman L, 1990 *Phys. Rev.* B **41** 907
[12] Ceperley D M and Alder B J 1980 *Phys. Rev. Lett.* **45** 566
[13] Mansur H S, Mansur A A P and González J C 2011 *Polymer* **52** 1045
[14] Ghoswami M, Ghosh R, Chakraborty G, Gupta K and Meikap A 2011 *Polym. Comp.* **32** 2017

Studies on the uv- visible and photoluminescent emission in nanocrystalline tungsten oxide

C L Veenas, L R Asitha, Vipin C Bose, A S Aiswarya Raj, G Madhu, and V Biju*
Department of Physics, University of Kerala, Kariavattom Campus
Thiruvananthapuram, KERALA, INDIA, Pin – 695581

E-mail: bijunano@gmail.com

Abstract. Nanocrystalline Tungsten Oxide (WO_3) samples having different average crystallite sizes 10nm, 11 nm, 22 nm and 26 nm were synthesized through controlled chemical precipitation method. The phase purity and average crystallite sizes were estimated from X-ray Diffraction (XRD) analysis. The vibrational characteristics of the samples were studied by Raman spectroscopy. The X-ray Photoelectron Spectroscopic (XPS) analysis of the samples revealed the presence of oxygen vacancies which was confirmed by the uv-visible (uv- vis) absorption and photoluminescence (PL) emission studies.

1. Introduction

Tungsten Oxide (WO_3) is one of the most sought after material for the material researchers and shows photochromic ,electrochromic and gaschromic properties[1]. WO_3 has been extensively studied in the recent years due to its wide spread applications in smart windows, non-emissive displays, optical signal processing, sensors, information storage media etc. The most stable WO_3 at room temperature has monoclinic structure. Because of its electrochromic and photochromic applications, most of the studies of this material is in the thin film form and studies on nanocrystalline samples of WO_3 in powder form is rather limited. Here an attempt has been made to synthesize nanocrystalline WO_3 in a fine powder form with different average crystalline sizes by employing a controlled chemical precipitation method.

2. Experimental

Nanocrystalline WO_3 samples were synthesized using sodium tungstate dihydrate (Na_2WO_4. $2H_2O$) and concentrated hydrochloric acid (HCl) as the starting materials and oxalic acid($H_2C_2O_4$) was used as the structure directing reagent. The thermogravimetric analysis (TGA) of the tungstic acid precursor revealed the decomposition temperature was at 250^0C. The precursor was decomposed at 250^0C (W1) and annealed at different temperatures viz; 350^0C (W2), 500^0C (W3) and 600^0C (W4) to get different average crystallite sizes. XRD pattern of the samples were recorded in the range $2\theta=10$ to 90^0 with a step size of 0.0170^0 using a Philips XPERT PRO diffractometer with Cu Kα radiation (λ = 1.54056 Å). Raman spectra of the samples were recorded in the region 50-1100 cm^{-1} using Brucker-FT Raman Spectrometer (RFS 100/s, Germany) using 1064 nm line of Nd:YAG laser. X-ray photoelectron spectroscopic (XPS) analysis was carried out using a Thermo Scientific MULTILAB 2000 X-ray Photoelectron Spectrometer under a vacuum of UHV 10^{-10} mbar using Twin anode M_g (300/400 W) X-ray source. A CARY 5E UV-VIS-NIR spectrometer was used to record the UV-Visible-near IR absorption spectra of the samples in the range 200- 1200 nm. The photoluminescence emission spectra of the samples dispersed in water was recorded at different excitation wavelengths viz. 250, 280 and 320 nm using a Flurolog III spectrophotometer. Both absorption and emission spectra are recorded at room temperature.

3. Results and Discussion

Figure 1 shows the XRD pattern of the four samples. The XRD pattern of the precursor material was shown in Figure 2a whose peaks at the 2θ values of 16.53^0 (020), 25.63^0(111), 30.47^0 (031), 34.13^0 (200), 34.93^0 (002), 38.88^0 (022), 45.86^0 (051), 49.64^0 (202), 52.68^0 (222), 56.15^0 (311) and 61.69^0 (331) corresponding to the peaks of tungsten oxide hydrate ($WO_3.H_2O$) having orthorhombic crystal structure which is in agreement with JCPDS -ICDD pattern no: 84-0886. The XRD pattern for samples W1, W2, W3 and W4 can be indexed with the JCPDS -ICDD pattern no: 83-0951 corresponding to WO_3 having monoclinic crystal symmetry with space group $P2_{1\text{-}n(14)}$. The average crystallite size of the samples were estimated from Debye-Scherrer equation after correction for instrumental broadening from the peaks corresponding to the planes (120), (202), (040) and are 10 nm, 11 nm, 22 nm and 26 nm for W1, W2, W3 and W4 respectively.

Figure 2a shows the raman spectrum of the precursor material ($WO_3.H_2O$) which shows mainly eleven vibrational bands in the range 50-1100 cm^{-1} centered at 963, 805, 696, 319, 281, 254, 211, 176, 115, 90 and 80 cm^{-1}. Tungstic acid has its main peaks at 806 and 710 cm^{-1} and these peaks are shifted to 805 and 696 $cm^{-1}$for the precursor material. The band around 963 cm^{-1} is assigned to the stretching mode of terminal W=O which is common for all types of tungsten oxide hydrates[2]. The raman spectrum of the samples W1, W2, W3 and W4 are quite similar, which clearly indicates that the WO_3 doesnot change the crystalline phase with the annealing temperature. The intense peaks centered at 807 and 717 cm^{-1} are the typical Raman peaks of the crystalline monoclinic WO_3 (m-phase), which corresponds to the stretching vibrations of the bridging oxygen. The sharp peaks at 276 and 321 cm^{-1} are assigned to the bending vibration of O-W-O[2].

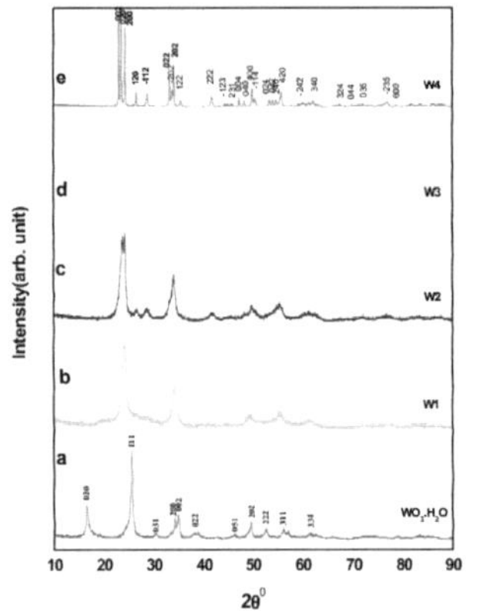

Figure 1. The XRD pattern of $WO_3.H_2O$, W1, W2, W3 & W4

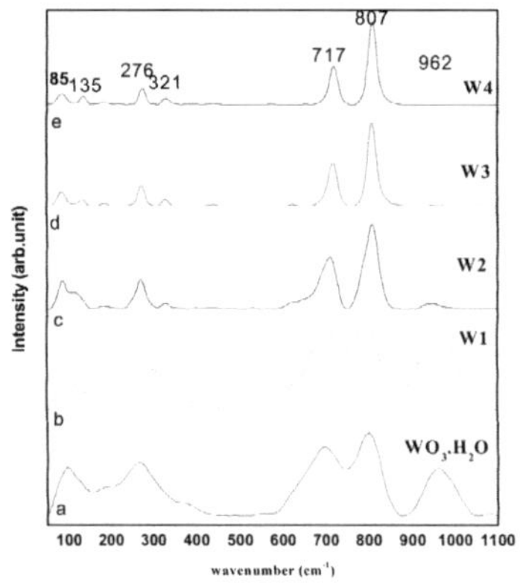

Figure 2. The Raman spectrum of $WO_3.H_2O$, W1, W2, W3 & W4.

The XPS spectra of W4f and O1s is recorded to obtain the possible valencies of tungsten (W) and the influence of –OH group in the binding energy of O^{2-} respectively. Figure 3a shows the XPS W4f core level spectrum for the sample W1 which can be deconvoluted into two doublets having binding energies at 35.5, 37.7, 36.1 and 38.3 eV. The binding energy at the positions 35.5 and 37.7 eV corresponding to the binding energy of $W4f_{7/2}$ and $W4f_{5/2}$ levels of tungsten atoms for W^{6+} state of oxide and to hydroxide which is due to the presence of adsorbed water on the surface respectively[3]. Figure 3b corresponds to the W4f core level spectrum of W^{6+} state having spin orbit separation of 2.1

eV[3]. Figure 3c corresponds to the XPS W4f core level spectra of W3 which can be deconvoluted into four peaks corresponds W^{6+} state and W^{5+} state. The deconvolution of the XPS spectra of the W4f core level spectra of W4 has binding energies at the position 36.1 eV corresponds to the $W4f_{7/2}$ of W^{6+} state and 35.1 eV corresponds to the $W4f_{7/2}$ of W^{5+} state which is slightly higher than the energy 34.8 eV which is related to the formation of oxygen vacancy at higher annealing temperatures and the presence of W^{4+} state[4].

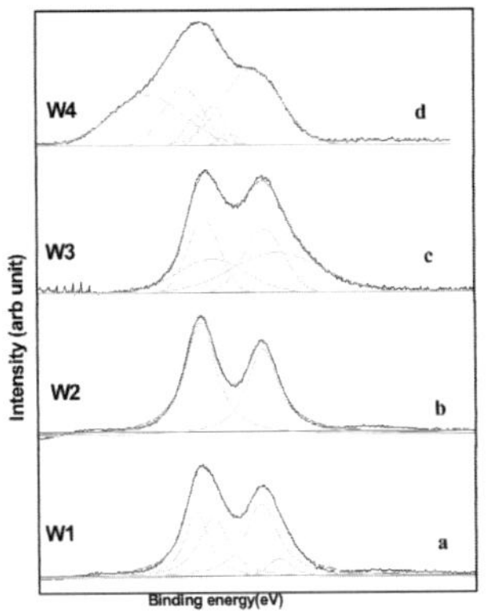

Figure 3. The XPS spectrum of W4f of W1, W2, W3 & W4.

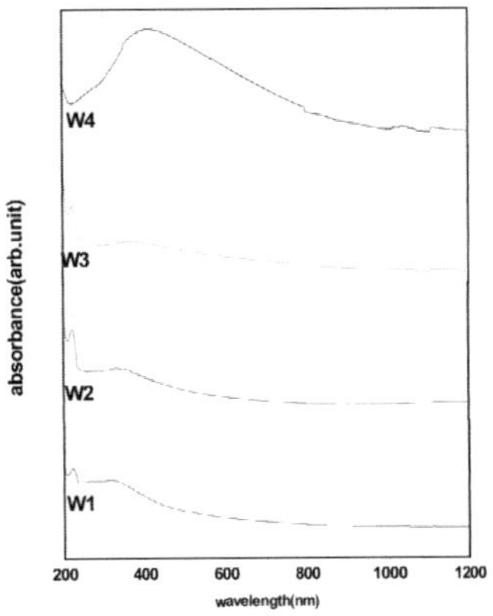

Figure4. The uv-visible absorption spectrum of W4f of W1,W2, W3 & W4.

The uv-visible absorption spectra of the samples are recorded in the wavelength range 200-1200 nm and is shown in the Figure 4. The optical band gap is calculated from the frequency dependent absorption coefficient (α) using the relation $(\alpha h\nu) = A(h\nu - E_g)^{1/2}$ where E_g is the direct band gap energy and hν is the photon energy[5]. The straight line behaviour of the $(h\nu - E_g)^2$ vs. hν for the samples confirms the direct band gap and the band gap value calculated for the sample W1 is 2.8 eV which is in agreement with the reported indirect band gap value of 2.7 eV. The optical band gap values obtained for the samples W2 and W3 are 4.16 and 4.34 eV which are found to be greater than the reported optical band gap value of tungsten oxide. They are found to increasing with increasing temperature up to 500°C and the band structure of WO_3 proposed by J B Goodenough is used to explain the results obtained from the uv-visible absorption spectrum for the samples W2 and W3[5]. The proposed band structure is in which the bands are constructed from molecular orbitals of oxygen and tungsten electronic states. The transitions corresponding to 298 nm (4.16 eV) and 286 nm (4.34) may be due to the transitions from valence band to the split d bands (e_g state). each part.

Figure 5 shows the room temperature photoluminescence (PL) spectra of the four samples for the excitation wavelength at 250 nm. All the samples have peaks at 352, 396, 426, 438, 53,468, 483 and 492 nm with maximum intense peak at 468 nm. The peaks at 453, 468, 483 and 492 nm are in agreement with the peaks same as that of the bulk phase WO_3 powder[6]. The wavelength of the peaks in the range 320-492 nm doesnot change with the annealing temperature. To study the dependence of emission spectrum on excitation wavelength ,the room temperature PL spectrum of the samples is recorded at different excitation wavelengths viz., 250, 280 and 320 nm. For all the four samples, the intensity of the peak at 382 nm increases for the excitation wavelength 280 nm and for the excitation wavelength increases to 320 nm, the intensity of the peak at 422 nm increases considerably. But the position of the bands in the range 390- 468 nm doesnot change with excitation wavelength which is attributed to the band-to-band transition in nanocrystalline WO_3. The uv- emission in the range 350-422 nm is attributed due to the presence of oxygen vacancies and Luo et al. described the defect emissions in WO_3 using the configuration-coordinate (CC) diagram for single defect center[6]. They have observed that the defect emissions from the high resonant state in the conduction band shows a red shift in the uv-emission with respect to the increasing in excitation wavelength. A similar kind of results is observed in the PL spectra of our samples under various excitations viz. 250, 280 and 320 nm and the uv emissions at 352, 382 and 422 nm is attributed due to the transition to valence band from the high lying defect state in the conduction band due to oxygen vacancies. As the excitation wavelength increases the intensity of uv emission at 422 nm reaches the maximum which may be due to the presence of increase in oxygen vacancy due to annealing[7].

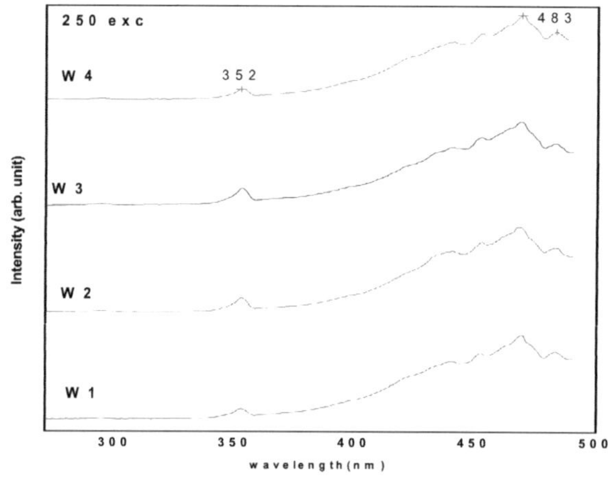

Figure 5. The photoluminescence spectrum of W1, W2, W3 & W4 under the excitation wavelength 250 nm

4.Conclusions

Nanocrystalline powder samples of tungsten oxide having average crystallite size in the range 10-27 nm are synthesized through controlled chemical precipitation method**.** The structural and vibrational characteristics of the samples are done by using XRD and Raman spectroscopy. The X-ray Photoelectron Spectroscopic (XPS) analysis of the samples reveales the presence of oxygen vacancies which was confirmed by the uv-visible absorption and photoluminescence (PL) emission studies.

5.Reference

[1] G A Niklasson, G. Granqvist 2007 *J. Mater. Chem.* **17** 127

[2] M F Daniel, B.Desbat, J C Lassegues, B Gerand, M Figlarz 1987 *J.Solid. State. Chem.* **67** 235

[3] G Leftheriotis, S Papaefthimiou, P Yianoulis, A Siokou, D Kefalas, 2003 *Appl. Surf. Sci.* **218** 275

[4] A P Shpak, A M Korduban, M M Medvedsky, V O Kandyba 2000 *Thin Solid Films* **358** 40

[5] S K Deb 1973 *Philo. Mag.* **27** 80

[6] J Y Luo, F L Zhao, L Gong, H J Chen, J Zhou, Z L Li, S Z Deng, N S Xu, 2007 *Appl. Phys. Lett.* **91** 093124

[7] K Lee, W S Seo, J T Park 2003 *J.Am.Chem.Soc.* **125** 3408

Study on the optical band gap and photoluminescence of $PbMoO_4$ nano powder synthesized by an auto igniting combustion technique.

S Vidya and J K Thomas*
Electronic Materials Research Laboratory, Department of Physics, Mar Ivanios College, Thiruvananthapuram- 695 015, Kerala, India

* Corresponding author e-mail: jkthomasemrl@yahoo.com

Abstract. Nano crystalline $PbMoO_4$ was synthesized through an auto-ignited combustion technique. The X-ray diffraction studies of $PbMoO_4$ nanoparticles reveals that the as-prepared powder itself is single phase and possess tetragonal structure. The average particle size of the as-prepared powder calculated using scherrer formula is 28nm. Fourier transform Infrared spectrum shows that the as prepared powder itself is phase pure with no formation of secondary phase .The optical band gap determined from UV-Visible absorption spectra is 3.20eV.Photoluminescence spectra of the samples shows blue emission.

1. Introduction

Synthesis of advanced ceramics materials as nanocrystals is one of the major fields in materials processing technology. The importance of nanomaterials is that its properties can be tuned by controlling the particle size. Molybdates with scheelite structure have received much attention owing to their remarkable characteristic properties in the field of optical devices. In the scheeelite structure $[MoO_4]^{2-}$ group adopts a tetrahedral co-ordination in space group $I4_1/_a$ and the cation is bonded to eight oxygen atoms forming an octahedron. Lead molybdate ($PbMoO_4$) is a well-known scheelite class of compound with various applications in numerous fields [1,2].

In this paper we report the synthesis of nanocrystalline phase pure $PbMoO_4$ through an auto ignited combustion technique for the first time. The structural characterization, optical band gap study and photo luminescent emission behavior are also presented.

2. Experimental

Nano crystalline $PbMoO_4$ was synthesized thgrough a modified combustion technique using MoO_3 and $Pb(NO_3)_2$ as starting material. The precursor solution was prepared by dissolving stoichiometric amount of $Pb(NO_3)_2$ in double distilled water and MoO_3 in ammonia solution. Citric acid was added to the solution as complexing agent. Oxidant to fuel ratio of the system was adjusted by adding $ConHNO_3$ and ammonium hydroxide. The precursor solution of pH ~7.0 was heated on a hot plate in a ventilated fume hood. The solution boils and dehydrates accompanied by foam. On persistent heating the foam ignites by itself giving nano powders of $PbMoO_4$.

Structure of the as-prepared powder was examined by powder X-ray diffraction (XRD) technique using a Bruker D-8 X-ray Diffractometer with Nickel filtered Cu Kα radiation. The infrared (IR) spectra of the samples were recorded in the range 400–4000 cm^{-1} on a Thermo-Nicolet Avatar 370 Fourier Transform Infrared (FTIR) Spectrometer using the KBr pellet method. The absorption spectra of the powder were measured using a Jasco U.V-Visible spectrophotometer. The photoluminescence spectra of the samples were measured using Flurolog@-3 Spectrofluorometer.

3. Results and discussions

The X-Ray diffraction pattern of as-prepared nano crystalline $PbMoO_4$ shown in figure 1.All the peaks are indexed for a perfect tetragonal structure with lattice parameters a=b=5.38 and c=12.075Å.The calculated lattice parameter values are well agreeing with the standard one reported in the JCPDS file no:44-1486. The average particle size calculated by Scherrer formula is 28nm. The as prepared powder itself is phase pure which implies that the phase formation was completed during the combustion process itself without the need of any calcination. Thus modified combustion technique in which citric acid is used as complexing agent and ammonia as fuel is a single step process unlike other conventional synthesis techniques such as solid state method in which high temperature calcinations is required for exact phase formation

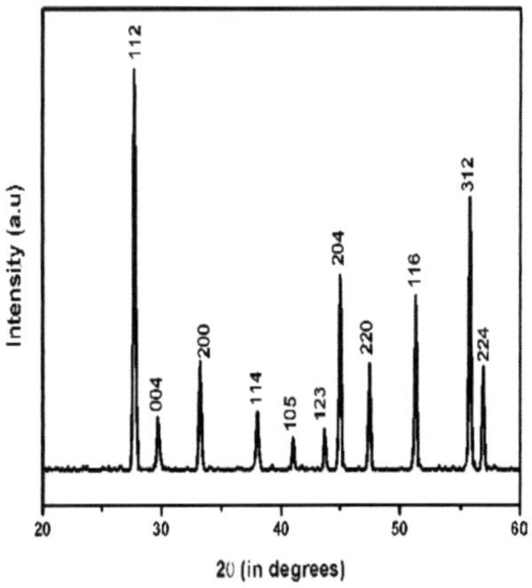

Figure 1. XRD pattern of nano $PbMoO_4$

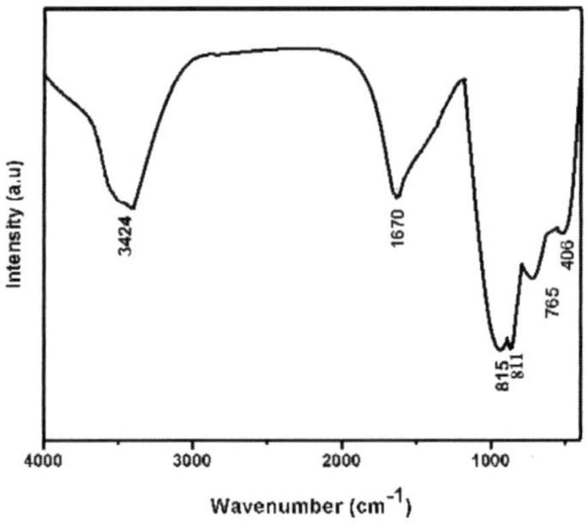

Figure 2. FT-IR spectrum of $PbMoO_4$

The phase purity of the as-prepared nano powder was studied using FT-IR spectra shown as figure 2. $PbMoO_4$ have tetrahedral symmetry (T_d) and hence the $F_2(\nu_3,\nu_4)$ modes are infrared active. In the FT-IR spectra the respective bands at 3424 and 1670cm^{-1} can be ascribed to O–H stretching vibration and H–O–H bending vibration of physically absorbed water on the sample surface. A strong absorption peak at 406cm^{-1} can be assigned to $F_2(\nu_4)$ vibration mode which represents the bending vibration of Mo–O. The antisymmetric stretching vibration $F_2(\nu_3)$ originating from the Mo–O in MoO_4^{2-} tetrahedron is corresponds to broad intense peak at 815cm^{-1}.Besides this strong bands, two other very weak bands can also be seen in the region 765cm^{-1} and 811cm^{-1} , attributed to Mo----O anti symmetric stretching vibrations. This occurred due to the fact that the combustion takes place at a relatively higher temperature, resulted in highly ordered crystal structure and well defined bands relating to Mo---O bonds. Thus vibrational analysis corroborates the XRD results.

The optical properties of nano crystalline $PbMoO_4$ were studied using UV-Visible spectra. Figure 3 shows the UV-Visible absorption spectrum. The sample absorbs heavily in the UV region and moderately in the visible region. Such materials found application in the field of UV filters and sensors. The optical band gap determined from the Tauc's plot given as figure 4 is 3.2eV which is exactly matching with the reported values [3].The band gap value suggest that PbMoO4 is a semiconductor. The absorption edge of the samples prepared by the combustion technique is blue-shifted compared to that of the sample prepared by the solid-state route which can be attributed to the nano nature of the sample.

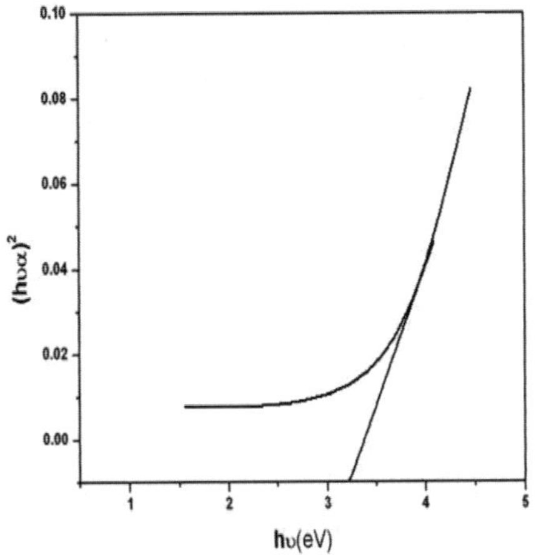

Figure 3. Tauc's plot of nano $PbMoO_4$

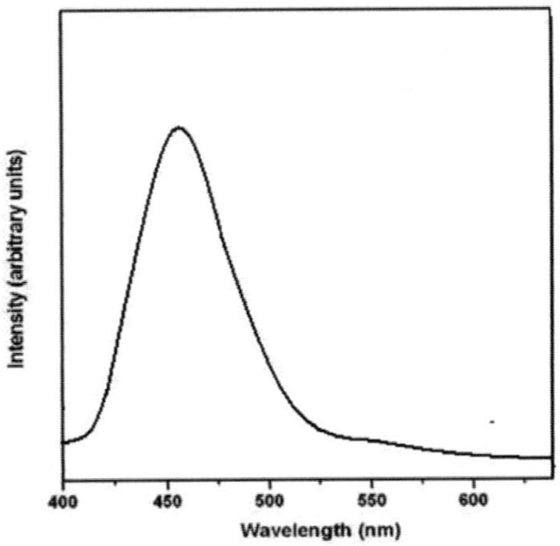

Figure 4. PL spectra of as prepared $PbMoO_4$

An intense blue PL emission at room temperature was observed when excited at 380nm and is shown in figure4. Generally MoO_4 tetrahedron unit is regarded as the reason for luminescent activity in scheelite compounds.$PbMoO_4$ is reported to show green and blue luminescence depending on structural distortions, particle size, morphology, processing temperature etc [4-6] .The PL spectra of $PbMoO_4$ shown in figure 4 has intense blue emission. Sczancoski et al [4] reported that the blue PL emissions are due to structural distortion in [MoO_4] and [PbO_8] clusters in the lattice. As our sample has no structural distortions the emission spectrum of the metal molybdates might be ascribed to the charge-transfer transitions within the [MoO_4] clusters. The blue PL emission of $PbMoO_4$ is assigned to $^1T_2$ to $^1A_1$ electronic transitions within the [MoO_4] tetrahedron groups.

4. Conclusions

Nano crystalline semiconducting $PbMoO_4$ was synthesized through a modified combustion process. The X-ray diffraction studies have shown that the nano powder was single phase and possess tetragonal structure with a lattice constant a=5.540 and c=12.795Å. The average particle size calculated from FWHM is 28nm. The FT-IR study also corroborates the XRD results. The average of band gap determined from the UV–Vis spectrum is 3.20 eV. The nano crystalline $PbMoO_4$ is found to be a photo luminescent material with strong blue emission due to MoO_4 clusters.

References

[1] Chesler R B Pinnow D A, Benson W W 1971 *Appl. Optics.* **10** 2562

[2] Shimodaira Y, Kato H, Kobayashi H, and Kudo A 2007 *Bull. Chem. Soc. Jpn.* **80** 885

[3] Bi Jinhong, Wu Ling, Zhang Yongfan, Zhaohui Li, Junqian Li, Xianzhi Fu 2009 *Appl Catal B: Environ.* **91** 135

[4] Sczancoski J C, Bomio M D R, Cavalcante L S, Joya M R, Pizani P S, Varela J A, Longo E, Siu Li M, and Andres J A 2009 *J.* Phys. Chem. C **113** 5812

[5] Xing G J , Liu R, Zhao C, Li Y L, Wang Y, Wu G M 2011 *Ceram. Int.* **37** 2951

[6] Groenink A and Blasse G 1980 *J. Solid State Chem.* **32** 9

IOP Conf. Series: Materials Science and Engineering **73** (2015) 012121 doi:10.1088/1757-899X/73/1/012121

Effect of Processing on Synthesis and Dielectric Properties of Lead free $(Bi_{0.98}R_{0.02})_{0.5}Na_{0.5}TiO_3$ Ceramics

Vijayeta Pal and R K Dwivedi*
Department of Physics & Material Science & Engineering, Jaypee Institute of Information Technology, Noida – 201307, UP, India

*Email: rk.dwivedi@jiit.ac.in

Abstract. In the present work, an effort has been made to synthesize $(Bi_{1-x}R_x)_{0.5}Na_{0.5}TiO_3$ (BRNT) system with compositions x = 0 & 0.02, where R = La, Nd and Gd by Semi–Wet Technique. Processing of the composition with x = 0 has been optimized by two precursor solution one by citric acid and other by ethylene glycol. The XRD patterns of these samples, prepared by ethylene glycol precursor solution, have shown pure phase of perovskite structure with a rhombohedral symmetry. The studies on structure, phase transitions and dielectric properties for all the samples have been carried out over the temperature range from RT to 450 °C at 100 kHz frequency. It has been observed that two phase transitions (i) ferroelectric to anti-ferroelectric and (ii) anti ferroelectric to paraelectric occur in all the samples. All samples exhibit a modified Curie–Weiss law above Tc. A linear fitting of the modified Curie- Weiss law to the experimental data shows diffuse type transition. The dielectric properties of BNT ceramics have been found to be improved with the substitution of rare earth elements.

1. Introduction

Lead oxide based ceramics with perovskite structure have been the subject of attraction for high performance sensors, actuators, transducers and other applications. In recent years many fields have expressed the need for actuating and sensing devices which can be used at higher temperatures (> 400°C) such as automotive, aerospace, and related industrial applications. Lead, in most of the cases, constitutes more than 60 % of the composition of these piezoelectric devices. Lead, which is volatile and known to be highly toxic, is released to the atmosphere during sintering, caused serious environmental and health problems [1]. This has lead to replace lead (Pb) in the field of piezoelectric ceramics. A lot of research has been carried out on lead free piezoceramic products in the last fifty years.

Pure BNT, discovered by Smolensky et al. in 1961, is a ferroelectric having Bi^{3+} and Na^+ complex ions on the A-site of ABO_3 type perovskite structure with a rhombohedral symmetry. Because of a large remanent polarization (P_r = 38 μC/cm2) at room temperature, BNT ceramic is considered as one of the promising candidates for lead free piezoelectric ceramics [2]. However, the poling of pure BNT ceramic is very difficult due to its high coercive field (E_c = 7.3 kV/mm). The pure BNT ceramic usually exhibits weak piezoelectric properties. To improve the electrical and piezoelectric properties of the material, a number of BNT based solid solutions and a number of processing methods have been studied extensively [3-4].

In the present work, a new lead free $(Bi_{1-x}R_x)_{0.5}Na_{0.5}TiO_3$ ceramic system where R = Nd, Gd and La have been investigated by the partial substitution of R^{3+} for Bi^{3+} on A-sites in $Bi_{0.5}Na_{0.5}TiO_3$ using ethylene glycol precursor by novel semi wet technique and its structural, phase transitions, dielectric properties and ferroelectric properties have been studied. To the best of my knowledge, the BRNT system has been synthesized and studied first time using semi wet technique in BNT System. This technique has been applied to make other systems to study their electrical properties [5].

2. Experimental

A novel Semi-wet technique was used to prepare pure BNT and $(Bi_{1-x}R_x)_{0.5}Na_{0.5}TiO_3$ (BRNT) system with compositions x = 0.02, where R = Nd, La and Gd, using analytical-grade metal oxides or nitrate powders of sigma Aldrich as raw materials such as Bi_2O_3 (99%), $NaNO_3$ (99%), Nd_2O_3 (99.9%), La_2O_3 (99.9%), Gd_2O_3 (99.9%), TiO_2 (99.9%) citric acid and ethylene glycol. In this process, appropriate amount of nitrates solutions of A-site cations and ethylene glycol/ citric acid were added with continuous stirring at appropriate temperature. An aqueous ammonia solution was added drop wise to adjust the pH of the solution (6-8) in the citric acid solution. Both the precursor solution was dehydrated by putting on heater at 80°C for 3 hrs to form a viscous gel which was placed in an oven at 150°C for overnight to combust the gel into ash powder. Here precursor solutions are expected to distribute the cations atomically homogeneously throughout the polymeric structure forming a polymeric complex. The ash, highly fine, homogeneous and highly reactive was mixed with appropriate amounts of TiO_2 powder thoroughly in ethanol using mortal pestle for 2 hrs followed by solid state route.

Pure BNT and BRNT powders were calcined at 750°C and 850-900°C for 2 hrs respectively. These calcined powders were again grounded and mixed thoroughly with a polyvinyl alcohol (PVA) binder solution and then pressed into the form of disk with 12 mm diameter and 1.5 mm thickness. The sintering of all these samples was carried out at 1150-1200°C. The crystalline structure of the sintered samples was examined using XRD analysis with Cu-Kα radiations (DX-1000). The dielectric constant (ε_r) and dielectric loss (tan δ) of BRNT ceramics at 100 kHz were measured as a function of temperature using LCR meter (Hioki 3522). P – E loop tracer (Marine india), was used to measure the polarization hysteresis at 50 Hz.

3. Result and Discussion

In the present work, preparation of pure $Bi_{0.5}Na_{0.5}TiO_3$ was optimized using two chemical precursors by citric acid and ethylene glycol. These samples are abbreviated as BNT-CA and BNT-EG [Figure 1 (a, b)]. It is observed from the XRD patterns that the sample prepared by ethylene glycol precursor solution, has shown better phase formation whereas BNT-CA has formed partially along with the presence of other phase identified as $Bi_2Ti_2O_7$ [6]. The experimental density observed in BNT-EG sample was 5.69 gm/ cm^2 which is 95 % of theoretical density. Therefore, a typical composition with x= 0.02 in the $(Bi_{1-x}R_x)_{0.5}Na_{0.5}TiO_3$ (BRNT) system, where R = La, Nd and Gd were prepared by semi wet technique using ethylene glycol precursor.

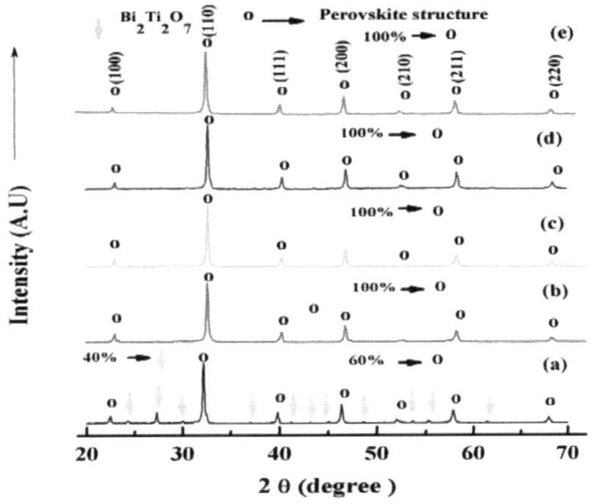

Figure 1. XRD patterns of BRNT ceramics (a) BNT-CA, (b) BNT-EG, (c) R = La, (c) R = Nd and (c) R =Gd

Figure 1(c-e) shows the XRD patterns of BRNT ceramics. All the samples exhibit single phase formation with a rhombohedral strcture.The substitution of La^{3+}, Nd^{3+}, and Gd^{3+} in BNT has significantly influenced the dielectric behaviour [Figure 2]. The highest value of ε_r (~1120) and lowest value of Tan δ (~ 0.040) are obtained with the addition of Gd in the BNT system at room temperature. The electrical parameters of BRNT ceramics at 100 kHz are tabulated in Table 1. Two dielectric anomalies are clearly shown in all the samples at temperature T1 and T2, termed as 'T_d' and 'T_m' respectively, which corresponds to dielectric transitions from ferroelectric (FE) to anti-ferroelectric (AFE) and anti-ferroelectric (AFE) to paraelectric (PE) respectively. In case of La doping, both phase transitions T_d and T_m shifts towards high temperature. This may be due to the partial replacement of A-site cation with larger ionic radius and larger amount, decreases the relative displacement of B-site cation with respect to the oxygen octahedral cage, hence, this increases transition temperature. Moreover, it has been observed that T_d shifts towards low temperature while T_m continuously increased with Nd and Gd for x = 0.02 in BRNT ceramics. The detailed explanation was given somewhere else [7-8].

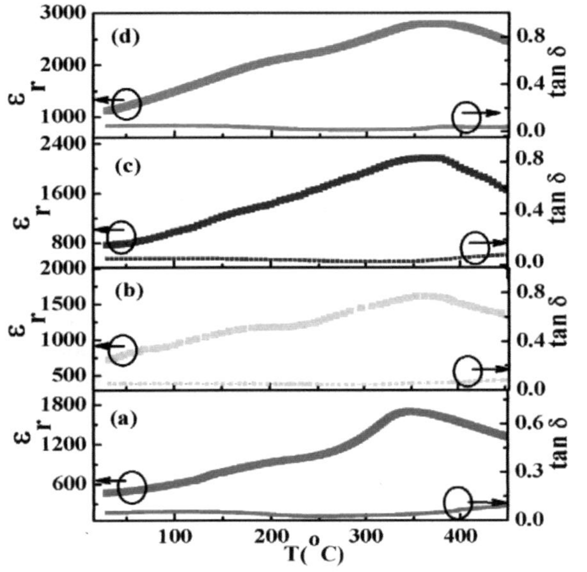

Figure 2. Temperature dependence of ε_r and tan δ of BRNT ceramics at 100 kHz (a) BNT-EG, (b) R = La, (c) R = Nd and (d) R = Gd.

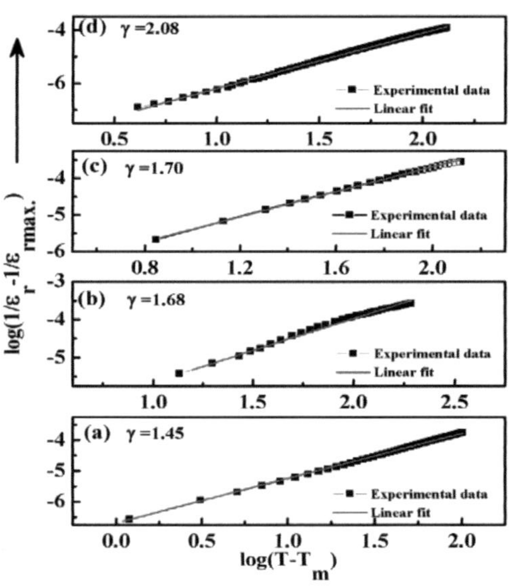

Figure 3. The log $(1/\varepsilon_r - 1/\varepsilon_{rmax})$ versus log $(T-T_m)$ of BRNT ceramics at 100 kHz (a) BNT-EG, (b) R = La, (c) R = Nd and (d) R = Gd.

The diffuseness in the phase transition can be described by the equation in relaxor ferroelectrics [9], where ε_{rmax} is the maximum value of dielectric constant at T_m, γ is the degree of diffuseness and C is the curie like coefficient. γ can have a value ranging from 1 for normal ferroelectric to 2 for an ideal relaxor ferroelectric. A linear relationship is observed in all the samples above T_m. It has been observed that γ of all ceramic samples is < 2, which attributes that the phase transition is diffused type [Figure 3].

Table 1. The electrical properties of all samples in BRNT ceramics.

$(Bi_{1-x}R_x)_{0.5}Na_{0.5}TiO_3$ (x = 0.02)	ε_r	Tan δ	T_d (°C)	T_m (°C)	γ	P_r (μC/cm²)	E_c (kV/cm)
R = 0	461	0.04	179	350	1.45	13.11	33.20
R = La	730	0.05	200	355	1.68	17.31	30.11
R = Nd	765	0.05	145	370	1.70	31.64	25.63
R = Gd	1120	0.04	160	370	2.08	27.39	26.03

4. Conclusion

In summary, ethylene glycol prepared BNT and BRNT samples have shown better crystallization and pure phase formation. The structural, phase transition and electrical properties of all the samples have been investigated. A linear fitting of the modified Curie- Weiss law to experimental data shows that all the samples have diffuse type phase transition. The highest value of dielectric constant (ε_r ~ 1120) and low value of dielectric loss (Tan δ ~ 0.04) is obtained for Gd doped sample at room temperature.

Acknowledgement

Financial support from DST & DRDO, Govt. of India is gratefully acknowledged and one of the authors Ms V. Pal is thankful to JIIT for providing teaching assistance ship to carry out her research work

References

[1] Jaffe B, Cook W.R and Jaffe H 1971 Academic Press, London Chapter 7
[2] Takennaka T, Maruyama K, Sakata K 1991 Jpn. J. Appl. Phys. **30** 2236
[3] Lin D, Xiao D, Zhu J and Yu P 2006 Appl. Phys. Lett. **88** 062901
[4] Hao J J, Wang X H, Chen R Z and Li L T 2005 Mater. Chem. Phys. **90** 282
[5] Pal V, Dwivedi R K and Thakur O P 2013 Advances in Materials Science and Engineering Volume **2013**, Article ID 125634, 7 pages
[6] Kitagawa K and Toyoda T 2003 J. Mater. Sci. **38** 2241
[7] Pal V, Dwivedi R K and Thakur O P 2014 Curr. Appl. Phys. **14** 99
[8] Pal V, Dwivedi R K and Thakur O P 2014 Mater. Res. Bull. **51** 189
[9] Uchino K and Nomura S 1982 ferroelectric Lett Sect **44** 55

Citrate mediated synthesis and tuning of luminescence in Eu^{3+} incorporated Gd_2O_3 nanophosphors

R G Abhilash Kumar and K G Gopchandran

Department of Optoelectronics, University of Kerala, Thiruvananthapuram – 695 581, Kerala, India

E-mail : gopchandran@yahoo.com

Abstract. $Gd_{1.9}Eu_{0.1}O_3$ nanophosphors were prepared successfully by a large-scale facile solution based citrate-metal complex controlled combustion method and was systematically studied by varying the citric acid to metal cation ratio. X-ray diffraction (XRD), scanning electron microscopy (SEM), transmission electron microscopy (TEM), photoluminescence (PL) measurements and radiative properties were done to evaluate the crystal structure, phase formation, phase composition, surface morphology, radiative and luminescent properties of the prepared nanophosphors. Photoluminescent emission intensity of the samples can be correlated with the amount of citric acid, improved crystallinity, uniform morphology, particle size, reduced defects, and proper diffusion of Eu^{3+} in to the crystal structure of Gd_2O_3. Higher asymmetricity results in intense red emission (612 nm) due to $^5D_0$-$^7F_2$ forced electric dipole transition and found that photoluminescence intensity is highest for the sample prepared with citric acid to metal cation ratio of 2:1. The existence of strong red emission from $Gd_{1.9}Eu_{0.1}O_3$ nanophosphor corresponding to $^5D_0$-$^7F_2$ transition (612 nm) of Eu^{3+} under UV light excitation make it a promising candidate for applications in bio assays, magnetic resonance imaging, deep uv LED's, solid state lighting, fluorescent lamps and flat panel displays.

1. Introduction

Phosphors, especially rare earth ion doped nanophosphors have been widely used in display devices, optical fibers, solid state lighting, amplifiers and in medical fields as imaging, biolabels, immunoassays etc. because they posses excellent luminescent properties as a result of their inner shell electronic transitions between the 4f-4f energy levels. The f-f transitions of trivalent rare-earth ions (except Ce^{3+} ion) is forbidden by the parity selection rule, so the emission of trivalent rare-earth ions (except Ce^{3+} ion) doped phosphors yield sharp lines in the optical spectra and a good color rendering index [1]. The rare earth sesquioxide gadolinium oxide (Gd_2O_3) isostructural to Y_2O_3 with high density (7.64 g/cm^3), superior chemical durability, superior colour purity, thermal stability, thermal conductivity and low phonon energy has been shown to be a good host for the luminescent rare earth ions and Eu^{3+} and Tb^{3+} ions are particularly used as efficient luminescent centres especially in display applications. Eu^{3+} doped Gd_2O_3 nanocrystals have shown very rich luminescent features due to the possible paths of activating the trivalent rare-earth ion Eu^{3+} through different sensitization processes, viz. the host Gd_2O_3 absorption, Eu-O charge transfer, the Gd^{3+} ion absorption, and the Eu^{3+} ion self-excitation. Due to a $^5D_0$-$^7F_2$ transition within europium, Gd_2O_3:Eu^{3+} shows red luminescence at 612 nm.

It should be noted that a morphology controlled synthesis by a single technique is of special importance, but remains as a challenge in the field of engineering of particles. The grain size, chemical homogeneity, morphology, agglomeration, and/or surface passivation are well-known factors which should influence phosphor efficiency [2]. Spherical shaped particles posses high packing fraction and also minimize light scattering for practical display applications. In the present study, we report a cost effective way to synthesis almost uniform spherical shaped $Gd_{1.9}Eu_{0.1}O_3$ nanophosphors by a large-scale and facile solution based citrate-metal complex controlled combustion method and was systematically studied by varying citric acid to metal cation ratio. In this synthesis method, citric acid acts as a chelating agent of metal cations via the hydroxyl and carboxyl groups. Polymerizing ability of citric acid is effectively utilized for the formation of metal-citrate complex in diethylene glycol medium, where the metal ions are uniformly distributed. Apart from that, citric acid is also performed as a fuel and caused controlled exothermic reaction with nitrates to result in the formation of nanophosphors.

2. Experimental

Materials used were Gadolinium oxide (Gd_2O_3, 99.99%, Aldrich), Europium oxide (Eu_2O_3, 99.99%, Aldrich), Conc. HNO_3 (70%), Polyethylene glycol 200 (PEG, 99%, Merck), Diethylene glycol (DEG, $C_4H_{10}O_3$, Merck, 99%) and Citric acid monohydrate ($C_6H_8O_7.H_2O$, A.R. grade) and were used without further purification. Stoichiometric amounts ($Gd_{1.9}Eu_{0.1}O_3$) of Eu_2O_3 and Gd_2O_3 were dissolved in concentrated nitric acid and deionized water ($1HNO_3$:$1H_2O$) to make their respective nitrate solution. The two solutions were then mixed under magnetic stirring to assure uniform mixing and then citric acid in DEG was added slowly into the prepared aqueous nitrate solution to chelate metal ions to form metal-citrate complex with continuous stirring. In order to study the effect of citric acid, molar ratio of citric acid to metal cation (CM) ratio is varied as 0.5:1, 1:1, 2:1 and 3:1. About 2 ml of PEG is also added to this complex solution as a mineraliser. The mixed solution was then maintained at ~100 ^{O}C with continuous stirring in a water bath for about 5 h to form a highly transparent viscous solution. The resulting viscous solution is then subjected to heat treatment in a muffle furnace at 180 ^{O}C for 1h and then the temperature is raised to 400 ^{O}C for 2h to result the greyish precursor. The precursor powder thus obtained were fully ground and was then subjected to sintering in a muffle furnace at 800 ^{O}C for 2h at a heating rate of 4 ^{O}C/min to obtain the $Gd_{1.9}Eu_{0.1}O_3$ nanophosphors.

Phase formation, phase analysis and microstructural characterizations of the phosphor samples were done on a X-ray diffractometer (Philips X'pert Pro) in the angular range (2θ) 10^0-70^0 with $CuK\alpha_1$ (λ =1.54056 Å) radiation employing X'celerator and monochromator at the diffracted beam side. The morphology of the samples was investigated using a Field Emission Scanning Electron Microscope (FEI-FEG, Sirion) operating at 5 KV in SE mode and by a transmission electron microscope (FEI Tecnai F20) operating at 200 KV equipped with a Gatan CCD camera. Jobin Yvon Horiba Fluorolog III (FL3-11) Spectrofluorometer equipped with a 450W Xenon lamp as excitation source and a photomultiplier tube in photon counting mode (Hamamatsu R928P) as detector is employed to record the excitation and emission photoluminescence spectra at room temperature.

3. Results and Discussion

3.1. Phase, Structure and Morphology Studies

X-ray diffraction (XRD) patterns of $Gd_{1.9}Eu_{0.1}O_3$ nanophosphor prepared with different citric acid to metal cation ratio (CM ratio) and heat treated at 800 ^{O}C for 2h is shown in figure 1. All the diffraction patterns observed can be indexed to cubic Gd_2O_3 (JCPDS File No. 43-1014, Ia3 (206) space group). No additional peaks of other phases have been found, which shows that Eu^{3+} has been effectively incorporated in to the Gd_2O_3 host lattice in all samples. It is found that the intensity of XRD peaks increase and width of peaks decrease with increase of CM ratio. This indicates an improvement of crystallinity and increase in crystallite size of the samples with increase of citric acid to metal cation ratio.

The crystallite size (D) of the phosphors were calculated from XRD patterns using the Scherrer equation

$$Crystallite\ size\ (D) = \frac{K\lambda}{\beta \mathrm{Cos}\theta} \tag{1}$$

where, K is a constant (=0.9), λ is the wavelength of X-rays (0.15406 nm), β is the diffracted full width at half maximum (FWHM) in radian, θ is the Bragg diffraction angle.

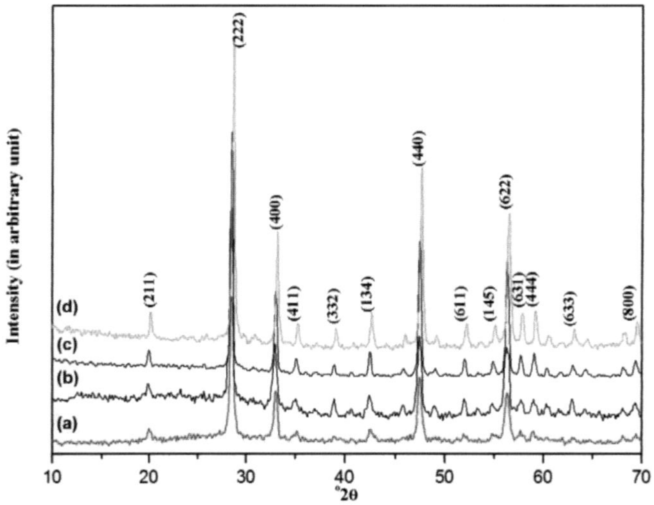

Figure 1. XRD patterns of $Gd_{1.9}Eu_{0.1}O_3$ nanophosphor prepared with CM ratio (a) 0.5:1, (b) 1:1, (c) 2:1 and (d) 3:1.

Figure 2 depicts the variation of lattice constant, cell volume and the evolution of crystallite size of the $Gd_{1.9}Eu_{0.1}O_3$ nanophosphor as function of CM ratio. Lattice constant obtained are in agreement with the standard JCPDS File No. 43-1014 (a = 10.813 Å). It can be seen that at lower CM ratio, there is no appreciable change in the crystallite size, but it increases more rapidly when CM ratio crosses 2:1. This is in agreement with the sharpening of X-ray diffraction peaks observed.

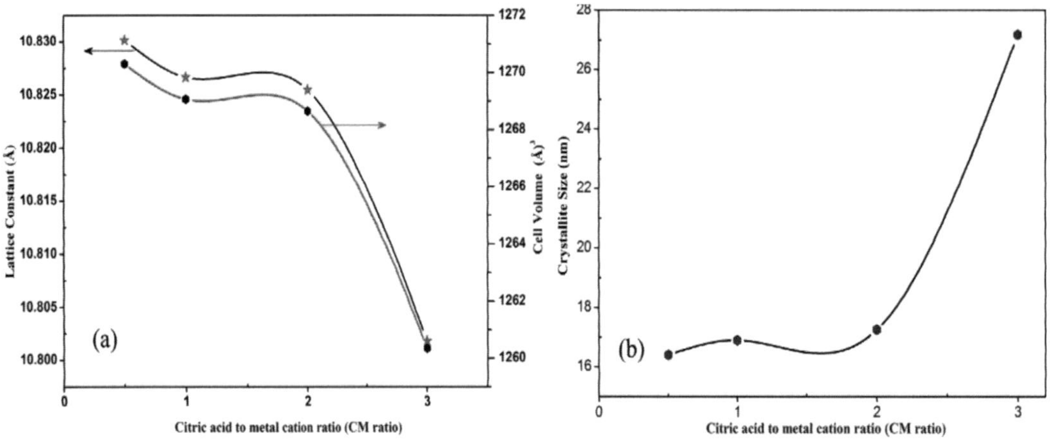

Figure 2. Variation of (a) lattice constant, cell volume and (b) the evolution of crystallite size of $Gd_{1.9}Eu_{0.1}O_3$ nanophosphor as function of CM ratio.

Figure 3. presents the typical scanning electron micrograph (SEM) and transmission electron micrograph (TEM) of the $Gd_{1.9}Eu_{0.1}O_3$ nanophosphor synthesized using the CM ratio of 2:1. From the SEM, it can be seen that almost spherical shaped morphology is observed. The size distribution of these nanoparticles is uniform with regular shape. TEM pattern reveals that the nanoparticles are monodisperse and uniform with almost spherical shape. The size of the nanoparticles measured from the TEM images was around 25 nm in diameter, which is consistent with XRD results. It is well known that spherical shaped grains are of greater importance because of their high packing density, lower scattering of light and better luminescence properties [3] for display applications.

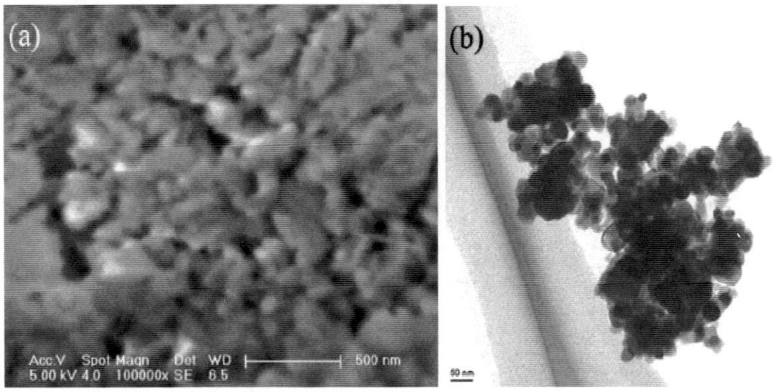

Figure 3. Typical (a) scanning electron micrograph and (b) transmission electron micrograph of the $Gd_{1.9}Eu_{0.1}O_3$ nanophosphor synthesized with a CM ratio of 2:1.

3.2. Photoluminescence Studies

The photoluminescent excitation spectrum consist of a strong broad band in the wavelength range 230-300 nm with a maximum at 265 nm and superimposed with weak lines. Strong broad band is due to the charge transfer (CT) electronic transition from the 2p orbital of O^{2-} to the empty 4f orbital of central Eu^{3+} ions [1.] Weak lines observed at 276, 279, 280 (^8S-^6I) and 314 nm (^8S-^6P) are related to internal f-f transitions of Gd^{3+} ion, indicates that there exists an energy transfer from Gd^{3+} to Eu^{3+} [4]. Photoluminescent (PL) emission spectra of $Gd_{1.9}Eu_{0.1}O_3$ nanophosphor (under 265 nm CT excitation) prepared by different CM ratio is shown in figure 4 (a). All the emission spectra have similar emission profiles and there are five groups of distinctive emission peaks between 550 and 720 nm are observed corresponding to the transitions from excited $^5D_0$ to ^7F_J (J =0, 1, 2, 3, 4) levels of Eu^{3+} ions [5] and are assigned to $^5D_0$-$^7F_0$ (581 nm), $^5D_0$-$^7F_1$ (588,593,599 nm), $^5D_0$-$^7F_2$ (612, 630 nm), $^5D_0$-$^7F_3$ (651 nm) and $^5D_0$-$^7F_4$ (707 nm) transitions, respectively. The strongest red emission peak at 612 nm corresponds to the hypersensitive forced electric dipole $^5D_0$-$^7F_2$ transition of Eu^{3+} ion with the selection rule $\Delta J = 2$ [1], induced by the lack of inversion symmetry and the $^5D_0$-$^7F_1$ is due to the magnetic dipole transition. Integrated photoluminescent intensity (PL) of the nanophosphors as a function of CM ratio is shown in figure 4 (b) and found that the highest emission intensity was observed when CM is 2:1.This may be due to the fact that high citric acid concentration brought about a higher carbon impurity left in the samples which is not favourable for luminescence while lower CM ratio accelerated the rate of hydrolysis which may yield in the formation of inhomogeneous nanoparticles. Thus the observed results showed that a CM ratio of 2:1 shows high luminescent intensity may be due to homogeneity, improvement of doping, good activation between host and activator, better crystallinity and lesser surface defects.

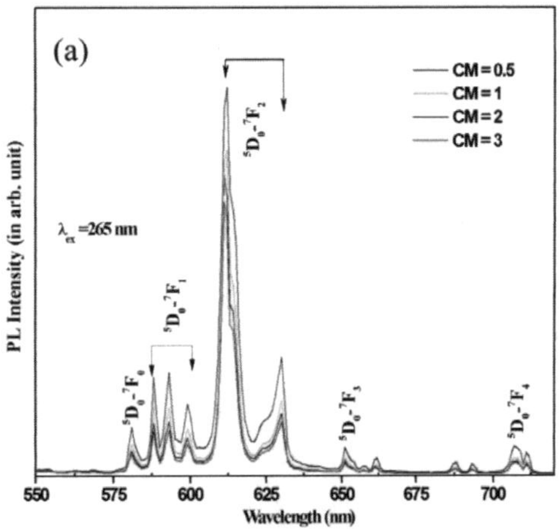

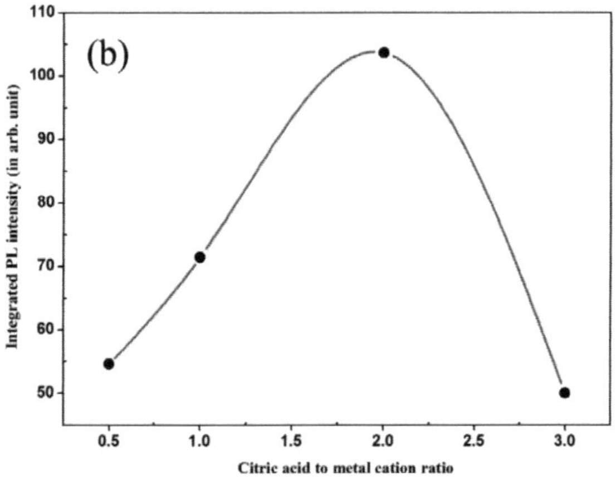

Figure 4. Photoluminescent (PL) emission spectra (a) and integrated PL intensity curve (b) of $Gd_{1.9}Eu_{0.1}O_3$ nanophosphor (under 265 nm excitation) as a function of CM ratio.

4. Conclusions

An efficient method for the synthesis of Eu^{3+} incorporated Gd_2O_3 nanophosphors using metal-citrate complex method is established and studied the effect of citric acid to metal cation ratio on the structure, morphology and luminescent properties of nanophosphors. Photoluminescent emission intensity of the samples can be correlated with the amount of citric acid, improved crystallinity, uniform morphology, particle size, reduced defects, and proper diffusion of Eu^{3+} in to the crystal structure of Gd_2O_3. Higher asymmetric environment around europium results in intense red emission (612 nm) due to $^5D_0$-$^7F_2$ forced electric dipole transition and found that photoluminescence intensity is highest for the sample prepared with a CM ratio of 2:1. The present study reveals that emission intensity can be tuned by suitably selecting the amount of citric acid during synthesis, which can be employed for other phosphor systems. The existence of strong red emission from $Gd_{1.9}Eu_{0.1}O_3$ nanophosphor corresponding to $^5D_0$-$^7F_2$ transition (612 nm) of Eu^{3+} make it a promising candidate for applications in bio assays, magnetic resonance imaging, deep uv LED's, solid state lighting, fluorescent lamps and flat panel displays.

Acknowledgements

R G Abhilash Kumar acknowledges University Grants Commission (UGC), India for the award of teacher fellowship under the faculty development programme.

References

[1] Blasse G and Grabmaier B C 1994 *Luminescent Materials* (Berlin : Springer-Verlag)
[2] Bredol M, Kynast U and Ronda C 1991 *Adv. Mater.* **7–8** 361
[3] Wang W N, Widiyastuti W, Ogi T, Lenggoro I W and Okuyama K 2007 *Chem. Mater.* **19** 1723-1730
[4] Li G Z, Yu M, Wang Z L, Lin J, Wang R S and Fang J 2006 *J. Nanosci. Nanotechnol.* **6** 1416-1422
[5] Xu L, Wei B, Zhang Z, Lu Z, Gao H and Zhang Y 2006 *Nanotechnology* **17** 4327-4331

Structural and optical properties of CuS thin films deposited by Thermal co-evaporation

A K Sahoo[1], P Mohanta[2] and A S Bhattacharyya [3]
[1]Department of Metallurgical Engineering and Materials Science, Indian Institute of Technology Bombay, Mumbai - 400076, India
[2]Center for Research in Nano-Technology and Science, Indian Institute of Technology Bombay, Mumbai - 400076, India
[3]School for Engineering and Technologies, Central University of Jharkhand, Ranchi, 835205, India

E-mail: ashishkumarsahoo443@gmail.com

Abstract. Copper sulfide (CuS) thin films with thickness 100, 150 and 200 nm have been deposited on glass substrates by thermal co-evaporation of Copper and Sulphur. The effect of CuS film thickness on the structural and optical properties have investigated and discussed. Structural and optical investigations of the films were carried out by X-ray diffraction, atomic force microscopy, high-resolution transmission electron microscopy and UV spectroscopy. XRD and selected area electron diffraction conforms that polycrystalline in nature with hexagonal crystal structure. AFM studies revealed a smooth surface morphology with root mean-square roughness values increases from 24 nm to 42 nm as the film thickness increase from 100 nm to 200 nm. AFM image showed that grain size increases with thickness of film increases and good agreement with the calculated from full width half maximum of the X-ray diffraction peak using Scherrer's formula and Williamson-Hall plot. The absorbance of the thin films were absorbed decreases with wavelength through UV-visible regions but showed a increasing in the near-infrared regions. The reflectance spectra also showed lower reflectance peak (25% to 32%) in visible region and high reflectance peak (49 % to 54 %) in near-infrared region. These high absorbance films made them for photo-thermal conversion of solar energy.

1. Introduction

The last few decades transition metal chalcogenides have been widely studied due to their excellent physical and chemical properties. Among them, copper sulfide is an important p-type semiconductor material which has great interest due to their importance in basic scientific research and potential applications for solid state solar cells, electro conductive coatings, lithium ion batteries, electrodes. The high absorbance of the films also used for photo-thermal conversion of solar energy. Generally copper sulfide can easily form a series of nonstoichiometric compounds, depends on the exact composition Cu_xS (x = 1 - 2) with a crystal structure varying from orthogonal to hexagonal. Copper sulfide thin films have been synthesized by several growth methods such as thermal evaporation [2], sputtering, spray pyrolysis [1], chemical bath deposition [3,5-7,9], electro deposition [8], atomic layer deposition and electrochemical methods. In most of the reports, CuS films were deposited by chemical roots techniques using harmful chemical and long deposition time for film preparation, these are main disadvantage of this technique. There are also few reports for the growth of CuS thin films by thermal co-evaporation of elemental constituents.

Here we report, the synthesis and characterization of CuS thin film have been deposited on glass substrates by thermal co-evaporation of copper and sulphur. The effect of CuS film thickness on the structural, morphological and optical properties have been investigated and discussed. The structural, morphological and optical properties of these samples were studied by using X-ray diffraction (XRD), high resolution transmission electron microscopy (HRTEM), atomic force microscopy (AFM) and UV–visible spectrophotometer.

2. Experimental

Copper sulfide based thin films were deposited on glass substrate by thermal co-evaporation of Cu and S with 99.999 % purity. A home-made thermal evaporation system was used for this purpose. The base vacuum was of ~5 x 10^{-6} mbar. The distance between the source and substrate was of about 20 cm. Keeping all other parameter constant changing the evaporation time, different thicknesses of films such as 100, 150 and 200 nm were deposited on glass substrate at the substrate temperature of 250 °C. Quartz crystal thickness monitor was mounted close to the substrate, to estimate the thickness and deposition rate. All these films thickness were also measured by surface profile meter (model Ambios XP-2). The effect of CuS film thicknesses on the structural, morphology and optical properties were investigated. Structural and composition analysis of these films were performed using a X'Pert PRO PANalytical X-ray powder diffractometer with Cuk$_\alpha$ radiation (λ = 1.54056 Å). The surface morphology of the grown films were studied by using AFM carried out in a contact mode with silicon nitride probes using Digital Instruments Nanoscope IV multimode SPM. Absorption and reflectance were carried out on Perkin-Elmer Lambda-950 UV – visible spectrophotometer. A JEOL model JEM-2100F HRTEM operated at 200 kV was used to record the bright field images and selected area electron diffraction pattern (SAED). For the TEM analysis, the CuS film was scratched by using sharp bleat and put into the beaker containing de-ionized water. The CuS film was found completely floating on the surface of de-ionized water. Then this film was transferred to carbon coated copper grid.

3. Results and Discussion

X-ray diffraction pattern have been carried out (see figure 1) for deposited films with different thickness of 100, 150 and 200 nm. It shows that a broad humps at 22.5° which corresponding to glass

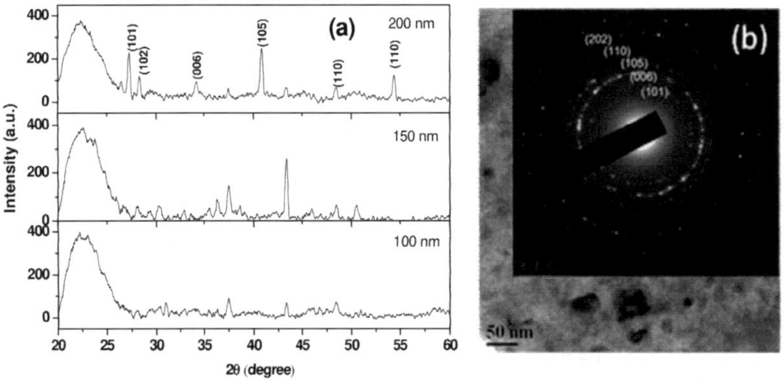

Figure 1. (a) X-ray diffraction patterns for CuS thin films of different thickness, (b) Bright field TEM image of 200 nm film and corresponding SAED pattern.

substrate. The intensities of (101), (102), (006), (105) and (110) XRD peaks are comparable with JCPDS (780876) (figure 1). We are not getting higher intensity peaks of CuS covellite phase in case of lower thickness films and additionally minor presence of Cu_xS phase. Thus, XRD result conform that deposited CuS films are polycrystalline and hexagonal. On the other hand, the crystalline size was also calculated from the full width half maximum (FWHM) of all these peaks using the Debye–Scherrer formula (1).

$$t = 0.9\lambda / \beta \cos\theta \tag{1}$$

where, λ is the X-ray wave length and θ is the diffraction Bragg angle. The calculated crystallite size ranged from about 28, 32 and 54 nm with increase the film thickness from 100 to 200 nm. We also used another procedure based on the Williamson–Hall equation to calculate strain and crystallite size of all deposited films. The Williamson–Hall equation (2) is

$$\beta \cos\theta = k\lambda (t)^{-1} + 2\varepsilon \sin\theta \tag{2}$$

where, β is the FWHM of the XRD all peaks, K is Scherrer constant, t is the crystalline size, λ is the wave length of the X-ray, ε is the lattice strain, and θ is the Bragg angle. In this method, βcosθ is plotted against 2sinθ. Using a linear extrapolation to this plot, the intercept gives the crystallite size and slope gives the strain. These crystallite values are comparable with the crystalline size which is calculated in Debye–Scherrer (see table 1). In order to study the crystal structure, transmission electron microscopy analysis was carried out on the 100 nm thick film. The electron diffraction pattern represent that the film is polycrystalline and the diffraction rings indexed as (101), (006), (105), (110), (202) (In figure 1(b)), which are properly match with the hexagonal covellite CuS film. This is in good agreement with the XRD result. Thus, TEM and XRD result conform that deposited films are hexagonal covellite CuS film.

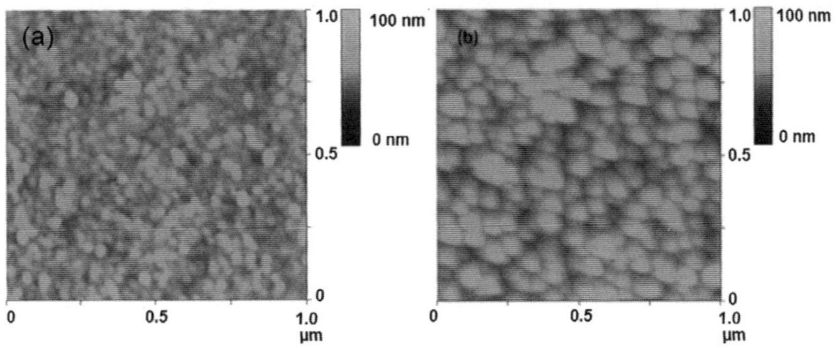

Figure 2. AFM images of CuS films of different thickness (a) 100 nm and (b) 200 nm.

AFM image have been carried out for the deposited CuS films of different thickness from 100 to 200 nm (Figure 2). It is clearly seen that grain size increases, values ranged from 32 nm to 61 nm for film with thickness increase from 100 nm to 200 nm, respectively. AFM image showed that grain size increases with thickness of film increases and good agreement with the calculated crystallite size from full width half maximum of the the XRD peak using Scherrer*'s* formula and Williamson-Hall plot (see table 1). Also we observed that the grain size values obtained from AFM images were slightly larger than those calculated from the XRD measurement. AFM studies also reveal a smooth surface morphology with root mean-square roughness values increases from 24 nm to 42 nm as the film thickness increase from 100 nm to 200 nm (see table 1).

Table 1. Comparison between crystalline size calculated from XRD using Debye–Scherrer formula, Williamson–Hall plot and AFM image for different thickness.

Film thickness (nm)	Debye–Scherrer crystalline size (nm)	Williamson–Hall crystalline size (nm)	Particle size from AFM (nm)
100	28	24	31
150	32	29	37
200	54	56	58

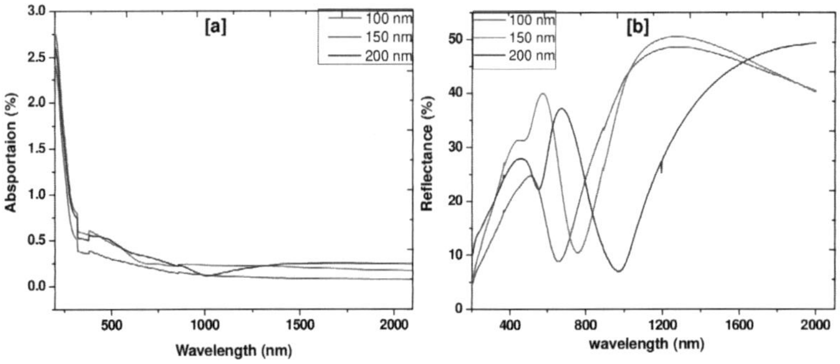

Figure 3. Optical (a) absorption (b) reflectance for deposited CuS thin films.

The optical absorption and reflectance measurement were carried out for all the deposited films (see figure 3). The absorbance of the thin films were absorbed decreases with wavelength through UV-visible regions but showed a increasing in the near-infrared regions, clearly absolved in case of 200 nm thick CuS film. The reflectance spectra also showed two prominent peak in all these deposited films, lower reflectance peak (~ 25 % to 32 %) in visible region and high reflectance peak (~ 49 % to 54 %) in near-infrared region.

4. Conclusions

CuS thin films with hexagonal have been successfully deposited on glass substrate by thermal co-evaporation of copper and sulphur. XRD and TEM results confirm that deposited films are hexagonal covellite CuS films. The increased in crystallite size are absorbed as film thickness increases from 100 nm to 200 nm and conformed from XRD and AFM studies. The deposited films also exhibit good optical properties, which make it for application in photo-thermal conversion of solar energy.

References

[1] Majhi S K, Mukherjee N, Dutta A K, Srivastava D N, Paul P, Karmakar B, Mondal A and Adhikary B 2011 *Mater. Chem. Phys.* **130** 392
[2] Bollero A, Grossberg M, Asenjo B and Gutelerrez M T 2009 *Surf. Coat. Technol.* **204** 593
[3] Hu H and Nair P K 1996 *Surf. Coat. Technol.* **81** 183
[4] Yuan K D, Wu J J, Liu M L, Zhang L L and Xu F F 2008 *Appl. Phys. Lett.* **93** 132106
[5] Güneri E, Kariper A 2012 *J. Alloys Compd.* **516** 20
[6] Castillo S J, Apolinar-Iribe A, León D A and Ruvalcaba-Cornejo C 2011 *J. Optoelectron. Adv. Mater.* **13** 1258
[7] Rodríguez-Lazcano Y , Martínez H, Calixto-Rodríguez M and Rodríguez A N 2009 *Thin Solid Films* **517** 5951
[8] Chen Y, Davoisne C, Tarascon J M and Guery C 2012 *J. Mater. Chem.* **22** 5295
[9] Xin M, Li K and Wang H 2009 *Appl. Surf. Sci.* **256** 1436

IOP Conf. Series: Materials Science and Engineering **73** (2015) 012124 doi:10.1088/1757-899X/73/1/012124

Microwave Dielectric and Magnetic Properties of Co-Zn Ferrites

A.R.Lamani[1] H.S.Jayanna[1] C.S.Naveen[1] M.P.Rajeeva[1] and G.D.Prasanna[2]

[1] Department of Physics, Kuvempu University Shankarghatta Karnataka, India

[2] PG Department of Physics, JSS College, Ooty Road, Mysore, Karnataka, India

E-mail: ashok1571972@gmail.com

Abstract. The dielectric permittivity, loss tangent, in the frequency range 1MHz to 1.8 GHz and hysteresis loop parameters at room temperature were studied on a series of Zn substituted cobalt ferrites with general formula$Co_{1-x}Zn_x$ Fe_2O_4 where (x=0.0, 0.2, 0.4, 0.6, 1.0.). The experimental results indicate that the Zn substitution affects all these parameters. The observed dispersion in dielectric permittivity with frequency is in accordance with Maxwell-Wagner model. The high temperature sintering is used to synthesize these materials via solid state reaction route and these samples were characterized by X-ray diffractometer (XRD), vibrating sample magnetometer (VSM). The saturation magnetizations (M_S) Hc and Mr of the Particles were measured at room temperature. Here for the smaller dopent concentration Ms increases with increasing in the Zn content this can explained on the basis of increased number of magnetic ions in the spinel lattice, at some point Ms decrease because of the difference between the magnetic moment of Fe^{2+} and Zn^{2+}, the magnetic moment of the A sub lattice will increases and the moment of the B sub lattice will decrease. The variation of crystalline shape ellipsoid is correlated with variation of dielectric constant.

1. Introduction

Spinel ferrites have potential application as electrical components such as memory and microwave devices due to their high resistivity and low loss eddy current [1-5]. In the recent years Co-Zn ferrites have attracted considerable attention for their interesting structural, magnetic and electrical properties. Non magnetic Zn doped to Co ferrites are of interest in fundamental understanding of the properties and variety of application such as radio frequency circuit, transform cores, antennas and read /write heads for high speed digital tape and in high quality filters [6-9]. The study of dielectric properties of ferrites offer much valuable information on the behavior localized electric charge carriers this leads to good explanation and understanding of the mechanism of dielectric behavior in ferrites. The lattice parameter [10] Mossbauer and DC electrical conductivity [11] have studied for Co-Zn ferrites but the attempt are not made for the above composition to study dielectric constant and loss in the frequency range 1Mhz -1.8Mhz and the magnetic properties at room temperature.

2. Experimental procedure

The mixed Co-Zn ferrites having the chemical formula $Co_{1-x}Zn_x Fe_2O_4$ (where x = 0.0 0.2, 0.4, 0.6, 1.0) were synthesized by ceramic technique using cobalt oxide, zinc oxide, ferric oxide (99.9% purity sigma Aldrich). The constituent chemicals were taken in stoichiometric proportion and mixed in the agate mortar for 4 hrs and then pre sintered in air in an electronically controlled furnace at 1200°C for 4 h using silica crucible. Then grinding process was repeated before pressing this powder in to disc shaped pellets of diameter of about 1 cm and the thickness ranging from 0.05 cm to 0.1cm using pallet maker. The pressed samples were annealed at 1000 °C to remove the binder (PVA) and to reduce the porosity of the samples. Later the samples were cooled to room temperature. The dielectric constant and loss of Zn substituted Co ferrite were measured in the frequency range from 1MHz to 1.8GHz using Agilent 4291B impedance analyzer. X-ray diffraction pattern of the powder compositions of Co-Zn ferrites shows spinel structure. The measurement of saturation magnetization and magnetic moments of the powder samples of Co-Zn ferrites at different compositions at room temperature using Lakeshore VSM model 7400 series at IISc Bangalore, with magnetic field strength of 10KG.The magnetic moment in terms of Bohr magneton were calculated.

3. Results and discussion

From the X-ray diffraction pattern as shown in the figure 1, shows spinel structure and the variation in the lattice constant is explained on the basis that the modification of the A-B interaction strength due to the change in Fe^{3+} concentration between A and B sites it is confirmed decrease in the lattice constant 8.41-8.34.The variation of dielectric constant and loss as shown in the figure 2 and 3, high value of the dielectric constant can be explained on the basis of Maxwell –Wagner model and Koops phenomenological theory [12, 13, 14].The space charge polarization is governed by the available free charges on the grain boundary and the conductivity of the samples. Koops proposed that the effect of grain boundaries is predominate at low frequencies. The thinner the grain boundary, higher the dielectric constant value. The observed decrease of ε' with increasing the frequency can be attributed to the fact that the electron exchange between Fe^{2+} and Fe^{3+} ions cannot fallow the change of the external applied field beyond certain frequency [15]. In the crystal structure of Co-Zn ferrite Zn^{2+} occupy tetrahedral sites(A), Co^{2+} ions have a tendency to enter in to B-sites,Fe^{2+} occupies octahedral sites (B) sites and Fe^{3+} occupies either tetrahedral or octahedral sites randomly [16,17].For the small dopant concentration, Ms increases with increasing in the Zn content. This result is expected if most Zn^{2+} ions enter the lattice interstitial sites instead of occupying Fe site ions in the lattice. Because the magnetic Zn^{2+} ions are dissolved in the spinel lattice, the net value of magnetic moment and also Ms may be increased by increased number of magnetic ions in the spinel lattice. The variation are as shown in the figure 4, 5, 6 and the data are tabulated in table1.

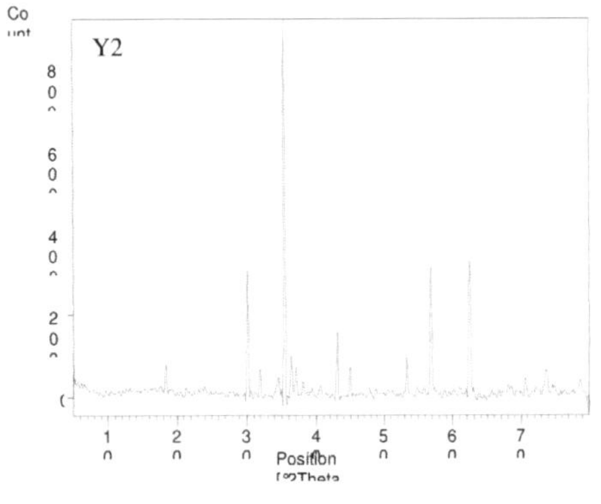

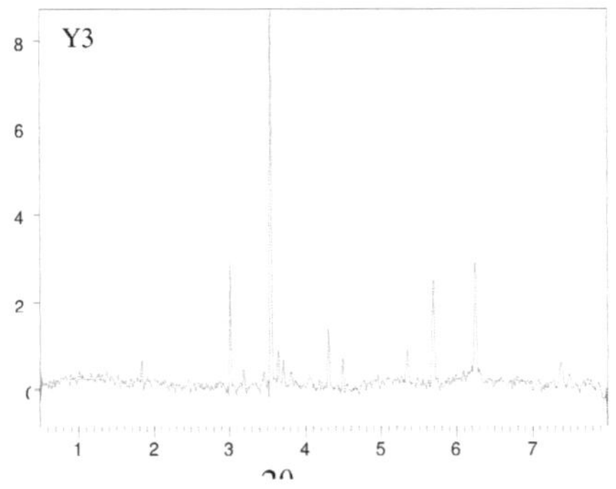

Figure 1 XRD pattern of Co-Zn ferrites for different concentration. (Y2,X=0.2,Y3,X=0.4)

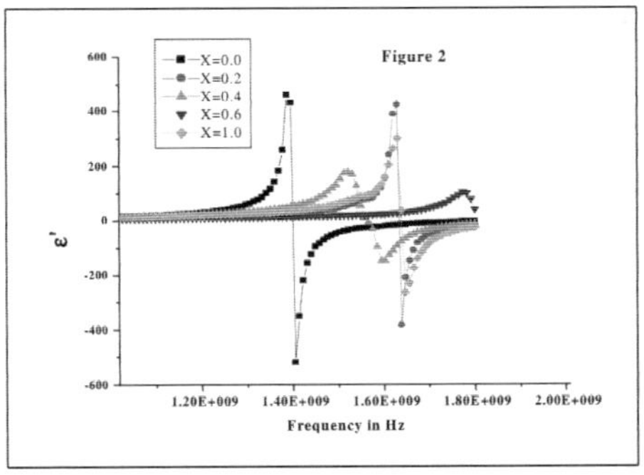

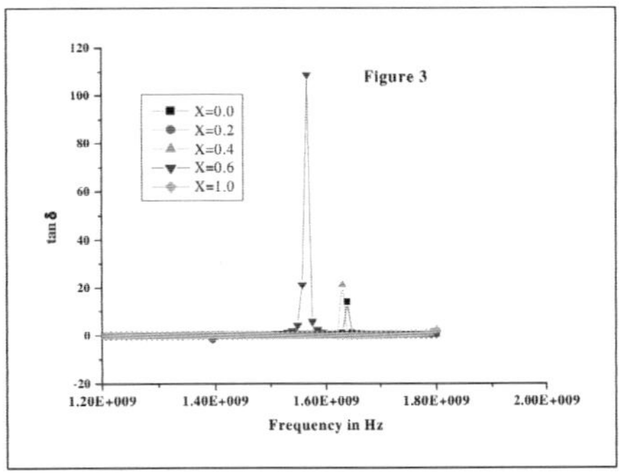

Fig 2 Permittivity of Co-Zn ferrites at different concentrations
Fig 3 Dielectric loss of Co-Zn ferrites at different concentrations

Further increases in the dopant concentration Ms decreases. A possible explanation of this decrease may be that Zn^{2+} entering the B-sites, maybe that some Fe^{3+} ions prevent to occupy this position, pushing the Fe^{3+} ions from B to A sites with further increasing dopant concentration as discussed by [18,19]. When Fe^{3+} ions at B-sites are replaced by magnetic Zn^{2+}, because of the difference between the magnetic moments of Fe^{3+} and Zn^{2+}, the magnetic moments of the A-sites sub lattice (M_A) will increases and the moment of the B-site sub lattice (M_B) will decreases. Therefore, the value of the net moment M being equal to M_B-M_A, will decreases. Explain the observed decrease.

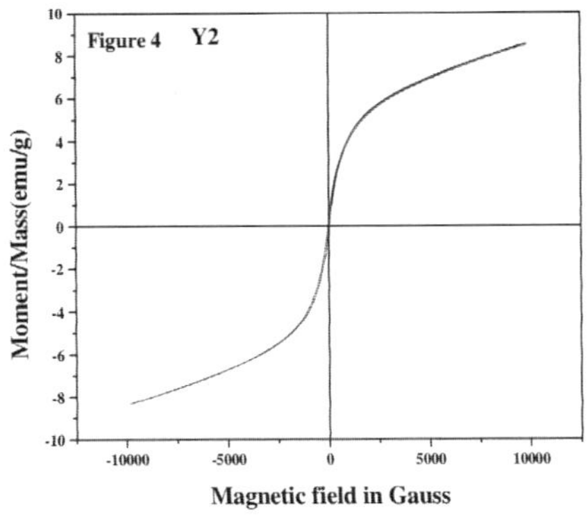

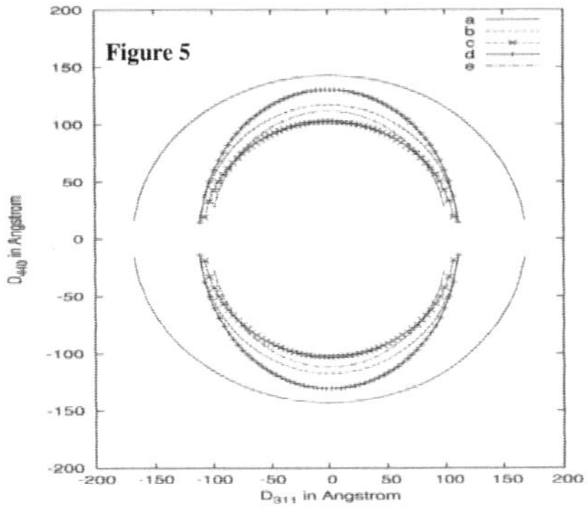

Figure 4: Hysteresis of Co-Zn ferrites at different concentration.
Figure 5: Crystalline shape ellipsoid of Co-Zn ferrites.

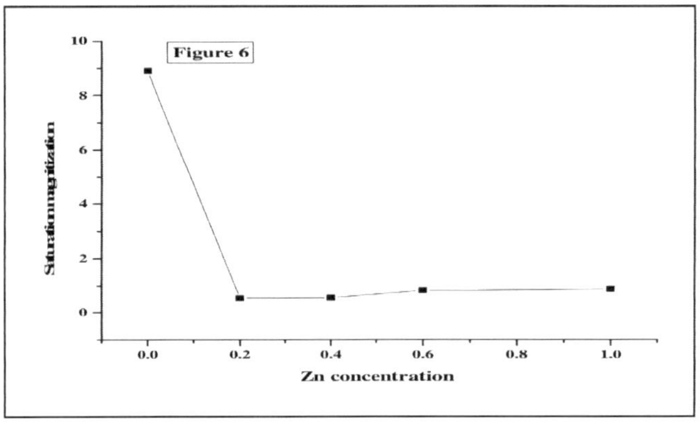

Figure 6: Variation of saturation magnetization in Co-Zn at different concentration.

Table1 X-ray and Magnetic parameter.

Sample	Lattice constant 'a' A^o	Saturation magnetization(Ms)	$n_B(\mu B)$	Mr	Hc Gauss
Y2	8.4174	0.54	0. 22	1.39	18.51
Y3	8.4121	0.56	0. 237	0.0745	27.75
Y4	8.3977	0.82	0. 35	0.0517	74.77
Y6	8.3411	0.88	0. 38	0.139	407.40

4. Conclusion:

From the x-ray diffraction studies shows that Co-Zn ferrite has spinel structure because of the presence of most prominent reflection (311) and (400).In ferrites, the mechanism of electrical conduction and dielectric polarization are similar. They could be explained on the basis of the hopping conduction mechanism (hopping of localized electrons between adjacent Fe^{2+} and Fe^{3+} ions at B sites). As such when the hopping frequency equal to that of the externally applied field a maximum loss tangent may be observed. Because the magnetic Zn^{2+} ions are dissolved in the spinel lattice, the net value of magnetic moment and also Ms may be increased by increased number of magnetic ions in the spinel lattice.When Fe^{3+} ions at B-sites are replaced by magnetic Zn^{2+}, because of the difference between the magnetic moments of Fe^{3+} and Zn^{2+}, the magnetic moments of the A-sites sub lattice (M_A) will increases and the moment of the B-site sub lattice (M_B) will decreases. From the crystalline shape ellipsoid is related to the dielectric constants and is maximum for x=0.0

References.

[1] S.C .Watanabe, S.P.Bamne, S.P.Tangsali, (2007) Mater.Chem.Phys.**103** 323.
[2]J.R.Dahn, U.Vomsacken, C.A.Micha. (1990) Solid State Ionics **44** 87.
[3] G.M.Aargentina.P.D.Babu, (1974) IEEE Trans. Microwave theory Tech.MTT 22
[4] P.P.Hankare, R.P.Patil, U.B.Sankpal, S.D.Jadhav, P.D.Lokhande, K.M.Jadhav,R.Sasikala, (2009) J.Solid State Chem. **182** 3217.
[5]Mazhar U, Rana Tahir Abbas, (2000) J.Mater.Chem.Phys. **65** 345

IOP Conf. Series: Materials Science and Engineering **73** (2015) 012125 doi:10.1088/1757-899X/73/1/012125

Growth and Magnetic characterization of 1D Permalloy Nanowires using self developed AAO Templates

A.K. Singh, G.G. Khan, B. Das, K. Mandal

Magnetism Laboratory, S.N Bose National Centre for Basic Sciences, Block-JD, Sector-III, Salt Lake, Kolkata-700098, India

E-mail ashuvishen@gmail.com & aksingh@bose.res.in

Abstract. 1D Permalloy refers to an alloy of Ni and Fe with 80% and 20% composition respectively. 1D Permalloy nanowires are particularly attractive because of their high permeability, low coercivity, near zero magnetostriction and high anisotropic magnetoresistance. Because of low magnetostriction of Permalloy shape anisotropy plays a very important role. As a result, the nanowires show unidirectional anisotropy along their length. Because of this property, they can be used in many applications such as recording head sensors, magnetic storage devices etc. In the present work 1D Permalloy nanowires arrays were fabricated into the pores of self engineered Anodic Aluminium Oxide (AAO) templates by a simple electrodeposition technique (EDT). By varying the Anodization voltage and the parameters of the electrolytic solutions we developed various AAO templates with different average pore diameters. We developed the 1D Permalloy NW's of different diameters depending on the pore size arrangement of AAO templates by varying the deposition conditions. Structural characterization of AAO templates and 1D Permalloy NW's was performed by Transmission and Scanning Electron Microscopy (TEM & SEM). XRD studies of 1D Permalloy NW's shows their fcc crystalline structure and the AAO template was found to be amorphous in nature. Magnetic studies show the 1D Permalloy NW's arrays to have obvious anisotropy, and the easy axis was found to be parallel to the nanowires axis. We performed the angular dependence measurement of 1D Permalloy NW's. When the applied magnetic field was parallel to the nanowires, the coercivity (H_c) and the maximum remanent ratio (M_r/M_s) were considerably higher than those while the magnetic field perpendicular to the nanowires. 1D Permalloy NW's developed in this work are expected to be utilize in magnetic memory and magnetic recording devices..

1. Introduction

In the past few years, one dimensional (1D) Permalloy nanostructure material such as nanowires, nanochains and nanotubes have attracted attention [1-3], due to their novel properties and potential applications in microelectronics and magnetic nanodevices. Permalloy refers to an alloy of Ni and Fe with 80% and 20% composition respectively [4]. There are various techniques [2-4], to fabricate these magnetic nanostructures. Among these methods, electro deposition technique (EDT) may be a promising candidate or competitive technique for the controlled preparation of high-quality 1D nanostructure and the simplicity of the required equipment. Porous anodic aluminum oxide template (AAO) is one of the templates utilized to fabricate various 1D nanomaterials [5,6]. This has been fuelled by their versatile applications in fields of electronics, magnetic [3, 7], energy storage [8] and biosensors [9]. Anodic aluminium oxide (AAO) template has remarkable properties such as pores uniform in diameter and length paralleling each other. 1D Permalloy magnetic nanowire arrays have the potential applications in super high density magnetic recording media and storage devices attract more attention in present. In this work, the 1D Permalloy nanowire arrays were fabricated into the nanometer scaled pores of AAO template using a simple electro deposition technique (EDT) and the magnetic properties of the nanowires have also been investigated as well as we report a simple and low-cost approach to fabricate large-area highly ordered nanoporous alumina films by two-step high-field anodization.

2. Experimental Details

In order to prepare alumina membrane in a more simple way, we first try to anodize a thin sheet of aluminum (Al) with the thickness of ~ 150 nm and dimension of ~ (5 × 10) mm using oxalic acid ($C_2H_2O_4$) electrolyte solution. The aluminum sheet has the purity of ~ 99.998% and purchased from Sigma Aldrich. The sheets are available with prior electro-polished. One side of the sheets is covered by a Bakelite structure so that only the other side of aluminum comes to the contact of electrolyte solution. Then a two-step anodizing method was employed to obtain the AAO template.

Table 1. Anodizing conditions and morphologic properties of AAO templates.

Sample No.	Anodizing Voltage (V)	First Anodization Time (min.)	Second Anodization Time (h)	Average Pore Diameter (nm)	Average Pore Distance (nm)
AAO#1	50	15	3	45	78
AAO#2	55	15	3	58	94
AAO#3	60	15	3	73	120
AAO#4	65	15	3	95	138

The concentration of oxalic acid is taken as 3 wt% as the voltage maintained at a constant value of ~ 50 to 65 V to achieve a current density of 200-500 Am^{-2}. The anodization is carried out at 11^0 C to dissipate the heat generated during the exothermic reaction of conversion of aluminum in to alumina.

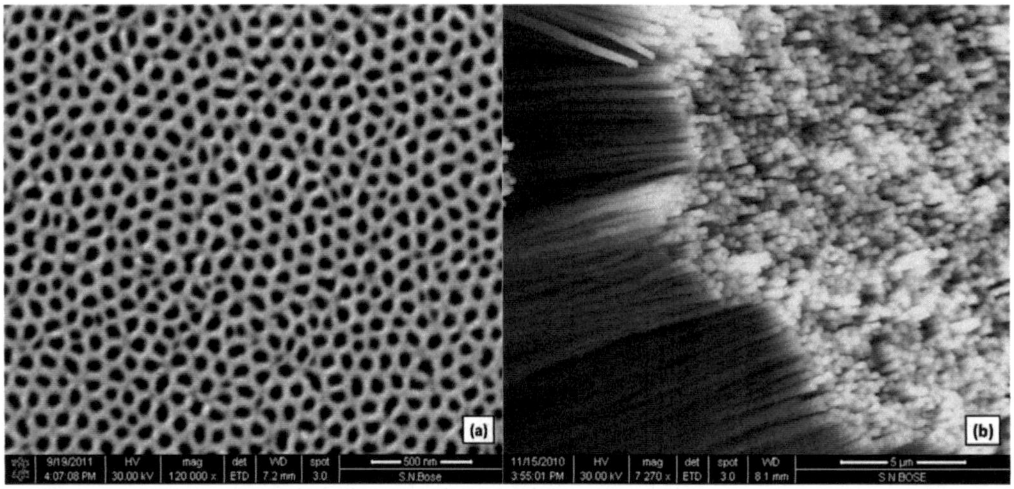

Figure 1. (a) Top view ESEM images of the AAO templates with 45 nm average pore diameter. **(b)** ESEM micrographs of NiFe NWs with 40 nm diameter.

The first anodization is carried out for 15 minutes. The thin porous alumina layer grown on the aluminium foil surface after the first-stage of anodization was removed by chemical etching using a mixture of 60wt% phosphoric (H_3PO_4), 28wt% sulphuric (H_2SO_4) and 12wt% nitric (HNO_3) acid solution. The etched Al foil was re- anodized for 3hours for the second time keeping all the anodization parameters same as that in the first-stage to obtain more uniform porous structure. Finally the hexagonally arranged nanopores structure is detached from the base aluminum using mercury chloride

($HgCl_2$) solution. The as prepared AAO template was dipped inside the 10wt% phosphoric acid (H_3PO_4) solution for 15min for pore widening and pore rounding of the self-organized nanoporous template. Consequently, AAO templates with regular pore arrangement were obtained. Table.1 shows the anodizing conditions for preparing AAO templates. AAO templates with different pore distributions were obtained by changing the anodizing voltage. To use it as an electrode, one side of it was coated with a conductive gold layer of 0.1 μm thickness by the thermal evaporation technique. All electrochemical deposition of nanowires was carried out using an AUTOLAB-30 potentiostat and a conventional three-electrode cell (20 cm^3 capacity). Electrolyte solution used for the permalloy NW deposition consists of 100 g/L $NiSO_4$, 20 g/L $FeSO_4$, 20 g/L ascorbic acid and 30 g/L H_3BO_3. The pH of the solution is maintained at 3.5. Permalloy nanowires were electrodeposited at room temperature. The electrodeposition time was 30 minutes. The length of the nanowires obtained was ~14 μm. For structural characterization of nanowires, the electrodeposited AAO templates were dissolved in 2M of NaOH solution at room temperature. Environmental scanning electron microscope (FEI Quanta-200 Mark-2) and Transmission electron microscope (FEI TECNAI TF20ST) were used to observe the nanowires and their surface morphology and to determine their length and diameter. The crystalline structure of FeNi nanowires were investigated by transmission electron microscopy. The phase of the samples has confirmed by x-ray diffraction (X'Pert Pro, Panalytical). Magnetic properties were studied using a Vibrating sample magnetometer (Lakeshore, model 7144) while the angle θ between the nanowires and applied field changed from 0 to 90°.

3. Results and Discussion

As shown in Fig. 1(b) & 2(a) is the ESEM and TEM image of the Permalloy nanowires prepared on AAO templates. The Permalloy nanowires have the diameter of about 40 & 70 nm respectively and the average length of about 14 μm. This figure also indicates the uniformity in length and diameter of the nanowires. The x-ray diffraction (XRD) pattern of the 70 nm sample, prepared by 30 min electrodeposition, is shown in Fig. 2(b) for better clarity of the peaks. The XRD peaks correspond to either face-centered-cubic FeNi as shown in this figure.

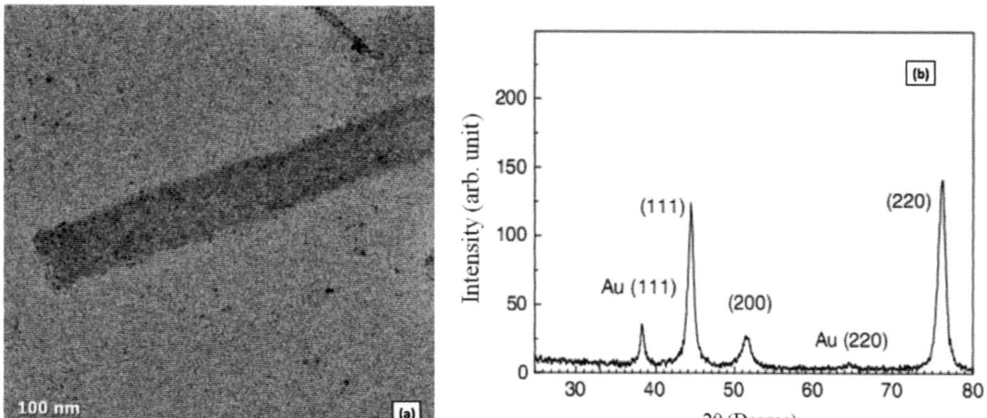

Figure 2.(a) TEM image of the Permalloy nanowire with the average diameter of 70 nm.
(b) XRD image of NiFe NW's.

The magnetic hysteresis loops with different θ, which means the angle between the applied magnetic field and the nanowires, were shown in Fig.3. The magnetic data, such as coercivities (H_c) and maximum remnant ratios (M_r/M_s) are shown in Table.2. For the Permalloy nanowires with the diameter of 40 & 70 nm, when the applied field parallel to the nanowires, θ = 0°, the H_c and M_r/M_s are 1307 Oe and 0.75. In contrast, when the magnetic field perpendicular to the nanowires, θ = 90°, the H_c and M_r/M_s are much lower to 258 Oe and 0.05, respectively. The magnetic measurement results of Permalloy nanowires with the diameter of 70 nm showed the same phenomenon. The H_c and M_r/M_s for Permalloy nanowires of 40 nm are a little higher than those for the nanowires of 70 nm because the larger aspect ratio. The change of the magnetic properties demonstrates that the Permalloy nanowires prepared in this work show obvious magnetic anisotropy and the easy axis are parallel to the nanowires.

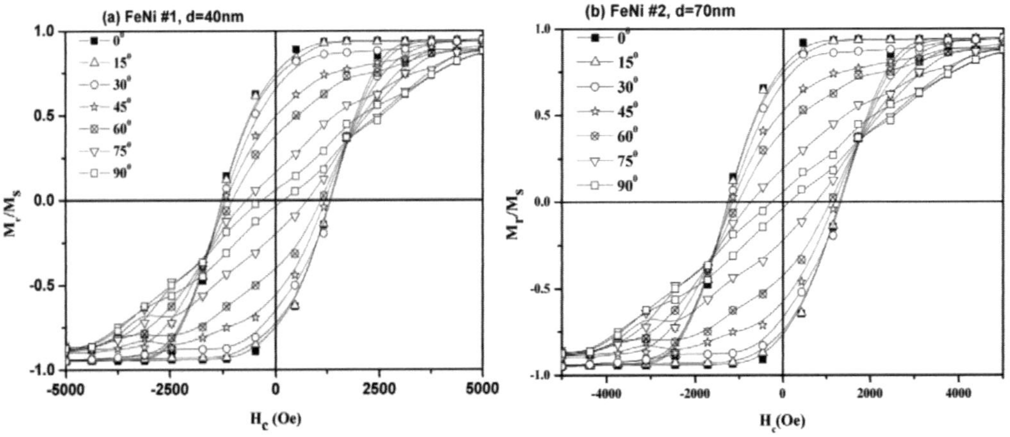

Figure 3. Hysteresis loops of the Permalloy nanowires arrays with different diameters **(a)** 40 nm and **(b)** 70 nm when the angle θ changed from 0° to 90°.

Table 2. H_c and M_r/M_s data for FeNi nanowires of different angle θ with applied magnetic field.

	$\theta(^{0})$	0	15	30	45	60	75	90
	H_c(Oe)	1307	1302	1298	1211	1072	727	258
FeNi 40nm	**M_r/M_s**	0.75	0.725	0.655	0.53	0.39	0.185	0.05
	H_c(Oe)	1290	1279	1266	1176	1056	713	196
FeNi 70nm	**M_r/M_s**	0.775	0.75	0.68	0.55	0.415	0.205	0.055

The magnetic anisotropic property is one important factor for the magnetic recording media materials and the maximum coercivity of the Permalloy nanowires, about 1300 Oe, is also proper for the magnetic recording technique.

4. Conclusion

Permalloy nanowires arrays were fabricated into the AAO template and the morphologic and magnetic properties of the nanowires were characterized. The AAO templates prepared by two-step anodizing method have highly ordered and high density pore arrangement, which is proper for the fabrication of 1D nanowire arrays. The Permalloy nanowires arrays show obviously magnetic anisotropy with the easy axis parallel to the nanowires. Permalloy nanowires have application in magnetic storage devices.

References

[1] Shen J et al 1997, J. *Physical Review B* **56**, 2340.
[2] Wernsdorfer W et al 1996, *Physical Review Letter* **77**, 1873.
[3] Whitney T et al 1993, *Science* **261**, 1316.
[4] Wegrowe J.E et al 1999, *Physical Review Letter* **82**, 3681.
[5] Yamaguchi A et al 2004, *Nature Materials* **3**, 337.
[6] Wang W et al 2004, J. *Journal of Applied Physics* **96**, 615.
[7] Nielsch K et al 2000, *Advanced Materials* **12**, 582.
[8] Che G. L et al 1998, *Nature* **393**, 346.
[9] Matsumoto F et al 2004, *Advanced Materials* **16**, 2105.

IOP Conf. Series: Materials Science and Engineering **73** (2015) 012126 doi:10.1088/1757-899X/73/1/012126

Combustion synthesis of $SrAl_4O_7$:Eu^{2+}, Dy^{3+}, Gd^{3+} long lasting phosphor

A N Yerpude[1] and S J Dhoble[1]
[1]Department of Physics, RTM Nagpur University, Nagpur -440033, India

E-mail: atulyerpude@gmail.com

Abstract. Long afterglow strontium aluminates $SrAl_4O_7$:Eu^{2+},Dy^{3+},Gd^{3+} phosphor was prepared by combustion method. The X-ray diffraction (XRD), Scanning electron microscope (SEM) and photoluminescence (PL) spectrum as well as the luminescence decay were used to characterize the long lasting phosphor. The emission spectrum of the phosphor shows broadened band at 495 nm assigned to the $4f^6 5d \rightarrow 4f^7$ ($^8S_{7/2}$) transition of Eu^{2+}. The present phosphor shows green persistent luminescence after removal of the UV light.

1. Introduction

Long-lasting phosphors are the materials that can absorb UV light and then gradually emit visible light after the light source is removed. The long-afterglow emission can also come from the trapping of electrons through tunneling, without having to pass through the conduction band [1].These materials are used for lighting, display, detection of high energy rays, multidimensional optical memory, luminous paints, emergency lighting, safe traffic, wall painting, films, artificial fibers, rubbers and textiles ceramic and imaging storage [2]. Divalent europium is the most significant one of rare-earth ions widely investigated and applied in the field of long-lasting phosphorescence (LLP) materials. The emission of Eu^{2+} can be varying from blue to red depending on the host lattice [3,4]. In the recent years, strontium aluminates phosphors activated by Eu^{2+} ions, new kinds of long persistent phosphors shows with better properties, have been invented. Due to their excellent properties such as high quantum efficiency, long persistence of phosphorescence, stability, as well as the fact that they are environmentally friendly[2, 5-7]. In this paper we report the $SrAl_4O_7$:Eu^{2+},Dy^{3+},Gd^{3+} phosphor prepared by combustion method. The prepared phosphor is well characterized by X ray diffraction, scanning electron microscope and photoluminescence and phosphorescence decay curves.

2. Experimental

The $SrAl_4O_7$:Eu^{2+},Dy^{3+},Gd^{3+} phosphor were synthesized by a gel-combustion method. The combustion synthesis was carried out at 550^{0}C by Merck analytical grade materials which are as follows $Sr(NO_3)_2$, $Al(NO_3)_3 .9H_2O$, Eu_2O_3 ,Dy_2O_3 , Gd_2O_3 and urea (NH_2CONH_2). Urea was used as fuel. Eu_2O_3, Dy_2O_3, Gd_2O_3 converted in to nitrate form by mixing of appropriate amount of dilute nitric acid. Metal nitrates were weighed stoichiometrically, dissolved in distilled water in a beaker and stirred in order to obtain a clear solution. The resulting colorless solution was further heated with stirring at 85–90 °C until a highly viscous wet gel was formed. The resulting solution was transferred into a china crucible, which was then introduced into a muffle furnace maintained at 550 °C. Then, spontaneous ignition occurred and underwent combustion with enormous swellings occurring. The whole process spent a few minutes and resulted into a foamy white powder. The voluminous and foamy combustion ash was milled to obtain the final $SrAl_4O_7:Eu^{2+},Dy^{3+},Gd^{3+}$ phosphor. The final product obtained is in the fluffy form, which is used for the further investigations.

3. Results and discussion

3.1. XRD phase analysis. The formation of crystalline phases in the samples prepared by the combustion method was confirmed by powder XRD measurement. Figure 1 show that the XRD pattern of the prepared $SrAl_4O_7$ phosphor is well matched with standard JCPDs (Joint Committee on Powder Diffraction Standards) file no 00-025-1208.

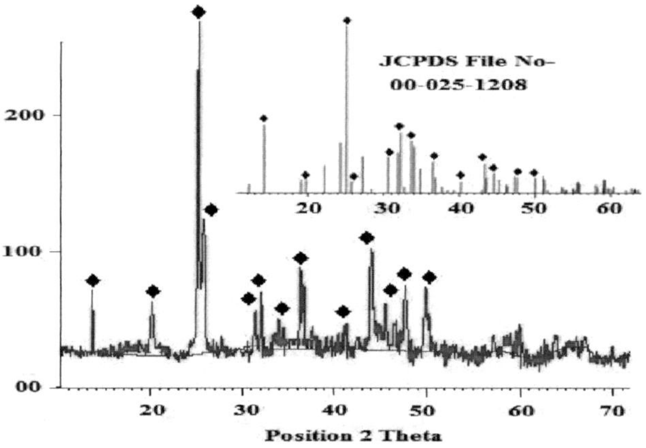

Figure 1. XRD of $SrAl_4O_7$ host.

3.2. Morphology of the $SrAl_4O_7$ phosphor. Scanning electron microscopy (SEM) study was carried out to investigate the surface morphology and crystallite sizes of the synthesized phosphor powder. Figure 2 shows the SEM morphology of the phosphor prepared by the combustion method at 550^0 C. The micrographs show that the crystallite sizes are varying from few microns to several tens of microns. The particles possess foamy like morphology formed from highly agglomerated crystallites nature.

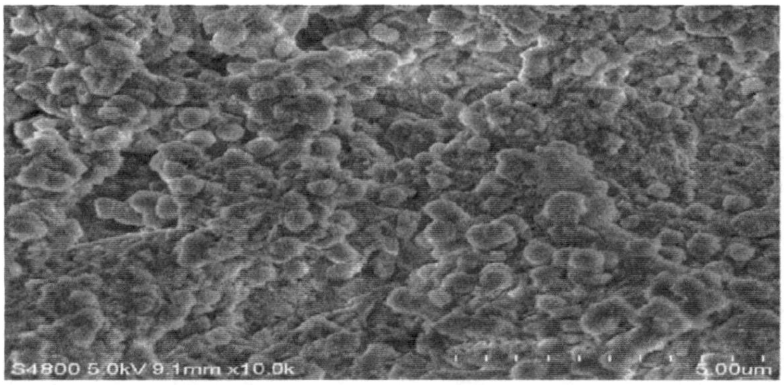

Figure 2. Morphology of the $SrAl_4O_7$ phosphor.

IOP Conf. Series: Materials Science and Engineering **73** (2015) 012126 doi:10.1088/1757-899X/73/1/012126

3.3 Photoluminescence and long lasting properties of $SrAl_4O_7:Eu^{2+},Dy^{3+},Gd^{3+}$ phosphor

Figure 3 and 4 shows the excitation and emission spectra of $SrAl_4O_7:Eu^{2+},Dy^{3+},Gd^{3+}$ phosphor. At room temperature, both the emission and excitation are broad bands. The excitation spectrum shows broad band with the peak at 365nm, corresponding to the crystal-field components of the 5d level in the excited $4f^6 5d$ configuration of the Eu^{2+} ions. The prepared phosphor exhibit a green emission broad band centered at 495 nm, which is attributed to the typical $4f^6 5d^1 \rightarrow 4f^7$ transition of Eu^{2+} ion. No special emission peaks of Eu^{3+} were observed in the spectra. This means that Eu^{3+} in the crystal matrix has been completely reduced to Eu^{2+}. It is worthy to note that, rare earth Dy^{3+}, Gd^{3+} as co-activator into $SrAl_4O_7:Eu^{2+}$ also gave the only green broadband emission and no characteristic emission of the co-activators themselves. The special Dy^{3+}, Gd^{3+} emission peaks are not present, indicating that it only serves as the trap, not as the luminescent center in the host. An important result of our present work is that we observe bright green long-lasting phosphorescence in $SrAl_4O_7:Eu^{2+},Dy^{3+},Gd^{3+}$. Figure 5 gives the decay curves of the long-lasting phosphorescence. The afterglow decay time was measured after the phosphors were irradiated by near UV lamp with a peaking wavelength of 365 nm for 2 minutes. When the light source was removed, the phosphors showed a rapid decay and subsequently an afterglow. The mechanism of the long persistence is due to the holes trapped-transported-detrapped process [8]. However, the detail of the mechanism is still unclear.

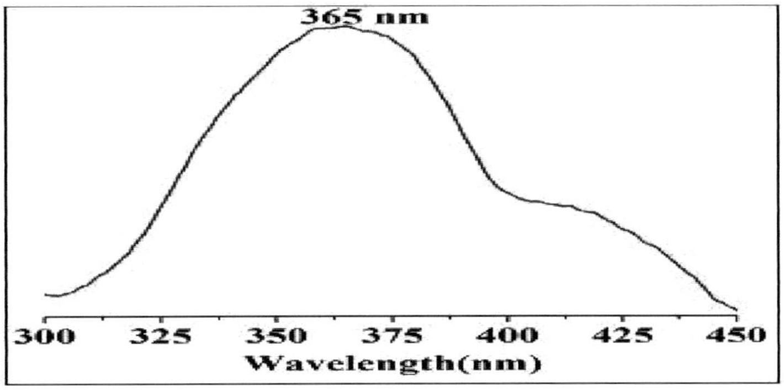

Figure 3. Excitation spectrum of $SrAl_4O_7:Eu^{2+},Dy^{3+},Gd^{3+}$ monitored at 495 nm emission.

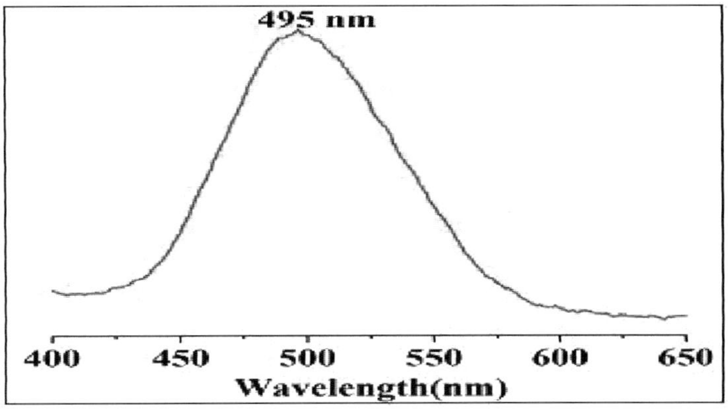

Figure 4. Emission spectrum of $SrAl_4O_7:Eu^{2+},Dy^{3+},Gd^{3+}$ under excitation 365 nm.

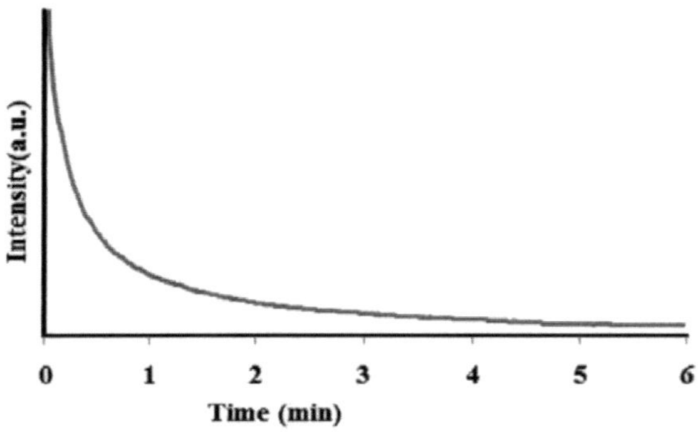

Figure 5. Phosphorescence decay curve of $SrAl_4O_7:Eu^{2+},Dy^{3+},Gd^{3+}$ phosphor.

4. Conclusions

In present work, a green long-lasting phosphor $SrAl_4O_7:Eu^{2+},Dy^{3+},Gd^{3+}$ was prepared by combustion method. The excitation and emission spectra are broadband with the peaks at 365nm and 495 nm, respectively. The green emission corresponds to transitions from $4f^65d^1$ to $4f^7$ of Eu^{2+} ion. Decay time of the long lasting phosphor indicated that the persistent phosphor has bright phosphorescence and maintains duration of several minutes. The SEM photograph indicated that the sample exhibited a rounded shaped particles uniformly attached with each other. Combustion method lowers the cost and can save energy and can be operated easily.

References

[1] Lei B, Li B, Zhang H, Zhang L, Cong Y and Li W 2007 *J. Electrochem. Soc.* **154** 623
[2] Matsuzawa T, Aoki Y, Takeuchi N and Murayama Y 1996 *J. Electrochem. Soc.* **143** 2670
[3] Yerpude A N and Dhoble S J 2012 *J. Lumin.* **132** 1781
[4] Yerpude A N and Dhoble S J 2012 *Luminescence* **27** 450
[5] Yang P, Lu M.K, Song C.F, Xu D, Yuan D.R, Xia G.M and Liu S.W 2002 *Inorg. Chem. Commun.* **5** 919
[6] Jia W Y, Yuan H B, Lu L Z, Liu H M and Yen W M 1998 *J. Lumin.* **76** 424
[7] Chang C, Li W, Huang X , Wang Z, Chen X, Qian X, Guo R, Ding Y and Mao D 2010 *J. Lumin.* **130** 347
[8] Zhang J. Y, Zhang Z. T, Wang T. M and Hao W C 2003 *Mater. Lett.* **57** 4315

Physical and optical studies on Li_2O-Na_2O-WO_3-B_2O_3 glasses

Avula Edukondalu[1,3*], B. Kavitha[1,3], Abdul Hameed[2] and K. Siva Kumar[1,3]

[1]Department of Physics, Osmania University,Hyderabad,India.

[2]Department of Physics, Nizam College, Hyderabad, India.

[3]Department of Physics, Osmania University College for Women, Hyderabad, India.

E-mail: kondalou@gmail.com (Edukondalu)

Abstract. Glasses with composition xLi_2O-$(30-x)Na_2O$-$10WO_3$-$60B_2O_3$ (where x= 0, 5, 10, 15, 20, 25 and 30 mol%) have been prepared using the melt quenching technique. In the present work, the mixed alkali effect (MAE) has been investigated in the above glass system through density and optical studies. The density of the present gasses varies non-linearly, the exhibiting the mixed alkali effect. From the optical absorption studies, the values of direct optical band gap, indirect optical band gap energy (E_o) and Urbach energy(ΔE) have been evaluated. The values of E_o and ΔE vary non-linearly with composition parameter, showing the mixed alkali effect.

1. Introduction

Optical absorption studies on amorphous materials yield important information regarding electronic and vibrational edges plus the contributions from impurities such as transition metal ions. Stevels [1] was the first to suggest that the intrinsic absorption edge of an oxide glass corresponded to the transition of valence electron of an oxygen ion in the glass network to an excited state. Although thermal vibrations are assumed to be responsible for the tail part of the optical absorption edge a number of glasses obeying Urbach rule for the absorption coefficient, whose physical origin is not well understood.

Tungsten oxide containing glasses is of great interest because they can exhibit unusual electrochromic or photosensitive properties related to the ability of tungsten atoms to exhibit various oxidation states (W^{4+}, W^{5+} or W^{6+}). The extent and ability of incorporating WO_3 in glasses strongly depends on the glass former, size of ions in the glass structure, mobility of the modifier cation, etc. Large amount of WO_3 is easily incorporated into germinate, tellurite, phosphate and borate glasses [2-5].

The present investigation is a part of an ongoing program to study the MAE in lithium sodium tungsten borate glasses. Therefore, in this work we have concentrated on the influence of mixed alkalis on physical and optical absorption properties of xLi_2O–$(30-x)Na_2O$–$10WO_3$– $60B_2O_3$ ($0 \leq x \leq 30$ mol%) glasses. The physical properties of the glasses have also been determined with respect to glass compositional parameter R_{Li} defined as R_{Li}= Li_2O (mol%)/(Li_2O+Na_2O) (mol%) which takes the values 0, 0.166, 0.332, 0.5, 0.666, 0.832 and 1.

2. Experimental

The xLi_2O–$(30-x)Na_2O$–$10WO_3$–$60B_2O_3$ ($0 \leq x \leq 30$ mol%) glass system was prepared through the standard melt quenching method. Details of the preparation and characterization of the glass specimens are given in ref. [6,7]. The optical absorption spectra of the glasses were recorded in the wavelength range 200–800 nm at room temperature using a Schimandzu 3100 UV-VIS-NIR spectrometer. The uncertainty in the observed wavelength was about ±1 nm.

3. Results and Discussion

3.1. Physical parameters

The molar volume (V_m) and oxygen packing density (O) are calculated using the measured values of density [8]. From table 1 it can be observed that the variation of density, molar volume and oxygen packing density as a function of compositional parameter is non-linear. This non-linear variation can be taken as a signature of the mixed alkali effect. The change in the molar volume is a result of the creation of non-bridging oxygens (NBOs), which will break the bonds and increase space in the glass network. This result indicates that the glass structure becomes loosely packed.

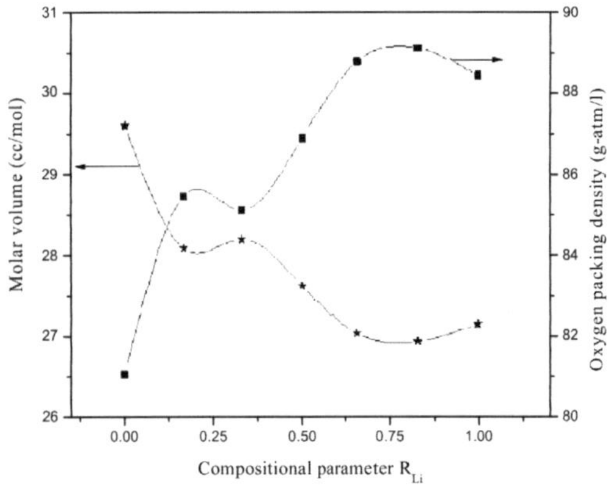

Figure. 1. Variation of V_m and OPD as function of compositional parameter R_{Li} .

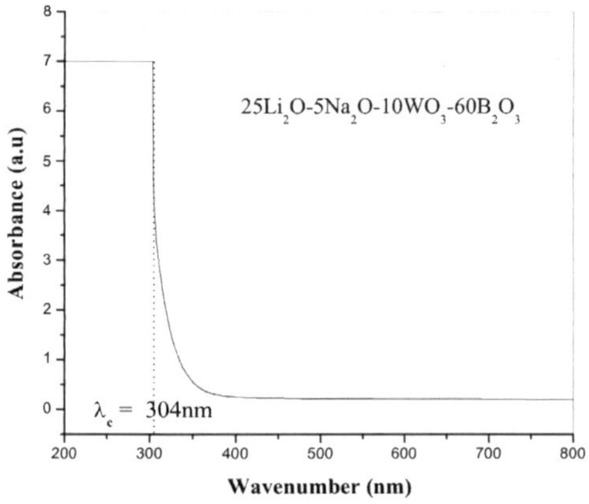

Figure. 2. Optical absorption spectra of x= 15 glasses sample.

3.2. Optical absorption

The optical absorption spectra of xLi_2O-$(30-x)Na_2O$-$10WO_3$-$60B_2O_3$ glass system is shown in Fig. 2. Other glass samples in the present series also show similar spectral behavior. The optical absorption co-efficient α(v), near the fundamental absorption edge of the curves in the above figure was determined from the relation

$$\alpha (v) = (1/d) \ln(I_0/I_t) = 2.303A/d \qquad (1)$$

where I_0 and I_t are the intensities of the incident and transmitted beams, respectively, and d is the thickness of the glass sample. The factor $\log(I_0/I_t)$ corresponds to absorbance. Davis and Mott and Tauc and Menth [9,10] relate this data to the optical band gap, E_{opt} through the following general relation proposed for amorphous materials

$(\alpha h\nu)^n = B(h\nu - E_{opt})$ (2)

where B is a constant related to the extent of the band tailing and hν is incident photon energy.

The index n determines the type of electronic transitions causing the absorption, and takes the values 1/2, 2, 2/3 and 1/3 for indirect allowed, direct allowed, direct forbidden and indirect forbidden transitions, respectively. According to the theory of electronic structure of amorphous materials [11] the absorption edge of the amorphous material is interpreted in terms of indirect allowed transitions. The values of optical band gap energy E_{opt} can be obtained by extrapolating the absorption coefficient to zero absorption in the $(\alpha h\nu)^n$ against photon energy hν plots as shown in Fig 3. The optical band gap energy, thus evaluated for the glass samples at different values of n are listed in Table 1. In the present glasses, the allowed indirect and direct band gap energy varies from 3.54 to 3.83 eV and 3.66 to 3.98 eV, respectively. Srinivasa Rao *et al* [12]. and Huang *et al* [5]. Observed a large band gap energies around 3.25 eV and 4.18 eV in $PbO\text{-}Bi_2O_3\text{-}As_2O_3\text{-}WO_3$ and $Li_2O\text{-}B_2O_3\text{-}WO_3$ glasses, respectively. It is observed that the band gap energies vary non-linearly with compositional parameter, indicating the existence of a mixed alkali effect.

The main feature of the absorption edge of amorphous materials is an exponential increase of absorption coefficient α(ν) with photon energy hν, which is given by the Urbach rule [13]

$\alpha(\nu) = C \exp(h\nu/\Delta E)$ (3)

where C is a constant, and ΔE is the Urbach energy which is a measure of band tailing. Fig. 4 plots the variation of ln (α) as function of photon energy hν. The values of Urbach energy (ΔE) were determined by taking the reciprocals of the slopes of the linear portion of ln(α) vs hν curves. The Urbach energy values of the present glass samples are presented in Table 1. The compositional dependence of Urbach energy ΔE is illustrated in Fig. 4. Urbach energy is varying non-linearly with a compositional parameter indicating the existence of a mixed alkali effect in present glasses.

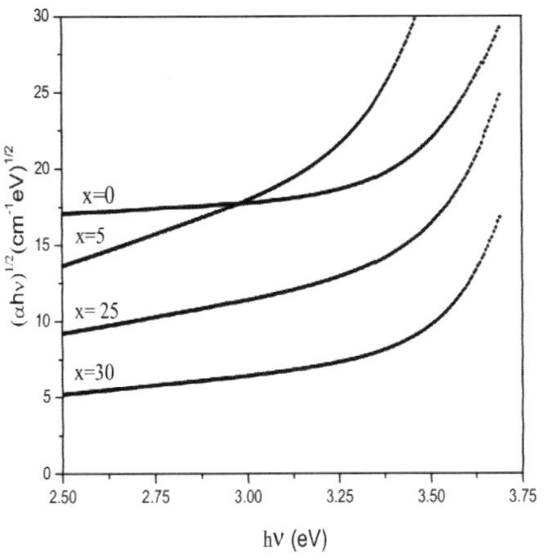

Figure 3. $(\alpha h\nu)^{1/2}$ versus hν, Tauc's plot for the present glass system.

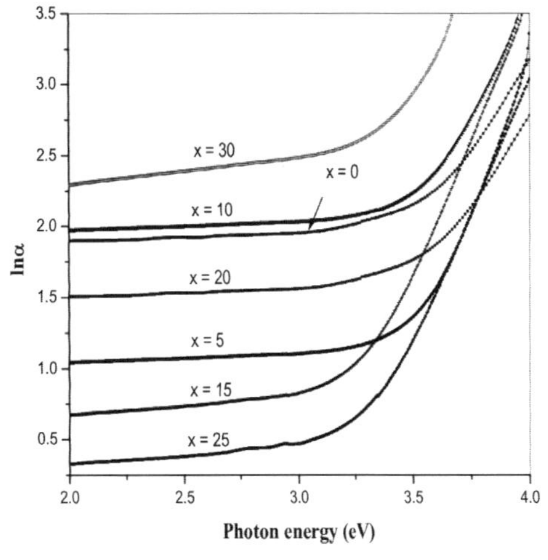

Figure 4. lnα versus hν, Tauc's plot for the present glass system.

Table 1. Density (ρ), Molar volume (V_m), Oxygen Packing Density and Optical parameters of the xLi_2O-$(30-x)Na_2O$-$10WO_3$-$60B_2O_3$ glass system

Parameters	x=0	x=5	x=10	x=15	x=20	x=25	x=30
ρ (g/cc) (±0.001)	2.821	2.918	2.849	2.851	2.854	2.805	2.724
V_m (cc/mol)(±0.001)	29.616	28.082	28.199	27.616	27.025	26.925	27.136
OPD (g-atm/l))(±0.001)	81.037	85.464	85.109	86.906	88.807	89.136	88.443
λ_c (nm) (±1)	338	311	287	313	307	304	313
Indirect E_o (eV)	3.41	3.80	3.83	3.62	3.57	3.72	3.73
Direct E_o (eV) (±0.05)	3.66	3.93	3.99	3.77	3.86	3.85	3.92
ΔE (eV) (±0.001)	0.285	0.342	0.394	0.266	0.288	0.265	0.235

4. Conclusions

Mixed alkali tungsten borate glasses in the form of xLi_2O–$(30-x)Na_2O$–$10WO_3$–$60B_2O_3$ ($0 \leq x \leq 30$) were prepared, and their physical and optical properties have been studied. The following conclusions were made:The physical properties like density, refractive Index and glass transition temperature were evaluated for the glass studied in the present work varies non-linearly with a compositional parameter which shows the MAE. From optical absorption edges the direct and indirect optical band gap energies and Urbach energies were evaluated and their non-linear behavior confirms the mixed alkali effect.

References

[1] Bih L, Abbas L, Mohdachi S and Nadiri A 2008 *J. Mol. Struct.* **891** 173.
[2] Bergman A G, Machavar Z N and Maltsev V T 1972 *Russ. J. Inorg. Chem.* **17** 3106.
[3] Shaltout I, Tang Y, Braunstein R and Abu-Elazm A M 1995 *J. Phys. Chem. Solids*, **56** 141.
[4] Carla C. de Araujo, Wenzel Strojek, Long Zhang, Hellmut Eckert, Gael Poirier, Sidney J. L. Ribeirob and Younes Messaddeq 2006 *J. Mater. Chem.* **16** 3277.
[5] Huang P N, Huang X H and Gan F X 1990 *Solid State Ionics* **44** 11.
[6] Edukondalu A , Purnima M, Srinivasu Ch, Sripathi T, Awasthi A M, Syed Rahman and Siva Kumar K 2012 *J. Non-Cryst. Solids* **358** 2581–2588.
[7] Edukondalu A, Sathe V, Syed Rahman, Siva Kumar K and Sreenivasu D 2014 *Vibrational Spectroscopy* **71** 91–97.
[8] Edukondalu A, Sathe V, Syed Rahman and Siva Kumar K 2014 *Physica B* **438** 120.
[9] Davis E A and Mott N F 1970 *Philos. Mag.* **22** 903.
[10] Tauc J and Menth A 1972 *J. Non-Cryst. Solids* **8** 569.
[11] Edukondalu A, Kavitha B, Samee M A, Shaik Kareem ahmmad, Syed Rahman and Siva kumar K 2013 *J. Alloys Compd.* **322** 324 -335.
[12] Rao N S, Ravi Kumar V and N. Veeraiah N 2011 *Physica B* **406**4494.
[13] Urbach F1953 *Phys. Rev.* **92** 1324.

Studies on PVA based nanocomposite Proton Exchange Membrane for Direct methanol fuel cell (DMFC) applications

P Bahavan Palani[1], R Kannan[1]*, S Rajashabala[2],S Rajendran[3],G.Velraj[4]

[1]Department of Physics, Anna University of Technology Madurai, Dindigul campus, Dindigul-624622

[2]School of Physics, Madurai Kamaraj University, Madurai-625021

[3]Department of Physics, Alagappa University, Karaikudi-3

[4]Department of Physics, Periyar University, Selam-2

* kkdikans@yahoo.com, bahavanpalani@gmail.com

Abstract. Different concentrations of Poly (vinyl alcohol)/Montmorillonite (PVA/MMT) based proton exchange membranes (PEMs) have been prepared by solution casting method. The structural and electrical properties of these composite membranes have been characterized by using X-ray diffraction (XRD), Fourier transform infrared spectroscopic (FTIR) and AC impedance spectroscopic methods. The conductivity of the PEMs has been estimated for the different concentration of MMT. Water/Methanol uptake measurement were also analyzed for the prepared PEMs and presented. The proton conductivity studies were carried out at room temperature with 100% of humidity.

1. Introduction

Two challenges for humanity in the next 20 years are the striking availability of fossil fuels and global warming that result from the ever increasing use of fossil fuels. One possible solution to these problems is to use hydrogen as energy carrier. Fuel cell is an alternative for traditional energy sources because it provides ultra clean environment. The preparation of proton exchange membrane for fuel cells has been active research for the past one decade. Direct methanol fuel cells (DMFC) have received lot of attention because of their high energy density and low emission of pollutants. Methanol crossover in DMFC causes a loss of fuel and in turn it reduces the performance of DMFC. These drawbacks gave way to find new materials to minimize the methanol crossover through PEMs while maintaining good proton conductivity. Nafion, a commercially available membrane has a high chemical and electrical resistance as well as high proton conductivity. However it has been found that 40% of methanol can be lost across the Nafion membrane in DMFCs [1, 2]. So Nafion is not an effective barrier to methanol crossover. Therefore, a PEM with significantly decreased methanol crossover need to be developed. Composite membranes have been prepared by introducing nano-sized inorganic materials, such as clay nanoparticles [3-8]. New cross-linked Nafion-PVA composite polymer electrolyte membrane has been prepared [Shao et al.]. It was found that the methanol permeability was reduced by about 48% [3]. Son et al. reported that poly (vinyl alcohol) based polymer electrolyte membranes containing polyrotaxane and the size of the polyrotaxanes play an important role in proton transport and methanol crossover in DMFC [4]. Yang has fabricated poly (vinyl alcohol)/montmorillonite/ poly (styrene sulfonic acid) proton conducting membranes for DMFCs [5]. It is noticed that, there is an enhancement of ionic conductivity due to the presence of –SO3H group in modified MMT fillers and PSSA polymer. Moreover this composite membrane exhibited a good electrochemical performance and it will be a potential candidate for DMFC in near future. In this work, PVA based nanocomposite proton exchange membranes were prepared by solution casting method and it is characterized for DMFC.

2. Experimental

2.1. Materials

Poly (vinyl alcohol) (Average MW~1,15,000) was purchased from Loba, India and it used as a host material. Montmorillonite K10 (Na+MMT) obtained from Hamada, India were used as recieved. Ion exchange resin obtained from Central Drug House (P) Ltd, India.

2.2. Preparation of protonated Montmorillonite (H^+MMT)

Montmorillonite (Na^+MMT) was dispersed in distilled water at room temperature. The dispersed solution was passed through the ion exchange resin which was taken in a burette for at least four times to ensure a complete exchange into H^+. Then the collected protonated solution was dried at room temperature for several hours. The dried protonated MMT was grinded well using a mortar. Now the nano size protonated powder is ready for membrane preparation.

2.3. Preparation of Polymer Electrolyte Membrane

The composite membranes were prepared with appropriate weight ratios of PVA and MMT fillers. The host polymer and MMT clay were mixed in distilled water by normal stirring for 12 hrs in order to obtain homogeneous solution. The resulting solution was poured onto a glass plate or Petri dish. It was allowed to evaporate at room temperature. After complete evaporation of the solvent, the composite polymer membranes were peeled off from the Petri dish or glass plate. The thickness of the prepared composite polymer membrane is around 0.08- 0.12 mm. The prepared membranes were subjected to characterization.

2.4. Preparation of Sulfonated PVA/ MMT Membranes

The sulfonated composite solid polymer membranes were prepared with different molar concentrations of sulfuric acid. The prepared composite membranes were immersed separately in 5M, 8M and 10M solutions for at least 24 hrs at room temperature.

3. Results and discussion

3.1. FTIR analysis

The FTIR technique is used to find out the presence of chemical groups in the composite polymer electrolyte membranes and they were recorded in the range of wave number from 4000 to 500 cm^{-1}. Figure.1 shows, the FTIR spectra from PVA/ MMT composite polymer electrolyte membranes. It shows strong characteristic peaks of the PVA polymer at 1440, 1711, 1145, 920 and 853cm^{-1}[6, 9, 10]. It can be seen that, the broad transmission peak of the O-H stretching bands at around 3600-3200 cm^{-1} due to the strong hydrogen bonding of (water) intramolecular or intermolecular type. The vibration peak of C-H band at 2945 cm^{-1} refers to the stretching of alkyl groups and the transmission peak at

1711.63 cm^{-1} arises due to the asymmetric stretching vibration of carbonyl (C=O) from the acetate group. Additionally, the strong peak at 1140.33 cm^{-1} is due to the C-O stretch, C-C stretching vibration. In addition, there were several vibrational peaks at 919.50 and 853.21 cm-1 were due to the C-C stretching.

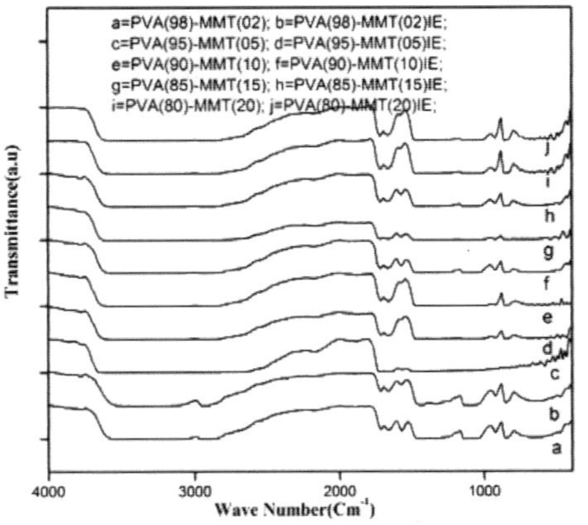

Figure 1. FTIR spectra of PVA/MMT composite PEMs.

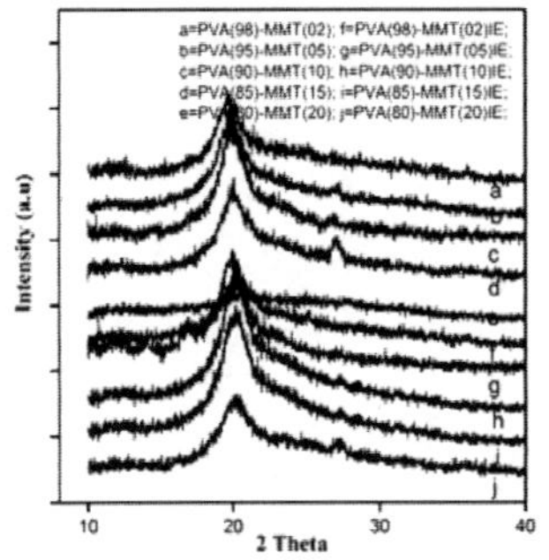

Figure 2. XRD patterns of PVA/MMT composite PEMs.

3.2. XRD analysis

The X-ray diffraction patterns were taken by X-ray diffractometer. Radical scan were recorded in the reflection scanning mode with 2θ being changed from 10° to 60°. Figure 2 Illustrates the diffraction pattern of PVA/MMT composite polymer membranes that were prepared by adding with varied amounts of Na+MMT and H+MMT fillers. The pure PVA polymer is well known to exhibit a semi-crystalline structure with a large crystalline peak at a 2Θ angle of 19-20°. The peak at 19.6° is correspond to the (101) plane in the pure PVA film [8]. It is observed that the same peak for PVA/20wt.% MMT composite membrane shows there is no crystalline peak in Figure. 2(e) because, the intensities of the (101) peak for PVA/MMT composite membranes were greatly reduced when the amount of modified MMT fillers increased. Under such circumstances, it was clearly observed that the amorphous phases existed in the PVA/MMT composite membrane. Therefore, the PVA/MMT composite membranes show good ionic conductivity for the higher concentration of nanoparticles. This is due to the more amorphous phases and more flexible of local PVA chain segmental motion in the PVA/MMT composite membranes.

3.3. Water /Methanol uptake analysis

The composite polymer membranes were kept at room temperature and weighed accurately with high accuracy electronic weighing balance. The selected membranes were taken for water and methanol uptake measurement. The uptake measurements of samples were calculated for different timing like 30, 60,120,180,240,300 and 600 seconds. The percentage of water uptake have been calculated by an equation

$$WA\% = \frac{W_{wet} - W_{dry}}{W_{wet}} \times 100 \qquad (1)$$

where, W_{wet} is the weight of the wet membrane; W_{dry} is the weight of the dry membrane.

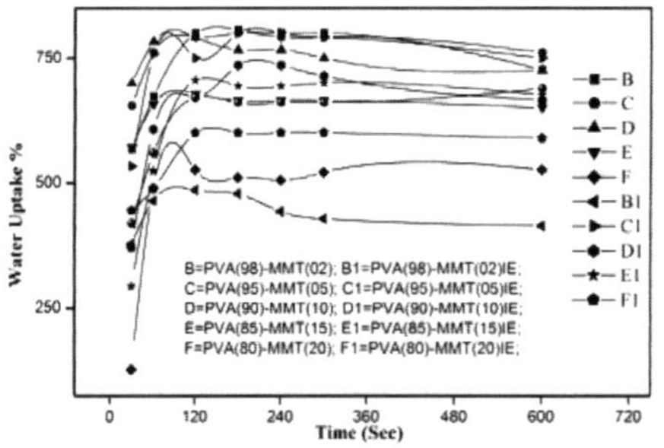

Figure 3. Variation of water uptake of PEMs as a function of time.

Figure 3 shows that, the percentage of water uptake with respect to time. The concentration of MMT increases the hydrophilicity of the polymer membranes decreases in weight percentage. Because, protonated MMT to enhance acidity and hydrophilicity. Further, the hydrophilic nature of the membranes depends on the number of available protonated particle in acid groups and their dissociation capability in water. It is observed that the low concentration of MMT leads to absorb more water molecule so that the weight percentage of membrane shows good affinity with water.

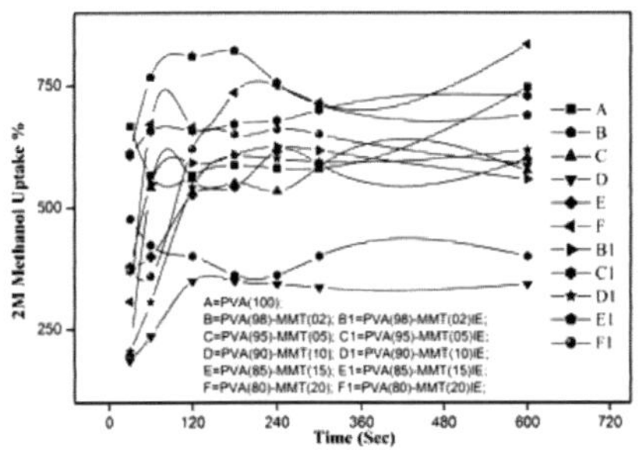

Figure 4. Variation of 2M% methanol uptake of PEMs as a function of time.

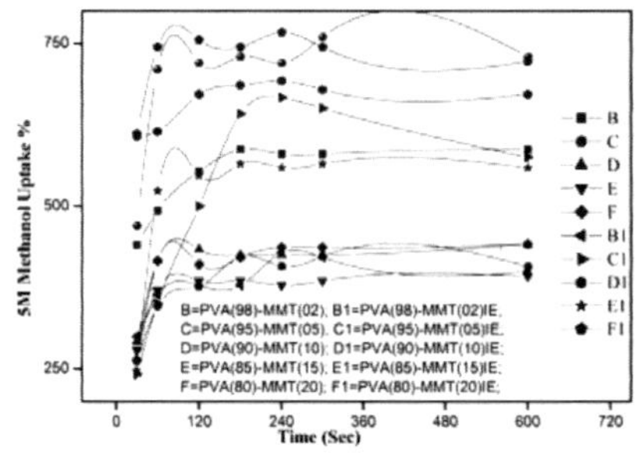

Figure 5. Variation of 5M% methanol uptake of PEMs as a function of time.

Similarly, the percentages of methanol uptake of PVA/MMT composite membranes are shown in Figure (4-5) for different molar concentrations like 2M and 5M. From these, by increasing amount of MMT concentration, the methanol uptakes of the composite polymer membranes were decreased. In different concentration of methanol solution, PVA (90)-MMT (10) with protonated membrane shows lower methanol uptake. After 300 seconds the methanol uptake gets saturated for all the prepared samples. Thus, the results show that, the protonated membranes are suitable for DMFC applications.

3.4. Proton Conductivity analysis

The conductivity measurements were made for the PVA/MMT composite polymer electrolyte membranes by an AC impedance Analyzer. The composite polymer membranes were immersed in 5M, 8M, and 10M of H_2SO_4 solutions for at least 24 hrs at room temperature before measuring the conductivity measurement. The PVA/MMT composite polymer electrolyte membranes were placed between stainless steel, with surface area about 2.99 cm^2. The proton conductivity of composite polymer electrolyte membranes was recorded at room temperature. The proton conductivity (σ) is given by

$$\sigma = \frac{l}{Rdw} \quad (2)$$

where, l is the distance between the two stainless steel electrodes; d and w are the thickness and width of composite PEMs; R is the measured resistance of the membranes.

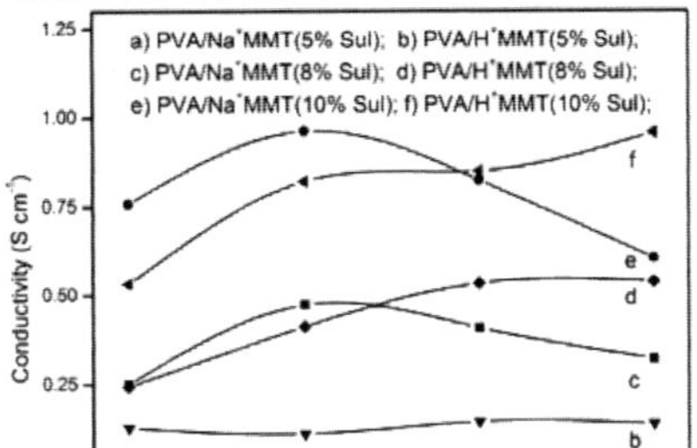

Figure 6. variation of proton conductivity as a function of MMT concentration.

In Figure 6 the MMT concentration vs. conductivity of PVA/Na^{+}MMT polymer membranes without sulfonation and with sulfonation (5M, 8M, 10M). It is observed that the MMT concentration increases with PVA, the conductivity values initially increased upto 10% of MMT, further it is decreased in conductivity at room temperature (Table.1). The proton conduction occur by the hoping of protons because of an acid-base complex is formed between the acidic sulfonic groups of the polymer electrolyte and basic group of the MMT compound. On the other hand, this mechanism is disfavored by the increase of MMT content in the membrane because more basic than acidic groups are present at the composite membranes without ion exchange. As a result PEMs became anionic.

Table 1. Proton conductivities of PEMs at varied amount of Na^{+}MMT and H^{+}MMT

	Conductivity S/Cm							
	without Ion Exchange resin				with Ion Exchange resin			
concentration of MMT	PVA	5% Sul-PVA	8% Sul-PVA	10% sul-PVA	PVA	5% Sul-PVA	8% Sul-PVA	10% sul-PVA
5	0.0019	0.0096	0.2510	0.7587	0.0017	0.1297	0.2439	0.5326
10	0.0022	0.1053	0.4730	0.9620	0.0021	0.1106	0.4114	0.8207
15	0.0017	0.1816	0.4071	0.8240	0.0016	0.1437	0.5322	0.9417
20	0.0027	0.1735	0.3203	0.6053	0.0006	0.1376	0.5371	0.9572

The conductivity of PVA/H^{+}MMT composite membranes with different molar concentration of sulfonation is another beneficial effect. It is noticed that the increasing amount of MMT fillers as well as the conductivity also increases. This may be due to protonated MMT particle with sulfonic acid groups. The increase in Proton conductivity wrap up by addition of H^{+}MMT fillers, is most likely due to the beneficial effect of the acidic $-SO_3H$ terminal groups. From these above observations it is noticed that the proton conductivity value of 0.9572 S cm^{-1}.

4. Conclusion

The Protonated and non-protonated MMT with PVA Membranes were carried out. The proton conductivity value around 10-1S/Cm of PVA (80)-MMT (20) obtained at room temperature with 100% humidity. In the structural studies, we obtained amorphous phase in the higher concentration of H+MMT. It shows that, the MMT fillers are dispersed well in the polymer matrix. The above conductivity result is higher than that of Yang et al. reported for the same material.

References

[1] Tricolo V, Carretta N, Bartolozzi M 2000 J. Electrochem. Soc. **147** 1286
[2] Kreuer K D 1996 Chem. Mater. **8** 610
[3] Shao Z G, Wang X, Hsing I M 2002 J. Membr. Sci. **210** 147
[4] Son J H, Kang Y S, Won J 2006 J.Membr. Sci. **281** 345-350
[5] Yang C C, Chiu S J, Kuo S C 2011 Curr. Appl Phys. **11** S229-S237
[6] Kadir M F Z, Aspanut Z, Majid S R, Arof A K 2011 Spectrochim. Acta, Part A **78** 1068-1074
[7] Yang C C, Wu G M 2009 Mater. Chem. Phys. **114** 948
[8] Yang C C, Chiu S J, Chien W C 2006 J. Power Sources **162** 21
[9] Mansur H S, Sadahira C M, Souza A N, Mansur A A P 2008 Mater. Sci. Eng., C **28** 539-548
[10] Mansur H S, Costa H S, 2008 Chem. Eng. J. **137** 72-83

Preparation and characterization of perovskite Cu doped $LaFeO_3$ semiconductor ceramics

Bandi Vittal Prasad[1], B. Venugopal Rao[1], K. Narsaiah[1], G. Narsinga Rao[2] , J.W.Chen[2] and D.Suresh Babu[1,*]

[1]Department of Physics, Nizam College (Osmania University), Hyderabad, India.
[2]Department of Physics, National Taiwan University, Taipei, Taiwan, R. O. C.,

[*] Corresponding author E-mail: s_devarasetty1956@yahoo.co.uk

Abstract. Orthorhombic structured polycrystalline Cu doped $LaFeO_3$ perovskite ceramics were prepared by sol-gel route and characterized by X-ray diffraction (XRD), differential thermal analysis (DTA), Scanning electron microscopy (SEM) and Fourier transform infrared spectroscopic (FT-IR) techniques. The complex dielectric properties for these ceramic samples investigated as function of temperature, in the range 80K - 300K, and frequency, in the range 20 Hz -1MHz. These samples exhibiting colossal dielectric constant value of 10^4 near room temperature at low frequencies. The response is similar to that observed for relaxor ferroelectrics. Impedance spectroscopy (IS) data analysis indicates the ceramics to be electrically heterogeneous semiconductors consisting of semiconducting grains with dielectric constant below 100. We conclude, therefore that grain boundary effects is the primary source for the high effective dielectric constant in Cu doped $LaFeO_3$ ceramics.

[1] Introduction

Materials exhibiting colossal values of the dielectric constant are the subject of intense research today owing to their ability to reduce the size of capacitive components, offering an opportunity to miniaturize electronic systems. In general, high intrinsic dielectric constant ϵ' ($_n$ ' >10^3) is exhibited exclusively by ferroelectric and relaxorferroelectrics. In recent years, materials exhibiting a so called giant dielectric constant have been reported [1-8]. One of the giant dielectric constant (ϵ') materials $CaCu_3Ti_4O12$ (CCTO) has recently attracted much attention. Right from the time this effect was first observed in CCTO, attention has been concentrated on materials with perovskite structure (ABO_3), due to the set of specific properties belonged to these materials.

Cu doped $LaFeO_3$ (LFCO) materials are semi conductive in nature at room temperature and conductive at high temperatures. In this paper we report details concerning the dielectric response of LFCO as a function of frequency and temperature along with the possible underlying mechanisms.

[2] Experimental

Polycrystalline powders of Cu doped $LaFeO_3$ (LFCO) were prepared by the citric acid route. Collected powders were annealed at 700 ^{0}C for 2 h. The prepared pellets were sintered at 1000 ^{0}C (1200 for $LaFeO_3$ sample) for 4 h.

IOP Conf. Series: Materials Science and Engineering **73** (2015) 012129 doi:10.1088/1757-899X/73/1/012129

[3] Results and discussions

The XRD patterns obtained at room temperature of the polycrystalline LFCO synthesized by citric acid method are shown in Figure 1. It is observed that the XRD spectrum is of a single phase LFCO. All the peaks could be indexed to orthorhombic cell associated with space group Pbnm. The lattice parameters were given in the Table1.

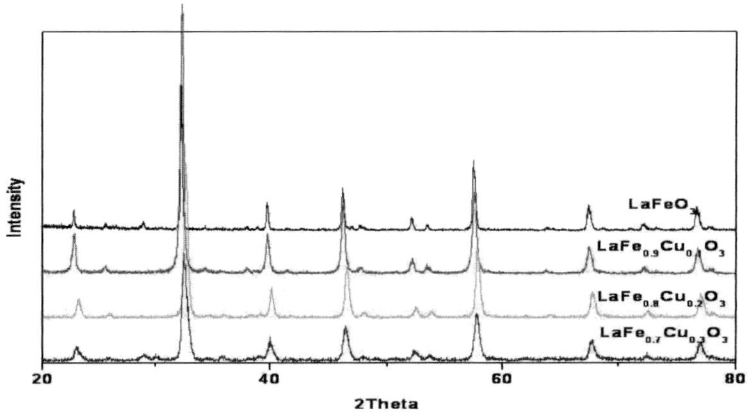

Figure1. The XRD patterns obtained at room temperature of the $LaFe_{1-x}Cu_xO_3$ (x= 0, 0.1, 0.2, 0.3) ceramics

Table 1. Lattice parameters of $LaFe_{1-x}Cu_xO_3$ (x= 0, 0.1, 0.2, 0.3)

	a (A^0)	**b** (A^0)	**c** (A^0)
$LaFeO_3$	5.576(3)	7.857(3)	5.550(2)
$LaFe_{0.9}Cu_{0.1}O_3$	5.554(4)	7.854(4)	5.554(2)
$LaFe_{0.8}Cu_{0.2}O_3$	5.567(3)	7.869(3)	5.563(2)
$LaFe_{0.7}Cu_{0.3}O_3$	5.571(4)	7.862(4)	5.560(2)

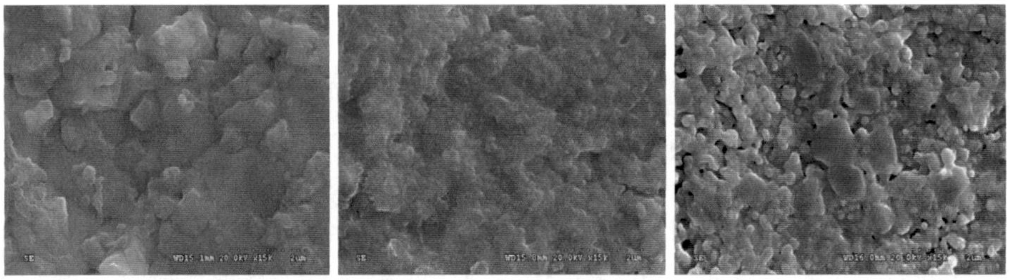

Figure 2. SEM micrograph of the $LaFe_{1-x}Cu_xO_3$ (x= 0, 0.1, 0.2, 0.3) ceramics pellets

IOP Conf. Series: Materials Science and Engineering **73** (2015) 012129 doi:10.1088/1757-899X/73/1/012129

SEM micrograph of the sintered LFCO pellets in air shown in Figure 2. The micrograph reveals that the pellet consists of small grains with considerable porosity.

The deferential thermal analysis has been performed for all the dried gel. The DTA curves for LFCO compounds are shown in Figure 3(a). Strong exothermic peaks on DTA curves are observed, which are due to vigorous combustion reactions of the organic compounds, where maximum weight loss (70%) seen in TG. Weight loss is also observed between 600 ^{0}C and 700 ^{0}C which corresponds to the exothermic peaks in DTA curves it corresponds to the decomposition of an amorphous Oxyhydroxycarbonate. Beyond this temperature range, the curves tend to be very flat, showing no intense decomposition.

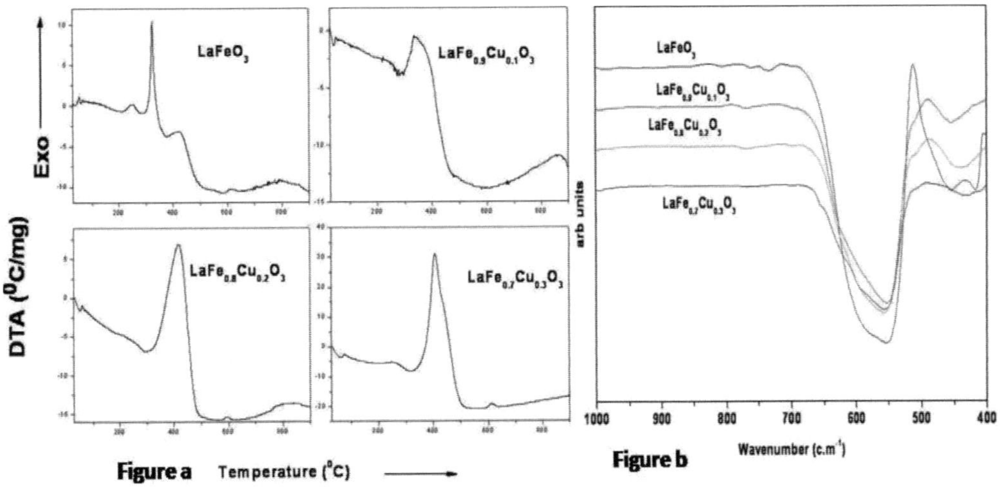

Figure 3. The DTA(Figure a.) and FTIR(Figure b.) curves of the $LaFe_{1-x}Cu_xO_3$ (x = 0, 0.1, 0.2, 0.3) ceremics

The Crystal structure of the rare earth ferrites is known to be of the distorted $GdFeO_3$ – type structure The two infrared-active vibrations are shown in the Figure 3(b). The bands around 556 cm^{-1} and 435 cm^{-1} corresponds to stretching mode v_s, which involves the internal motion of the Fe-O bond and the bending mode v_b, which is sensitive to a change in the Fe-O-Fe bond angle

The temperature dependence of the dielectric constant (ϵ') and dielectric loss (tan δ) for $LaFe_{1-x}Cu_xO_3$ (x = 0, 0.2, 0.3) samples are shown in Figure 4. The room temperature dielectric constant attains a value ~ 10^4 at 1 kHz. Two ϵ' plateaus at higher and low temperature are clearly seen Figure 4. Between two plateaus the dielectric constant (ϵ') increases sharply by a factor of ~100, accompanied by a peak in the dielectric loss (tan δ) at the temperature T_P. The peak positions (value of T_P) shifted to high temperatures with increase in Cu composition, and with increase in measuring frequency. It can also be seen that there is an exponential like increasing background in tan δ curve.

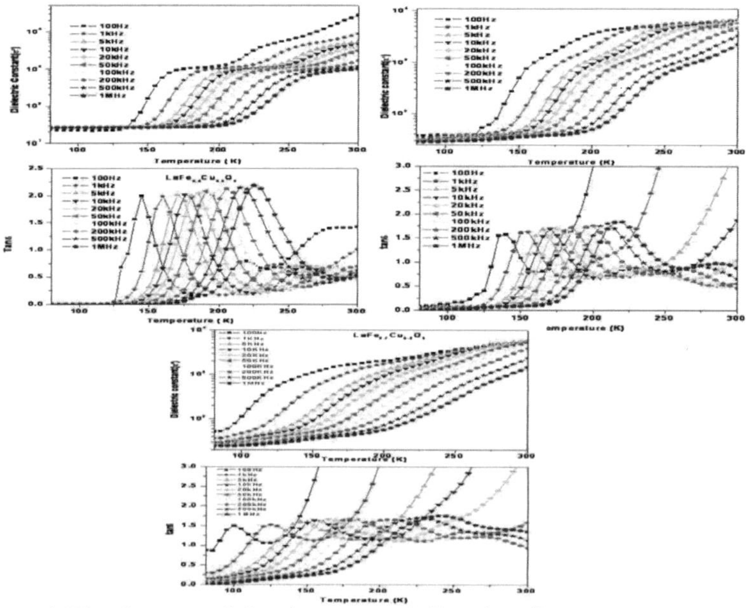

Figure 4. Plots between dielectric constant (ε') and tanδ versus temperature for $LaFe_{1-x}Cu_xO_3$ (for x= 0, 0.2, 0.3)

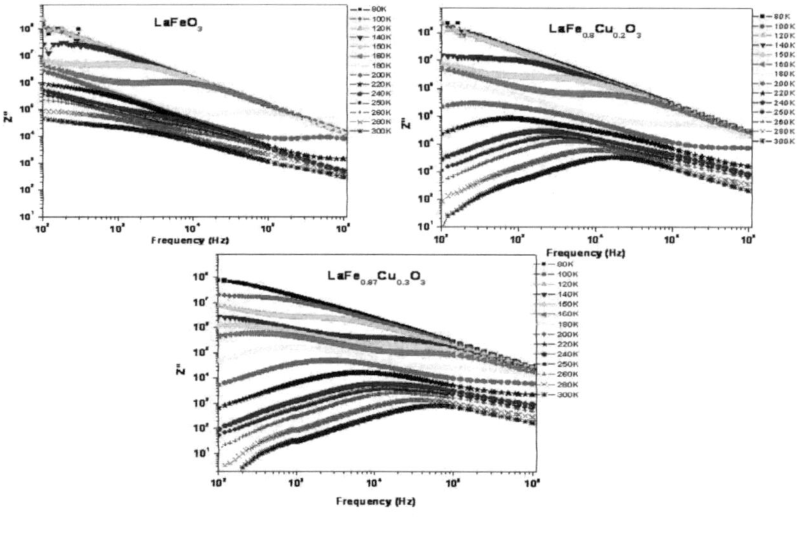

Figure 5. Imaginary part of impedance versus frequency at different temperatures for the $LaFe_{1-x}Cu_xO_3$ (for x= 0, 0.2, 0.3) ceramic samples

To clarify reason for the high dielectric constant we performed impedance analysis. Figure 5. shows Z'' vs frequency for different temperatures. The higher values of Z'' at lower frequencies and at lower temperatures and the merging of Z'' curves at higher frequency end indicate the presence of space charge polarization in these samples. we can observe a weak peak at low frequencies and the peak are shifted to higher frequencies as temperature increases and the peak are moved out of measuring window as temperature is more then 200K. As temperature increases further a strong peak is observed at lower frequencies for LC02 and LC03 samples which is moving to higher frequencies as temperature increases with decrease in intensity, indicating the both electric responses are thermally activated. The broadness of the peak indicates presence of multiple relaxations in the material [9-10]. Peak width at half maxima of Z'' curves is more than a decade (> 1.14 decades) indicating non-Debye relaxation behavior.

1. Conclusions

$LaFeO_3$ samples with Cu doping are prepared through chemical route and single-phase formation is confirmed by XRD. SEM micrographs pour size and number of pours increases with Cu and grain size decreases. The temperature dependent dielectric constant for $LaFe_{1-x}Cu_xO3$ (x = 0, 0.2, 0.3) resembles to that of CaCu3Ti4O12, widely studied in recent year. The observed temperature and frequency dependent high dielectric constant in semiconducting $LaFe_{1-x}Cu_xO3$ (x = 0, 0.2, 0.3) ceramics showing non Debye relaxation and is dominated by grain boundary response.

References

[4] M.A.Subramanian, D.Li,N.Duan, B.A.Reisner, and A. W. Sleight, 2000 *J.Solid State Chem.* **151** 323.
[5] T. B. Adams, D. C. Sinclair, and A. R. West, 2002 *Adv. Mater.* (Weinheim, Ger.) **14**, 1321.
[6] D. C. Sinclair, T. B. Adams, F. D. Morrison, and A. R. West 2002, *Appl. Phys. Lett.* **80**, 2153.
[7] M. Li, A. Feteira, and D. C. Sinclair 2005, *J. Appl. Phys.* **98**, 084101.
[8] A. P. Ramirez, M. A. Subramanian, M. Gardel, G. Blumberg, D. Li, T. Vogt, and S. M. Shapiro, 2000 *Solid State Commun.* **115**, 217.
[9] Bandi Vittal Prasad, G. Narsinga Rao, J. W. Chen, D.Suresh Babu, 2011 *Mater. Chem. Phy.* **126**, 918.
[10] Bandi Vittal Prasad, G. Narsinga Rao, J. W. Chen, D.Suresh Babu,2011 *Mater. Res. Bull.* **46**, 1670.
[11] Bandi Vittal Prasad, G. Narsinga Rao, J. W. Chen, D.Suresh Babu,2012 *solid state sci.* **14**, 225.
[12] A. K. Jonscher, 1983 *Dielectric Relaxation in Solids*, (Chelsea Dielectrics London).
[13] J. R. Macdonald 1987 *Impendance Spectroscopy* (Wiley New York)

IOP Conf. Series: Materials Science and Engineering **73** (2015) 012130 doi:10.1088/1757-899X/73/1/012130

Physical, Optical absorption and EPR studies on fluoro-bismuthate glasses

Ch. Srinivasu[a], M.A. Samee[b], A. Edukondalu[b], C. Laxmi kanth[c] and Syed Rahman[b]*

[a]Department of Physics & Chemistry, MGIT, Gandipet, Hyderabad - 500075

[b]Department of Physics, Osmania University, Hyderabad, INDIA-500 007

[c]Shool of Physical sciences, University of Dodoma, Tanzania

Email: cnumgit14@gmail.com

ABSTRACT: Glasses of the xLiF-(50-x)Li2O-20SrO-30Bi2O3 system, with $0 \leq x \leq 20$ mole % were studied by EPR and Optical measurements. The changes in both density and molar volume indicate structural modifications occur due to addition of LiF. The glass transition temperatures are observed to decrease with an increase in LiF content in the compositions. The local structure around Cu^{2+} ions has been examined by means of electron paramagnetic resonance and optical absorption measurements. It is observed that the spin-Hamiltonian parameters calculated from the EPR spectra are influenced by the glass composition. The Cu^{2+} ions are in well-defined axial sites but subjected to small distortion leading to the broadening of the spectra. The spin-Hamiltonian parameter values indicate that the ground state of Cu2+ is and the site symmetry around Cu2+ ions is tetragonally distorted octahedral. The optical absorption spectra exhibited a broad band corresponding to the d-d transition bands of Cu2+ ion. By correlating EPR and optical absorption data, the bond parameters are evaluated.

1. INTRODUCTION

Bismuthate glasses have been the subject of renewed interest because of their manifold possible applications [1, 2]. These glasses were found to be an efficient γ-ray absorber [3] and also considered for use in scintillation detectors for high energy physics [4, 5]. Studies of EPR and optical properties of transition metal ions in glasses have made it possible not only to interpret the energy levels involved in the observed transitions, but also to comment on the chemical and structural environments about the metal ion center. The presence of transition metal ions (TMIs) in glasses is detected by optical absorption and electron paramagnetic resonance (EPR) studies by identifying the relevant absorption band positions due to TMIs. Copper can be present in glasses as the Cu^{2+} ion (d^9), the Cu^+ ion (d^{10}), and under some circumstances as metallic Cu0, the latter occurring in the so-called copper-ruby glass as a colloidal suspension [7]. The data obtained from EPR and optical absorption can be correlated to obtain information regarding the bond parameters which determine the metal-ligand bond nature in the glasses. There are reports from our laboratory regarding EPR and optical absorption studies of Cu2+ containing heavy metal oxide glasses [8, 9].

2. EXPERIMENTAL

Pure and 1 mole % copper doped xLiF-(50-x) Li2O-20SrO-30Bi2O3 glasses, ($0 \leq x \leq 20$) are prepared by conventional melt quench technique. Details of the preparation and characterizationof the present glassesare given in ref. [10]. The glass transition temperature, Tg, was measured in all samples using a temperature modulated differential scanning calorimeter (TA Instruments, DSC 2910). All samples were heated at the standard rate of 10 oC min-1 in aluminum pans. The optical absorption spectra of the present glass samples were recorded at room temperature using a double beam Shimadzu spectrometer (model UV-3100) in the wavelength range 400-800 nm. The uncertainty in the observed wave length is about ±1 nm. The room temperature EPR spectra of copper doped powdered glass samples were recorded using a JEOL-1X- EPR spectrometer in the range 2200 G – 4200 G operating in the X-band and employing a field modulation of 100 kHz. DPPH was used as the standard *g* marker for the determination of magnetic field.

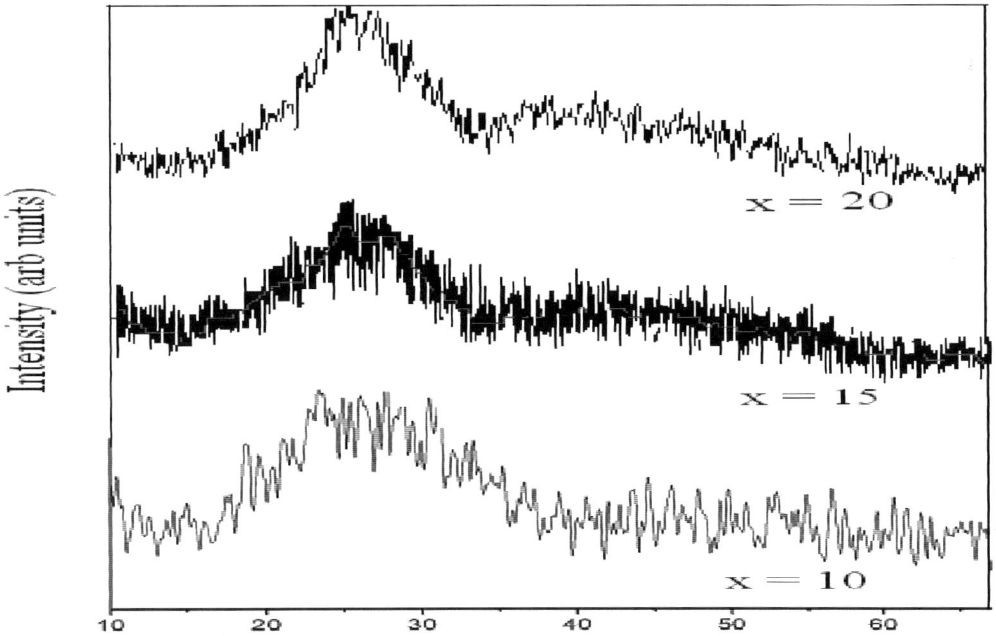

Fig. 1. X-ray diffraction pattern for different glasses xLiF-(50-x)Li2O-20SrO-30Bi2O3 system.

IOP Conf. Series: Materials Science and Engineering **73** (2015) 012130 doi:10.1088/1757-899X/73/1/012130

3. RESULTS & DISCUSSIONS

3.1 Physical properties

The amorphous nature of all the samples was confirmed by the absence of Bragg's peak in X-ray diffraction pattern as shown in figure 1. The variation of measured densities (ρ) of the present glasses along with evaluated values of molar volume (V_m) as a function of LiF. It is clear that the density of these glasses increases where as molar volume decreases with increase in LiF content. Since the fluorine ion has nearly the same radius (0.136 nm) as that of oxygen ion (0.140 nm), one can expect that the probability of fluorine ions to substitute oxygen is much higher or it may occupy interstitial position.

3.2 Optical absorption

Optical absorption spectra of the pure glasses are presented in figure 2. A distinct cutoff (λ_c) was observed for each glass sample. It can be seen that λ_c decreases with LiF content.

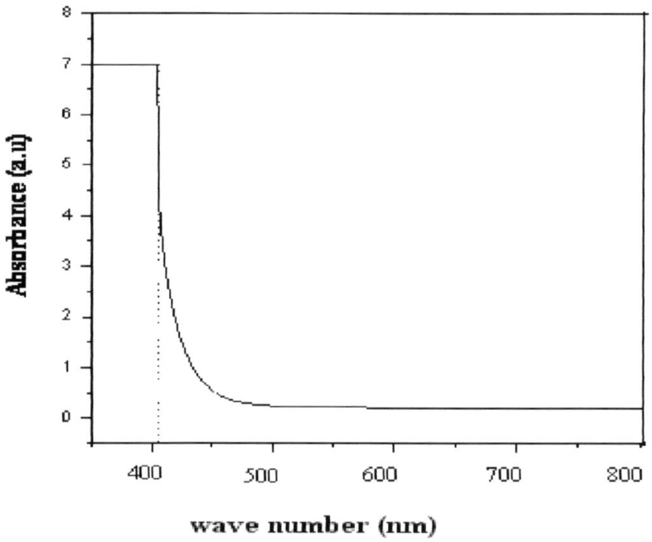

Fig. 2 Optical absorption spectra of a typical glass sample.

3.3. EPR and Optical absorption spectra

The EPR spectra of 1 mole% of Cu^{2+} doped samples is shown in figure 3. The spectra closely resembles that of Cu^{2+} in most oxide glasses, which can easily be recognized on the basis of four-line hyperfine splitting due to ^{63}Cu, ^{62}Cu (I=3/2), but the isotope splittings were not resolved owing to nearly identical nuclear moments. The hyperfine features were observed on the parallel of the spectrum where as the fourth line must have been overlapped

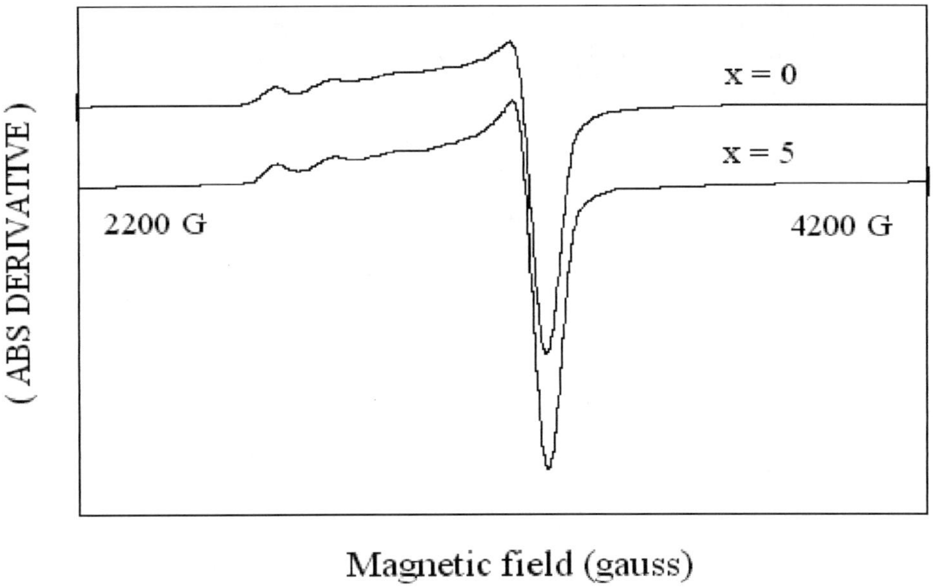

Fig. 3 EPR spectra of 1 mole% of Cu2+ doped glasses.

on perpendicular features of the spectrum. The spectra were analysed using the spin-Hamiltonian

$$H = g_{||}\beta H_z S_z + g_{\perp}\beta (H_x S_x + H_y S_y) + A_{||} I_z S_z + A_{\perp}(I_x S_x + I_y S_y) \quad (1)$$

where the symbols have their usual meaning. The values of $A_{||}$ and $A_{\perp}$ were calculated using [11] the following equations

$$H_{||}(-3/2)-H_{||}(+3/2) = 3A_{||} \quad \text{and} \quad H_{\perp}(-3/2)-H_{\perp}(+3/2) = 3A_{\perp} \quad (2)$$

The spin-Hamiltonian parameters are presented in table 1. From the *g* values it is observed that $g_{||} > g_{\perp} > 2.0023$ (=g_e). Therefore it is concluded that Cu^{2+} has an octahedral environment along one of the cube axis, and the ground state of Cu^{2+} is $d_{x^2-y^2}$. The anisotropic hyperfine structure is due to John-Teller distortion [11]. The changes in the spin-Hamiltonian parameters with varying concentration of lithium ions are due to the change in electron cloud density at the Cu^{2+} ion as the Li_2O enters the glass network. The optical absorption spectra of 1 mole % Cu^{2+} doped samples is shown in figure 4. These spectra show a single optical absorption bands which corresponds to Cu^{2+} and are influenced by host structure. The spectra associated with the Cu^{2+} ion are *d–d* transitions which can be described in terms of the ligand field theory [12]. In a regular octahedral field, the $3d^9$ configuration would result in a degenerate ground state (^2E_g). In glasses it is assumed that due to disordered vitreous state no site is perfectly cubic. Therefore, tetragonal distortions are endemic to the vitreous state which leads to splitting of energy levels. It is observed that the elongated structures are usually more energetically favoured than the compressed ones [13]. But only a single optical absorption maximum was observed in most of the cases [14, 15]. This single optical band was interpreted as the overlap of all the three transitions. Hence in the present investigation the observed asymmetric band around 13,907 cm^{-1} is due to the overlap of $^2B_{1g} \rightarrow ^2A_{1g}$ and $^2B_{1g} \rightarrow ^2B_{2g}$ transitions. Most of the authors [16, 17] assigned the observed optical peak to the $^2B_{1g} \rightarrow ^2B_{2g}$ transition (ΔE_{xy}) and have used this value in the evaluation of the bond parameters.

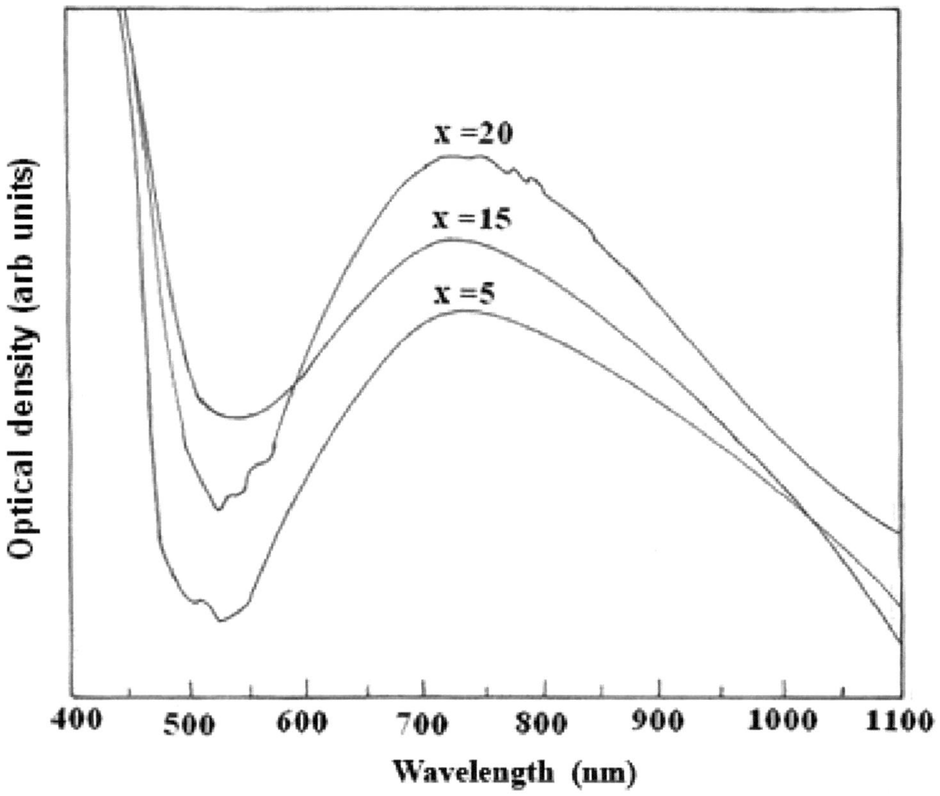

Fig. 4 Optical absorption spectra of LiF-Li2O-SrO-Bi2O3 glasses.

4. Conclusions

The density of these glasses increases where as molar volume decreases with increase in LiF With the increase of LiF content the optical band gap energy decreases.From the observed values of g|| > g⊥ > ge (2.0023) it is concluded that the ground state of Cu2+ ions in all the samples under present study is orbital (2B1g state) and the site symmetry around Cu2+ ion is tetragonally distorted octahedral sites. The values of α^2 and $\beta1^2$ indicate that the in-plane sigma bonding is covalent where as in-plane and out-of-plane π bonding are ionic in nature.

REFERENCES

[1] Oprea I, Hesse H and Betzler K, 2004 *Opt. Mater.* 26 235.
[2] Eldin Metwalli E, 2003 *J. non-Cryst. Solids* 317 221.
[3] Bishay A and Maghrabi C, 1969 *Phys. Chem. Glasses*, 10(1) 1.
[4] Edukondalu A , Purnima M, Srinivasu Ch, Sripathi T, Awasthi A M, Syed Rahman and Siva Kumar K 2012 *J. Non-Cryst. Solids* 358 2581–2588.
[5] Edukondalu A, Sathe V, Syed Rahman, Siva Kumar K and Sreenivasu D 2014 *Vibrational Spectroscopy* 71 91–97.
[6] Edukondalu A, Sathe V, Syed Rahman and Siva Kumar K 2014 *Physica B* 438 120.
[7] Tauc J and Menth A 1972 *J. Non-Cryst. Solids* 8 569.
[8] Edukondalu A, Kavitha B, Samee M A, Shaik Kareem ahmmad, Syed Rahman and Siva kumar K 2013 *J. Alloys Compd.* 322 324 -335.
[9] Srinivasa Rao, Purnima M, Bale S, Siva Kumar K and Syed Rahman, 2005 *Bull. Mater. Sci.* 28 589.
[10] Srinivasa Rao, Purnima M, Bale S, Siva Kumar K and Syed Rahman, 2006 *Bull. Mater. Sci.* 29 365.
[11] Bale S and Syed Rahman, 2008 *Opt. Mater.* 31 333.
[12] Bates T, Mackenzie J D, 1962 *Modren aspects of the vitreous state* 2 195.
[13] Jorgensen C K, 1955 *Acta Chem. Scand.* 9 1362.
[14] Orgel E L, 1955 *J. Chem. Phys* 23 1004.
[15] C.J. Ballhausen, 1962 *Introduction to Ligand Field Theory, McGraw-Hill, NY,.*
[16] A. H. Maki, B. R. Mc Garvey, J. Chem. Phys. 29 (1958) 31.
[17] Kivelson D and Neiman R, 1961 *J. Chem. Phys* 35 145.

Structural and Optical properties of HNO_3 etched Ag thin films

Desapogu Rajesh and C S Sunandana

School of Physics, University of Hyderabad

sunandana@gmail.com; rajesh.desapogu@gmail.com

Abstract. Structural and optical properties, elemental analysis of Ag thin films before and after etching were investigate by using XRD, UV-Vis spectrophotometer, Raman spectroscopy and FESEM. These films were found to exhibit a strong SERS effect and excellent thermal stability. The HNO_3 roughened silver film seems to be a more routinely preparable substrate system for SERS study. The HNO_3 roughened Ag thin films exhibits excellent stability for enhancement, because the acid roughened Ag films features a strong enhancement factor better stability than pure Ag thin films. Hence, the method will be useful in the development of plasmon-based analytical devices, specifically SERS-based biosensors.

1. Introduction

Surface-enhanced Raman scattering (SERS) is a sensitive tool for exploring metal adsorbate interactions and reactivity of adsorbed species [1-10]. Since Fleischmann et al.'s initial study [1], SERS has been subjected to extensively theoretical and experimental investigations. Surface Plasmon resonance, which is associated with collective electron resonance induced by incident light on a rough metal surface, is one of the electromagnetic effects contributing largely to the SERS phenomenon. In addition to their easy preparation and chemical stability, this kind of monolayer can find practical applications such as in the area of highly resistant surface modifications. It also provides insight into the structure and molecular packing of monolayer and multilayer organic films that is now a growing area of technique and materials sciences [11]. Knowledge of the structure of the thin films is the key to understanding the relationships between microscopic structure and macroscopic physical and chemical properties. Surface-enhanced Raman scattering (SERS), is a important analytical tool which allows the study of molecules adsorbed on a nanostructured rough surface or nanoparticles of noble (Ag, Cu, Au) metals [12,13] SERS is widely used for obtaining information about the vibrational spectrum of different chemical and biological substances. Development of the SERS technique provides the possibility to obtain Raman spectra from a very low amount of substance, even from a single molecule [14–18]. The SERS effect leads to a large enhancement factor relative to normal Raman scattering. The enhancement factor depends not only on the particular noble metal and excitation wavelength, but also on the size and shape of the nanostructures and their spatial locations. The method of metal nanostructure synthesis is crucial for these parameters. The SERS effect relates to the localized surface plasmon resonance (SPR) of metal nanoparticles, which is manifested in the absorption spectrum as an intensive and broad band. The SERS-active substrates for broad applications in the field of molecular detection and characterization. Surface enhancement depends critically on the creation of surface roughness, the HNO_3 etched roughened silver on which the remarkable phenomena. Herein we aim to report a study and optical properties on the etching rate dependence of the surface-induced morphology of silver thin films.

2. Experimental part

Series of Ag thin films have been obtained by thermal evaporation in a high-vacuum chamber with base pressure 10^{-6} mbar. The silver wire used for evaporation was purchased from Aldrich Chemical (99.9% pure) and used as such. The substrates were borosilicate glass for structural, optical studies. All thin films were grown at room temperature (25 ^{0}C) and low deposition rates (<0.01 nm/Sec). The studied Ag thin films samples were 60nm thick on the average. The Ag films etched with diluted HNO3 at different etched times. Silver films of 60nm thickness were immersed into diluted HNO_3 and distilled water (1:10) at room temperature, for about 20sec. This procedure creates a roughness future on a 2 or 3nm scale, as shown by the FESEM graphs. After etching, the silver films was thoroughly rinsed with distilled water and dried in air. This procedure is similar to the etching method for formation mechanism of γ-AgI thin films by sunandana and senthil kumar [19]. The pure, etched Ag films structural properties studied by X-Ray diffraction (XRD) and Field emission scanning electron microscopy (FESEM) techniques were used to evaluate the microstructure of the samples. The optical absorption of Ag thin films was determined by JASCO optical Spectrophotometer. The optical absorption was measured in the UV-Visible range 300nm to 800nm.

3. Results & Discussion

3.1 XRD

From fig1 x-ray diffraction, it was observed that the no crystal phases present in Ag thermal evaporated films but a broad huge peak observed at 30 degrees. Thus 60nm thick Ag films appear by and large quasi amorphous and it is observed no crystalline phases are present. The quasi-amorphous XRD pattern reflects the nanocrystaline nature of the Ag Nanoparticles. The Ag film was etched with diluted $HONO_3$ for 20 sec; upon which a high intensity broader peak with a larger FWHM was observed. These results indicate increase in the particle size and a decrease in density of particles which are in agreement with FESEM results.

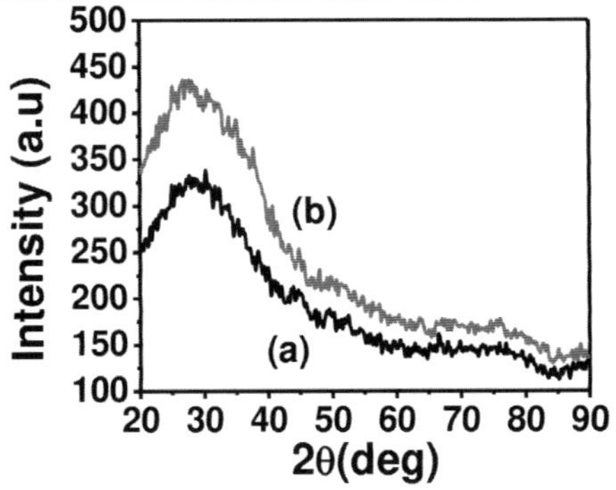

Figure 1. XRD Pattern of 60nm Ag film (a) Pure (b) 20 sec etched Ag film

3.2 Optical properties

UV-VIS absorption spectra have been proved to be quite sensitive to the formation of silver because silver nanoparticles exhibit an intense absorption peak due to the surface plasmon excitation. Fig. 2

shows that the intensity of SPR band and its full width at half maxima (FWHM) increases with etching time. There was no obvious change in peak position for etching, except for the increase of absorbance. These signatures of absorption spectrum are in accordance with the increase size of Ag nanoparticles. It is obvious that with the progress of etching reaction more and more Ag+ are reduced and this results in the increase size of silver nanocrystals through diffusion controlled mechanism. This tremendous increment in absorption spectrum useful to chemical sensors, detectors as well as some sensitive spectroscopic measurements. With the increasing size of Ag nanoparticles, multiple transitions of surface plasmon become more prominent and this is reflected on the increase in absorption intensity of etched Ag thin films. These broad (FWHM) SPR peaks mainly depend on the etching rate, thickness of the film, dielectric constant and preparation of the sample. One major reason for SPR broadening is electron surface scattering, which may be enhanced for very small clusters. For smaller particles, electrons reach the surface faster and scatter quickly, losing the coherence of collective oscillation. In this case thickness, etching rate is playing important role, based on this we are tuning the optical, structural properties.

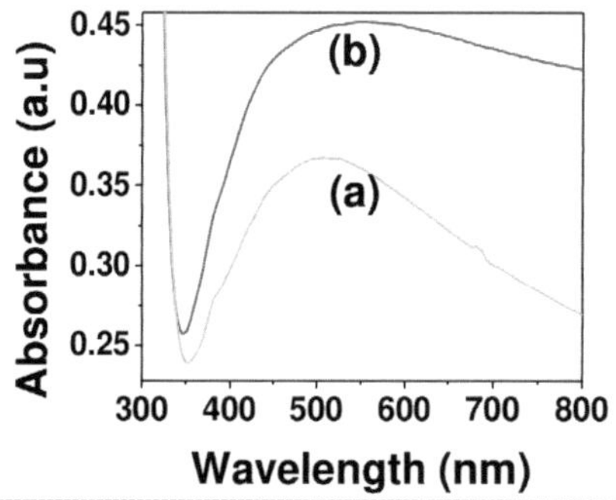

Figure 2. Absorbance spectrum of Ag films (a) Pure Ag (b) Ag film 20 sec etched with diluted HNO_3

3.3 FESEM

The morphology of Ag thin films was examined by FESEM. Figure 3a shows the FESEM image of as deposited Ag thin film have uniform distribution with small nano agglomeration particles. In the case of etched Ag thin films were completely isolated after etching due to their agglomeration; the occurrence of separation, agglomeration on chemically etched silver films in atmosphere can be explained on consideration of the surface diffusion of Ag atoms. The etching rate is quick using strong acid solution because the consumed etchant can be resupplied rapidly in the strong acid. From FESEM results the average Ag nanoclusters sizes for unetched, etched Ag thin films are estimated as 15±5, 60±10nm respectively.

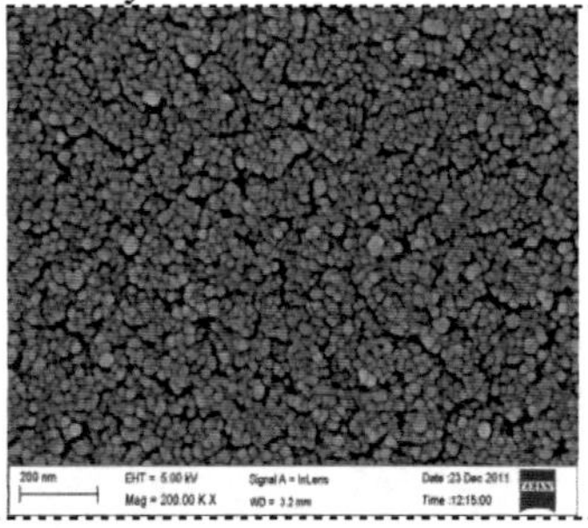

Figure 3.a. FESEM image shows complete agglomeration in 60nm Ag thin films

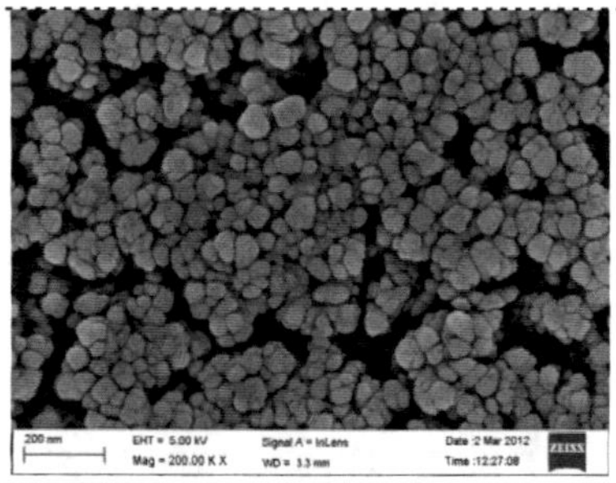

Figure 2. FESEM image of 60nm Ag film etched for 20sec etched with diluted HNO_3 in atmosphere

	showing Iceland separation and agglomeration.

3.4 Raman

A shape and well-resolved Raman line was observed at 1548cm^{-1} in unetched film which upon HNO_3 etching became sharper and more intense and also shifted to 1587 cm^{-1}, charactaristics of Ag [20]. An excellent SERS spectrum was obtained from HNO_3 – roughened Ag films as shown in the fig 4. The HNO3 roughened Ag thin films exhibits excellent stability for enhancement, because the acid roughened Ag films features a string enhancement factor better stability than pure Ag thin films. This preliminary result shows that a careful optimization of film thickness and etching time would lead to SERS quality films. Studies in that direction are in progress. In summary, we found that very stable and optically tunable; SERS-active Ag films can be reproducibly fabricated simply by Ag film etched with diluted HNO_3. Hence, the method will be useful in the development of plasmon-based analytical devices, specifically SERS-based biosensors [21].

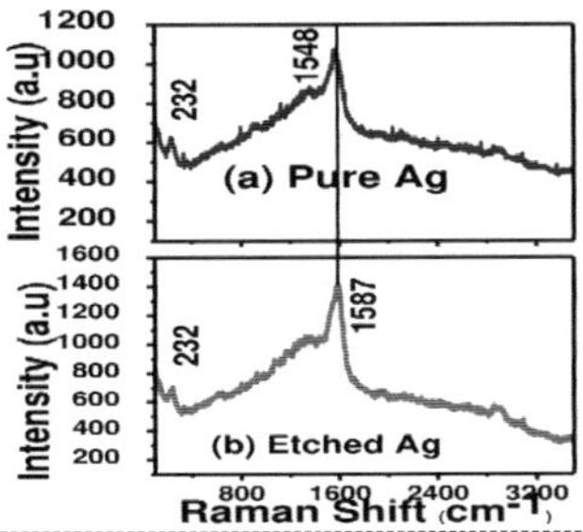

Figure 3. SERS Spectrum of 60nm Ag thin film (a) Pure Ag (b) HNO_3 etched Ag film

4. Conclusion

Structural and optical properties, elemental analysis of Ag thin films before and after etching were investigate by using XRD, UV-Vis spectrophotometer and FESEM, Raman. UV-Vis spectra exhibited Surface Plasmon absorption characteristic of thin films and nano particle ensembles, respectively. Based on the etching mechanism, the etching rate should be change the morphology of the silver thin films was observed by using FESM. Thermal evaporated Ag films were found to exhibit a strong SERS effect and excellent thermal stability. The HNO_3 roughened silver film seems to be a more routinely preparable substrate system for SERS study. The HNO_3 roughened Ag thin films exhibits excellent stability for enhancement, because the acid roughened Ag films features a strong enhancement factor better stability than pure Ag thin films.

Acknowledgement:

DR thanks to University of Hyderabad for providing UGC-BSR fellowship.

References:

[1] Fleischmann M, Hendra D J and McQuillan A J 1947 *Chem. Phys. Lett.* **26** 163
[2] Jeanmaire D L and Van Duyne P R 1977 J. *Electroanal. Chem.* **84** 1
[3] Albrecht M G and Creighton J A 1977 J. *Am. Chem. Soc.* **99** 5215

[4] Van Duyne P R and Moore C B (Ed.), *Chemical and Biochemical Applications of Lasers,* Academic
Press, New York, 1979, **4** 101
[5] Furtak T E and Chang R K (Eds.) 1982 *Surface Enhanced Raman Scattering, Plenum Press*, New York,
[6] Moskovits M, 1985 *Rev. Mod. Phys.* **57** 783
[7] Otto A, Mrozek I, Grabhorn H and Akemann W 1992 *J. Phys.: Condens. Matter* **4** 1143.
[8] Vo-Dinh T and Halevi P (Ed.) 1995 *Photonic Probes of Surfaces*, Elsevier, Amsterdam, **1** 67.
[9] Cotton T M, Clark R J H and Hester R.E (Eds.) 1993 *Spectroscopy of Surface*, Wiley, Chichester, 91.
[10] Campion A 1998 *Chem. Soc. Rev.* **4** 241.
[11] Swalen J D, Allara D L, Andrade J D, Chandross E A and Garoff S 1987 Langmuir **3** 932.
[12] Moskovits M 2005 *J. Raman Spectrosc.* **36** 485
[13] Kneipp H, Kneipp K, Itzkan I, Dasari R R and Feld M S 2002 *J. Phys.: Condens. Matter.* ***14,*** R597.
[14] Leopold N and Lendl B 2003 *J. Phys. Chem. B* **107** 5723
[15] Wang Y, Chen H, Dong S and Wang E 2007 *J. Raman Spectrosc.* **38** 515
[16] Kneipp K, Wang Y, Kneipp H, Perelman L T, Itzkan I, Dasari R R and Feld M S 1997 *Phys. Re*V. *Lett.* **78** 1667
[17] Kneipp K, Kneipp H, Kartha V B, Manjharam R and Deinum G 1998 *Phys. Re*V. *E* **57** R6281.
[18] Goulet P J G, Pieczonka N P W and Aroca R F 2005 *J. Raman Spectrosc.* **36**, 574
[19] Senthil Kumar P and Sunandana C S 1999 *Thin Solid films* **357** 111
[20] Yun L, Gi X and Jian D 1993 *Applied Surface Science* **68** 485
[21] Hyoung K P, Jae Keun Y and Kwan K 2006 *Langmuir* **22** 1626

Structural and optical response of mixed phase CdS nanocrystals to Ni^{2+} ions

G Murali, D Amaranatha Reddy, B Poornaprakash, R P Vijayalakshmi* and B K Reddy
Department of Physics, Sri Venkateswara University, Tirupati 517502, India

E-mail: vijayaraguru@gmail.com

Abstract. Cadmium sulfide (CdS) semiconductor nanocrystals doped with Ni^{2+} have been synthesized via a solution-based method and employing 2-mercaptoehonal as a capping agent. The obtained nanocrystals were characterized using X-ray diffraction (XRD) and Diffuse reflectance spectroscopy studies (DRS). XRD results showed that the undoped CdS nanocrystals are in mixed phase of cubic and hexagonal, where as the doped CdS nanocrystals in hexagonal phase. DRS results supported the mixed phase of undoped nanocrystals by showing two absorption edges at 363 nm and 428 nm. The absorption edge at 363 nm can be attributed to cubic phase while the other can be attributed to hexagonal phase of the sample. But for Ni^{2+} doped CdS nanocrystals only one absorption edge at 468 nm was observed. The observed red shift in the band edge of the nickel doped CdS nanocrystals is due to sp-d exchange interactions between the band electrons and the localized d-electrons of the Ni^{2+} ions substituting Cd^{2+} ions in the doped sample.

1. Introduction

Recent progress in the synthesis and characterization of nanocrystals has been driven by the need to understand the novel physical properties of nanocrystals and their potential application in nanoscale electronic and optoelectronic devices [1-6]. A variety of synthetic methods are exist to prepare the nanocrystals [7-13] and in solution based chemical methods a capping agent was added to the reaction medium inorder to avoid agglomeration and to provide surface passivation to the nanocrystals [14].The large surface area of a nanoparticle and its immediate environment, such as capping agents and solvent molecules, have strong effects on the particle properties [15, 16].

Semiconductor nanophosphors, especially sulfides, doped with transition metal ions and rare earth ions, have been studied extensively because of their excellent optical properties. CdS is one of the most important II-VI semiconductors, with a wide band gap of 2.42 eV at room temperature, is used in light-emitting diodes, solar cells, and other optoelectronic devices [17-19]. CdS has three types of crystal structures namely hexagonal wurtzite, cubic zinc blende and high pressure rock-salt phase. Hexagonal phase has been observed in both the bulk and nanocrystalline CdS while cubic and rock-salt phases are observed only in nanocrystalline CdS [20]. The quantum size effect of CdS nanocrystals occurs when the crystallite diameter is comparable to or below 5-6 nm, the size of the exciton [22]

The doping of CdS with suitable transition metal impurities such as Mn, Cu, Ni and Co and rare earth metals such as Er and Tb opens up the possibility of forming a new class of nanophosphors, whose structural and optical properties differ from the corresponding host nanomaterials. In the present paper, we attempt to elucidate the effects of Ni doping on the structural and optical properties of CdS nanocrystals synthesized at a temperature 80 ^{0}C by chemical co-precipitation method.

2. Experimental

For synthesis of CdS:Ni (0 and 2%) nanocrystals, 0.2 M cadmium acetate (Cd $(CH_3COO)_2{\cdot}2H_2O$) solution was mixed with varying concentrations (0 and 2%) of $Ni(CH_3COO)_2$ solution with constant magnetic stirring. Then 0.2 M Na_2S solution was added to the mixture. An appropriate amount of 2-mercaptoethanol (1 ml) was also added to the reaction medium to stabilize the particle surfaces. Then reaction temperature was fixed at 80 ^{0}C and the solution was stirred for about 3 hours. The final product was filtered out and washed with double distilled water several times and then dried at 100 ^{0}C for 12 h, to remove both water and organic capping agent and other by products formed during the reaction process. The dried flakes were crushed in a pestle-motor till very fine powder was obtained.

X-ray diffraction (XRD) studies were carried out to determine the structure of the CdS host and to determine sizes of the crystallites, using a Seifert 3003 TT X-ray Diffractometer, with Cu-Kα radiation with a wavelength of 1.542 A^0. Diffraction patterns were recorded over a range of 15 to 70°. Diffuse reflectance spectra of the samples were recorded at room temperature using Jasco V 570 UV-Vis spectrophotometer in the wavelength range 200–850 nm.

3. Results and discussion

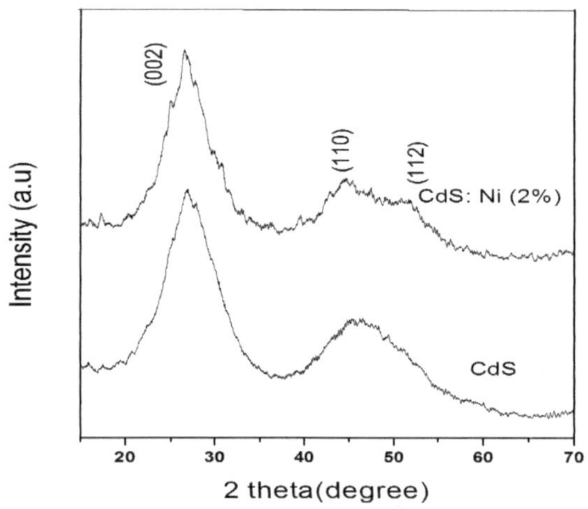

Figure 1. XRD spectra of Undoped and Ni doped CdS nanocrystals

Figure 1 shows the X-ray diffraction patterns of undoped and 2% Ni-doped CdS nanocrystals. The XRD spectrum from un doped CdS apparently exhibits only two broad peaks, but a more careful study of the pattern suggests that the peak at higher angle is a overlap of multiple peaks of cubic and wurtzite XRD patterns. Hence, undoped CdS can be considered as crystallized in both zinc blende and wurtzite structure. The mixed phase could either arise due to twinning or faulting within a single cubic particle or due to domains of the cubic and hexagonal phase coexisting within the same particle, sharing their respective close-packed planes at the domain boundaries. For Ni doped CdS nanocrystals, the diffraction peaks positioned at 2θ values 26.48^0, 43.93^0 and 51.94^0 were observed and are indexed as (002), (110) and (112) planes corresponding to that of hexagonal CdS (JCPDS# 02-0549). The average crystallite size has been calculated using the scherrer's formula [23]

$$D=0.89\lambda/\beta\cos\theta$$

Where, D is the mean crystallite size, λ the X-ray wavelength, β the FWHM of the diffraction peak and θ the diffraction angle. The crystallite size of CdS and Ni doped CdS nanocrystals are found to be lie in the range of 1.2-1.8 nm

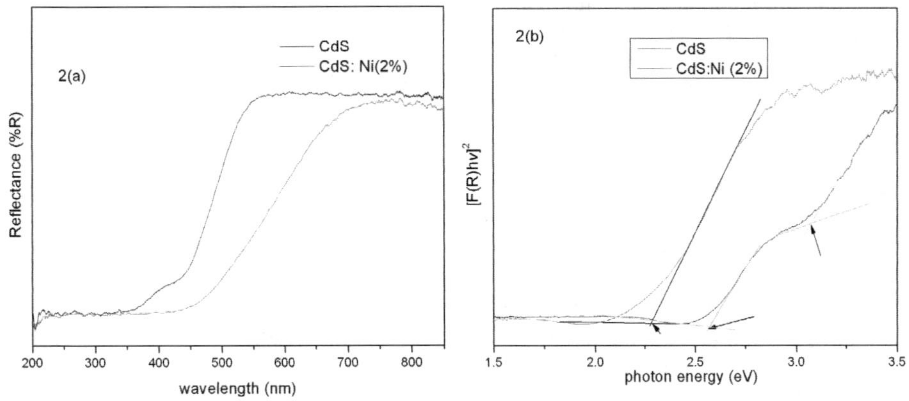

Figure 2. (a) Diffuse reflectance spectra of undoped and Ni doped CdS nanocrystals (b) Kubelka–Munk plots and band gap energy estimation for the undoped and Ni doped CdS nanocrystals.

The band gap of the present samples was estimated from diffuse–reflectance spectra by plotting the square of the Kubelka–Munk function($F(R)^2$) versus energy and extrapolating the linear part of the curve to $F(R)^2 = 0$, as shown in figure 2(b). DRS results supported the mixed phase of undoped nanocrystals with showing two absorption edges at 363 nm and 428 nm with corresponding energy gaps 3.07 eV and 2.56 eV respectively**.** Since the band gap of the bulk hexagonal phase (2.5 eV) is smaller than the band gap for the 'bulk' cubic phase (3.5 eV), the smaller of the two band gaps in CdS was assigned to the hexagonal phase while the larger to the cubic phase. But for Ni^{2+} (2%) doped CdS nanocrystals, only one absorption edge at 468 nm with energy gap value 2.27 eV was observed. The observed red shift in the band edge of the Ni^{2+} doped CdS nanocrystals is due to the increasing sp-d exchange interactions between the band electrons and the localized d-electrons of the Ni^{2+} ions substituting Cd^{2+} ions in the doped sample.

4. Conclusions

Undoped and Ni doped CdS nanocrystals were prepared using chemical co-precipitation method at 80^0C. XRD spectra showed that the undoped CdS NCs are in mixed phase (cubic and hexagonal) and with Ni doping the structure was transformed to hexagonal phase. DRS results supported the mixed phase of undoped nanocrystals with showing two absorption edges at 363 nm and 428 nm. The energy gap of hexagonal and cubic phase of cds nanocrystals are 2.56 eV and 3.07 eV respectively. Ni doping results in decrease of band gap energy to 2.27 eV.

Acknowledgements The authors are highly grateful to the University Grants Commission, New Delhi, India, for providing the financial support.

References

[1] Park S J, Kim S, Lee S, Khim Z G, Char K and Hyeon T 2000 *J. Am. Chem. Soc.* **122** 8581.
[2] Peng X, Manna L, Yang W, Wickham J, Scher E, Kadavanich A and Alivisatons A P 2000 *Nature* **404** 59.
[3] Sapra S and Sarma D D 2004 *Phys. ReV. B* **69** 125304.

[4] Viswanatha R, Sapra S, Saha- Dasgupta T and Sarma D D 2005 *Phys.ReV. B* **72** 045333.
[5] Peng Z A and Peng X 2001 *J. Am. Chem. Soc.* **123** 183.
[6] Nanda J, Sapra S, Sarma, D D, Chandrasekharan N and Hodes, G 2000 *Chem. Mater.* **12** 1018.
[7] Singh V and Chauhan P 2009 *J. Phys. Chem. Solids* **70** 1074.
[8] Phuruangrat A, Thongtem T and Thongtem S 2009 *J. Exp. Nanosci.* **4** 47.
[9] Lalayan A A 2005 *Appl. Surf. Sci.* **248** 209.
[10] Jinxin Z, Gaoling Z and Gaorong H 2007 Front *Chem Chin* **2** 98.
[11] Marandi M, Taghavinia N, Iraji A and Mahadavi S M 2005 *Nanotechnology* **16** 334.
[12] Tong H and Zhu Y J 2006 *Nanotechnology* **17** 845.
[13] Murali G, Amaranatha Reddy D, PoornaPrakash B, Vijayalakshmi R P, Reddy B K, Venugopal R 2012 *Physica B* **407** 2084.
[14] Tang H, Mi Y, Hui Z, Xia M and Deren Y 2005 *Mater. Lett.* **59** 1024.
[15] Murali G, Amaranatha Reddy D, PoornaPrakash B, Vijayalakshmi R P 2012 *Cryst. Res. Technol.* **47** 202.
[16] Seker F, Meeker K, Kuech T F and Ellis A B 2000 *Chem. Rev.* **100** 2505.
[17] Mcclean I P and Thomas C B 1992 *Semicond. Sci. Technol.* **7** 1394.
[18] Deshmukh I P, Holikatti S G and Hankare P P 1994 *J.Phys.D* **27** 1786.
[19] Banerjee R, Jayakrishnan R and Ayyub P 2006 *J. Phys.: Condens. Matter* **12** 2377.
[20] Banerjee R, Jayakrishnan R and Ayyub P 2000 *J. Phys.: Condens.Matter* **12** 10647.
[21] Liu W, Jin C, Jia C, Yao L, Cai W and Li X 2004 *Chem.Lett.* **33** 228.
[22] Lazell M and O'Brien P 1999 *J.Mater.Chem.* **9** 1381.
[23] Hemant S, Mukesh C and Dhanajay B, 2009 *Mater. Lett.* **63** 767.

Structural and Optical properties of ZnO thin films grown on various substrates by RF magnetron sputtering

G.Anil Kumar[1], M.V.Ramana Reddy[1] and Katta Narasimha Reddy[2]

[1]Department of Physics, Osmania University, Hyderabad,A.P. India.

[2]Department of Physics, Mahatma Gandhi University, Nalgonda,A.P. India.
E-mail: anilhcu@gmail.com

Abstract. ZnO thin films were deposited on glass, quartz and silicon substrates under the same growth condition by r.f. magnetron sputtering technique using a high purity (99.99%) ZnO target of 2-in. diameter and 3 mm thickness in an Argon atmosphere with sputtering power of 50W and sputtering pressure of $2x10^{-2}$ Torr. A systematic study has been made of the influence of substrate on the film structural properties. Crystalline properties of ZnO films as a function of deposited substrate were investigated using X-ray diffraction. XRD analysis revealed that the deposited films were polycrystalline in nature with strong preferential orientation of grains along the c-axis. The micro structural parameters, such as the lattice constant, crystallite size, stress and strain are calculated. The effect of substrate on the deposited films was discussed. All the films present a high transmittance of above 90% in the wavelength range of the visible spectrum and sharp absorption edge near 380 nm.

1.Introduction

Recently, ZnO film has gained much more interest due to its potential use in many applications, ranging from UV light emitters [1], varistors [2], transparent high-power electronics [3], surface acoustic wave devices [4], piezoelectric transducers , and chemical and gas sensing [5], which attributed to its wide and direct band gap (about 3.37 eV), excellent chemical and thermal stability, and a larger exciton binding energy (_60 meV). Moreover, ZnO has novel and very important optical properties.The elements doped ZnO films have emerged as excellent candidates for potential applications as transparent electrodes in optoelectronic devices [6]. So far, various preparation methods have been used to deposit the ZnO films, such as: RF magnetron sputtering [7], spray pyrolysis , chemical vapor deposition(CVD) [8], pulsed laser deposition [9], and sol–gel process [10]. Among these means, RF Magnetron sputtering is preferred due its reproducibility and for getting highly oriented films. This physical vapor deposition is used to produce films with the desired optoelectronic and nanostructure properties by precisely controlling the parameters like Ar-O_2 flux deposition pressure, target-substrate distance, RF power and substrate temperature. In the present study, we have made an attempt to produce highly textured ZnO nanocrystalline films in the absence of oxygen environment on glass, quartz and silicon substrates using RF magnetron sputtering and studied the effect of substrates on the structural and optical properties.

2.Experimental details

2.1.ZnO thin film deposition

ZnO thin films were grown on ultrasonically cleaned glass, quartz and silicon substrates by using RF magnetron sputtering. The sputtering system is capable of creating an ultimate vacuum of 5 x 10^{-5} Torr. A power supply operated at a crystal-controlled frequency of 13.56 MHz. The sputtering chamber was pumped with turbo molecular pump and rotary pump combination. The pressure in the sputter chamber was measured using digital Pirani and Penning gauge combination. The magnetron target assembly was mounted on top of the sputter chamber such that the sputtering could be done by sputter down configuration. A high purity (99.99%) ZnO target of 2-in. diameter and 3 mm thickness was used for the sputtering. Pure argon was used as sputtering gas .The flow rate of argon gas was controlled by Tylan mass flow controller. The sputtering chamber was pumped down to the vacuum of 5 x 10^{-5} Torr before the introduction of Argon gas. The target was pre-sputtered for 15 min to remove any impurity on the surface of the target. A shutter was placed between the target and substrate to control period of coating, The target was water-cooled and the substrate was kept at room temperature initially. During the film deposition, the substrate temperature was increased due to the bombardments of the sputtering ions and the final substrate temperature was nearly 40–60 ^{0}C. To measure the deposition rate and film thickness, a vibrating quartz crystal was used. The sputtering conditions maintained during the growth of ZnO films are given in Table 1.

Table 1:

Deposition parameters maintained during the deposition of ZnO films by RF magnetron sputtering .

Sputtering target	ZnO (99.99%) 2-in. diameter and 3 mm thickness
Target to substrate distance	70 mm
Substrates	glass,quartz and silicon
Ar gas flow rate	80 SCCM
Ultimate pressure	5x10^{-5}Torr
Sputtering pressure	1.2x10^{-2} Torr
Substrate temperature	Room Temperature
Sputtering power	50 W
Thickness	100nm

The crystalline structure of the films was studied by X-ray diffraction measurements obtained with a Philips X'Pert X-Ray Diffractometer (XRD) in the 2θ range of 20°- 80° using CuK_α radiation of wavelength λ=1.5406 Å at room temperature. X-ray tube was operated at 40 kV (Voltage) and 30 mA (current) with scanning speed of 0.5°/min. Optical transmittance spectra of films were recorded with a SHIMAZDU 3100 UV-Vis spectrometer over wavelength range from 200 to 800 nm.

3.Results and discussion

3.1.Structural properties

XRD patterns of ZnO films grown on glass, quartz and silicon substrates at RT are shown in Fig. 1. It is found that XRD patterns of ZnO thin films show the same form, that is only a (002) diffraction peak is detected in the films. This indicates that ZnO films prepared by sputtering deposition show a good c-axis orientation perpendicular to the substrate. The diffraction peak angle of the films grown on glass, quartz and silicon substrates at RT appears at 34.01^0, 34.17^0 and 34.31^0, respectively. Compared with the (002) peak position of ZnO powder (2θ = 34.42^0), the diffraction angle of the films decreases, which results in the increase of c-axis value. This indicates that ZnO films suffer compressive stress along the interfaces. The full-width at half-maximum (FWHM), determined from the XRD spectra of ZnO films prepared on glass,quartz and silicon substrates, are 0.1952,0.1968 and 0.144, respectively.The smaller FWHM value indicates that the ZnO films are more textured along the c-axis. The peak intensity of ZnO films deposited on Si substrates is higher than that of those prepared on glass and quartz substrates. However, a slight increase of crystalline size is observed for the sample deposited on Si substrate.

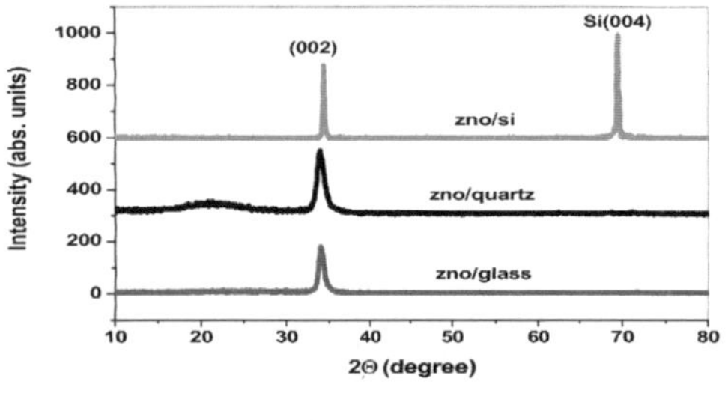

Fig.1. XRD patterns of ZnO films deposited on different substrates.

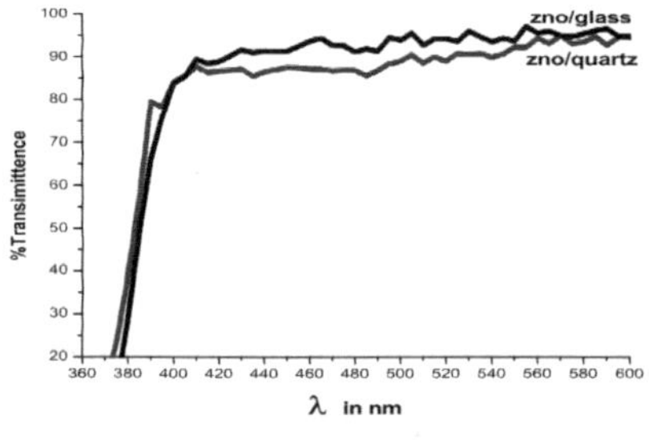

Fig. 2. Optical transmittance spectra of ZnO films.

The average crystallite sizes of the films grown on glass, quartz and silicon substrates have been calculated using the Scherrer's formula:

$$D = 0.94\lambda/(\beta \cos \theta) \qquad (1)$$

Where λ,θ and β are X-ray wavelength, the Bragg's diffraction angle and the full width at half maxima (FWHM) of the peak corresponding to the ''θ'' value.
Dislocation density (δ) and Number of crystallites per unit area (N) have been calculated using the crystallite size and thickness of the film [11-13].
The average uniform strain (e_{zz}) in the lattice along the c-axis in the randomly oriented ZnO films grown on glass, quartz and silicon substrates have been estimated from the lattice parameters using the following expression [14],

$$e_{zz}=(d-d_0)/d_0 \qquad (2)$$

Where ''d'' is the lattice parameter of the ZnO film calculated from (0 0 2) peak of XRD pattern and the "d_0'' is the lattice parameter for the ZnO bulk.
For hexagonal crystals, the stress (σ) in the plane of the film can be calculated using the biaxial strain model [15]:

$$\sigma = (2C_{13} - (C_{11}+C_{12})\, C_{33}/C_{13}).e_{zz} \qquad (3)$$

here C_{ij} are elastic stiffness constants (C_{11} = 2.1 x 10^{11} N/m^2 , C_{33} = 2.1x 10^{11}N/m^2, C_{12} = 1.2 x 10^{11} N/m^2, and C_{13} = 1.05 x10^{11} N/m^2). This yields the following numerical relation for the stress derived from the change in the 'd' lattice parameter:

$$\sigma\ (N\ m^{-2}) = -\ 4.5\times10^{11}.e_{zz} \qquad (4)$$

The various structural parameters for ZnO thin films grown on glass, quartz and silicon substrates were calculated and are represented in Table 2.

Table 2: Structural information of ZnO thin films grown on glass, quartz and silicon substrates.

sample	2θ(^{0})	Lattice parameter (nm)	FWHM (mrad)	Grain size (nm)	Strain (10^{-3})	Stress (GPa)	Dislocation density (10^{15}lines/m^2)	Number of grains/unit area(10^{16} m^{-2})
ZnO/glass	34.01	0.5271	3.405	42.58	12.29	-5.53	0.5515	0.1295
ZnO/quartz	34.17	0.5240	3.433	42.25	6.33	-2.84	0.5602	0.1325
ZnO/silicon	34.31	0.5222	2.512	57.76	2.88	-1.29	0.2997	0.0518

From the Table 2. it has been observed that the lattice constant for the thin film deposited on the glass, quartz and silicon substrates decreases . The change in lattice constant for the thin films deposited over the bulk clearly suggests that the film's grains are under stress and that may be due to a change in the nature and concentration of the native imperfections. This causes either elongation (tensile) or compression of the lattice parameters. The density of film is, therefore, expected to change in accordance with changes in the lattice constant.The average internal stress for the ZnO films deposited at RT is found to be compressional in nature. The negative sign indicates that the films are in a state of compressive stress. The total stress in the film commonly consists of two components. One is the intrinsic stress introduced by impurities, defects and lattice distortions in the crystal, and the other is the extrinsic stress introduced by the lattice mismatch and thermal expansion coefficient mismatch between the film and substrate. The extrinsic stress in the thin films normally relaxes if the thickness of the film is larger [16]. In the present case, thickness of all the films is 100 nm. Thus, the extrinsic stress will not be present and the total estimated stress values seem to be dominantly intrinsic. In present case of sputtering deposition, this intrinsic stress arises due to the bombardment of the energetic species. The films deposited at lower temperature exhibits strong compressive stress. In the sputtering process, the highly energetic species could be implanted below the ZnO surface, which is likely to induce high intrinsic stresses by creating the defects [17].The microstrain (e) , the dislocation density (δ) and , the number of crystallites(N) are observed to exhibit a slow decreasing trend from glass to Si substrate. The decrease in e , δ and N may be due to the movement of interstitial Zn atoms from inside the crystallites to the grain boundary, which leads to a reduction in the concentration of lattice imperfections.

3.2. Optical properties

Fig. 2 shows the optical transmittance spectra of ZnO films grown on glass, quartz and silicon substrates.All the films present a high transmittance of above 90% in the wavelength range of the visible spectrum and sharp absorption edge near 380 nm. The absorption edge was shifted towards lower wavelength with the quartz substrate.The increase in transmittance of the films was related to an increase in grain size of the films.

4.Conclusion

Highly oriented and smooth ZnO thin films have been grown on glass,quartz and Si substrates using the RF magnetron sputtering technique. The influence of the substrate on the crystalline and oriented properties of ZnO films was investigated. Each film showed a preferred orientation along the (0 0 2) plane. The average internal stress developed in the films were compressional in nature. All the films present a high transmittance of above 90% in the wavelength range of the visible spectrum and sharp absorption edge near 380 nm.

Acknowledgments

The authors thank UGC-SAP(DSA-II) for providing necessary facilities. One of the authors (GAK) thank UGC,New Delhi for awarding the JRF under the UGC scheme of RFSMS.One of us (MVRR) thank OU-DST-PURSE Programme,Osmania University, Hyderabad, for providing financial assistance to carry out this work.

References

[1] D.C. Look, R.L. Jones, J.R. Sizelove, N.Y. Garces, N.C. Giles, L.E. Halliburton, Phys. Status Solidi A 195 (1) (2003) 171.

[2] D. Xu, L.Y. Shi, Z.H. Wu, Q.D. Zhong, X.X. Wu, J. Eur. Ceram. Soc. 29 (2009) 1789.

[3] E. Bellingeri, D. Marré, L. Pellegrino, I. Pallecchi, G. Canu, M. Vignolo, C. Bernini, A.S. Siri, Superlattices Microstruct. 38(2005) 446.

[4] H. Ohta,K. Kawamura, M.Orita, M.Hirano,N.Sarukura,H.Hosono, Appl. Phys. Lett.77 (2000) 475.

[5] N.Han,L.Y.Chai,Q.Wang,Y.J.Tian,P.Y.Deng,Y.F Chen, Sensor. Actuat. B. Chem. 147 (2010) 525.

[6] H.Y.Liu,VAvrutin,N. Izyumskaya, Ü. Özgür,H. Morkoc, Superlattices Microstruct. 48 (2011) 458.

[7] K.B. Sundaram, A. Khan, Thin Solid Films 295 (1997) 87.

[8] K. Haga, P.S. Wijesena, H. Watanabe, Appl. Surf. Sci. 169–170(2001) 504–507.

[9] J.N. Zeng, J.K. Low, Z.M. Ren, T. Liew, Y.F. Lu, Appl. Surf.Sci. 197–198 (2002) 362–367.

[10] Y. Natsume, H. Sakata, Mater. Chem. Phys. 78 (2002) 170–176.

[11] V.Bilgin, S. Kose, F. Atay, I. Akyur, Mater. Chem. Phys. **94**, 103-108 (2005).

[12] J.B. Seon, S. Lee, J.M. Kim, H.D. Jeong, Chem. Mater. **21** (2009) 604-611.

[13] M.B. Ortuño López, J.J . Valenzuela-Jáuregui, M. Sotelo-Lerma, A. Mendoza-Galván, R. Ramirez-Bon, Thin Solid Films **429**, 34-39 (2003).

[14] T. Minami, H. Sato, S. Takata, N. Ogawa, T. Mouri, Jpn. J. Appl. Phys. 3 (1992) L1106–L1109.

[15] S. Maniv, W.D. Westwood, E. Colombini, J. Vac. Sci. Technol. 20 (1982)162.

[16] T.B. Hur, Y.H. Hwang, H.K. Kim, I.J. Lee, J. Appl. Phys. 99 (2006) 064308.

[17] P. Patsalas, C. Gravalidis, S. Logothetidis, J. Appl. Phys. 96 (2004) 6234– 6246.

IOP Conf. Series: Materials Science and Engineering **73** (2015) 012134 doi:10.1088/1757-899X/73/1/012134

Electrical and Magneto-electronic Properties of p-$La_{0.7}Ca_{0.3}MnO_3$/ $SrTiO_3$/n-Si Heterostructure for Spintronics Application

J. Panda, S. K. Giri, and T. K. Nath*
Department of Physics and Meteorology, Indian Institute of Technology, Kharagpur, West Bengal 721302, India

Abstract. An experimental study of p-$La_{0.7}Ca_{0.3}MnO_3$/ $SrTiO_3$/n-Si heterostructure in which the LCMO and Si are separated by a thin interfacial $SrTiO_3$ (STO) layer with typical thickness of about 15 nm, has been in situ fabricated with the Pulsed laser deposition technique. The $La_{0.7}Ca_{0.3}MnO_3$ (LCMO) film of about 70 nm has been grown on STO at substrate (n-Si) temperature of 800°C in 0.5 mbar oxygen pressure. The junction exhibits good rectifying behavior over the temperature range of 10 - 300 K. Electrical properties of p-$La_{0.7}Ca_{0.3}MnO_3$/STO/n-Si MOS like heterostructure exhibits nonlinear I–V characteristics in a wide temperature range. The heterostructure also exhibits MOS-diode like behavior with all type of possible current flow mechanisms (such as thermionic emission, tunneling, recombination, degeneration, etc.) through the heterojunction. The ideality factor, reverse saturation current, series resistances and turn-on voltages of the heterojunction have been estimated at different operating temperatures. The junction magnetoresistance (JMR) properties of p-LCMO/STO/n-Si heterostructure have been studied over the temperature range of 100-300 K. The current–voltage characteristics at all temperatures conclusively show the high sensitivity junction magnetoresistance (JMR ~ 48% at 150 K and ~ 30% at 300 K) under the external 7 T magnetic field. The JMR is positive and strongly depends on temperature at an applied forward bias voltage (3 V). The relation between JMR and external magnetic field is found to be of ($\Delta\rho/\rho \approx \alpha H^{\beta}$) type having both α and β temperature dependent. We attribute the emergence of positive JMR to the quantum mechanical tunneling transport mechanism along with the modification of barrier height across the heterojunction.

1. Introduction

The field of spintronics which employs the electron spin along with the electron charge to design devices with novel functionalities is the focus of much current research interest. Utilization of the spin degree of freedom provides opportunities to create devices with superior performance compared with those that use only electron charge [1, 2]. One class of materials that is promising for use in spintronic devices is the 3*d* transition-metal complex oxides, which owing to their highly correlated *d* electrons, exhibit a wide variety of electronic and magnetic properties [3, 4]. The research on manganite materials ($R_{1-x}A_xMnO_3$, where R is a trivalent rare earth and A is a divalent alkaline ion) for their integration into spintronics devices has recently received much attention. The half-metallicity of these materials makes them proper electrodes in magnetic tunnel junctions (MTJ's). Generation of spin polarization usually means creating a non-equilibrium spin population which can be achieved in several ways. Recently, a lot of research studies have been focused on perovskite based oxide semiconductors and semimetals heterojunctions with different semiconductors [5-7]. It is still an open question to model the interface between different perovskite-type oxides. Previous studies reveal that the transport process in manganite heterostructures can be approximated by p-n junctions or Schottky type junctions [8, 9]. P.L. Lang et al. also demonstrated that $La_{0.67}Ca_{0.33}MnO_3/SiO_2/Si$ type p-n heterojunctions show Space Charge Limit (SCL) diode characteristics with the presence of the SiO_2 layer [10]. The most of the studies focused on the transport process based on current voltage characteristics. Generally the current-voltage characteristic is guided by tunneling of conduction electrons between the junctions though the presence of edge leakage which makes the transport mechanism complicated. As a result there is difficulty in the extraction of actual barrier height.

In present work we have fabricated p-LCMO/STO/n-Si heterojunction by pulsed laser deposition technique. We have explicitly studied the junction current–voltage (*I –V*) characteristics of p-LCMO/STO/n-Si junction without and with 7 T external magnetic field. The JMR properties of the same heterojunction also studied under the external magnetic field up to 7 T. The origin of JMR has been explained using standard spin injection theory.

2. Experimental procedures

At first we have prepared the LCMO target using the "prypophoric reaction" route. Final sintering of the LCMO target was carried out at 1200 °C for 24 h. $La_{0.7}Ca_{0.3}MnO_3$ film on n-Si (100) substrate was grown by Pulsed Laser Deposition process. The substrate was thoroughly cleaned in ultrasonic bath using acetone and alcohol followed by chemical cleaning with mixture of H_2O_2 and H_2SO_4 (1:1). The oxide layer was etched by 10 % HF. The film was deposited on <100> Si substrate using following steps. (1) The substrate was first heated at 800 ºC and STO film was deposited at 0.5 mbr O_2 pressure. (2) Then LCMO film was deposited at the same physical condition. The laser pulse of energy 300 mJ was applied on the STO and LCMO targets for 20 min at a frequency of 10 Hz and the target - substrate distance was kept at 4 cm. (3) finally, the film was then sintered at 800 ºC and 0.5 mbar O_2 pressure for 1 h. A cryogen free superconducting magnet (8 T) fitted with a closed cycle helium refrigerator system was used to measure the electrical transport properties of the LCMO/STO/n-Si heterojunction. The junction magnetoresistance properties of p-LCMO/STO/n-Si heterostructure were measured at different temperatures in the range of 100 K-300 K. The ohmic contact on the back side of n-Si semiconductor was made by high purity indium and other side contact on LCMO film was made by silver paste.

3. Results and discussion

The I-V characteristics of p-LCMO/STO/n-Si heterojunction have been shown in Fig. 1(a) at different temperatures (150 K, 200 K, 250 K and 300 K) in absence of external magnetic field. The heterojunction shows very good rectifying diode like behavior in all temperatures.

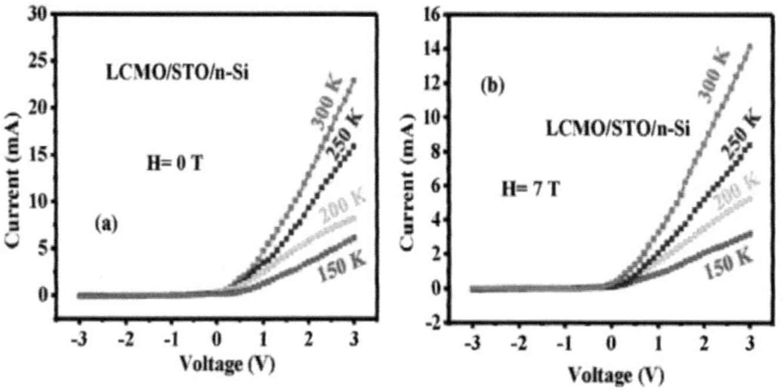

Figure 1. (a) and (b) The I-V characteristics of LCMO/STO/n-Si heterojunction at different temperatures (150 K, 200 K, 250 K and 300 K) with uot and with external magnetic field 7 T, respectively.

The junction current is found to increase with increasing temperature which is generally expected in p-n junction characteristics. Figure 1(b) shows the I-V characteristics of same heterojunction at same temperatures with external magnetic field 7 T. At low forward voltage the current increased exponentially which has been usually observed in diodes and attributed to recombination, degeneration or tunneling mechanism. The detailed comparative study of I-V without and with applied 7 T magnetic field measured at different temperatures (150 K and 300 K) for LCMO/STO/n-Si heterojunction have been shown in Fig. 2. The current across the junction decreases with increasing the magnetic field at a 3 V forward bias voltage. The junction I-V characteristics of this heterojunction can be described using modified diode equation [11]

$$I = I_0 \exp\left(\frac{e(V - V_0 - IR_s)}{\eta kT} \right) \tag{1}$$

where I_0 is the reverse saturation current. V_0, η and R_s are turn on voltage, ideality factor and junction series resistance, respectively. The best fit parameters (η, I_0, R_s and V_0) for LCMO/STO/n-Si heterojunction have been evaluated using equation (1) at different temperatures.

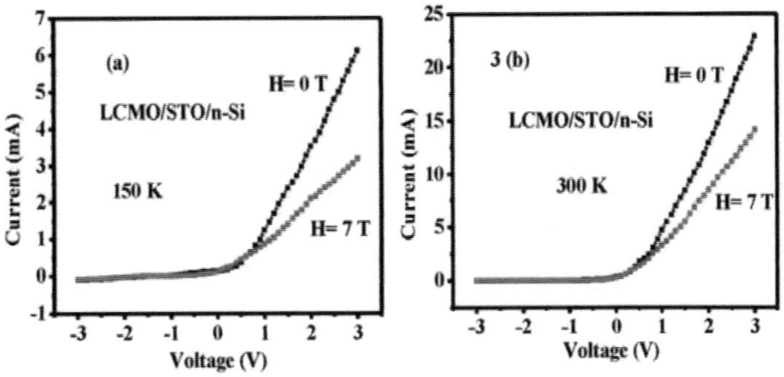

Figure 2. (a) and (b) The detailed comparative I-V study for LCMO/STO/n-Si heterojunction at 150 K and 300 K without and with external magnetic field 7 T, respectively.

The JMR (JMR=(($R_{H=6T}$ − $R_{H=0T}$)/$R_{H=0T}$)*100)properties (up to 6 T magnetic field) of the same heterojunction have been studied at different temperatures (150 K and 300 K) under the bias voltage of 3 V, which has been shown in Fig. 3. The junction resistance is increases with increasing of magnetic field which leads to positive JMR. It shows that the JMR behavior follows a simple empirical relation with magnetic field as [12],

$$JMR = \alpha H^{\beta} \tag{2}$$

where, α, β and are coefficients. The best fit parameters (α =16.08, 4.2 and β =0.56, 0.97 at 150 K and 300 K, respectively) for LCMO/STO/n-Si heterojunction have been evaluated using equation (2). The coefficient β is lower than one at both temperatures, showing non-linear magnetic field dependence of positive JMR of the junction. It shows the positive JMR properties for all the temperatures and the magnitude of positive JMR is also temperature dependent. The % JMR at a fixed bias voltage of 3V has been found to decrease sharply with the increase in temperature. The origin of positive JMR in this heterojunction can be best explained by the spin injection theory of magnetic p-n junction where p-Si is nonmagnetic semiconductor. Spin injection generally occurs in n-side and spin extraction generally occurs in p-side. Both spin injection and spin extraction becomes large with applied magnetic field. The current through the magnetic p-n junction depends on magnetic field because of the spin orbit splitting of magnetic materials. The magnetically active electrons in LCMO cause non-equilibrium polarized conduction electrons by the application of external magnetic field. This non-equilibrium population at magnetic n-side enhances the spin injection through the space-charge region [13]. It causes a high magnetoresistance (positive junction MR) according to standard spin injection theory [14]. Highest positive JMR for LCMO/STO/n-Si heterojunction has been found to be 48 % at 150K (30 % at 300K) at a fixed applied bias voltage of 3 V. Ferromagnetic metel (F) – non magnetic semiconductor (N) tunneling is the process, where the electrons tunnel through a barrier in the presence of a high electric field.

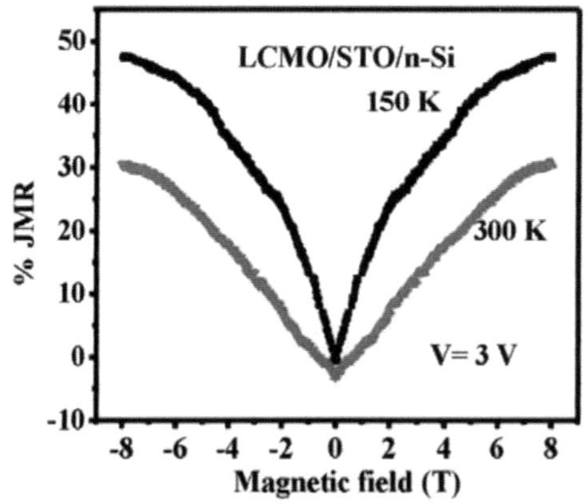

Figure 3. The JMR properties for LCMO/STO/n-Si heterojunction at different temperatures (150 K and 300 K) at a fixed bias voltage of 3 V (up to 7 T magnetic field).

This quantum-mechanical tunneling process is an important mechanism for thin barriers as those in metal– semiconductor junctions on highly doped semiconductors [15-18]. The dominating current transport mechanism is tunneling. Hence, it can be concluded that the appearance of positive JMR is due to tunneling of electron through the heterojunction.

4. Conclusions

In conclusion, the I-V properties of LCMO/STO/n-Si heterojunction have been studied at several temperatures. The heterojunction shows good rectifying behavior at all temperatures. The JMR behavior of the same heterojunction has been studied which shows that the high value positive JMR (48 %) observed at 150 K. The % JMR at a fixed bias voltage of 3V has been found to decrease sharply with the increase in temperature. The junction magnetoresistive property has been explained using standard spin injection mechanism in magnetic p-n junction. The properties of these heterojunction are very useful in the area of spintronics devices.

Acknowledgement
Authors would like to acknowledge gratefully the financial support of Department Science and Technology (DST), New Delhi, India, through Project No. : IR/S2/PU-04/2006.

References

[1] S. A. Wolf, D. D. Awschalom, R. A. Buhrman, J. M. Daughton, S. von Molnar, M. L. Roukes, A. Y. Chtchelkanova, and D. M. Treger, *Science* **294**, 1488 (2001).
[2] I. Zutic, J. Fabian, and S. Das Sarma, *Rev. Mod. Phys.* **76**, 323 (2004).
[3] J. Cibert, J. F. Bobo, and U. Luders, C. R. *Phys.* **6**, 977 (2005).
[4] C. H. Ahn, A. Bhattacharya, M. Di Ventra, J. N. Eckstein, C. D. Frisbie, M. E. Gershenson, A. M. Goldman, I. H. Inoue, J.Mannhart, A. J. Millis, A. F. Morpurgo, D. Natelson, and J. M. Triscone, *Rev. Mod. Phys.* **78**, 1185 (2006).
[5] J. Xing, K. Zhao, G.Z. Liu, M. He, K.J. Jin, H.B. Lu, *J. Phys. D. Appl. Phys.* **40,** 5892 (2007).
[6] K. Zhao, K.J. Jin, H. Lu, Y. Huang, Q. Zhou, M. He, Z. Chen, Y. Zhou, G. Yang, *Appl. Phys. Lett.* **88,** 141914 (2006).

[7] K. Lord, D. Hunter, T.M. Williams, A.K. Pradhan, *Appl. Phys. Lett.* **89**, 052116 (2006).
[8] Z. Lue, J. Gao, A.B. Djuri_sic, C.T. Yip, G.B. Zhang, *Appl. Phys. Lett.* **92,** 182501 (2008).
[9] K. Wang, X.D. Hun, Z. Zhang, T. Li, M. Zhang, H. Yan, *J. Appl. Phys.* **103,** 014315 (2008).
[10] P.L. Lang, Y.G. Zhao, C.M. Xiong, P. Wang, J. Li, D.N. Zheng, *J. Appl. Phys.* **100,** 053909 (2006).
[11] May S J and Wessels B W *Appl. Phys. Lett.* **88,** 072105 (2006).
[12] Ilie A and Equer B *Phys. Rev.* B **57,** 15349 (1998).
[13] [15] Mark Johnson, Magnetoelectronics, *academic press (Elsevier)*, (2005).
[14] S. Chattopadhyay and T. K. Nath, *J.Appl. Phys.* **108**, 083904 (2010).
[15] Vincent G, Chantre A and Bois D *J. Appl. Phys.* **50,** 5484 (1979).
[16] Buchwald W R and Johnson N M *J. Appl. Phys.* **64,** 958 (1988).
[17] Auret F D, Goodman S A and Meyer W E 1995 *Semicond. Sci.Technol.* **10,** (1376).
[18] Martin P A, Streetman B G and Hess K *J. Appl. Phys.* **52,** 7409 (1981).

Electronic and magnetic properties of CdI_2-type MX_2 (M = V, Nb; X = Al, Ga and In) compounds

G Jaiganesh[1] **and G Kalpana**

Department of Physics, Anna University, Chennai - 600 025, India

[1]E-mail address: jai_spectrum@yahoo.com

Abstract. First-principle calculations within the density-functional theory is used to investigate the possibility of existence of ferromagnetism in hypothetical CdI_2-type MX_2 (M = V, Nb; X = Al, Ga and In) compounds. The TBLMTO-ASA program is used for this purpose. Both spin-polarization and non-spin-polarization calculations are carried out for each compound. The spin-polarization calculation shows that VX_2 (X = Al, Ga and In) compounds are ferromagnets whereas NbX_2 (X = Al, Ga and In) compounds are non-magnets because no effective polarization of the energy states occurs at their equilibrium volume. However, the NbX_2 (X = Al, Ga and In) compounds become ferromagnets under large volume expansion. Ferromagnetism in these compounds is predicted for the first time. The variation of lattice constant with magnetic moment for VX_2 (X = Al, Ga and In) compounds are also analysed. The magnetism arises mainly from the cation *d*-like states. This result is similar to other transition-metal based pnictide and chalcogenide compounds. The ground-state properties like equilibrium lattice parameters, bulk modulus, heat of formation and total and partial magnetic moments are calculated.

1. Introduction

Several layered transition-metal compounds have interesting magnetic properties [1, 2]. In this work, we have systematically studied the electronic band structure calculations for the hypothetical layered CdI_2-type MX_2 (M = V, Nb; X = Al, Ga and In) compounds in the ferromagnetic and nonmagnetic phases. To the best of our knowledge, there are no experimental and theoretical studies on these compounds. The motivation of the present work is to investigate the ferromagnetism in these compounds by performing first principles electronic structure calculations by means of TBLMTO method within the density functional formalism. Ferromagnetism in these compounds is predicted for the first time. The obtained results are compared with other transition-metal ferromagnets [3].

2. Crystal Structure and Method of Calculations

Within the framework of DFT, the electronic structure calculations are performed using the TBLMTO method. The potential and charge density inside each atomic sphere are assumed to have spherical symmetry. This method has been described well in the literature [4-6]. In the CdI_2-type structure (space group: P-3m1, No.: 164), the M and X atoms occupy the (0, 0, 0) and (0.33, 0.66, z) positions. The LMTO method gives accurate results only for close packed structures. Hence, an empty sphere is placed at the appropriate position, E: (0, 0, 0.82). For the titled compounds, the cell parameter values are optimized using the experimental cell parameter values of CdI_2-type Vanadium di halide compounds [2].

3. Result and discussions

Spin-polarization (Ferromagnetic phase) and non-spin-polarization (Nonmagnetic phase) calculations were carried out for the titled compounds in the hypothetical CdI_2-type structure in a manner similar to our earlier work [7, 8]. The total energy as a function of molecular volume is calculated for all these compounds. As an example, the result for CdI_2-type VAl_2 and $NbAl_2$ compounds are shown in figure 1. From the total energy calculations, it is observed that VX_2 compounds have lower energies in ferromagnetic phase compared with non-magnetic phase whereas NbX_2 compounds does not occur effective spin-polarization. Heat of formation or formation energy is calculated in order to study the stability of phase. The calculated ground-state properties and heat of formation of these compounds are listed in Table 1.

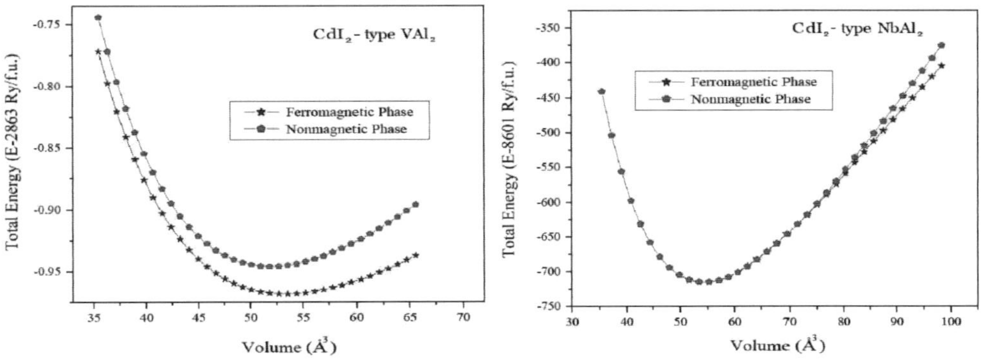

Figure 1. Calculated total energy per formula unit vs. Volume (Å^3).

The calculated spin-dependent density of states for VAl_2 compound at their equilibrium volume is shown in Figure 2. This compound exhibit ferromagnetic property with small magnetic moment and both the spins have metallic nature. In this compound, the states which are lying around -10 eV are mainly due to the anion s-like states. The states lying above this are mainly contributed by anion p-like states and cation s- and d-like states. The hybridized states of anion p-like and cation d- and s-like states lie close to the Fermi level. The density of states profiles of majority spin and minority spin for these compounds are slightly different. This may be one of the reasons why there is small magnetic moment induced in the anion p-like states. The overall density of states profiles for VGa_2 and VIn_2 compounds are similar. The calculated total magnetic moments for these compounds are given in Table 2 along with partial magnetic moments at each atomic site.

Table 1. Calculated equilibrium lattice constants (a, c) in Å, bulk modulus (B_0) in GPa and heat of formation (H) in kJ/mol. in FM phase. The values in the parenthesis are for NM phase.

Compounds	Lattice constants				B_0		H	
	a		c					
VAl_2	3.389	(3.332)	5.492	(5.399)	471.57	(508.25)	124.891	(132.532)
VGa_2	3.473	(3.442)	5.637	(5.587)	399.71	(412.01)	120.987	(129.106)
VIn_2	3.561	(3.517)	5.779	(5.707)	313.53	(355.32)	91.286	(104.337)
$NbAl_2$	3.445		5.594		585.96		98.631	
$NbGa_2$	3.517		5.707		503.55		89.343	
$NbIn_2$	3.621		5.878		484.64		68.731	

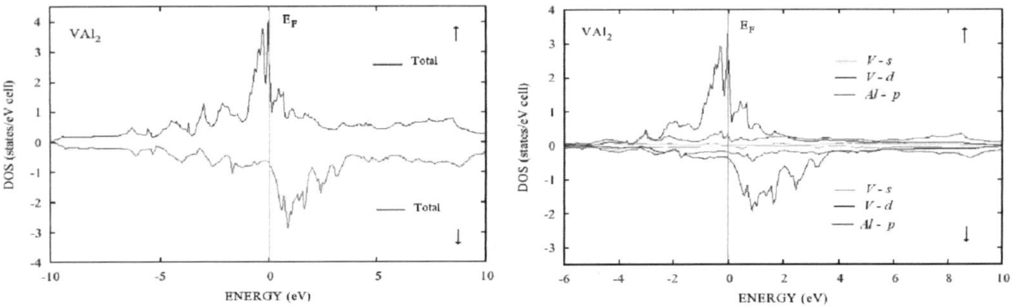

Figure 2. Spin-dependent total (left) and partial (right) DOS's of VAl_2 compound at their equilibrium volume.

From the table, it can be observed that the main contribution to the magnetic moment comes from the transition-metal atom. From the Figure 3, it can be seen that the magnetic moment for VX_2 compounds is slowly decreasing with reducing the cell volumes, finally it becomes non-magnetic.

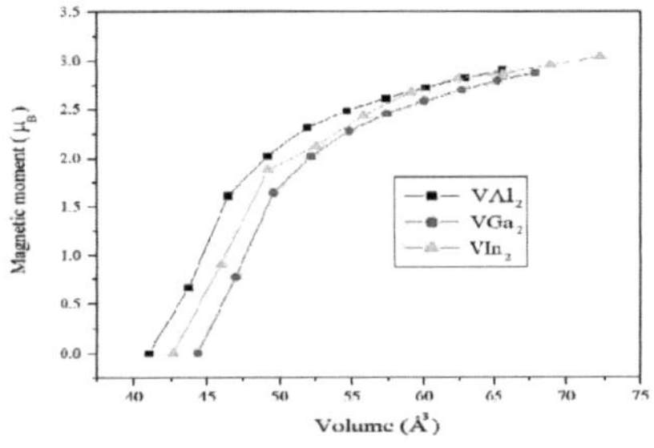

Figure 3. Volume vs. magnetic moment of VX_2 compounds.

Table 2. Calculated magnetic moments (μ) in μ_B.

	Magnetic moments			
	Cation	**Anion**	**Empty site**	**Total**
VAl_2	2.3302	0.1315	0.0197	2.4814
VGa_2	2.1606	0.1001	0.0149	2.2756
VIn_2	2.7203	0.0868	0.0111	2.8182

Both spin-polarization and non-spin-polarization calculations are carried out for NbX_2 compounds. The overall density of states profiles for these compounds at their equilibrium volume are similar to majority spin density of states of VX_2 compounds. These compounds exhibit non-magnetic behavior. In these compounds, both the spin bands are metallic. For expanded volume, these compounds exhibit ferromagnetic property with small magnetic moment. Both spins have metallic character. The difference between majority spin and minority spin states is that, around the Fermi level, the hybridized states of anion p-like and cation d-like states shifts towards the lower energy and lies below the Fermi level in the spin up states. With increasing volume, the cation d-electrons transfer from minority spin to majority spin. The spin-dependent total and partial density of states for $NbAl_2$ compound at their critical or expanded volume is given in Figure 4. The critical lattice constants and magnetic moments for NbX_2 compounds are given in Table 4.

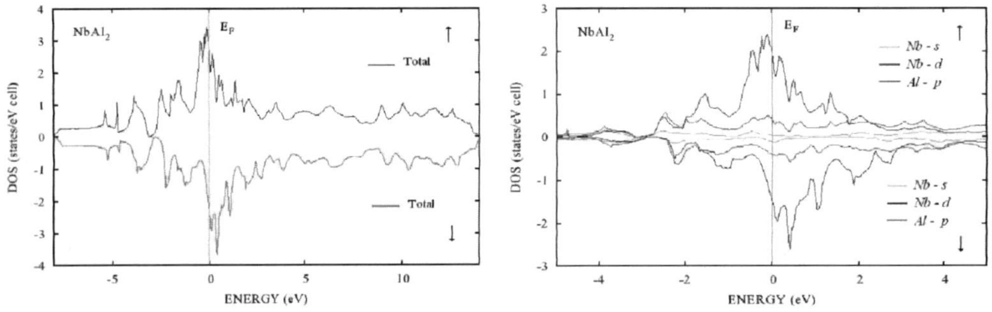

Figure 4. Spin-dependent total (left) and partial (right) DOS's of $NbAl_2$ at their expanded volume.

Table 4. Calculated critical lattice constants (a_{cri}, c_{cri}) in Å, magnetic moments (μ) in μ_B for NbX_2 compounds.

	a_{cri}	c_{cri}	Magnetic moments			
			Cation	**Anion**	**Empty**	**Total**
$NbAl_2$	3.836	6.216	1.3335	0.1604	0.0134	1.5075
$NbGa_2$	3.775	6.127	1.2672	0.1391	0.0149	1.4212
$NbIn_2$	3.879	6.215	0.1547	0.0110	0.0006	0.1663

4. Summary and Conclusion

To summarize, the electronic band structure calculations have been carried out for titled compounds in hypothetical CdI_2-type structure in order to find ferromagnetism. From the total energy calculations, one can observe easily that VX_2 (X = Al, Ga and In) compounds have lower energies in ferromagnetic phase compared with nonmagnetic phase whereas NbX_2 (X = Al, Ga and In) compounds are nonmagnets at their equilibrium volume. In these compounds, the magnetism arises mainly from the cation d-like states. VX_2 (X = Al, Ga and In) compounds exhibit ferromagnetic property with small magnetic moment at their equilibrium volume. However, NbX_2 (X = Al, Ga and In) compounds exhibit ferromagnetism under large volume expansion. In these compounds, unit cell volume plays an important role in exhibiting ferromagnetism. This result is similar to other transition-metal based group V compounds [3, 9].

Acknowledgement
The first author (**G. J.**) gratefully acknowledges the financial support from University Grant Commission (UGC), India.

References

[1] Hirakawa K, Ikeda H, Kadowaki H and Ubukoshi K 1983 *J. of the Phys. Soc. of Japan* **52** 2882
[2] Hirakawa K, Kadowaki H and Ubukoshi K 1983 *J. of the Phys. Soc. of Japan* **52** 1814
[3] Jaiganesh G *et al* 2010 *Computational Materials Science* **49** 112
[4] H L Skriver 1984 *The LMTO method*, Springer–Verlag, Heidelberg
[5] Hohenberg P and Kohn W 1964 *Phys. Rev.* **136** B864
[6] Kohn W and Sham L J 1965 *Phys. Rev.* **140** A1133
[7] Jaiganesh G *et al* 2007 *Phys. Stat. Sol. (b)* **244** 4643
[8] Jaiganesh G *et al* 2008 *Metals, Materials and Processes* **20** 183
[9] Picozzi S, Continenza A and Freeman A J 2002 *Phys. Rev. B* **66** 094421

IOP Conf. Series: Materials Science and Engineering **73** (2015) 012136 doi:10.1088/1757-899X/73/1/012136

Growth and birefringence studies of semi organic non-linear optical LHB single crystal

V Jayaramakrishnan[1], T Prasanyaa[2], M Haris[2] and G Bhoopathi[1]

[1]Department of Physics, PSG College of Arts and Science, Coimbatore - 641 014, Tamilnadu, India
[2]Department of Physics, Karunya University, Coimbatore - 641 114, Tamilnadu, India

E-mail: jayaram.si@gmail.com

Abstract. In the last few decades nonlinear optical materials are getting attention in the field of optical data storage, telecommunication, second harmonic generation (SHG) and optical signal processing, etc. In the present work we are reporting the single crystal growth of L-Histidine with hydro-bromic acid. The L-Histidine bromide (LHB) single crystals have been harvested from the solution in a span of 34 days by adopting slow cooling solution growth technique. The grown crystals have been subjected to powder X-ray diffraction studies to identify the cell parameters and structure. The crystalline perfection has been defined by rocking curve (HRXRD) analysis. Optical transmission spectra reveal the optical properties of the grown crystals. The Modified channel spectrum (MCS) method has been adopted for the study of spectral dependence of linear birefringence over the wavelength range 480–620 nm. The second harmonic generation efficiency was tested by using Kurtz and Perry method, keeping KDP as reference.

1. Introduction

The investigations on materials with potential for large second-order NLO applications are very much emphasized in view of the key-role played by them in the frequency conversion process required for the production of new laser output covering ultraviolet and blue-green region. Recently extensive research is going on materials with the capable of efficient frequency conversion of IR radiation to visible and UV wavelengths. Especially, those materials, which can generate highly efficient second harmonic blue light using laser diodes, are of great interest for applications including optical computing, optical information processing, optical disk data storage, medical diagnostics, etc [1]. Organic amino acids are interesting materials formed by weak van der Waal hydrogen bonds; they contain a proton donor carboxyl acid (–COO) group and the proton acceptor amino (–NH_2) group. Due to this dipolar nature, amino acids have physical properties that make them ideal candidates for NLO applications[2]. Amino acids are widely utilized because they not only contain chiral carbon atoms directing to the crystallization process with the non-centrosymmetric space group, but also possess zwitter ionic nature, which is favouring for crystal hardness [3]. In specific, semi organic combinations render many entertaining structures and bonding patterns for the molecular engineering of novel materials [4].

L-histidine bromide (LHB) is one such semi organic NLO single crystal with molecular formula of $C_6H_{12}N_3O_3Br$ was reported by Reena and Sagayaraj [5]. Structural, vibrational and theoretical studies of LHB single crystals have explained by Ahamad *et al* [6]. From the structural point of view, LHB crystallizes in the non-centrosymmetric space group ($P2_12_12_1$) of the orthorhombic system, whose unit cell parameters are a = 7.0530 Å , b = 9.0409 Å and c = 15.2758 Å [7]. The structure of LHB crystal is found to be isostructural with that of L-histidine hydrochloride. Reena *et al.*, Ahamad *et al.*, and other researchers [5-10] have reported on the studies of L-histidine bromide crystals. Ramajothi and Dhanuskodi [11] have reported on the growth of LHB and reported about SHG conversion efficiency and laser damage threshold were measured using a Nd:YAG laser (1064 nm). Joema *et al.*, [12] have studied the effect of doping urea and ammonium sulphate into LHB crystals. In the present work, the LHB single crystals was successfully synthesized, growth, structural and optical properties were studied, which are mandatory in view of NLO applications and the results obtained are reported here discussed in detail.

2. Experimental details

2.1. Synthesis and crystal growth

The starting material are obtained from Sigma-Aldrich (L-histidine) and Merck (Hydrobromic acid), India. For the entire synthesis and growth processes Millipore water resistivity of 18.2 MΩ.cm was used. The LHB has been synthesized by dissolving L-histidine ($NH_2C_6H_7N_2O_2$) and hydrobromic acid (HBr) in the stoichiometry ratio of 1:1 respectively. After the synthesis of raw material, the saturated solution was prepared (60 rpm) at 30 °C, which was kept in jacketed vessel under controlled temperature water bath for slow cooling technique. The temperature of the saturated solution was raised by 5 °C in order to achieve a homogeneous mixture of solute in solvent. The growth solution of LHB was maintained at the saturation temperature (30 °C) for 2 days and then temperature reduction has done in the solution in order to employ slow cooling technique. The solution was initially cooled at a rate of 0.1 °C /day and subsequently 0.2 °C /day as the growth progresses. The crystals shown in figure 1 were harvested when they attained an optimal size and shape in the growth vessels. Optically transparent and well defined single crystal of dimension 9 × 10 × 7 mm^3 was harvested from the solution in a span of 34 days.

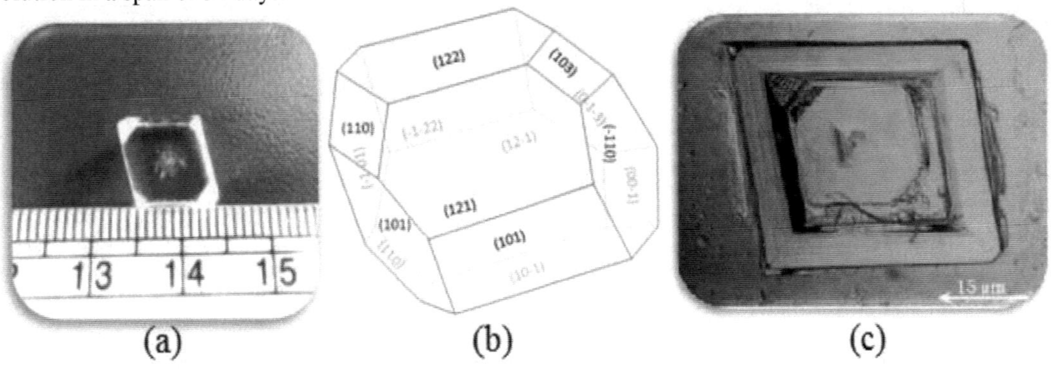

Figure 1. (a) Photograph of grown LHB single crystal (b) morphology of LHB single crystal and (c) micro crystallization image of LHB in the solution.

2.2. Characterizations

Well defined single crystals of good transparency were collected and crused properly into the powder for powder x-ray diffration analysis, which was measured by an advanced D8 powder X-ray diffractometer. The rocking curves of the crystals for the (200) diffraction planes were recorded in symmetrical Bragg geometry using the (121) natural facets by performing a x scan with triple-axis geometry. The monochromated X-ray beam incident on the specimen was obtained using a hybrid two-bounce Ge(220) monochromator with a parabolic multilayer mirror assembly. The diffracted beam from the specimen was detected using a scintillation detector with a triple-axis three bounce Ge(220) analyzer [13]. A Perkin Elmer spectrum 1 Fourier transform infrared spectrometer was employed to determine the infrared spectrum at room temperature in the range of 4000–450 cm^{-1}. The sample was prepared by pressing the crystal powder with KBr to a pellet form. The UV-Vis transmittance spectrum was recorded with a PerkinElmer Lambda 35 spectrometer in the range 200-800 nm. SHG test for the grown crystals was performed by powder technique of Kurtz and Perry. The modified channel spectrum method (MCS) has been employed for dispersion of birefringence behavior of the grown specimen.

3. Results and Discussion

3.1. XRD analysis

The structure and lattice dimensions of the grown LHB crystal have been found from the powder X-ray diffraction analysis and the recorded spectrum is shown in Figure 2. (a). The well defined prominent Bragg's peaks were indexed by using TREOR programme. The calculated lattice parameters are a =7.0162 Å, b =9.0213 Å, c =15.2747 Å, V= 972.42 $Å^3$ and the results are in close agreement with those reported by previous researchers [7, 10]. The narrow, sharp and high intensity peaks reveal that the grown crystals were of high degree of crystallinity and the pattern exhibits the single phase of parent orthorhombic phase. However, the microscopic defect cannot be reflected in the XRD analysis.

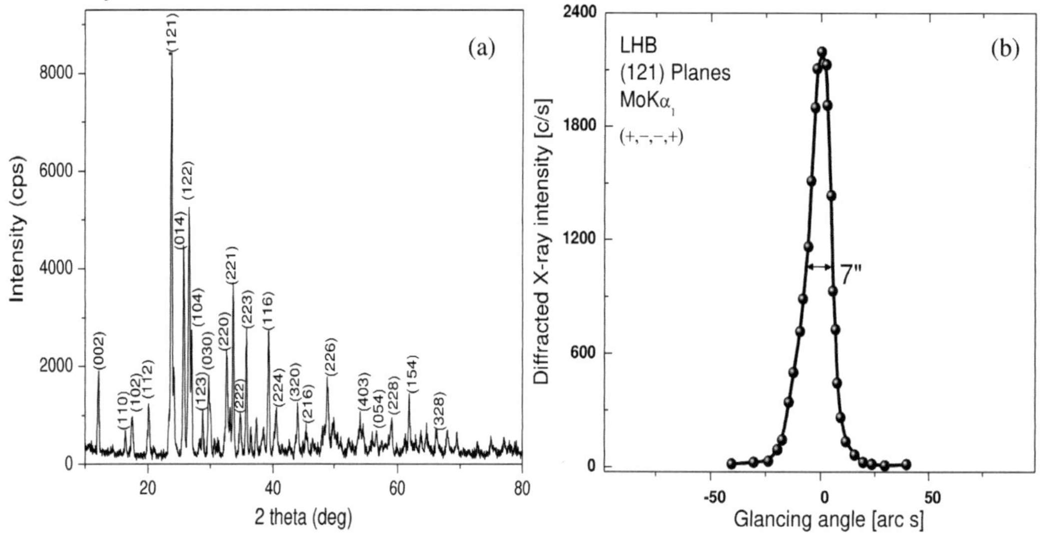

Figure 2. (a) Powder X-ray diffraction Indexed spectrum and (b) HRXRD curve of LHB crystal

3.2. HRXRD analysis

The crystalline perfection of the grown single crystals was characterized by HRXRD by employing a multicrystal X-ray diffractometer. Figure 2. (b) shows the high-resolution DC recorded for a slow cooling grown LHB single crystal using (121) diffracting planes in symmetrical Bragg geometry by employing the multi crystal X-ray diffractometer with $MoK\alpha_1$ radiation. As seen in the Figure 2. (b) the DC is very sharp without any satellite peaks which may otherwise be observed either due to internal structural grain boundaries [14] or due to epitaxial layer which may sometimes be formed in the crystals grown from solution [15]. The full width at half maximum (FWHM) of the diffraction curves is 7 arc sec, which is close to that expected from the plane wave theory of dynamical X-ray diffraction [16]. The single sharp diffraction curve with low FWHM indicates that the crystalline perfection is extremely good. The slow cooling grown LHB specimen is a nearly perfect single crystal without any internal structural grain boundaries. High quality and crystalline perfection contributes to attractiveness of LHB crystals in future device fabrications.

3.3 FTIR spectral analysis

Functional groups present in the grown LHB crystal were analyzed by FTIR analysis and the spectrum is portrayed in the Figure 3. The protonation of COOH and NH_2 occurs, which gives rise to hydrogen bonding interaction between NH_3^+ and COO^-. This confirms that the amino and imidazole are protonated to counter balance the negative charge of carboxylate and bromide ions [12]. There is a broad envelope in the wave number region from 3500 cm^{-1} to 2000 cm^{-1} which is due to superimposed O-H stretch of COOH and N-H stretch of NH_3^+ vibrations. This occurs in the given spectra at 3444 cm^{-1}. This absorption is due to the multiple fine structures in the low wave number regions and it extends up to 2000 cm^{-1}.

Generally in the overtone region, there is a strong band near 1700 cm^{-1}, which is due to the combination of asymmetrical NH_3^+ bending vibration and the torsional oscillations of NH_3^+ group. In the present work, this corresponds to a wave number of 1984 cm^{-1}. The peaks at 1602 cm^{-1} and 1492 cm^{-1} are due to asymmetric and symmetric bending vibrations of NH_3^+ respectively. The asymmetric and symmetric stretching modes of carboxylate group are observed at 1573 cm^{-1} and 1411 cm^{-1} respectively. The strong peak at 1637 cm^{-1} is due to the C=O stretching modes of vibration. The shift of –C = O stretch to near 1000 cm^{-1} supports unprotonation of the carboxyl group, thus confirming the stoichiometry of the LHB crystal, Histidine : HBr, 1:1.

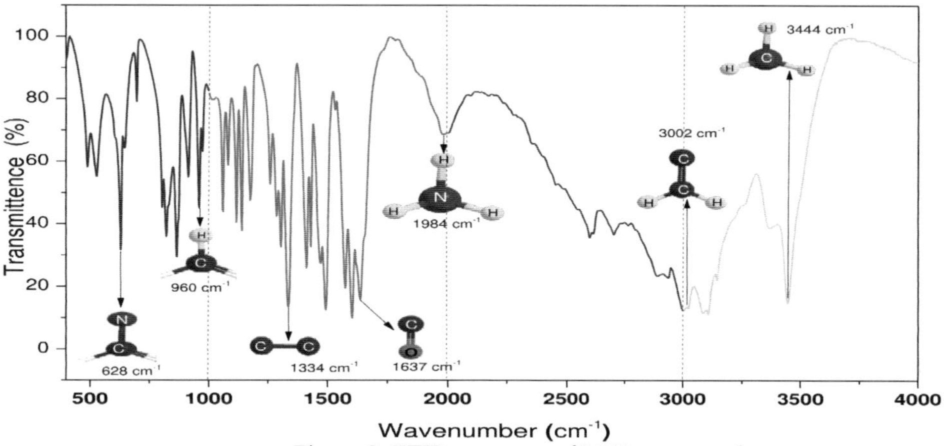

Figure 3. FTIR spectrum of LHB compound

The CH_2 vibrations, which generally lie just below 3002cm^{-1} is clearly resolved in the given spectrum [9]. The strong peak at 1334 cm^{-1} is attributed to C-C stretching vibrations. There are absorption bands at 1286 cm^{-1}, 1175 cm^{-1} and 1082 cm^{-1} which corresponds to NH bending. The bands from 600 $cm^{-1}$to 1000 cm^{-1} are assigned to CH bending. The torsional oscillation of NH_3^+ occurs nearly at 526 cm^{-1}.There is a very strong peak at 1139 cm^{-1} which corresponds to CH deformation. The peak at 809 cm^{-1} clearly indicates bromide ion bonding with carbon atom. From the spectral analysis it is confirmed that the amino and imidazole group are protonated and counter balance the negative charge of the carboxylate and bromide ions functionality. In the FTIR spectrum all the peaks are not only sharp, but also and highly intense, which indicates grown LHB crystals are lesser defects.

3.4 Linear and nonlinear optical analysis

UV-Visible transmission spectrum for LHB crystal of 6.14×10^{-3} m thickness was taken. It was observed that the crystal has good transmission in the entire visible region. Optical transmission spectrum is (Figure 4. (a)) indicates that the LHB crystal have shorter cut off wavelength of 231 nm and it is close agreement with already reported work[11]. The spectrum suggests that the single transition at near UV region in the grown LHB single crystal. The appearance of higher transmittance about 83% of grown LHB crystal (t=6.14× 10^{-3} m) shows not only very low defect concentration but also higher transmission in the entire visible region. The large transmission window in UV and entire visible region facilitates, it to be a potential material for second and third harmonic generation of Nd:YAG laser fundamental and for the generation of the higher harmonics of the GaAlAs laser diode emitting the fundamental in the vicinity of 800 nm for the achievement of blue lasers. No appreciable absorption was detected in the visible region of the spectrum and thereby confirming the advantages of the crystal grown in low and homogeneous stirring rate. This is an intrinsic property of all the amino acids, which illustrates that the crystals have good optical obtain properties.

The grown single crystal of LHB was properly powdered and the samples were sieved out for a uniform particle size of 150 μm and used for the SHG Kurtz-Perry experiment. The acquired value from the plot is (Figure 4. (b)) clearly indicates that in the grown LHB sample SHG output is higher in efficiency (1.8 times) when compared to that of the standard KDP. Generally, SHG is always lower for the defective sectors compared to the more perfect sectors. In the grown LHB crystal SHG conversion efficiency is higher than the earlier works [9,11]. This higher efficiency of LHB may be due to better crystalline perfection was achieved by stirring rate (60 rpm), which increases the bond energy and exhibit higher SHG output during present experiment.

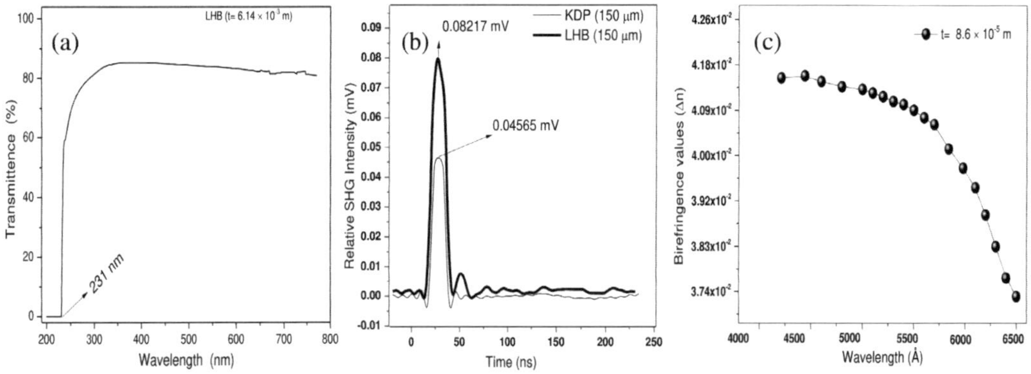

Figure 4. (a) UV-Vis transmission spectrum (b) SHG output of the grown LHB and (c) dispersion of birefringence versus wavelength of LHB crystal

3.5 *Birefringence spectrometry analysis*

Birefringence measurement is a precise technique to investigate the optical homogeneity in NLO crystals and the value of birefringence is an important parameter to calculate phase matching angles [17]. The semi organic LHB crystal grown from slow cooling technique was subjected to MCS method in order to obtain accurate birefringence values. The carefully cut and polished LHB crystal with high transparency and without cracks was subjected for this study at room temperature. The effect of crystal, it produce two linearly polarized component rays whose planes of vibration are polarized perpendicular to each other and will encounter two waves, which have amplitudes of $1/\sqrt{2}$ equal to that of the incoming wave (neglecting surface reflections). When light passes through the LHB, the extraordinary and ordinary waves have a wavelength of λ/n_e and λ/n_o respectively. The light wave exit from the flat surface of the SUT has 180° phase difference between two polarized components. The values of the birefringence have been calculated by finding the absolute fringe order for particular wavelength and computed using the relation: $\Delta n = k\lambda/t$, where λ is the wavelength in nm, t, the thickness of the crystal in mm and k, the fringe order [18]. The graph showing the variation of birefringence (Δn) against wavelength (λ) is presented in Figure 4. (c). . The crystal possesses low birefringence values and it lies between 4.1310×10^{-2} and 3.3640×10^{-2} for a thickness of 8.6×10^{-5} m and the values are in good agreement with the already published article [8]. The experimental result shows that the birefringence decreases with increasing wavelength of entire visible region, and so the grown LHB single crystal is classified as optically positive at room temperature.

4. Conclusion

The research presented in this paper was intended to grow good quality single crystal of LHB by the slow cooling technique with homogenioues stirring of 60 rpm. The grown LHB single crystal possesses not only bulky prismatic morphology but also highly transparent and non hygroscopic. The orthorhombic crystal structure and cell parameters of the LHB were confirmed by powder X-ray diffraction analysis. HRXRD analysis experiment suggested that the LHB crystal is nearly perfect single crystal without any grain boundaries. The UV–Vis spectrum showed good transparency in the UV and Visible region with a lower cut off wavelength of 231 nm and there is no appreciable absorption was detected in the visible region of the spectrum. The SHG relative efficiency of LHB crystal was found to be 1.8 times higher than that of KDP. The crystals showed very low variation in the value of birefringence over a wide wavelength range. The highly perfect LHB crystals obtained from homogeneous stirring rate can be utilized in potential blue laser and NLO related device fabrications.

Acknowledgement
One of the authors (V.J) is grateful to Department of Collegiate Education (DCE), Government of Tamilnadu and University Grants Commision (UGC) for granting the financial support for this work through the Ph.D Scholarship Scheme (2009-2012) and Major Research Project (File No: 38-167/2009-SR; dt. 24-12-2009).

References

[1] Jiang M H and Fang Q 1999 *Adv. Mater.* **11** 1147
[2] Prasanyaa T, Haris M, Jayaramakrishnan V, Amgalan M and Mathivanan V 2013 *Phys. Scr.* **88** 045403
[3] Bhat M N and Dharmaprakash S M 2002 *J. Crystal Growth* **236** 376
[4] Prasanyaa T, Jayaramakrishnan V and Haris M 2013 *Spectrochim. Acta A* **104** 110
[5] Reena I and Sagayaraj P 2003 *J. Crystal Growth* **249** 557
[6] Ahmed A B, Feki H, Abid Y, Boughzala H and Mlayah A 2008 *J. Mol. Str.* **888** 180.
[7] Reena I, Preema T C, Anand D P, Palanichamy M and Sagayaraj P 2005 *Mat. Chem. Phy.* **93** 272.
[8] Vijayan N, Bhagavannarayana G, Nagarajan K and Upadhyaya V 2009 *Mat. Chem. Phy.* **115** 656.
[9] Anandan P, Parthipan G, Saravanan T, MohanKumar R, Bhagavannarayana G, Jayavel R 2010 *Physica B* **405** 4951.
[10] Rajendran K V, Jayaraman D, Jayavel R and Ramasamy P 2003 *J. Crystal Growth* **254** 461.
[11] Ramajothi J and Dhanuskodi S 2007 *Spectro. Acta Part A* **68** 1213.
[12] Joema S E, Perumal S, Ramalingam S and Selvarajan P 2011 *J. Res. Scie. Tech.* **3** 63.
[13] Krishnan P, Gayathri K, Bhagavannarayana G, Jayaramakrishnan V, Gunasekaran S and Anbalagan G 2013 *Spectrochim. Acta A* **112** 152
[14] Lal K and Bhagavannarayana G 1989 *J. Appl. Cryst.* **22** 209
[15] Bhagavannarayana G, Ananthamurthy R V, Budakoti G C, Kumar B and Bartwal K S 2005 *J. Appl. Cryst.* **38** 768
[16] Betterman B W and Cole H 1964 *Rev. Mod. Phys.* **36** 681
[17] Moovendaran K, Jayaramakrishnan V and Natarajan S 2013 Phot. and Optoele. **3** 9
[18] Bhoopathi G, Jayaramakrishnan V, Ravikumar K, Prasanyaa T and Karthikeyan S 2013 Mater. Sci. Poland **31** 1

IOP Conf. Series: Materials Science and Engineering **73** (2015) 012137 doi:10.1088/1757-899X/73/1/012137

Synthesis and Structural Characterization of Niobium Doped Lead-Telluride Glass-Ceramics

M Sathish[1] and B Eraiah[*]
[1]Govt First Grade College, Department of physics, Doddaballapur-561203, INDIA
[*]Department of physics, Bangalore University, Bangalore-560056, INDIA

E-mail:sathishphy79@gmail.com, eraiah@rediffmail.com

Abstract. The basic glasses with composition (70-x) TeO_2-30PbO-xNb_2O_5 (where x=0.1 mol % and 0.2 mol %) were prepared by melt quenching method and heat treated at 280^{0}C for 30 min. The samples becoming glass ceramics was confirmed by SEM. The XRD parameters such as crystallite size of these glass ceramics decreases as increase the impurity and is the order of 184-109A^0. However, micro strain (ε) and dislocation density (δ) increases. Glass transition and thermal stability estimated from DSC measurements and it has been found that both increase with increasing of impurity. Infrared Absorption spectra were measured for TeO_2 glass and glass ceramic doped with Nb_2O_5. The recorded bands attributed to the different modes of vibration and stretching of Te-O band. Optical Absorption spectra of TeO_2-PbO-Nb_2O_5 system shows that the absorption edge has a tail extending towards the lower energies and shifts towards for higher energies for rare earths-doped glass-ceramics. The degree of the edge shift was found to depend on the structural rearrangement and the relative concentrations of the glass basic units. The general appearance of the absorption spectra of these rare earth doped TeO_2 glasses are similar to the spectra observed for other glasses doped with the same kind of rare earth oxides.

1. Introduction

Tellurium oxide glass and glass ceramics have been considered as the best materials for use in memories, laser hosts and non linear optical devices like optical amplifiers and optical filters in view of their high refractive index, low phonon energy, high dielectric constant, good infrared transmission and large third order non-linear susceptibility[1-3]. It is important to study the thermal behaviour and the crystallization mechanisms of these glasses because transparent glass ceramic materials, especially once manufactured via controlled crystallization ,are recently perceived as a new type of non linear optical materials because of their extremely fine grain sizes, many different techniques are to be used to find a correlation between physical characteristics, structural characterizations and the atomic arrangement observed in these glasses such as XRD , SEM , DSC and UV-Visible.

2. Experimental Procedure

2.1 Sample Preparation

The basic glasses with composition (70-x) TeO_2-30PbO-xNb_2O_5 (where x=0.1mol % and 0.2 mol %) were prepared by melt quenching method. The starting materials, Nb_2O_5, PbO and TeO_2 taken in the appropriate proportion weighed in digital electronic balance and mixed thoroughly in a porcelain crucible, then heated in an electrical furnace by the temperature gradually the mixture was melt at a temperature around 850^{0}C. Thus obtained homogeneous melt was quenched between two brass plates. Prepared samples were annealed at 280^{0}C for 30 minutes and during which the base glass transfers into glass ceramic composite.

[*]Corresponding author Email:eraiah@rediffmail.com

3. Results and discussion

3.1 X-Ray diffraction

A Typical XRD patterns of niobium doped lead telluride glass and glass ceramics as shown in figure (1). It observed that prepared glasses at room temperature possess broad peak indicating amorphous structure. The presence of intense peak at $2\theta = 29^0$ in (111) direction for heat treated samples reveals that these glasses are crystalline in nature. The 2θ values observed in the XRD and those of JCPDS (01-0582) data were found in fair agreement between them. Doped glasses were found to exhibit two diffraction peaks associated with (111) and (110), of which the intensity of (111) orientation is predominant.

Table 1. XRD parameters of glass ceramics.

Sample name	d(A^0)	D(A^0)	$\varepsilon\ 10^{-3}(lin^{-2}\ m^{-4})$	$\delta\ 10^{15}\ (lin/\ m^2)$
0.1 mol%	2.92	184	1.876	2.93
0.2 mol%	3.02	109	3.175	8.389

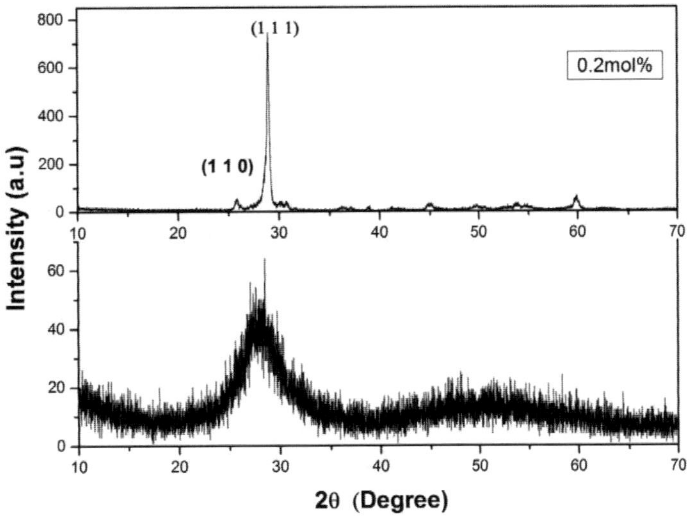

Figure 1. A typical XRD spectra for 0.2 mol % doped glass and glass-ceramic

3.2 Scanning electron microscopy

SEM investigations were performed on glass and glass ceramics. Figure (2) show Typical SEM photographs of the samples undergo translation from amorphous to partial crystallization. It appears that the process of annealing with impurity decreases the number of grain boundaries. As a result, an improved microstructure and morphology are created. Average grain size was found to be 49.2μm for 0.1mol% and 58.1μm for 0.2mol%. Grain growth in similar materials is well known and reported for solid state re-crystallization method.

IOP Conf. Series: Materials Science and Engineering **73** (2015) 012137 doi:10.1088/1757-899X/73/1/012137

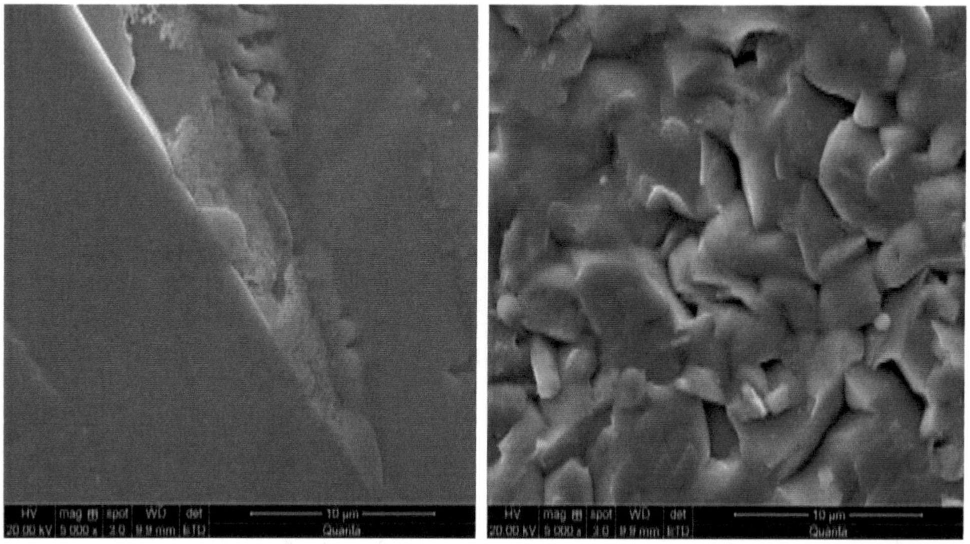

Figure 2. A typical SEM photographs of 0.1 mol % and 0.2 mol % doped glass-ceramics.

3.3 Differential scanning calorimetric Studies

A typical DSC trace of glass is as shown in figure (3) and its clear that the 0.1mol% and 0.2mol% impurity sample contains only one endothermic hump corresponds to glass transition T_g and two endothermic peaks confined to crystallization temperatures T_{C1} and T_{C2} , however , for annealed glass ceramic powders recorded at the same heating rate of 10^0C/min .Of the two thermal events observed, the first one for 0.1 mol %, appearing as an endothermic hump, is the glass transition and exothermic hump was crystallization peak. Secondly for 0.2 mol % consists of two endothermic peaks consider to melting peaks and annealing makes glass transition temperature to decrease from 300^0c to 275^0c. The reason for increase in the T_g is that Nb-O-Te and Nb-O-Nb linkages are increasing as shown in table 2, requiring higher temperature for relaxation [4]. The difference between the onset crystallization temperature and glass transition temperature (T_X-T_g) has been frequently been quoted as a rough indicator of glass stability (ΔT) as depicted in the table 2. It represents the temperature interval during which nucleation takes place [5].

Table 2. Variation of Transition temperature T_g with impurities

Glass composition (mol%)			T_g in ^{0}C	T_c in ^{0} C	$\Delta T = T_x - T_g$ in ^{0}C
TeO_2	PbO	Nb_2O_5			
69.9	30	0.1	298	350	50
69.8	30	0.2	300	360	62

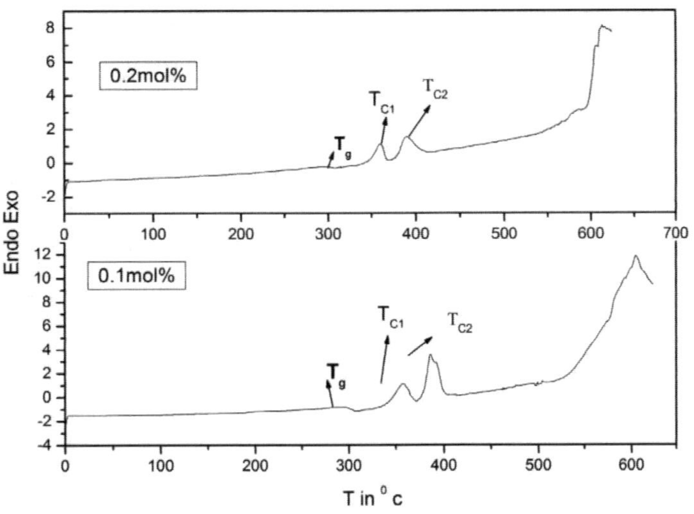

Figure 3. A typical DSC spectra for 0.1 mol % and 0.2 mol % doped glass.

3.4 Optical absorption

Figure4 (A) shows the typical UV absorption spectra of glass. It has been found that the UV of TeO_2 edge shifts to shorter wavelengths with increasing the Nb_2O_5 content in the binary telluride glasses. In this study, the position of fundamental absorption edge shifts to higher energy (shorter wavelength) with increasing Nb_2O_5 in the system. The shift of the ultraviolet absorption band to shorter wavelengths corresponds to transitions from NBO with bound exited electron less tightly than the bridging oxygen [6]. The general appearance of the absorption spectra of the present samples is similar as found by burger et al and other glasses doped with same kind of rare earth oxides, such as P_2O_5-Pr_6O_{11} glass and vanadium-phosphate glasses doped with different rare earth oxides [7-8].

It can be seen that there exists a linear dependence of $(hv)^2$ in the photon energy as shown in figure 4(B). This suggests that at higher photon energies the transitions occurring in the present glass and glass ceramics are of indirect type. In the present work the addition of rare earths to TeO_2 is to produce replacement of the O-Te-O ,and/or O-Te-O bands in the glass network and to be replaced by RE-Te-O,RE-O-Te and/or O-RE-O bands which is reflected in the absorption spectra by a significant shifting of the absorption edge are most likely to related to structural rearrangement of the glass and the relative concentration of the various fundamental units.

The optical band gap for glass with impurity 0.1mol% and 0.2mol% is the order of 1.70eV and 2eV,wherese as the annealed samples of the order of 1.73eV and 2.04eV respectively. The absorption edge depends on the oxygen bond strength in the glass forming network. However, by doping Nb_2O_5 the optical band gap is constituted by the host materials PbO does not alter much the band gap picture of glass but increases slightly in the case of glass ceramics.

IOP Conf. Series: Materials Science and Engineering **73** (2015) 012137 doi:10.1088/1757-899X/73/1/012137

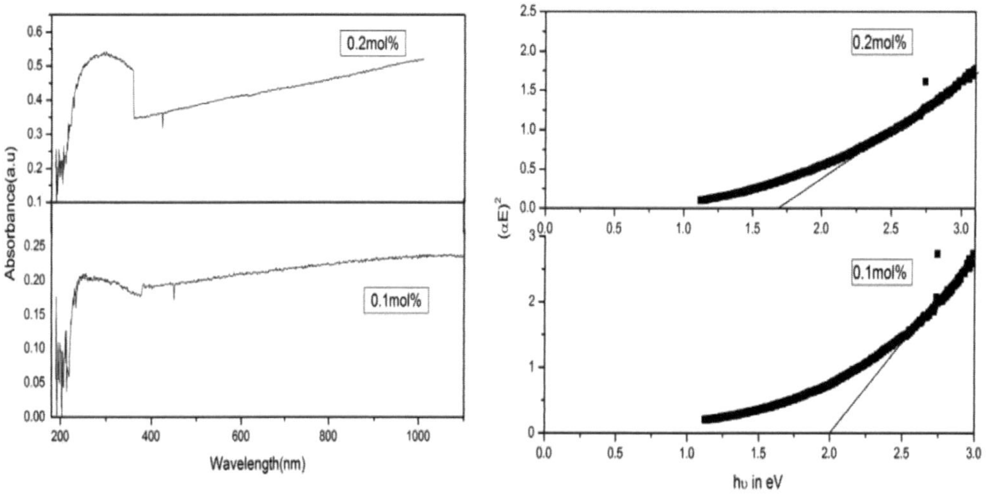

Figure 4(A). UV-Visible spectra of doped glass. **(B).** A Typical $(\alpha E)^2$ as a function of energy for the composition (70-x) TeO_2-30PbO-xNb_2O_5 glass ceramic samples with 0.0mol% and 0.2mol% impurity.

3.5 Infrared spectroscopy

Figure (5) shows the Typical IR spectra of the TeO_2 glasses doped with the rare earths oxides of the type Nb_2O_5 in the spectral range 400-4000 cm^{-1}. For all glass and glass ceramics presented in this work, the spectra showed bands from 640 cm^{-1} to 778 cm^{-1}.

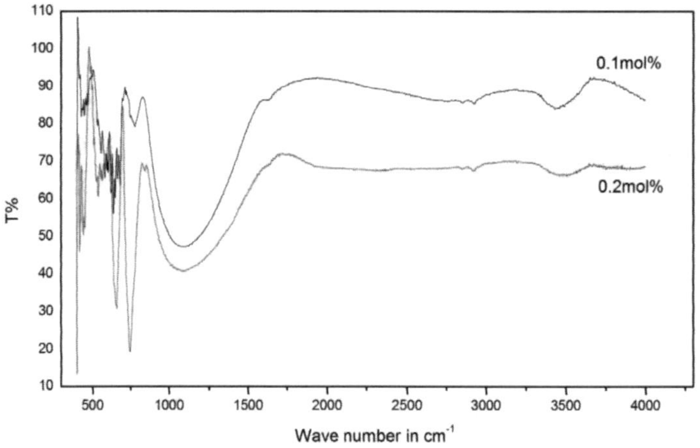

Figure 5. A typical IR spectra for 0.1 mol % and 0.2 mol % doped glass.

The main IR band at 640 cm^{-1} attributed to symmetrical vibration oscillations of Te-O bands. Rare earths connected to the chains of TeO_4 groups are identified on the basis of simultaneous presence of the bands at 660 cm^{-1}-625 cm^{-1}. For rare earths-doped samples new bands were detected from 418 cm^{-1} to 594 cm^{-1}, respectively. These bands attributed to O-Nb-O stretching vibrations. The spectra of all the studied samples show the typical broadening of the observed bands. In the most cases, they are similar to the spectra of the crystalline phases after heat treatment. This is the direct proof for the similarities of structural units and of the short range order. For this reason, interpretations of spectra of

TeO_2 based samples based on the conclusions drawn for their crystalline phases. This is assumed that this is the result of particular structure of trigonal by pyramids of TeO_4 groups and presence of a free electron pair in one of the broadening positions [9]. It is clear that beyond the existence of an average deformation of the network and long range disorder in the system [10]. By analogy with the crystalline telluride [9], it may be accepted that with the introduction of other oxides in the telluride samples. As new structural arrangement of the back side of four coordinated tellurium ions favour the stronger polarizability of the free electron pair in certain directions. As a result, parts of the TeO_4 groups are transformed into TeO_3. 1000 cm^{-1} and 1100 cm^{-1} are attributed to **Pb-O** the asymmetric stretching vibrations [$\mathbf{PbO_n}$] tetrahedral structural units [11-13]. 816cm^{-1} assigned to vibration of **Nb** and its neighboring non bridging system and to bridging to **Nb-O** group motion [14-15].

4. Conclusions

It is observed that the grain size decreases which intern increases the strain and dislocation density. But there are no noticeable peaks in XRD indicating presence of impurity Nb_2O_5 into the samples. Due to the heat treatment grain growth will takes place which gives the partial crystallinity and good surface morphology. These results good agreement with XRD characteristics.SEM reveals that samples are translated to partial crystalline state. It is clear that glass stability increases as the Nb_2O_5 content increases and decrease of glass transition temperature in annealed sample. Quantitative justification of some of the bands for both binary TeO_2 rare earth doped system was done. Optical observation characteristic of niobium doped lead telluride glass and glass ceramic were studied. The absorption spectra of rare earth ions in the glass differ from that of rare earth crystals and broadening of the absorption bands is due to the multiplicity of rare earth sites in the system. The results show that optical band gap values increase on addition of impurity to glass. However, upon heat treatment the band gaps were observed to increase.

References

[1] Vijay Prakash G, Narayana Rao D and Bhatnagar A K 2001 *Solid.State.Commun.* **39** 119
[2] Waber M J, Meyers J D and Blackbum D H 1981 *J.Appl.Phy.* **52** 2944
[3] Vogel E M, Weber M J and Krol D M 1991 *Phys Chem Glasses.* **32** 231
[4] Sene f f, Martinelli J.R and Gomes 2004 *J.Non-Cryst. Solids.* **30** 348
[5] J Chinna Babu and V Chandra Mouli 2009 *Indian J pure & App Phy.* **47** 241
[6] Harini, R R Hogarth C A and Ahamed M M 1984 *J.Mater.Sci.Lett.* **3** 843
[7] Sidek,H A A, S Chow, Z A Talib and S.A.Halim 2004 *Turk.J.Phys.* **28** 65
[8] Khan M N, Harini and Ahamed M M 1985 *J.Mater.Sci.* **20** 2207
[9] Lindqvist O 1968 *Acta .Chem.Scand.* **22** 977
[10] Dimitriv Y and Dimitrow V 1978 *Mater.Res.Bull.* **13** 1071
[11] Mandal S, S Hazara and Ghosh A 1994 *J.Mater.Sci.Lett.* **13** 1054
[12] Hazara S and Ghosh A 1995 *J.Mater.Res.* **10 (9)** 2374
[13] Milanova M, Lardonova R and Kostov K L 2009 *J.Non.Cryst.Solids.* **3536** 379
[14] Senthil Murgan G and Ohishi Y 2005 *Phy.Chem.Glasses.* **46(4)** 359
[15] Takyuki Komatsu, Hiroharu Mhari and Kazumasa Malusita 1991 *J.Non-Cryst Solids.* **135** 105

International Conference on Materials Science and Technology (ICMST 2012) IOP Publishing
IOP Conf. Series: Materials Science and Engineering **73** (2015) 012138 doi:10.1088/1757-899X/73/1/012138

Theoretical and Experimental studies on nonlinear optical L-phenylalaninium maleate

M. Victor Antony Raj and J. Madhavan*
Department of Physics, Loyola College, Chennai -34

E-mail: jmadhavang@yahoo.com

Abstract. Good quality single crystals of L-phenylalaninium maleate (LPM) were grown successfully by slow evaporation technique. The grown crystal was subjected to powder X-ray diffraction study at room temperature. From the powder XRD studies it was observed that the crystal belongs to monoclinic crystal system having non-centrosymmetry with $P2_1$ space group. Quantum chemical Density functional theoretical (DFT) computations were performed and the optimized structure parameters of LPM molecule calculated by DFT (B3LYP) level with 6-31G (d,p) basis set were obtained. The optical absorption spectrum is used to study various linear optical parameters as a function of incident photon energy. It is found that LPM is a suitable material for NLO applications. HOMO-LUMO calculations were performed and the energy band gap was obtained.

1. Introduction

Nonlinear optical chromopores have attracted great interest in recent years and their compounds are important for the development of high-speed optical modulators and switches. In this context, researchers have been focusing on the search for materials with large optical nonlinearities and for their application in the fields of high speed data transmission, processing and storage [1-3]. Considering the significance and influence of Aminoacids in NLO materials, in this paper, we study the molecular and nonlinear optical properties of L-Phynelalaninium maleate(LPM) using ab initio and density functional theory (DFT) calculations.

2. Experimental Procedure

High purity L-phenylalanine (Merck 99%) and maleic acid (AR grade) were taken in 1:1 molar ratio and dissolved in deionized water. The reaction is as follows,

$$C_6H_5CH_2CH\,(NH_2) + C_4H_4O_4 \longrightarrow C_6H_5CH_2CH\,(NH_3)^+\,C_4H_3O_4^-$$

The photograph of as grown single crystals of LPM crystal is shown in Figure 1.

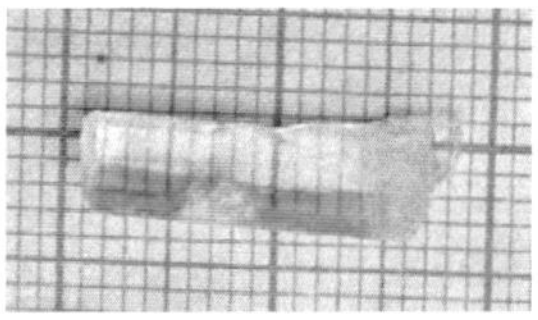

Figure 1 Photograph of as grown LPM single crystal

3. Computational details

Quantum chemical Density functional theoretical (DFT) computations were performed using closed-shell Becke–Lee–Yang–Parr hybrid exchange-correlation three-parameter functional (B3LYP) in combination with 6-31G (d,p) basis set to derive the complete geometry optimizations and normal-mode analysis on isolated entities. Above said task was achieved using Gaussian03W [4].

4. Results and Discussion

4.1 Powder X-ray Diffraction Analysis

The grown single crystal of LPM was subjected to powder X-ray diffraction. The powder form of the above-mentioned crystal was taken for analysis using a Rich powder diffractometer - RICH SIEFERT & CO with a scan speed of 1°/min.

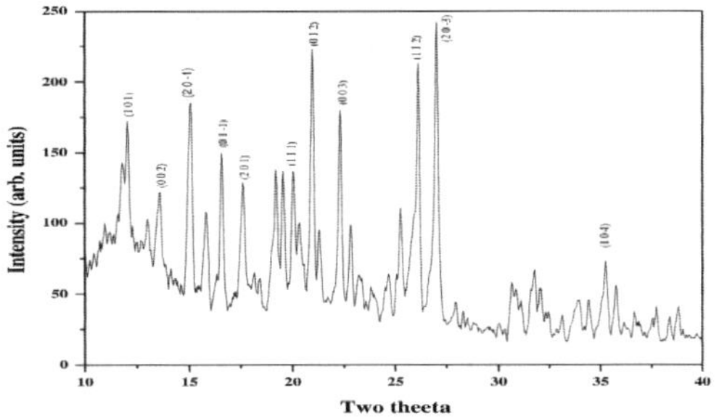

Figure. 2 Powder XRD pattern of LPM

The resulting powder X-ray diffraction pattern is shown in Figure 2. The obtained two-theta values were used for indexing by using the XRDA software package. It was confirmed that the crystal belongs to the monoclinic crystal system having non-centrosymmetry with $P2_1$ space group. Lattice parameters have been determined as a=11.056 Å, b=5.332 Å, c=11.471 Å, α=90.00°, β=101.070°, γ=90.00°.

4.2 Molecular geometry

The optimized molecular structure of the isolated LPM molecule calculated using DFT theory at B3LYP/6-31 level is shown in Figure 3.The optimized geometrical parameters such as bond length and bond angles are given in Table 1.

Table 1. Selected bond length and Bond angles of LPM molecule

S.No	Bondlength(Å)		Bond angles(°)	
1	C_1-C_6	1.39483	C_1-C_6-C_5	119.999
2.	C_1-C_2	1.39516	C_6-C_1-C_2	119.998
3.	C_2-C_3	1.39471	C_1-C_2-C_3	120.008
4.	C_3-C_4	1.39543	C_2-C_3-C_4	119.994
5.	C_4-C_5	1.39482	C_4-C_5-C_6	119.994
6.	C_5-C_6	1.39514	C_4-C_3-C_{12}	119.993
7.	C_3-C_{12}	1.54000	C_3-C_{12}-C_{15}	109.471
8.	C_{12}-C_{15}	1.54000	C_{15}-C_{21}-O_{22}	119.99999
9.	C_{15}-N_{17}	1.47000	O_{23}-C_{21}-O_{22}	119.99996
10	C_{15}-C_{21}	1.54000	C_2-C_3-C_{12}	120.01277
11.	C_{21}-O_{23}	1.43000	H_{19}- N_{17}-H_{18}	109.47119
12	C_{21}-O_{22}	1.43000	C_{21}-C_{15}-H_{16}	109.47122
13.	N_{17}-H_{19}	1.00000	C_{12}-C_{15}-C_{21}	109.47119
14.	O_{22}-H_{24}	1.25000	C_{12}-C_{15}-N_{17}	109.47123

IOP Conf. Series: Materials Science and Engineering **73** (2015) 012138 doi:10.1088/1757-899X/73/1/012138

4.3HOMO-LOMO Analysis

The energies of the highest occupied molecular orbital (HOMO) and the lowest unoccupied molecular orbital (LUMO) are computed at B3LYP/6-31G(d, p) level. HOMO and LUMO orbitals are shown in Figure 4. Generally, the energy values of LUMO, HOMO and their energy gap reflect the chemical activity of the molecule. The energies of the HOMO and LUMO, based on the optimized structure are computed as -0.10814 and -0.05570 a.u, respectively. The HOMO–LUMO energy gap is -0.05244 a.u. The calculated HOMO and LUMO energies clearly show that charge transfer occurs within the molecule.

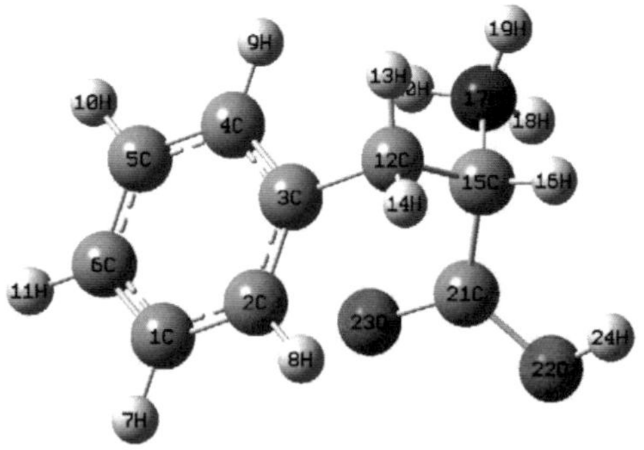

Figure-3. Atomic numbering system adapted for ab initio computations of LPM molecule.

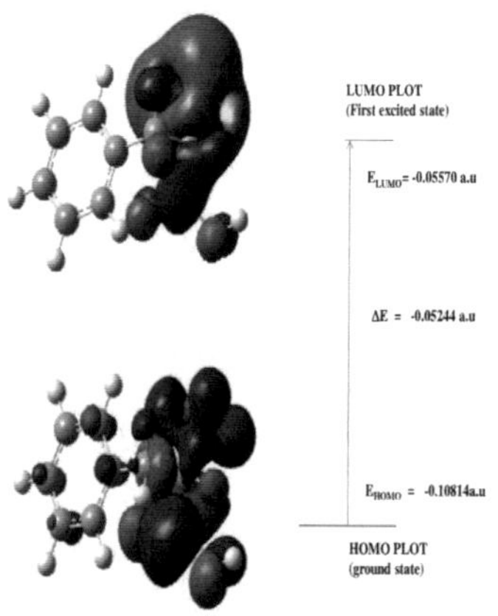

Figure 4. HOMO – LUMO plot of LPM molecule.

4.4 Optical Properties of LPM

The optical properties of a material are important, as they provide information on the electronic band structures, localized states and types of optical transitions. To determine the transmission range and hence to know the suitability of LPM single crystals for optical applications, the UV-Vis-NIR spectra (Figure 5) were recorded in the range of 200–1200 nm. The spectrum reveals that the LPM compound has a wide transparency window in the region 240–1000 nm (~80%) and shows no remarkable absorption in the visible region. Figure 6 and Figure 7 shows the plot of extinction coefficient and refractive index against incident photon energy. It's found that extinction coefficient remains constant for incident photon energy of 1.2 eV to 5.5 eV. The variation of refractive index with incident photon energy shows that initially the refractive index decreases with increasing photon energy then becomes constant.

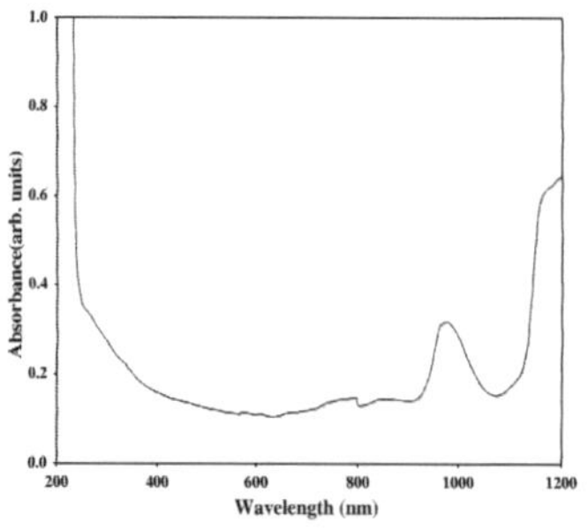

Figure 5. Optical absorption spectrum of LPM

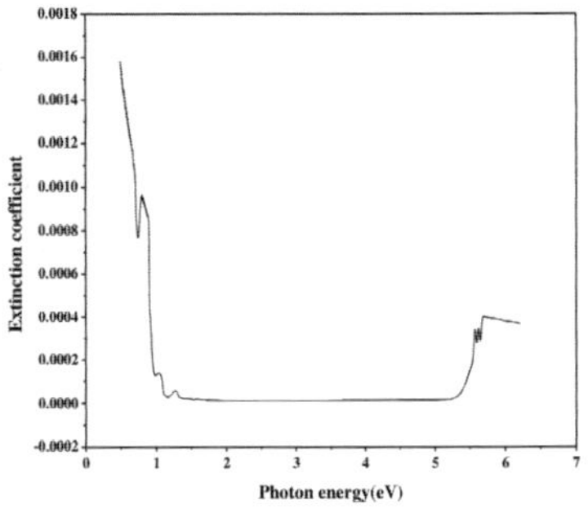

Figure 6. Extinction coefficient Vs Incident Photon Energy of LPM

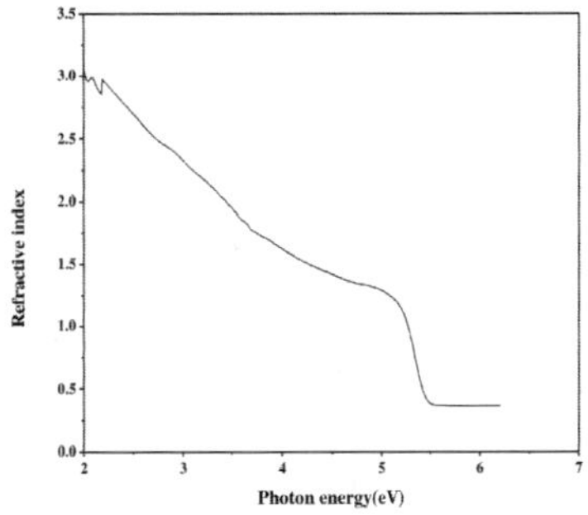

Figure 7. Refractive index Vs Incidient photon energy of LPM

5. Conclusions

Optically good quality single crystals of LPM were grown by slow solvent evaporation technique. The grown crystal is subjected to Powder X-ray diffraction analysis and it is found that the crystal belongs to the monoclinic system with $P2_1$ space group. Density functional theory (DFT) computations using (B3LYP) level with 6-31 G (d,p) basis set gives optimized structure parameters of LPM molecule. It is understood that hydrogen bonds are prevelant in this material and the optimized C–C bond lengths of LPM fall in the range 1.39471–1.54000 A°. Molecular energy gap of LPM was found as -0.05244 au by HOMO-LUMO analysis. Optical absorption spectrum was recorded for the given crystal and it is found that it has minimum absorption between 240 -1000 nm. Linear optical constants like extinction coefficient and refractive index are calculated and related with incident photon energy. This optical analysis reveals the NLO behavior of the material.

6. References

[1]. Senthil. S, Pari. S, Sagayaraj. P, Madhavan. J, *Physica B*, 404 (2009) 295.
[2]. Sankar. D, Praveen Kumar. P, Madhavan J, *Physica B*, 405(2010) 1233.
[3]. Madhavan J., Aruna S., Anuradha A., Premanand D., Vetha Potheher I., Thamizharasan K and Sagayaraj P. (2006c) , *Optical Materials*, Vol. 29, pp. 1211-1216.
[4]. Frisch M.J, Trucks G.W, Schlegel H.B, Scuseria G.E, Robb M.A, J.R. Cheeseman J.R, Montgomery Jr J.A, Vreven T, Kudin K.N, Burant J.C, Millam J.M, Iyengar S.S, Tomasi J, Barone V, Mennucci B, Cossi M, Scalmani G,.Rega N, Petersson G.A, Nakatsuji H, Hada M, Ehara M, Toyota K, Fukuda R, Hasegawa J, Ishida M, Nakajima T, Honda Y, Kitao O, Nakai H, Klene M, Li X, Knox J.E, Hratchian H.P, Cross J.B, Adamo C, Jaramillo J, Gomperts R, Stratmann R.E, Yazyev O, Austin A.J, Cammi R, Pomelli C, Ochterski J.W, Ayala P.Y, Morokuma K, Voth G.A, Salvador P, Dannenberg J.J, Zakrzewski V.G, Dapprich S, Daniels A.D, Strain M.C, Farkas O, Malick D.K, Rabuck A.D, Raghavachari K, Foresman J.B, Ortiz J.V, Cui Q, Baboul A.G, Clifford S, Cioslowski J, Stefanov B.B, Liu G, A.Liashenko, Piskorz P, Komaromi I, Martin R.L, Fox D.J, Keith T.,.Al-Laham M.A, Peng C.Y, Nanayakkara A, Challacombe M, Gill P.M.W, Johnson B, Chen W, Wong M.W, Gonzalez C, Pople J.A, Gaussian 03, Revision C.02, Gaussian Inc.,Wallingford, CT, 2004.

IOP Conf. Series: Materials Science and Engineering **73** (2015) 012139 doi:10.1088/1757-899X/73/1/012139

HYDROGEN AS FUEL CARRIER IN PEM FUELCELL FOR AUTOMOBILE APPLICATIONS

Mudassir Ali. Sk[1], K. Venkateswara Rao[2], Jagirdar V. Ramana Rao[3]

[1,2,3] Center for Nano Science and Technology, Institute of Science and Technology
Jawaharlal Nehru Technological University Hyderabad
Kukatpally, Hyderabad, 500085-INDIA

E-mail: smali.nanot@gmail.com

Abstract. The present work focuses the application of nanostructured materials for storing of hydrogen in different carbon materials by physisorption method. To market a hydrogen-fuel cell vehicle as competitively as the present internal combustion engine vehicles, there is a need for materials that can store a minimum of 6.5wt% of hydrogen. Carbon materials are being heavily investigated because of their promise to offer an economical solution to the challenge of safe storage of large hydrogen quantities. Hydrogen is important as a new source of energy for automotive applications. It is clear that the key challenge in developing this technology is hydrogen storage. Combustion of fossil fuels and their overuse is at present a serious concern as it is creates severe air pollution and global environmental problems; like global warming, acid rains, ozone depletion in stratosphere etc. This necessitated the search for possible alternative sources of energy. Though there are a number of primary energy sources available, such as thermonuclear energy, solar energy, wind energy, hydropower, geothermal energy etc, in contrast to the fossil fuels in most cases, these new primary energy sources cannot be used directly and thus they must be converted into fuels, that is to say, a new energy carrier is needed. Hydrogen fuel cells are two to three times more efficient than combustion engines. As they become more widely available, they will reduce dependence on fossil fuels. In a fuel cell, hydrogen and oxygen are combined in an electrochemical reaction that produces electricity and, as a byproduct, water.

1.Introduction

Nanomaterials have attracted great interest in recent years because of the unusual mechanical, electrical, electronic, optical, magnetic and surface properties. The high surface/volume ratio of these materials has significant implications with respect to energy storage. Both the high surface area and the opportunity for nanomaterial consolidation are key attributes of this new class of materials for hydrogen storage devices. Recent investigations have shown that nanoscale materials may offer advantages if certain physical and chemical effects related to the nanoscale can be used efficiently. The present review focuses the application of nanostructured materials for storing of hydrogen in different carbon materials by physisorption method. To market a hydrogen-fuel cell vehicle as competitively as the present internal combustion engine vehicles, there is a need for materials that can store a minimum of 6.5wt% of hydrogen. Carbon materials are being heavily investigated because of their promise to offer an economical solution to the challenge of safe storage of large hydrogen quantities. Hydrogen is important as a new source of energy for automotive applications. It is clear that the key challenge in developing this technology is hydrogen storage. Hydrogen is clean and renewable energy carrier and an Hydrogen energy system is expected to progressively replace fossil fuels in future. Due to compressed gas, containers in vehicles have less volume than the classic ones. A clear picture of the true capacity of nanotubes for hydrogen storage is still being developed by efforts in experiment and theory. Because CNT received directly after synthesis, generally, have closed caps, methods for chemical activation are required to achieve their full potential. These selected carbon nanomaterials are characterized by the different techniques, for structural characterization and crystallite size determination XRD is used. SEM for the morphological study and FTIR for the spectral data of carbon nanomaterials were used. The surface areas of Selected Carbon nanomaterials were measured using BET.

2.Hydrogen Adsorption Principle

In hydrogen adsorption studies, the hydrogen concentration absorbed by Carbon Nano materials can be determined by various techniques such as the gravimetric, electrolytic and the gasometric methods. From the gravimetric methods, the hydrogen concentration present in the material can be obtained by the change the mass of the material before and after hydrogen adsorption. The sample forms one of the electrodes in the case of electrolytic method, where hydrogen is given by the electrolyte (e.g., KOH) when a known current passed through the circuit. In gasometric method, the hydrogen concentration present in the material can be obtained from volumetric or pressure reduction technique. In volumetric technique, the measurement of the change in the volume gives the hydrogen concentration of a material at a constant pressure and temperature. In the pressure reduction technique, hydrogen concentration is calculated from the observed pressure change before and after hydrogen adsorption at a constant calibrated volume and at a constant temperature. The hydrogen adsorption capacity of selected carbons is expressed in terms of wt%. It is defined as

$$\text{Weight Percent (Wt\%)} = \frac{\text{Weight of the hydrogen molecule}}{\text{Weight of the material}} * 100$$

3.Experimental Methods

A test of five different carbon samples, such as Commercial CNTs, Carbon Black, Activated Carbon, Exfoliated Graphite, Graphitic Flakes are assembled. Experiments have been carried on these samples at room temperature of 297 Kelvin (24^{0}C) using hydrogen gas to be adsorbed. The parameters taken while conducting the adsorption experiment have been studied and maintained in a proper manner so that the measurement is accurate. Different temperatures and pressure ranges are taken to study the possibilities of hydrogen intake for selected carbon materials.

4.Results and Discussion

The data of hydrogen adsorption on different carbon materials produced by arc-discharge at different experimental conditions with a constant amount (0.1 gram) was carried out. Selected samples were examined for their ability to adsorb H_2 at pressures up to 100 millibars. Surface area for the selected samples is shown in Table 1. Hydrogen adsorption isotherms of selected samples of selected carbons are shown in Figure 6.

Table 1. BET Surface Area of Selected Carbon Nano Materials

Sample	**BET Surface area (m^2/g)**
Activated Carbon	1050
Carbon Black	70
Graphite Flakes	30

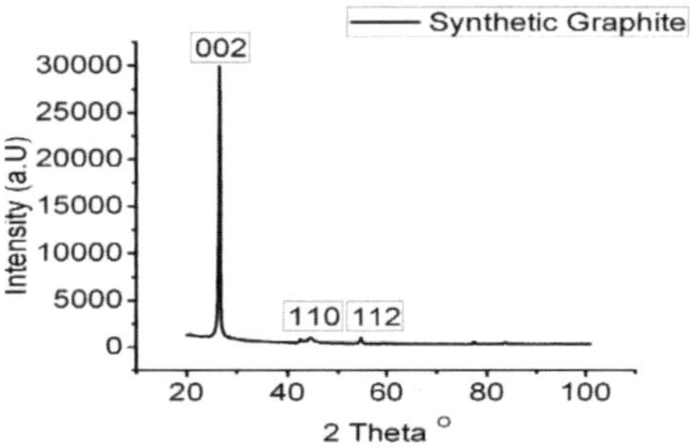

Figure 1. X-Ray Diffraction of Synthetic Graphite, the peaks are identified for their lattice structure and the orientation

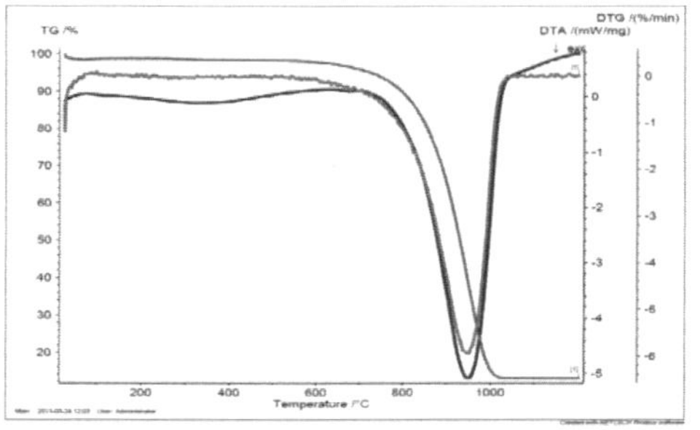

Figure 2. TG/DTA/DTG of Graphitic Flakes, There is a drastic weight loss with almost 55 % of weight loss can be seen.

Figure 2., the green line indicates the TG analysis, shows which, the weight is gradually decreases at approximately 750°C resulting in drastic weight loss, which can be attributed to burning of graphite flakes from 740°C - 1020°C.

Figure 3. SEM of Exfoliated Graphite, has domain of multilayer structure with many diamond shaped pores

When heating at 350°C, an apparent volume change of the exfoliated graphite is observed. It can be seen in Figure 3., that Exfoliated Graphite is worm-like or accordion-like and graphite layers are held together at their edges. It is clear that the exfoliated domain has a multilayer structure with many diamond shaped pores. The particles have an average diameter range between 50-300 nm and lengths upto 10-60 μm.

Figure 4. SEM of Graphite Flakes,

Figure 5. SEM of Carbon Black. Based on the variable thickness of the particles

A study conducted by SEM is conclusive to characterize the texture (elemental particles and their aggregating) as well as the structure of the carbon black Figure 6. It shows that the arc-discharge plasma process is a versatile process for growing a wide range of carbon black.

IOP Conf. Series: Materials Science and Engineering **73** (2015) 012139 doi:10.1088/1757-899X/73/1/012139

- From the results, it can be noticed that the hydrogen adsorption capacities of different carbon samples in the range of 0.8 wt%-4.0 wt%.
- In the case of Graphitic flakes, the sample showed the adsorption cpacity of 4.4 wt% at high pressures (100 millibars).

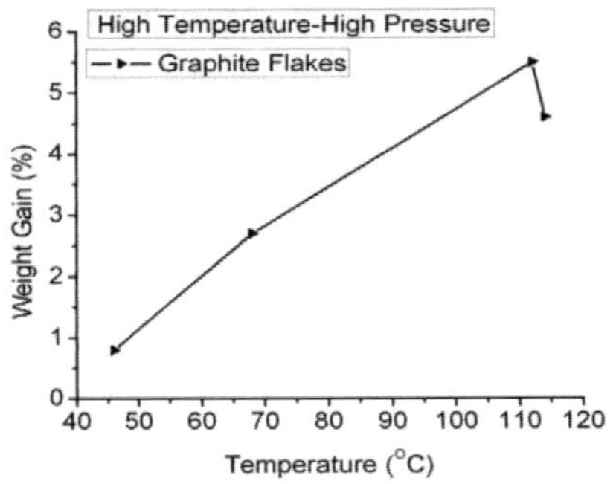

Figure 6. Adsorption isotherm of Graphite Flakes

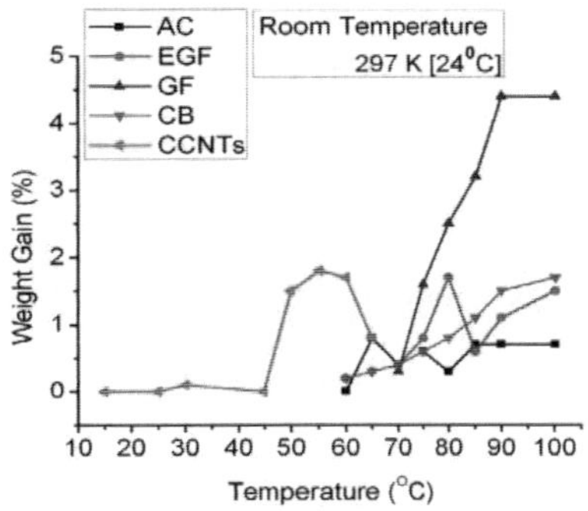

Figure 7. Adsorption Isotherms of H_2 on Selected Carbon

Figure.7, shows the uptake of hydrogen begins at 46°C giving rise to 0.8 wt% and continuesly shoots up to 68°C to 2.7 wt%, and at 112°C to 5.5 wt% respectively. At temperature 114°C the uptake gradually slows down to 4.6 wt%. The maximum hydrogen adsorbed can be noticed to be 5.5 wt% at 112°C. Isotherm of graphite flakes, the adsorption of gas suddenly decreased to 5.9 wt% which is almost equal to 6 wt% at 37°C, due to reason that flake graphite has a distinctly flaky or platy morphology.

5.Conclusion

Hydrogen storage in carbon single-wall nanotubes has become the focus of numerous research groups in the world. However, obtaining activated SWNT hydrogen storage materials with highly reproducible adsorption capacities has not yet been achieved. One reason for this may be that hydrogen storage is only optimized for a very specific and narrow distribution of SWNTs of distinct types and diameters.

References

[1] Dillon AC, Jones KM, Bekkedahl TA, Kiang CH, Bethune DS, Heben MJ. 1997 *Nature* **9** 386.
[2] Wu XB, Chen P, Lin J, Tan KL. *Int. J. Hydrogen Energy.* 2000 **5** 261.
[3] Wang QK, Zhu CC, Liu WH, Wu T. *Int. J. Hydrogen Energy.* 2002 **500** 497.
[4] Zuttel A, Sudan P, Mauron P, Kiyobayashi T, Emmenegger C, Schlapbach L. *Int. J. Hydrogen Energy.* 2002 **12** 203.
[5] Smith MR, Bittner EW, Bockrath BC. 2002 *Preprints Symp. Am. Chem. Soc. Div. Fuel Chem.* **5** 784.
[6] Hou PX, Xu ST, Ying Z, Yang QH, Liu C, Cheng HM. 2003 *J. Carbon* **6** 2471.
[7] Ma Y, Xia Y, Zhao M, Wang R, Mei L. 2001 *Phys. Rev.* B **63** 115422.

Acknowledgement

Author would like to thank Center for Nano Science and Technology, Institute of Science and Technology, Jawaharlal Nehru Technological University Hyderabad, Center for Environmental Science, Institute of Science and Technology, Jawaharlal Nehru Technological University Hyderabad for supporting me and arranging me the Laboratory instruments to carry on my Project work and Authors would like to thank, Department of Carbon Materials, The International Research Center for Powder Metallurgy and New Materials, (ARCI) Balapur Village, Hyderabad, an autonomous R&D center, Government of INDIA, Department of Science and Technology (DST), for providing the Carbon Materials and detailed Characterization analysis.

Electronic properties of armchair graphene nanoribbons doped with Cobalt atoms

Neeraj K. Jaiswal and Pankaj Srivastava

Nanomaterials Research Group,
Computational Nanoscience and Technology Lab (CNTL),
ABV-Indian Institute of Information Technology and Management, Gwalior, India

E-mail address: neerajkjaiswal@gmail.com

Abstract. We have investigated the electronic properties of Co doped armchair graphene nanoribbons (AGNR) within the framework of density functional theory. We have considered Co as an adsorbed atom as well as substitutional dopant in AGNR. Adsorption is found energetically favorable over substitutional doping. The calculated formation energy suggests edge as the most energetically favorable site for Co atoms in AGNR. It is observed that for all the considered sites, Co atoms effectively modify the electronic band structures of AGNR and introduce metallicity. Moreover, substitutional Co-doping induces a large number of conducting channels crossing the Fermi energy and thereby enhancing the metallicity in AGNR as compared to Co-adsorption. The observed highly metallic nanoribbons are crucial for nano interconnects and other technological applications.

1. Introduction

Graphene is the one atom thick 2-D sheet of C atoms exhibiting extraordinary electronic, physical and chemical properties. Initially it was assumed that 2-D materials do not exists in real physical world. The successful synthesis of graphene by K. S. Novoselov and A. K. Geim in 2004 [1] worked as the formal invitation for the research community to work with graphene. Research on graphene was further accelerated by the development of various experimental (synthesis and characterization) techniques which made it possible to produce graphene in controlled number of layers [2], large area synthesis of monolayer graphene [3] and production of graphene nanoribbons (GNR) with sharp edges [4]. GNR are quasi 1-D strips of graphene presenting promising properties for various applications such as field effect transistors [5, 6], spin filters [7], p-n junction diodes [8] and negative differential resistance devices [9]. Depending up on the shape of edges, GNR are known as either zigzag graphene nanoribbons (ZGNR) or armchair graphene nanoribbons (AGNR) [10]. The most striking application of GNR has been shown in electrical, chemical and bio-sensors [11-13]. Recently, Wang et al demonstrated single atom substitutional doping of monolayer graphene [14]. They used metal atoms to deposit in the mono and di-vacancies of single layer graphene. Efforts have been made to alter the band gap of GNR with metallic impurities [15, 16]. It is reported in previous theoretical investigations that transition metal (TM) impurities can induce a magnetic moment in GNR [17, 18]. Further, TM impurity can exist in graphene/GNR either as a substitutional dopant or in the form of adsorbed adatoms. Efforts have been done to study both types of TM impurities in GNR [21-23]. However, both types of TM impurities have been investigated individually but still, there is lack of research comparing the adsorption of TM impurities with that of substitutional doping in nanoribbons. Therefore, we study the adsorption of Co atoms in AGNR and compared the results with substitutional Co impurities in AGNR.

2. Computational Method

In the present work, we perform first principles calculations within the density functional theory (DFT) framework. We use supercell model to probe the stability and electronic properties of adsorbed as well as substitutionally doped Co atoms in AGNR. Vacuum space of 10 Å was used to eliminate the Coulomb interaction of ribbons with their periodic images. All the considered structures under consideration were optimized by using local density approximation (LDA) for the exchange correlation functional as proposed by Perdew and Zunger [19]. For structure relaxation and obtaining the ground state of the system, we used DFT based ATK-VNL simulation package as provided by quantumwise [20]. The convergence process was continued in a self consistent manner until the maximum force on every atom (including the impurity atoms) reaches to less than 0.05eV/Å. The *k-point* sampling of *1x1x100* k-points was selected with an energy cutoff value of 100 Rydberg.

3. Results and discussions

We select AGNR having width=7 as the representative ribbon and considerd four configurations to compare the adsorption and substitutional doping of Co atoms. These configurations are edge adsorption (EA), centre adsorption (CA), edge substitutional doping (ES) and centre substitutional doping (CS). The considered doping sites and the rule adopted to define the ribbon width are presented in Figure 1.

3.1. Stability analysis

To examine the energetic favourability and stability of various considered AGNR structures, we have calculated formation energy (FE) and binding energy (BE) and summarized the findings in Table 1. It is noticed that adsorption of Co on AGNR is energetically more favorable than its substitutional doping as indicated by large negative FE [Table 1]. Moreover, edge is the most preferred site for adsorption of Co atoms in AGNR. The difference between FE of ES and CS is much appreciable than the FE difference of EA and CA configurations. Previously, Cocchi *et al.*, also found edge as the most preferred site for Co impurities in GNR [21]. Analysis of BE suggests that substitutional Co atoms have stronger binding with AGNR whereas adsorbed Co impurities exhibit a rather loose binding which is also supported by the positive BE. We observed that Co is most strongly bound with AGNR when present as a substitutional dopant at the centre of the ribbon which is similar to the behavior of Ni substitutional impurity in AGNR [22]. From the optimized bond lengths ($d_{C\text{-}Co}$) [Table 1], we can infer that substitutional Co possesses the shorter bond length which also points towards the maximum stability. The larger bond lengths of adsorbed Co atom are according to the weaker binding of these configurations as compared to substitutionally doped Co atoms.

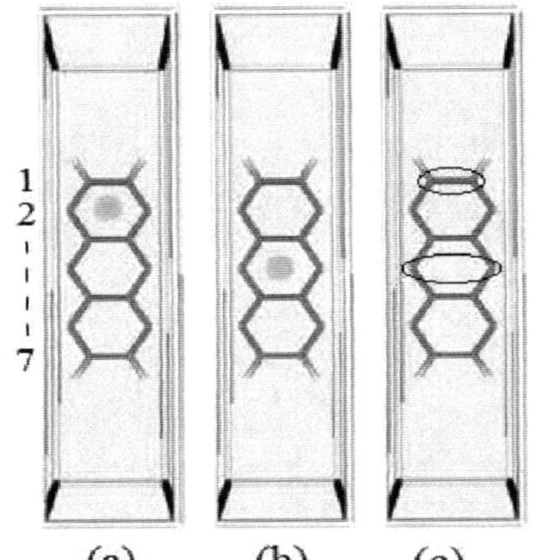

Figure 1. The schematic diagram of supercells depicting the considered adsorption and substitutional sites. Pink sphere represent the position of Co atom in (a) edge Co-adsorption (EA) and (b) centre Co-adsorption (CA). (c) The ellipses at the edge and the centre of supercell respectively represent the position of Co atoms for edge substitutional (ES) and centre substitutional (CS) doping.

Table 1. The calculated formation energy (FE), binding energy (BE) and the optimized bond length between C-Co

Configuration	FE (eV)	BE (eV)	d_{C-Co} (Å)
EA	-20.43	1.42	1.98
CA	-19.89	1.96	1.98
ES	-5.52	-2.85	1.80
CS	-3.18	-3.30	1.81

3.2 Electronic Structures

The band structures of considered configurations are depicted in Figure 2 (a)-(c). Only the band structures of EA and CS configurations have been shown since these are the most stable sites for adsorption and substitution of Co atoms respectively. The band structure of pristine AGNR has been presented for the comparison. The behavior of CA (ES) was also metallic similar to EA (CS). It is observed that band structure of pristine ribbon is significantly disturbed due to the presence of Co impurity. For EA, there exists a flat dispersive band at the Fermi level which is due to localized *3d* electrons of Co [Figure 2(b)]. For CS, we noticed maximum perturbation in the band structure [Figure 2 (c)]. Strong hybridization between C-*2p* and Co-*3d* orbital causes the origin of a number of additional bands across the Fermi level which makes this configuration highly metallic.

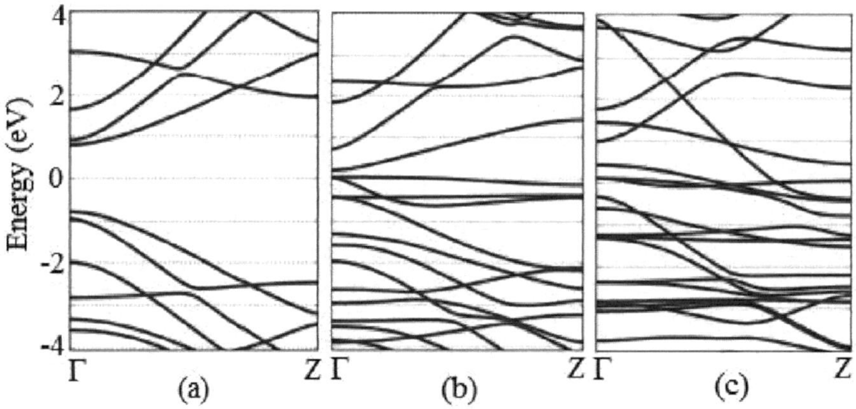

Figure 2. The band structure of (a) Pristine (b) edge Co-adsorbed and (c) centre Co doped AGNR. The Fermi energy is set at 0 eV.

Similar additional bands across the Fermi level were also observed for ES configuration (to be concise, that band structure is not shown here). Therefore, we can conclude from the above observations that both types of Co impurities (adsorption and substitutional) cause metallicity in AGNR which were otherwise semiconducting. This induced metallicity sustains irrespective of the doping site. We also noticed that metallicity is much enhanced when Co is present as the substitutional dopant rather than the adsorbed atom.

4. Conclusion

We probed the stability and electronic properties of AGNR with adsorbed/substitutional Co impurities by means of DFT using LDA. It is revealed that the adsorption of Co impurity on AGNR is energetically more favorable over the substitutional doping. However, the substitutional doping of Co results in stronger binding than that of adsorption. The electronic band structure of the ribbons was significantly altered and a semiconductor to metallic transition was observed in all the considered structures. An enhanced metallicity is obtained in AGNR irrespective of the doping site. The observed high metallicity is crucial for making metallic contacts, interconnects in nanoelectronic and various other technological applications.

Acknowledgments
We thank the Computational Nanoscience and Technology Laboratory (CNTL), ABV- Indian Institute of Information Technology and Management, Gwalior for providing the computational resources and infrastructural facilities.

References
[1] Novoselov K S, Jiang D, Schedin F, Booth T J, Khotkevich V V, Morozov S V and Geim A K 2004 *Science* **306** 666
[2] Osvath Z, Darabont A, Incze P N, Horvath E, Horvath Z E and Biro L P 2007 *Carbon* **45** 3022
[3] Lee B J, Yu H Y and Jeong G H 2010 *Nanoscale Res. Lett.* **5** 1768
[4] Li X, Wang X, Zhang L, Lee S and Dai H 2008 *Science* **319** 1229
[5] Liang X, Fu Z and Chou S Y 2007 *Nano Lett.* **7** 3840
[6] Meric I, Han M Y, Young A F, Ozyilmaz B, Kim P and Shepard K L 2008 *Nat. Nanotech.* **3** 654
[7] Zeng J, Chen K Q, He J, Zhang X J and Sun C Q 2011 *J. Phys. Chem.* C **115** 25072
[8] Abanin D A and Levitov L S 2007 *Science* **317** 641
[9] Ren H, Li Q X, Luo Y and Yang J 2009 *Appl. Phys. Lett.* **94** 173110
[10] Son Y W, Cohen M L and Louie S G 2006 *Phys. Rev. Lett.* **97** 216803
[11] Schedin F, Geim A K, Morozov S V, Hill E W, Blake P, Katsnelson M I and Novoselov K S 2007 *Nat. Mater.* **6** 652
[12] Fowler J D, Allen M J, Tung V C, Yang Y, Kaner R B and Weiller B H 2009 *ACS Nano* **3** 301
[13] Myung S, Solanki A, Kim C, Park J, Kim K S and Lee K B 2011 *Adv. Mater.* **23** 2221
[14] Wang H, Wang Q, Cheng Y, Li K, Yao Y, Zhang Q, Dong C, Wang P, Schwingenschlögl U, Yang W and Zhang X X 2012 *Nano Lett.* **12** 141
[15] Kan E J, Xiang H J, Yang J and Hou J G 2007 *J. Chem. Phys.* **127** 164706
[16] Jaiswal N K and Srivastava P 2011 *Physica E* **44** 75
[17] Gorjizadeh N and Kawazoe Y 2008 *Mater. Trans.* **49** 2445
[18] Longo R C, Carrete J and Gallego L J 2011 *J. Chem. Phys.* **134** 024704
[19] Perdew J P and Zunger A 1981 *Phys. Rev.* B **23** 5048
[20] *www.quantumwise.com*
[21] Cocchi C, Prezzi D, Calzolari A and Molinari E 2010 *J. Chem. Phys.* **133** 124703
[22] Jaiswal N K and Srivastava P 2011 *Solid State Commun.* **151** 1490
[23] Jaiswal N K and Srivastava P 2011 *AIP Conf. Proc.* **1349** (Part A) 283

Raman and Dielectric Studies on Lead free $(K_{0.5}Na_{0.5})$ NbO_3 Piezoelectric Ceramics

P. Mahesh[1], **D. Pamu**[1]
[1]Department of Physics, Indian Institute of Technology Guwahati, Guwahati-39, India

E-mail: pamu@iitg.ernet.in

Abstract. The present study demonstrates the preparation of $(K_{0.5}Na_{0.5})NbO_3$ (KNN) ceramics at low temperatures by using mechanochemical synthesis process. The effect of calcination temperature on structure, and dielectric properties of KNN ceramics have been studied systematically. It is found that both the dielectric constant and tanδ of KNN ceramics as a function of temperature exhibited two sharp phase transitions indicating orthorhombic to tetragonal (184°C) and ferroelectric tetragonal to paraelectric cubic phases (385°C). It was observed that the ceramics calcined at 700°C and sintered at 1000°C shows the high Curie temperature, high dielectric permittivity and low dielectric loss. Further with increase in calcination temperature, the ferroelectric tetragonal to paraelectric cubic phase shifted to lower temperatures. Raman spectra of KNN ceramics were obtained and the three characteristic Raman peaks in the spectra of KNN are related to the internal vibrations of the NbO_6 octahedron. The effect of processing parameters on dielectric properties and phase transitions studied systematically.

1. Introduction

Lead-free piezoelectric materials have been studied due to the environmental and biological advantages and almost equivalent piezoelectric properties to PZT [1, 2]. Despite its excellent piezoelectric properties, one of the major disadvantages of PZT is potential lead volatility at higher temperatures. It is known that lead poisoning can result in adverse effects on health such as reduced IQ, slowed body growth, and kidney damage [3].

$(K_{0.5}Na_{0.5})NbO_3$ (KNN)-based piezoelectric materials have drawn great attention due to their competitive piezoelectric properties among lead-free ceramics and broad operation temperature range. $(K_{0.5}Na_{0.5})NbO_3$ (KNN) is currently considered as a potential candidate of lead-free piezoelectric materials. KNN is a solid solution of ferroelectric $KNbO_3$ and antiferroelectric $NaNbO_3$ [4]. KNN exhibits an orthorhombic structure at room temperature. Saito et al. reported that most of the piezoelectric properties are comparable to those of PZT [1]. Pure KNN-ceramics have difficulty becoming fully dense using ordinary sintering methods. This is because the Na_2O and K_2O easily evaporate at high temperatures. However, densification of KNN ceramics is reported to be difficult at low temperatures. The present study demonstrates the preparation of KNN ceramics at low temperatures by using solid state reaction process. The effect of calcination temperature on structure, and dielectric properties of KNN ceramics as a function of temperature have been studied systematically.

[1] Corresponding Author: Dr, D. Pamu, pamu@iitg.ernet.in.

2. Experimental Procedure

Samples of KNN ceramics were prepared by conventional solid-state reaction method from individual high purity powders (>99.99%) of K_2CO_3, Na_2CO_3, and Nb_2O_5. The starting materials were mixed in accordance with desired stoichiometry of the KNN ceramics. A planetary ball mill (Fritsch GmbH, Germany) was used to prepare these powders. The powders were mixed with proponol and the mixed powders were dried and calcined at different temperatures (600°C, 700°C, 800°C and 900°C) for three hours. The calcined powders again grounded for 5 hours and are uniaxially pressed into cylindrical samples of diameter 13mm and height 2mm isostatically. These pellets were sintered at temperature 1000°C for 5hours. The phase purity of the calcined powders and the sintered pellets was identified by X-Ray Diffractometer (Bruker D8). The dielectric constant and dielectric loss were measured using LCR meter (Wayne Kerr Electronics Pvt. Ltd., Model 1J43100). A PID temperature controller is used to control the temperature of the heating assembly up to 500°C. The Raman spectra of the samples were collected using a LABRAM HR800 Raman spectrometric analyzer developed by JOBIN YVON with a He-Ne source.

3. Results and Discussions

Figure 1 shows the XRD patterns of the KNN ceramics, calcined at different temperatures and sintered at 1000°C for 5 hours. It is observed that all the samples exhibited single orthorhombic phase and the sample calcined at 600°C exhibited the secondary phase peaks but the single phase has been confirmed at 650°C. Chang et al [5] reported the single phase for the KNN ceramics at 850°C. This shows the importance of mixing the initial mixing reagents by high energy milling in reducing the initial particle size and uniform mixing for the low temperature synthesis of KNN ceramics. The basic principle of processing temperature by high energy milling is that this synthesis not only makes the material finer but also includes structural changes, phase transformations and even solid-state reactions among the solid reagents. These physicochemical changes occur due to the efficient transformation of the mechanical energy of the grinding media to the material particles and the intensive mechanochemical force during the milling process. The smaller particle sizes would accelerate the chemical reaction between the initial reagents and also reduces the reaction temperature. It is also observed that as the calcination temperature increases the peaks are broadened and peak position shifted to lower side. The average crystallite sizes of the KNN ceramics are in the range of 60nm to 20nm.

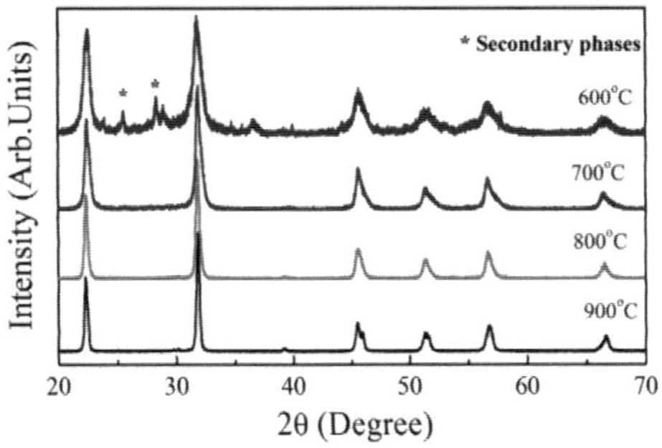

Figure 1. XRD patterns of the KNN ceramics calcined at different temperatures and sintered at 1000°C for 5 hours.

Figure 2. SEM image of the KNN ceramics calcined at (a) 600°C, (b) 700°C and sintered at 1000°C for 5hrs.

The SEM micrograph of the KNN samples sintered at 1000°C for 5 hours is shown in Figure 2. The surface morphology shows the rectangular size of the grains with dense packing with an average grain size of 1.2μm. The densities of the KNN ceramics were in the range of 84-90%.

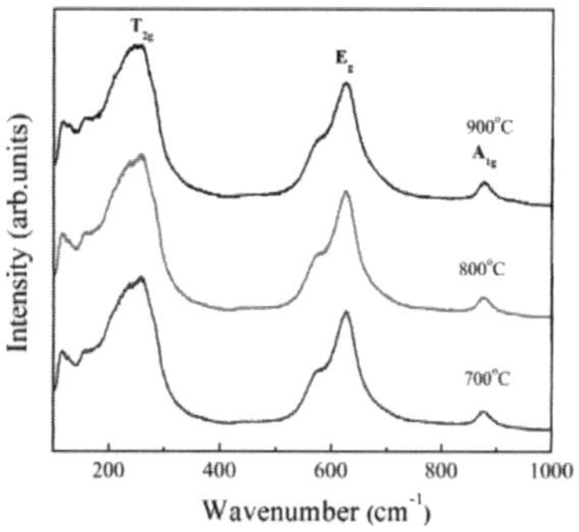

Figure 3. Raman spectra for the KNN ceramics calcined at different temperatures and sintered at 1000°C for 5 hours.

The Raman spectroscopy is an extremely sensitive tool to study at a local scale the structural deformations of perovskites, which are induced both by the tilting of octahedra and by the cationic displacements. These structural modifications induce large changes in internal modes associated with NbO_6 octahedron and thus a modification of the Raman spectra. Figure 3 shows the Raman spectra of KNN powders recorded at room temperature. The spectra measured are similar to those reported earlier [6, 7]. The Raman spectrum can be described by means of internal vibrations of the octahedral NbO_6 molecular group. The free NbO_6 group exhibits cubic O_h symmetry, whose dynamics is characterized by 15 internal vibrational degrees of freedom or six normal vibrations ν_i:
$\Gamma_{vib} = A_{1g}(Raman) + E_g(Raman) + 2T_{1u}(IR) + T_{2g}(Raman) + T_{2u}(silent)$.

The ν_1 vibration is related to symmetric A_{1g} type mode, which represents the Nb-O bonds vibration, ν_2 means doubly degenerated E_g vibrations of the O-Nb-O bonds and ν_3 is triply degenerated T_{1u} vibration of the Nb-O bonds. The ν_1, ν_2, ν_3 vibrations are the stretching modes of the NbO_6 octahedral group. The other three normal vibrational modes: triply degenerated ν_4 of T_{1u} symmetry of the O-Nb-O bonds, triply degenerated ν_5 with T_{2g} symmetry of the O-Nb-O and triply degenerated ν_6 with T_{2u} symmetry of the O-Nb-O bonds are all the bending vibration modes. Thus there are three characteristic Raman peaks (ν_1=870 cm^{-1}, ν_2=620 cm^{-1} and ν_5=250 cm^{-1}) in the spectra of KNN which are related to the internal vibrations of the NbO_6 octahedron. The broad Raman bands peaking at 250 and 620 cm^{-1} in the cubic phase shown in Figure 3 are a clear signal of disorder, for Raman activity in cubic perovskites is symmetry forbidden. These features are common in the cubic phase of pervoskite ferroelectrics and they signal local static and/or dynamic disorder reminiscent of the distortions in ferroelectric phases [7].

Figure 4 (a) shows the temperature dependence of dielectric constant of KNN ceramics calcined at different temperatures and sintered for 5 hours. It is observed that two sharp phase transitions are observed in dielectric constant versus temperature curve corresponding to the phase from orthorhombic to tetragonal (184°C) and tetragonal to cubic (385°C) respectively. The phase transition around 184°C is the ferroelectric phase transition temperature whereas the transition around 385°C is the ferroelectric to paraelectric transition temperature which is also called as the Curie temperature (T_C). It is observed that the KNN sample calcined at 600°C exhibited lowest Curie temperature, low dielectric constant and high dielectric loss attributed to the presence of secondary phases on the other hand KNN sample calcined at 700°C exhibited the highest dielectric constant and lowest loss. Further with increasing the calcination temperature dielectric properties as well as Curie temperature decreases, this may due to the volatilization of alkali elements. The dielectric anomaly, near the T_C, can be explained by the fact that soft mode frequency softens, giving rise to large value of dielectric constant [10]. The variation in tanδ as a function of temperature for the KNN samples calcined at different temperatures and sintered at 1000°C was shown in Figure 4(b).

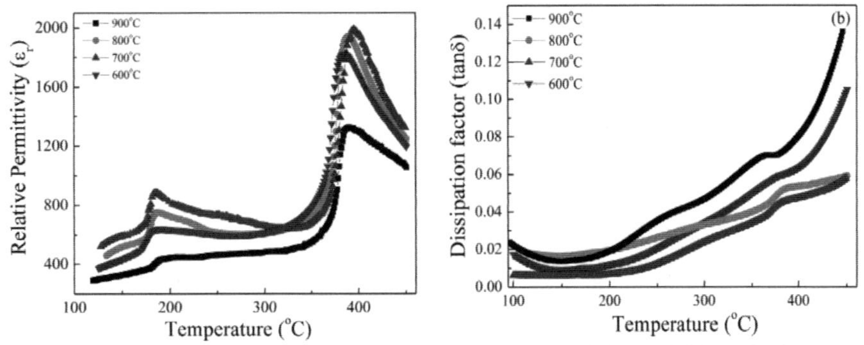

Figure 4. Temperature Variation of (a) relative dielectric permittivity and (b) tanδ of KNN ceramics calcined at different temperatures and sintered at 1000°C.

The variation in tanδ as a function of temperature is also similar to the behaviour of the dielectric constant. In the present study it is interesting to note that the observed tanδ values are much lower and can be attributed to the higher densities and uniform microstructure and the obtained tanδ values were in the range of 0.006 to 0.14. The higher loss in the case of the sample calcined at 600°C may be due to the secondary phases which are identified by XRD patterns. It is well known that the dielectric loss is heavily influenced by the secondary phases. In the present study the obtained lower dielectric properties compared to the reported values in the literature and can be attributed to the lower densities and the smaller grain sizes [5]

4. Conclusions

In conclusion, KNN ceramics are prepared by solid state reaction method. The effect of calcination temperature on structure and dielectric properties of KNN ceramics are investigated. The phase pure KNN ceramics were formed at 700°C itself with the initial small particles size. The Phase transitions from orthorhombic to tetragonal (184°C) and ferroelectric tetragonal to paraelectric cubic (385°C) phase transitions are observed in the present study. The KNN ceramics calcined at 700°C and sintered at 1000°C exhibited the highest dielectric constant and lowest dielectric loss. The obtained lower dielectric constant values can be attributed to the lower densities and the smaller grain sizes

Acknowledgements
Authors acknowledge BRNS and DRDO for the financial support.

References

[1] Saito Y, Takao H, Tani T, Nonoyama T, Takatori K, Homma T, Nagaya T and Nakamura M 2004 *Nature* **432** 84.

[2] Cross E 2004 *Nature* **432** 24.
[3] Woolf A D, Goldman R and Bellinger D C 2007 *Pediatr Clin North Am.* **54** 271.
[4] Zhang H, Jiang S and Kajiyoshi K 2010 *J. Am. Ceram. Soc.* **93** 1957.
[5] Chang Y, Yang Z, Chao X, Zhang R and Li X 2007 *Mater. Lett.* **61** 785.
[6] Połomska M, Hilczer B, Kosec M and Malic B 2008 *Ferroelectrics* **369** 149.
[7] Trodahl H J, Klein N, Damjanovic D, Setter N, Ludbrook B, Rytz D, and Kuball M 2008 *Appl. Phys. Lett.* **93** 262901.
[8] Pytte E 1972 *Phys. Rev.* **B 5** 3758.
[9] Cochran W 1960 *Adv. Phys.* **9** 387.
[10] Kandari A S, Bhandari A, Bourai A and Panwar N S 2009 *Ferroelectrics* **386** 139.

IOP Conf. Series: Materials Science and Engineering **73** (2015) 012142 doi:10.1088/1757-899X/73/1/012142

Electronic and magneto transport property of $Pr_{0.7}Sr_{0.3}MnO_3$ system in nanometric gran size modulation

Proloy T. Das[1], **J. Panda, A. Taraphder and T. K. Nath**
1 Department of Physics and Meteorology, Indian Institute of Technology Kharagpur, Kharagpur - 721302, India.

E-mail: dasproloy@phy.iitkgp.ernet.in; tnath@phy.iitkgp.ernet.in

Abstract. Grain size effect on electronic- and magnto-transport properties of hole doped manganite $Pr_{0.7}Sr_{0.3}MnO_3$ (PSMO) in nanometric dimension has been thoroughly investigated under external magnetic field range from zero to ± 8 Tesla down to 4 K. Metal-insulator transition temperature (T_P) which is observed for all the PSMO nanometric samples, differs from the ferromagnetic to paramagnetic transition temperature (T_C) in the nanometric regime compare to its reported bulk counterpart mainly due to the enhanced surface disorder effect. Temperature dependent low field complex magneto- impedance (LFMI) behaviour has also been studied in presence of external magnetic field range of 0 - 800 Oe down to 77 K.

1. Introduction

The perovskite manganites with generic formula $Re_{1-x}R_xMnO_3$, where Re is trivalent rare earth cation (e.g. La, Pr, Nd etc.) and R is divalent alkaline earth cation (e.g. Ca, Sr, Ba, Pb etc.) have drawn considerable attention, especially following the discovery of their negative colossal magneto-resistance (CMR) [1-3] effect. In these materials subtle balance between charge, spin, lattice and orbital degrees of freedom leads to a variety of phases with fascinating properties like charge ordering (CO) and orbital ordering (OO) [4,5], metal - insulator (M-I) transition [6,7], ferromagnetic insulator (FMI), canted anti-ferromagnetic insulator (CAFI) [8] etc. In the bulk polycrystalline or single crystal manganite system the double exchange mechanism strongly correlates the electronic transport property with its magnetic property. In nanometric grain size modulation, a drastic change in physical properties mainly arises due to finite size effect. A number of reports have already been published [9-11], but there is no such details systematic study on PSMO system in nanometric size modualtion. Investigation on electronic- and magneto-transport properties of a series of chemically synthesized PSMO nanoparticles is the main focus of this paper.

2. Experimental details

A series of nanometric PSMO samples with different particle size (Φ) were synthesized by chemical 'pyrophoric' reaction process [12] making solution of Pr_6O_{11}, $Sr(NO_3)_2$ and Mn $(CH_3COO)_2$ in required stoichiometric ratios. The burnt ash like powder was calcined at 850°C for 5 hours followed by three different sintering temperatures (850^0 C, 950^0C and 1050^0C) to obtain samples with different particle size. X-ray diffraction (Philips PW1710) study was carried out with Cu-K_α radiation (λ = 1.5418 Å) to investigate the proper phase formation in the samples. High field magneto-transport and electronic-transport study of PSMO samples were carried out (down to 2 K) by a standard four probe contact method using a high field low temperature facilities (8 T CFM VTI, CRYOGENIC Ltd., U.K.). The low-field (up to 5000 Oe) magneto- transport (down to 77 K) measurement was also carried out using a homemade set up. Temperature and frequency dependent (down to 77 K) low field magneto impedance (LFMI) measurements were studied using a 1000 Oe solenoid coil magnet along with a high frequency HIOKI 3532-50 LCR Hi-TESTER meter. A high precision temperature controller (model 331, Lakeshore) was used to measure and control the temperature (within ± 10 mK resolution) for all the measurements.

3. Results and Discussions

The XRD patterns in Fig. 1(a) including its inset confirm the single phase of PSMO sample with orthorhombic structure (Pbnm space group). The average crystallite sizes of the samples are found to be 28 nm (PSMO 850), 37 nm (PSMO 950), and 42 nm (PSMO 1050). The log-normal distribution of particle size of PSMO 850 sample gives an average particle size of 28 nm. From the high resolution TEM (HRTEM) micrograph the estimated inter planer lattice spacing (d) is ~ (3.501 ±0.001) Å, which corresponds to the (111) plane of the crystal. From the field dependent MR study of various temperatures under 8 Tesla magnetic field, we have observed maximum 85% change in MR for PSMO 850 sample, 83.5% change in MR for PSMO 950 (as shown in Fig. 1(b)) near T_P, and a 82.27% change in MR for PSMO 1050 at 150 K (not shown here), respectively. Almost ~ 44% change in MR has been found with respect to the MR value obtained at lowest temperature (4 K) for all of PSMO nanometric samples at 8 Tesla. To find out the spin polarized tunnelling contribution (MR_{SPT}) over total magneto-resistance ($MR = MR_{SPT} + MR_{INT}$ for the PSMO polycrystalline nanoparticles, we have used a generic model proposed by the Raychoudhuri *et al.* [13].

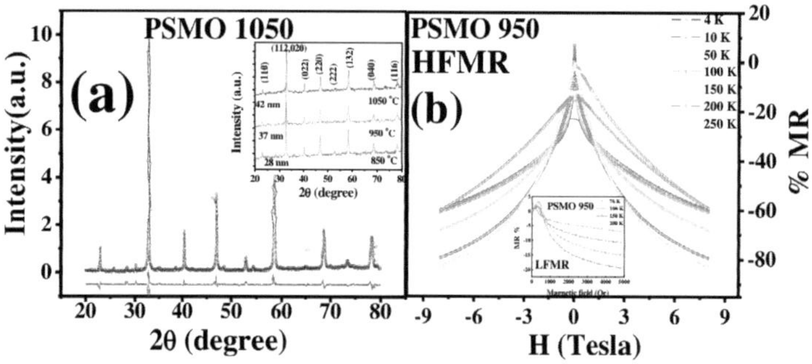

Figure 1 (a) Rietveld refinement of HRXRD of PSMO 1050 sample. The inset shows normal XRD of PSMO 850, 950, 1050 samples; (b) HFMR and LFMR variation of PSMO 950 sample at different temperatures in the magnetic field range of 0 to ± 8 Tesla.

From the fitting of experimental LFMR data, the calculated spin polarized tunnelling (SPT) contribution as well as the intrinsic contribution (INT) in magneto-resistance data (as tabulated in Table - 1) indicates that MR_{SPT} increases in PSMO 1050 sample up to 1.53% with respect to PSMO 850 sample, whereas MR_{INT} decreases up to 2.38% in PSMO 1050 sample with respect to PSMO 850 sample. The expected values of MR and calculated values of MR match very well for each nanometric samples. High field magneto-resistance is related to the contribution from spin polarized inter granular tunneling of electrons across the grain boundaries and spin dependant scattering across the grain surfaces due to field induced-depression of local magnetic disorder near the grain boundary. A strong temperature (down to 77 K) and frequency (up to 2 MHz) dependence low field complex magneto impedance (LFMI) [$= \{Z(\omega, H, T) - Z(\omega, 0, T)\}/ Z(\omega, 0, T)$] have been observed for all PSMO nanometric samples [14]. Figure 2(a) depicts a temperature dependent MI of PSMO 1050 sample whereas inset of Fig. 2(a) shows frequency dependent MI change of PSMO 1050 sample. PSMO 1050 sample shows maximum percentage of MI change (~ - 45%) at 250 K. Figure 2(b) shows a combined plot of % MI change for all PSMO nanometric samples as a function of magnetic field at 77 K and at a frequency of 1 MHz. Maximum -35 % MI change has been found for PSMO1050 sample and only -6% change in MI has been found for PSMO 850 sample.

Table 1 LFMR values and estimated various contributions of MR of PSMO samples for different particle size.

Crystallite size	MR_{SPT} (%)	MR_{INT} (%)	MR_{EXP} (%)	MR (%)
PSMO 850	-8.54	-11.54	-20.08	-20.3
PSMO 950	-6.95	-12.21	-19.16	-19.4
PSMO 1050	-21.63	-2.63	-15.85	-16.1

In PSMO 950 sample, the maximum change in MI, have been found -21% at 2 MHz frequency (is not shown here) by sweeping the dc magnetic field up to 800 Oe. PSMO 850 sample depicts maximum -8% change in MI at lowest attainable temperature (77 K) at 1 MHz whereas no MI is observed at room temperature. Frequency dependent MI of PSMO 850 sample at 77 K shows maximum -15% change in MI at higher frequency (2 MHz) rather than 100 KHz. In higher frequency range, most likely the skin penetration depth becomes smaller and charge carrier conduction process along the surface region of the PSMO samples increases due to skin depth effect of electrodynamics, which can produce a moderate value of MI [15]. Figure 3(a) shows resistivity plot of PSMO 850 at different magnetic fields of zero, 1, and 8 Tesla. MR variation as a function of temperatures for samples PSMO 850 at 1, and 8 Tesla magnetic fields is shown in the inset of Fig. 3(a). We have observed the metal-insulator transition around ~ 180 K for PSMO 850 and PSMO 950 nanometric samples whereas in case of PSMO 1050 (not shown here), it was found near 250 K. The temperature dependent resistivity plots of

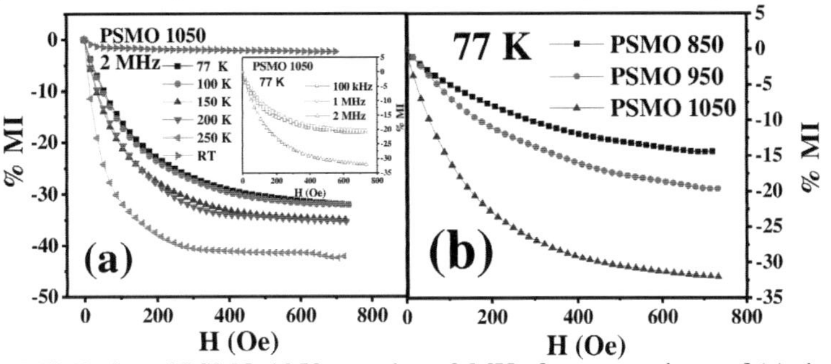

Figure 2 (a)LFMI plot of PSMO 1050 sample at 2 MHz frequency; inset of (a) shows frequency dependent LFMI behavior of PSMO 1050 sample at 77 K, and (b) Combined LFMI behavior of PSMO 850, 950, and 1050 samples at 77 K and 1 MHz frequency.

all PSMO samples under without magnetic field is shown in Fig. 3(b). It is clear from the Fig. 3 (b) that the peak resistivity value monotonically decreases (~ 13.91%) with size reduction at 0 Tesla magnetic fields. We have observed that lowest MR value of all nanometric samples appear around room temperature which corroborates the resistivity data of PSMO sample. It is clear that MIT transition temperatures are suppressed with increasing of magnetic fields. Shifting of T_P towards the lower temperature with variation of particle size is most likely due to the grain surface effect on electronic transport. Low temperature resistivity upturn has been observed below 50 K for all PSMO samples. The upturn of resistivity at low temperatures regime for all the nanometric samples may

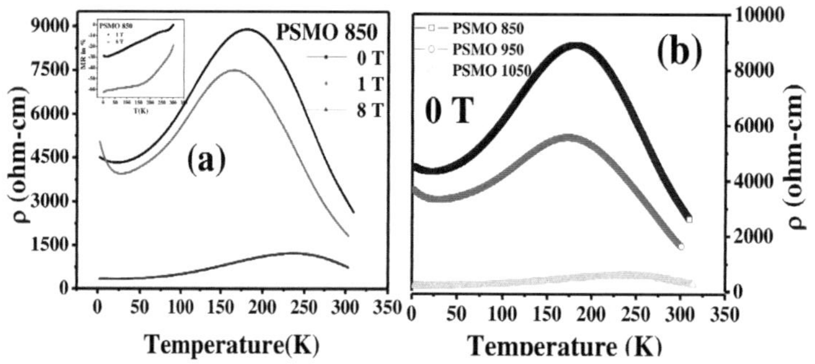

Figure 3 (a) Resistivity plot as a function of temperature of PSMO 850 sample at 0, 1, and 8 Tesla respectively; Inset of (a), % of MR variation of PSMO 850, sample at 1 and 8 Tesla respectively, (b) Combined resistivity plot of PSMO 850, 950 and 1050 samples at 0 Tesla.

arises due to various reasons, namely Coulomb blockade charging effect, Kondo-like mechanism, high spin disorder increment of impurity scattering of electrons in magnetic inhomogeneous matrix, however it has been found that electron-electronic interaction mechanism [16] and weak localization can explain the resistivity upturn properly for these kind of hole doped PSMO nanomanganites.

4. Conclusion

In summary chemically synthesized PSMO nanoparticles exhibit a strong dependency on electronic and magneto- transport behaviors in nanometric grain size modulation. The HFMR in these nanoparticles arises mainly due to the spin polarized tunneling of conduction electrons between the adjacent ferromagnetic grains, which is enhanced with the decrease of grain size. We have observed relatively high MI values at higher frequency regime than low frequency regime. It confirms that the surface conduction increases in the PSMO manganite with decreases of the skin depth. The MIT transition temperatures (T_P) slightly differ with grain size variation. Nano size induced grain boundary effect is believed to be responsible for this modification of T_P.

5. Acknowledgement

One of the authors, Poloy T. Das wishing to acknowledge Council of Scientific and Industrial Research (CSIR) for the financial support through award no.: 09/081 (1077)/ 2010-EMR-I.

6. References

[1] R von Helmolt et al 1993 *Phys. Rev. Lett.* vol. **71** 2331
[2] Tokura Y 2000 *Colossal magneto resistive oxides* (Gordon and Breach Science Publishers)
[3] Dagotto E et al 2001 *Phys. Rep.* **344** 153
[4] Tomioka Y et al 1995 *Phys. Rev. Lett.* **74** 5108
[5] Kuwahara H et al 1995 *Science* **270** 961
[6] Imada M et al 1998 *Rev. Mod. Phys.* **70** 1216
[7] Rao C N R and Raveau B 1998 *Colossal Magnetoresistance, Charge Ordering and Related Properties of Manganese Oxides* (World Scientific, Singapore)
[8] Shannon R D 1976 *Acta Cryst.* **A 32** 751
[9] Sen V et al 2007 *J. All. Comp.* **439** 205
[10] Boujelben W et al 2001 J. Soild State Chem. **156** 68
[11] Chand U et al 2011 *Archives of Applied Science Research* **3** 1
[12] Dey P and Nath T K 2006 *Phys. Rev. B* **73** 214425
[13] Raychaudhuri P et al 1998 *J. App. Phys.* **84** 2048
[14] Qin H et al 2002 *J. App. Phys.* **91** 12
[15] Yelon A et al 1996 *Appl. Phys. Lett.* **69** 20
[16] Kumar D et al 2002 *Phys. Rev. B* **65** 094407

IOP Conf. Series: Materials Science and Engineering **73** (2015) 012143 doi:10.1088/1757-899X/73/1/012143

Silver nanocluster films on ITO coated glass as novel substrates for the detection of molecules using Surface Enhanced Raman Scattering (SERS)

Raju Botta[1], **G. Upender, C. Bansal**
School of Physics, University of Hyderabad, Hyderabad-500 046, India.

E-mail: raazbotta@gmail.com

Abstract. A novel surface enhanced Raman scattering (SERS) substrates were prepared by a nanocluster deposition system. Silver nanoclusters were deposited on the indium tin oxide (ITO) coated glass slides. These films were annealed at 300°C for two hours to obtain the required size distribution for the desirable optical properties. The surface morphology of the films was examined using filed emission scanning electron microscope (FESEM) and atomic force microscope (AFM). The surface plasmon resonance (SPR) of the pre annealed and the annealed Ag deposited ITO coated glass substrates were measured using UV-Vis spectrophotometer. The SERS studies were carried out on these substrates with Methylene Blue (MB) as a test molecule with 10^{-4} to 10^{-8} M concentrations using 514.5 nm and 632.81 nm laser excitation wavelengths. The results show that Ag nanoclusters of average particle size of ~ 150 nm act as very good SERS substrates for the detection 10^{-6} M of MB. Both the electromagnetic (EM) and chemical enhancement could be contributed to the SERS signal enhancement of MB molecule.

1. Introduction

Surface Enhanced Raman Scattering (SERS) is a spectroscopic tool for the detection of molecules adsorbed on the metal nanoparticles. SERS is a known powerful technique which can be used in analytical, biological, explosives detections [1] and medical diagnostics. SERS technique is the only one which can combainely give single molecule detection and chemical fingerprint [5]. Two effects are responsible for these enhancements one is Electromagnetic (EM) enhancement which is due to the excitation of localized surface plasmon resonance (LSPR) at the metal nanoparticle structures. And second one is chemical enhancement (CHEM) which is a contribution between the probed molecule and the substrate. The LSPR is depends on the size and shape of the nanoparticles, for which optimization is required. In our previous report we optimize the nanocluster size [2]. In this present study we use Methylene Blue (MB) as probed molecule. Optical absorption by UV- Vis spectrometer and surface characterization was carried using and Field Emission Scanning Electron Microscope (FESEM) and Atomic Force Microscope (AFM).

[1] To whom any correspondence should be addressed.

2. Experimental

2.1 Preparation and characterization of Ag nanocluster films/substrates

ITO coated glass were used to preparare substrates for Ag cluster deposition. The surface contamination was removed by the sonication. Subsequently dried all these substrates using nitrogen gas before putting in the vacuum chamber. Silver nanoclusters (Ag NC's) were deposited on these ITO coated substrates at a base pressure of 2 x10^{-6} Torr using cluster deposition system (NANODEP 60 from Oxford Applied Research, UK). Magnetron power was maintained at ~ 70 W. The Ag nanoclusters were formed while they traversed towards the substrates. A detailed description of cluster deposition is found elsewhere [3]. We annealed these substrates at 300 °C/ 2 hrs under Ar environment to grow nanocluster sizes. Optical absorption of these substrates at room temperature were carried using UV-Vis spectrophotometer in the range from 300-800 nm. Surface morphology of Ag NC's on ITO coated glass (Ag NC/ITO glass) was carried using field emission scanning electron microscope (FESEM).

2.2. Surface Enhanced Raman Scattering studies

SERS spectra of MB adsorbed on Ag NC/ITO glass was recorded using micro-Raman spectrometer(LabRam HR) using laser excitation wavelengths of 514.5 and 632.8 nm at room temperature. The wide range of concentrations of MB was prepared from 10^{-4} to 10^{-8} M. For each test we took 10 micro moles of MB and dropped on the annealed Ag NC/ITO glass substrates and dried within the edges of the substrate at room temperature.

3. Results and discussions

Surface roughness of ITO coated glass is around 10 nm which was shown in figure 1B. Surface morphology of Ag NC/ITO glass were carried out by FESEM and FESEM image of annealed Ag NC/ITO glass substrate was shown in figure 1A.We calculate the average nanocluster sizes of Ag NC/ITO glass substrate which turn out ~ 150 nm. Surface plasmon resonance (SPR) of Ag NC/ITO glass substrates was turnout to be at ~ 590 as shown in figure 2. In this figure there are two maximum absorption peaks, one at ~ 360 nm is due to smaller size nanoclusters i.e. < 100 nm and second one at ~ 590 nm is due to bigger size nanoclusters i.e. > 100 nm. The laser excitation (633 nm) was chosen in such a way that the excitation wavelength of falls within the SPR absorption band of the nanoclusters and is also in resonance with the MB (~ 590 nm) absorption.

In this present study SERS measurements carried out with the laser excitation line of 514 and 633 nm. From our measurements we observe that the SERS spectrum from laser excitation wavelength of 633nm is better than the 514 nm. It is may be due to the SPR of Ag NC/ITO glass is close to the laser excitation wavelength (633 nm) and also the absorption band of MB is close to the laser excitation wavelength and SPR of Ag NC/ITO glass. Fig.4 shows SERS spectra of 10^{-4} M concentration of MB adsorbed on the Ag NC/ITO glass substrate. Peaks appeared at ~ 445, ~ 479, ~ 601, ~ 673, 773, ~ 806, ~ 1036, ~ 1228, ~ 1392, ~ 1429 and ~ 1623 cm^{-1} are characteristic peaks of MB [4]. The spectral assignments of these peaks are found elsewhere [4]. In the present data we got very good reproducible SERS spectra which is shown in figure 4. SERS measurements were taken on different locations over the Ag nanoclustes on ITO coated glass substrates.

We got very good SERS signals of MB even at 10^{-8} molar concentration. We compare the intensities of characteristic peaks with varying concentrations. Peak intensities of 479, 1036, 1429 and 1623 cm^{-1} were estimated by applying the Lorentzian shape to the spectral line [2] and resulting intensities were shown in figure 5.

Figure 5 shows the comparison of the Raman lines of 479, 1036, 1429 and 1623 cm^{-1} at different molar concentrations (i.e. 10^{-4}, 10^{-6} and 10^{-8}) of MB. Intensities of Raman line from 10^{-4} to 10^{-8} molar concentrations are linearly varying with concentration. This gives us confidence in the

IOP Conf. Series: Materials Science and Engineering **73** (2015) 012143 doi:10.1088/1757-899X/73/1/012143

efficacy of our measurements.

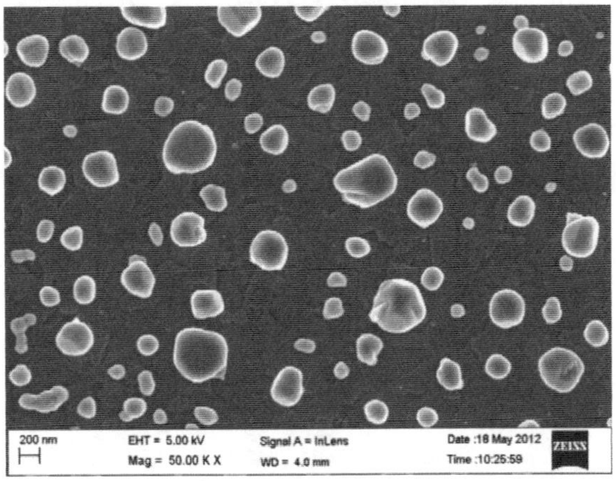

Figure 1A . FESEM image of annealed Ag nanoclusters on ITO coated glass substrate

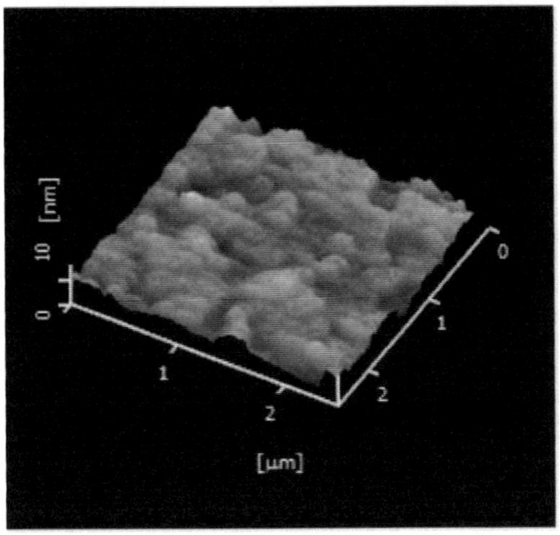

Figure 1B. AFM image of ITO coated glass substrate, surface roughness was around 10 nm

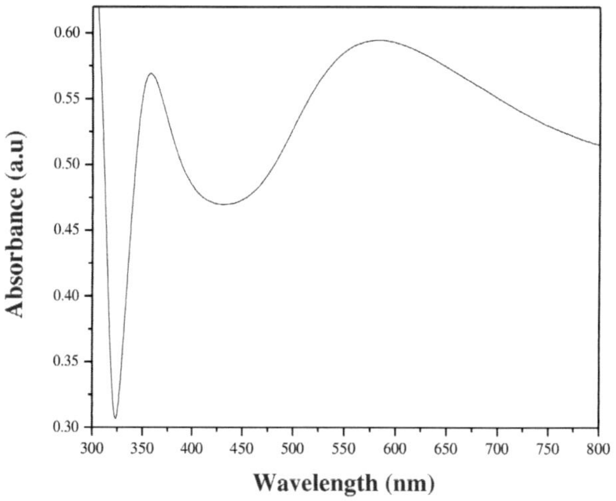

Figure 2 UV-Vis spectrum of annealed AgNC's on ITO coated on glass substrate

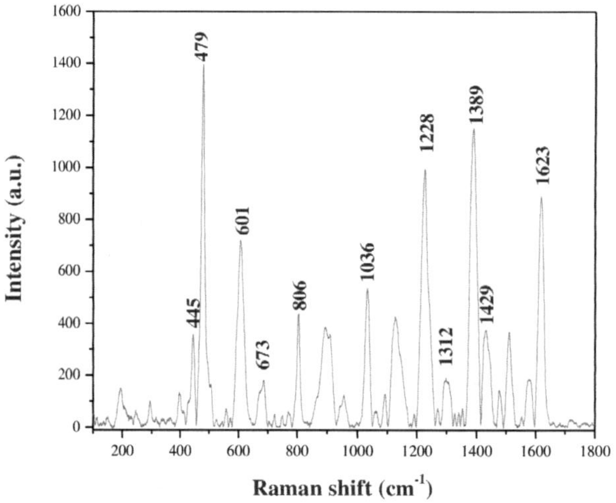

Figure 3 SERS spectra 10^{-4} M MB adsorbed Ag nanoclusters on coated on glass substrateas with laser excitation wavelength of 632.8 nm.

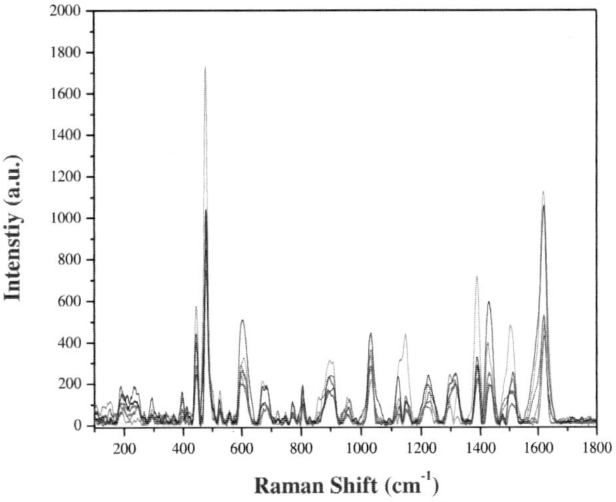

Figure 4 Reproducible SERS spectra of 10^{-4} M MB

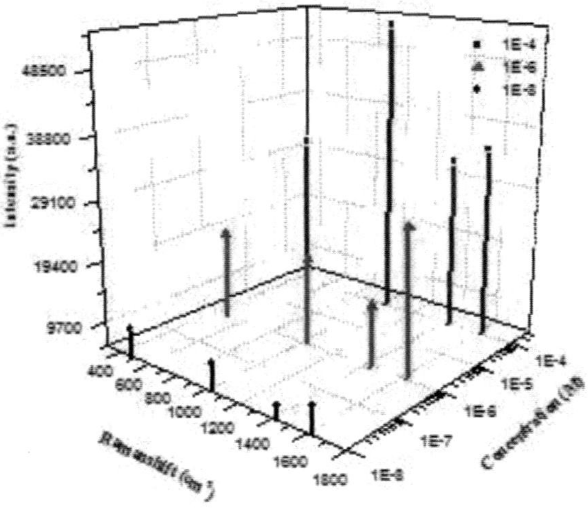

Figure 5 Comparison of Raman lines at 479, 1036, 1429 and 1623cm^{-1} of MB at different molar concentrations (10^{-4}, 10^{-6} and 10^{-8} molar concentration) with laser excitation wavelength of 632.8 nm.

4. ConclusionsIn

Novel SERS substrates were prepared using plasma inert gas phase condensation technique based cluster deposition system. AgNC on ITO coated glass substrates prepared using this technique were found to be good SERS-active substrates for detection of molecules. Very low concentrations of MB were detected in the present study. This technique can be used for trace analysis in analytical chemistry. Limit of detection of MB was 10 nM concentration.

Acknowledges:
One of the authors Raju Botta acknowledges to UGC-BSR (RFSMS) fellowship for financial support.

IOP Conf. Series: Materials Science and Engineering **73** (2015) 012143 doi:10.1088/1757-899X/73/1/012143

References

[1] Samuel S. R. Dasary, A. K. Singh, D. Senapati, H. Yu, and P. Chandra Ray 2009 *J. Am. Chem. Soc.* **131** 13806.

[2] G. Upender, R. Satyavathi, B. Raju, Shadak Ali, D. Narayana Rao and C. Bansal 2011 *Chem. Phy. Lett.* **511** 309.

[3] I. Shyjumon, M. Gopinadham, C. A. Helm, B. M. Smirnov, R. Hippler 2006 *Thin Solid films* **500** 41.

[4] S.Sarkar, S Pande, S. Jana, A. K. Sinha, M. Pradhan, M. Basu, J. Chowdhury, T. Pal 2008 *J. Phys. Chem. C* **112** 17862.

[5] K. Kneipp, Y. Wang, H. Kneipp, L.T. Perelman, I. Itzkan, R. Dasari, M.S. Feld 1997 *Phys. Rev. Lett.* **78** 1667.

IOP Conf. Series: Materials Science and Engineering **73** (2015) 012144 doi:10.1088/1757-899X/73/1/012144

Novel Analytical model for current-voltage characteristics and transconductance of undoped AlGaN/GaN MISHFETs and performance comparison with different high-k dielectrics

Rangababu Ganta

[1]Department of Electrical and Electronics Communication Engineering,IIT Kharagpur, Kharagpur -721302, India

E-mail: `rangababu.ganta@gmail.com`

Saptarsi Ghosh

[2]Advanced Technology Development Center, IIT Kharagpur, Kharagpur -721302, India

E-mail: `saptarsi123@gmail.com`

Dhrubes Biswas1[1,2]

E-mail: `dbiswas@ece.iitkgp.ernet.in`

Abstract. In this paper an improved charge controlled model of undoped ,lattice mismatched AlGaN / GaN MISHFET is presented .This model gives an accurate estimation of the 2-DEG sheet carrier concentration at the interface contributed entirely by spontaneous and piezoelectric polarization .The dependence of this sheet carrier density on Al composition and the insulator thickness is investigated in detail. Based on these developments an analytical current - voltage model incorporating Source/Drain parasitic resistance is presented.Also, the non-linear behavior of saturated Drain current is properly formulated and computation of transconductance is carried out for the same. Even without any intentional doping the proposed model exhibits fairly encouraging values of output current and conductance.

1. Introduction

With the impulsive rise of mobile computing and communication sector, high power RF devices are destined to play significant role in consumer electronics. Wide band gap semiconductors such as GaN are strong contender for high power electronic and optoelectronic circuits. Larger direct band gap (3.4 eV), as compared to Silicon (1.1 eV) and GaAs (1.4 eV) leads to higher values of critical fields and breakdown voltages. Recently GaN/AlGaN based heterostructure devices are grabbing much attention of both material and device community due to excellent electron transport properties and high sheet carrier densities. In GaN/AlGaN based systems a significant amount of charge concentration is found to originate at the interface without any intentional doping in the AlGaN layer. This carrier charge confinement results from both piezoelectric and spontaneous polarization in the GaN/AlGaN layers. The omission of delta doping in the AlGaN layer leads to simpler and economic fabrication technologies. But such HFETs are susceptible to high gate leakage due to the moderate Schottky barrier height of the gate metal (Au/Ni). For this purpose nowadays the device structure further modified , as an insulator layer with high permittivity is grown between the gate metal and the AlGaN layer [1, 2]. This ensures lower gate leakage in terms of body currents which in turn reduces low power loss in OFF state and also serves the purpose of passivating the AlGaN layer. Now reliable and predictive analytical models of such metal-insulator-semiconductor heterostructure FETs are needed along with their fast development in circuit implementation. But till date only a few of such models have been developed and there exists a great void to be filled.

In this paper we present a threshold voltage based analytical model to compute the I-V characteristics of undoped AlGaN/GaN MISHFET structures. In our model we take into account the nonlinear relationship of carrier concentration with Fermi level. As natural continuation of our earlier work we have formulated the drain current in the saturation region also with non zero sloped behaviour. Therefore this present model accurately predicts the current in weak, moderate and strong inversion regions. Transconductance, which effectively determines the device cut-off frequency, is also found out analytically. All the calculations are repeated for different dielectric materials for device performance optimization.

2. Compact MISHFET Model

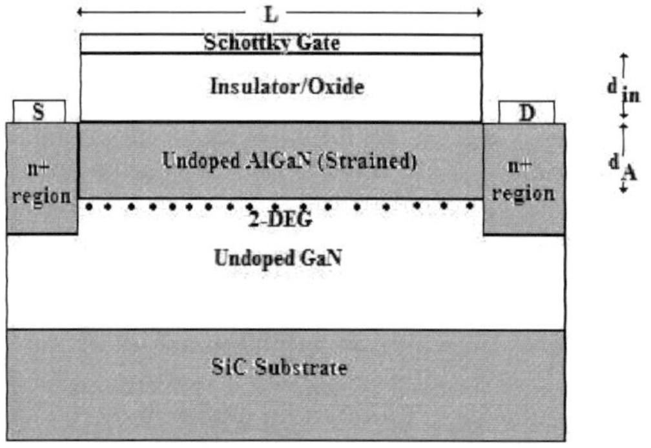

Figure 1. Schematic of an undoped AlGaN/GaN MISHFET

In an undoped MISHFET structure the free charge concentration at the AlGaN/GaN interface is entirely comprised of those originating from polarization (both spontaneous & piezoelectric).A generalized layered structure of the same is as shown in Figure 1.The active GaN mono layers are grown on stress reliving buffer layers of the same material .For a metamorphic growth of this nature, the substrate chosen is generally Sic owing to its high thermal conductivity for high temperature application .Up on this layers of undoped $Al_xGa_{1-x}N$ (strained), insulator and the Schottky gate metal are grown respectively .For a n-channel device the Source and Drain are heavily doped.

2.1. Current-Voltage Characteristics

The threshold voltage of the undoped MISHFET structure is given by [3]:

$$V_{th} = \Phi_b - \frac{\Delta Ec}{q} - \frac{\sigma_{pz}(m)}{s}(da + \frac{\in stin}{\in in}) + \frac{Ef}{q} + k1 \tag{1}$$

Where $\sigma_{pz(m)}$ is the charge induced due to spontaneous and piezoeletric polarization, Ec is the conduction band discontinuity of the heterostructure, Ef is the fermi level distance from the bottom of the conduction band. It is to be noted Ef and charge density ,Ns are interrelated and have to solved simultaneously .Now for mobility in GaN we follow the model given by [4] and proceed as per [5].Before saturation the boundary conditions are

at Source end

$$Vs = I_D R_S \tag{2}$$

at Drain end

$$V_L = V_{DS} - I_D R_D \tag{3}$$

by appropriate substitution we get

$$\frac{I_D}{\lambda E_C}(1 + pe_0 + e_0^4) = (e_0 + qe_0^4)(V_{GS} - V_{TH} - I_D R_S) \tag{4}$$

$$\frac{I_D}{\lambda E_C}(1 + pe_L + e_L^4) = (e_L + qe_L^4)(V_{GS} - V_{TH} - V_{DS} - I_D R_D) \tag{5}$$

$$\frac{I_D}{\lambda E_C}(\frac{e_L}{2e_0} + \frac{e_0}{2e_L} + p(e_L - e_0))$$
$$= (V_{GS} - V_{TH})(e_L - e_0) + e_0(V_{DS} - I_D R_D) - e_L I_D R_S - e_0 e_L E_C L \tag{6}$$

Where λ is the body effect factor, ex = E(x)/Ec, Ec being the critical electrical field. By simultaneously solving the above equations, we obtain current in the linear region. In the saturation region if l be the point of pinch off then.

$$V_l = V_{DSSAT} \tag{7}$$

$$V_L = V_{DS} - I_{DSAT} R_D \tag{8}$$

Hence we acquire the following relations

$$\frac{I_{DSAT}}{\lambda E_C}(1 + pe_0 + e_0^4) = (e_0 + qe_0^4)(V_{GS} - V_{TH} - I_{DSAT} R_S) \tag{9}$$

$$\frac{I_{DSAT}}{\lambda E_C}(2+p) = (1+q)(V_{GS} - V_{TH} - V_{DS} - I_{DSAT}R_D) \tag{10}$$

$$\frac{I_{DSAT}}{\lambda E_C}(\frac{1}{2e_0} + \frac{e_0}{2} + p(1-e_0))$$
$$= (V_{GS} - V_{TH})(1-e_0) + e_0(V_{DS} - I_{DSAT}R_D) - I_{DSAT}R_S - e_0 E_C L \tag{11}$$

$$V_l + \Psi E_C \sinh(L - \frac{Ll}{\Psi}) = V_{DS} - I_{DSAT}R_D \tag{12}$$

Similarly by solving the above we get current in saturation region.

2.2. *Transconductance*

Differentiating saturation drain current with respect to gate to source voltage, the transconductance in the saturation region can be computed.

3. Results & Discussion

The present model evaluates both the d.c current-voltage characteristics and small signal gain of undoped AlGaN/GaN MISHFET.

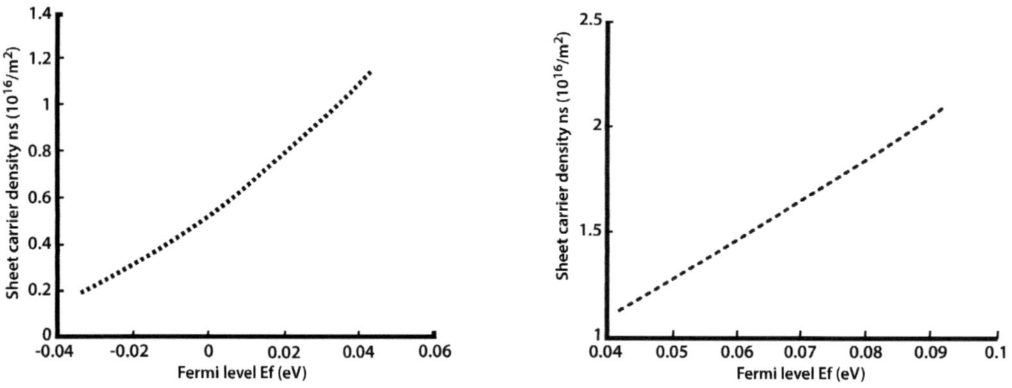

(*a*) (*b*)

Figure 2. Variation of 2-DEG density with Fermi level for Al mole fraction of(a).15 and (b).3 in a V_{GS} sweep of -2 to 2V

The variation of sheet carrier density with Fermi level is plotted in Figure 2. From these graphs we obtain a comparative analysis by changing the aluminium content. Evidently an increment in the aluminium mole fraction induces higher polarized charges. Figure 3.(a) & (b) shows the variation of drain current with drain to source voltage for different gate to source biases. As the Gate overdrive voltage increases so does the current .From the figures it is clear that the present analysis correctly incorporates the channel length modulation effect in the saturation region. An observed increase in the drain current is which obviously originates from a higher value of sheet carrier density.

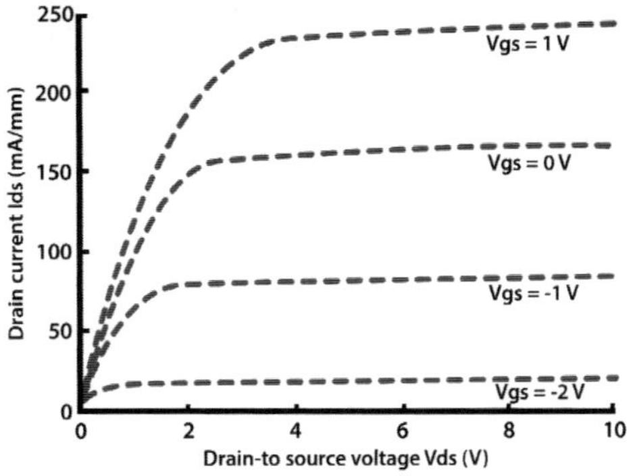

(a)

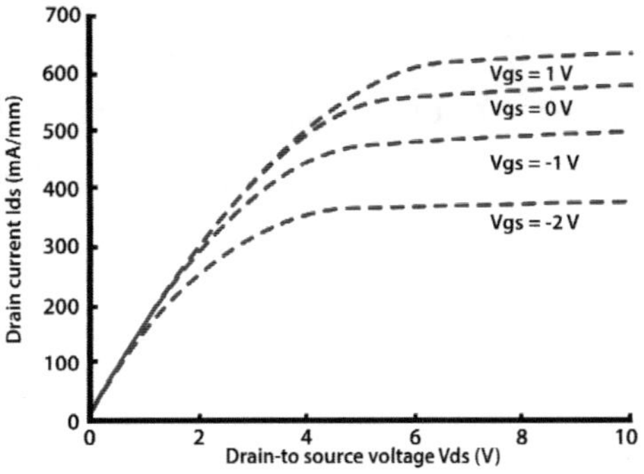

(b)

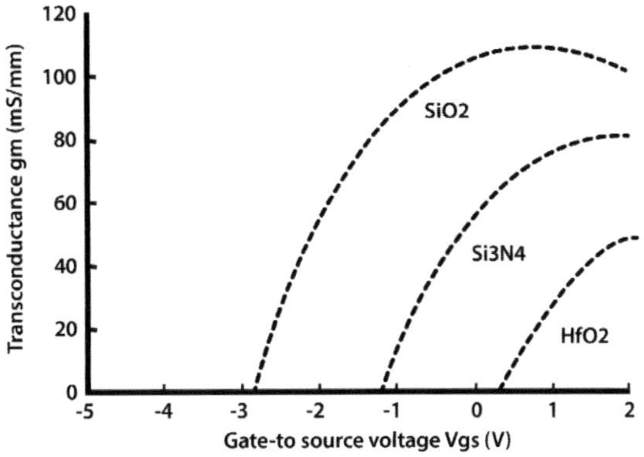

(c)

Figure 3. Variation of drain current with drain to source bias for SiO2 as the gate insulator for Al mole fraction of (a).15 and (b).3. Variation of transconductance with gate to source voltage for different insulators is plotted in (c).

4. Conclusion

The proposed model gives a fair representation of undoped MISHFETs. It can be concluded that even without any intentional doping such structures can operate under normal biases and may be considered as a substitute of doped structures.

References

[1] Takehiko Nomura, Hiroshi Kambayashi, Mitsuru Masuda,Sonomi Ishii, Nariaki Ikeda, Jiang Lee, and Seikoh Yoshida, *2006 IEEE Transactions on Electron Devices 53.*

[2] K. H. Lee, C. Chang and S. J. Chang, *2011 Applied Physics Letters 99.*

[3] Saptarsi Ghosh,Servin Rathi, Partha Mukhopadhyay, Sanjay kumar Jana,and Dhrubes Biswas, *International Conference on Advances in Engineering and Technology ICAET 2012,Conference Proceedings.*

[4] F.Schwierz,*2005 Solid State Electron. 49 889-95.*

[5] Miao Li and Yan Wang,*2008 IEEE Transactions on Electron Devices 55 261-67.*

IOP Conf. Series: Materials Science and Engineering **73** (2015) 012145 doi:10.1088/1757-899X/73/1/012145

Studies on electrical and optical properties of PVP:KIO_4 complexed polymer electrolyte films

M Ravi, K Kiran Kumar and V.V.R.Narasimha Rao

Sri Venkateswara University, Department of Physics, Andhra Pradesh, Tirupati-517502, India.

E-mail: m.raviphd@gmail.com

Abstract. Solid polymer electrolytes based on poly (vinyl pyrrolidone) (PVP) complexed with potassium periodate (KIO_4) salt at different weight percent ratios were prepared using solution-cast technique. X- ray diffraction (XRD) results revealed that the amorphous nature of PVP polymer matrix increased with the increase of KIO_4 salt concentration. Electrical conductivity was measured with an AC impedance analyzer in the frequency and temperature range 1 Hz-1 MHz and 303 K-373 K respectively. The maximum ionic conductivity $1.421\times10^{-4}\,Scm^{-1}$ was obtained for 15 wt% KIO_4 doped polymer electrolyte at room temperature. The variation of ac conductivity with frequency obeyed Jonscher power law. Optical absorption studies were performed in the wavelength range 200-600 nm and the absorption edge, direct band gap and indirect band gap values were evaluated.

1. Introduction

In recent years, solid polymer electrolytes have drawn significant interest due to their theoretical as well as practical applications and hence have been extensively investigated for commercial applications such as high energy density batteries, electrochromic display devices, electrochemical sensors and super capacitors [1-3]. The polymeric electrolytes have many advantages due to their favorable electrical, optical and mechanical properties, ease of fabrication in thin films and ability to form effective electrode-electrolyte contacts. Because of the mechanical stability and suitable conductivity of solid polymer electrolytes, many physicists and chemists have devoted their efforts to study and improve the electrical properties of polymers by doping them with metal salts (ions) [4]. Several types of polymers such as (PMMA),(PVC), (PVDF) and (PEO) etc have been studied to see their potential as electrolytes . Among these polymers, poly (vinyl pyrrolidone) (PVP) deserves a special attention because of its good electrical properties, easy processability, moderate electrical conductivity and charge transport mechanism [5]. The PVP is amorphous polymer, which can permit faster ionic mobility compared to other semi-crystalline polymers, and due to the presence of carbonyl group (C=O) in the side chains of PVP, it forms a variety of complexes with various inorganic salts. It interacts well with several types of ions and enhances the number of free ions in the system. PVP is a potential material having a very high dielectric strength, low scattering loss, good charge storage capacity and dopant dependant electrical properties.

2. Experimental

Thin films (~120 μm) of poly (vinyl pyrrolidone) and various compositions of complexed films of PVP with KIO_4 salt were prepared by solution cast technique [6]. The X-ray diffraction patterns of pure PVP, KIO_4 and KIO_4 complexed PVP polymer electrolytes have been recorded by SEIFERT 3003RT X-ray diffractometer with CuK_α radiation (λ= 1.54178 A° operated at 40 KV and 30 mA), in the 2θ range 5-60°. The electrical conductivity measurement of polymer films was carried out by AC impedance spectroscopic technique using computer controlled PSM 1700 (Phase Sensitive Multimeter) meter over the frequency and temperature range 1Hz-1MHz and 303-373K respectively. Optical absorption spectra were recorded at room temperature in the wave length range 200-600 nm using PERKIN ELMER UV-VIS-NIR (model 3100) spectrophotometer.

3. Results and Discussion

3. 1 XRD analysis: The XRD profiles of pure PVP, KIO_4 salt and PVP: KIO_4 systems are shown in Fig.1

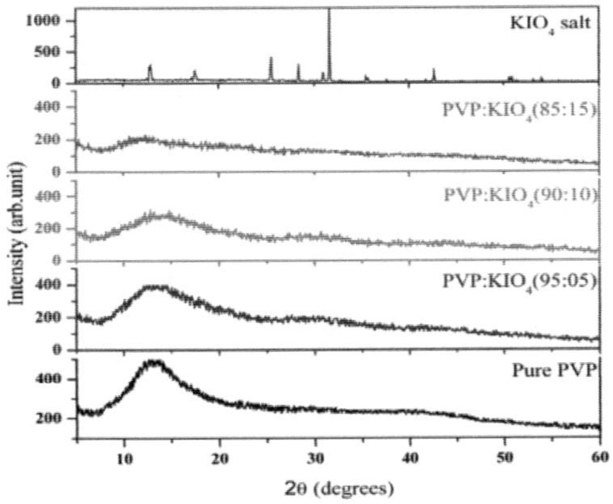

Figure 1 XRD profile of KIO_4 salt doped PVP PVP polymer electrolytes

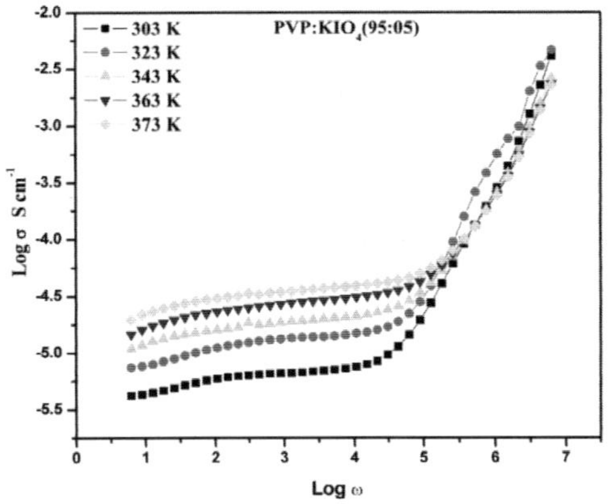

Figure 2 Logσ Vs Logω for PVP: KIO_4 (95:05) polymer electrolytes at different temperatures

The XRD profile of KIO_4 salt shows intensive sharp peaks which indicate the crystalline nature of the salt. These peaks are found to be absent in the PVP: KIO_4 systems indicating the complete dissociation of salt in the polymer matrix. For pure PVP a broad peak was observed at 2θ region of $11-18^o$ which can be attributed to the amorphous nature of pure PVP**.** This peak is broadened and its relative intensity decreases with increase of KIO_4 salt concentration, which indicates the increase in the amorphous nature of the host polymer. The increase in the amorphous nature causes a reduction in the energy barrier to the segmental motion of the polymer electrolyte. This amorphous nature results in greater ionic diffusivity and high ionic conductivity, which can be observed in amorphous polymers having flexible back-bone [7, 8].

3. 2 Conductance spectra: The variation of conductivity as a function of frequency at different temperatures for PVP: KIO_4 (95:05) is shown in Fig 2. It is evident from the figure that in the measured frequency range two distinct regions namely, low frequency plateau region and high frequency dispersion region were observed. The low frequency plateau region which describes the electrode-electrolyte interface phenomena is attributed to the space charge polarization at the blocking electrode and is associated with the dc conductivity$(\sigma_{dc})$of the complexed polymer electrolyte. In the high frequency dispersion region the ionic conductivity increases with increasing frequency which is associated with the ac conductivity of the polymer electrolyte. However, at high frequencies, the conductivity variation with temperature is much less than that at low frequencies. DC conductivity values were obtained by taking the intercept of logσ at zero frequency are listed in table 1.

The total conductivity of polymer electrolytes may be expressed using Jonscher's universal power law equation [9]

$$\sigma(\omega) = \sigma_{dc} + A\omega^{\alpha} \tag{1}$$

where α is the temperature dependent frequency exponent, A, temperature dependent parameter , (σ_{dc}) the dc conductivity and $\omega = 2\pi f$ is the angular frequency.

According to the jump relaxation model reported by Funke [10], the power law exponent α relates the back hop rate to the site relaxation time:

$$\alpha = \text{(back hop rate)/(site relaxation time)}$$

If α is less than one, the backward hopping is slower than the site relaxation; it results in translational motion of the K^+ ion. The conductivity spectra of the polymer electrolyte were fitted to eq (1) using a nonlinear least-square fitting procedure and the evaluated parameter α is in the range from 0.55-0.86, i.e., less than one (<1) for PVP: KIO_4 (95:05) polymer electrolyte and found to decrease with increase in temperature for all KIO_4 salt concentrations.

3. 3 Optical Properties: The absorption coefficient α was calculated from the absorbance (A). After correction for reflection, α was determined using the following relation:

$$\alpha = \frac{2.303}{x}\log(I/I_0) = \frac{2.303}{x}A$$

where I_0 and I is the intensities of incident and transmitted radiation respectively, x is the thickness of the sample and A corresponds to $\log(I/I_0)$

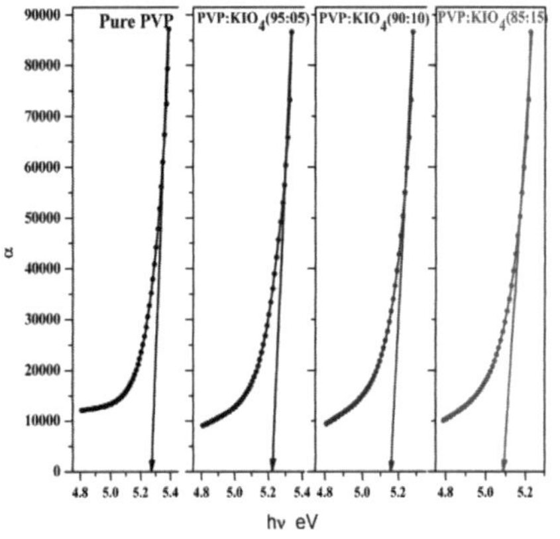

Figure 3 α Vs hν plots of PVP:KIO_4 polymer electrolyte films

Table 1. Conductivity, Band edge, Band gap values of PVP:KIO_4 polymer films.

Sample	σ at 303 K	Band edge(eV)	Band Gap (eV) Direct	Indirect
Pure PVP	7.421×10^{-8}	5.29	5.32	5.12
5wt% KIO_4	6.661×10^{-6}	5.21	5.28	5.08
10wt%KIO_4	3.046×10^{-5}	5.18	5.19	5.01
15wt%KIO_4	1.421×10^{-4}	5.12	5.10	4.91

Davis and Shalliday [11] reported that near the fundamental band edge, both direct and indirect transition occur and can be observed by plotting $\alpha^{1/2}$ and α^2 as a function of energy hν (ν is the frequency of incident light , and h is Planck's constant). The position of the absorption edge was determined by extrapolation of the linear portion of α-hυ curves (Fig.3) to the zero absorption value.

When a direct band gap exists, α has the following dependence on the energy of the incident photon [11-13]. $\alpha h\upsilon = C(h\upsilon - E_g)^{1/2}$

where E_g is the band gap and $C = 4\pi\sigma_o / nc\Delta E$ is a constant dependent on the specimen structure. For indirect transitions, which require phonon assistance, α has the following dependence on hυ [11-13]

$$\alpha h\upsilon = A(h\upsilon - E_g + E_p)^2 + B(h\upsilon - E_g - E_p)^2$$

where E_p is the energy of the photon associated with the transition and A and B are constants depending on the band structure.The direct and indirect band gap values were obtained from the plots of $(\alpha h\upsilon)^2 - h\upsilon$ and $(\alpha h\upsilon)^{1/2} - h\upsilon$ as shown in Figure 4-5.

The values of the absorption edge, direct band gap, and indirect band gap are tabulated in table 1. It is clear form table 1 that the direct and indirect band gaps showed a decrease on doping with KIO_4

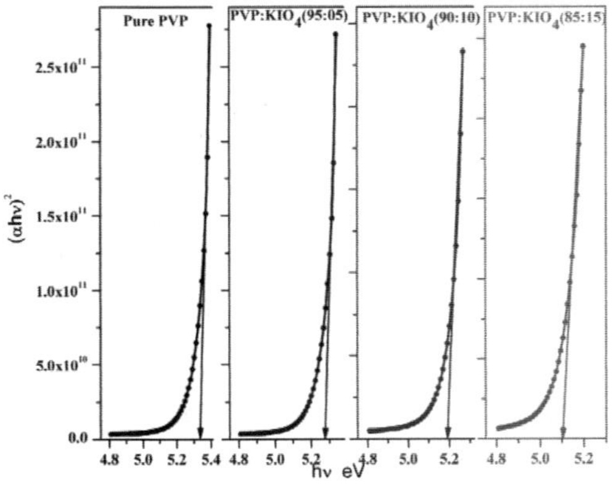

Figure 4 $(\alpha h\nu)^2$ Vs hv plots of PVP:KIO_4 polymer electrolyte films

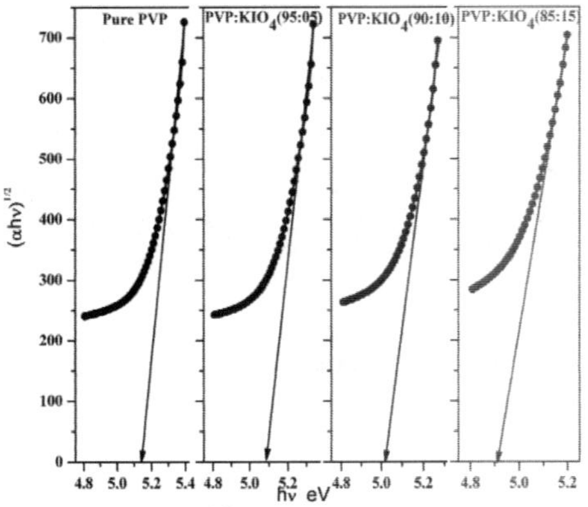

Figure 5 $(\alpha h\nu)^{1/2}$ Vs hv plots of PVP:KIO_4 polymer electrolyte films

4. Conclusions

The XRD patterns of the PVP: KIO_4 electrolyte suggests an increase in the degree of amorphousity with increasing KIO_4 salt concentrations in the polymer matrix. The variation of ac conductivity with frequency obeys Jonscher power law. The optical absorption edge and bad gap (both direct and indirect) showed a decreasing trend with an increased concentration of the dopant.

Acknowledgement: One of the authors Mr.M. Ravi acknowledges the UNIVERSITY GRANTS COMMISSION, NEW DELHI for providing financial support under **Research Fellowship for Meritorious Students (RFSMS)** scheme to carry out this work.

REFERENCES

[1] Gray, F. M 1997 *Polymer electrolytes* (The Royal society of Chemistry, England)
[2] Amrtha Bhide, Hariharn K 2006 *J. Power sources* **159** 1450
[3] Mac Callum J R and Vincent C A 1987 *Polymer electrolyte review* (Elsevier, London)
[4] Subba Reddy C V, Jin A P, Zhu Q Y, Mai L Q, Chen W 2006 *Eur. Phys. J. E* **19** 471.
[5] Armand M B, Chabagno M J, Duclot M 1998 *Int. Conf. Solid Electrolytes Ext Abst.* University of St Andrews, Scotland 826
[6] Ravi M, Pavani Y, Kiran Kumar K, Bhavani S, Sharma A.K and Narasunga Rao V.V.R 2011 *Mater. Chem. Phys.***130** 442
[7] Morita M, Araki F, Yoshimoto N, Ishikawa M and Tsutsumi H 2000 *Solid State Ionics* **136** 1167
[8] Baskaran R and Sevasekarapadian S 2004 *Ionics* **10** 129.
[9] Jonscher A K 1978 *J. Mater. Sci.* **13** 553.
[10] Funke K, Roling B and Lange M 1998 *Solid State Ionics* **105** 195.
[11] Davis D S and Shalliday T S 1960 *Phys. Rev.* **118** 1020.
[12] Thutupalli G M and Tomlin S G 1976 *J. Phys. D App. Phys.* **9** 1639.
[13] Pankoue J I 1971 *Optical processes in semiconductors* (Dover Publication, New York)

Effect of addition of $BaTiO_3$ nano particles on the electrical transport properties of YBCO superconductor

P P Rejith, S Vidya and J K Thomas*

Electronic Materials Research Laboratory, Department of Physics, Mar Ivanios College, Thiruvananthapuram- 695 015, Kerala, India

*e-mail: jkthomasemrl@yahoo.com

Abstract. The flux pinning properties of YBCO bulk superconductors synthesized by conventional solid state route and are added with different weight% (x=0, 0.5, 1, 2,3, 5) of nano $BaTiO_3$ which are prepared by a modified combustion route are studied systematically. The phase analysis of the samples was done by using X-ray diffraction and scanning electron microscopy. Temperature-resistivity measurements, magnetic field dependence of critical current density (J_c-B Characteristics) and flux pinning force calculations were done at 77 K. From the SEM images the microstructure of the sample showed a relative uniform distribution of the nano-particles within the specimen. We found that, even though the transition temperature (T_c) does not change considerably with the $BaTiO_3$ addition, both the critical current density (J_c) and flux pinning force (F_p) increased systematically up to 2 wt% $BaTiO_3$ in the composite, in the presence of magnetic field ranging between 0 and 0.6 T .The J_c value in 2 wt% $BaTiO_3$ added sample showed at least 250% larger than that of the pure YBCO. Also the flux pinning force calculated for the 2 wt% $BaTiO_3$ added is found to be enhanced more than 9 times that of pure YBCO. These observations suggest that the $BaTiO_3$ addition to the Y-123-compounds improve the electrical connection between superconducting grains to result in the increase in J_c.

1. Introduction

$YBa_2Cu_3O_{7-\delta}$ (YBCO) is one of the most widely studied high temperature superconductor, because of its promising and attractive aspects of applications, which include energy storage systems, current limiters, magnetic bearings etc. Improvement of critical current of YBCO materials is attained by the creation of additional defects acting as pinning centers by irradiation, incorporation of nanoparticles and doping [1-3]. Flux pinning by non-superconducting inclusions and crystal defects are also widely studied for YBCO. The main effects of these methods were the inclusion of artificial pinning centers and the introduction of secondary phases related to the additives which results in the enhancement of J_c. Here we report the results of our studies on the variation in the electrical transport properties such as T_c, J_c, and flux pinning force and also the microstructure analysis of YBCO added with different wt% of nanoparticles of $BaTiO_3$ prepared using an auto igniting combustion technique.

2. Experimental

The $YBa_2Cu_3O_{7-\delta}$ superconductor was prepared through conventional solid state route by thoroughly mixing high purity Y_2O_3, $BaCO_3$, and CuO in the stoichiometric ratio. The mixture was calcined at 930°C for a total of 72 hours with two intermediate wet grindings. The nano particles of $BaTiO_3$ were prepared through an auto igniting combustion process. For the typical synthesis, aqueous solution containing ions of Ba and Ti was prepared by dissolving high purity $Ba(NO_3)_2$ in distilled water and $C_{12}H_{28}O_4Ti$ in ethanol. Citric acid was then added as the complexing agent. Amount of citric acid was calculated based on total valence of the oxidizing and the reducing agents for maximum release of energy during combustion [4]. Oxidant/fuel ratio of the system was adjusted by adding nitric acid and ammonium hydroxide. The solution containing the precursor mixture at a pH of ~7.0 was heated using a hot plate at ~250°C, which ignites giving nano powder of phase pure $BaTiO_3$. Different wt% of 15nm sized $BaTiO_3$ thus obtained is then added to YBCO, mixed thoroughly, pelletized and sintered at 980±5°C for 24 hours with a heating rate of 3°C/minute. Then the samples was then cooled to 550°C in 8 hours and kept 24 hours for oxygenation.

The structural characterization of $YBa_2Cu_3O_{7-\delta}$ (Y0), nano crystalline $BaTiO_3$ added $YBa_2Cu_3O_{7-\delta}$ samples were examined by powder X-ray diffraction (XRD) technique using a Bruker D-8 X-ray Diffractometer. The surface morphology of the sintered samples was studied using Scanning Electron Microscopy (SEM) (JEOL JSM 6390 LV). The critical transition temperature (T_c) of both pure and $BaTiO_3$ added samples were measured using standard four probe technique using Keithley source meter 2440 and nano voltmeter 2182A along with lakeshore temperature controller equipped with PT111 platinum sensor. The in-field transport critical current (J_c-B) measurements were done using a magnetic field set up fabricated indigenously using bipolar electromagnets with required value of current which can generate a maximum field upto 1 Tesla at liquid nitrogen temperature using the four probe method with the standard 1μV/cm criterion

Hereafter we denote Y0, YT0.5, YT1, YT2, YT3, YT5 for 0, 0.5, 1, 2, 3, 5 wt% respectively for nanopowders of $BaTiO_3$ addition in YBCO.

3. Results and Discussion

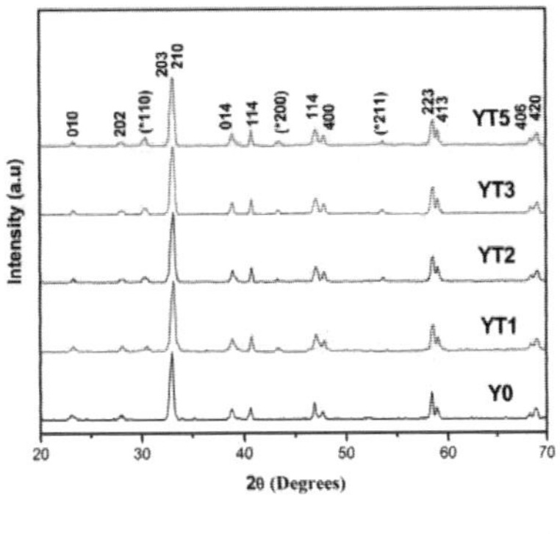

Figure 1 XRD Patterns of different wt% of $BaTiO_3$ added YBCO

Figure 1shows the XRD pattern of pure and 1, 2, 3, and 5 wt% of nano $BaTiO_3$ added YBCO. In the XRD pattern of the YBCO–$BaTiO_3$ composite samples, the peaks corresponding to cubic $BaTiO_3$ could be observed at 2θ= 31.1, 42.9 and 55.14. Even upto 5 wt % of nano $BaTiO_3$ was added, the peak position of YBCO remains unaltered without any shift.

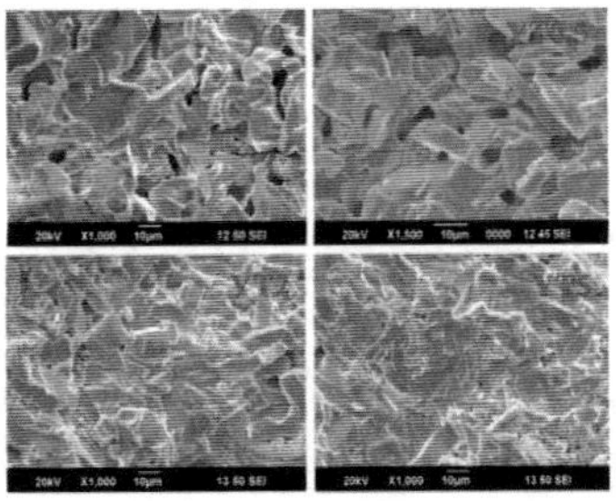

Figure 2 SEM images of the fracture surface of the samples

The SEM micrographs of the fractured surface of sintered YBCO- $BaTiO_3$ composites are shown in figure 2.The microstructure of the samples seems to improve with the increasing content of $BaTiO_3$ in YBCO. It is clearly seen that pure YBCO sample is porous with significant amount of voids of micron size, resulted a low sintered density. It is also clear from the micrographs that the porosity lowers as the percentage of addition of nanocrystalline $BaTiO_3$ in YBCO increases.

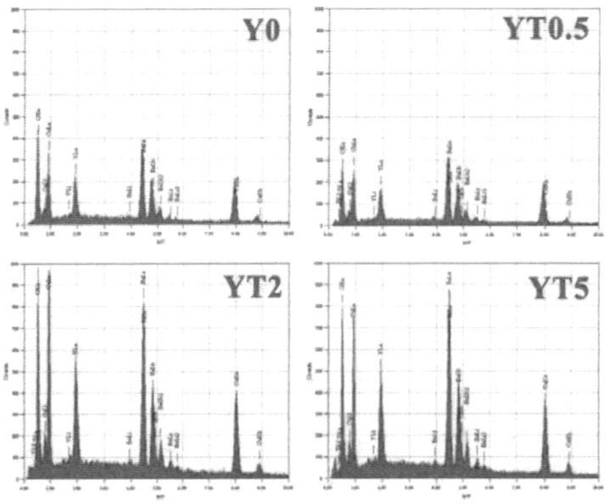

Figure 3 EDAX spectrums of the YBCO-$BaTiO_3$ composites

The EDAX analysis for pure and $BaTiO_3$ added samples are shown in figure 3. The spectra indicate the presence of $BaTiO_3$ in the YBCO compound in its respective stoichiometric ratios.

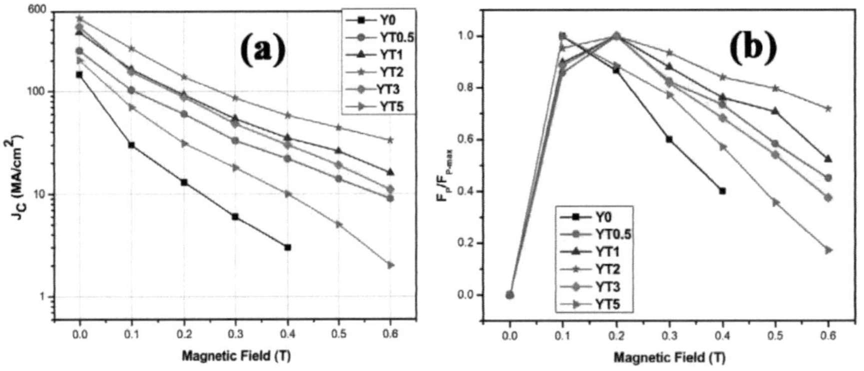

Figure 4(a) Magnetic field dependence of transport J_c **(b)** Variation of the normalized pinning force density ($F_p/F_{p\text{-}max}$) with applied field

Magnetic field dependence of transport J_c (J_c-B characteristics) of the samples at 77 K is shown in figure 4 (a). The J_c-B characteristics of $BaTiO_3$ added samples are found to be much better than that of the pure sample. The samples YT0.5, YT1 and YT2 display good J_c-B behavior in the graph, of which the YT2 exhibit the best. J_c increases appreciably with increasing concentration $BaTiO_3$ from 147 MA/m^2 for Y0 to an optimum value of 516 MA/m^2 for YT2 sample and then decreases for further increasing concentration of nano additives. The obtained value for J_c of YT0.5 (251 MA/m^2), YT1 (378 MA/m^2), YT2 (516 MA/m^2) are 70%, 150%, 250% respectively, more than that of the pure YBCO (147MA/m^2) sample. These characteristics show that addition of $BaTiO_3$ in optimum concentration (x =2 wt %) in YBCO enhances the self-field J_c, and gives significant improvement in the J_c–B characteristics of the superconductor.

Using the volume pinning force density, $F_p = J_c \times B$, one can investigate the flux pinning strength of the system [5-6]. The normalized volume pinning force density ($F_p/F_{p\text{-}max}$) calculated from the data of J_c-B characteristics as a function of the applied field is plotted in figure 4(b). It is seen that addition of $BaTiO_3$ shifts the peak positions of $F_p/F_{p\text{-}max}$ towards higher fields. This confirms that the flux pinning strength of nano $BaTiO_3$ added samples increases significantly compared to the pure YBCO. Table 1 summarizes all the results obtained for pure and different weight percentages of $BaTiO_3$ added YBCO samples.

Table1. Transport properties of YBCO with different weight percentages of $BaTiO_3$

Sample	Sintering temperature (°C)	Sintered Density (gcm^{-3})	$T_{c\text{-}onset}$ (K)	$T_{c\text{-}zero}$ (K)	Room Temp resistivity (μΩm)	Self field J_c (MA/m^2)	$F_{p\text{-}max}$ (x 10^5 N/m^3)
Y0	985	5.53	98	92	9.3	147	30
YT0.5	983	5.68	102	92	19.9	251	120
YT1	982	5.82	103	92	30.7	378	184
YT2	980	5.88	103	91.5	46.8	516	276
YT3	978	5.86	102	91	58.2	428	176
YT5	975	5.84	99	90	74	202	70

The obtained results show that with the $BaTiO_3$ addition, the vortex pinning increases. The F_p for the composite sample, YT2 is 9 times higher than the value of pure YBCO. Such a higher value of F_p in the composite sample could be the result of $BaTiO_3$ particles acting as artificially generated pinning centers. Thus, in future possible applications, the addition of $BaTiO_3$ to Y-123 can be effective in enhancing the superconducting properties such as critical current density and pinning force in a relatively low magnetic field region below 1 T.

In our studies we conclude that the change in the intrinsic parameters and the structure defect density induced by the addition of $BaTiO_3$ could be the reason for enhanced flux pinning. The observed improvement in J_c over a small range of $BaTiO_3$ content is direct evidence for the presence of effective flux pinning. The $BaTiO_3$ particles stay around outside of the grain might also act like catalyst to improve the structural quality at the grain boundaries. This in turn can causes the increase of inter-grain contact surface and increase the number of flux pining centers at the inter-grain regions and results in the enhancement of J_c even in the presence of higher magnetic field.

4. Conclusions

In conclusion, we synthesized pure YBCO and YBCO with different weight % of nano $BaTiO_3$ composite samples and systematically investigated its effect as artificially generated pinning centers. There is not much variation in transition temperatures (T_c) of pure and composite samples was observed while a significant enhancement in the value of J_c and F_p for an optimum 2 weight percentage $BaTiO_3$ added YBCO samples, relative to pure YBCO is observed. About 250% of enhancement in the value of J_c is achieved in the composite sample when compared to pure YBCO, over the range of applied magnetic field. The introduction of $BaTiO_3$ nanoparticles in YBCO resulted in the increase in pinning force density from 30×10^5 N/m^3 to 276×10^5 N/m^3 at 77 K

References

[1] Paulius L M, Almasan C C and Maple M B 1993 *Phys. Rev. B* **47** 11627
[2] Skakle J M S 1998 *Mater. Sci. Eng. R* **23** 1
[3] Harada T and Yoshida K 2002 *Physica C* **383** 48
[4] Patil K C 1993 *Bull. Mater. Sci.* **16** 533
[5] Koblischka M R, van Dalen A J J, Higuchi T, Yoo S I and Murakami M 1998 *Phys Rev. B* **58** 2863
[6] Kramer E J 1973 *J. Appl. Phys.* **44** 1360

IOP Conf. Series: Materials Science and Engineering **73** (2015) 012147 doi:10.1088/1757-899X/73/1/012147

Synthesis of $AgInS_2$ semiconductor nano crystals by solvothermal method

S Sugan and R Dhanasekaran
Crystal Growth Centre, Anna University, Chennai-600 025, India.

E-mail: rdhanasekaran@annauniv.edu

Abstract. A solvothermal route was proposed to synthesis nanocrystalline silver indium sulfide ($AgInS_2$) using ethylenediamine as a solvent. The reaction was carried out in an autoclave in the temperature of 200 °C with AgCl, Indium (In) and sulfur (S) as reactants. Powder X-ray diffraction studies show that the synthesized nano crystals belongs to orthorhombic phase $AgInS_2$ with cell parameters a= 7.001 Å, b= 8.278 Å, and c= 6.698 Å. High resolution scanning electron microscope (HR-SEM) and Energy Dispersive X-ray (EDX) analyses have been carried out for the synthesized nano crystals to find the surface morphology and composition respectively. The band gap of the synthesized nano crystals was determined from UV-Vis spectrum and photoluminescence spectrum. The melting point of the $AgInS_2$ nano crystals was found to be 956°C using differential thermal analysis (DTA).

1. Introduction

$AgInS_2$ is a semiconductor compound of the I–III–VI_2 groups, which crystallizes in chalcopyrite and orthorhombic ordered phases. It might be used in light emitting diodes, nonlinear optics and solar cells [1]. Semiconductor materials, such as $AgInS_2$, $CuInS_2$ and $CuInGaSe_2$ have been broadly investigated for photovoltaic device applications [2, 3]. So far, substantial attempts were made to the preparation of $AgInS_2$ thin films and single crystals [4]. As a potential different, $AgInS_2$ based chalcopyrite semiconductor is also likely to make a good solar-cell absorber layer because of its large absorption coefficient [5]. To realize large amounts of size-controlled semiconductor nanoparticles, chemical synthesis methods are suitable, and are mainly well developed for type II–VI semiconductor nanoparticles. Indeed, the optical properties of CdS nanoparticles have been studied for a long time [6]. However, conventional binary-compound semiconductors contain toxic elements such as Cd, Hg, Pb, Te, Se and As. Hence, nontoxic and semiconducting materials are required for the perspective of control of hazardous substances, such as I–III–VI_2 chalcopyrite semiconductors mainly $CuInS_2$, $AgGaS_2$, $AgInS_2$ etc, are the promising and alternative materials. Much interest has been decided on chalcopyrite semiconductors as an applicant for solar cell materials, because they have direct band gaps with energies well matched to the solar spectrum, high absorption coefficients and high environmental stability [7, 8]. In the present work solvothermal method has been adopted for the synthesis of $AgInS_2$ nano crystals using ethylenediamine as a solvent. We discuss the formation of $AgInS_2$ nano crystals, optical and thermal properties.

2. Experimental

2.1. Synthesis of $AgInS_2$ nano crystals

Analytical grade of Silver chloride (AgCl), Indium (In) and Sulphur (S) were selected as the reactants. In a characteristic procedure 1:1:2 mole ratio of AgCl, In and S were put into a stainless steel autoclave of 20 ml capacity. The autoclave was then filled with ethylenediamine up to 85 % of the total volume. The autoclave was maintained at 200 °C for 8 hrs and allowed to cool to room temperature. The precipitate was separated by filtration and then washed sequentially with carbon disulfide, absolute ethanol and distilled water to remove residual impurities such as S and Cl. A dark brown color $AgInS_2$ powder was obtained after drying at 80 °C for 2 hrs. The final products were collected and subjected to further investigation by using powder X-ray diffractometer (SEI FERT) JSO DEBYEFLEX 2002 model with $CuK\alpha_1$ radiation (1.541 Å). The accelerating voltage and the applied current were 40 kV and 30 mA respectively. The surface morphology of as synthesized samples was investigated using FEI Quanta FEG 200 HR-SEM. The chemical compositions of the samples were studied using EDX system connected with a HR-SEM operating at an accelerating voltage of 30 keV. UV-Vis absorption spectrum has been carried out using Cary 5E high resolution spectrophotometer. The emission spectrum was recorded using JOBIN - YUON Flurolog-3-11spectroflurometer and DTA analysis was carried out by using Netzsch STA 409 c/cd thermal analyzer at the heating rate of 10°C/min in the Nitrogen atmosphere.

3. Results and discussion

3.1. X- ray diffraction

Figure 1 shows the XRD pattern of as synthesized $AgInS_2$ nano crystals by solvothermal method. All the diffraction peaks could be indexed to the orthorhombic $AgInS_2$ phase with cell parameters a= 7.001Å, b= 8.278 Å and c= 6.698 Å in agreement with reported data in the literature (JCPDS, 25-1328). The XRD peak intensity of (002) plane was relatively higher than the other planes. Whereas, no characteristic peaks of other impurities such as In and Ag_2S were observed. The broad peak indicates polycrystalline behaviour of the particles. From this pattern, the scherrer's equation gives the crystalline size 21.31 nm of (002) plane.

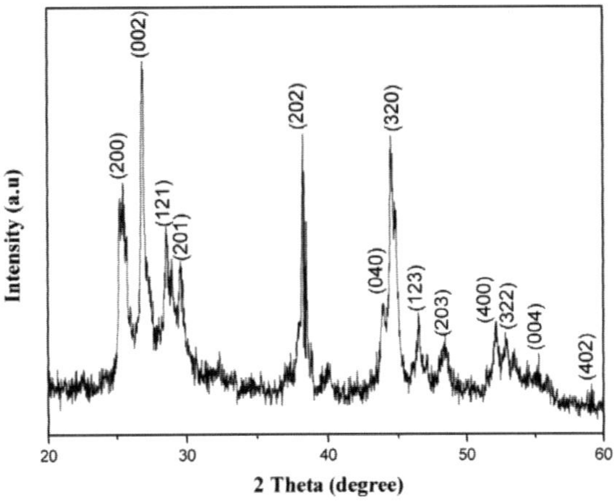

Figure 1. XRD pattern of $AgInS_2$.

3.2. Surface morphology analysis

The surface morphology of as synthesized nano crystalline $AgInS_2$ are shown in figure 2. It can be seen that nano crystalline $AgInS_2$ consist of agglomerate like flower structure. The diameter of the particles is in the range around 20-30 nm in accord with the XRD result.

Figure 2. HR-SEM images of $AgInS_2$.

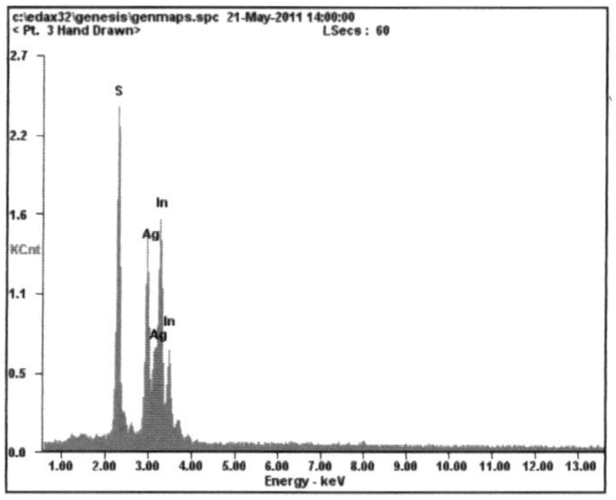

Figure 3. EDX spectrum of $AgInS_2$.

3.3. Composition analysis

Figure 3 Shows the typical EDX spectrum of as synthesized nano crystalline $AgInS_2$. The results showed that the sample was composed of only Silver (Ag), Indium (In), Sulfur (S) elements are in the following molar ratio 24.85: 26.37: 48.85, which is more close to the stoichiometric composition. No apparent impurities like chloride and elemental sulfide could be detected in the samples.

3.4. UV-Vis absorption spectrum

Figure 4 shows the UV-Vis absorption spectrum of as synthesized $AgInS_2$ nano crystals and the absorption edge was observed at approximately 590 nm. Inset of the figure 4 shows $(\alpha h\nu)^2$ versus photon energy examined by absorption spectrum, with α and $h\nu$ being the absorption coefficient and photon energy respectively. Band energy (E_g) 2.08 eV can be calculated using the following equation $(\alpha h\nu)^2 = (h\nu - E_g)$. As compared with bulk band gap of 1.98 eV the band gap of obtained nano crystals was widened approximately 0.1 eV due to quantum size effect [9].

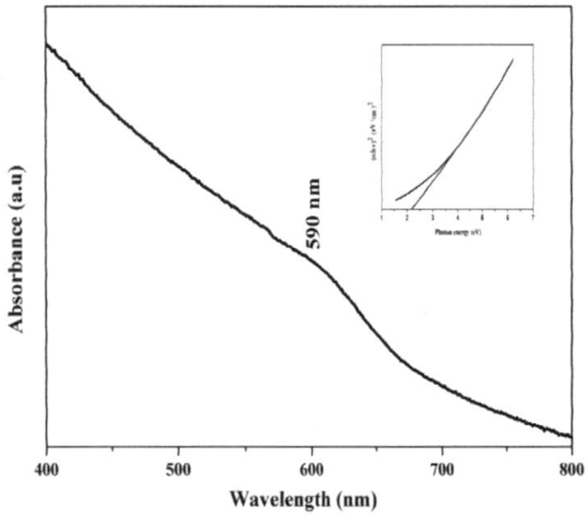

Figure 4. UV-Vis absorption spectrum.

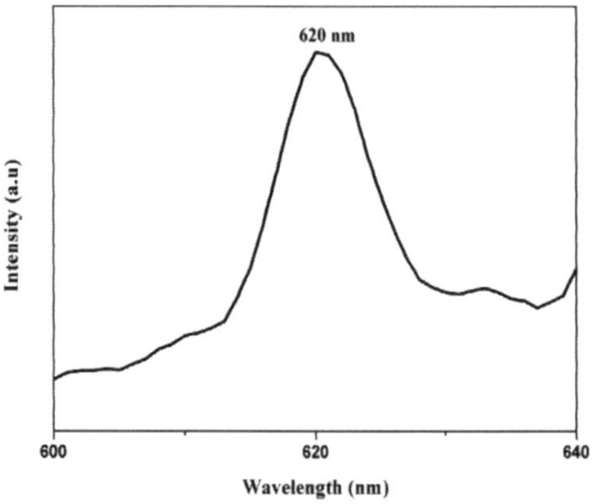

Figure 5. Photoluminescence spectrum.

3.5. Photoluminescence (PL) spectrum

The photoluminescence spectrum of the synthesized nano crystals were obtained at the excitation wavelength of 590 nm using xenon as a source and it is shown in the Figure 5. An emission peak located around 620 nm (1.99 eV) were approximately 0.09 eV smaller than the optical band gap it is due to the electron hole recombination pair.

3.6. *Differential thermal analysis (DTA)*

As synthesized $AgInS_2$ nano crystalline powder was carried out using DTA in nitrogen atmosphere with a gas flow rate of 50 cm^3/min. The heating rate was 10 °C/min. Al_2O_3 was used as a standard to measure the absolute value of temperature reading. DTA curve of $AgInS_2$ indicates only one sharp endothermic peak at 956 °C corresponding to the melting point of the material.

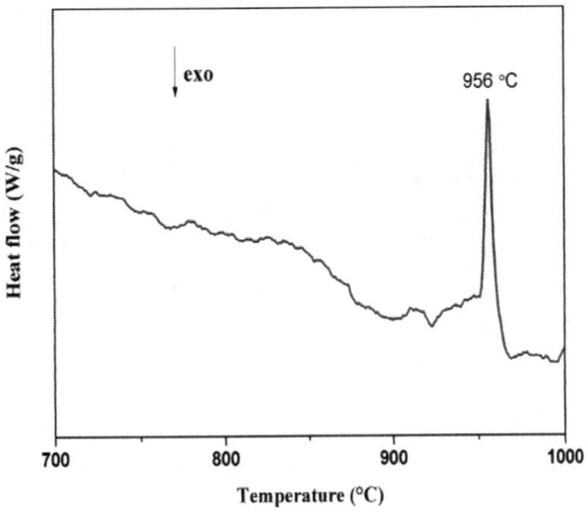

Figure 6. Differential thermal analysis of $AgInS_2$.

4. Conclusions

Ternary $AgInS_2$ nano crystals were synthesized by solvothermal method using ethylenediamine as a solvent. The formation of nano crystals were confirmed with powder XRD studies and the cell parameter shows that these nano crystals possess orthorhombic system. HR-SEM analysis indicates the as synthesized nano crystals consist of flower like structure. The stoichiometric composition of the synthesized $AgInS_2$ nano crystals was determined using EDX analysis. UV-Vis absorption spectrum of $AgInS_2$ nano crystals show an absorption peak around 590 nm. The emission band appears at 620 nm (1.99 eV) were determined from PL spectrum is due to the electron hole recombination pair. The melting point of the $AgInS_2$ was found to be 956°C using DTA.

Acknowledgements

This work is supported by the Council of Scientific and Industrial Research (CSIR), India.

References

[1] Shay J L and Wernick J H 1975 Ternary Chalcopyrite Semiconductors Growth Electronic Properties and Applications (*Pergamon Press, New York*)
[2] Wang D, Zheng W, Hao C, Peng Q and Li Y 2008 *Chem. Commun.* **22** 2556
[3] Nanu M, Schoonman J and Goossens A 2005 *Nano Lett.* **5** 1716
[4] Cheng K W and Wang S C 2009 *Mater. Chem. Phys.* **115** 14
[5] Aissa Z, Ben Nasrallah T, Amlouk M, Bernede J C and Belgacem S 2006 *Sol. Energy Mater. Sol. Cells.* **90** 1137
[6] Eychmuller A 2000 *J. Phys. Chem.* B. **104** 6514
[7] Bailey S G and Flood D J 1998 *Prog. Photovolt. Res. Appl.* **6** 1
[8] Schock H W and Noufi R 2000 *Prog. Photovolt. Res. Appl.* **8** 151
[9] Chen H, Yu S M, Shin D W and Yoo J B 2010 *Nanoscale Res Lett.* **5** 217

IOP Conf. Series: Materials Science and Engineering **73** (2015) 012148 doi:10.1088/1757-899X/73/1/012148

Size-Induced Exchange Bias Effect and Griffiths Phase in Electron Doped $Sm_{0.09}Ca_{0.91}MnO_3$ manganites

S K Giri[1,2] and T K Nath[1]
[1]Department of Physics and Meteorology, Indian Institute of Technology, Kharagpur, West Bengal 721302, India

Email: samir@phy.iitkgp.ernet.in

Abstract. We report here the magnetic properties of electron-doped $Sm_{1-x}Ca_xMnO_3$ manganites with the doping level of x=0.91. Exchange bias effect has been observed in $Sm_{0.09}Ca_{0.91}MnO_3$ nanomanagnites system and can be tuned by the strength of cooling magnetic field (H_{cool}). The values of exchange bias parameter i.e. exchange bias fields (H_E), coercivity (H_C), remanence asymmetry (M_E) and magnetic coercivity (M_C) are found to strongly depend on H_{cool}. The larger effective magnetic moments and deviation of inverse susceptibility (χ^{-1}) from Curie-Weiss law indicate the possible existence of Griffiths phase (GP). The effect of size reduction on exchange bias effect and GP is addressed here. The enhancement of exchange bias effect and GP has been argued to be due to the modification of the phase separated state on size reduction.

1. Introduction

The effect of size reduction on the magnetic properties of electron doped manganites is a topic of considerable interest which exhibits various fundamental phenomena such as colossal magnetoresistance (CMR), exchange bias, Griffiths phase and first-order ferromagnetic transition etc. [1- 4]. The main identification of the presence of exchange bias (EB) effect is (1) horizontal shift of the field-cooled (FC) magnetic hysteresis loop (2) enhancement of magnetic coercivity as compared to the zero-field-cooled (ZFC) case (3) the existence of the magnetic training effect (TE). One of another interesting behavior near and above the ferromagnetic (FM) transition temperature T_C focuses on the performance of ferromagnetic cluster [1]. This often leads to Griffiths phase (GP) [4] which is

[2] To whom any correspondence should be addressed.

originally proposed for randomly diluted Ising ferromagnet. The Griffiths phase means the existence of short range ferromagnetic (FM) clusters in paramagnetic regime in the temperature range $T_C \leq T \leq T_G$. Where T_G is Griffiths temperature at which the FM clusters start to nucleate [5] and T_C is the ferromagnetic Curie temperature.

Here we report the magnetic properties of $Sm_{0.09}Ca_{0.91}MnO_3$ electron doped manganites showing the exchange bias effect below T_N in a field cooled magnetic state and the FM interaction dominates the inhomogeneous phase above T_C i.e. the Griffiths singularity in the PM regime. Here we have reported only two particle sizes; one is S750 (nano) and other one is S1150 (bulk) sample. Our results suggest that the enhancement of exchange bias effect and GP has been argued to be due to the modification of the phase separated state on size reduction.

2. Results and Discussion

2.1. Dc Magnetization

Temperature dependent zero field cooled (ZFC) and field cooled (FC) magnetization curve of S750 sample are shown in the figure 1 (a). The Curie temperature (T_C) of the sample is estimated from the dM/dT vs. T plot (not shown here). The interesting behavior is that the bifurcation between the curves (ZFC and FC) starts much above T_C i.e. from 113 K (inset of figure 1 (a)). But for a typical ferromagnet, the bifurcation starts from T_C. This feature clearly demonstrates the onset of short range ferromagnetic ordering above T_C. The dc magnetization plot of S1150 sample is shown in the figure 1 (b). It is found that the T_C (~ 107 K) value of S1150 sample is lower than that of S750 (~110 K) sample and no anomaly between the ZFC and FC curves of S1150 sample is observed. So, we conclude that T_C of $Sm_{0.09}Ca_{0.91}MnO_3$ manganites increases with decreasing the particle size.

2.2. Griffiths phase singularity

The downturn nature of inverse magnetic susceptibility (χ^{-1}) as a function of temperature above T_C is considered as a hall mark of Griffiths phase (GP) singularity [5]. The temperature dependence inverse dc susceptibility of S750 sample at 50 Oe dc magnetic field has been plotted in figure 2 (a). At higher temperature, χ^{-1} varies almost linearly with T following the Curie-Weiss law. But with decreasing the temperature, a sharp downturn is observed below a characteristic temperature. The onset of this downturn is denoted as Griffiths temperature (T_G) i.e. the temperature where χ^{-1} deviates from Curie –Weiss law and below this temperature FM clusters emerge in the PM matrix as described in Griffiths phase system [5]. Generally, GP is univocally characterized by the $\chi^{-1}(T)$ exponent (λ) lower than unity, as obtained from the following relation: [6]

$$\chi^{-1} = \left(T - T_C^{Rand}\right)^{1-\lambda}, \; 0 \leq \lambda \leq 1 \qquad (1)$$

where T_C^{Rand} is the critical temperature of the random ferromagnet where susceptibility tend to diverge. In figure 2 (a), the red line shows the fitted data with equation (1). The fitted results shows $\lambda = 0.72$ and $T_C^{Rand} = 115$ K. The value of T_C^{Rand} is close to the value of T_C. In the paramagnetic regime, the value of λ is expected to be zero. Here $\lambda = 0.72$, this concludes that there is a finite size ferromagnetic clusters in the paramagnetic regime above T_C where susceptibility deviates from the CW behavior. For S1150 sample, the inverse magnetic susceptibility (χ^{-1}) vs. temperature plot at 50 Oe dc magnetic field is shown in the figure 2(b). In this case χ^{-1} varies linearly with T following the Curie-Weiss law from T_C to higher temperature and GP like properties are not observed.

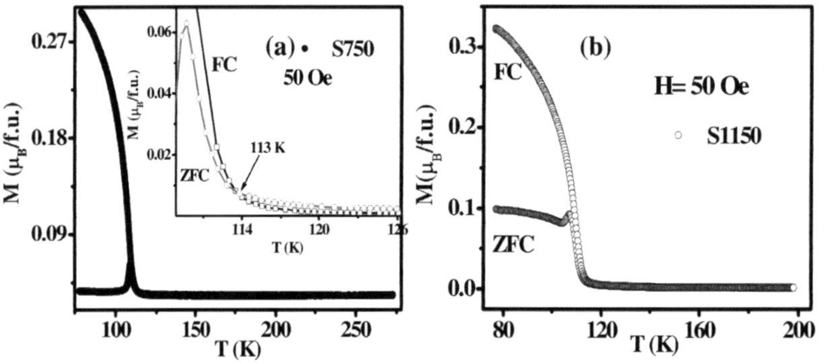

Figure 1. (a) M vs. T plot of S750 sample. Inset shows its zoomed view of a selected portion. (b) M vs. T plot of S1150 sample.

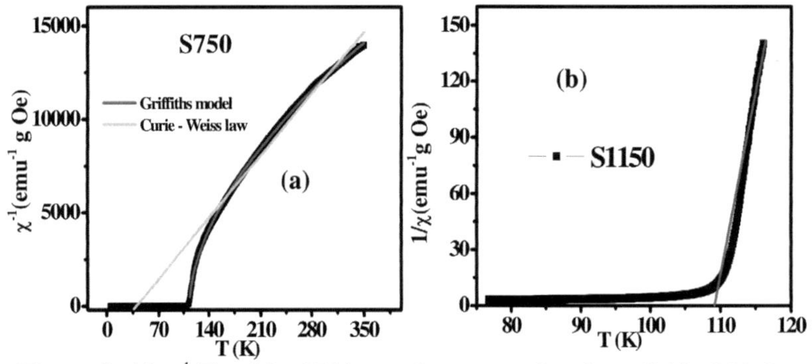

Figure 2. (a) χ^{-1}(T) of the S750 sample measured under a field of 50 Oe. The green line is the fitting result following the CW law and the red solid line is the fitting result following the Griffiths model. (b) χ^{-1}(T) of S1150 sample measured under a field of 50 Oe.

2.3. Exchange bias effect

We have observed the shift in magnetic hysteresis loop when the S750 sample is cooled in a static dc magnetic field. The shift is almost absent when the cooling magnetic field is almost zero. A typical magnetic hysteresis loop at 5 K after cooling the sample with 10 kOe magnetic field is shown in the inset of figure 3 (a). We have calculated the Exchange Bias (EB) parameter by using the standard definition [7]. The value of H_E and H_C is consistent with the value of reported other electron doped nanomanganites [8]. We have studied the magnetic cooling field (H_{cool}) dependence of exchange bias parameter. Figure 3 (a) shows the variation of exchange bias parameter H_E with different cooling field at 5 K and hysteresis is measured with ± 5T magnetic field. At first, H_E increases with H_{cool} up to the cooling field = 10 kOe. Again it shows the decreasing trend with further increasing of H_{cool}. The value of M_E also increases up to 10 kOe magnetic field and then decreases with further increase of H_{cool} as shown in the figure 3 (b). It is found that the value of H_E decreases up to ~ 11 % for the increase of cooling field from 10 to 70 kOe. This suggests that the growth of FM cluster at the high cooling field could contribute partly to the decay of H_E. Thus we conclude that at the high cooling field, not only the alignment of degree of FM cluster increases but the size of FM cluster also increases. Again for S1150 sample, we have also studied the magnetic hysteresis loop after the field cooled with 10 kOe magnetic field at 5 K (inset of figure 3 (b)). But there is no loop shift with the cooling field. So we conclude that the EB effect enhances in the nanomanganites S750 sample.

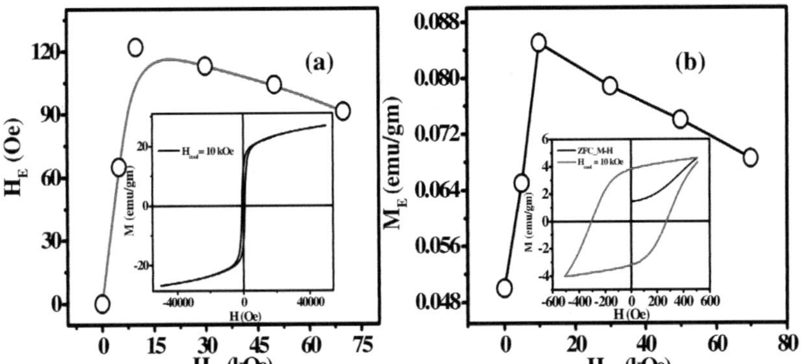

Figure 3. (a) H_E vs. H_{cool} of S750 sample. Inset shows the M-H loop of S750 sample after FC with 10 kOe magnetic field (b) M_E vs. H_{cool} of S750 sample. Inset shows M-H loop of S1150 sample after FC with H_{cool} = 0 and 10 kOe.

3. Conclusions

In summary, a systematic investigation on the magnetic properties of the electron-doped $Sm_{0.09}Ca_{0.91}MnO_3$ manganites has been investigated. A distinct signature of Griffiths phase was found in inverse magnetic susceptibility above T_C only for S750 sample. We have also observed the intrinsic interface EB effect in nanomanganite system with disordered glassy interfaces. Thus for $Sm_{0.09}Ca_{0.91}MnO_3$ manganites, the Griffiths phase and EB effect gradually enhances with decreasing the particle size and we have explained it on the basis of modification of phase separated state on size reduction.

References

[1] Dagotto E, 2002 Nanoscale Phase Separation and Colossal Magnetoresistance, Springer, Berlin
[2] Tokura Y 2006 *Rep. Prog. Phys.* **69** 797
[3] Meiklejohn W H and Bean C P 1956, *Phys. Rev.* **102** 1413
[4] Griffiths R B 1969 *Phys. Rev. Lett.* **23** 17
[5] Tong P, Kim Bongju, Daeyoung Kwon, Qian T, Lee Sung-Ik, Cheong S-W, and Kim Bog G 2008 *Phys. Rev. B* **77** 184432
[6] Castro Neto A H, Castilla G, and Jones B A 1998 *Phys. Rev. Lett.* **81** 3531.
[7] Giri S K and Nath T K 2011 *AIP Advances* **1** 032110
[8] Huang X H, Ding J F, Zhang G Q, Hou Y, Yao Y P, and Li X G 2008 *Phys. Rev. B* **78** 224408

IOP Conf. Series: Materials Science and Engineering **73** (2015) 012149 doi:10.1088/1757-899X/73/1/012149

Effect of Annealing Temperature on the Structural and Optical Properties of Zinc Oxide (ZnO) Thin Films Prepared by Spin Coating Process

Sandeep Sanjeev and Dhananjaya Kekuda
Department of Sciences, Manipal University, Manipal-576 104, Karnataka, INDIA

E-mail: sandeepssanjeev@gmail.com
Telephone: 09844488973

ABSTRACT: ZnO thin films were deposited onto the glass substrates by spin coating method. Zinc acetate dihydrate, 2-methoxyethanol and monoethanolamine (MEA) were used as starting materials for the thin film preparation. Films were obtained by spin coating at 4000 rpm for 30 sec at room temperature and were annealed at different temperatures ranging from 200°C to 400°C. The effect of annealing temperature on the structural and optical properties of the ZnO thin films was investigated. Surface morphology of thin films was studied using atomic force microscope (AFM). The optical absorbance and transmittance measurements were recorded by using a single beam spectrophotometer in the wavelength range 300 nm to 900 nm. Evaluated optical band gap value agrees approximately with that of bulk ZnO. It is observed that band gap decreases as the annealing temperature is increased from 200°C to 400°C.

1. Introduction

Zinc oxide is an inexpensive n-type semiconductor having direct band gap of 3.3 eV which crystallizes in hexagonal wurtzite structure. Due to large exciton binding energy of 60 meV, they have potential applications in Optoelectronic devices such as in solar cells Light emitting diodes (LED) Zinc oxide thin films are applied in Thin Film Transistors (TFT). Thin films of Zinc oxide can be prepared by various techniques; among them are thermal evaporation, Sputtering, Chemical Vapor Deposition (CVD), Spin coating and Spray pyrolysis. Apart from doping, to increase the functionality of ZnO thin film, the effect of preparation conditions on the properties have to be considered for its effective technological applications. In the present work we have used spin coating process for film preparation. Zinc acetate dihydrate was used as the starting material. The Sol-gel process has the advantages of controllability of compositions, simplicity in processing and is cost effective (Shakti, 2010). We have studied the effect of annealing temperature on the structural and optical properties of zinc oxide (ZnO) thin films. Surface morphology of thin films was studied using atomic force microscope (AFM). UV-VIS spectrometry, were used for optical characterization.

2. Experimental details

ZnO thin films were prepared on glass substrates by spin coating method. Zinc acetate dihydrate (Zn $(CH_3COO)_2 . 2H_2O$) was used as a starting material, 2-methoxyethanol and monoethanolamine (MEA) were used as a solvent and stabilizer respectively for the solution preparation. Zinc acetate dihydrate was first dissolved in 2-methoxyethanol and then MEA were added, at room temperature. The molar ratio of MEA to zinc acetate dihydrate was maintained at 1.0 and the concentration of zinc acetate dihydrate was 0.3 M. The solution were stirred at 80°C for 2 hours, thus obtained clear solution was then allowed to cool to room temperature. After the filtration coating was made. The solution dropped onto the pre-cleaned glass substrate, which were rotated at 4000 rpm for 30 sec using spin coater at room temperature. After depositing by spin coating, the films were dried at 150°C for 10 min over a hot plate to evaporate the solvent. The process from coating to drying was repeated to obtain the workable thickness of the film. Thus obtained multilayer films annealed in muffle furnace at 200°C, 300°C and 400°C for 1hr were prepared to study the effect of annealing.

Surface morphology of thin films was studied using atomic force microscope (AFM). The optical absorbance and transmittance measurements were recorded by using a single beam spectrophotometer in the wavelength range 300 nm to 900 nm.

3. Results and discussion

3.1 Surface Morphology Of The ZnO Thin Film

Surface morphology of thin films was studied using atomic force microscope (AFM). AFM images of the thin films annealed at 250°C, 350 °C and 400°C are shown in fig.1, fig.2 and fig.3 respectively. The thin film prepared by spin coting is uniform; this was observed by AFM images taken at two different regions of the same film. By comparing the image of thin films annealed at 250°C, 350°C and 400°C it is observed that film roughness increases as increase in annealing temperature. This is due to increase in grain size with increase in annealing temperature.

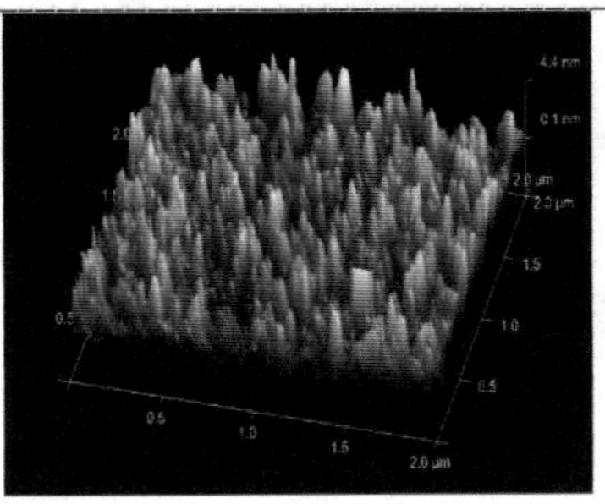

Fig.1 AFM image of ZnO thin film annealed at 250°C

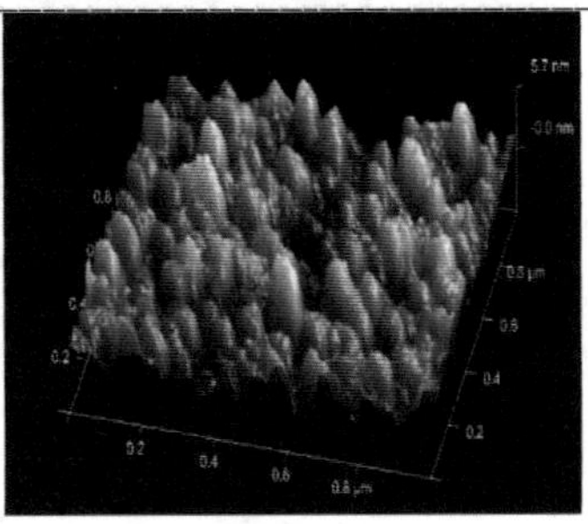

Fig.2 AFM image of ZnO thin film annealed at 350°C

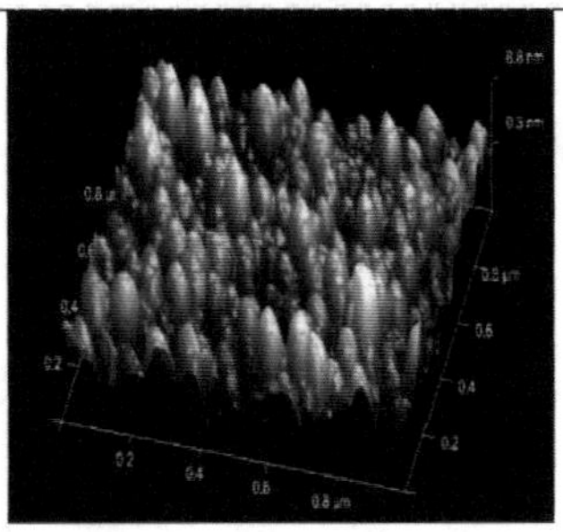

Fig.3 AFM image of ZnO thin film annealed at 400°C

3.2 Optical Properties of The ZnO Thin Films

The optical absorbance and transmittance measurements were recorded by using a single beam spectrophotometer in the wavelength range 300 nm to 900 nm. Absorption and transmission spectra are shown in Fig.4 and Fig.5 respectively, for ZnO thin films annealed at different temperatures. Figures show the increase in absorption of photon energy as increase in annealing temperature. Transmission increases with increases in wavelength and decreases with increase in annealing temperature. The transmission of the film is above 80% in this visible region.

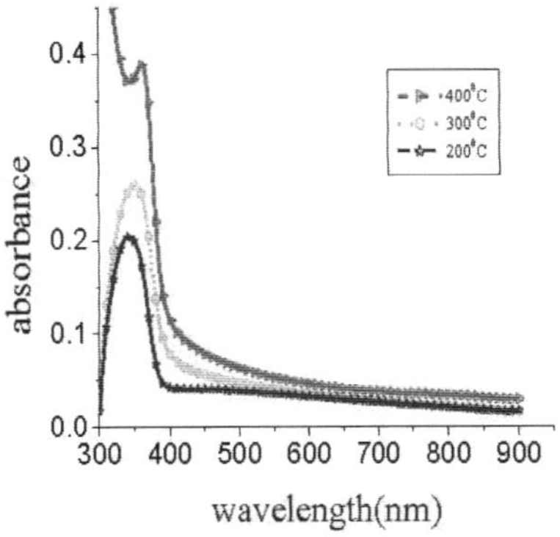

Fig.4 Absorption spectrum of zinc oxide thin films annealed at different temperatures.

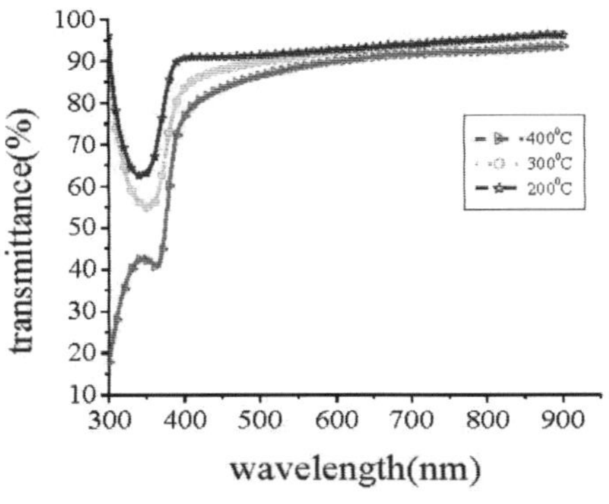

Fig.5 Transmittance spectrum of zinc oxide thin films annealed at different temperatures.

Refractive index of ZnO thin film was calculated from equation (1) (Hussein et al., 2011)

$$n = [1 + (R)^{0.5} / 1 - (R)^{0.5}] \quad (1)$$

The reflectance R was calculated using the following equation (2)

$$R = 1 - [T / e^{-A}]^{0.5} \quad (2)$$

The variation of refractive index with wavelength is shown in fig.6. From fig.6 it is observed that refractive index decreases as wavelength increased.
The absorption coefficient of thin film is calculated from the expression (3) (Kumar et al., 2011)

$$\alpha = 2.303(A/t) \quad (3)$$

Where, A is absorbance and t is the thickness. Shakti (2010) notes that the absorption coefficient α and the extinction coefficient k are related by the formula (4) (Shaaban E.R., 2006).

$$k = \alpha\lambda/4\pi \quad (4)$$

Where, λ is the wavelength. The variation of extinction coefficient with wavelength is shown in Fig.7.

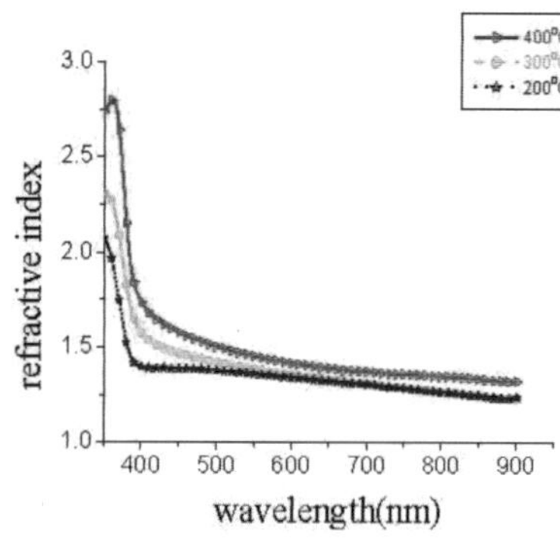

Fig.6 plot of refractive index versus wavelength for zinc oxide thin films annealed at different temperatures

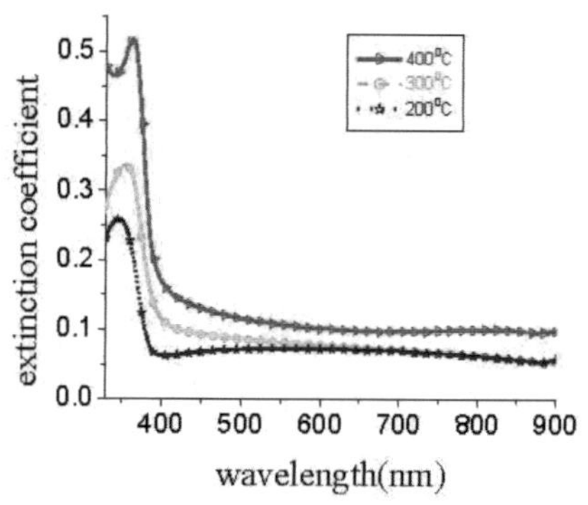

Fig.7 Plot of extinction coefficient versus wavelength for zinc oxide thin films annealed at different temperatures

It is well known that ZnO is a direct-gap semiconductor. The Optical band gap energy E_g and absorption coefficient α is related by the equation (5) (Tauc, 1970)

$$(\alpha h\nu)^2 = h\nu - E_g \quad (5)$$

Where, α is absorption coefficient and $h\nu$ is the incident photon energy. The variation of $(\alpha h\nu)^2$ with $h\nu$ is plotted, for the ZnO thin films, were annealed at different temperature, which is shown in Fig.8 The band gap energy E_g of the films was evaluated from the intercept of the linear portion of the each curve for different annealing temperature with the $h\nu$ in X-axis. The value of Band gap energy E_g as calculated above agrees nearly with band gap of bulk ZnO (3.37 eV). It is observed that band gap value decreases from 3.28eV to 3.16eV as the annealing temperature is increased from 200°C to 400°C. The effect annealing time on optical band gap were studied by annealing the thin film for different time intervals. The variation of E_g with annealing time is shown in fig.9 it is observed that band gap value decreases linearly as the annealing time interval increases.

IOP Conf. Series: Materials Science and Engineering **73** (2015) 012149 doi:10.1088/1757-899X/73/1/012149

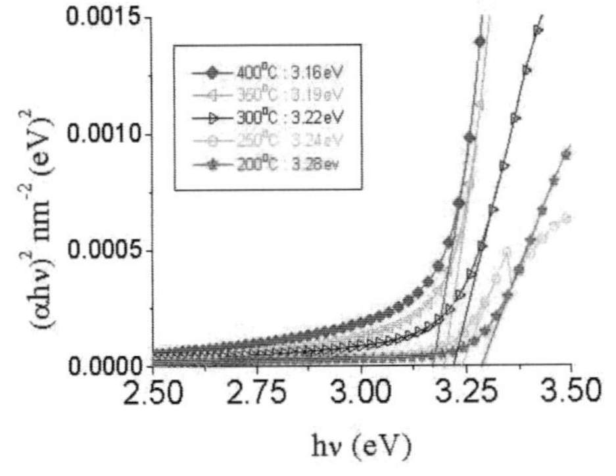

Fig.8 Plot of $(\alpha hv)^2$ versus hv for zinc oxide thin films for zinc oxide thin films annealed at different temperature

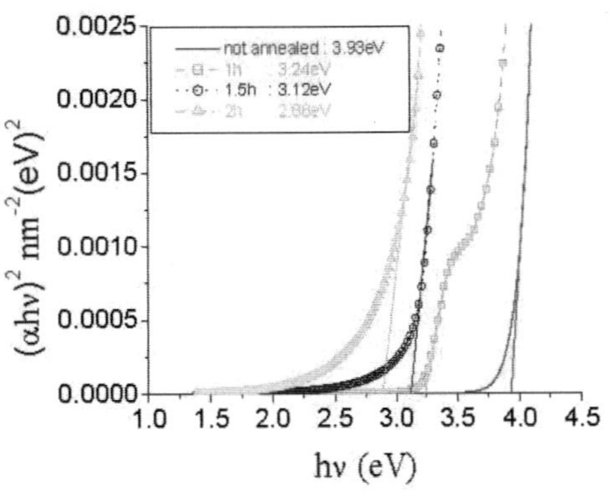

Fig.9 Plot of $(\alpha hv)^2$ versus hv for 1Mzinc oxide thin films annealed at 300°C for different time intervals

4. CONCLUSIONS

Zinc Oxide thin films prepared on glass substrate by spin coating process and annealed at different temperatures (200°C to 400°C). It is investigated that annealing temperature affects the structural and optical properties of the ZnO thin film. The transmission spectrums of the films were recorded by UV-VIS Spectrophotometer. The films showed high transparency (>80%) in the visible region. The refractive indices, extinction coefficients were calculated. The refractive index and extinction coefficient showed some variation with rise in annealing temperature of ZnO film. The Optical energy band gap values were evaluated by plot of $(\alpha hv)^2$ versus hv. The value of band gaps agrees approximately with that of bulk ZnO. Band gap value decreases as the annealing temperature increases. AFM image of the samples results the increase in roughness as annealing temperature increases, this is possibly because of increase in grain size, so that band gap energy Eg decreases.

5. ACKNOWLEDGEMENTS

Authors are grateful to Innovation Centre and Center for Atomic and Molecular Physics (CAMP), Manipal Institute of Technology,Manipal University for helping to do the project.

6. REFERENCES

Hussein [1]H.F.,Ghufran Mohammad Shabeeb[2], S.Sh. Hashim[3]. Preparation ZnO Thin Film by using Sol-gel-processed and determination of thickness and study optical properties. J. Mater. Environ. Sci. 2 (4) (2011) 423-426
ISSN : 2028-2508.http://www.jmaterenvironsci.com

ILICAN*, S., Y. CAGLAR, M. CAGLAR, 2008.Preparation and characterization of ZnO thin films deposited by sol-gel spin coating method. JOURNAL OF OPTOELECTRONICS AND ADVANCED MATERIALS Vol. 10, No. 10. 2578 – 2583.

Kumar1 *,K. Balachandraand P. Raji2, 2011. SYNTHESIS AND CHARACTERIZATION OF NANO ZINC OXIDE BY SOL GEL SPIN COATING. Recent Research in Science and Technology 2011, 3(3): 48-52
ISSN: 2076-5061 www.recent-science.com

Shakti, Nanda and P. S. Gupta, 2010.Structural and Optical Properties of Sol-gel Prepared ZnO Thin Film. Applied Physics Research Vol. 2, No. 1. 19-28 www.ccsenet.org/apr

Tauc, J.,1970. The optical properties of solids (North-Holland, Amsterdam, 1970).

IOP Conf. Series: Materials Science and Engineering **73** (2015) 012150 doi:10.1088/1757-899X/73/1/012150

FTIR and Optical Properties of various Se-S-Zn Chalcogenide Glasses

Sanjay, N. Kishore, R. S. Kundu, S. Dhankhar

Department of Applied Physics, Guru Jambheshwar University of Science and Technology, Hisar- 125001 (INDIA)

sanjay2000angira@yahoo.co.in

Abstract. Se-S-Zn chalcogenide glasses are prepared by melt-quenching technique. The amorphous nature of the glasses is ascertained by XRD patterns. In this work, the structural changes in these glasses are investigated by FTIR- spectra. An absorption band at $500cm^{-1}$ found in each glass is assigned due to the presence of Se_8 ring. The optical band gap is observed to decrease with increase in Zn content. This decrease has been explained on the basis of metallic bonding due to incorporation of Zn. The Urbach's energy is used to characterize the degree of disorder in amorphous solids. SEM has been used to prove the formation of nanosized particles.

1. Introduction

Recently, a great deal of work has been reported in the field of chalcogenide glasses due to their wide range of applications both in scientific and technological fields [1] such as in solid state devices, infrared optical fibres. These glasses exhibit unique IR transmission and electrical properties that make them useful for several applications such as threshold switching, memory switching, inorganic photoreceptors, detection through lenses and optical waveguide [2]. Amorphous Se is found to be useful as photo receptor in Xerox-graphy [2]. In case of Selenium (Se), each atom needs two neighbours to satisfy the valence requirements. This can be achieved either by the formation of small molecules (Se_8) or linear polymeric chain (Se_n) [3].The properties of amorphous Se (a-Se) and the effect of alloying S into a-Se have been studied by various workers [3]. These studies indicate that when S is incorporated to amorphous Se, it is dissolved in the Se chains to satisfy its coordination requirements and to form a cross-linked structure which retards the crystallization probability. The addition of third element in Se-S glass converts it into an interesting material and new promising properties of the material are expected [4]. The addition of Zinc (Zn) to Se-S is expected to create the compositional and configurational disorder in the material. Optical properties of multicomponent chalcogenide glasses have attracted great attention because of their high transmittance in IR spectral region [5].In our research paper, the optical properties of various compositions of Se-S-Zn chalcogenide glasses in terms of optical band gap and Urbach's energy are reported. The FTIR spectra of present glasses are also discussed.

2. Experimental details

Chalcogenide glasses $Se_{(72-x)}S_{28}Zn_x$ (x = 0, 2, 4) are prepared by the melt-quenching technique. High purity (99.999%) constituent elements in appropriate proportions are weighed using electronic balance and put into a quartz ampoule (length ~12cm, inner diameter ~14mm and outer diameter ~16mm). The ampoule containing chemicals (10g) is sealed in a vacuum of 10^{-3} Torr by using rotary pump and heated in a furnace where the temperature is raised at a rate of 4 ^{0}C per minute up to 600 ^{0}C and thereafter temperature is raised at a slower rate of 1.5 ^{0}C per minute up to the temperature in the range 875-925 ^{0}C depending upon the composition of the sample. The ampoule is kept at this temperature for 15 hours. The ampoule is agitated frequently in order to intermix the constituents to ensure homogenization of the melt. The molten sample in quartz ampoule is rapidly quenched in ice-cooled water to get the glassy state. The amorphous nature of the samples is ascertained by XRD patterns as recorded by using Rigaku X-ray diffractometer. The SEM model JSM-Joel 6510 is used to study the microstructure of the glasses. The IR transmission spectra of these glasses in KBr pellets are recorded using Perkin Elmer Spectrophotometer in the wave number range 400-4000cm^{-1}. The optical transmission spectra of all the samples are recorded using UV/VIS/NIR Spectrophotometer (Cary 5000) in the wavelength range 200-800nm.

3. Results and discussion

3.1. X-ray diffraction and SEM studies

XRD patterns of various glasses are found to be similar and typical XRD patterns for sample with x = 2 and 4 are shown in Fig.1. Perusal of this figure shows that the samples exhibit the glassy (amorphous) behaviour with a broad hump. No peak corresponding to Zn is observed which indicates that Zn has completely entered the glass matrix. The microstructure of the glasses is identified by scanning electron microscopy (SEM) studies. The Fig.2 shows the typical SEM micrograph of studied glasses for x = 2 (at%) of Zn which is found to be a glassy (amorphous) nature of the sample.

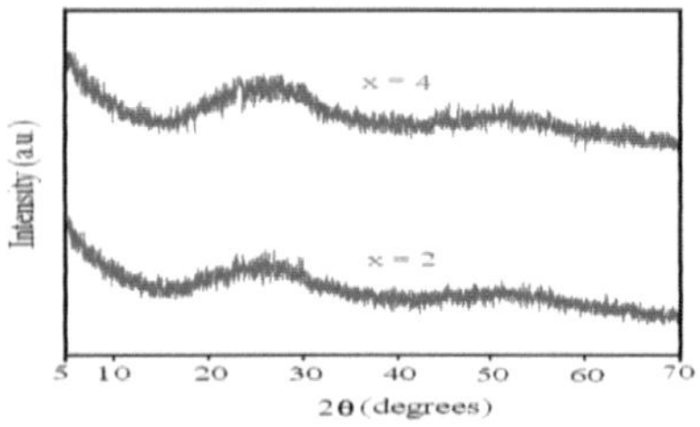

Fig.1 X-ray diffraction patterns of $Se_{(72-x)}S_{28}Zn_x$ chalcogenide glasses: for x = 2, 4 (at %).

Fig.2 Typical SEM of $Se_{(72-x)}S_{28}Zn_x$ chalcogenide glasses: for x = 2 (at %).

3.2. Fourier Transform Infrared (FTIR) Spectra

FTIR transmission study provides valuable information about the atomic configuration of glasses. The spectra of $Se_{(72-x)}S_{28}Zn_x$ (x = 0, 2, 4) chalcogenide glasses are shown in Fig.3(a-c). FTIR spectra of samples with x= 2 and 4 exhibit (except $Se_{72}S_{28}$) four absorption bands:

i) band at about 3500 cm^{-1} ii) band at about 1500 cm^{-1} iii) band at about 1000 cm^{-1}
iv) band at about 500 cm^{-1}

The intensity of all these bands is observed to increase with increase in Zn concentration (from x = 2 to x = 4). This may be due to the occurs of hetero-polar bonds like Zn-S, Zn-Se or Zn-S-Se. FTIR spectra of $Se_{72}S_{28}$ and $Se_{68}S_{28}Zn_4$ show some additional absorption peaks in region between 2500-3000cm^{-1} and 1500-2000cm^{-1} in mid-IR region. These absorption peaks may arise due to presence of H_2O and other impurity contents in glassy samples [6]. However, no such types of absorption peaks are observed in remaining glass sample. An absorption band observed at around 500cm^{-1} is assigned to the presence of Se_8 ring [7]. This absorption band is absent in $Se_{72}S_{28}$ glass. Another absorption band appearing at about 1500cm^{-1} is assigned to the presence of Zn-Se-Zn. Similar results have been reported by Goyal and Maan [8]. The absorption band at about 1000cm^{-1} is assigned to vibrational mode associated with mixed vibrations due to hetero –polar bonds like Zn-Se, Zn-S and/or Zn-Se-S [9]. From the energy point of view, the hetero-polar bonds are preferred compared to homo-polar bonds. Due to chemical ordering, the features such as the change in slope or kink occur for the various

IOP Conf. Series: Materials Science and Engineering **73** (2015) 012150 doi:10.1088/1757-899X/73/1/012150

properties and therefore, the glass structure is made of cross-linked structural units consisting of hetero-polar bonds only. Each constituent is coordinated by 8-n atoms, where n is the number of electrons in the outer shell (valence shell) and result in neglecting the other types of the bonds like dangling bonds and the other valence defects [10]. As the amount of Zn goes on increasing formation of hetero-polar Se-Zn bonds starts increasing at the expense of homo-polar bonds like Se-Se, S-S and Zn-Zn bonds [11]. So formation of Se-Zn and/or S-Zn bonds reduces the average energy of the system which consequently supports the decrease in optical band gap [12].

Fig.3 (a) FTIR spectrum of $Se_{(72-x)}S_{28}Zn_x$ (x = 0) glasses.

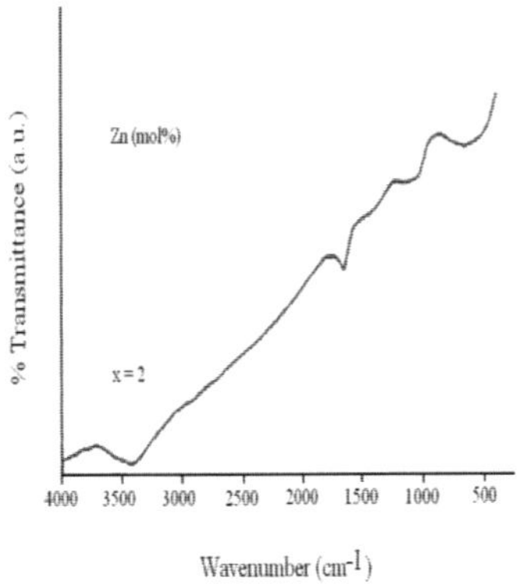

Fig.3 (b) FTIR spectrum of $Se_{(72-x)}S_{28}Zn_x$ (x = 2) glasses.

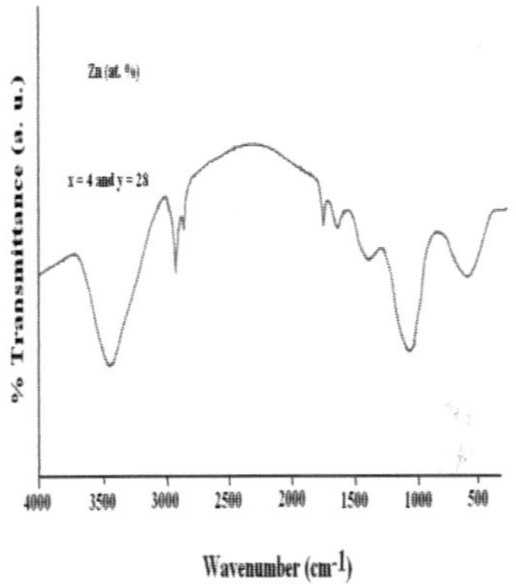

Fig.3 (c) FTIR spectrum of $Se_{(72-x)}S_{28}Zn_x$ (x = 4) glasses.

3.3. UV-Vis absorption spectra

The band gap of glassy systems is obtained with the help of evaluated values of absorption coefficient (α) by using Tauc's relation [13].

$$\alpha h\upsilon = B(h\upsilon - E_g)^r \qquad (1)$$

where B is the slope of Tauc edge called band tailing parameter that depends on the width of localized states [14] in the band gap (i.e. transition probability). E_g is the optical band gap of the investigated glasses and r is a number which characterizes the transition process. Plots of $(\alpha hv)^{1/2}$ versus photon energy (hv) (r = 2) for $Se_{(72-x)}S_{28}Zn_x$ (x = 0, 2, 4) glasses are shown in Fig.4. Value of E_g is determined by extrapolation of the linear part of the plots of $(\alpha hv)^{1/2}$ versus photon energy (hv) and it intercepts with the abscissa used value of E_g (at α = 0, E_g = hv). The values of optical energy band gap (E_g) lie between 1.98- 2.81eV and it shows that the optical energy band gap is decreasing with increase in Zn doping. It can be suggested, following Kitao et al. [15], that the compositional change in the host material could exist by itself. This effect is due to the change in bond angles and/or bond lengths. Present results are consistent with the previous work dealing with the decrease of the energy gap with composition [16].The decrease in optical band gap may be correlated with the electronegativity difference of the elements involved. According to Kastner et al. [17], the valence band in chalcogenide glasses is constituted by lone pair p-orbital contributed by the chalcogen atoms. These lone pair electrons will have higher values of energy adjacent to electropositive atoms than those of electronegative atoms [16]. Therefore, the addition of electropositive elements to the alloy may raise the energy of lone pair states which is further responsible for the broadening of valence band inside the forbidden gap. The electrogetivites of Se, S and Zn are 2.55, 2.58 and 1.65, respectively. From these electrogetivity values, it is clear that Zn is less electronegative than Se and S, so the substitution of Zn for Se and/or S may raise the energy of some lone pair states and hence broaden the valence band. This leads to band tailing and hence shrinking of the band gap thereby E_g value decreases with Zn content [16]. Further, the value of Urbach's energy 'ΔE' has been calculated from the reciprocal of the slopes of the linear portion of curves between ln{α(υ)}and hυ i.e. Urbach's plots. In general, the materials with larger Urbach's energy would have greater tendency to convert weak bonds into defects. The value of 'ΔE' for the present glass system lies in the range 2.30- 3.21 eV which is increasing with Zn content, suggesting the glassy nature of our system.

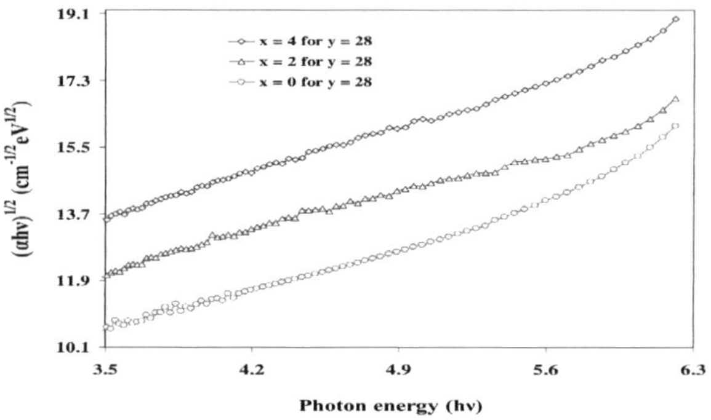

Fig.4 Tauc's plots for various $Se_{(72-x)}S_{28}Zn_x$ glassy alloys.

4. Conclusions

From XRD patterns, it is confirmed that the given samples are amorphous in nature. FTIR spectra suggets the formation/exhistence of the heteropolar bonds like Se-Zn, Se-S at the expense of homopolar bonds like Se-Se, Zn-Zn, S-S with increase in Zn content. The optical absorption in the given glass systems seems to be of indirect type and the optical band gap is found to decrease from 2.81 to 1.98 eV with the addition of Zn content. Urbach's energy increases confirming thereby the amorphous nature of the studied glasses.

5. Acknowledgements

Authors are highly thankful to DST-FIST, DRDO and UGC New Delhi for financial assistance. Authors are also thankful to Dr. B.K. Shrivastva, Dr. N.S.Saxena, Sh. K.B. Razdan for providing lab. facility (i.e. sealing of ampoule) for preparing these glasses.

References

[1] S.Faheem Naqvi, Deepika, N.S. Saxena, Kananbala Sharma, D. Bhandari, J.Alloys Compd. 506 (2010) 956.
S.Faheem Naqvi, Deepika, N.S. Saxena, Kananbala Sharma, D. Bhandari, J.Alloys Compd. 506 (2010) 956.
[2] Kedar Singh, Abhay Kumar Singh, N.S. Saxena, Current Applied Physics 8 (2008) 159.
[3] G. Lucovsy, Mater. Res. Bull. 4 (1996) 505.
[4] N.Mehta, S.K.Arrahari, A.Kumar, Mater.Lett. 61 (3) (2007) 837.
[5] A.V.Kolobov, J. Tominaya,J.Optoelectron. Adv. Mater. 4 (2002) 679.
[6] Abhay Kumar Singh, Neeraj Mehta, Kedar Singh, Physica B 404 (2009) 3470.
[7] Parikshit Sharma, V. S. Rangra, Pankaj Sharma, S. C. Katyal, J. Alloys Compd. 480 (2009) 934.
[8] D. R. Goyal, A. S. Maan, J. Non-Cryst. Solids 183 (1995) 182.
[9] Q. H. Nie, G. X. Wang, X. S. Wang, S. X. Dai, S. W. Deng, T. F. Xu, X. Shen, Opt. Commun. 283 (2010) 4004.
[10] K. tanaka, Phys. Rev. B 39 (1989) 1270.
[11] S. S. Fouad, A. H. Ammar, M. Abo-Ghazala, Physica B 229 (1997) 249.
[12] P. Sharma, V. S. Rangra, P. Sharma, S. C. Katyal, J. Phys.D: Appl. Phys. 41 (2008) 225.
[13] J.Tauc, in: J.Tauc(Ed), Amorphous and Liquid Semiconductors, Plenum Press, New York, 1979, p.159.
[14] J.Tauc, The Optical Properties of Solids, North-Holland, Amsterdam, 1970.
[15] M.Kitao, H.Akao, T.IshiKawa, S.Yamada, Phys.Stat.Sol.A 64 (1981) 493.
[16] M.A. Majeed Khan, M. Wasi Khan, M. Husain, M. Zulfequar, J. Alloys Compd. 486(2009) 876.
[17] M.Kastner, D.Adler, H.Fritzsche, Phys.Rev.Lett. 37 (1976) 1504.

IOP Conf. Series: Materials Science and Engineering **73** (2015) 012151 doi:10.1088/1757-899X/73/1/012151

Strain Effects on Band Structure of Wurtzite InGaN/GaN Quantum Well on Si Substrate

S. K. Jana[1], S. Ghosh[1], S. M. Dinara[1], T. D. Das[1] and D. Biswas[2]*

[1]*Advanced Technology Development Centre, Indian Institute of Technology, Kharagpur, India, 721302*
[2]*Dept. of Electronics & Electrical Communication Engineering, Indian Institute of Technology, Kharagpur, India, 721302*
[*]*E-mail: skjiit@gmail.com*

Abstract Heteroepitaxial crystal growth of InGaN/GaN Quantum Well (QW) on Si substrate is necessary for the development of visible and ultraviolet (UV) Light Emitting Diodes (LEDs). In this work we accumulate the efforts that have been made to focus the analysis of strain effects on the band structure of GaN quantum well systems on Si substrate due to its potential application. We limit this investigation on the effect of pseudomorphic GaN quantum wells on Si substrate grown at different GaN planes. It is observed that the deformation potential constants and anisotropic strain separate Heavy Hole (HH) and Light Hole (LH) band and eventually making LH the top most bands.

1. Introduction The development of LED devices on Si substrate using InGaN/GaN is quite challenging because of significant lattice mismatch (almost 16%, leading to strain in the GaN layers) and thermal mismatch (approximately 54%) between Si and GaN. Presently the InGaN/GaN QW Light Emitting Diodes are fabricated on SiC, Sapphire (α- Al_2O_3) or Si substrate. The GaN, InN and their ternary alloy materials have a wurtzite crystal structure and a direct energy bandgap. At room temperature the band gap of GaN and InN are 3.44eV and 1.89eV respectively [1]. The band gap of the ternary alloy $In_xGa_{1-x}N$ can be interpolated using Vegard's law [2]. This band gap depends on the mole fraction of In. In that case a bowing parameter is also taken into account. GaN/InGaN based blue and green light emitting diodes (LED) are very important for shorter wavelength application [3-5]. The optical properties of the LED devices depend significantly on the epi-layer quality. Different types of substrates are used for growing InGaN/GaN quantum well layer such as 6H-SiC, Sapphire, and Si. Our main focus is to investigate devices on Si substrate because of its obvious benefits [6]. In the pseudomorphic growth, strain is generated in the epitaxial layers. This strain modifies the band structure [7-10]. We have investigated this phenomenon.

2. Effect of strain on valance band spectra in pseudomorphic InGaN/GaN quantum wells

The idea of quantum well first took shape when Esaki and Tsu [11, 12] suggested in 1969 that a heterostructure sandwich of ultrathin layer of two semiconductors with different band gaps should exhibit some novel useful properties. The band-edge potential varies from layer to layer due to different band gaps and a periodically varying potential is produced in the structure for layer thickness of the order of few nano meters. The band structure of the quantum well is modified by this strain. The effect of strain on the band structure of semiconductor was first presented by Bir and Pikus [13]. The unstrained lattice

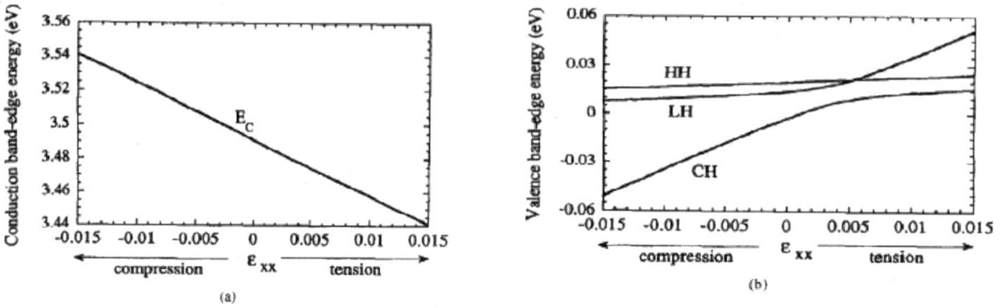

Figure 1. The band edges of (a) the conduction band and (b) the HH, the LH, and the CH bands as a function of the compressive and tensile for a GaN crystal. The reference energy is set at $\mathbf{E_v^0}$of an unstrained GaN material[14].

vectors (x_i) and the strained lattice vectors $(x_i^{'})$ are related through the relation $x_i^{'j} = x_i^j + {}_j\, \varepsilon_{ij}\; x_i^j$, where ε_{ij} denote the strain tensor components. A model for the electronic band structure of wurtzite strained quantum well was presented in [14] by introducing strain effect on **k.p** band structure. This theoretical model provides a foundation of wurtzite- strained quantum well LED. The band edge energy of the conduction band E_c , the HH, the LH, and the CH bands are shown in Fig.1.

The same band structure of strained semiconductor was modified with the introduction of spin- orbit split-off bands coupling by C.Y.P. Chao et al. [15]. As a result the [6×6] Hamiltoninian at the band edge at K=0, is takes a new form. The most important results are obtained from the series expansion of E up to the second order of K near the band edges given by:

$$E_{HH}(k) \approx E_{HH}(0) - \left(\frac{\hbar^2}{2m_0}\right)\left[(\gamma_1+\gamma_2)K_t^2 + (\gamma_1 - 2\gamma_2)K_z^2\right] \qquad (1)$$

$$E_{LH}(k) \approx E_{LH}(0) - \left(\frac{\hbar^2}{2m_0}\right)\left[(\gamma_1 - f_+\gamma_2)K_t^2 + (\gamma_1 + 2f_+\gamma_2)K_z^2\right] \qquad (2)$$

$$E_{SO}(k) \approx E_{SO}(0) - \left(\frac{\hbar^2}{2m_0}\right)\left[(\gamma_1 - f_-\gamma_2)K_t^2 + (\gamma_1 + 2f_-\gamma_2)K_z^2\right] \qquad (3)$$

Where f_+ and f_- are dimensionless, strain- dependent factors.

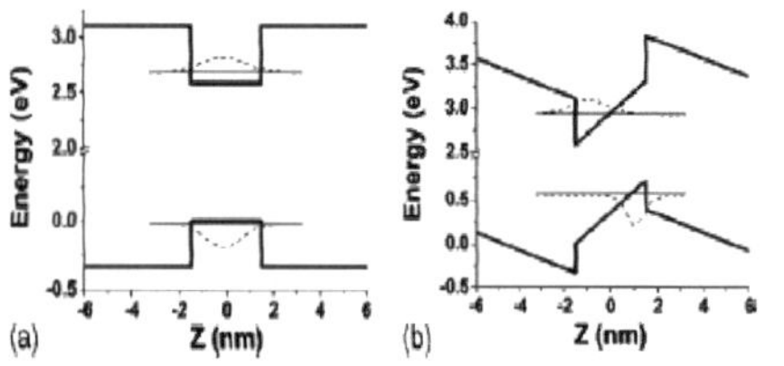

Figure 2. Shows the valance sub band dispersion spectrum of $In_{0.16}Ga_{0.84}N/GaN$ QW structure (a) without and (b) with the internal field L_w =3nm and L_b=7nm **[19]**.

A further improvisation on this theoretical study based on the RSP Hamiltonian uses Envelop function analysis which successfully explains the origin of piezoelectric effects [16]. The piezoelectric polarization P_{piezo} is generally formulated using Vegards's interpolation formula [17]. For valance band they have used [6×6] Hamiltonian in block diagonal forms [18]. Fig. 2. shows the valance sub band dispersion spectrum of $In_{0.16}Ga_{0.84}\,N/GaN$ QWstructure caused by internal polarization field. In this case the blue energy shift of all states is caused by the indium surface segregation effect. Here the energy shift is 20meV for the valence band amounting to a total blue shift of 70mev. The effects of strain on the valence band structure of $(11\bar{2}2)$ semipolar InGaN grown on GaN were analyzed by **k.p** method also by Qimin Yan et al [20] which showed that the anisotropic strain the c-plane and shear strain are the most important factors for decoding the ordering of the two top most valence bands. The shear strain deformation potential D_6 is often calculated for GaN and InN using density functional theory with the Heyd – Scuseria – Emzerhof hybrid function. [21]

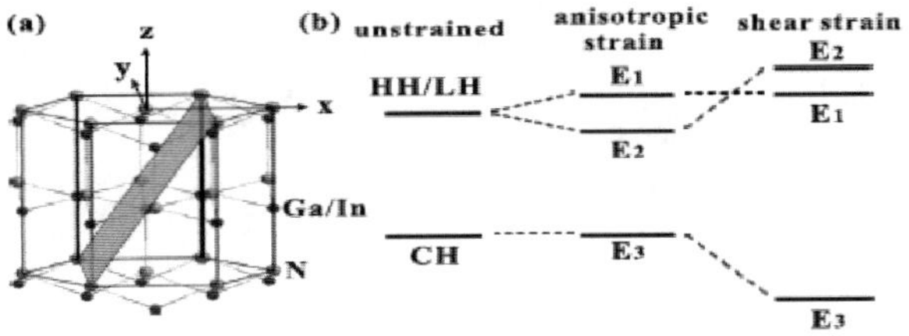

Figure 3. (a) the normal coordinate system of the wurtzite GaN crystal. The $(11\bar{2}2)$ plane is shown in gray. (b) Schematic illustration of strain effects on the band structure in $(11\bar{2}2)$ semipolar InGaN. In an unstrained alloy, the heavy hole (HH) and light hole (LH) bands are doubly degenerate [20].

Under shear strain, the E_2 band goes up may switch order with the E_1 band [20]. It also gives a measure of the Energy separation ΔE_{23} between valence bands E_2 and E_3 . It is seen that InGaN/GaN MQWs structure suffers from anisotropic in-plane compressive strain, breaking the original valence band states [22] .The peak energy shifts results from the splitting valence sub-bands that rises with increase of in-plane strain [23]. Conversely reduction of strain leads to the reduction of the polarization ratio [24, 25]. Due to strain, the VB states are strongly modified affecting both the energies as well as the polarization selection rules for the transitions. Hamiltonian for the strain dependence of the VB can be modeled by the matrix given by [26] yielding three distinct VB maxima with energies E_j^v. The parameter values thus obtained are $\alpha_\perp = \alpha_\parallel = -44.5eV$, $D_1 = -41.4eV$, $D_2 = -33.3eV$, $D_4 = -4.1eV$, and $D_5 = -4.7eV$. The resulting inter band hydrostatic deformation potentials are $\alpha - D_1 = -3.1eV$ and $\alpha - D_2 = -11.2eV$. The calculation of reference [25] gave slightly modified values of deformation potential at $D_3 = 9.4eV$, and $D_4 = -4.7eV$. The diagonalization of the traditional [6×6] VB Hamiltonian was reduced to give three distinct sub-band dispersion relations, where each sub-band is a Kramers degenerate pair [27]. At Γ point, where k=0, the VB Hamiltonian is automatically blocked diagonalized as:

$$H(k=0,\varepsilon) = \begin{pmatrix} H^U & 0 \\ 0 & H^L \end{pmatrix}, \text{ where } H^U = \begin{bmatrix} F_0 & -K_0^* & 0 \\ -K_0 & G_0 & \sqrt{2}\,\Delta_3 \\ 0 & \sqrt{2}\,\Delta_3 & \lambda_0 \end{bmatrix}$$

Where, H^L is the complex conjugate of H^U. References [28] and [29] further reformed the Hamiltonians and their interband optical matrix elements for three important crystal orientation; namely c-plane, m-plane, and a-plane by using multiband effective mass theory. It is to be noted that strain in any form has a degrading effect on the performance and as a preventive measure usage of pre-strained layer has been suggested for partial relaxation resulting in improvement of the quantum efficiencies of the LED [30].

3. Conclusion

In summary we have analyzed the various theories for band structure modification of InGaN/GaN quantum wells due to strain. It is seen that generation of the piezoelectric field is the primary cause for alteration of valance band structure in the quantum well.

Acknowledgment
We thankfully acknowledge the financial support received from DST, INDIA (Sanction No. 100/IFD/196/2010-11, dated: 03/06/10). We also thank Mr. N.N.Halder for his valuable help.

References

[1] Seoung-Hwan Park, Doyeol Ahn, S. L. chuang," Electronic and optical Properties of a-and m- plane wurtzite InGaN-GaN quantum Wells" IEEE J. of Quantum Electronics, Vol. 43, No.12, (2007)

[2] D. Xiao and k.W. Kim," Envelope-function analysis of wurtzite InGaN/GaN quantum well light emitting diodes: JA P, Vol-96 No. 1(2004)
[3] S.Nakamura et al. " High brightness blue, green and yellow light emmiting diodes with quantum well structure" Jpn. J. Appl. Phys., Vol.34, PP. L797-L799, July 1995.s,
[4] I. Akasaki and H. Amano, "Crystal growth and conductivity control of group III –nitride Semiconductors and their applications to short wave length light emitter," Jpn. J. Appl . phys., Vol 36, PP. 5393-5408,sept.1997.
[5] W.C. Lai, et al, "InGaN/AlInGaN light emitting diodes." IEEE photo. Technology lett, Vol 13, PP. 559, 2001
[6] Hiroo Yonezu, Yuzo Furukawa and Akihiro Wakahara," III-V epitaxy on Si for photonics applications" Journal of Cryastal Growth 310, 4757-4762 (2008)
[7] S.L. chuang and C.S. Chang," k.p method for strained wurtzite semiconductors" Physical Review B Vol. 54, Num 4 July(1996)
[8] K. Domen, K. Horino, a. Kuramata, and T. Tanahashi," Optical Gain for wurtzite GaN with Anisotropic strain in c Plane," IEEE Journal of Quantum Electronics Vol. 3. No.2. April 1997.
[9] Yee Chia Yeo et al. " Electronic Band Structure and Optical Gain Spectra of Strained Wurtzite GaN-AlGaN quantum - well Lasers," IEE J. Quantum Electronics Vol.34, NO. 3, March 1998.
[10] S. L. Chuang," Efficient band – structure calculation of strained quantum wells" Physical review B, Vol. 43, No.12 April, 1991
[11] L. Esaki and R. Tsu, IBM Internal Res. Rep, RC 2418, Mach 26 (1969)
[12]L. Esaki and R. Tsu, IBM J. Res Dev. 14,61 (1970)
[13] G.L. Bir, and G. E. Pikus," symmetry and strain induced Effects in Semiconductor" wiley, New york, 1974.
[14] S.L Chuang " IEEE Journal of quantum Electronics, Vol 32 N0-10, October 1996.
[15] C.Y.P.Chao and S.L. Chuang " Spin-orbit-coupling effects on the valence band structure of strained Semiconductor quantum wells " Phys. Rev. B 46, 4110-41(1992)
[16] Mykhailo V. Klymenko, Oleksiy V. Shulika," Band structure of InGaN/GaN Quntum Wells under Influence of Internal Fields and Indium surface Segregation" LFNM 2010 ICLFONM 12-14 September, 2010 Ukrain.
[17] V. Fiorentini, F. Bernardini , O. Ambacher, " Evidence for non linear macroscopic polarization in III-V nitride alloy heterostructute, " Appl. Phys. Lett, Vol.80,pp.1204-1206, 2002.
[18] S.L. Chuang, C.S. Chang," A band structure model of strained quantum well wurtzite semiconductors" Semicond sci, Technol. Vol 12, PP. 252-263, 1997
[19] D. Xiao and k.W. Kim," Envelope-function analysis of wurtzite InGaN/GaN quantum well light emitting diodes: JA P, Vol-96 No. 1(2004)
[20] Qimin Yan et al., Appl. Phys. Lett. 97, 181102 (2010)
[21] J. Heyd at al.," J. chem. Physics 124, 219906(2006)
[22] Huei-Min Huang at al, J. Lightwave Technology, Vol. 29, No. 18, SEP. 15, 2011.
[23] V.Y Davydov et al." Raman and Photoluminescence Structure of biaxial strain in GaN epitaxial layers grown on 6H-SiC" J. Appl. Phys. Vol. 82, PP. 5097(1997)
[24] B. Liu et all," Anisotropic crystallographic properties, strain, and their effects on band structure of m-plane GaN on $LiAl0_2$
[25] Deyi Fu et al. JAP Vol. 106, 023714 (2009)
[26] Sandip Ghosh, P. Waltereit, O. Brandt, H.T. Grahn,and K.H. Ploog, Phys. Rev. B, Vol. 65, 075202
[27] S.L. Chuang and C.S. Chang, Phys, Rev. B 54, 2491(1996)
[28] S.L. Chuang and C.S. Chang "The K.P methode for Strained Wurtzite Semiconductors," Phys-Rev. B. Vol. 54, PP. 2491-2504, !996
[29] S.L Chauang "Optical gain of strained wurtzite GaN quantum-well layers," IEEE J-Quantum Electronics Vol 32, no. 10, PP. 1791-1800 Oct.1996
[30] Jae-Ho song et al." Effects of prestrained layers on piezoelectric field and indium incorporation in InGaN/GaN Quantum wells" 2009 OSA/ CLEO/IQEC.

IOP Conf. Series: Materials Science and Engineering **73** (2015) 012152 doi:10.1088/1757-899X/73/1/012152

Effect of Laser Peening without Coating on 316L austenitic stainless steel

S Sathyajith[1,3] and S kalainathan[2]
[1]Department of Physics MVJ College of Engineering, Bangalore,India
[2] School of Advanced Sciences, VIT University, Vellore-63201, India

[2]sathyajithsathyanesan@msn.com

Abstract. Laser Peening without Coating (LPwC) is an innovative surface modification technique used for the in-suit preventive maintenance of nuclear reactor components using frequency doubled (green) laser. The advantage of LPwC is that the laser required for this technique is in milli joule range and the processes can perform in aqueous environment. This paper discussed the effect of LPwC on 316L austenitic stainless steel using low energy Nd: YAG laser with various laser pulse density. The base specimen and laser peened specimen were subjected to surface residual stress, surface morphology, micro hardness and potentiodynamic polarization studies. The laser peened surface exhibit significant improvement in surface compressive residual stress. The depth profile of micro hardness revealed higher strain hardening on laser peened specimens. Though corrosion potential reported an anodic shift,current density is found to be increased after LPwC for the specimen peened with higher pulse density.

1. Introduction

Enhancement of surface properties of metals and alloys is an important methodology used to improve the durability, fatigue strength, wear resistance and corrosion resistance of structural components. Laser shot peening is one of such surface enhancement technique which is found to be effective in preventing foreign object damage of aircraft engine fan blades, preventing crack initiation of turbine blades and recently used as a preventive maintenance technique against stress corrosion cracking of nuclear plants[1,2,3]. Thae last two decade experience an intense research in the versatility of laser shot peening under various application sectors and the results are promising compare to conventional shot peening technique[2] .The laser peening technique was evolved as a fully mechanical treatment technique using high power pulsed laser[2]. But after the discovery of new alternative technique by Mukai etal ,low power lasers also become a choice for laser shot peening[3].This alternative method was termed, Laser Peening Without Coating(LPwC).This note worthy advancement in laser shot peening increase the scope of laser peening technology and the method was successfully implemented as a preventive maintenance strategy against stress corrosion cracking in Japanese nuclear power plant[4].

Laser peening without coating encouraged the feasibility of using low energy laser for laser peening application. The authors already reported the successfulness of using low energy laser for laser peening application [5,6,7]. This present work focused on the effect of laser peening without coating on 316L SS using low energy Nd:YAG laser. The usage of IR radiation has advantages and disadvantages with respect to frequency doubles (Green) laser. The transparency of green light in water is higher compare to IR, however dielectric breakdown threshold will reduce with respect to the decrease of wave length, this in effect cause decrease of peak power pressure [2].

2. Experimental

Low carbon stainless steel 316L was selected for present investigation which has wide application as valve pump, jet engine parts, digesters parts, power plant components due to its weldability and high corrosion resistance strength. The specimen was placed on an XY computer controllable translation stage consists of two perpendicularly fixed translation stages. The movement on each axis was controlled using RS485 daisy chain of two servo motors. Laser source was Nd:YAG laser operating at pulse width 10 ns,5 Hz repetition rate and 1064 nm fundamental wavelength. The experimental parameters are tabulated in table.1.

Table1. Experimental parameters

Specimen	Beam diameter (mm)	Laser Fluence (GW/cm^2)	Pulse density ($pulses/mm^2$)
LP-1	1	2.5	9
LP-2	1	2.5	22
LP-3	1	2.5	32

After laser peening the surface roughness of the specimens were measured using Mitutoyo profilometer of resolution 0.01 μm. Surface residual stress was evaluated based on X-ray diffraction $\sin^2\psi$ method [8,9,10]. Mn –Kα line was used as X-ray source (20 kV, 4 mA), diffraction angle was 151.2° with 2 mm beam diameter. Microstructure analysis was conducted by taking cross section of the laser peened specimen across the direction of laser peening. The sample was electrochemically etched using 10% oxalic acid solution with 5.6 V potential and etching time 1.3 minute. Vickers micro hardness studies were conducted to evaluate the work hardening after laser peening. Depth profiles of the specimens were made with 50 g load and 10 s hold time. Potentiodynamic polarization study was performed to investigate the corrosion behaviour in 3.5% NaCl solution using a potentiostat (Gill A.C, ACM make).

3. Result and Discussion

3.1 Surface Roughness and Surface Residual stress. The surface roughness values of laser peened and unpeened specimens are represented in table 2. The direct laser ablation lead to an increase of surface roughness compare to untreated surface. The average roughness value (Ra) indicates an increase with increase of laser pulse density. However in respect to conventional shot peening increase in roughness was appeared to be less [2].

Table 2. Surface Roughness

Specimen	Ra (μm)	Ry (μm)	Rq (μm)
Base material	0.41	3.61	0.55
LP-1	0.69	6.71	1
LP-2	1.04	7.6	1.34
LP-3	1.42	10.18	1.78

Ra: Average roughness
Ry: Maximum Height of the Profile
Rq: Root mean square roughness

Table 3. Surface residual stress

Specimen	σ_x (MPa)
Base material	-12.6
LP-1	-273.1
LP-2	-180.8
LP-3	-165.4

The surface residual stress of unpeened and laser peened specimens are tabulated in table 3. From the residual stress result it can be confirmed that laser peening without coating induces higher compressive residual stress whilst the increase of laser pulse density reduces the magnitude of surface compressive stress. Increase of thermal influence with increase of pulse density will be the reason for decrease of magnitude of stress.

3.2 Microstructure analysis. The microstructure of unpeened and laser peened specimens are presented in figure 1. There is no significant difference identified before and after laser peening. No

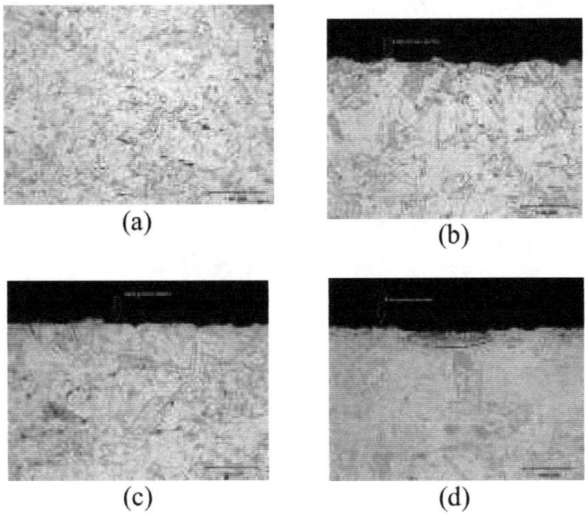

Figure 1. Micro structure of laser peened specimens (a)Un peened (b)LP-1 (c)LP-2 (d)LP-3

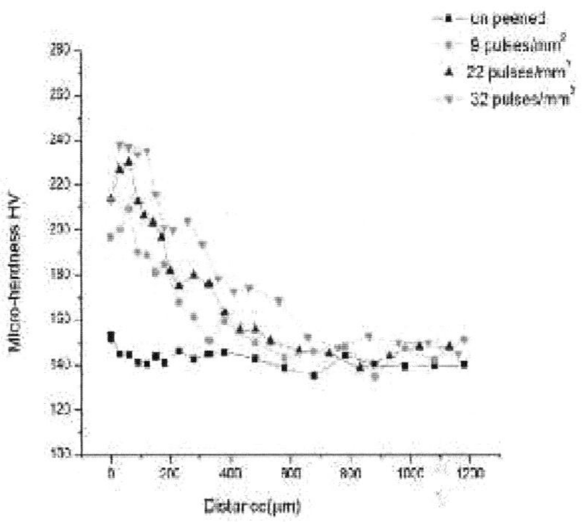

Figure 2. Depth profile of micro hardness

Table 4 Corrosion potential and corrosion current density

Specimen	I_{corr}(mA/cm^2)	E_{corr}(mV)
Untreated	2.3793	-406.71
LP-1	2.3035	-369.42
LP-2	2.4827	-384.78
LP-3	2.6339	-351.87

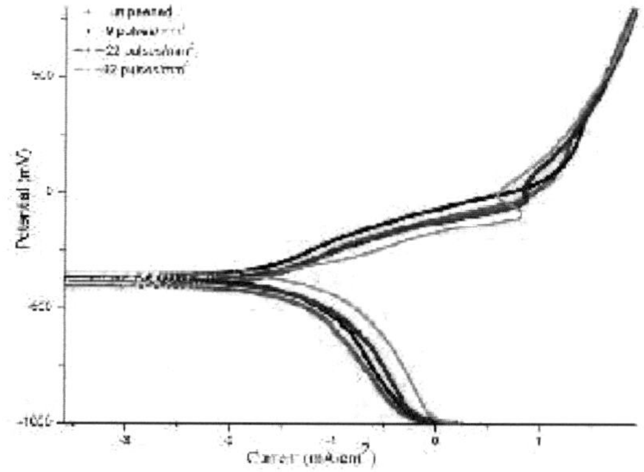

Figure 3.Potentiodynamic polarization plot

AUTHOR INDEX

AUTHOR INDEX

AUTHOR INDEX

AUTHOR INDEX

Institute of Physics
Dirac House, Temple Back
Bristol BS1 6BE UK

CURRAN ASSOCIATES INC
proceedings.com

6th International Conference on Inertial Fusion Sciences and Applications 2009

Journal of Physics: Conference Series Volume 244

San Francisco, California, USA
6-11 September 2009

ISBN: 978-1-61738-932-0
ISSN: 1742-6588

IOP Institute of Physics